MOLECULAR BIOLOGY
of ASSEMBLIES
and MACHINES

MOLECULAR BIOLOGY
of ASSEMBLIES
and MACHINES

Alasdair C. Steven

Wolfgang Baumeister

Louise N. Johnson

Richard N. Perham

Garland Science

Taylor & Francis Group

NEW YORK AND LONDON

Garland Science

Vice President: Denise Schanck
Senior Editor: Summers Scholl
Editorial Assistants: Michael Roberts and William Sudry
Developmental Editor: Mary Purton
Production Editors and Layout: Georgina Lucas and
EJ Publishing Services
Illustrator: Nigel Orme
Cover Design: Matthew McClements, Blink Studio, Ltd
Copyeditors: Bruce Goatly and Sally Livitt
Proofreader: Sally Huish
Indexer: Liza Furnival, Medical Indexing Limited
Permissions Coordinator: Sheri Gilbert

About the Authors

Alasdair C. Steven trained as a theoretical physicist (PhD Cambridge), before migrating into biology as a postdoctoral fellow at the Biozentrum, University of Basel. As a principal investigator, his research is aimed at elucidating structure–function–assembly–dynamics relationships of macromolecular complexes, such as viruses, energy-dependent proteases, and amyloid filaments. He has developed novel methods for image reconstruction and analysis of electron micrographs, and 'hybrid approaches' in which cryo-electron microscopy is systematically combined with other experimental methods.

Wolfgang Baumeister received his PhD in biophysics from the University of Düsseldorf where he remained, apart from a year as Heisenberg Scholar at the University of Cambridge, before moving to the Max-Planck-Institute of Biochemistry in Martinsried. There he is head of the Department of Structural Molecular Biology and an Honorary Professor of Physics at the Technical University of Munich. Baumeister's laboratory pioneered the development of cryo-electron tomography enabling structural biology studies *in situ*. His work has shaped the understanding of the structure and function of the cellular machinery of protein degradation, most notably the proteasome.

Dame Louise N. Johnson was a leader in X-ray crystallography who contributed to solving the first (lysozyme) and second (RNase) enzymes to be described by this technique. She received her PhD from University College London, spent a postdoctoral year at Yale, and made her career at the University of Oxford, where she was Professor of Structural Biology from 1990–2007. She also served as Director of Life Sciences at Diamond Light Source. Johnson's laboratory determined the structures of protein kinases, regulatory proteins of the cell cycle, and glycogen metabolism—yielding insights as to the origin of diseases and drug design.

Richard N. Perham was a protein biochemist with broad-ranging interests. After national service in the Royal Navy he embarked on an academic career (PhD from Cambridge; postdoctoral fellowship at Yale). He then returned to Cambridge eventually becoming Chair of the Department of Biochemistry and Master of St John's College. Among the systems he studied were tobacco mosaic virus assembly, bacteriorhodopsin, and multienzyme complexes including pyruvate dehydrogenase. Perham also made influential contributions to synthetic biology and commercial applications of phage display.

Library of Congress Cataloging-in-Publication Data
Names: Steven, Alasdair C., author. | Baumeister, W. (Wolfgang), 1946- , author. | Johnson, L. N., author. | Perham, R. N., author.
Title: Molecular biology of assemblies and machines / Alasdair C. Steven, Wolfgang Baumeister, Louise N. Johnson, Richard N. Perham.
Description: New York, NY : Garland Science, [2016] | Includes bibliographical references.
Identifiers: LCCN 2015042845 | ISBN 9780815341666 (alk. paper)
Subjects: | MESH: Macromolecular Substances--metabolism. | Biochemical Processes. | Biophysical Processes. | Cell Physiological Processes. | Macromolecular Substances--chemistry.
Classification: LCC QP171 | NLM QU 45 | DDC 572/.4--dc23
LC record available at http://lccn.loc.gov/2015042845

Published by Garland Science, Taylor & Francis Group, LLC, an informa business,
711 Third Avenue, New York, NY 10017, USA, and 3 Park Square, Milton Park, Abingdon, OX14 4RN, UK.

Printed in the United States of America
15 14 13 12 11 10 9 8 7 6 5 4 3 2 1

Garland Science
Taylor & Francis Group

Visit our website at http://www.garlandscience.com

FOREWORD

Structural biology has become a popular subject. Scientific journals use beautiful pictures of protein folds to decorate their covers. Even advertisers in the popular press resort to pictures of DNA or alpha helices to demonstrate the good scientific pedigree of their products. In *Molecular Biology of Assemblies and Machines*, biology is explained in terms of these well-known structural motifs. The book covers almost every basic biological topic at a level that will allow the inquisitive reader to quickly absorb the fundamentals in terms of the currently available structural information. The four authors bring together extensive knowledge of biology, chemistry, biochemistry, physics, and mathematics from their diverse backgrounds. Together they emphasize the mechanistic basics of biology and provide insights that are often missing in books with more restricted interests. These will stimulate the reader to look further for greater detail elsewhere.

This book is by no means the first one that views biology from a structural point of view. One of the earliest books of this genre was written by Dickerson and Geis in 1969. Others followed, including those by Schulz and Schirmer (1980), Brändén and Tooze (1991), and Perutz (1992). But these books tend to concentrate on specific biological topics that have seen significant structural achievements. The present book reverses this approach by surveying the complete biological landscape in terms of the functions required by a living cell. Then, currently available structural information is used to illuminate the mechanisms of the biological machines that are required to maintain life.

The first protein structure was determined in the 1950s for sperm whale myoglobin, which has a molecular mass of 17 kDa. As technology advanced, it became possible to solve much larger complexes. For instance, some 20 years later, viruses with icosahedral symmetry built of 180 identical subunits, each as big or bigger than myoglobin, were first described and more recently, much larger viruses have been described in comparable detail. Many of the molecular machines covered in this book have multiple copies of many different subunits—as for instance the well-known and much studied bacteriophage T4. This is to be expected as many biological functions are complex and require the coordinated efforts of different specialized subunits. The larger the complex (for instance, a ribosome), the more multifaceted will be the functions required of it; also, the less likely it is that every assembled unit will have the exact same structure. Variants can arise from functionally relevant conformational changes, but also from mistakes occurring during assembly, much as they do in the assembling of houses from bricks. It is often difficult to exactly repeat an experiment performed with biological macromolecules; indeed that is why the science of biology is often considered to be an inexact science. Thus the wonderful structural tools developed for relatively small, homogeneous molecules like myoglobin fade into obsolescence as investigations are extended to larger assemblies that individually have small differences but yet function similarly. Eventually structural investigations at near-atomic resolution must address whole cells. In this case the assumption of structural uniformity that has brought us to the present state of knowledge will need to be completely discarded. This book is being published at a time when major changes are happening in the technology and direction of structural biology, whereby objects of interest each have their own individual structure and yet are functionally similar.

Michael G. Rossmann

PREFACE

We were motivated to undertake this book project based on the dawning realization—already evident more than 10 years ago—that virtually all major biological activities are carried out not by individual macromolecules, but by assemblies thereof, ranging from several to very many components. These assemblies can be either transient or long-lived and they may integrate multiple reactions or interconnect signaling pathways. This is the realm of mesobiology. Among the assemblies, we distinguish an important subset with unmistakably machine-like properties, while others can be viewed as 'smart biomaterials.' It is assemblies that generate energy; replicate and repair DNA (the master-molecules of life); transcribe it into RNA; synthesize proteins and assist in their folding; dispose of proteins that are defective or no longer needed; control the cell cycle; communicate within and between cells; transport material between compartments; and perform many other vital functions. Our goal, then, became to give a systematic account of their structures and provide a mechanistic understanding in the light of the structures. It was clear at the outset that this was a moving target: the extent to which they are understood varies widely from system to system, and new assemblies are still being discovered on a regular basis. With that *caveat*, we are as up-to-date as possible and our emphasis has been on structural principles rather than an encyclopedic compilation.

This textbook is intended to be accessible to advanced undergraduates and graduate students throughout the life sciences, and should also appeal to constituencies in physics, chemistry, and engineering—especially in the machine-like aspects. Some familiarity with the basics of protein structure and cell biology and in biochemistry and genetics will be helpful, but Chapter 1 of this volume is intended to provide some background and set the stage. The subsequent chapters should then enable the reader to move further into the original literature on given topics.

While there is some emphasis on the assemblies of eukaryotic cells, those of bacteria, archaea, and even viruses, have not been neglected. In many cases, the assemblies of prokaryotes afford valuable insights as simpler and experimentally tractable versions of their eukaryotic counterparts.

We also mention, where possible, how many diseases have their roots in malfunctions of assemblies or in mutations in their components. Aberrant assembly is the basis of sickle cell disease; here a single mutation in hemoglobin causes filaments to assemble that distort the red blood cells, impeding the circulation and thus eliciting the symptoms of this disease. More recently, it has been recognized that aberrant assembly of another kind of protein filament—amyloids, which are misfolded insoluble aggregates of some cellular proteins—is deeply implicated in neurodegenerative diseases. One of these, Alzheimer's, is emerging as the major medical challenge of our time. Mechanistically in another direction, aberrant phosphorylation by mutated kinases and their impact on signaling underlie many cancers.

A wealth of relevant information—graphic and otherwise—is already available on the Internet. We see these resources as complementary to rather than as an alternative to what our textbook has to offer. It will also be useful for the reader to have some familiarity with a user-friendly molecular graphics program—for example, Pymol (www.pymol.org), Proteopedia (proteopedia. org), or Chimera (http://www.cgl.ucsf.edu/)—to display, manipulate, and otherwise familiarize themselves with the three-dimensional organization of assemblies of interest.

On didactics, we envisage that this textbook can provide a basis for many different college courses, particularly those focused on the structures and functions of biological macromolecules. Possible courses include but are not limited to: the biophysics or biochemistry of macromolecules; macromolecular structure, function, and dynamics; macromolecular assemblies and/or machines. Given the density of material covered, the text can also be tailored to suit concentrated advanced courses. A one-semester course might typically cover Chapter 1 plus three or four complementary and more focused chapters. For instance, a course on structural genetics might include Chapters 2 (chromatin), 3 (DNA replication), and 4 (DNA repair); one addressing biomolecular synthesis might be based on Chapters 5 (transcription), 6 (protein synthesis), 7 (protein degradation), and 8 (mechanisms of self-assembly); communications, motility and transport might be covered with Chapters 10, 14, and 16; the enzymology and catalytic properties of multiprotein complexes could be based on Chapters 9, 15, and 17; and the assemblies that govern some cellular behaviors might go for Chapters 12, 13, and 17. Many other combinations of chapters should also work well together.

As Lord Rutherford famously put it "*All science is either physics or stamp-collecting.*" In fact, we may now detect a convergence between physics and structural biology. The synchrotron beam lines and electron microscopes that made possible the structure determinations that are defining mesobiology had their origins in physics. Indeed as this discipline moves closer to maturity it may be seen as evolving toward a physics of complex systems. However, science—like beauty—is in the eye of the beholder. The structures of macromolecular assemblies such as the ribosome, the proteasome, and complex viruses stand out in their visual beauty and high value—just like the 'Penny Blacks' and 'Cape of Good Hope triangulars' in the eyes of philatelists.

Finally, we close by acknowledging the vital roles played by our coauthors, Louise Johnson and Richard Perham, who have not lived to see completion of our project. It has been a joy and a privilege to work with them. We owe much to Garland's indefatigable Summers Scholl, to Mary Purton for expert developmental editing, and to Nigel Orme for elegant and insightful artwork. We thank Umar Salam for his support and help during the book's development. We affectionately thank Ray Steven, Gudrun Baumeister, and Nancy Lane Perham for encouragement and support throughout this project.

RESOURCES FOR INSTRUCTORS AND STUDENTS

The teaching and learning resources for instructors and students are available online. The instructor's resources on the Garland Science website are password-protected and available only to adopting instructors. The student resources on the Garland Science website are available to everyone. We hope these resources will enhance student learning and make it easier for instructors to prepare dynamic lectures and activities for the classroom.

Instructor Resources

Instructor Resources are available on the Garland Science Instructor Resource Center, located at www.garlandscience.com/instructors. The website provides access not only to the teaching resources for this book but also to all other Garland Science textbooks. Adopting instructors can obtain access to the site from their sales representative or by emailing science@garland.com.

Art of Molecular Biology of Assemblies and Machines

The images from the book are available in two convenient formats: PowerPoint® and JPEG. They have been optimized for display on a computer. Figures are searchable by figure number, by figure name, or by keywords used in the figure legend from the book.

Student Resources

The resources for students are available on the *Molecular Biology of Assemblies and Machines* Student Website, located at www.garlandscience.com and selecting the 'Student' tab.

Flashcards

Each chapter contains flashcards, built into the student website, that allow students to review key terms from the text.

Glossary

The comprehensive glossary of key terms from the book is online and can be searched or browsed.

Online Appendices

Appendix I: *A Field Guide to Protein Folds* illustrates the structure and topology of commonly encountered protein folds.
Appendix II: *Methods of Structure Determination* describes the main methods currently used in the practice of structural biology organized by computational approaches, electron microscopy, nuclear magnetic resonance, and X-ray crystallography. Each entry lists key references for further study.

ACKNOWLEDGMENTS

The authors and publisher of *Molecular Biology of Assemblies and Machines* specially acknowledge and thank the following people for their help during the development of the book. Those who provided contributions to chapters are listed first (*), followed by readers (†), and reviewers (‡).

Chapter 1:

* Neil Rzechorzek (University of Cambridge), ‡ Robert Barber (University of Wisconsin, Parkside), Jijie Chai (Tsinghua University), James J. Champoux (University of Washington), Qintong Li (Sichuan University), Eric May (University of Connecticut), Jesus Perez-Gil (Complutense University of Madrid), Jose Rizo-Rey (University of Texas Southwestern Medical Center), Rajan Sankaranarayanan (CSIR-Centre for Cellular and Molecular Biology), Sen-Fang Sui (Tsinghua University), Kunchithapadam Swaminathan (National University of Singapore), Hongwei Wang (Tsinghua University), Zhisong Wang (National University of Singapore)

Chapter 2:

* Bradley Cairns (University of Utah School of Medicine), Jeff Hansen (Colorado State University), Karolin Luger (University of Colorado, Boulder), Patrick Varga-Weisz (The Babraham Institute), † Andrew Travers (MRC Laboratory of Molecular Biology), ‡ Pui Shing Ho (Colorado State University), Hengbin Wang (University of Alabama, Birmingham)

Chapter 3:

* Stephen D. Bell (Indiana University at Bloomington), David Jeruzalmi (City University of New York), Anthony Maxwell (John Innes Centre), ‡ Deepak Bastia (Medical University of South Carolina), Sue Cotterill (St George's College, University of London), Agnieszka Gambus (University of Birmingham), Stephen Kearsey (University of Oxford), Kenneth N. Kreuzer (Duke University), Linda Reha-Krantz (University of Alberta), Phoebe A. Rice (University of Chicago)

Chapter 4:

* Tom Ellenberger (Washington University of St Louis School of Medicine), Luca Pellegrini (University of Cambridge), Phoebe A. Rice (University of Chicago), Neil Rzechorzek (University of Cambridge), ‡ Li Fan (University of California, Riverside), Youri Pavlov (University of Nebraska Medical Center), John Tainer (The Scripps Research Institute), Hongwei Wang (Tsinghua University), Mike Williamson (University of Sheffield), Yanbin Zhang (University of Miami School of Medicine)

Chapter 5:

* Patrick Cramer (Max Planck Institute for Biophysical Chemistry), Reinhard Lührmann (Max Planck Institute for Biophysical Chemistry), Cindy L. Will (Max Planck Institute for Biophysical Chemistry), † Kiyoshi Nagai (MRC Laboratory of Molecular Biology), ‡ Robert Barber (University of Wisconsin, Parkside), Sonja Baumli (University of Oxford), Elena Conti (Max Planck Institute for Biochemistry), Zhi John Lu (Tsinghua University),

Leticia Márquez-Magaña (San Francisco State University), Daniel Reines (Emory University School of Medicine), Hongwei Wang (Tsinghua University), Mike Williamson (University of Sheffield)

Chapter 6:

* Anders Liljas (Lund University), Elizabeth Villa (University of California, San Diego), † Andreas Bracher (Max Planck Institute for Biochemistry), Venkatraman Ramakrishnan (MRC Laboratory of Molecular Biology), Daniel Wilson (Ludwig Maximilians Universität München), ‡ Scott C. Blanchard (Weill Cornell Medical College), Jane Dyson (The Scripps Research Institute), Katrina Forest (University of Wisconsin, Madison), Kurt Frederick (Ohio State University), Richard Jackson (University of Cambridge), Daniel Kaganovich (Hebrew University), Harry Noller (University of California, Santa Cruz), Peter Tompa (Flanders Institute of Biotechnology)

Chapter 7:

† Gregory Effantin (Institut de Biologie Structurale, Grenoble), ‡ Robert Barber (University of Wisconsin, Parkside), Raymond Deshaies (California Institute of Technology), Fred Gorelick (Yale University), Jörg Martin (Max Planck Institute for Developmental Biology), Mike Williamson (University of Sheffield), Zhaohui Xu (University of Michigan)

Chapter 8:

‡ Timothy S. Baker (University of California, San Diego), Jay Brown (University of Virginia), Peter Prevelige (University of Alabama, Birmingham), Peter Stockley (University of Leeds), Nicola Stonehouse (University of Leeds), Stephen Stray (University of Mississippi Medical Center), John A. Tainer (The Scripps Research Institute)

Chapter 9:

* Edward Bayer (Weizmann Institute of Science, Israel), Raphael Lamed (Tel Aviv University), Neil Rzechorzek (University of Cambridge), ‡ Robert Barber (University of Wisconsin, Parkside), Katrina Forest (University of Wisconsin, Madison), Marco W. Fraaije (University of Groningen), Perry A. Frey (University of Wisconsin, Madison), Eric May (University of Connecticut), Edith Miles (National Institutes of Health), Jesus Perez-Gil (Complutense University of Madrid)

Chapter 10:

* Martin Beck (European Molecular Biology Laboratory), Elena Conti (Max Planck Institute of Biochemistry), Oliver Daumke (Max Delbrück Center for Molecular Medicine), Jenny Hinshaw (National Institutes of Health), Tomas Kirchhausen (Harvard Medical School), Jose Rizo-Rey (University of Texas Southwestern

Medical Center), Natalie Strynadka (University of British Columbia), Thomas Spreter (University of British Columbia), Elizabeth Villa (University of California, San Diego), † Stuart T. Endo-Streeter (Duke University), Gabriel Waksman (Institute of Structural and Molecular Biology, University of London), ‡ Kenneth D. Belanger (Colgate University), Yuh Min Chook (University of Texas Southwestern Medical Center), Graeme Henderson (University of Bristol), Eamonn Kelly (University of Bristol), Robert Kretsinger (University of Virginia), Qingzhen Liu (Wuhan University), Danton H. O'Day (University of Toronto, Mississauga), Brian Reilly (Texas Tech University), Karen K. Resendes (Westminster College, New Wilmington), David Stephens (University of Bristol), Mike Williamson (University of Sheffield)

Chapter 11:
* Josephine C. Adams (University of Bristol), Jürgen Engel (University of Basel), Maria K. Hoellerer (University of Oxford), Martin E. M. Noble (Newcastle University), Gina E. Sosinsky (University of California, San Diego), William I. Weis (Stanford University School of Medicine), † Gabriel Waksman (Institute of Structural and Molecular Biology, University of London), ‡ Clive R. Bagshaw (University of Leicester), Iain Campbell (University of Oxford), Matthias M. Falk (Lehigh University), Peter J. Koch (University of Colorado, Denver)

Chapter 12:
* David Barford (MRC Laboratory of Molecular Biology), † Jane Endicott (Newcastle University), John Heath (University of Birmingham), Carl-Henrik Heldin (Uppsala University), Brenda Schulman (St Jude Children's Research Hospital, Memphis), Stephen Sprang (University of Montana), Susan Taylor (University of California, San Diego), Nick Tonks (Cold Spring Harbor Laboratory), ‡ Josephine C. Adams (University of Bristol), Ye-Guang Chen (Tsinghua University), Christopher Garcia (Stanford University School of Medicine), Graeme Henderson (University of Bristol), Eamonn Kelly (University of Bristol), Robin Ketteler (University College London), John Kuriyan (University of California, Berkeley), Qingzhen Liu (Wuhan University), David Stephens (University of Bristol), John J. G. Tesmer (University of Michigan), Qi Wang (University of California, Berkeley), Mike Williamson (University of Sheffield)

Chapter 13:
* Anthony A. Hyman (Max Planck Institute of Molecular Cell Biology and Genetics), Andrea Musacchio (Max Planck Institute of Molecular Physiology), Laurence Pelletier (Mount Sinai Hospital), Stefan Riedl (Sanford-Burnham Medical Research Institute), † Kim Nasmyth (University of Oxford), ‡ Clive R. Bagshaw (University of California, Santa Cruz), Solomon Bhupanapadu Sunkesula (Brody School of Medicine at East Carolina University), Robert Friis (University of Bern), Stephan Geley (Innsbruck Medical University), Leonard R. Johnson (University of Tennessee College of Medicine), Brett D. Keiper (Brody School of Medicine at East Carolina University), Qingzhen Liu (Wuhan University), David Morgan (University of California, San Francisco), David Stephens (University of Bristol), Zhaohui Xu (University of Michigan)

Chapter 14:
* Linda A. Amos (MRC Laboratory of Molecular Biology), Dennis Bray (University of Cambridge), Roger Craig (University of Massachusetts Medical School), Kenneth C. Holmes (Max Planck Institute for Medical Research), Dennis R. Thomas (Max Planck Institute for Biochemistry), ‡ Clive R. Bagshaw (University of California, Santa Cruz), Jonathon Howard (Yale University), Hugh Huxley (Brandeis University), Junmin Pan (Tsinghua University), Mary E. Porter (University of Minnesota, Twin Cities), Ivan Rayment (University of Wisconsin, Madison), Karl-Gösta Sundqvist (Karolinska Institute), Ron Vale (University of California, San Francisco), Zhisong Wang (National University of Singapore)

Chapter 15:
* Gary Cecchini (The Veteran's Health Research Institute), Richard J. Cogdell (University of Glasgow), Ryota Iino (National Institutes of Natural Sciences, Japan), Hiroyuki Noji (University of Tokyo), ‡ Christoph van Ballmoos (Stockholm University), Giorgio Lenaz (University of Bologna), Thomas Meier (Max Planck Institute of Biophysics), Peter Rich (University College London), Alastair Stewart (Victor Chang Cardiac Research Institute), Zhisong Wang (National University of Singapore), Joachim Weber (Texas Tech University)

Chapter 16:
* Konstantinos Beis (Imperial College London), Alex Cameron (Diamond Light Source), Thomas Sørensen (Diamond Light Source), Robert M. Stroud (University of California, San Francisco), Nigel Unwin (MRC Laboratory of Molecular Biology), † Frances Ashcroft (University of Oxford), ‡ Isaiah T. Arkin (Hebrew University), Chris Miller (Brandeis University), Mark Sansom (University of Oxford), G. Graham Shipley (Boston University), John J. G. Tesmer (University of Michigan), Gareth R. Tibbs (Columbia University), Giovanni Zifarelli (Italian National Research Council)

Chapter 17:
* Piet Gros (University of Utrecht), David A. Shore (The Scripps Research Institute), Robyn L. Stanfield (The Scripps Research Institute), Ian A. Wilson (The Scripps Research Institute), † Nick Gay (University of Cambridge), Susan Lea (University of Oxford), Anton van der Merve (University of Oxford), Bob Sim (University of Oxford), ‡ Qingzhen Liu (Wuhan University), Mary Jane Niles (University of San Francisco), Brian Reilly (Texas Tech University), David Stephens (University of Bristol), Ronald Paul Taylor (University of Virginia School of Medicine), Zhaohui Xu (University of Michigan)

Appendix I, A Field Guide to Protein Folds:
* Nick Furnham (London School of Hygiene & Tropical Medicine)

Appendix II, Methods of Structure Determination:
* Bauke W. Dijkstra (University of Groningen), Angela M. Gronenborn (University of Pittsburgh), Tatyana Polenova (University of Delaware), Andrej Sali (University of California, San Francisco)

CONTENTS

SPECIAL FEATURES

DETAILED CONTENTS

Life Processes are Driven by Macromolecular Assemblies and Machines

Life on Earth depends on two key features. One is the presence of water to enable the chemical reactions on which life depends. The second is a system for capturing information and passing it on to succeeding generations. It is now clear that both features relate to the properties of macromolecular complexes. Water as a medium is paramount for them to assume their native functional structures and to support sufficient mobility of the complexes and their substrates in a fluid phase. A subset of these complexes is responsible for information storage, expression, and replication.

For many years now, scientists have sought to understand life processes by studying individually the many and varied macromolecules that make up a cell—the smallest viable unit of life. This reductionist approach of isolating macromolecules for structural and functional investigation has, in fact, been enormously successful. The atomic resolution structures of DNA and proteins such as myoglobin, hemoglobin, and lysozyme in the 1950s and 1960s were triumphs of the then-emerging discipline of structural molecular biology and provided a solid foundation for understanding their mechanisms of action. Nowadays, the myriad solved structures in the Protein Data Bank (PDB, http://www.rcsb.org) as well as the smaller but growing Electron Microscopy Data Bank (EMDB, http://www.ebi.ac.uk/pdbe/emdb/) attest to the success of this approach and afford invaluable resources for molecular biologists, biochemists, and life scientists generally.

In spite of these achievements, awareness has grown in recent years that only rarely can a complex biological function be attributed to an individual macromolecule. Rather, most cellular functions are performed by assemblies of several different species of macromolecules (proteins, RNA, and DNA, as appropriate) acting in concert, and the ability of an assembly to function depends critically on its structural integrity. In general, the structure and function of an assembly cannot be inferred from the properties of its components alone: the scope of an assembly extends far beyond the sum of its parts. It is evident that close coordination of components can greatly improve the speed and efficiency of its operation and make possible reactions of staggering complexity (**Figure 1**). Many assemblies are endowed with explicitly machine-like properties and they are the central theme of this book. Other assemblies have primarily structural roles as 'smart' biomaterials but exhibit a similar diversity in their composition and sophistication in how they are regulated.

In a seminal review published in 1998, Bruce Alberts wrote: "we now know that nearly every major process in a cell is carried out by assemblies of 10 or more protein molecules. And, as it carries out its biological functions, each of these protein assemblies interacts with several other large complexes of proteins. Indeed, the entire cell can be viewed as a factory that contains an elaborate network of interlocking assembly lines, each of which is composed of a set of large protein machines."

Assemblies are held together by weak (noncovalent) interactions and they vary widely in stability and longevity. The methods used to break open cells to allow the isolation of their contents may break down labile assemblies, and removal from the crowded environment of a cell interior may have a similar effect. Complexes robust enough to withstand the rigors of purification can be studied structurally with the traditional methods—primarily X-ray crystallography and cryo-electron microscopy. However, labile complexes can be of no less importance, and fleeting interactions can be crucially important to the cell. Fragile complexes may be amenable to a 'divide and conquer' approach in which individual protein subunits or subcomplexes are solved at high resolution by X-ray crystallography or nuclear magnetic resonance (NMR) spectroscopy or computational methods, and then

Figure 1 Charlie Chaplin and friend ponder the workings of a machine. (From Modern Times. With permission from Roy Export S.A.S.)

fitted together to match the structure of the complex as determined by electron microscopy (EM). Until recently, EM structures have been at lower resolution but technical advances are now yielding—in favorable cases—structures of complexes and subcomplexes at resolutions comparable to those achieved in X-ray crystallography. On a broader front, labile and transient complexes may be pursued by 'hybrid' approaches that also integrate information from methods such as chemical cross-linking, mass spectrometry, hydrodynamic analysis, small-angle X-ray scattering, predictive bioinformatics, fluorescence techniques, and electron paramagnetic resonance spectroscopy.

Almost all major cellular functions are carried out by assemblies and it appears that almost the entire content of a cell (other than water and ions) consists of large structures that can be sedimented by centrifugation. However, it is difficult to obtain reliable estimates for the numbers of assemblies—stable, unstable, or transient—in any given cell, even one as small as a bacterium. Interaction maps derived from large-scale proteomics experiments provide a glimpse of the abundance of some complexes, but most remain ill defined because it is notoriously difficult to discriminate between direct physical interactions and indirect functional links, between binary and multiple interactions, and between transient ('kiss and run') and stable interactions. It is possible, even likely, that many metabolic or signaling pathways are organized in ways that are difficult to observe, by nonrandom interactions mediated by weak intermolecular forces. To elucidate them—and to fully relate findings made *in vitro* with what happens inside a cell—will require the further development and integration of methods such as electron tomography and 'super-resolution' light microscopy that allow structural studies *in situ*; that is, in unperturbed cellular environments. Thus, structural biology and cell biology need to converge to achieve a deeper understanding of the multifaceted organization of cells. That is the challenge for the emerging discipline of mesobiology: to rationalize the biological activity of macromolecular assemblies and machines.

References

Alberts B (1998) The cell as a collection of protein machines: preparing the next generation of molecular biologists. *Cell* 92:291–294.

Ball P (2003) Portrait of a molecule. *Nature* 421:421–422.

Campbell ID (2008) Structure of the living cell. *Phil Trans Roy Soc B* 363:2379–2391.

Hartwell LH, Hopfield JL, Leibler S & Murray AW (1999) From molecular to modular cell biology. *Nature* 402:C47–C52 (6761 Suppl.).

Gierasch LM & Gershenson A (2009) Post-reductionist protein science, or putting Humpty Dumpty back together again. *Nat Chem Biol* 5:774–777.

Sali A, Glaeser R, Earnest T & Baumeister W (2003) From words to literature in structural proteomics. *Nature* 422:216–225.

The Machines and Assemblies of Life

The mechanics of life—indeed life itself—depends on processes operating in that crucial level in the structural hierarchy that lies between individual macromolecules and cells or organelles: the realm of macromolecular assemblies. This book describes the assemblies that function in particular biological contexts. Overall we consider one class—macromolecular machines, viewed as assemblies that have an overtly mechanical aspect to their biological activities and, like macroscopic machines, do mechanical work—alongside other kinds of assembly with different properties. The latter assemblies nevertheless exhibit similar complexity in their structures and interactions and in the sophistication with which their synthesis is regulated. Chapter 1 sets the scene, surveying key concepts relating to the interactions that define macromolecular structures, govern their folding, and orchestrate their assembly. We start at the near-atomic scale, discussing interactions between molecules, and work up to the cellular level. En route, we dwell on the symmetries that underlie the architecture of many assemblies, and on the dynamic properties essential to their functioning: assemblies are far from static structures! Recognizing the fundamental need for communication within and between assembly systems, we examine the essentials of information storage, retrieval, and transmission, and, finally, consider assemblies as they operate in the crowded interiors of cells, and their place in the timeline of evolution.

1.1 EXPRESSION OF THE GENETIC BLUEPRINT

The 'Central Dogma' of molecular biology formulated by Francis Crick in 1958 described a one-way flow of sequence information between the three major classes of biopolymers (DNA, RNA, and protein), later paraphrased by Marshall Nirenberg as "DNA makes RNA makes protein." *Transcription* and *translation* are the processes by which cells read out and then express the information stored in genes, which are segments of chromosomal DNA. Transcription transfers the information encoded in a gene into an RNA nucleotide sequence—a messenger RNA (mRNA)—that is complementary to one of the DNA strands. In translation, the RNA nucleotide sequence is rendered into a sequence of amino acids, a polypeptide chain. So what makes DNA—whose replication is essential for the maintenance and transmission of genetic information and involves the unwinding of the double helix and the synthesis of two complementary strands (**Figure 1.1**)? In fact, DNA and RNA on their own make very little. We now know that all the steps of information storage, copying, and transfer are accomplished by elaborate macromolecular machines. These are mostly multi-subunit protein complexes but some also have RNA components. None has a DNA component.

Transcription is mediated by *RNA polymerases* and steered by other proteins called *transcription factors*. In bacteria and archaea, the processes of transcription and translation are not spatially segregated within the cell, and the mRNA is used directly. In *eukaryotes*, transcription takes place in the nucleus and the mRNA is exported into the cytoplasm for translation; first, however, the primary transcript (pre-mRNA) is modified by protein 'capping' factors and in many cases is processed further by splicing. In splicing, RNA segments derived from coding regions of DNA (*exons*) are fused together and the intervening noncoding segments (*introns*) are excised (**Figure 1.2**). Splicing is performed by RNA/protein complexes called *spliceosomes*. Combining exons in different ways (alternative splicing) expands the repertoire of mRNAs derived from a single gene. The mature mRNA then associates with a *ribosome*, a large and highly conserved RNA/protein complex, which synthesizes the corresponding polypeptide chain. For the protein to become functional, the polypeptide chain

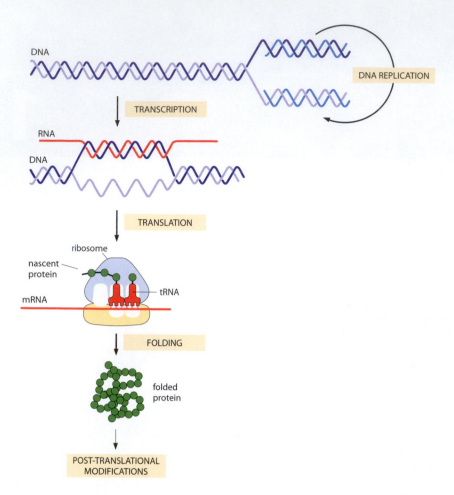

Figure 1.1 The flow of genetic information from chromosomal DNA to a natively folded protein.

must fold into a specific three-dimensional structure, a spontaneous process that in many cases is facilitated by other proteins, called *molecular chaperones.*

The flow of information is not perfect and not always in one direction

DNA replication is carried out by another protein machine, the *replisome*, which consists of *DNA polymerases* and associated proteins. The fidelity of replication is high (error rate $\sim10^{-9}$ per base pair) but it is not a perfect process: this would be thermodynamically impossible; moreover, the ability to evolve—essential to life—depends on the occasional error. The error rate for mRNA synthesis or translation is much higher, $\sim10^{-4}$, which is permissible because mRNA and proteins are not information storage molecules and, if faulty, can be replaced.

It has turned out that there are exceptions to this unidirectional flow of information from DNA to RNA to protein. One important exception is **reverse transcription**, the transfer of information from single-stranded RNA to double-stranded DNA. This is the mechanism employed by retroviruses (hence the term *retro*-virus) to replicate their RNA genomes, and by eukaryotic cells in synthesizing **telomeres** to protect the ends of their chromosomal DNA. Many other viruses also have RNA, not DNA, as their genetic material and they reproduce it by direct RNA-dependent RNA replication. The RNA polymerase machines responsible are generally encoded by viral genes, unlike many other functions for which the virus relies on resources provided by the host cell.

RNAs that act in a regulatory way are also turning out to be of major importance in eukaryotes. A multicellular organism consists of cells of many different kinds, and yet the chromosomal DNA is exactly the same in each cell. For a given organism, the differences in size, shape, and function of its various cells reflect the differing combinations of genes that are expressed in them. A simple eukaryote, the nematode worm *Caenorhabditis elegans,* and

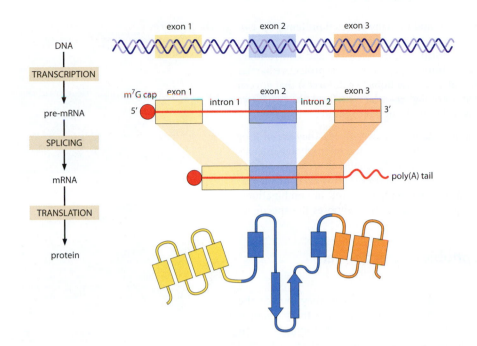

Figure 1.2 Transcription and translation of a eukaryotic gene with three exons.

a more complex one, *Homo sapiens*, both have ~20,000 genes but they differ vastly in cell number (~10^3 for the nematode, in contrast with ~10^{14} for humans) and cell differentiation. In human DNA, less than 2% of its 3 billion base pairs encode proteins, while much of the remaining noncoding DNA is transcribed into RNAs that have crucial roles in regulating gene expression and the translation of mRNAs. With the nematode, the fraction of protein-coding DNA is much higher, at ~33%. Most probably, it is differences in the complexity of this regulatory RNA world that underlie the huge differences in anatomical and physiological complexity between the worm and the human.

Nor should the effects of chemical modification of DNA and proteins be overlooked. In vertebrates, chemical modification of genomic DNA can affect gene expression without changing the underlying nucleotide sequence. DNA methylation, for example, can lock genes into a silent state. If heritable, effects of this kind are termed *epigenetic* and can have important roles in controlling expression of the genetic blueprint, such as X-chromosome inactivation in females or the preferred expression of maternally or paternally derived genes.

Most eukaryotic proteins undergo *post-translational* changes to their amino acid sequences that modify or modulate their activities. These include selective cleavage by proteases, the self-excision of peptide segments (inteins), and the covalent attachment of polypeptide 'markers,' as in *ubiquitylation*. In both prokaryotes and eukaryotes, an extensive range of post-translational modifications expands the set of amino acid side chains far beyond the 20 specified by DNA codons, without requiring genetic changes. Reversible phosphorylation is of particular importance in regulating biological activity in eukaryotes.

A glaring exception to the classical Central Dogma is given by *prions*, the agents of transmissible neuropathic diseases (spongiform encephalopothies) in mammals or of metabolic regulation in fungi. A prion is an infectious aggregate of protein in a fibrillar non-native conformation (termed *amyloid*) that serves as a template for recruiting the native protein into the aggregate, changing its conformation as it does so. In the replication of a prion, there is no participation of DNA or RNA other than in synthesis of the original proteins; here, protein makes protein. Prions differ from other amyloids in their infectivity, whereby the prion is somehow transferred into another previously uninfected cell or organism.

1.2   WEAK FORCES AND MOLECULAR INTERACTIONS

Covalent bonds are required for the synthesis of individual macromolecules; phosphodiester bonds and peptide bonds are of particular importance in nucleic acids and proteins respectively. In a few situations they also contribute to protein folding, stability, and assembly. Examples include the disulfide bridges found in many secreted proteins, the iso-peptide cross-links formed between pairs of Lys and Asn residues in certain viruses and in some bacterial *pili*, and the isopeptide bond formation between Lys and Gln residues catalyzed by transglutaminase enzymes in maintaining the epidermis, the protective barrier of skin. With these few exceptions, it is noncovalent interactions (**weak forces**) that govern protein folding and oligomerization, the assembly of higher-order structures, the structures of nucleic acids and membranes, and every aspect of molecular recognition.

The importance of these noncovalent interactions cannot be overstated. They are of four kinds: electrostatics, hydrogen bonds, van der Waals interactions, and hydrophobic interactions. All must operate in the context of water as medium. Individually, all are much weaker (~8–38 kJ/mol) than a typical covalent bond (~350–450 kJ/mol). However, in macro-molecular assemblies, their effect is amplified by their being deployed in combinations, forming complementary **interaction patches** whose organization imposes the specificity of an interaction.

All weak forces other than hydrophobic interactions are electrostatic in origin

Electrostatic interactions can be attractive or repulsive. With proteins, they include the attraction between a negatively charged residue (Asp or Glu) and a positively charged residue (Arg or Lys) in what is often called a 'salt bridge,' after the paired ions, Na^+ and Cl^-, in common salt. The electrostatic potential is a long-range effect, proportional to $1/r$ (**Figure 1.3A**). The force exerted between the charges (which is the derivative of the potential) falls off as $1/r^2$. Fully calibrated, the potential energy per mole for point charges q_1 and q_2 is given by

$$U = q_1 q_2 N_A / 4\pi\varepsilon_0\,\varepsilon r$$

where N_A is Avogadro's number (6.02×10^{23}/mol); ε_0 is the permittivity of free space, $8.85 \times 10^{12}\ C^2/(m^2\,N)$; ε is the dielectric constant of the medium; $1\ J = 1\ N \times m$; 1 elementary charge (that is, + on a proton or – on an electron) = 1.60×10^{-19} coulomb (C). In solution, this interaction is weakened by the high dielectric constant (80) of water and the solvation of charged groups.

Lys and Arg side chains are not fully interchangeable. The N^6-NH_3^+ group of Lys is essen-tially a point positive charge, whereas the delocalization of the positive charge over the guanidinium group of Arg enables it to impose directionality in its interaction with a delo-calized negative charge. One example is the interaction with the negatively charged carboxyl group (**Figure 1.4A**); another is that with a phosphate ($-OPO_3^{2-}$) group, which explains why Arg is so often found in binding sites, especially those for nucleotide substrates. The surface charges on a protein molecule generate an electrostatic field, and this influences the orientation of potential interaction partners, as exemplified by the field surrounding the enzyme superoxide dismutase (Figure 1.4B).

Another type of electrostatic interaction is the **hydrogen bond** (H-bond), formed when the hydrogen atom attached to an electronegative atom approaches another electronegative atom and there is a partial sharing of the proton. In crystal structures of proteins, where hydrogen atom positions are not usually determined, an H-bond is assumed if two electro-negative atoms (the donor D and acceptor A) are closer than 3.3 Å and the enclosed angle is >90° (**Figure 1.5A**). In proteins, most H-bonds are of the type N–H⋯O or O–H⋯O, and

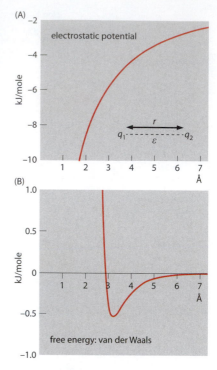

(A) electrostatic potential

(B) free energy: van der Waals

Figure 1.3 Distance dependence of electrostatic and van der Waals interactions. (A) Electrostatics. The potential energy per mole for point charges q_1 and q_2, distance r apart. This curve shows the field associated with an attractive interaction between oppositely charged residues in water. At $r = 3$ Å, $U = -5.7$ kJ/mole. In a protein interior, where $\varepsilon \approx 4$, $\Delta U = -115$ kJ/mol for this interaction. (B) A van der Waals energy curve. Energy is plotted as a function of separation of atoms (r), showing the net sum between repulsive energy ($1/r^{12}$ dependence) and attractive energy ($1/r^6$ dependence) with a minimum at 3.2 Å, corresponding to an energy $E_m = -0.54$ kJ/mol. (Adapted from G.E. Schulz and R.H. Schirmer, Principles of Protein Structure. New York: Springer-Verlag, 1979.)

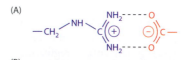

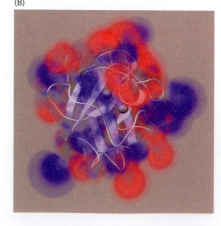

Figure 1.4 Directionality in electrostatic interactions. (A) Bidentate interaction between delocalized charges over the guanidinium group of arginine (blue) and a carboxyl group (red) restricts the relative orientation of the two charges. (B) The enzyme Cu–Zn superoxide dismutase (SOD) functions as a homodimer which, through its electrostatic field, attracts the negatively charged superoxide, $O_2^{-\cdot}$, a reactive oxygen species (see Section 15.2), to the positively charged metal ions in its catalytic sites, where it is converted to the less toxic hydrogen peroxide. A monomer of bovine SOD is shown, with its backbone depicted as a white ribbon diagram and the Cu^{2+} and Zn^{2+} in the catalytic site as spheres. The surrounding electrostatic field of the protein subunit is displayed as a volume-rendered cloud with positive potential in blue and negative potential in red. (PDB 2SOD) (B, courtesy of Art Olson.)

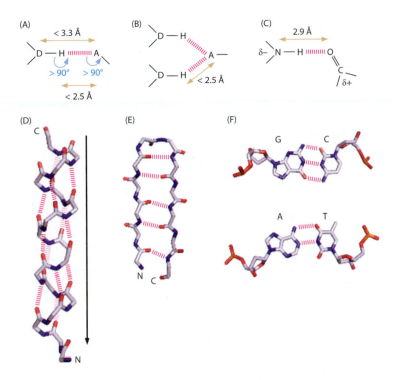

Figure 1.5 H-bond interactions. The bonds are marked by ladders of red dashes. In (A), D is the donor atom and A is the acceptor atom. The standard geometrical criteria for identifying an H-bond in a protein are indicated. (B) Bifurcated H-bond with two sharing donors. (C) The H-bond between peptide groups, showing the partial electric charges ($\delta-$ and $\delta+$) on the atoms. (D) The main chain conformation of an α helix in which the peptide NH of residue $i + 4$ forms an H-bond with the C=O of residue i. Alignment of the peptide dipoles generates an overall dipole for the helix (arrow) with a small positive charge at the N-terminal end. Side chains are not shown. (E) The main chain conformation of a two-stranded anti-parallel β sheet (that is, a β hairpin) with H-bonds between the strands. (F) Watson–Crick H-bonds between G base-paired with C, and A base-paired with T.

less often N–H···N (Figure 1.5C), and almost all (~98%) potential donors and acceptors are found with partners. Sometimes one donor has two acceptors or *vice versa*, generating a bifurcated H-bond (Figure 1.5B). H-bonds have energies of –12 to –38 kJ/mol, depending on the local environment, and they are directional in character. The bond energy is highest when the two atoms and the proton lie on a straight line; a 30° departure from linearity reduces it by about 30%. Two crucial roles (among many) of H-bonding lie in the interactions between main-chain atoms that stabilize the secondary structure of proteins, namely the α helices, β sheets, and turns (Figure 1.5D–E), and in the base pairing of nucleotides in duplex DNA (Figure 1.5F).

Charged (and uncharged) groups also engage in other electrostatic-based interactions. Even nonpolar groups (meaning, here, a group lacking a permanent dipole) develop transient dipoles that interact with similarly fluctuating dipoles in neighboring groups. These interactions, known collectively as **van der Waals interactions**, are short-range and represent a balance between mutual repulsion of the respective electron clouds as atoms come into close contact, for which the energy has a $1/r^{12}$ dependence, and attraction from dipole/dipole interactions whose energy falls off as $1/r^6$. The equilibrium distance is ~3.2 Å (see Figure 1.3B). The net forces are individually weak: for example, a methylene group in a crystalline hydrocarbon has a van der Waals energy of ~–8.4 kJ/mol. However, their influence is enhanced by multiplicity, as in the core of a folded protein, the central region of a *lipid bilayer*, or the base stacking that stabilizes duplex DNA and RNA.

Hydrophobic interactions drive the folding and assembly of macromolecules

The H_2O molecule is unusual. It has a permanent dipole moment and can participate in four H-bonds, two as donor and two as acceptor. Thus water molecules have fluctuating structures based on a tetrahedral H-bonded arrangement: typically, each H_2O participates in three or four H-bonds at any one time. The dual donor/acceptor property means that water molecules often engage in H-bonded networks linking different parts of a macromolecule (**Figure 1.6**). In bulk water, the free energy change when an H-bond formed is small because it is accompanied by rupture of the H-bonds that each previously made with other water molecules. Some macromolecules have 'buried' water molecules that form an integral part of their structure separate from the bulk water. A buried water/protein H-bond contributes ~–2.5 kJ/mol to the stability of a protein.

(A)

(B)

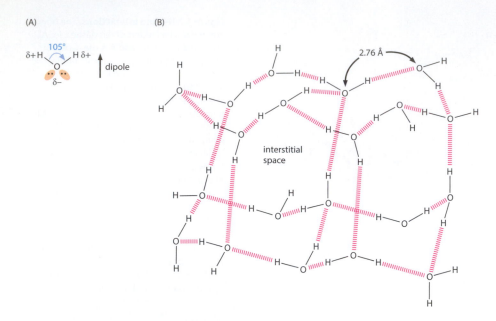

Figure 1.6 Water as a polar solvent.
(A) The dipole of a water molecule. The oxygen *sp*³ orbitals are shown as yellow ellipses containing the lone pair of electrons that act as acceptors in H-bonds. (B) A possible arrangement of H-bonds (red dash ladders) for a network of water molecules in solution.

However, electrostatic-based forces do not explain why nonpolar amino acids tend to congregate in the interior of proteins, leaving most charged residues on the surface and exposed to the solvent. Buried charged residues, when they occur, are almost always neutralized by pairing with complementary charges. Uncharged polar groups are found both on the surface and in the interior (**Figure 1.7**); in the latter, they usually engage in H-bonding. Thus, the interior of a folded protein is a hydrophobic milieu where nonpolar groups engage in van der Waals interactions with other nonpolar groups. This partitioning is due to **hydrophobic effects.**

The presence of a nonpolar group in water is energetically unfavorable on account of its effect on the local organization of water molecules, which no longer assume fluctuating tetrahedral structures but form a clathrate (cage-like) structure around the nonpolar group. For example, it takes a minimum of 20 water molecules to make a clathrate around CH_4, and their partial ordering reduces the overall number of degrees of freedom. The corresponding decrease in **entropy** (an energy-related measure of disorder) is relieved if the nonpolar group associates with another nonpolar group, releasing water molecules from the interface. This more than compensates for the loss of configurational entropy from the association of the two nonpolar molecules.

Hydrophobicity may be measured in terms of the free energy change when a nonpolar group is transferred from water into a nonpolar solvent. Some examples are given in **Table 1.1**. Hydrophobic interactions are nondirectional and they are relatively long-range, making them early contributors to protein folding and assembly. The lack of directionality means that a single hydrophobic group cannot give rise to specificity, but the way in which hydrophobic residues are distributed over an *interaction patch* does confer this property (see Figure 1.7).

(A)

(B)

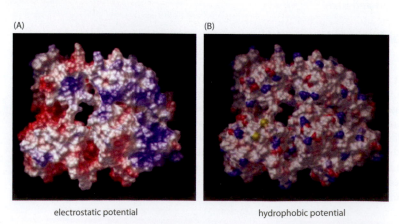

electrostatic potential   hydrophobic potential

Figure 1.7 Mapping distributions of charge and hydrophobicity over a protein surface. The images were rendered with PMV (Python Molecular Viewer). (A) The electrostatic potential (positive, blue; negative, red) of a tyrosine kinase was computed with APBS (Adaptive Poisson–Bolzmann Solver). (PDB 1GCF) (B) Hydrophobic potential at the molecular surface shown in a coloring scheme in which all carbon atoms (hydrophobic) are white, while polar atoms are desaturated red and blue, and charged atoms are saturated red and blue. This gives a mixed-color solvent-excluded surface with charged and polar atoms distributed among the white (hydrophobic) carbon atoms. The small yellow patches are sulfur atoms in Cys residues. (From S.I. O'Donoghue et al., *Nat. Methods Suppl.* 7:S42–S55, 2010. With permission from Macmillan Publishers Ltd.)

Table 1.1 Free energy changes on the transfer of hydrophobic groups from water to n-octanol.	
Group (relative to H atom)	**Free energy change (kJ/mol)**
-CH_3	−2.9
-CH_2CH_3	−5.7
-$(CH_2)_2CH_3$	−8.6
-CH_2-phenyl	−15.0
-OH	6.6
(Adapted from A. Fersht, Structure and Mechanism in Protein Science. New York: WH Freeman, 1999.)	

Intrinsic membrane proteins represent a special case in that they have transmembrane domains consisting mainly of amino acids with hydrophobic side chains that are embedded in the hydrophobic phase of a lipid bilayer. The main chain atoms of these segments of polypeptide chain compensate for their polar nature by H-bonding, most commonly in the form of α helices and occasionally in β sheets.

The energy balance in folding and assembly has both enthalpic and entropic contributions

The energetic exchanges in transitions involving biological (and other) macromolecules are related by the equation

$$\Delta G = \Delta H - T\Delta S$$

where **ΔG** is the change in **Gibbs free energy**, ΔH is the change in **enthalpy** (the heat given up or taken in by the system), ΔS is the change in entropy, and T is the absolute temperature (in kelvins). In general, if a system is not already at a free energy minimum, it will spontaneously shift toward a lower (more negative) free energy state. Note, however, that ΔG tells nothing about kinetics, in other words the rate at which a new equilibrium will be reached. In folding and assembly, the enthalpic term (ΔH) with contributions from charge/charge interactions, H-bonds, and van der Waals interactions is generally negative (favorable). The entropic term ($-T\Delta S$) may be positive or negative, depending on whether the loss of configurational entropy on entering a more ordered state exceeds the gain in entropy from liberating water molecules. When ΔH is close to zero, a positive entropy contribution (making $-T\Delta S$ negative) is decisive. This is the basis of entropy-driven processes, which include the assembly of *microtubules* (Section 14.5), *tobacco mosaic virus* (Section 8.2), or *collagen fibrils* (Section 11.7). Ordering processes of this kind have the counter-intuitive property of being promoted by a rise in temperature, which increases the value of $-T\Delta S$: (normally, raising the temperature promotes kinetic disorder, leading to dissociation).

Size and topography matter for interaction patches

The concept of **solvent-accessible surface** was introduced by B-K. Lee and Frederic Richards as a way of quantifying how much surface is buried when proteins associate. It defines the accessible surface area (ASA) over which a protein and solvent make contact, as marked out by computationally rolling a spherical solvent molecule over the van der Waals surface of the protein (**Figure 1.8A**). The probe radius is usually set to 1.4 Å for water. Later, the molecular surface area (MSA) of a protein was defined as the area covered by an inward-facing probe rolled over the van der Waals surface (Figure 1.8B). ASA areas tend to be somewhat larger (**Table 1.2**). The buried surface area (BSA) when macromolecules X and Y form a complex is given by

$$BSA = ASA(X) + ASA(Y) - ASA(XY)$$

A similar equation applies when MSAs are used instead. The BSA is roughly twice the area of an interaction patch at the X/Y interface because approximately the same amount of exposed surface is lost from both X and Y (it is exactly the same amount if XY is a symmetric

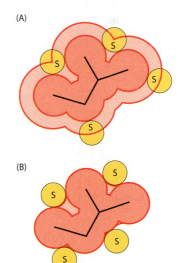

(A)

(B)

Figure 1.8 Accessible surface area (ASA) and molecular surface area (MSA). In both cases, a solvent probe, S, is rolled computationally over the van der Waals surface of the molecule. In (A), its center defines the ASA. In (B), its point of contact with the van der Waals surface defines the MSA. (Adapted from I. Tunon et al., *Protein Eng.* 5:715-716, 1992. With permission from Oxford University Press.)

Table 1.2 Solvent-accessible surface areas (ASAs) and molecular surface areas (MSAs) for the Cdk2/cyclin A heterodimer.

	Cdk2 (X)	Cyclin A (Y)	Cdk2/cyclin A (XY)	BSA (X + Y − XY)
ASA ($Å^2$)	14,396	12,300	23,346	3350
MSA ($Å^2$)	11,733	10,308	19,158	2884

ASAs were calculated with Areaimol from the CCP4 suite of programs. MSAs were calculated with Aesop. (Courtesy of Martin Noble.)

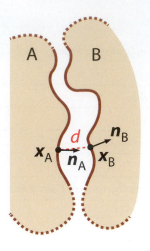

Figure 1.9 Shape complementarity of paired molecular surfaces. The measure S_c is defined in the text. Bold symbols denote vectors. The solid lines mark the regions on molecules A and B that form the interface under consideration. For point x_A, x_B is the nearest point on molecule B. The scalar product $n_A.n_B$ is the cosine of the angle α between the two unit vectors n_A and n_B, and cannot exceed 1. Similarly, the exponential decay according to the distance d between x_A and x_B cannot exceed 1 and only achieves this value if the two surfaces make direct contact ($d = 0$). (Adapted from M.C. Lawrence, CCP4 Newsletter 39.)

dimer). In another convention used by some investigators (but not here), the areas defined above are halved and the redefined BSAs are similar in size to the interaction patch.

Atoms in the core regions of proteins are inaccessible to solvent. Crystal structures reveal that they are packed as tightly as atoms in an organic solid, such as a small-molecule crystal. Amino acid residues in the core tend to pack in a precise jigsaw fit, although some proteins contain cavities large enough to accommodate one or more water molecules (see above). Surface residues are less tightly packed and some proteins have surface-exposed regions that are flexible or even completely disordered.

The surface area buried between protein subunits can be used as a guideline to discriminate between crystal packing and a physiologically meaningful interaction. Crystal contacts are usually smaller and less complementary. An analysis of crystalline proteins known to form monomers or dimers in solution gave a success rate of ~85% when 1712 $Å^2$ was taken as a cut-off value for ΔASA. The Protein Interfaces Surfaces and Assemblies (PISA) program predicts quaternary structures on the basis of calculated interactions and thermodynamic parameters and has an estimated success rate of ~90%. Paired interaction patches also have complementary topographies. A convenient measure of **complementarity**, S_c, averages the following quantity over all points on both surfaces:

$$S_c = \exp(-wd^2)\cos\alpha$$

where d is the distance between x_A, a point on surface A, and x_B the nearest point on surface B; α is the angle between the unit vector normal to surface A at x_A and the unit vector normal to surface B at x_B, and w is a free parameter (**Figure 1.9**). The higher the value of S_c, the better the fit. A perfect fit has the maximum value $S_c = 1$. Values of 0.65–0.75 are typical for genuine interactions. Favorable shape complementarity is helpful when assessing interfaces, real and potential, but although it is conducive to the formation of hydrophobic and van der Waals interactions, it does not necessarily point to a physical interaction.

As noted above, complementary interaction patches are quite large. Although they include some hydrophobic sites, they also involve charge pairing and H-bonds. An interaction patch of 1000 $Å^2$ or so cannot be entirely hydrophobic, because the molecule bearing it would be liable to nonspecific aggregation and therefore insoluble. Two examples of complementary pairs of interaction patches are shown in **Figure 1.10**. However, interactions within an assembly are more complex in that most subunits have binding interfaces with more than one other subunit: for example, subunits in viral capsids make contact with between four and seven neighbors (see Chapter 8).

A certain minimum strength of interaction is required for specificity

Interactions are gauged in terms of their **affinity** or its inverse, the **dissociation constant** (K_d), defined by

$$K_d = [A][B]/[AB]$$

where [A] is the molar concentration of component A, and so on. Affinities are related to free energies of association by

$$K_d = \exp(-\Delta G/RT)$$

or, rearranging,

$$\Delta G = -RT\ln K_d$$

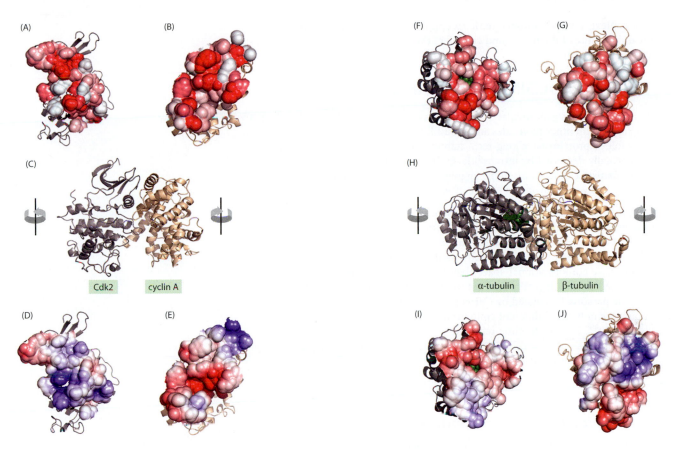

(A)   (B)   (F)   (G)

(C)   (H)

Cdk2   cyclin A   α-tubulin   β-tubulin

(D)   (E)   (I)   (J)

R is the gas constant. If solutions of [A] and [B], both at 2 μM (a typical concentration for *in vitro* experiments with 100 kDa proteins), are mixed, and they have a K_d of 1 μM, then at equilibrium there are equal concentrations of A and B, that is 1 μM, free and as AB. This does not state a strong preference for complex formation. If, in contrast, the K_d is 1 nM, then about 98% of both A and B are as AB; in other words, association is strongly favored. Given the large areas and chemical complexity of their surfaces, there is an enormous number of ways in which two macromolecules can come together, and many of them represent attractive interactions. Whether or not an association is meaningful depends on the K_d. A micromolar K_d is considered borderline, but a nanomolar K_d is taken to indicate a stable complex. In antigen–antibody binding, for example, the K_d is typically in the nanomolar to picomolar range (Section 17.4).

Cooperativity enhances stability in multi-subunit complexes

In assemblies, most subunits interact with more than one neighbor. The net ΔASA for one subunit, considered as an interaction between it and the rest of the complex, can be taken as the sum of all the ΔASAs involved. For this reason, complexes tend to be more stable than the individually weak interactions would suggest. The combined effect of several interactions may be described in terms of their **avidity**. For simplicity, let us consider a two-component system (C, D) in which there is more than one interface between C and D, each with its own $K_d^{(i)}$, $i = 1, 2, \ldots$. In principle, their combined effect is given by the product of these dissociation constants:

$$K_d^{tot} = \Pi\, K_d^{(i)}$$

However, it can also be considered in terms of a single 'effective' interaction whose dissociation constant, K_d^{eff}, represents the avidity of the reaction. Engagement at one interface may affect the readiness of binding to take place at another and this gives rise to the concept of **cooperativity**, described by parameter α:

$$\alpha = K_d^{tot}/K_d^{eff}$$

If $\alpha = 1$, the interaction is non-cooperative; if $\alpha > 1$, there is positive cooperativity (synergy); and if $\alpha < 1$, there is negative cooperativity (interference). Cooperativity is also important

Figure 1.10 Interaction patches at inter-subunit interfaces of two heterodimeric proteins. (A–E) Cdk2/cyclin A. (PDB 1FIN) A side-view ribbon diagram is shown in (C). (A, B) and (D, E) show the paired interaction patches as surface renderings color-coded to convey different properties. Their areas are given in Table 1.2. To present the interaction patches *en face*, the subunits were rotated 90° as shown. Other visible parts are shown as ribbons. (A, B) are coded for hydrophobicity: stronger red means more hydrophobic. Complementarity is expressed in terms of the extent to which the red regions observe mirror symmetry across the vertical axis. (D, E) are coded for electrostatic potential: blue, positive; red, negative; white, uncharged. Complementarity is expressed in terms of red regions being reflected in blue regions across the vertical axis. Greater, although not total, complementarity is observed in terms of hydrophobicity. As the buried surface areas are much larger than the minimum required for a meaningful interaction (see main text), such departures from complementarity can be accommodated. (F–J) α–β Tubulin. (PDB 1JFF) The presentation is as in (A–E). The buried surface area at the interface is 3780 Å², and about half that for each interaction patch. The bound nucleotide (GTP) is green. The hydrophobicity complementarity is greater than in (A–E) but still not perfect.

in a different context in which an oligomeric receptor molecule has multiple binding sites for a ligand and binding of the first ligand can affect the affinity for a second ligand and so on, as discussed in Section 1.7.

1.3   PROTEIN FOLDING AND STABILITY

Most proteins must adopt a particular three-dimensional structure that allows them to interact specifically with other molecules and to fulfill their biological functions. Moreover, only correctly folded proteins have long-term stability in cellular environments. Misfolded proteins are normally degraded by intracellular proteases, but those that escape this quality-control mechanism can form insoluble aggregates, for example the fibrillar assemblies (*amyloids*) associated with neurodegenerative diseases (Section 6.5).

How a given protein reaches its folded state after synthesis on the ribosome is an important issue often referred to as the 'protein folding problem.' Christian Anfinsen's classic experiment (**Figure 1.11**) showed that the linear sequence of amino acids in a polypeptide chain contains all the information needed to specify its three-dimensional structure. Moreover, despite the immense conformational space that a polypeptide chain can sample, it must reach its native fold on a biologically relevant timescale. The magnitude of this problem is illustrated by the paradox first stated by Cyrus Levinthal: if, with a protein of 100 aa, each residue can assume only three different configurations, the number of possible conformations would be ~10^{49}. Even if the time required for switching from one conformation to another is very small, say 10^{-11} seconds, then a random search of conformational space would take 10^{29} years (the estimated age of the universe is only ~14×10^9 years)! This clearly rules out any random-search process. In reality, proteins fold on a timescale of seconds or less.

Protein folding follows pathways populated with intermediates

A common feature of most hypotheses aimed at resolving the Levinthal paradox is some sort of folding pathway, with the formation of structured elements occurring early in the process. These elements form in parallel, not sequentially, and subsequent sampling of conformational space involves fewer, larger, components; both considerations accelerate the folding process. The 'hydrophobic collapse' model assumes that folding starts with a rapid condensation of the polypeptide chain into a dynamic, flexible form with the beginnings of

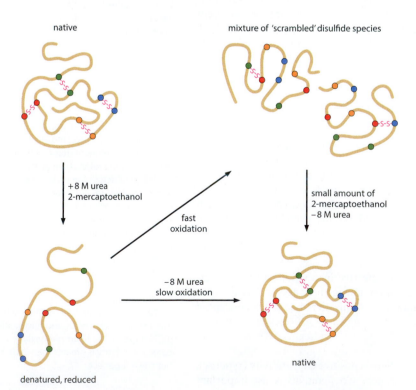

Figure 1.11 The Anfinsen experiment demonstrating protein unfolding and spontaneous refolding. Ribonuclease, an enzyme of 124 aa with four disulfide bridges, is exposed to 8 M urea to denature the protein, and 2-mercaptoethanol to cleave the disulfide bridges. This renders the enzyme inactive. When the cysteines are allowed to reoxidize rapidly in the absence of 2-mercaptoethanol under denaturing conditions, the protein adopts a mixture of conformations with randomly formed disulfide bridges and remains inactive. However, if the urea is removed slowly, disulfide exchange occurs and the enzyme refolds into a conformationally homogeneous fully active species. The four disulfide bridges are correctly formed out of the 105 ($7 \times 5 \times 3 \times 1$) possible pairings of cysteines.

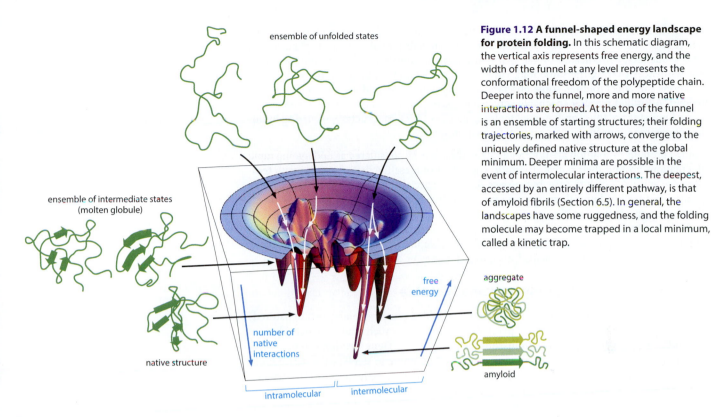

ensemble of unfolded states

ensemble of intermediate states
(molten globule)

native structure

free
energy

number of
native
interactions

intramolecular   intermolecular

aggregate

amyloid

Figure 1.12 A funnel-shaped energy landscape for protein folding. In this schematic diagram, the vertical axis represents free energy, and the width of the funnel at any level represents the conformational freedom of the polypeptide chain. Deeper into the funnel, more and more native interactions are formed. At the top of the funnel is an ensemble of starting structures; their folding trajectories, marked with arrows, converge to the uniquely defined native structure at the global minimum. Deeper minima are possible in the event of intermolecular interactions. The deepest, accessed by an entirely different pathway, is that of amyloid fibrils (Section 6.5). In general, the landscapes have some ruggedness, and the folding molecule may become trapped in a local minimum, called a kinetic trap.

a hydrophobic core, a so-called **molten globule**. Whether hydrophobic collapse or secondary structure formation occurs first is still an open question.

There is now ample evidence that folding proceeds in stages and that there is no single, directed pathway. The process is determined by the protein's **free energy landscape**, which can be represented graphically as a funnel (**Figure 1.12**). In this view of folding, the non-native protein exists as a large ensemble of different structures. The molecules move spontaneously down free energy gradients, passing through multiple partly folded states until they converge on a unique global minimum. On the way, more and more native interactions are established. In an alternative folding pathway, fibrillar aggregates called *amyloid* may be formed. Most amyloids that have been studied are the result of misfolding to a state that is a lower free energy minimum than the native fold. However, there are some proteins for which amyloid is their native state (see Section 6.5).

For proteins that fold via a two-state transition (denatured/native), the free energy funnel has no significant valleys or barriers (that is, no local minima). Such a scenario is rare and largely confined to small proteins (see below). More commonly, the energy landscape is rugged and the folding protein must navigate through local minima. This ruggedness reflects the temporary formation of folding intermediates that depend on multiple weak interactions not necessarily found in the end state (the native fold). These local minima represent kinetic traps that slow the folding process. If a local minimum is deep, the corresponding intermediate is longer-lived and the likelihood of off-pathway events such as aggregation increases. A full characterization of all the non-native species (unfolded, transitional states, and partly folded intermediates) poses a formidable challenge that has yet to be solved.

Larger proteins (longer than ~200 aa) do not usually behave as a single folding unit; many consist of two or more regions whose folds are recognized as they appear in several, even many, other proteins. These regions, called **domains**, vary in size between ~50 and ~250 aa. Some are capable of fully independent folding if separately expressed, or they may be excised proteolytically in folded form from the parent protein: such domains make few, if any, contacts with the other part(s) of the protein. Other domains make many stabilizing interactions with other parts in their parent protein and are essential contributors to its hydrophobic core. They would leave unacceptably large amounts of hydrophobic surface exposed in the absence of the rest of the protein: in this sense, they fold semi-autonomously. Many domains are responsible for particular functions, such as the **Rossmann fold** that binds NAD$^+$/NADP in oxidoreductases or the SH2 domain that binds phosphorylated Ser residues in other proteins. Domains may tolerate substantial changes to their amino acid sequences, as indicated by the many instances of domains with substantially different sequences sharing the same fold. However, single amino acid substitutions in crucial locations can affect folding, some observed as temperature- or cold-sensitive mutants (see below). A limited number (probably somewhat above 1000) of domain folds is thought to exist. Domains are thus fundamental building blocks of protein structure, recurring in different combinations in the vast array of known proteins.

Folded proteins in solution are not rigid; rather, they may be considered as undergoing rapid fluctuations around their native structure. The extent of these fluctuations varies with temperature and from case to case, but surface regions, especially loops and projections, are in general less well ordered than core regions. Multidomain proteins sometimes have disordered segments that act as hinges or extended flexible linkers (see, for example, Section 9.4), and some proteins have sizable surface-exposed regions (in extreme cases, the entire molecule) that remain unfolded in the absence of a suitable interaction partner. These regions, which are said to be *natively unfolded*, are covered in greater detail in Sections 1.5 and 6.4.

Protein structures are only marginally stable

It has long been known that the native structures of folded proteins can be lost (denatured) by heating or exposure to extremes of pH or denaturing agents such as urea or guanidinium chloride, or detergents such as sodium dodecylsulfate (SDS). **Denaturation** can be understood as a disruption of the multiple weak forces that combine to stabilize the native folds. Denatured proteins are highly protonated or deprotonated (at extremes of pH) or coated with denaturants, and thus tend to have extended shapes. Indeed, the extended form of SDS-denatured proteins saturated with bound negatively charged dodecylsulfate ions is the basis for their separation by size in SDS polyacrylamide-gel electrophoresis (SDS-PAGE). Unfolding can be partial and, in some cases, reversible on removal of the denaturing agent, but it generally involves a cooperative loss of structure as the weak forces are collectively disrupted. The denatured protein is best regarded as an ensemble of unfolded structures, for which the term **random coil** is sometimes used. However, it may also refer to specific conformations that lack regular secondary structure elements: that is, α helices and β strands.

Denaturation exposes hydrophobic groups that are normally confined within the protein core, and these will tend to interact, causing nonspecific aggregation. Binding of chemical denaturants such as urea keeps the denatured protein in solution, but thermal denaturation or exposure to low pH or organic solvents leaves the exposed hydrophobic groups free to interact and this can lead to precipitation.

The free energy of stabilization (ΔG) of folded proteins is small, usually in the range of −21 to −42 kJ/mol at 25°C. This is distributed over many weak interactions but is energetically equivalent to just two or three H-bonds. It has been argued that this property of marginal stability is important in facilitating functionally relevant conformational changes and unfolding prior to degradation and recycling. Others have argued that, provided that a protein achieves a certain threshold of stability, further increases are evolutionarily neutral and are not selected for.

Protein stability correlates with size and other factors such as covalent cross-links

Data from studies of the reversible folding of 63 globular proteins have shown that thermal denaturation depends more on the length of the polypeptide chain than on its amino acid composition or its secondary or tertiary structure. The principal contributions to stability come from interactions in the hydrophobic core, whose volume increases with the cube of the radius (r^3), whereas the surface area increases only with the square (r^2). On this basis, it is possible to assess the entire proteomes of organisms whose genomes have been sequenced and arrive at an estimate of the average stability. For *E. coli* proteins at 37°C, the average ΔG of folding comes to ~30 kJ/mol; for *S. cerevisiae* it is ~37 kJ/mol, and for *C. elegans* it is ~34 kJ/mol.

It is widely accepted that a minimum length of ~50 aa is required to generate a sufficient hydrophobic core to sustain a folded structure, although there are a few exceptions.

The peripheral subunit-binding domain (PSBD, 33 aa in its structured region) of 2-oxo acid dehydrogenase complexes is one of the smallest proteins to have a stable fold (see Figure 9.4B). Its two short parallel α helices, separated by a short 3_{10} helix and a loop, enclose a close-packed hydrophobic core; folding takes place on an ultrafast (microseconds) timescale with a ΔG of ~–8 to –17 kJ/mol at 25°C in an apparent two-state transition. Its single robust folding pathway suggests the presence of a single nucleation-competent motif (or *foldon*).

Another factor that affects stability is covalent cross-linking (see Section 1.2), notably disulfide bridges, which are largely confined to extracellular (secreted) proteins such as the protease chymotrypsin. The oxygen-rich extracellular environment is much harsher than the reducing environment found inside cells. A disulfide bridge restricts the conformational space open to an otherwise unfolded polypeptide chain and thus stabilizes folds by limiting the increase in entropy from disorder. Experiments indicate that a disulfide bridge can contribute up to ~350 kJ/mol to the free energy of stabilization. For this reason, disulfide-bridged proteases such as subtilisin have found wide application in washing powders, for example, and the biotechnology industry seeks to render many proteins less prone to denaturation by engineering new disulfide bridges into them.

Many cellular proteins denature collectively under thermal stress

Based on the spread of protein sizes in its proteome, the range of protein stabilities in any one organism is predicted to be broad (**Figure 1.13**). About 650 (15%) of the 4300 proteins of *E. coli* have predicted ΔG values of denaturation of <17 kJ/mol, so that they are only marginally stable at the ambient temperature of 37°C. Temperatures that are substantially higher should lead to catastrophic levels of denaturation, with 1% of *E. coli* proteins predicted to denature at 47°C, rising to 50% at 54°C. These temperatures coincide with the range beyond which *E. coli* cells cannot survive, suggesting that collective denaturation of the less stable proteins is responsible for halting cell function. Other calculations suggest that similar behavior is widespread.

In the account given above, no attention was paid to the fact that many proteins have multiple domains. This should result in lower stabilities as estimated on the basis of domain length rather than total polypeptide chain length. Nor does it recognize that many proteins are stabilized by incorporation into multi-subunit assemblies. These qualifications notwithstanding, changes of just a few degrees can shift thermal equilibria substantially: for example, it is estimated that raising the temperature from 37°C to 41°C destabilizes the average protein by almost 20%. To combat thermal perturbation, cells mobilize their *heat-shock response* systems, up-regulating chaperone proteins that promote the refolding of thermally perturbed proteins (Section 6.3).

In temperature-sensitive mutants, minor changes to the primary structure of a protein that folds correctly and is biologically active at one temperature—say, 30°C—render it inactive and presumably denatured at a slightly higher temperature—say, 37°C. Such mutations have proved invaluable in unraveling many biochemical and assembly pathways. Conversely, cold-sensitive mutants function at the higher but not at the lower temperature, presumably because the mutated protein remains stuck in a kinetic trap until released by additional thermal (enthalpic) energy.

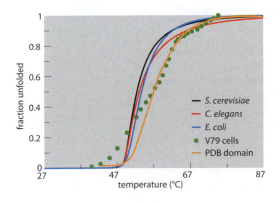

Figure 1.13 Fractions of proteins in the proteomes of *E. coli*, *S. cerevisiae*, and *C. elegans* unfolded as a function of temperature. The blue, black, and red curves are based on the predicted distributions of domain lengths in the respective proteomes of a bacterium, a fungus, and a nematode worm. The fraction of denatured proteins for mammalian V79 cells (green points) was measured by differential scanning calorimetry. The yellow curve is the prediction based on domain-length distribution obtained from structures in the Protein Data Bank. (Adapted from K. Ghosh and K. Dill, *Biophys. J.* 99:3996–4002, 2010. With permission from Elsevier.)

Proteins from thermophilic organisms are not very different from mesophilic homologs

Unlike organisms that live at temperatures between 0°C and 40°C (mesophiles), thermophiles flourish only at higher temperatures (up to 80°C) and hyperthermophiles thrive in even hotter environments (80–120°C), such as hot springs or deep sea vents. Many thermophiles and hyperthermophiles are archaea (Section 1.9), but some bacteria have adapted to life at high temperatures. Nevertheless, regardless of preferred growth temperature, these organisms all have similar proteomes, but proteins from thermophiles are substantially more heat-resistant than their mesophile counterparts. Indeed, the ability of the *Taq* polymerase from the bacterium *Thermus aquaticus* to function at temperatures above those that denature mesophile polymerases and cause mesophile DNAs to melt (strand separation) is the basis of the polymerase chain reaction (PCR).

The structures of proteins from thermophiles, esteemed by crystallographers because of their amenability to crystallization, are generally similar to homologs from mesophiles. The differences reside mostly in their having tighter surface loops and more stabilizing electrostatic interactions (notably, salt bridges) at the protein surface. These changes do not have to increase the free energy of folding very much to be effective. If a mesophile protein at, say 100°C (373 K), denatures in 2 seconds and the thermophile homolog survives 1000 times as long (~30 min), that is potentially long enough for a typical bacterial cell division. The change in free energy of folding to achieve this extra stability, $\Delta\Delta G \approx RT\ln 1000$, or $2 \times 373 \times 2.3 \times 3 = 21$ kJ/mol, is roughly equal to a couple of H-bonds. So it is perhaps not surprising that it is hard to recognize the subtle underlying changes in weak forces distributed over the entire protein molecule.

1.4 SELF-ASSEMBLY AND SYMMETRY

Just as the primary structures of protein subunits specify their secondary structures (α helices and β sheets) and tertiary structures (three-dimensional folds), they also underlie the quaternary structures (oligomeric and polymeric) of the machines and assemblies that permeate biology. In many cases, just one kind of protein is involved (**homomeric assembly**), but in others several different polypeptide chains co-assemble (**heteromeric assembly**), and other components (such as nucleic acids or lipids) may also be incorporated. The exquisite selectivity exercised in the choice of components to be incorporated into any given complex is a striking feature of all assembly processes. Nonetheless, it is possible to codify the various types of quaternary structure that are encountered and to see how some might have been evolutionarily favored.

Many assembly processes can be recreated from purified components—for example, in the reversible denaturation of oligomeric enzymes, whereby refolding of the subunits and reassembly of the oligomers take place upon removal of the denaturant. Another classic example is the reassembly of tobacco mosaic virus from its RNA and coat protein (see Section 8.3). We call this **self-assembly**, and the fact that it is spontaneous indicates that, like protein folding, it is driven by an overall decrease in Gibbs free energy. In other systems, assembly may require other proteins (assembly factors or scaffolding proteins) that are not retained in the final structure. Alternatively, an input of energy (deriving ultimately from the hydrolysis of ATP) may be needed.

Most proteins form symmetrical oligomers with two or more subunits

Bioinformatics, structural and functional genomics, and protein structure databases all concur that the majority of proteins exist as oligomers. Of the complexes found in *E. coli* cells, it is estimated that about half are homomeric. Dimers and tetramers are common, but there are also many larger assemblies. Assemblies with odd numbers of subunits are much rarer than those with even numbers. Inter-subunit interfaces in oligomers are generally more highly conserved across species than is the rest of the protein surface. Given that these interfaces are largely hydrophobic, it is tempting to imagine that dimers arose by mutations that formed a hydrophobic interaction patch on the surface of a monomer. If the interaction were face-to-face (an **isologous** interaction), a symmetrical dimer would result (**Figure 1.14**). A second face-to-face interaction generated in the same way would give a

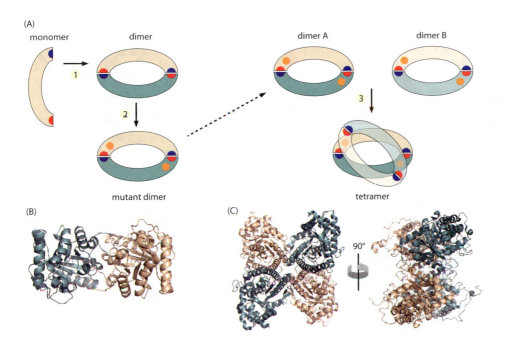

(A)

monomer   dimer   dimer A   dimer B

mutant dimer   tetramer

(B)   (C)

90°

Figure 1.14 Formation of symmetrical homodimers and homotetramers. (A) In step 1, a monomer (in wheat) with complementary binding patches (red and blue) interacts with an identical monomer (in teal), creating a dimer with a 2-fold axis perpendicular to the plane of the page. In step 2, mutation creates a second complementary binding patch (orange) on each monomer and, in step 3, two such dimers interact across a 2-fold axis, giving a tetramer. (B) A symmetrical homodimer of chicken muscle triose phosphate isomerase. (PDB 8TIM) (C) Two views of the symmetrical homotetramer of rabbit muscle fructose-1,6-bisphosphate aldolase. (PDB 3B8D) In (B) and (C), the subunits are colored as in (A).

tetramer, but the emergence of a third interaction patch on each subunit in the tetramer could not generate a closed octamer with all potential binding surfaces satisfied internally. Interaction surfaces would be left open for head-to-tail interactions of one tetramer with another, with no limit on the number. This would probably be selected against, which may explain the prevalence of dimers and tetramers with 2-fold axes.

Symmetrical interactions of this kind create closed dimers and tetramers and these have several potential advantages. In a dimeric enzyme, two active sites can be generated across the 2-fold axis, with contributions from both subunits to each one. In addition, in a symmetrical homo-oligomer the effect of any mutation is multiplied and this can enhance the sensitivity to evolutionary selection, as in the properties of *cooperativity* and *allostery*, which are almost always associated with multiple binding sites for substrates and/or ligands (Section 1.7).

Symmetry defines a set of larger structures composed of multiple copies of identical subunits

Over and above dimers and tetramers, there are many much larger structures among the assemblies and machines that this book addresses. Some are homomeric; others are heteromeric. If only one type of subunit is used, the simplest arrangements have the subunits packed such that the bonding pattern is the same for each; in other words, they are *equivalent*. (Departures from equivalence are discussed in Section 8.2.) The final structure is then necessarily a symmetrical one. Since proteins are chiral—that is, different from their mirror images—reflections are forbidden and the only symmetry operations permitted are rotations and translations. Thus all structures must belong to one of the line, point, plane, or space groups. Plane groups are relevant to biological membranes and space groups to protein crystals, but here we confine the discussion to finite structures defined by line and point groups.

Line and cyclic point group symmetries generate helices and rings

If protein subunits bind head-to-tail in identical fashion, every subunit will have exactly the same environment except those at the two ends, an arrangement known as **line symmetry** (**Figure 1.15A**). In general, this gives rise to a helical structure, which may be further stabilized by axial interactions between subunits in successive turns. (A simple linear stack is rare in biology.) Some protein helices are essentially straight, reflecting rigidity, whereas others are more flexible. There is no restriction on the number of subunits per turn, nor does it have to be integral.

one extreme, steps in the photocycles of light-harvesting complexes, involving electron transfers or the *cis–trans* isomerization of *chromophores*, are in the femtosecond (10^{-12} s) range. On the other hand, changes that entail global movements of a macromolecular machine, involving many or all of its components, usually take place in milliseconds or, in some cases, much longer. For example, the assembly of a clathrin-coated pit (Section 10.2) can take 30–60 seconds, and its final pinching off perhaps 1 second. The times characteristic of some of the machine-based processes described in this book are summarized in **Table 1.3**.

Table 1.3 Times characteristic of movements of some macromolecular machines.

Movement	Time (ms)	Book section
Photocycles of light-harvesting complexes	10^{-9}	15.3
One revolution of a bacterial flagellum[a]	0.8–3.2	14.11
Viral DNA packaging motor translocates one base pair[b]	Minimum times: $1.5(\lambda) - 6(\phi 29)$	8.5
DNA polymerase of phage T7 adds one base pair to a replicating DNA molecule[c]	3–4	3.2
Passage of a water molecule through an aquaporin channel[d]	8–50	16.4
Turnover number of a typical multienzyme complex (~100/s)	10	9.1–9.3
RNA polymerase adds one nucleotide to an mRNA[e]	10 (Pol II); 15–150 (Pol I)	5.2
Beating cycle of a sperm flagellum	25–33	14.8
One revolution of ATP synthase (F_1F_o ATPase)[f]	30	15.4
Translocation step by a motor protein[g]	0.2–400	14.3, 14.7
Chaperone-assisted protease ClpXP; 0.1 ms for scission; 4–15 sec for a substrate translocation step[h]	$0.1–10^4$	7.2
Ribosome adds one amino acid to a nascent polypeptide chain	50 (bacterial)–500 (eukaryotic)	6.2
A lamellipodium advances by 10 nm[i]	300	14.9
Translocon transfers one aa of polypeptide chain across a membrane	$10^2–10^3$	10.2

[a]300 Hz (driven by proton gradient)–1200 Hz (driven by gradient of Na^+).

[b]Maximum rates achieved early in packaging by bacteriophages λ and $\phi 29$. Rates become slower as packaging proceeds on account of mounting back-pressure. Two base pairs are translocated for each ATP hydrolyzed. (D.N. Fuller et al., *J. Mol. Biol.* 373:1113–1122, 2007.)

[c]In elongation phase, replication having been initiated. Bases are added in separate processes to the two strands of a replicating duplex at approximately equal rates. There are two main steps: polymerization and proof-checking. (K.A. Johnson, *Biochim. Biophys. Acta* 1804:1041–1048, 2010.)

[d]Transport of glycerol (through a glycerol-specific channel) is slower than water transport.

[e]Elongation phase, assuming that transcription has been initiated and is proceeding processively. Transcription also involves other, slower, steps such as promoter recognition and binding, and engagement with transcription factors. (G.L. Hager, J.G. McNally and T. Misteli, *Mol. Cell* 35:741–753, 2009.)

[f]Three molecules of ATP synthesized per revolution. Each synthesis involves multiple substeps.

[g]For successive steps in processive motion along a cytoskeletal filament. They correspond to translocation speeds of 0.2–60 μm/s for myosins (0.2–50 ms per cycle), and 0.02–2 μm/s for kinesins (4–400 ms per cycle). (L.C. Sweeney and A. Houdusse *Phil. Trans. R. Soc. B* 359:1829–1841, 2004.)

[h]Assumes that the protease complex has already bound substrate and initiated translocation or cleavage steps. There may be 50–100 ATP hydrolyzed to achieve a translocation step. (S.R. Barkow et al., *Chem. Biol.* 16:605–612, 2009.)

[i]2 μm/min for leading edge. (J.I. Lim et al., *Exp. Cell Res.* 316:2027–2041, 2010.)

The structural changes involved in these transitions are of several kinds: the addition or detachment of protein subunits; rigid-body movements (rotations and translations) of individual domains or subcomplexes; order–disorder transitions; and refolding (**Figure 1.20**). There is also strong evidence that complexes (and individual proteins) undergo rapid (picoseconds to milliseconds) thermal fluctuations or '*breathing*' motions whereby they fluctuate about their ground state, transiently exposing surfaces that are normally buried and/or unfolding peripheral motifs.

The experimental methods used to probe the dynamic behavior of macromolecules and complexes may be placed in two classes: ensemble methods and 'single-molecule' methods. Ensemble methods measure the net signal from numerous contributors. This property amplifies the signal to a detectable level but raises the problem that when a transition of interest is induced, the contributors do not necessarily proceed in synchrony, so that the signal recorded is a temporally smeared time-average. This problem is overcome in 'single-molecule' approaches that interrogate the complex of interest, one copy at a time, but the resulting signals are necessarily very weak. Nevertheless, by combining data from multiple measurements, it is possible to strengthen the signal as well as to assess its stochastic variability. In 'single-molecule' experiments, it is often necessary to adjust the experimental conditions so as to slow the transition down from its natural rate *in vivo*, to allow measurements to be made. It may also be necessary to modify the complex chemically to render the moving parts detectable.

Ensemble methods measure the net signal from numerous contributors

In time-resolved X-ray diffraction, the specimen is illuminated with an intense synchrotron-derived beam of X-rays (see **Methods**). A transition is initiated—for instance, by an electrical stimulus or a flash of light—and diffraction is monitored on a fast read-out detector. Millisecond time resolution is possible with current devices. One may distinguish two kinds of experiment, depending on the specimen and the portion of the diffraction pattern that is monitored. The smaller this portion, the more rapid the read-out that is possible.

Time-resolved fiber diffraction studies have been pursued mainly in the context of muscle fiber activation. The specimen has components that exhibit two forms of helical symmetry so that the signal is confined to two sets of discrete reflections ('layer-lines') in the diffraction pattern. One set relates to the actin filaments and the other to the myosin filaments. An example is shown in Figure 14.24C,D. To interpret these data, a mathematical model of the interacting structures is formulated and its parameters are fitted to reproduce the observed changes in diffraction intensity and their time course.

In SAXS (small-angle X-ray scattering) experiments, the specimen consists of many randomly oriented molecules that may diffuse rotationally and translationally during the measurement. As a result, the diffraction pattern is rotationally symmetric. In this situation, it may be sufficient to sample the diffraction intensity on a single radial line, although a higher signal-to-noise ratio is obtained if a full two-dimensional pattern is recorded and the data are averaged around each ring. SAXS diffraction patterns have a relatively low information content but can yield valuable information about kinetics, particularly if other information is available about the structures of components. As an example, an analysis of the R→T transition of aspartate transcarbamoylase, which takes place within a tenth of a second, is given in **Figure 1.21**.

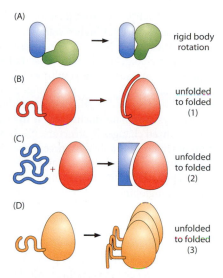

Figure 1.20 Different kinds of conformational change. (A) Rigid-body rotations of domains or subunits. (B–D) Association-induced folding of initially unfolded motifs, subunits, and domains. (B) The motif folds on binding to the folded core of the same protein. (C) An unfolded subunit (blue) folds on binding to another already folded subunit (red). (D) An initially unfolded domain polymerizes into an amyloid fibril decorated with folded globular domains of the same protein.

Figure 1.21 Kinetics of T→R transition of aspartate transcarbamylase monitored by SAXS. ATCase catalyzes the reaction of carbamoyl phosphate (CP) with l-aspartate (Asp) to form *N*-carbamoyl-L-aspartate and inorganic phosphate (P_i) (see Figure 1.31). The *E. coli* protein has a catalytic core consisting of a dimer of homotrimers of 34 kDa subunits plus three regulatory dimers of 17 kDa subunits (see Figure 1.15C). The enzyme is regulated by binding nucleotide triphosphates. Crystal structures are known for both the low-activity (no substrates) T state and the high-activity R state. In the T→R transition, the catalytic trimers move apart by 11 Å and rotate by 5°. (A) SAXS patterns recorded 38 ms (blue curve), 380 ms (orange curve), and 3800 ms (purple curve) after mixing 0.75 mM ATCase with 50 mM substrates (CP and L-Asp). The pattern from the enzyme with CP and D-Asp at 3800 ms (red curve) is a T-state control. The green curve represents the sum of 0.33 × (T-state curve) + 0.67 × (R-state curve). Spatial frequency $s = 2\sin\theta/\lambda$, where 2θ is the scattering angle and λ is the X-ray wavelength (1.38 Å). (B) Time courses of the quaternary structure change (red curve) and control (green curve), as monitored by scattering in the frequency band 0.085–0.152 Å^{-1}. Inset: the first 300 ms with a curve (two exponentials) fitted to the data points. (Adapted from J.M. West, J.R. Xia and E.R. Kantrowitz, *J. Mol. Biol.* 384:206–218, 2008. With permission from Elsevier.)

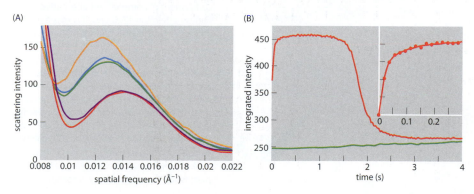

(A)

scattering intensity

spatial frequency (Å⁻¹)

(B)

integrated intensity

time (s)

Time-resolved X-ray crystallography (see **Methods**). Although crystallography has been the primary source of high-resolution information on static structures, it is less well suited to dynamic studies. Nevertheless, several avenues have been explored. They typically involve inducing a change in a protein crystal of known structure—for instance, by a flash of laser light—then measuring the resulting changes in diffraction intensities and interpreting them in structural terms. To collect enough photons in a fraction of a second, very intense X-ray sources are required, risking radiation damage. It is essential that the unit cell parameters of the crystal remain unchanged in the transition, which limits applicability to very small and localized changes in structure. One conceptually appealing approach uses Laue diffraction in which a broadband source is used instead of the usual monochromatic X-ray beam. Reflecting the formidable technical difficulties of these experiments, there have been only a few applications to proteins, for instance a study of the dissociation of carbon monoxide from hemoglobin.

Nuclear magnetic resonance (NMR) spectroscopy (see **Methods**) is based on the ability of a nucleus with a spin of ½ (for example, ^{1}H, ^{13}C, ^{15}N, and ^{31}P) to adopt two different orientations in a magnetic field. The distribution of nuclei between the two states can be changed by subjecting them to a short pulse of radiation with a frequency commensurate with the energy difference between them. Monitoring the magnetic signals in the subsequent decay can yield dynamic information about the orientation and spacing of the nuclei, which provide restraints that can be turned into structural information.

A molecule in solution is free to tumble; however, tumbling slows with increasing size, causing broadening and overlapping of signals. Thus, the determination of protein structures by solution NMR has been restricted to molecules of no more than about 40 kDa. It is therefore of limited applicability to molecular machines but has been invaluable in determining the structures of components that could not be crystallized and in studies of the conformationally flexible parts of larger proteins that are invisible to X-ray crystallography. One of the earliest examples was the detection and identification of the Ala/Pro-rich linkers required for catalytic activity in a pyruvate dehydrogenase multienzyme complex (Section 9.4). Another example is the detection of mobility on a timescale of seconds in the N-terminal motifs that gate the α ring of subunits in the 20S proteasome (Section 7.3).

In time-resolved **FRET** (**fluorescence resonance energy transfer**), two fluorophores—a donor and an acceptor—conjugated to different components of a complex shift when the transition is induced (**Figure 1.22A**). Excitation of the donor is followed by a transfer of energy to the acceptor if they are sufficiently close. The acceptor then emits light at its given wavelength. This emission signal is monitored in time. Each donor/acceptor pair has a characteristic spacing (the Förster radius), the spacing at which FRET efficiency is 50% and falls off thereafter as the sixth power. An example of time-resolved FRET is given in Figure 1.22B, C for an investigation of the weak ($K_d \approx 10\ \mu M$) interaction in which a 210 kDa complex is formed between the complement protein, C3b and a regulating factor (see Section 17.3).

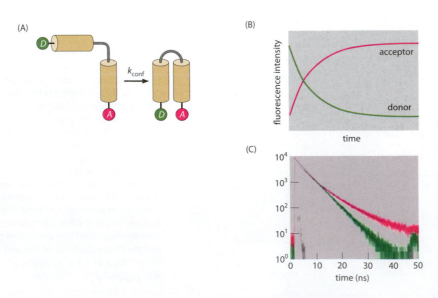

Figure 1.22 Kinetics of a conformational change from time-resolved FRET. (A) The distance between donor (*D*) and acceptor (*A*) fluorophores attached to different sites changes in a conformational transition. The transition may be initiated by rapid mixing with a suitable ligand. (B) The decrease in donor fluorescence and the accompanying increase in acceptor fluorescence report on a shortening of the interfluorophore distance. Fluorescence decay is generally used to monitor the kinetics. (C) Time-resolved fluorescence-decay curves for the binding of C3bAx488 to FH1–4Ax555. C3b is a key component of complement alternative pathway (Section 17.3) and FH1–4 is a four-domain, functionally competent, fragment of Factor H which helps regulate that pathway. The donor dye Ax488 (Alexa Fluor 488) was attached to a unique site in C3b and the acceptor Ax555 was attached to a Cys residue substituted for Asn102 in FH1–4. Donor-only decay is in green and donor/acceptor decay in pink. (B, adapted from D. Klostermeier and D.P. Millar, *Methods* 23:240–254, 2001. With permission from Elsevier; C, from I.C. Pechtl et al., *Protein Sci.* 20:2102–2112, 2011. With permission from John Wiley & Sons.)

Figure 1.23 Time-resolved observation of kinesin-1 stepping by optical trapping. Kinesin-1 is a dimeric motor protein that moves along microtubules in 8 nm steps, driven by ATP hydrolysis (Section 14.7). Its stepping is highly processive. The steps can be forward (toward the plus end of the microtubule) or backward. These plots record the stepping histories in three experiments with individual molecules of the mutant, Kin6AA, which has a six-residue insertion in its neck-linker region and is more liable than wild-type kinesin-1 to take backward steps. The plots were smoothed by median filtering. Against an applied force of 3 pN most steps are forward, but at 7 pN all steps are backward. The time between successive steps varies stochastically. Data of this kind were used to formulate a five-state model of stepping kinetics. (From B.E. Clancy et al., *Nat. Struct. Mol. Biol.* 18:1020–1027, 2011. With permission from Macmillan Publishers Ltd.)

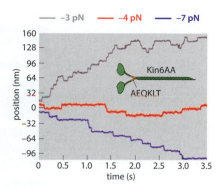

Cysteines were introduced at potentially informative positions in the proteins as sites for fluorophore attachment.

'Single-molecule' methods interrogate macromolecules one at a time

In force measurements by optical trapping, two reference points on the complex of interest are attached to the experimental apparatus either by antibodies or by covalent attachment to a glass bead, and an optical trap (Figure 14.17) is connected to each. The reaction is induced and the apparatus measures temporal changes in force exerted by the complex on the sensor. An example of using this approach to characterize the discrete steps made by a motor protein moving along microtubules is shown in **Figure 1.23**.

In time-resolved atomic force microscopy (AFM) (see **Methods**), a very fine tip is scanned across the surface of a specimen, mapping its height above a flat substrate. The technique has been refined to the point that dynamic events involving mobile components of a complex anchored on a substrate can be observed at video rates (~0.03 seconds per frame). An example of this approach has been visualization of the 36 nm 'hand over hand' steps taken by the motor protein myosin V (Section 14.3) moving along an actin filament (**Figure 1.24**).

There are several ways in which fluorescent probes can be used to track movements of macromolecular machines. In one, a fluorescent dye (or better, multiple copies of the dye) is conjugated to the particle of interest and its progress into and through a cell is followed at video rate in live cell imaging with a camera of sufficient sensitivity and resolution. Thus, movement of the particle of interest—viewed as a dot—can be charted in a biologically relevant context (**Figure 1.25**). In another application, rotational stepping by the F_1F_o ATPase has been visualized by attaching a fluorescently labeled actin filament to the rotary motor and observing its movements, the filament being long enough for its rotation to be observed by light microscopy (Section 15.4).

Although cryo-EM (see **Methods**) visualizes individual molecules, it is not possible to record the same moving molecule at successive time points. Rather, in time-resolved cryo-electron microscopy (also called four-dimensional EM), dynamics are visualized by comparing images of molecules captured in successive states; in this sense, it is an ensemble method that depends on being able to distinguish between different conformational states. This source of variability has to be disentangled from that arising from the differing

Figure 1.24 Video-rate imaging of myosin V stepping by atomic force microscopy. (A) Diagram of a myosin V construct with two heads and a coiled-coil linker (heavy meromyosin, HMM) moving along an actin filament. L-head, leading head; T-head, trailing head. (B) Successive AFM images taken at 146.7 ms per frame, showing the movement of M-V-HMM in 1 mM ATP. Pixel brightness conveys the height of the sampled surface above the substrate. The white arrow indicates the coiled-coil tail tilted toward the minus end of actin. Vertical dashed lines mark the centers of mass of the motor domains. The color-coded squares denote the states captured in each frame. The 'foot-stomp' is a head movement in which the head detaches and reattaches without translocating. (Adapted from N. Kodera et al., *Nature* 468:72–76, 2010. With permission from Macmillan Publishers Ltd.)

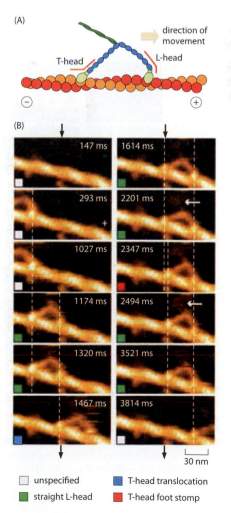

anterograde motion ⟶

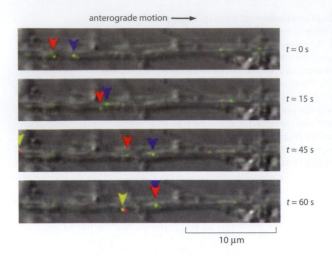

t = 0 s

t = 15 s

t = 45 s

t = 60 s

10 μm

Figure 1.25 Intracellular transport of herpesvirus capsids monitored by time-resolved fluorescence microscopy. Anterograde transport (that is, outward from the cell body toward the synaptic terminal) of pseudorabies virus (PRV) capsids along a neuronal axon imaged by confocal microscopy. PRV is a porcine virus belonging to the same subfamily as herpes simplex virus. The time-series images are merged overlays of differential interference contrast and fluorescence images with GFP and red fluorescent protein (RFP). Blue, red, and yellow arrowheads track the movement of fluorescent dots—mostly individual capsids, but see below—through the field of view. Red and blue arrowheads highlight yellow dots, which contain both GFP and RFP fused to a small capsid protein of which there are ~900 copies per capsid. The yellow arrowhead marks a predominantly red dot and a predominantly green dot transported together that eventually separate later, not shown. (From M.G. Lyman et al., *J. Virol.* 81:11363–11371, 2007. With permission from American Society for Microbiology.)

orientations that the molecule presents in projection. For this to be done reliably, the differences between the states must be substantial, and the states manageably few in number. Images are classified by computational techniques. The images in each class are averaged to reduce noise or reconstructed to obtain a three-dimensional structure. Data are collected at successive time points after initiating the transition, and the time courses with which the various conformers wax and wane are determined. Two examples of this approach have been to visualize the maturation of viral capsids (**Figure 1.26**) and to describe the power-stroke of the motor protein *dynein*.

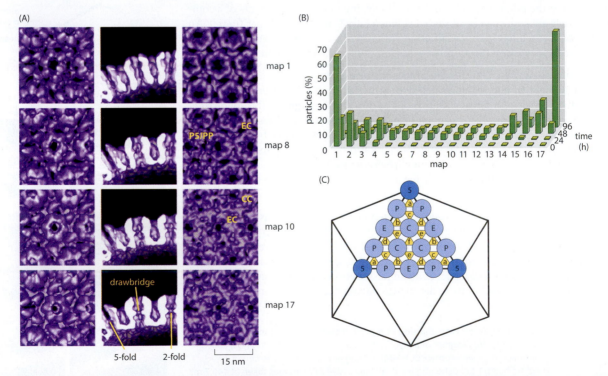

Figure 1.26 Successive states in the maturation of herpes simplex virus capsid captured by time-resolved cryo-electron microscopy. Micrographs of the maturing capsids were recorded at four time points out to 96 hours. (*In vivo* maturation is much faster but is likely to proceed along the same pathway.) Capsid images were sorted into 17 classes, and a three-dimensional structure was calculated for each class. (A) Details from maps 1 (earliest precursor state), 8, 10, and 17 (final mature state). There is an axial channel through each capsomer, illustrated for a peripentonal

hexon in the cutaway views. The 'drawbridge' domain moves into a position in which it almost closes off this channel in the mature state. (B) Kinetics of maturation. The populations of each class were determined for each time point. (C) Mapping of the various capsomers: edge hexons (E), central hexons (C), peripentonal hexon (P), pentons (5), and triplexes (yellow, a–e). In (A), EC is the interface between an edge hexon and a central hexon, and so on. (Adapted from J.B. Heymann et al., *Nat. Struct. Biol.* 10:334–341, 2003. With permission from Macmillan Publishers Ltd.)

Molecular dynamics models the motions of crystal structures in the presence of a force field

To explore the full dynamic behavior of macromolecules, one enters the realm of computational biology (see **Methods**). Here, 'molecular dynamics' (MD) studies are typically based on numerical simulation of movements taking place, starting from a known crystal structure in the presence of a mathematically defined force field. An all-atom simulation will involve many hundreds, if not thousands, of variables (the coordinates of each atom), each considered as a function of time. The scope of an MD experiment depends on their number as well as the duration of the simulation (now often as long as 1 µs) as well as the reliability of the force field employed. These are complex calculations. Because most assemblies and machines discussed in this book are at least an order of magnitude larger than individual macromolecules, and their motions are slower—in the millisecond range or slower—they are tackled instead in coarse-grained approaches, in which the elements used to describe the dynamic structure are larger than single atoms; examples are amino acids, secondary structure elements, domains, and even protein subunits. Molecular dynamics has been particularly effective in analyzing the transport of solutes through membranes via channels or transporters (see Chapter 16). Indeed, the interior of a cell has been portrayed as being in a state of constant motion.

1.6   CATALYSIS

Most cells grow and replicate in an aqueous environment (about 70% of the weight of a cell is water), at or near neutral pH, and at temperatures no higher than 40°C. These conditions, not least water as solvent, are far from those that a chemist would expect to use in the laboratory. In biology, reactions require catalysis. Enzymes are catalysts that speed up the rate at which chemical reactions reach equilibrium. They cannot alter the free energy change (ΔG) of a reaction and thus its equilibrium position; rather, they accelerate the forward and back reactions equally. The rate enhancements achieved by enzymes are vast—up to 10^{17}-fold in one or two cases—and our understanding of their ability to do this continues to grow but is still incomplete.

All enzymes are proteins, apart from a few important exceptions where RNA acts catalytically, as in the ribosome (Section 6.2). In uncatalyzed chemical reactions, the rate is normally approximately doubled for a 10°C rise in temperature ($Q_{10} \approx 2$). However, enzymes function best at or near the growth temperature of the host organism and cease to function beyond their characteristic denaturation temperatures. An enzyme also usually works best on a given substrate over a particular pH range, centered on its optimum pH. This is usually near-neutral, except for enzymes that function in special environments, such as pepsin, a protease that operates at the low pH (<2) of the stomach, or enzymes of some extremophiles that live in conditions of low or high pH.

Enzymes form highly specific but transient complexes with their substrates

The network of biochemical reactions—the metabolic map—varies from organism to organism but in all cases is highly complex (see Nicolson maps cited in References 1.6). In general, each reaction is catalyzed by a dedicated enzyme that exhibits exquisite specificity for its substrate and can distinguish even between **stereoisomers** (**Box 1.1**). Biological chemistry is almost always restricted to particular stereoisomers, such as the L-series of amino acids and the D-series of sugars. The 'wrong' isomers are normally not recognized by the enzymes concerned; in some important instances, organisms may have taken evolutionary advantage of this. For example, the presence of D-amino acids in some bacterial antibiotics makes them resistant to degradative defense systems. In the pharmaceutical industry, the presence of the 'wrong' stereoisomer can be of crucial importance, for example in the disastrous effects of the drug thalidomide (see Box 1.1).

Enzymes follow **saturation kinetics**: as the substrate concentration is increased, the rate of the reaction catalyzed also rises, but it levels off at high substrate concentrations (**Figure 1.27**). This, and their exacting specificity, led to the concept of an enzyme **active site** and an enzyme–substrate complex being formed as a key step in catalysis. The **lock-and-key hypothesis** of Emil Fischer more than a century ago postulated a perfect fit between

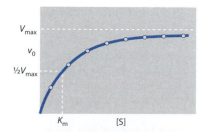

Figure 1.27 Plot of the initial velocity, v_0, against substrate concentration [S] for an enzyme-catalyzed reaction. The Michaelis constant, K_m, is the substrate concentration at which v_0 is half the maximal value, V_{max}, achieved at saturating substrate concentration.

Box 1.1   Isomerism

There are three sorts of isomerism important in biology. **Structural isomerism** is the existence of two or more molecules made of the same atoms but with different bond arrangements, for example the sugars glucose and fructose ($C_6H_{12}O_6$), The presence of a double bond in a molecule introduces the possibility of **geometric isomerism**: *cis* and *trans* forms, different molecules with alternative dispositions of the substituents across the double bond. The third sort is stereoisomerism, in which a molecule contains a carbon atom with four different substituents, a so-called **chiral** carbon. There is no internal plane of symmetry and the molecule can exist in two mirror-image forms called **enantiomers**.

The original naming of all stereoisomers was based on glyceraldehyde, which contains one chiral carbon (**Figure 1.1.1**). Enantiomers rotate plane-polarized light in opposite directions; *levo* (–) to the left and *dextro* (+) to the right. All compounds with an absolute configuration of atoms around a chiral carbon atom that can be related to D-(+)-glyceraldehyde are part of the D-series, and conversely for the L-series. The designations D and L do not indicate whether the molecule is *dextro*- or *levo*-rotatory. Another naming convention is the *R/S* notation, but we use D- and L- in this book.

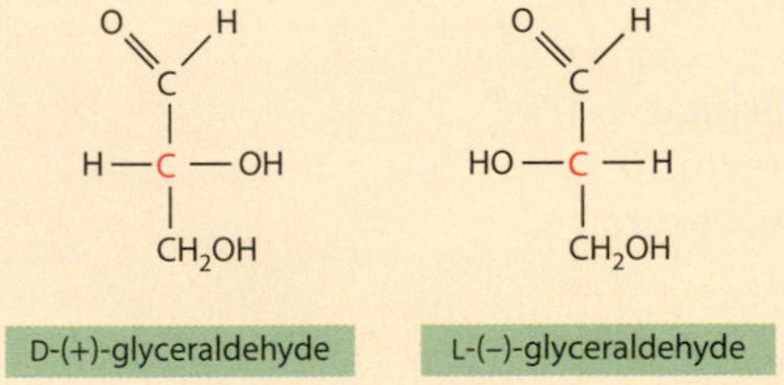

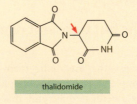

Figure 1.1.1 Glyceraldehyde exists in two enantiomeric forms. The chiral carbon atom is in red. The two forms are mirror images in an imaginary plane between them.

D-(+)-glyceraldehyde   L-(–)-glyceraldehyde

If a molecule contains more than one chiral carbon it will exhibit **diastereoisomerism**; there will be as many chemically different compounds as there are chiral carbons, each of which exists as a pair of enantiomers. For example, the four-carbon sugars threose and erythrose (two chiral carbons) are diastereoisomers but not enantiomers; in other words, they are not mirror images of one another.

Both exist in D- and L-forms, which are enantiomeric. The forms depicted in **Figure 1.1.2** have the same configuration about

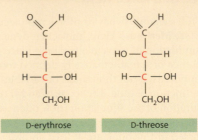

D-erythrose   D-threose

Figure 1.1.2 D-Erythrose and D-threose. The chiral carbon atoms are in red.

the bottom chiral carbon atom as D-glyceraldehyde (see Figure 1.1.1) and are therefore designated D-erythrose and D-threose. Note that D-erythrose and D-threose are *levo*-rotatory, whereas D-glyceraldehyde is *dextro*-rotatory.

Stereoisomerism is of major importance in biology. Proteins almost always consist of only L-amino acids, and carbohydrates of D-sugars. Macromolecules are asymmetric and therefore likely to recognize only one enantiomer of any given compound. Failure to take account of this can be dangerous. Thalidomide was sold as a medication effective against morning sickness for pregnant women. It contains a chiral carbon atom and so exists in D- and L-forms (**Figure 1.1.3**). The commercial preparation was a **racemic mixture**—it contained equal amounts of the two enantiomers, one of which induced developmental abnormalities and led to a tragic outbreak of birth defects. It is thought that this isomer may be able to insert into the major groove of DNA at GC-rich regions, interfering with gene transcription and thus inhibiting the formation of blood vessels. Unfortunately the enantiomers are able to interconvert *in vivo* and administering only one enantiomer means that both enantiomers will still be present in the serum. More recently, thalidomide has found favor again, in carefully controlled application, for such diseases as AIDS, leprosy, lupus, and tuberculosis.

thalidomide

Figure 1.1.3 The structure of thalidomide. The chiral carbon is arrowed.

the enzyme and its substrate. Today we speak instead of molecular complementarity. This accommodates the possibility of conformational change on the part of the protein when it comes into contact with the substrate, or the selection of a particularly efficacious conformation of the enzyme from an ensemble of related structures.

Enzyme kinetics are governed by a few equations

For the single-substrate reaction

$$E + S \underset{k_{-1}}{\overset{k_1}{\leftrightarrow}} ES \overset{k_2}{\rightarrow} E + P$$

where S is the substrate and P the product, if $[E_0]$ is the total concentration of the enzyme, the initial rate, v, of the forward reaction is given by

$$v = \frac{k_2[E_0][S]}{[S] + (k_2 + k_{-1})/k_1}$$

and the **Michaelis constant**, K_m, by

$$K_m = (k_2 + k_{-1})/k_1$$

an equation first derived by George Briggs and J.B.S. Haldane in 1925. This followed earlier work of Leonor Michaelis and Maud Menten in 1913, who derived a similar equation but assumed that k_2 was very small compared with k_{-1}. If $k_2 << k_{-1}$, $K_m \approx k_{-1}/k_1$, which is the same as K_s, the dissociation constant of the enzyme–substrate complex. The **catalytic rate constant** for the forward reaction, conventionally known as k_{cat}, equals k_2. It is sometimes referred to as the **turnover number**, because it reflects the number of substrate molecules converted to product per second at the enzyme active site. From the equation above, K_m can be defined as the substrate concentration at which half-maximal velocity ($\frac{1}{2}V_{max}$) is achieved (see Figure 1.27). At low substrate concentrations, $v = (k_{cat}/K_m)[E_0][S]$, which implies that k_{cat}/K_m can be regarded as an apparent second-order rate constant. Thus it is a measure of the specificity of the enzyme for its particular substrate—hence its other name of **specificity constant**.

In many instances, $k_2 << k_{-1}$ as Michaelis and Menten assumed. In such cases, $K_m \approx k_{-1}/k_1$ (see above) and gives a direct measure of the affinity ($k_1/k_{-1} = 1/K_s$) of the enzyme for its substrate. However, the assumption should be tested for any given enzyme. If $k_2 >> k_{-1}$, $k_{cat}/K_m \approx k_1$, the rate constant for the formation of the enzyme–substrate complex. k_1 is typically about $10^8/(s\ M)$ for enzymes and their substrates, and a measurement of k_{cat}/K_m approaching this value is a strong indication that Briggs–Haldane rather than Michaelis–Menten kinetics are being followed. With some enzymes, additional intermediates occur on the reaction pathway, and then $K_m < K_s$.

A key feature of enzyme catalysis is the tight binding of the transition state

In a typical chemical reaction, the reactants pass through one or more **transition states** on the way to product, and the energy required to reach the highest point in the profile is the **activation energy** that must be surmounted if the reaction is to go to completion (**Figure 1.28A**). In the transition state, chemical bonds are being made and broken, whereas the depressions in an energy profile are populated by intermediates in which bonds are fully formed. The rate of the overall reaction is the concentration of the transition state multiplied by the rate constant for its decomposition. In an enzyme-catalyzed reaction, the activation energy is lowered (Figure 1.28B) and the rate of reaction is thereby enhanced.

Following earlier work by J.B.S. Haldane, Linus Pauling suggested that an enzyme speeds up a reaction by stabilizing the transition state, binding it more tightly than the substrate and thereby lowering the activation energy. Specificity, represented by the formation of an ES complex, and catalysis, represented by the transition state, are linked, but the crucial feature is that the enzyme and transition state should adopt complementary structures. The results of protein engineering experiments designed to dissect the contributions made by individual amino acid side chains to the catalytic mechanisms of particular enzymes are consistent with this idea. Moreover, if a molecule resembling the transition state of a chemical reaction is used as an antigen, **catalytic antibodies** can be elicited that are able to catalyze, albeit sometimes only poorly, the reaction represented by the transition state analog.

(A)

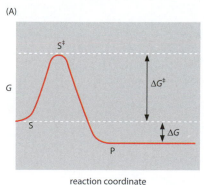

(B)

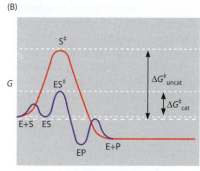

reaction coordinate

Figure 1.28 Free energy profiles of uncatalyzed and enzyme-catalyzed reactions. G, Gibbs free energy; S, substrate (reactant); E, enzyme; P, product. (A) A simple uncatalyzed chemical reaction passes through the transition state, $S^{\ddagger}$. $\Delta G^{\ddagger}$ is the free energy of activation (that is, the difference in free energy between reactant and the transition state). ΔG represents the difference in free energy between reactant and product. (B) In a simple enzyme-catalyzed reaction (blue curve), $ES^{\ddagger}$ is the transition state and ES and EP represent the enzyme–substrate and enzyme–product complexes. ΔG is unchanged. $\Delta G^{\ddagger}_{uncat}$ is the free energy of activation for the uncatalyzed reaction, and $\Delta G^{\ddagger}_{cat}$ is the free energy of activation for the enzyme-catalyzed reaction. The difference, $\Delta\Delta G^{\ddagger}$, is a measure of the efficacy of the catalyst.

Enzyme-catalyzed reactions usually proceed through the direct making or breaking of bonds over distances of up to 2 Å (see below). However, in some electron transfer proteins (Chapter 15), *redox centers* are as much as 15–20 Å apart. In such enzymes, electron transfer proceeds by *quantum mechanical tunneling*; that is, the electron tunnels through a transition state barrier that classically it could not surmount. The rate of electron transfer is rapid, much faster than the millisecond turnover times that are characteristic of most enzymes, and can readily accommodate the picosecond reaction times common in photosynthesis. Some enzymes involved in C–H transfers also appear to operate by quantum mechanical tunneling of the hydrogen, though because of its much higher mass (1800× that of an electron) the tunneling distance is limited to at most 1 Å. In contrast with conventional transition state theory, tunneling is more a function of barrier width than of barrier height (Chapter 15).

Enzymes generate catalytic rate enhancements in multiple ways

An enzyme creates a 'pre-organized environment' that promotes catalysis. It selectively binds substrates and orients them advantageously in the active site, creating a microenvironment that facilitates reactions that would be difficult to achieve in an aqueous environment. It positions catalytically important amino acid side chains to facilitate acid–base catalysis and/or other features of an efficient catalytic mechanism. It also promotes a high effective concentration of intramolecular species that overcomes the loss of entropy that is unavoidable in intermolecular reactions. It makes possible covalent catalysis in which relevant groups (for example phosphate) can be transiently lodged on the protein, and it binds **cofactors**. Cofactors can be metal ions, **coenzymes** (such as NAD^+ or $NADP^+$, which shuttle as carriers between oxidoreductases), and **prosthetic groups**, which differ from coenzymes in being bound very tightly or even covalently to the parent protein. The chemical mechanism of the protease chymotrypsin (**Figure 1.29**) illustrates some of these features.

Enzymes are not static structures. In addition to allowing the enzyme to adopt a structure complementary to the transition state, conformational flexibility may be required to allow substrates to enter and products to leave, and to close off the active site from the aqueous environment. If water molecules are expelled, there will be a gain in entropy that promotes catalysis. Groups essential to catalysis can be moved into position as the reaction proceeds, and high precision (<1 Å) in their placing is usually necessary.

Most enzymes operate on a millisecond timescale or faster. NMR spectroscopy and other techniques have shown that proteins undergo multiple reversible motions away from their average structures. The timescales involved vary from seconds to milliseconds for the rearrangement of subunits or domains down to nanoseconds to picoseconds for local movements of side chains. How coupled ensembles of fluctuating conformations contribute to the rate enhancement of reactions as enzymes progress through their catalytic cycles is not yet settled, but the concept of an energy landscape like that observed in protein folding (see Figure 1.12) is gaining acceptance. Nonetheless, the conclusion remains that the catalytic power of enzymes resides in a lowering of the activation energy and an increase in the generalized transmission coefficient that results from dynamic crossing and recrossing of the activation barrier.

Enzymes can be inhibited reversibly and irreversibly

Enzymes can be inhibited in various ways. **Competitive inhibitors** closely resemble the substrate in structure and can bind reversibly at the active site, thereby blocking the reaction. They do not affect the value of V_{max}, because saturating concentrations of substrate will swamp a finite concentration of inhibitor. However, the K_m for the substrate will apparently increase, as the enzyme requires a higher concentration of substrate to reach $\frac{1}{2}V_{max}$. In another type of reversible inhibition, a transition state analog (a molecule that resembles the transition state but cannot be acted on by the enzyme) binds more tightly to the active site than does the substrate itself. Transition state analogs have been developed for several enzymes and have helped in the elucidation of catalytic mechanisms and in structure-based drug design.

A third and different form of inhibition, generally irreversible, is covalent modification of the enzyme. For example, a reagent may preferentially react with a protein functional group that displays unusual reactivity. If this group is in or near the active site, the catalytic reaction may be impaired. In a related form of inhibition, **active site-directed**

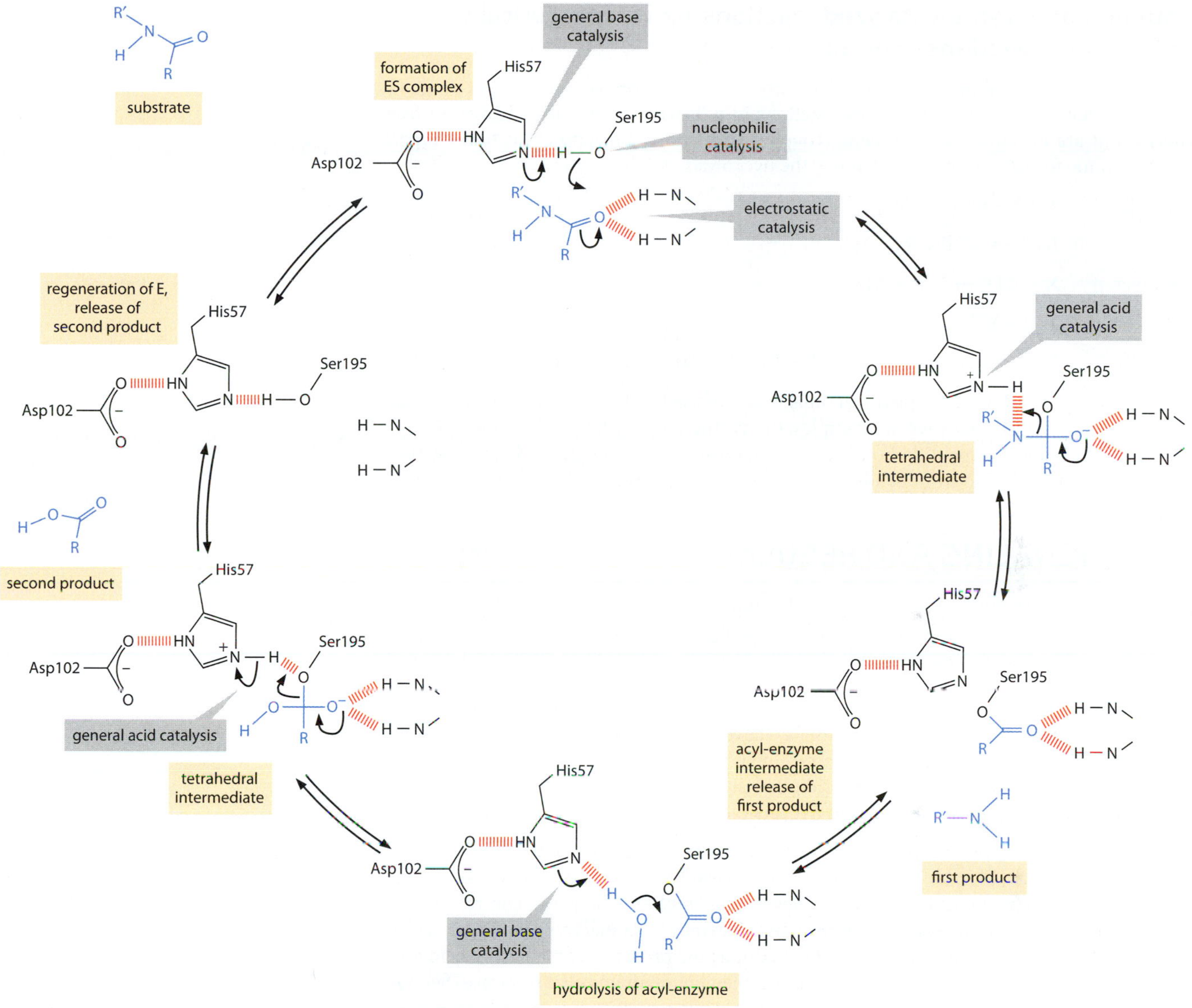

Figure 1.29 Catalytic mechanism of the pancreatic protease chymotrypsin. Substrate specificity is provided by the bulky hydrophobic side chain of the amino acid (Phe, Tyr, Trp) on the N-terminal side of the scissile peptide bond, which enters a narrow hydrophobic pocket located close to the catalytically important groups of the enzyme. The catalytic reaction can be considered as comprising two halves. In the first, the negative charge on the carboxyl group of Asp102 acts to stabilize a positive charge on the imidazole ring of His57, to which it is H-bonded. This makes it easier for the hydrogen of the hydroxyl group in the side chain of Ser195 to be donated to the uncharged imidazole ring (general base catalysis), thereby promoting nucleophilic attack of Ser195 on the >C=O group of the scissile peptide bond of the bound peptide substrate. The trigonal carbon of the >C=O is held in the correct position for this attack by two fractionally charged >NH groups (Gly193 and Ser195) in a so-called oxyanion hole provided by the peptide backbone of the enzyme (electrostatic catalysis). This generates the tetrahedral intermediate in which the newly formed >C–O⁻ moves deeper into the oxyanion hole, where it forms H-bonds with the NH groups of Gly193 and Ser195 (stabilization of the transition state). This adduct in turn collapses to the acyl-enzyme intermediate, facilitated by the donation of a proton from the positively charged imidazole ring of His57 to the R'NH– leaving group (general acid catalysis), and release of the first product, R'NH₂. In the second half of the reaction, a water molecule interacts with the now uncharged imidazole ring of the Asp102-His57 pair. This helps activate it (general base catalysis) as a nucleophile for attack on the acyl-enzyme, thereby forming another tetrahedral intermediate. Collapse of this tetrahedral intermediate, with the return of a proton from the positively charged imidazole ring (general acid catalysis), regenerates the hydroxyl group of Ser195 and thus the starting enzyme, which is accompanied by release of the second product, RCOOH. The numbering of residues is that of the inactive precursor (zymogen), bovine chymotrypsinogen. (Adapted from A.R. Fersht, Structure and Mechanism in Protein Science: A Guide to Enzyme Catalysis and Protein Folding. W.H. Freeman, 1998.)

inhibitors—molecules that resemble the substrate but also carry a 'chemical warhead'—are targeted at the active site. Selective reaction of the warhead with a nearby functional group can irreversibly inhibit the enzyme. The warhead might be an alkylating agent, or an electrophile in search of a protein nucleophile, or even a highly reactive species generated *in situ* from a benign substituent by a laser flash (such as a nitrene from an azide).

Coupling of enzyme-catalyzed reactions allows energetically unfavorable reactions to occur

A reaction will proceed in a given direction only if it has a favorable free energy change (a negative value of ΔG). However, many reactions in cells have unfavorable changes in free energy and are brought about by being coupled with another reaction that has a larger, negative value for ΔG—often, for example, the hydrolysis of ATP.

Typically, $X + Y \rightarrow X\text{-}Y, \quad \Delta G_1 > 0$

$ATP \rightarrow ADP + P_i, \quad \Delta G_2 < 0$

However, if $X + ATP \rightarrow X\text{-}P + ADP$

and $X\text{-}P + Y \rightarrow X\text{-}Y + P_i$

(overall) $X + Y + ATP \rightarrow X\text{-}Y + ADP + P_i, \quad \Delta G < 0 \text{ if } \Delta G_2 > \Delta G1$

Facilitating the catalysis of coupled reactions is one of the key features of many molecular machines; to give just three diverse examples, the synthesis of DNA by polymerases (Chapter 3), the formation of peptide bonds by the ribosome (Section 6.2), and carboxylation reactions catalyzed by biotin-dependent multienzyme complexes (Section 9.4).

1.7   SIGNALING AND REGULATORY MECHANISMS

Biological processes must be coordinated and regulated if cells are to function in an orderly way. Signals are of crucial importance, whether coming from outside the cell or sent between or within compartments inside the cell. The fate of a cell (or indeed an organism) can rest on the speed and efficiency with which signals are transmitted and acted upon. Growth factors and hormones—chemical messengers such as insulin and epinephrine (adrenaline) that circulate in the blood of higher animals—have long been known to transmit signals across the cell membrane, generating intracellular responses. Many molecular machines and assemblies are key players in these and other signaling processes. Their activity can be regulated in several ways, some slow and irreversible, others faster and reversible.

Cells are able to turn on and off the synthesis of a given protein by controlling the transcription of its gene (Chapter 5). However, this process is slow (minutes to hours) and the protein remains active until it decays or is degraded. Another widespread control mechanism is the action of proteases to generate active enzymes from inactive precursors, called **pro-enzymes** or **zymogens**. This is how, for example, the proteases responsible for initiating apoptosis (Section 13.6) or for activating complement in an immune response (Section 17.3) are generated. Again, though, the relevant proteins remain active until degraded. In contrast, a protein can undergo a conformational change on the millisecond timescale in response to binding a ligand or to post-translational modification. This mechanism underlies many rapid signaling processes and can be envisaged as having conferred great evolutionary advantage.

Ligand-induced conformational change and cooperativity are widespread methods of controlling biological activity

A general feature of signal transduction is that of ligand-induced conformational change (**Figure 1.30**). The ligand can be small—an ion or small molecule (a *chemical messenger*)— or large, such as another protein. Binding of the ligand to a protein induces a conformational

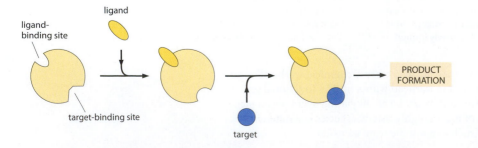

Figure 1.30 Ligand-induced conformational change. Binding of ligand at one site induces a change at the remote active site on the same protein molecule, thereby enabling it to bind the target and process it.

change that enables the protein to recognize its target molecule and act on it to initiate the desired biological activity. Activity can thus be turned on or off in a fraction of a second. The effect of a ligand-induced conformational change might be simple and direct, such as stimulating an enzyme to bind its substrate and initiate a catalytic reaction, but it can be more elaborate. For example, the binding of lactose to the *lac* repressor induces a conformational change that causes the repressor to dissociate from the *lac* operator sequence in *E. coli* DNA, allowing the transcription of genes in the *lac* operon; equally, the binding of insulin to its transmembrane receptor (Section 16.2) or of growth factors to G-protein-coupled receptors (Section 12.2), sets in train intracellular *protein kinase* cascades that amplify and transmit signals to the nucleus or other targets (Chapter 12). Many more examples recur throughout this book.

Protein oligomers offer the possibility of a ligand-induced conformational change in one subunit being transmitted to a neighboring subunit. If the oligomer has two conformational states, it might then be locked into one or the other state by ligand binding. These possibilities are manifested in cooperativity (Section 1.2); that is, one event can make another similar event either more or less likely to follow. Cooperativity is generally a property of oligomers and commonly of symmetrical oligomers. It is a widespread method of transmitting signals through macromolecular complexes, as in the oxygenation of the hemoglobin $\alpha_2\beta_2$ tetramer whereby the binding of one O_2 molecule makes it easier for a second to bind, and so on up to a maximum of four.

Allosteric proteins are regulated by a special form of cooperativity

Consider a metabolic pathway of the kind

$$A \rightarrow B \rightarrow C \rightarrow D \rightarrow E \rightarrow F \rightarrow G \rightarrow H$$

In the biosynthesis of pyrimidine nucleotides, the first $(A \rightarrow B)$ of seven steps is the formation of *N*-carbamoylaspartate from aspartate and carbamoyl phosphate, catalyzed by aspartate transcarbamoylase (ATCase) (**Figure 1.31**). ATCase is reversibly inhibited by CTP, the end product (H) of the pathway. Thus, if the concentration of CTP in the cell rises, it down-regulates its own biosynthesis (and conversely, if the concentration of CTP falls). This process is termed **feedback inhibition**. CTP has a structure totally different from either of the substrates for ATCase and is therefore unlikely to act as a competitive inhibitor. ATCase is a complicated oligomeric enzyme (see Figure 1.15C). Substrates are bound and acted upon at a site in a catalytic subunit, whereas CTP binds to a site in a regulatory subunit yet is able to exert its inhibitory effect on a catalytic site more than 50 Å away. A time course of activation is shown in Figure 1.21. In contrast, the presence of ATP stimulates the activity of ATCase, and a high concentration of purine nucleotides in the cell therefore acts as a signal to ATCase to increase the production of the complementary pyrimidine nucleotides for nucleic acid synthesis. ATP, like CTP, is quite different in structure from aspartate and carbamoyl phosphate and binds at a different site, also remote from the catalytic site.

Pyrimidine biosynthesis is just one of many examples of this kind of metabolic control. The fact that the inhibitory or stimulatory molecules are structurally different from the substrates in such enzymes led Jacques Monod, Jeffries Wyman, and Jean-Pierre Changeux to introduce the term **allostery** (Greek *allos*, other, and *stereos*, shape). ATCase is an extreme example of the principle of separate but communicating binding sites for substrates, allosteric inhibitors, and allosteric activators in that its catalytic and regulatory sites are, unusually, in different subunits (**Figure 1.32**).

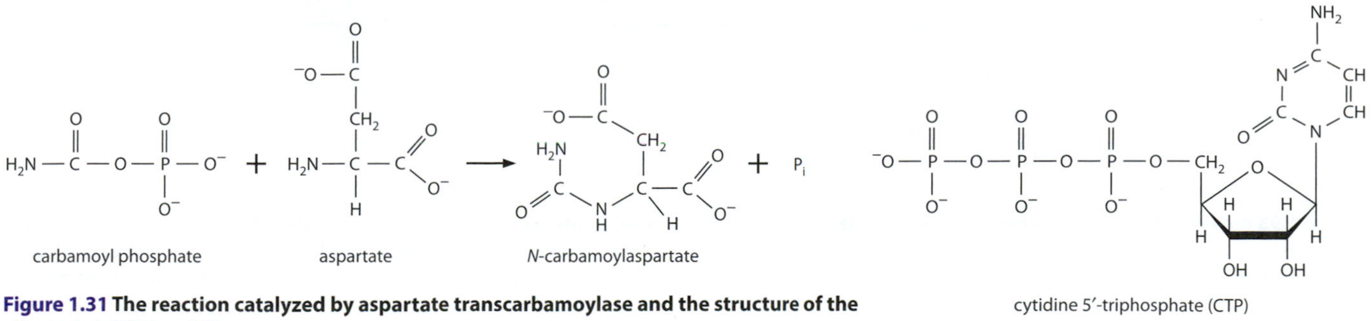

carbamoyl phosphate   aspartate   *N*-carbamoylaspartate

cytidine 5′-triphosphate (CTP)

Figure 1.31 The reaction catalyzed by aspartate transcarbamoylase and the structure of the allosteric inhibitor, CTP.

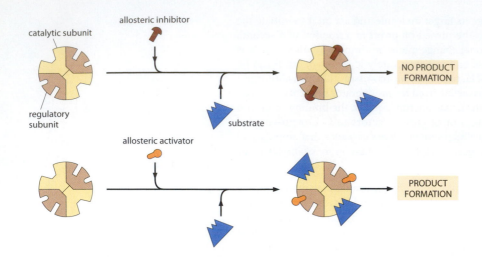

Figure 1.32 Allosteric regulation of a multi-subunit enzyme. This model enzyme is a hetero-oligomer of regulatory and catalytic subunits. Binding of an effector molecule (inhibitor or activator) to the regulatory subunits is transmitted to, and controls the activity of, the catalytic subunits.

Allosteric enzymes do not follow Michaelis–Menten kinetics

A plot of velocity against substrate concentration for an allosteric enzyme is an S-shaped (sigmoid) curve (**Figure 1.33A**), not the hyperbola typically observed for unregulated enzymes. This is the result of positive cooperativity. A sigmoidal v/[S] plot means that a small change in substrate concentration produces a bigger change in catalytic rate than it does with a hyperbolic Michaelis–Menten v/[S] plot. Moreover, the effect can be further amplified by allosteric inhibitors or activators (Figure 1.33B). Interactions involving identical molecules (such as the cooperative binding of substrate to an enzyme or of O_2 to hemoglobin) are described as **homotropic**, and those involving allosteric effectors (inhibitors and activators) as **heterotropic**.

Allostery is mediated by protein/protein interactions and conformational changes

Most allosteric proteins are small oligomers, and the Monod–Wyman–Changeux (MWC) theory explained cooperativity in allosteric enzymes and hemoglobin on the basis of the following assumptions (**Figure 1.34A**):

*   The quaternary structure is symmetrical (or pseudo-symmetrical, as in $\alpha_2\beta_2$ hemoglobin).
*   The oligomers are in dynamic equilibrium between two conformations, a 'tense' (T) state and a 'relaxed' (R) state.
*   In the absence of any substrate or signaling molecule, at equilibrium the concentration of the T state is higher than that of the R state.
*   The T state has a lower affinity for ligands, and hence lower catalytic activity, than the R state.
*   Binding of a substrate or allosteric activator shifts the equilibrium in the direction of the R state, whereas an allosteric inhibitor shifts it toward the T state.

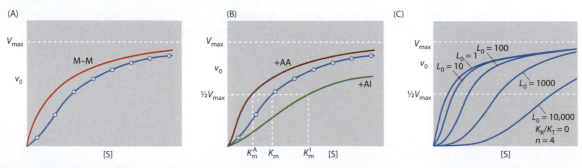

Figure 1.33 Kinetics of allosteric enzymes. (A) Plot of the initial velocity, v_0, against substrate concentration [S] for an allosteric enzyme (blue) and an enzyme obeying Michaelis–Menten (M–M) kinetics (red). (B) Plot of v_0 against [S] for an allosteric enzyme in the presence of an allosteric activator (AA) or an allosteric inhibitor (AI). The K_m for substrate S decreases in the presence of the activator and increases in the presence of the inhibitor. This is typical for the class of allosteric enzymes in which the value of V_{max} remains unchanged. (C) Effect of differing values of L_0 for an allosteric tetramer ($n = 4$) with $K_R/K_T = 0$.

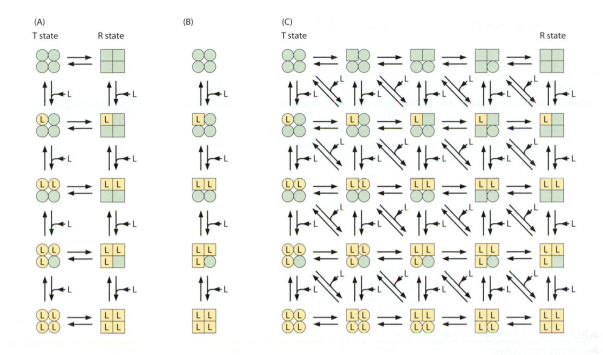

The sigmoid binding curve of an allosteric protein can then be calculated from a few parameters: a constant $L_0 = [T]/[R]$, the ratio of the T and R states in the absence of any ligand binding, and K_R and K_T, the dissociation constants for substrate binding to the R and T states, respectively. In comparison with the Michaelis–Menten equation, the $v/[S]$ relationship becomes

$$v = \frac{V([S]/K_R)(1 + [S]/K_R)^{n-1}}{L_0(1 + [S]/K_T)^n + (1 + [S]/K_R)^n}$$

For most allosteric enzymes K_R/K_T is in the range 0.01–0.04 (meaning that, to a good first approximation, the $[S]/K_T$ term can be neglected) and L_0 ranges from ~60 to 10^4 (see Figure 1.33C).

If an allosteric activator displaces the R/T equilibrium so that the R state predominates, cooperativity in substrate binding is no longer observed and the enzyme follows Michaelis–Menten kinetics. This has been verified experimentally for several enzymes and is beautifully clear for ATCase: if separated from the regulatory dimers, the catalytic trimers exhibit straightforward Michaelis–Menten kinetics.

An alternative model for allostery was proposed by Daniel Koshland, George Nemethy, and David Filmer (KNF) in which, in the absence of substrate, the oligomer exists in a single state constrained by subunit/subunit interactions (Figure 1.34B). Symmetry is not conserved, and the conformation of each subunit changes when it binds substrate. If, as a result, this makes it easier for a neighboring subunit to switch to the active conformation, there is positive cooperativity; if it is made more difficult, negative cooperativity is observed. The KNF model can readily be extended to incorporate heterotropic interactions by regulatory activator or inhibitor molecules. Unlike the MWC theory, the KNF theory is able to explain negative cooperativity and is also compatible with 'half-of-the-sites reactivity,' in which only half of the active sites in the oligomer are active at any one time.

If the subunit/subunit constraints are sufficiently tight, once one subunit is in the R state, all others must take up the R state, and the KNF and MWC theories coalesce. Indeed, Manfred Eigen pointed out that they can both be accommodated in a general overall framework (Figure 1.34C) in which the MWC model populates only the extreme states, whereas the KNF model populates the states found on the diagonal. There are many examples of positive cooperativity; negative cooperativity, although less common, is observed in machines such as the GroEL/GroES chaperone (Section 6.3), the E1 components of pyruvate dehydrogenase multienzyme complexes (Section 9.4), and ATP synthase (Section 15.4). However, allosteric complexes often show elements of both MWC and KNF theory.

Figure 1.34 Models of allostery. The models are illustrated for a tetrameric protein, L, a ligand (substrate or allosteric modulator); circles represent T-state subunits (low affinity for L) and squares represent R-state subunits (high affinity for L). (A) The MWC model. The conformational changes in the protein subunits are concerted; no tetramer contains both R- and T-state subunits at any one time. (B) The KNF model. Conformational changes are not concerted and each subunit can be in either the T-state or the R-state. In both models, increasing the concentration of L causes more subunits to switch into the R-state. (C) The generalized model of allostery in which the MWC model is the two outer columns and the KNF model is the diagonal from top left to bottom right.

Reversible covalent modification controls the activities of some proteins

Chemical modifications afford another way to rapidly regulate the activity of a target protein. In eukaryotes, post-translational modifications of proteins are many and varied, but phosphorylation (catalyzed by kinases) and dephosphorylation (catalyzed by phosphatases) are the most common (Chapter 12). It is important that the modification and its reversal are carried out by different enzymes, so that a signal used to turn on the targeted activity differs from the one that turns it off. The sites of phosphorylation are the hydroxyl groups of serine, threonine, or tyrosine residues. The hydroxyl group, although polar, carries no charge at neutral pH (the pK_a of the OH group in serine or threonine is around pH 14; that of the OH in tyrosine is about pH 10.5). The phosphorylated side chain, however, is negatively charged, which can have the effect of promoting a conformational change in the modified protein that modulates its activity.

The widespread importance of phosphorylation and dephosphorylation as a signaling and control mechanism is reflected in the fact that 518 putative protein kinase genes have been identified in the human genome; of these, 244 map to disease loci. Some 150 different phosphatases, >100 of which are phosphotyrosine-specific, catalyze dephosphorylation. As many as one-third of human intracellular proteins may be subject to regulatory phosphorylation.

Homeostasis is an important aspect of response to environmental change

The concept of the **rate-limiting step** has been a key feature for understanding how a sequence of reactions in a metabolic pathway is regulated. However, **metabolic control analysis** has demonstrated, counter-intuitively perhaps, that each enzyme contributes appreciably to the overall flux control (**Box 1.2**). In response to a signal, an increase or decrease in the flux through the pathway is accompanied by smaller changes to the concentrations of individual metabolites. It appears that maintenance of stable levels of metabolite concentrations (**homeostasis**, from the Greek *homoios*, same; *stasis*, standing still) is a key feature of metabolic regulation, as indeed it has long been noted to be in the hormonal control of blood glucose or pH. Homeostasis also appears to be a key feature in the maintenance of high concentrations of macromolecules in cells experiencing changes in their

---

Box 1.2 Metabolic control analysis

Metabolic control analysis posits that each step in a metabolic pathway contributes in some degree to the control of flux through the pathway. Each enzyme is assigned a *flux control coefficient*, and the sum of the control coefficients for all the enzymes in the pathway must come to 1.0. The higher its flux control coefficient, the more significant will be the part played by a given enzyme in the control process. Likewise, for each metabolite there is a concentration control coefficient derived from the effects that each enzyme in the pathway has on its concentration. The 'connectivity theorem' states that the overall sum of the concentration control coefficients should be 0 for each metabolite.

According to theory, a change in the activity of a single enzyme is likely to produce a larger change in the concentrations of metabolites in a pathway than in the flux through the pathway. Moreover, changing the activity of enzymes catalyzing reactions that are at or near equilibrium is unlikely to generate large effects on either metabolite concentrations or metabolic flux. Raising the flux through a metabolic pathway is rarely achieved by controlling the activity of a single enzyme, which would require its flux control efficient to be close to 1.0.

This explains why the overexpression of genes encoding what were thought to be rate-limiting enzymes has been observed to have little or no effect on the flux through the relevant metabolic pathway *in vivo*.

The usual outcome of a signal in a metabolic pathway is an increase or decrease in the flux through the pathway, but this is accompanied by smaller changes in the concentrations of individual metabolites. It appears that maintenance of stable levels of metabolite concentrations is a major factor in metabolic regulation. In turn, it has been argued that feedback inhibition of allosteric enzymes catalyzing supposedly rate-limiting reactions has more to do with metabolite homeostasis than with flux changes. Metabolic control analysis predicts that if the activities of all the enzymes in a pathway are modified by the same numerical factor, the metabolic flux will be modified identically without change to the concentrations of metabolic intermediates. The effect of the feedback inhibition in a pathway of the kind described above will be to lower the flux control coefficient of the target enzyme, which therefore means that control must pass to subsequent steps.

environments (Section 1.8). We are still only in the early stages of a quantitative understanding of the complexity of signaling and the regulation of interconnected systems, but
homeostasis is increasingly viewed as an important property.

1.8 MACROMOLECULAR CROWDING

Up to this point, we have described the properties of macromolecular assemblies mainly
in terms of their structures and those of their components as determined by X-ray crystallography and other techniques, and their functional attributes as established by biochemical
and genetic approaches. In general, the specimens studied are in dilute solution. Most *in
vitro* experiments with proteins are carried out at concentrations of <1 mg/ml; for example,
enzyme kinetic studies usually use only a few μg per ml; at the upper end, proteins crystallize from solutions at 5–40 mg/ml. *In vivo* the situation is quite different: the interior of a
cell is a very crowded environment (**Figure 1.35**). The total concentration of protein inside
an *E. coli* cell is 200–300 mg/ml and that of RNA is 75–150 mg/ml, giving a total macromolecular concentration of ~300–400 mg/ml. Eukaryotic cells are similarly congested, and
this congestion has major consequences for biological activity within a cell.

In crowded situations, the concentrations of macromolecules can be so high that they
account for a significant fraction of the total volume, which is thus unavailable to other
macromolecules. This is termed **macromolecular crowding** (or sometimes, the **excluded
volume effect**). The concentrations given above imply that 25–30% of the volume in living
cells is excluded. For comparison, the densest possible packing of uniform spheres (cubic or
hexagonal close packing) occupies ~74% of total volume, as shown by Carl Friedrich Gauss
in 1831. With random packing, this number falls to ~64%. In protein crystals, the fractional
volume occupied by proteins is typically ~55%, but values as low as ~25% are observed. It
appears that lattice contacts in protein crystals do not alter the overall conformations of the
molecules, although they may cause some local perturbations. On this basis, we assume
that the effects of molecular crowding are not due to any major perturbation of protein
structure.

Advances in mass spectrometry have made it possible to examine the concentrations of
individual proteins in cells. For example, in the pathogenic spirochete *Leptospira interrogans*, with a total protein concentration of ~250 mg/ml, more than 2200 proteins were identified; their abundances ranged from 40 or fewer to 40,000 copies per cell. (A cellular quota
of 40,000 copies of a 25 kDa protein corresponds to a concentration of ~20 μg/ml.) Proteins

**Figure 1.35 Dense packing of
macromolecules and organelles in a
eukaryotic cell.** An electron tomogram
is shown for a 400 nm-thick section of
a cultured rat cell. The specimen was
prepared by high-pressure freezing, freeze
substitution, and plastic embedding. A tilt
series of 80 micrographs was collected and
used to calculate the tomogram, which
was then 'segmented' by an observer who
manually traced out the various subcellular
components in each slice. The segmented
tomogram is viewed *en face* from one side.
The plough-shaped feature is the Golgi
apparatus with seven cisternae. Color
coding: yellow, endoplasmic reticulum;
blue, membrane-bound ribosomes; orange,
free ribosomes; bright green, microtubules;
bright blue, dense core vesicles; white,
clathrin-negative vesicles; bright red, clathrin-
positive compartments and vesicles; purple,
clathrin-negative compartments and vesicles;
dark green, mitochondria. (From B.J. Marsh,
Biochim. Biophys. Acta 1744:273–292, 2005.
With permission from Elsevier.)

involved in protein synthesis and quality control and proteins needed for motility were the most abundant. Other studies indicate that similar protein concentrations and abundances are widespread in prokaryotes. In response to changes in conditions, protein compositions can change. For example, when the spirochete is exposed to an antibiotic, it responds by expressing the genes for a small number of normally unrepresented proteins (of unknown function), but its total protein concentration remains constant. This homeostatic behavior resembles the response of cells in keeping concentrations of individual metabolites constant even when the flux through a metabolic pathway changes (Section 1.7).

Molecular crowding affects reaction rates, protein folding, assembly, and stability

The effective concentration of a macromolecule in any reaction (a property defined as its **thermodynamic activity**) is higher than its actual physical concentration. The **activity coefficient** is defined as the ratio of the two concentrations. At low concentrations, as in most experiments *in vitro*, the activity coefficient has a value very close to 1, but at high concentrations (crowding), it is well above 1. This has several important kinetic and thermodynamic consequences, among them effects on diffusion rates, reaction rates, and assembly (association or dissociation).

Studies of bacterial DNA replication were among the first *in vitro* experiments to be carried out under crowding conditions, when it was discovered in Arthur Kornberg's laboratory that it was necessary to add a synthetic polymer, polyethylene glycol (PEG), to the reaction mixture for DNA synthesis to occur. For reasons described below, this was probably due to PEG-induced molecular crowding, promoting the assembly of a particular complex at the origin of replication. It is important to bear such considerations in mind when attempting to relate the properties of a macromolecular system studied *in vitro* to the situation *in vivo*.

Crowding promotes the binding of macromolecules to one another, but specificity in these interactions depends on their structures. When macromolecules associate, there is a reduction in excluded volume and the association constants for macromolecular complexes are increased in comparison with their values in dilute solution. The degree of these changes depends on the sizes, shapes, and binding affinities of the macromolecules.

If another molecule is put into a solution of a globular protein at 300 mg/ml, its activity coefficient increases appreciably for molecules above 1 kDa and steeply for those above 10 kDa. For example, it has been estimated that for a 40 kDa protein in a monomer–dimer equilibrium, the association constant at the overall protein concentration found inside an *E. coli* cell would be 8–40 times higher than in dilute solution. If it were a monomer–tetramer equilibrium, the association constant would increase by a factor of 10^3–10^5. Thus, it appears that crowding effects are essentially confined to macromolecules, and small molecules are largely unaffected.

In intracellular protein folding, crowding favors the condensation of newly synthesized polypeptide chains into molten globule-like conformations (Section 1.3); in parallel, however, crowding also enhances the tendency of partly folded chains to come together in aberrant aggregates. Macromolecular machines called *chaperones* safeguard nascent polypeptide chains by sequestering them during the folding process (Section 6.3).

The increase in activity coefficients at high concentrations of reactants is not linear with respect to concentration. For example, the activity coefficient of hemoglobin at 200 mg/ml is ~10, and at 300 mg/ml it is ~100. The high concentration of hemoglobin in the red blood cell (~350 mg/ml) underlies the disastrous pathology of sickle-cell disease (**Box 1.3**). In contrast, the crystallin proteins of the eye lens, essential for transparency, do not aggregate to the point of scattering light, which would impair transparency, despite total protein concentrations of >500 mg/ml. Crystallins remain unreplaced throughout life, and their extraordinary stability and durability have been attributed to their high concentration.

Macromolecular crowding affects diffusion rates

Diffusion of macromolecules in cells generally follows Fick's laws (**Box 1.4**). The high concentration of protein might be expected to reduce mobility for all sorts of molecules. However, data obtained by single-particle tracking (SPT) using selectively labeled proteins or by fluorescence recovery after photobleaching (FRAP—the return of fluorescent protein

Box 1.3   Sickle-cell anemia and aberrant polymerization of a mutant hemoglobin

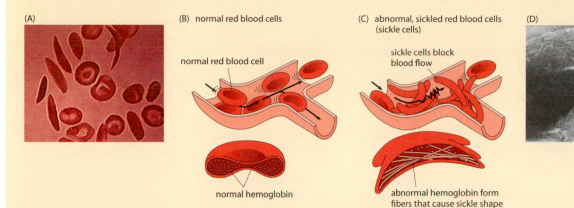

(A)

(B) normal red blood cells

normal red blood cell

normal hemoglobin

(C) abnormal, sickled red blood cells (sickle cells)

sickle cells block blood flow

abnormal hemoglobin form fibers that cause sickle shape

(D)

1 µm

Figure 1.3.1 Capillary blocking by red blood cells distorted by fibers of a mutant hemoglobin. (A) Scanning electron micrograph of normal erythrocytes (flattened concave disks) and sickled (elongated crescent) erythrocytes. (B,C) Schematic diagrams of (B) normal erythrocytes which are pliable and pass easily along narrow capillaries, and (C) sickled erythrocytes, which are stiffer and block capillaries. (D) Electron micrograph of a lysed sickled erythrocyte, negatively stained, showing fibers spilling out of the cell. (A, from the U.S. National Library of Medicine; B and C, adapted, courtesy of NIH Gov. NHLBI Diseases; D, from R. Josephs, H.S. Jarosch and S.J. Edelstein, *J. Mol. Biol.* 102:409–426, 1976. With permission from Elsevier.)

Hemoglobin (Hb) is an $\alpha_2\beta_2$ tetramer, whose α and β subunits (each 17 kDa) are homologous with similar folds. Hb serves as an oxygen carrier in red blood cells (erythrocytes). In humans, the mature erythrocyte has no nucleus or other organelle; its cytosol is essentially a highly concentrated solution of Hb.

Sickle-cell anemia is a debilitating and often fatal disease, in which the erythrocytes are deformed ('sickled'; see **Figure 1.3.1**). Sickled erythrocytes are fragile, and their rupture causes hemolytic anemia. They are also more rigid than normal erythrocytes and obstruct capillaries, restricting blood flow, causing pain and other symptoms. In 1945, Linus Pauling proposed that sickling is caused by a mutation in the Hb gene. In 1956, Vernon Ingram showed that there was a Glu→Val substitution at position 6 in the β subunit of sickle-cell Hb. This was the first instance of a disease being attributed to an amino acid replacement in a protein, a discovery that marked the birth of 'molecular medicine.'

Sickling is due to the deoxy-form of hemoglobin S (HbS) aggregating into filaments whose growth distorts the erythrocyte into an elongated shape (**Figure 1.3.2**). The hydrophobic side chain of Val6 protruding from one HbS tetramer can fit into a hydrophobic pocket on another tetramer. Repetition of this interaction generates paired strands of tetramers in which the two strands are half-staggered and related by a 2-fold screw axis. Seven such pairs twist about a common axis to make 14-stranded fibers with a diameter of ~24 nm and a mean helical pitch of ~270 nm. In wild-type Hb, the charged Glu side chain at position

$\beta6$ prevents any such interaction, and the surface pocket is absent in oxy-Hb or oxy-HbS because of the conformational change that accompanies oxygenation. However, when the erythrocyte enters a capillary, the Hb becomes deoxygenated and the surface pocket forms. This allows the deoxy-HbS tetramers to fibrillize, causing sickling, and the pathological symptoms ensue. Oxygenation causes the HbS fibers to redissolve. Much effort has gone into seeking treatments that can inhibit fiber formation, thus far without major success.

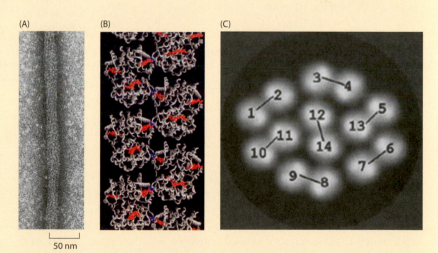

(A)

50 nm

(B)

(C)

1 2
3 4
12
13 5
11
10 14
6
9 8
7

Figure 1.3.2 Molecular packing in HbS fibers. (A) Electron micrograph of a negatively stained fiber. (B) Two-stranded fibers of HbS tetramers are present in the crystal structure. The backbones of the HbS subunits are in white; the heme groups in red, and the mutant βVal6 residues in blue. (PDB 2HBS) (C) Cross section of a fiber model reconstructed from the electron micrographs, showing the 14 strands (numbered 1–14) of HbS tetramers. The strands are paired as shown by the connecting lines. The fiber is elliptical in cross section, ~24 nm in diameter on the major axis. (A and C, from Z. Wang et al., *J. Struct. Biol.* 131:197–209, 2000. With permission from Elsevier; B, from D.J. Harrington, K. Adachi and W.E. Royer, *J. Mol. Biol.* 272:398–407, 1997. With permission from Elsevier.)

Box 1.4   Diffusion: Fick's laws

Molecules in solution are in constant motion, a random process called normal **diffusion**, synonymous with Brownian motion. The primary determinants of diffusion are the size and shape of the molecule and the viscosity of the solvent. If there is a concentration gradient, the molecules will diffuse from regions of higher concentration to lower concentration (in other words, their motion is entropy-driven).

Fick's First Law states that the net flux of diffusing substance, J (mol/(m^2 s)), in one dimension is given by

$$J = -D d\phi/dx$$

where D is the diffusion coefficient (in m^2/s), ϕ is the local concentration of the diffusing molecule (in mol/m^3), and x is distance (in m).

In three dimensions, this becomes

$$\boldsymbol{J} = -D\nabla\phi$$

where $\boldsymbol{J}$ is the flux vector and ∇ represents the vectorial derivative.

Fick's Second Law gives the change in concentration gradient with time:

$$d\phi/dt = D d^2\phi/dx^2$$

or, in three dimensions,

$$\partial\phi/\partial t = D\nabla^2\phi$$

Fick's laws were derived in 1855. In 1851, Stokes put forward his analysis of a small sphere moving with velocity v through a viscous medium in laminar flow (smooth flow without turbulence). He derived the equation

$$F = -6\pi\eta r v,$$

where F is the frictional force (viscous drag), η is the viscosity, and r is the radius of the sphere.

In 1905, Einstein analyzed Brownian motion in terms of the kinetic theory of heat and showed that, for a particle under such conditions,

$$D = RT/Nf_0$$

where R is the universal gas constant, T is the absolute temperature, N is Avogadro's number and f_0 is the solvent friction coefficient. This simplifies to

$$D = kT/f_0$$

where k = Boltzmann's constant. f_0 for a small spherical particle is, as Stokes had formulated, $6\pi\eta r$.

Biological macromolecules are not perfect spheres. This causes departures from the Einstein theory, and the value of the frictional coefficient, f, is greater than that of f_0 for the equivalent sphere. The extent to which the frictional ratio, f/f_0, is greater than 1.0 gives an estimate of how well a macromolecule can be modeled instead as, say, a prolate ellipsoid or a cylinder. For *tobacco mosaic virus*, f/f_0 has a value close to 2.0, reflecting its rod-like structure (axial ratio ~17:1). However, if a macromolecule or assembly has a high value of f/f_0 but electron microscopy or some other method of shape determination indicates that it is roughly spherical, it follows that the macromolecule or assembly carries unusually large amounts of bound water. The 2-oxo acid dehydrogenase multienzyme complexes described in Section 9.4 are a good example of this. For most globular proteins, the amount of bound water is ~0.35 g/g protein, and the value of f/f_0 is typically ~1.3, reflecting both the departure from perfect sphericity and their typical hydration.

At the macroscopic level, the importance of water is becoming apparent in some surprising ways. Given the strong crowding effects of intracellular proteins, small changes in cellular water cause large changes in activity coefficients. In older people, there is a significant loss of cellular and tissue water; a diminution in the occupiable volume available to proteins may contribute to the age-related onset of protein misfolding and aggregation diseases.

On the basis of Fick's laws and Einstein's work, the root mean square displacement (RMSD) $<r>$ of a solute particle in time t can be calculated from the equation

$$<r> = \sqrt{(6Dt)}.$$

For diffusion in one or two dimensions, the factor 6 is replaced by 2 or 4 respectively.

---

into a volume bleached by a laser flash) show only modest slowing (threefold to fourfold) for small solutes in the eukaryotic cytosol compared with their rates in water. With larger proteins, the effects are more pronounced: hemoglobin (68 kDa) in an erythrocyte diffuses six times more slowly than in dilute solution, and green fluorescent protein (GFP, 27 kDa) recombinantly expressed in the *E. coli* cytoplasm is slowed 11-fold. In a typical FRAP experiment, the bacterial chemotaxis signaling protein CheY (14 kDa; Section 14.12) fused with a fluorescent protein yielded a diffusion coefficient of 2 μm^2/s (**Figure 1.36**). Other measurements using fluorescent correlation spectroscopy gave a value of 4.6 μm^2/s and the diffusion coefficient of CheY alone has been estimated at 10 μm^2/s. Thus, it would take ~50 ms ($r^2/2D$; Box 1.4) for a CheY molecule to diffuse 1 μm from a pole of an *E. coli* cell to the flagellar motors along the sides of the cell (see Figure 14.73). This is consistent with the timescale of the chemotactic response.

Figure 1.36 Time course of protein diffusion in an *E. coli* cell. Simulation *in silico* of the return of fluorescence attributed to CheY tagged with yellow fluorescent protein (CheY–YFP) after photobleaching. The results were fitted to a single exponential curve on the assumption that the labeled molecules were homogeneous and in an isotropic environment, yielding a diffusion coefficient of 2 μm²/s. (From M.A. DePristo et al., *Prog. Biophys. Mol. Biol.* 100:25–32, 2009. With permission from Elsevier.)

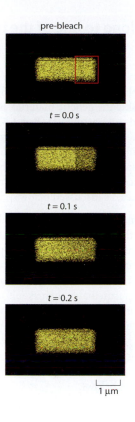

pre-bleach

t = 0.0 s

t = 0.1 s

t = 0.2 s

1 μm

In a cell, the central assumption of a continuous fluid of uniform viscosity across the cytoplasm or within an organelle does not normally hold. High concentrations of solutes and the presence of cytoskeletal elements, membranes, or other organelles affect the process, turning it into **anomalous diffusion**. Nevertheless, in many instances, solutes do observe the equations of normal diffusion.

Other factors may come into play; for example, the cytoskeleton restricts diffusion. For some complexes (such as dextrans above 500 kDa) or for linear DNAs, diffusion through the eukaryotic cytosol has been found to be greatly slowed. Within organelles, effects on diffusion may be smaller than theoretical calculations have suggested. Thus in the mitochondrial matrix, the diffusion of GFP is slowed down only about 3-fold, which has been attributed to the clustering of matrix enzymes into membrane-associated complexes (Chapter 15), thereby increasing the space through which other proteins can diffuse. However, in all cells and compartments, any diffusion-limited process stands to be affected; and the bigger the molecule, the larger the effect.

Models of crowded intracellular environments can now be built

It is now possible, conceptually and technically, to perform realistic computational modeling of the densely crowded cellular interior. For example, a simulation has been performed of the *E. coli* cytoplasm, based on 100 macromolecules representing 50 of the most abundant constituents of *E. coli* (45 proteins, the 50S and 30S ribosomal subunits and three typical tRNAs) plus 8 molecules of GFP, at a total concentration of 275 mg/ml. The structures of the chosen molecules were taken from the Protein Data Bank, and their relative abundances were derived from proteome measurements. Together they account for ~85% by weight of the cytoplasmic proteome. Their diffusion was calculated under conditions of Brownian motion, allowing for steric exclusion and electrostatic and short-range hydrophobic interactions. The process was calibrated by monitoring the GFP molecules, for which estimates of the translational diffusion coefficient (7.7–9.0 μm²/s) *in vivo* had been obtained by FRAP. In a 15 μs simulation, the GFP molecules were found to move, on average, about six molecular diameters (320 Å), encountering ~80 different neighbors in doing so. A still image is shown at **Figure 1.37**. Larger molecules moved more slowly, a ribosome hardly moving at all but still experiencing encounters with many more mobile neighbors.

1.9   CELLULAR COMPARTMENTATION AND EVOLUTION

As long ago as the 1930s, Frederick Gowland Hopkins made clear that it is wrong to perceive the cell as "just a bag of enzymes" but it is only now that the complexities are becoming recognized and, to some extent, amenable to investigation. The cell interior is not a homogeneous medium in which macromolecules are free to diffuse: in eukaryotic cells, much of the cytosol is occupied by organelles and crisscrossed by the filamentous network of the cytoskeleton. This leads to the confinement of many cytosolic macromolecules. Some attach (by electrostatics) to surfaces of organelles or the cytoskeleton, and others are anchored (by hydrophobic interactions) in membranes. Even in a prokaryotic cell, there exist microenvironments that differ significantly from the general cytosol.

Each kind of organelle represents a chemically distinct environment populated by a particular set of macromolecules. There is nevertheless communication and exchange of materials between organelles. Proteins are conveyed in vesicles that bleb off the donor compartment and fuse into the membrane that encloses the acceptor compartment. Some of these vesicles have protein coats, and transport is effected mainly by motor proteins traveling along cytoskeletal filaments.

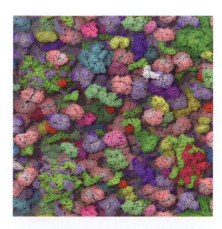

Figure 1.37 A dynamic model of the *E. coli* cytoplasm. Shown is a rendering of the model cytoplasm after a simulation of Brownian motion. RNA is depicted in green and yellow. (From S.R. McGuffee and A.H. Elcock, *PLoS Comp. Biol.* 6:e1000694, 2010.)

All cells belong to one of the three Urkingdoms: Archaea, Bacteria, Eukarya

Until the 1980s, organisms were classified as either **prokaryotes** or **eukaryotes**, a division based on morphological criteria. The defining feature of eukaryotic cells was the presence of a nucleus housing the genetic material; in addition, they possessed other organelles (membrane-enclosed compartments) (**Box 1.5**) that provide the chemical environments required for specific reactions or separate concurrent metabolic processes. Prokaryotes, which were mostly unicellular and small (1–5 μm), lacked a nucleus; intracellular organelles were found in only a few species such as photosynthetic bacteria. These defining properties remain in effect today except that prokaryotes are now recognized as falling into two kingdoms, bacteria and archaea, as discussed below.

The fundamental similarities at the molecular level shared by all present-day organisms make it likely that life as we know it has emerged on Earth only once. Most probably, all extant organisms are descended from a **progenote**, a hypothetical entity with an organizational level simpler than that of prokaryotic cells (**Figure 1.38**). Given that the subcellular organization of even the simplest prokaryotes is already quite complex, there must have been simpler intermediates such as self-replicating RNA-like molecules contained in semipermeable vesicular structures from which the ability to assemble amino acids into proteins subsequently evolved. At the level of the progenote, 'genes' and the proteins encoded by them were probably small and the accuracy of transcription and translation was probably low. The progenote is distinct from and precedes the **last universal common ancestor** (LUCA), which lies at the base of phylogenetic trees and is thought to have been a prokaryote-like cell (see below).

The idea of a Tree of Life goes back to Charles Darwin (**Figure 1.39**), but nowadays phylogenetic trees are based on molecular criteria rather than comparative anatomy or morphology. Protein or gene sequences and, increasingly, whole genomes are used to construct the trees (**Figure 1.40**). In the 1970s, it started to become apparent that prokaryotes can be partitioned into two fundamentally different life forms. In particular, a comparison of ribosomal RNA sequences indicated the existence of two distinct groups, archaea and bacteria. Carl Woese proposed that all organisms belong to one of three kingdoms, later renamed **Archaea**, **Bacteria**, and **Eukarya** (see Figure 1.38). This reclassification was challenging but is now supported by multiple lines of evidence. Nevertheless, it is still useful in some contexts to retain the term 'prokaryote' for non-eukaryotes; in other words, archaea or bacteria.

Bacteria have an open compartment, the nucleoid, and a membrane-delimited compartment, the periplasm

The absence of organelles in most prokaryotes does not mean that their cytosol lacks organization. The genome, usually in the form of a single circular chromosome, is confined to a region, the **nucleoid**, which can be detected by light microscopy using DNA-specific stains. Bacteria belonging to the phylum *Planctomycetes* are an exception in that they have membrane-enclosed compartments; their nucleoid is enclosed by a membrane and is therefore analogous to the eukaryotic nucleus. Moreover, some bacteria possess microcompartments enclosed by polyhedral protein shells that contain specific enzymes; an example in cyanobacteria is the **carboxysome**, which houses the CO_2-fixing enzyme RuBisCo (ribulose-1,5-bisphosphate carboxylase).

In addition to membrane-enclosed compartments, cells have 'open' compartments, regions where certain components congregate. The bacterial nucleoid affords one example. (Eukaryotes also have open compartments such as the *nucleolus*—a site for ribosome synthesis within the nucleus.) Open compartments are likely to have some sort of scaffold. In the cytosol, this role is fulfilled in at least some cases by the cytoskeleton. For a long time it was held that prokaryotes do not have a cytoskeleton, but recently genomic sequences, molecular structures, and advanced imaging have shown that homologs or analogs of the major eukaryotic cytoskeletal proteins are found in prokaryotes. Gram-negative bacteria also have a clearly defined compartment distinct from their cytosol, the **periplasm** or **periplasmic space** situated between the plasma membrane (also called the inner membrane) and the outer membrane.

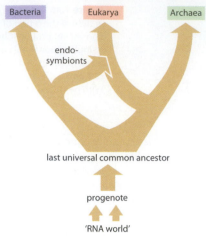

Figure 1.38 A likely scenario for the early evolution of cellular life. After the RNA world, DNA gradually replaced RNA for information storage, and proteins supplanted RNA in most aspects of biocatalysis. The LUCA is held to have been a prokaryote, whereas the progenote represents an earlier and simpler life form. (Adapted from J.P. Gogarten and L. Olendzenski, in Encyclopedia of Molecular Biology [T. Creighton, ed.]. Wiley On-line Library.)

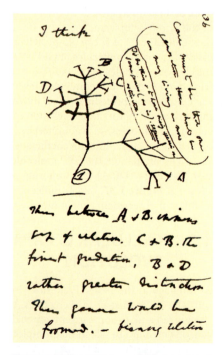

Figure 1.39 Darwin's tree of life. Annotated sketch from his notebook B (1837). Right bubble: "Case must be that in one generation then, there should be as many living as now". Second bubble: "To do this & to have many species in the same genus requires extinction." Bottom: "Thus between A & B, immense gap of relation. C & B, the finest gradation. B & D, rather greater distinction. Thus genera would be formed – bearing relation". (Transcripts courtesy of the American Museum of Natural History.)

Box 1.5   Major compartments in eukaryotic cells

Chloroplasts: Organelles found in plant cells and some algae, enclosed by a double membrane. Their interior (*stroma*) houses a complex system of stacked *thylakoid membranes*, known as *grana*. Chloroplasts, ~5 μm across, are the site of photosynthesis.

Cytosol: Main part of the cytoplasm, excluding the nucleus and other organelles. Typically it constitutes about 50% of the cell volume. It is the major site of metabolic activity.

Endoplasmic reticulum (**ER**): Convoluted compartment where lipids, secretory proteins, and membrane proteins are synthesized. Part of it, the rough ER, is densely studded with ribosomes on the cytosolic surface. Together with the cisternae of the Golgi apparatus, it occupies 15–20% of the cell volume.

Endosomes: Organelles containing materials newly ingested into the cell (by **endocytosis**) and destined to be fused into lysosomes where their protein cargo is degraded.

Golgi apparatus (**Golgi complex**): A stacked system of flattened vesicles (cisternae); the part proximal to the ER is the *cis* face, and the part distal to it is the *trans* face. Here, proteins from the ER are modified (for example, by glycosylation) and sorted for secretion from the cell or delivery to other compartments.

Lysosomes: Vesicular structures containing enzymes that degrade foreign materials taken into the cell by endocytosis. These enzymes are fully active only at the acidic pH (~2) within a lysosome.

Mitochondria: Organelles 1–2 μm long by ~0.5 μm wide. The inner mitochondrial membrane has deep infoldings called **cristae**. Mitochondria occupy up to 20% of the cellular volume. In aerobic metabolism, they produce most of the ATP.

Nucleus: Surrounded by the nuclear envelope, a double membrane, the nucleus occupies 5–10% of the cell volume. The outer nuclear membrane is continuous with the ER.

Peroxisomes: These are small vesicles harboring enzymes that carry out oxidative reactions. Some use molecular oxygen to oxidize organic substrates, and this generates toxic hydrogen peroxide (H_2O_2), which is decomposed into H_2O and O_2 by catalase.

Mitochondria and chloroplasts are of particular relevance to this book through their roles in generating the energy needed to support life. The main properties of mitochondria are summarized here; chloroplasts are covered in Chapter 15. Eukaryotic cells typically contain 1000–2000 mitochondria, which are delimited by an outer and an inner membrane. The inner membrane encloses a compartment called the **mitochondrial matrix** (**Figure 1.5.1**). Its surface area is enlarged by the cristae, whose number correlates with the respiratory intensity of the cell or tissue; for example, there are many more in heart mitochondria than in those of skeletal muscle. The sides of the inner membrane facing the matrix and inter-membrane space are termed the N and P sides respectively, because the **membrane potential** is negative on the inside (matrix) and positive on the outside (inter-membrane space). The inner membrane is essentially impermeable to ions and polar molecules of all kinds but allows the passage of O_2 and CO_2. In contrast, the outer membrane is largely permeable to ions and small molecules (<5–10 kDa) because it contains protein pores, *mitochondrial porins*, among them the voltage-dependent anion channel (VDAC).

A mitochondrion contains from one to many copies of a circular DNA, which varies widely in size between species and encodes a variety of RNAs (tRNAs, rRNAs, etc.) and proteins. The existence of circular mitochondrial DNA and the presence of bacterial-type ribosomes, together with many other similarities to bacterial cells, have led to the *endosymbiont theory* (see the main text) that mitochondria and chloroplasts arose from the engulfment of bacteria by primordial eukaryotic cells.

Figure 1.5.1 The structure of a mitochondrion. (A) Electron micrograph of a thin section of a mitochondrion. (B) Schematic diagram of a mitochondrion, showing the different compartments contained within the outer membrane. The respiratory chain complexes and ATP synthases (Chapter 15) are arranged vectorially in the inner membrane. The ATP synthases span the inner membrane and cluster as dimers (depicted as yellow spheres) along the sharp folds in the cristae. The matrix contains many enzymes of oxidative metabolism, including those involved in the citric acid cycle and fatty acid degradation (Chapter 9), plus ribosomes and molecules of mitochondrial DNA. (A, from B. Alberts et al., Molecular Biology of the Cell, 6th ed. New York: Garland Science, 2015.)

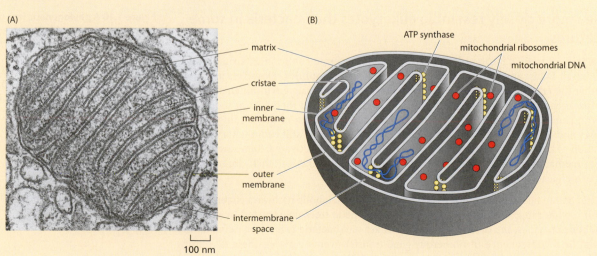

(A)

100 nm

(B)

matrix
cristae
inner membrane
outer membrane
intermembrane space

ATP synthase
mitochondrial ribosomes
mitochondrial DNA

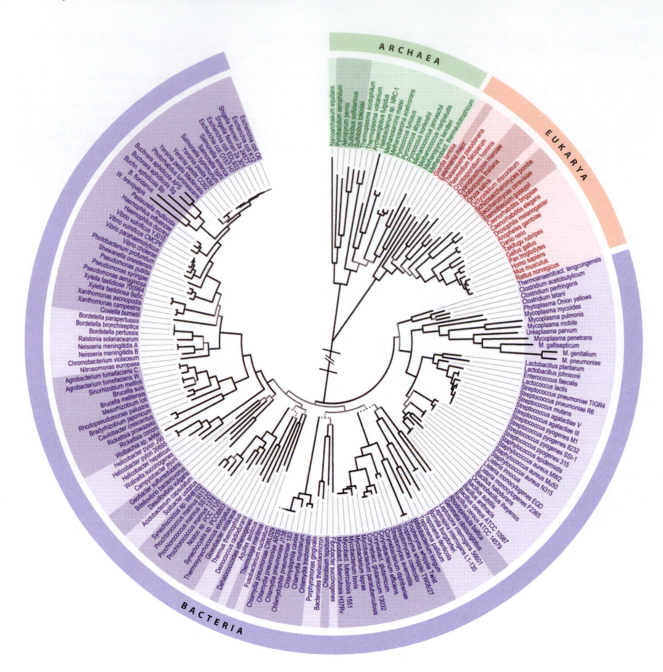

Archaea more closely resemble eukaryotes than bacteria in some key features

The first archaea to be described happened to inhabit extreme environments such as hot springs or salt lakes, and so they were perceived as **extremophiles**. However, it is now known that the great majority of archaeal species inhabit moderate environments such as the oceans or the soil. It has been estimated that archaea account for ~40% of the Earth's biomass and are major players in its carbon and nitrogen cycles. Our understanding of their physiology and their contribution to the planet's global ecology is still rudimentary: only a tiny fraction of known archaeal species have been isolated and cultivated. Some archaea, in particular methanogens, inhabit the guts of humans and ruminants, but there are no examples yet of archaeal pathogens or parasites.

In archaea, and to a lesser extent in bacteria, a variety of metabolic pathways enables them to exploit a wide range of energy sources. In phylogenetic trees constructed by comparing metabolic enzymes, the division between archaea and bacteria is not well defined; this appears to be a consequence of numerous gene transfer events between these two kingdoms. However, the machinery of gene expression (DNA replication, transcription and

Figure 1.40 A phylogenetic tree derived from the genomes of fully sequenced organisms. The representatives of Archaea are shown in green, Eukarya in red, and Bacteria in blue. (Adapted from F.D. Ciccarelli et al., *Science* 311:1283–1297, 2006. With permission from American Association for the Advancement of Science.)

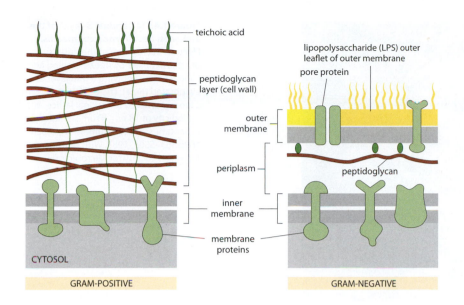

Figure 1.41 The cell envelope architectures of Gram-positive and Gram-negative bacteria. The dye introduced by H.C. Gram in 1864 binds to the thick peptidoglycan layers of Gram-positive bacteria, producing a purple color. Gram-negative bacteria, with less peptidoglycan, bind less stain and are colored pink. The green tendrils in the Gram-positive cell wall are the polyglyceridephosphate chains of lipoteichoic acid molecules. (Adapted from B. Alberts et al., Molecular Biology of the Cell, 6th ed. New York: Garland Science, 2015.)

translation), although fundamentally similar, exhibits distinct differences. In particular, the replication machinery of archaea resembles that of eukaryotes, whereas archaeal ribosomes superficially resemble bacterial ribosomes but at the level of ribosomal RNA and protein structure are more closely related to eukaryotic ribosomes.

Major differences exist between archaeal and bacterial cell envelopes

Almost all bacteria possess a peptidoglycan layer as part of their cell envelope. In Gram-positive bacteria this layer is thick (20–100 nm) and covers the plasma membrane; in Gram-negative bacteria it is thin (<10 nm) and located in the periplasm (**Figure 1.41**). **Peptidoglycan**, also called **murein**, is a sheetlike polymer formed by linear chains of two amino sugars, N-acetyl glucosamine and N-acetyl muramic acid, linked by a β-(1,4) glycosidic bond and cross-linked by short (4–5 aa) peptides. Archaeal cells lack peptidoglycan layers, but a few species have an analog called pseudopeptidoglycan or pseudomurein (**Figure 1.42**). It is different in chemical composition (N-acetylglucosamine and N-acetyltalosaminuronic acid linked by a β-(1,3) glycosidic bond), but otherwise resembles bacterial peptidoglycans. In many archaea, the plasma membrane is reinforced and protected by regularly arrayed (glyco)proteins forming surface layers (**S-layers**) (**Figure 1.43** and **Figure 1.44**). Some S-layers have been implicated in determining cell shape. S-layers unrelated in primary structure have also been found in some bacteria, most of them Gram-positive.

Whereas the membranes of bacteria and eukarya are composed mainly of glycerol-ester lipids, archaeal membranes are made of glycerol-ether lipids. Typically, lipids are amphipathic with a hydrophilic 'head group' and hydrophobic 'tails.' In eukaryotic and bacterial phospholipids, a tail is made of long fatty acid chains, but archaeal hydrophobic tails

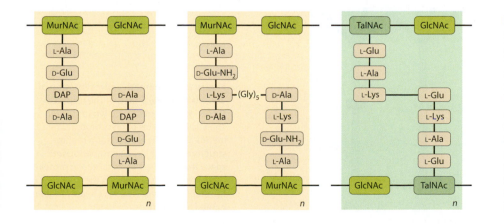

Figure 1.42 Cross-linking patterns of bacterial murein and archaeal pseudomurein. In peptidoglycan (murein), the sugar derivatives N-acetylmuramic acid (MurNAc) and N-acetylglucosamine (GlcNAc) form a linear backbone cross-linked by short peptides. n denotes the number of units in the backbone (many). Some archaea have an evolutionarily unrelated cell envelope component referred to as pseudopeptidoglycan or pseudomurein (right). Here GlcNAc and acetyltalosaminuronic acid (TalANAc) alternate; unlike murein, pseudomurein lacks D-amino acids.

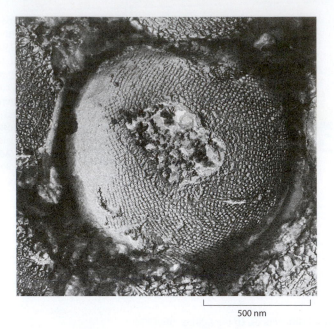

500 nm

Figure 1.43 Tetragonal S-layer of the archaeon *Desulfurococcus mobilis*. Electron micrograph of the outer surface of a cell exposed by freeze-etching and contrasted by applying a thin layer of evaporated metal. The cell is covered with a loosely ordered tetragonal lattice with a periodicity of 18 nm. (From I. Wildhaber, U. Santarius and W. Baumeister, *J. Bacteriol.* 169:5563–5568, 1987. With permission from American Society for Microbiology.)

are isoprenoid in nature, often containing methyl branches or cyclopentane rings. Some archaea, in particular extreme thermophiles, contain tetraether lipids that have two polar headgroups with their 2,3-*sn*-glycerol moieties covalently linked by the C_{40} isoprenoid chains. These tetraether lipids can form membranous monolayers. The presence of tetraether lipids and of isoprenoid tails, often with cyclopropane or cyclohexane rings and ether bonds replacing ester bonds, are all thought to increase membrane stability, enabling the organisms to tolerate extremes of temperature, salt concentration, or pH (**Figure 1.45**).

Eukaryotic cell organelles probably arose by engulfing bacteria

Most scenarios for the origin of eukaryotic cells accept the **endosymbiont theory** advocated by Lynn Margulis. This proposes that their organelles arose from the internalization of one organism by another, followed by their symbiotic association (see Figure 1.38). The supporting arguments are strong for mitochondria and chloroplasts: both replicate by a process similar to the binary fission of bacteria; both are surrounded by two membranes; in both cases their ribosomes more closely resemble those of bacteria than those of eukaryotes, particularly for chloroplasts; and both have circular genomes distinct from the linear DNA of nuclear chromosomes. However, their genomes are much smaller than those of bacteria: human mitochondrial DNA (16,569 bp) has only 37 genes (encoding two ribosomal RNAs, 22 tRNAs, and 13 proteins). Chloroplast DNA is generally larger; for example, that of *Arabidopsis* (154,478 bp) has about 120 genes (encoding 4 ribosomal RNAs, 37 tRNAs, and potentially 87 proteins). Most of the present-day mitochondrial and chloroplast proteins

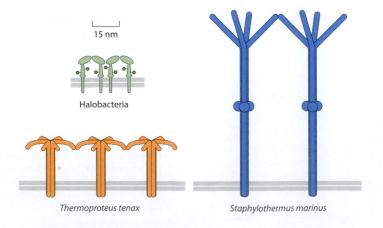

15 nm

Halobacteria

Thermoproteus tenax

Staphylothermus marinus

Figure 1.44 Schematic representation of various archaeal S-layers. The respective repeating elements are drawn to scale. The S-layer proteins or glycoproteins are membrane-anchored and form rooflike structures delimiting a quasi-periplasmic space that may harbor extracellular proteins such as a protease bound to the long stalk of the *Staphylothermus marinus* S-layer protein named tetrabrachion.

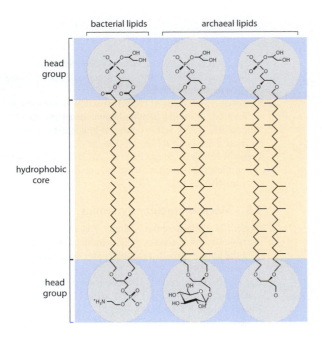

bacterial lipids   archaeal lipids

head group

hydrophobic core

head group

Figure 1.45 Comparison of bacterial and archaeal lipids. In bacteria and eukarya, the glycerol moiety is ester-linked to an *sn*-glycerol 3-phosphate backbone, whereas in archaea the isoprenoid side chains are ether-linked to an *sn*-glycerol 1-phosphate moiety. The bilayer-forming lipids in bacteria are phosphatidylglycerol (upper) and phosphatidylethanolamine (lower). The archaeal lipid can either be a C_{40} tetraether forming a continuous monolayer (left) or C_{20} diether lipids forming a bilayer (right). In bacterial lipids, the hydrophobic core is formed by the fatty acid hydrocarbon chains, whereas in archaea it consists of isoprenoids, often with methyl branches. (Adapted from S.-V. Albers and B.H. Meyer, *Nat. Rev. Microbiol.* 9:414–426, 2011. With permission from Macmillan Publishers Ltd.)

are encoded in nuclear DNA, and it appears that, during evolution, many genes were transferred from organelle to nucleus and the organellar DNAs shrank in size. These events are thought to have happened more than 1.5 billion years ago, when substantial amounts of oxygen entered the atmosphere from the activity of bacterial cells engaged in photosynthesis (Section 15.3).

Phylogenetic analysis based on genome sequences supports the endosymbiont theory and suggests that the bacterial progenitor of mitochondria was an ancestor of *Rickettsia prowazekii*, which has $>10^6$ bp of DNA and is a cause of lice-borne typhus, and that chloroplasts arose from the engulfment of an O_2-producing photosynthetic bacterium resembling a modern cyanobacterium.

The origin of the first cell with a nucleus remains enigmatic. Possibly, it reflects another fusion event, involving bacterial hosts related to the *Planctomycetes* (see above) and archaea as symbionts. The common ancestor of plants and animals was probably a eukaryotic cell that had mitochondria but not chloroplasts, and it is likely that plants and animals evolved multicellularity independently.

Higher eukaryotes have similar numbers of genes as lower eukaryotes but many more regulatory elements

The genomes of archaea and bacteria are similar in size. The smallest (for example, that of *Mycoplasma genitatum*) encode about 500 proteins, the largest (for example, that of *Bacillus cereus*) about 5000, in the same range as the genomes of some single-celled eukaryotes such as yeast (*S. cerevisiae*). Multicellular higher eukaryotes have between 13,600 genes (the fly *Drosophila melanogaster*) and 28,000 genes (the plant *Arabidopsis thaliana*). The human genome contains about 20,000 protein-coding genes—about the same in number (and with largely similar functions) as those of the nematode worm *C. elegans*. However, this worm has only 10^3 cells and $\sim 10^8$ bp in its chromosomal DNA, compared with $\sim 10^{14}$ cells and $\sim 3.2 \times 10^9$ bp for an adult human. In general, the ratio of non-protein-coding (formerly referred to, misleadingly, as 'junk' DNA) to protein-coding DNA increases with developmental complexity; it is approximately 3:1 in *C. elegans* and 50:1 in humans. Almost all of the genomes of humans and worms are now known to be transcribed, generating vast numbers of non-protein-coding RNAs, and there is a growing awareness that these form the basis of complex regulatory systems that underlie cellular differentiation and development. Thus the role of RNA is by no means confined to that of a messenger for protein synthesis and serving as components (structural and catalytic) of the ribosome (Section 6.2).

REFERENCES

1.1   Expression of the genetic blueprint

Ahlquist P (2002) RNA-dependent RNA polymerases, viruses, and RNA silencing. *Science* 296:1270–1273.

Crick F (1970) Central dogma of molecular biology. *Nature* 227:561–563.

Kempner ES & Miller JH (2003) The molecular biology of *Euglena gracilis*. XV. Recovery from centrifugation-induced stratification. *Cell Motil Cytoskel* 56:219–224.

Mattick JS (2011) The central role of RNA in human development and cognition. *FEBS Lett* 585:1600–1616.

Nandkumar J & Cech T (2013) Finding the end: recruitment of telomerase to telomeres. *Nat Rev Mol Cell Biol* 14:69–82.

1.2   Weak forces and molecular interactions

Frank RAW, Pratap JV, Pei XY et al (2005) The molecular origins of specificity in the assembly of a multienzyme complex. *Structure* 13:1119–1130.

Janin J, Bahadur RP & Chakrabarti P (2008) Protein–protein interaction and quaternary structure. *Q Rev Biophys* 41:133–180.

Jung H-I, Bowden SJ, Cooper A et al (2002) Thermodynamic analysis of the binding of component enzymes in the assembly of the pyruvate dehydrogenase multienzyme complex of *Bacillus stearothermophilus*. *Protein Sci* 11:1091–1100.

Kuriyan J, Konforti B & Wemmer D (2012) The Molecules of Life: Physical and Chemical Principles. Garland Science.

Lee B & Richards FM (1971) The interpretation of protein structures: estimation of static accessibility. *J Mol Biol* 55:379–400.

McDonald IK & Thornton JM (1994) Satisfying hydrogen bonding potential in proteins. *J Mol Biol* 238:777–793.

PISA may be accessed at www.ebi.ac.uk/pdbe/pisa/

Zhang Q, Sanner M & Olson AJ (2008) Shape complementarity of protein–protein complexes at multiple resolutions. *Proteins* 75:453–467.

1.3   Protein folding and stability

Brockwell DJ & Radford SE (2007) Intermediates: ubiquitous species on folding energy landscapes? *Curr Opin Struct Biol* 17:30–37.

Dill KA, Ozkan SB, Shell, MS et al (2008) The protein folding problem. *Annu Rev Biophys* 37:289–316.

Dobson CM (2003) Protein folding and misfolding. *Nature* 426:884–890.

Hartl U & Hayer-Hartl M (2009) Converging concepts of protein folding *in vitro* and *in vivo*. *Nat Struct Mol Biol* 16:564–558.

Kalia Y, Brocklehurst SM, Hipps DS et al (1993) The high-resolution structure of the peripheral subunit-binding domain of dihydrolipoamide acetyltransferase from the pyruvate dehydrogenase multienzyme complex of *Bacillus stearothermophilus*. *J Mol Biol* 230:323–341.

Karshikoff A & Ladenstein R (1998) Proteins from thermophilic and mesophilic organisms essentially do not differ in packing. *Prot Eng* 11:867–872.

Neuweiller H, Sharpe TD, Rutherford TJ et al (2009) The folding mechanism of BBL: plasticity of transition-state structure observed within an ultrafast folding protein family. *J Mol Biol* 390:1060–1073.

Pauling L (1960) The Nature of the Chemical Bond. Cornell University Press.

Privalov PL & Gill SJ (1988) Stability of protein structure and hydrophobic interaction. *Adv Prot Chem* 39:191–234.

1.4   Self-assembly and symmetry

Cansu S & Doruker P (2008) Dimerization affects collective dynamics of triosephosphate isomerase. *Biochemistry* 47:1358–1368.

Chan WW-C, Kaiser C, Salvo JM et al (1974) Formation of dissociated enzyme subunits by chemical treatment during renaturation. *J Mol Biol* 87:847–852.

Engström A & Strandberg B (eds) (1969) Symmetry and Function of Biological Systems at the Macromolecular Level: Nobel Symposium 11. Wiley Interscience.

Goodsell DS & Olson AJ (2000) Structural symmetry and protein function. *Annu Rev Biophys Biomol Struct* 29:105–153.

Perham RN (1975) Self-assembly of biological macromolecules. *Phil Trans R Soc Lond B* 272:123–136.

Perica T, Marsh JA, Sousa FL et al (2012) The emergence of protein complexes: quaternary structure, dynamics and allostery. *Biochem Soc Trans* 40:475–491.

1.5   Macromolecular dynamics

Baxter RH, Ponomarenko N, Srajer V et al (2004) Time-resolved crystallographic studies of light-induced structural changes in the photosynthetic reaction center. *Proc Natl Acad Sci USA* 101:5982–5987.

Kay LE (2011) Solution NMR spectroscopy of supra-molecular systems, why bother? A methyl-TROSY view. *J Magn Reson* 210:159–170.

Khalili-Araghi F, Gumbart J, Wen PC et al (2009) Molecular dynamics simulations of membrane channels and transporters. *Curr Opin Struct Biol* 19:128–137.

Ma BC, Tsai J, Haliloglu T & Nussinov R (2011) Dynamic allostery: linkers are not merely flexible. *Structure* 19:907–917.

Perrais D & Merrifield CJ (2005) Dynamics of endocytic vesicle creation. *Dev Cell* 9:581–592.

Torchia DA (2011) Dynamics of biomolecules from picoseconds to seconds at atomic resolution. *J Magn Reson* 212:1–10.

Tsuruta H & Irving TC (2008) Experimental approaches for solution X-ray scattering and fiber diffraction. *Curr Opin Struct Biol* 18:601–608.

Zhou J, Ha KS, La Porta A et al (2011) Applied force provides insight into transcriptional pausing and its modulation by transcription factor NusA. *Mol Cell* 44:635–646.

1.6   Catalysis

Cornish-Bowden A (2012) Fundamentals of Enzyme Kinetics, 4th ed. Wiley-VCH.

Glowacki DR, Harvey JN & Mulholland AJ (2012) Taking Ockham's razor to enzyme dynamics and catalysis. *Nat Chem* 4:169–176.

Hammes GG, Benkovic SJ & Hammes-Schiffer S (2011) Flexibility, diversity, and cooperativity: pillars of enzyme catalysis. *Biochemistry* 50:10422–10430.

Hay S & Scrutton NS (2012) Good vibrations in enzyme-catalysed reactions. *Nat Chem* 4:161–168.

Maps of metabolic pathways may be accessed at www.iubmb-nicholson.org/chart.html

Moser CC, Chobot SE, Page CC et al (2008) Distance metrics for heme protein electron tunneling. *Biochim Biophys Acta* 1777:1032–1037.

Klinman JP (2014) The power of integrating kinetic isotope effects into the formalism of the Michaelis–Menten equation. *FEBS J* 281:489–497.

1.7   Signaling and regulatory mechanisms

Alonso A, Sasin J, Bottini N et al (2004) Protein tyrosine phosphatases in the human genome. *Cell* 117:699–711.

Fell DA (1992) Metabolic control analysis: a survey of its theoretical and experimental development. *Biochem J* 286:313–330.

Hofmeyr J-H & Cornish-Bowden A (2000) Regulating the cellular economy of supply and demand. *FEBS Lett* 476:47–51.

Johnson SA & Hunter T (2005) Kinomics: methods for deciphering the kinome. *Nat Methods* 2:17–25.

Koshland DE Jr, Nemethy G & Filmer D (1966) Comparison of experimental binding data and theoretical models in proteins containing subunits. *Biochemistry* 5:365–385.

Monod J, Wyman J & Changeux JP (1965) On the nature of allosteric transitions: a plausible model. *J Mol Biol* 12:88–118.

Swain JF & Gierasch LM (2006) The changing landscape of protein allostery. *Curr Opin Struct Biol* 16:102–108.

1.8 Macromolecular crowding

Bloomfield V, Dalton WO & Van Holde KE (1967) Frictional coefficients of multisubunit structures. I. Theory. *Biopolymers* 5:135–148.

Dix JA & Verkman AS (2008) Crowding effects on diffusion in solutions and cells. *Annu Rev Biophys* 37:247–263.

Doster W & Longeville S (2007) Microscopic diffusion and hydrodynamic interactions of hemoglobin in red blood cells. *Biophys J* 93:1360–1368.

Malmström J, Beck M, Schmidt A et al (2009) Proteome-wide cellular protein concentration of the human pathogen *Leptospira interrogans*. *Nature* 460:762–766.

McGuffee SR & Elcock AH (2010) Cytoplasm full energy model. doi:10.1371/journal.pcbi.1000694.s014.

Minton AP (2006) How can biochemical reactions within cells differ from those in test tubes? *J Cell Sci* 119:2863–2869.

Schachman HK (1959) Ultracentrifugation in Biochemistry. Academic Press.

Zhou H-X, Rivas G & Minton AP (2008) Macromolecular crowding and confinement: biochemical, biophysical, and potential physiological consequences. *Annu Rev Biophys* 37:375–397.

1.9 Cellular compartmentation and evolution

Archibald JM (2008) The eocyte hypothesis and the origin of eukaryotic cells. *Proc Natl Acad Sci USA* 105:20049–20050.

Ciccarelli FD, Doerks T, von Mering C et al (2006) Toward automatic reconstruction of a highly resolved tree of life. *Science* 311:1283–1287.

Cox CJ, Foster PG, Hirt RP, Harris SR & Embley TM (2008) The archaebacterial origin of eukaryotes. *Proc Natl Acad Sci USA* 105:20356–20361.

McInerney JO, Martin WF, Koonin EV et al (2011) Planctomycetes and eukaryotes: a case of analogy not homology. *BioEssays* 33:810–817.

Woese CR, Kandler O & Wheelis ML (1990) Towards a natural system of organisms: proposal for the domains Archaea, Bacteria, and Eucarya. *Proc Natl Acad Sci USA* 87:4576–4579.

DBD

conformational
change

DBD

new remodeling
cycle

DBD

DNA loop
propagation
around the 2nd half
of the nucleosome

anchor

DNA loop
formation

anchor

anchor

anchor

Tr

Tr

Tr

Tr

Chromatin

2

2.1   INTRODUCTION

The study of **chromatin** had its beginnings in the late nineteenth century when scientists started using microscopy to investigate cell division and biochemistry to analyze the chemical composition of nuclear material. In fact, the term chromatin was first introduced around this time to describe "that substance in the cell nucleus which is readily stained." In the 1930s, light microscopy revealed two types of chromatin, a highly condensed form called *heterochromatin* and *euchromatin*, a less dense form. However, it was not until the advent of chromatography and gel electrophoresis in the 1960s that it became possible to characterize the protein components of chromatin, leading to the identification of four major classes of core *histones* (H2A, H2B, H3, H4) and a fifth class of linker histones.

Early models suggested that chromatin consisted of supercoiled DNA coated by histones. Nuclease digestion of chromatin (unlike that of free DNA) produces a regularly spaced 'laddered' pattern in gel electrophoresis, indicating that chromatin has a simple, repeating substructure. Filaments decorated by repeating subunits ('beads on a string') were also observed in electron micrographs of chromosomal material and purified chromatin preparations. The appearance of these structures was very sensitive to ionic conditions; lowering the ionic strength led to uncoiling, while increasing the ionic strength resulted in compaction and higher order structures (**Figure 2.1**).

In the mid-1970s several groups succeeded in isolating stable nucleoprotein particles that could be characterized by analytical ultracentrifugation. These particles were presumably the 'beads' observed in electron micrographs, which we now call *nucleosomes*. In 1974 a structural model based on biochemical evidence was introduced in which two copies each of histones H2A, H2B, H3, and H4 form an octameric structure. The (H3–H4)$_2$ tetramer was proposed to coil ~200 base pairs (bp) of DNA around its periphery, while binding of the

(A)   (B)   (C)

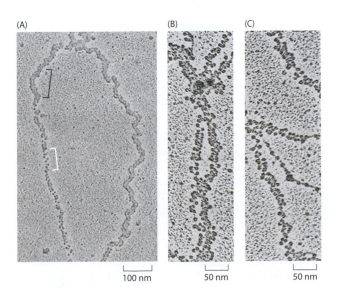

100 nm   50 nm   50 nm

Figure 2.1 Electron micrographs of metal-shadowed chromatin fibers released from metaphase cells of chicken cell line by detergent lysis. (A) The fibers illustrate the transition from a loosely packed ribbon to a compact fiber (regions in white and black brackets). (B) Regions of the fiber showing face-to-face stacking of nucleosomes. (C) Parallel row arrangements coexist with regions where the face-to-face interaction is lost. (From C.L.F. Woodcock, L.L.Y. Frado and J.B. Rattner, *J. Cell Biol.* 99:42–52, 1984. With permission from Rockefeller University Press.)

H2A–H2B dimers stabilized the octamer–DNA complexes. The DNA was no longer seen as being coated by histones but as being wrapped around an octameric histone core, forming a contiguous repetitive array separated by histone-free and nuclease-accessible 'linker' DNA. Although essentially correct, the model overestimated the amount of histone-bound DNA and underestimated the amount of linker DNA between the octameric particles. Ultimately, this model was validated by high-resolution crystal structures. While the nucleosome structure is now understood in great detail, higher order chromatin structures beyond the level of the nucleosome remain controversial. The heterogeneity and absence of long-range order make high-resolution structural studies a daunting challenge.

For a long time, chromatin was thought to have primarily a structural role, allowing compaction of DNA. Indeed, in a human cell more than 2 meters of DNA must be organized and packed into a nucleus that is only a few micrometers in diameter (**Box 2.1**). More recently, the dynamic aspects of chromatin structure have come into focus. This raises the question of how histone–histone and histone–DNA contacts can be both tight and dynamic. Given that nucleosomes partially occlude the DNA, how is DNA accessed during replication, repair, transcription, and other processes? How are nucleosomes assembled and disassembled, and how are histones replaced or exchanged? All these activities involve molecular machines, which are the focus of this chapter. *Chromatin remodeling complexes* hydrolyze ATP to restructure or mobilize nucleosomes, which allows regulated exposure of DNA in chromatin (Section 2.3). The dynamics of chromatin structure are governed in part by post-translational modifications of histones (see also Section 2.4). It has recently become clear that covalent histone modifications have crucial functions in gene regulation and other chromatin-templated processes. These modifications may constitute a *histone code*, and *epigenetic* information is thought to extend the genetic message beyond the sequence of DNA bases. But first, this chapter will look at the structure of nucleosomes and other higher order types of chromatin, and the role of histone chaperones in assembling nucleosomes.

2.2   NUCLEOSOMES AND HIGHER ORDER CHROMATIN STRUCTURES

The core histones have a two-domain organization and structure

The four core classes of **histones** (H2A, H2B, H3, and H4) are highly conserved proteins. Only 2 of the 102 amino acids are different between the H4 of plants and mammals. The core histones of all eukaryotes are small (11–15 kDa), highly positively charged proteins that consist of two structurally and functionally separable domains (**Figure 2.2**). The N-terminal domain (NTD) or tail domain consists of ~20–35 residues. Most of the minor sequence variability occurs in the NTDs, which serve as essential determinants of chromatin condensation (see below). The next ~80–100 residues are mostly α-helical and contain the 'histone fold' domains and extensions. These interact with other histones and nucleosomal DNA. In addition, histone H2A also has an unstructured C-terminal tail domain (CTD).

The histone fold domain comprises three α helices (α1–3) connected by two short loops (L1–2). The central (28 aa) helix is flanked by two short (10–14 aa) helices, to form a shallow U-shaped cleft. Despite a low degree of sequence similarity between them, the H2A, H2B, H3, and H4 histone fold domains are structurally very similar. Additionally, each histone has unique secondary structural elements (called histone fold extensions), which are either involved in histone octamer stabilization (see below) or define the surface of the histone octamer.

The core histone NTDs are disordered and not visible in the crystal structure of the nucleosome or the histone octamer. These domains are deficient in hydrophobic amino acids, and approximately one-third of the residues are lysines or arginines, many of which are sites of specific post-translational modifications. In both their unmodified and modified states, the core histone NTDs can bind DNA and various proteins. Upon interaction with these macromolecules, the intrinsically disordered NTDs become structured (see Section 6.4 for a further discussion of natively unfolded protein domains).

Many archaea, but not bacteria, have a small set of histone-like proteins that consist only of the minimal histone fold motif. These proteins bind DNA as dimers or tetramers but do not form nucleosome-like structures. Archaeal histone-like proteins do not contain NTDs or

Box 2.1   The hierarchical condensation of DNA in the eukaryotic nucleus

To fit into the nucleus, DNA needs to be compacted (**Figure 2.1.1**). Compaction proceeds in a hierarchical manner and is effected by the interaction of the DNA with histone proteins. The first level of compaction is the formation of nucleosomes. The organization of DNA into 1.7 superhelical turns around the histone octamer to form nucleosomes provides a compaction of ~7-fold. The histone octamer forms a cylinder 9 nm in diameter and 5 nm in height. The human genome is organized in the form of ~32 million nucleosomes, which occupy ~10% of a 6 μm diameter nucleus or ~20% when including the associated DNA. Mediated by histone N-terminal tails and linker histone H1, the 'beads-on-a-string' nucleosomal arrays locally fold into more compact chromatin fibers. Folding occurs through a series of steps to ultimately produce a highly compacted structure with a diameter of ~30 nm. Formation of the 30 nm fiber produces an additional 50–100-fold compaction. In all likelihood, chromatin folding is governed by several modes of nucleosome–nucleosome interactions and is stabilized by chromatin architectural proteins and linker histone H1. In one model for a moderately folded array the 30 nm fiber is further coiled to form *chromonema fibers* (~300 nm); a meandering structure with loops emanating from a central backbone, or arrays of domains formed by intermingling fibers and radial loops. Chromonema fibers condense further into a 700 nm diameter interphase chromosome. The total degree of compaction achieved in going from naked DNA to the mitotic chromosomal fiber is ~10,000-fold. This allows up to 2 meters of chromosomal DNA to fit into the eukaryotic nucleus (**Figure 2.1.2**).

Figure 2.1.2 The many levels of chromatin packing. Each DNA molecule has been packaged into a mitotic chromosome that is 10,000-fold shorter than its extended length. (Adapted from B. Alberts et al., Molecular Biology of the Cell, 6th ed. New York: Garland Science, 2015.)

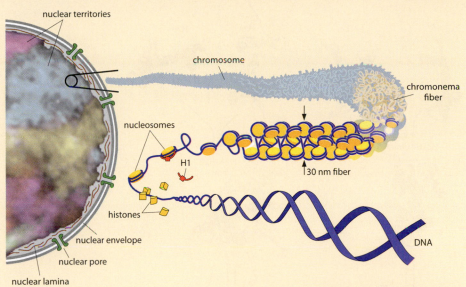

Figure 2.1.1 Major structures in DNA compaction. Within nuclei, chromosomes occupy discrete territories as first inferred from microlaser experiments inducing local genome damage and later visualized by light microscopy using chromosome-specific fluorescent probes. This spatial organization is emerging as a crucial aspect of gene regulation and genome stability.

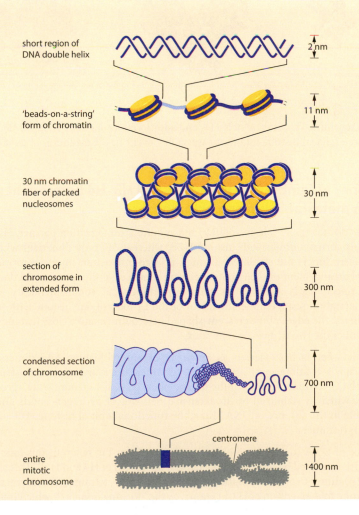

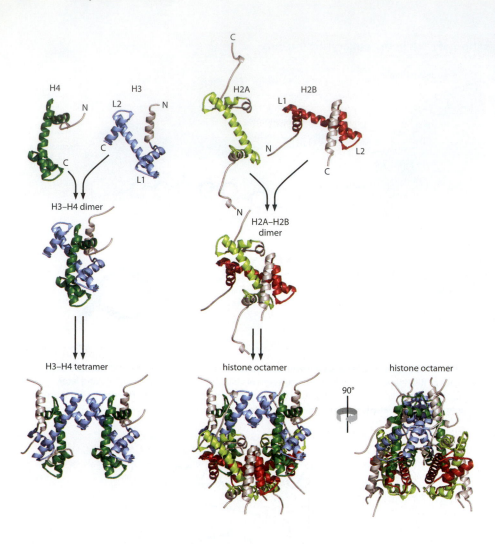

Figure 2.2 Structures of the four core histones H3, H4, H2A, and H2B and their assembly intermediates in histone octamer formation.

histone fold extensions. Thus, fusion of NTDs and additional secondary structural elements to the primitive archaeal histone fold motif was a key event in the evolution of eukaryotic chromatin.

Core histones assemble into H2A–H2B and H3–H4 heterodimers and form a metastable histone octamer

Single histones do not stably fold in solution but require a dimerization partner. Histone H2A exclusively forms a dimer with histone H2B, whereas H3 only dimerizes with H4. These tight heterodimers are stabilized through a large interaction interface formed by the antiparallel arrangement of the two long central α helices of the histone fold domains (Figure 2.2). Selectivity is achieved by extensive amino acid complementarity at the interface. Dimerization results in the alignment of the L1 loop of one histone with the L2 loop of the other at the edges of the crescent-shaped heterodimer, and a close juxtaposition of the N termini of the two α1 helices at its apex.

The histone octamer is best viewed as an arrangement of histone fold dimers (see Figure 2.2). Two H3–H4 dimers are arranged in a tetramer via a four-helix bundle structure composed of the C-terminal regions of the α2 and short α3 helices from two H3 proteins. This interface, which is stabilized by an extensive network of hydrogen bonds and salt bridges, is stable in the absence of DNA. Thus, the functional unit of histones H3 and H4 is an (H3–H4)$_2$ tetramer.

In the presence of DNA (or at high ionic strength), a similar but heterologous four-helix bundle structure is formed by residues of the α2 and α3 helices of H4 and H2B. This tethers one (H2A–H2B) dimer to one 'arm' of the W-shaped (H3–H4)$_2$ tetramer. The interaction is aided by a histone fold extension of H2A (known as the 'docking domain'), which interacts with the other arm of the (H3–H4)$_2$ tetramer to form a short β sheet shared between H4 and H2A. Only one small contact area exists between the H2A–H2B dimers within a histone octamer. It is formed by the L1 loop of H2A. Thus, each of the four histone fold dimers in the histone octamer is connected with its three partners via an intricate interaction network. The histone octamer is stable only above 2 M KCl or NaCl; the high salt condition shields the many basic residues that are normally neutralized by binding to DNA within the nucleosome. Histone octamers have been crystallized under these high salt conditions, and the resulting structures are very similar to those of the octamer within the nucleosome under more physiological salt conditions.

A nucleosome is 147 bp of DNA wrapped around a histone octamer

The structure of the **nucleosome** was determined by X-ray crystallography (see **Methods**). Each histone octamer interacts with 147 bp of DNA. Tight contacts between the rigid framework of the histone octamer and the DNA at 14 independent sites, 12 of which are formed by the antiparallel pairing of the histone folds, contort the DNA into the shape of a flat, left-handed superhelix. The DNA binds within a basic groove on the wedgelike surface of the histone octamer (**Figure 2.3**). Specifically, the central ~60 bp of DNA are organized by the (H3–H4)$_2$ tetramer, with the next ~30 bp on each side contacting the H2A–H2B dimers. The final ~10 bp on each side interact with the αN helix of histone H3, which is held in place by the H2A docking domain.

Contacts between histones and DNA are observed only at the minor grooves, as they face the histone octamer. The DNA bases are less exposed in the minor groove, and this minimizes base-specific contacts between DNA and the histone octamer (**Figure 2.4**). Indeed, the majority of the ~120 noncovalent interactions at the 14 individual interaction interfaces are hydrogen bonds between the amide of the protein main chain and the phosphodiester backbone of the DNA.

The rigid, primarily α-helical histone octamer is much better suited to enforce the significant distortion of the DNA in the form of a DNA superhelix (**Box 2.2**). Three to five such interactions exist at each site; they are reinforced by a varying number of hydrogen bonds with basic side chains. Numerous water-mediated contacts are also found. Not all of the 14 interfaces exhibit the same number of interactions. The central ~80 bp are bound more tightly than the rest of the DNA, which is consistent with the dynamic character of the peripheral DNA regions. This has important implications for nucleosome assembly and disassembly.

Nucleosomal DNA is a highly distorted superhelix

Nucleosomal DNA is distorted into 1.65 turns of a flat, left-handed superhelix, with a radius of 42 Å and a pitch of 26 Å. Increasing the resolution of the nucleosome structure to 2 Å revealed a wealth of detail on the parameters of nucleosomal DNA, which are unique

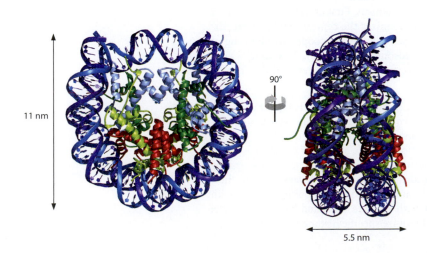

Figure 2.3 Structure of the nucleosome. Core view from in front (down the superhelical DNA axis) and side. Histones H3, H4, H2A, and H2B are shown in light blue, dark green, light green, and red, respectively. The structure contains 147 bp of DNA and two copies of each of the four core histone proteins. (PDB 1AOI) (From K. Luger et al., *Nature* 389:251–260, 1997.)

1 ('end')   146 ('end')

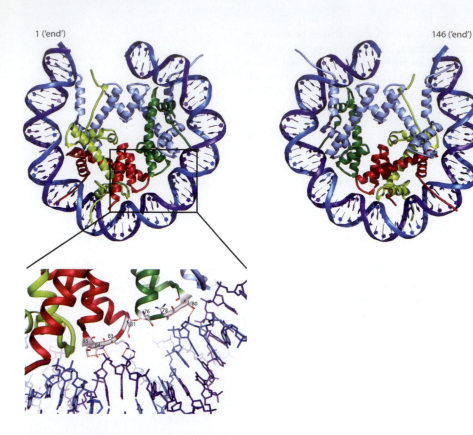

Figure 2.4 Core nucleosome structures. Structures viewed from the 1 ('end') and 146 ('end') and detailing the interactions between core histones and the DNA.

yet within the characteristic range of B-form DNA. However, as a consequence of super-helix formation, the major and minor grooves of the DNA that face the histone octamer are compressed and deep, whereas they are more extended and flattened at the outside (see Figure 2.4). The geometry of the superhelix brings into alignment major (and minor) grooves that are separated by ~80 bp of linear DNA. The DNA regions that are exposed on the outside of the nucleosome are surprisingly flexible and mobile, as demonstrated by high crystallographic B-factors. It is conceivable that stretches of up to 6 bp can be recognized by sequence-specific DNA-binding proteins, even in the context of a nucleosome. This has been demonstrated for small-molecule DNA-binding ligands, which bind nucleosomal DNA with high affinity (see **Box 2.3**).

Nucleosomes are assembled sequentially with the help of histone chaperones

Because of the instability of the histone octamer under physiological conditions, the nucleosome is not assembled by binding of an intact histone octamer to DNA. Rather, nucleosome assembly occurs via a sequential step-wise mechanism. First an $(H3–H4)_2$ tetramer is deposited onto DNA to form a 'tetrasome' that organizes the central ~60 bp. Next, the first H2A–H2B dimer is added to form a 'hexasome,' followed by cooperative binding of the second H2A–H2B dimer to complete the nucleosome (**Figure 2.5**). This same assembly mechanism operates *in vivo* and *in vitro* with very minor differences, and thus can be thought of as intrinsic to the nucleosome.

Due to their extremely basic nature, soluble H2A–H2B dimers and $(H3–H4)_2$ tetramers are toxic to the cell. Consequently, they are bound *in vivo* to members of a diverse class of proteins termed **histone chaperones**. Histone chaperones (not to be confused with the chaperones involved in protein folding, discussed in Section 6.3) are a structurally diverse group of proteins that, together with dedicated chromatin assembly and spacing factors, assemble nucleosomes through replication-dependent or -independent pathways. True to their name, histone chaperones ensure that histones do not engage in improper interaction with nucleosomes and DNA. Replication-dependent pathways are regulated by the cell cycle; they are closely coupled to histone synthesis and require specific histone post-translational modifications (see below).

Box 2.2 DNA structure

The physicochemical properties of DNA strongly influence its manipulation and packaging by protein assemblies. Structurally, DNA is a right-handed double helix in which the base sequences on the two antiparallel strands are complementary. Physically, DNA behaves as a conformationally flexible and dynamic polymer. These physical properties are strongly dependent on the DNA sequence and, in particular, on the interactions between adjacent base pairs (base steps). The melting temperatures (temperature at which the strands separate) of individual base steps differ substantially as also do the stacking energies (related to melting temperatures), which are a measure of the interaction between adjacent base pairs in a sequence. The sequence heterogeneity of DNA thus determines which sequences are more easily melted—and so can promote the initiation of transcription or DNA replication—and which sequences are stiffer or more flexible. For example, TpA with the weak stacking characteristic of pyrimidine-purine steps, is the base step with lowest melting temperature and is often preferentially enriched at sites where strand separation is initiated, such as the –10 hexamer, TATAAT, in bacterial promoters. Conversely the base step GpC, a purine-pyrimidine step with stronger stacking, has the highest melting temperature.

Not only does sequence heterogeneity influence the probability of strand separation, it also directly affects the DNA stiffness and hence DNA packaging. When free in solution, A/T-rich sequences are (with the exception of poly dA.dT) more flexible, allowing the DNA chain to adopt a greater range of configurations.

When wrapped around a histone octamer, the axis of the DNA polymer is bent, on average, by ~45°/double-helical turn. In any bent DNA, but especially on the surface of the octamer, the DNA grooves (both major and minor) on the outer face of the DNA are much wider than those on the inner face (**Figure 2.2.1**). Importantly the conformational preferences of individual base steps are strongly sequence dependent. A/T-rich sequences on average prefer to adopt a narrow minor groove and G/C-rich sequences a wide minor groove. In part this is a consequence of the steric effects of the protruding exocyclic groups on the bases; in A-T base pairs, the 5-methyl group of thymine is exposed in the major groove while in G/C base pairs the 2-amino group of cytosine is exposed in the minor groove. In addition, while A/T-rich sequences on average stack on each other across the longitudinal axis of the molecule, the stacking of G/C base pairs is displaced away from the axis (**Figure 2.2.2**) This displacement is coupled to a change in the conformation of sugar–phosphate backbone (technically from a BI to a BII conformation) associated with a widening of the minor groove. It is thus energetically favorable for G/C-rich sequences to adopt a wider minor groove and less favorable for otherwise more flexible A/T-rich sequences. As a consequence of these factors, optimal DNA-binding sites for the histone octamer consist of alternating short A/T- and G/C-rich sequences in phase with the helical repeat. This sequence organization allows bending in a preferred direction (anisotropic bendability). A further factor influencing protein-induced DNA wrapping is the asymmetric charge neutralization by basic amino acids at the interface of the protein and DNA.

Although the function of DNA is often discussed in the context of its double-helical character, DNA is conformationally dynamic and can assume other structures in a sequence-dependent manner. For example, there is transient base-pair separation, possibly involving the formation of bubbles or the flipping out of bases. Among other possible alternative structures are: DNA cruciforms, which form at palindromic sequences; right-handed Z-DNA, which can form at certain G/C-rich sequences; G-quadruplexes found at telomeres where the DNA is single-stranded and also in other internal positions where the DNA is double-stranded; and triple-stranded H-DNA formed at certain repeating sequences. These structures are generally less energetically favorable than the canonical double helix but may be formed *in vivo* and *in vitro* in negatively supercoiled DNA (see Box 2.3).

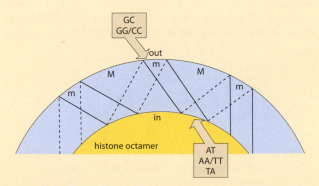

Figure 2.2.1 Sequence-dependent bending preferences of DNA wrapped around the histone octamer. The preferred base steps occurring where the minor groove points out away from and in toward the histone octamer are indicated. (Adapted from A. Travers and A. Klug, *Proc. R. Soc. Lond. B* 317:537–561, 1987.)

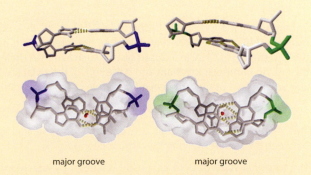

major groove   major groove

Figure 2.2.2 The stacking of adjacent base pairs. In AA/TT (left) the center of the base pairs coincides with the helical axis while in GC (right) the base pairs are displaced toward the major groove. The BI and BII sugar-phosphate backbone conformations are shown in blue and green, respectively. (Adapted from C. Oguey, N. Foloppe and B. Harmann, *PLoS One* e15931, 2010.)

Box 2.3 DNA topology

The application of torque to a DNA molecule that is not free to rotate about its longitudinal axis can alter its torsional state to form a more energetic 'supercoiled' form. This supercoiling can be 'positive'—overwinding—in the same right-handed sense as the DNA double helix or 'negative'—underwinding—in the opposite sense (**Figure 2.3.1**). Changing the torsional state can be reflected in one or both of two ways. The twist between adjacent base pairs may be altered or the additional torsion may induce a coiling, or writhing, of the DNA chain itself. This induced coiling can take two forms (Figure 2.3.1). The toroidal form is a single simple coil and is exemplified by the path of the DNA duplex in a nucleosome. By contrast, in the plectonemic form the DNA duplexes form an interwound double helix; this is usually the form adopted by DNA in solution. Supercoiling can also result in the formation of knotted structures within and between DNA molecules (see Chapter 4). Supercoiled DNA represents a higher energy level of DNA than the relaxed form not under torsional strain and consequently supercoiling directly affects other physical parameters of DNA: for example positive and negative superhelicity, respectively, lower and raise the melting temperature.

Figure 2.3.1 Configurations of negatively supercoiled DNA. (A) Toroidal. The path of the DNA is a left-handed spiral. (B) Plectonemic. Two right-handed coils of DNA duplexes form a double-helical structure. For positively supercoiled DNA the sense of the coiling in both forms is reversed. (Adapted from Cozzarelli et al. (1990) In DNA topology and its biological effects, pp. 139–184. With permission from the copyright holder, Cold Spring Harbor Laboratory Press.)

A fundamental topological property of DNA is its **linking number** (**Lk**), a parameter that describes the intertwining of the two strands of DNA. In a closed system such as a DNA circle, the linking number is invariant and is the sum of the **twist** (Tw) and **writhe** (Wr) of the supercoiled molecule:

$$Lk = Tw + Wr$$

This relationship implies that the partition of supercoiling between Tw and Wr can change without changing Lk.

The torsional state of DNA can also be altered by enzymatic manipulation using energy from ATP. This manipulation of supercoiling can take two forms. First, an enzyme, termed a *topoisomerase*, can bind to a single site and directly change the coiling (see Chapter 4). Second, coiling can be changed by the movement of a protein or protein complex, such as RNA polymerase, along the DNA, under particular constraints. An example of the first case is DNA gyrase, a bacterial topoisomerase that introduces negative supercoiling into DNA, thus facilitating both compaction and strand separation at biologically important DNA sequences. In this example the energy level of DNA is raised. Other topoisomerases can reverse this effect, so relaxing DNA.

While topoisomerases can establish and maintain an equilibrium state of supercoiling, the processes of DNA replication and transcription generate transient changes in **DNA supercoiling** following the translocation of the protein complexes along the DNA. These transients arise as a direct consequence of the double-helical structure of DNA. When proteins such as RNA polymerase move along DNA they do not track linearly along the molecule but instead rotate along a helical path, following one or other of the grooves. Many protein complexes are much more bulky than DNA and their freedom to rotate around the DNA is constrained by molecular crowding. In some cases the polymerizing enzymes may even be restrained in a fixed spatial position by physical attachment to extensive structures such as membranes. Under these circumstances, provided that the rotation of the DNA molecule is also constrained, torsional strain is generated such that the DNA is overwound downstream of the advancing enzyme and underwound upstream (**Figure 2.3.2**) and is thus overall topologically neutral—there is no net change in Lk. This principle, first proposed by Liu and Wang, plays a key role in the genetic organization of chromosomes.

In both the eukaryotic nucleus and the bacterial nucleoid, DNA is usually packaged in a negatively supercoiled form. The total superhelical content is often described as being partitioned between 'constrained' superhelicity where the DNA is protein-bound and so unavailable to facilitate, for example, transcription initiation, and 'unconstrained' where the supercoils are present in free DNA. In the context of functioning molecular machines acting on DNA, it is important to recognize that the superhelical density of a closed domain is not homogeneously distributed. The values reported in the literature are averaged over a whole domain but in practice there is considerable variation within a domain. Since superhelicity changes the local structure of DNA, the localized superhelical density

Figure 2.3.2 Induction of superhelicity in a constrained DNA domain by a fixed elongating DNA translocase. Negative superhelicity is generated upstream of the translocating enzyme (R) and positive superhelicity downstream. In eukaryotic chromatin the downstream positive superhelicity potentially destabilizes the wrapping of negatively supercoiled DNA on nucleosomes, while the upstream negative superhelicity facilitates DNA wrapping on the histone octamer. (Adapted from L.F. Liu and J.C. Wang, *Proc. Natl Acad. Sci. USA* 84:7024–7027, 1987.)

Box 2.3 *continued*

is sequence-dependent. Sequences with low average melting temperatures will melt and so change their linking number under the influence of negative superhelicity more readily than those of higher average melting temperature. Similarly the superhelicity, both positive and negative, generated by translocase movement is highest close to the functioning complex and gradually decreases by diffusion and the action of topoisomerases with increasing distance from the complex. An important consequence of the sequence-dependent variation of superhelical density is that the effect of superhelicity—whether untwisting or writhing—will depend on the length and context of sequences sensitive to superhelical deformation. For example, negative superhelicity will more likely facilitate the untwisting, and eventual melting, rather than the writhing, of short sequences with a low melting temperature

immediately flanked by sequences of higher melting temperature. Conversely, for longer sequences of lower than average melting temperature, writhing will be favored because such sequences are also, on average, more elastic. Variation in the response to superhelicity explains why the sequences with lowest average melting temperatures in fly and yeast chromosomes are not located in regions where the DNA is untwisted or melted (for example, promoters) but at the 3' end of genes where they provide binding sites for topoisomerase II (see Chapter 4), potentially enabling the relaxation of positive superhelicity generated downstream of an actively transcribing polymerase. Similarly, the regions in proximity to the Ter sites for bacterial replication are usually the least stable in bacterial chromosomes.

The structure of the nucleosome is intrinsically dynamic

Nucleosomes are often portrayed as 'monolithic' immovable objects that stand in the path of DNA and RNA polymerases. However, it is now well established that nucleosomes are intrinsically dynamic. The interactions between DNA and protein are strongest near the center (nucleosomal dyad), and weaker toward the periphery. This has recently been confirmed by single-molecule studies. In a mononucleosome, the ends of the DNA are in rapid equilibrium between the bound and unbound state, thereby facilitating the binding of transcription factors or other proteins that require access to nucleosomal DNA.

Partial unraveling of histone–DNA contacts facilitates the spontaneous or chaperone-mediated removal of H2A–H2B dimers from the nucleosome to form hexasomes or tetrasomes. This occurs by disrupting the dimer/tetramer interface. Nucleosome disassembly (whether partial or complete) followed by re-assembly are important steps during gene activation and

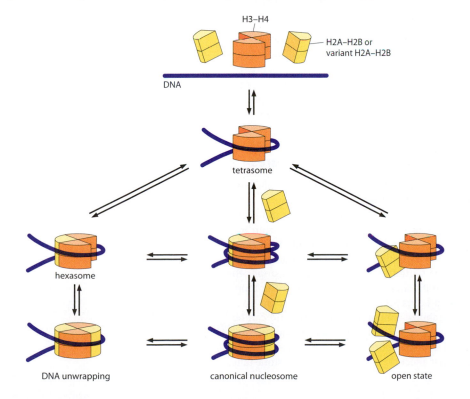

Figure 2.5 Structural states of the nucleosome. The interchangeable states include the tetrasome, which is formed by wrapping ~80 bp DNA around a (H3–H4)$_2$ tetramer. Hexasomes (lacking one H2A–H2B heterodimer) are intermediates in the assembly/disassembly of canonical nucleosomes. (Adapted from K. Luger, M.L. Dechassa and D.J. Tremethick, *Nat. Rev. Mol. Cell Biol.* 13:436–447, 2012. With permission from Macmillan Publishers Ltd.)

transcription. Dictated by the modular architecture of the nucleosome, the outer ~30 bp of DNA have to unravel to allow dissociation of the H2A–H2B dimers, followed by dissociation of the (H3–H4)$_2$ tetramer from the DNA. These processes are facilitated by ATP-dependent chromatin remodeling factors (see also Section 2.3), often in conjunction with histone chaperones. The stability of nucleosomes is also affected by the incorporation of histone variants or the presence of certain post-translational modifications.

DNA sequence directs specific positioning of nucleosomes *in vitro* and *in vivo*

As discussed above, histones make only limited direct contacts with the bases of the DNA, consistent with their role as global organizers of DNA. Nevertheless, it has long been known that *in vitro* nucleosomes form preferentially on certain sequences. Nucleosomes are considered to be positioned if their specific locations on the DNA sequence are fixed. This is likely a consequence of the propensity of certain DNA sequences to conform more readily to the left-handed helical pitch dictated by the histone (H3–H4)$_2$ tetramer. Nucleosome positioning *in vivo* is an important regulatory aspect that underlies the competition with regulatory proteins, dynamic histone exchange, remodeler dependence, and the utilization of a given origin of replication during cell division.

Nucleosomal arrays form higher order structures that differ in their degree of condensation

On a given DNA molecule, contiguous nucleosomes are connected by 'linker' DNA and form conformationally dynamic nucleosomal arrays. The folding of these arrays is reminiscent of polypeptide folding, although with fewer topological possibilities. In the completely unfolded (or fully extended) form referred to as 'beads on a string' there are no nucleosome–nucleosome interactions. Condensation of nucleosomal arrays to form higher order structures is driven by nucleosome–nucleosome interactions. Proper chromatin folding requires arrays with consistent nucleosome spacing (or repeat length), an attribute accomplished by chromatin **remodelers** specialized in chromatin assembly (see also Section 2.3).

As in protein folding, nucleosomal arrays proceed through folding intermediates before achieving their final folded state. The first step involves interaction of neighboring nucleosomes causing a moderate shortening of the array. The next step involves additional compaction and stabilizing interactions between nucleosomes further apart to form what is commonly referred to as the '30 nm fiber' (see Box 2.1).

When studied as short fragments *in vitro*, nucleosomal arrays not only fold but also reversibly and cooperatively self-associate to form oligomeric chromatin structures. This transition involves nucleosome–nucleosome interactions in *trans* and occurs independently of the folded state of the nucleosomal array; unfolded arrays, moderately folded arrays, and terminally folded 30 nm fibers are all capable of forming condensed arrays under physiological conditions. It remains to be seen to what extent the structures observed *in vitro* mimic the *in vivo* situation.

Linker histones stabilize condensed 30 nm chromatin structures

Linker histones (for example, H1, H5) comprise a family of nucleosome-binding proteins that stabilize folded chromatin states. In most eukaryotic chromatin, they are found at an average stoichiometry of ~0.8 linker histone per nucleosome. The ratio is higher in tightly packed heterochromatin and lower in less dense euchromatin (see below). Apart from the core histones themselves, linker histones are the most abundant proteins found in chromatin. They are structurally unrelated to the core histones, with a three-domain organization. The globular domain is a winged helix motif (see **Guide**) that targets linker histones to nucleosomes by binding both high- and low-affinity sites on the nucleosome. The long, intrinsically disordered CTD binds linker DNA and thus protects an additional 20 bp from nuclease digestion. This domain is required to stabilize condensed chromatin and can be specifically phosphorylated and acetylated. Although little is known about the shorter, disordered NTD, it can be specifically methylated and phosphorylated. These post-translational modifications impart electrostatic effects that affect the association of the histone tails with the negatively charged DNA duplex.

Although model studies have shown that linker histones are not necessary for 30 nm fiber formation, they help to stabilize fibers. For example, under conditions where nucleosomal arrays equilibrate between the beads-on-a-string and 30 nm fiber conformations, linker histone-containing chromatin fibers exist solely in the 30 nm fiber state. Binding of linker histones to nucleosomal arrays stabilizes the entering and exiting nucleosomal DNA and causes the linker DNA to form an apposed stem motif. This likely affects the conformation of the 30 nm fiber.

Proteins that bind nucleosomes and in the process antagonize linker histone H1 action would generally destabilize condensed chromatin fibers. The high mobility group nucleosome-binding (HMGN) members of the HMG superfamily are one such group of small, disordered non-histone proteins. They can be thought of as remodelers of the 30 nm fiber. Accordingly, the HMGN family of proteins is involved in many cellular processes, including regulation of gene expression, DNA repair, and cellular differentiation.

The structure of the 30 nm fiber remains unsettled

Whereas there is universal agreement that nucleosomal arrays and linker histone-bound chromatin can form 30 nm fibers under physiological conditions, the internal organization of the 30 nm fiber remains controversial. Several models have been proposed, in part due to the fact that the chromatin samples structurally characterized in early studies were compositionally heterogeneous.

The 'superbead' model was proposed in 1976 based mainly on electron microscopy (EM) (see **Methods**) images from many laboratories showing a segmented, beaded 30 nm fiber structure. Superbeads were routinely isolated from sucrose density gradients after partial nuclease digestion of chromatin and were reported to yield approximately spherical particles with a diameter of 30 nm. While there is strong evidence that beaded chromatin suprastructures exist under certain conditions, a major problem with the superbead model is that there was never agreement about the packed structure of the superbeads and estimates of the number of nucleosomes in each bead ranged from 6 to 48.

As an alternative to superbeads, 30 nm fiber models based on a continuous, closely packed helical arrangement of nucleosomes were proposed. In 1976, EM and early X-ray diffraction studies led to a one-start 'solenoid.' According to this model, the faces of the nucleosome are approximately parallel to the axis of the fiber. Adjacent nucleosomes are in contact (N with both n+1 and n+6), linker DNA segments are bent, and the nucleosomal DNA is continuously supercoiled. This one-start solenoid was conceptually pleasing and quickly gained popularity. However, in the early 1980s, several groups proposed alternatives. In these models the chain of nucleosomes first forms a two-start zigzag in which nucleosomes n and n+2 are in close proximity, the linker DNA is straight, and the zigzag chain may twist into a helical ribbon (**Figure 2.6**).

The 30 nm fiber has a heteromorphic structure dependent on nucleosome repeat length and packing order

Structural studies with reconstituted model nucleosome arrays suggest that nucleosome repeat length influences the resulting 30 nm fiber structure. By digesting model 12-mer nucleosomal arrays containing a 177 bp repeat length with restriction enzymes specific for linker DNA, it was shown that folded, cross-linked nucleosomal arrays yielded two 'stacks.' This indicates a fundamental two-start organization. The crystal structure of an intermediately folded tetranucleosome with a slightly shorter repeat length (167 bp) directly confirms a two-start organization. However, a tetranucleosome is too short to form a 30 nm fiber, and thus no structural information beyond the intermediate folded state was obtained. While it must be derived from a two-start zigzag, the structural details of the terminally folded 30 nm fiber remain unknown. The studies with 167 and 177 bp repeat length model systems are direct and definitive, and seemingly solve a fundamental question regarding the backbone organization of higher order chromatin structure. Recent cryo-EM (see **Methods**) structures of 30 nm fibers reconstituted in the presence of linker histone H1 and with different nucleosome repeat lengths (12 × 177 bp and 12 × 187 bp) show basically a two-start (double-helical) zigzag configuration for nucleosome arrangement. The structures show left-handed twist of the repeating tetranucleosomes, within which the four nucleosomes

(A) 'one-start' helix
(solenoid)

(C) 'two-start' helix
(zigzag)

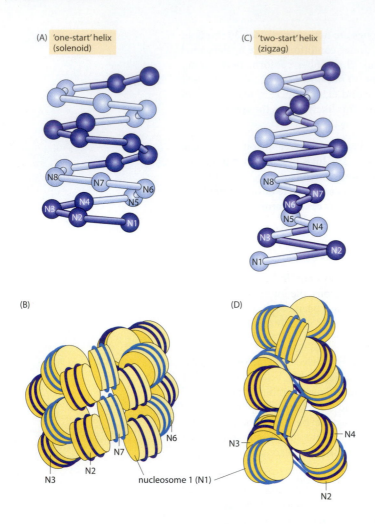

(B)

(D)

Figure 2.6 Two models for chromatin secondary structure. The solenoid model (A, B) is characterized by interactions between consecutive nucleosomes (n, n+1). In the zigzag model (C, D) alternate nucleosomes interact (n, n+2). Alternative nucleosomes are numbered from N1 to N8. In the solenoid model the 30 nm fiber is an interdigitated one-start helix in which a nucleosome interacts with its fifth and sixth neighbor. In the zigzag model, the chromatin fiber is a two-start helix in which the nucleosomes are arranged in a zigzag manner such that a nucleosome binds to its second neighbor. (Adapted from G. Li and D. Reinberg, *Curr. Opin. Genet. Dev.* 21:175–186, 2011. With permission from Elsevier.)

zigzag back and forth with a straight linker DNA. Histone H1, which is asymmetrically bound, appears to have a key role in determining this configuration (**Figure 2.7**).

Can 30 years of seemingly disparate data on the structure of the 30 nm fiber be reconciled? Recently, 12-mer model systems with variable nucleosome repeat lengths have been created, and their folding has been analyzed by mesoscopic computer modeling and EM-assisted nucleosome capture techniques. The latter method identified nearest neighbor nucleosome–nucleosome interactions in the folded fiber. These studies also provided evidence for a two-start heteromorphic 30 nm fiber in which both n±1 and n±2 interactions were detected; some of the linker DNA segments were straight, while others were bent. Thus, it seems likely that there is no single 30 nm structure. A chain of nucleosomes may be able to form one-start solenoids, two-start zigzags, and superbead structures depending on specific local conditions. These structures may even coexist within one 30 nm fiber. Unlike idealized model systems, native chromatin has variable linker DNA lengths, is compositionally heterogeneous, and can be interrupted by nucleosome-free regions. Thus, the multiple structures observed by EM may reflect the ability of the 30 nm fiber to accommodate natural compositional and configurational heterogeneity of chromatin while maintaining its compact packing.

Higher order folding of nucleosomal arrays is regulated by cations and the core histone N-terminal tail domains

In solution, nucleosomal arrays are in equilibrium between partially folded, folded, and condensed structures. Many factors influence this equilibrium. Inorganic (for example Na^+, Mg^{2+}) and organic (for example polyamines, linker histones) cations bind to the DNA of nucleosomal arrays, neutralizing backbone negative charge, and thereby facilitating close approach of nucleosomes in the condensed states. Nucleosomal arrays with longer linker DNA lengths require a greater concentration of salts to achieve maximal folding than arrays

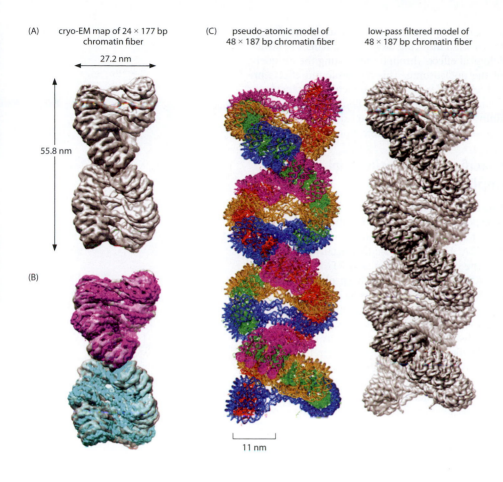

(A) cryo-EM map of 24 × 177 bp chromatin fiber

27.2 nm

55.8 nm

(B)

(C) pseudo-atomic model of 48 × 187 bp chromatin fiber

low-pass filtered model of 48 × 187 bp chromatin fiber

11 nm

Figure 2.7 30 nm chromatin fiber structural models. (A) Cryo-EM (electron microscopy) map of the 30 nm fiber reconstituted on 24 × 177 bp DNA. The length and diameter of the fiber are indicated. (B) The structure of 24 × 177 bp fibers was docked from two copies of 12 × 177 bp fibers. (C) A pseudo-atomic model (left) and its corresponding density map low-pass filtered to 11 Å (right) built by stacking the EM structures of dodecanucleosomal 30 nm fibers with 187 bp on top of each other to generate a continuous fiber. The linker histone H1 was omitted for clarity. (From Feng Song et al., *Science* 344:376–380, 2014.)

with shorter lengths of linker DNA. Thus, as discussed above, the nucleosome repeat length will have some influence on the condensation equilibrium and perhaps also on the actual structure of folded nucleosomal arrays.

Formation of both the folded and oligomeric condensed states of nucleosomal arrays is dependent on the core histone NTDs. Early experiments removed large portions of all the NTDs and showed that folding and self-association were abolished, even in the presence of high divalent salt concentrations. Studies with recombinant nucleosomal arrays indicate that the H4 NTD plays the dominant role in the folding transitions, but all four core histone NTDs contribute non-redundant functions during self-association. The NTDs are mobile during the condensation process, and their position in condensed nucleosomal arrays is not the same as in single nucleosomes. An emerging determinant of nucleosomal array folding and self-association is the nucleosome surface, which is highly charged and contoured. The H4 NTD has been proposed to interact with charged patches on the surfaces of adjacent nucleosomes while mediating nucleosomal array folding, whereas other regions of the surface have been shown to be repressive to condensation. The role of the nucleosome surface in array condensation is just beginning to be explored experimentally.

Core histone isoforms have variant sequences and functions

Heterogeneity in the chromatin fiber is introduced by histone variants, which impart different structural properties to chromatin and are, therefore, a critical component of genome regulation. All eukaryotic organisms contain histone variants, which have one or a few amino acid differences and also specific expression patterns. These 'special histones for special occasions' may be stably associated with specialized chromatin domains; for example CENP-A, a variant H3 histone, is an essential component of centromeres and macroH2A is enriched in the inactive X chromosome of female mammals. Histone variants may also be associated with specific gene activation states, or they may serve important functions in DNA repair.

Structural analyses of nucleosomes containing histone variants show little structural change but subtle differences in stability. Thus, rather than inducing large-scale changes in structure, histone variants likely exert their biological effects through modulating the uniquely charged and shaped surface landscape of the nucleosome. This has profound effects on nucleosome–nucleosome interactions and higher order chromatin structure. For instance, macroH2A may reinforce chromosome condensation, whereas H2A.Bbd is believed to help unfold nucleosomal arrays. Since many amino acid substitutions also exist in the tails, histone variants exhibit diverging patterns of post-translational modification.

Core histones undergo many specific post-translational modifications with structural implications

Post-translational modification (*PTM*) of histones was first reported in 1964, but did not receive much attention until the enzymes responsible for these modifications were discovered. Specific conserved PTMs of the core histone residues have now been documented in the histone fold domains, and the CTD of H2A. However, those that occur in the NTDs have been the most widely studied. Every type of known amino acid PTM (for example acetylation, methylation, phosphorylation, ADP-ribosylation) has been observed in histone proteins, and hundreds of specific PTMs have been mapped using chromatin immunoprecipitation (ChIP) assays. However, the function of only a small number of histone PTMs has been determined. Without exception, PTMs on histones are reversible, and the enzymes that add and remove them are themselves tightly regulated. In many cases, histone modifications are interdependent; for example, the addition of a modification to a particular residue depends on the presence (or absence) of another modification nearby. Recent advances in mass spectrometry and sequencing have provided a wealth of information on the modification state of nucleosomes. *In vivo*, the patterns of PTM accumulation across the genome have been mapped for multiple modifications, and these patterns are variable within a cell population.

Virtually all attention has focused on the biological effects of histone PTMs (see Section 2.4), and far less is known about the structural consequences. From the limited knowledge available, it seems likely that most (if not all) histone PTMs exert their effects by changing the properties of the malleable and disordered NTDs, allowing them to assume distinct structures and engage in many different macromolecular interactions depending on the particular PTM pattern. Modifications in the structured regions of histones are likely to change the molecular surface of the nucleosome, perhaps affecting nucleosome–nucleosome interactions or interactions with histone chaperones and chromatin remodelers, all of which could influence higher order chromatin structure. Other PTMs are likely to alter histone–histone or histone–DNA interactions, although the effects are expected to be subtle at most. How individual histone PTMs on the same nucleosome or on neighboring nucleosomes synergize to invoke a plethora of finely tuned structural effects is a matter of intense investigation.

At least some of the PTMs on the core histone NTDs have been shown to affect the condensation equilibrium. Early model studies showed that bulk acetylation of the NTDs disrupted higher order chromatin structures. More recent studies have examined specific PTMs. For example, acetylation of K16 on the H4 NTD is as effective at disrupting folding and self-association as is removal of the entire NTD. In contrast, trimethylation at H4 K20 enhances higher order chromatin structure formation, by pushing the equilibrium to the more condensed conformational states. Given the large number of documented NTD PTMs, much more work will be necessary to fully understand the role that specific NTD modifications have on nucleosomal array condensation.

Chromatin architectural proteins are essential for higher order chromatin structures

So far the focus of higher order folding has been on the intrinsic formation and stabilization of condensed chromatin structures, as influenced by salts, DNA, and histone proteins. An emerging, but far from complete, story relating to higher order chromatin structure involves chromatin architectural proteins. These proteins bind to nucleosomal arrays and mediate assembly of higher order chromatin structures that differ from those formed via the intrinsic pathway. The first proteins found to fulfill such a role were yeast Sir3 and avian MENT, followed shortly thereafter by human MeCP2 and a *Drosophila* Polycomb group complex. Each of these proteins, when mixed with beads-on-a-string nucleosomal arrays under low salt conditions, induces formation of unique condensed chromatin structures. For example, MeCP2 assembles an undulating higher order chromatin structure in which the nucleosomes are arranged mostly edge-to-edge rather than face-to-face. Each Polycomb group complex compacts three nucleosomes and essentially wraps the nucleosomal array around itself. HP1-α is a unique heterochromatin-associated protein that preferentially binds to intrinsically folded nucleosomal arrays and rearranges their condensed structures. Many chromatin architectural proteins also mediate the *in vitro* assembly of individual arrays into novel supramolecular structures.

Genomic chromatin is a heterogeneous and complex macromolecular assembly

Model system studies have advanced our understanding of nucleosomal array and chromatin fiber behavior *in vitro*. Due to the paucity of *in vivo* studies of higher order chromatin architecture, we can only extrapolate how model system properties may translate to the structure of genomic chromatin. Genomic regions in which the nucleosomal array is uninterrupted and which contain at least one linker histone per nucleosome will likely be folded into a stable 30 nm fiber of some sort. H1-rich regions of the genome are associated with **heterochromatin**, a term historically used to describe highly electron-dense chromatin in the nucleus (especially at the nuclear periphery). This includes constitutive heterochromatin that is usually formed in all cell types (for example around the centromeres in mammalian cells) and facultative heterochromatin that is formed only under specific conditions (for example in specific cell types or developmental stages and the inactive X chromosome in female mammalian cells). **Euchromatin** is the term used to describe genomic chromatin of a less electron-dense form than heterochromatin. Whereas heterochromatin is associated with transcriptional repression, the genes packaged into euchromatin can be constitutively active or transcriptionally regulated (in either direction). Euchromatin has lower concentrations of linker histones, consistent with its less condensed and more active state.

Given what we know about the structure and function of the chromatin fiber, the terms heterochromatin and euchromatin are much too general. In addition to linker histones, there are many chromatin architectural proteins and specific PTMs associated with highly condensed chromatin. This strongly suggests that there is more than one specific type and conformation of heterochromatin. The precise structure of any given region of euchromatin will be determined by the nucleosome repeat length, the location of nucleosome-free regions in the array, the PTM status of the core histone NTDs, and the presence of linker histones or chromatin architectural proteins. In other words, for both heterochromatin and euchromatin, the specific primary structure of the chromatin region in question will dictate the condensed chromatin structures that are formed under physiological conditions. Correspondingly, the number of specific higher order structures that exist in genomic chromatin is likely to be quite large.

Elucidating chromosomal architecture beyond the 30 nm fiber remains a challenge

Structural characterization of condensed chromatin beyond the 30 nm fiber is more based on inference than on precise experimentally determined structures. Interphase chromosomes are generally 200–400 nm in diameter. EM studies of isolated human interphase chromosomes have shown that in order to achieve this level of compaction, the 30 nm fiber is first twisted into 60–80 nm fibers. These further condense into 130 nm diameter chromosomal domains termed **chromonema fibers** (see Box 2.1) which are then condensed even more to form the G1 chromatids (see also Section 13.2). Interestingly, the higher order structures observed in interphase chromosomes are only present under isolation conditions including the same concentrations of Mg^{2+} or polyamines that induce self-association of model nucleosomal arrays. If low salt buffers are used for isolation, only 30 nm fibers are observed. This led to the conclusion that self-association of model nucleosomal arrays and chromatin fibers is an *in vitro* manifestation of a mechanism responsible for stabilizing condensed structures at the chromosomal level. A popular notion is that 30 nm chromatin fibers are organized into large 'radial loops' by protein-mediated interactions between two widely separated regions of the fiber and the nuclear matrix. Indeed, interphase chromosomes are segregated into specific territories of the nucleus. A better understanding of large-scale chromosomal structure and organization remains a challenge for the future.

2.3   REMODELING COMPLEXES

DNA-binding proteins help orchestrate processes such as transcription and replication. However, to access DNA they have to compete with nucleosomes. Packaging by nucleosomes and histone chaperones condenses and organizes the genome but can impede access, as most proteins cannot bind DNA at sites located within the boundary of a nucleosome. Thus, occlusion and competition is one mode whereby nucleosomes contribute to gene repression. To manage the access to nucleosomal DNA, eukaryotic cells have evolved a set of **chromatin remodeling complexes** (termed *remodelers*) that alter the presence, position, or structure of nucleosomes. This enables proper chromatin assembly, as well as rapid and regulated access of DNA-binding proteins to the genome.

Remodelers regulate DNA exposure in chromatin

Remodelers impact nucleosomes via four modes (**Figure 2.8**). (1) Histone octamers can be moved to a new position (sliding) thus exposing DNA previously within the nucleosome. (2) Nucleosomes can also be ejected, exposing a larger region of DNA. (3) H2A–H2B dimer removal leaves only the central $(H3–H4)_2$ tetramer, which exposes flanking DNA and destabilizes the nucleosome. (4) Alternatively, resident H2A–H2B dimers can be exchanged for those containing H2B and the histone variant H2A.Z, which creates a specialized and unstable nucleosome with unique sequence and modification features. Notably, both sliding and ejection require the breakage of all 14 histone–DNA contacts (though they are reformed following sliding), and dimer replacement involves breaking six contacts and the dimer/tetramer interface. These thermodynamic barriers underlie the slow movement and dissociation rates of nucleosomes *in vitro*, and pose energetic obstacles that remodelers must overcome.

All remodelers share three basic properties: (1) an affinity for the nucleosome that involves both DNA and histone epitopes, which may be influenced by covalent histone modifications; (2) a similar catalytic ATPase domain that functions as a DNA-translocating motor to break histone–DNA contacts; and (3) domains and/or proteins that regulate the ATPase domain. Together, these shared properties allow nucleosome selection and remodeling, with specialized domains and proteins influencing which modes of remodeling will be conducted. The majority of remodelers form large multiprotein complexes, but there are some that function as monomers (see below).

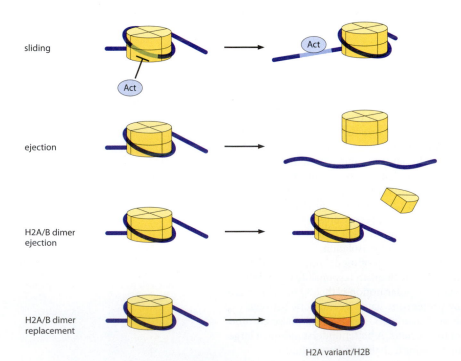

sliding

ejection

H2A/B dimer ejection

H2A/B dimer replacement

H2A variant/H2B

Figure 2.8 Modes of nucleosome remodeling. Remodelers enable access to nucleosomal DNA through sliding, ejection, or H2A–H2B dimer removal. Sliding is a common property of SWI/SNF, ISWI, and CHD family remodelers. However, ejection is accomplished primarily by SWI/SNF family remodelers. The SWR1 complex is unique in its ability to replace H2A–H2B dimers with those containing the histone variant H2A.Z, whereas INO80-related complexes may perform the opposite function. Act denotes a regulatory protein or activator (pale blue) with a binding site on the nucleosome.

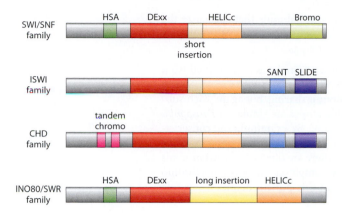

Figure 2.9 Schematic of remodeler family ATPase structures. All remodeler families contain an ATPase domain that is split into two RecA-like parts: DExx (red) and HELICc (orange). Each remodeler family is distinguished by the unique domains residing within, or adjacent to, the ATPase domain. Remodelers of the SWI/SNF, ISWI, and CHD families each have a distinctive short insertion (tan) within the ATPase domain, whereas remodelers of INO80/SWR1 family contain a long insertion (yellow). Each family is further defined by distinct combinations of flanking domains: bromodomain (light green) and HSA (helicase-SANT) domain (dark green) for SWI/SNF family, SANT-SLIDE module (blue) for ISWI family, tandem chromodomains (magenta) for the CHD family, and HSA domain (dark green) for the INO80/SWR1 family.

Remodelers can be separated into four families defined by their composition and activities

All remodelers contain a central catalytic ATPase subunit, which contains a 'split DExx/HELICc' ATPase domain that has two RecA-like lobes (see below), a demonstrated helicase/translocase motor motif (see also Box 3.1). However, each catalytic subunit also contains additional domains that are unique in function and protein association, which allow their classification into four families: SWI/SNF, ISWI, CHD/NURD/Mi-2, and INO80/SWR1 (**Figure 2.9**). Most eukaryotes contain at least one remodeling complex from each of these four families. Within each family, the non-catalytic subunits provide important regulatory and targeting functions. For example, SWI/SNF family remodelers all contain actin-related proteins that help regulate the ATPase activity and bromodomain-containing proteins that bind acetylated lysines in histones and other proteins.

Many organisms build multiple alternative complexes within each remodeler family. The themes of assembling these alternative complexes can be illustrated by examining the SWI/SNF family. One common theme involves the utilization of paralogs of the remodeler subunits (including the ATPase subunit) to build two or more alternative remodeling complexes (**Figure 2.10**). For example, the SWI/SNF and RSC complexes in *Saccharomyces cerevisiae*, both SWI/SNF family members, are built largely from separate paralogs and have only two

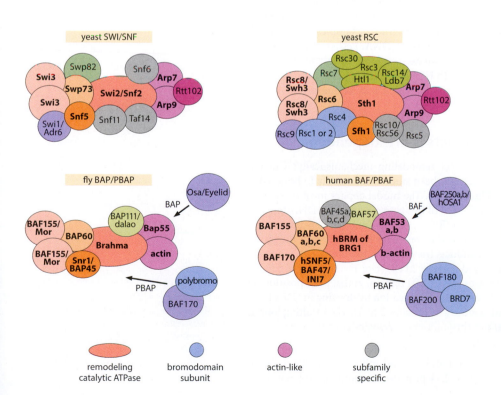

Figure 2.10 Composition of SWI/SNF remodelers in yeast, flies, and humans. Subunits with similar colors within a complex indicate functional modules, and identical colors between organisms denote related subunits. Bold font indicates subunits that are conserved in all SWI/SNF remodelers. A consistent feature of SWI/SNF family remodelers is a dimer of actin-related proteins (Arps, purple). In metazoans, this dimer consists of actin itself and an actin-related protein.

actin-related proteins in common. They also contain unique subunits that specialize the complexes for particular tasks. The assembly of these distinct subunits into the complex relies on unique domains present on only one of the two paralogs.

The concept of building two main separate SWI/SNF family complexes is conserved from yeast to man. Human cells assemble the related BAF (SWI/SNFa) and PBAF (SWI/SNFb) complexes using a combination of gene paralogs and subunit sharing, but again, each sub-complex also contains unique polypeptides. The assembly modes are similar in *Drosophila melanogaster*, although flies rely on a single ATPase subunit. By examining the conservation of composition (proteins and domains) within remodeling complexes, one can arrive at a set of 'core' members that define the family. Likewise there are often conserved proteins that define an alternative assembly; for SWI/SNF family complexes, there is a polybromo ortholog in most tested eukaryotes and in each organism this protein defines a separate SWI/SNF assembly.

Interestingly, combinations of particular paralogs can create additional assemblies that are utilized in particular cell types or developmental contexts to impart additional specialized functions. For instance, unique SWI/SNF family complexes are built during neuronal development in mice, such that alternative actin-related protein paralogs form modules that help guide neuronal progenitors through developmental decisions and transitions. Finally, although the majority of remodeler families form large complexes, those in the CHD family function as monomers. However, roles in large complexes are well documented for a subset (Mi-2). The likely determinant for monomer sufficiency is whether the requirements for proper nucleosome selection and remodeling regulation *in vivo* can be accommodated within the catalytic subunit itself.

Remodelers have specialized as well as common properties

Despite sharing properties that allow proper nucleosome selection and remodeling, remodelers are equipped for particular tasks by attendant subunits. ISWI remodelers promote chromatin assembly and order by equally spacing nucleosomes on a DNA template, a property shared with a subset of CHD remodelers. For ISWI and CHD remodelers, their catalytic subunits alone can conduct chromatin assembly and ordered nucleosome spacing, with attendant subunits influencing the efficiency of the reaction and the precise spacing between nucleosomes. In contrast, SWI/SNF remodelers randomize the positions of nucleosomes on DNA arrays, a function consistent with providing access of DNA-binding proteins to chromatin. Notably, SWI/SNF family remodelers can eject nucleosomes, whereas ISWI family remodelers lack this activity, and dimer ejection has been observed with SWI/SNF remodelers but with only a subset of ISWI remodelers. These examples reinforce the notion that SWI/SNF family remodelers provide access to the DNA. Perhaps the most specialized function of remodelers involves histone variant exchange. Here, only SWR1 family remodelers efficiently replace H2A–H2B dimers with H2A.Z–H2B dimers *in vitro*, whereas the INO80 remodeler can conduct the reverse reaction, which may be important for the recovery from DNA repair.

The disruption of histone–DNA contacts is ATP-dependent

How does ATP hydrolysis by the remodeler affect remodeling mechanistically? Current models favor a DNA translocation model, which has been studied extensively in both SWI/SNF family and ISWI family remodelers (**Figure 2.11**). The model applies most directly to ISWI and can be summarized by considering the use of two active domains: a DNA-binding domain (DBD) and a DNA translocation domain (Tr). First, the remodeler anchors to the nucleosome using histone-binding motifs. The DBD then binds the DNA at or near the proximal linker, whereas the ATPase domain that conducts translocation binds DNA at a location inside the nucleosome, two turns from the dyad (Figure 2.11B, state 1). After undergoing a conformational change, the ATPase/translocase domain (Tr) remains at that fixed position on the octamer, from which it conducts directional DNA translocation by drawing in DNA from the linker and pumping it toward the dyad (Figure 2.11B, states 2 to 3). The resulting loop is then propagated in an energy-neutral manner throughout the remainder of the nucleosome (Figure 2.11B, states 3 to 4).

How are DNA translocation and loop formation regulated and utilized to derive different remodeling outcomes? The mechanism in Figure 2.11 provides an intuitive explanation of

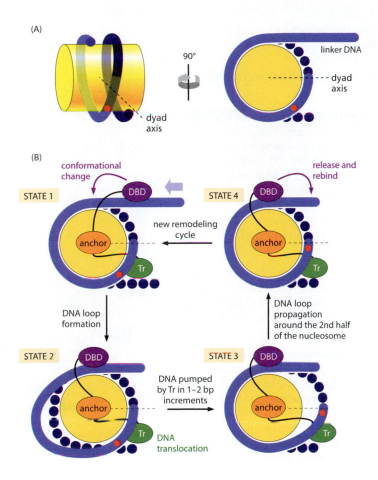

(A)

90°

dyad
axis

linker DNA

dyad
axis

(B)

conformational
change

STATE 1

DBD

anchor

Tr

new remodeling
cycle

DNA loop
formation

STATE 2

DBD

anchor

Tr   DNA
      translocation

DNA pumped
by Tr in 1–2 bp
increments

STATE 3

DBD

anchor

Tr

DNA loop
propagation
around the 2nd half
of the nucleosome

release and
rebind

STATE 4

DBD

anchor

Tr

Figure 2.11 Model of DNA movement during a remodeling event.
(A) Left, a nucleosome side view emphasizing the left-handed wrapping of DNA (light to dark blue) around the histone octamer (yellow transparent cylinder). DNA color changes from light to dark blue when passing the nucleosomal dyad axis. Right, the nucleosome is rotated 90° and depicted in two dimensions. After passing the dyad axis, the second turn of DNA is presented in dark blue dots instead of light blue to reinforce the perspective of DNA wrapping. The red dot provides a reference point on the DNA, useful for following the translocation of DNA along the octamer surface. (B) Successive steps (states 1–4) occurring during a remodeling event. The concerted action of a DNA-binding domain (DBD) located on the linker DNA and a translocation domain (Tr) located near the dyad generates a small DNA loop that propagates along the nucleosome surface. The remodeler undergoes a conformational change in its DBD when the DNA loop is generated (states 1 to 2), followed by the translocation of the DNA through the Tr domain toward the dyad (states 2 to 3). The DNA loop continues its propagation around the nucleosome surface by one-dimensional diffusion, breaking histone–DNA contacts at the leading edge of the loop while replacing those contacts at the lagging edge. Loop propagation continues to the distal linker, resulting in nucleosome repositioning (states 3 to 4). The remodeler resets its conformation with its original binding contacts and is ready for a new remodeling cycle (states 4 to 1).

sliding but must be expanded to account for the other modes of remodeling. For nucleosome ejection, a large DNA loop on the surface might enable histone chaperones to remove the octamer while histone–DNA contacts are disrupted. For dimer ejection, a small DNA loop might be restricted to the position on the nucleosome surface where H2A–H2B dimers reside, facilitating their loss. Likewise, the dimer replacement function conducted by SWR1 family complexes may first involve a similar mechanism as that for dimer loss, followed by H2A.Z–H2B dimer insertion using the dedicated chaperone Chz1. An alternative mechanism for dimer or octamer ejection involves DNA translocation occurring on the remodeler-bound nucleosome, which removes DNA from the adjacent nucleosome by unspooling. This could result in the ejection of the adjacent dimer or the entire octamer, depending on the amount of DNA removed from the adjacent nucleosome. These possible mechanisms and their implications are important areas of future study.

Remodeler regulation depends on the interplay with histone post-translational modifications

In principle, histone PTMs could affect remodeler activity, remodeler targeting, or both. For instance, the ISWI ATPase is activated by a highly basic region on the histone H4 tail (residues 17–19), but not when this tail is acetylated (H4K16ac). This PTM does not affect the overall affinity of the remodeler for the nucleosome. Rather it reduces the extent of DNA engagement within the DNA translocase domain and lowers ATPase activity. This example demonstrates that the histone tail can regulate remodeler activity instead of affinity. However, H4 acetylation increases the efficiency of remodeling by the yeast RSC complex, likely due to the recognition of histone acetylation by the multiple bromodomains residing on SWI/SNF family complexes. In this case, affinity is increased, but a role for acetylation in regulating ATPase activity has not been determined. For CHD remodelers, the N terminus of their catalytic ATPase contains two tandem chromodomains, which bind methylated lysine residues and may be utilized for remodeler targeting or regulation. Interestingly, omission of the N terminus (including the chromodomains) leads to activation of the CHD ATPase domains, suggesting an autoinhibition mechanism that may be relieved by the binding of the chromodomains to histone PTMs.

Structural models inform how remodelers engage and remodel nucleosomes

Biochemical studies on SWI/SNF family remodelers suggest that a single nucleosome is the primary unit of remodeling, which commonly involves a complex comprising one nucleosome and one remodeler. This notion is supported by three EM structures of the yeast SWI/SNF family complex RSC, along with two of metazoan SWI/SNF (PBAF) complexes. One such RSC structure is depicted in **Figure 2.12A**, with a nucleosome modeled into the deep single pocket. This pocket accommodates a nucleosome snugly but with no steric clashes, strongly supporting the 1:1 stoichiometry. The exact position and orientation of the nucleosome are speculative. Since the pocket is slightly larger than the nucleosome, a DNA loop involving a few base pairs could be accommodated, in keeping with the models proposed in Figure 2.11.

Biochemical studies with certain ISWI complexes also suggest a 1:1 stoichiometry, while other ISWI complexes (such as ACF) operate as a dimer. Recent EM work reveals two ACF complexes bound in symmetrically equivalent positions on either side of the nucleosome (Figure 2.12B). They probably alternate action, pumping DNA in opposite directions to move the octamer back and forth along the DNA.

Presently, the field lacks a high-resolution structure of a remodeler complex. However, high-resolution structures (either alone or in association with DNA) are available for monomeric Rad54-related proteins, which are SNF2 family DNA translocases utilized in DNA repair (recombination) that serve as excellent structural models for remodeler translocases. Furthermore, a high-resolution structure of a CHD monomer remodeler (*S. cerevisiae* Chd1) has been solved, revealing a DNA translocase domain similar to Rad54 with chromodomains present at the N terminus. This may prevent access of DNA to the translocase domain, in keeping with the autoregulation mechanism for CHD remodelers described above.

2.4 EPIGENETIC MECHANISMS

Every cell in our body has the same blueprint, its genome; however, all the different cell types are characterized by distinct gene expression patterns. These patterns are inherited from parent to daughter cell during the development of a cell lineage. **Epigenetics** (epi is from the Greek επί—meaning over or above) is the study of heritable changes in phenotype caused by mechanisms that alter gene expression without changing the underlying DNA sequence. Epigenetic changes include histone variant incorporation, histone PTMs, DNA methylation, and actions of noncoding RNAs. Epigenetic regulation enables differential expression of a given genome, such that cells can differentiate into a wide range of specialized types. Unlike the underlying DNA sequence of the genome, the epigenome is dynamic (**Figure 2.13**).

In addition to the roles they play in differentiation and development, epigenetic mechanisms are also important in unicellular organisms. In yeast and protozoa, for example, epigenetic regulation helps in the maintenance of *centromeres* (the part of a chromosome that links sister chromatids) and *telomeres* (regions of repetitive nucleotide sequences at each end of a chromatid), in the organisms' adaptation to the environment, and in simple differentiation processes (for example sporulation). While epigenetic inheritance of phenotypes

(A)
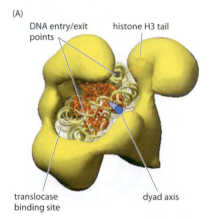

DNA entry/exit points   histone H3 tail

translocase binding site   dyad axis

(B)
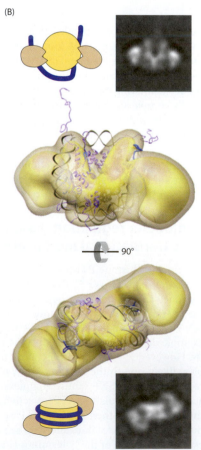

90°

Figure 2.12 EM models of SWI/SNF and ISWI remodelers.
(A) There are three similar electron microscopy (EM) structures of the yeast SWI/SNF family complex RSC (Remodels the Structure of Chromatin) with the nucleosome modeled into the pocket. The pocket is slightly larger than the nucleosome dimension and fits without steric clashes. The orientation of the nucleosome and the position of the labeled features are speculative, but they are consistent with biochemical evidence. (B) EM structure of the ISWI family ACF dimer bound to a nucleosome. (A, from A.E. Leschziner et al., *Proc. Natl Acad. Sci. USA* 104:4913–4918, 2007. National Academy of Sciences, USA. Reprinted with permission.)

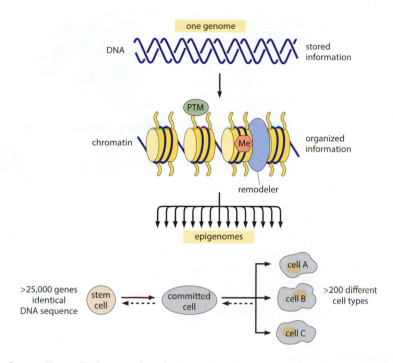

one genome

DNA — stored information

PTM

chromatin — organized information

Me

remodeler

epigenomes

>25,000 genes identical DNA sequence

stem cell → committed cell →

cell A

cell B   >200 different cell types

cell C

Figure 2.13 DNA versus chromatin. The genome is represented by an invariant DNA sequence (blue double helix) of an individual. The epigenome is represented by the overall heritable chromatin state, which indexes the entire genome in any given cell. It varies according to cell type and in response to internal and external signals. Epigenome diversification occurs during development in multicellular organisms as differentiation proceeds from a single stem cell (the zygote) to more committed cells. Reversal of differentiation or transdifferentiation requires the reprogramming of the cell's epigenome. (Adapted from C.D. Allis, T. Jenuwein, and D. Reinberg (eds). Epigenetics. Cold Spring Harbor, NY: Cold Spring Harbor Laboratory Press, 2007.)

occurs from cell to cell, the transfer of epigenetic patterns can also occur in a transgenerational manner, from organism to organism. This is well established in plants and can lead to non-Mendelian inheritance of traits. Recently, transgenerational inheritance has also been observed in animals.

Epigenetic mechanisms commonly employ 'readers' and 'writers' of epigenetic marks that interact with each other (sometimes directly) for mark propagation. The inheritance of epigenetic marks involves 'readers' and 'writers' that often interact, as will be discussed in the examples of histone PTMs and DNA methylation below.

Histone post-translational modifications are carriers of epigenetic information

Covalent PTMs of histones can be transmitted to daughter cells as epigenetic marks. This is supported experimentally by studies in yeast, where single histone residues can be mutated. However, the role of histone PTMs as carriers of epigenetic information in mammals has been largely inferred from the histone-modifying activity of many factors important for epigenetic pathways and the presence of the corresponding histone PTM within that epigenetic domain. For instance, EZH2 is a Polycomb group protein with histone methyltransferase activity that is involved in inactivation of one X chromosome in XX females (see below). Specifically, EZH2 (the 'writer') trimethylates lysine 27 of histone H3, and the inactive X chromosome is decorated by the corresponding H3K27me3 'mark.'

Special protein domains recognize specifically modified histone residues

Histone PTMs create binding sites for modification-specific protein modules (readers). Usually, these binding modules are small domains (50–150 aa) containing pockets that accommodate a given post-translationally modified histone residue. Bromodomains (see **Guide**) interact with acetylated histones, whereas chromodomains and PHD (plant homeo domain) finger motifs bind methylated histones.

As an example, the chromodomain-containing heterochromatin protein 1 (HP1) family recognizes a specific epigenetic histone PTM. By binding di- and trimethylated lysine 9 of histone H3 tails (H3K9me2/3) with high specificity, the HP1 family mediates gene silencing and heterochromatin formation in organisms as diverse as fission yeast, *Drosophila*, and mammals. X-ray crystallographic and NMR (nuclear magnetic resonance) (see **Methods**) spectroscopic studies show that the histone tail binds as an extended β strand in a groove on one face of the chromodomain, completing a three-stranded β sheet between two chromodomain β regions and the coplanar tail. The chromodomain–tail interaction requires that the tail is methylated at lysine 9 (**Figure 2.14**). The modified residue is bound within a conserved aromatic pocket (Tyr24, Trp45, and Tyr48 in *Drosophila*), which forms a three-walled cage around the methyl groups via van der Waals and cation-π interactions. The remaining tail residues are bound through backbone hydrogen bonds and complementary surface interaction from their side chains. Therefore, recognition of H3K9me3 occurs only when presented in the correct context of the surrounding histone tail. Consequently, the HP1 chromodomain does not interact with H3K4me3, a PTM linked to transcriptional activation, because the sequence context of this PTM site is different to the one around H3K9me2/3. HP1 proteins also bind the H3K9me3 methyltransferase SUVAR39, which binds the H3K9me3 mark through its own chromodomain. The interaction between the methyltransferase and HP1 proteins seemingly aids the propagation of this heterochromatic mark in some organisms.

The Polycomb protein is another chromodomain-containing factor that assembles into repressive chromatin complexes, which are important for the heritable silencing of genes during development. The chromodomain of Polycomb interacts specifically with H3K27me3. Structural analysis of the Polycomb chromodomain shows that it is related to the HP1 chromodomain and interacts with H3K27me3 in a similar manner. However, a longer groove allows the binding of five additional residues and creates specificity to K27 methylation. Despite the previous examples, chromodomain recognition of histone lysine methylation is not always linked with repressive chromatin; the human nucleosome remodeling factor CHD1, whose chromodomain binds H3K4me3, is involved in active transcription. **Figure 2.15** summarizes some of the known interactions between chromatin factors, their specific recognition domains, and histone PTMs.

The histone code hypothesis suggests that the PTM pattern of a nucleosome acts as a 'barcode'

Many site-specific histone PTMs are often mutually exclusive at a given residue (for example a lysine residue may either be methylated or acetylated). The identification of such PTMs and the corresponding factors that bind to them led to the **histone code** hypothesis. This hypothesis proposes an epigenetic marking system using different combinations of histone PTM patterns to regulate functional outputs of eukaryotic genomes. Indeed, several binding factors require the PTM of distinct histone residues for function, as if the PTM pattern of a nucleosome creates a sort of 'barcode' (**Figure 2.16**)

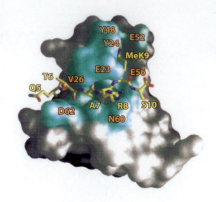

Figure 2.14 Side chain interactions between the H3K9me2 tail and the *Drosophila* chromodomain. (From S.A. Jacobs and S. Khorasanizadeh, *Science* 295:2080–2083, 2002. With permission from AAAS.)

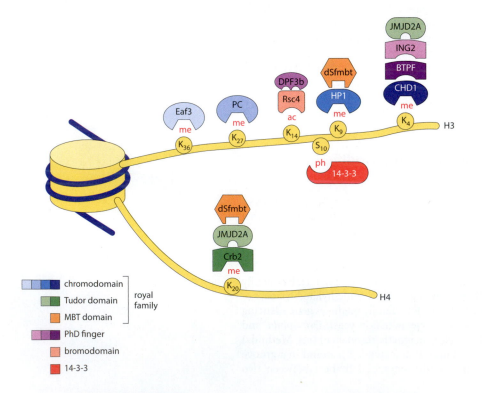

Figure 2.15 A cartoon summarizing some of the known post-translational modifications (PTMs) on histones H3 and H4 and their known interaction domains. (Adapted from A.J. Bannister and T. Kouzarides, *Cell Res.* 21:381–395, 2011. With permission from Macmillan Publishers Ltd.)

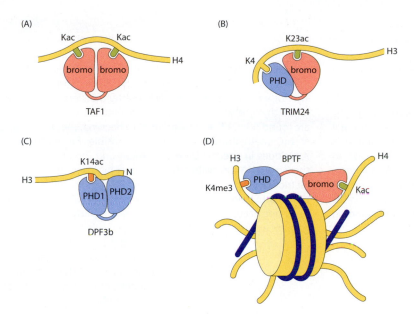

Figure 2.16 Models of combinatorial readout of two histone post-translational modifications (PTMs) by paired chromatin-associated modules. (A) The TAF1 double bromodomain binds two acetylated lysines (Kac) on the H4 histone tail. (B) The TRIM24 PHD-bromo cassette binds unmethylated H3K4 and H3K23ac PTM pairs. (C) The DPF3b double PHD binds both the N-terminal H3 and the H3K14ac mark. (D) The BPTF PHD-bromo cassette binds H3K4me3 and H4Kac paired histone marks on a single nucleosome. (Adapted from Z. Wang and D.J. Patel, *J. Biol. Chem.* 286:18363–18368, 2011.)

While it has not been shown that histone PTMs alone can act as instructive or self-propagating determinants of epigenetic outcomes, it is clear that they are intimately involved in various epigenetic processes. Histone post-translational modifications are carriers of epigenetic information.

Despite the accepted role of histone PTMs as a source of epigenetic information, it remains unclear how these patterns are passed from one cell generation to the next. A few models have been proposed. Random segregation of histones from a pool of new and old (with the latter retaining their PTM status) would dilute the marks and disrupt their location within the genome. Therefore, this model would only effectively transmit those PTMs that were highly repetitive. Alternatively, segregation could be semi-conservative, with an equal distribution of new and old histones within a replicated nucleosome. Although this model would produce a template from which PTM patterns could be copied heritably, it is not known how such ordered histone segregation would occur.

Methylation of CpG islands is another epigenetic marker that results in widespread gene silencing

In vertebrate genomes, only about 3% of the C nucleotides are methylated, but the majority of C residues that occur in CpG dinucleotide sequences are methylated (**Box 2.4**). The 5mC residues can spontaneously deaminate to thymine, more than the repair system (thymine DNA-glycosylase, Chapter 4) can normally cope with. However, the genomes are punctuated by regions ~1 kbp long, rich in CpG sequences. These **CpG islands** (estimated to be ~25,000 in human somatic cell DNA) overlap the promoter regions upstream of many genes and most remain hypomethylated during early developmental stages. However, some become methylated during normal development and this affects gene regulation by occluding binding sites for certain transcription factors, while creating sites for 5mC-binding proteins that mediate gene repression. There is good evidence that methylation of CpG islands acts as an epigenetic marker and may be a more widespread means of gene silencing in somatic cells.

Replication of CpG methylation during embryogenesis and development is an important characteristic of epigenetic inheritance in mammals. DNA methyltransferase 1 (Dnmt1) mediates the maintenance of DNA methylation patterns from parent to daughter cells, by methylating newly replicated CpG sites occurring opposite those methylated on the mother strand. However, the methylation 'writer' DNMT1 requires the 'reader' cofactor UHRF1 (also known as ICBP90 and Np95), which interacts preferentially with hemi-methylated CpG sites and recruits Dnmt1 to the replication fork. Genetic deletion of UHRF1 in mice leads to a catastrophic loss of DNA methylation and, consequently, a loss of transcriptional silencing by heterochromatin and imprinting.

Box 2.4 DNA methylation

The bases in DNA can undergo chemical modification, the most common form of which is methylation. The modification is catalyzed by enzymes called DNA methyltransferases, which form three different families each with a different specificity. As in most biological methylations, the source of the methyl group is *S*-adenosyl methionine (SAM) and the reaction is illustrated in **Figure 2.4.1**. The methyl groups in the modified bases project into the major groove of B-DNA and do not interfere with base pairing, but importantly they are able to recruit and interact with DNA-binding proteins that in vertebrates can affect gene expression, notably methyl-CpG-binding domain proteins.

The enzyme that generates 5-methylcytosine (5mC) is found in some lower eukaryotes, most higher plants and in animals. The other two forms of DNA methyltransferase generate N^4-methylcytosine and N^6-methyladenine, respectively, and are found primarily in prokaryotes. In bacteria, the N^6-mA- and N^4-mC methyltransferases are responsible for methylating A residues in specific base sequences and thereby rendering the host DNA insensitive to digestion by the protective host restriction enzymes. In addition, in *Escherichia coli* it is clear that methylation of all A residues found in GATC sequences does not occur until shortly after DNA replication. This time gap allows freshly synthesized DNA strands to be transiently distinguished from their template strands, of crucial importance to the system of *strand-directed mismatch DNA repair* (Chapter 4).

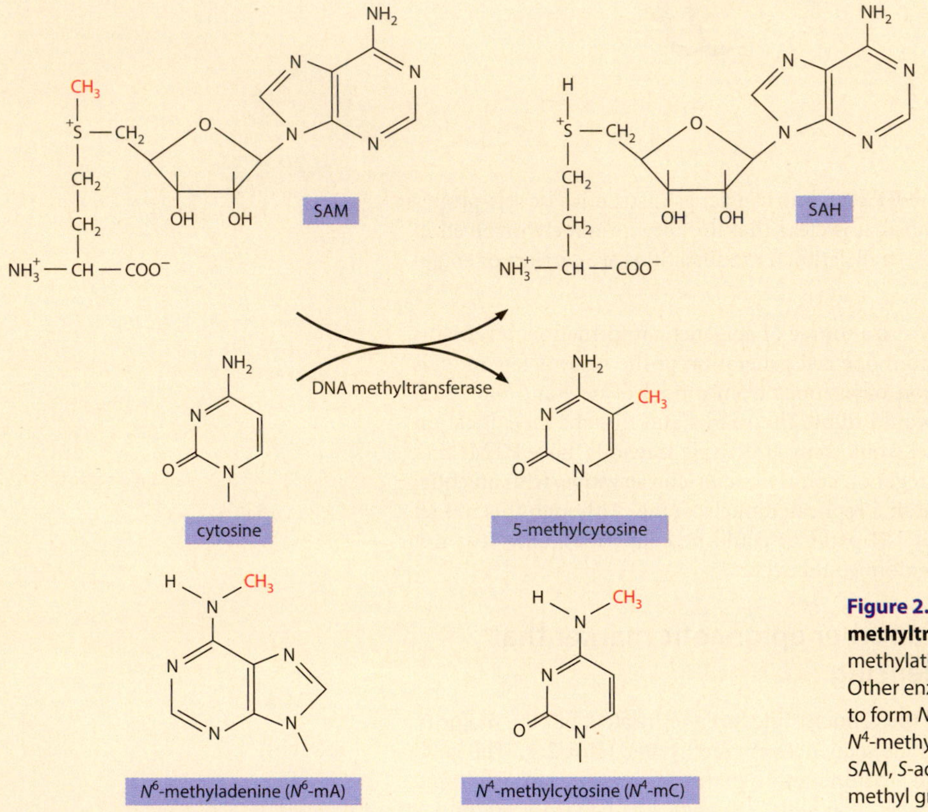

Figure 2.4.1 Methylation of DNA by methyltransferases. The reaction shown is the methylation of cytosine to 5-methylcytosine. Other enzymes are able to methylate A residues to form N^6-methyladenine or C residues to N^4-methylcytosine. SAH, *S*-adenosylhomocysteine; SAM, *S*-adenosylmethionine. The transferred methyl group is in red.

The preferential affinity of UHRF1 for hemi-methylated over symmetrically methylated DNA is accomplished via its SRA (Set and RING-associated) domain. X-ray crystallographic studies show that the SRA domain of the human or mouse UHRF1 protein folds into a globular structure, formed largely by the packing of two twisted β sheets (**Figure 2.17**). A concave surface faces the DNA and a tight-fitting binding pocket accommodates the 5mC that is flipped out of the duplex DNA, binding it by van der Waals interactions. This cavity would be unfavorably hydrated with a nonmethylated cytosine ligand, and it would not bind any of the other DNA bases well. Two loops flanking the concave surface reach through the gap generated by the base flipping of the 5mC and contact the DNA in the major and minor grooves. These 'fingers' read the other three bases of the CpG duplex, conferring specificity for the CpG dinucleotide as well as discriminating against methylation on the cytosine of the complementary strand. Therefore, multiple specific contacts to the methylated and nonmethylated CpG bases explain the selectivity of the SRA domain for hemi-methylated DNA over both nonmethylated and fully methylated DNA. Upon recognition of the base-flipped hemi-methylated CpG site by the SRA domain of UHRF1, the maintenance methyltransferase Dnmt1 is recruited. Modeling suggests that a coordinated transfer of the hemi-methylated DNA to Dnmt1 occurs upon flipping out of the target cytosine for methylation (**Figure 2.18**).

DNA methylation is normally absent from the DNA of zygotes and the methylation pattern is subsequently set up by *de novo* methyltransferases. It is generally irreversible and can be inherited during cell division because so-called maintenance methyltransferases preferentially methylate CpG sequences in the DNA strand opposite already methylated CpG sequences (in reverse orientation) on the opposing DNA strand. Without this latter activity the DNA methylation pattern would gradually be lost over time. Disruption of normal DNA methylation is widely recognized as a hallmark of neoplastic cells and frequently observed in inherited diseases.

However, DNA methylation is not a universal mechanism for gene silencing in eukaryotes. Lower invertebrates such as *D. melanogaster* and *Caenorhabditis elegans* or yeasts (*S. cerevisiae*) exhibit little or no DNA methylation, but the genomes of many plants (such as *Arabidopsis thaliana*) are heavily methylated. In plants DNA methylation can be induced by RNA molecules and it may be that it represents some form of defense mechanism against RNA viruses or transposons.

X-chromosome inactivation and imprinting are important epigenetic phenomena in mammalian cells

An important paradigm for mammalian epigenetics is **X-inactivation**: the random and subsequently heritable silencing of one of the two X chromosomes in female somatic lineages, initiated during early development. This ensures that XX females express the same number of X-chromosome gene products as XY males (known as dosage compensation). Condensed heterochromatin prevents the expression of most genes on the inactive X chromosome. The establishment and propagation of this repression involve most parts of the chromosome and are accomplished through the combined effects of several epigenetic mechanisms. For instance, the inactive X chromosome is coated with the noncoding RNA Xist (X-inactive specific transcript), DNA methylated, modified at specific histone PTM sites, and enriched in histone variant macroH2A.

Another phenomenon linked to epigenetics is **imprinting**, whereby differential regulation of a gene or gene cluster leads to expression of only one of the two parental alleles which is heritable in all cell types. One example of this is the maternally imprinted (and thus silent) insulin-like growth factor 2 gene, resulting in the exclusive expression of the paternal allele. As some alleles are maternally imprinted and others paternally, both parental genomes are required in mammals, as confirmed by the failed development of embryos engineered to contain two maternal or two paternal genomes. Imprinting is controlled by *cis*-acting DNA elements called imprint control elements (ICEs). DNA methylation plays a critical role in this process; differential methylation of the parental ICEs is established in the germ line before (or during) gametogenesis and is maintained after fertilization in somatic tissues. Histone PTMs and chromosome folding also contribute to the maintenance of genomic imprinting.

The study of epigenetic mechanisms is a rapidly developing field, where many new principles will emerge in the years to come. Structural studies of the factors that are involved in epigenetic processes will be indispensible for a better understanding.

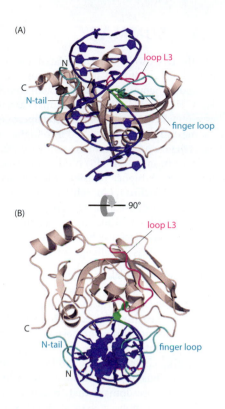

Figure 2.17 Structure of the UHRF1 SRA hemi-methylated CpG DNA complex from mouse. SRA (Set and RING-associated; gray), loops penetrating into the DNA through the gap generated by the base flipping of the 5mC (magenta), DNA (blue), flipped-out 5mC (green). (PDB 2ZKD, 2ZKE) (From K. Arita et al., *Nature* 455:818–821, 2008. With permission from Macmillan Publishers Ltd.)

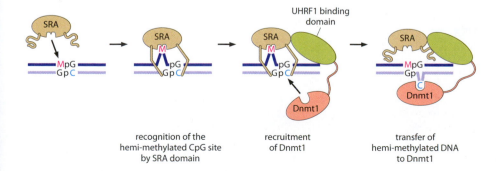

recognition of the hemi-methylated CpG site by SRA domain

recruitment of Dnmt1

transfer of hemi-methylated DNA to Dnmt1

Figure 2.18 Schematic model showing cooperative action by UHRF1 and Dnmt for maintenance of CpG methylation. (Adapted from K. Arita et al., *Nature* 455:818–821, 2008. With permission from Macmillan Publishers Ltd.)

2.5   SUMMARY

For the DNA to be accommodated within the nucleus, it needs to be compacted. The core particle of chromatin is the nucleosome, which is formed by 147 bp of DNA wrapped around an octameric histone protein complex. There are four families of core histones, with H2A–H2B and H3–H4 heterodimers assembling to form the histone octamer. Nucleosome assembly is assisted by histone chaperones. The basic 'beads-on-a-string' array of nucleosomes is further compacted into higher order structures of a 30 nm fiber. However, their exact *in vivo* packing remains controversial, and even its very existence has been challenged. Linker histones (H1, H5) bind to the DNA between nucleosomes and help to stabilize 30 nm fibers. A higher density of linker histones is normally associated with the electron-dense heterochromatin, which is associated with transcriptional repression, whereas the less electron-dense and transcriptionally active euchromatin has a lower concentration of linker histones. Chromatin architectural proteins, such as the *Drosophila* Polycomb group complex, have recently been shown to mediate assembly of higher order chromatin structures. It remains a challenge for the future to elucidate their molecular architecture.

Chromatin remodeling complexes move, reconfigure, or eject nucleosomes to provide the chromatin fluidity needed for chromosomal processes such as gene regulation, DNA repair, replication, and recombination. There are four families of chromatin remodeling complexes, each endorsed with particular subunit compositions to perform specialized modes of remodeling. Furthermore, within each family, the subunits can assemble in more than one configuration, in part through the use of paralogs. All remodelers hydrolyze ATP, converting chemical energy to force, as needed to alter the relatively stable histone–DNA contacts. Thus, in spite of their compositional differences, all remodelers contain a highly similar ATPase domain that functions as an ATP-dependent DNA translocase. For each remodeler, the non-catalytic attendant subunits provide important targeting and regulatory functions, which likely involve the 'reading' of histone PTMs and domains.

Genome expression is also regulated in a heritable way by epigenetic changes that do not alter the underlying DNA sequence. These epigenetic changes include variant histones, histone PTMS, DNA methylation, and the action of noncoding RNAs. Epigenetic mechanisms often involve both 'writers' and 'readers' of epigenetic marks. Specific histone PTMs are recognized by domains such as the bromodomains interacting with acetylated histones or the chromodomains with methylated histones. The sequence context of the PTM is also important; for example the chromodomain-containing heterochromatin protein 1 (HP1) family recognizes only methylated lysines in the tail of histone H3. The 'reader' HPI also binds the 'writer' methyltransferase, with this interaction aiding the propagation of this epigenetic mark. The histone code hypothesis proposes an epigenetic marking system, with different combinations of histone PTM patterns regulating the functional output of eukaryotic genomes. However, it remains unclear how these patterns are passed from one cell generation to the next. DNA methylation particularly of CpG islands (regions rich in CpG dinucleotides) is another epigenetic marker that results in widespread gene silencing. A combination of epigenetic markers including the coating with the noncoding RNA Xist, results in the random and heritable silencing of one the two X chromosomes in female somatic cells, thus ensuring equivalence in the expression of X-chromosome gene products in both females and XY males (dosage compensation). Another epigenetic phenomenon is imprinting, whereby specific genes or gene clusters from one parent are silenced in all cell types.

Sections of this chapter are based on the following contributions

2.2   Karolin Luger, University of Colorado, Boulder, USA and Jeffrey C. Hansen, Colorado State University, Fort Collins, USA.

2.3   Bradley R. Cairns, University of Utah School of Medicine, USA.

2.4   Patrick Varga-Weisz, The Babraham Institute, Cambridge, UK.

REFERENCES

2.2 Nucleosomes and higher order chromatin structures

Clark RF & Felsenfeld G (1971) Structure of chromatin. *Nat New Biol* 229:101–106.

Fan JY, Gordon F, Luger K et al (2002) The essential histone variant H2A.Z regulates the equilibrium between different chromatin conformational states. *Nat Struct Biol* 9:172–176.

Ghirlando R & Felsenfeld G (2008) Hydrodynamic studies on defined heterochromatin fragments support a 30-nm fiber having six nucleosomes per turn. *J Mol Biol* 376:1417–1425.

Hall MA, Shundrovsky A, Bai L et al (2009) High-resolution dynamic mapping of histone-DNA interactions in a nucleosome. *Nat Struct Mol Biol* 16:124–129.

Hart CM & Laemmli UK (1998) Facilitation of chromatin dynamics by SARs. *Curr Opin Genet Dev* 8:519–525.

Kornberg RD & Lorch Y (2007) Chromatin rules. *Nat Struct Mol Biol* 14:986–988.

Kruithof M, Chien FT, Routh A et al (2009) Single-molecule force spectroscopy reveals a highly compliant helical folding for the 30-nm chromatin fiber. *Nat Struct Mol Biol* 16:534–540.

Luger K (2001) Nucleosomes: Structure and Function. Encyclopedia of Life Sciences eLS. John Wiley & Sons. doi: 10.1038/npg.els.0001155.

Luger K, Mäder AW, Richmond RK et al (1997) Crystal structure of the nucleosome core particle at 2.8 Å resolution. *Nature* 389:251–260.

Moudrianakis EN & Arents G (1993) Structure of the histone octamer core of the nucleosome and its potential interactions with DNA. *Cold Spring Harb Symp Quant Biol* 58:273–279.

Olins DE & Olins AL (2003) Chromatin history: our view from the bridge. *Nat Rev Mol Cell Biol* 4:809–814.

Ris H & Kubai DF (1970) Chromosome structure. *Annu Rev Genet* 4:263–294.

Schalch T, Duda S, Sargent DF & Richmond TJ (2005) X-ray structure of a tetranucleosome and its implications for the chromatin fibre. *Nature* 436:138–141.

Segal E, Fondufe-Mittendorf Y, Chen L et al (2006) A genomic code for nucleosome positioning. *Nature* 442:772–778.

Song F, Chen P, Sun D et al (2014) Cryo-EM study of the chromatin fiber reveals a double helix twisted by tetranucleosomal units. *Science* 344:376–380.

Thoma F, Koller T & Klug A (1979) Involvement of histone H1 in the organization of the nucleosome and of the salt-dependent superstructures of chromatin. *J Cell Biol* 83:403–427.

Tremethick DJ (2007) Higher-order structures of chromatin: the elusive 30 nm fiber. *Cell* 128:651–654.

Van Holde KE (1988) Chromatin. Series in Molecular Biology (Rich A ed) Springer-Verlag.

Wolffe A (1998) Chromatin: Structure and Function, 3rd ed. Academic Press.

Woodcock CL & Ghosh RP (2010) Chromatin higher-order structure and dynamics. *Cold Spring Harb Perspect Biol* 2:a000596.

Woodcock CLF, Frado L-LY & Rattner JB (1984) The higher-order structure of chromatin: evidence for a helical ribbon arrangement. *J Cell Biol* 99:42–52.

2.3 Remodeling complexes

Asturias FJ, Chung WH, Kornberg RD & Lorch Y (2002) Structural analysis of the RSC chromatin-remodeling complex. *Proc Natl Acad Sci USA* 99:13477–13480.

Chaban Y, Ezeokonkwo C, Chung WH et al (2008) Structure of a RSC-nucleosome complex and insights into remodeling. *Nat Struct Mol Biol* 15:1272–1277.

Clapier C & Cairns BR (2009) The biology of chromatin remodeling complexes. *Annu Rev Biochem* 78:273–304.

Ho L & Crabtree GR (2010) Chromatin remodelling during development. *Nature* 463:474–484.

Leschziner AE, Saha A, Wittmeyer J et al (2007) Conformational flexibility in the chromatin remodeler RSC observed by electron microscopy and the orthogonal tilt reconstruction method. *Proc Natl Acad Sci USA* 104:4913–4918.

Liu LF & Wang JC (1987) Supercoiling of the DNA template during transcription. *Proc Natl Acad Sci USA* 84:7024–7027.

Skiniotis G, Moazed D & Walz T (2007) Acetylated histone tail peptides induce structural rearrangements in the RSC chromatin remodeling complex. *J Biol Chem* 282:20804–20808.

2.4 Epigenetic mechanisms

Arita K, Ariyoshi M, Tochio H et al (2008) Recognition of hemi-methylated DNA by the SRA protein UHRF1 by a base-flipping mechanism. *Nature* 455:818–821.

Avvakumov GV, Walker JR, Xue S et al (2008) Structural basis for recognition of hemimethylated DNA by the SRA domain of human UHRF1. *Nature* 455:822–825.

Bannister AJ & Kouzarides T (2011) Regulation of chromatin by histone modifications. *Cell Res* 21:381–395.

Deaton AM & Bird A (2011) CpG islands and the regulation of transcription. *Genes Dev* 25:1010–1022.

Fischle W, Wang Y, Jacobs SA et al (2003) Molecular basis for the discrimination of repressive methyl-lysine marks in histone H3 by Polycomb and HP1 chromodomains. *Genes Dev* 17:1870–1881.

Hashimoto H, Horton JR, Zhang X et al (2008) The SRA domain of UHRF1 flips 5-methylcytosine out of the DNA helix. *Nature* 455:826–829.

Koerner MV & Barlow DP (2010) Genomic imprinting—an epigenetic generegulatory model. *Curr Opin Genet Dev* 20:164–170.

Maison C & Almouzni G (2004) HP1 and the dynamics of heterochromatin maintenance. *Nat Rev Mol Cell Biol* 5:296–304.

Min J, Zhang Y & Xu RM (2003) Structural basis for specific binding of Polycomb chromodomain to histone H3 methylated at Lys 27. *Genes Dev* 17:1823–1828.

Murray IA, Clark TA, Morgan RD et al (2012) The methylomes of six bacteria. *Nucleic Acids Res* 40:11450–11462.

Nielsen PR, Nietlispach D, Mott HR et al (2002) Structure of the HP1 chromodomain bound to histone H3 methylated at lysine 9. *Nature* 416:103–107.

Wutz A (2011) Gene silencing in X-chromosome inactivation: advances in understanding facultative heterochromatin formation. *Nat Rev Genet* 12:542–553.

DNA Replication

3.1   INTRODUCTION

Francois Jacob observed that "Everything in a living being is centered on reproduction. A bacterium … what destiny can they dream of other than forming two bacteria?" Central to this anthropomorphic simplification is the process of replication of DNA, the information-bearing molecule in the cell. DNA is replicated at the moment when a number of internal and external factors align. The overall framework for understanding DNA replication was laid down in 1953 with elucidation of the double-helical structure of DNA and the idea of complementary base pairing. Indeed, at the time, James Watson and Francis Crick fore-shadowed decades of studies to understand associated mechanisms when they wrote one of the most memorable sentences in any scientific paper, "It has not escaped our notice that the specific pairing we have postulated immediately suggests a possible copying mechanism for the genetic material". The clear implication was that the two strands of the double helix could separate to expose nucleobases in each strand, which would then act as templates for the generation of two exact daughter molecules (**Figure 3.1A**). This concept of **DNA replication**, now known as semi-conservative replication, was borne out experimentally soon afterwards by Herbert Taylor, Philip Woods, and Walter Hughes, and independently by Matthew Meselson and Franklin Stahl. These researchers used orthogonal approaches to show that the replication of the parent duplex gave rise to two molecules of daughter DNA in each of which one strand was parental and the other (complementary) strand was newly synthesized.

Although conceptually simple, semi-conservative replication of DNA requires a complex array of protein machines, brought together in space and time, to ensure the faithful propagation of the genetic blueprint; regulation of these machines is now also understood to be an important feature. All organisms replicate their DNA by what is essentially a common mechanism, a process that can be subdivided into three key stages—initiation, elongation, and termination—based on a common structural element, the **replication fork**. The replication fork was first noted in electron microscopy (see **Methods**) analysis of replication

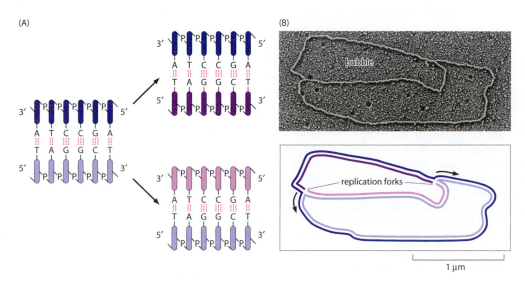

(A)

(B)

bubble

replication forks

1 µm

Figure 3.1 Semi-conservative replication of DNA, with two replication forks moving in opposite directions. (A) Each strand of the DNA double helix (dark blue and light blue) is used as a template, generating two exact copies (purple and lilac) of the double helix. An A/T base pair is held together by two H-bonds (pink dots); a G/C base pair is held together by three H-bonds. (B) Upper, electron micrograph of the circular *E. coli* chromosome (4.6 × 10^6 bp) being replicated, with a bubble that grows in size as the DNA replication process proceeds. Lower, schematic representation, with two replication forks moving in opposite directions from a fixed starting point on the chromosome, as indicated by the arrows. Color scheme as in (A). (B, Courtesy of Jerome Vinograd.)

of the circular chromosome of *Escherichia coli*, which revealed that replication began from a single point, and that a 'bubble'-shaped structure (Figure 3.1B) grew in size as replication proceeded. These, and other genetic analyses, implied participation of two replication forks moving in opposite directions at ~500–1000 bp per second until they met up at a diametrically opposite point on the circular chromosome. This bidirectionality has profound consequences for the mechanism of DNA replication.

Details of the molecular mechanisms associated with DNA replication have emerged from many insightful studies in phage, bacteria, archaea, and eukaryotes. The basic processes of DNA replication are conserved throughout evolution, although a number of exceptions to the normative view have been described. In this chapter we present this normative view, focusing on the molecular machines in the sequence in which they appear in our story. Analyses from bacteria, where core mechanisms are easier to access experimentally, are integrated with studies in archaea and eukaryotes where elaborations and regulatory complexity are of great interest. Where possible, notable exceptions to the normative view will be highlighted.

3.2   INITIATION AND ELONGATION

DNA replication starts at special sites on the chromosome called origins of replication. Origins provide binding sites for multiprotein complexes (also known as pre-replicative complexes) that carry out several important tasks, including: (1) melting the duplex at the origin; (2) assembling the replicative helicase, which will travel ahead of the replication fork during DNA synthesis; and (3) regulating initiations.

The logic for initiating replication at defined rather than random sites reflects a regulatory priority to coordinate DNA synthesis with other events in the cell (cell size, metabolic status, etc.). Control is manifest by regulating melting of the duplex; separating the two strands of duplex DNA is required for assembly and operation of the synthesis machinery. In addition to this control of access to the duplex, *DNA polymerases* typically can only initiate synthesis when the duplex has been melted and a short **RNA primer**, which has been synthesized by a specialized RNA polymerase (primase), is provided.

Origins vary in length (100–500 bp). In bacteria, archaea, and yeast, origins are defined by sequence. By contrast, in some eukaryotes (some fungi and metazoans), origins are defined by context, with the result that any DNA sequence can serve as an origin.

A series of multiprotein complexes are recruited to the bacterial origin of DNA replication

Studies in a large number of organisms show that a series of multiprotein complexes are recruited to the origin just before the initiation of DNA synthesis to implement the tasks listed above. Initiation proteins are either recruited to dsDNA-binding sites at the origin and associate into an oligomer (bacterial DnaA) or arrive as a pre-formed oligomer (eukaryotic ORC). The resulting multiprotein–DNA complex then recruits additional factors that mature into a pair of replication forks.

Bacterial chromosomes, irrespective of whether they are linear or circular, are replicated from a single origin, termed *oriC*. In *E. coli,* the *oriC* element (**Figure 3.2A**) spans ~270 bp; the right-most 200 bp contains at least 10 asymmetric '9-mer' DnaA-binding sites (termed R1–5, I-sites). DnaA monomers bind to these sites and associate into a large nucleoprotein assembly. The left-most segment of *OriC* contains three '13-mer' bp sites, and is terminated by an AT-rich region, termed the DNA unwinding element (DUE); this segment harbors the site of initial unwinding. The AT-rich DUEs require less energy to melt because there are fewer H-bonds (two per AT bp, compared with three per GC bp) holding the chains together (see Figure 3.1A). The '13-mers' appear to interact with the ATP form of DnaA after melting has occurred. Melted DNA at the origin enables the subsequent recruitment of additional proteins required for replication. The polarity of the 9-mer sites, their relative spacing, and distance to the 13-mers are important. Origin DNA melted by the DnaA nucleoprotein complex serves as entry point for the DnaB replicative helicase; entry requires the DnaC loader/helicase inhibitor protein. The DnaA nucleoprotein assembly forms in the presence of ADP or ATP, but it is the ATP form that is required for melting of the origin; melted DNA is then captured within the oligomeric DnaA assembly. Other proteins such

(A)

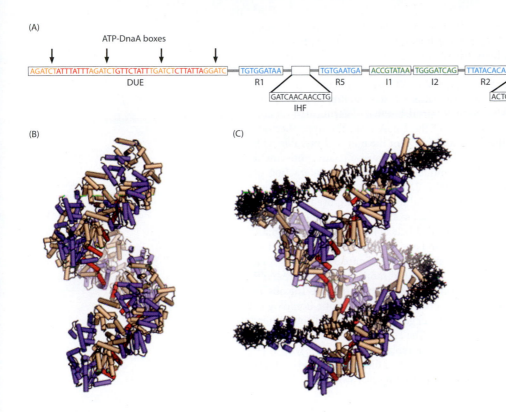

ATP-DnaA boxes

AGATCTATTTATTTAGATCTGTTCTATTTGATCTCTTATTAGGATC | TGTGGATAA | | | TGTGAATGA | ACCGTATAA | TGGGATCAG | TTATACACA | | | TTTGGATAA | TTGATCCAA | TTATCCACA

DUE   R1   R5   I1   I2   R2   R3   I3   R4

GATCAACAACCTG
IHF

ACTCAAAAACTGAAC
Fis

(B)   (C)

Figure 3.2 The initiation of DNA replication in bacteria. (A) The origin of replication (*oriC*) in *E. coli*. DUE, DNA unwinding element; R1–R5, I1–I3, and ATP-DnaA boxes, binding sites for DnaA; IHF and Fis, binding sites for architectural proteins IHF and Fis, respectively. (B) A helical filament composed of four truncated DnaA molecules (AAA+ module and HTH DNA-binding domain) in complex with the ATP analog AMPPCP. The AAA+ module is depicted in beige with initiator sequence motif (ISM) insert in the core domain shown in red; the helix-turn-helix (HTH) domain is in blue. (C) Hypothetical model of DnaA helical filament with HTH domains having undergone an outward rigid-body rotation that permits DNA (depicted in black) to be wrapped around the outside. (PDB 2HCB) (Adapted from K.E. Duderstadt and J.M. Berger, *Crit. Rev. Biochem. Mol. Biol* 43:163–187, 2008. With permission from Informa Healthcare.)

as the integration host factor (IHF) and inversion stimulation factor (Fis) also have binding sites in *oriC* and modulate the interaction with DnaA during initiation; both proteins promote origin complex formation by modifying the structure of DNA. The total concentration of DnaA remains unchanged throughout the bacterial cell cycle, but the amount of DnaA in the ATP-bound form rises just before initiation of DNA replication (the requirement for ATP by DnaA is not universal). This, in turn, stimulates saturation of the R1–R5, I-sites and DnaA boxes with DnaA-ATP.

A number of regulatory of switches and circuits have been described for control of replication initiation. For example, in *E. coli*, initiation is controlled via two broad strategies: (1) regulating the activity of DnaA, and (2) controlling access of DnaA to origin DNA. The *E. coli* DiaA protein influences the timing of replication initiation by forming a tetramer-based assembly with DnaA. Once initiation has taken place, a regulatory circuit consisting of the Hda and *sliding clamp* proteins (associated with the elongating replication fork) operates to inactivate DnaA for additional initiations by stimulating hydrolysis of ATP. Inclusion of the sliding clamp in this circuit ensures that inactivation takes place only after establishment of an active replication fork. Very little is known about how DnaA is reactivated for another round of replication. Acidic phospholipids have been implicated in this process, but the mechanism is not well understood. Recently, two chromosomal elements have been identified that serve to reactivate DnaA. Other strategies for regulating DnaA activity include transcriptional control of protein levels, and the presence of sites on the chromosome (the data locus) that attract DnaA and titrate it away from the replication origin.

A second broad approach to regulating initiation centers on controlling access to origin DNA. In *E. coli*, this strategy exploits the fact that the parent chromosome is fully methylated on GATC sequences by the Dam methylase. Semi-conservative replication of DNA produces molecules that are hemi-methylated; binding of the SeqA protein to GATC sequences within the origin serves to inhibit additional initiations. Sequestration of the origin by SeqA is reversed (to enable initiations) by an incompletely understood mechanism.

The active form of DnaA is an ATP-dependent oligomer

As noted above, the form of DnaA that is active for replication initiation is an ATP-dependent oligomer. Many structural and biochemical studies have provided insight into this oligomer and how it might function in replication. These studies imply an architecture that consists of four subdomains: (1) an N-terminal domain that codes for contacts to the replicative helicase DnaB, described below; (2) a flexible linker; (3) the AAA+ ATPase module (core and lid subdomains); and (4) a helix-turn-helix (HTH) dsDNA-binding domain (see **Guide**), which acts as the primary determinant of the binding of DnaA to the replication origin. DnaA is a member of the AAA+ family of ATPases, which are important participants in many and various molecular machines (**Box 3.1**, see also Box 7.2). As yet no structure for a full-length DnaA (52 kDa) has been obtained, but that of a truncated version (containing the AAA+ module and HTH domain) of the DnaA from the hyperthermophile *Aquifex aeolicus* has been solved in the presence of ADP or the non-hydrolyzable ATP analog, AMPPCP (Figure 3.2B). The binding of ATP appears to cause a relative movement of lid and core subdomains in the AAA+ ATPase module, transmitted by interaction of the sensor II Arg with the γ-phosphate of the ATP (Box 3.1). This conformational change makes it possible for another DnaA-ATP protomer to bind to a newly exposed surface on the first protomer, and the assembly proceeds not into the hexamer more commonly found with AAA+ ATPase modules, but into an 8_1 right-handed helical filament. Notably, DnaA is a member of the 'initiator clade' of AAA+ ATPases, and one of the distinguishing features of this clade is the presence of the initiator sequence motif (ISM) (Figure 3.1.4 in Box 3.1). In the crystal structure of DnaA spanning domains 3 and 4, this motif prevents DnaA-ATP from forming planar oligomers and directs formation of the 8_1 filament instead. The observation that in the structure of the origin recognition complex (the eukaryotic homolog to DnaA), these motifs form a nearly planar arrangement suggests that additional conformers of DnaA, yet to be described, may play functional roles. The current view suggests that the 'active' version of DnaA is the ATP-bound form, and the replacement of ADP with ATP in the AAA+ module promotes assembly of the protomer into a filament capable of binding dsDNA. In this instance, the AAA+ ATPase module is acting as a switch which, when turned on, initiates DNA replication at *oriC*.

The AAA+ module and HTH domain in DnaA are separated by a linker region including an extended α helix and, in model building, if the HTH domains in the DnaA filament are repositioned, it becomes possible to wind the *oriC* DNA around the outside of the filament (Figure 3.2C). Wrapping of origin DNA around the exterior of the DnaA oligomer appears to be a key step in melting the DUE, perhaps topologically by constraining positive DNA supercoils (see Section 3.4); oligomerization of DnaA also creates a binding site for melted single-stranded DNA (ssDNA). The precise architecture of the DnaA-origin DNA complex, and concomitantly its precise role in the mechanism of initiation, remain to be established.

DNA replication in eukaryotes proceeds from multiple origins

Although conserved throughout evolution, the molecular mechanisms and associated protein machines of DNA replication required the introduction of new features to handle the greater complexities of eukaryotic chromosomes. Two aspects of eukaryotic genomes demand special attention. First, they are composed of multiple chromosomes, not just the single one found in most bacteria, and second, they are considerably larger than those in bacteria (*E. coli*, 4.6 Mbp; human, two copies of 3080 Mbp, arranged in 23 pairs of chromosomes). Both of these aspects are addressed by initiating replication from multiple origins, rather than just one, as in bacteria. Chromosomal DNA in all eukaryotes has multiple origins of replication per chromosome, but they are less well understood than the *oriC* region in *E. coli* DNA (Figure 3.2). The best studied eukaryotic origins are from the budding yeast *Saccharomyces cerevisiae*, where they range in length from 150 to 200 bp and are called the **autonomously replicating sequence** (**ARS**). ARS origins can contain several motifs, termed A, B1, B2, and B3. The A and B1 elements serve as the major point of contact between origin recognition complex (below) and origin DNA. Origins defined by nucleotide sequence are also present in mitochondrial DNA and some DNA viruses. In contrast to budding yeast, origins in other fungi and metazoans are not specified by nucleotide sequence. Rather, origin sequences appear to be located in intergenic regions of the DNA free of *nucleosomes*. The density of origins varies between species; the chromosomal DNA of yeast (~12 Mbp) contains several hundred (an average of one every 40–50 kbp) whereas that of humans (~3000 Mbp) has ~30,000 (an average of one roughly every 100 kbp). Electron micrographs of DNA in rapidly dividing cleavage nuclei reveal the existence of multiple replication forks and 'bubbles' in operation (**Figure 3.3A**). At higher resolution, two small regions representing ssDNA about 200 bases long, can be seen on diametrically opposite parts of the replication bubble (Figure 3.3B). These indicate

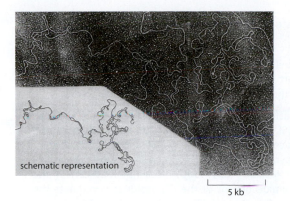

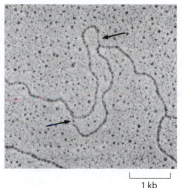

schematic representation

5 kb

1 kb

Figure 3.3 Fragment of replicating chromosomal DNA from lysed cleavage nuclei of *D. melanogaster* cells. (A) Electron micrograph showing 119 kbp of DNA containing a total of 23 replication 'bubbles'. A schematic representation is at bottom left. (B) Two replication forks in a replication bubble. The arrows point to short regions of ssDNA, on opposite sides of the DNA in the bubble, consistent with the forks moving in opposite directions. (From H.J. Kriegstein and D.S. Hogness, *Proc. Natl Acad. Sci. USA* 71:135–139, 1974. With permission from the authors.)

bidirectional movement of the replication forks, as in bacteria, and their relatively small size correlates with the much shorter lengths of *Okazaki fragments* in eukaryotes than bacteria (see below). However, in eukaryotes generally, only a fraction of the possible origins are observed to be active and giving rise to bubbles in S phase of the cell cycle. This might be a fail-safe mechanism to offer an additional supply of origins in the event of replication forks stalling in the replication of these large genomes.

The large size of the eukaryotic chromosomes is a second aspect that is efficiently handled by the presence of multiple origins. For example, if the largest chromosome (62,000 kbp, equivalent to 2.1 cm) in *Drosophila melanogaster*, was serviced by a single origin of replication, the replication fork would have to attain a rate of ≥ 18,000 kbp/min to accomplish DNA replication in the observed interphase period of 3.4 minutes. (Notably, such a short interphase provides an upper limit for fork rate, a rate only observed in certain cells and at specific developmental stages. Nevertheless, this example serves to establish a striking contrast between eukaryotic and bacterial DNA replication.) In fact, the measured velocity of replication forks in eukaryotes *in vivo* is typically ~20 or more times slower than that of a replication fork in *E. coli* (~60 kbp/min). Initiation of replication from multiple sites is an adaptation to the requirement for large genomes in eukaryotes.

The use of multiple origins in eukaryotic DNA replication raises the question of the manner of their coordination. Eukaryotic DNA replication is further complicated by the fact eukaryotic chromosomes are contained within an organelle, the nucleus, and packaged by histones to form *chromatin*. These questions and complexities influence chromosome biology, and are active areas of investigation.

Archaea, although they resemble bacteria in size and many other properties (Section 1.9), also exhibit some notable differences. Archaea also have multiple (up to three) DNA replication origins, which vary between and within organisms and display some sequence specificity. Archaeal DNA replication, in particular, much more closely resembles that of eukaryotes, although it requires a significantly smaller set of proteins.

The origin recognition complex is the homolog of the DnaA oligomer in eukaryotes

Despite their variability, eukaryotic origins of replication become activated by interaction with a single set of initiator proteins that associate into a complex called the **origin recognition complex (ORC)**. This six-subunit protein assembly was first identified in *S. cerevisiae*, where the proteins were designated Orc1–6 in descending order of size (914 aa in Orc1, down to 426 aa in Orc6). Orc1–5 are homologous proteins, all containing an AAA+ ATPase module of the initiator clade (the same as DnaA in *E. coli*, Box 3.1) and a winged-helix (WH) domain involved in DNA binding. By contrast, DnaA has an HTH domain. Orc6 has no obvious relation to the others and is poorly conserved between yeast and metazoan eukaryotes. In addition to the proteins that make up the ORC, another protein, termed Cdc6 (513 aa), partners with the ORC to initiate replication (**Figure 3.4A**). Like the Orc1–5 proteins, the Cdc6 protein also contains an AAA+ ATPase module and a WH domain; Cdc6 binds ATP and Cdc6/ATP binds to the ORC and enhances its specificity for DNA. Assembly of the ORC begins a cascade of protein-binding events that, as we will see below, leads to establishment of a pair of replication forks at each origin. Notably, in

Box 3.1   ATPases are important components of a wide variety of protein machines

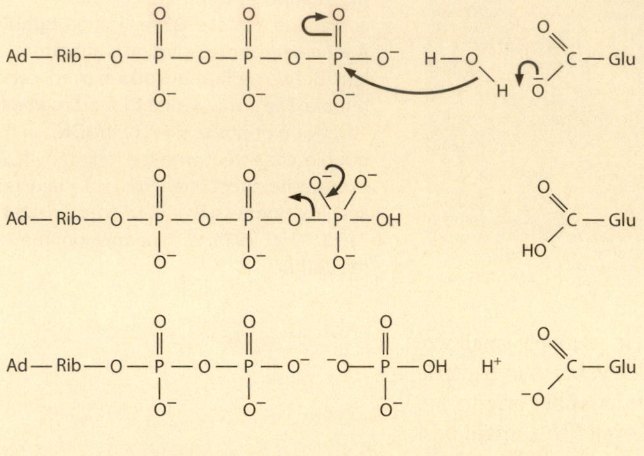

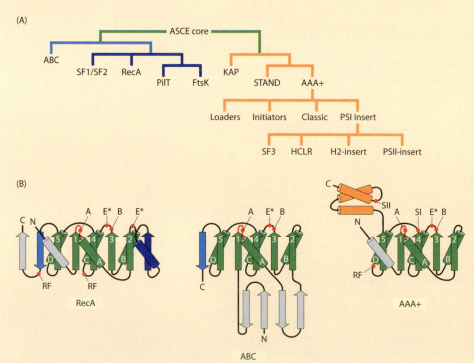

Figure 3.1.1 Mechanism of hydrolysis of ATP by ASCE ATPases. A water molecule activated by a conserved Glu residue takes part in an in-line associative attack on the β-γ phosphoanhydride bond of the ATP. Ad, adenine; Rib, ribose. The formation of a pentacoordinate intermediate is followed by its collapse to ADP and Pi.

Much of the genome (~10–20%) in any prokaryote or eukaryote is taken up with genes encoding proteins that use nucleoside triphosphates (NTPs) as substrates. These proteins engage in a wide variety of activities and most bring about the hydrolysis of the NTP (usually, ATP) to NDP and Pi (**Figure 3.1.1**) The free energy associated with this reaction can be used to drive a coupled reaction with an adverse (positive) free energy change, as described in Chapter 1. Often this is by kinases that generate phosphorylated intermediates. In contrast, NTPases of the kind

described below harness this free energy more directly to drive another sort of reaction, one that typically involves the performing of mechanical work on a macromolecular client.

On the basis of phylogenetic analysis of the F1-ATPase (Chapter 15), myosin (Chapter 14), and other kinases, John Walker and colleagues identified two sequence motifs as common to enzymes that bind ATP. Crystal structures have shown that in the Walker-A (or **P-loop**) motif (G/AXXXXGKT/S), the backbone amide groups of the Gly residues coordinate the phosphates of the bound ATP, the Lys residue interacts with the β-γ phosphates, and the T/S residue coordinates a bound Mg^{2+}. In the Walker-B motif (ΦΦΦΦD/E, where Φ is any hydrophobic amino acid), the D/E residue further coordinates the Mg^{2+}. A large subset of P-loop NTPases, the **ASCE** (**additional strand conserved E**) superfamily, has an additional β strand in the core fold, plus a conserved Glu residue essential for catalysis.

ASCE ATPases possess a common topology of five β strands sandwiched between α helices, a Rossmann-fold, but distinct families (clades) can be picked out (**Figure 3.1.2A**). The folds of three examples are given in Figure 3.1.2B. In addition to the Walker motifs and conserved Glu, they have other important features. Thus, AAA+ ATPases (see Box 7.2) and most members of the RecA family (Section 4.4) contain an Arg side chain (RF, or Arg finger), which interacts with the phosphates of the bound ATP. In the AAA+ family, the RF is located at the end of helix D before β-strand 5, the so-called second region of homology (SRH); but in the RecA family, its position in the primary structure can vary. No Arg finger is found in the ABC family (Section 16.5). Another conserved feature in AAA+ ATPases and many RecA ATPases is a polar residue (often Asn) at the

Figure 3.1.2 Classification of ASCE ATPases. (A) A representation of the different subgroups (clades), based on sequence alignments and topology diagrams. The branches are colored according to their relationship to the ASCE core. (B) Topology diagrams to illustrate RecA, ABC, and AAA+ types. The core fold (α helices A–D; β strands 1–5) is shown in green, with additional β strands and α helices in light and dark blue, and those found in only some clade members in gray. The 'lid' characteristic of AAA+ ATPases is in yellow. The P-loop between strand-1 and helix-A is in red. A, Walker-A motif; B, Walker-B motif; E*, conserved Glu; RF, Arg finger; SI, sensor I; SII, sensor II. Alternative positions of RF and E* in RecA are indicated. (Adapted from J.P. Erzberger and J.M. Berger, *Annu. Rev. Biophys. Biomol. Struct.* 35:93–114, 2006; and L.M. Iyer et al., *Nucleic Acids Res.* 32:5260–5279, 2004.)

Box 3.1 *continued*

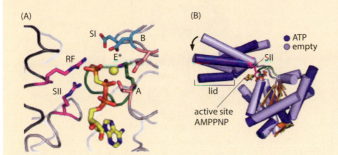

(A)

(B)

Figure 3.1.3 Active site and conformational change of AAA+ ATPases. (A) Active site of DnaA from *Aquifex aeolicus* (initiator clade). Catalytic residues are indicated. For one subunit, the backbone is in black and that of its neighbor in pale gray. AMPPC, an unhydrolyzable ATP analog, is shown in stick representation. Mg^{2+} is a yellow ball. (PDB 2HCB). (B) Movement of the lid domain relative to the core domain during the catalytic cycle in a subunit of the *E. coli* protease ClpY/ HslU (HCLR clade) – see Section 7.3. Loops have been removed for clarity. The β strands of the core domain are in yellow. (PDB 1DO2). (From N.D. Thomsen and J.M. Berger, *Mol. Microbiology.* 69:1071-1090, 2008. With permission from John Wiley & Sons.)

end of strand β4, which acts as a 'sensor' (S-I) to inform about the presence or absence of the γ-phosphate of bound nucleotide. AAA+ ATPases have an α-helical 'lid' domain, appended at the C-terminal end of the core fold (Figure 3.1.2B), and this houses an additional sensor, S-II, usually Arg, which plays a key role in a movement of the lid and core domains during the catalytic cycle.

In ASCE ATPases the ATP-binding site lies between two neighboring protomers and the enzymes are active only as oligomers. Many RecA and AAA+ ATPases are ringlike hexamers, in which binding of ATP induces a conformational change in one of the protomers. This remodels its interface with the neighboring protomer and the effect is amplified by similar changes at the other ATP-binding sites around the hexamer. An important mediator is the Arg finger, which extends from the back of the core domain of a given protomer into the ATP-binding site formed with its neighbor (**Figure 3.1.3A**). In RecA-like ATPases, the β sheet may twist a little during catalysis, whereas in AAA+ ATPases the core is essentially unchanged but there is a realignment of the lid—essentially a rigid-body rotation of up to

20°—with which S-II from the lid domain of the same protomer is associated.(Figure 3.1.3B). These rearrangements of the oligomer form the basis of the mechanical work that RecA-like and AAA+ ATPases perform.

AAA+ ATPase modules (200–250 aa) can be subdivided into seven clades on the basis of their fold topology (Figure 3.1.2A). Three examples associated with DNA replication are illustrated in **Figure 3.1.4**. The SF3 clade (superfamily 3 helicases) is unique in lacking a lid domain but instead has an additional helix bundle. The grafting of accessory domains on to the core further defines the different clades.

The rate at which the bound ATP is hydrolyzed varies. A low rate means that the conformational change that accompanies ATP binding is essentially an on-off switch, reset when the ATP is hydrolyzed and the products released. In contrast, rapid hydrolysis of bound ATP ensures that the conformational changes recur at a high rate, a property suited for a molecular motor.

AAA+ ATPases conduct many different biological processes and are discussed further in Box 7.2.

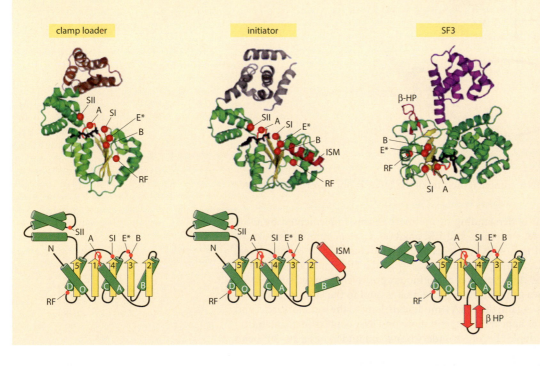

Figure 3.1.4 Structures of AAA+ ATPase modules from three different clades. The lid and core domains are depicted in green (α helices, O and A–D) and yellow (β strands, 1–5). Additional activity-specific domains are in purple or slate. Bound nucleotide is in black. Beneath each structure is a topology diagram in the same color code, with clade-specific changes highlighted in red. ISM, initiator specific motif; β HP= β hairpin. This 'clamp loader' is the eukaryotic RFC-B (PDB 1SXJ), the 'initiator' is *E. coli* DnaA (PDB 2HCB), and the SF3 protein is the SV40 helicase (PDB 1SVM). (Adapted from K.E. Duderstadt and J.M. Berger, *Crit. Rev. Biochem. Mol. Biol.* 43:163–187, 2008. With permission from Informa Healthcare.)

(A)

(B)

CDK

ORC loading

CDK
DDK

check-
point

helicase
loading

(C)

helicase
activation

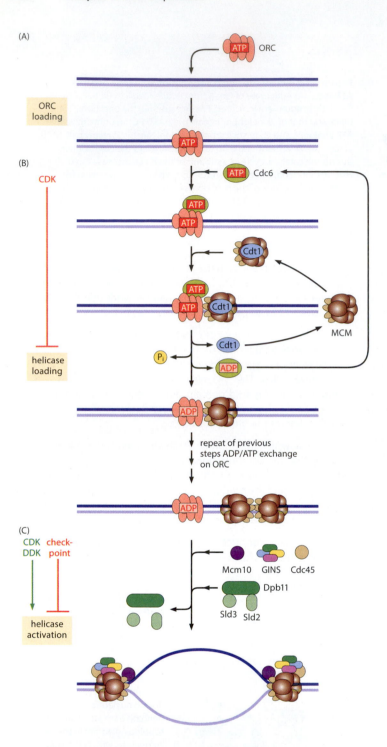

repeat of previous
steps ADP/ATP exchange
on ORC

Mcm10   GINS   Cdc45

Dpb11

Sld3   Sld2

(D)

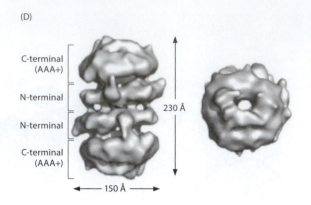

C-terminal
(AAA+)

N-terminal

N-terminal

C-terminal
(AAA+)

230 Å

150 Å

Figure 3.4 Schematic pathway of assembly of ORC and MCM at an origin of replication. (A) Binding of ORC. (B) Loading of inactive mini-chromosome maintenance (MCM) in G1 phase. (C) Activation of MCM in S phase. (D) Electron microscopic reconstruction of the double hexamer of *S. cerevisiae* MCM, showing the head-to-head arrangement. (A–C, adapted from D. Remus and J.F. Diffley, *Curr. Opin. Cell Biol.* 21:771–777, 2009, and M. L. Bell and S. D. DePamphilis, Genome Duplication. New York: Garland Science, 2011; D, from D. Remus et al., *Cell* 139:719–730, 2009. With permission from Elsevier.)

contrast to the DnaA-origin DNA complex, melting of the DNA is not observed until later in the assembly process.

The beginnings of a structural understanding of the mechanisms implemented by ORC have been obtained from crystal structures and EM analyses. Crystallographic studies of the archaeal (*Sulfolobus solfataricus*) ORC proteins bound to archaeal DNA origins reveal that the proteins Orc1-1 and Orc1-3 bind to the same face of the DNA and, although they share only a small interface (360 Å^2), they offer a continuous positively charged surface that covers 28 bp of DNA. In archaea generally, the WH domains bind to the DNA and, unusually, so do the ATPase domains. Again, the AAA+ modules belong to the initiator clade, and the binding of the ORC proteins appears to depend more on DNA context (deformability) than base sequence. The DNA duplex becomes underwound and undergoes a 20–30° bend

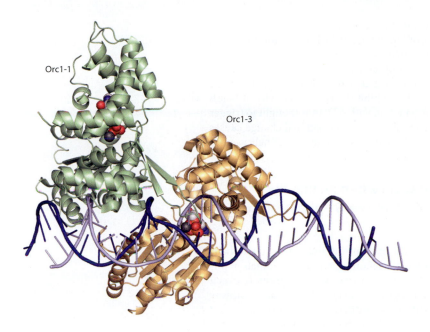

Figure 3.5 Archaeal ORC proteins Orc1-1 and Orc1-3 of *S. sulfataricus* bound to a replication origin in archaeal DNA. The proteins were crystallized in complex with a 33 bp fragment of DNA. Both contain AAA+ ATPase modules (with bound ADP) and WH domains. ADP molecules are shown in sphere representation and Mg^{2+} as green spheres. (PDB 2QBY)

(**Figure 3.5**). Electron microscopic studies of yeast and *Drosophila* ORC complexes have revealed an overall architecture which, in combination with a recent crystal structure of *Drosophila* ORC, will push forward our still incomplete understanding of what ORC does at the eukaryotic origin.

Bacterial DnaA melts DNA at the origin to enable loading of the DnaB replicative helicase

Binding of active DnaA to *oriC* produces unwinding of, at most, a small and localized region of the genomic DNA but, for replication of the whole chromosome, its entire DNA content must be unwound and copied. The unwinding requires a dedicated enzyme, the **replicative helicase**, DnaB, which is recruited by DnaA and assembled at the origin of replication. DnaB of *E. coli* (a hexamer of 52 kDa subunits) is recruited in the form of a complex with a further protein, the helicase loader DnaC—there are six DnaC subunits bound to each hexameric DnaB. DnaC is another AAA+ protein and closely related structurally to the initiator protein, DnaA. Indeed, in protein/protein interaction studies DnaA can interact with DnaC via their respective AAA+ domains when ATP is present. Moreover, the N-terminal domains of DnaB can interact directly with the N-terminal domains of DnaA. All this suggests that DnaC can recruit DnaB to the end of a DnaA filament that has DNA bound on the outside.

These findings offer a plausible model for the assembly of a bidirectional replication fork (**Figure 3.6**), based on the idea that there may be two distinct modes of recruiting DnaB/DnaC to DNA. Biochemical studies have revealed that DNA melts on one side of the DnaA-binding site and that replication is initiated on both of the exposed DNA strands. The DnaB

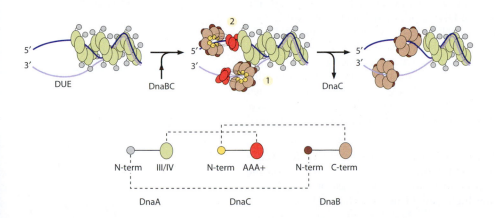

Figure 3.6 A model for the loading of two replicative helicases facing in opposite directions at *E. coli oriC*. Left, DnaA molecules assemble at *oriC* to make the filament to which DNA binds, causing unwinding of the DUE (DNA unwinding element). Middle, the DnaBC complex assembles on the 'bottom' DNA strand (1) by direct interaction of the N-terminal domains of DnaB with the N-terminal domains of DnaA, and on the 'top' DNA strand (2) by interaction of AAA+ modules of DnaC with the end of the DnaA filament. Right, the binding of DnaC to the ssDNA triggers hydrolysis of its bound ATP to ADP, which causes it to dissociate and leave a DnaB hexamer bound to each ssDNA strand. The DnaB hexamers face in opposite directions ready to unwind DNA further in the 5′–3′ direction in each instance as part of a replication fork. The critical interactions between domains of DnaA, DnaC, and DnaB are indicated schematically below. (Adapted from M.L. Mott et al., *Cell* 135:623–634, 2008. With permission from Elsevier.)

helicase (see below) has a fixed direction of movement (5′ to 3′) and, given the anti-parallel nature of the two strands of DNA, two hexamers of DnaB need to be loaded onto the DNA, but facing in opposite directions. Thus it is possible that a hexamer of DnaB is loaded on the 'top' (5′–3′ strand in Figure 3.6) facing in one direction by virtue of DnaC/DnaA interactions, whereas on the 'bottom' (5′–3′) strand DnaB is loaded facing the other way by virtue of DnaB/DnaA interactions. DnaC has been found to inhibit the helicase activity of DnaB but, on loading of the DnaC/DnaB complex, its interaction with ssDNA is thought to trigger hydrolysis of the ATP bound by DnaC. The accompanying conformational change causes DnaC to dissociate from DnaB, thereby liberating the active form of the DnaB helicase. All is then set to initiate replication of both strands of the DNA.

Eukaryotic DNA is licensed for replication when an inactive replicative helicase is loaded at the ORC in the G1 phase of the cell cycle

Eukaryotes have evolved considerably more complex mechanisms and machinery to handle the large size and segmented nature of eukaryotic genomes, and to coordinate replication with other cellular events. Despite the differences in size and complexity between eukaryotic genomes, the assembly of replication forks follows essentially the same pathway in humans as in yeast. These mechanisms consist of a coordinated series of intricate and highly regulated biochemical events, known as the cell division cycle, and they govern the growth and proliferation of eukaryotic cells. A complete treatment of the cell division cycle is beyond the scope of this work, and the reader is referred to appropriate literature for additional information.

One of the central events in the cell cycle is the replication of chromosomal DNA, which must occur accurately for continued progress through the cycle. Elaborate timing and quality controls ensure that each daughter cell acquires a single, functional copy of the genome and also sets the stage for the next round of chromosomal duplication. Cells prepare for the next round of DNA replication by licensing replication origins for activity. As noted above, ORC binds to origin DNA and recruits the Cdc6 protein, which in turn recruits the **minichromosome maintenance (MCM)** assembly which arrives at origin DNA complexed to the Cdt1 protein. There is no evidence of local DNA melting of the kind induced by DnaA in *E. coli* to accompany the binding of ORC/Cdc6 at the eukaryotic origin; rupture of base pairing only begins once the replicative helicase is fully assembled and activated. The MCM assembly is recruited to the origin in an inactive form; once additional factors are recruited in a later stage, it will become the engine of the replicative helicase at the fork. The MCM assembly consists of six distinct but related polypeptides (termed Mcm2–7), each of which contains an AAA+ ATPase module preceded by an additional unique N-terminal domain. Analysis of the loaded MCM assembly by electron microscopy (Figure 3.4D) reveals a structure with overall dimensions ~230 × 150 Å that has been interpreted to represent two MCM hexamers forming a back-to-back pair held together by interactions between their N-terminal domains (that is, the constituent hexamers face in opposite directions). Biochemical analysis reveals that the MCM double hexamer can slide passively on duplex DNA, but displays no helicase activity at this stage. It is only once additional proteins are recruited to the origin (notably Mcm10, Cdc45, and the GINS complex) in S phase, accompanied by the activity of two S phase kinases CDK and DDK (Cdc7 kinase-Dbf4 regulatory protein), that the helicase can begin to unwind duplex DNA as part of a replication fork (Figure 3.4D). This ensures that DNA replication from the many origins is confined to the S phase and occurs only once per cell cycle. However, more than one double hexamer can be loaded at the same origin and it has been suggested that any surplus could be a source of pre-loaded helicases ready to restore activity at a stalled replication fork, obviating the need to relicense origins in the face of mechanisms that otherwise ensure this occurs only once per cell cycle. Note that mitochondrial DNA, which also has to be copied before cell division, is made throughout interphase. The ensemble of proteins (ORC, Cdc6, MCM, Cdt1) that are deposited sequentially during the G1 phase of the cell cycle are termed the **pre-replicative complex (pre-RC)**.

The assembly and activity of the pre-RC are governed by two kinases—CDK (cyclin-dependent kinase) and DDK (DBF4-dependent kinase)—whose activity depends on the phase of the cell cycle. Activity of both kinases is low during the G1 phase of the cell cycle, but rises sharply during S phase, and remains high until the next G1 phase. Although CDK and DDK have a similar cell cycle-dependent activity profile, they have distinct phosphoryl-ation substrates, and concomitantly, different regulatory roles. CDK phosphorylates distinct sites on the ORC ensemble, the Cdc6 protein, and the MCM complex. The effect of these modifications is to inhibit loading of the MCM helicase to DNA. CDK also phosphorylates the Sld2 and Sld3 proteins (see below), and this has the effect of activating the helicase. DDK phosphorylates the MCM4 and MCM6 subunits of the MCM complex, which is important for origin activation. The cell cycle-dependent activity profile of CDK and DDK, their sub-strate preferences, and functional consequences of phosphorylation lead to the conclusion that pre-RCs can only assemble during G1, and become activated during S phase. During G1, when CDK activity is low, pre-RCs can assemble, until origins fire and the cells enter S phase, at which point assembly of additional pre-RCs is restricted to the next pass through the cell cycle. The activity of both CDK and DDK rises on entry into S phase, and this enables firing of the origin (DDK) and activation of the helicase (CDK). Together, the role played by these kinases represents one of the components which ensures that chromosomes are replicated once per cell cycle. A simple way to envision the cell cycle-dependent activ-ity of origin complexes involves the concept of licensing. Through the interplay of protein complexes and kinases, origins are licensed for replication during the G1 phase. However, the license to replicate is withdrawn during S phase and is withheld until late in M phase of the next pass through the cycle.

Helicases use the energy of nucleotide binding and hydrolysis to unwind duplex DNA

Replicative DNA helicases are ATPases that require NTP (normally ATP) hydrolysis to progress along the DNA. In bacteria, bacteriophages, and archaea, they are homohexamers; in eukaryotes, a heterohexamer. In principle, a helicase could move along a single strand of DNA in either the 5′ to 3′ or the 3′ to 5′ direction and indeed both sorts of helicases have been discovered. In bacterial DNA replication, the DnaB helicase moves in the 5′ to 3′ direction, and so operates on the *lagging strand*. By contrast, the eukaryotic CMG (Cdc45, MCM2-7, GINS, below) replicative helicase travels in the 3′ to 5′ direction and, consequently, must operate on the *leading strand. In vitro*, the DnaB helicase translocates along dsDNA in an ATP-dependent manner, the physiological role of such an activity remains obscure.

The replicative helicase of bacteria, DnaB, is a hexamer of 52 kDa subunits; each subunit has an N-terminal domain (NTD, ~150 aa), which is a helical bundle and hairpin, a flexible linker of ~30 aa, and a C-terminal domain (CTD, ~270 aa), which adopts the RecA fold (see Box 3.1). The NTD interacts with the bacterial primase, DnaG, in the *replisome* (see below). A number of important insights have come from the biochemical and structural analysis of the related helicase gp4 of *E. coli* bacteriophage T7. It too is a hexamer of 52 kDa subunits but it can be cleaved by limited proteolysis into two different activities: the N-terminal 245 aa constitute the T7 primase, the C-terminal 295 aa contain the helicase; and the two parts are connected by a linker region of 26 aa. A crystal structure of the hexameric form of the helicase domain (residues 241–566) reveals it to be toroidal in shape (**Figure 3.7A**). The central hole is lined by positively charged loops, three contributed by each monomer, which can interact with the ssDNA that is able to pass through it (the hole is too small for

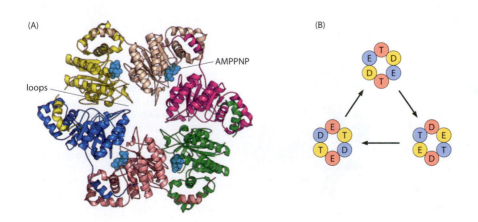

Figure 3.7 The helicase domain of bacteriophage T7 gp4. (A) Structure of the helicase with each of the six subunits in a different color. Four molecules of AMPPNP (blue) are bound between subunits. The ssDNA at a replication fork passes through the central hole, where it makes contact with three loops contributed by each subunit. (B) A possible reaction mechanism in which the six subunits successively adopt three slightly different conformations, as seen in (A). Each of the three is shown in a different color and designated: T, NTP-bound; D, nucleotide hydrolyzed (that is NDP + P$_i$); E (empty, that is nucleotide hydrolyzed and released). (PDB 1E0J) (Adapted from M.R. Singleton, M.S. Dillingham and D.B. Wigley, *Annu. Rev. Biochem.* 76:23–50, 2007.)

duplex DNA). The NTP (gp4 is unusual in preferring dTTP to ATP) is bound at active sites between subunits but, despite the subunits being identical, they adopt three subtly different conformations in the hexamer. These follow each other sequentially around the ring and can be correlated with three classes of NTP-binding site: a high-affinity site, a low-affinity site, and an empty site. It has been proposed that the high- and low-affinity sites correspond to dTTP- and dTDP-bound forms, respectively. Thus, a sequence of nucleotide binding, hydrolysis, and release, with accompanying conformational changes (see Box 3.1 for the mechanism of RecA-like ATPases), can propagate around the hexamer (Figure 3.7B). This is similar to the binding-change mechanism of the F_1F_0-ATP synthase described in Chapter 15, but different in that in the ATP synthase the ring is an $\alpha_3\beta_3$ hexamer in which only the β subunits are catalytically active. In gp4, the observation of four bound nucleotides per hexamer indicates that four NTP sites are involved per cycle, with two sites unoccupied.

Precisely how the binding of NTP, its hydrolysis and release of the products, is harnessed to movement of the helicase along the ssDNA remains unclear. It seems likely that changes in the positions of the basic loops within the central channel of the helicase and abutting the DNA will accompany the NTP hydrolysis. There is evidence that the loops track the spiral of the ssDNA backbone, and there need be little rotation of the DNA relative to the protein as the helicase is driven forward by NTP hydrolysis. This could be important in reducing the problems of DNA supercoiling during replication (Section 3.4). In no instance so far has the step size (that is the number of base pairs unwound per NTP hydrolyzed) been unequivocally determined. However, the size of the gp4 toroid is compatible with biochemical estimates of 2–3 bp, although this may differ between helicases.

The MCM helicase is activated in S phase of the cell cycle

How the MCM helicase is activated in S phase is poorly understood. An active *D. melanogaster* MCM helicase can be reconstituted *in vitro* from Mcm2–7 and two other proteins, Cdc45 and the GINS complex. The GINS complex is an elongated heterotetramer of homologous 22 kDa proteins (Sld5, Psf1, Psf2, and Psf3; the name GINS derives from Go, Ichi, Ni, and San, the numbers five, one, two, and three in Japanese), predominantly α-helical, but with surface patches of unstructured regions that may provide binding sites for other proteins (**Figure 3.8**). A central pore runs through the complex, ~10 Å in diameter at its narrowest point and sufficient to encircle ssDNA, although whether it does so is unknown. The N-terminal 16 aa of Psf3 form a loop that partially blocks the entrance to the pore and may control accessibility to it. The biochemical evidence is that the GINS complex binds in particular to the Mcm4 subunit and that a monomeric Cdc45 (566 aa in humans), the hexameric MCM, and tetrameric GINS exhibit mutual interactions that bring them together, probably accompanied by some conformational remodeling of the MCM, to make a CMG complex that has an increased affinity for DNA and exhibits a previously latent helicase activity. *In vivo* activation of helicase activity also depends critically on the action of two S phase kinases, CDK and DDK. Phosphorylation of the N-terminal region of Mcm4 by DDK is required for binding of Cdc45, and DDK also phosphorylates the N-terminal regions of Mcm2 and Mcm6. Although these phosphorylation events activate helicase activity, in combination with the phosphorylation of Cdc6 and ORC, they also inhibit assembly of additional pre-RC entities, and in doing so ensure that replication is restricted to once and only once per cell cycle (Figure 3.4C).

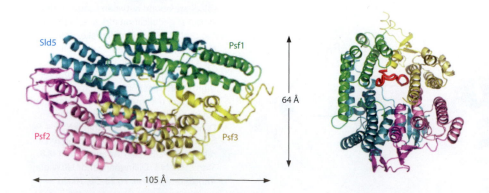

Figure 3.8 Structure of the tetrameric GINS complex. Two orthogonal views of the human GINS complex are shown. Left, viewed from the side. Right, viewed across an end, showing the central pore and the N-terminal 16 aa of Psf3 as a loop, in red, in place at the entrance to the pore. (PDB 2Q9Q) (From Y.P. Chang et al., *Proc. Natl Acad. Sci. USA* 104:12685–12690, 2007. Reprinted with permission.)

The replicative helicases of eukaryotes and archaea track on DNA in a 3′ to 5′ direction

The hexameric helicases of eukaryotes and their viruses, together with those of archaea, differ significantly from DnaB of *E. coli* in tracking along the DNA in the 3′ to 5′ direction rather than 5′ to 3′. Moreover, they are based on AAA+ ATPase modules, whereas the ATPases of DnaB and bacteriophage T7 gp4 are of the RecA type (Box 3.1). The structure of Mcm2–7 from electron microscopy (Figure 3.4D), and its ability before activation to slide passively on linear fragments of DNA, are consistent with its encircling dsDNA. Thus, if it is to act like *E. coli* DnaB (Figure 3.6) or T7 gp4 (Figure 3.7), albeit migrating 3′ to 5′, it must undergo some further remodeling in the activation process and there must be some melting of the DNA that enables the helicase to change so as to encircle only the 3′–5′ strand as it leaves the origin.

Papillomavirus E1 helicase tracks on ssDNA in the 3′ to 5′ direction

A precedent for a 3′–5′ helicase of this kind is the E1 helicase of papillomavirus, which has AAA+ ATPase domains of the SF3 clade (Box 3.1). The papillomavirus genome is a small circular dsDNA and, with the assistance of another viral protein E2, E1 is capable of assembly as a double hexamer at an origin of replication, held together by a DNA-binding and oligomerization domain on the N-terminal side of each AAA+ ATPase module (**Figure 3.9A**). Local melting of the DNA must also occur, as with *E. coli* DnaA/DnaB, for each hexamer then to surround ssDNA.

The crystal structure of bovine papillomavirus E1 (residues 306–577, which includes the AAA+ ATPase module and DNA-binding domain) in complex with a 13-base dT fragment and ADP and Mg^{2+} ions, shows the six subunits encircling ssDNA (Figure 3.9B). The ssDNA passes through the central channel, which narrows to 13 Å in diameter, too small for dsDNA (20 Å). Hairpins projecting from the six protein subunits (see Figure 3.7) track the sugar–phosphate backbone of the ssDNA in a right-handed helical arrangement, like steps on a spiral staircase. From each subunit, the N_6 amino group of Lys506 interacts with one DNA phosphate and the main chain amide proton of His507 forms an H-bond with the phosphate of an adjacent nucleotide (Figure 3.9C). The N-terminal oligomerization domains are pointing towards the 5′ end of the DNA, with the AAA+ ATPase domains

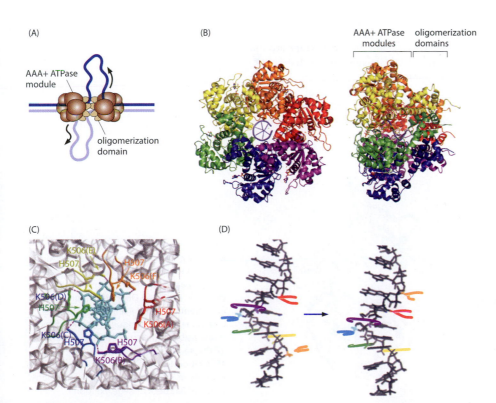

(A)

AAA+ ATPase module

oligomerization domain

(B)

AAA+ ATPase modules   oligomerization domains

(C)

(D)

Figure 3.9 The E1 helicase of bovine papillomavirus. (A) Schematic model of the double hexameric rings of the E1 helicase on dsDNA. The two rings are held together by their collars of DNA-binding/oligomerization domains. Each ring is tracking 3′ to 5′ on ssDNA and the two loops of ssDNA of increasing size being generated are extruded from the sides. (B) Two orthogonal views of a single hexamer with the subunits depicted in different (rainbow) colors. In the left-hand view, the collar of oligomerization domains lies in front of the ring of AAA+ ATPase modules. The strand of ssDNA (oligo dT) tracks through the central hole, which is not big enough to accommodate dsDNA. (C) ssDNA strand in the central hole, with Lys506 (K506) and His507 (H507) in the loops protruding from each subunit. (D) Spiral staircase model of DNA translocation, the loop from a given subunit remaining associated with a unique nucleotide as it escorts it from the top to the bottom of the hole through the hexameric helicase and bound ATP is hydrolyzed. (PDB 2GXA) (D, adapted from K.D. Raney, *Nat. Struct. Mol. Biol.* 13:671–672, 2006. With permission from Macmillan Publishers Ltd.)

therefore on the 3′-terminal side. Although ADP is bound at each of the six subunit inter-faces, the subunits are found to adopt three different conformations, interpreted as ATP-binding, ADP-binding, and empty. The ATP-binding sites are associated with subunit loops at the upper (5′ end) of the spiral staircase, the ADP-binding sites with loops in the middle, and the empty sites with loops at the bottom.

A plausible mechanism of helicase action can be inferred, based on each loop remaining associated with a unique nucleotide and migrating downwards through a cycle of ATP hydrolysis and ADP release (see Figure 3.7B), pulling the ssDNA with it. At the bottom, the now empty site binds ATP and the loop moves back to the top of the staircase to repeat the process (Figure 3.9D). In this 'escort mechanism,' six ATPs are therefore hydrolyzed per full cycle, and six nucleotides of ssDNA are translated through the helicase in the 3′ to 5′ direction. The steadily extending ssDNA strands emerge from the sides of the double hexamer (Figure 3.9A). The collar of oligomerization domains acts rather like a β clamp round the ssDNA, conferring processivity, as well as holding the two hexameric helicases together. The helicase (large T antigen, Tag) of virus SV40 (Chapter 8) is likely to have a similar mechanism. However, it has also been suggested that MCM may differ from DnaB and E1 in that it remains encircling dsDNA, as loaded. The helicase would then have to pump dsDNA towards another protein or proteins (for example Cdc45 and/or GINS), which would act like a 'ploughshare' to participate in strand separation. A precedent for this sort of mechanism can be found in bacterial Holliday junction branch migration, in which two RuvB hexamers pump dsDNA and a tetramer of RuvA acts as the separation 'pin' and coordinates reannealing of the strands (Section 4.4). The AAA+ ATPase of MCM is of the PSII-insertion clade, whereas that of RuvB is of the HCLR clade (Box 3.1).

Single-stranded DNA is bound by a protective protein before it enters the DNA polymerase

The two strands of ssDNA generated by the advancing helicase are prevented from rean-nealing and from hydrolysis or other damage before they encounter their relevant poly-merase by interaction with a **single-stranded DNA-binding protein**. The *E. coli* SSB is a tetramer of a 177 aa protein, of which the N-terminal 112 aa constitute the DNA-binding domain and the remainder (aa 113–177) appear to be disordered, even in the presence of bound ssDNA. SSB can bind ssDNA in two modes: one occludes on average 35 nucleo-tides ((SSB)$_{35}$), and the other on average 65 nucleotides ((SSB)$_{65}$). The crystal structure of the tetrameric ssDNA-binding domain bound to deoxycytidylate DNA strands is shown in **Figure 3.10**.

The binding domain takes the form of an OB fold (see **Guide**) and the DNA wraps around the outside of the tetramer, interacting mostly through both its sugar–phosphate back-bone and its bases with Lys or Arg side chains. *In vitro*, at low salt concentrations and high protein:DNA density, the complex is in the (SSB)$_{35}$ mode, in which the DNA strand is not bound to one of the four protein subunits, but distributes itself over one and partly over another two. At higher salt concentrations and lower protein concentrations, the complex

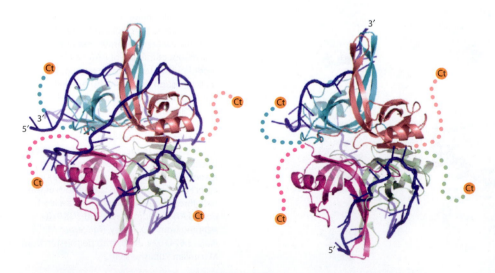

Figure 3.10 Models of *E. coli* SSB interacting with single-stranded DNA. The DNA in stick representation is depicted in blue. Each of the four SSB monomers (residues 1–112, an OB fold) is depicted in a different color and each has a disordered C-terminal tail (Ct, residues 113–177), which is not visible in the crystal structure. Left, (SSB)$_{65}$ mode; right, (SSB)$_{35}$ mode. (PDB 1EYG) (Courtesy of Dr Gabriel Waksman, Institute of Structural and Molecular Biology, UCL and Birkbeck College, London.)

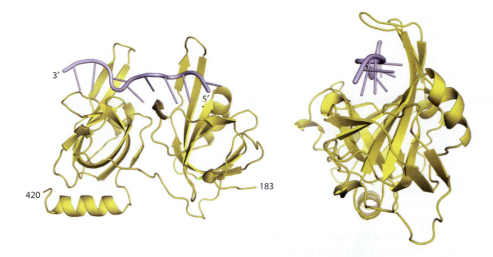

Figure 3.11 DNA-binding domain of RPA. Two orthogonal views of a fragment of ssDNA (dC8) depicted in blue in complex with the ssDNA-binding region (residues 183–420, containing two successive OB folds) of the RPA70 component of human RPA. Domain I (at right) represents Ile198–Asp291; domain II (at left) represents Ile305–Asp402. (PDB 1JMC)

is in the $(SSB)_{65}$ mode, in which all four protein subunits contribute to the interaction with the DNA. Cooperative inter-tetramer protein/protein interactions in the $(SSB)_{35}$ mode can lead to long clusters along the DNA strand. No particular biological function has yet been ascribed to one or other of the DNA-binding modes. The disordered C-terminal tail (rich in Pro, Gly, and Gln residues) has been identified as a partner in the interaction of SSB with the χ subunit of the *E. coli clamp loader* complex and the primase, key parts of the replisome in association with DNA Pol III. This is described in more detail below.

Bacterial SSBs are almost invariably tetramers, but the corresponding bacteriophage proteins are normally monomers (T4, gp32) or dimers (T7, gp2.5) and both also have one OB fold in the monomer.

The eukaryotic ssDNA-binding protein RPA also helps organize many other proteins in the replication fork

RPA is a highly conserved heterotrimeric protein (subunits ~70, 32, and 14 kDa). The largest subunit (RPA70) contains the DNA-binding region, with four successive homologous OB folds connected by flexible unstructured linker regions. Two of the OB folds (those principally involved in ssDNA binding) are shown in **Figure 3.11**. The DNA is tightly bound ($K_d \sim 10^{-9}$ M) in an extended form in a channel that runs across the two domains. RPA32 and RPA14 each also contain an OB fold, making six in all in RPA. The OB folds are similar to those in *E. coli* SSB or bacteriophage T7 gp2.5 (see **Figure 3.12**) and RPA fulfils a similar function in the replication forks of archaeal and eukaryotic cells. It binds to and helps organize more than 20 other proteins involved in DNA replication, including DNA Pol α and RFC.

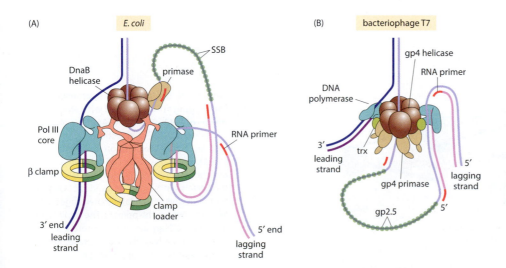

Figure 3.12 Schematic diagrams of replisomes in *E. coli* and bacteriophage T7. (A) In the *E. coli* replisome, the two τ subunits of a τ2γδδ′ clamp loader are shown binding to the α subunits of the two Pol III polymerases and to the DnaB helicase. With a τ3δδ′ clamp loader, there would be three Pol III polymerases bound (one on the leading strand and two perhaps on the lagging strand). There are three primases bound per DnaB helicase (only one shown here) and the interaction of the SSB with the clamp loader via the χψ heterodimer is also omitted for clarity. (B) Bacteriophage T7 DNA being replicated in *E. coli*. The DNA polymerase comprises gp5 and the *E. coli* host protein, thioredoxin (trx). A third DNA polymerase/thioredoxin may also be bound by the helicase. gp2.5, single-stranded DNA-binding protein; gp4, covalently linked helicase and primase (RNA polymerase domain and smaller Zn^{2+}-binding domain). Again, some interactions have been omitted for clarity. Neither model is drawn to scale. (A, adapted from N.Y. Yao and M. O'Donnell, *Curr. Opin. Cell Biol.* 21:336–343, 2009. With permission from Elsevier. B, adapted from J.B. Lee et al., *Nature* 439:621–624, 2006. With permission from Macmillan Publishers Ltd.)

The RNA primers for DNA polymerases are synthesized by primase, a special polymerase

DNA polymerases are unable to initiate the synthesis of new DNA strands. This is puzzling, as RNA polymerases can generate new mRNA strands in transcription (Chapter 5) and RNA-dependent replicases can replicate viral RNAs without primers, all while using essentially the same chemistry. However, several different ways of priming the new DNA strand have been identified. In the case of some viruses, such as adenovirus, the side chain OH of a Ser residue in a protein is used to make a phosphodiester bond with the first dNTP, leaving its 3′-OH free to react with the second dNTP, catalyzed by the viral DNA polymerase in the usual way. In mitochondrial DNA synthesis, a mitochondrial RNA polymerase creates RNA primers with a free 3′-OH group. But the most widespread mechanism of priming is that a special RNA polymerase, **primase** (or **DNA primase**), part of the replisome, synthesizes RNA **primers**. This dedicated RNA polymerase synthesizes short pieces of RNA (up to 60 nt depending on the species) that are used to prime DNA synthesis by the DNA polymerase (**Figure 3.13**). In principle, only a single RNA primer is required, at the origin of replication, for leading strand synthesis; in practice, evidence suggests that leading strand synthesis can be reprimed, for example, downstream of DNA damage. This situation stands in contrast to the lagging strand where a new primer is needed for synthesis of each Okazaki fragment. In bacteria, the newly synthesized DNA in an Okazaki fragment is 1–2 kb long, but only 100–200 nucleotides in archaea and eukaryotes.

Primases make primers of defined length but exhibit low fidelity in the copying process

Primases exhibit low fidelity in the copying process, but make primers of prescribed length. This has led Irving Lehman to joke that primase is "An enzyme that can count but can't read". Isolated primases have notably lower turnover numbers than the DNA polymerases but, on binding to the helicase-binding domain of DnaG, the affinity for ssDNA is greatly increased, as is the rate (1000-fold) of primer synthesis. The association of primase with helicase also limits the length of the primers synthesized to ~10–14 nucleotides. (T7 gp4 synthesizes primers of four or five nucleotides.) This ability to count the number of nucleotides added has been attributed, at least in part, to interaction of the Zn^{2+}-binding domain of one DnaG monomer with the polymerase domain of another *in trans*, which limits the amount of template that can be copied. Moreover, primer synthesis does not occur randomly on the template. In the bacterial systems, most commonly a trinucleotide serves as starting point (**Table 3.1**). In the trinucleotide 3′-XN_1N_2-5′, X serves to direct the primase to a suitable site and is not replicated and N_1 is normally a pyrimidine base. Eukaryotic primases are much more tolerant about template specificity.

RNA primers introduced as part of the replication process must later be removed before the Okazaki fragments, which have been synthesized discontinuously, are ligated together, and the resulting gap filled in. RNase-H, a nonspecific endonuclease that degrades RNA in an RNA/DNA duplex, can help to achieve this but is mechanistically incapable of cleaving off

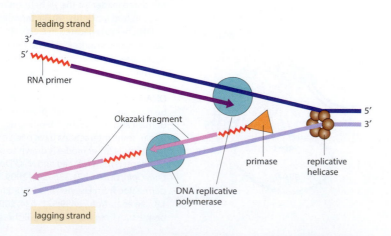

Figure 3.13 Discontinuous synthesis on the lagging strand primed with short pieces of RNA. A single RNA primer initiates synthesis in the 5′–3′ direction on the leading strand. Synthesis on the lagging strand, which must also be in the 5′–3′ direction, requires multiple re-initiations and the DNA is synthesized as a series of Okazaki fragments, which are later joined together after removal of the primers. The replicative helicase is advancing on the parental 5′–3′ strand but the two DNA polymerases are shown as moving in opposite directions.

Table 3.1 Some template initiator sequences used by primases

Organism	Template initiator sequence 3′–5′	Primer synthesized 5′–3′
E. coli	3′-GTC	pppAG-
B. stearothermophilus	3′-ATPy	pppAPu-
Bacteriophage T7	3′-CTG	pppAC-
Bacteriophage T4	3′-TTG	pppAC-
Human	3′-PyNN	pppPuPu-

N, any nucleotide; Pu, purine; Py, pyrimidine.

the last ribonucleotide from the DNA strand. The essential enzyme is DNA Pol I which has a 5′–3′ exonuclease activity that is capable of sequentially removing all the ribonucleotides of the primer. This is the true biological function of Pol I, not a polymerase as such. Also, the lower fidelity of DNA primase (~1 error per 10^2 nucleotide additions) than a replicative DNA polymerase means that the RNA primer sequence may contain copying errors; the process of RNA primer by excision and new DNA synthesis offers an opportunity to correct such errors if they have occurred. The final result is a complete and accurate copy of the original double-stranded DNA.

DNA is copied by DNA polymerases

The first enzyme capable of replicating DNA (**DNA polymerase**) was discovered by Arthur Kornberg and colleagues in E. coli, and is now known as DNA polymerase I (Pol I). Because mutations in the gene encoding this enzyme were later found not to be lethal, it turned out that Pol I is not the enzyme that copies DNA as the helicase unwinds it. Further experiments identified this as Pol III. Instead, Pol I has a key role in another part of the replication process and, as described in Chapter 4, in some areas of DNA repair. DNA polymerases are now seen as a diverse set of enzymes that fall into seven distinct superfamilies (A, B, C, D, X, Y, and the reverse transcriptases), based on sequence analysis. Polymerases in these superfamilies can also be classified, based on whether they primarily function in DNA replication or repair. Replicative DNA polymerases are found in family A (mitochondria, bacteriophage T7), family B (human DNA pols alpha, delta, and epsilon; viral DNA polymerases—herpesvirus, vaccinia virus, adenovirus; bacteriophage T4 and phi29 DNA polymerases), family C (bacterial DNA pols), or family D (archaeal DNA pols). The enzymes that act in DNA repair are in family A (DNA polymerase I), family X (human DNA pol beta), or family Y (translesion DNA pols that can replicate past DNA damage).

DNA polymerases copy an ssDNA template using dNTPs as substrate. The newly formed DNA strand is synthesized in the direction 5′–3′, dNTPs being added one at a time in a sequence complementary to the base sequence on the strand (3′–5′) being copied (**Figure 3.14A**). The side product, pyrophosphate, can be hydrolyzed by a suitable phosphatase to two molecules of P_i; thus the addition of one nucleotide to the growing strand utilizes the free energy of hydrolysis of two phosphoanhydride bonds. All DNA polymerases also have a common architecture, consisting of three domains arranged in the form of a right hand: fingers, palm, and thumb (**Figure 3.15**). The residues involved in catalysis are located in the palm domain, while the thumb grips the newly formed duplex DNA, which lies in a cleft. The fingers grasp the template ahead of the active site and additionally facilitate binding and positioning of the incoming dNTP; they also harbor a sequence that binds to the sliding clamp responsible for ensuring that the polymerase remains anchored to the DNA (see below). Two acidic amino acids, usually Asp, in the palm domain and located at the bottom of the cleft, coordinate two metal ions (usually Mg^{2+}) that are crucial to the mechanism (Figure 3.14B) but, depending on the family, a third Asp can be involved.

Based on the fold of the core regions in their palm domains, there are two structural classes of DNA polymerase (**Figure 3.16**): the 'classic' fold displayed by four of the six families of

(A)

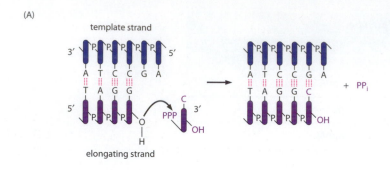

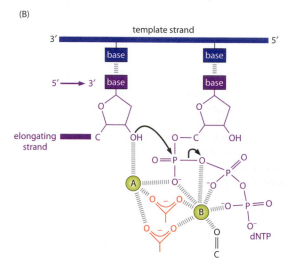

(B)

Figure 3.14 Mechanism of DNA polymerases. (A) Schematic diagram showing base complementarity of new DNA strand growing in the 5′–3′ direction, with attack of the free 3′-hydroxyl group of its terminal deoxyribose on the α-phosphate of the incoming dNTP. The vertical dashed lines between the bases on opposing strands represent the H-bonds (two for A/T, three for G/C) between complementary base pairs. (B) The two-metal mechanism in which metal ions are coordinated by two Asp side chains (purple). Metal ion A helps lower the pK_a of the deoxyribose 3′-OH group, activating it for inline nucleophilic attack on the α-phosphate of the incoming NTP. A pentacoordinate intermediate is formed (cf. mechanism of AAA+ ATPase described in Box 3.1). Metal ion B coordinates with and orients the triphosphate of the dNTP, and facilitates collapse of the intermediate and release of the pyrophosphate to complete the reaction.

DNA polymerases, and the βNT fold, named for the Pol β nucleotidyltransferase, a eukaryotic polymerase whose function lies in DNA repair (Chapter 4) is restricted to the family C and family X enzymes. The lack of discernible sequence homology between the classic and βNT folds has been taken to indicate that they may have arisen by convergent evolution towards a common mechanism.

DNA polymerases contain a 3′ to 5′ exonuclease as well as a 5′ to 3′ polymerase activity

Much of the early structural and mechanistic work was done on *E. coli* Pol I. Limited proteolysis cleaved the single polypeptide chain (~100 kDa) into two separate parts: residues 1–323, which was found to contain 5′–3′ exonuclease activity and a larger (Klenow) fragment, residues 324–928, which contained the DNA polymerase active site and 3′–5′ exonuclease activity. From this and from later studies on other DNA polymerases (prokaryotic, archaeal, and eukaryotic), some other important general features have emerged, as follows.

The 3′–5′ exonuclease activity is part of a proofreading mechanism to ensure high fidelity of replication (see below) and its active site is some distance away (~25 Å) from that of the polymerase.

- The cleft is lined with positive charges and the protein contacts only the sugar–phosphate backbone of the DNA along its minor groove, thereby ensuring that it is not sequence-specific.
- Kinetic experiments and comparisons of structures of different polymerases binding DNA and dNTP in different complexes indicate that the fingers domain moves rapidly to bury the incoming dNTP, which, if correctly base-paired with the template, allows closure of the active site.
- Conformational flexibility in the polymerase active site may be such that the active conformation is only achieved when substrate has bound.

The 5′–3′ exonuclease activity is found only in Pol I, in keeping with its true role in DNA replication.

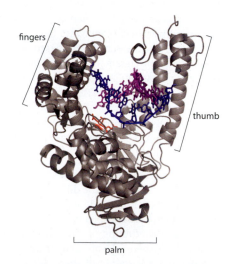

Figure 3.15 Architecture of a typical DNA polymerase. The structure of residues 290–831 of the replicative DNA polymerase of the Gram-negative bacterium *Thermus aquaticus*, viewed as a right hand with palm at the bottom, thumb up to the right, and partly closed fingers to the left. The DNA in the cleft is shown with template strand (3′–5′) in blue and elongating strand (5′–3′) in purple. The two catalytically important Asp residues at the bottom of the cleft are shown in orange. (PDB 1TAU)

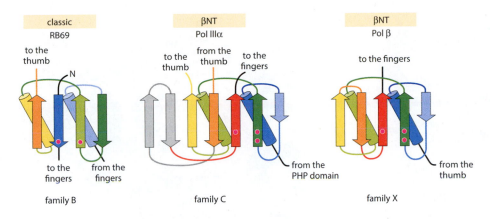

Figure 3.16 Two different topologies of the core regions of the palm domains of DNA polymerases. Left, the classic fold, exemplified here by the family B DNA polymerase of bacteriophage RB69; middle, the βNT fold of the family C DNA polymerase Pol IIIα of *E. coli*; right, the βNT fold of the family X DNA polymerase Pol β. Secondary structure is colored from the N-terminal end (blue) to the C-terminal end (red). The red circles indicate the locations of the catalytic acidic residues (two in the classic fold, plus a third in the βNT fold). The extra β strands in Pol IIIα compared with Pol β are shown in gray. (Figure adapted from S. Bailey, R.A. Wing and T.A. Steitz , *Cell* 126:893–904, 2006. With permission from Elsevier.)

The replicative polymerase of *E. coli*, Pol III holoenzyme (~800 kD) is composed of at least 15 proteins. It can be considered as having three components: (1) the holoenzyme core, (2) the gamma complex clamp loader (below), and (3) the beta sliding clamp (below). The holoenzyme core is composed of three subunits, the polymerase (α), which harbors DNA synthesis activity; a separate 3′–5′ proofreading exonuclease (ε); and the third subunit (θ), with no identified function as yet. All of the bacterial replicative polymerases (family C) follow the same pattern, but with interesting variations. This can be seen clearly in the structure of the replicative polymerase, Pol C, from the Gram-positive organism *Geobacillus kaustophilus* (**Figure 3.17**) and a comparison with the replicative polymerase of the Gram-negative bacterium *Thermus aquaticus* (Figure 3.15). Residues 828–1293 form the core (fingers, palm, and thumb). At the N-terminal end is an oligonucleotide/oligosaccharide-binding (OB) domain followed by a polymerase-histidinol phosphatase (PHP) domain. A PHP domain is in all family C polymerases but rarely in others. It is named for its homology to yeast histidinol phosphatase, an enzyme that catalyzes metal-dependent hydrolysis of phosphoester bonds. This has led to the idea that it may, at least in some instances, be involved with hydrolyzing the pyrophosphate released at each step of the polymerase reaction cycle. Inserted in the PHP domain is a sequence identifiable as a (Zn^{2+}-dependent) 3′–5′ exonuclease (deleted from the Pol C polypeptide before crystallization to prevent degradation of the DNA/primer co-crystallized with it and a dGTP molecule). In Pol C, the OB domain is located where it could bind the template strand 15–20 nt ahead of the polymerase active site. At the C-terminal end is the duplex binding which helps bind the newly formed duplex DNA against the thumb, and it also forms a crevice with the fingers through which the template strand is guided into the polymerase active site.

The Pol C polymerase active site in ternary complex with a DNA/primer template and dGTP is shown in Figure 3.17A. The DNA duplex is in the B-form except at the 3′ end of the growing strand in the polymerase active site where the deoxyribose is in the C3′ *endo* conformation, which would position the 3′-OH group for inline attack on the αP of the dGTP.

Most of these features are reproduced in other family C DNA polymerases, but with occasional differences and variations. For example, in the *T. aquaticus* Pol IIIα, as in Pol C, the PHP domain contains a proofreading 3′–5′ exonuclease, whereas the same domain in *E. coli* Pol IIIα appears to have lost that activity and proofreading is confined to the ε subunit. The OB domain of Pol C is in the N-terminal region of the protein whereas those of *T. aquaticus* and *E. coli* Pol IIIα are located towards the C-terminal end (Figure 3.17B) and appear on the opposite side of the polymerase domain, but may still fulfill the same function in binding the ssDNA template.

The high fidelity of replicative polymerases arises from multiple sources

The copying of genomic DNA is accomplished by the cell with a mutation rate of one for every billion base pairs copied (1×10^{-9}). This astonishingly low error rate arises from three broad sources: (1) Watson–Crick base pairing and DNA polymerase selectivity, (2) proofreading by the 3′–5′ exonuclease, and (3) mismatch repair (covered in Chapter 4). Replicative DNA polymerases exhibit an error rate of 10^{-6}–10^{-7} by combining Watson–Crick

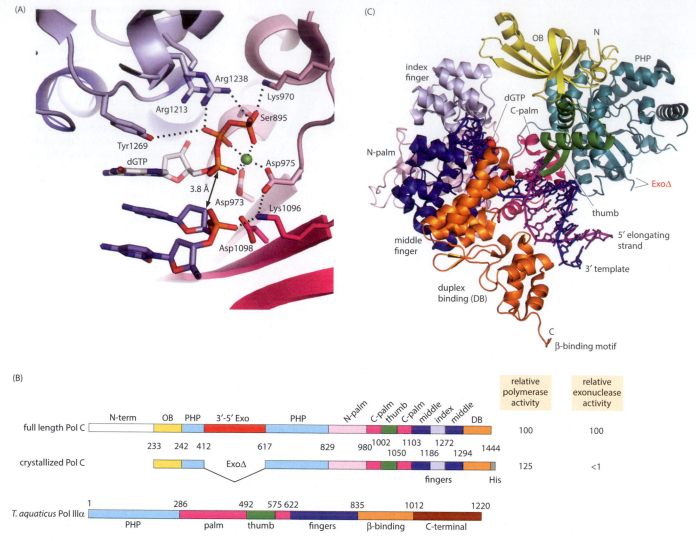

Figure 3.17 Structure of the replicative polymerase Pol C from *G. kaustophilus*. (A) Close-up of active site, with key residues highlighted. Metal B (Mg²⁺) is shown as a green sphere; no metal ion is visible in the A site, possibly because the elongating strand was prepared without a 3′-OH group to prevent catalysis. If present, the 3′-OH group of the elongating strand would be 3.8 Å from the α-phosphate of the dGTP, indicated by the black double-headed arrow. (B) Domain structure of the polypeptide chain of Pol C in comparison with that of Pol IIIα of *T. aquaticus*. The C-terminal region of *T. aquaticus* Pol IIIα contains an OB domain (yellow) like that near the N terminus of Pol C. Before Pol C was crystallized, the 3′–5′ exonuclease domain was deleted, together with the poorly conserved N-terminal domain that does not contribute to the polymerase activity. The His-tag used in its purification remains. (C) Structure of Pol C with DNA template (blue) and elongating strand (purple) bound in the active site, together with dGTP (ball and stick representation) as incoming NTP. Domains are color-coded as in (B). The region (ExoΔ) where the deleted 3′–5′ exonuclease domain would be is indicated. (PDB 3F2B, 3F2C, and 3F2D) (Adapted from R.J. Evans et al., *Proc. Natl Acad. Sci .USA* 105:20695–20700, 2008. Reprinted with permission.)

pairing, polymerase architecture, and proofreading; mismatch repair operates after replication has been completed.

The high fidelity with which DNA polymerases copy the DNA template strand stems from three features of the enzymatic process. Although the four Watson–Crick base pairs adopt very similar sizes and shapes, which are distinct from those of various mismatched pairs, free energy calculations reveal that the energetic cost of incorporating the 'wrong' base is very low (0.2–4 cal/mol); this translates into an error rate of at most 10^{-2}. This intrinsic error rate arising from the four Watson–Crick base pairs is further reduced by a series of architectural features found in high-fidelity DNA polymerases. At each nucleotide addition, arrival of the correct dNTP promotes closure of the fingers subdomain on the nascent base pair (A/T and G/C base pairs, if correctly made, are the same size) to form a pocket from residues from the finger and palm domains that fits the base pairs snugly. Correct insertion also enables formation of hydrogen bonds between the polymerase (palm subdomain) and acceptors in the minor groove of the nascent base pair; these hydrogen bond donors are found in approximately the same position in all four Watson–Crick base pairs. In addition, hydrogen bonds to the newly formed eight base pairs enable the polymerase to continue monitoring fidelity. The correct staging of the above two events leads to optimal positioning of the active site aspartate residues around the 3′-OH group for inline attack on the alpha phosphate of the incoming dNTP. Arrival of the incorrect dNTP increases the probability that the conformational change of the fingers to close the snug pocket will not occur, that the minor groove contacts will not form, and/or that the alpha phosphate of the dNTP will not be positioned appropriately for chemistry. In such a case, the dNTP may be rejected. The above interplay between the geometry of the Watson–Crick base pair and the architecture of the DNA polymerase produces an error rate of 10^{-5}–10^{-6}.

Fidelity of the final duplex is further increased by the action of the 3′–5′ exonuclease. Although the error rate of the DNA polymerase is quite low, misincorporation can still take place. When this happens, distortions in the duplex due to the mismatch destabilize some contacts between the polymerase and the newly formed duplex. In Pol A and Pol C, residues of the palm domain contact the minor groove of the newly synthesized duplex DNA; this enables the sensing of a mismatch up to 8 bp after its incorporation. The close match between the position of hydrogen bond donors and acceptors in the case of correct incorporation leads to stability of the complex on DNA. However, the presence of a misincorporated base pair results in weakening of binding of the DNA, and triggers displacement of the 3′ end of the elongating chain from the polymerase active site to the physically remote 3′–5′ exonuclease active site (Figure 3.17C), a distance of some 25 Å. Indeed, the exonuclease activity is in a different subunit (ε subunit) of Pol III in *E. coli* and other Gram-negative bacteria. The exonuclease chews back along the elongating chain until the incorrectly base-paired nucleotide that escaped detection in the polymerase active site has been removed, the trimmed DNA relocates to the polymerase active site, and polymerase activity can then resume.

Biochemical experiments indicate that polymerases, such as Pol IIIα of *E. coli*, have an error rate of ~1 in 10^5 steps of NTP incorporation. With the 3′–5′ exonuclease activity functioning, either in the same polypeptide chain in Pol C or in the ε subunit of Pol III of *E. coli*, the fidelity is further improved, the error rate falling to ~1 in 10^6–10^7. In some Gram-negative bacteria, such as *Thermus thermophilus*, the PHP domain of the polymerase chain has also retained its 3′–5′ exonuclease activity, and this may add to the overall proofreading ability in such instances. Note that without some albeit low level of mistakes going uncorrected, genetic variation and natural selection would be impossible.

DNA replication is continuous on one strand but not the other

DNA polymerases synthesize a DNA strand only in a 5′ to 3′ direction, copying the parental strand in the 3′ to 5′ direction at the replication fork. Given that the two strands in double-stranded DNA are arranged in anti-parallel configuration, how can a polymerase copy the parental strand that runs in the opposite direction? In 1968, Reiji Okazaki discovered that the answer lies in one strand, the **leading strand**, being synthesized continuously in a 5′ to 3′ direction, whereas the other strand, the **lagging strand**, is copied off the parental DNA strand, also in the 5′ to 3′ direction, but as a series of short segments. A series of processing reactions enables maturation of the so-called **Okazaki fragments** (Figure 3.13) into the final product.

Numerous enzymes interact at a replication fork and function as a concerted giant assembly

Many enzymes participate in the synthesis of new duplex DNA, of which only a few are depicted schematically in Figure 3.13, and they are marshaled at a replication fork to act in a concerted way. Together they constitute a giant assembly termed a **replisome**. Much of what we know about the bacterial replisome has come from studies of *E. coli* and the bacteriophages T7 and T4 which infect *E. coli* but encode their own replication machineries. Archaeal and eukaryotic replication forks have many corresponding features but are considerably more complex and somewhat less well understood.

The components of the replisomes of *E. coli* and bacteriophages T4 and T7 are set out in **Table 3.2**. The principal components of all replisomes are: a DNA helicase; a DNA polymerase and associated sliding clamp; a DNA primase; and a single-stranded DNA-binding protein. The DNA helicase advances the replication fork; in *E. coli* and the bacteriophages T7

Table 3.2 Components of the replisomes of *E. coli* and bacteriophages T4 and T7

Component	*E. coli*	T4*	T7
Helicase	DnaB	gp41	gp4
Replicative polymerase	Pol III (α, ε, θ)	gp43	gp5
Primase	DnaG	gp61	gp4
Single-stranded DNA-binding protein	SSB	gp32	gp2.5
Clamp (processivity factor)	β	gp45	Thioredoxin†
Clamp loader	τ₂γ δδ'χφ	gp44/62 (4:1)	None

*Proteins identified as gp (gene product) are encoded by phage genes given the designated numbers.
†The protein thioredoxin is encoded by the host *E. coli* genome.

and T4, the active form is a hexamer. DNA polymerases have very high catalytic rates, being capable of adding ~1000 nt per second to the elongating chain, but they are relatively inefficient, extending the elongating chain by only a few nucleotides per binding event. Indeed, Pol III can be regarded as essentially distributive, that is to say it falls off the template strand after adding only one or a few nucleotides. Its processivity, that is, its ability to remain in continuous contact with the template strand, adding nucleotides to the elongating strand, is enhanced by tight but noncovalent interaction of the DNA polymerase with the clamp or processivity factor. In the case of bacteriophage T4 and *E. coli*, the clamp is a toroidal complex that requires a dedicated machine, the clamp loader, to assemble it onto DNA. The DNA polymerase of bacteriophage T7 uses an *E. coli* host protein, thioredoxin, for the purpose.

DNA is synthesized simultaneously on leading and lagging strands

As the replicative helicase moves along the 5′ to 3′ strand of the DNA duplex, melting the DNA ahead of the replication fork (Figure 3.13), the two DNA polymerases, one operating on what has become the leading strand and the other on the lagging strand, are carried with it (see below). This simultaneous synthesis of DNA on both strands requires that the lagging strand must form a loop of increasing size with its associated Okazaki fragments (Figure 3.18). As the preceding primer/Okazaki fragment is drawn towards its polymerase

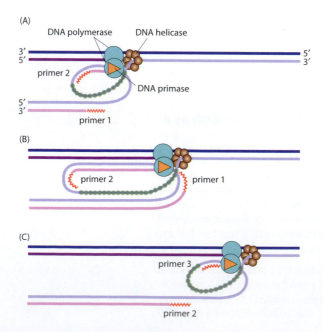

Figure 3.18 Schematic model of two Pol III molecules in an *E. coli* replisome coordinating DNA synthesis on leading and lagging strands. DNA helicase loaded on the 5′–3′ lagging strand unwinds the DNA ahead of the replication fork. The two DNA polymerases are linked together. (A) The upper DNA polymerase synthesizes DNA continuously (dark purple strand) on the 3′–5′ leading strand of the parent DNA. The lower DNA polymerase, associated with DNA primase, synthesizes Okazaki fragments (light purple) by copying the 5′–3′ lagging strand. The first Okazaki fragment has been synthesized from primer 1, and primer 2 is being used to synthesize a second Okazaki fragment. Single-stranded DNA is protected by interaction with multiple copies of SSB (green balls). (B) As the replication fork progresses, the lagging strand, together with its growing Okazaki fragment, balloons out. Eventually primer 1/Okazaki fragment 1 is drawn into collision with the polymerase. (C) The collision causes the polymerase to relinquish its grip on the lagging strand, a new primer (primer 3) is synthesized by the primase and the DNA polymerase then rebinds to the lagging strand and begins synthesizing another Okazaki fragment.

Table 3.3 Activities of DNA polymerases in eukaryotes				
Enzyme	Subunit composition (kDa)	Function	Processivity (nt per event)	3′–5′ proofreading
Pol α	166*, 68, 58†, 48†	Primase/polymerase	10–30	No
Pol δ	125*, 68, 50, 12	Leading strand polymerase	500–600	Yes
Pol ε	261*, 59, 17, 12	Lagging strand polymerase	~400	Yes

The molecular masses are those of the human polymerases.
*Polymerase activity and 3′–5′ exonuclease activity where present.
†Primase.

by the elongation of the Okazaki fragment undergoing synthesis, it will eventually bump into the polymerase. At this point the polymerase must let go of the lagging strand template and allow primase to synthesize another RNA primer, thereby also releasing the loop (the 'trombone mechanism'). The DNA polymerase will then resume DNA synthesis from the primer to make the next Okazaki fragment and another loop will form and expand until it is released in turn.

Three different DNA polymerases are required for DNA replication and operate differently on the leading and lagging strands

All eukaryotes possess three DNA polymerases (Pol α, Pol δ, and Pol ε), all of which belong to family B and are required for DNA replication (**Table 3.3**). The heterotetrameric Pol α, in line with its low processivity (10–30 nucleotides) and lack of 3′–5′ proofreading activity, is concerned only with the initiation of RNA-primed DNA synthesis on both leading and lagging strands. The 166 kDa subunit houses the polymerase activity, the 68 kDa subunit has a regulatory role (it is reversibly phosphorylated at different stages of the cell cycle), and the two smallest subunits constitute the primase activity required to make the RNA primers. Pol α synthesizes an RNA primer of about 10 ribonucleotides and then continues to add deoxyribonucleotides until the new RNA/DNA fragment has reached about 30 nt in length, Pol α is then released and synthesis is transferred to Pol ε (leading strand) or Pol δ (lagging strand). These two polymerases are also both heterotetramers in which the largest subunit supplies the polymerase activity but also has a 3′–5′ exonuclease proofreading action. Neither polymerase has primase activity but both exhibit high processivity, enhanced by interaction with the eukaryotic sliding clamp (PCNA).

The association of Pol ε with the leading strand and Pol δ with the lagging strand was elegantly established in *S. cerevisiae* by replacing the wild-type gene for the catalytic subunit of Pol δ with one that encoded a subunit with a higher error rate and observing that mutations were principally to be found on the lagging strand. Relatively little is yet known about the structure of eukaryotic primases, though the initiation of RNA primer synthesis is rate-limiting in lagging strand synthesis, as in bacteria. Whereas *E. coli* requires its primase to initiate synthesis about 5000 times in the replication of its chromosome, humans require about 30,000,000 such events, and eukaryotic primases are predictably more tolerant in template specificity (Table 3.1). To replicate their DNA, archaea probably use a DNA polymerase of family B, which also has a 3′–5′ proofreading activity, and also interacts with a sliding clamp.

The processivity of DNA polymerases is enhanced by a sliding clamp

For the synthesis of DNA to be efficient, the DNA polymerase has to retain its hold on the strand that it is copying and catalyze many consecutive rounds of nucleotide addition. In *E. coli* this is achieved by physical association of the polymerase with a sliding clamp. Pol III interacts with β clamp, which raises the processivity from a very few to ~5000 nucleotides added per encounter. The structure of the β clamp in complex with a 14-mer/10-mer fragment of DNA representing a primed site is shown in **Figure 3.19**. The two identical

(A)

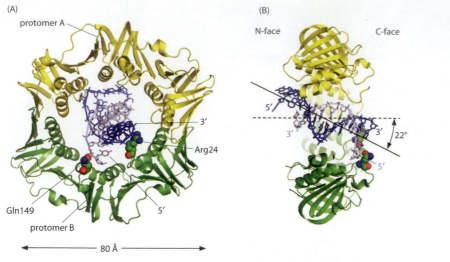

(B)

(C)

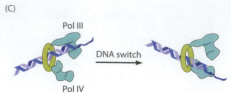

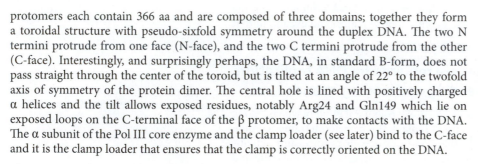

Figure 3.19 Structure of the β clamp of *E. coli* DNA polymerase Pol III in complex with primed DNA. (A) The complex viewed end on. The two protomers are shown in different colors. The DNA passes through the central hole (~35 Å diameter) of the toroidal protein; the template strand being copied is shown in dark blue, complementary strand in light blue. (B) Cut-away side view, showing the 22° tilt of the DNA duplex with respect to the twofold axis of the protein dimer. Arg24 and Gln149 are shown in ball and stick representation. (C) There are two possible orientations of the tilted DNA with respect to two polymerases that can be bound to the C-terminal face of the clamp. This suggests how activity might be switched from the catalytic site of one polymerase to that of a different polymerase. (PDB 3BEP) (C, adapted from R.E. Georgescu et al., *Cell* 132:43–54, 2008. With permission from Elsevier.)

protomers each contain 366 aa and are composed of three domains; together they form a toroidal structure with pseudo-sixfold symmetry around the duplex DNA. The two N termini protrude from one face (N-face), and the two C termini protrude from the other (C-face). Interestingly, and surprisingly perhaps, the DNA, in standard B-form, does not pass straight through the center of the toroid, but is tilted at an angle of 22° to the twofold axis of symmetry of the protein dimer. The central hole is lined with positively charged α helices and the tilt allows exposed residues, notably Arg24 and Gln149 which lie on exposed loops on the C-terminal face of the β protomer, to make contacts with the DNA. The α subunit of the Pol III core enzyme and the clamp loader (see later) bind to the C-face and it is the clamp loader that ensures that the clamp is correctly oriented on the DNA.

The presence of two identical protomers in the clamp implies the presence of two identical binding sites for polymerases and there is experimental evidence with Pol III and Pol IV that the clamp can indeed bind two different polymerases at once. Arg24 and Gln149 in each protomer are offset with respect to the central axis, and thus two possible orientations of tilted DNA also exist, depending on which protomer houses the Arg24 and Gln149. This leads to a model for how it may be possible to switch between two different polymerases and introduce a new catalytic activity (Figure 3.19C), an essential feature of the DNA repair process (see Chapter 4).

The direct interaction of the β clamp with DNA means that it cannot slide freely—even though the noncovalent bonds must be transient and readily broken—and single-molecule experiments indicate that it diffuses along DNA (D ~10^{-14} m²/s) 100–1000 times more slowly than if it were free. Nonetheless, this makes the clamp still capable of moving faster than the polymerase bound to it can incorporate nucleotides and it does not diminish the replication rate.

DNA Pol ε and Pol δ, together with many other proteins, associate with a sliding clamp on the DNA

Eukaryotes (and archaea) contain a protein entitled PCNA (proliferating cell nuclear antigen), named for an antigen that was detected in the nuclei of cells during the DNA synthesis phase of the cell cycle. PCNA resembles the β clamp of *E. coli* in both structure and function, despite little sequence similarity. Instead of two identical subunits each with three similar domains, as in the β clamp, PCNA has three identical subunits (30 kDa) each with two similar domains (in some archaea the three subunits are different but related) arranged to generate sixfold pseudosymmetry (**Figure 3.20**). The axis of the DNA passing through the central hole is tilted at 40° with respect to the axis of the PCNA ring. This echoes the 22° tilting of the DNA in the *E. coli* β clamp (Figure 3.19), but is substantially more pronounced. Although there are defining contacts between the negatively charged DNA backbone and positively charged amino acid side chains projecting into the central hole, the interaction between DNA and PCNA is relatively limited, as it has to be if the clamp is to slide reasonably freely.

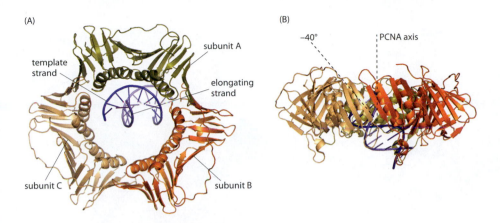

(A)

template strand

subunit A

elongating strand

subunit C

subunit B

(B)

−40°

PCNA axis

Figure 3.20 The sliding clamp of eukaryotes bound to DNA. Two orthogonal views of *S. cerevisiae* PCNA bound to a fragment of dsDNA (template strand 5′-TTTTATACGATGGG-3′; elongating strand 5′-CCCATCGTAT-3′). (A) Cross section with DNA in the central hole. The three monomers are depicted in different colors (for crystallization purposes, the three PCNA monomers were covalently joined together by means of two 11 aa linkers). (B) Side view showing the axis of the dsDNA tilted at 40° to that of the PCNA ring. (PDB 3K4X)

In addition to conferring high processivity, PCNA is central to the DNA replication process in its ability to bind a wide range of partner proteins (helicases, polymerases, ligases, etc.). The partner proteins typically have a short PCNA-interacting-protein (PIP) motif of consensus sequence QXX(I/L/M)XXF(F/Y) in their N- or C-terminal regions, which binds to a surface region of the connecting peptide that links the two domains of a PCNA monomer together. Thus, the PCNA trimer can, in principle, bind up to three partner proteins at any one time and act as a platform for assembling the DNA replication machinery in one place.

Bacteriophage T7 replicates its DNA without a ring-shaped sliding clamp. Rather, its DNA polymerase (gp5) achieves processivity by complexing with a host protein thioredoxin (**Figure 3.21**). One likely consequence of this interaction is that a 76 aa loop at the tip of the thumb region of the polymerase comes to bend round to encircle the DNA. The polymerase/thioredoxin complex is then highly processive (~1000 nucleotides added per DNA-binding event).

The β clamp and DNA polymerase are loaded onto the RNA-primed DNA by an ATP-dependent clamp loader

As described above, the replicative helicase, advancing ahead of the replication fork, binds to primase (**Figure 3.22**). Before entering the primase, the ssDNA generated by the helicase

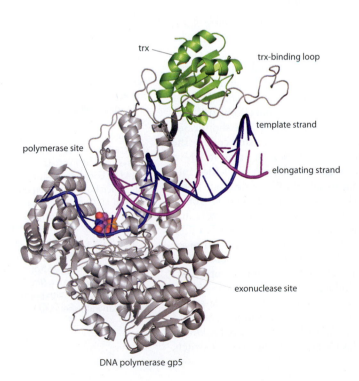

trx

trx-binding loop

polymerase site

template strand

elongating strand

exonuclease site

DNA polymerase gp5

Figure 3.21 The *E. coli* host protein thioredoxin acts as processivity factor for bacteriophage T7 DNA polymerase. The DNA polymerase is shown in complex with a 21 bp duplex DNA with a single-stranded four-base extension at the 5′ end of the template strand. The incoming nucleoside triphosphate (2′3′-dideoxyCTP) to be added to the elongating strand is shown in sphere representation. The 3′–5′ exonuclease activity was eliminated by deleting six residues (118–123) from the exonuclease domain. The proofreading exonuclease site is ~30 Å from the polymerase site. trx, thioredoxin. (PDB 1T8E)

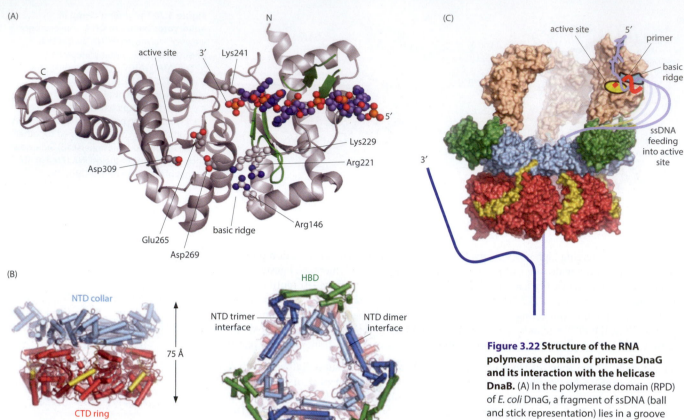

(A)

(B)

(C)

Figure 3.22 Structure of the RNA polymerase domain of primase DnaG and its interaction with the helicase DnaB. (A) In the polymerase domain (RPD) of *E. coli* DnaG, a fragment of ssDNA (ball and stick representation) lies in a groove between two β hairpins (green; upper β2/β3, lower β5/β6). Negatively charged residues (stick representation) in the active site are separated from a positively charged region designated as a basic ridge. (B) Structure of *G. stearothermophilus* DnaB and its interaction with the helix-binding domain (HBD) of DnaG. Left, the six N-terminal domains (NTDs; residues 1–152, blue) of DnaB form a ring above the six C-terminal domains (CTDs; residues 186–454, red) with the intervening linker regions in yellow. Right, a top view of the complex of DnaB with the HBD of DnaG. Three HBDs (green) are bound at the NTD trimer interfaces and make contact only with the NTDs. (PDB 3B39 and PDB 2R6A) (C) Docking of three DnaG RPDs (depicted in beige), as illustrated in (A), onto the complex of DnaB and DnaG HBD, illustrated in (B). The growing RNA primer is depicted in red and the Zn-binding domain of DnaG has been omitted for clarity. ssDNA would be complexed with multiple copies of SSB, which have also been omitted for clarity. (A, from J.E. Corn, J.G. Pelton and J.M. Berger, *Nat. Struct. Mol. Biol.* 15:163–169, 2008. With permission from Macmillan Publishers Ltd. B, from S. Bailey, W.K. Eliason and T.A. Steitz, *Science* 318:459–463, 2007. With permission from AAAS. C, adapted from J.E. Corn, J.G. Pelton and J.M. Berger, *Nat. Struct. Mol. Biol.* 15:163–169, 2008. With permission from Macmillan Publishers Ltd.)

is protected by binding to SSB, the disordered C-terminal region of which (Figure 3.10) binds to another machine, the **clamp loader** (or γ complex). The clamp loader is responsible for assembling the β clamp round the DNA (Figure 3.19) and in the correct orientation for its associated DNA polymerase to synthesize DNA in the 5′ – 3′ direction.

Below, we make the simplifying assumption that the bacterial clamp loader operates alone; this is not the case, in fact, as the clamp loader forms the organizing core of DNA polymerase III holoenzyme. Such an organizing role may also extend to the eukaryotic clamp loader (RFC), although this is far from established. In *E. coli*, the clamp loader is a pentamer (**Figure 3.23**), most likely containing two copies of a τ subunit, one γ subunit, and one copy each of structurally related δ and δ′ subunits, but there is increasing evidence that the most favored structure *in vivo* will be one with three τ subunits and no γ subunit (see below). The τ (71 kDa) and γ (47 kDa) proteins are encoded by the same gene (*dnaX*), the shorter one being generated by a translational frameshift in the gene that introduces a stop codon. The C-terminal region (~46 aa) of γ appears to be a conformationally flexible linker to which the 24 kDa C-terminal extension is attached in τ. The extension itself can be divided into two domains, the inner of which binds to the DnaB helicase and the C-terminal one to the Pol III α subunit. However, most structural and mechanistic evidence has come from a variant assembly with the simpler composition γ₃δδ′. The two N-terminal domains (1 and 2) of each subunit constitute an AAA+ ATPase module (clamp loader clade, see Box 3.1), whereas the C-terminal domain (3) self-associates to create the pentameric 'collar.' However, only the γ subunits (or τ subunits if present) are active ATPases, as the δ and δ′ subunits do not have competent ATP-binding sites. There is biochemical and electron microscopy (EM) evidence that the binding of ATP to the clamp loader induces a stable interaction with an open clamp and that interaction with a primer-template DNA (double-strand/single-strand junction with a 5′ overhang) promotes ATP hydrolysis, release of the clamp loader, and deposition of a closed clamp round the DNA (Figure 3.23A. As described above for DnaA, the AAA+ ATPase modules in the clamp loader complex change conformation as a switch, rather than as a motor.

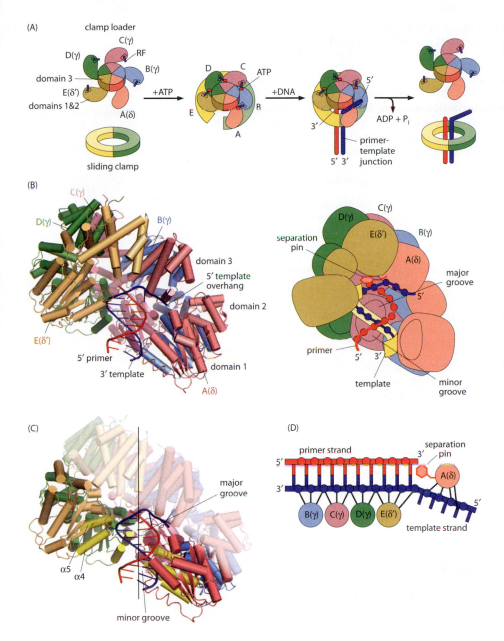

Figure 3.23 The *E. coli* clamp loader complex bound to primer–template DNA. (A) Schematic mechanism of the cycle of clamp loading onto DNA. The five subunits of the clamp loader are labeled A–E; in each subunit domains 1 and 2 make up an AAA+ ATPase module and domain 3 forms the collar. RF, arginine finger in an AAA+ ATPase module. ATP is bound only between subunits B/C, C/D, and D/E. The template DNA strand is in blue, the primer in red. (B) Left, structure of the clamp loader bound to a primer–template DNA; right, schematic representation of the structure. The five protein subunits are colored as in (A). The protein contacts with the DNA are largely confined to the template strand, in the region highlighted in yellow. (C) The clamp loader subunits form a spiral tracking the DNA template strand. The N termini of helices α4 and α5 (yellow) in subunits B–E interact with the DNA phosphate groups. Subunit A is more detached. The axis of the DNA is indicated by the vertical line. (D) Subunits B–E interact with dinucleotide steps of the template strand and Tyr316 in domain 3 (collar) of subunit A(δ) caps the 3′ end of the primer. (PDB 3GLF) (Adapted from K.R. Simonetta et al., *Cell* 137:659–671, 2009. With permission from Elsevier.)

The clamp loader forms an ATP-dependent spiral structure round the primer–template junction

The structure of a clamp loader bound to a synthetic primer–template junction (10 bp of DNA with a 5-nucleotide 5′ overhang) and an ATP analog (ADP.BeF3) is shown in Figure 3.23B. Overall, the clamp loader forms a spiral (like a 'lock-washer') around the DNA, with the 5′ overhang of the template strand protruding from a gap in the spiral (the clamp loader can thus be envisaged as a twisted hexameric helicase that lacks one subunit, creating the gap). The protein/DNA contacts are mostly with the template strand. The three γ subunits and the δ′ subunit (B–E in Figure 3.23A and B) track the template strand of the duplex portion of the DNA in dinucleotide steps, with the positively charged helix dipoles at the N-terminal ends of helices α4 and α5 in each subunit interacting with the negatively charged phosphates on the DNA backbone (Figure 3.23C). The δ subunit (subunit A) is responsible for an important contact with the primer strand at the end of the duplex portion: Tyr316 in its collar domain (domain 3) stacks against the base of the nucleotide at the 3′ end of the primer strand, thereby capping it and causing the template strand to bend sharply to exit the complex (Figure 3.23D). Model building suggests that this would be true irrespective of whether the primer is RNA or DNA. This primer capping mechanism is akin to that of the UvrD helicase involved in DNA repair, in which an aromatic residue acts as a 'separation pin' to help strand separation (Chapter 4), and helps explain the specificity for primer–template junction sites.

Subunits B–E in the spiral are related to neighbors by a rise of ~7.3 Å and a rotation of ~60° around a central axis, which coincides with that of the duplex DNA (Figure 3.23C). In B-DNA, the corresponding values for dinucleotide steps are ~6.8 Å and ~72°, which indicates that the DNA becomes slightly underwound as a result of its interactions with the clamp loader. Binding of the clamp loader to DNA serves to introduce rotational symmetry into the arrangement of its subunits and, in so doing, optimally repositions key residues, among them the arginine fingers (Box 3.1) of the AAA+ ATPase modules (for example Arg149 of the γ subunit), rendering the three competent ATPase sites active.

There is currently little information about how binding of ATP to the clamp loader is coupled to opening of the β clamp. The δ subunit has been assigned a major part in this process—and has been termed the 'wrench' in consequence. No specific role for the δ′ subunit has been proposed as yet.

Comparison with the structure of the corresponding eukaryotic clamp loader/sliding clamp complex (RFC-PCNA, below) suggests that hydrolysis of the bound ATP causes the clamp loader to undergo a conformational change, revert to its nonspiral structure and dissociate from the β clamp, leaving the latter installed on the DNA duplex. The structure of the *E. coli* clamp loader bound to DNA (and a nonhydrolyzable analog of ATP) shown in Figure 3.23 may thus reflect the structure at the point of ATP hydrolysis.

In eukaryotes, PCNA is loaded onto DNA by a clamp loader called replication factor-C (RFC), which closely resembles that of *E. coli*. It too has five subunits (A–E), which accommodate AAA+ ATPase modules of the same 'clamp loader' clade (Box 3.1), and these adopt the form of a rising right-handed spiral, perceived as a hexamer with one subunit missing. By means of a similar cycle of ATP binding and hydrolysis, RFC can bind round the DNA at a primer-template junction, prise open a PCNA sliding clamp, install it on the DNA and then release it to bind to the polymerase. (In archaea, subunits B–E are identical.) The structure of the RFC of *S. cerevisiae* bound to a PCNA is shown in **Figure 3.24**. In this form, the

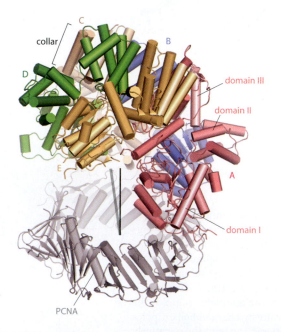

Figure 3.24 Structure of the clamp loader of *S. cerevisiae* bound to its PCNA. The five subunits (A–E) of the RFC are shown in different colors (cf. *E. coli* clamp loader in Figure 3.23). RFC-A is a truncated version (residues 295–785) and the bound nucleotide in each subunit is nonhydrolyzable ATP-γS. Domains I and II constitute the AAA+ ATPase module in each subunit, and the five copies of domain III, one from each subunit, interact to form the collar. The three subunits of PCNA are shown in gray. The screw axis of the RFC is indicated in black, and the rotation axis of PCNA is indicated in gray. (PDB 1SXJ)

PCNA ring is planar and makes contact with only three RFC subunits, but if it is forced into a spiral like that of the RFC, it can make contact with all five subunits. As with the *E. coli* clamp loader, the primer–template junction would be capped within the RFC, conferring specificity for loading PCNA rings at such junctions.

The clamp loader is bound to SSB by a heterodimer of χ and ψ proteins

Upstream of newly synthesized Okazaki fragments, SSB binds to and protects the ssDNA, and binds to the clamp loader by means of a heterodimer, χ (147 aa) and ψ (137 aa). The collar of the clamp loader binds tightly to the ψ protein in a 1:1 stoichiometry, whereas the χ protein binds to the last 26 or so residues of the intrinsically disordered C-terminal tail segment (residues 113–177) of SSB. The ψ protein stimulates DNA-dependent ATP hydrolysis by the clamp loader. The structure of the *E. coli* clamp loader bound to a synthetic peptide representing residues 2–28 (identified biochemically as the critical region, K_d ~7 nM) at the N-terminal end of the ψ protein, is shown in **Figure 3.25**. Based on the known structures of the ψχ heterodimer and the tetrameric SSB bound to DNA, it is then possible to put together a model for the binding of the SSB to the clamp loader at a replication fork (Figure 3.25).

DNA Pol I and DNA ligase are required to fill in and close the gaps between Okazaki fragments on the lagging strand

As we saw above, when DNA Pol III has finished DNA replication, RNA primers must be removed and the gaps filled in. Two enzymes are needed to do this: DNA polymerase Pol I, with its 5′–3′ editing function to degrade the RNA primer (with or without help from RNase H) and its polymerase function to replace it with the correct DNA; and **DNA ligase** (in *E. coli*, LigA) to make the phosphodiester bonds required to join up the pieces. DNA ligases were discovered in extracts of *E. coli* and mammalian cells in the late 1960s. Their ability to join pieces of DNA together was of major importance in making recombinant DNA technology possible. The ligases have an unusual three-step mechanism (**Figure 3.26A**) in which a specific Lys enters into a phosphoamide bond with AMP. In bacteria, the AMP comes from cleavage of NAD^+ and release of nicotinamide; in archaea and eukaryotes, it derives from cleavage of ATP and release of PP_i. The AMP is then transferred to the phosphorylated 5′ end of the DNA at the nick remaining after Pol I has done its work, which is followed by nucleophilic attack of the 3′-OH group of the adjoining strand and generation of the proper phosphodiester bond. A similar mechanism is employed by RNA ligases and

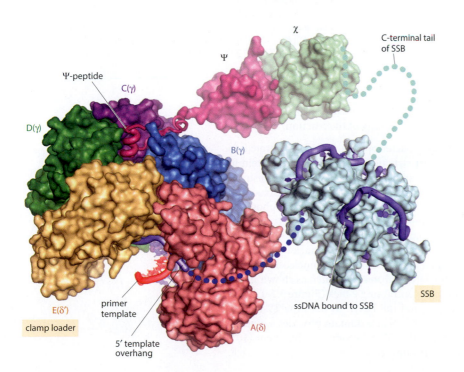

Figure 3.25 The clamp loader is linked to SSB by the ψχ heterodimer. The ψ peptide (residues 2–28 of the ψ protein) is depicted in dark pink, bound across the collar domains of subunits B, C, and D in the clamp loader. The disordered C-terminal region of one of the four subunits in the SSB (Figure 3.10) binds to the χ protein. The ssDNA of the template strand is shown leaving the SSB and entering the clamp loader as part of the primer–template duplex. The color coding of the clamp loader and primer–template DNA is the same as in Figure 3.23. (PDB 3GLH, clamp-loader/ψ peptide complex; PDB 1EM8, ψχ heterodimer.) (Adapted from K.R. Simonetta et al., *Cell* 137:659–671, 2009. With permission from Elsevier.)

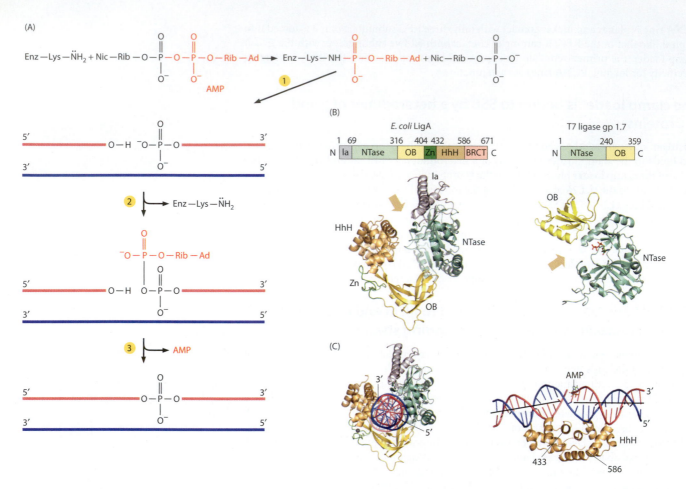

Figure 3.26 Structure and mechanism of DNA ligases.
(A) The three-step mechanism: step 1, reaction of a Lys residue in the active site with NAD⁺ (or ATP); step 2, binding to nicked DNA of lagging strand (template strand in blue and fragments to be joined in salmon pink) followed by transfer of AMP from the active site Lys to the 5′-phosphate end of a DNA fragment; step 3, nucleophilic attack of 3′-OH group of adjacent DNA fragment and formation of phosphodiester bond. Ad, adenine; Nic, nicotinamide; NMN, nicotinamide mononucleotide; Rib, ribose. (B) The DNA ligases of *E. coli* and bacteriophage T7. The arrows indicate the gap at which the DNA substrate enters to make contact with the active site. The BRCT domain is disordered and not visible in the crystal structure of LigA. The C-terminal 10 amino acids of the T7 ligase are also disordered. (C) *E. coli* LigA in complex with a 26 bp duplex DNA, colored as in (A), with a centrally located nick. Left, end-on view showing the various protein domains encircling the DNA. Right, a view at about 90° around a vertical axis. The HhH domain (residues 433–586) binds symmetrically across the nick in the DNA, with the 5′ end of the nick having already become adenylated. Four helix-turn-helix motifs crucially abut the DNA. A bend of 10° in the DNA is observed. (PDB 2OWO (LigA) and 1A0I (T7)) (C, adapted from J. Nanadkumar, P.A. Nair and S. Shuman, *Mol Cell* 26:257–271, 2007. With permission from Elsevier.)

by the GTP-dependent mRNA capping enzyme (see Chapter 5). The degradation of ATP or NAD⁺ contributes to the overall negative (favorable) free energy of the reaction.

The structures of the DNA ligases of *E. coli* and bacteriophage T7 are illustrated in Figure 3.26B. Both are monomers and have modular structures that enable the protein to encircle the dsDNA substrate. The much simpler T7 ligase (40 kDa) uses ATP as substrate; the more elaborate *E. coli* LigA (77 kDa) is NAD⁺-dependent. Comparison of the structures of various bacterial LigA fragments with just NAD⁺ or NMN bound or in the phospho-amide state, indicate that substantial domain movements accompany the various stages of the enzymatic reaction. LigA has an N-terminal domain Ia (unique to NAD⁺-dependent ligases) that binds the NMN moiety of NAD⁺ and sets up the NAD⁺ for attack by the active site Lys residue; domain Ia and the OB domain then pivot away from the nucleotidyl trans-ferase (NTase) domain, making it possible for the DNA substrate to enter, as shown in the complex with a 26 bp fragment of DNA containing a central single nick (Figure 3.26C). The NTase and OB domains form a C-shaped clamp round the DNA (as does the T7 ligase), interacting with its minor groove near the nick, and the NTase domain provides a binding pocket for the adenine of the 5′-AMPP at the nick. The DNA becomes slightly bent (10°) and is locally distorted towards an RNA-like A-form immediately either side of the nick, widening both major and minor grooves. The 3′-OH at the nick is suitably poised for attack 3.2 Å from the target P atom, with the AMP leaving group apically oriented. The HhH domain (a tandem repeat of two (HhH)$_2$ modules) binds on the other side of the substrate with approximate twofold symmetry across the nick. The Zn finger domain bridges the OB and HhH domains, having perhaps just a stabilizing structural role. The C-terminal BRCT (BRCA1 C-terminal) domain (see **Guide**) is intrinsically disordered and its role, if any, is unclear. Two metal ions are also likely to be involved in catalysis, but they have been excluded from the proteins during crystallization to prevent a catalytic cycle taking place and are not included in the discussion.

The replisome is held together by an array of protein/protein interactions

The various components that comprise a replisome make multiple and overlapping protein/protein contacts. The bacterial replicative helicase is traveling only in the 5′–3′ direction but both DNA strands are being copied simultaneously, though by different mechanisms on the leading and lagging strands. Central to this is the clamp loader which, among the other things it does, holds the two Pol III cores together. A schematic model of the *E. coli* replisome (a holoenzyme of ~800 kDa), reflecting the various protein/protein interactions described above, is shown in Figure 3.12.

A more dynamic view of the constructs than is depicted in Figure 3.12 is now emerging. For example, with T4, T7, and *E. coli* replisomes assembled *in vitro* (the last with three τ subunits rather than a τ2γ configuration in the clamp loader), three polymerase molecules can interact with the helicase. In its simplest formulation, two of these would be active at any one time, one on each of the leading and lagging strands, with the third carried effectively as a 'spare.' The τ3 replisome may be the normal form *in vivo*—the composition is still uncertain. It has been suggested that the third polymerase can engage with the lagging strand on a new primer before the previous Okazaki fragment has been completed (there is some EM evidence in support of this) and it may also be that it could be invoked to circumvent lesions that stall a loaded polymerase. How replisomes can skip over DNA lesions (translesion polymerases) is described in Chapter 4.

A specialized set of proteins is required for RNA primer excision

As we saw with bacteria, in eukaryotes the RNA primer, which enables the process of DNA synthesis to start, must be removed; indeed, the whole fragment synthesized by DNA Pol α is removed. RNase H, as in *E. coli*, can play a part in the removal of the RNA primers that is required before Okazaki fragments can be ligated, but the main means lies with a set of two specialized nucleases. In the dominant pathway, Pol δ synthesizing an Okazaki fragment on the lagging strand bumps into the downstream Okazaki fragment and displaces a short 'flap' (2–10 nucleotides) at the latter's 5′ end, thereby exposing it to nuclease activity. At the same time, it typically leaves an overhang of one nucleotide at the 3′ end of the Okazaki fragment it has been synthesizing. At this point Pol δ is replaced by an endonuclease termed flap endonuclease-1 (FEN1), whose preferred substrate is duplex DNA with a protruding 5′ flap at an internal position (**Figure 3.27**). The exposed oligonucleotide flap is excised, leaving a gap with a free 3′-OH group on one side and a phosphorylated 5′-OH on the 5′ side. DNA ligase can then join the two ends together.

FEN1 (human 380 aa) contains a PIP motif in its long (~45 aa) C-terminal tail and its structure in complex with a PCNA ring demonstrates that the binding is due to

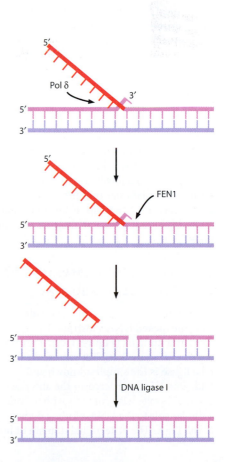

Figure 3.27 Schematic diagram of primer removal by FEN1. A flap (about 10 nucleotides of RNA/DNA, in red) at the 5′ end of the downstream Okazaki fragment is displaced by the action of DNA Pol δ as it completes the synthesis of the current Okazaki fragment. This creates a substrate for FEN1 (flap endonuclease-1), which cuts the flap off at the Y junction with duplex DNA, leaving a phosphodiester bond between the two Okazaki fragments to be created by the action of DNA ligase I. The PCNA (proliferating cell nuclear antigen) ring to which DNA Pol δ and FEN1 are bound is omitted for clarity. (Adapted from M. L. Bell and S. D. DePamphilis, Genome Duplication. New York: Garland Science, 2011.)

(A)

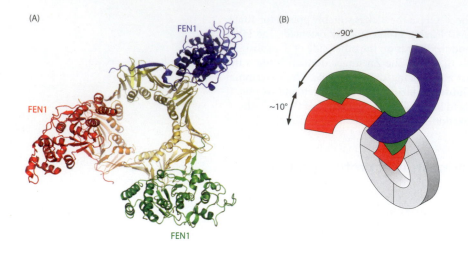

(B)

Figure 3.28 PCNA binds three molecules of FEN1. (A) The three PCNA subunits are depicted as in Figure 3.20. Each FEN1 molecule (red, blue, and green) is bound to a subunit of PCNA by means of the PIP motif in its C-terminal tail plus additional protein/protein interactions. The three FEN1 molecules are in different orientations, with two (red and green) closer to the central hole in PCNA through which the DNA passes. (B) Schematic diagram of a FEN1 molecule on PCNA, in each of the three orientations in (A). The DNA would be expected to pass through PCNA at an angle (Figure 3.20) and the three orientations shown may represent possible orientations through which any one FEN1 molecule would go in approaching the DNA to remove the primer flap. (PDB 1UL1) (Adapted from S. Sakurai et al., *EMBO J.* 24:683–693, 2005.)

interactions with short segments of FEN1 as well as a region with a more extensive interface (**Figure 3.28A**). Of the three FEN1 molecules bound to the PCNA, each is in a slightly different conformation and held in a different orientation to the ring. The nuclease active site is in the N-terminal core domain (residues 1–332, a homolog of the 5′–3′ exonuclease domain of *E. coli* DNA Pol I) and utilizes two metal ions held in clusters of acidic side chains to promote nucleophilic attack on the scissile phosphodiester bond in the target substrate. The linker region between the nuclease core domain and C-terminal PCNA-binding tail appears to act as a 'hinge' allowing the nuclease domain of one FEN1 molecule to shift towards the DNA running through the PCNA ring, while the other two are held back, perhaps awaiting their turn to act (Figure 3.28B).

FEN1 cannot handle long flaps. If flaps of 30 nucleotides or more are formed, the ssDNA may attract the ssDNA-binding protein RPA (replication protein A). RPA coats ssDNA strands and inhibits FEN1 activity but, in turn, attracts another nuclease, Dna2. Dna2 was originally identified in mutant yeast strains and has the ability to cleave a long segment of the RNA/DNA primer flap, reducing it to a short (~5–7 nucleotides) RPA-free flap that is susceptible to the action of FEN1. Dna2 is of crucial importance in yeasts but less so in mammals.

As we have seen above, primase is not highly accurate in its copying and Pol α has no 3′–5′ exonuclease proofreading activity. The proofreading ability of Pol δ may help in correcting any mistakes. Likewise, primer removal by FEN1/Dna2 will take with it any fault in the DNA region of the primer synthesized by Pol α, leaving Pol δ to copy the template strand correctly before ligation seals the gap. If a mistake occurs beyond the point reached by flap displacement, it will have to be repaired by other mechanisms, such as mismatch repair (Chapter 4). To allow ligase to bind to PCNA, FEN1 must dissociate and this is dependent on post-translational modification: FEN1 methylated at Arg192 can bind to PCNA but, after cleavage of the flap, it becomes demethylated. Phosphorylation at Ser187 by a cell cycle-dependent kinase (Cdk2) ensues and this, in turn, causes FEN1 to dissociate from the PCNA.

Eukaryotic DNA ligases resemble *E. coli* LigA but use ATP rather than NAD⁺ as co-substrate

Eukaryotic DNA ligases can vary widely in complexity but all resemble *E. coli* LigA in encircling the nicked DNA substrate. This is evident in the structures of the ligases from chlorella virus and humans (**Figure 3.29**). They use the same three-step mechanism as the bacterial enzymes (Figure 3.26) but require ATP rather than NAD⁺ in the activation step. The chlorella ligase is the simplest known and contains a 30-residue insertion (the 'latch') in its OB fold, which is disordered in the absence of DNA but closes the ring to envelope the DNA and makes extensive contacts with its minor groove opposite the nick. In this regard, it plays a similar role to the DNA-binding domain (DBD) of human ligase I and the HhH domain of *E. coli* LigA. The N-terminal domain of human ligase I (not in the crystal structure shown

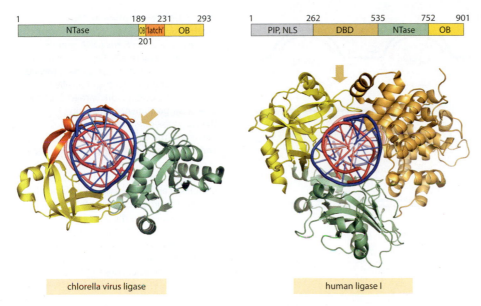

chlorella virus ligase

human ligase I

Figure 3.29 DNA ligases from chlorella virus and humans. The ligases are shown in complex with an encircling nicked duplex DNA, colored as in Figure 3.26. That of humans (right) has an N-terminal domain that contains the PIP (PCNA-interacting-protein) motif and a nuclear localization signal (PIP, NLS) but this was not present in the protein crystallized. The arrows show where the ring opens to admit the DNA substrate. (PDB 2Q2T (chlorella virus) and 1X9N (human))

in Figure 3.29) contains the PIP motif that is responsible for binding to PCNA (see above) and a *nuclear localization signal*.

3.3   TERMINATION OF DNA REPLICATION

DNA replication has to stop as well as to start. With the circular chromosomes of bacteria, replication forks traveling at about the same speed in opposite directions from an origin of replication will eventually collide halfway round. In eukaryotes, there is the problem of how to halt replication at the ends of the linear DNA in chromosomes.

Bacterial chromosomes contain termination sites for DNA replication

In *E. coli*, the evidence is that DNA replication is terminated in a region approximately diametrically opposite *oriC* and which contains four termination sites (none of which is absolute). These termination sites (*TerA*, *TerB*, *TerC*, and *TerD*) are 23 bp motifs centered round a consensus sequence of 5′-AGNATGTTGTAACTAA-3′, and each acts as a replication fork barrier, able to block a replication fork traveling in one direction rather than the other. They do this by binding Tus (termination utilization substance, a 36 kDa protein) with high affinity ($K_d \sim 10^{-13}$ M). One face of Tus is able to interact with DnaB in the replication fork tracking towards it in the 5′–3′ direction on the lagging strand. This causes DnaB to dissociate from the fork, thereby arresting replication. However, a fork tracking 5′–3′ on the other strand finds Tus in the opposite orientation, no interaction with DnaB occurs, and the fork is therefore allowed to pass through (**Figure 3.30A**). It is likely that two barriers exist in each orientation so that if a replication fork manages to pass through one, it is arrested by the other.

Termination of replication inevitably follows collision of replication forks. The DNA that lies between the forks will be shortening and eventually will be fully unwound ahead of the oncoming helicases. In the circular DNA of *E. coli*, the 3′ end of the new leading strand being synthesized at each fork eventually approaches the primer at the 5′ end of the nascent (lagging) strand being synthesized at the other fork, the primers are removed and the ssDNA gaps are filled in. At this point, the two circular chromosomes are complete and can separate.

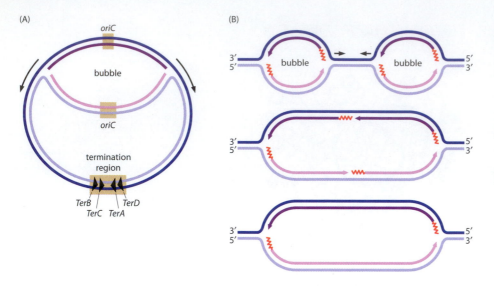

(A)

(B)

Figure 3.30 Termination of DNA replication. (A) Replication of a circular chromosome, as in *E. coli* (DNA strands colored as in Figure 3.1). Replication begins at *oriC* and proceeds bidirectionally towards the termination region diametrically opposite. Replication fork barriers *TerD* and *TerA* allow passage of a replication fork in the clockwise direction; *TerB* and *TerC* allow passage in the counterclockwise direction. Forks coming from the opposite direction in each instance are arrested. (B) Three stages in the fusion of two adjacent replication bubbles in eukaryotic DNA. The 3′ end of the nascent strand on each template strand in a bubble approaches the 5′ end of the nascent strand in the neighboring bubble (top), the bubbles coalesce (middle), the primer is removed, and the ssDNA gap is filled (bottom). Primers are depicted in red.

The linear DNA in eukaryotic chromosomes is replicated by a special mechanism that also protects its ends

In eukaryotes the effect of replication fork collision will be to join two replication bubbles and make them into one longer one, a process that must recur many times until all the replication bubbles have been joined (Figure 3.30B). Replication fork barriers exist also in eukaryotes, although they are infrequently used, as expected if the multiple replication bubbles are to coalesce. Instead, their function appears to be in preventing replication forks from colliding with the molecular machines that transcribe DNA into RNA (Chapter 5) and have been best characterized in the transcription of rRNA genes. This still leaves the problems of the ends of linear DNA.

DNA polymerases synthesize DNA only in the 5′–3′ direction and need a primer. Therefore the 3′ ends of linear DNA cannot be copied into daughter DNA and, without intervention, the chromosomal DNA would shorten a little each time it is replicated. The effect would be cumulative and the end result disastrous for chromosomal integrity. In addition, eukaryotes have evolved crucially important mechanisms that detect and repair the exposed ends of broken chromosomal DNA (Chapter 4), and such double-strand breaks are potentially indistinguishable from the natural ends of intact chromosomal DNA. Both problems are overcome by the existence of telomeres at the ends of chromosomal DNA.

Telomeric DNA comprises multiple tandem repeats (up to several hundred kilobase pairs of dsDNA) of a 6–8 nucleotide motif (TTAGGG in mammals, TTGGGG in *Tetrahymena*) at the 3′ end of the DNA. This G/T-rich region base pairs with a complementary C/A-rich region on the opposite strand, but extends beyond it as an ssDNA overhang up to several hundred nucleotides long. In EM, large lasso-like structures (t-loops) are observed, equivalent to thousands of TTAGGG (TG) repeats and thought to be formed by the overhang displacing normal base pairing in the double-stranded DNA region (**Figure 3.31A**). The telomeres are protected by interaction with a collection of proteins, the nature of which varies widely between organisms, and which also distinguish the natural ends of chromosomes from unwanted dsDNA breaks. The **shelterin** complex in humans comprises six proteins (protection of telomere Pot1 and Tpp1, which bind to ssDNA repeats, and the dsDNA telomere-binding proteins Trf1 and Trf2, Rap1 and Tin2). A yet more complicated structure capable of adoption by telomeric DNA because of its high G content is the **G-quadruplex**. This is a four-stranded structure generated by the stacking of G-tetrads formed from four G nucleotides held together by Hoogsteen (rather than canonical Watson–Crick) base pairing. It is stabilized by a monovalent metal ion such as Na^+ or K^+ coordinating the guanine >C=O groups at the center of the tetrad (Figure 3.31B). G-Quadruplexes differ in terms of the orientation (parallel or anti-parallel) of the strands, the conformation (*syn* or *anti*) of the glycosidic bond, and the arrangement of the connecting loops. A set of G-quadruplex structures formed from human telomeric DNA *in vitro* is depicted schematically in Figure 3.31C. The existence and role, if any, of quadruplex DNA

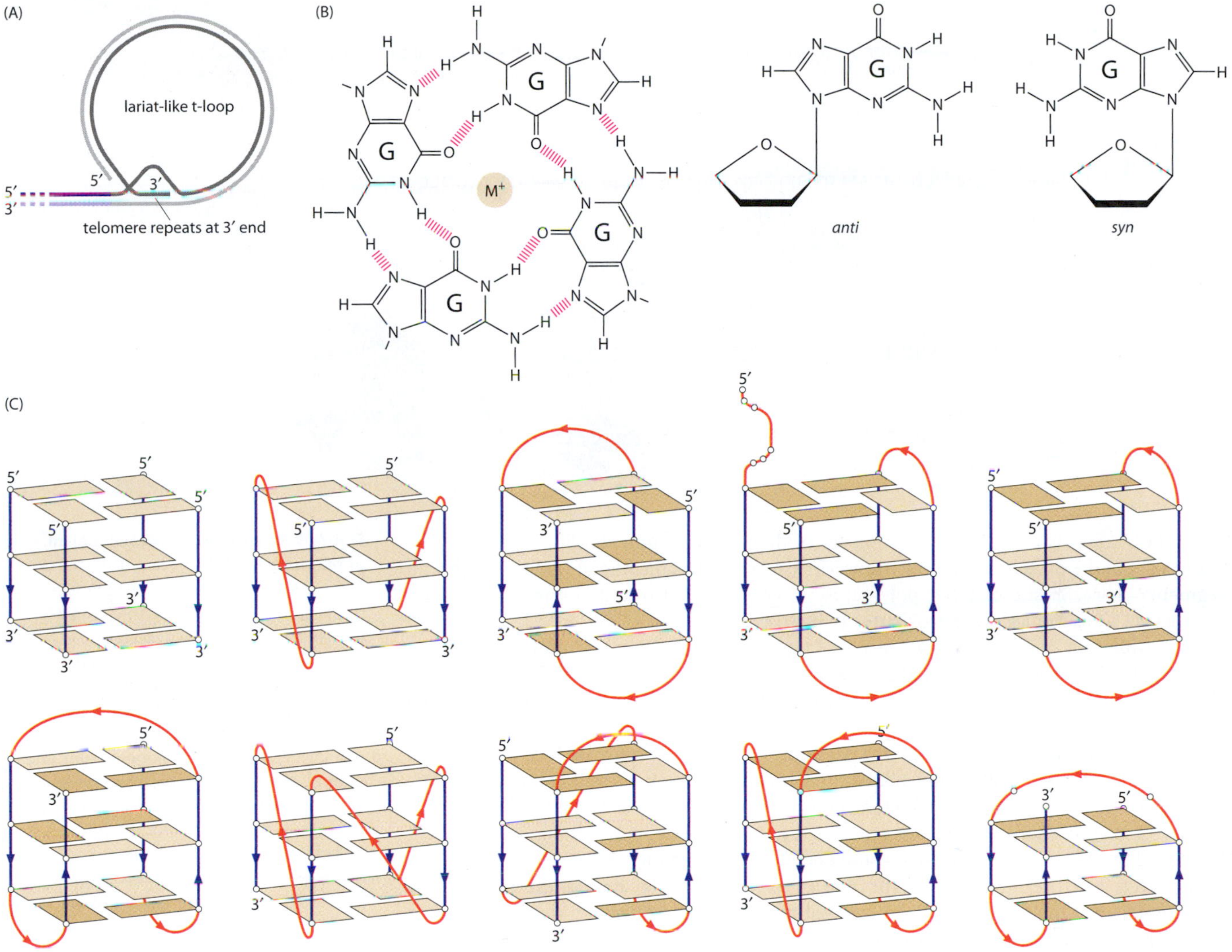

Figure 3.31 Structures adopted by telomeric DNA. (A) Formation of t-loops by the 3′-ssDNA overhang displacing the 5′–3′ strand in the upstream dsDNA region and base pairing with the 3′–5′ strand. (B) Structure of a G-tetrad. Left, G-tetrad stabilized by Hoogsteen base-pairing and a monovalent metal ion; right, *syn* and *anti* forms of G-nucleoside. (C) Different forms of G-tetrad stacking for human telomeric DNA. DNA strands, blue and loops, red. Arrows indicate direction of strands. *Syn* conformations, pink; *anti* conformations, beige. (B and C, adapted from A.T. Phan, *FEBS J.* 277:1107–1117, 2010. With permission from John Wiley & Sons.)

in telomeres *in vivo* has yet to be established, but suggestions include end protection, an indicator of DNA damage, and recruitment of the enzyme, telomerase, which is responsible for maintaining the telomeric DNA.

Telomeric DNA is synthesized and maintained by telomerase, a specialized polymerase that utilizes RNA as a template

Telomeric DNA would be shortened with each round of cell division but for an enzyme, **telomerase**, which synthesizes fresh TG repeats and maintains the telomere at an appropriate length. The enzyme, first discovered by Carol Greider and Elizabeth Blackburn in extracts of *Tetrahymena*, contains an ssRNA (telomerase RNA, TER) used as a template for the synthesis of the TG repeats, a catalytic protein subunit, telomerase reverse transcriptase (TERT), and various species-specific accessory proteins. For example, in human telomerase, the proteins dyskerin, pontin, and reptin appear to form a scaffold that recruits and stabilizes TER and then dissociates from the ribonucleoprotein. The TER component differs widely among species: that of *Tetrahymena thermophila* is only 159 nt, that of humans is 451 nt, whereas *S. cerevisiae* has a TER of 1167 nt. Telomerase is recruited to the telomeres, for example by proteins of the shelterin complex in humans, which leads to partial pairing of the RNA template with the 3′ end of the ssDNA overhang. This positions the terminal 3′-OH in the active site of TERT, which then synthesizes one TG repeat, after which the RNA/DNA hybrid temporarily dissociates and then reforms with the new 3′-OH repositioned on the template ready for a further round of synthesis (**Figure 3.32A**), a process designated 'repeat addition processivity.'

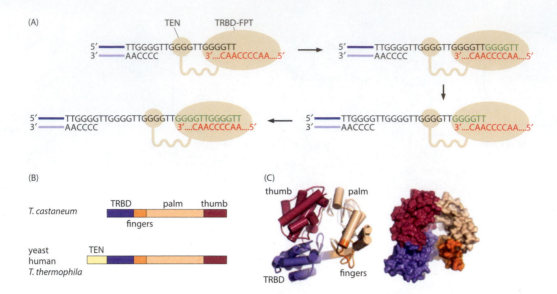

Figure 3.32 Structure and mechanism of telomerase. (A) Extension of 3′ strand overhang of telomere DNA by repetitive synthesis of TG repeats. TERT (telomerase reverse transcriptase) is depicted schematically in beige. The N-terminal TEN (telomerase essential N-terminal) domain is linked to the TRBD (telomerase RNA-binding domain) by a flexible linker region; F, P, and T are the fingers, palm, and thumb segments of a typical polymerase. The region of the *T. thermophila* TER used as template is depicted in red. The TG repeats successively synthesized are depicted in green. (B) Domain maps of TERTs from two species. (C) Two representations of the structure of TERT from *T. castaneum*, with the various segments colored as in (B). (PDB 3DU5 and 3DU) (B and C, adapted from A.J. Gillis, A.P. Schuller and E. Skordalakes, *Nature* 455:633–637, 2008. With permission from Macmillan Publishers Ltd.)

TERT (127 kDa) is based on a common set of domains or segments (Figure 3.32B). The major part of the protein is a reverse transcriptase, which has the fingers, palm, and thumb segments characteristic of DNA polymerases (see Figure 3.15) and is homologous with the reverse transcriptases of retroviruses (Chapter 8). This part is preceded by a telomerase RNA-binding domain (TRBD). In most organisms, there is an additional N-terminal domain, TEN (telomerase essential N-terminal), attached by a long flexible linker to the TRBD. A crystal structure of the TERT from the flour beetle *Tribolium castaneum*, which lacks the TEN domain, shows that the other four segments adopt a ringlike structure (Figure 3.32C) and resembles retroviral reverse transcriptases and family B DNA polymerases (Figure 3.16). Various catalytically important motifs line the ring, including the three invariant Asp residues (251, 343, and 344) that would participate in a two-metal mechanism (see Figure 3.14). The interior dimensions of the ring (26 Å across by 21 Å long) are sufficient to host a DNA/RNA hybrid of 7–8 bp, and could position the incoming 3′ end of the DNA in the active site. The TRBD is thought to help to facilitate the location of the 5′ end of the RNA template at the entrance to a groove that runs across the interior of the ring. During the catalytic cycle, the RNA/DNA hybrid in telomerase is maintained at a constant length by the melting of H-bonds at the distal end as new bonds are formed with the addition of new nucleotides at the 3′ end of the DNA strand. The structural and mechanistic basis of repeat addition processivity and a precise role for the TEN domain have yet to be elucidated.

After the G-rich 3′-ssDNA overhang has been extended, it can be used as the template for DNA Pol α/primase to fill in the vacant space at the 5′ end of the C-rich 5′–3′ strand. Telomerase is active during S phase in germline and stem cells where it serves to maintain telomere length, but is repressed in most somatic cells. Only 20–50 copies of telomerase are present in human cells and the shelterin complex plays some part in recruiting them to the shortest telomeres. Gradual loss of multiple TG repeats ultimately leads to activation of a p53-dependent DNA damage response and to cell senescence (which may have a relationship with the aging process of the organism) or apoptosis (Chapter 13). Activation of telomerase is thus crucially important in disease and perhaps aging, and more than 85% of human cancer cells constitutively express the gene for TERT.

Replicons and factories

It takes many proteins to bring about the replication of a bacterial genome, even more (>60) for an archaeal or eukaryotic genome. A single unit of DNA replication, the **replicon**, comprising a replication origin, a specific protein initiator, and the replication forks traveling in opposite directions, is now reasonably well understood. In the simpler bacterial systems, the replisome model (Figure 3.12) suffices, but in eukaryotes we have many such origins of replication and there are indications that their replication forks come together to form localized 'factories.' These can be visualized by staining cells with an antibody against the base analog 5-bromouracil (BrdU) after allowing new DNA to be synthesized with the analog present. About 1000 factories appear at the beginning of S phase in mammalian cells, each factory being estimated to contain between 20 and 200 replication forks. The vast assembly of molecular machines, all contributing different activities, that together make possible DNA replication, displays some organization but does not form a highly ordered whole. Components come and go according to the stage of the process, in concert with the cell cycle. The temporal control is perhaps as remarkable as the molecular machinery of the process itself.

3.4   DNA TOPOLOGY IN REPLICATION

The description of DNA replication in bacteria, archaea, and eukaryotes has thus far been simplified by neglecting the topological problems that are associated with dsDNA as a double-helical structure. This property of DNA imposes topological constraints as it is condensed into chromosomes, unwound and replicated, repaired, and transcribed into RNA. In the processes of DNA replication, the DNA is not normally free to rotate about its own axis, being closed circular (as in *E. coli*) or too big or tied up in chromosomes (as in eukaryotes) Thus, every turn (10.5 bp, right-handed) of DNA unwound by a helicase must be compensated for by the introduction of a left-handed (positive) superhelical turn appearing ahead of the replication fork and leaving intertwined duplex DNA (precatenanes) behind (**Figure 3.33A**). The positive supercoils will accumulate and their presence will ultimately force DNA replication to a halt unless they are relieved. A further problem is that when the

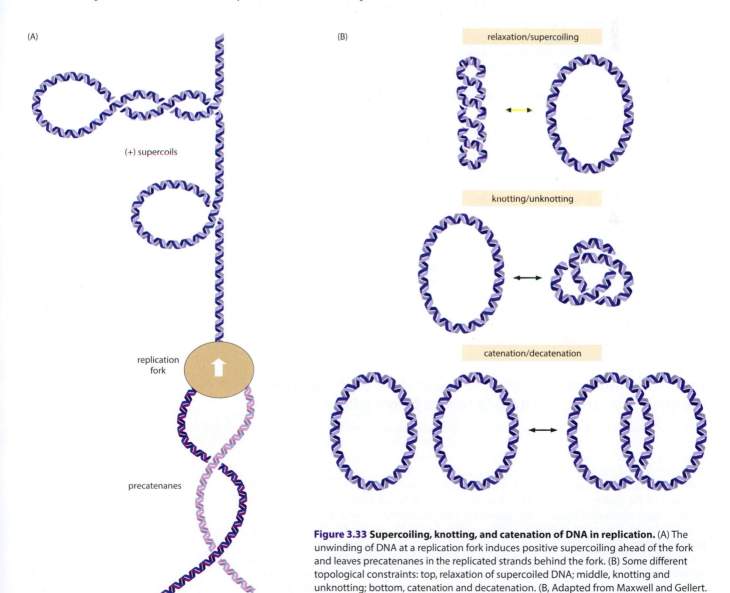

Figure 3.33 Supercoiling, knotting, and catenation of DNA in replication. (A) The unwinding of DNA at a replication fork induces positive supercoiling ahead of the fork and leaves precatenanes in the replicated strands behind the fork. (B) Some different topological constraints: top, relaxation of supercoiled DNA; middle, knotting and unknotting; bottom, catenation and decatenation. (B, Adapted from Maxwell and Gellert. (1986). *Adv. Prot. Chem.* 38: 69–107. With permission from Elsevier.)

replication of a circular DNA chromosome is terminated (as described above), it leaves the two daughter chromosomes linked together (catenated). Difficulties such as these occasioned some early skepticism of semi-conservative replication of DNA; the hydrodynamic and energetic problems appeared insuperable, although the biochemical evidence was increasingly incontrovertible.

DNA topoisomerases overcome topological constraints in DNA replication

The answer to the problems of DNA topology during replication awaited the discovery of a class of enzymes, the **topoisomerases**, which can resolve supercoiling, catenation, and knotting (Figure 3.33B). Topoisomerases are able to change the linking number (Lk) of DNA molecules (see Box 2.3), either by changes in twist (Tw, the number of times one strand wraps round the other in duplex DNA) or more usually in writhe (Wr, the coiling of the DNA axis in space, sometimes equated to supercoiling). Unsupercoiled DNA will have a 'ground state' linking number Lk_m ($\approx N/10.5$, where N is the number of bp and 10.5 = number of bp/turn in dsDNA *in vivo*). The great compaction of DNA in bacterial nucleoids or eukaryotic chromosomes is achieved by changing its linking number, chiefly by changes in Wr. Bacterial and archaeal circular chromosomes are normally stored with an excess of negative supercoils (see below), which is an underwound state that has the favorable consequence of making it easier for the replicative helicase to melt/unwind the right-handed DNA ahead of the replication fork through its propensity to insert positive supercoils. Moreover, the change in Lk_m associated with the excess of negative supercoils in the parent chromosome stores free energy in the DNA and this is harnessed, at least in part, by topoisomerases in their catalysis of topological changes.

DNA topoisomerases fall into two classes with similar but different mechanisms

To achieve its effect, a topoisomerase must transiently cleave one or both strands of the duplex DNA. Two mechanisms have been encountered: one is of an enzyme (type I) that cuts one of the strands of the duplex, allowing the unbroken strand to pass through the gap (strand passage) or to swivel about the phosphodiester bond that remains intact opposite the nick (controlled rotation); the other (type II) is simultaneous cleavage of both strands, thereby allowing another dsDNA to pass through the cut (strand passage again) before both cuts are resealed. All organisms contain at least one type I and one type II enzyme.

It turns out that in both types, there is transient covalent DNA–protein linkage in the form of a mixed anhydride bond between a terminal phosphate group on the nicked DNA and the phenolic -OH group of a specific Tyr residue in the enzyme active site (**Figure 3.34A**). Each enzyme type can be further subdivided into two subtypes (A and B) on the basis of mechanism and sequence similarities; moreover, different topoisomerases carry out different functions *in vivo* (**Table 3.4**). Type I enzymes are all monomeric (single-strand cleavage), whereas type II enzymes are all dimers or heterotetramers (consistent with double-strand cleavage).

Type IA topoisomerases cleave one strand of the duplex and pass the other strand through the gap created before religation

Type IA topoisomerases are found in bacteria, archaea, and eukaryotes. They operate by formation of a transient covalent link between a protein Tyr -OH group and the 5′-phosphate of the DNA strand at the cleavage point, followed by passage of the opposing strand through the newly created gap before it is resealed (Figure 3.34B). The proteins are padlock-shaped, with four distinct domains surrounding a large (27 Å) hole through the center. A crystal structure of the catalytic core (67 kDa, residues 2–595) of *E. coli* topo I complexed with a 13-mer oligonucleotide (**Figure 3.35**) shows the ssDNA cleaved but with the 5′-phosphate of the 3′-terminal product in covalent linkage with Tyr319 in domain III. The 5′-terminal product is bound in a deep groove across domains I, III, and IV, and four phosphate groups (positions +1, −1, −2, and −4) interact with four Arg side chains (Arg321, 493, 195, and 507). Specificity for phosphodiester bond cleavage is due to a cytosine base at the −4 position relative to the cleavage site being uniquely able to fit snugly into a pocket with a stacking interaction with Tyr177. Allied to this is a network of H-bonds and ionic interactions involving

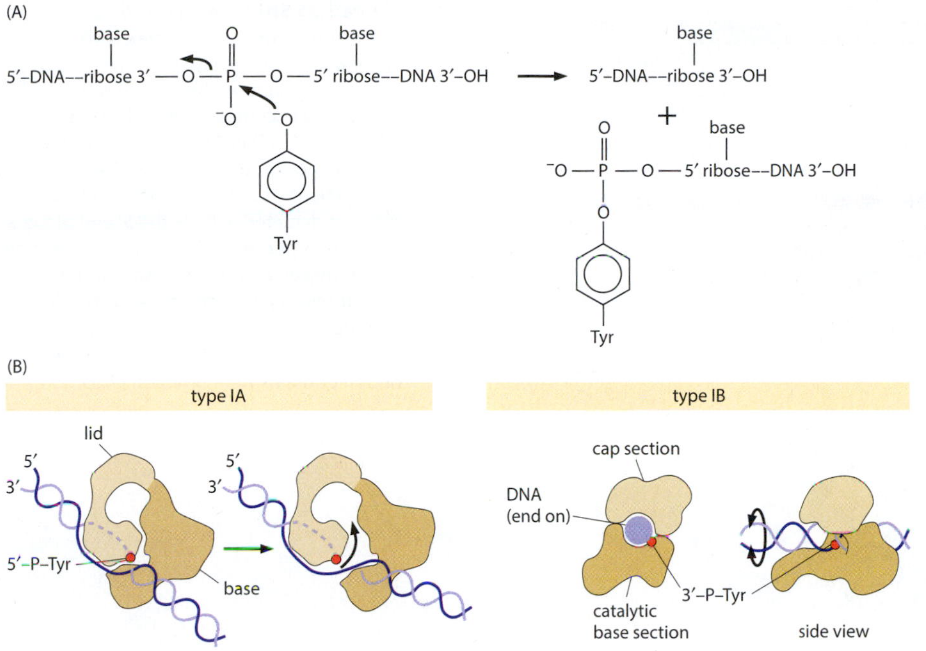

(A)

(B)

Figure 3.34 Cleavage of DNA strands by topoisomerases and mechanisms of type I enzymes. (A) DNA strand cleavage by attack of a Tyr -OH group on a phosphodiester bond, generating a mixed anhydride with the Tyr. In this instance, the leaving group is the 3′-OH group of the deoxyribose of the 5′ fragment and the 3′ fragment is covalently attached at its 5′ end. The reaction can be reversed to reconstitute the phosphodiester bond. The participation of a general acid (see Box 3.1) is omitted for clarity. In other instances, the leaving group is the 5′-OH group of the deoxyribose of the 3′ fragment, and a covalent bond is formed with the 3′-OH group of the 5′-terminal fragment. (B) Type IA topoisomerases use the 5′-phosphoTyr attachment and a strand passage mechanism, as indicated by the black arrow. Type IB topoisomerases use a 3′-phosphoTyr attachment and a controlled rotation mechanism, as indicated by the double-headed black arrow. (B and C, adapted from A.D. Bates and A. Maxwell, DNA Topology. OUP, 2005. With permission from Oxford University Press.)

Glu115, Arg168, and Asp172 and the deoxyribose rings at positions −2 and −3, which also help orient the DNA chain for cleavage. A triad of acidic residues (Asp111, Asp113, and Glu115) are involved in divalent metal coordination, and Glu9—aided by His365—may serve as the general acid in the attack of Tyr319 on the scissile phosphate.

A plausible mechanism for type IA topoisomerases is that a susceptible region of dsDNA melts, and the exposed 3′–5′ single strand binds across the lid segment, accompanied by remodeling of the active site. Once the binding groove is filled, cleavage occurs and the 5′-phosphate bond is formed with Tyr319. Movements of the protein domains (domain I has a much higher temperature factor in the cleaved complex state than in the apoenzyme) opens a gate into the interior, the unpaired partner strand enters, and the gate then closes. This leaves the uncleaved strand on the other side. The cleaved strand can now be religated by a reverse of the cleavage reaction, and a second gate opening and closing allows the unpaired strand to exit. The newly restored dsDNA has undergone a change in Lk of 1, and the enzyme is now reset for another catalytic cycle. The Zn ribbon domains may contribute to stabilizing the structure with ssDNA, although their number (from zero to five) and their retention of the tetracysteine motif required to bind Zn^{2+} varies between bacterial species.

Table 3.4 Classification of DNA topoisomerases

Type	Examples	5′- or 3′-phosphate link	Mechanism	Relax supercoils	Add supercoils	Catenate or decatenate
IA	*E. coli* topo I, III	5′	Strand passage	(−) only		
	Archaeal reverse gyrase			(−)	ATP (+)	
IB	Human topo I	3′	Controlled rotation	(−)	ATP (+)	
IIA	*E. coli* gyrase	5′	Strand passage	(+), (−)	ATP (−)	Inefficient
	E. coli topo IV					Efficient
	Human topo II					Efficient
IIB	Archaeal and plant topo VI	5′	Strand passage	(+), (−)		Efficient

(+), (−) indicate the direction of supercoils introduced. ATP is required to add supercoils where shown. There is also a type IC enzyme, topo V, but it is thus far restricted to the hyperthermophilic archaeon *Methanopyrus kandleri*. It resembles class IB enzymes in employing controlled rotation as a mechanism.

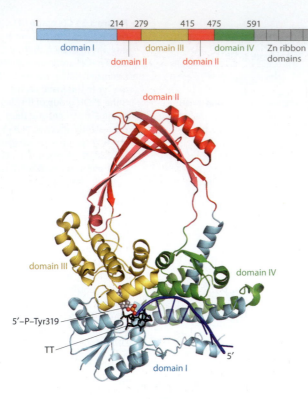

domain II

domain III

domain IV

5′–P–Tyr319

TT

5′

domain I

Figure 3.35 Structure of *E. coli* topoisomerase IA in complex with ssDNA. The domain layout is shown above the structure. The protein was crystallized in complex with a 13-base (5′AATGCGCT-TTGGG3′, hyphen indicates cleavage point) strand of DNA. The noncovalently bound 5′ fragment of this strand is depicted in blue; the 5′ end of the 3′ fragment represented by the TT dinucleotide (black) makes a covalent bond with Tyr139. The protein crystallized carried a D111N mutation, which inhibits the rejoining of the two parts of the DNA fragment, and the Zn ribbon domains were omitted. (PDB 3PX7)

Note that both new ends of the DNA at the cleavage site are carefully protected, one by covalent linkage to a Tyr residue, the other by tight binding in the ssDNA groove. This ensures that they do not become frayed or otherwise damaged and are in the right position for the religation reaction. Moreover, because a segment of ssDNA is necessary to initiate the cleavage, underwound (that is negatively supercoiled) dsDNA will be the preferred substrate, as observed *in vivo*.

A related form of type IA topoisomerase, reverse gyrase, is found only in hyperthermophilic archaea. The C-terminal half of the chain is typical of the topoisomerase IA structure, but the N-terminal half includes two RecA-like helicase domains (SF2 group, see Box 3.1). Reverse gyrase binds and hydrolyzes ATP but displays no helicase activity. The energy derived from ATP breakdown is required for its biological activity: the insertion of positive (+) supercoils. Positive supercoiling diminishes the propensity of dsDNA to unwind, and this may be beneficial in maintaining the integrity of the archaeal genome during hyperthermophilic growth.

Type IB topoisomerases cleave one strand of the duplex and allow part of the duplex DNA to undergo controlled rotation before religation

Type IB topoisomerases are largely if not entirely confined to eukaryotes (and poxviruses). Their mechanism entails cleavage of one strand of the substrate dsDNA, attachment of the 3′-OH end of the 5′ fragment through its phosphate group to a Tyr residue, and then a controlled rotation about the intact phosphodiester bond in the other strand (Figure 3.34B). The structure of the human topo IB is shown in **Figure 3.36**. The DNA is tightly clamped by the encircling domains of the protein, which line the hole with positive charges. One strand of the dsDNA is cleaved and its 3′ end anchored in covalent linkage to the active site Tyr, with Arg488, Arg590, and His632 all thought to help stabilize the transition state and Lys532 acting as general acid. The clamp seen in the DNA-bound structure must open for the DNA to enter and ~10 bp of DNA are enveloped by the protein, sealed in by the 'lips' that extend from domains I and III.

The part of the DNA projecting out of the hole is left free to rotate, and does so, at least in part, by harnessing the free energy of the release of (+) or (−) supercoiling. To allow

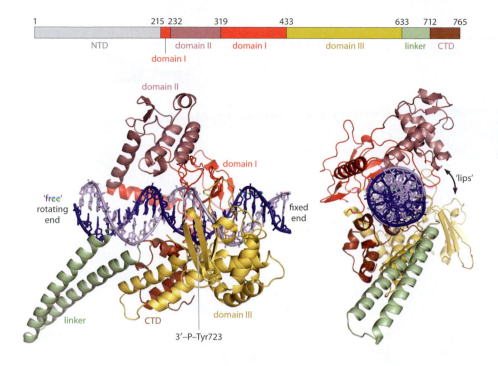

Figure 3.36 Structure of human topoisomerase IB in complex with dsDNA. The domain layout is shown above the structure. The N-terminal domain (NTD) is protease-sensitive (possibly disordered). It does not participate in the reaction, at least *in vitro*, but contains nuclear localization signals. The remainder of the protein (residues 175–765) was crystallized in complex with a 22 bp fragment of DNA. Left, side view of structure. Right, 90° rotation to show DNA end on. One strand is cleaved, becoming anchored by a covalent bond between its new 3′-phosphate and Tyr723. The portion of DNA projecting beyond the protein is now free to swivel/rotate, as indicated, before the original DNA phosphodiester bond is restored. (PDB 1A36)

the enzyme to reseal the broken strand, the two ends must align again in the active site. This could be a temporary halt imposed at some intermediate stage by friction within the enzyme or by the DNA becoming fully relaxed. Helices projecting out from domain III present a positively charged surface that could interact with the rotating DNA; the linker region may also be important in that a functional enzyme created without linkers has substantially altered religation kinetics. When the dsDNA has been reconstituted, the protein clamp must open to allow it to exit the enzyme. This mechanism explains how topo IB can relax (+) or (−) supercoiling and generate multiple integral changes in Lk, with experiments typically indicating ΔLk of ± 5.

Type II topoisomerases cut both strands of dsDNA and pass one segment of DNA through the gap created

Type II topoisomerases work by means of a strand passage mechanism, but one in which both strands of a segment (G segment) of the target DNA are cleaved and another segment (T segment) is passed through the gap thereby created (**Figure 3.37A**). For such a mechanism, ΔLk = 2 for each turnover. Even though the supercoiling of the DNA may favor relaxation, the enzymes also use the free energy of hydrolysis associated with the breakdown of two molecules of ATP to drive the reaction, and while type II enzymes relax supercoils, DNA gyrase can also create them. As mentioned earlier, all type II topoisomerases are dimers or tetramers and they are the most widely found topoisomerases throughout all three domains of life. DNA gyrase of *E. coli* (a GyrA$_2$GyrB$_2$ heterotetramer) is one of the most intensively studied, categorized as a type IIA enzyme to distinguish it and related enzymes from the more recently discovered topo VI of bacteria, plants, and algae, which has some significant differences and has been designated type IIB. Another widely studied type IIA enzyme is the topo II of *S. cerevisiae*, a homodimer.

Gyrase is encoded by two genes, *gyrA* and *gyrB* (**Figure 3.38**). The monomer of topo II of *S. cerevisiae* and the gyrase B and A proteins are homologous and can be closely aligned, but the yeast enzyme lacks the CTD at the end of Gyr A. Structural and biochemical studies of these and other type II enzymes, in DNA-free and DNA-bound forms, have led to a growing understanding of their molecular mechanism. GyrB (90 kDa) comprises three domains: the N-terminal domain is an ATPase, a member of the GHKL superfamily (so called because it is found in gyrase, the molecular chaperone Hsp90, CheA-type histidine kinases, and the DNA mismatch repair enzyme, MutL), followed by a transducer domain which is thought to transmit signals to GyrA through the third (C-terminal) domain (~47 kDa) of GyrB. The structure of the GHKL and transducer region (~43 kDa) is depicted in

(A)

(B)

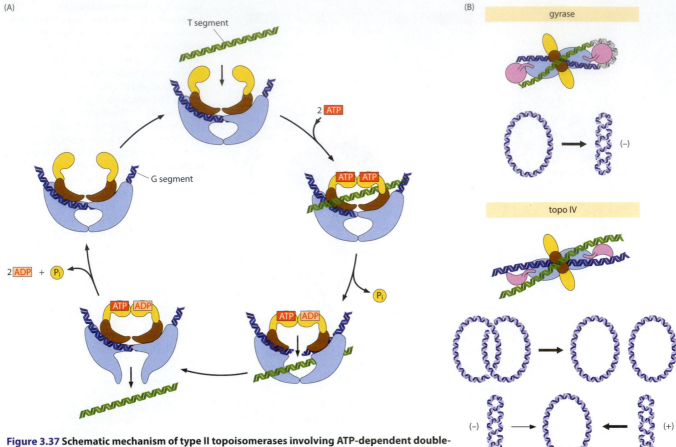

Figure 3.37 Schematic mechanism of type II topoisomerases involving ATP-dependent double-strand cleavage of DNA and passage of another duplex through the gap. (A) The topoisomerase GyrA dimer (blue) binds a G segment of DNA and bends it, and the GyrB dimer (yellow and brown) binds another T segment. The binding of two ATP molecules causes the gate through which the T segment entered to close. One of the ATP molecules is hydrolyzed, and both strands of the G segment are severed, allowing the T segment to pass through the gap created into a cavity in the GyrA dimer. The GyrA gate opens and the T segment exits, whereupon the remaining ATP bound is hydrolyzed and the products are released, resetting the enzyme to bind another T segment. $\Delta Lk = 2$. (B) The binding of dsDNA to two different type II topoisomerases, gyrase and topo IV, showing how their C-terminal domains (CTD, lilac) bind DNA differently. This causes gyrase to favor the insertion of negative (−) supercoils into its target DNA, whereas topo IV much prefers to decatenate DNA circles and relax negatively and especially positively (+) supercoiled DNA. (B, adapted from K.C. Neuman, *Proc. Natl Acad. Sci. USA* 107:22363–22364, 2010.)

Figure 3.38. The entry gate (N gate) for the T segment of DNA can be identified where the two GHKL motifs come together across the twofold axis of the dimer, and the ATP-binding sites reside. Binding of ATP induces the dimerization, and the N gate is closed by the 15 aa extension (a 'strap') from the N terminus of each GHKL domain The strap in each subunit, in which the hydrophobic Ile10 plays an important part, binds across the twofold axis to its partner, sequestering the DNA segment in a cavity ~24 Å across, big enough to accept a DNA duplex.

The G segment of DNA binds at the DNA gate in the GyrA dimer (Figure 3.38), where Tyr122 in a winged-helix domain (WHD) in each monomer (Tyr782 in the yeast topo II) forms the 5′ P-Tyr bond in the cleavage reaction with the DNA. The conserved residues Arg121 (from the same monomer) and His80, Arg32, and Lys42 (from the opposite monomer) form a cluster of positive charges in the active site, and Arg46 and Arg47 are available to anchor the noncovalently linked 3′ end of the cleaved DNA. The G segment (~26 bp in the binding site) is substantially bent (as much as 150° overall in the yeast enzyme) on binding, adopting an A-form in the center but retaining B-form at the ends, and is cleaved with a four-base overhang on each side. The hydrolysis of the bound ATP molecules is asynchronous, in that one ATP is hydrolyzed at this stage, one P_i is released, the WHDs separate

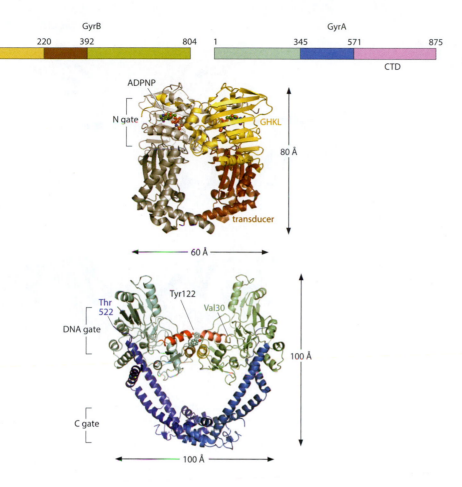

Figure 3.38 Structure of the major catalytic fragments of GyrB and GyrA of *E. coli* DNA gyrase. The color coding is as in Figure 3.37.

(along with other domain movements), and the T segment passes through into a cavity ~34 Å across, a little bigger than that in GyrB. This allows the G segment to be resealed and the DNA gate to close again, ensuring unidirectionality. The molecular details of this mechanism, and the involvement of the transducer domain, are as yet poorly understood.

The presence of the remaining bound ATP may ensure that the N gate remains closed at least until the T segment has passed through the DNA gate and, maybe, until the catalytic cycle has been completed. DNA binding transmits structural changes to the C gate, associated with the opening of the gap that allows the T segment to escape. The second bound ATP is hydrolyzed, and two ADP and a P_i are released, causing the N gate to reopen, ready to receive another T segment for the process to begin again.

The C-terminal domains of bacterial topoisomerase IIA enzymes impose their specific biological functions

In *E. coli* and most other bacteria, there is a second type II topoisomerase, topo IV. Gyrase and topo IV are homologous, have the same basic structure (topo IV is a $ParC_2ParE_2$ heterotetramer) and follow the same mechanistic pathway, but they have different activities *in vivo*. The function of gyrase is chiefly to introduce negative supercoils into DNA, which are important to the condensation of the bacterial chromosome (see above). Topo IV, on the other hand, is associated with the removal of positive supercoils and decatenation. These differences can be traced back to the CTD in the GyrA protein and its counterpart in topo IV ParC, which are involved in binding the DNA that is acted on by their respective enzymes.

The isolated CTD of gyrase is able to bind DNA and bend it by 180° whereas that of topo IV bends it much less. A crystal structure of the ParC dimer shows the CTD attached to the core part of gyrase by a linker region (-PSEP-) close to the DNA gate (**Figure 3.39A**). In this state of this enzyme, the N gate is open wider than the same gate in GyrA and the two catalytic Tyr residues are 33 Å apart. The CTD in gyrase may be further displaced from

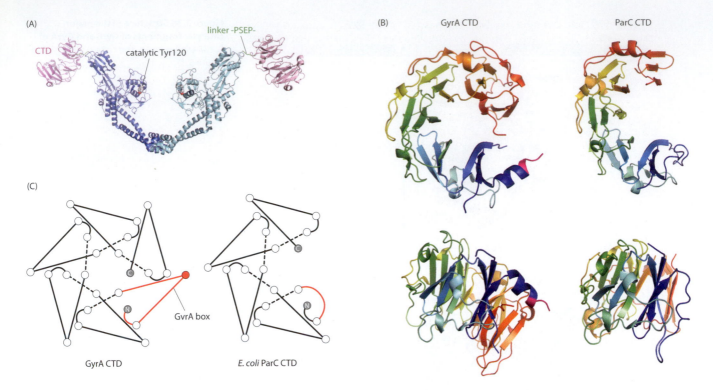

Figure 3.39 Structure of the C-terminal domains (CTDs) of gyrase and topoisomerase IV of *E. coli*. (A) Structure of ParC dimer of *E. coli* topo IV. The core (catalytic) part is depicted in blue and the CTDs are in lilac. The linker region that attaches a CTD to the core part is in green. (B) Upper, plan view of the 'pinwheel' CTDs from GyrA and ParC. Each protein is color-coded from indigo (N terminus) to red (C terminus) and the six blades in GyrA and four in ParC essentially coincide with the different rainbow colors. In GyrA, the first four residues (QRRG) of the Gyr Box sequence motif in the N-terminal blade are picked out in red; the remainder (GKG) are missing because residues Gly563–Glu575 are disordered. The truncated N-terminal blade in ParC can be seen, as can the lack of blade 6. Lower, side views in which the spiral nature of the GyrA CTD is evident, in contrast with the flatter shape of the ParC CTD. (C) A schematic view of the structures in (B), illustrating the lack of the sixth blade and the GyrA motif in the truncated N-terminal blade in the ParC CTD. The blade containing the GyrA box is depicted in red. (PDB 1ZVU (ParC dimer), 1ZI0 (GyrA CTD), 1ZVT (ParC CTD)) (Adapted from K.D. Corbett et al., *J. Mol. Biol.* 351:545–561, 2005. With permission from Elsevier.).

the core and more flexibly attached, possibly reaching down to the C gate. The structure of the gyrase CTD of *E coli* is shown in Figure 3.39B. It has a pinwheel structure, with six repeating 'blades', each comprising a four-stranded anti-parallel β sheet. The ring is not flat, but the successive blades are arranged in a rising right-handed spiral (lock-washer) arrangement. The outer edge is rich in positively charged side chains, obviously suitable for binding a DNA wound round in the same spiral. Biochemical experiments indicate that a conserved motif (-QRRGGKG-, the GyrA box) located in the most N-terminal of the GyrA CTD blades is of major importance for the supercoiling activity of gyrase, presumably affecting the way the DNA is bound. Other bacterial CTDs are very similar, but the rings may be flat, and differ also in the number of blades and whether they are whole or partial (Figure 3.39C). The ParC of *E. coli* topo IV is a typical ParC in that it lacks the GyrA box and has only four intact blades.

A likely mechanism, based on these structures and the known biological outcomes, is that gyrase operates by a CTD interacting with a G segment of DNA as it juts out from the DNA gate, bending the DNA back round and upwards towards the N gate. Assuming some flexibility in the linker region attaching it to the core enzyme, the CTD may accompany the bent DNA as the T segment passes through the N gate and exits from the C gate (Figure 3.37B). This imposes a left-handed cross-over geometry in the DNA, consistent with the known outcome of introducing negative (–) supercoils and a ΔLk of –2. With topo IV, in contrast, the difference in the CTD means it is unable to bend the DNA round and back but, being also less flexibly attached, associates with the G strand as it juts out from the core enzyme. The T segment therefore enters the N gate in *trans*, which makes decatenations possible and promotes relaxation of positive (+) rather than negative (–) supercoils (Figure 3.37B). The importance of the CTD and GyrA box has been emphasized by the conversion of a topo IV-like topoisomerase from *A. aeolicus* to a gyrase by replacing its CTD with one that contains an intact GyrA box.

Topoisomerases have become popular targets for cytotoxic and therapeutic agents. For example, the fluoroquinolones, such as ciprofloxacin, stabilize the enzyme-bound dsDNA break in bacterial gyrase and are famously antibacterial. The anti-cancer agent camptothecin stabilizes the cleavage complex of human topo I, and etoposide acts in the same way on human topo II. Novobiocin inhibits gyrase by blocking the binding of ATP. In general, topoisomerase inhibitors are good chemotherapeutic drugs because of their unique mechanism in stabilizing the covalent complex between protein and DNA.

3.5 SUMMARY

DNA is replicated by copying each parental strand to create a complementary daughter strand, thereby creating two new double helices in place of one. This process of semi-conservative replication is catalyzed by a multicomponent molecular machine known as a replisome. Replication begins at a single starting point (*OriC*) on bacterial circular DNA but has multiple origins on archaeal and the much larger eukaryotic chromosomes. Eukaryotic DNA replication is highly regulated so that each origin of replication needs to be licensed to initiate; copying of DNA is restricted to once per cell cycle.

Helicases are ATPases that propagate the unwinding of DNA away from the origin of replication (5′–3′ in bacteria, 3′–5′ in archaea and eukaryotes). This forms a structure known as a replication fork, with each strand of parental DNA separated and ready for copying by DNA polymerases that form part of the replisome. DNA polymerases work in the 5′ to 3′ direction, synthesizing continuously on the leading strand, but discontinuously on the lagging strand, generating a succession of Okazaki fragments. The processivity of DNA polymerases is enhanced by a protein sliding clamp, which is assembled around the DNA by a clamp loader, and their high fidelity of copying is enhanced by a 3′–5′ proofreading activity. DNA polymerases require a primer, which is synthesized by a further enzyme, primase. These primers are then removed and the gaps created between the Okazaki fragments filled in by RNase H/DNA Pol I (in bacteria) and by RNase H/DNA Pol δ/FEN1 (in eukaryotes); DNA ligases join the pieces together. Bacterial circular DNA replication terminates at a region diametrically opposite the origin of replication.

Linear DNA molecules in eukaryotic chromosomes have a special structure, a telomere, at their 3′ ends, comprising multiple repeats of a short G/T-rich motif with an ssDNA overhang. To avoid continual shortening, telomeric DNA is synthesized and maintained by telomerase, a polymerase that uses an RNA template. Topoisomerases act to relieve the topological constraints implicit in DNA replication (superhelical turns introduced when helicases unwind a helical dsDNA, concatenations in replicated circular DNA, and so on). Type I enzymes temporarily break one strand of the DNA, allowing either the unbroken strand to pass through the gap or a controlled rotation of dsDNA around the breakpoint. Type II topoisomerases break both strands, allowing dsDNA through the cut before it is resealed. In eukaryotes, individual replicons (units of DNA replication) come together to form factories of between 20 and 200 replication forks, with about 1000 factories in mammalian cells at the beginning of S phase.

Sections of this chapter are based on the following contributions

3.2   Stephen D. Bell, Indiana University at Bloomington, USA.
3.4   Anthony Maxwell, John Innes Centre, Norwich, UK.

Expert revision of this chapter provided by

David Jeruzalmi, City College of New York, New York, USA.

REFERENCES

General

Bates AD & Maxwell A (2005) *DNA Topology*. Oxford University Press.

DePamphilis ML & Bell SD (2011) *Genome Duplication*. Garland Science.

Duderstadt KE & Berger JM (2008) AAA+ ATPases in the initiation of DNA replication. *Crit Rev Biochem Mol Biol* 43:163–187.

McHenry CS (2003) Chromosomal replicases as asymmetric dimers: studies of subunit arrangement and functional consequences. *Mol Microbiol* 49:1157–1165.

Meselson M & Stahl FW (1958) The replication of DNA in *Escherichia coli*. *Proc Natl Acad Sci USA* 44:671–682.

Taylor JH, Woods PS & Hughes WL (1957) The organization and duplication of chromosomes as revealed by autoradiographic studies using tritium-labeled thymidine. *Proc Natl Acad Sci USA* 43: 122–128.

Watson JD & Crick FHC (1953) Molecular structure of nucleic acids; a structure for deoxyribose nucleic acid. *Nature* 171:737–738.

Yao NY & O'Donnell M (2009) Replisome structure and conformational dynamics underlie fork progression past obstacles. *Curr Opin Cell Biol* 21:336–343.

3.2   Initiation and elongation

Bailey S, Eliason WK & Steitz TA (2007) Structure of hexameric DnaB helicase and its complex with a domain of DnaG primase. *Science* 318:459–462.

Bailey S, Wing RA & Steitz TA (2006) The structure of *T. aquaticus* DNA polymerase III is distinct from eukaryotic replicative DNA polymerases. *Cell* 126:893–904.

Bowman A, Ward R, Wiechens N et al (2011) The histone chaperones Nap1 and Vps75 bind histones H3 and H4 in a tetrameric conformation. *Mol Cell* 41:398–408.

Corn JE, Pelton JG & Berger JM (2008) Identification of a DNA primase tracking site redefines the geometry of primer synthesis. *Nat Struct Mol Biol* 15:163–169.

Cunningham Dueber EL, Corn JE, Bell SD & Berger JM (2007) Replication origin recognition and deformation by a heterodimeric archaeal Orc1 complex. *Science* 317:1210–1213.

Das C, Tyler JK & Churchill MEA (2010) The histone shuffle: histone chaperones in an energetic dance. *Trends Biochem Sci* 35:476–489.

Ellenberger T & Tomkinson AE (2008) Eukaryotic DNA ligases: structural and functional insights. *Annu Rev Biochem* 77:313–338.

Enemark EJ & Joshua-Tor L (2006) Mechanism of DNA translocation in a replicative hexameric helicase. *Nature* 442:270–275.

Evans RJ, Davies DR, Bullard JM et al (2008) Structure of PolC reveals unique DNA binding and fidelity determinants. *Proc Natl Acad Sci USA* 105:20695–20700.

Georgescu RE, Kim S-S, Yurieva O et al (2008) Structure of a sliding clamp on DNA. *Cell* 132:43–54.

Guo Z, Zheng L, Xu H et al (2010) Methylation of FEN1 suppresses nearby phosphorylation and facilitates PCNA binding. *Nat Chem Biol* 6:766–773.

Hamdan SM, Loparo JJ, Takahashi M et al (2009) Dynamics of DNA replication loops reveal temporal control of lagging-strand synthesis. *Nature* 457:336–339.

Ilves I, Petojevic T, Pesavento JJ & Botchan MR (2010) Activation of the MCM2-7 helicase by association with Cdc45 and GINS proteins. *Mol Cell* 37:247–258.

Lee J-B, Hite RK, Hamdan SM et al (2006) DNA primase acts as a molecular brake in DNA replication. *Nature* 439:621–624.

McNally R, Bowman GD, Goedken ER et al (2010) Analysis of the role of PCNA-DNA contacts during clamp loading. *BMC Struct Biol* 10:3.

Nossal NG, Makhov AM, Chastain PD et al (2007) Architecture of the bacteriophage T4 replication complex revealed with nanoscale biopointers. *J Biol Chem* 282:1098–1108.

Oyama T, Ishino S, Fujino S et al (2011) Architectures of archaeal GINS complexes: essential DNA replication factors. *BMC Biol* 9:28.

Pascal JM (2008) DNA and RNA ligases: structural variations and shared mechanisms. *Curr Opin Struct Biol* 18:96–105.

Pietroni, P & von Hippel PH (2008) Multiple ATP binding is required to stabilize the "activated" (clamp open) clamp loader of the T4 DNA replication complex. *J Biol Chem* 283:28338–28353.

Remus D & Diffley JFX (2009) Eukaryotic DNA replication: lock and load, then fire. *Curr Opin Cell Biol* 21:1–7.

Remus D, Beuron F, Tolun G et al (2009) Concerted loading of Mcm2-7 double hexamers around DNA during DNA replication origin licensing. *Cell* 139:719–730.

Shereda RD, Kozlov AG, Lohman TM et al (2008) SSB as an organizer/mobilize of genome maintenance complexes. *Crit Rev Biochem Mol Biol* 43:298–318.

Simonetta KR, Kazmirski SL, Goedken ER et al (2009) The mechanism of ATP-dependent primer-template recognition by a clamp loader complex. *Cell* 137:659–671.

Singleton MR, Dillingham MS & Wigley DB (2007) Structure and mechanism of helicases and nucleic acid translocases. *Annu Rev Biochem* 76:23–50.

Spiering MM, Nelson SW & Benkovic SJ (2008) Repetitive lagging strand DNA synthesis by the bacteriophage T4 replisome. *Mol BioSyst* 4:1070–1074.

Thomsen ND & Berger JM (2008) Structural frameworks for considering microbial protein and nucleic-acid dependent motor ATPases. *Mol Microbiol* 69:1071–1090.

Wigley DB (2008) ORC proteins: marking the start. *Curr Opin Struct Biol* 19:72–78.

Wing RA, Bailey S & Steitz RA (2008) Insights into the replisome from the structure of a ternary complex of the DNA polymerase III α-subunit. *J Mol Biol* 382:859–869.

Zegerman P & Diffley JFX (2010) Checkpoint-dependent inhibition of DNA replication initiation by Sld3 and Dbf4 phosphorylation. *Nature* 467:474–478.

Zheng L & Shen B (2011) Okazaki fragment maturation: nucleases take centre stage. *J Mol Cell Biol* 3:23–30.

3.3   Termination of DNA replication

Mason M, Schuller A & Skordalakes E (2011) Telomerase structure function. *Curr Opin Struct Biol* 21:92–100.

Phan AT (2010) Human telomeric G-quadruplex: structures of DNA and RNA sequences. *FEBS J* 277:1107–1117.

Wyatt HDM, West SC & Beattie TL (2010) InTERTpreting telomerase structure and function. *Nucleic Acids Res* 38:5609–5622.

3.4   DNA topology in replication

Corbett KD, Schoeffler AJ, Thomsen ND & Berger JM (2005) The structural basis for substrate specificity in DNA topoisomerase IV. *J Mol Biol* 351:545–561.

Fu G, Wu J, Lui W et al (2009) Crystal structure of DNA gyrase B' domain sheds lights on the mechanism for T-segment navigation. *Nucleic Acids Res* 37:5908–5916.

Schoeffler AJ & Berger JM (2008) DNA topoisomerases: harnessing and constraining energy to govern chromosome topology. *Q Rev Biophys* 41:41–101.

Tretter EM, Lerman JC & Berger JM (2011) A naturally chimeric Type IIA topoisomerase in *Aquifex aeolicus* highlights an evolutionary path for the emergence of functional paralogs. *Proc Natl Acad Sci USA* 107:22055–22059.

Wang JC (2002) Cellular roles of DNA topoisomerases: a molecular perspective. *Nat Rev Mol Cell Biol* 3:430–440.

Zhang Z, Cheng B & Tse-Dinh YC (2011) Crystal structure of a covalent intermediate in DNA cleavage and rejoining by *Escherichia coli* DNA topoisomerase I. *Proc Natl Acad Sci USA* 108:6939–6944.

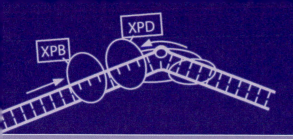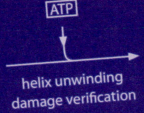

helix unwinding
damage verification

DNA Repair and Recombination

4

4.1   INTRODUCTION

In Chapter 3, we saw the complexity of the molecular machines deployed by a cell to ensure the faithful replication of its DNA. Even so, occasional mistakes (changes in DNA base sequence) slip through and these need to be detected and corrected, if possible. DNA can also be damaged by reactive metabolites or environmental toxicants and such lesions—depending on their severity—could stall the next round of DNA replication, disrupt cell division, or lead to cancer and other diseases. In this chapter, we look at the ways in which cells undertake DNA repair. Considerable resources are expended for this purpose: several percent of the coding capacity of the genome are devoted to the enzymes required to detect and correct lesions and large amounts of ATP are consumed in the process. If the correction fails, there will be a **mutation**, that is, a change of base sequence (substitution, insertion, deletion, etc.). It has been estimated that the genome of a typical human cell gets damaged over 10,000 times every day, but less than 1 in 1000 of these lesions gives rise to a permanent mutation.

As a chemical entity, DNA is relatively stable but even inside the cell it is subject to damage in a variety of ways: from reactive oxygen species generated as by-products of aerobic metabolism (Chapter 15); by aberrant methylation by the methyl group donor S-adenosyl-methionine; by spontaneous hydrolysis of the N-C^1 glycosidic bonds between purine bases and their deoxyribose residues (depurination, which happens about 5000 times per day in a human cell); by deamination by DNA editing enzymes; by spontaneous loss of exocyclic amino groups of bases (deamination, which includes the conversion of cytosine to uracil, the latter occurring about 100 times per day in a human cell). DNA can also be damaged by exogenous factors, including ultraviolet (UV) light; ionizing radiation such as γ-rays; and chemicals in the environment (for example, nitrous acid from nitrite, which deaminates bases and is widely used as a meat preservative, and polycyclic aromatic amines and hydro-carbons, which form bulky adducts and are found in cigarette smoke and exhaust fumes). Some examples of the damage that can be incurred are depicted in **Figure 4.1**.

In a **point mutation** (for example, one base pair substituted with another), we can distinguish transitions (one purine or pyrimidine is replaced by another purine or pyrimidine, for example, A/T replaced with G/C) from transversions (one purine or pyrimidine is replaced by a pyrimidine or purine, respectively, for example, A/T replaced with C/G). Altered/unnatural base pairings caused by DNA damage (such as 8-oxoG) and mismatches left after DNA replication (natural mispairing of A with C and G with T between rare tautomeric forms) can lead to mutations, as shown in **Figure 4.2**. The mutation may pass unnoticed (a silent mutation) if the effect on a gene or gene product is negligible, helped by the existence of multiple codons for some amino acids. Rarely, there may be some advantage conferred, without which there would be no evolution by mutation and selection. More commonly, however, a mutation will be deleterious: there is a strong correlation between the accumulation of mutations and disease, including cancer in mammals.

Lesions can occur in one or both strands of duplex DNA and are repaired by five different enzyme systems

The existence of the two complementary strands in a DNA double helix makes it easy to see how damage to one strand might be repaired by reference to an intact complementary strand. But, in some instances, both strands of a duplex DNA are compromised. This is

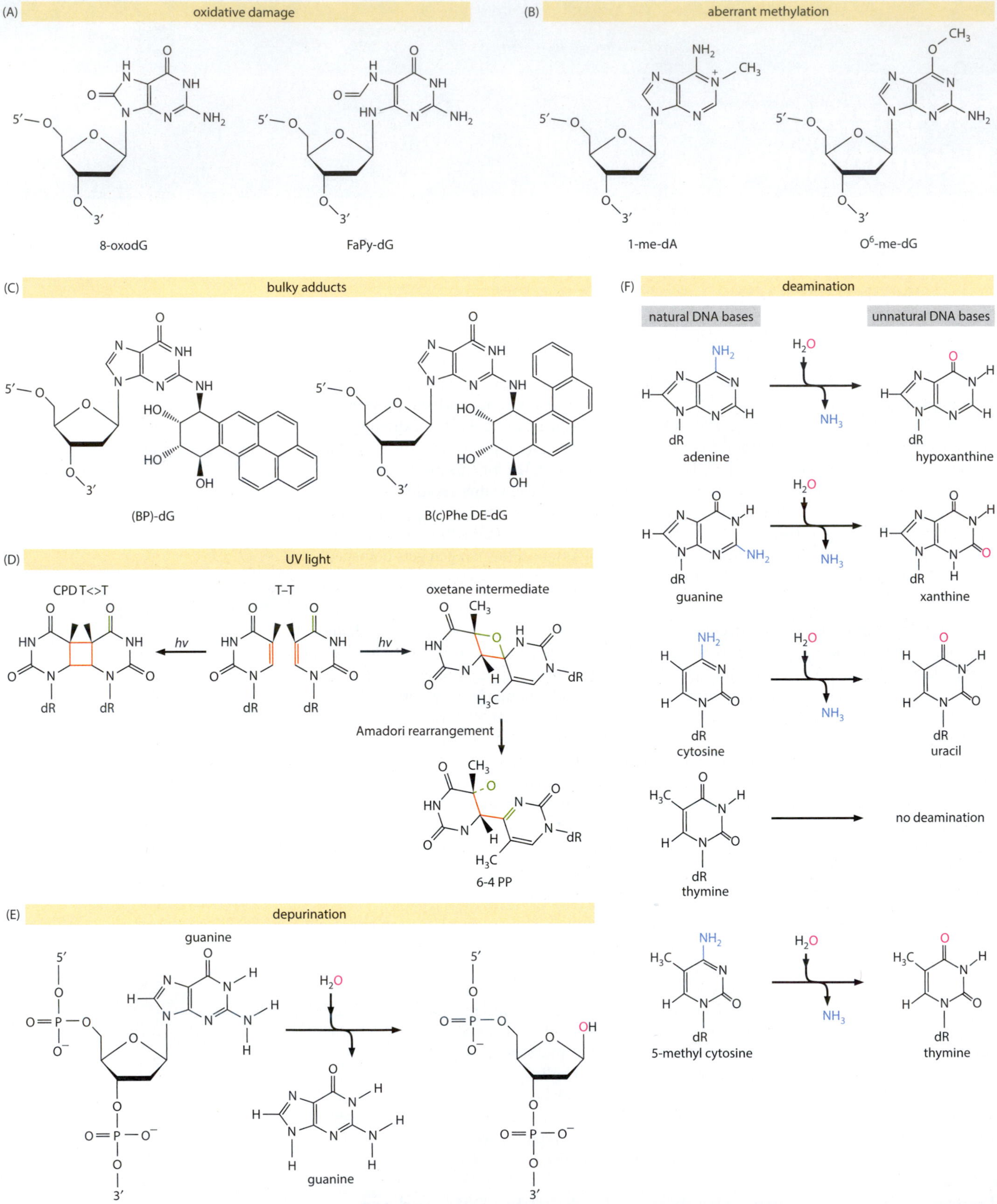

(A)
oxidative damage

8-oxodG

FaPy-dG

(B)
aberrant methylation

1-me-dA

O^6-me-dG

(C)
bulky adducts

(BP)-dG

B(c)Phe DE-dG

(D)
UV light

CPD T<>T

T–T

oxetane intermediate

Amadori rearrangement

6-4 PP

(E)
depurination

guanine

guanine

(F)
deamination

natural DNA bases

unnatural DNA bases

dR
adenine

H_2O

NH_3

dR
hypoxanthine

dR
guanine

H_2O

NH_3

dR
xanthine

dR
cytosine

H_2O

NH_3

dR
uracil

dR
thymine

no deamination

dR
5-methyl cytosine

H_2O

NH_3

dR
thymine

Figure 4.1 (*left*) Some examples of damage that can be incurred by DNA bases. (A) Bases damaged by reactive oxygen species. G has the lowest redox potential and is the prime target. 8-oxodG (8-oxo 7,8-dihydroguanine); FaPy-dG, 2,6-diamino 4-hydroxy 5-formamidopyrimidine. (B) Bases damaged by aberrant methylation: 1-me-dA (1-methyl-dA); O^6-me-dG (O^6-methyl-dG). (C) Bases damaged by bulky environmental adducts: [BP]-dG, (+)-*trans-anti*-benzo[a]pyrene-N^2-dG; B[c]Phe DE-dG, (-)-1R-2S-epoxy-3S-4R-dihydroxy 1,2,3,4-tetrahydrobenzo(c)phenanthrene-N^2-dG. (D) Formation of UV-induced lesions between adjacent pyrimidine bases. Bonds involved in the photochemical reactions between two adjacent thymines are highlighted in red and green. (E) Depurination, spontaneous hydrolysis of the glycosidic bond linking the purine base (A or G) to the 1-position of its deoxyribose. (F) Deamination, spontaneous hydrolysis of the exocyclic amino groups of purine and pyrimidine bases. dR, deoxyribose in N-glycosidic linkage. (D, adapted from L.O. Essen, *Curr. Opin. Struct. Biol.* 16:51–59, 2006. With permission from Elsevier.)

particularly dangerous; the complementary strand is no longer an intact template, and for a DNA double-strand break, the cell faces the additional problem of recognizing the ends of the broken DNA as different from natural ends of chromosomal DNA (Chapter 3).

Different forms of DNA damage are detected and repaired by specialized enzyme systems. Depending on the lesion, the process can involve a single protein or an assembly of up to 30–40 subunits working in concerted fashion. Systems that repair single-strand lesions can either chemically reverse the damage done to the affected base(s) or excise the damaged strand and use the complementary strand as a template to replace the missing nucleotides by gap-filling. Many of the spontaneous modifications to DNA are corrected by **base-excision repair** (BER), involving selective removal of the damaged base and its replacement with an undamaged one. The repair of bulkier modifications of bases, which induce local distortions of the double helix, is catalyzed by the enzymes of **nucleotide-excision repair** (NER), whereby a short segment containing the defective nucleotide is removed and replaced. Repair of mismatched bases arising from errors in DNA replication is catalyzed by the enzymes of **mismatch repair** (MMR), which are remarkable for their ability to sense a mismatch, identify the incorrect base against the parentally correct one, and re-introduce the correct partner by replacing a segment of the affected strand of the double helix. We examine these pathways in more detail below.

Finding the damaged base(s) in the midst of a large excess of unaffected DNA presents the cell with an extraordinary challenge. It appears that many DNA repair proteins recognize the site of DNA damage by exploiting the locally decreased stability of the double helix, caused by partial or complete loss of Watson–Crick base pairing and diminished base stacking. In other instances, a static distortion of the double helix creates a shape that is specifically recognized by the cognate repair enzyme. After initial recognition, there follows a series of actions to expose the damaged DNA to enzymatic activities that vary with the lesion and the type of repair pathway invoked. In dsDNA, the bases are essentially hidden within the double helix and a common feature of the repair systems is exposure of the damaged base by its being 'flipped out.'

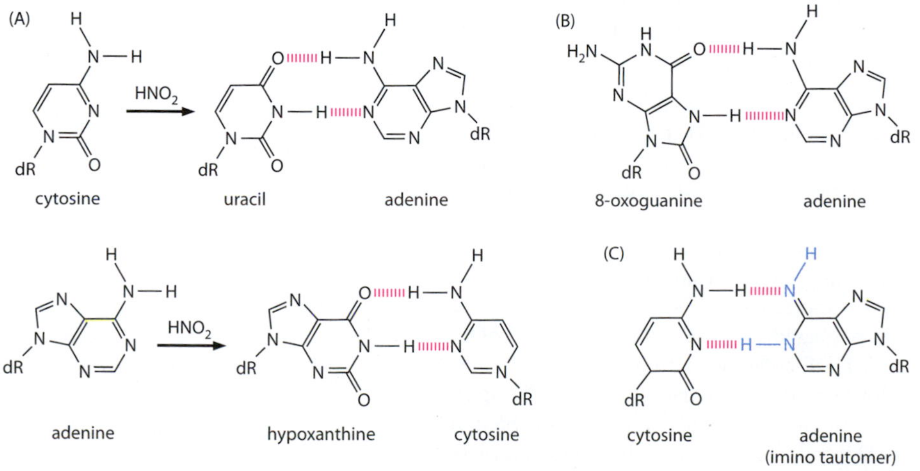

Figure 4.2 DNA damage and errors in DNA replication lead to point mutations. (A) Unnatural DNA bases formed by spontaneous deamination (Figure 4.1) can generate point mutations. Hypoxanthine (deamination product of A) can base-pair with C (A/T goes to C/G) and uracil (deamination product of C) can base-pair with A (C/G goes to T/A). Thymine (deamination product of 5-methyl-C) can base-pair with A (C/G goes to T/A) (not shown). (B) 8-OxoG, a damaged base generated by reactive oxygen species, can form a Hoogsteen base pair with A. This resembles a T/A base pair sufficiently closely to pass scrutiny by the DNA polymerase, generating a G/C to T/A transversion and explaining the highly mutagenic effect of 8-oxoG. (C) A natural but very rare tautomeric form of A (the transfer of a proton from the exocyclic amino group at position 6 to the ring N^1, shown in blue) can form an unnatural but acceptable base pair with C, generating a spontaneous A/T to G/C transition. dR, deoxyribose in N-glycosidic linkage.

Double-strand breaks are detected and repaired by the alternative processes of **nonhomologous end joining** (**NHEJ**) and **homologous recombination** (**HR**). HR has additional roles in transferring DNA strands between homologous DNA molecules during reactivation of stalled replication forks (Chapter 3) and in meiotic cell division (Chapter 13). HR should be distinguished from a second type of recombination, **site-specific recombination**, which allows certain nucleotide sequences (called **mobile genetic elements**) to move about the genome—a remarkable process that over evolutionary history has been responsible for much of what we now see as important rearrangements of DNA (Section 4.5). It is further manifested in the way in which some viruses move their genomes in and out of the chromosomes of their host cells and, in vertebrates, in the generation of antibody diversity (Chapter 17).

HR repairs DNA double-strand breaks by an error-free process that utilizes a homologous chromosome as a template for repair. This is the primary mechanism for repairing double-strand breaks in dividing cells, which produce two copies of the genome during the S phase of the cell cycle. In nondividing cells such as neurons, which have only one copy of each chromosome, DNA breaks are fixed by the error-prone NHEJ pathway. NHEJ consists of resecting the broken DNA ends and pasting them back together with some loss of genetic information at the site of the repair.

4.2   DIRECT REVERSAL OF DAMAGE IN ONE STRAND OF DUPLEX DNA

Direct repair of damaged bases is catalyzed by two important classes of enzymes: the photolyases that reverse the damage inflicted by UV light, and the enzymes that reverse methylation modifications/damage caused by alkylating agents.

Lesions induced by UV light can be repaired directly

DNA is susceptible to damage by ultraviolet light (200–300 nm) in the sun's rays, notably the formation of cross-linked dimers between neighboring pyrimidines (T and/or C but mainly T) on the same strand. The major product (75%) is a cyclobutane pyrimidine dimer (CPD), designated T< >T in the case of thymine, and the minor product (25%) is a pyrimidine-pyrimidone photoproduct (6-4 PP, see Figure 4.1D). Both lesions can be chemically reversed by **DNA photolyases**, which rely on photoactivation by blue light in the visible region (350–450 nm). Photolyases are monomeric proteins (55–65 kDa) that occur widely in prokaryotes and eukaryotes, but not in placental mammals which instead utilize the NER pathway for the repair of UV photolesions (see below). Photolyases contain two noncovalently bound prosthetic groups, one of which is always FADH$^-$ (a $2e$/1H$^+$-reduced form of FAD—see Chapter 15), and the second, either methylenetetrahydrofolate (MTHF) with three to six Glu residues in a polyGlu tail, as in *Escherichia coli* CPD photolyase, or 8-hydroxy-7,8-didemethyl-5-deazariboflavin (8-HDF), as in *Anacystis nidulans* photolyases (**Figure 4.3A**). The proposed reaction scheme of the *E. coli* CPD photolyase, inferred from femtosecond spectroscopy and site-directed mutagenesis, is consistent with an electron transfer radical mechanism (Figure 4.3B) in which the reduced FAD cofactor donates an electron to the pyrimidine dimer, forming a flavin radical that is subsequently reduced by an electron donated by the DNA substrate. The cyclobutane pyrimidine dimer is split into two successive steps, the overall reaction taking less than 700 ps.

The crystal structure of the homologous photolyase from *A. nidulans* bound to a 14-residue fragment of DNA containing a synthetic CPD analog (**Figure 4.4**) shows that the pyrimidine photodimer has been flipped out of the DNA double helix into a hydrophobic pocket, where it packs against Trp residues in a π-π stacking arrangement. Outside the T-dimer contact site, the DNA is essentially in the B-type conformation and the predominant interactions are with the sugar–phosphate backbone of the DNA, as expected of an enzyme that shows no base sequence specificity. In B-DNA, a single CPD lesion induces bending by 20–30° but in the active site of the photolyase this increases to ~50°, consistent with the enzyme recognizing diminished stability of the damaged DNA. Only the FADH$^-$ cofactor is essential for catalysis, with the MTHF or 8-HDF cofactors acting as photoantennae (Chapter 15) that pass on energy to FADH$^-$. However, the two cofactors are too far apart for efficient direct electron transfer and electron tunneling (Chapter 15) must be invoked. It is proposed that the electron moves from the FAD isoalloxazine ring to the adenine moiety

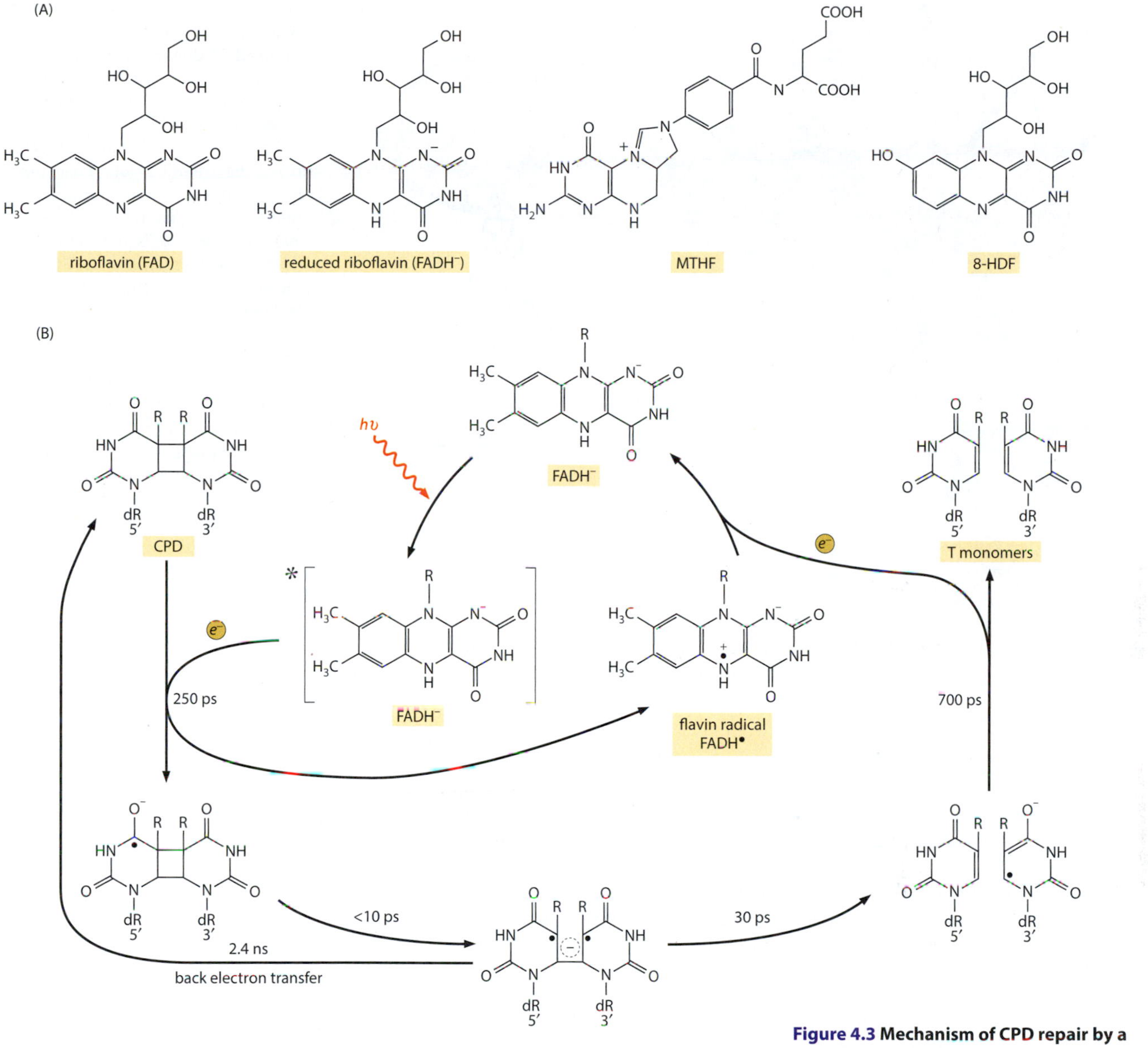

(A)

riboflavin (FAD)   reduced riboflavin (FADH⁻)   MTHF   8-HDF

(B)

Figure 4.3 Mechanism of CPD repair by a specific photolyase.
(A) Co-factors involved in the photocycles of CPD and 6-4 PP repair. FAD, flavin adenine dinucleotide (only the isoalloxazine ring and ribitol sugar of the riboflavin moiety are shown); FADH⁻, two-electron reduced form of FAD (2e, 1H⁺); MTHF, methenyltetrahydrofolate; 8-HDF, 8-hydroxy 7,8-didemethyl 5-deazariboflavin.
(B) Photocycle and kinetics of the reaction catalyzed by the CPD photolyase of *E. coli*. The co-factor FADH⁻ is initially photoactivated (by direct blue light, or exciton energy transfer from MTHF or 8-HDF) to an activated state *FADH⁻, which transfers an electron to the cyclobutane substrate and becomes a flavin radical FADH•, and this is then converted back to FADH⁻ by return of an electron as the two T monomers are generated. The forward reactions take place on the picosecond timescale, and a much slower (nanosecond) back electron transfer can be detected.

and thence to the 5′ side of the CPD to initiate the splitting of the cyclobutane ring. After breakage of both C-C bonds, the electron remains on the 3′ side and tunnels back via the adenine moiety again.

A crystal structure and kinetic analysis of the functionally distinct but related (6-4 PP) photolyase, which specifically repairs the pyrimidine-pyrimidone photolesion, indicate a generally similar mechanism. It turns out that there is a remarkable structural and mechanistic relatedness of the photolyases to the cryptochromes. The latter proteins are not involved in DNA repair but act as photoreceptors that transduce blue light signals, and take part in a wide range of biological processes such as flowering in plants and the circadian clock in mammals.

Some aberrant methylations can also be repaired directly

Three different mechanisms for restoring DNA damaged by alkylation have been discovered. Direct demethylation of O^6-me-dG and O^4-me-dT (Figure 4.1A) is catalyzed by O^6-alkylguanine-DNA-alkyltransferase (AGT), whereas removal of the methyl group of 1-me-dA, 1-me-dG, 3-me-dC, and 3-me-dT is brought about by the alpha-ketoglutarate-dependent dioxygenase, AlkB. O^6-me-dG in DNA preferentially base-pairs with T during replication and thus causes G/C to A/T transitions. AGT and AlkB first locate the lesion and

(A)

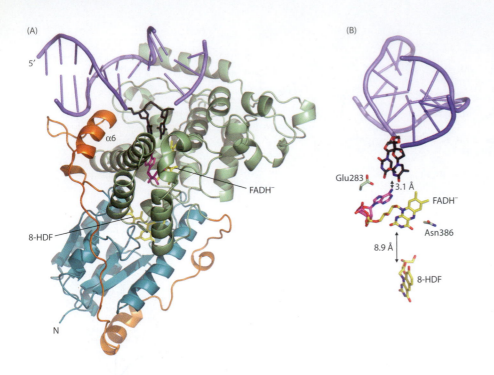

(B)

Figure 4.4 Structure of a complex of DNA CPD photolyase with DNA containing a repaired T-T dimer. (A) The photolyase from *A. nidulans* in complex with a 14-residue DNA fragment with the T-dimer (in black) flipped out of the double helix. An N-terminal (α/β) domain (residues 1–131 in *E. coli*) is shown in teal, the C-terminal (α) domain (residues 204–471 in *E. coli*) in green, and a long interdomain region (residues 132–203 in *E. coli*) in orange. A helix in the interdomain region is clearly seen with its N-terminal end apposed to the sugar–phosphate backbone of the DNA. The photoantenna 8-HDF (*A. nidulans* photolyase uses 8-HDF, rather than the MTFH of the *E. coli* enzyme) is bound at the interface between the domains, whereas the FADH⁻ is bound in the C-terminal domain. The AMP moiety of FADH⁻ is in magenta, the riboflavin in yellow. (B) The arrangement of the co-factors. Two critical conserved residues adjacent to the FADH⁻, Glu283 (Glu274 in *E. coli*) and Asn386 (Asn378 in *E. coli*), are included. This view is rotated relative to that shown in (A)—see the DNA. (PDB 1TEZ)

then flip the offending base out of the double helix for direct chemical removal of the methyl group. The third process for repair of alkylation damage is catalyzed by lesion-specific BER glycosylases, as described below.

Human AGT functions by transferring the methyl group from the damaged base to the thiolate of Cys145 in the enzyme active site (**Figure 4.5A**). Human AGT binds DNA containing a mispaired O^6-me-dG without major conformational change. The nucleophilic Cys145 is adjacent to a helix-turn-helix (HTH) motif (see **Guide**) in the C-terminal part of the protein (Figure 4.5B), ready for in-line attack on the O^6-Me bond, <3 Å away (compare with cysteine proteases in Box 7.1). The HTH motif, which in most known instances binds in the major groove of DNA, is unusual in that its second helix (Ala127-Gly136) binds within the minor groove. This type of contact minimizes sequence-specific interactions, thereby facilitating action on different DNA molecules. The minor groove is forced wider, the DNA (essentially B-form) bends ~15° away from the protein, and the offending O^6-me-dG is flipped out. Arg128 of the HTH helps to stabilize the bend by projecting into the minor groove, where it stacks in the space vacated by the flipped-out base and interacts with the 'orphaned' C on the opposite strand. Steric repulsion with the side chain of Tyr114 aids flipping out by inducing rotation in the 3′-phosphate bond of the target base. The thioether linkage of Cys145 with the methyl group abstracted from the DNA is irreversible, and AGT

(A)

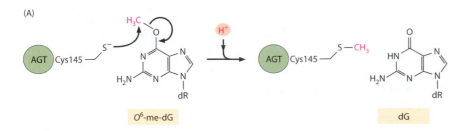

(B)

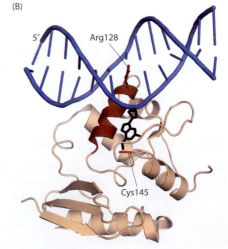

Figure 4.5 Direct demethylation of O^6-me-guanosine by AGT. (A) The methyl group on the O^6 of O^6-me-dG is removed by direct nucleophilic attack of the thiolate group of a Cys residue in the protein (Cys145 in human AGT). (B) Structure of human AGT (lacking the C-terminal 30 residues) bound to a 13 bp fragment of dsDNA with an O^6-me-dG in the middle. The 5′ end of the damaged DNA strand (cartoon representation in blue) is indicated. The flipped-out O^6-me-dG is depicted in black, Cys145 (replaced by Ser145 to inactivate the enzyme for crystallization) is in red, and the second helix of the helix-turn-helix motif in the protein that makes contact with the DNA in its minor groove in a darker brown. (PDB 1T38)

can thus be used only once. It is an indication of the importance of DNA repair that a 22 kDa protein is synthesized to make just one correction before being discarded.

E. coli AlkB and related human enzymes remove the methyl group by an oxidative process requiring dioxygen, an Fe^{2+} cofactor, and reducing power provided by the oxidative decarboxylation of 2-oxoglutarate (**Figure 4.6A**). Human paralogs of AlkB, including ABH2 (29 kDa), repair both DNA and RNA damaged by alkylation. They all have a jelly roll fold (see **Guide**), as found in all Fe^{2+}/2-oxoglutarate-dependent oxygenases. In both *E. coli* AlkB and human ABH2 the modified base is flipped out into the enzyme active site for repair but the enzymes differ in how they interact with dsDNA containing the 1-me-dA lesion (Figure 4.6B and C). In ABH2, a finger residue (Phe102) is intercalated into the space vacated by the flipped-out base (compare Arg128 in human AGT). In contrast, AlkB differs from these other repair enzymes by 'squeezing' together the bases that flank the flipped base. They form a one base-wide stack (3.4 Å separation) on the base from the strand opposite the flipped-out 1-me-dA, thereby maintaining the length of normal duplex DNA, whereas the preceding base is forced out of the helix and becomes disordered. The squeezing of the bases is accompanied by the deoxyriboses of the two nucleosides adopting unusual conformations: one (3′ side) is C3′ *endo*, as in A-form DNA, and the other (5′ side) is twisted by ~180 Å and held in this inverted position by local interaction with residues (Thr51-Pro52-Gly53) of the protein. AlkB and ABH2 also differ in that AlkB interacts almost exclusively with the damaged strand and the interactions are confined to the flanking regions of the flipped-out base, whereas ABH2 makes interactions with both strands of the DNA, including an interaction with an RKK motif located in a protruding loop some distance away from the lesion site. The observed interactions with one or both strands of the lesioned DNA help to explain the preference of AlkB to undertake repair of ssDNA, whereas ABH2 prefers dsDNA.

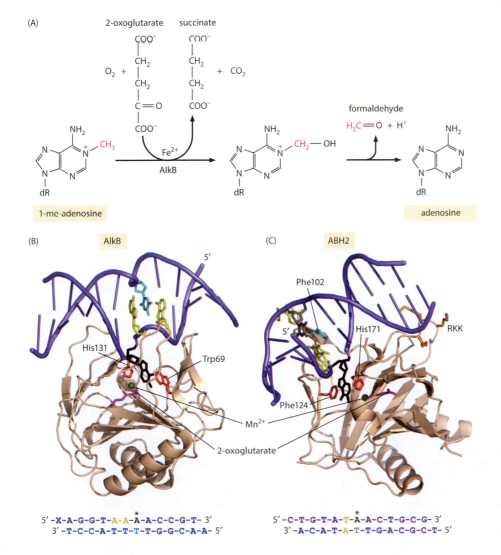

(A) 2-oxoglutarate   succinate

1-me-adenosine

formaldehyde

adenosine

(B) AlkB

(C) ABH2

5′- X–A–G–G–T–A–A–A̲*–A–C–C–G–T– 3′
3′ -T–C–C–A–T–T–T–T–G–G–C–A–A– 5′

5′ -C–T–G–T–A–T–A̲*–A–C–T–G–C–G- 3′
3′ -A–C–A–T–A–T–T–G–A–C–G–C–T- 5′

Figure 4.6 Direct demethylation of 1-me-adenosine by AlkB and ABH2. (A) The methyl group on 1-me-dA is removed by a process of coupled oxidative decarboxylation. The methyl group undergoing removal is in red. (B and C) Structures of AlkB (lacking N-terminal 11 aa) and ABH2 (lacking N-terminal 55 aa) in complex with a 13 bp fragment of dsDNA (cartoon representation in blue) containing a 1-me-A lesion in the middle. The 5′ end of the damaged strand is indicated in each case. The DNA sequence is given below each structure, key bases are colored as in the structures, and the 1-me-A is marked with an asterisk. The 1-me-A (shown in black) is flipped out, the 2-oxoglutarate co-substrate is in magenta, and the Mn^{2+} is in green. In AlkB, the orphaned base (dT7, in cyan) stacks between the dA6 and dA7 (both in yellow) coming from the damaged strand and the flipped-out 1-me-dA8 interacts with Trp69 and His131. dT8 of the undamaged strand is invisible, having been forced out of the helix and thus disordered. In ABH2, Phe124 and His171 cushion the flipped-out 1-me-A (position 7 in the damaged strand of DNA in the sequence depicted below the structure) and Phe102 (shown in cyan) is inserted into the vacant space in the DNA duplex thus created. The orphaned dT8, gray in the undamaged strand, forms a partial base pair with the T6 (in yellow) on the 5′ side of the damaged A. The dA9 (also yellow) on the undamaged strand and originally base-paired with T6 on the damaged strand can still be seen but has been forced out of the double helix. The RKK motif interacting with the sugar–phosphate backbone of the undamaged strand is in orange. (PDB 3BIE (AlkB), 3BUC (ABH2))

4.3   TEMPLATED REPAIR OF LESIONS AFFECTING ONE STRAND OF THE DNA DUPLEX

The most widespread methods of repairing a lesion in one strand of dsDNA utilize the complementary strand as a template against which the damaged base can be replaced. As in the direct reversal of the chemical modification of a base described above, removal of a damaged base begins with flipping it out of the DNA helix, after which the repair can be completed by copying the complementary, undamaged strand. For larger, helix-distorting lesions recognized by the NER pathway, an oligonucleotide containing the damaged base is excised from DNA and the resulting gap is filled by DNA polymerase.

Base-excision repair is initiated by DNA glycosylases

The enzymes responsible for initiating base-excision repair are the DNA **glycosylases**, which recognize the site of the lesion and then catalyze cleavage of the glycosidic bond between the damaged base and the deoxyribose of the sugar–phosphate backbone. This creates an apurinic or apyrimidinic site (an **abasic** or **AP site**), a situation that also occurs in the spontaneous hydrolysis of the glycosidic bonds in DNA (Figure 4.1E). DNA glycosylases are ubiquitous and many have been characterized; some are highly selective for a specific alteration, like the uracil DNA glycosylase (UDG) which removes the uracil resulting from the deamination of cytosine (100–500 occurrences per day in a human cell), whereas others have a broader specificity for several, chemically related base modifications. The BER pathway repairs many of the most common types of spontaneous DNA damage, and is responsible for repairing more than 10,000 damaged bases per cell per day.

DNA glycosylases that recognize different types of damage exhibit a conserved mechanism of 'base flipping' to expose damaged nucleotides to the enzyme active site. Two examples are illustrated in **Figure 4.7A, B**, human UDG (25 kDa, one of at least four homologs in humans) and 8-oxoguanine DNA glycosylase (OGG1), 38 kDa, a member of the so-called

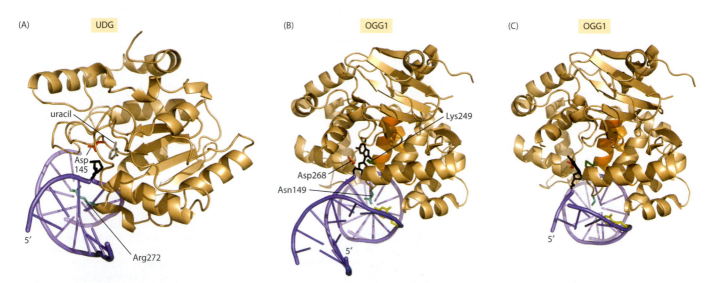

Figure 4.7 Structures of uracil DNA glycosylase and 8-oxoguanine glycosylase. (A) The structure of a human UDG in complex with a 10-base fragment of dsDNA (in blue) with a central U/G mismatch. The 5′ end of the damaged strand is indicated. The aberrant U (depicted in black) is flipped out into a deep active site in the enzyme and the glycosidic bond to the deoxyribose has been cleaved (that is, an enzyme–product complex). The U-containing sugar–phosphate backbone binds along an active site groove in UDG presenting complementary positive charges. Arg272 (cyan) projects into the DNA helix from across the minor groove, occupying the space vacated by the flipped base. Asp145 (red) may activate a water molecule for direct in-line attack on the N-C^1 glycosidic bond. (PDB 4SKN) (B) In the human OGG1 (residues 12–327 representing the core catalytic region), the 15-base dsDNA (blue) is bound in a largely electrically neutral channel, but a large number of α helices have their

N termini (positively charged dipole) oriented towards the DNA. The protein–DNA interaction is confined to the 8-oxoG-containing strand, and a helix-hairpin-helix (HhH) motif (Glu234-Ala258, orange) binding on the 3′ side of the lesion appears to play a major role in positioning the DNA with the 8-oxoG (black) flipped into the active site. The orphan C on the complementary strand is in yellow The DNA is kinked by 70° and the side chain of Asn149 (cyan) occupies the space vacated by the flipped base. The nucleophile for cleavage of the N-C^1 glycosidic bond appears to be the N^6-NH$_2$ of Lys249 (green, actually Gln249 in the protein crystallized, to render the enzyme inactive), unprotonated by its interaction with nearby Asp268 (red). (PDB 1EBM) (C) Human OGG1 is captured flipping unmodified G from a normal G/C base pair, but the flipped base is located differently in a pre-recognition pocket distinct from the active site, in which only 8-oxoG can be stably bound. (PDB 1YQK)

HhH-GPD superfamily (helix-hairpin-helix followed by a Gly/Pro-rich loop and conserved Asp). Interaction of the DNA with the glycosylase induces a 180° torsion around the sugar-phosphate backbone, which helps flip the damaged nucleotide out of the double helix into the active site of the enzyme. This is sometimes accompanied by DNA bending, by as much as 70° in OGG1, mediated by non-sequence-specific interactions with the sugar–phosphate backbone, although the overall B-DNA conformation is generally maintained. The flipped base makes extensive interactions within the active site that dictate specificity for modified bases and offset the energetic cost of distorting the DNA structure by contributions from hydrogen bonding and stacking with aromatic amino acids. A stabilizing side chain (Arg272 in UDG and Asn149 in OGG1) is inserted into the space in the double helix previously occupied by the flipped base. An undamaged DNA base is poorly retained within the active site, thereby limiting the excision of normal bases from DNA.

DNA glycosylases search for lesions in DNA by transient encounters, both passive and active

BER deals with modified bases that cause little if any distortion of the DNA double helix. How then are such lesions recognized? In the case of human UDG, structural and biochemical studies indicate that the process is essentially passive on the part of the enzyme and depends on thermally driven base-pair dynamics. Spontaneous transient expulsion of pyrimidines occurs, and the glycosylase binds the exposed base (U or T from U/A or T/A base pairs respectively) in a cryptic site. At this stage, the expelled base has undergone only a small part (30°) of the torsional distortion (180°) required to move it into the catalytic site. The 'wrong' base, U, can make the specific interactions that allow it to proceed further along the reaction pathway, whereas T has a bulky 5-methyl group that does not fit in the enzyme active site. The spontaneous rate of opening for U/T base pairs in DNA is estimated to be 1400/s at 24°C, fast enough to account for the observed rate of uracil flipping by the UDG enzyme (<700/s) at the same temperature.

In contrast, OGG1 locates damaged bases by a process of facilitated diffusion. Single-molecule assays show the enzyme as binding and then moving in either direction by Brownian motion over kilobase lengths of DNA, with a one-dimensional diffusion constant (D) of 4.8 $\pm$ 1.1 $\times$ 10^6 bp^2/s ($D = <x^2>/2t$, described in Box 1.4). This rapid sliding of OGG1 on DNA is affected very little by salt concentration and thus appears to be relevant for physiological conditions. The structure of OGG1 in complex with DNA containing a normal G/C base pair indicates that the DNA is substantially bent (80°, more than the 70° in the complex with a DNA fragment containing an 8-oxoG/A base pair) and has expelled the G (Figure 4.7C). However, the extra-helical G is not in the active site pocket (where it might become erroneously excised) but in a more exposed exo-site, ~5 Å away. Recognition of 8-oxoG does not depend on its 8-oxo moiety but instead on a crucial H-bond between the backbone carbonyl of Gly42 in the active site and the N7 proton found on 8-oxoG but not on other purines. No such bond can be made by G, which causes G to be rejected whereas 8-oxoG is stably bound in the active site and then excised. Comparable experiments with MutM, the bacterial counterpart of OGG1, reveal that when the protein encounters the 8-oxoG lesion it bends the DNA even while the 8-oxoG is still intra-helical. This facilitates expulsion of the base by causing a loss in stacking interactions with its neighbors, further aided by interactions with two invading side chains, a Phe and a Met. Another invading Arg residue competes with the base pairing to the complementary base on the opposing strand, promoting disruption of the target pair. In each of these examples, the DNA double helix is interrogated for damaged bases in a step-wise process, without a need to completely expose every nucleotide. The base-flipping mechanism favors damaged bases over undamaged DNA because most base modifications weaken base-pairing interactions that pose a barrier to base flipping, and because modified bases are accommodated by a binding pocket in the enzyme that discriminates against binding of normal bases.

Replacement of the excised base requires additional enzymes

BER doesn't end with excision of the damaged base. Moreover, AP sites are also generated by the spontaneous hydrolysis of glycosidic bonds (Figure 4.1E). The AP site must be dealt with efficiently because it will stall replication forks and disrupt topoisomerase activity. The site lacks an N-glycosidic bond at the deoxyribose C^1, which means that the sugar ring can open to expose a chemically reactive free aldehyde group.

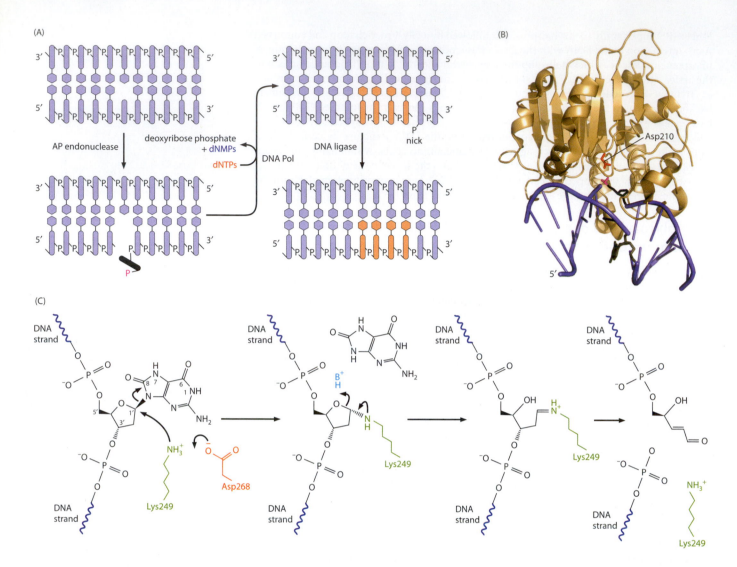

Figure 4.8 Replacement of bases removed by accidental hydrolysis of glycosidic bonds or by action of specific glycosylases. (A) AP endonuclease (APE1) cleaves the sugar–phosphate backbone of the AP site-containing strand; the dangling 5′phosphodeoxyribose (magenta and black) and nucleotides to the 3′ side are removed by the 5′–3′ exonuclease action of DNA polymerase and replaced using the complementary strand as template; and the nick is sealed with a DNA ligase. (B) Structure of human APE1 (residues 40–318) in complex with an AP-containing 11-base fragment of DNA. The new 5′-phosphodeoxyribose is in magenta and black, and the orphan base on the complementary strand is in dark gray. The DNA is kinked by 35° and Asp210 activates a water molecule to act as the nucleophile in the sugar–phosphate bond cleavage. (PDB 1DE8) (C) The reaction catalyzed by OGG1 (Figure 4.7), showing the nucleophile as an unprotonated Lys249 residue, aided by a general acid BH⁺. This leads also to cleavage of the sugar-phosphate backbone at the AP site, as opposed to requiring a separate AP endonuclease.

The AP site is repaired by enzymes that restore the structure of DNA. First, the AP endonuclease (APE1 in humans) cuts the DNA strand on the 5′ side of the site, creating a free 3′-OH and a 5′-deoxyribose phosphate on the other side of a one nucleotide-wide gap (**Figure 4.8A**). After the damaged base has been excised, APE1 kinks the baseless part of the DNA by 35°, deforming it in the process (Figure 4.8B). The phosphodiester bond on the 5′ side is cleaved by nucleophilic attack from a water molecule activated by a suitably positioned unprotonated Asp210 (Figure 4.8B), creating a single nucleotide gap that can be filled in by DNA polymerase (see Figure 3.1.1 in Box 3.1). The 3′-OH group is used by DNA Pol β (DNA Pol I in *E. coli*) to introduce one nucleotide into the DNA, so-called 'short patch repair.' In an alternative mechanism known as 'long patch repair,' human DNA polymerase ε inserts several nucleotides, displacing the downstream DNA strand to create a 5′ flap that is subsequently excised by the flap endonuclease FEN1. In the final step of BER, the nicked DNA strand is ligated together by DNA ligase III (short patch repair) or by DNA ligase I (long patch repair) (Figure 4.8A). Many of the DNA intermediates in BER and other repair pathways are reactive or mutagenic, so there is an impetus to coordinate the steps of the repair process to avoid a loss of genetic information or untoward cellular responses to toxic DNA intermediates. The coordination of multiple enzymatic activities during DNA repair has been difficult to verify experimentally, but is the proposed function of 'scaffold proteins' like XRCC1 (X-ray cross complementation protein 1) that bind to multiple DNA repair factors at sites of damage.

The DNA glycosylase OGG1 represents a class of bifunctional glycosylases that, having hydrolyzed the glycosidic bond to the damaged base, go on to cleave the abasic strand at the adjoining 3'-phosphate by virtue of their β-lyase activity (Figure 4.8C). In this instance, the repair process requires an additional 'clean up' step to remove the 3'-phosphate, primarily by polynucleotide kinase/phosphatase (PNPK), so that the DNA ends are ready for DNA polymerase and ligase to complete the job as before.

Nucleotide-excision repair deals with bulky lesions

The NER pathway, found in all cells, responds to bulky DNA adducts (Figure 4.1C) that cause a significant local distortion in the double helix. Among these NER substrates are the lesions caused by UV light, cigarette smoke, and other environmental risk factors, which are prominent causes of cancer. Recognition of the lesion is followed by an orderly recruitment of multiple repair proteins; the proteins differ between bacteria and eukaryotes, but the mechanisms are similar. In the much simpler bacterial system, exemplified by *E. coli*, three proteins form a complex with unusual endonuclease activity. Two copies of UvrA (103 kDa), which has intrinsic DNA-binding capacity, come together and then recruit a UvrB subunit (74 kDa). The UvrA$_2$/UvrB$_1$ complex recognizes the presence of a bulky lesion in the DNA and UvrA then dissociates, leaving UvrB bound tightly to the DNA. ATP hydrolysis by UvrA is required for this process. UvrB in turn recruits an endonuclease, UvrC (67 kDa), which sequentially catalyzes two incisions in the strand carrying the damaged base, one on either side of the lesion. Irrespective of the nature of the lesion, the first cut is at the fourth or fifth phosphodiester bond on the 3' side of the lesion, and the second is typically at the eighth phosphodiester bond on the 5' side (**Figure 4.9**). UvrC and the cleaved oligonucleotide (12–13 residues) containing the damaged base are then displaced by a helicase (UvrD, helicase II). UvrB remains bound and recruits DNA Pol I, which displaces UvrB as it fills in the gap in the DNA. The remaining nick in the DNA is sealed by DNA ligase and the excised oligonucleotide is degraded.

The NER process in eukaryotes is more complicated, requiring at least 18 enzymes and other proteins, and the excised oligonucleotide containing the damage is ~25–30 residues long, as described below. Compared with BER, NER consumes more energy in that oligonucleotides are excised and degraded and new DNA has to be synthesized, but NER is of pivotal importance in placental mammals, which do not contain DNA photolyases for the repair of UV-induced CPD dimers. Thus, the onus on NER is apparent. Numerous hereditary diseases are caused by mutations in proteins of the NER pathway and other defects in DNA repair. Some examples are listed in **Table 4.1**.

UvrABC interact and function sequentially in bacterial NER

UvrA is a member of the ABC superfamily of ATPases, described as membrane transporters in Section 16.5. Each subunit in the dimer contains two ABC modules, designated nucleotide-binding domains (NBDs) I and II respectively, separated by a flexible linker. Each NBD has an ATP-binding domain and a 'signature' domain, as in other ABC ATPases, but NBD-I also has two unusual domains: a UvrB-binding domain and an 'insertion' domain (**Figure 4.10A**). In all, four ATP molecules can be bound per dimer, each with the usual Walker A and B motifs. However, the ATP molecules are not bound at dimer interfaces as in other ABC ATPases, and their role in dimer formation and the NER process is not well understood. Curiously too, the UvrA monomer binds three Zn^{2+} ions, not just the two expected (one per NBD), and the Zn-binding motifs do not adopt the 'Zn-finger' structure. Their role is likely to be structural, and Zn modules 2 and 3 both coordinate their Zn^{2+} ions with four Cys residues.

A clue to the interaction between UvrA and UvrB came with the solving of the crystal structure of a complex of two excised domains from UvrA (residues 131–245) and UvrB (residues 149–250) (Figure 4.10B). The 'docking domains' of both UvrA and UvrB contain long regions (residues 154–199 and 157–250 respectively), which are disordered in the separate

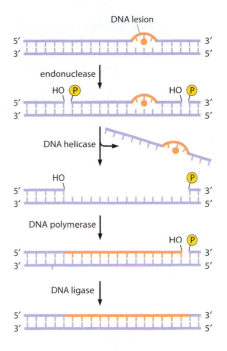

Figure 4.9 Nucleotide-excision repair in bacteria. Steps involved in nucleotide-excision repair in *E. coli*. The pathway in eukaryotes is similar but involves many more proteins and excises a much larger oligonucleotide containing the damaged base(s).

Table 4.1 Human hereditary diseases with defective cellular responses to DNA damage

Human disease	Gene(s)	Principal response to defect	Principal clinical features
Xeroderma pigmentosum (XP)	*XPA-XPG, XPV*	Nucleotide-excision repair (NER); translesion DNA synthesis	Dermatitis, freckling, skin cancer, sometimes neurological defects
Ataxia telangiectasia (AT)	*ATM*	Repair of DNA strand breaks	Cerebellar ataxia, defective immune function, neurological problems, predisposition to hematolymphoid cancer
Nijmegen breakage syndrome (NBS)	*NBSI*	Repair of DNA strand breaks	Developmental abnormalities, growth retardation, cancer predisposition
LIG4 syndrome	*LIG4*	Repair of DNA strand breaks	Defective immune function, neurological problems, predisposition to hematolymphoid cancer
Bloom syndrome (BS)	*BLM*	Resolution of stalled replication/transcription intermediates	Dwarfism, immunodeficiency, cancer predisposition
Werner syndrome (WS)	*WRN*	Resolution of stalled replication/transcription intermediates	Premature aging, cancer predisposition
Hereditary nonpolyposis colon cancer (HNPCC)	*MLHI, MSH2, MSH6, PMSI, PMS2, MLH3, EXOI*	Mismatch repair	Colon and other cancers

(From E.C. Friedberg et al. DNA Repair and Mutagenesis, 2nd ed. ASM Press, 2005.)

proteins but become ordered in the complex, helping to create an interface that is dominated by polar interactions. The likely DNA interaction surface, suggested by the structure and the results of site-directed mutagenesis, is depicted by a blue box in Figure 4.10C. The length of DNA required to interact with the modeled UvrA$_2$/UvrB$_1$ complex (~43 bp) would be consistent with DNA footprinting experiments. There are two binding sites for UvrB on the UvrA dimer, but the stoichiometry of A$_2$B$_1$ appears best to fit experimental results. UvrB

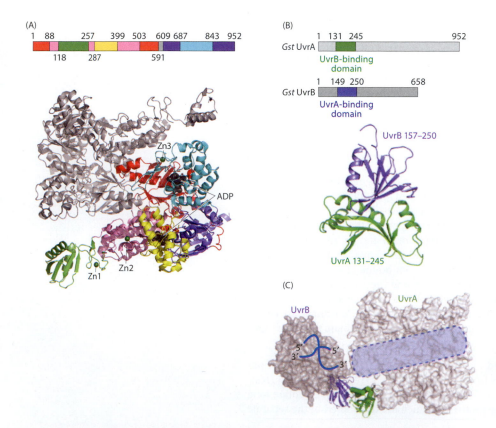

Figure 4.10 Structures and interaction of UvrA and UvrB. (A) Domain map and structure of *Geobacillus stearothermophilus* UvrA. In one subunit, the domains are colored, and the other subunit in gray. ATP-binding domains I, red, II dark blue; signature domains I, pink; II, cyan; UvrB-binding, green; insertion, yellow; and linker, gray. The Zn^{2+} are numbered by module. The two ADP molecules bound per subunit (dark gray) are shown in space-filling format. (PDB 2R6F) (B) Structure of the interaction domains from UvrA (green) and UvrB (blue) in a 1:1 complex. Only the residues indicated form the ordered structure. (PDB 3FPN) (C) Model depicting how UvrA and UvrB might dock onto one another in an A$_2$B$_1$ complex. A likely binding surface for DNA is in blue. (Adapted from D. Pakotiprapha et al., *J. Biol. Chem.* 284:12837–12844, 2009. With permission from the American Society for Biochemistry and Molecular Biology.)

has recognizable motifs characteristic of helicases, but appears to function not as a true helicase but rather as a helix destabilizer that facilitates the incisions to be made by UvrC.

UvrC has two endonuclease domains, one in each half of the protein, each with its distinctive and independent active site. The N-terminal site is responsible for the 3′ incision, whereas the C-terminal site catalyzes the 5′ incision. The C-terminal endonuclease is structurally similar to RNase H (Section 3.2) despite a lack of sequence similarity and, further towards the C terminus, there are two HhH motifs linked by a short helix and held together by a hydrophobic core. The HhH domain is commonly observed to interact in a non-sequence-specific manner with the sugar–phosphate backbone across the minor groove of DNA (see, for example, BER above) and this, supported by site-directed mutagenesis results, suggests that they are involved in the binding and positioning of UvrC in a pre-incision complex. The endonuclease reaction depends on a water molecule activated as a nucleophile to attack a phosphodiester bond, with catalytic assistance from two metal ions coordinated by conserved Asp residues, again reminiscent of the mechanism of polymerases and ATPases (Figure 3.1.1, Box 3.1, and Figure 3.4).

Many proteins act sequentially in eukaryotic NER

In eukaryotes, 18 or more proteins, coming and going in transient complexes, are required to catalyze NER. Many of these proteins were originally identified in studies of the genetic basis of diseases like xeroderma pigmentosum and Cockayne disease which result in developmental defects, particularly in the nervous system, and skin cancer. These symptoms highlight the importance of NER for DNA repair in post-mitotic cells such as neurons, and for the repair of UV photodimers in skin exposed to sunlight. This scientific legacy resulted in a complicated naming scheme for NER proteins that includes acronyms based on hereditary diseases (XP from xeroderma pigmentosum, CS from Cockayne syndrome) and genetic screens (ERCC from excision repair cross-complementing genes, and so on). A schematic outline of the NER pathway is shown in **Figure 4.11**. The protein XPC recognizes a wide range of lesions, initiating an orderly recruitment of repair proteins. The 10-subunit TFIIH complex, also involved in transcription (Section 5.3) provides two ATP-dependent helicases (XPB, unwinding 3′ → 5′; XPD, unwinding 5′ → 3′), suggesting that TFIIH acts to open up the DNA structure as a 20 bp 'bubble' round the lesion that is stabilized by the ssDNA-binding protein RPA bound to the undamaged strand.

Instead of a bifunctional UvrC as in *E. coli*, two separate endonucleases make the incisions into the damaged strand: ERCC1-XPF on the 5′ and XPG 3′ on the side of the lesion. An oligonucleotide of ~30 residues is released, and PCNA, RPA, DNA Pol δ or ε, and DNA ligase I are required to restore the correct double helix. Like DNA replication in eukaryotes, NER also requires that the DNA be dislodged temporarily from its close association with histones in the nucleosomes of chromatin.

Interaction of damaged DNA with XPC (Rad4) is the first step in eukaryotic NER

XPC (yeast ortholog, Rad4) is found with a partner protein (RAD23B, yeast ortholog Rad23), whose function may be to protect XPC from ubiquitin-mediated degradation (Section 7.4). RAD23B has an XPC-binding domain (R4BD), a ubiquitin-like domain (UBL), and two ubiquitin-associated domains (UBA1 and UBA2) (**Figure 4.12A**). A crystal structure of Rad4 in complex with a 24 bp duplex DNA containing a CTD T-T lesion (see above) and mismatch plus a large fragment of Rad23 containing the UBL and UBA2 domains, is shown in Figure 4.12B. Rad4 consists of an N-terminal α/β domain of 310 aa, within which is a 45 aa segment with a *transglutaminase* fold, followed by three related β-hairpin-containing domains (BHDs) each of 50–90 aa.

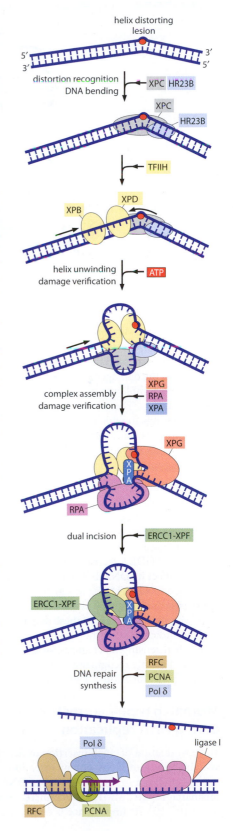

Figure 4.11 Nucleotide-excision repair in eukaryotes. The lesion is recognized by XPC/HR23B and a succession of proteins is accreted to unwind the helix locally (XPB, XPD) creating a 20 bp 'bubble,' protect the exposed ssDNA (RPA), make two incisions in the damaged strand (XPG, XPF), discard the oligonucleotide (24–32 residues) containing the damaged base, and fill in the gap using the undamaged strand as template (RFC, PCNA, Pol δ, and DNA ligase I). (Adapted from L. C. Gillett and O.D. Schärer, *Chem. Rev.* 106:253–276, 2006. With permission from ACS.)

(A)

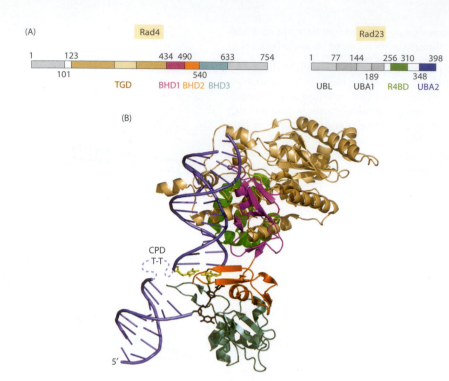

(B)

Figure 4.12 Damaged base flipping by NER protein Rad4 (XPC). (A) Domain maps of Rad4 and Rad23. The domains shown in gray were not included in the proteins crystallized. (B) Crystal structure of Rad4 (residues 101–632) in complex with Rad23 (domains R4BD and UBA2, only) and a 24 bp fragment of DNA with T and a CPD lesion at nucleotide positions 14, 15, and 16, respectively (damaged strand), and T mismatches at all three complementary positions in the undamaged strand. The two T bases opposite the CPD lesion are in gray and the neighboring mismatched T residues are in yellow. Rad23 UBA2 was not visible and assumed to be disordered. The CPD lesion was also invisible, and presumably disordered. (PDB 2QSG)

As with other DNA repair enzymes, the bound DNA is kinked (~42°) and the damaged bases are flipped out, but in this instance they are left disordered and exposed to solvent and thus accessible for processing by the downstream proteins. An important feature is the way the β hairpin of BHD3 inserts through the double helix, the inference being that it helps to flip the damaged bases out into the solvent. The opposing DNA strand and the two undamaged complementary bases are flipped out into a groove provided by the BHD2 and BHD3 domains. The relative lack of specificity in the binding is consistent with the wide range of lesions to which NER is directed. It appears that XPC recognizes lesions passively by sampling the conformational state of the DNA, and actively by probing the stability of the double helix against insertion of the BHD3 hairpin cooperatively to facilitate flipping out of the damaged bases.

In mammalian cells, a second protein complex detects UV-damaged DNA and recruits NER proteins to sites of damage. Within minutes after UV irradiation of cells, the DDB1/DDB2 (DNA damage-binding proteins 1 and 2) complex translocates from the cytoplasm to the nucleus where it binds tightly to damaged DNA. DDB1/DDB2 is physically associated with the cullin family ubiquitin ligase CUL4-RBX1, which ubiquitinates XPC and DDB2. Ubiquitination of DDB2 decreases its affinity for damaged DNA, whereas ubiquitination of XPC does not affect its DNA-binding activity. A plausible model for the respective roles of XPC/RAD23 and DDB1/DDB2 in sensing DNA damage is that DDB1/DDB2 is the initial damage sensor that hands off the lesioned DNA to XPC following ubiquitylation of both proteins. Mutations of either of these damage sensors can cause sensitivity to UV-induced DNA damage, indicating that they are not redundant systems. The added complexity in mammals of this dual sensing mechanism for UV lesions is not fully understood and may reflect the need to efficiently recognize a variety of UV-induced modifications while strongly discriminating against the futile repair of unmodified DNA.

Mismatch repair corrects mispaired bases that are left uncorrected during DNA replication

The introduction of small insertion or deletion loops (IDLs) of up to three or four residues and the failure of DNA polymerases to correct misincorporated bases is a rare but significant problem in the replication of DNA. This post-replicative repair is the province of **mismatch repair** (MMR), which increases the overall fidelity of DNA replication by a factor of 10^2–10^3. Its importance is emphasized by the predisposition of humans to sporadic tumors and, most notably, hereditary non-polyposis colorectal cancer, if their MMR genes are incapacitated by mutation (see Table 4.1). The process of MMR has been conserved from bacteria to eukaryotes but only in bacteria has the mechanism been elucidated by which the parental DNA strand is identified vis-à-vis the strand containing a base substitution that is to be repaired.

In *E. coli*, the initial recognition of a lesion is followed by sequential recruitment of several proteins, as in NER (**Figure 4.13**). The ATP-dependent protein MutS (a dimer of 94 kDa subunits) recognizes a broad range of mispaired or up to four unpaired bases. It clamps onto the DNA and this is followed by ATP-dependent binding of the MutL ATPase (a dimer of 68 kDa subunits). Bidirectional scanning of the DNA ensues, driven by ATP hydrolysis. The dsDNA is drawn through into a growing loop until a 'strand-discrimination signal' is encountered in the form of a hemi-methylated GATC tetranucleotide palindrome. Newly synthesized DNA is unmethylated and remains so until it is acted upon by a specific Dam methyltransferase (see Box 2.4). Thus, in a mismatch, the base in the methylated strand can be assumed to be the correct one, and the opposing base (or loops) targeted for repair. At this point, an endonuclease, MutH (26 kDa), is recruited and activated to make a nick on the

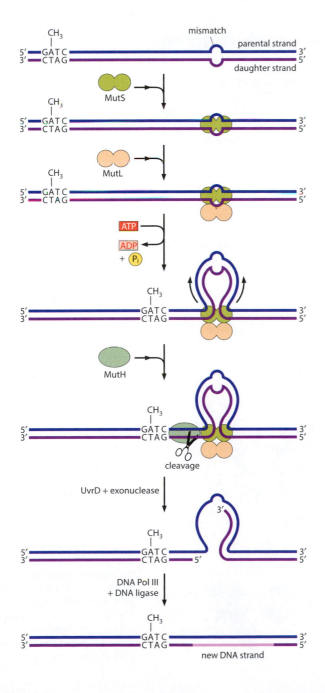

Figure 4.13 Mismatch repair in *E. coli*.
A mismatch is detected by the dimeric MutS, which recruits MutL and scanning begins on either side of the lesion. When a strand recognition signal (a hemi-methylated GATC sequence) is encountered in the parental DNA strand (in this instance the first one is found on the 5′ side of the lesion), the endonuclease MutH makes an incision in the daughter strand. A helicase (UvrD— see Figure 4.9) and exonuclease excise nucleotides of the daughter strand back beyond the lesion. DNA Pol III and DNA ligase, with appropriate accessory proteins, fill in the gap and seal the nick.

5′ side of the unmethylated GATC sequence. The helicase UvrD (Figure 4.9) unwinds the DNA towards the mismatch and the defective daughter strand, including the mismatched base, is removed by an exonuclease (3′ → 5′ in Figure 4.13). The gap thereby created is filled by DNA Pol III and DNA ligase, aided by SSB (Chapter 3). Note that the first strand-discrimination signal might instead be encountered in scanning on the other side of the lesion from that shown in Figure 4.13, in which case a helicase and exonuclease operating 5′ → 3′ would respectively unwind the DNA and degrade the daughter strand.

Either way, long tracts of the damaged strand may be removed, up to 1000 bp from the lesion. In terms of the large amounts of ATP consumed, both in DNA strand removal and resynthesis, this makes MMR an energetically costly form of DNA repair. How the parental DNA strand is recognized in organisms other than *E. coli* is as yet unclear. MutH is found only in Gram-negative bacteria and strand recognition does not involve hemi-methylated GATC sequences; rather, it may be a search for nicks in newly synthesized DNA strands. In principle, unprocessed *Okazaki fragments* in the lagging strand (see Chapter 3) could designate the newly synthesized DNA strand to be repaired, but this mechanism would not be applicable to mismatches occurring during synthesis of the leading strand of the replication fork, which occurs in a continuous manner.

MutS forms an asymmetric homodimer and mismatch recognition involves interaction with only one subunit

The MutS subunit has six domains (**Figure 4.14A**). Crystal structures of bacterial MutS binding to DNA that contains a mismatch show the dimer forming a clamp around it (Figure 4.14B). The positively charged clamp domains from both subunits contact the DNA sugar-phosphate backbone in a sequence-independent manner. The ATPase domain belongs to the ABC transporter family (as does UvrC in NER). The bound DNA is sharply kinked (~60°) at the mismatch, and the local bases are distorted, with the sugar puckering changed from C2′-*endo* (typical of B-form DNA) to C3′-*endo* (characteristic of A-form DNA). Interestingly, the dimer is asymmetric and only one of the two mismatch-recognition domains makes intimate contact with the mismatch in the bound DNA. This domain is on the side of the minor groove of the DNA widened by the kinking and distortion, whereas the other recognition domain makes limited contact with the DNA backbone elsewhere. In contrast to the BER and NER proteins discussed above, MutS does not flip the misincorporated base targeted for excision out of the double helix.

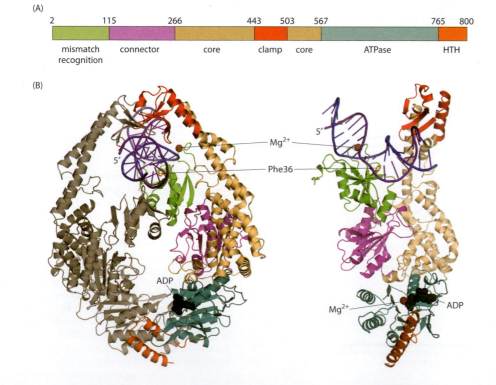

(A)

| 2 | 115 | 266 | 443 | 503 | 567 | 765 | 800 |

mismatch recognition | connector | core | clamp | core | ATPase | HTH

(B)

Figure 4.14 MutS dimer bound to a fragment of DNA with a G/T mismatch. (A) Domain map of *E. coli* MutS, residues 2–800; the C-terminal 53 aa are omitted. (B) Structure of the MutS dimer (residues 2–800) bound to a 30 bp fragment of dsDNA with a G/T mismatch at position 9. The domains in the mismatch-binding monomer are colored as in (A); the other subunit is in gray. The 5′ end of the damaged strand (purple) is indicated. The bound ADP molecule in each subunit is in dark gray in space-filling format. (PDB 1E3M)

Two potential ATP-binding sites are buried in the dimer interface of MutS, each with an ordered loop contributed *in trans* from the opposite subunit, but in the crystal structure, electron density is found for an ADP and Mg^{2+} only in the mismatch-binding subunit of the asymmetric dimer. This is consistent with biochemical data indicating that the two ATPase sites are non-equivalent and that alternating active sites might be a feature of MutS. Eukaryotic MutS homologs, such as MutSα (MSH2/MSH6) and MutSβ (MSH2/MSH3), are heterodimers, formalizing the asymmetry seen in the bacterial MutS homodimers. For MutS to bind to DNA, the clamp domains, disordered in crystal structures of MutS without DNA, provide an entry point for the DNA. Once DNA is bound, the clamp domains become ordered as they encircle the DNA. Single-molecule measurements in a light microscope with yeast MSH2/MSH6 proteins have demonstrated that they can slide along tethered, extended DNA by one-dimensional diffusion triggered by an ATP replacement of bound ADP but without the need for ATP hydrolysis. The search for mismatches might rest on continual testing of local instability in the DNA double helix and attempts by MutS to wedge a Phe residue (Figure 4.14) into the base stack. The diffusion coefficient is ~0.25 $\mu m^2/s$ and the DNA in *Saccharomyces cerevisiae* (2.4×10^7 bp) is 8160 μm in length. Assuming that there are ~1000 molecules of MSH2 per cell, it would take 2.9 min to scan the whole genome for mismatches if the DNA were naked and there was no impediment to MSH movement. It is of interest that MSH3 and MSH6 both bind to PCNA via PCNA-binding motifs in their N-terminal regions and the heterodimers of which they are part co-localize with DNA replication factories in S phase *in vivo*. Association with a *replisome* (DNA-replicating complex) would have the MMR proteins ideally placed to scan newly synthesized DNA that is free of nucleosomes and other interfering proteins. Understanding of mismatch recognition by MutS is now well advanced, whereas the mechanism for recruiting the next enzyme in the pathway, MutL, has yet to be worked out.

4.4   REPAIR OF DOUBLE-STRAND BREAKS

A double-strand break (DSB) affecting both phosphodiester backbones of dsDNA may occur through exposure to ionizing radiation, to free radicals generated in aerobic metabolism, or to anti-cancer drugs such as camptothecin, an inhibitor of topoisomerases. DSBs may also occur when DNA replication forks stall at sites of damage. DSBs are especially dangerous and difficult for the cell to repair. The ends of the break need to be detected and kept together and protected until a repair can be effected, to avoid nucleolytic degradation or aberrant fusion with other chromosomes. DSBs are also potent activators of cell cycle checkpoints (see Chapter 13) that arrest cell growth until repair of the DSB is complete; failure to do so could lead to catastrophic problems at mitosis.

The two pathways for DSB repair—**nonhomologous end joining** (**NHEJ**) and **homologous recombination** (**HR**)—differ mainly in the fidelity of maintaining sequence information at the break site. In NHEJ and its variant **microhomology-mediated end joining** (**MMEJ**), the broken DNA ends are ligated back together after resection of any damaged nucleotides. The trimming of damaged DNA ends can lead to a loss of sequence information. In contrast, in HR-mediated repair of DSBs, the identical sister chromatid or homologous chromosomal DNA is used as a template to reconstitute the original sequence or a close facsimile thereof across the site of the break (**Figure 4.15**). Until recently, HR was regarded as the predominant form of DSB repair in bacterial cells, but bacterial NHEJ has since been documented. In eukaryotes, NHEJ operates throughout the cell cycle, but is particularly important in quiescent cells such as neurons and in dividing cells during G1 and early S phase (see Chapter 13), before DNA replication has occurred. HR becomes important in late S and G2 phase, when sister chromatids are available to act as DNA templates for recombination.

As with DNA replication, DNA repair in eukaryotes has to take place in the context of intimate association of the DNA with histones and other proteins in chromatin. For the most part, we describe only the activity on bare DNA, but a growing body of evidence shows that chromatin structure and remodeling activities play significant roles in controlling access to DNA during many DNA repair processes.

Recognition of the broken DNA ends is the first step in DSB repair

The recognition of DSBs initiates the repair processes and, in eukaryotes, triggers the G2/M cell cycle checkpoint (see Chapter 13) to ensure that repair is complete before cell division

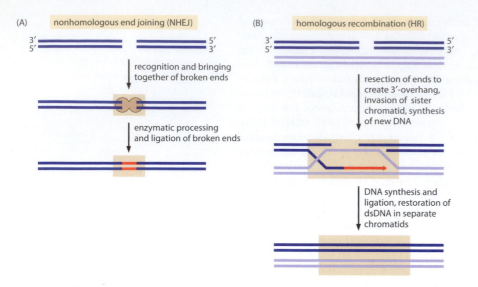

(A) nonhomologous end joining (NHEJ)

recognition and bringing together of broken ends

enzymatic processing and ligation of broken ends

(B) homologous recombination (HR)

resection of ends to create 3'-overhang, invasion of sister chromatid, synthesis of new DNA

DNA synthesis and ligation, restoration of dsDNA in separate chromatids

Figure 4.15 Schematic outlines of nonhomologous end joining and homologous recombination. (A) In NHEJ, the two broken DNA ends are brought together, processed, and ligated (participating proteins are in beige). (B) HR involves resection to create a long 3' overhang that invades the sister chromatid, followed by a complex mechanism of copying from the unbroken DNA in the sister chromatid. Participating proteins are omitted for clarity. In (A) and (B) the regions of DNA involved in processing and synthesis are highlighted. In NHEJ, newly synthesized DNA at the site of end joining (red) may be different from that in the original dsDNA; in HR, new DNA synthesis begins (red) with invasion of a 3' overhang and the breaks in both strands are repaired with complete fidelity to the original.

takes place. The key molecular machine for recognition of DSBs is the MR (Mre11-Rad50) complex, which exists in all three kingdoms of life as a heterotetramer, M_2R_2. In eukaryotes, the MR complex acquires an additional protein, Nbs1, creating the MRN complex ($M_2R_2N_2$). The MR complex holds DNA ends together via short- and long-range interactions while taking part in their remodeling, which varies with the particular DNA substrate presented. Often, the ends of broken DNA strands require enzymatic 'tidying up' before they can be rejoined. This generally means resection of the ends to create 3' overhangs. Moreover, the exposed DNA ends need to be protected from nonspecific degradation by nucleases and kept together in preparation for rejoining. The MR and MRN complexes not only have intrinsic exonuclease activity but can also interact with some other proteins, notably factors such as DNA endonuclease RBBP8 (CtBP-interacting protein, CtIP), ataxia telangiectasia mutated serine-protein kinase (ATM), ataxia telangiectasia and Rad3-related protein homolog serine/threonine-protein kinase (ATR), and mediator of the DNA-damage checkpoint 1 (MDC1). The MR complex participates in most if not all processes that involve DNA ends, including NHEJ, MMEJ, and HR forms of DSB repair, checkpoint signaling, the restart of stalled replication forks, meiotic recombination, and telomere maintenance (**Figure 4.16**).

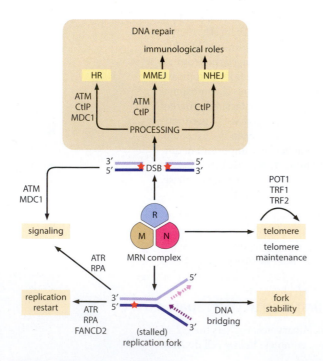

DNA repair

immunological roles

HR   MMEJ   NHEJ

ATM   ATM
CtIP   CtIP   CtIP
MDC1

PROCESSING

DSB

ATM
MDC1

signaling

ATR
RPA

MRN complex

POT1
TRF1
TRF2

telomere

telomere maintenance

replication restart

ATR
RPA
FANCD2

(stalled) replication fork

DNA bridging

fork stability

Figure 4.16 The MRN complex plays a central role in various DNA repair processes. The interaction of MRN with additional proteins, such as ATM, ATR, and CtIP, makes it a key sensor in detecting double-strand breaks (DSBs), designating different repair pathways (HR, microhomology mediated-end-joining, NHEJ), facilitating restart at stalled replication forks and triggering appropriate signaling pathways. Sites of DNA damage are indicated with red stars. It is also involved in telomere maintenance (Chapter 3). (Adapted from G.J. Williams, S.P. Lees-Miller and J.A. Tainer, *DNA Repair* 9:1299–1306, 2010. With permission from Elsevier)

The Mre11 component of the MRN complex is a dimeric nuclease that can bind DNA ends

Mre11 (meiotic recombination 11 homolog) is a multidomain, calcineurin-like phosphoesterase with multiple activities: Mg^{2+}-dependent 3′–5′ dsDNA exonuclease, ssDNA endonuclease, and DNA hairpin-opener (**Figure 4.17A**). Mre11 functions as a dimer that binds symmetrically to the broken ends of B-form dsDNA or asymmetrically to branched DNA such as a stalled replication fork (Figure 4.17B, C). In the dimer, the phosphoesterase and capping domains of each subunit bind to a DNA end, primarily contacting the minor groove and sugar–phosphate backbone in a sequence-independent manner. When bound to branched DNA, one Mre11 subunit engages the DNA and the other one undergoes a conformational change that blocks its DNA-binding activity by a rotation of the capping domain. The alternative conformation in the symmetrical form of Mre11 bound to a DSB might serve to activate ATM signaling, whereas an asymmetric Mre11 could activate ATR signaling in response to a stalled replication fork (Figure 4.16). Mutational analysis suggests that the 3′–5′ exonuclease activity of Mre11 is not critical for HR repair, but that the endonuclease activity is required to resect damaged DNA ends. Other enzymes bound to Mre11 in the MRN complex, among them CtIP, assist in resecting the DNA in a 5′ → 3′ manner as an early step in HR (see below).

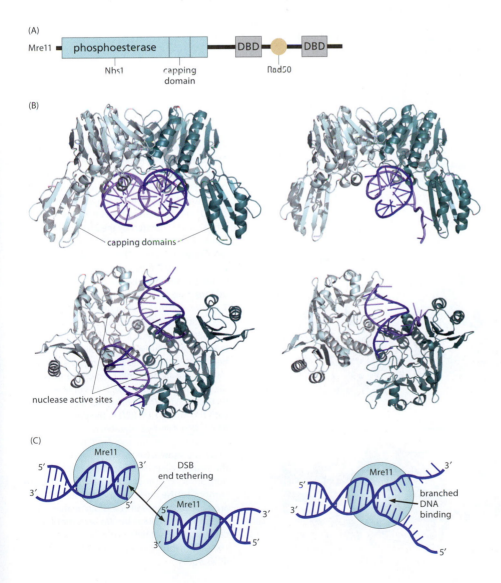

Figure 4.17 The Mre11 component of the MR complex. (A) Domain map of Mre1: the long C-terminal region contains DNA-binding domains (DBD) and a Rad50-binding motif. (B) Views of the dimeric *P. furiosus* Mre11 phosphoesterase domains (residues 1–342). Left, two DNA fragments are bound; right, a one-ended break mimicking a replication fork is bound. To ensure unidirectional binding, the DNA fragment 5′-CACAAGCTTTTGCTTGTGAC-3′ which adopts a hairpin shape with a 4T loop (underlined) was used in the crystallization of the two-ended break. In contrast, the fragment 5′-CGCGCACAAGCTTTTGCTTGTGGATA-3′, which also forms a hairpin shape with a 4T loop but has an open non-base-paired end, was used for single-end break crystallization. In the lower images, the structures are rotated by 90° about a horizontal axis. For clarity, one subunit is depicted in a lighter shade and the 4T loops are omitted. (C) Diagram of Mre11 dimers binding (left) two separate DNA ends or (right) a branched DNA. (PDB 3DSC and 3DSD) (C, adapted from R.S. Williams et al., *Cell* 135:97–109, 2008. With permission from Elsevier)

Rad50 is an ATPase that interacts with Mre11 and undergoes reversible ATP-dependent dimerization

Rad50 is a member of the ABC family of ATPases (see Box 3.1) that partners Mre11 in the MR complex. The Rad50 polypeptide contains a remarkably long tract of heptad repeats that separates the N-terminal end of the ABC ATPase housing the Walker A motif from the C-terminal end housing the Walker B motif (**Figure 4.18A**). The heptads, nearly 1000 aa in human Rad50 and 600 in the *Pyrococcus furiosus* protein, form an anti-parallel coiled coil with the Zn hook motif (see below) at its tip. The flexibility of the coiled coil allows the N- and C-terminal lobes of the Rad50 to come together for enzymatic activity. A crystal structure of the *Thermotoga maritima* Mre11 dimer complexed with two Rad50 ATPases containing truncated coiled coils shows the M_2R_2 in an 'open' conformation with an angle of ~120° between the two coiled coils (Figure 4.18B). A helix-loop-helix segment from the C-terminal end of the Mre11 phosphoesterase domain is flexibly linked and binds to the coiled coil motif of Rad50 adjacent to the ATPase domains.

ATP binding at the dimer interface of Rad50 activates the MR complex, bringing together the ATPase domains to facilitate interactions of conserved, functionally important residues at the interface (Figure 4.18C). ATP-dependent activation of Rad50 involves a substantial rotation of subunits that decreases the angle between the coiled coils. Cycles of ATP binding, hydrolysis, and release may control the conformational changes that drive the process of DNA loading, remodeling, and discharge from the MR complex.

The Zn hook projecting from the tip of the Rad50 coiled coil exposes a Zn^{2+}-binding, C-X-X-C motif that serves to dimerize two Rad50 subunits (**Figure 4.19A**). In the dimer, the two ATPases are tethered by an exceedingly long coiled coil and can adopt widely varying conformations separated by as much as ~1200 Å, which would be consistent with a role in bridging between two sister chromatids during DSB repair (Figure 4.19B).

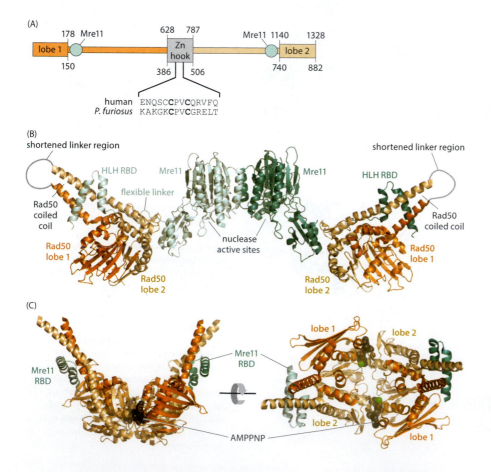

Figure 4.18 The Rad50 component of the MR complex. (A) Domain map of Rad50 subunits of eukaryotes (human numbering on top) and archaea (*P. furiosus* numbering on bottom). They differ in the length of the coiled-coil region, depicted in orange and wheat, and separated by the Zn hook region. Mre11-binding sites are indicated. The Cys-X-X-Cys sequences of the Zn hook regions of human and *P. furiosus* Rad50 chains are included for comparison. (B) Structure of a complex between the *T. maritima* Mre11 (full length 385 aa) and a reconstituted Rad50 [residues 1–190 (lobe 1) and 686–852 (lobe 2) joined by a synthetic linker (GlyGlyAlaGlyGlyAlaGlyGly)]. The helix-loop-helix Rad50-binding domain (HLH RBD) and its flexible linker region comprise the C-terminal residues 343–385 of Mre11. (PDB 3QG5) (C) *T. maritima* Rad50 complexed with the Mre11 HLH RBD in the presence of AMPPNP, a non-hydrolyzable ATP analog. Two Mg^{2+} (green) and the two AMPPNP molecules (gray) are shown. On the right-hand side, the structure has been rotated 90° about a horizontal axis. The two ATPase active sites are composite: the three phosphates of AMPPMP are bound by the Walker A motif of lobe 1 in one subunit, whereas the Walker B motif and catalytically essential Glu798 (see Box 3.1) come from lobe 2 in the other subunit. (PDB 3QF7)

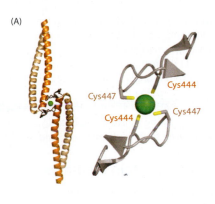

(A)

Figure 4.19 Coiled-coil regions hold the Rad50 ATPase domains together. (A) Crystal structure of the hook holding two *P. furiosus* Rad50 chains together. The two Cys residues in a loop at the tip of each of two coiled-coil regions (residues 396–506) form a tetrahedral adduct with a Zn²⁺ (green). (PDB 1L8D) (B) Schematic structure of Rad50 dimers in two orientations, held together by Zn hooks. The Mre11-binding sites are indicated with blue dotted circles. In the presence of ATP, two ATPase domains can come together direct from a single dimer (left) or from two adjacent dimers (right). The rotation of ATPase lobe 2 required for this to happen is indicated. In the extended form (right), the spacing of the ATPases (up to ~700 Å in *P. furiosus* and ~1200 Å in eukaryotes) is dictated by the length of the coiled-coil regions. The electron micrograph is of the human (M₂R₂)₂ heterotetramer in the extended conformation (~1200 Å between the two catalytic heads). cc, coiled coil; hd, catalytic head; hk, hook. (Schematic diagram redrawn from Williams et al., *DNA Repair* 9:1299–1306, 2010. With permission from Elsevier. Electron micrograph from Hopfner et al., *Nature* 418:562–566, 2002. With permission from Macmillan Publishers Ltd.)

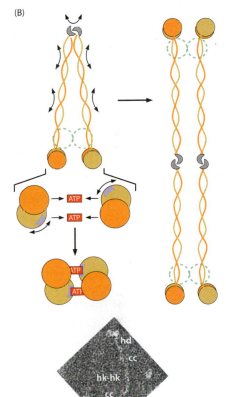

(B)

The Nbs1 component helps link the MRN complex to the appropriate DSB response

The third component of the vertebrate MRN complex is Nbs1, the product of the gene mutated in Nijmegen breakage syndrome 1 (called Xrs2 in yeast). Nbs1 does not directly participate in the DSB repair process but instead links MR complex activity to many different proteins that are in turn regulated by phosphorylation. Nbs1 contains several protein interaction domains in its N-terminal half, including an FHA (forkhead-associated) domain followed by two BRCT domains (**Figure 4.20A**) (see **Guide**). A protease-sensitive peptide links these domains to a C-terminal elongated chain with binding sites for Mre11 and the ATM kinase. A crystal structure of the N-terminal half of Nbs1 from *Schizosaccharomyces pombe* reveals an elongated structure in which the FHA domain and first BRCT domain make extensive contact and are less intimately associated with the second BRCT domain (Figure 4.20B). The FHA and BRCT domains function independently, interacting with phosphothreonine- and phosphoserine-containing proteins respectively. The interaction of the FHA domain with a pThr-containing peptide representing Ctp1 is illustrated in Figure 4.20B.

The modular structure of Nbs1 enables it to bind a variety of proteins: among them, CtIP (Sae2 in *S. cerevisiae*, Ctp1 in *S. pombe*), MDC1, ATR, ATM, Mre11, and the WRN (3′–5′) helicase, each with different cellular consequences for DSB repair (Figure 4.16). Nbs1 in the MRN complex bound to a DSB activates the ATM checkpoint kinase and is itself phosphorylated by ATM. ATM also phosphorylates the tails of histones H2AX adjacent to a DSB. In mammals, phosphorylated H2AX attracts MDC1 which like Nbs1 contains an N-terminal FHA domain followed by two BRCT domains. MDC1 binds phospho-H2AX through its two BRCT domains, and recruits more MRN-ATM complexes via interactions with Nbs1 to amplify and sustain the S phase checkpoint signal in response to DSBs. Lower eukaryotes lack MDC1 and other pathways may exist to reinforce checkpoint activation.

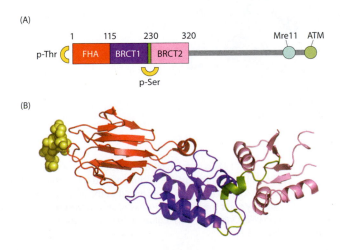

Figure 4.20 The Nbs1 component of the MRN complex. (A) Domain map. The sites for binding pThr- and pSer-containing proteins are marked, as are the binding sites for Mre11 and ATM in the elongated flexible C-terminal half of the protein. (B) Structure of the N-terminal half of Nbs1 from *S. pombe*, complexed with a small peptide (KKIQELD(pS)T(pT)DEDEI) representing the FHA-binding site on Ctp1. The FHA (residues 1–114), BRCT1 (residues 115–210), and BRCT2 (residues 230–320) domains, and the linker are colored as in (A). The Ctp1 peptide is shown as yellow spheres (only the residues close to the pThr exhibit electron density). (PDB 3HUF)

Many post-translational modifications of the MRN complex have been documented, including >50 *in vivo* phosphorylation sites and a few methylations and acetylations. These findings presage an intricate network for regulating MRN activity, although the functional effects of these post-translational modifications are not fully characterized. However, it is clear that there is a multiplicity of protein–protein interactions between Mre11, Rad50, and Nbs1 themselves and with partner proteins, with a corresponding multiplicity of states and conformational changes. This enables the MRN complex to act as a flexible sensor, effector, and signaling platform, able to initiate and engage in a wide array of responses to a DSB, according to the circumstance and need of the cell at any given time.

The Ku protein mediates DSB repair by the NHEJ pathway

Ku is another dimeric protein (Ku70/Ku80 in mammals) that recognizes the broken ends of DSBs and participates in recruiting other NHEJ factors, such as DNA ligase IV, to the break. The Ku heterodimer holds the fractured DNA ends together while they are tidied up by nucleases and polymerases ready for ligation. The structure of human Ku70/Ku80 complexed with a DNA fragment representing a dsDNA end (**Figure 4.21**) shows a quasi-symmetrical heterodimer despite the low sequence identity (<20%) of Ku70 and Ku80. Each subunit has an N-terminal α/β domain followed by a β-barrel domain containing a ~70-residue loop that completes a ring around the DNA cradled against the rest of the protein. Correspondingly, Ku binds tightly to DNA (K_d ~0.2 nM) via sequence-independent interactions confined to the sugar–phosphate backbone from the minor groove. However, the Ku ring binds to DNA in a preferred orientation with Ku70 proximal and Ku80 distal to the free DNA end. This polarity may reflect an inherent asymmetry in the ring and/or the electrostatic repulsion of DNA by a highly acidic N-terminal region in Ku70, which is disordered and may orient the ring vis-à-vis Ku80's DNA-binding interface. Ku-bound DNA ends are brought together in a 'presynaptic complex' that enables NHEJ activity. Ku is an efficient 5'-dRP/AP lyase that can incise DNA at terminal (5'-dRP) and near-terminal AP sites, which frequently accompany DSBs. Ku's lyase activity resembles that of the short patch BER/SSBR enzymes discussed above, nicking DNA on the 3' side of an abasic site. Inhibition of Ku's lyase activity interferes with NHEJ, indicating that processing of DNA ends before ligation is of crucial importance.

DNA-bound Ku recruits other factors essential for NHEJ

Once bound to DNA, the Ku heterodimer, perhaps as a result of a conformational change, recruits other proteins involved in NHEJ. The DNA-dependent protein kinase catalytic

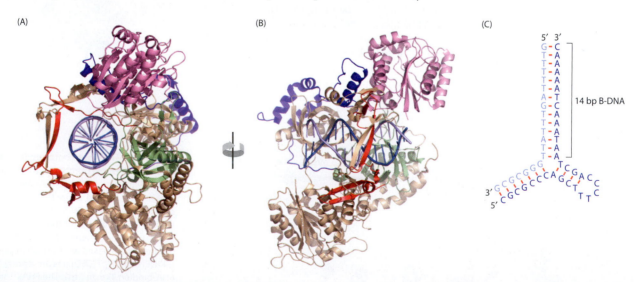

Figure 4.21 Structure of Ku heterodimer bound to a broken DNA end. (A) The domains in the Ku70 subunit are depicted as follows: N-terminal (residues 34–251), dark pink; β barrel (252–276 and 342–442), green; DNA-enclosing loop (277–341), red; C-terminal arm (443–534), blue. The Ku80 subunit is in wheat and the DNA in dark and light blue. (B) The same as in (A) but rotated 90° about a vertical axis. A C-terminal 19 kDa domain of Ku80 responsible for binding DNA-PKcs was omitted. (C) The dsDNA fragment used for crystallization has a large hairpin-looped region. This ensures that the end representing the DSB can only enter the Ku ring one way round, in the preferred orientation as *in vivo*. Only the 14 bp region of B-DNA is in the structure in (A). (PDB 1JEY)

(A)

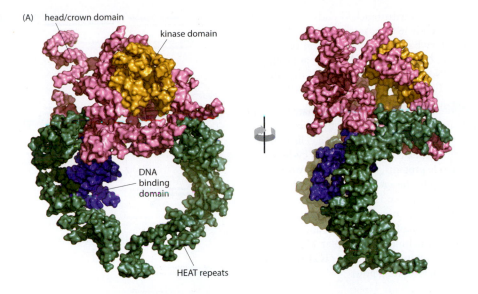

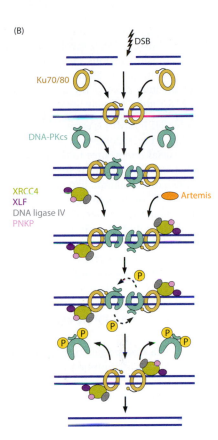

(B)

Figure 4.22 DNA-PKcs and a schematic outline of nonhomologous end joining (NHEJ). (A) Orthogonal views of a low-resolution structure of human DNA-PKcs showing the ring composed of HEAT repeats (in teal), the head/crown region (pink), the kinase domain (orange), and putative DNA-binding domain (blue). (PDB 1 KEJ) (B) Schematic outline of NJEH. The recruitment of DNA-PKcs moves the Ku rings inwards, allowing additional proteins to be recruited to resect the broken ends of the DNA. Among these proteins are possibly Artemis, PNKP/XRCC4, and MRN (MRN and DNA polymerases are not shown, for clarity). Autophosphorylation of DNA-PKcs which opens it up to release the DNA is depicted in yellow. The final step, ligation of the processed DNA ends, is catalyzed by a complex of ligase IV-XRCC4-XLF. (Redrawn after T.A. Dobbs, J.A. Tainer and S.P. Lees-Miller, *DNA Repair* 9:1307–1314, 2010. With permission from Elsevier.)

subunit (DNA PKcs) binds to the C-terminal region of the Ku80 subunit (absent from the crystal structure in Figure 4.21), activating the serine/threonine kinase activity of DNA PKcs that contributes to H2AX phosphorylation and checkpoint signaling (see above). DNA PKcs is a large (~4100 aa) horseshoe-shaped protein containing a long stretch of helix-turn-helix HEAT repeats (see **Guide**). The curved structure of DNA PKcs may enable it to encircle DNA (**Figure 4.22A**) by an opening and closing of the gap. A small globular domain, also formed of HEAT repeats, is a likely DNA-binding site, with the kinase domain occupying a 'head-crown' region opposite the gap (Figure 4.22A).

The role of DNA PKcs autophosphorylation in promoting NHEJ is unclear, but it may serve to remodel the repair complex by pushing the Ku ring inward, away from the broken DNA end, to create space for other repair factors (Figure 4.22B). The identities of the participating factors will depend on the nature of the DNA break (incompatible 5′ overhang, 3′ overhang, abnormal hairpin, etc.). The MRN complex is one of the important factors recruited to DSBs in association with NHEJ; others are the endonuclease Artemis activated by DNA PKcs and polynucleotide kinase/phosphatase (PNKP). Incompatible overhangs at DNA ends may be filled in by DNA polymerases μ and λ, both of which contain BRCT domains at their N termini that bind to Ku70/Ku80. The final step of NHEJ, ligation of the processed DNA ends, is carried out by the dedicated DNA ligase IV/XRCC4/XLF complex.

In bacterial NHEJ, no equivalent of DNA PKcs has been identified, but a counterpart of Ku is known plus a dedicated DNA ligase (LigD) of unusual complexity, containing an ATP (not NAD$^+$)-dependent ligase domain, a polymerase domain, and a phosphoesterase domain. Much more remains to be discovered.

MMEJ is a variant of NHEJ

DSBs with irregular structures, such as covalent attachment of polypeptides after treatment of patients with camptothecin, or the existence of hairpins in the DNA, pose challenges for

Ku binding and conventional NHEJ. The repair of DSBs with blocked ends can be completed by MMEJ, which involves resection of both ends to expose single-stranded overhangs with short regions (5–20 nt) of complementary base sequence that can participate in base-pairing interactions between the ends of a DSB (**Figure 4.23**). After annealing of complementary strands, the noncomplementary DNA flaps are removed, and ligation follows. MRN and CtIP play important, but as yet not fully understood, roles in MMEJ.

As with NHEJ, MMEJ is error-prone because of potential loss of an indeterminate length of DNA by end resection before ligation of the new junction. Nevertheless, the two processes are of crucial importance to the repair of DNA, particularly in nondividing cells that lack the alternative HR pathway for repairing DSBs during the S phase of the cell cycle. Given that much of the mammalian genome does not encode proteins, the 'scars' left by the repair process (2000 or more in the genome of a human being by the age of 70) appear to be a lesser problem than the disaster of leaving broken DNA ends unmended. Moreover, NHEJ is put to good use in diversification of the antibody repertoire, rejoining DSBs deliberately created by the RAG1/RAG2 protein complex in the recombination of multiple V, D, and J gene segments, and in the process of immunoglobulin class switching (see Chapter 17). However, the ability to restore the original DNA sequence in rejoining broken ends is also of crucial importance, and it is to homologous recombination that we turn now.

HR operates by DNA strand exchange in all three kingdoms of life

The ability to retrieve genetic information by the exchange of strands between homologous DNA molecules is common to bacteria, archaea, and eukaryotes and must have arisen early in evolution. In addition to repairing DSBs, strand exchange is the basis of reactivating replication forks that have stalled as a consequence of DNA damage and for reshuffling genetic material during *meiosis* (Chapter 13). As in NHEJ, recruitment of the MRN complex may help tether the broken DNA ends with respect to the sister chromatid that is to template HR repair (Figure 4.19B). In eukaryotes, the MRN complex serves to recruit the ATM checkpoint-activated kinase (Figure 4.16), leading to phosphorylation of H2AX histones in chromatin near the breakpoint and the activation of the cell cycle checkpoint. However, in contrast to NHEJ and MMEJ, which involve a limited amount of DNA end resection, in HR there is extensive resection to create long 3′ overhangs required for invasion of the sister chromatid. The DNA end resection step is a tipping point between the NHEJ and HR pathways for repair of DSBs. Extensive resection of DSBs in the HR pathway requires the coordinated activity of specialized nucleases and helicases, the best characterized of which is the bacterial RecBCD complex described below. In eukaryotes (using *S. cerevisiae* as model), the experimental evidence is that the helicase Sgs1 and two exonucleases, Dna2 and Exo1, are involved. The new ssDNA in the 3′ overhang is protected by interaction with the relevant ssDNA-binding protein (RPA in eukaryotes).

The 3′ overhang formed by resection invades the sister chromatid DNA and anneals with a complementary DNA strand that will serve as a template for extension of the invading strand by DNA polymerase. Thus, the homologous sister chromatid provides the sequence that will be joined to the broken ends of the DSB. Following strand invasion and DNA synthesis, the repair can take place in several ways, notably **synthesis-dependent strand annealing (SDSA)** and **double-strand break repair (DSBR)** (**Figure 4.24**). In SDSA, DNA synthesis and ligation repair the DSB, and the DNA in the sister chromatid is returned unchanged. In DSBR, both ends of the break invade the sister chromatid, creating a double crossover structure that can be resolved either with or without physical exchange (crossover) of the DNA segments flanking the crossovers. Mitotic DSBR normally operates without crossover, causing only a local change in the sequences around the break. However, programmed HR in meiosis is responsible for the multiple crossovers that generate genetic diversity.

DNA strand exchange is catalyzed by filaments of a RecA family ATPase

DNA strand exchange during HR is promoted by a highly conserved ATP-dependent **recombinase** (~40 kDa) called RecA in bacteria, RadA in archaea, and Rad51 in eukaryotes. Its core is an ATPase domain typical of the RecA class of proteins that undergo conformational changes coupled with ATP hydrolysis to perform mechanical work on a DNA substrate (see Box 3.1). The ability of the RecA, RadA, and Rad51 recombinases to assemble

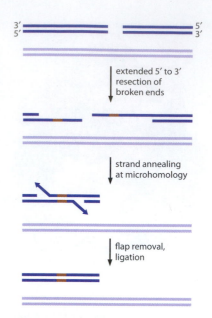

Figure 4.23 Schematic outline of microhomology-mediated end joining (MMEJ). The dsDNA ends are resected 5′ → 3′ until a suitable microhomology (5–20 residues, in brown) between one strand and the other is revealed. Any DNA flap this creates is removed and the new juxtaposed ends are ligated. There will be a loss in DNA sequence around the end-joining site, whereas the sister chromatid is unaffected and remains full length.

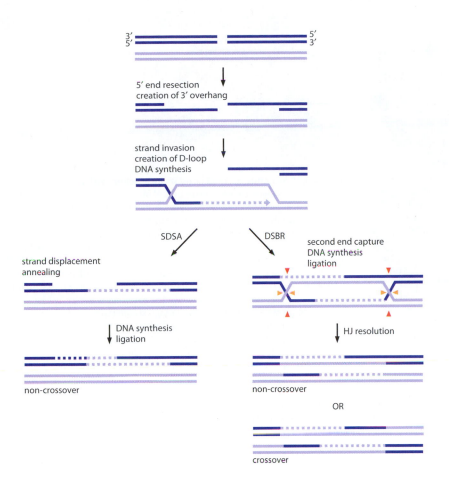

Figure 4.24 Schematic outline of homology-directed repair. A double-strand break (DSB) is resected to create long 3′ overhangs, one of which invades the sister chromatid and becomes base-paired with the complementary sequence. The displaced strand forms a D-loop and a DNA polymerase begins to synthesize new DNA (in direction of arrowhead). In SDSA (left), the D-loop migrates and polymerases synthesize new DNA (shown as dashed, and color-coded according to which complementary strand is being copied). The repaired DNA strands are ligated. In DSB repair (DSBR) (right), capture of the second end creates another Holliday junction (HJ), new DNA (dashes) is synthesized, and the completed DNA strands are ligated. Resolution of the HJs can happen in two ways: one without crossover (yellow arrowhead at both HJs), and one with crossover (yellow arrowhead at one HJ and red arrowhead at the other). (Redrawn from F P. Sung and H. Klein, *Nat. Rev. Mol. Cell Biol.* 7:739–750, 2006. With permission from Macmillan Publishers Ltd.)

into rings or helical assemblies is crucial for the remodeling of DNA. The Rad51 recombinase displaces the ssDNA-binding protein from the 3′ overhang of ssDNA with the help of mediator proteins that load Rad51 onto DNA. This generates a helical recombinase/DNA filament in the form of a right-handed solenoid. The *S. cerevisiae* Rad51 filament (**Figure 4.25A**) has approximately six subunits per turn, with the ATPase active sites in the interfaces between adjacent protomers. The pitch is 130 Å, at the high end of the 90–120 Å range reported for recombinase filaments. Electron microscopy (see **Methods**) of human Rad51/DNA filaments reveals a similar helical structure, but with a pitch of ~90 Å (Figure 4.25B). Association of protomers in the filament is guided by a short hydrophobic peptide external to the ATPase domain of one protomer, forming an additional strand on the edge of a β sheet of a neighboring protomer (Figure 4.25C).

Single-molecule studies have shown that as few as four or five recombinase protomers can cooperatively initiate formation of nucleoprotein filaments with ssDNA, followed by rapid extension of the filament at rates of 3–10 nm/s. Protomers associate and dissociate at both ends of the filament, but a higher binding rate at the 3′ end means that net overall growth is in the 5′ → 3′ direction. RecA/Rad51 filaments adopt two conformations. In the presence of DNA and a non-hydrolyzable analog of ATP, it is narrower and extended, with a pitch of 91–97 Å, ~6.2 RecA protomers per turn, ~3 nucleotides per protomer, and an average rise of ~5.1 Å per base in the DNA (compared with 3.4 Å per base in B-form DNA). This is thought to be the active form that catalyzes strand exchange. In the absence of DNA, the filament is wider and compressed, with a pitch of ~83 Å. These numbers can be regarded as averages, with differences between individual protomers possible within a filament. Structural adaptability is likely to be important for catalyzing the DNA strand exchange reaction.

A synaptic complex between a RecA nucleoprotein filament and sister chromatid DNA promotes fidelity in strand exchange

DNA strand exchange can occur *in vitro* by incubating the recombinase with substrate DNAs in the presence of ATP and Mg^{2+}. The details are unclear, but the recombinase filament wrapped around the ssDNA promotes invasion of the target dsDNA by the ssDNA

(A)

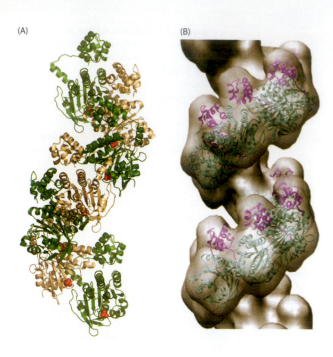

(B)

(C)

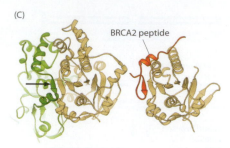

BRCA2 peptide

Figure 4.25 Structure of Rad51 and its interaction with BRCA2. (A) In this crystal, the *S. cerevisiae* Rad51 forms helical filaments with a pitch of 130 Å; here, seven adjacent ATPase protomers (alternating green and wheat) in one such helix are shown. The filament has approximate sixfold screw symmetry. The red balls represent sulfate ions binding in place of phosphate in the ATPase active sites between adjacent protomers. For crystallization, a potentially disordered N-terminal segment of 79 aa (unique to yeast) was removed. (PDB 1SZP) (B) Electron microscopy reconstruction of the Rad51/DNA filament shown as a gray transparent surface with the Rad51 subunit crystal structure from (A) fitted in (the core is in cyan and the N-terminal domain in magenta). (C) Formation of Rad51 filament involves a β strand (arrowed) from one subunit (green) adding to the edge of a β sheet in a neighboring subunit (wheat). A peptide region (red) from the BRCA2 protein (human) can mimic this interaction and prevent filament formation. (PDB 1NOW) (B, from D. Lucarelli et al., *J. Mol. Biol.* 391:269–274, 2009. With permission from Elsevier.")

to create a three-stranded joint molecule by a process known as **synapsis**. The search for a region of homology then aligns the invading ssDNA with the complementary strand of the target DNA in a strand exchange reaction powered by ATP hydrolysis (**Figure 4.26A**).

Crystal structures of the bacterial RecA recombinase in complexes with ssDNA and dsDNA have given snapshots of pre- and postsynaptic states of an HR nucleoprotein filament (Figure 4.26B, C). In the ssDNA–RecA complex, each RecA subunit binds exactly three nucleotides (a nucleotide triplet) and the ssDNA is helically arranged around the filament

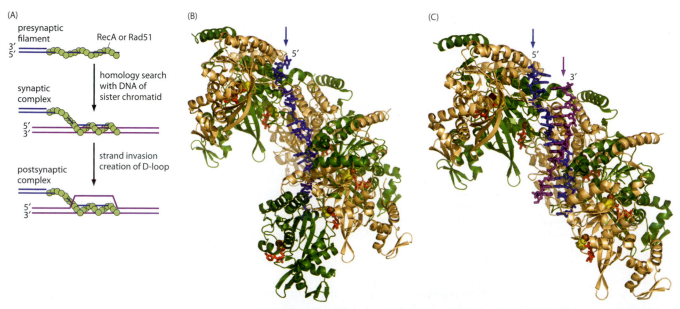

Figure 4.26 Presynaptic and postsynaptic structures of RecA filaments with ssDNA and dsDNA. (A) Schematic model of the formation of a synaptic complex and creation of a D-loop. (B) The presynaptic complex of an *E. coli* RecA$_6$-(ADP-AlF$_4$-Mg^{2+})$_6$ filament with an oligonucleotide, (dT)$_{18}$ in blue, representing an ssDNA 3′ overhang. RecA subunits alternate in green and wheat. For crystallization, the RecA filament was generated from expression of a gene encoding a linear array of six RecA subunits connected by 14-aa linkers. The N-terminal

RecA subunit is at the top. The ADP(red)-AlF$_4$-Mg^{2+} (yellow) molecules lie in the active sites between adjacent RecA subunits. Only 15 of the 18 nucleotides are ordered. (C) A putative postsynaptic complex generated by soaking crystals of a RecA$_5$-(ADP-AlF$_4$-Mg^{2+})$_5$-(dT)$_{15}$ filament in a solution containing the complementary (dA)$_{12}$. The (dA)$_{12}$ strand is in purple; otherwise the color code is as in (B). The (dT)$_{15}$ and (dA)$_{12}$ strands have 13 and 10 ordered nucleotides, respectively. (PDB 3CMU and 3 CMX)

axis, with 18.5 nucleotides per turn (pitch 94 Å) and an average rise per base of 5.08 Å. The DNA is thus stretched, as expected (see above), but the stretching is found to be far from uniform: each nucleotide triplet is arranged more like a strand in B-form DNA (two of the three bases stacking with a spacing of 3.5–4.2 Å), whereas there is a much longer (7.2 Å) axial rise coupled with a negative twist between the third base of one triplet and the first base of the succeeding one. The sugar–phosphate backbone of the ssDNA is embedded in the protein subunits of the filament, which helps to compensate for the lack of base stacking in the inter-triplet gaps. In contrast, the bases in the triplets are exposed, poised to begin homology searching a heterologous DNA strand by means of Watson–Crick hydrogen-bonding interactions.

The structure of a RecA–dsDNA complex supports this idea (Figure 4.26C). The DNA strand mimicking that of a newly exchanged strand from the sister chromatid dsDNA is anti-parallel, held in place by Watson–Crick hydrogen bonds and making very few contacts with the RecA protein. The filament parameters are not much changed but the gap between adjacent triplets (now as base pairs) has increased to 8.4 Å. The limited contact between the exchanged strand and the RecA protein filament could be important in ensuring the fidelity of HR, in that homology search and strand exchange are dominated by base-pairing interactions, and the RecA nucleoprotein filament templating a B-form DNA helix that fosters Watson–Crick base pairing and helps to exclude erroneous forms (for example, mismatched Hoogsteen).

Coordinated activities of helicases and nucleases generate the long 3′ overhangs for strand invasion

The bacterial (E. coli) RecBCD complex is a particularly well-characterized example of a molecular machine catalyzing DNA end resection for HR. It operates in conjunction with a **Chi (crossover hot spot instigator)** sequence in the DNA, about 1000 of which are found in the E. coli chromosome. RecB and RecD are both ATP-dependent helicases belonging to the SF1 structural class, which has two RecA-like ATPase domains per subunit (see Box 3.1). However, crucially, they display opposite polarity of translocation on DNA strands: RecB (subclass SF1A) moves in the 3′ → 5′ direction, whereas RecD (subclass SF1B) moves in the 5′ → 3′ direction. RecB protein also has a single nuclease domain as its C-terminal part and RecC resembles it in having an N-terminal SF1A helicase domain and a C-terminal nuclease domain. However, the nuclease domains in both RecB and RecC have critical active site residues substituted and are catalytically inactive. The 330 kDa RecBCD heterotrimer is illustrated in **Figure 4.27A**.

With its RecB and RecD helicase motors operating with opposite polarities on complementary DNA strands, RecBCD unwinds dsDNA with great speed (1–2 kb/s) and processivity (up to 30 kb). ATP hydrolysis drives translocation on DNA, dragging the DNA strands through the complex and forcing them apart at a 'pin' region contributed by the catalytically

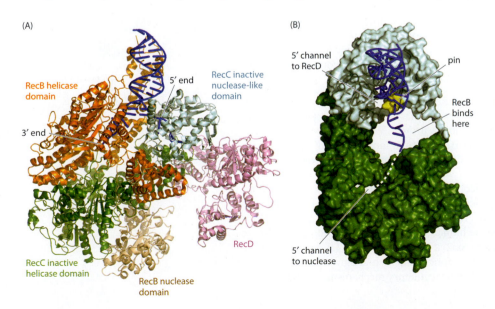

(A)

RecB helicase domain

5′ end

RecC inactive nuclease-like domain

3′ end

RecD

RecC inactive helicase domain

RecB nuclease domain

(B)

5′ channel to RecD

pin

RecB binds here

5′ channel to nuclease

Figure 4.27 Structure of the *E. coli* RecBCD complex. (A) Intact RecBCD crystallized with a 43-mer oligodeoxyribonucleotide (blue) which forms a five-base hairpin loop at one end and a 19 bp ds region with blunt end at the other. At the blunt end, 4 bp have been unwound, creating a forked structure with separate 5′ and 3′ ends. (B) RecC structure from the complex, omitting RecB- and RecD-associated density. The 5′ and 3′ ends of the DNA are heading towards different tunnels through RecC, separated by a 'pin' region (Asn-Met-Met-Val-Arg, residues 1078–1082) in yellow. RecC is shown in surface representation, color-coded as in (A).

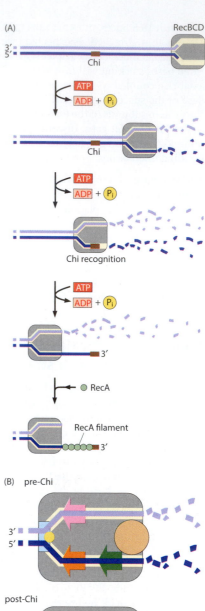

Figure 4.28 Schematic model of the catalytic activity of the RecBCD complex. (A) The helicases in RecBCD work their way along both DNA strands and the nuclease domain of RecB degrades the two emerging strands, discarding the products, until the complex encounters a Chi sequence in the DNA. Degradation of the 3′-terminal strand then ceases, while that of the 5′-terminal strand continues. RecBCD helps load RecA onto the 3′-ssDNA thus formed to create a presynaptic recombinase filament. (B) Arrangement of RecBCD protein domains and their contributions to catalytic activity. Arrowheads represent helicases, color-coded as in Figure 4.27 (that of RecC is inactive but recognizes the Chi sequence). The large wheat-colored circle represents the nuclease domain of RecB, and the small yellow circle represents the pin region in RecC where the DNA strands are separated. (Redrawn from J.T. Yeeles and M.S. Dillingham, *DNA Repair* 9:276–285, 2010. With permission from Elsevier.")

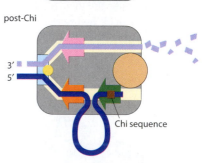

inactive RecC subunit (Figure 4.27B). Downstream of the pin, the 3′ ssDNA tail is acted upon by RecB whereas the 5′ tail is dealt with by RecD. The strands converge again at the nuclease domain in the RecB subunit, where they are intermittently chopped off by the RecB nuclease (**Figure 4.28A**) as the DNA exits the complex. This coordinated activity, often called a dual motor–single nuclease mechanism, continues until RecBCD encounters a Chi sequence (5′-GCTGGTGG-3′ in *E. coli* but different in other bacteria) on the 5′–3′ strand. Interaction of the Chi sequence with the inactive helicase domain of RecC brings about a major change in the enzymatic behavior of the RecBCD complex (Figure 4.28B). The complex pauses at Chi and then resumes translocation but at reduced speed and without digestion of the 5′–3′ strand following the encounter with the Chi sequence. In contrast, the frequency of cleavage events on the 3′–5′ strand is increased after Chi recognition, generating the 3′ tail of ssDNA that will be a substrate for the RecA recombinase. The RecBCD complex eventually disengages from the 3′ tail and RecA loads on the ssDNA, generating a nucleoprotein filament for the HR-catalyzed strand exchange reaction.

DNA end resection and the loading of RecA onto the 3′ ssDNA overhangs are integrated activities of the bacterial RecBCD complex. In archaea, yeast, and other eukaryotes, a growing number of helicases and nucleases have been identified with roles in resecting DNA ends before HR. The MRN complex (see above) is involved, acting probably in conjunction with the Sae2/CtIP nuclease, together with extensive resection promoted by Sgs1 (an SF2 helicase), Exo1 (a 5′–3′ exonuclease), and Dna2 (a 5′–3′ helicase-endonuclease). However, no equivalent of the Chi sequence has been discovered.

Assembly and disassembly of the recombinase nucleoprotein filament are tightly regulated

The RecA and Rad51 recombinases are essential DNA repair proteins with potentially lethal consequences for the cell if they are not properly regulated. All stages of the strand exchange reaction promoted by the RecA/Rad51/RadA family of recombinases are regulated by interactions with proteins known as recombination mediators that control nucleoprotein filament assembly and disassembly. Unlike the very high degree of sequence and structural conservation observed for the recombinases, the prokaryotic and eukaryotic mediator proteins appear unrelated and appear to have evolved largely independently.

In bacteria, the RecBCD complex serves to load RecA onto ssDNA, creating a nucleoprotein filament. In higher eukaryotes, the Rad51 recombinase is loaded onto DNA by a specific mediator protein, the breast cancer susceptibility protein BRCA2. A repeated sequence motif (BRC, ~30 aa) located in the middle of BRCA2 functions as a binding epitope for Rad51, mimicking the β strand of Rad51 that contributes inter-subunit interactions in the nucleoprotein filament (Figure 4.25C). In this way, BRCA2 can sequester Rad51 monomers and deliver them to the growing nucleoprotein filament. The biological importance of BRCA2 is underlined by the high-risk susceptibility to breast, ovarian, and other cancers associated with certain mutations in the BRC repeats of BRCA2 that weaken the affinity for Rad51. In *S. cerevisiae*, the Sgs1 helicase (the human counterpart is the Bloom's syndrome protein, BLM) appears to modulate HR, and Srs2, a protein with an ATP-dependent 3′–5′ helicase activity, can dismantle the Rad51 presynaptic filament, highlighting additional layers of regulation in the HR pathway. The analogous regulatory mechanisms for mammalian HR are an active area of research.

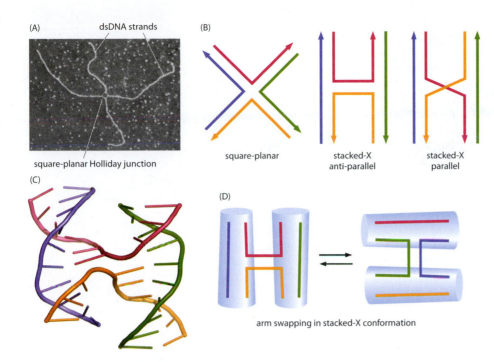

(A) dsDNA strands

square-planar Holliday junction

(B) square-planar   stacked-X anti-parallel   stacked-X parallel

(C)

(D) arm swapping in stacked-X conformation

Figure 4.29 Holliday junctions (HJs).
(A) Electron micrograph of a recombination intermediate HJ found in *E. coli* colicin plasmid DNA. (B) Schematic drawings of HJ DNA in square-planar and stacked-X arrangements. (C) Crystal structure of a fragment of DNA, d(CCGGTACCGG), forming a stacked-X HJ. (PDB 1DCW) (D) Reversible rearrangement of arms in antiparallel stacked-X conformation. (Adapted from H. Potter and D. Dressler, *Proc. Natl Acad. Sci. USA* 73:3000–3004, 1976.)

Four-way (Holliday) junctions are key intermediates in DSB repair and meiosis

The four-way junction, also called **Holliday junction** (**HJ**) in recognition of its proposer (Robin Holliday, in 1964), was originally postulated as a structural intermediate in the process of genetic exchange by meiotic recombination. In an HJ, two DNA strands in two double helices have exchanged partners and the resulting crossover junction (**Figure 4.29A**) displays important structural and dynamic features. In low salt solution, HRs adopt a planar cruciform structure, in which each arm points to the corner of a square (square-planar conformation), thereby minimizing electrostatic repulsion between the phosphodiester backbones. The square-planar conformation is also seen in structures of HJ DNAs bound to enzymes that remodel or resolve HJ structures. In the presence of divalent cations such as Mg^{2+} or a higher concentration of monovalent cations, a protein-free HJ folds to a more compact, stacked structure (stacked-X) featuring the pairwise, co-axial stacking of helical arms into two semi-continuous duplexes (Figure 4.29B), which are related by an ~50° right-handed twist of their axes. In principle, two stacked-X conformations are possible, related by a 180° rotation of one duplex around an axis through the midpoint of the two duplexes; however, only the anti-parallel stacked-X structure is observed in solution (Figure 4.29B, C). The arms of the anti-parallel stacked-X structure can swap stacking partners, leading to two different conformations that undergo rapid inter-conversion via the square-planar intermediate (Figure 4.29D). In addition, the crossover point of the HJ can translocate by a process termed **branch migration**, so long as the flanking DNA sequences are homologous and can base-pair with one another. Branch migration is an isoenergetic reaction, but in practice is catalyzed by helicase-like remodeling enzymes that move the crossover point in a directed manner. HJ remodeling during HR can influence the genetic outcome of DSBR (Figure 4.24), as the isomers can be resolved by the nicking endonucleases to create either patched or spliced gene products.

Branch migration is driven by the RuvAB complex

The best-studied system of branch migration is that of the RuvA and RuvB proteins in *E. coli*. RuvA is a DNA-binding protein that specifically recognizes an HJ, whereas RuvB is a helicase of the AAA+ ATPase family. A tetramer of RuvA subunits forms a large, shallow concave surface, positively charged overall, to which the square-planar HJ DNA can adhere (**Figure 4.30A–C**). Fourfold symmetry is retained in the RuvA-HJ complex, with 8 bp of B-DNA in each arm covered by a RuvA monomer. The two helix-loop-helix motifs in the middle domain (II) of RuvA make interactions with the phosphodiester backbone of the minor groove of the DNA. A long β hairpin with a pair of acidic residues (Glu55, Asp56) at

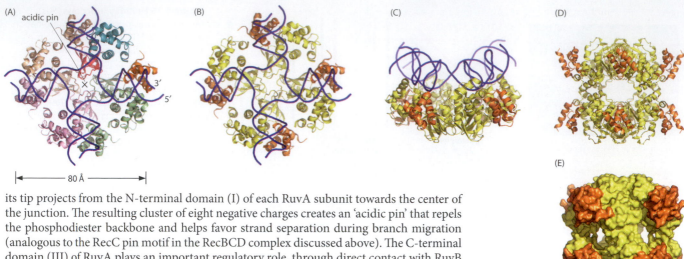

its tip projects from the N-terminal domain (I) of each RuvA subunit towards the center of the junction. The resulting cluster of eight negative charges creates an 'acidic pin' that repels the phosphodiester backbone and helps favor strand separation during branch migration (analogous to the RecC pin motif in the RecBCD complex discussed above). The C-terminal domain (III) of RuvA plays an important regulatory role, through direct contact with RuvB (see below). Two RuvA tetramers can associate face-to-face in an octameric assembly with a large internal cavity shaped as two perpendicular channels into which the four arms of an HJ in the square-planar conformation could fit (Figure 4.30D, E).

A combination of biochemical, electron microscopy, and X-ray crystallographic studies has led to a model for HJ remodeling and branch migration catalyzed by the RuvAB complex (**Figure 4.31**). Two hexamers of the RuvB helicase in opposing orientations encircle opposite arms of the HJ as these emerge from the octameric RuvA complex. Each hexamer is held by interactions of two RuvB protomers with domains III of two RuvA subunits, leaving four copies of RuvA domain III without a partner. Branch migration is thought to derive from the concerted action of the two RuvB helicases, driven by ATP hydrolysis, acting to

Figure 4.30 Structure of RuvA. (A) The *E. coli* RuvA tetramer with a Holliday junction (HJ) DNA bound. Three subunits are in different colors (green, pink, wheat), while the fourth subunit has domain I (residues 1–64) in red; domain II (residues 65–150) in teal; domain III (residues 155–203) in orange. Residues 151–154 are disordered. The four arms of the HJ DNA are in blue. (PDB 1C7Y) (B and C) Orthogonal views of the same structure as in (A) but with all four subunits yellow (domains I and II) and orange (domain III). (D) Octameric form of RuvA from *Mycobacterium leprae*, with two RuvA tetramers face-to-face, color-coded as in (B). (E) Surface representation of the structure in (D), viewed across an edge. (PDB 1BVS)

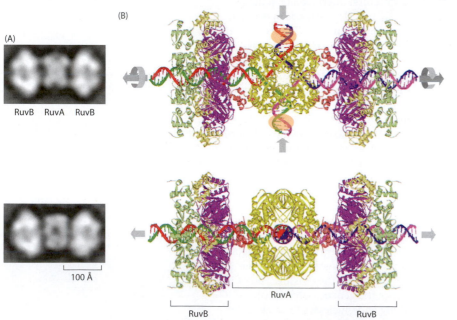

Figure 4.31 Branch migration catalyzed by RuvAB. (A) Averaged electron micrographs showing two side views of a *T. thermophilus* RuvAB-HJ DNA ternary complex, contrasted by negative staining. The views are related by a 90° rotation around the *x* axis. The Holliday junction (HJ), which has 36 bp on the long arm and 13 bp on the short arm, is not well visible. (B) Corresponding side views of a model constructed from crystal structures of RuvA and an RuvA domain III-RuvB complex. Two hexameric rings of RuvB (domain N, purple; domain M, pale yellow; domain C, green) pull two spirally rotating DNA duplexes out from the HJ, on opposite sides of the complex (arrows). RuvA, a double tetramer (domains I and II, yellow; domains III, orange) is depicted with the other two duplexes of the HJ entering the complex at right angles to the two exits (gray arrows in upper panel). A RuvB hexamer has six binding sites for RuvA, of which only two can be occupied, due to the 6:4 symmetry mismatch. Binding of two domains III elicits a conformational change in RuvB, enhancing its ATP-dependent helicase activity. (A, from K. Yamada et al., *Mol. Cell* 10:671–681 2002. With permission from Elsevier. B, adapted from K. Yamada et al., Mol. Cell 10:671-681, 2002. With permission from Elsevier.)

pull the opposing arms of the HJ through and across the pin region in RuvA, separating the DNA strands and pushing recombined DNA products out through the ends of the complex. Other helicases similarly encircling and acting on dsDNA in DNA replication are described in Chapter 3.

HJs are resolved by specific endonucleases or dissolved by a helicase/topoisomerase

After the process of branch migration, the cruciform DNA can be resolved by HJ-specific nucleases (resolvases), which introduce symmetrical single-strand cleavages of the DNA near the branch point of the junction (Figure 4.24). This is followed by joining of the new ends by DNA ligase, generating two separate molecules of duplex DNA and concluding the process of HR. It is likely that these nucleases (for example, RuvC in *E. coli*) displace one RuvA tetramer to gain access to the DNA. The structures of two HJ resolvases, from bacteriophages T7 (endo I) and T4 (endo VII), have been determined in complex with HJ DNA. They exhibit substantial differences in HJ recognition and handling, but the HJ bound by both enzymes is in an unusual conformation (**Figure 4.32**), that is intermediate between the square-planar and stacked-X conformations observed for free HJs in solution.

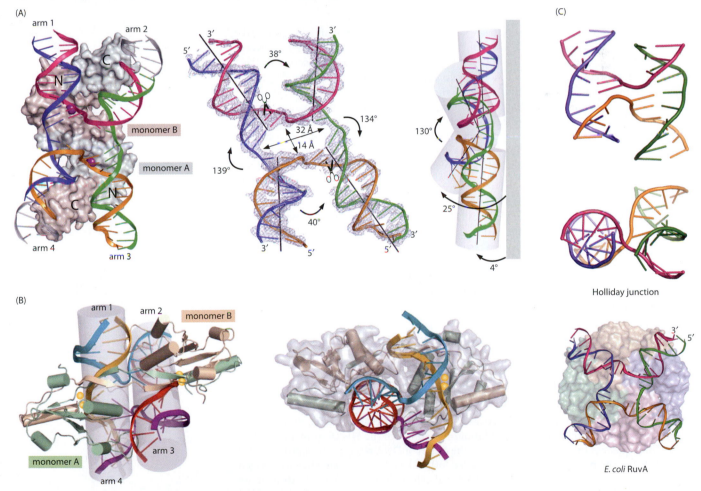

Figure 4.32 Structures of two Holliday junction (HJ) resolvases complexed with synthetic HJs. (A) Bacteriophage T4 endonuclease VII-HJ complex. The two protein subunits (molecular surface representation) are in pale lilac and pale blue, with N and C termini marked. The four DNA strands are color-coded blue and green for non-exchanging strands, magenta and orange for exchanging strands (symmetry-related DNAs are in gray). In the second and third panels, angles subtended by the various arms with each other and with a flat surface are given, Mg^{2+} ions at the two active sites are depicted as purple balls, and the positions of single-strand cleavage in arms 1 and 3 are marked with scissors. (PDB 2QNF) (B) Bacteriophage T7 endonuclease I-HJ complex. The two protein subunits are in green and beige. The two Ca^{2+} ions (replacing the catalytically active Mg^{2+}) in each active site are represented as yellow balls. The four DNA strands in the HJ are color-coded red and gold for non-exchanging strands, blue and purple for exchanging strands. The single-strand cleavages occur in arms 1 and 3 in the gold and red strands at the scissile bonds each 1 bp away from the crossover, adjacent to the Ca^{2+}. (PDB 2PFJ) (C) For comparison, two orthogonal views of an anti-parallel stacked-X form of HJ-DNA on RuvA, color-coded as in (A). (A, after C. Biertümpfel, W. Yang and D. Suck, *Nature* 449:616–620, 2007; B, after J.M. Hadden et al., *Nature* 449:621–624, 2007.)

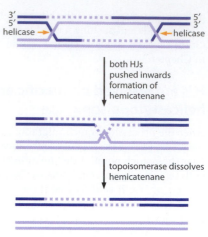

Figure 4.33 Dissolution of double Holliday junction (HJ) structure by action of helicase and topoisomerase. After branch migration, the two HJs (see Figure 4.24) can be pushed together by an ATP-dependent helicase (e.g. the BLM helicase) until they meet and form a hemicatenane. This can be acted upon by topoisomerase III, re-forming the duplex DNA molecules in the sister chromatids without any crossing over.

Endo VII is a symmetrical head-to-tail dimer of intertwined chains (157 aa) folded into an S shape. Its shallow, concave structure provides a large positively charged surface tightly molded to the DNA, encroaching from the minor groove side of the junction point and thus consistent with the lack of sequence specificity. The HJ bound to endo VII is anti-parallel, its four arms in the B-DNA conformation and Watson–Crick base-paired throughout. However, conventional pairwise co-axial stacking of the arms is lost and the angles between adjacent arms are ~40° and ~135°, giving the junction an 'open-H' appearance (Figure 4.32A). At the opened up crossover point, the phosphodiester backbones of the exchanging strands are sharply kinked, whereas the non-exchanging strands are held further apart. This positions the exchanging strands ready for symmetrical cleavage in the two active sites of the nuclease, with the cuts (not necessarily simultaneous) two nucleotides 3′ to the branch point (see also Figure 4.24).

In contrast, the HJ in complex with the bacteriophage T7 endo I dimer maintains a stacked-X conformation, with pairs of DNA arms essentially co-axial. However, the angle between the axes is ~−80°, a counterclockwise rotation by 130° from that of +50° observed for the free junction. Moreover, the two α/β domains in each protein subunit have undergone a major rearrangement in binding the HJ DNA, thereby creating two almost perpendicular channels, ~30 Å long, to grip the DNA duplexes in this unusual conformation (Figure 4.32B). It is likely that endo I recognizes a conformer of the stacked-X junction that is transiently populated in solution, stabilizing this HJ structure in the protein–DNA complex. The two active sites of the endo I dimer are positioned for symmetrical cleavage of the HJ, in this instance at the phosphodiester bonds between the first and second nucleotides of the continuous strands on the 5′ side of the crossover.

In some instances—notably, mitotic HR in eukaryotes—crossover resolution of HJs is suppressed in favor of a non-crossover product (**Figure 4.33**) by the activity of family RecQ helicases. The BLM helicase (counterpart of Sgs1 in yeast) can push the two HJs inwards and the *hemicatenane* thus formed is then resolved by the action of topoisomerase III. There are no crossover products, which has the merit of preventing chromosome rearrangements and preserving chromosome integrity in mitosis. Proportionate use of this mechanism is also important in ensuring that the appropriate number of crossovers occurs in HR in the first meiotic division.

Homology repair also restarts stalled replication forks

DNA replication can be stalled disastrously by damaged DNA upstream of a replication fork, including a nick or gap, or an AP site. Stalled replication forks can be restarted at the point of breakdown by several different processes, including an error-prone **translesion synthesis (TLS)** described below. The mechanisms of replication fork restart resemble HR in DSB repair, beginning with the creation of a long 3′ overhang and continuing with strand invasion and either SDSA or resolution/dissolution of the HJs (**Figure 4.34**). Most of the Chi sequences in bacteria face towards the origin of DNA replication, appropriate for RecBCD to initiate recombination from a stalled fork. It has been estimated that a damaged replication fork occurs at least once in a bacterial cell division and perhaps 10 or more times in a eukaryotic cell cycle. Indeed it has been suggested that the major function of HR (except in meiosis in eukaryotes) may be the rescue of stalled replication forks rather than DSB repair.

Having restored the branched structure of the DNA, the cell has to re-assemble a functional replication fork. In *E. coli,* this requires a complex of seven proteins, which includes the DnaB, DnaC, and DnaG proteins sufficient to assemble a replisome at *oriC* (Chapter 3) but additionally containing a 3′–5′ helicase (PriA) and three assembly factors (PriB, PriC, and DnaT). This combination of proteins was originally discovered as a requirement for the replication of bacteriophage φX174 DNA, but is now known as the **replication restart primosome**. In eukaryotes, the DNA damage response is not limited to the period

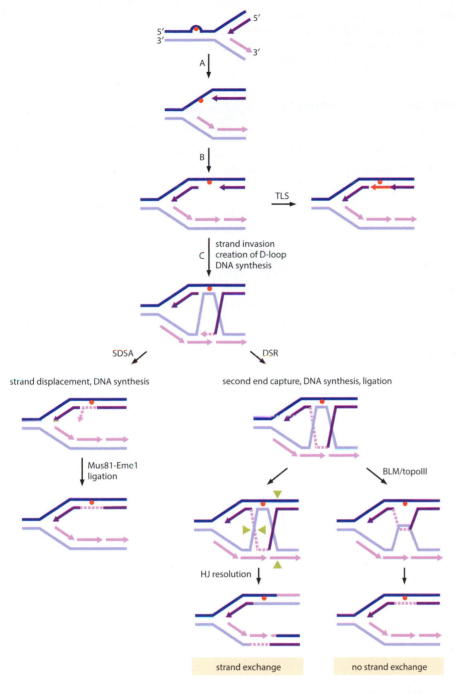

Figure 4.34 Repair of blocked or collapsed replication fork. A replication fork meets a damage site upstream (red circle) through which it cannot pass, pauses, and restarts but leaves a gap (steps A and B). This can be rectified in some circumstances by a translesion synthetase (TLS) that can synthesize a copy strand with mistakes incorporated. Alternatively, strand invasion can be followed by recombinational repair (step C), involving either (left) SDSA or (right) double-strand repair (DSR) based on second end capture. Srs2-dependent SDSA proceeds without strand exchange as in Figure 4.24, but in archaea/eukaryotes may generate a surplus 3′ flap, which is removed by an endonuclease (Mus81-Eme1) of the XPF/Mus81 family. In DSR, one option is for the double HJs to be resolved by a nicking endonuclease (yellow arrowheads), with strand exchange also as in Figure 4.24; another is for it to be dissolved by the separate BLM/topoIII activity without strand exchange, as shown in more detail in Figure 4.33. (After A. Ciccia, N. McDonald and S.C. West, *Annu. Rev. Biochem.* 77:259–287, 2008.)

of chromosomal DNA replication but operates throughout interphase. The key players are two checkpoint signaling kinases, ATM and ATR, as shown in Figure 4.16, and the effector kinases, Chk2 and Chk1, that target Cdc25, a phosphatase and regulator of cyclin-dependent kinases that control progression through the cell cycle. This is discussed further in Chapter 13.

Damage tolerance and error-prone translesion DNA synthesis

All cells possess a 'fall-back' system to deal with DNA lesions that are not readily repaired by processes described above. This situation could be the result of a plethora of mutations being incurred, possibly from a drastic change in environment. In these circumstances, special DNA polymerases can continue replication past the lesion, at the cost of a higher error rate during normal DNA synthesis, perhaps 10^{-2}–10^{-3}, about two orders of magnitude worse than regular DNA polymerases even without their intrinsic proofreading activity (Chapter 3). Among the first **translesion synthases (TLSs)** to be defined was Rad30 from *S. cerevisiae*, designated Pol η as it was the seventh eukaryotic DNA polymerase to be described. It is capable of reading past a T-T cyclobutane dimer and efficiently inserting two A residues in the growing polynucleotide chain. A form of the disease xeroderma pigmentosum is due to defects in the human Pol η, presumably because of the increased difficulty in dealing with a lesion that is induced by UV light. The human TLS Pol ι (the ninth eukaryotic DNA polymerase described) also bypasses the T-T dimer but does so inefficiently. These and other TLSs share some sequence similarity and have been assigned to a separate Y family of polymerases. Database searches have identified Y family members in organisms representing all three kingdoms of life. For example, there are two Y family polymerases (Pol IV and Pol V) in *E. coli* and four (Pol η, Pol ι, Pol κ, and Rev1) in humans.

Structures of Y family polymerases with and without bound substrates indicate that the active sites of these enzymes are preformed and not subject to a substrate-induced fit, and are more solvent-exposed and less constraining than those of high-fidelity polymerases, enabling the incorporation of nucleotides with fewer constraints on base pairing with the template strand. These properties are consistent with enzymes exhibiting a low fidelity of DNA-templated DNA synthesis. Thus, *E. coli* Pol IV can bypass G residues with large adducts but allows −1 frameshifts and missense mutations in copying undamaged DNA. Along with many other proteins including replicative polymerases, the Y family polymerases bind to the sliding clamps (β clamps in *E. coli*, PCNA in eukaryotes) through B/PIP interaction motifs. Different polymerases can bind simultaneously to PCNA, for example, *E. coli* Pol III and Pol IV, perhaps as part of a handover mechanism between synthesis and repair.

Access of TLSs to undamaged DNA must be strictly controlled. How this is achieved is poorly understood but may involve protein–protein interaction with RecA, post-translational modification, and ubiquitylation and/or SUMOylation and proteasomal degradation (see Section 7.4). In addition, the activity of TLS polymerases is limited by their distributive activity, which provides an opportunity for a processive, high-fidelity polymerase to resume replication once the lesion is bypassed. In *E. coli*, the TLSs are part of a multifactor stress response to extensive DNA damage, commonly known as the **SOS response** and based on the controlled expression of 43 genes encoding proteins such as Pol IV, Pol V, SSB, UvrA and B, and RecA. These genes are normally repressed by the protein LexA but formation of the RecA nucleoprotein filament causes LexA to cleave itself, thereby releasing the SOS genes from repression. Following DNA repair and disassembly of the RecA nucleoprotein filament, newly synthesized LexA resumes repression of the SOS genes.

4.5   SITE-SPECIFIC DNA RECOMBINATION AND DNA TRANSPOSITION

HR is used in DNA repair to bring about rearrangements between homologous DNAs that possess a high level of sequence similarity, thereby restoring the original sequence to the broken DNA. The order of genes in the DNA molecules remains unchanged. A different type of recombination, called **site-specific DNA recombination and DNA transposition**, is catalyzed by a different set of molecular machines that manipulate DNA strands that have only limited sequence similarity. Originally discovered by Barbara McClintock in her study of so-called 'jumping genes' in maize, the biological consequences are numerous and important, including the alteration of gene order, the introduction of new genes, insertion of viral genomes into host chromosomes and their corresponding removal, altered gene expression, and the dissemination of antibiotic resistance. These systems have also provided biotechnology with many useful tools, such as *Cre* recombinase and transposon-insertion mutagenesis. They also underlie some important diseases, where inappropriate DNA transposition has been associated with hemophilia, porphyria, and Duchenne muscular dystrophy.

In conservative site-specific DNA recombination, the cutting and religating of DNA at specific sites creates inversions, deletions, or insertions, depending on the relative orientation of these sites. This is usually a very tidy process, and proceeds with no net loss or gain of nucleotides (**Figure 4.35A**). Such reactions are catalyzed by members of two large enzyme families, referred to as serine and tyrosine recombinases based on the nature of key residues in their active sites. In contrast, **DNA transposition** is the movement of defined DNA elements from one location in a genome to another (Figure 4.35B). The enzymes that catalyze such reactions, termed **transposases**, are usually encoded by the

(A)

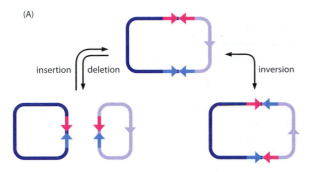

(B)

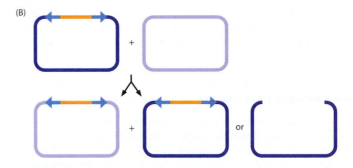

Figure 4.35 Schematic illustration of site-specific recombination and transposition. (A) Site-specific recombination. A recombinase recognizes specific inverted sequence repeats (mid-blue and magenta arrows) on a plasmid or chromosome duplex DNA (represented as dark blue and light blue lines; not drawn to scale). The DNA is then cut within those sites, rearranged, and religated. Depending on how the sites are aligned, the gray segment can be either inverted or deleted/inserted. (B) DNA transposition. In the starting duplex DNA (dark blue), the transposon (yellow) carries a gene for a transposase but may also carry many other genes. The transposase recognizes the ends (mid-blue arrows) of the transposon, and catalyzes insertion of the transposon into a new location in duplex DNA (light-blue). The original host replicon (dark blue) may retain a copy of the transposon, or be left with double-strand breaks to be repaired by the host organism. Depending on the mechanism of insertion, the new copy may be flanked by direct repeats of a few nucleotides.

mobile DNA element itself and interact specifically with its ends. The simplest elements, **insertion sequences (IS)**, consist of just a transposase gene flanked by its binding sites. However, larger elements (**transposons**) can carry many additional genes, such as antibiotic resistance determinants, and can even have other mobile elements nested within them.

Site-specific recombination and transposition overlap and some, but not all, site-specific recombinases function as transposases (**Table 4.2**). Mobile DNA elements are genetic opportunists that have co-opted various bits of cellular machinery to serve their purposes, and an ever-increasing number of mechanisms by which transposition can occur is being discovered. This has led to a confusing nomenclature. For example, the DDE family of DNA transposases is mechanistically and structurally related to HIV integrase, which inserts the HIV genome into the human one. In contrast, bacteriophage λ integrase, which stitches a viral genome into that of its *E. coli* host, belongs to the tyrosine recombinase family, some members of which also act as transposases. To clarify the situation, there has been a gradual shift toward including both the active site motif and the function when classifying DNA recombinases (for example, 'serine integrase').

Tyrosine recombinases cut one strand of each partner duplex at a time

Tyrosine and serine recombinases both cleave DNA through the nucleophilic attack of an active site hydroxyl group provided by a tyrosine or serine residue respectively, on the phosphodiester backbone. This generates a covalent protein–DNA intermediate, which can be viewed as an ester that stores the free energy of hydrolysis of the broken DNA phosphodiester bond. There is thus no need for ATP in religating the recombinant DNA. Unlike other phosphotransferases (for example, DNA polymerases, Chapter 3), the tyrosine and serine recombinases require no metal ions in their active sites; instead, highly conserved positively charged side chains surround the scissile phosphate. In all other respects, the strand exchange mechanisms and the primary and tertiary structures of the two enzyme families are unrelated.

The most widely known members of the tyrosine recombinase family are Cre and Flp, because of their usefulness in the genetic manipulation of other organisms, and bacteriophage λ integrase, owing to the latter's historical significance in understanding site-specific recombination. In tyrosine recombinases, one strand of each partner duplex is cut and religated before the other is attacked, thus avoiding DSB formation (**Figure 4.36A**). Each DNA recombination site contains a pair of specific binding sites for the recombinase, arranged as an inverted repeat separated by a central spacer region of 6–8 bp, depending on the enzyme. Two recombination sites are brought together by a tetramer of recombinase subunits in which the conformation of pairs of protomers alternates between an active and an inactive state. Asymmetric bends are forced on the spacers. In the first step, one strand in each duplex is cleaved at one end of the spacer as the active site Tyr hydroxyl group attacks the scissile phosphate. This generates a phosphotyrosine linkage between the protein and the deoxyribose 3'-phosphate. The newly created DNA 5' ends then exchange places, and their 5'-hydroxyl groups attack the opposing 3'-phosphotyrosine linkages. The Tyr hydroxyls are thus regenerated and the DNA strands are religated, generating an HJ intermediate. Isomerization of the complex then swaps the pairs of active and inactive protomers and the HJ is resolved by a similar set of reactions involving the other strand of each duplex to yield the recombinant DNA product.

Most tyrosine recombinases display little or no sequence specificity for the spacer region but the sequences of the two partner sites must match, and they must be aligned in antiparallel orientations for the recombination reaction to go to completion. Otherwise, the product DNA would contain mismatches and the reverse reaction, to reform the substrates, would be favored.

Table 4.2 Families and functions of recombinases and transposases

Function	Family			
	Phosphotyrosine intermediate	**Phosphoserine intermediate**	**DDE, Mg^{2+}; no covalent intermediate**	**HUH phosphotyrosine intermediate**
Bacteriophage or viral integrase	λ integrase (integration and excision of bacteriophage λ DNA into and from *E. coli* chromosome)	Bxb1, fC31; integrases	Prototype foamy virus (PFV) integrase (insertion of dsDNA copy of retroviral RNA into host chromosome)	
DNA invertase	Flp (2 μm plasmid of *S. cerevisiae*, analogous to Cre/Lox in bacteriophage P1)	Hin (inversion of a 900 bp segment of DNA in the *Salmonella* genome, alternating expression of genes encoding flagellin)	Piv	
DNA resolvase	Cre (cyclization of linear genome and resolution of dimeric genomes at Lox sites of bacteriophage P1 DNA replicated in *E. coli*); XerC/D (resolution of replicated chromosomes and ColE1 plasmids in *E. coli*)	Sin (resolution of dimeric multidrug resistance plasmids in *Staphylococcus aureus*); γδ (resolvase encoded by γδ transposon, member of prokaryotic Tn3 family in *E. coli*)		
Integron integrase	IntI (*Vibrio cholerae*, integration and excision of antibiotic resistance gene cassettes)			
Telomere resolvase	TelK (resolution of dimeric *Klebsiella oxytoca* bacteriophage φKO2 DNA into two linear chromosomes with hairpin ends)			
Transposase	CTnDOT	IS607, SCCmec	γδ (transposase encoded by γδ transposon, member of prokaryotic Tn3 family in *E. coli*); Tn5 (transposase encoded by prokaryotic transposon Tn5, member of the IS4 family of mobile elements); Mos1 (transposase encoded by eukaryotic *mariner* transposon *Mos1*); Tn7, μ, γδ	ISHp608
Related proteins	Type IB topoisomerases (Section 3.4)		RNase H; some HJ resolvases (Section 4.3)	Rolling circle replication proteins

Notes: Enzymes in red are described in the text. Transposases are named according to the transposon encoding them.

(A)

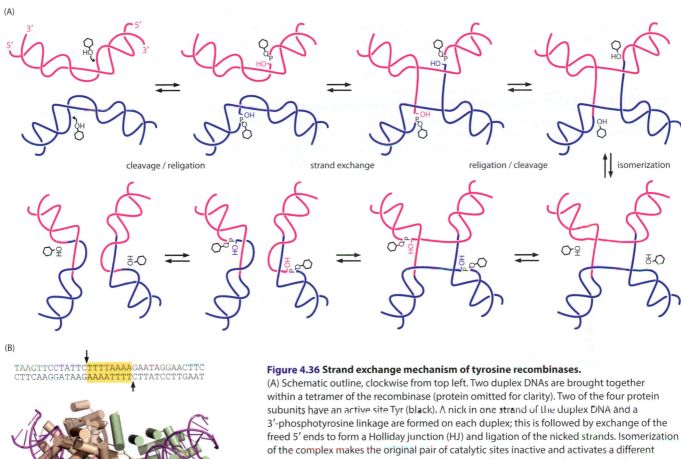

(B)

```
TAAGTTCCTATTCTTTTAAAAGAATAGGAACTTC
CTTCAAGGATAAGAAAATTTTCTTATCCTTGAAT
```

Tyr343

Figure 4.36 Strand exchange mechanism of tyrosine recombinases.
(A) Schematic outline, clockwise from top left. Two duplex DNAs are brought together within a tetramer of the recombinase (protein omitted for clarity). Two of the four protein subunits have an active site Tyr (black). A nick in one strand of the duplex DNA and a 3′-phosphotyrosine linkage are formed on each duplex; this is followed by exchange of the freed 5′ ends to form a Holliday junction (HJ) and ligation of the nicked strands. Isomerization of the complex makes the original pair of catalytic sites inactive and activates a different pair. A repetition of the original series of reactions then completes the process. All steps are freely reversible. (B) Structure of *S. cerevisiae* Flp (423 aa) in complex with a nicked HJ, corresponding to the third schematic DNA structure in (A). The synthetic DNA used in the crystallization has inverted repeats separated by an 8 bp spacer (in yellow box) and the sites of nicking and creation of transient Tyr 3′-phosphodiester bonds are indicated by arrows. Alternate protomers of Flp are colored wheat and green, recombining DNA in blue and magenta, as in (A). In this view, the catalytic domains lie behind the DNA and additional N-terminal domains are located in front. In Flp, unlike most other tyrosine recombinases, the helix that houses the nucleophilic Tyr (Tyr343) contributes *in trans* to the active site of a neighboring protomer. In this structure, the Tyr343 residues on helices of wheat subunits pack into green protomers, and the corresponding Tyr343 helices on green subunits are disordered. (PBD 1M6X)

Formation and resolution of the HJ DNA intermediate requires little movement in the recombinase tetramer

A crystal structure of the *S. cerevisiae* Flp recombinase in complex with its substrate DNA shows the HJ intermediate bound with each of its four arms in contact with one of the four protein subunits in the tetramer in a pseudo-square-planar conformation (Figure 4.36B). Two of the subunits have active catalytic sites, whereas the other two have their nucleophilic Tyr residues displaced and rendered inactive, consistent with alternating 'half-of-the-sites' activity. The approximate fourfold symmetry of the recombinase–DNA complex matches that of the HJ intermediate, which allows recombination to proceed with only a few nucleotides at each free 5′ end undergoing significant motion. The reaction cycle evidently proceeds with little movement on the part of the protein. Similar fourfold symmetry is displayed by RuvB, which facilitates the migration of HJ branch points after homologous recombination (see Figure 4.30). Half-of-the-sites reactivity is a conserved feature of tyrosine recombinases, but there may be significant differences in the way it is achieved in, for example, Cre and bacteriophage λ integrase.

The net change in phosphoester bond energy in tyrosine recombinase activity is zero, making the reaction completely reversible. In the case of Flp, which inverts a segment of the yeast 2 µm plasmid, the lack of directionality appears to be acceptable. Other tyrosine recombinases have acquired a variety of accessory proteins and DNA sites that bias their reactions in one direction over the other, in accordance with their biological activity.

The catalytic domains of tyrosine recombinases are related to type IB topoisomerases (Chapter 3) and to telomere resolvases (see below). All transiently nick DNA by displacing a 5′-OH at a scissile phosphate with an active site Tyr and, despite minimal sequence similarity, their catalytic domains have the same fold. Most members of both families have at least one additional DNA-binding domain N-terminal to the catalytic one. However, the folds of the additional domains vary widely, implying that these enzymes evolved in a modular fashion. In most tyrosine recombinases, both the catalytic domain and the binding domain next to it participate in recognizing the binding site sequence and, as in Figure 4.36B, wrap around the substrate DNA.

Bacteriophage λ integrase makes use of ancillary DNA-bending proteins

The integration and excision of bacteriophage λ DNA into and out of the *E. coli* chromosome was the first site-specific DNA recombination reaction to be studied *in vitro* with purified components. Nonetheless, its complicated regulatory system is still not completely understood. A key feature is that the **λ integrase** recognizes two related but different DNA sequences: *attP* in the phage DNA and *attB* in the bacterial DNA (**Figure 4.37A**). These

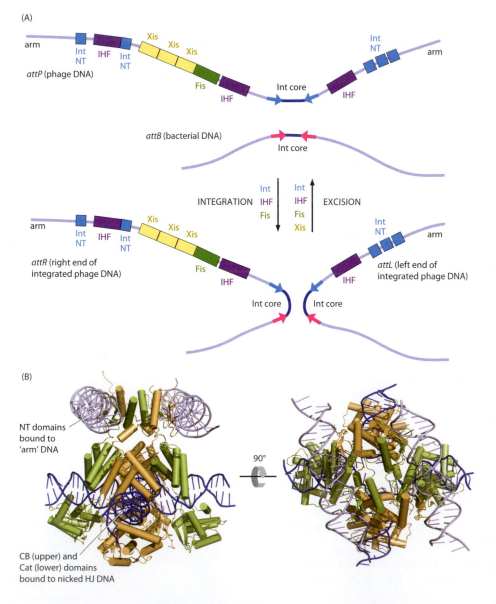

Figure 4.37 Bacteriophage λ DNA integration and excision. (A) Arrangement of binding sites for integrase and accessory proteins on partner DNAs undergoing recombination; core regions of DNA, are in dark blue; inverted repeats, as paired mid-blue and magenta arrows. Integration into the *E. coli* chromosome occurs when the attachment site on the phage DNA (*attP*) is bound by the host-encoded DNA-bending proteins Fis and IHF, as well as the integrase core domain (Int core) and N-terminal domains (Int NT). This highly folded complex catalyzes the integration of phage DNA into the simpler attachment site (*attB*) on the bacterial DNA. Excision of bacteriophage DNA from the host chromosome is aided by expression of the phage-encoded Xis protein, which helps bring together the *attL* and *attR* sites in a different highly folded complex. (B) Crystal structure of a bacteriophage λ integrase (356 aa) tetramer. Alternate subunits are colored wheat and green. The NT domains (residues 1–63) bind to sites on the DNA duplex 'arms' and are linked to the Int core domains by α-helical regions (residues 64–74). The Int core comprises central binding (CB) and Cat domains (residues 75–175 and 176–356, respectively), which are bound to a nicked Holliday junction that mimics the central steps of the strand exchange reaction (Figure 4.36). (PDB 1Z1G)

Figure 4.38 Bending of arm DNA by ancillary proteins in bacteriophage λ DNA integration and excision. (A) Structure of three Xis subunits (yellow) bound in tandem array to a regulatory DNA element in λ *attR*. A nick in the DNA used to form the crystal is visible. The Xis subunits (69 aa) adopt winged-helix conformations. They bind cooperatively head-to-tail via non-sequence-specific interactions with the DNA, cumulatively inducing a bending of the DNA by ~72°. (PDB 2IEF) (B) Structure of the Fis homodimer bound to a 27 bp fragment of DNA harboring a 15 bp AT-rich Fis-binding consensus sequence. The two 98 aa subunits are in different shades of green. The C-terminal helix in each subunit forms a recognition helix in a helix-turn-helix (HTH) motif and the two helices insert into adjacent major grooves of the DNA. This is made possible by bending of the DNA by ~65° and accompanying compression of the minor groove in the center of the site. (PDB 3JR9) (C) Structure of the IHF heterodimer (purple; α subunit 99 aa, β subunit 94 aa) bound to a cognate DNA site in λ *attL*. A nick in the DNA used to form the crystal is visible. The DNA is bent by ~160°, but is close to planar with a dihedral angle of only 10–15°. (PDB 1IHF)

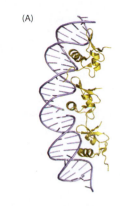

(A)

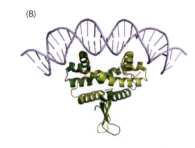

(B)

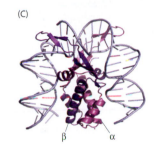

(C)

recombination sites include not only 'core' inverted repeats to which the core-binding (CB) and C-terminal catalytic (Cat) domains of the integrase bind, but also sites for three different DNA-bending proteins. Fis and IHF encoded by the *E. coli* host promote integration of λ DNA into the *E. coli* chromosome, whereas the phage-encoded Xis protein helps trigger excision 10^6-fold, while also inhibiting integration. In addition, 'arm' sites bind to an N-terminal domain of the integrase. Depending on the cellular concentrations of the DNA-bending proteins, different tightly folded complexes are formed that preferentially perform integration or excision. Exactly how these large complexes bias the reaction is unclear, but it is likely that the initial complex is set up like a coiled spring that is more relaxed in the product state, thereby driving an otherwise balanced reaction in the forward direction.

The structure of a λ integrase tetramer simultaneously bound to 'arm' and 'core' site DNAs is shown in Figure 4.37B. The accessory proteins Xis (**Figure 4.38A**) and Fis (Figure 4.38B) bend their target DNAs by ~70°, and IHF (Figure 4.38C) induces almost a U-turn. An armlike loop projects from each subunit of the IHF heterodimer (despite sharing modest (<30%) sequence identity, the α and β subunits have almost identical folds) and interacts exclusively with the minor groove of the DNA. Most of the bending takes place at two very large kinks 9 bp apart, where a proline residue at the tip of a projecting loop intercalates between base pairs for which stacking is disrupted, and the bent DNA is stabilized by interactions with the rest of the protein. These structures of λ integrase and a set of accessory proteins point to the way in which together they can bend the relevant *att* sites and catalyze insertion or excision of the λ DNA. How they might be assembled *in vivo* has yet to be determined.

Unusual tyrosine recombinases act as resolvases of DNA replicon dimers

When a stalled bacterial replication fork is repaired by HR, the HJ intermediate can be resolved in two ways: one in which the two daughter circular DNAs are generated, and another that produces one continuous circle of dimeric DNA (**Figure 4.39**). The dimeric DNA must be resolved before the cell divides. For the *E. coli* chromosome, this is accomplished by two proteins, XerC and XerD, acting on a site termed *dif*, located near the replication terminus.

XerC and XerD are unusual tyrosine recombinases in that they act as a heterotetramer rather than homotetramer. There is a specific binding site for XerC on one side of the spacer in the target site, and a different one for XerD on the other. Regulation and directionality are provided by interaction of XerD with a cell division-associated DNA translocase, FtsK. In the absence of FtsK, XerD is inactive and XerC catalyzes strand cleavage and exchange, but the reaction cannot proceed beyond the HJ intermediate and this can only be resolved back to the substrate. In the presence of FtsK, however, XerD is activated and catalyzes strand cleavage and exchange before XerC catalyzes a second set of strand exchanges to complete the reaction. XerC and XerD are widespread in bacteria. Certain plasmids, such as ColE1, also take advantage of the system to resolve their dimers but, rather than using FtsK to enforce directionality, they exploit a DNA-bending protein, pepA, to bias the reaction.

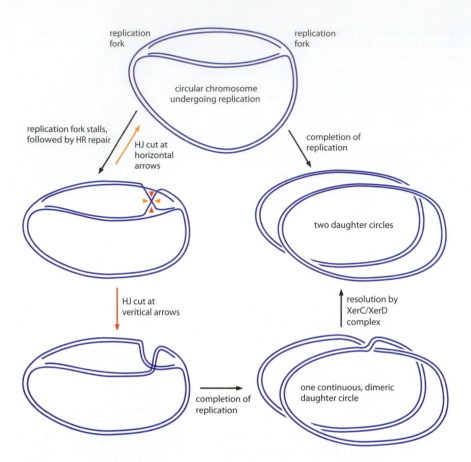

Figure 4.39 Formation and resolution of dimeric replicons. During replication of a circular chromosome or plasmid, HR-mediated repair of a stalled replication fork can lead to the final product being one large, dimeric circle rather than two daughter circles. The dimer can be resolved into the proper products by site-specific recombinases (for the *E. coli* chromosome, XerC and XerD).

Integron integrases are tyrosine recombinases that recognize hairpinned ssDNA substrates

Hairpinned DNA can also act as substrate for specialized tyrosine recombinases. **Integrons** are DNA elements that consist of strings of gene cassettes, often encoding antibiotic resistance, and a gene for a tyrosine recombinase (IntI) that can individually mobilize those cassettes. The inverted repeats for the **integron integrases**, unlike those for integrases such as Cre and Flp, are imperfect; one (*attI*) adjacent to the *IntI* gene, and others (*attC*) in individual gene cassettes, show poor sequence conservation and vary in length (57–141 bp). IntI acts on the *attI* and an *attC* site for integration, and on two *attC* sites for excision. Single-stranded regions of DNA are generated during bacterial conjugation or plasmid transformation and this enables the *att* sites to fold into hairpins that the integron integrase is able to recognize.

The structure of an integron integrase in complex with a hairpin ssDNA (**Figure 4.40**) illustrates how the enzyme is able to cope with the variability in the *attC* sites. The integron integrase resembles other tyrosine recombinases in that the protein subunit has an N-terminal domain with two helix-turn-helix (HTH) motifs, but the catalytic C-terminal domain differs in possessing a characteristic insertion of about 20 residues. Substrate recognition is dependent on two conserved nucleotides that bulge out of the hairpin, thereby helping to locate the enzyme on the DNA backbone and mediate the higher-order assembly of the synaptic complex. This allows for sequence-dependent binding at *attI* and sequence-degenerate binding at *attC*. However, in contrast with the flipped-out bases seen in complexes of BER enzymes with damaged DNA (Section 4.3), in the *att* sites the two bases have no partners on the other strand and are presumably already extra-helical.

Hairpin telomeres at the end of linear chromosomes are resolved by specialized tyrosine recombinases

In bacteria with circular genomes, the problem of protecting the ends of the chromosome in DNA replication does not arise. In DNA replication in eukaryotes, protection of the ends

(A)

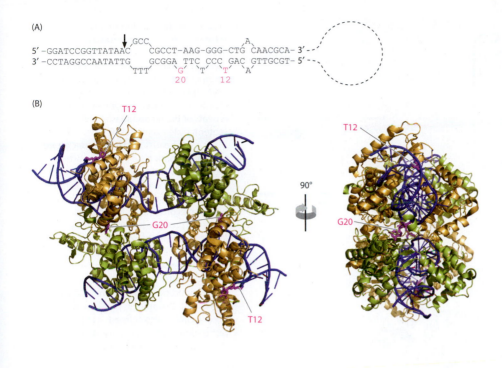

(B)

Figure 4.40 Recognition of *att* sites in ssDNA by an integron integrase. (A) The folded ssDNA substrate that is recognized by an integron integrase (IntI) from *Vibrio cholerae*. Extra-helical nucleotides T12 and G20 (magenta) are key features for site recognition. The site of cutting by nucleophilic attack of Tyr302 is arrowed. (B) Structure of a tetramer of IntI synapsing two such sites. In each subunit the N-terminal domain (residues 1–85) has two HTH motifs; the C-terminal domain (105–320) resembles that of other tyrosine recombinases apart from an insertion (residues 192–210) characteristic of integron integrases. In the left-hand image, the N-terminal domains are visible above the DNA, and the conserved catalytic domains are beneath the DNA. Each T12 (magenta) is intercalated between two conserved His residues (240, 241) on the same protomer (non-attacking, depicted in wheat) that binds the duplex DNA from which it protrudes. In contrast, each G20 (magenta) stabilizes synapsis by docking into a hydrophobic pocket on a different protomer (attacking, depicted in green) that is bound to the other duplex. (PDB 2A3V)

of linear chromosomes is achieved by the presence of telomeres and telomerase (Chapter 3). Certain bacteria and viruses, such as the plant pathogen, *Agrobacterium tumefaciens*, and the linear plasmid prophage of *Borrelia* (the causative agent of Lyme disease), also have linear chromosomes, but instead of the complexity of eukaryotic telomeres, their chromosome ends are protected by the elegant and simple expedient of hairpin formation. Replication of such hairpin-ended linear chromosomes produces dimeric circles (**Figure 4.41A**), which are resolved by a specialized tyrosine recombinase called **protelomerase**. The substrate DNA contains an inverted repeat of the sequence normally found at the chromosome end; this is bound by two copies of the resolvase, which nick at both ends of the spacer (Figure 4.41B). The free 5′ ends then double back on themselves to attack the 3′-phosphotyrosine linkage on the opposite subunit, creating a DSB with covalently closed hairpin ends.

The structure of a protelomerase in complex with its substrate DNA (Figure 4.41C) reveals that each monomer engulfs one-half of the recombination site in the dimer, bending the DNA by ~73°, but unlike λ integrase (Figure 4.37), not imposing a sharp kink that breaks the internal twofold symmetry. As in other tyrosine recombinases, the monomer contains an N-terminal DNA-binding domain joined to the catalytic domain by a long α-helical linker, but differs in having an additional C-terminal DNA-binding domain. It also has an insert in the N-terminal domain that interacts with the opposing subunit and diverts the path of the DNA across the dimer interface. There are few interactions with the central 6 bp that go on to form the hairpins at the ends of the chromosomes.

Serine recombinases overlap tyrosine recombinases in biological function but differ evolutionarily and mechanistically

Serine and tyrosine recombinases almost completely overlap in terms of biological function (Table 4.2) but are mechanistically and evolutionarily unrelated. Unlike tyrosine recombinases, serine recombinases introduce DSBs in their substrates, and their complexes with DNA substrates undergo large internal movements to accomplish DNA strand exchange. They have a very modular architecture. The catalytic domain, usually at the N-terminal end, has little, if any, DNA sequence specificity and the most extensively studied family members have been found to recognize their cognate sites primarily through a C-terminal HTH domain. This modularity allows for domain swaps with other binding modules of different sequence specificities, which greatly expands the potential of these enzymes for use as genetic tools.

A generalized pathway for serine recombinase-mediated DNA strand exchange is shown in **Figure 4.42A**. An active site Ser in each of two protomers introduces a break in each

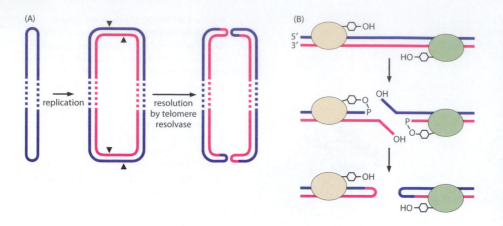

(C)

N-terminal domains

catalytic domains

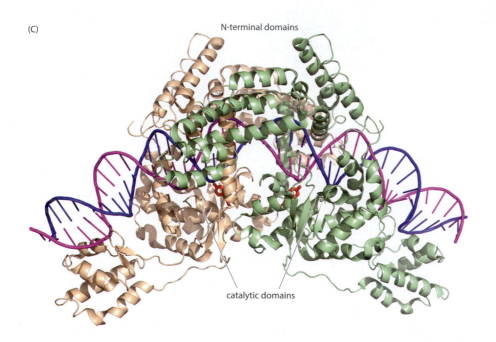

Figure 4.41 Resolution of hairpin telomeres by protelomerase K. (A) Schematic view of a circular replicon (in blue) with hairpinned ends and a recombinase binding site near each end. DNA replication (new DNA in magenta) creates a dimeric circle, with inverted repeats of the recombinase binding site (potential cleavage sites depicted by black arrowheads). Action of a specialized tyrosine recombinase at each inverted repeat resolves the dimeric chromosome into two linear ones, each with a hairpinned end. (B) Schematic mechanism of a dimeric tyrosine recombinase cleaving an inverted repeat to produce two hairpin ends. (C) Structure of a hairpin telomere resolvase (TelK) from *Klebsiella oxytoca* bacteriophage φKO2 in complex with a 44 bp synthetic DNA substrate. One TelK protomer (residues 1–538, omitting the C-terminal 102 aa) is in wheat, the other in pale green, as in (B): the DNA is also colored as in (B). The nucleophilic Tyr residues (Tyr 425) are shown as orange sticks. (PDB 2V6E)

strand of a target DNA duplex, generating 5-phosphoserine links to the protein and liberating two free 3′ ends. This can be seen in structures of the serine γδ resolvase encoded by the γδ transposon (Figure 4.42B). The resolvase is needed to 'clean up' after transposition of this mobile element, which results in fusion of the donor and acceptor replicons, with a copy of the transposon at each junction. It acts at specific sites within the transposons, and resolves the 'co-integrate' into two daughter circles, each containing a copy of the transposon. Each crossover site within the transposon is bound by a resolvase dimer, and two resolvase dimers come together to form an active tetramer. Tetramerization is mediated by the catalytic domains of the protomers, and the DNA segments lie on the periphery of the complex. This contrasts with the tyrosine recombinases, in which the center of the tetramer is occupied by DNA.

After the DSBs are introduced, a remarkable rotation of 180° by two subunits of γδ resolvase relative to the other two is thought to occur. They carry their broken DNA ends with them, which thus aligns them with new partners (Figure 4.42B). Religation is the direct reversal of the cleavage reaction: the free 3′-OH groups attack the opposite protein–DNA linkages, displacing the OH groups of the two serine residues, but the DNAs being joined have changed partners. This subunit rotation mechanism is supported by substantial biochemical and topological data, and is further supported by the structure of the active tetramer. The central interface is remarkably flat, which would facilitate smooth rotation, yet is quite hydrophobic, which would make complete dissociation unfavorable.

(A)

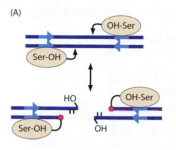

Figure 4.42 Double-strand break (DSB) and subunit rotation in serine recombinases. (A) Schematic mechanism of DSB formation catalyzed by a serine recombinase. One crossover site is shown, bound by two subunits of the tetrameric enzyme. The two active site Ser residues attack the scissile phosphates, creating DSBs in the DNA; at the breakpoint in each strand, a new 3'-hydroxyl is generated whereas the 5' end (red) is linked by a phosphodiester bond to the protein. (B) Strand exchange illustrated with γδ resolvase structures. Two dimers (left) are activated by additional factors (or in this case, by mutations) to form a tetramer (center panels). DNA-binding domains (DBDs) in the form of HTH motifs help hold the DNA duplexes in place and the active-site Ser residues (Ser10, one is labeled) cut the DNA and become covalently attached to the 5' ends at the breakpoint. The scissile 5'-phosphate (labeled) is in yellow. Subunit rotation then re-aligns the DNAs for religation in a recombinant configuration. Note that tetramerization requires large conformational changes and remodeling of the dimer interface. (PDB 1GDT and 1ZR4)

(B)

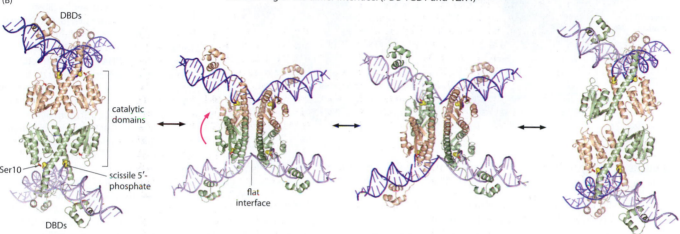

Serine recombinases can be regulated by means of accessory proteins

The reaction pathway shown in Figure 4.42 is fully reversible, with a ΔG° close to zero. Thermal energy is enough to drive subunit rotation, but it cannot drive the reaction in one direction over the other. Furthermore, the crossover sites are locally symmetric, yet they must be aligned in the correct relative orientation to produce deletions rather than inversions. Additional proteins are required, which bind at accessory sites and bring the two recombination sites together in an intertwined synaptic complex that traps three supercoiling nodes, illustrated schematically in **Figure 4.43**. The resolvase is only active in the context of the complex with DNA, which seems to trigger the dimer-to-tetramer conformational change shown in the first step of Figure 4.42B (the tetramer structure shown in the center panels was determined using a constitutively active mutant).

This synaptic complex accomplishes several things: it aligns the crossover sites properly; it limits recombination to intra- rather than inter-molecular reactions, since it is much easier to trap the three DNA crossings within one supercoiled circle rather than between two circles; and it harnesses supercoiling as an energy source. Starting from this geometry, the change in *linking number* for the reaction (see Box 2.3) is +4: the 180° rotation rewinds each duplex by a half-turn, and introduces a new crossing of one strand over the other, the hand of which corresponds to one positive supercoil. The two crossings trapped in the catenated product no longer count as supercoiling strain. The energy released by this change in

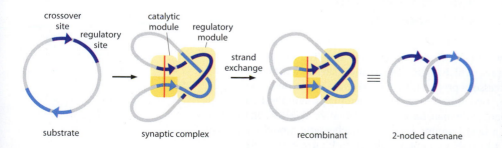

Figure 4.43 Formation of a serine resolvase synaptic complex with a defined topology. Interactions among proteins bound to asymmetric regulatory sites dictate the formation of a synaptic complex with a specific topology. This aligns the crossover sites correctly, and strand exchange (central arrow) is driven forward by a favorable change in the linking number. The catenated product circles are separated later by a type II topoisomerase. (After K.W. Mouw et al., *Mol. Cell* 30:145–155, 2008. With permission from Elsevier.)

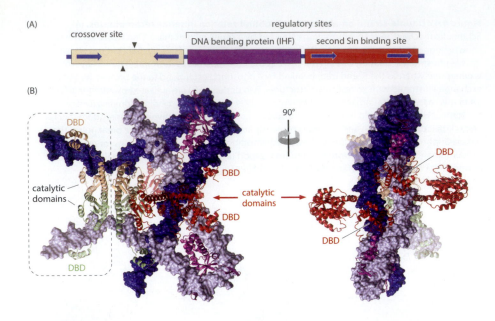

(A)

crossover site

regulatory sites

DNA bending protein (IHF)   second Sin binding site

(B)

90°

DBD

DBD

DBD

catalytic
domains

catalytic
domains

DBD

DBD

DBD

DBD

Figure 4.44 Model of the full synaptic complex for Sin resolvase. Two crossover and regulatory sites are brought together in the full synaptic complex. (A) Each full site DNA includes two binding sites for Sin dimers. These are the crossover site and a regulatory second Sin-binding site, separated by a site for the DNA-bending protein, IHF (see Figure 4.38). The individual Sin-binding sites (blue arrows) are arranged as inverted repeats in the crossover site (wheat), where DNA cleavage and rejoining occur (at vertical arrows), and as direct repeats in the regulatory site. (B) Orthogonal views of a structural model of the synaptic complex, constructed by docking three separate protein–DNA co-crystal structures: an activated γδ resolvase tetramer (two protomers in wheat and two in pale green, as in Figure 4.42) bound to crossover site DNA (in dark blue and gray), IHF (purple) bound to its cognate site, and Sin (dark red) bound to its second, regulatory site. Approximately one turn of modeled B-form DNA was added to the right end of each DNA to show its overall trajectory. During strand exchange, the portion of the complex within the dotted rectangle would rotate 180° relative to the rest. The DNA-binding domains (DBDs) are labeled. Note that the Sin recombinase dimers (dark red) bound to the regulatory site interact through their DBDs, whereas those of γδ (wheat/green) bound at the crossover site interact through their catalytic domains. (Courtesy of Phoebe Rice.)

linking number depends on plasmid size and the starting density of supercoiling, but under physiologically reasonable assumptions is similar to that for an ATP hydrolysis.

The best model for a full synaptic complex has been developed for Sin, a serine resolvase encoded by certain plasmids of *Staphylococcus aureus*, and so closely related to γδ resolvase (~30% identical in sequence, with very similar biochemistry) that the activated, crossover site-bound tetramers must closely resemble each other. Sin resolves plasmid dimers to monomers to be partitioned between daughter cells, similar to the role played by the tyrosine recombinases XerC and XerD for the *E. coli* chromosome (Figure 4.39). In addition to the crossover site, each recombination site includes a binding site for a DNA-bending protein (IHF), and an 'accessory' binding site for a second dimer of Sin (**Figure 4.44**). The enzyme dimers bound to accessory sites at each recombination site interact to form a tetramer, which remains catalytically inactive but plays a key part in organizing the complex by interacting also with dimers bound at the crossover sites. This generates a compact synaptic complex that brings the two crossover sites into close proximity and correctly aligned for the resolution reaction.

The accessory Sin-binding site, which is required for activation of DNA cleavage at the crossover site, is unusual in that the individual half-sites are arranged as a direct repeat, rather than inverted (Figure 4.44A). The dimeric protein can accommodate this because there is a flexible linker between its catalytic and DNA-binding domains. To hold together the central crossing of the synaptic complex, the dimers bound to the accessory sites come together to form a tetramer, mediated by contacts between the DNA-binding domains. The direct repeat geometry of the half-sites allows identical and synergistic protein–protein contacts between both pairs of DNA-binding domains. Thus only accessory site-bound dimers can generate a stable tetramer mediated by DNA-binding domains. This is very different from the catalytic domain-mediated tetramer seen for activated γδ resolvase bound at the crossover sites. Sin emerges as a remarkably versatile protein able to form two completely different tetramers in different circumstances.

Accessory proteins can direct a serine recombinase to catalyze inversion rather than deletion of a DNA segment

Serine recombinases that catalyze inversion of a DNA segment rather than its deletion are closely related in sequence and domain structure to the serine resolvases. The difference lies in the topology of the synaptic complex, which aligns the crossover sites in a different relative orientation and is dictated by accessory proteins. The best-studied serine invertase is Hin, which requires the participation of two dimers of the DNA-bending protein Fis. Hin is responsible for inverting a DNA segment in *Salmonella*, causing alternative flagellin gene expression as a mechanism of evading a host immune response. A schematic model for the Hin 'invertasome' and its action on a Fis-DNA complex is shown in **Figure 4.45**. Note

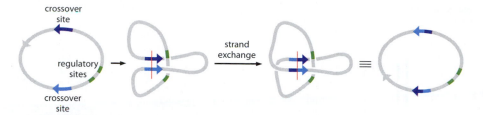

Figure 4.45 A serine recombinase can be directed to catalyze inversion. Dark blue and mid-blue arrows represent binding sites for Hin dimers on duplex DNA (light blue). Binding sites for Fis dimers that regulate the reaction are in green (see Figure 4.37). The full synaptic complex traps two supercoiling nodes and, after strand exchange (middle arrow), the segment of DNA between the two crossover sites has been inverted.

how this differs from the serine recombinase synaptic complex that specifies resolution in Figure 4.43.

The same catalytic module is used for different purposes in the large serine recombinase subfamily, which includes certain bacteriophage integrases and DNA transposases. These enzymes contain a large DNA-binding domain spanning hundreds of amino acids, rather than the small HTH domain of the serine resolvases and invertases. The structure of this domain is unknown, but it appears to play a role in synapsing the two crossover sites.

Another family of transposases and retroviral integrases is defined by a DDE motif in the active site

Members of this third large family of enzymes (DDE = Asp–Asp–Glu) are associated with diverse mobile DNA elements in all three kingdoms of life. Additionally, the recombinase that assembles antibody genes in V(D)J recombination (Chapter 17), is probably derived from a DDE transposase. The catalytic domains of these enzymes bear structural and mechanistic similarities to non-sequence-specific hydrolytic nucleases such as RNaseH and the HJ resolvase RuvC (Section 4.4), and their active sites contain conserved Asp and Glu carboxylate groups that bind catalytically important Mg^{2+} ions. Unlike the nucleases, they use not just water as a nucleophile but also the 3'-OH of a DNA end, and this is key to their ability to insert one DNA segment into another. Members of this family have a variety of additional domains (and sometimes subunits) that mediate sequence-specific DNA binding and, in some cases, provide regulation and/or other catalytic activities.

Two key chemical steps are common to the reactions catalyzed by these enzymes (**Figure 4.46**). In the first, water attacks the scissile phosphate at the end of the DNA element, displacing its 3'-OH group, and in the second, the 3'-OH attacks a phosphate group in a target DNA, creating a new connectivity. Both reactions are catalyzed by the same active site, and they usually occur within a large complex that brings together both ends of the element. For some elements, such as the γδ transposon, the transposase catalyzes only these two reactions. The resulting branched DNA products are converted into replication forks that use the two strands of the transposon as templates, resulting in the co-integrate that is resolved by a serine resolvase (see Figure 4.42B). Many other transposases excise the element from its previous host DNA before inserting it into a new one, and apparently rely on the DSB repair pathways of the host cell to clean up after their action.

Most enzymes belonging to the DDE group display little or no sequence specificity for the target DNA, but some have mechanisms for homing in on certain locations or structures in the genomes of their hosts. Their lack of specificity makes some of them good tools for generating libraries of random insertions, among other uses. Unlike the serine and tyrosine recombinases, the initial cleavage reaction is essentially irreversible. Controlling when and under what circumstances it can occur is thus particularly important. The second reaction simply exchanges one phosphodiester bond for another, raising a different question: what makes it go forward? This is best understood in the case of bacteriophage mu, which uses a massive burst of transposition to replicate: Mu transposase–DNA complexes become increasingly stable as the reactions progress, and the final one is so stable that it blocks replication unless the product is released by action of the ClpXP unfoldase/protease (Section 7.3).

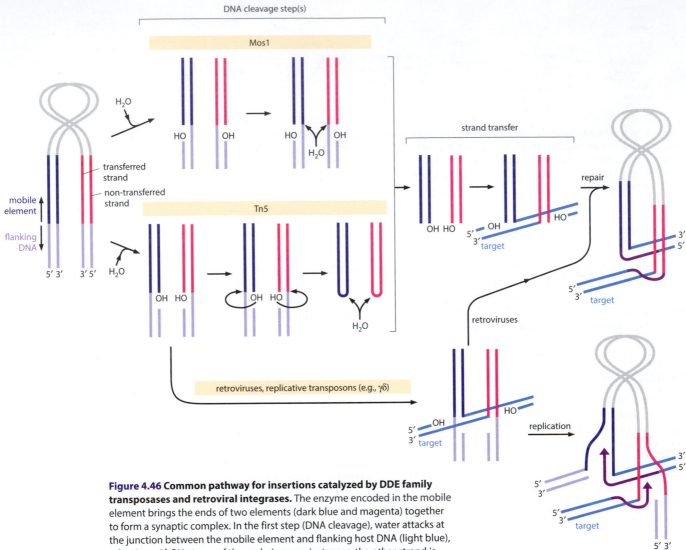

Figure 4.46 Common pathway for insertions catalyzed by DDE family transposases and retroviral integrases. The enzyme encoded in the mobile element brings the ends of two elements (dark blue and magenta) together to form a synaptic complex. In the first step (DNA cleavage), water attacks at the junction between the mobile element and flanking host DNA (light blue), releasing a 3′-OH at one of the ends. In some instances, the other strand is also cleaved to create a double-strand break (DSB). One method of doing this, exemplified by Mos1 transposase, is for the enzyme to catalyze a second attack of water, this time on the opposite DNA strand. Another method, exemplified by Tn5 transposase, is for the 3′-OH that was the leaving group in the first step to carry out a nucleophilic attack on a phosphate on the opposite strand, creating a hairpinned end. Water now attacks again, releasing a 3′-OH at the element end. In all cases, the next step (strand transfer) is the attack of the 3′-OH at the end of the mobile element on both strands of a target DNA (mid-blue). Depending on the element, the distance between scissile phosphates on the target is 2–9 bp. For transposons that make DSBs and for retroviruses, where the 'flanking host' DNA is only a 2 nt extension, the final step is repair of the small gaps in the target DNA by host enzymes (new DNA shown in purple). For replicative transposons (such as γδ), the displaced 3′-OH of the target DNA is used as a primer for replication of the entire transposon.

DDE transposases and retroviral integrases are diverse but share some structural features

Structures of two transposase–DNA complexes (Mos1 and Tn5) show a dimer of the transposase interacting with two transposon ends (**Figure 4.47A, B**). Each transposase subunit interacts with both DNA segments: its catalytic domain binds the end of one DNA, while its sequence-specific DNA binding domain(s) bind the other DNA. This *trans* arrangement ensures coordination of the chemical reactions at the two transposon ends. Both of the structurally characterized DDE transposases have HTH domains N-terminal to the catalytic

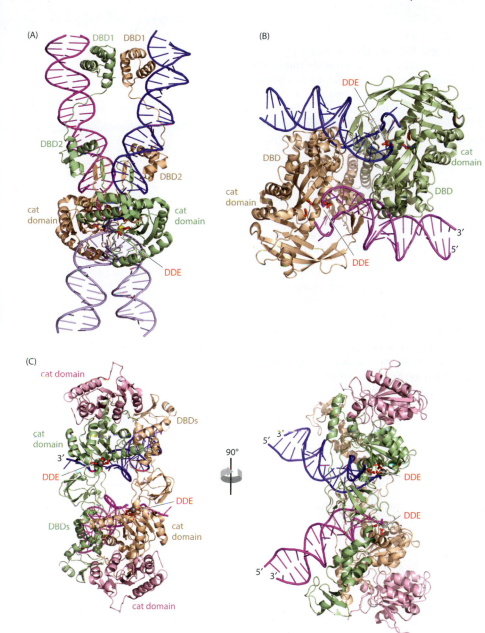

Figure 4.47 Structures of DDE family members in complex with DNA. In all of them, one protomer is depicted in wheat and the other in pale green. The catalytic domains (cat) and DNA-binding domains (DBDs) are labeled. The active site DDE residues are highlighted in red. (A) Mos1 transposase (DDD: Asp156, Asp249, Asp284); (B) Tn5 transposase (DDE: Asp97, Asp188, Glu326); (C) PFV integrase (DDE: Asp128, Asp185, Glu221). The catalytic domains of two additional protomers found in this complex are in pink. (PDB 3HOS, 3ECP, and 3L2R)

domain, but the relative position in the two structures is very different. In the Mos1-DNA complex, there are two small DNA-binding domains, and these mediate synapsis between two nearly parallel DNA segments. Although HTH domains are not usually thought of as protein–protein interaction motifs, recall that the DNA-binding domains of the serine resolvase, Sin, play a similar role (see Figure 4.44). In the Tn5-DNA complex, the DNA-binding domains are more peripheral, and an additional C-terminal segment forms critical protein–protein contacts. The transposition mechanisms of Mos1 and Tn5 have major differences in detail, that of Tn5 passing through a hairpin-ended intermediate, as illustrated in Figure 4.46. The catalytic domain of Tn5 transposase contains an ~90 aa insertion that promotes this intermediate, in part by providing a hydrophobic pocket to stabilize a base that is flipped out of the helical stack during hairpin formation (labeled G20 in Figure 4.40).

The structure of the integrase of a retrovirus, prototypic foamy virus (PFV), in complex with two viral DNA ends (Figure 4.47C) shows each 3′ end docked into the catalytic domain of one protomer but bound by the DNA-binding domains of the opposite protomer. Retroviral integration requires a tetramer, and the catalytic domains of two additional protomers appear to buttress interactions between the two functional protomers without contacting the DNA.

These similarities are remarkable, given the evolutionary breadth covered by these three structures: Tn5 is from *E. coli*, Mos1 from insects, and PFV infects mammals. However, there are also striking differences, most notably in the overall organization of the complexes: the protein–protein contacts are mediated by different domains, the DNA ends approach one another at different angles, and the catalytic domains they dock into are in different relative orientations. The latter probably reflects different spacings of the attack sites on the target DNA (2 bp for Mos1, 9 bp for Tn5, and 4 bp for the retroviral integrase). This is reminiscent of restriction enzymes, where the same catalytic module dimerizes very differently to produce different overhangs after cleavage.

Mobile elements are responsible for a large proportion of important evolutionary changes in genomes

The widespread importance of site-specific recombination (conservative and transpositional) is clear. The systems described above are some of those for which considerable structural information is available.

In its various guises, site-directed recombination has also caused many changes in the genomes of bacteria, archaea, and eukaryotes. We have seen how retroviruses can move themselves in and out of chromosomes. Retroviral-like **retrotransposons** constitute a large family of transposons present in a wide range of organisms, from yeasts, to insects, to mammals, and utilize the same mechanism. A classic example is the *Ty1* element in yeast, which is transcribed to an mRNA copy that encodes a reverse transcriptase that in turn generates a dsDNA copy of the entire element. The dsDNA is then integrated into the yeast genome by the action of an integrase encoded in the *Ty1* DNA. The resemblance to a retrovirus is striking, but *Ty1* lacks a protein coat and cannot leave the cell within which it operates. Another category of elements is that of the non-retroviral retrotransposons, occurring as repeated DNA sequences that make up a large proportion of many vertebrate chromosomes. In humans, most of these are immobile but some retain the ability to move. They do so by a totally different mechanism, which requires an endonuclease to nick the target DNA and an RNA-mediated process that again requires a reverse transcriptase. This falls outside the scope of this chapter, but it should be noted that some 40% of the human genome, including the abundant *Alu* repeats that constitute more than 10%, derives from mobile elements that depend on such RNA-mediated transposition.

4.6   SUMMARY

Although DNA replication is a highly efficient process (error rate $\sim 10^{-9}$), DNA can suffer damage through the occasional error in replication or from chemical or physical insults. Such damage can generate mutations or cause cell death, and it contributes to many human diseases, including cancers. Accordingly, cells make a heavy investment in DNA repair systems that reduce the overall error rate per cell doubling by 2–3 orders of magnitude.

Some types of damage affecting one DNA strand can be repaired by direct chemical reversal, whereas others are repaired by excision and replacement of the damaged nucleotides. Repair enzymes recognize damaged bases/nucleobases by a local decrease in stability or distortion of the double helix. The efficiency of damage recognition can be enhanced by the enzyme sliding along the DNA and detecting the modified DNA before flipping the damaged base out of the double helix to facilitate the repair, accompanied by bending or kinking the DNA. Short (3–4 residues) insertions or deletions can be fixed by removal and replacement of the affected DNA segment (*mismatch repair*). Damage affecting both strands, such as interstrand crosslinks or double strand breaks, poses a larger challenge for repair. These lesions require the coordinated activities of multienzyme machines that may include a specialized lesion-bypass polymerase capable of synthesizing through the damaged region.

DNA double-strand breaks need to be distinguished from the normal ends of chromosomes, and the ends thus created need to be protected from loss of nucleotides and fusion with the wrong chromosomes. In *nonhomologous end joining* (NHEJ), the DNA ends are remodeled to excise damaged nucleotides then rejoined, often with some loss of genetic information at the junction. By contrast, *homologous recombination* (HR) uses identical or homologous chromosomal DNA as a template to reconstruct the exact original sequence. In NHEJ, the MR complex (Mre11-Rad50) is able to hold DNA ends together and also takes part in their remodeling. The Ku protein recognizes ends for NHEJ and then recruits other proteins, including a DNA-dependent protein kinase (DNA PKcs), that stimulate repair activities.

In *recombination*, DNA tracts are exchanged between two interacting chromosomes. In prokaryotes, it is involved in DNA repair; in eukaryotes, it is also associated with crossover during meiosis. In HR, there is extensive processing of the DNA ends by nucleases and helicases, for example the bacterial RecBCD complex. This generates a 3′ ssDNA overhang that will recombine with an homologous dsDNA for templated DNA synthesis. DNA strand exchanges are promoted by the ATP-dependent recombinase RecA (bacteria and archaea) and Rad51 (eukaryotes). In complex with single-stranded DNA, these proteins polymerize into helical filaments in which the DNA in extended form is lined up with an homologous target dsDNA to form a synapse before strand exchange occurs. At an intermediate stage in recombination when two duplexes have formed a synapse but the product duplexes have not yet been separated, a four-way junction known as a *Holliday junction* (HJ) is formed. The HJ is finally resolved by single-strand nicks near the branch point and ligation of the new ends.

If other repair mechanisms fail, the cell may fall back on translesion synthesis in which a special kind of DNA polymerase synthesizes through the damaged region, albeit with relatively low fidelity. In addition to its role in DNA repair, translesion synthesis can restart stalled replication forks.

Whereas HR takes place between DNA molecules with long tracts of sequence homology, exchanges between short sites of homology are mediated by the machines of *site-specific recombination (SSR)*. They can bring about DNA inversions, deletions or insertions, as catalyzed by a structurally diverse and phylogenetically widespread group of enzymes. Site-specific recombinases use active site hydroxyls or divalent metals to catalyze DNA cleavage and ligation, while accessory proteins bind to DNA and bend it into a compact form that promotes the recombination reaction.

In *transposition*, DNA cutting and exchange processes related to those employed in SSR move defined DNA elements from one site to another within a genome. The element can be a single gene (an insertion sequence) or multiple genes (a transposon). These enzymes have provided useful tools for biotechnology, including the Cre recombinase and transposon-insertion mutagenesis. Related to the transposases that perform these operations are retroviral integrases that insert viral genomes into the host cell chromosome.

Sections of this chapter are based on the following contributions

4.1–4.4   Luca Pellegrini, University of Cambridge, UK.
4.5   Phoebe A. Rice, University of Chicago, USA.

Expert revision of this chapter was provided by

Tom Ellenberger, Washington University of St Louis School of Medicine, St Louis, USA.

REFERENCES

4.1   Introduction

Friedberg EC (2003) DNA damage and repair. *Nature* 421:436–440.

Friedberg EC (2014) Master molecule, heal thyself. *J Biol Chem* 289:13691–13700.

Jackson SP & Bartek J (2009) The DNA-damage response in human biology and disease. *Nature* 461:1071–1078.

Kunkel TA (2004) DNA replication fidelity. *J Biol Chem* 279:16895–16898.

Lindahl T (2013) My journey to DNA repair. *Genomics Proteomics Bioinformatics* 11:2–7.

Perry JJ, Cotner-Gohara E, Ellenberger T & Tainer JA (2010) Structural dynamics in DNA damage signaling and repair. *Curr Opin Struct Biol* 20:283–294.

4.2   Direct reversal of damage in one strand of duplex DNA

Brettel K & Byrdin M (2010) Reaction mechanisms of DNA photolyase. *Curr Opin Struct Biol* 20:693–701.

Liu Z, Tan C, Guo X et al (2011) Dynamics and mechanism of cyclobutane pyrimidine dimer repair by DNA photolyase. *Proc Natl Acad Sci USA* 108:14831–14836.

Mees A, Klar, T, Gnau P et al (2004) Crystal structure of a photolyase bound to a CPD-like DNA lesion after in situ repair. *Science* 306:1789–1793.

Müller M & Carell T (2009) Structural biology of DNA photolyases and cryptochromes. *Curr Opin Struct Biol* 19:277–285.

Sancar A (2003) Structure and function of DNA photolyase and cryptochrome blue-light photoreceptors. *Chem Rev* 103:2203–2237.

Yi C & He C (2013) DNA repair by reversal of DNA damage. *Cold Spring Harb Perspect Biol* 5:a012575–a012575.

4.3 Templated repair of lesions affecting one strand of the DNA duplex

Blainey PC, van Oijen AM, Banerjee A et al (2006) A base-excision DNA-repair protein finds intrahelical lesion bases by fast sliding in contact with DNA. *Proc Natl Acad Sci USA* 103:5752–5757.

Fromme JC & Verdine GL (2004) Base excision repair. *Adv Protein Chem* 69:1–41.

Giller LCJ & Schärer OD (2006) Molecular mechanisms of mammalian global genome nucleotide excision repair. *Chem Rev* 106:253–276.

Hsieh P & Yamane K (2008) DNA mismatch repair: molecular mechanism, cancer, and ageing. *Mech Ageing Dev* 129:391–407.

Iyer R, Pluciennik A, Burdett V & Modrich P (2006) DNA mismatch repair: functions and mechanisms. *Chem Rev* 106:302–323.

Jiricny J (2006) The multifaceted mismatch-repair system. *Nat Rev Mol Cell Biol* 7:335–346.

Krokan HE & Bjoras M (2013) Base excision repair. *Cold Spring Harb Perspect Biol* 5:a012583–a012583 .

Min J-H & Pavletich NP (2007) Recognition of DNA damage by the Rad4 nucleotide excision repair protein. *Nature* 449:570–576.

Mol CD, Izumi T, Mitra S & Tainer JA (2000) DNA-bound structures and mutants reveal abasic DNA binding by APE1 DNA repair and coordination. *Nature* 403:451–456.

Pakotiprapha D, Lui Y, Verdine GL & Jeruzalmi D (2009) A structural model for the damage-sensing complex in bacterial nucleotide excision repair. *J Biol Chem* 284:12837–12844.

Prasad R, Shock DD, Beard WA & Wilson SH (2010) Substrate channeling in mammalian base excision repair pathways: passing the baton. *J Biol Chem* 285:40479–40488.

Qi Y, Spong MC, Nam K et al (2009) Encounter and extrusion of an intrahelical lesion by a DNA repair enzyme. *Nature* 462:762–768.

Reyes GX, Schmidt TT, Kolodner RD & Hombauer H (2015) New insights into the mechanism of DNA mismatch repair. *Chromosoma* (in press doi:10.1007/s00412-015-0514-0).

Sixma TK (2001) DNA mismatch repair: MutS structures bound to mismatches. *Curr Opin Struct Biol* 11:47–52.

Wallace SS (2014) Base excision repair: a critical player in many games. *DNA Repair (Amst)* 19:14–26.

4.4 Repair of double-strand breaks

Brisset NC, Pitcher RS, Juarez R et al (2007) Structure of a NHEJ polymerase-mediated DNA synaptic complex. *Science* 318:456–459.

Chen Z, Yang H & Pavletich NP (2008) Mechanism of homologous recombination from the RecA–ssDNA/dsDNA structures. *Nature* 453:489–496.

Cohn MA & D'Andrea AD (2008) Chromatin recruitment of DNA repair proteins: lessons from the fanconi anemia and double-strand break repair pathways. *Mol Cell* 32:306–312.

Conway AB, Lynch TW, Zhang Y et al (2004) Crystal structure of a Rad51 filament. *Nature Struct Mol Biol* 11:791–796.

Dillingham MS & Kowalczykowski SC (2008) RecBCD enzyme and the repair of double-stranded DNA breaks. *Microbiol Mol Biol Rev* 72:642–671.

Dobbs TA, Tainer JA & Lees-Miller SP (2010) A structural model for regulation of NHEJ by DNA-PKcs autophosphorylation. *DNA Repair* 9:1307–1314.

Goodarzi AA & Jeggo PA (2013) The repair and signaling responses to DNA double-strand breaks. *Adv Genet* 82:1–45.

Hiom K (2010) Coping with DNA double strand breaks. *DNA Repair* 9:1256–1263.

Hopfner K-P (2009) DNA double-strand breaks come into focus. *Cell* 139:25–27.

Joo C, McKinney SA, Nakamura M et al (2006) Real-time observation of RecA filament dynamics with single monomer resolution. *Cell* 126:515–527.

Krejci L, Altmannova V, Spirek M & Zhao X (2012) Homologous recombination and its regulation. *Nucleic Acids Res* 40:5795–5818.

Kuzminov A (1999) Recombinational repair of DNA damage in *Escherichia coli* and bacteriophage lambda. *Microbiol Mol Biol Rev* 63:751–813.

Lammens K, Bemeleit DJ, Möcke C et al (2011) The Mre11:Rad50 structure shows an ATP-dependent molecular clamp in DNA double-strand break repair. *Cell* 145:54–66.

Li X & Heyer W-D (2008) Homologous recombination in DNA repair and DNA damage tolerance. *Cell Res* 18:99–113.

Lieber MR (2010) The mechanism of double-strand DNA break repair by the nonhomologous DNA end-joining pathway. *Annu Rev Biochem* 79:181–211.

Zahn KE, Wallace SS & Doublié S (2011) DNA polymerases provide a canon of strategies for translesion synthesis past oxidatively generated lesions. *Curr Opin Struct Biol* 21:358–369.

4.5 Site-specific DNA recombination and DNA transposition

Gellert M & Nash H (1987) Communication between segments of DNA during site-specific recombination. *Nature* 325:401–404.

Grindley NDF, Whiteson KL & Rice PA (2006) Mechanisms of site-specific recombination. *Annu Rev Biochem* 75:567–605.

Hickman AB & Dyda F (2015) Mechanisms of DNA transposition. *Microbiol Spectr* 3 (doi: 10.1128).

Jung YD, Ahn K, Kim YJ et al (2013) Retroelements: molecular features and implications for disease. *Genes Genet Syst* 88:31–43.

Kim MS, Lapkouski M, Yang W & Gellert M (2015) Crystal structure of the V(D)J recombinase RAG1-RAG2. *Nature* 518:507–511 (doi: 10.1038/nature14174).

Krishnan L & Engelman A (2012) Retroviral integrase proteins and HIV-1 DNA integration. *J Biol Chem* 287:40858–40866.

Landy A (2015) The λ integrase site-specific recombination pathway. *Microbiol Spectr* 3 (doi: 10.1128).

Rutherford K & Van Duyne GD (2014) The ins and outs of serine integrase site-specific recombination. *Curr Opin Struct Biol* 24:125-131.

Schneider S, Schorr S & Carell T (2009) Crystal structure analysis of DNA lesion repair and tolerance mechanisms. *Curr Opin Struct Biol* 19:87–95.

Takata M, Sasaki MS, Sonoda E et al (1998) Homologous recombination and non-homologous end-joining pathways of DNA double-strand break repair have overlapping roles in the maintenance of chromosomal integrity in vertebrate cells. *EMBO J* 17:5497–5508.

Van Duyne GD (2015) Cre recombinase. *Microbiol Spectr* 3 (doi:1128/microbiolspec).

Webb CJ, Wu Y & Zakian VA (2013) DNA repair at telomeres: keeping the ends intact. *Cold Spring Harb Perspect Biol* a012666.

prokaryote

TTGA −35

TATA box −10

transcription

DNA   ∼ −40 to −100

upstream regulatory sequences

σ factor

+1 start

RNA polymerase

Transcription

5

5.1   INTRODUCTION

Gene **transcription** is the first step in decoding genetic information. During transcription, information in the DNA sequence is copied to RNA. Differential transcription of genes determines the nature and the properties of a cell. Transcription is the major stage at which gene expression is controlled. The key enzymes of transcription are the DNA-dependent **RNA polymerases**, molecular machines that transcribe selected regions of DNA into RNA. In prokaryotes, the DNA is ready to be transcribed. In eukaryotes, the DNA is wrapped around histone proteins to form *nucleosomes*, which are organized into higher-order structures called chromatin (Chapter 2). To provide the RNA polymerases with access to the DNA, eukaryotic cells contain chromatin remodeling and modification complexes that use ATP hydrolysis to move or restructure nucleosomes (Section 2.3).

In all organisms, transcription depends on specific DNA sequence motifs in the **promoter** regions. Recognition of promoters by **general transcription factors** directs the polymerase to the transcription start site. In prokaryotes, the general transcription factor **sigma (σ)** recognizes promoter elements located 10 and 35 base pairs (bp) upstream of the transcription start site, the −10 and −35 elements (**Figure 5.1A**). In eukaryotes, the situation is more complex (Figure 5.1B). The best-studied eukaryotic promoter motif is the **TATA box** with the consensus DNA sequence TATAA/TAA/T. The TATA box is recognized by the TATA box binding protein (TBP), a subunit of the general transcription factor TFIID. TFIID and other general transcription factors (TFIIA, TFIIB, TFIIE, TFIIF, and TFIIH; together with RNA polymerase II—Pol II) form the **pre-initiation complex** (**PIC**) on promoter DNA (Section 5.3). The co-activator protein complex, **Mediator**, stabilizes the PIC and communicates signals from gene-specific transcription factors, which bind to sites that are usually upstream from the start site and can act positively (activators) or negatively (repressors).

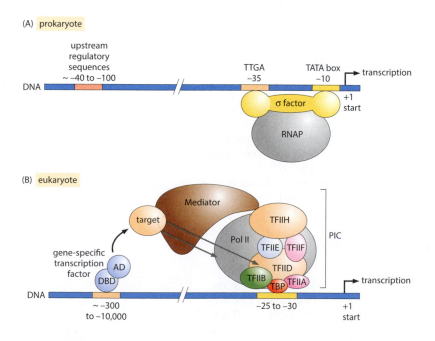

Figure 5.1 Initiation of transcription by RNA polymerase. (A) The prokaryotic system. The σ factor recruits RNA polymerase (RNAP) to the promoter site through recognition of the TATA box and the TTGA sequences. Sites for binding gene-specific regulatory transcription factors are further upstream. Numbers denote positions in the DNA sequence relative to the start site at +1. (B) Eukaryotic RNA polymerase II (Pol II). Pol II is assembled into the pre-initiation complex (PIC), which includes the six general transcription factor components of TFII. TFIID binds to the TATA box promoter through its TATA box binding protein (TBP). Gene-specific transcription factors bind to specific DNA sequences, which may be far from the promoter site, through their DNA binding domain (DBD), while their activation domain (AD) interacts with a target protein that may be part of the Mediator complex or a separate protein.

In prokaryotes, gene-specific transcription factor-binding sites are usually not far from the start site (typically, between −40 and −140 bp) and can overlap promoter motifs. Activators and repressors can bind to upstream regulatory sites, allowing gene expression to be controlled in response to changes in the environment. In eukaryotes, gene-specific transcription factor-binding sites tend to be remote from the promoter. Transcription factors can bind additively or cooperatively to achieve specific transcription patterns. Hence, a small number of regulators and signals can bring about a large and diverse repertoire of gene regulatory events. The *Saccharomyces cerevisiae* genome encodes almost 300 transcription factors, whereas in the human genome there may be several thousand.

Cells synthesize several types of RNA. Many genes are transcribed to messenger RNAs (mRNAs) to make proteins, but others (more than 10% in *S. cerevisiae;* the majority in humans) are transcribed to RNA as the final product. These include ribosomal RNAs (rRNAs) and transfer RNAs (tRNAs) which are key players in translation (Section 6.2), and RNAs that help to regulate gene expression, viz. small nuclear RNAs (snRNAs), small nucleolar RNAs (snoRNAs), and microRNAs (miRNAs), or serve as components of the *spliceosome* (see Section 5.4).

Bacterial and archaeal cells have a single RNA polymerase that makes mRNAs, rRNAs, and tRNAs. In contrast, eukaryotic cells employ three different RNA polymerases: of these, Pol I transcribes the abundant rRNAs; Pol II transcribes mRNAs but also snRNAs, snoRNAs and miRNAs; and Pol III transcribes small noncoding RNAs including tRNAs and 5S rRNA. Plants have two additional RNA polymerases, Pol IV and Pol V, which are related to Pol II. Viruses encode their own RNA polymerases which can be RNA-dependent or DNA-dependent, depending on the nature of the viral genome. In addition, the mitochondrial genome is transcribed by a single-subunit RNA polymerase that is related to RNA polymerases of bacteriophages.

In eukaryotes, primary RNA transcripts are processed before they exit from the nucleus. Precursor mRNA undergoes three major processing events that are linked to transcription: first, the addition of a cap structure at the RNA 5′ end; second, the generation of a poly(A) tail of about 200 adenine nucleotides at the 3′ end (Section 5.2); and third, the splicing out of the noncoding regions (*introns*) from the pre-mRNA (Section 5.4). Ribosomal RNAs and tRNAs are neither capped nor polyadenylated, but both are substantially modified (Section 6.2). Nearly 10% of tRNA nucleotides are modified enzymatically by reactions that include deamination of adenine to inosine, dimethylation of guanine, and modification of uridine to dihydrouridine or to thiouridine.

In this chapter, we describe the structural basis for RNA polymerase mechanisms, focusing on Pol II, for which most structural information is available (Section 5.2). We discuss assembly of the pre-initiation complex and its role in regulating and directing Pol II to the transcription start site (Section 5.3). Also included is a description of post-transcriptional events whereby the spliceosome machinery selectively removes introns (Section 5.4) and the exosome degrades unwanted RNA (Section 5.5).

5.2   RNA POLYMERASE II (POL II) AND THE ELONGATION COMPLEX

RNA polymerases were discovered in *Escherichia coli*, mammalian cells, and plants during 1959–1960. The first detailed insights into their architecture and mechanism were obtained in about 2000, when crystal structures were solved for Pol II from the yeast *S. cerevisiae*, and for the bacterial RNA polymerase (RNAP) from *Thermus aquaticus*. To initiate transcription, RNA polymerase is directed to the start site by gene-specific and general transcription factors. The resulting pre-initiation complex with duplex DNA is then converted to an open complex in which the DNA strands are locally separated to form a transcription bubble, exposing the bases that are used as a template for RNA synthesis (**Figure 5.2**).

RNA polymerases are multi-subunit enzymes that share a conserved core

Bacterial and archaeal RNA polymerases have 5 and 11 subunits respectively, whereas the three eukaryotic polymerases, Pol I, II, and III, are more complex, with 14, 12, and 17 subunits respectively, giving total molecular masses of 589, 514, and 693 kDa (**Table 5.1**). There

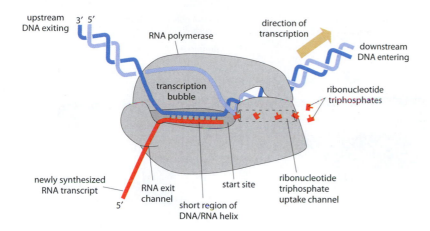

Figure 5.2 Transcription elongation by RNA polymerase. RNA polymerase associates with DNA (blue), and the double helix is unwound locally to form a transcription bubble. The polymerase adds nucleotides (small T-shapes in orange) to the growing RNA chain (red), using the exposed single-strand DNA (dark blue) as template. (Adapted from B. Alberts et al., Molecular Biology of the Cell, 6th ed. New York: Garland Science, 2015.)

is a common core of five homologous subunits (**Figure 5.3**). The two large subunits (Rpb1 and Rpb2 in Pol II; β' and β in bacteria) form the central mass and lie on opposite sides of a positively charged active center cleft. They are anchored by two small subunits that participate in polymerase assembly (Rpb3/Rpb11 heterodimer in eukaryotes; α homodimer in bacteria). A fifth core subunit (Rpb6 in eukaryotes; ω in bacteria) stabilizes the largest

Table 5.1	**RNA polymerase subunits.**				
Bacteria	**Archaea**	**Eukaryotes**			**Class**[a]
		Pol I	**Pol II**[b]	**Pol III**	
β'	A' + A''	A190	Rpb1	C160	Core
β	B (B' + B'')	A135	Rpb2	C128	Core
α	D	AC40	Rpb3	AC40	Core
α	L	AC19	Rpb11	AC19	Core
ω	K	Rpb6 (ABC23)	Rpb6	Rpb6	Core
–	H	Rpb5 (ABC27)	Rpb5	Rpb5	Common
–	–	Rpb8 (ABC14.5)	Rpb8	Rpb8	Common
–	X	A12.2	Rpb9	C11	Common
–	N	Rpb10 (ABC10a)	Rpb10	Rpb10	Common
–	P	Rpb12 (ABC10b)	Rpb12	Rpb12	Common
–	F	A14	Rpb4	C17	Rpb4/7 subcomplex
–	E	A43	Rpb7	C25	
		A49	(Tfg1/Rap74)	C37	TFIIF-like subcomplex[c]
		A34.5	(Tfg2/Rap30)	C53	
		–	–	C82	Pol III-specific subcomplex
		–	–	C34	
		–	–	C31	

[a]Core: sequence partly homologous in all RNA polymerases. Common: shared by all eukaryotic RNA polymerases.
[b]Rpb derives from an early name for RNA polymerase II: RNA polymerase B.
[c]The two subunits in Pol I and Pol III form heterodimers that are predicted to resemble part of the Pol II initiation/elongation factor TFIIF, which is composed of subunits Tfg1, Tfg2, and Tfg3 in *S. cerevisiae*, and of subunits Rap74 and Rap30 in humans.

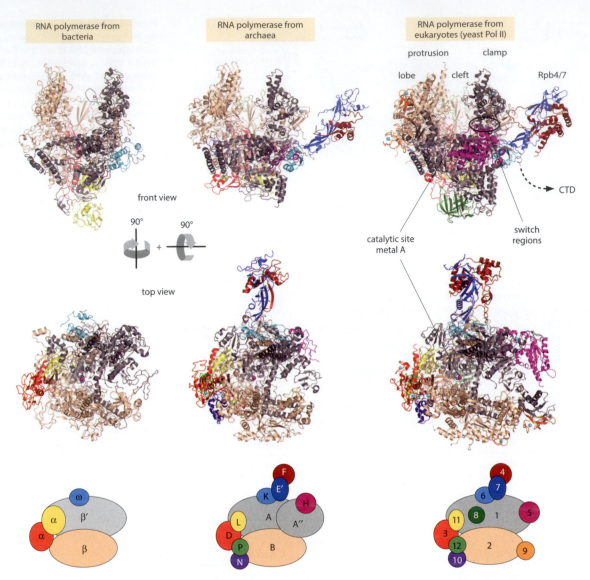

RNA polymerase from bacteria

RNA polymerase from archaea

RNA polymerase from eukaryotes (yeast Pol II)

protrusion   clamp

lobe   cleft   Rpb4/7

front view

90°   90°

top view

CTD

catalytic site metal A   switch regions

Figure 5.3 Three RNA polymerase structures. The color code and schematic representations of the subunit arrangements are shown in the bottom panel, matching the view shown in the middle panel. Views in the top and middle rows are related by two 90° rotations. Bacteria, *Thermus thermophilus* (PDB 1IW7); archaea, *Sulfolobus solfataricus* (PDB 2PMZ); eukaryote, *S. cerevisiae* (PDB 1I50 and 1WCM).

subunit. Prokaryotic and eukaryotic polymerases differ in their peripheral features where interactions with other factors take place.

The catalytic site is located on the floor of the cleft, near the center of the complex (see Figure 5.3). The Rpb1 side of the cleft forms a mobile clamp, connected to the body of the polymerase by mobile switch regions (discussed below). The clamp is closed when the polymerase is bound to DNA and RNA. The Rpb2 side of the cleft consists of the lobe and protrusion domains. Rpb2 also forms a protein 'wall' (called the 'flap' in the bacterial enzyme) that blocks the end of the cleft. In Pol I, II, and III and in the archaeal enzyme, the core is extended by five common subunits, namely Rpb5, Rpb8, Rpb9, Rpb10, and Rpb12 (see Table 5.1). Pol II contains, in addition, the peripherally placed heterodimer, Rpb4/7.

The repeat-containing C-terminal domain (CTD) is a unique feature of Rpb1, the largest Pol II subunit: it emerges from Rpb1 below the Rpb4/7 subcomplex and is mobile (see Figure 5.3). The CTD consists of heptapeptide repeats of the sequence YS^{2}PTS^{5}PS. There are 52 repeats in the human polymerase and 26 in the yeast polymerase. To start transcription, the CTD is phosphorylated on Ser5 and, later, on Ser2 during transcription elongation (see Section 5.3).

The elongation complex binds template DNA, nucleotide triphosphates, and newly synthesized RNA

Transcription elongation begins with binding of a nucleoside triphosphate (NTP) to the **elongation complex** (**EC**) formed by the polymerase, DNA, and RNA (see Figure 5.2). The transcription bubble contains a short heteroduplex formed between the DNA template strand and the RNA product strand. The NTP is added to the 3′ end of the growing RNA strand, reading off the template strand of DNA and releasing a pyrophosphate ion (**Figure 5.4**). Translocation of DNA and RNA then frees the substrate site for binding the next NTP. As transcription proceeds, the RNA emerges from the catalytic site.

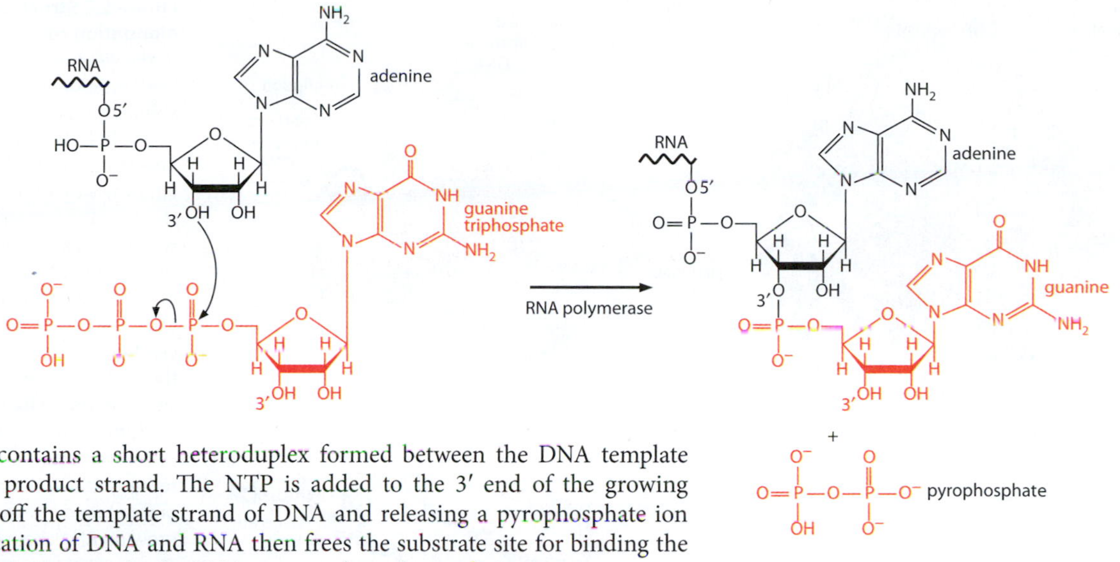

The mechanism of RNA elongation was elucidated in studies of Pol II/nucleic acid complexes. The structure of core Pol II transcribing a minimal double-stranded DNA fragment with a single-stranded extension revealed downstream DNA entering the cleft, and an 8–9 bp DNA/RNA hybrid in the active center. Subsequent studies of EC structures, using hybrid duplexes designed to mimic the natural arrangement of nucleic acids in the catalytic site, revealed the exact location of the downstream DNA and several nucleotides upstream of the hybrid. Thus, conserved surfaces and protein loops unwind downstream DNA and separate the RNA product from the DNA template at the end of the hybrid (**Figure 5.5A,B**).

Nucleotide selection is coupled to catalysis

Crystal structures of Pol II ECs with bound NTP substrates showed how the polymerase selects the correct NTP and how it incorporates a nucleotide into RNA. The NTP was observed bound at the insertion site, the site that is occupied during catalysis, and also at an overlapping, slightly different site, suggesting an inactive NTP-bound pre-insertion state of the enzyme. In both positions, the NTP forms a Watson–Crick base pair with the templating DNA. In the pre-insertion state, the NTP is bound to an open catalytic site, whereas in the insertion state it is bound to a closed catalytic site.

Crystallographic and biochemical studies identified two important elements of RNA polymerases, the bridge helix (Rpb1 residues 810–848) and the trigger loop, a helix–loop–helix motif (Rpb1 residues 1070–1100) (see Figure 5.5A,B). The complementary Watson–Crick hydrogen bonds from the NTP to the template DNA explain selectivity, while the contacts from the NTP 2′-OH partly explain the discrimination in favor of ribonucleotides and against deoxynucleotides (Figure 5.5C,D). Folding of the trigger loop leads to closure of the catalytic site, delivery of the NTP to the insertion site, and catalysis.

Catalytic nucleotide incorporation follows a two-metal-ion mechanism, as in DNA polymerases (Chapter 3). The catalytic site contains a tightly bound metal ion (metal A) and a second one (metal B) that generally binds with the NTP (see Figure 5.5D). Metal A is held by three invariant Asp side chains (not shown in Figure 5.5D) and contacts the RNA 3′-OH, whereas metal B binds to the NTP triphosphate moiety. In a proposal for the catalytic mechanism, His1085 from the trigger loop assists the withdrawal of electrons from the β-phosphate and facilitates attack by the RNA ribose 3′-OH on the α-phosphate of the NTP (see Figure 5.5D).

Nucleotide addition and translocation require a dynamic polymerase catalytic site

The first step of the catalytic cycle involves selection of the correct NTP (**Figure 5.6**, states 1–4). NTP selection begins with the sampling of nucleotides in the pre-insertion state, in which Watson–Crick interactions between the NTP and the DNA template are

Figure 5.4 RNA polymerase synthesizes RNA chains in the 5′ → 3′ direction. An RNA chain is shown with an adenine nucleotide at the 3′ end. RNA polymerase catalyzes the attack by the 3′-OH on the α-phosphate of an NTP, here guanosine triphosphate, with elimination of pyrophosphate to form an RNA chain with a guanine nucleotide added at the 3′ end. The reaction is repeated with additional NTPs to extend the growing RNA chain.

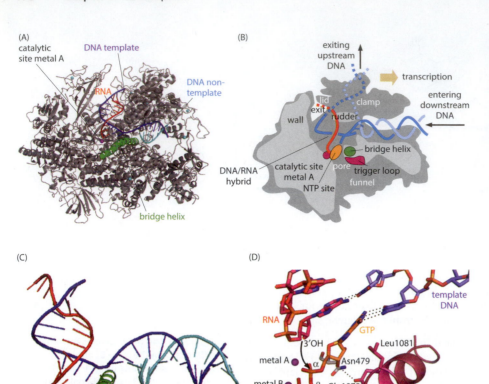

(A) catalytic site metal A

DNA template

RNA

DNA non-template

bridge helix

(B) exiting upstream DNA

transcription

entering downstream DNA

lid   clamp
exit   rudder
wall   bridge helix

DNA/RNA hybrid

catalytic site metal A

pore   trigger loop

NTP site   funnel

(C)

metal A
metal B

NTP

trigger loop/helices

bridge helix

(D) template DNA

RNA

GTP

3'OH   Leu1081

metal A   α   Asn479

metal B   β   Gln1078

Arg1020   γ   Arg766

Lys752   trigger loop

His1085

Figure 5.5 Structure of the Pol II elongation complex. (A) The view is related to the views in the top row in Figure 5.2 by a 90° rotation around a vertical axis. All Pol II subunits are in gray; DNA is shown in dark and light blue and the newly synthesized RNA in red. (PDB 2E2H) (B) A schematic cutaway view of the elongation complex, showing the transcription bubble. The NTP is shown as an orange ellipse. The catalytic site connects to a pore ('secondary channel' in the bacterial enzyme) in the floor of the cleft. The pore widens towards the outside, creating a funnel that allows access of the NTP. (C) Close-up view of the catalytic site. The trigger loop is in the closed conformation. (D) The nucleotide addition cycle. The catalytic insertion site with a representative NTP (GTP) (carbon atoms orange) ready to be incorporated into the growing RNA chain (carbon atoms red). The guanine base recognizes the cytosine of the template DNA (carbon atoms blue) through Watson–Crick hydrogen bonds. There is a nonpolar contact to Leu1081 from the trigger loop. Two residues from Rpb1 (Asn479 and Gln1078) are close to, but just outside, direct H-bonding distance to the ribose 2'-hydroxyl and contribute specificity for ribonucleotides. The basic group Lys752 (from Rpb1) is directed toward the β-phosphate and Arg766 and Arg1020 (from Rpb2) contact the γ-phosphate. His1085 from the trigger loop is in a position to contact the β-phosphate, but this residue is not always well ordered. The 3'-OH of the RNA is poised to attack the α-phosphate of the NTP to form a phosphate ester bond with release of pyrophosphate. The metal positions are shown, but their contacts are omitted for simplicity. (PDB 2E2J) (Adapted from R.D. Kornberg, *Proc. Natl Acad. Sci. USA* 104:12955–12961, 2009.)

established (state 2). The correct NTP is then delivered to the insertion site through a shift to the closed state of the trigger loop, and additional contacts discriminate against deoxy-NTPs, to prevent synthesis of DNA (state 3). Subsequent catalysis leads to nucleotide incorporation into the nascent RNA chain and the release of pyrophosphate (state 4), with the trigger loop shifting back to its open conformation (state 5).

During translocation, the bridge helix and the trigger loop cooperate to form a ratchet that underlies the translocation of the DNA/RNA duplex to the next template position after nucleotide incorporation (see Figure 5.6, states 5–7). Data from the yeast polymerase with the inhibitor amanitin (**Box 5.1**) have suggested a conserved two-step mechanism of translocation via a trigger-loop-stabilized EC intermediate with an altered structure of the bridge helix. During translocation step 1 (see Figure 5.6, states 6 and 7), the hybrid DNA/RNA product moves from the pre-translocation to the post-translocation position, and the downstream DNA translocates until the next DNA template base reaches a pre-templating position above the bridge helix. During translocation step 2, the DNA template base twists by 90° to reach its templating position in the active center (register +1).

A *Brownian ratchet* model has been proposed for the origin of the conformational changes that drive translocation. In this model, reversible diffusion of the polymerase along the DNA template in its pre- and post-translocation states is directionally rectified by NTP binding and hydrolysis, leading to unidirectional motion. In single-molecule measurements with *E. coli* RNAP, the polymerase moved in discrete steps of 3.7 ± 0.6 Å, close to the distance between base pairs in B-form DNA (3.4 Å). Analyses of velocity versus force at different NTP concentrations were consistent with the Brownian ratchet model.

Inhibitors of RNA polymerase have helped to define mechanisms

Information on translocation has come from binding studies with α-amanitin, a potent inhibitor of Pol II (see Box 5.1). Amanitin binds in the Pol II funnel and contacts the trigger loop and bridge helix, thus restricting these elements and interfering with translocation

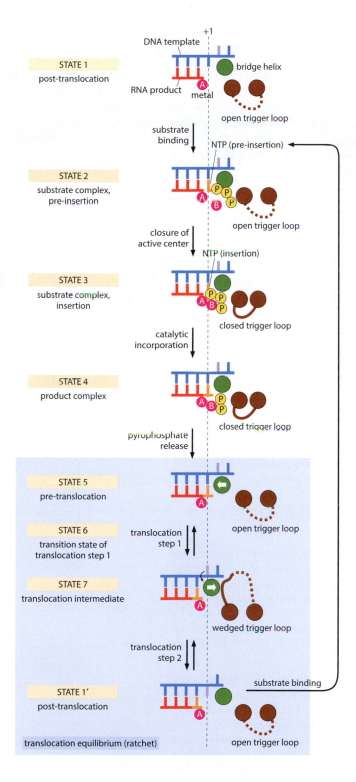

Figure 5.6 Schematic representation of the nucleotide addition cycle and elongation mechanisms. The two metal ions, A and B, are represented by pink circles. PPP is the triphosphate of the NTP (orange) that is being added to the growing RNA. The vertical dashed line marks the register +1 on the DNA template, and the next template base is in pale blue. States 1–4 indicate the steps for the nucleotide addition. States 5–7 indicate the translocation mechanism. The conformational changes of the trigger loop and bridge helix are indicated. (Adapted from F. Brueckner and P. Cramer, *Nat. Struct. Mol. Biol.* 15:811–818, 2008. With permission from Macmillan Publishers Ltd.)

(**Figure 5.7**). The inhibitor rifampicin (see Box 5.1) binds tightly to bacterial RNAP in a pocket in the β subunit (equivalent to Rbp2 in yeast) at the DNA/RNA hybrid-binding region, more than 12 Å from the catalytic site. Rifampicin acts by blocking the path of the elongating RNA when the transcript reaches 2–3 nucleotides (nt) in length. Streptolydigin (see Box 5.1) also inhibits bacterial polymerases. Structural studies showed that streptolydigin inhibits both initiation and elongation by binding at a site on the β′ subunit (equivalent to Rpb1 in yeast). This site is adjacent to the bridge helix but distinct from the NTP site. Streptolydigin acts by interfering with closure of the catalytic site. Its mechanism is thus similar to that of α-amanitin. However, superposition of the yeast and bacterial polymerases shows that the two binding sites are about 20 Å apart and at opposite ends of the bridge

Box 5.1   Inhibitors of RNA polymerase

The mushroom toxin α-amanitin (**Figure 5.1.1**), a cyclic octa-peptide from the death cap fungus *Amanita phalloides*, inhibits eukaryotic Pol II most potently and also Pol III but not Pol I.

Inhibitors of bacterial RNAP have been developed as antibacterial drugs. The broad-spectrum antibiotic rifampicin is used for anti-tuberculosis therapy. It binds to a pocket in the β subunit that is conserved among different bacteria but not

between bacterial and eukaryotic polymerases and hence is suitable as an antibiotic without harming the human host. Streptolydigin also inhibits the bacterial polymerase but not the eukaryotic one. It is effective against bacterial polymerases *in vitro* but its affinity is relatively low ($K_i \approx 10\ \mu M$) and bacterial membranes are not permeable to the compound. Hence it is not in clinical use.

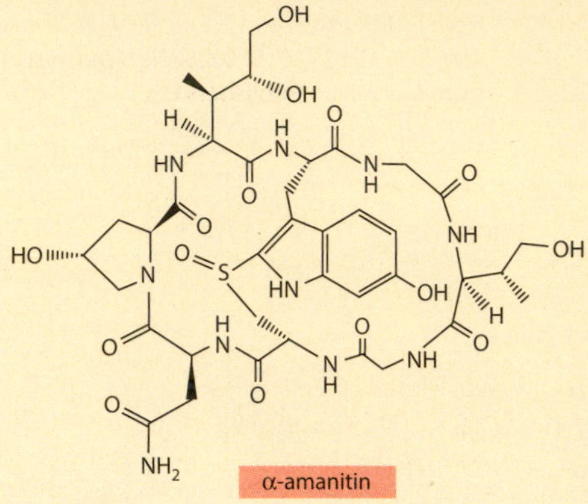

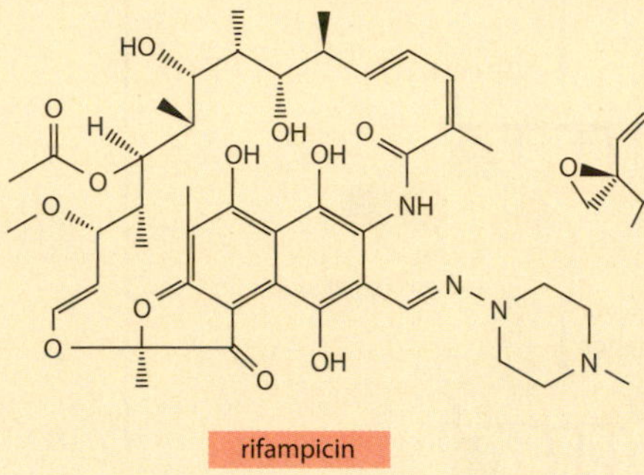

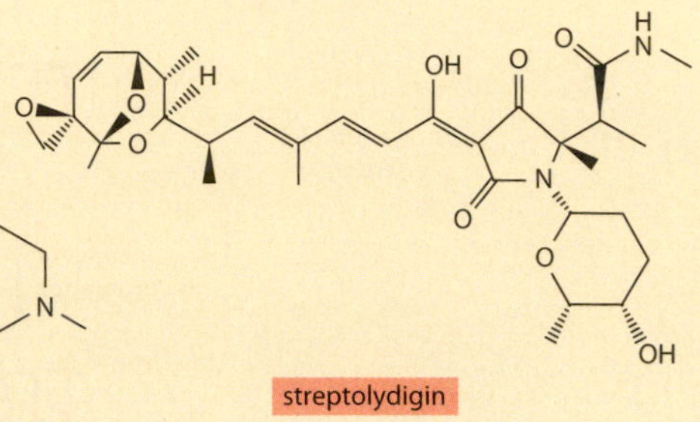

Figure 5.1.1 Chemical structures of three RNA polymerase inhibitors.

α-amanitin   rifampicin   streptolydigin

helix. Likewise, the streptolydigin and rifampicin sites are 12 Å apart in the bacterial polymerase and are non-overlapping. Thus, yeast and bacterial polymerases show three distinct sites that accommodate different inhibitors derived from natural products.

The polymerase can overcome natural obstacles to transcription elongation

During elongation, the polymerase must overcome intrinsic DNA arrest sites. These regions are often A/T-rich, leading to weaker DNA/RNA hybrids than C/G-rich regions. At arrest sites, the polymerase moves backward, resulting in threading of the 3′ end of the RNA into the pore (see Figure 5.5B). Backtracking by a few residues leads to pausing and is reversible, but extensive backward movement leads to arrest from which recovery is only possible by cleaving the RNA transcript in the polymerase active site. Backtracking predominates when forward movement is impeded, for example by damage in the template DNA or by nucleotide misincorporation in the RNA.

The arrested polymerase can be rescued by the RNA-cleavage stimulatory factor TFIIS, which creates a new RNA 3′ end at the active site from which transcription can resume (**Figure 5.8**). TFIIS enhances the intrinsic RNA nuclease activity of Pol II. It has the following components: domain I, an N-terminal four-helix bundle; domain II, a three-helix bundle; and domain III, a C-terminal domain with a zinc-ribbon fold (see **Guide**) and a protruding β-hairpin. Domain II and the linker between domains II and III are required for Pol II binding, whereas domain III is essential for stimulation of RNA cleavage. The mechanism of TFIIS function was determined from structures of Pol II with TFIIS, a Pol II EC with TFIIS, and a complex of Pol II with backtracked RNA. TFIIS changes the polymerase conformation and inserts its hairpin loop into the polymerase pore, presenting two acidic residues (Asp290 and Glu291) to the catalytic site (see Figure 5.8). The Pol II catalytic site is 'tunable'; that is, it can catalyze both RNA synthesis and RNA hydrolysis, of which the latter is enhanced by TFIIS. Bacterial RNAP uses an analogous mechanism of release from transcriptional arrest, but the corresponding bacterial cleavage stimulatory factors, GreA and GreB, are unrelated in structure to TFIIS.

Single-molecule dual-trap optical-tweezer measurements with Pol II have further elucidated these events. Template DNA (9.8 kbp) was tethered between two polystyrene beads

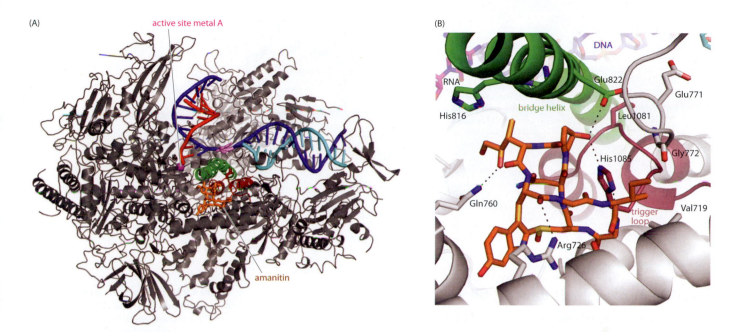

(A)   active site metal A

amanitin

Figure 5.7 Structure of the Pol II elongation complex inhibited by α-amanitin. (A) The complete Pol II structure with DNA (dark and light blue), RNA (red), and amanitin (orange) bound and the bridge helix (green) and the trigger loop (maroon) indicated. (B) The molecular contacts of amanitin with Pol II residues in the trigger loop and bridge helix. H-bonds are marked with dashes. Some Pol II residues (763–769), which also contact amanitin, have been removed for clarity. DNA and RNA can be seen at the top of the figure. The orientation is similar to that in (A). (PDB 2VUM)

held in place by optical traps. Upon addition of NTPs, Pol II began transcription; this shortened the tether between the two beads and increased the force (**Figure 5.9A**). The average pause-free velocity was 12.2 ± 4.5 nt per second (s) and frequent short (< 20 s) pauses were observed. Pol II ceased to transcribe above a force of 7.5 ± 2 pN. The presence of TFIIS rescued backtracked enzymes in ~ 25% of runs and allowed transcription to proceed up to a force of 16.9 ± 3.4 pN, thus more than doubling the force limit by allowing the enzyme to switch between backtracking and active transcription (Figure 5.9B).

RNA synthesis has a higher error rate than DNA synthesis

RNA polymerases synthesize RNA at rates of 20–70 nt/s and many transcripts can generally be synthesized in less than 1 hour. The error rate is about 1 in 10^4 nt. This is higher than the error rate of 1 in 10^7 for DNA polymerase (Chapter 3). Errors in DNA replication, in which the genetic material is copied and used by the next generation of cells, are more serious than errors in mRNA synthesis, in which many mRNA transcripts are made and then destroyed and the information is not passed on to the next generation. RNA polymerases use a simple proofreading mechanism; if an incorrect nucleotide is added to the growing RNA chain, the polymerase stalls, backtracks, and the wrong nucleotide is excised by the hydrolytic RNA nuclease activity of the polymerase. An incorrect nucleotide is added about 500 times more slowly than a correct one and the addition of another nucleotide after an incorrect one is slow. This gives time for the intrinsic nuclease activity to remove the incorrect nucleotide. The nuclease activity is slow but can be enhanced strongly by association of TFIIS.

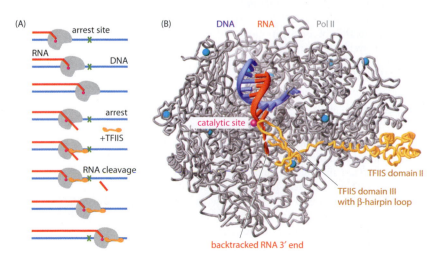

Figure 5.8 TFIIS rescues arrested Pol II. (A) The mechanism of rescue involves backtracking and TFIIS-induced mRNA cleavage. (B) Structure of the Pol II/TFIIS complex. Structural Zn^{2+} are shown as cyan spheres. The view is similar to that in Figure 5.5. (PDB 1PQV)

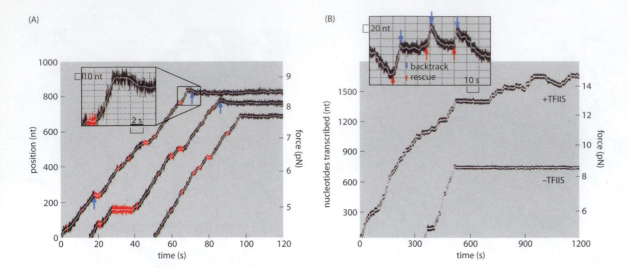

Figure 5.9 Single-molecule transcription measurements. (A) The DNA template position (nt transcribed) (left-hand axis) and force (right-hand axis) are plotted against time. Yeast Pol II exhibits elongation (black), pausing (red), backtracking (blue arrows and inset) and arrest. (B) TFIIS can rescue backtracked transcription and can promote transcription up to a higher force. In (A) the more detailed record inserted at top left is a blow-up of the boxed region, whereas in (B) the more detailed record (inserted panel) is from a different experiment. (From E.A. Galburt et al., *Nature* 446:820–823, 2007. With permission from Macmillan Publishers Ltd.)

Small RNAs can inhibit transcription

Transcription may be inhibited by noncoding RNAs that bind to and inhibit RNA polymerases. For example, in mice, a small noncoding RNA that is transcribed by Pol III binds Pol II and inhibits the transcription of protein-coding genes. Structural studies of yeast Pol II with a 33-mer RNA inhibitor showed the RNA as a double helix in the catalytic cleft. Comparison with the elongation complex showed that the RNA site overlapped the downstream DNA-binding site. The structure suggests that, in the free Pol II enzyme, the RNA inhibitor blocks DNA from entering the catalytic cleft but, once the elongation complex has been formed, the nucleic acids in the cleft block binding of the RNA inhibitor. Thus, the RNA inhibits transcription initiation but not elongation.

Messenger RNA is protected at the 5′ end by a cap structure

The first modification to mRNA occurs co-transcriptionally after 20–30 nt have been synthesized. A RNA 5′ phosphatase hydrolyzes a phosphate from the first 5′-nucleotide triphosphate. A guanylyltransferase then catalyzes the addition of GMP, derived from GTP, via an unusual 5′–5′ triphosphate link. Finally, a methyltransferase methylates the N7 position of the transferred GMP moiety (**Figure 5.10**). All three enzymes are localized at the Pol II CTD (C-terminal domain) after it has become phosphorylated on Ser5 by TFIIH during transcription initiation (Section 5.3). The cap structure is recognized by the cap-binding complex, which protects the mRNA from 5′–3′ exonucleases.

Termination is closely associated with 3′-polyadenylation of pre-mRNA

The stable polymerase elongation complex must be disrupted during termination. Different mechanisms are used in prokaryotes and eukaryotes. Prokaryotes employ a termination system in which a GC-rich region in the RNA promotes the formation of a hairpin loop. This RNA sequence is followed by a poly(U) tract, which forms weaker contacts with the DNA because U/A base pairs contain two hydrogen bonds compared with the three made by G/C pairs. These two RNA factors, the hairpin and the poly(U) tract, appear sufficient to dissociate RNAP. The process may be assisted by an accessory protein, Rho, which is a homohexamer with RNA/DNA helicase and RNA-dependent ATPase activity. Rho binds to nascent RNA and translocates 5′ → 3′ along RNA, using energy from ATP hydrolysis.

In eukaryotes, termination is intimately linked to the 3′-polyadenylation machinery. The formation of the mRNA poly(A) tail is directed by sequences present in the nascent RNA and catalyzed by the **polyadenylation** machinery, consisting of at least six multimeric protein factors (**Figure 5.11**). Before the addition of poly(A), the mRNA is cleaved. In most mRNAs, cleavage occurs at a CA dinucleotide that is about 10–30 nt downstream from a highly conserved AAUAAA motif and less than 30 nt upstream from a GU-rich or U-rich motif. These RNA sequences bind two protein complexes, CstF (cleavage stimulation factor) and CPSF (cleavage and polyadenylation specificity factor), which are recruited to

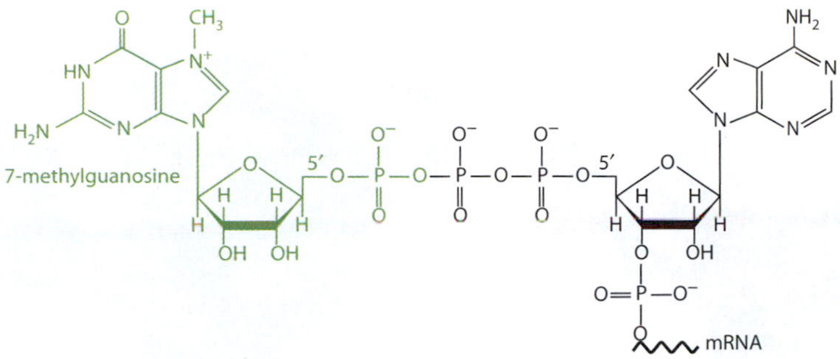

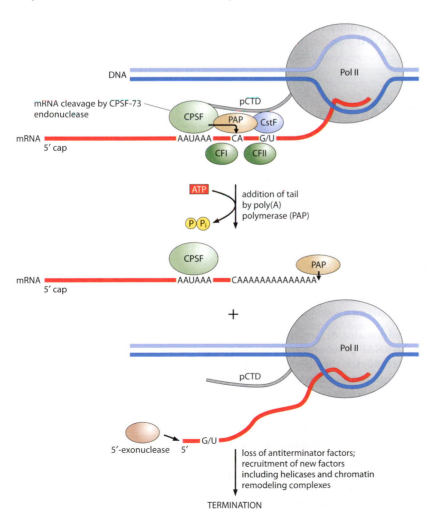

Figure 5.10 Capping of mRNA transcripts. The 5′ end of eukaryotic mRNA is capped by a 7-methylguanosine (green) linked by a 5′-to-5′ triphosphate bridge to the first 5′ nucleotide of the mRNA, which is usually an adenine (as shown here) or a guanine.

the phosphorylated Pol II CTD (discussed in Section 5.3), and assemble with the cleavage stimulatory factors CFI and CFII. The endoribonuclease that catalyzes RNA cleavage after the CA motif is the CPSF-73 subunit of CPSF, a metallo-β-lactamase protein. After mRNA cleavage, the poly(A) polymerase (PAP) adds ~200 A residues to the 3′ end of the mRNA.

After these events, Pol II continues to transcribe and eventually terminates by a process that is not completely understood. Pol II transcription proceeds beyond the poly(A) site by more than 1.5 kb in human genes, but in yeast, termination occurs close to the poly(A) site. The transcribed extra piece of RNA lacks a 5′ cap and is rapidly degraded by 5′–3′ exonucleases.

Pol I and Pol III are similar in structure to Pol II but synthesize different RNAs

The number of core subunits in Pol I, Pol II, and Pol III is conserved (see Table 5.1), and the catalytic centers are similar in all three enzymes. Pol I and Pol III also contain distant

Figure 5.11 Transcription termination. Diagram showing polyadenylation at the 3′ end of mRNA, involving RNA cleavage followed by poly(A) synthesis. After cleavage of the mRNA by CPSF-73 endonuclease (a subunit of CPSF) and polyadenylation of the cleaved mRNA, transcription continues and is eventually terminated by digestion of the unprotected 5′ end of RNA, the loss of antiterminator factors, and recruitment of other factors. (Adapted from N.J. Proudfoot et al., *Cell* 108:501–512, 2002. With permission from Elsevier.)

They can be defined by the presence of a TATA box (TATAA/TAA/T), situated about 30 bp upstream of the start site in higher eukaryotes (30–90 bp in the yeast *S. cerevisiae*). Eukaryotic promoters often lack a TATA box and/or contain other elements that facilitate recognition. These include BRE (TFIIB-recognition element), Inr (initiator element; TFIID TAF1/TAF2 recognition), and DPE (downstream promoter element; TFIID TAF6/TAF9 recognition).

TFIID acts as a scaffold for the assembly of other general transcription factors

PIC assembly begins with DNA binding by TFIID, a 700 kDa complex composed of the TATA box binding protein (TBP) and a set of TBP-associated factors (TAFs). TBP binds tightly ($K_d \approx 1$ nM) to the TATA box, through its C-terminal domain (180 aa). TBP has a saddle-like shape and recognizes the minor groove of DNA (**Figure 5.14**). Binding of TBP causes a pronounced bend in the DNA, exploiting the greater ease of distortion of AT-rich regions (see Figure 5.14 inset). Recognition of the TATAAAA sequence is primarily by four Phe residues, two of which intercalate between the first two base pairs and two between the last two base pairs of the TATAAAA box. Despite the distortion, base pairing is preserved and there is no apparent strain in the DNA, because partial unwinding is compensated for by right-handed supercoiling. The distorted DNA structure signals the location of an active promoter for the recruitment of other proteins.

The TFIID TAFs vary in number and function in different species. Several gene-specific transcription factors recognize elements of TFIID directly as well as their particular DNA sequences, thus providing a connection between the gene-specific transcription factor and the PIC without the Mediator complex (discussed below). Nine of the 13 conserved TAFs have a histone fold domain, but they are unlikely to mimic the role of nucleosomal histones because many of the nucleosomal DNA-interacting residues are not conserved. There is a further connection between TAFs and histones. TAF1, a 250 kDa protein, has acetyltransferase activity for histones H3 and H4, and its bromodomain (see **Guide**) recognizes specific acetylated Lys residues. This interaction might enhance TFIID binding to acetylated nucleosomes at promoters previously modified by activator-recruited histone acetyltransferases.

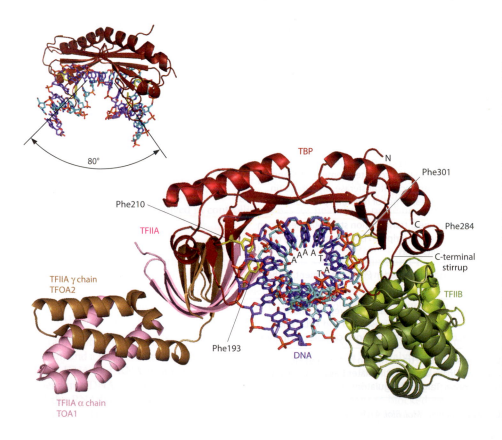

Figure 5.14 Assembly of the eukaryotic TBP/TFIIB/TFIIA/TATA box preinitiation complex. Main panel: a composite showing TBP bound to a TATA box DNA element with the TFIIB core domain from the TBP/DNA/TFIIBc structure and TFIIA from the TBP/DNA/TFIIA structure juxtaposed. The symmetry between the two TBP domains (red; C-terminal residues 155–333) is apparent. The 15 bp DNA duplex is represented with carbon atoms on one strand in dark blue and the other in cyan. The TATAAAA sequence on the dark blue strand is labeled. The four Phe residues (two from each TBP domain) that contact the DNA are in yellow. TFIIB (114–300) is shown in green for the N-terminal cyclin box domain, and in lime green for the C-terminal cyclin box domain. TFIIA was crystallized from two fragments of the α chain; the N-terminal 58 residues and the C-terminal 76 residues (PDB TOA1) and the complete γ chain (122 residues; PDB TOA2). The inset at upper left shows the TBP/DNA complex in a different view, to demonstrate the 80° bend in the DNA on binding TBP. (PDB 1CDW [TBP/DNA], 1VOL [TFIIB/TBP/DNA], and 1NVP [TFIIA/TBP/DNA])

TFIIA and TFIIB help stabilize the TBP/TATA complex

In the second stage of PIC assembly, the TFIID/DNA complex recruits TFIIA and TFIIB, which stabilize it. TFIIA may not be essential but it does function to overcome transcriptional inhibition. It binds to one side of the TBP/DNA complex (see Figure 5.14). TFIIB consists of an N-terminal zinc-ribbon domain connected by a flexible region to a C-terminal core domain (TFIIBc). The crystal structure of the TFIIBc/TBP/TATA-element ternary complex showed that the core TFIIBc protein contains two all α-helical cyclin-box domains (see **Guide**) that resemble those of the cell cycle regulator *cyclin A* (Section 13.3). The ternary complex is formed by TFIIBc clamping the acidic C-terminal 'stirrup' of TBP in its cleft, and interacting with other elements of TBP and the phosphoribose backbone of the TATA box (see Figure 5.14). TFIIBc binds to TBP on the opposite side to that which binds TFIIA. In addition to stabilizing the TBP/DNA complex, TFIIBc binding enhances the polarity of the quasi-symmetric TATA element recognition.

TFIIB recruits the promoter complex to Pol II

TFIIB bridges between promoter DNA and Pol II and sets the spacer distance to the transcription start site. The crystal structure of Pol II in complex with TFIIB revealed that TFIIB consists of an N-terminal zinc-ribbon domain, followed by a B-reader helix and loop, a B-linker strand, and a B-linker helix that connects to TFIIBc (**Figure 5.15A**). TFIIB binds to Pol II via its zinc-ribbon domain, which binds the Pol II dock domain. The B-reader binds at the upstream region of the DNA/RNA hybrid near the RNA exit tunnel, and the B-linker region contacts the clamp. The core TFIIBc N-terminal cyclin box domain binds on the Pol II wall (Figure 5.15B). With knowledge of the structure of TFIIBc/TBP/TATA box DNA complex (see Figure 5.14), these elements could be added and a model of a closed promoter complex constructed in which the DNA was extended in both directions (see Figure 5.15B). In the closed complex, TFIIB recruits promoter DNA to Pol II, using its contacts from the zinc-ribbon domain to Pol II and to the TBP/DNA complex.

To model an open promoter complex, it was assumed that the DNA is melted ~20 bp downstream of the TATA box with the help of TFIIB, to allow the emerging template strand to slip into the template tunnel, as observed in the elongation complex (Figure 5.15C). In the proposed model, 34 nt of DNA connect the TATA box to the transcription start site. The DNA region between the TATA box and the transcription start site serves as a spacer. After open complex formation, yeast Pol II scans the DNA for an initiator sequence motif that defines the transcription start site. The TFIIB reader contributes to recognition of the

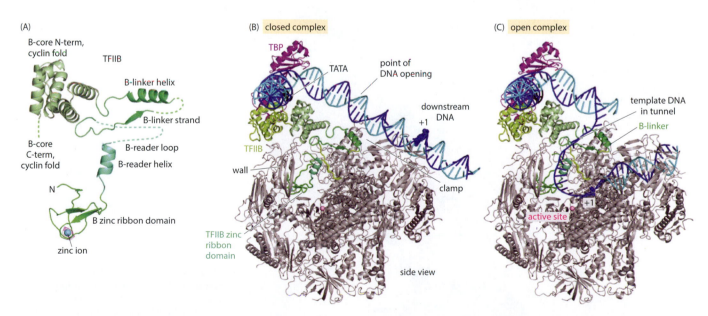

Figure 5.15 TFIIB in complex with Pol II. (A) The fold of TFIIB, showing the organization of the structure. (B) The closed Pol II structure with TFIIB, DNA, and the TBP superposed. The nucleotide that would bind in the +1 site in the open complex is shown as a space-filling model.

(C) Model of the open complex in which template DNA has melted to contact the catalytic site and interact with TFIIB. (From D. Kostrewa et al., *Nature* 462:323–330, 2009. With permission from Macmillan Publishers Ltd.)

initiator element. In the transition from initiation to elongation, the first two NTPs are then positioned opposite a conserved initiator dinucleotide motif and RNA synthesis begins. RNA growth after the first few nucleotides leads to clashes with the TFIIB reader loop, and short RNA fragments may be ejected (abortive transcription). Further growth of RNA and rewinding of upstream DNA trigger the release of TFIIB and formation of the elongation complex.

There are some similarities in the initiation mechanisms of TFIIB/Pol II and σ^{70}/RNAP, despite there being little sequence or structural similarity in the respective transcription factors. The TFIIB zinc ribbon, linker, and N-terminal cyclin box domains bind to polymerase surfaces similar to those that are bound by σ^{70} domains σ_4, σ_2, and σ_3 respectively, and the TFIIB reader loop corresponds to the loop between σ_3 and σ_4 (see Figure 5.13). However, the courses of the polypeptide chains of TFIIB and σ^{70} show opposite directionality and the domain structures differ, indicating that the two features in common result from convergent evolution.

TFIIH contains enzymes that unwind DNA and phosphorylate the Pol II CTD

In eukaryotes, transcription initiation also requires TFIIE, TFIIF, and TFIIH. TFIIF binds to Pol II above the Rpb2 side of the cleft, whereas TFIIE binds to the clamp and TFIIH is located near the polymerase jaws (**Figure 5.16**). TFIIE and TFIIH are apparently the last factors to enter the PIC. TFIIH has 10 subunits, including two DNA helicases, a DNA-dependent ATPase, and a Pol II CTD kinase. The helicase activity (of its subunits XPB and XPD) uses energy from ATP hydrolysis to unwind the DNA at the transcription start site (helicases are discussed in Chapters 3 and 4). This allows the polymerase to gain access to the template strand of DNA. Although the location of TFIIH on Pol II is known only approximately, its placement at a downstream DNA site (but remaining capable of reaching upstream) would permit a role in DNA negative supercoiling, enhancing transient opening of the DNA double helix.

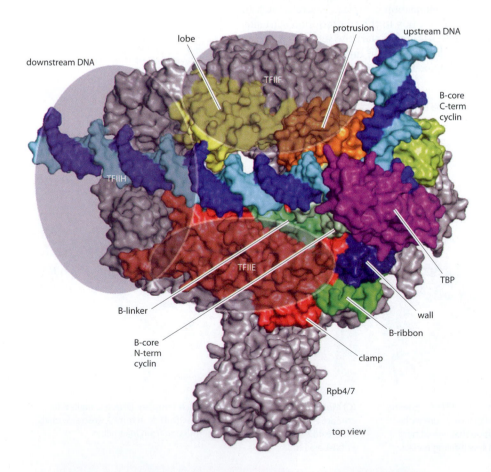

Figure 5.16 Model of the Pol II initiation complex with general transcription factor positions. Pol II is shown as a surface representation in gray. The locations of TFIIE, TFIIF, and TFIIH are indicated. (From D. Kostrewa et al., *Nature* 462:323–330, 2009. With permission from Macmillan Publishers Ltd.)

The TFIIH kinase activity is contributed by the heterotrimer CDK7/cyclin H/Mat1, which phosphorylates the Pol II CTD heptapeptide sequence (YSPTS^{5}PS) at Ser5. CDK7/cyclin H is also the master cyclin-dependent protein kinase that phosphorylates and activates the cyclin-dependent protein kinases (CDKs) that control the cell cycle (Section 13.2).

Assembly of the PIC promotes the transcription of short RNA transcripts, after which transcription pauses. The pause allows time for the ^{7}MeG capping of the 5′ end of the nascent mRNA by recruitment of the capping enzymes to the Ser5-phosphorylated CTD (Section 5.2). The Ser5-phosphorylated CTD also recruits a histone methyltransferase and splicing enzymes (Section 5.4).

Transcription elongation requires protein kinase activity of Cdk9/cyclin T (P-TEFb)

The protein kinase Cdk9/cyclin T, also known as positive transcription elongation factor (P-TEFb), regulates the transition from initiation to elongation. *In vivo*, elongation is inhibited by two proteins: the negative elongation factor (NELF) and the DRB-sensitivity-inducing factor (DSIF). Phosphorylation of NELF and DSIF by Cdk9/cyclin T results in relief from repression of transcription elongation. Phospho-NELF dissociates from the complex, whereas phospho-DSIF remains with the complex but changes conformation to allow activation (**Figure 5.17**). Cdk9/cyclin T also phosphorylates the Pol II CTD repeats YS^{2}PTSPS on Ser2. This phosphorylation allows the CTD to recruit the RNA 3′-polyadenylation (Section 5.2) and other RNA-processing factors, hence coupling transcription and RNA processing. After these phosphorylations, Pol II enters the elongation phase of transcription. Cdk9/cyclin T is itself under regulation; it is inhibited by association with 7SK RNA (an snRNA) in complex with proteins HEXIM and LARP7, and this inhibition may be relieved by the protein Brd4.

Gene-specific transcription factors regulate transcription

In prokaryotes, gene transcription in response to nutritional status or environmental conditions is controlled by repressors and activators that bind close to or within ~140 bp of the promoter sites. The affinity of repressors or activators for DNA is mediated via their DNA-binding domains. This affinity is modulated by activator domains that bind ligands associated with nutritional status or the environment. In this way, bacteria are able to respond to environmental conditions. There is a wealth of structural information on the association of these regulatory transcription factors with their cognate DNA and several protein-DNA binding motifs have been identified.

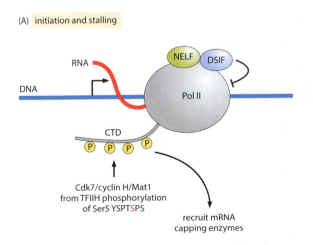
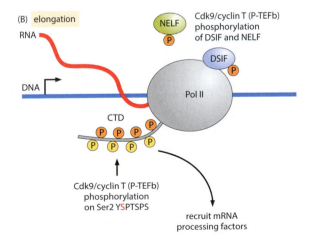

Figure 5.17 Regulation of Pol II transcription by Cdk7/cyclin H/Mat1 and P-TEFb (Cdk9/cyclin T). (A) On initiation of transcription, the Pol II CTD is phosphorylated at Ser5 by Cdk7 from the Cdk7/cyclin H/Mat1 complex, part of the TFIIH general transcription factor. Transcription elongation is blocked by the action of NELF (negative elongation factor) and DSIF (5,6-dichloro-1-β-D-ribofuranosylbenzimidazole (DRB) sensitivity-inducing factor). Pol II produces short RNA chains and then stalls, giving time for the mRNA 5′-capping enzymes to act. (B) The change to transcription elongation is promoted by the action of Cdk9/cyclin T (P-TEFb), which phosphorylates NELF and DSIF, resulting in relief from transcription repression. Cdk9/cyclin T also phosphorylates the Pol II CTD at Ser2. The CTD phosphorylated Pol II recruits mRNA processing factors.

Although eukaryotic cells also have transcription factors that respond to environment (for example, oxygen tension or hormonal stimulation), most of their transcription factors are involved in promoting cell development and differentiation at the right time. They recognize gene-specific sequences, usually upstream of the start site. Recognition regions may be close to the start site (promoter-proximal); others may be tens of kilobase pairs away. Communication between the gene-specific transcription factors, the general transcription factors, and Pol II is mediated by co-activators such as the Mediator complex (see Figure 5.1B). Gene-specific transcription factors often have three regions: a DNA-binding domain, a ligand recognition and/or regulatory region, and an activator domain that provides the connection to the polymerase through other proteins.

In humans, there are ~3000 gene-specific transcription factors and ~20,000 protein-coding genes. Thus, not every different type or set of genes in eukaryotes requires its own transcription factor. Gene-specific transcription factors can act in combination and create a regulatory complex that allows a cell to respond to multiple external signals. The transcriptional control elements also include **enhancers**, silencers, and insulator elements on the DNA. These can be located as much as 10–50 kbp upstream or downstream of the TATA box.

The **enhanceosome** for interferon-β (IFN-β) is an example of a compact enhancer with clustered DNA sites. Expression of the virus-inducible IFN-β gene requires multiple transcription factors: the heterodimer ATF-2/c-Jun, the interferon response factors IRF-3 and IRF-7, and NFκB (p50/RelA). These proteins bind cooperatively to a nucleosome-free region of the promoter, spanning the nucleotides from –102 to –47 bp relative to the transcription start site. Assembly of the enhanceosome leads to the recruitment of enzymes that catalyze nucleosome acetylation and chromatin remodeling (see Figure 5.1B). These activities result in repositioning of the nucleosome that covers the TATA box and allow the TBP and RNA polymerase machinery to access the promoter. A composite model based on structures of individual complexes shows how the binding of eight transcription factor DNA-binding domains to specific DNA sites upstream of the transcription start site creates a continuous recognition surface with extensive overlap of individual binding sites (**Figure 5.18**).

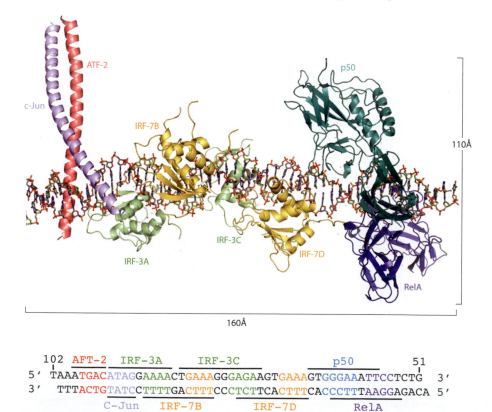

Figure 5.18 The IFN-β enhanceosome. A composite model of the enhanceosome, which comprises eight transcription factors, shown in different colors with their core DNA-binding sites colored accordingly. The DNA is bent around the four IRF domains. (From D.D. Panne et al., *Cell* 129:1111–1123, 2007. With permission from Elsevier.)

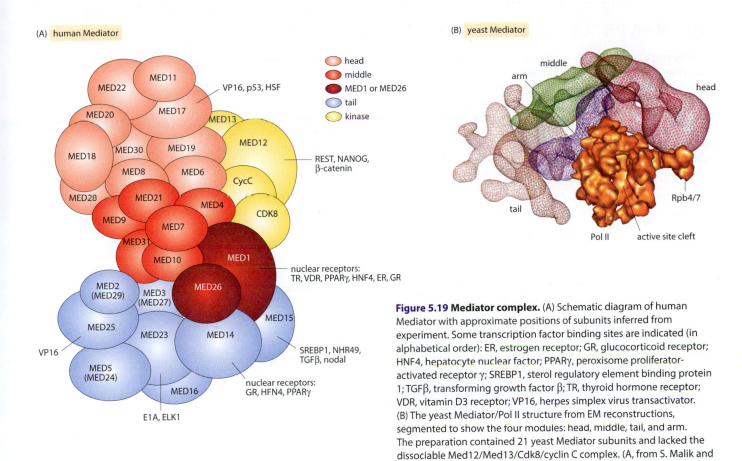

(A) human Mediator

(B) yeast Mediator

head
middle
MED1 or MED26
tail
kinase

VP16, p53, HSF

REST, NANOG,
β-catenin

nuclear receptors:
TR, VDR, PPARγ, HNF4, ER, GR

SREBP1, NHR49,
TGFβ, nodal

nuclear receptors:
GR, HFN4, PPARγ

E1A, ELK1

VP16

middle

arm

head

tail

Pol II   active site cleft

Rpb4/7

Figure 5.19 Mediator complex. (A) Schematic diagram of human Mediator with approximate positions of subunits inferred from experiment. Some transcription factor binding sites are indicated (in alphabetical order): ER, estrogen receptor; GR, glucocorticoid receptor; HNF4, hepatocyte nuclear factor; PPARγ, peroxisome proliferator-activated receptor γ; SREBP1, sterol regulatory element binding protein 1; TGFβ, transforming growth factor β; TR, thyroid hormone receptor; VDR, vitamin D3 receptor; VP16, herpes simplex virus transactivator. (B) The yeast Mediator/Pol II structure from EM reconstructions, segmented to show the four modules: head, middle, tail, and arm. The preparation contained 21 yeast Mediator subunits and lacked the dissociable Med12/Med13/Cdk8/cyclin C complex. (A, from S. Malik and R.G. Roeder, *Nat. Rev. Genet.* 11:761–772, 2010. With permission from Macmillan Publishers Ltd.; B, from G. Cai et al., *Structure* 17:559–567, 2009. With permission from Elsevier.)

The Mediator complex links gene-specific transcription factors to the PIC

Mediator is a ~1.2 MDa multiprotein complex that is required for transcription from all Pol II promoters. Mediator functions as a bridge to convey information from gene-specific regulatory proteins to the basal Pol II machinery. Mediator is recruited to promoters by gene-specific transcription factors and interacts with the PIC. It functions in transcriptional activation and in the repression and stabilization of the PIC (see Figure 5.1B). Human Mediator is composed of 26 core subunits, each named MED followed by a number (**Figure 5.19A**). EM studies have identified 'head,' 'middle,' and 'tail' modules in Mediator (Figure 5.19B). In addition, there is a regulatory module, the kinase module, consisting of the kinase Cdk8/cyclin C and two further subunits, MED12 and MED13, also involved in regulation. Many Mediator subunits are less conserved than the general transcription factors, suggesting that they may serve as species-specific regulatory factor targets.

Transcription factors have been shown to bind through their activation domains to at least seven Mediator subunits (see Figure 5.19A). For example, MED1, the largest of the core subunits (168 kDa), is targeted by several ligand-inducible nuclear hormone receptors, including the human thyroid hormone receptor (TR). MED15 (87 kDa) is targeted by the transcriptional activator SREBP1, required for the transcription of genes encoding enzymes involved in lipid homeostasis, and by the transcription factor TGFβ, a multifunctional protein that controls proliferation, differentiation, and other functions in many cell types (Chapter 12). Mediator binding to Pol II is enhanced by transcription factors and *vice versa*. Cryo-EM (see **Methods**) studies of Mediator from *S. cerevisiae* have revealed that the head module binds to the Pol II surface around the RNA exit tunnel and the Rpb4–Rpb7 stalk (Figure 5.19B), whereas the middle module extends toward the Pol II foot domain and the tail module faces upstream DNA, where transcription activators bind.

5.4   RNA PROCESSING: THE SPLICEOSOME

In eukaryotes, most protein-coding genes are interrupted by noncoding sequences known as **introns**. These genes are transcribed into precursor mRNAs (pre-mRNAs) that undergo splicing to remove the introns. The coding sequences (**exons**) are ligated together to form mRNA. Whereas some transcripts are constitutively spliced—that is, each mRNA produced from a given pre-mRNA is the same—many are alternatively spliced to generate different mRNAs from a single pre-mRNA species. Alternative splicing is prevalent in higher eukaryotes and it increases the number of distinct proteins expressed from a single gene. Most exons are less than 200 nt in length, whereas introns can vary from ~50 to more than 10,000 nt.

A large, dynamic protein/RNA complex, termed the **spliceosome**, performs the splicing. To date, two spliceosomes differing in composition have been characterized. The U2-dependent (major) spliceosome is found in all eukaryotes and catalyzes the removal of U2-type introns, the most commonly encountered class of introns. (U2 stands for uridine-rich small nuclear ribonucleoprotein particle, number 2; see below). The less abundant U12-dependent (minor) spliceosome splices out the rare U12-type class of pre-mRNA introns and is present in only a subset of eukaryotes. Here we focus on the structure and function of the more widely studied U2-dependent spliceosome.

The pre-mRNA introns are removed in a two-step process

Sequence analysis of introns has shown that nearly all U2-type introns begin at the 5′ end with GU, and end at the 3′ end with AG (**Figure 5.20A**). Introns also have an important internal branch site, which is located about 20–40 nt upstream of the 3′ splice site (3′ss) and always contains an adenine. In higher eukaryotes, this branch site is followed by a polypyrimidine tract (Y in Figure 5.20A).

In the first step of intron removal, the 2′-OH group of the A in the intron branch point promotes a nucleophilic attack on the 5′ss, resulting in cleavage of the pre-mRNA at this site. Concomitantly, the 5′ end of the intron is ligated, via a 2′–5′ phosphodiester bond, to the branchpoint A, generating a lariat-like structure. In the second step, the 3′-OH group of the 5′ exon carries out a nucleophilic attack on the 3′ss. Upon cleavage of the latter, the 5′ and 3′ exons are ligated together, forming the mRNA, and the intron is released (see Figure 5.20B).

Assembly and disassembly of the spliceosome proceed in stepwise fashion

The active conformation of the intron is generated by the binding of the spliceosome to the pre-mRNA. Spliceosomes are assembled onto each intron by the stepwise interaction of four small nuclear ribonucleoprotein particles (snRNPs) and numerous non-snRNP proteins. The uridine (U)-rich snRNPs U1, U2, U4/U6, and U5 are the main building-blocks of the U2-dependent spliceosome.

snRNPs consist of an snRNA molecule (or two RNA molecules for the U4/U6 snRNP) that range in length from 106 to 187 nt in humans. With the exception of U6, each has a unique 2,2,7-trimethylguanosine cap at its 5′ end. The snRNAs are bound by seven Sm proteins (B/B′, D1, D2, D3, E, F, and G) plus several particle-specific proteins (**Figure 5.21**), except for U6, whose snRNA is bound by seven homologous LSm (Like-Sm) proteins (LSm2–8). Sm proteins were first discovered as an autoantigen in patients with Systemic Lupus Erythematosus (SLE) and named for Stephanie Smith, an SLE sufferer. They are distinguished by a conserved seven-residue sequence motif and bind U-rich RNAs; an example of an Sm structure is shown below in Figure 5.28C.) The U4/U6.U5 tri-snRNP is formed by association of the U4/U6 and U5 snRNPs, accompanied by loss of the U5-52K protein and recruitment of three additional proteins (Snu66, Sad1, and 27K).

The simplest assembly of spliceosomes occurs on introns that are relatively short (< 200 nt) or on pre-mRNA that contains a single intron (**Figure 5.22**). First, the U1 snRNP interacts with the 5′ss of the pre-mRNA to form the so-called E complex. The U1 snRNP contains a sequence near its 5′ end that is complementary to the short consensus sequence at the 5′ end of introns. The U2 snRNP then stably associates with the branch site, generating the A complex. The U2 RNA has a sequence that is largely complementary to the consensus

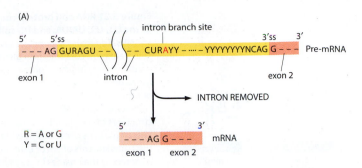

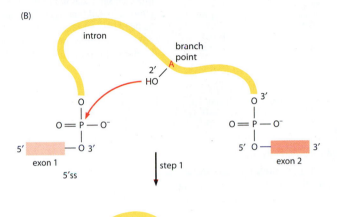

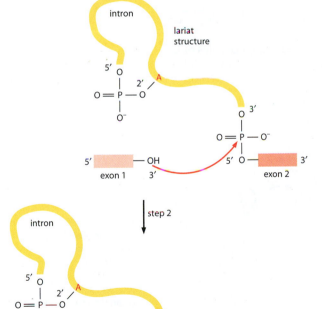

Figure 5.20 Pre-mRNA splicing reactions.
(A) Consensus nucleotide sequences in a pre-mRNA molecule signal the start and end of most introns in humans. An A residue (red) indicates the branch point of the lariat produced by splicing. The distance between the 5'ss and the branch point usually exceeds the distance between the 3'ss and the branch point. (B) The splicing of exons in pre-mRNA occurs by two transesterification reactions. In the first step, the 2'-hydroxyl of the branchpoint adenosine (A, red) attacks the 3' phosphorus of exon 1. In the second step, the released 3'-hydroxyl of exon 1 attacks the 5' phosphorus of exon 2. (A, adapted from H. Lodish et al., Molecular Cell Biology, 4th ed. New York: WH Freeman, 2000; B, adapted from B. Alberts et al., Molecular Biology of the Cell, 6th ed. New York: Garland Science, 2015.)

sequence flanking the branch point of the pre-mRNA. Subsequent recruitment of U5 and U4/U6 snRNPs as the pre-assembled U4/U6.U5 tri-snRNP yields the B complex, which does not yet have an active site. After major conformational and compositional rearrangements, including the release of U1 and U4, an activated complex termed Bact is formed.

The catalytically activated B* complex is formed from Bact by the action of Prp2, a DExD/H-box RNA-dependent ATPase/helicase (**Box 5.2**). B* catalyzes the first step of splicing (5'ss

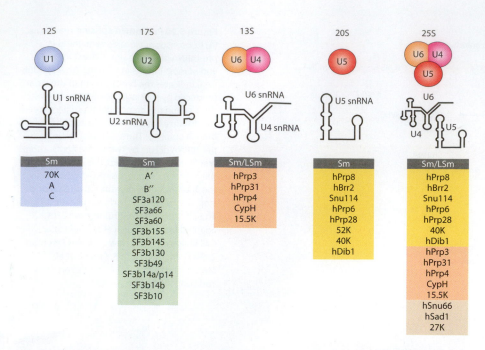

Figure 5.21 RNA and protein composition of the U1, U2, U4/U6, and U5 spliceosomal snRNPs. Top row: the numbers give the sedimentation coefficients in glycerol gradient centrifugation for the predominant forms of human snRNPs. Each snRNP has one or more (in the case of U4/U6) uridine-rich snRNA molecules. Middle row: the secondary structures of these snRNAs. Bottom row: Lists of proteins contained in addition to the complements of Sm or LSm proteins, as denoted by their names in gray fill at top. The U4/U6.U5 tri-snRNP has two Sm sets and one LSm set. (Adapted from C.L. Will and R. Lührmann, *Cold Spring Harb. Perspect. Biol.* 3:a0037 [doi: 10.1101/cshperspect.a0037], 2011. With permission from the copyright holder, Cold Spring Harbor Laboratory Press.)

cleavage and ligation of the intron's 5′ end to branchpoint A) to generate the C complex. After additional rearrangements, which are currently not well understood, the C complex catalyzes the second step of splicing. The spliceosome subsequently disassembles, releasing the mRNA and excised intron, and the snRNPs are recycled for additional rounds of splicing (see Figure 5.22).

Most mammalian pre-mRNAs contain multiple introns. When an intron length exceeds ~200 nt (and most do), formation of the initial splicing complex occurs across the exon, a process called exon definition. Splicing then involves more complex rearrangements of U1 snRNP and U2 snRNP and the recruitment of auxiliary splicing factors to RNA sequences known as exon splicing enhancers. Many of these factors are SR proteins. Members of this

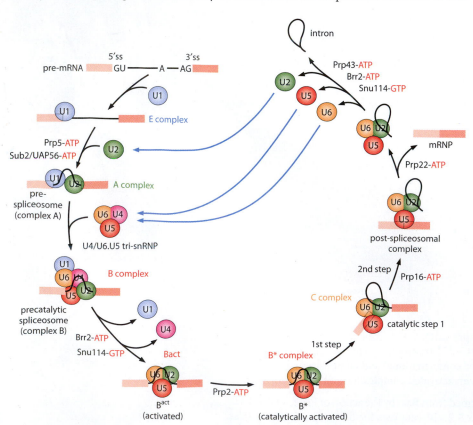

Figure 5.22 Assembly and subsequent disassembly of the spliceosome. The stepwise association and dissociation of the spliceosomal snRNPs (indicated by colored circles) with the pre-mRNA across an intron. The 5′ and 3′ exons of the pre-mRNA are indicated by red boxes, and the intron by a black line. The DExH/D-box ATPase/helicase proteins (Prp5, UAP56, Prp28, Brr2, Prp2, Prp16, Prp22, Prp43) and the GTPase Snu114 are shown at the step (or steps) of splicing at which they act. Pr, pre-mRNA-processing factor. Prp28, Brr2, and Snu114 are part of the U5 snRNP. The other enzymes are recruited individually. After their release from the spliceosome, the snRNPs are recycled and take part in new rounds of splicing. (Adapted from C.L. Will and R. Lührmann, *Cold Spring Harb. Perspect. Biol.* 3:a0037, 2011. doi: 10.1101/cshperspect.a0037. With permission from the copyright holder, Cold Spring Harbor Laboratory Press.)

family contain a domain rich in Arg-Ser (RS) dipeptides and they promote both cross-exon and cross-intron interactions between snRNPs, thereby facilitating spliceosome assembly.

The marking of exon and intron boundaries and assembly of the spliceosome on a pre-mRNA molecule are concomitant with RNA polymerase activity, but the actual process of intron removal can occur later. Introns are not necessarily removed in the order in which they occur. Hence, although spliceosome assembly is co-transcriptional, some splicing reactions occur post-transcriptionally.

A dynamic spliceosomal RNA/RNA interaction network is formed during splicing

An RNA/RNA interaction network, involving both the snRNAs and the pre-mRNA, has a central role in juxtaposing the reactive groups of the intron (**Figure 5.23A**). Initially, U1 snRNA forms base pairs with the 5′ss, and U2 forms base pairs with the branch site. The U1 snRNA contains a sequence 3′-CAUUCA-5′ near its 5′ end that is complementary (with one mismatch) to the *S. cerevisiae* 5′ss 5′-GUAUGU-3′. Upon formation of Bact, U6/U4.U5 tri-snRNP addition, and loss of U1 (Figure 5.23B), nucleotides of the U6 snRNA are essential components of the spliceosome's active site, but they are initially delivered to the spliceosome in an inactive form, base-paired to U4. This ensures that the pre-mRNA is not cleaved prematurely. During catalytic activation, U4/U6 base-pairing interactions are disrupted, and the U6 snRNA forms base pairs with intron nucleotides at the 5′ss, displacing U1 in the process. U6 also forms short duplexes with U2 and an intramolecular U6 stem-loop (U6-ISL) that is involved in catalytic metal ion binding. These RNA structures involving U2 and U6 snRNA have crucial roles in the catalytic core of the spliceosome, with U6 nucleotides directly involved in pre-mRNA splicing. U5 interactions have a key role in tethering the 5′ exon to the spliceosome after cleavage at the 5′ss during step 1 of splicing and later for correctly positioning the exons for the second step of splicing.

The remodeling of RNA/RNA, RNA/protein, and protein/protein interactions during spliceosome assembly requires rearrangements and conformational changes, regulated to

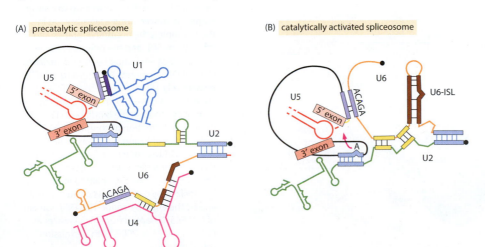

(A) precatalytic spliceosome

(B) catalytically activated spliceosome

Figure 5.23 Rearrangements in the spliceosome RNA/RNA interaction network during catalytic activation. (A) The precatalytic spliceosome B complex. Full-length U1, U2, U4, and U6 snRNA, and only loop 1 of the U5 snRNA (not all drawn to scale) are shown schematically. The pre-mRNA is shown with the 5′ and 3′ exons as red boxes connected by the intron as a black curve. Note that the RNA network is depicted before the action of ATPase Prp28, because the U1/5′ss interaction is still intact. (B) The catalytically activated spliceosome Bact. The U4/U6 interaction has been disrupted and the U6 snRNA base pairs with the 5′ss, displacing the U1 snRNA. The 5′ end of each snRNA is indicated by a black dot. (Adapted from M.C. Wahl et al., *Cell* 136:701–718, 2009. With permission from Elsevier.)

ensure that they occur at the correct stage of the splicing process. Major participants in spliceosome rearrangements are the DExD/H-box RNA-dependent ATPases/helicases. At least two of these helicases, Prp28 and Brr2, are integral components of the U5 snRNP. Both proteins catalyze RNP rearrangements that lead to the catalytic activation of the spliceosome. Prp28 facilitates the exchange of the U1 snRNP with U6 at the 5'ss (in moving from the B to the Bact complex in Figure 5.22), and Brr2 catalyzes the subsequent unwinding of the U4/U6 duplex that allows U6 to form base pairs with U2 in transforming the B complex to Bact (see Figures 5.22 and 5.23).

Prp8, the largest spliceosomal protein (2335 aa in humans) is a component of the U5 snRNP, interacting with Brr2 and regulating its helicase activity. The U5-associated GTPase Snu114 (see Figure 5.22) that acts during the B-to-Bact conversion also modulates Brr2 activity. Like Brr2, Snu114 interacts with Prp8, suggesting that Prp28 coordinates the control of Brr2 by Snu114. The three proteins form a stable complex that can be isolated by salt-induced dissociation of the U5 snRNP. They represent a molecular motor that drives RNP rearrangements required for catalytic activation of the spliceosome.

Splice-site recognition in the E and A complexes involves the coordinated action of RNA and protein

A major task of the spliceosome is the recognition and pairing of the correct 5' and 3' splice sites. During spliceosome assembly, the splice sites and branch site are recognized multiple times by proteins and snRNAs, and both kinds of molecules contribute to ensuring the precision of the splicing reaction. Many functionally important binary interactions within the spliceosome are weak, but the overall stability of functional complexes is enhanced by combining multiple weak interactions (see Section 1.2). These principles are well illustrated during the early stages of spliceosome assembly of complex A and complex E.

Multiple recognition events occur at the 5'ss. The U1 snRNA, as part of the U1 snRNP, engages in a base-pairing interaction with the 5'ss as described above. This interaction is stabilized by the U1-associated 70K and C proteins, and also by members of the SR protein family (see above) that interact with both the pre-mRNA and the U1-70K protein.

At the branch site and 3'ss, three proteins—SF1/BBP (splicing factor 1/branchpoint binding protein; 68 kDa) and U2 Auxiliary Factors U2AF65 (65 kDa) and U2AF35 (35 kDa)—bind to the intron branchpoint sequence, the polypyrimidine tract, and the 3'ss AG dinucleotide respectively (**Figure 5.24A**). SF1/BBP interacts with the branch site through its KH domain (**Box 5.3**) and also with U2AF65, which contains an N-terminal RS domain followed by three RNA recognition motifs (RRMs) (see Box 5.3). RRM1 and RRM2 bind RNA, and RRM3 mediates the interaction with SF1/BBP. During the subsequent formation of the A

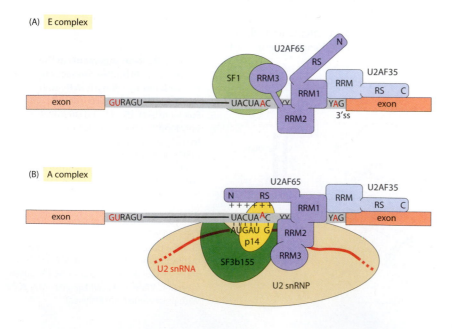

Figure 5.24 Molecular interactions at the branch site and the 3'ss during the early stages of spliceosome assembly. (A) In the E complex, the pre-mRNA branch site is bound by SF1, and the polypyrimidine tract and 3'ss are bound by the U2 auxiliary factor (U2AF) subunits U2AF65 and U2AF35 respectively. U2AF65 binds both SF1 and U2AF35. Exons are shown in red, and introns in gray. (B) Upon stable binding of U2 snRNP during the formation of complex A, SF1 is displaced, allowing the U2-associated protein p14 to contact the branch site and U2AF65 to interact with U2 protein SF3b155. The U2/branch site base-pairing interaction is stabilized by components of the U2 snRNP and by the arginine-serine-rich (RS) domain of U2AF65. (Adapted from M.C. Wahl et al., *Cell* 136:701–718, 2009. With permission from Elsevier.)

Box 5.3 KH, S1, and RRM RNA-binding domains

KH, S1, and RRM domains (see **Guide**) are RNA-binding domains that have been recognized as components of many proteins. This box shows the canonical binding modes for RNA, but variations exist. There are also RRM domains that bind proteins and not RNA. The fold is indicative of function, but not all function in the same way.

The 70-aa KH domain was originally identified in a human protein called heterogeneous nuclear ribonucleoprotein (hnRNP) K, which has three KH domains. The KH domain binds single-stranded polynucleotides with dissociation constants ranging from 10^{-6} to 10^{-9} M. The KH domain comprises a three-stranded β sheet packed against three α helices (**Figure 5.3.1A**). There is a conserved GxxG sequence in the loop between the two helices of the βααβ core. Single-stranded RNA binds with the bases docked against a hydrophobic surface and the phosphate groups exposed to the solvent. The RNA is gripped between the GxxG loop and a variable loop that connects β2 and β3. The surface and interacting residues can provide a sequence-specific binding region.

The S1 RNA-binding domain was originally identified in ribosomal protein S1. The ~70 aa domain consists of a five-stranded antiparallel β barrel with a Greek key motif (see **Guide**). The RNA-binding site is formed by conserved residues on one face of the barrel and adjacent loops. The structure is similar to the cold shock domain (CSD) (see **Guide**) from the protein that regulates response to cold shock in bacteria and binds both DNA and RNA. The same topology is seen in the OB domain (oligonucleotide-oligosaccharide-binding domain; see **Guide**) (Figure 5.3.1B).

The RNA recognition motif (RRM) is the most abundant RNA-binding module and is found in proteins involved in pre-mRNA splicing and mRNA translation, in which it recognizes the poly(A) tails or the cap of eukaryotic mRNAs. It can bind RNA specifically and tightly, as in the U1A protein, a component of the U1 small nuclear ribonucleoprotein particle (U1 snRNP), in which snRNA binds with a K_d of 10^{-10} M. In other proteins that bind RNA transiently, the affinity is much lower. The fold comprises a four-stranded antiparallel β sheet with two α helices (αA and αB) arranged in an αββαββ topology. In some proteins, the structure is extended with further helices or a fifth β strand. The RRM domains are characterized by 6-residue and 8-residue consensus sequences (RNP1 and RNP2 respectively), which carry exposed aromatic residues from the two central β strands. The major interactions of RNA are with the β sheet, but the loops are also important in determining specificity (Figure 5.3.1C,D).

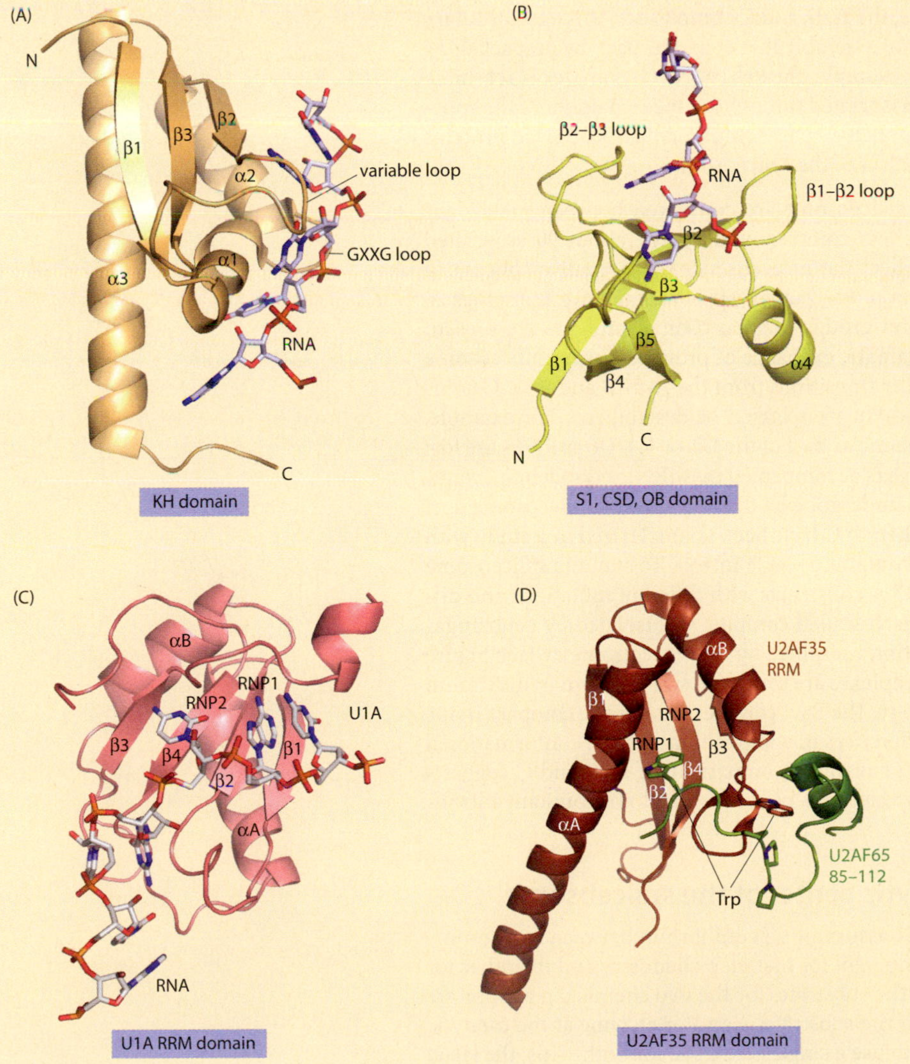

(A) KH domain

(B) S1, CSD, OB domain

(C) U1A RRM domain

(D) U2AF35 RRM domain

Figure 5.3.1 Structures of the KH, S1, and OB domains. (A) On the KH domain, RNA docks against α helices and the variable loop. (PDB 1EC6) (B) In this S1 structure, the five-stranded β sheet is interrupted between strands β2 and β3 by helix α4. This region is a loop structure in other S1 domains. RNA docks against the β2 and β3 strands and is gripped by the β1–β2 loop on one side and by the β2–β3 loop on the other side. (PDB 2A8V) (C) RRM domain of U1A bound to a 21-nt RNA hairpin (only eight nt are shown). The single-stranded RNA binds to the surface of the β sheet, and the AUUGCAC sequence of the loop interacts with the RNP1 and RNP2 motifs. (PDB 1URN) (D) RRM domain of U2AF35 from the spliceosome bound to a Pro-rich motif from U2AF65. This is an example of a protein/protein interaction of an RRM domain. Here, the binding is from the opposite side of the β sheet to that shown for the U1A protein, and the main recognition motifs are a Trp residue from U1AF65 and one from U1AF35 that dock 'tongue-in-groove.' (PDB 1JMT)

complex, the U2 snRNA engages in a base-pairing interaction with the branch site, displacing SF1/BBP. The branch-site sequence 5'-UACUAAC-3' is paired with the U2 snRNA sequence 3'-AUGAUG-5', with the crucial A not paired and looped out, leaving it available for its role in catalysis (Figure 5.24B). Heteromeric protein complexes within the U2 snRNP, namely subunits SF3a and SF3b, contact the pre-mRNA at or near the branch site and stabilize the U2/branch-site interaction. Most RNA/RNA interactions in the spliceosome involve only very short base-paired regions and so proteins must provide additional interactions. In a subsequent step, the branch-site A associates with the U2-associated p14 protein (at least in higher eukaryotes), and U2AF65 interacts with the U2-associated SF3b155 protein (Figure 5.24B). At later stages of splicing, the U2AF and SF3b proteins appear to dissociate, and new RNA/protein and protein/protein interactions involved in 3'ss recognition are formed.

These interaction networks have been established by multiple techniques including cross-linking assays, the use of recombinant proteins or fragments thereof, band-shift assays to study RNA/protein interactions, and, in some cases, structural studies of the binding partners. There are X-ray or NMR (see **Methods**) structures for several components, including U2AF RRM1 and RRM2 domains with RNA oligonucleotides, and the core U2AF65/U2AF35 heterodimer. They show how the first RRM domain of U2AF35 is a versatile recognition module and recognizes a Pro-rich sequence from U2AF65 (residues 85–112) via reciprocal 'tongue-in-groove' interactions involving Trp residues.

The spliceosome has a complex and dynamic protein composition

Characterization of purified human spliceosomal complexes by mass spectrometry has revealed that ~50 proteins associate with the snRNPs and over 125 proteins with the individual assembly intermediates (for example, the A, B, and C complexes). This extraordinary complexity poses a challenge for the order of assembly that is met in part by prepackaging many proteins in stable subcomplexes. For example, the Prp19 (NTC) complex is recruited during B-complex formation and remains associated throughout the remainder of the splicing cycle (not shown in Figure 5.22, for simplicity). This complex consists of seven proteins in humans and has a role in catalytic activation of the spliceosome.

The major pathways for assembly of the spliceosome are conserved between metazoans and yeast, but the latter spliceosomes are less complex, with 'only' about 90 associated proteins (**Figure 5.25**). This appears to reflect the near-absence of alternative splicing in *S. cerevisiae*. Nearly all of the core proteins of the human spliceosome have homologs in *S. cerevisiae* spliceosomes. Although several groups of spliceosomal proteins are present throughout the splicing cycle, there is a dramatic exchange of proteins during spliceosome assembly and catalytic activation. During the transitions from the A-to-B and B-to-C complexes, many proteins are recruited and others are released or destabilized. For example, essentially all proteins entering the spliceosome as part of the U1 or U4/U6 snRNPs are lost during catalytic activation. Protein exchange is accompanied by snRNP remodeling events. The changes in composition and structure undergone by the *S. cerevisiae* spliceosome as it converts from the B complex to Bact and then to C have been characterized in a study with the pre-mRNA for actin. This pre-mRNA contains a single intron. To stall the spliceosome at successive stages, three different pre-mRNA constructs with different mutations and different concentrations of ATP were used. Each stalled complex was isolated by centrifugation and affinity selection; and its composition examined by mass spectrometry (see Figure 5.25) and its structure by EM. The three complexes are similar in size (~40 nm in maximum dimension) but quite different in morphology. The Bact complex has a more compact shape than the B complex, consistent with the differences in composition and the conformational remodeling that is inferred to accompany activation. Comparison of Bact and C suggests that further structural changes accompany cleavage of the 5'ss and concomitant intron-lariat formation.

A two-state model for the catalytic center of the spliceosome

Most experimental data can be explained by assuming an equilibrium between two conformational states of the spliceosome, one competent for first-step chemistry and the other for second-step chemistry (see Figure 5.20B). The substrates for the two chemical reactions are different and hence some rearrangement of the substrate(s) and/or enzyme at the catalytic center is required. If the spliceosome were to use a single active site for both steps, the lariat intermediate formed in the first step would need to be displaced to allow positioning of the

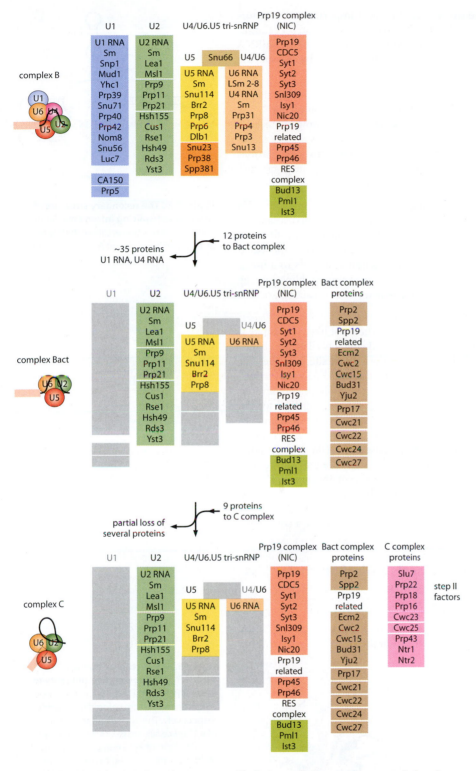

Figure 5.25 Compositional dynamics of the *S. cerevisiae* spliceosomal B, Bact, and C complexes. Proteins (yeast nomenclature) are grouped according to snRNP association, function, or association with a given spliceosomal complex. Gray boxes indicate proteins that have left a complex. Catalytic activation (B-to-Bact transition) is accompanied by the loss of ~35 proteins, including all U1 and U4/U6-associated proteins, whereas ~12 proteins (brown) are recruited. Nine proteins (pink) are recruited during the Bact-to-C transition. (From P. Fabrizio et al., *Mol. Cell* 36:593–608, 2009. With permission from Elsevier.)

3′ss for the second step. The spliceosome is likely to toggle between open and closed states during the catalytic phase. The ATPases Prp2 and Prp16 are needed to activate the spliceosome for the first and second catalytic steps, respectively (see Figure 5.22). These factors probably facilitate conformational changes in the catalytic center and thus also in positioning the substrates.

The spliceosome appears to act mostly as a ribozyme

The organization of the spliceosome's active site and the respective contributions of RNA and protein to splicing catalysis are not entirely clear. Evidence indicates that the spliceosome is likely to be a ribozyme, in which case the snRNAs are the most important contributors to the catalysis of both splicing steps. Consistent with RNA-based catalysis, step 1 requires the presence of a metal ion. Several intermolecular structures formed by the pre-mRNA and the U2, U5, and U6 snRNAs share similarities with structures formed by self-splicing group II introns.

Group II introns are protein-free RNA transcripts containing an intron sequence that *in vitro* slowly undergoes self-splicing in the presence of a high concentration of Mg^{2+}. The intron sequence contains an A whose 2′-OH acts as an attacking group resulting in the lariat structure, as in pre-mRNA processing. All group II introns fold into a secondary structure that brings the reactive elements together (**Figure 5.26**). The mechanistic similarities between self-splicing group II introns and spliceosomal splicing have led to the suggestion that snRNAs function analogously to the stem-loops of group II introns.

The U6 and U2 snRNAs are key participants in splicing. Specific mutations in U2 and U6 affect catalysis, and a nucleotide in a region of U6, whose structure mimics that of a catalytically active structure in group II introns, binds a divalent metal ion needed for the first step of splicing. Furthermore, base-paired fragments of the U6 and U2 RNAs, when incubated with two short RNA oligonucleotides, perform a two-step reaction that leads to a linear RNA product containing part of each oligonucleotide. Thus, a reaction resembling splicing can be catalyzed by U2 and U6. However, the kinetics and efficiency of this reaction are slow and low, implying that important cofactors are missing.

Several observations suggest that proteins may assist catalysis. The most likely candidate is Prp8, which is located in the spliceosome core as part of U5 and contacts all of the chemically reactive regions of the pre-mRNA including the 5′ss, the branch site, and the 3′ss. Significantly, a C-terminal region of Prp8 that is known to interact with the 5′ss has a ribonuclease H (RNase H)-like fold (**Figure 5.27**). Functional RNase H domains contain four carboxylates at their catalytic site that coordinate two Mg^{2+} ions that are critical for the catalysis of RNA cleavage (discussed in Section 5.5). The Prp8 RNase H-like domain contains an incomplete set of metal-binding residues and does not bind Mg^{2+}. Experiments based on cross-linking with ultraviolet radiation have identified a region (the 3_{10} helix) that contacts the 5′ss in complex B spliceosomes (see Figure 5.27). A unique part of Prp8, its β-hairpin, is reminiscent of the β-hairpin protrusions of ribosomal proteins such as S10 and L22, which stabilize the ribosomal RNA structure (Section 6.2). The data are consistent with a role for Prp8 in structuring the RNA network at the heart of the spliceosome and positioning the 5′ss for catalysis. Thus, the catalytic site of the spliceosome may be a composite, consisting of both RNA and protein.

A crystal structure of the U1 snRNP suggests a mechanism for 5′ss recognition

U1, the simplest of the snRNP complexes, recognizes the 5′ss of the pre-mRNA and initiates spliceosomal assembly. The U1 snRNA forms a four-stem-loop (SL) structure (**Figure 5.28A**). Its 5′ end base pairs with a short sequence at the 5′ss of pre-mRNA in a recognition

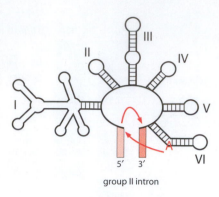

Figure 5.26 The secondary structure of Group II self-splicing intron RNA. Steps 1 and 2 of the reaction are indicated by red arrows. The branchpoint A residue is in red.

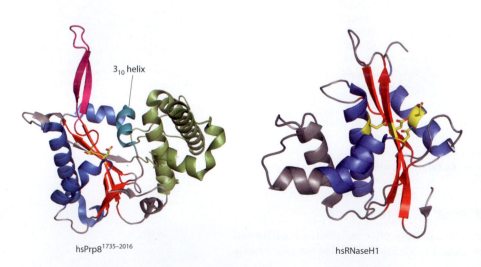

hsPrp8$^{1735-2016}$ hsRNaseH1

Figure 5.27 Structure of human Prp8$^{1735-2016}$, showing the similarity to human RNase H. The β sheet and α helices common to both are colored red and blue respectively. The 3_{10} helix that is the locus of a 5′ splice site cross-link is colored cyan. The β hairpin of Prp8 is in pink. Two Asp residues corresponding to Asp1781 and Asp1782 at the catalytic site of RNase H are in yellow with stick representation. In RNase H, the four catalytic residues Asp145, Glu186, Asp210, and Asp274 are in yellow. For crystallography, Asp210 was mutated to Asn. Regions colored green (hsPrp8) and gray (hsRNaseH) differ between the two enzymes. (PDB 3E9L [Prb8] and 2QKK [RNaseH]) (Adapted from V. Pena et al. *EMBO J.* 27:2929–2940, 2008. With permission from John Wiley & Sons.)

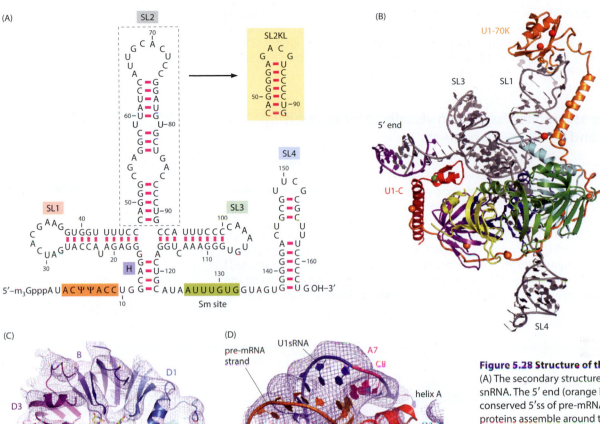

Figure 5.28 Structure of the U1 snRNP.
(A) The secondary structure of human U1 snRNA. The 5′ end (orange box) binds to the conserved 5′ss of pre-mRNA. The seven Sm proteins assemble around the Sm site (green box). In the form crystallized, stem-loop 2 (SL2) was replaced by a shorter loop SL2KL. Ψ is pseudouridine. (B) Crystal structure of the U1 snRNP. RNA is in gray. SL2KL is not visible in this view. The 5′ end of a neighboring snRNA (purple) base-pairs with the 5′ end of U1 snRNA. Orange balls mark the positions of selenomethionine residues introduced into U1-70K to assist in interpreting the electron density. (C) Interaction of the seven Sm proteins with the Sm site of the snRNA. The snRNA nucleotides are shown in stick representation. The molecular envelope from the density map is superimposed. (D) Interaction of protein U1-C (red) with the RNA duplex between the 5′ end of U1 snRNA and the putative pre-mRNA. (A, from A.J. Newman and K. Nagai, *Curr. Opin. Struct. Biol.* 20:82–89, 2010. With permission from Elsevier; B–D, from D.A. Pomeranz et al., *Nature* 458:475–480, 2009. With permission from Macmillan Publishers Ltd.)

event that is dependent on protein U1-C. A significant advance has come from a crystal structure of a U1 snRNP containing recombinant proteins (the seven Sm proteins [B/B′, D1, D2, D3, E, F, and G] and the U1-specific U1-70K and U1-C) together with an *in vitro*-transcribed snRNA that was modified to facilitate crystallization (the third U1 protein, U1-A, was missing from this complex, but the U1 snRNP is active in a splicing assay without it). Although the resolution of these data was relatively low, interpretation was aided by prior knowledge of structures for part of the RNA and some of the proteins.

Human U1 snRNA forms four stem-loop structures: SL1, SL2, SL3, and SL4. The three-dimensional structure comprises a four-helix junction with two coaxially stacked helices (SL1/SL2 and SL3/H-helix) followed by a single-stranded Sm-binding region and a 3′ SL4 stem-loop (Figure 5.28A,B). Each of the seven Sm proteins, whose structures comprise an N-terminal α helix and a five-stranded β sheet, bind the conserved Sm site of the U1 snRNA, forming a ring (Figure 5.28C). The individual nucleotides of the motif 5′-AUUUGUG-3′ probably interact with the Sm proteins E, G, D3, B, D1, D2, and F, in that order. The U1-70K protein has an unusual structure and winds around the snRNP interacting with U1-C protein at its N-terminal region and with SL1 at its C-terminal region (see Figure 5.28B).

In the crystal lattice, the 5′ end of the U1 snRNA was found base-paired with an adjacent U1 snRNA, and this packing arrangement gave clues to how the U1 snRNA might recognize the 5′ss of pre-mRNA during the first stage of spliceosome assembly. A helix (helix A) and a loop of the zinc-finger domain (see **Guide**) of the U1-C protein binds across the minor groove of this duplex, consistent with a role for U1-C in stabilizing the U1/5′ss duplex (Figure 5.28D). The invariant GU dinucleotide motif of the intron base-pairs with C8 and A7 of U1 snRNA respectively, and U1-C may contribute to a discrimination of nucleotides at these two positions. In a canonical pairing, the sequence 5′-G**GU**AAGU-3′ of the intron (where the conserved GU is in bold) would bind 3′ → 5′ to the sequence 5′-m3GpppAUAC$\Psi\Psi$A^{7}C^{8}C-3′ of the 5′ end of U1 snRNA, where mGppp is the RNA cap and Ψ is pseudouridine.

Spliceosome assemblies have been visualized by electron microscopy and labeling experiments

Several snRNP units of the spliceosome have been visualized in EM reconstructions. By fitting the EM density maps of the U5 and U4/U6 snRNPs into that of the U4/U6.U5 tri-snRNP, it was possible to determine how these components interact in the tetrahedrally shaped tri-snRNP (**Figure 5.29A**). In other EM studies, labels were introduced genetically into some spliceosomal proteins and their positions identified by difference mapping. In this way, proteins Prp8, Brr2, and Snu114 were localized within the *S. cerevisiae* U4/U6.U5 tri-snRNP (Figure 5.29B). Unwinding of the U4/U6 snRNA duplex requires several components of the U5 portion of the tri-snRNP, including the RNA helicase Brr2, the RNase H domain-containing Prp8, and the GTPase Snu114. The projection structure (a 2D image) showed a modular organization comprising three protruding domains that contact one

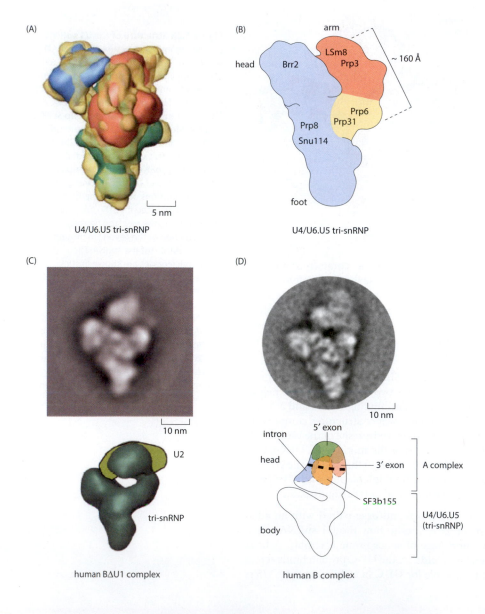

Figure 5.29 EM reconstructions of snRNP complexes. (A) Reconstruction of the human tri-snRNP complex, cryo-negatively stained. Color code: yellow, tri-snRNP overall; green, U5 body; blue, U5 head; red, U4/U6 di-snRNP. (B) Locations of functionally important proteins mapped on a diagrammatic projection of the yeast tri-snRNP: U5 (pale blue); U4/U6 (red); linker region (yellow). Protein names are placed to mark the locations of their C termini, as determined by antibody labeling EM. (C) The human BΔU1 complex. Upper, an averaged projection of the negatively stained complex. Lower, a three-dimensional reconstruction. The triangular body resembles the tri-snRNP complex in shape while the head adopts variable positions. (D) The human B complex. Protein A-coated colloidal gold and antibodies were used to locate the engineered binding sites in the pre-mRNA substrate. Antibodies against the U2-associated SF3b155 protein were used to map its position. Upper, an averaged projection of the negatively stained complex. Lower, interpretive sketch showing regions where the 5′ exon, 3′ exon, intron, and SF3b155 protein were mapped by immuno-EM, and the likely locations of the A complex (U1/U2) and the tri-snRNP (U4/U6.U5). (A, from B. Sander et al., *Mol. Cell* 24:267–278, 2006. With permission from Elsevier; B, from I. Häcker et al., *Nat. Struct. Mol. Biol.* 15:1206–1212, 2008; C, from D. Boehringer et al., *Nat. Struct. Mol. Biol.* 11:463–468, 2004. B and C, with permission from Macmillan Publishers Ltd; D, from E. Wolf et al., *EMBO J.* 28:2283–2292, 2009. With permission from John Wiley & Sons.)

another in the central portion. By generating genetically tagged tri-snRNP proteins, it was shown that U4/U6 snRNP forms a domain termed the arm. The head domain harbors Brr2, while Prp8 and the GTPase Snu114 are located centrally. The head and arm adopt variable relative positions.

The human precatalytic BΔU1 spliceosome, which contains the U2, U4/U6, and U5 snRNPs but lacks the U1 snRNP and several other non-snRNP proteins (a result of the preparation methods used), has a triangular body with a globular head that assumes various conformations (Figure 5.29C). Comparison of the human U4/U6.U5 tri-snRNP with BΔU1 suggested that the tri-snRNP resides in the lower part of the triangular body of BΔU1 and hence that the U2 snRNP should be located in the head. This is consistent with immuno-EM studies that mapped the U2-associated SF3b155 protein to the head region of intact human B complexes.

The positions of the intron and 5′ and 3′ exons of a model pre-mRNA in the intact human B complex have been determined by immuno-labeling of RNA tags inserted at specific sites. They are located in specific regions of the head complex together with the U2 snRNP core protein SF3b155, which contacts the pre-mRNA on both sides of the branch site (Figure 5.29D).

Splicing enhancers and silencers regulate alternative splicing

A hallmark of the spliceosome is its inherent flexibility, a characteristic that lends itself to regulation. The mRNA ultimately produced is determined by which 5′ and 3′ splice sites are recognized and subsequently paired by the spliceosome. The inclusion or exclusion of particular sets of exons can lead to multiple mRNAs being generated from one pre-mRNA species by alternative splicing. Several patterns of alternative splicing have been documented (**Figure 5.30**). For some pre-mRNAs, two or more alternatively spliced mRNAs (in a given ratio) are generated in all cell types. For many others, the choice of alternative splice sites is regulated in a tissue-specific or development-specific manner, or in response to changes in the cell's environment. Alternative splicing can lead to both quantitative and qualitative changes in protein expression. An example of alternative splicing is given in **Box 5.4** for tau, a microtubule-associated protein implicated in several neurodegenerative diseases.

Splice site recognition and selection in higher eukaryotes are influenced by several sequence elements of the pre-mRNA. Because most splice site consensus sequences are relatively degenerate, such splice sites alone are not capable of directing spliceosome assembly. The recognition and selection of splice sites are influenced, in most cases, by flanking pre-mRNA regulatory sequences—so-called intronic and exonic splicing enhancers or silencers (ISEs, ISSs, ESEs, or ESSs)—that modulate splice site usage. Most splicing regulatory proteins (that is, *trans*-acting alternative splicing factors) are members of the SR or heterogeneous nuclear ribonucleoprotein (hnRNP) protein families. SR proteins typically bind ESEs, often activating use of adjacent splice sites. In contrast, ESSs are typically bound by hnRNP proteins, which are characterized by RRM- and KH-type RNA-binding domains (see Box 5.3). Many alternative splice-site choices are determined by the combined actions of positive and negative regulatory proteins and the contribution of splicing enhancer and silencer sequences (in other words, combinatorial control). In most cases, regulatory events affect early stages of spliceosome assembly; however, they sometimes occur very late in splicing. Indeed, there are likely to be checkpoints throughout the splicing cycle at which decisions can be made that alter the outcome of the splicing process.

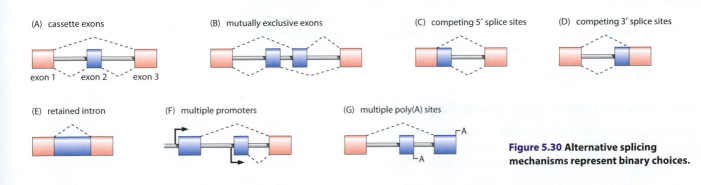

(A) cassette exons

exon 1   exon 2   exon 3

(B) mutually exclusive exons

(C) competing 5′ splice sites

(D) competing 3′ splice sites

(E) retained intron

(F) multiple promoters

(G) multiple poly(A) sites

Figure 5.30 Alternative splicing mechanisms represent binary choices.

Box 5.4 Alternative splicing and the tau protein

Tau is a microtubule-binding protein that assists in the assembly and stabilization of microtubules (see Section 14.6). It is abundant in the central nervous system, where it is expressed predominantly in axons. Tau proteins are encoded by a single gene that encodes 16 exons on Chromosome 17. In the normal adult human brain, six isoforms are produced by alternative RNA splicing. They differ in the presence or absence of a 29 or 58 aa insert in the N-terminal region of the protein and the inclusion or not of a 31 aa repeat, encoded by exon 10, in the C-terminal region (**Figure 5.4.1**). Exclusion of exon 10 leads to the production of three isoforms, each containing three repeats (3R-tau) of the microtubule-binding region of tau; inclusion of exon 10 leads to a further three isoforms, each containing four repeats (4R-tau). In the human brain, the ratio of 3R-tau to 4R-tau isoforms is ~ 1.

Tau is the major component of intracellular fibrillar deposits, which define several neurodegenerative diseases, including Alzheimer's disease, progressive supranuclear palsy (PSP), and the inherited frontotemporal dementia with Parkinsonism linked to Chromosome 17 (FTDP-17) disease. In the diseased state, hyperphosphorylated forms of tau aggregate to form insoluble fibrils; in certain diseases (PSP and FTDP-17), these are composed solely of the 4R-tau isoforms, suggesting that disease has some correlation with alternative splicing mechanisms. The mechanisms by which these changes in the ratio of 3R-tau to 4R-tau lead to neuronal dysfunction and cell death is unclear. It is possible that an overproduction of 4R-tau may lead to an excess of free tau in the cytoplasm, leading to its hyperphosphorylation and assembly into fibrils.

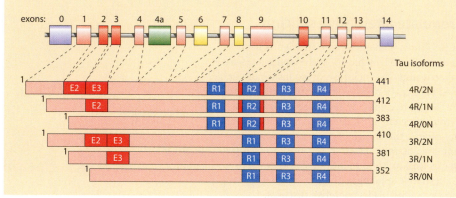

Figure 5.4.1 The human tau gene and the six isoforms generated by alternative splicing. Alternative splicing of exons E2, E3, and E10 (red boxes) produces the six isoforms. The blue boxes (R1–R4) depict the 18 aa microtubule-binding repeats. The exons and introns are not drawn to scale. (From V.M. Lee et al., *Annu. Rev. Neurosci.* 24:1121–1159, 2001.)

5.5　THE EXOSOME

The **exosome** is an RNA-degrading machine that is important for RNA quality control. (The same term, exosome, is also used for a different structure, alternatively known as extracellular vesicle, which is not discussed further.) Not all RNA that is synthesized is used. In the nucleus, excised introns and defective RNAs are degraded. In the cytoplasm, RNAs are degraded at the end of their useful lives. Messenger RNA molecules reaching the cytoplasm undergo progressive deadenylation at rates that are specific for each mRNA species. In vertebrates, mRNA half-lives range from a few minutes (for the transcription factor c-Fos, for example) to a few hours (for most mRNAs), to several days (for globin mRNA, for example). RNA lifetimes in other species vary and correlate with the generation lifetime of the species. Ribosomal RNAs, packaged in the ribosome, have considerably longer lifetimes than mRNA.

There are three major classes of intracellular RNA-degrading enzymes: endonucleases that cleave RNA internally; 5′-exonucleases that hydrolyze RNA from the 5′ end; and 3′-exonucleases that degrade RNA from the 3′ end. Most organisms have a plethora of ribonucleases (RNases), with partial redundancy. The exosome has 3′-exonuclease activity and degrades mRNA, snoRNAs, snRNAs, and rRNAs. This molecular machine was first described in 1997 in *S. cerevisiae* in the context of ribosomal RNA processing; hence, many of its subunits have the Rrp (ribosomal RNA processing) nomenclature. It is a major contributor to RNA degradation in all three kingdoms of life.

Exosomes are based on a hexameric ring structure

Exosomes increase in complexity from bacteria, where they are simple homo-oligomers, to archaea and eukaryotes, where they constitute larger hetero-oligomeric assemblies (**Table 5.3**). All have a central hexameric core to which additional subunits and cofactors have

Table 5.3 Exosome components.

Component	Domains	Bacteria	Archaea	Eukaryotes (*S. cerevisiae* and human)	Mass of human protein (kDa)
Exosome core	RNase PH	(RNase PH) × 6 or (PNPase) × 3	Rrp41	Rrp41	26
			Rrp42	Rrp45	49
			Rrp41	Rrp46	25
			Rrp42	Rrp43	30
			Rrp41	Mtr3	28
			Rrp42	Rrp42	32
	S1 and KH domains		(Rrp4) × 3	Rrp4	33
			or	Rrp40	30
			(Csl4) × 3	Csl4	21
Exonuclease			–	Rrp44	109
			–	Rrp6	101

(Adapted from M. Schmid and T.H. Jensen, *Trends Biochem. Sci.* 33:501–510, 2008.)

been added during evolution. In bacteria, RNA degradation is catalyzed by the enzymes RNase PH (phosphorolysis) and PNPase (polynucleotide phosphorylase). The prototypic RNase PII has a single exonuclease site. RNase PII subunits assemble into a homohexameric ring (**Figure 5.31A**). The PNPase subunit has two RNase PH domains, of which only the C-terminal one has exonuclease activity (Figure 5.31B). They are followed by two RNA-binding domains, S1 and KH, in the C-terminal region. (KH and S1 domains are described in Box 5.3.) Three PNPase molecules are arranged in a pseudohexameric ring complex, composed of the six RNase PH domains with the catalytic sites facing the center of the complex and the RNA-binding domains on the top of the ring.

The exosomes of other organisms have more elaborate arrangements of subunits. The archaeal exosome has a core built from two RNase PH homologs, Rrp41 and Rrp42 (Figure 5.31C). Only Rrp41 has catalytic residues, but both Rrp41 and Rrp42 are required to form an active site. The subunits are arranged in a hexameric ring with three copies of the catalytically active Rrp41/Rrp42 dimer. One or other of the RNA-binding proteins Rrp4 and Csl4 forms a trimeric cap on top of the ring, completing a nine-subunit complex. The eukaryotic exosome also contains six RNase PH-like subunits and three RNA recognition subunits (see Table 5.3 and Figure 5.31D). However the eukaryotic nine-subunit exosome is inactive and requires a tenth subunit, Rrp44, and/or an eleventh subunit, Rrp6, for exonuclease activity.

The exosome has processive 3′ → 5′ exoribonuclease activity

All exosomes cleave RNA from the 3′ end, one nucleotide at a time. The reaction is processive, whereby the next nucleotide on the substrate RNA shifts into the site vacated when the cleaved nucleotide has diffused away. In contrast to DNA polymerase, whose processivity is facilitated by the clamp loader (Chapter 3), the mechanisms for translocation of nucleotides in the exosome involves the recognition of RNA at a remote site (see below). Exosome

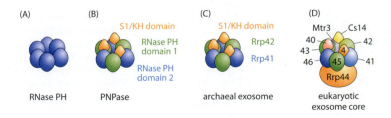

Figure 5.31 Schematic representations of exosomes. The exosomes are from bacteria (A, B), archaea (C), and eukaryotes (D). (Adapted from E. Lorentzen and E. Conti, *Cell* 125:651–654, 2006. With permission from Elsevier.)

activity results in complete degradation of single-stranded mRNA substrates and in partial degradation of structured RNAs.

Bacterial and archaeal exosomes use a phosphorolytic mechanism. They promote the nucleophilic attack of inorganic phosphate on the 3′-phosphorus, leading to cleavage of the phosphodiester bond and the release of products, namely nucleoside 5′-diphosphate and an RNA shortened by one nucleotide (**Figure 5.32A**). The eukaryotic exosome subunit Rrp44 promotes 3′–5′ exoribonuclease activity, probably by a two-metal-ion mechanism (as discussed for RNA polymerase in Section 5.2 and for DNA polymerase in Chapter 3), leading to a nucleoside monophosphate product (Figure 5.32B). These mechanisms may be contrasted with the mechanism for the extracellular endonuclease RNase A, in which the reaction proceeds through a 2′-3′ cyclic intermediate and is promoted by two His residues, one acting as an acid and the other as a base.

The archaeal exosome uses RNA-binding proteins to target RNA to the catalytic site

The structure of the 270 kDa exosome from the hypothermophilic archaeon *Sulfolobus solfataricus* is shown in **Figure 5.33**. The RNase PH subunits Rrp41 and Rrp42 are arranged in a pseudo-hexameric ring with three RNA-binding subunits (either Rrp4, as shown, or Csl4) on top. Rrp4 has three domains: an N-terminal six-stranded β-sheet domain (NT), followed by an S1 domain and a C-terminal KH domain. The S1 domains are located in the center and form a pore on top of the Rrp41/Rrp42 hexamer (see Figure 5.33A, B).

The RNA substrate is first recruited by the trimeric Rrp4 cap and is then channeled through the central hole of the hexameric ring to reach the catalytic sites of the RNase subunits. In the archaeal exosome crystal structure, RNA is observed bound at two sites. Site I is at the entrance to the central pore formed by the Rrp4 S1 domains, where one RNA nucleotide is

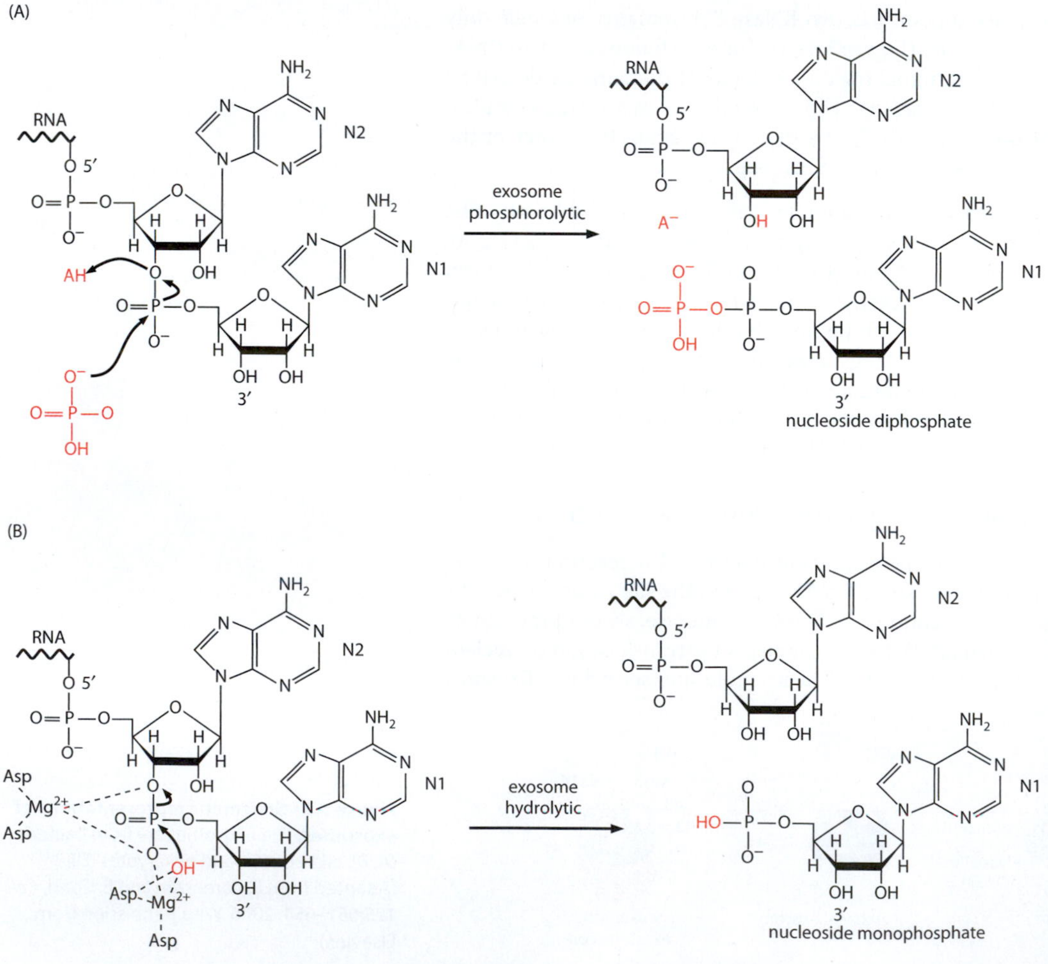

(A)

(B)

Figure 5.32 The phosphorolytic and hydrolytic reactions catalyzed by the exosome. Adenine is shown for illustration. (A) For the bacterial and archaeal exosomes, the phosphorolytic reaction is promoted by attack of inorganic phosphate, leading to the formation of a nucleoside diphosphate. AH is an acidic group poised to protonate the 3′ oxygen of the nucleotide in the N2 position. (B) In a two-metal mechanism, observed for the hydrolytic mechanism of RNase H and other enzymes, including the eukaryotic exosome protein Rrp4, the hydrolytic reaction proceeds by one metal ion activating a water molecule to produce a hydroxyl nucleophile, with the other metal ion poised to stabilize the pentavalent phosphoryl intermediate. Each metal is ligated by two conserved Asp residues. An acidic residue (possibly one of the conserved Asp residues) is needed to protonate the 3′ oxygen after cleavage of the ester bond.

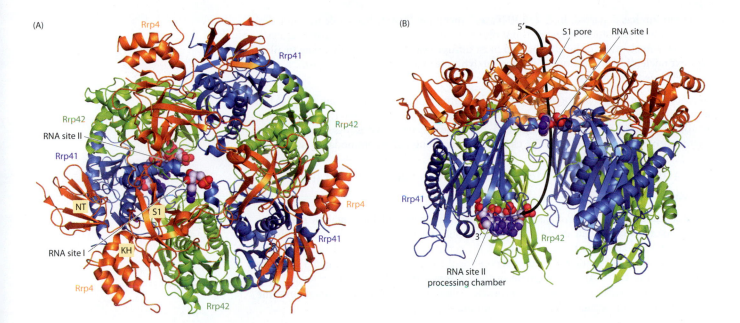

Figure 5.33 Structure of the *Sulfolobus solfataricus* archaeal exosome. (A) A view along the 3-fold axis of symmetry. The N-terminal (NT), KH, and S1 domains of one Rrp4 subunit are labeled. The Rrp4 subunits form a trimer capping the Rrp41/Rrp42 ring. A single RNA nucleotide, shown as spheres, is bound at the recruitment site, site I, at the pore of one Rrp4 S1 domain. Four RNA nucleotides are shown as spheres bound at the catalytic site, RNA site II, located between the Rrp41 and Rrp42 subunits. For simplicity, RNA binding to only one Rrp41/Rrp42/Rrp4 complex is shown. (PDB 2JEA) (B) Side view with the front Rrp42 (green subunit) removed to show the channel (black arrow) from the S1 pore to the processing chamber through which a single-stranded RNA molecule is threaded to reach the catalytic site.

observed bound (see Figure 5.33A,B). Site II is at the Rrp41 catalytic site, where there are four nucleotides bound with the 3′ end directed into the catalytic site and the 5′ end close to the central hole. Site I acts as a recruitment site and there is a direct path from there to the catalytic site II (see Figure 5.33B).

At the catalytic site, the RNA is recognized mainly by electrostatic interactions between the phosphate backbone and a ladder of Arg residues from both the Rrp41 and Rrp42 subunits (**Figure 5.34A**). There are no sequence-specific interactions with the bases, and this observation explains why the exosome is a general nonspecific RNA-degrading enzyme. Interactions of the exosome with the 2′-OH of the ribose of the nucleotides in the first (N1) and fourth (N4) positions allow the exosome to discriminate between RNA and DNA. Such discrimination is important in the nucleus, where both DNA and RNA are present but only unwanted RNA must be degraded. Residues at the catalytic sites are conserved between the bacterial and archaeal exosomes, suggesting a common catalytic mechanism. The inorganic phosphate is assumed to bind to a position occupied by a Cl⁻ in the crystal structure, where it is located close to the phosphate of the N1 nucleotide. The two negatively charged phosphate groups interact with the positively charged Arg99 and Arg139 from the Rrp41 subunit (Figure 5.34B).

Cleavage of the phospho-ester bond between the N1 and N2 nucleotides requires the donation of a proton to the 3′ oxygen of the N2 ribose. Residue Asp182 is positioned to play this role (AH in Figure 5.32A). The reaction probably goes through a pentavalent transition state formed at the 3′ phosphate, because the attacking group comes in before the

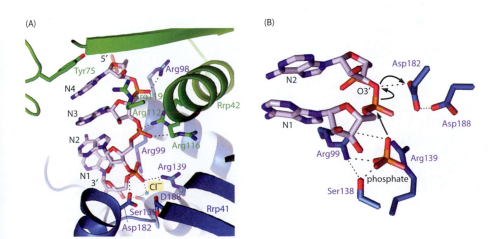

Figure 5.34 Details of RNA bound to the archaeal exosome. (A) At the catalytic site between Rrp41 and Rrp42 subunits, four nucleotides are bound, labeled N1–N4 from the 3′ end to the 5′ end. Residues that are important for binding and catalysis are labeled. A Cl⁻ at the catalytic site is labeled. (PDB 2C38) (B) Close-up view of the catalytic site, showing a postulated inorganic phosphate ion replacing the Cl⁻. The phosphorolytic catalytic mechanism could proceed by attack of the phosphate on the 3′ phosphate group of the RNA and general acid attack promoted by Asp182 on the 3′ oxygen of the ribose in N2.

leaving group has departed. In *E. coli* PNPase, a metal ion (Mn^{2+}) has a role in stabilizing the pentavalent transition state, but is not observed in the archaeal enzyme. After catalysis, the excised nucleotide in the N1 position must diffuse away from the catalytic site to allow the next nucleotide to translocate into the catalytic site in a processive mechanism. A side-channel close to the N1 site provides an escape route.

To reach the catalytic site, the RNA is threaded through a pore wide enough (8–10 Å) to accommodate single-stranded RNA but not double-stranded or structured RNA (see Figure 5.33B). Structured RNAs are not degraded, but larger unstructured RNAs are reduced to fragments of similar size (~9 nt). The channel is long enough to accommodate 9 nt.

There are similarities in mechanisms of the exosome and the proteasome

The archaeal exosome performs its processive degradation at a shielded catalytic site that is accessible only to unfolded single-stranded RNA substrates that are first located by the S1 domain of Rrp4 and then threaded through the central channel. This steric sequestration of the catalytic site is similar to that observed for the Clp proteases (Section 7.3) and the proteasome (Section 7.4). Confinement of the exosome's degradative activity to unfolded substrates is also reminiscent of those proteases whose peptidase components require the participation of partnering ATP-dependent unfoldases.

The human exosome core is similar in structure to the bacterial and archaeal exosomes but has no phosphorolytic catalytic activity

The eukaryotic exosome is assembled around a nine-subunit core (see Table 5.3). The six RNase PH-like subunits share 20–30% sequence identity with the *E. coli* RNase PH and PNPase PH domains. They are arranged as dimers Rrp41/Rrp45, Rrp46/Rrp43, and Mtr3/Rrp42 in a heterohexameric ring, with the first subunit of each pair equivalent to archaeal Rrp41 and the second subunit equivalent to Rrp42 (**Figure 5.35**). The eukaryotic exosome core also has subunits with RNA-binding domains, which are needed to stabilize the complex. Csl4 binds to Rrp43 and Mtr3, Rrp4 binds to Rrp42, and Rrp41 and Rrp40 bind to Rrp45 and Rrp46. The Rrp45 subunit has an extended C-terminal helix that wraps around the neighboring Rrp46/Rrp43 dimer (see Figure 5.35). The subunits make specific interactions with their partners so that the exosome can self-assemble with each of its nine subunits in the correct position.

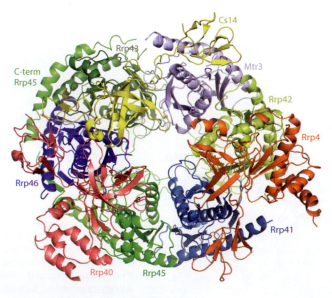

human exosome

Figure 5.35 The human exosome. Structure showing the six RNase PH-domain proteins (Rrp41, Rrp42, Mtr3, Rrp43, Rrp46, and Rrp45) and the three KH/S1 domain proteins (Rrp4, Rrp40, and Csl4). (PDB 2NN6)

Surprisingly, the eukaryotic exosome core has no exonuclease activity, despite containing six proteins with RNase PH domains. The lack of catalysis can be explained by substitutions affecting crucial amino acids at the catalytic sites. The Rrp41/Rrp45 pair has most of the RNA-binding Arg residues conserved, but the ones around the catalytic site, namely the archaeal equivalents of Arg99, Arg139, Asp182, and Asp188, are different. In place of the catalytic Asp182 there is a Glu residue. Glutamic acid can function as a general acid as efficiently as Asp, but in the human exosome, the larger side chain results in an incorrect orientation and a shift relative to the RNA substrate.

Additional subunits provide nuclease hydrolytic activity

Two further subunits, Rrp44 and Rrp6, associate with the eukaryotic exosome. Both have hydrolytic RNase activity. These subunits have no counterparts in prokaryotic or archaeal exosomes. Rrp6 is a metal-dependent exoribonuclease with non-processive 3′–5′ hydrolytic exoribonuclease activity, similar to the bacterial enzyme RNase D. Rrp44 co-purifies with the exosome and exhibits processive 3′–5′ RNase activity. Rrp44 has an N-terminal PIN domain with endoribonucleolytic activity, and a central exonuclease domain, which is similar to two bacterial enzymes, RNase II and RNase R, and is flanked by two oligonucleotide-oligosaccharide-binding (OB) folds at the N terminus (CSD1 and CSD2) and one OB-fold at the C-terminus (S1) (**Figure 5.36A**). The OB fold is based on a five-stranded β-sheet (see Box 5.3). PIN domains, which were first identified as part of the type IV pili (PilT N-terminus; see Section 11.8), are ~22 kDa nucleases present in all three kingdoms of life and are typically found in toxin/antitoxin systems in which toxicity is associated with RNase activity.

The Rrp44 crystal structure, in which the enzyme was inactivated by replacing the catalytic Asp with an Asn residue, showed the three OB folds present on one side of the molecule (Figure 5.36B). Nine nucleotides were observed bound at the catalytic site. The three nucleotides at the 5′ end interact with residues from both the nuclease domain and the CSD1 domain, and form an RNA recruitment site, whereas residues at the 3′ end reach into the catalytic site of the nuclease (see Figure 5.36B). At the catalytic site, four conserved Asp residues are arranged with similar geometry to the Asp residues of RNase II and RNase H (see Figure 5.27). By analogy, the reaction probably goes through a two-metal mechanism, as shown in Figure 5.32B, but this remains to be confirmed.

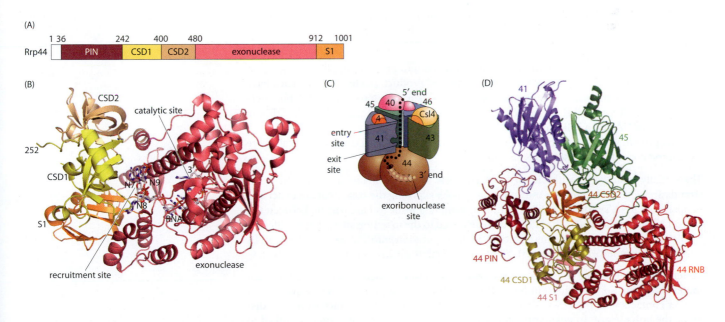

Figure 5.36 Rrp44 is the tenth subunit of the eukaryotic exosome and confers hydrolytic RNase activity. (A) Domain map of Rrp44. (B) Crystal structure of yeast Rrp44 (lacking the PIN domain). There are nine RNA nucleotides bound. The 3′ end is buried in the catalytic site, and the three nucleotides at the 5′ end (N7, N8, and N9) contact the recruitment site in the CSD1 domain. (PDB 2VNU) (C) Schematic representation of the yeast 10-subunit exosome with Rrp44 binding to the underside of the exosome core. Subunits Mtr3, Rrp43, and Csl4 on the upper side have been omitted for clarity. The path of the RNA, which has been probed by mutagenesis studies, is shown as a black dotted line. (D) The complex between human Rrp41/Rrp45 and Rrp44. (PDB 2WP8) (Adapted from F. Bonneau et al., *Cell* 139:547–549, 2009. With permission from Elsevier.)

Biochemical and crystallographic studies support the notion that the eukaryotic exosome core serves as a scaffold that threads RNA to the Rrp44 nuclease and has an active role in binding and selecting RNA substrates. RNase protection assays identified a 31–33 nt path that spans the nine-subunit exosome core to the catalytic site of Rrp44 (Figure 5.36C). Thus, the exosome core retains functional similarities to the archaeal exosome core in channeling the RNA, with conserved positively charged residues at the crucial entry and exit sites.

Rrp44 binds only to the Rrp41/Rrp45 dimer of the exosome core, as demonstrated in pull-down assays with tagged proteins. A crystal structure of the *S. cerevisiae* Rrp41/Rrp45/Rrp44 complex (Figure 5.36D) showed that the catalytic site of the Rrp44 PIN domain is directed to the solvent and away from the exosome core, so that it is able to use its endonuclease activity to nick loops in RNAs between secondary structure elements. In the absence of RNA, access to the exonuclease catalytic site of Rrp44 is blocked. It is assumed that conformational changes take place as RNA is threaded through the exosome core to allow access to the catalytic site. Rrp44 is able to process structured RNAs, provided that there is a 3′ tail long enough to allow the RNA to be threaded to the exonuclease site. Because the single-stranded RNA is degraded one nucleotide at a time, translocation would lead to disruption of the RNA secondary structure at the entrance to the exosome core, thus allowing unwinding of the RNA helix, promoted by multiple contacts between Rrp44 and RNA.

The exosome also co-purifies with several cofactors involved in quality control of mRNA turnover, nuclear RNAs, ribosome maturation, 3′ processing of noncoding RNAs, and gene silencing by RNA-mediated interference (RNAi). These multi-protein complexes that contain polymerases, helicases, and RNA chaperones are targets for future structural studies.

5.6   SUMMARY

In transcription, genetic information in the DNA sequence is copied to RNA by RNA polymerase machines. The tracts to be transcribed are designated by promoter sequences. Promoters are recognized by general transcription factors that recruit the polymerase to the start site. In prokaryotes, promoter elements are located 10 bp (the TATA box) and 35 bp upstream of the start site. In eukaryotes, the TATA box is recognized by a subunit of general transcription factor TFIID. TFIID and other general transcription factors combine with RNA polymerase II to form the pre-initiation complex (PIC) on promoter DNA. The Mediator complex stabilizes the PIC and communicates signals from gene-specific transcription factors, which mainly bind upstream from the start site and can act positively (activators) or negatively (repressors).

Many genes are transcribed to mRNAs to be translated into proteins, but for others—such as genes coding for tRNA and rRNA—RNA is the end product. Bacteria and archaea have a single RNA polymerase that synthesizes all RNAs. Eukaryotes have three RNA polymerases: Pol I for rRNAs, Pol II for mRNAs and small regulatory RNAs, and Pol III for tRNAs plus the 5S rRNA and a spliceosomal RNA. RNA polymerases have from 5 (bacterial) to 17 (Pol III) protein subunits, with cumulative masses in the 500–700 kDa range. They all share a common core of five homologous subunits.

The initially bound PIC with closed DNA converts to an open state in which the DNA strands are separated to form a transcription bubble, exposing the bases that are to act as a template for RNA synthesis. NTP selection by sampling leads to base pairing between the correct NTP and the DNA. The NTP is then delivered to the insertion site; catalysis, effected by a two-metal-ion mechanism, results in its incorporation into the nascent RNA. RNA polymerases synthesize at rates of 20–70 nt/s with a relatively high error rate (~1 in 10^4).

To terminate transcription, prokaryotes employ a system in which a GC-rich region promotes the formation of a hairpin loop. This is followed by a poly(U) tract, which forms weaker contacts with the DNA. Together, the hairpin and the poly(U) tract are able to dislodge the polymerase. In eukaryotes, termination is a more complicated process linked to the 3′-poly(A) machinery.

In eukaryotes, primary transcripts (pre-mRNAs) undergo three processing events:,(i) addition of a cap structure at the 5′ end; (ii) the generation of a poly(A) tail at the 3′ end; and (iii) the splicing out of the noncoding regions (introns). rRNA and tRNA are neither capped nor polyadenylated but undergo substantial chemical modifications.

Most eukaryotic pre-mRNAs have introns that are spliced out, leaving the coding sequences (exons) ligated together. In alternative splicing of transcripts with multiple introns, different combinations of retained exons generate different proteins. Splicing is performed by the spliceosome, a large dynamic protein/RNA complex. Spliceosomes assemble on each intron by sequential binding of four small nuclear ribonucleoprotein particles (snRNPs) and many other proteins. (Here, 'small' refers to the RNA, not the particle.) The five principal classes of U2 snRNPs have one to three U-rich snRNA molecules ranging from 106 to 187 nt in humans plus 10 to 36 protein subunits. Each snRNA is capped at its 5′ end. Although some spliceosomal proteins are present throughout the splicing cycle, there are major changes in protein composition during spliceosome assembly and catalytic activation. Splicing is a multi-step process involving sequential base-pairing interactions between the mRNA and the snRNAs.

To eliminate defective RNAs, excised introns, or mRNAs developmentally programmed for destruction (mRNA half-lives are typically minutes to days), cells have three major classes of RNA-degrading enzymes: endonucleases, 5′-exonucleases, and 3′-exonucleases. The exosome is an RNA-degrading machine with 3′-exonuclease activity that degrades mRNAs, snRNAs and rRNAs. Exosomes are based on a hexameric ring structure to which additional protein subunits have been added during evolution. RNA degradation is catalyzed by RNase PH (phosphorolysis) and PNPase (polynucleotide phosphorylase) activities. The basic PH subunit assembles as homohexamers. The PNPase subunit has two PH domains, followed by two RNA-binding domains. Three PNPase subunits give a pseudo-hexamer of PH domains, with the RNA-binding domains at one end. The active sites for RNA degradation are sequestered inside the ring.

Sections of this chapter are based on the following contributions

5.2   Patrick Cramer, Max Planck Institute for Biophysical Chemistry, Göttingen, Germany.

5.4   Cindy L. Will and Reinhard Luhrmann, Max Planck Institute for Biophysical Chemistry, Göttingen, Germany.

REFERENCES

5.1   Introduction

Alberts B et al (2015) The Molecular Biology of the Cell, 6th ed. Garland Science.

Lodish H et al (2012) Molecular Cell Biology, 7th ed. WH Freeman.

5.2   RNA polymerase

Abbondanzieri EA, Greenleaf WJ, Schaevitz JW et al (2005) Direct observation of base-pair stepping by RNA polymerase. *Nature* 438:460–465.

Brueckner F, Ortiz J & Cramer P (2009) A movie of the RNA polymerase nucleotide addition cycle. *Curr Opin Struct Biol* 19:294–299.

Cramer P, Armache KJ, Baumli S et al (2008) Structure of eukaryotic RNA polymerases. *Annu Rev Biophys* 37:337–352.

Cramer P, Bushnell DA & Kornberg RD (2001) Structural basis of transcription: RNA polymerase II at 2.8 Å resolution. *Science* 292:1863–1876.

Gnatt AL, Cramer P, Fu J et al (2001) Structural basis of transcription: an RNA polymerase II elongation complex at 3.3 Å resolution. *Science* 292:1876–1882.

Hahn S (2004) Structure and mechanism of the RNA polymerase II transcription machinery. *Nat Struct Mol Biol* 11:394–403.

Herbert KM, Greenleaf WJ & Block SM (2008) Single molecule studies of RNA polymerase: motoring along. *Annu Rev Biochem* 77:149–176.

Kettenberger H, Armache KJ & Cramer P (2003) Architecture of the RNA polymerase II–TFIIS complex and implications for mRNA cleavage. *Cell* 114:347–357.

Kornberg R (2007) The molecular basis of eukaryotic transcription (Nobel lecture) *Angew Chem Int Ed* 46:6956–6965.

Kostrewa D, Zeller ME, Armache KF et al (2009) RNA polymerase II–TFIIB structure and mechanism of transcription initiation. *Nature* 462:323–330.

Kuhn CD, Geiger SR, Baumli S et al (2007) Functional architecture of RNA polymerase I. *Cell* 131:1260–1272.

Murakami KS & Darst SA (2003) Bacterial RNA polymerase: the whole story. *Curr Opin Struct Biol* 13:31–39.

Richard P & Manley JL (2009) Transcription termination by nuclear RNA polymerases. *Genes Dev* 23:1247–1269.

Sainsbury S, Bernecky C & Cramer P (2015) Structural basis of transcription initiation by RNA polymerase II. *Nat Rev Mol Cell Biol* 16:129–143.

Shilatifard A, Conaway RC & Conaway JW (2003) The RNA polymerase II elongation complex. *Annu Rev Biochem* 72:693–715.

Svetlov V & Nudler E (2008) Jamming the ratchet of transcription. *Nat Struct Mol Biol* 15:777–779.

Wang D, Bushnell DA, Huang X et al (2009) Structural basis of transcription: backtracked RNA polymerase II at 3.4 Å resolution. *Science* 324:1203–1206.

Wang D, Bushnell DA, Westover KD et al (2006) Structural basis of transcription: role of the trigger loop in substrate specificity and catalysis. *Cell* 127:941–954.

5.3   The pre-initiation complex

Cheung AC & Cramer P (2012) A movie of RNA polymerase II transcription. *Cell* 149:1431–1437.

Green MR (2005) Eukaryotic transcription activation: right on target. *Mol Cell* 18:399–402.

Hirose Y & Ohkuma Y (2007) Phosphorylation of the C-terminal domain of RNA polymerase II plays central roles in the integrated events of eukaryotic gene expression. *J Biochem* 141:601–608.

Kouzarides T (2007) Chromatin modifications and their function. *Cell* 128:693–705.

Larivière L, Seizl M & Cramer P (2012) A structural perspective on Mediator function. *Curr Opin Cell Biol* 24:305–313.

Levine M & Tjian R (2003) Transcription regulation and animal diversity. *Nature* 424:147–151.

Malik S & Roeder RC (2010) The metazoan Mediator co-activator complex as an integrative hub for transcription regulation. *Nat Rev Genet* 11:761–772.

Panne D (2008) The enhanceosome. *Curr Opin Struct Biol* 18:236–242.

Plaschka C, Larivière L, Wenzeck L et al (2015) Architecture of the RNA polymerase II–Mediator core initiation complex. *Nature* 518:376–380.

Thomas MC & Chiang CM (2006) General transcription machinery and general cofactors. *Crit Rev Biochem Mol Biol* 41:105–178.

Tsai KL, Tomomori-Sato C, Sato S et al (2014) Subunit architecture and functional modular rearrangements of the transcriptional Mediator complex. *Cell* 157:1430–1444. Erratum, *Cell* 158:463 (2014).

Wang X, Sun Q, Ding Z et al (2014) Redefining the modular organization of the core Mediator complex. *Cell Res* 24:796–808.

5.4   The spliceosome

Fabrizio P, Dannenberg J, Dube P et al (2009) The evolutionarily conserved core design of the catalytic activation step of the *S. cerevisiae* spliceosome. *Mol Cell* 36:593–608.

Fica SM, Tuttle N, Novak T et al (2013) RNA catalyses nuclear pre-mRNA splicing. *Nature* 503:229–234.

Galej WP, Nguyen TH, Newman AJ et al (2014) Structural studies on the spliceosome: zooming into the heart of the machine. *Curr Opin Struct Biol* 25:57–66.

Lührmann R & Stark H (2009) Structural mapping of spliceosomes by electron microscopy. *Curr Opin Struct Biol* 19:96–102.

Matlin AJ, Clark F & Smith CW (2005) Understanding alternative splicing: towards a cellular code. *Nat Rev Mol Cell Biol* 6:386–398.

Pena V, Rozov A, Fabrizio P et al (2008) Structure and function of an RNase H domain at the heart of the spliceosome. *EMBO J* 27:2929–2940.

Pomeranz Krummel DA, Oubridge C, Leung AK et al (2009) Crystal structure of human spliceosomal U1 snRNP at 5.5 Å resolution. *Nature* 458:475–480.

Ritchie DB, Schellenberg MJ, MacMillan AM (2009) Spliceosome structure: piece by piece. *Biochim Biophys Acta* 1789:624–633.

Smith DJ, Query CC & Konarska MM (2008) "Nought may endure but mutability": spliceosome dynamics and the regulation of splicing. *Mol Cell* 30:657–666.

Wahl MC, Will CL & Lührmann R (2009) The spliceosome: design principles of a dynamic RNP machine. *Cell* 136:701–718.

Will CL & Lührmann R (2006) Spliceosome structure and function. In The RNA World (Gesteland RF, Cech TR & Atkins JF eds), pp 369–400. Cold Spring Harbor Laboratory Press.

Wolf E, Kastner B, Deckert J et al (2009) Exon, intron and splice site locations in the spliceosomal B complex. *EMBO J* 28:2283–2292.

5.5   The exosome

Bonneau F, Basquin J, Ebert J et al (2009) The *S. cerevisiae* exosome functions as a macromolecular cage to channel RNA substrates for degradation. *Cell* 139:547–559.

Houseley J & Tollervey D (2009) The many pathways of RNA degradation. *Cell* 136:763–776.

Lorentzen E & Conti E (2005) Structural basis of 3′ end RNA recognition and exoribonucleolytic cleavage by an exosome RNase PH core. *Mol Cell* 20:473–481.

Lorentzen E, Basquin J & Conti E (2008) Structural organization of the RNA-degrading exosome. *Curr Opin Struct Biol* 18:709–713.

Lorentzen E, Basquin J, Tomecki R et al (2008) Structure of the active subunit of the *S. cerevisiae* exosome core, Rrp44: diverse modes of substrate recruitment in the RNase nuclease family. *Mol Cell* 29:717–728.

Lorentzen E, Dziembowski A, Lindner D et al (2007) RNA channeling by the archaeal exosome. *EMBO Rep* 8:470–476.

Nurmohamed S, Vaidialingam B, Callaghan AJ et al (2009) Crystal structure of *Escherichia coli* polynucleotide phosphorylase core bound to RNase E, RNA and manganese: implications for catalytic mechanism and RNA degradosome assembly. *J Mol Biol* 389:17–33.

Schmid M & Jensen TH (2008) The exosome: a multipurpose RNA-decay machine. *Trends Biochem Sci* 33:501–510.

Symmons MF, Jones GH & Luisi BF (2000) A duplicated fold is the structural basis for polynucleotide phosphorylase catalytic activity, processivity, and regulation. *Structure* 8:1215–1226.

Protein Synthesis and Folding

6.1 INTRODUCTION

As outlined in Chapter 1, the genetic code relates the information stored in DNA codons to the sequence of amino acids in polypeptide chains. The process of translation from one language to the other is mediated by the ribosome, the quintessential machine of protein synthesis (Section 6.2). The ribosome is a large particle (2.5–4.2 MDa, depending on origin), built from two subunits, called large (L) and small (S). (Here 'subunit' refers to a complex of multiple proteins and RNA molecules rather than, as elsewhere in this book, a single polypeptide chain.) All ribosomes comprise three RNA molecules plus many (50–80) protein subunits. The overall organization is similar in all three kingdoms of life, reflecting ancient evolutionary origins. In sum, the ribosome binds an mRNA molecule and polypeptide synthesis ensues in four stages: initiation, elongation, termination, and recycling. At each stage, additional protein factors associate transiently with the ribosome–mRNA complex.

To become functional, a newly synthesized polypeptide chain must fold into a specific three-dimensional conformation. Usually, this 'native' conformation corresponds to the global minimum of the free energy landscape (Section 1.3). Folding is governed by thermodynamics, and all the information needed resides in the amino acid sequence. Folding can commence during protein synthesis—that is, co-translationally—but cannot be completed before the whole polypeptide chain has been synthesized. With a protein that consists of multiple independently folding domains, N-terminal domains fold before distal regions have been synthesized. Folding may stall in intermediate states and there may be non-productive interactions with other proteins. To avoid such outcomes, many proteins rely on specialized helper proteins, named *chaperones*, to fold correctly and to avoid aggregation (Section 6.3).

There is growing awareness that some proteins or substantial portions thereof exist natively in an unfolded state (Section 6.4). Many of them serve as flexible adaptors that promote specific interactions with other proteins and fold only when they engage with interaction partners. Their incidence and importance have been underestimated, probably because the most powerful methods that structural biology has at its disposal can study only defined structures but not ensembles of variable molecular species.

Cells have developed mechanisms to cope with proteins that fail to fold correctly, either by subjecting them to the unfolding–refolding machinery and/or by eliminating them through proteolysis (Chapter 7). In an alternative scenario, misfolded molecules may form intracellular aggregates, also called *inclusion bodies*, which resist proteolytic recycling. Inclusion bodies are often found when genes encoding recombinant proteins are overexpressed. In some cases, the misfolded proteins are polymerized in fibrils, called amyloids (Section 6.5). Amyloid fibrils of different proteins have in common structural features that endow them with extraordinary stability. Their accumulation has been correlated with a number of diseases. More recently, a growing number of proteins have been identified for which amyloid structures are the native conformation.

Prions are endogenous cellular proteins existing in a non-native conformation that behave as infectious agents (Section 6.6). Most prions are amyloids distinguished by their infectivity; in other words, they can be transmitted from one cell to another, where they propagate their non-native conformation in a domino-like manner.

6.2   THE RIBOSOME

The recipe for making a protein is encoded in its **messenger RNA** (mRNA), a single-stranded RNA molecule that is read in triplets of bases called codons. The genetic code serves as the dictionary that matches each codon to one of the 20 essential amino acids. **Transfer RNA** molecules (tRNAs) serve as the physical bridge between the two languages, each carrying on one side an anticodon and on the other side the corresponding amino acid. At the heart of the system lies the **ribosome**, which hosts all the components of the translation machinery and carries out translation. The ribosome recognizes a start codon in the mRNA and subsequently adds amino acids to the nascent protein until a stop codon is encountered, at which point the newly synthesized molecule is released. Additional participants in translation are the **tRNA synthetases** that charge each tRNA with the appropriate amino acid, and ancillary proteins called initiation, elongation, and **termination factors** that promote different stages of translation. In addition, there is a set of proteins and RNA molecules that prevent the translation machinery from taking non-productive actions such as translation of a fragmented mRNA. The overall process is outlined in **Figure 6.1**.

The translation system is, in its essentials, common to all living organisms. It comprises a large number of proteins and RNA molecules that are highly conserved across the three kingdoms of life. The account given here focuses on widely shared features of the translation machinery, exemplified primarily by the bacterial system. However, differences between bacterial, archaeal, eukaryotic, and organellar ribosomes are noted when relevant. The numberings of nucleotides and ribosomal proteins refer to those of *Escherichia coli*.

tRNAs are adaptor molecules between genes and proteins

When an understanding of protein synthesis was emerging in the late 1950s, Francis Crick postulated that there should be small adaptor molecules that could both read the base sequence of the mRNA and deliver the corresponding amino acids to the site of protein synthesis. He further suggested that these molecules could be small RNAs. Paul Zamecnik and Mahlon Hoagland, working on fractionation of cells and analysis of small 'junk' RNA, found that it was chemically associated with amino acids, in line with Crick's hypothesis. Further analysis of this RNA, later called 'transfer RNA,' gradually led to the recognition of one or more tRNAs specific for each amino acid and elucidation of their ~76 nucleotide-long sequences, their cloverleaf secondary structure, and L-shaped molecular structure

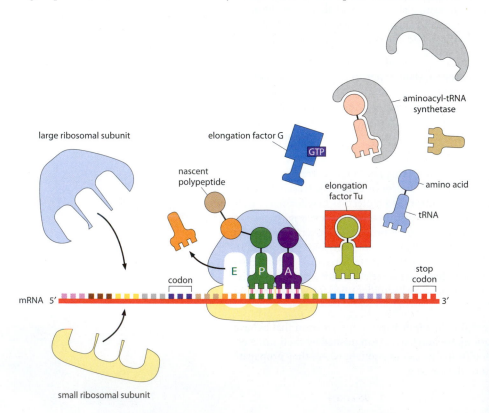

Figure 6.1 The key players in the translation process. At its center is the ribosome, composed of the large and small subunits. The ribosome uses an mRNA molecule as a template for making a protein. The tRNAs are each loaded with the correct amino acid by tRNA synthetases (gray). Amino acids are delivered to the ribosome by tRNAs, which enter at the A site and bind to the cognate codon in the mRNA. The attached amino acid is then linked to the growing polypeptide chain at its C-terminal end in the P site, and the now naked tRNA exits via the E site. The process is carried out with the assistance of elongation factors (red and blue) and is repeated, one amino acid at a time, until the stop codon near the 3' end of the mRNA is reached. The completed polypeptide chain is then released and the ribosome disassembles.

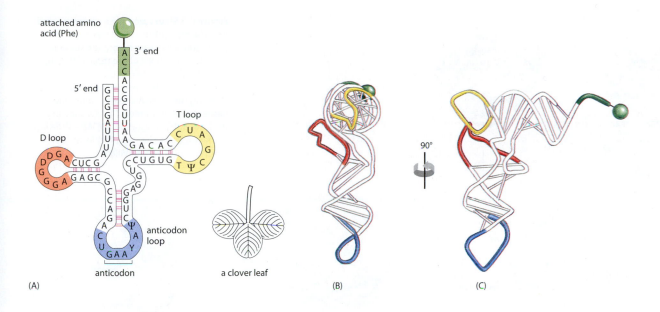

(A)

(B)

(C)

5′ GCGGAUUUAGCUCAGDDGGGAGAGCGCCAGACUGAAYAΨCUGGAGGUCCUGUGTΨCGAUCCACAGAAUUCGCACCA 3′

(D)

anticodon

(**Figure 6.2**). A remarkable feature of tRNA is the fact that the two functional sites are at opposite ends of the molecule. The anticodon recognizes a codon in the mRNA through interaction with the anticodon stem-loop (ASL) at one end of the L. The attachment site for the amino acid is at the CCA end (3′ end) of the acceptor stem at the opposite end of the L, ~75 Å away from the anticodon. Typically, ~10% of the bases in tRNAs are chemically modified. These altered bases serve to enhance codon recognition or to facilitate attachment to the correct amino acid.

Figure 6.2 **tRNAs are the adaptor molecules of translation.** (A) tRNAs can be drawn in two dimensions as a cloverleaf, in which the stems are the four base-paired helices. (B, C) tRNA adopts an L-shaped structure. The ASL (anticodon stem loop) is at the bottom (blue) and the amino acid is attached in ester linkage with the 3′-terminal A in the CCA at the top (green). (D) Nucleotide sequence of the molecule, color-coded to match (A–C). tRNAs contain unusual bases produced by post-synthesis modifications. This tRNA contains ψ (pseudouridine) and D (dihydrouridine) derived from uracil. (Adapted from B. Alberts et al., Molecular Biology of the Cell, 6th ed. New York: Garland Science, 2015.)

tRNA synthetases charge tRNAs with their cognate amino acids

The large distance between the functional sites of a tRNA presents a challenge for the molecules interacting with it, namely the ribosome and the **aminoacyl-tRNA synthetases** (aaRSs). The aaRSs recognize the cognate tRNAs and 'charge' them with the appropriate amino acids, yielding aminoacylated tRNAs (aa-tRNAs).

tRNA synthetases belong to two evolutionarily unrelated classes, each of which charges 10 different tRNAs. The initial step for both classes is activation of the amino acid by linking it to the AMP part of an ATP molecule. Next, the amino acid is attached to the tRNA. Class I aaRSs charge the tRNA on the 2′ OH of the terminal ribose, and class II aaRSs charge the tRNA on the 3′ OH. In the former case, the aminoacyl group transfers rapidly to the 3′ position, where the activated ester bond with the amino acid is present in all tRNAs.

The ribosome is an ancient molecular machine that catalyzes protein synthesis

In 1955, using electron microscopy (see **Methods**), the cell biologist George Palade discovered ribosomes and correctly identified their function as that of making proteins. Ribosomes are very abundant; in bacteria, they account for up to 50% of the cell's dry weight. They invariably have two subunits, the small and the large (SSU and LSU respectively; see Figure 6.1 and **Figure 6.3**).

Both ribosomal subunits vary in size, depending on their source. Their naming is based on their sedimentation rates during ultracentrifugation, in Svedberg units (S). Intact bacterial ribosomes sediment at 70S and their subunits at 30S and 50S respectively (see Figure 6.3). Eukaryotic ribosomes are larger, at 80S, with subunits of 40S and 60S. The ribosomes with the lowest sedimentation rates come from mammalian mitochondria (for example, those of the protozoan *Leishmania tarentolae* are 50S, with subunits of 30S and 40S).

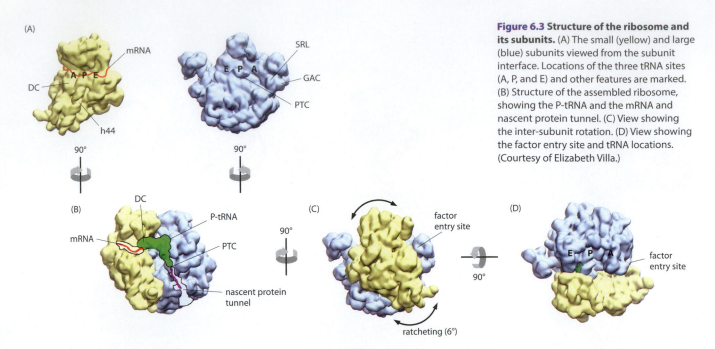

Figure 6.3 Structure of the ribosome and its subunits. (A) The small (yellow) and large (blue) subunits viewed from the subunit interface. Locations of the three tRNA sites (A, P, and E) and other features are marked. (B) Structure of the assembled ribosome, showing the P-tRNA and the mRNA and nascent protein tunnel. (C) View showing the inter-subunit rotation. (D) View showing the factor entry site and tRNA locations. (Courtesy of Elizabeth Villa.)

Organelles such as chloroplasts and mitochondria have their own translation systems with unique ribosomes, tRNAs, and associated protein factors; and some organisms have specialized ribosomes for translating a subset of mRNAs. Mitochondrial ribosomes have smaller ribosomal RNAs and many more proteins, most of them encoded in the nucleus. They are lower in mass, and they sediment more slowly than other ribosomes. Nevertheless all ribosomes share an evolutionarily conserved core that contains all the functional sites (**Table 6.1** and **Figure 6.4**).

Bacterial ribosomes contain three ribosomal RNA molecules (rRNAs). The 16S rRNA is in the small subunit, and the 23S and 5S rRNAs are in the large subunit. In eukaryotes, rRNAs are somewhat larger and often there are more than three of them (an SSU of 18S; LSUs of 5.8S, 25S, and 5S). Many mitochondrial ribosomes lack a 5S rRNA and their two main rRNAs can be as small as 9S and 12S respectively. Once the sequences of rRNAs from numerous species had been determined, it was recognized that the sequences of the SSU rRNA could be used to derive evolutionary relationships. In fact, the Tree of Life discussed in Section 1.9 was constructed on the basis of analyses of this kind.

Sequence alignments have also been used to identify tracts of complementary base pairing in different parts of the rRNAs. These tracts are capable of forming double-stranded RNAs

Table 6.1 **Composition of bacterial and eukaryotic ribosomes and the common core.**					
The common core	Bacteria (*Thermus thermophilus* or *E. coli*)	Archaea (*Pyrococcus furiosus*)	Lower eukaryotes (*S. cerevisiae*)	Higher eukaryotes (*Homo sapiens*)	Mitochondria (*H. sapiens*)
2.0 MDa	2.3 MDa	2.5 MDa	3.3 MDa	4.3 MDa	2.6 MDa
34 proteins	54 proteins	64 proteins	79 proteins	80 proteins	78 proteins
3 rRNAs	3 rRNAs	3 rRNAs	4 rRNAs	4 rRNAs	2 rRNAs
Large subunit: 19 proteins; 23S rRNA: 2843 bases; 5S: 121 bases	Large subunit (50S): 33 proteins; 23S rRNA: 2904 bases; 5S rRNA: 121 bases	Large subunit (50S): 39 proteins; 23S rRNA: 3049 bases; 5S rRNA: 126 bases	Large subunit (60S): 46 proteins; 5.8S rRNA: 158 bases; 25S rRNA: 3396 bases; 5S rRNA: 121 bases	Large subunit (60S): 47 proteins; 5.8S rRNA: 156 bases; 28S rRNA: 5034 bases; 5S rRNA: 121 bases	Large subunit (39S): 48 proteins; 16S rRNA: 1558 bases
Small subunit: 15 proteins; 16S rRNA: 1458 bases	Small subunit (30S): 21 proteins; 16S rRNA: 1542 bases	Small subunit (30S): 25 proteins; 16S rRNA: 1495 bases	Small subunit (40S): 33 proteins; 18S rRNA: 1800 bases	Small subunit (40S): 33 proteins; 18S rRNA: 1870 bases	Small subunit (28S): 30 proteins; 12S rRNA: 954 bases

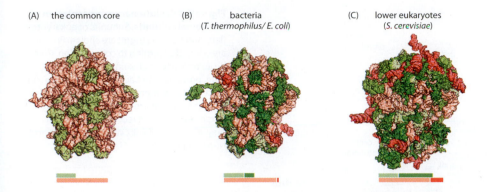

(A)   the common core

(B)   bacteria
(*T. thermophilus/ E. coli*)

(C)   lower eukaryotes
(*S. cerevisiae*)

Figure 6.4 Bacterial and eukaryotic ribosomes share a conserved core. The core is composed of RNA (light red) and proteins (light green). In addition to the core, ribosomes in each domain of life contain their own set of proteins, extensions and insertions in conserved proteins (both in dark green), and extension segments in the ribosomal RNAs (dark red). (Adapted from S. Melnikov et al., *Nature Struct. Mol. Biol.*, 19:560–567, 2012. With permission from Macmillan Publishers Ltd.)

that are referred to as 'helices.' RNA helices of the 16S RNA and 23S RNA are denoted **h** and **H** respectively, followed by a number. Overall, the 16S rRNA comprises four domains and the 23S rRNA comprises six (**Figure 6.5**). The additional RNA segments in eukaryotic ribosomes, referred to as **expansion segments**, reside on the surface of the ribosome surrounding the evolutionarily conserved core (see Figure 6.4). The functions of most expansion segments remain unknown.

In archaeal and bacterial ribosomes, the total protein mass is only about half of the RNA mass. In eukaryotes the protein content is somewhat higher, and in mitochondria the ratio is inverted, with up to three times more protein than RNA. It is likely that this additional protein content was acquired to compensate for the diminished RNA molecules. Some ribosomal proteins are conserved in all three kingdoms of life, whereas others are specific to bacteria, archaea, or eukarya. The protein subunits are named according to the ribosomal subunit to which they belong (S and L) and a number.

X-ray crystallography and electron microscopy have revealed the structure of ribosomes

X-ray crystallography (see **Methods**) has provided accounts of ribosomal subunits from bacteria and archaea, as well as complete bacterial ribosomes, at atomic resolution. In

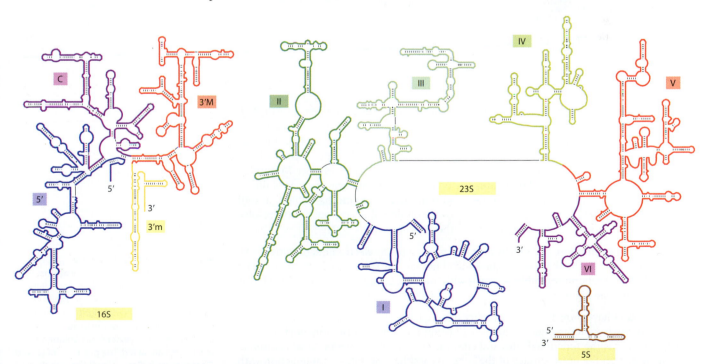

Figure 6.5 Secondary structures of the rRNA molecules from *E. coli*. At the left, the 16S RNA with its four domains: 5′, C, 3′M, and 3′m. Helix h44, the long yellow helix, is involved in the decoding process. At the right, the 5S and 23S RNAs. The 23S RNA has six domains. The peptidyl transfer site is associated with domain V. (Adapted from M.M. Yusupov et al., *Science* 292:883–896, 2001.)

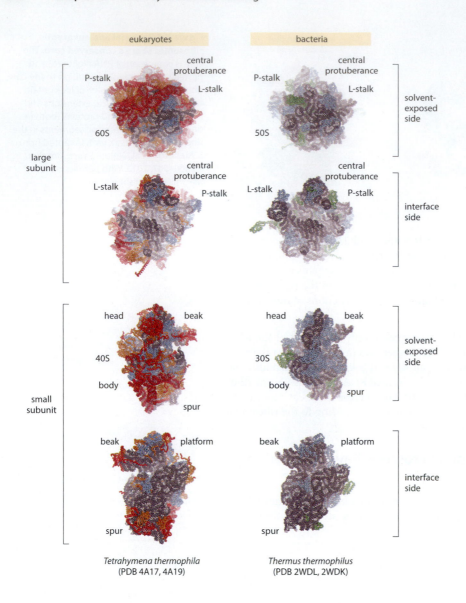

eukaryotes

bacteria

P-stalk

central protuberance

L-stalk

60S

large subunit

central protuberance

L-stalk

P-stalk

solvent-exposed side

interface side

P-stalk

central protuberance

L-stalk

50S

central protuberance

L-stalk

P-stalk

head   beak

40S

body

spur

small subunit

beak   platform

spur

head   beak

30S

body

spur

beak   platform

spur

solvent-exposed side

interface side

Tetrahymena thermophila
(PDB 4A17, 4A19)

Thermus thermophilus
(PDB 2WDL, 2WDK)

Figure 6.6 Evolutionary representation of ribosomal subunits. Subunits of eukaryotes (left) and bacteria (right) are shown. Proteins are colored according to conservation. Core proteins found in all kingdoms are in light blue, proteins with bacterial homologs are in gold, proteins or protein extensions unique to bacteria are in green, and those unique to eukaryotes are in red. RNA is shown in gray. (Courtesy of Felix Voigts-Hoffmann and Nenad Ban.)

addition to 'empty' ribosomes, a range of functional complexes have been investigated. Recently, structures of eukaryotic ribosomes have been added (see Figure 6.4 and **Figure 6.6**). Crystallization requires that certain ribosomal surfaces make contacts that often do not exist under natural conditions, and this prevents detailed insight into the interactions between factors and the ribosome. Cryo-electron microscopy (cryo-EM) (see **Methods**), not so constrained, has provided structures of ribosomes complexed with a range of different factors, in various conformational states, and from diverse organisms. Although until recently cryo-EM structures were of relatively low resolution, it is now possible to resolve comparable detail to many crystal structures of the ribosome.

The classical views of ribosomal structure were first revealed by electron microscopy (EM). The small subunit has the shape of a right-handed mitten. The side of the large subunit that binds to the small subunit resembles a crown with three protuberances, and is referred to as the crown view (see Figure 6.3).

X-ray structures have shown that the interface between the two subunits is mostly devoid of proteins. The two subunits interact through 12 bridges composed primarily of rRNA. The interface contains the functional sites and sites for binding other factors. Most ribosomal proteins reside on the outer surface of the subunits and have an intimate interaction with the rRNAs (see Figure 6.6). Many of the negative charges on the phosphates of the RNA are neutralized by positive charges on N- or C-terminal tails of the proteins or on loops that extend far from the small globular domains (**Figure 6.7**).

Figure 6.7 (*right*) **Evolutionary representation of ribosomal proteins.** Eukaryotic proteins are colored according to conservation. Protein cores found in all kingdoms are depicted in light blue, proteins with archaeal homologs are in gold, and proteins or protein extensions unique to eukaryotes are in red. The positions of N- and C-termini are indicated. Zn^{2+} ions are shown as green spheres. (Adapted from S. Klinge et al., *Science* 334:941–948, 2011 and J. Rabl et al., *Science* 331:730–736, 2011.)

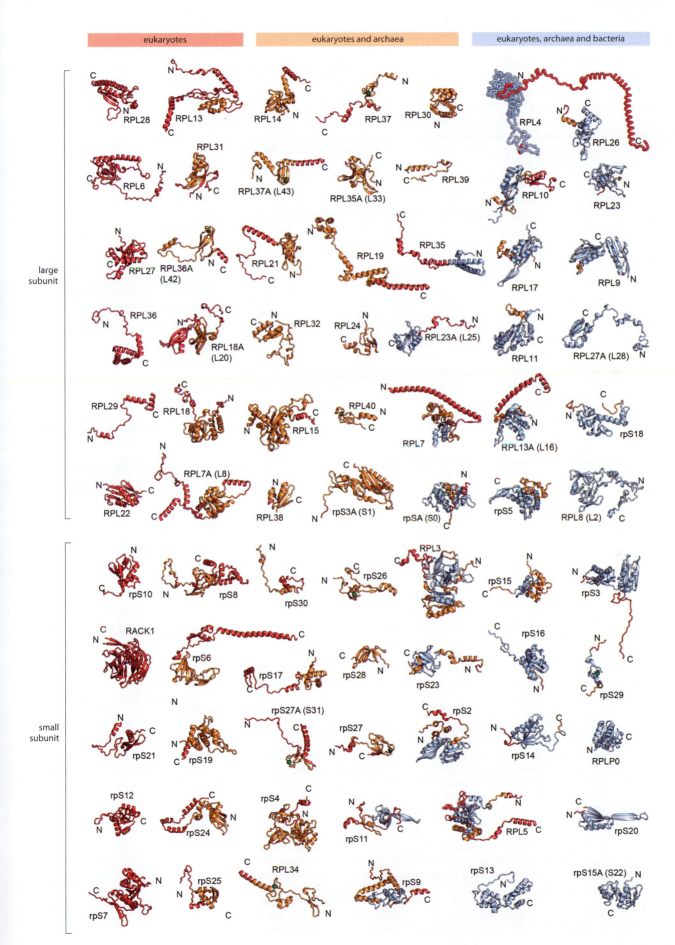

The ribosome is a ribozyme

The mRNA binds to the small subunit around the neck between the head domain and the body (see Figure 6.3). The small subunit contains the **decoding center** (**DC**), where the codon and the anticodon are probed for complementarity. The large subunit has the **peptidyl transfer center** (**PTC**), where amino acids are covalently added to the nascent protein. Evidently, the tRNAs and the ribosome have co-evolved, ensuring that the large distance between the anticodon of the tRNA and the attached amino acid is matched by keeping the ribosomal functional sites equally far apart on different subunits. Helix h44 of the small subunit (see Figures 6.3 and 6.5) is connected to the decoding center and makes essential contacts with the large subunit.

The ribosome has three binding sites for tRNAs (**Figure 6.8**). Each subunit contains A and P sites for aminoacyl-tRNA and peptidyl-tRNA respectively, and the E or exit site. The binding of tRNAs to these sites is independent in each subunit. When a tRNA binds to the same site in both subunits, it leads to the 'classical' A/A, P/P, and E/E conformations, while tRNAs bound to different sites lead to the 'hybrid' conformations.

The initial binding of the aminoacyl-tRNA to the ribosome occurs when the tRNA is in complex with elongation factor Tu (EF-Tu) and GTP and is positioned in the hybrid A/T conformation, in which the anticodon is located in the A site in contact with the codon, while the acceptor stem and amino acid are still bound to EF-Tu (T site). Once the codon–anticodon is recognized as a correct match, the tRNA is installed in the A site (A/A state). At this point, the C-terminal amino acid of the nascent polypeptide is attached to the tRNA in the P site. To add the incoming amino acid, peptide bond formation occurs, in which the nascent chain is transferred to the tRNA in the A site. This tRNA, along with the mRNA, must be moved on to the P site for synthesis to continue. This is achieved by the spontaneous formation of hybrid states called A/P and P/E, in which the tRNAs move along to their new sites in the 50S subunit, while the anticodon stem loops in the 30S subunit remain in their previous location. Eventually, the tRNA with the nascent chain is fully moved to the P site (P/P state), accompanying a large conformational shift called translocation, and the deacylated tRNA is moved into the E site (E/E state) before being released from the ribosome.

The mRNA is bound to the small subunit between the head and the body, facing the large subunit (see Figure 6.3B). The codon to be read is exposed in the A site (see Figure 6.3A). This decoding site is primarily composed of rRNA, but a minor part of protein S12 is close by. Base pairing between mRNA and tRNA is maintained in the P site.

The site for peptidyl transfer (the PTC) is situated on the large subunit and is also built primarily from rRNA. Thus, the ribosome is a *ribozyme* (that is, an RNA with enzymatic activity). The PTC is at the entrance to a tunnel in the large subunit, which is 80–100 Å long and through which the N-terminal end of the growing polypeptide exits from the site of synthesis (see Figure 6.3B). The tunnel walls are composed mainly of rRNA, but proteins L4 and L22 line parts of the tunnel. The tunnel exit on the external surface of the large subunit has several proteins around it: L17, L19, L22, L23, L24, L29, and L32. Several protein factors involved in the processing and folding of the emerging protein bind here. They include

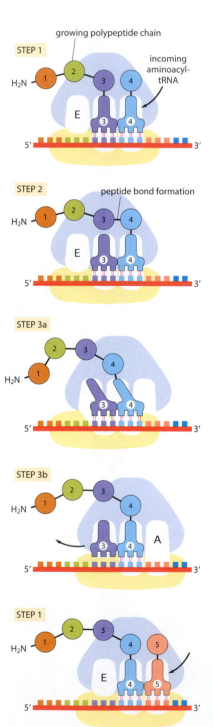

Figure 6.8 Positions of the tRNAs during the elongation cycle. The three-step cycle shown is repeated over and over during the synthesis of a protein. An aminoacyl-tRNA molecule binds to a vacant A site on the ribosome in STEP 1, a new peptide bond is formed in STEP 2, and the mRNA moves by a distance corresponding to three nucleotides through the small-subunit chain in STEP 3, ejecting the spent tRNA molecule and 'resetting' the ribosome so that the next aminoacyl-tRNA molecule can bind. This process is called translocation. The movement of the 30S subunit relative to the 50S subunit in STEP 3a results from conformational changes in the ribosome. As indicated, the mRNA is translated in the 5′ → 3′ direction, and the N-terminal end of a protein is made first, with each cycle adding one amino acid to the C terminus of the nascent protein. The position at which the growing polypeptide chain is attached to a tRNA does not change during the elongation cycle: it is always linked to the tRNA present in the P site of the 50S subunit. Notably, the ribosome can perform all elongation steps without the help of elongation factors, as depicted. Generally, EF-G binds the ribosome in STEP 3a, stabilizing the conformation of the rotated subunits.

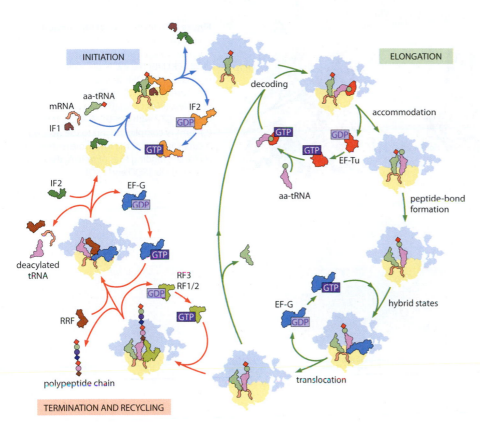

Figure 6.9 Overview of bacterial translation. There are four stages, namely: initiation, elongation, termination, and recycling. For simplicity, not all intermediate steps are shown. Yellow, 30S subunit; blue, 50S subunit. aa-tRNA, aminoacyl-tRNA; A site, aa-tRNA-binding site; E site, exit site for tRNA; EF, elongation factor; IF, initiation factor; P-tRNA, peptidyl-tRNA; P site, P-tRNA binding site; PTC, peptidyltransferase center; RRF, ribosome recycling factor; RF, release factor.

peptide deformylase, methionine aminopeptidase, signal recognition particle, the protein-conducting channel for transport through membranes, and the *trigger factor* chaperone (see Section 6.3).

Translation factors enhance the efficiency of protein synthesis

Ribosomes can translate unaided, albeit at a rate too slow to meet the needs of living organisms. To speed up the process, translation factors act as catalysts at each step (**Figure 6.9**). Depending on the environment and the context, the rates achieved range between 2 and 22 aa/s *in vivo*, with eukaryotic synthesis being considerably slower than bacterial. In addition, these factors increase the accuracy and fidelity of translation.

Several translation factors have GTPase activity; in fact, a GTPase participates in each step of translation. They all have G domains, mostly at the N-terminus. The G domains are related to small GTPases, such as p21-ras, but translational GTPases have additional domains. The domain that follows the G domain, domain II (**Figure 6.10A**), is common to all. Like most G proteins, translational GTPases function as molecular switches. When GDP is bound they are in the OFF state, and with GTP bound they are in the ON state. For further information on the structure and mechanism of G proteins, see Section 12.2.

The G domain has several functionally important and conserved features. The P-loop interacts with the phosphates of the G nucleotides. Like the P-loop, switches 1 and 2 undergo characteristic conformational changes during nucleotide binding and hydrolysis. Another important component is the Mg^{2+} ion that can bind to the β and γ phosphates as well as to the switch 1 loop. Most GTPases need a G-nucleotide exchange factor (GEF) to exchange the product for a new substrate. Among translational GTPases, only elongation factor Tu (EF-Tu) has a typical GEF elongation factor in EF-Ts (Figure 6.10C).

GTPases are inactive as enzymes until a GTPase-activating protein (GAP) induces them to hydrolyze their bound GTP. In translational GTPases, the residue that interacts with the hydrolytic water molecule is typically a His, whereas many other G proteins use a Gln. For them, the ribosome acts as the GAP with binding sites at the inter-subunit interface. The G domain and domain II always bind to the same sites, respectively the GTPase-associated center of the large subunit (GAC; see Figure 6.3A) and the small subunit. GTPase activation

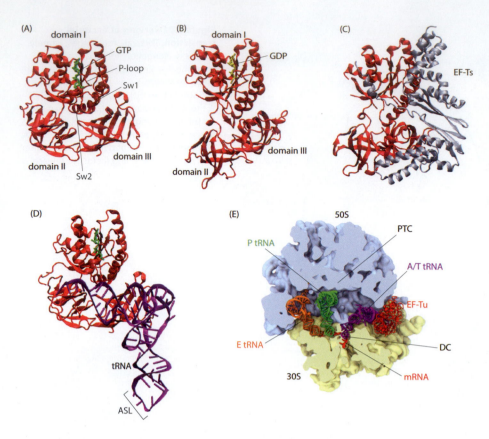

Figure 6.10 Structures of elongation factor Tu. (A) The molecule with GTP bound is in the ON state and has an 'open' structure. (PDB 1EFT) (B) The molecule with GDP bound is in the OFF state and has a compact structure. (PDB 1TUI) (C) The EF-Tu–EF-Ts complex. (PDB 1XB2) (D) Ternary complex of EF-Tu*GTP with Cys-tRNA. (PDB 1OB2) (E) Ternary complex bound to the ribosome. A cutaway view of the ribosome is shown to reveal the binding sites of mRNA, tRNA and factors. (PDB 2XQD and 2XQE)

is carried out by interaction of the catalytic residue of the GTPase and the sarcin–ricin loop (SRL; see Figure 6.3A) in the large subunit.

Translation occurs in four stages: initiation, elongation, termination, and recycling

Protein synthesis is carried out in four stages (see Figure 6.9). In the first stage, initiation, the mRNA is positioned in the ribosome to ensure that synthesis starts at the appropriate codon and in the correct reading frame. This produces the 70S initiation complex—a 70S ribosome programmed with the start codon of the mRNA and the initiator-tRNA located in the P site. In the elongation stage, amino acids are added sequentially to the nascent protein. During each elongation cycle, the tRNA moves through its three binding sites (A → P → E) on the ribosome, adopting so-called classical and hybrid states (see Figure 6.8). Elongation involves establishing the correct match between mRNA and tRNA, followed by peptide bond formation. Then, the ribosome undergoes a large conformational change as it translocates the mRNA and tRNAs by one codon, rendering the A site free and ready for the next incoming aminoacyl-tRNA. When a stop codon in the mRNA enters the A site, the termination and recycling stages begin. The stop signal triggers hydrolysis of the peptidyl-tRNA bond, and the translated polypeptide chain is released. The post-termination ribosome complexes are then split into subunits, which recycle in the next round of translation.

Initiation factors bind to the small subunit

Translation starts with the initiation phase, in which the ribosomal subunits are separate. The SSU must recruit an mRNA and set the reading frame. A preinitiation complex is first formed between the 30S subunit, several initiation factors, and the initiator tRNA. Bacteria have three **initiation factors**: IF1, IF2, and IF3 (**Figure 6.11**); only IF2 is a GTPase. Eukaryotes have corresponding factors, together with many additional ones.

IF3, which has a dumbbell shape with two domains (IF3N and IF3C) separated by a linker, has multiple roles. During initiation, it directs the initiator tRNA into the P site. After termination, IF3 promotes the dissociation of the 70S ribosome during recycling to prevent premature association of the two subunits. Results from cryo-EM and chemical footprinting

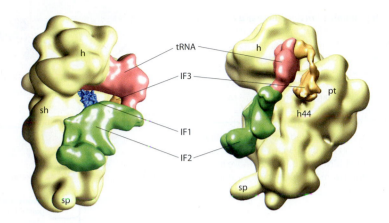

Figure 6.11 Structure of the 30S initiation complex. A cryo-EM map of the complete 30S initiation complex is shown in two orientations. The map is segmented into the 30S subunit (yellow), IF2 (green), fMet-tRNA[fMet] (red), and IF3 (orange), with an atomic model of IF1 fitted in (blue). Landmarks on the 30S subunit are h, head; sp, spur; sh, shoulder; pt, platform; h44, helix 44 from the 16S rRNA. (Adapted from P. Julián et al., *PLoS Biol.* 9(7):e1001095, 2011.)

suggest that IF3C binds to the platform, while IF3N probably binds the E-site region and may contact the elbow of the initiator tRNA. The binding of IF3C to helices h23 and h24 precludes the formation of inter-subunit bridge B2b, hindering subunit association.

IF1 is a small protein (9 kDa) with an OB-fold (see **Guide**) that binds to the A site in the decoding center (see Figure 6.11). Its precise role is still unclear, but it is known to bind cooperatively with IF2 and to sterically occlude the A site from initiator tRNA. It has also been suggested to affect the kinetics of subunit stabilization.

IF2 is the GTPase associated with initiation and is central to this process. In eukaryotes there is a factor called eIF2 that delivers the initiator tRNA to the small subunit. There is no counterpart in bacteria. The eukaryotic ortholog of IF2 is eIF5B: both proteins are involved in the association of the small and large subunits. IF2 promotes, recognizes, and stabilizes the binding of initiator tRNA, discriminating against the binding of elongator tRNAs. This is done through the direct interaction of IF2 with the fMet group in the initiator tRNA. By stabilizing this tRNA, IF2 indirectly influences mRNA binding by codon/anticodon interactions. Like other translational GTPases, interaction with the large subunit induces GTP hydrolysis and subsequent dissociation of IF2-GDP from the ribosome.

The subunits associate to form the 70S ribosome during initiation

In bacteria, initiation factors IF1–IF3 and initiator tRNA bind to the small subunit to form the 30S initiation complex (30S IC) (see Figure 6.9). The initiation factors were previously thought to bind independently, but recent studies have suggested that there may be several pathways for initiation, each with a preferred order of binding. Presumably, IF3 and IF2 bind to the 'empty' small subunit, with IF3 preventing premature association with the large subunit and accelerating the formation of the preinitiation complex. IF1 binds to the A site, probably preventing the tRNA from binding there instead of at the P site. Binding of mRNA around the neck of the small subunit occurs at any time, independently of the factors present in the 30S preinitiation complex. Instead, its binding depends on the mRNA concentration.

For many bacterial mRNAs, positioning of the mRNA is guided by base pairing between the so-called Shine–Dalgarno sequence at the 3′ end of the 16S RNA and a complementary sequence in the mRNA that places the initiator codon in the P site of the PTC. The timing of initiator tRNA binding remains unclear. Once the 30S IC has formed, IF2 promotes its association with the large subunit. Cryo-EM structures have been obtained for the 30S and 70S initiation complexes. The C-terminal domain of IF2 recognizes the formyl-Met bound to the initiator tRNA (fMet-tRNA). Domain II interacts with the 30S subunit and domain III with the large subunit. The G domain interacts with the GTPase activating center of the large subunit. Regions in the EM map corresponding to IF1 and IF3 were identified separately. IF2 has a large surface of interaction with the 50S subunit. This factor may promote association of the two subunits by being a well-fitting partner between them, or it may ensure that initiator tRNA is properly positioned before LSU joining. Upon hydrolysis of the GTP bound to IF2, the factors dissociate, permitting the elongation phase to begin.

In eukaryotes, the start codon might be hundreds of bases away from the point of mRNA attachment, and its location on the ribosome requires an entirely different mechanism,

Figure 6.12 Schematic view of the steps involved in decoding and peptidyl transfer. (A) The ternary complex of EF-Tu*GTP with an aminoacyl-tRNA binds to the A/T site. At this stage the tRNA is tightly bound to EF-Tu, and its shape is altered so that the anticodon can reach the codon in the decoding center. (B) The match between codon and anticodon is checked in a process called selection. Ternary complexes that are not a match dissociate. If the interaction is correct (cognate codon and anticodon), EF-Tu can hydrolyze its GTP to GDP and change its conformation, reducing its affinity for tRNA. (C) EF-Tu leaves and the tRNA swings into the A site, retaining the codon/anticodon interaction. Incorrect matches have low stability and can be discarded in a second fidelity check called proofreading. (D) Peptidyl transfer can occur in the PTC. (E) EF-G binds to the ribosome and helps the peptidyl-tRNA in the A site to move into the P site. (F) The preceding tRNA is discarded from the E site and the ribosome is now ready for the next round of elongation.

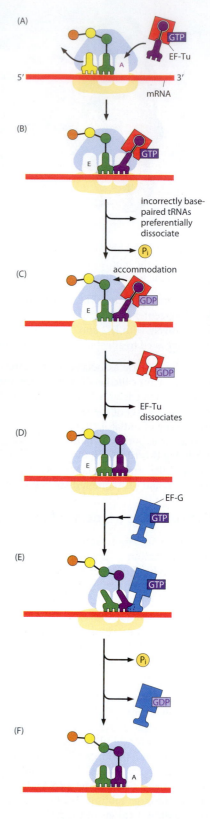

namely scanning along the mRNA from the 5-methyl-G cap that is placed at the 5′ end of the mRNA during processing.

Elongation factors escort tRNAs into and within the ribosome

Bacteria have three **elongation factors** (see Figure 6.10). EF-Tu and EF-G are both GTPases, whereas EF-Ts is a GEF. In eukaryotes, EF1a is the EF-Tu analog, and EF2 corresponds to EF-G. EF-Tu can bind all aminoacylated tRNAs except the initiator tRNA. EF-Tu, tRNA, and GTP form a ternary complex that can deliver tRNA to the empty A site of the ribosome (**Figure 6.12A**). EF-Tu has three domains, whose arrangement depends on whether GDP or GTP is bound. The one with GDP bound, the OFF conformation, is distinctly open (see Figure 6.12A). The ON conformation, with GTP bound, is more compact (Figure 6.12B) and is what binds tRNA in the ternary complex. EF-Tu interacts primarily with the acceptor stem, the T stem, and the CCA end of the tRNA (Figure 6.12D). The amino acid of the aminoacyl-tRNA is tucked into a pocket between the domains. The anticodon of the tRNA is on the other side of the complex from EF-Tu.

The complex of EF-Ts and EF-Tu is shown in Figure 6.10C. EF-Ts destabilizes the Mg^{2+} ion in EF-Tu that binds to the β phosphate of the GDP. This effect, together with a conformational change of the P-loop, results in nucleotide release. The excess GTP in the cell leads to a constant recharging with GTP, leaving EF-Tu in the ON state.

EF-G is the translocase of the ribosome. Translocation, accompanied by a large conformational change in the 70S ribosome, advances the tRNAs to their next positions in the small subunit. The 50S binding sites were occupied upon peptide bond formation when the tRNAs entered hybrid states (Figure 6.12E). EF-G is also a G protein. It has five domains and the N-terminal G domain has an extension that forms yet another domain called G′ (**Figure 6.13A**). Remarkably, EF-G mimics the shape of the ternary EF-Tu complex. Domains III–V correspond spatially to the tRNA (compare Figure 6.10D,E with Figure 6.13A,B). The binding of these two factors to the ribosome overlaps to a large extent.

Amino acids are added to the nascent protein during elongation

The initiation process leaves a ribosome with an empty A site and an aminoacylated initiator tRNA in the P site. In elongation, as each codon is read, various aminoacyl-tRNAs in complex with EF-Tu and GTP enter the A site but leave unless a cognate or near-cognate codon/anticodon interaction is identified (see Figure 6.12A). This induces conformational changes in the ribosome that trigger GTP hydrolysis by EF-Tu, and EF-Tu loses affinity for the tRNA, releasing it and leaving the ribosome (see Figure 6.12D). The tRNA then swings into the A site, with the aminoacyl end reaching the peptidyl transferase site of the 50S subunit (see Figure 6.12D).

Peptide bond formation, involving deacylation of the P-site tRNA and the transfer of the peptide chain to the A-site tRNA, follows spontaneously. This state of the ribosome is highly dynamic, and the tRNAs oscillate into so-called A/P and P/E hybrid states (Figure 6.12E). Next, translocation of the tRNAs occurs, a process catalyzed by EF-G (Figure 6.12F). Binding of EF-G to the ribosome locks the tRNAs in hybrid states, and the ensuing translocation reaction shifts the peptidyl-tRNA from the A/P hybrid state to the P site (P/P state) and the deacylated tRNA from the P/E to the E site (Figure 6.12F), thereby freeing the A site to bind

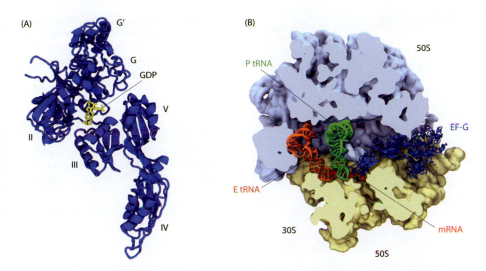

Figure 6.13 Structure and binding of EF-G. (A) EF-G complexed with GDP. Domain IV interacts with the decoding site on the ribosome. The shape of EF-G mimics that of the ternary complex (EF-Tu–tRNA–GTP). (PDB 1EFG) (B) EF-G bound to the ribosome in the post-translocational state (GDP-bound). (PDB 2WRL and 2WRK)

the next incoming aminoacyl-tRNA (see Figure 6.8). Thus, during an elongation cycle, the ribosome is thought to undergo a transition from the dynamic pre-translocation state to the more stable post-translocation state.

tRNA needs to deform to be probed for codon–anticodon complementarity

The beginning of elongation requires a ribosome with a sense codon (as opposed to a nonsense or stop codon) in an empty A site and an initiator fMet-tRNA or a nascent peptide in the P site. The abundant protein EF-Tu in its GTP-bound ON state makes ternary complexes with all aminoacyl-tRNAs for the standard 20 amino acids. In the recognition step of elongation, ternary complexes test their anticodons against the codon in the decoding site of the small subunit. When bound to EF-Tu, tRNA cannot bind directly into the A site (classical A/A state). Instead, the tRNA comes in at an angle from its position in the A site; that is, the A/T hybrid state (see Figure 6.8). In this conformation, the anticodon stem loop is in the A site matching the mRNA, and the CCA end is buried inside EF-Tu, far from the peptidyl transfer site. This causes a deformation between the anticodon stem and the D stem (see Figure 6.2) that presumably stores elastic energy along the tRNA. (Each of the three arms comprises a loop and a stem.)

The ribosome decodes the signal

The correct (cognate) match of codon and anticodon is based on Watson–Crick base pairing, recognized as such by the appropriate distance and orientation of the sugar moieties. The 16S RNA participates in checking for the correct anticodon (**Figure 6.14**). Adenine 1493 stabilizes Watson–Crick base pairs with the first base of the codon through H-bonds (Figure 6.14C). Similarly, A1492 and G530 of the 16S RNA stabilize Watson–Crick base pairs with the second base of the codon (Figure 6.14D). The third base pair of the codon/anticodon interaction is less rigorously checked, hence its description as the 'wobble' base pair (Figure 6.14E). The recognition also involves H-bonding between the minor-groove edges of the nucleotides in the decoding site (A1492, A1493, and G530) and the minor-groove edges of the paired codon–anticodon nucleotides, as well as a close steric fit. When there is a proper match between codon and anticodon, conformational changes are induced that result in a signal being transmitted to EF-Tu and inducing GTP hydrolysis, in a process called selection.

Crystal and cryo-EM structures have shown that several conformational changes occur upon cognate codon recognition in the ribosome (codon–anticodon monitoring, domain closure, disruption of inter-subunit bridge B8); the tRNA (in the A/T state, 3′-end distortion) and EF-Tu (β-loop interaction with 30S shoulder, G-domain shift) are essential for GTP hydrolysis. These conformations properly position the GTPase center of EF-Tu for activation by the sarcin–ricin loop (SRL) of the large subunit that is presumed to act as a GAP by activating His84 in EF-Tu. This mechanism of GTPase activation by the SRL is likely to be similar in all translational GTPases. After GTP hydrolysis, EF-Tu adopts the

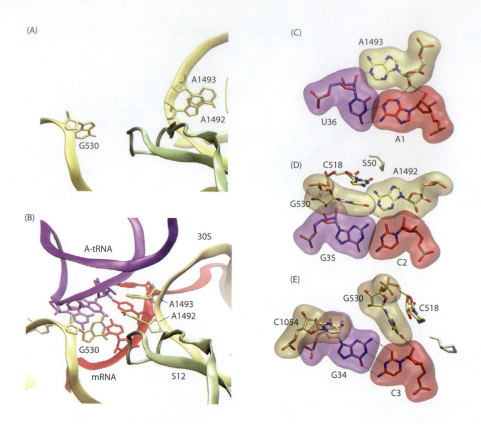

Figure 6.14 The decoding site. (A) In the ribosome without an A-site tRNA, A1492 and A1493 are stacked in h44. (B) When an aminoacyl-tRNA binds to its cognate codon in the mRNA in the A site, A1492, A1493, and G530 change conformation to interact with the minor groove of the mRNA–tRNA minihelix. Yellow, parts of the small subunit; red, mRNA; purple, tRNA. (C–E) The interactions between the codon bases, the anticodon bases, and nucleotides A1493, A1492, and G530 are thought to stabilize the Watson–Crick base pairs in positions 1 and 2.

GDP-bound conformation, which introduces strain in its interaction with the ribosome, resulting in a lowered binding affinity for the tRNA. Subsequently, EF-Tu leaves and the tRNA springs into its normal conformation. Since the ASL is stably positioned in the 30S A site, the tRNA swings into the A/A site in a step called accommodation (see Figures 6.9 and 6.12C).

Peptide bond formation occurs rapidly and spontaneously

The next step in elongation is peptidyl transfer. In this main chemical event in translation, the α-amino group of the aminoacyl-tRNA makes a nucleophilic attack on the ester carbonyl of the peptidyl-tRNA to form a new peptide bond (**Figure 6.15A**). With the tRNA accommodated, the CCA ends of the P- and A-site tRNAs are in close proximity, and they pair with bases in the P- and A-loops of the large subunit respectively (Figure 6.15B). The enzymatic active site, composed only of RNA, is configured so as to present the substrates in optimal orientation and distance, and peptide bond formation is rapid and spontaneous. The ribosome accelerates the rate of peptide bond formation at least 10^5-fold. However, no group from the ribosome has any dominant role in the catalysis. In fact, a comparison of the rates of peptide bond formation by the ribosome and by a ribosome-free model system suggested that the ribosome facilitates the reaction solely by entropic effects, which include substrate positioning, shielding from bulk solvent, or organization of the active site.

The attacking α-amino group of the aminoacyl-tRNA makes important H-bonds to the 2′ hydroxyl of the terminal ribose of the P-site tRNA and to the N3 atom in A2451. This network of H-bonds suggests a proton shuttling mechanism from the α-amino group into the 2′ OH in A76 to its 3′ O leaving group (Figure 6.15C). Peptidyl transfer leads to a peptidyl-tRNA in the A site; that is, the nascent chain is attached to the A-site tRNA (see Figure 6.8). To proceed, this tRNA has to be translocated to the P site.

The ribosome rocks during translocation

In translocation, the tRNAs and mRNA move by one position (one codon) at a time across the ribosome. Once peptidyl transfer has occurred, the P-site tRNA can move into a P/E hybrid site. This means that its CCA end moves into the E site, allowing the CCA end of the A-site tRNA to move into the A/P site. In doing so, the single-stranded CCA end rotates through 180° to form H-bonds with the P-loop instead of the A-loop. This orientation of

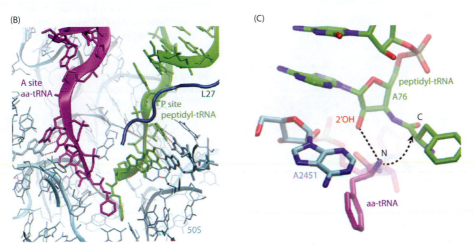

Figure 6.15 The peptidyl-transfer site.
(A) Schematic drawing of the reaction. Ade, adenine. The NH_2 group of the incoming aminoacyl-tRNA in the A site makes a nucleophilic attack on the ester bond between the C-terminal amino acid of the nascent polypeptide and the 3′ OH group of the A residue at the CCA end of the tRNA in the P site. (B) Binding of tRNAs to the PTC. (C) The nucleophilic NH_2 group is positioned by interaction with the 2′ OH of A76 of the peptidyl-tRNA and N3 of A2451, as part of an extensive network of hydrogen bonds. (Adapted from T.M. Schmeing and V. Ramakrishnan, *Nature* 461:1234–1242, 2009. With permission from Macmillan Publishers Ltd.)

the substrate is needed for the next step in translation. The hybrid states occur in tandem with a 'ratcheting' motion of the ribosome, in which the 30S subunit rotates through 6° with respect to the 50S subunit (see Figure 6.3C). In this equilibrium between unratcheted and ratcheted states of the ribosome and between classical and hybrid tRNA states, EF-G binds near the A site, stabilizing the ratcheted and hybrid states. EF-G binds to the ribosome with the G domain at the GAC, catalyzing movement from hybrid to post-translocational states. The rate of these conformational changes is enhanced by GTP hydrolysis. The peptidyl tRNA moves from the hybrid A/P site to the P site. At the same time, the deacylated tRNA in the P/E site translocates to the E site and the mRNA advances by one codon. After translocation, EF-G dissociates and the ribosome is ready for a new cycle of elongation (see Figure 6.12).

Release and recycling factors recognize and execute the end of the cycle

In most organisms, the genetic code has three stop codons: UAA, UAG, and UGA. There is no tRNA with a corresponding anticodon and, in bacteria, when a stop codon is exposed in the decoding part of the A site, one of the closely related release factors 1 and 2 recognizes the stop codon. The factor binds and catalyzes hydrolysis of the acyl bond linking the nascent polypeptide chain to the tRNA in the P site, releasing the newly synthesized protein from the ribosome. RF1 and RF2 both recognize UAA, whereas UAG and UGA are only recognized by RF1 and RF2 respectively. The sequence Gly-Gly-Gln (GGQ) is responsible for hydrolysis of the peptidyl-tRNA bond. The specificity of the stop codons is realized through several contacts including those made by different tripeptides in RF1 and RF2: Pro-Ala/Val-Thr or P(A/V)T, and Ser-Pro-Phe or SPF respectively. The crystal structure of RF2 (**Figure 6.16A**) shows a compact shape, with the GGQ and SPF motifs only 23 Å apart (compare with 75 Å between the decoding and PTC centers). However, cryo-EM and crystallographic structures of ribosome-bound factors have shown that RFs undergo large conformational changes, with the GGQ ending up in the peptidyl transfer site and the decoding peptides at the stop codons, as is needed (Figure 6.16B,C).

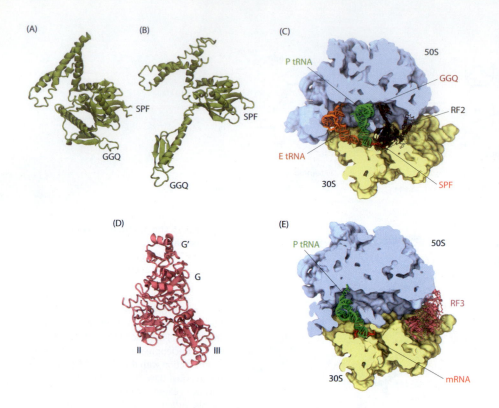

Figure 6.16 Structures of the release factors. (A) The crystal structure of isolated RF2 shows a compact shape. (PDB 1GQE) (B) RF2 in its ribosome-bound conformation exhibits a large conformational change, after which these two functional tripeptides are placed in their active sites. The SPF sequence (P(A/V)T in RF1) binds in the decoding area at the stop codon, while the GGQ sequence is active to release the polypeptide from the tRNA in the P site in the PTC. (PDB 2X9R and 2X9S) (C) RF2 as in (B), but shown bound to the intact 70S ribosome. (PDB 2X9R and 2X9S) (D) Structure of RF3, the GTPase involved in the release step, bound to a non-hydrolyzable GTP analog. The G′ domain differs in structure from that of EF-G. (PDB 3ZVO and 3ZVOP) (E) Structure of the 70S ribosome with RF3-GDPCP (RF3-guanosyl-methylene-triphosphate) bound in the fully rotated conformation. (PDB 3ZVO and 3ZVOP)

RF3 catalyzes the release of RF1 and RF2 from the ribosome. RF3 is a GTPase with three domains. As with EF-G, the G domain is extended with a G′ domain (Figure 6.16D,E). Eukaryotes have only one release factor, eRF1, which recognizes all three stop codons. Another factor, eRF3, stimulates peptide release by eRF1 in a GTP-dependent manner.

After peptide hydrolysis, the ribosome with a deacylated tRNA in the P site must be disassembled. This process is brought about by the **ribosome recycling factor (RRF)**, which is assisted by EF-G in separating the two subunits so that the mRNA and the deacylated tRNA can dissociate and the subunits can engage with other mRNAs for a new round of translation (**Figure 6.17**).

A protein is born and the cycle begins again

When a stop codon on the mRNA is reached, there is normally no cognate tRNA to bind to it: instead, a release factor binds with high affinity. Unbound release factors adopt a closed conformation (see Figure 6.16A). However, when bound to the ribosome, RF1 and RF2

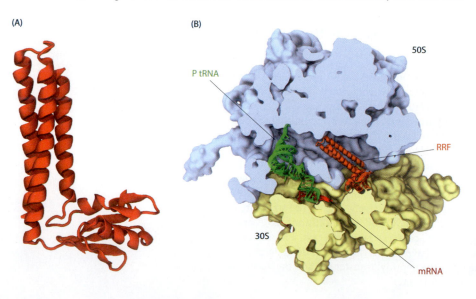

Figure 6.17 Structure of the ribosome-recycling factor, RRF. (A) The factor has an L-shaped structure like a tRNA but binds to the ribosome in an orientation substantially different from that of a tRNA. (B) Structure of the recycling complex. (PDB 4GD1 and 3R8S)

adopt extended conformations in which the peptide motifs—P(A/V)T and SPF—that bind to the stop codon can reach the A site, while the GGQ sequence that participates in hydrolyzing the protein from the tRNA is inserted into the PTC (see Figure 6.16C). Crystallographic analyses of the two factors when bound to the ribosome suggest that the Gln coordinates the putative catalytic water. Furthermore, studies have shown that Gln has a role in discriminating between water and other nucleophiles. After hydrolysis, RF3 removes RF1 or RF2 from the ribosome in a process powered by GTP hydrolysis.

After termination, ribosomes are left with a deacylated tRNA in the P site and an mRNA that has reached its 3′ end. This state is identified by RRF, which binds across both the A and P sites (see Figure 6.17). In some still not fully understood process, RRF and EF-G catalyze the dissociation of the ribosomal subunits in conjunction with GTP hydrolysis.

Proteins can be translocated into and across membranes

Whereas some newly synthesized proteins are released directly into the cytosol, others are designated for translocation into or across lipid membranes. For integral membrane proteins, the translating ribosome is targeted to the membrane, and translation and export take place concomitantly. Targeting is carried out by the **signal recognition particle** (**SRP**), an evolutionarily conserved ribonucleoprotein particle that recognizes signal sequences at the N-termini of nascent polypeptide chains emerging from the ribosome (**Figure 6.18A**) and docks them into the translocon system in the membrane via the SRP receptor (SR).

The **translocon system** interacts directly with the ribosome after hand-off from the SR (Figure 6.18B). Post-translational targeting to the translocon is carried out by SecB. The core translocon that carries out co-translational and post-translational translocation is a conserved heterotrimer in both prokaryotes (SecYEG) and eukaryotes (Sec61). Its largest subunit, SecY or Sec61α, forms a pore in the membrane through which the protein is threaded (Section 10.7). During translocation, the pore is 'unplugged' by the displacement of a single short transmembrane helix of SecY or Sec61α, and the pore widens by the movement of an intramolecular hinge at the distal part of SecY or Sec61α. These changes not only allow proteins to pass through the pore but they also enable transmembrane helices to exit from the translocon laterally as they become integrated into the lipid bilayer.

There has been a long-standing debate as to the copy number of the core translocon in the native complex. Although most structural studies using solubilized or reconstituted SecYEG–Sec61 complexes support the binding of a single core translocon to the ribosome, FRET measurements suggest that the core translocon may oligomerize in a native membrane environment. This is consistent with crystal contacts observed in two-dimensional crystals of the SecYEG complex and the results of cross-linking experiments with SecYEG complexes. Both ideas are reconciled in a model in which complexes of ribosome and nascent chain bind to monomeric translocon complexes and induce their oligomerization.

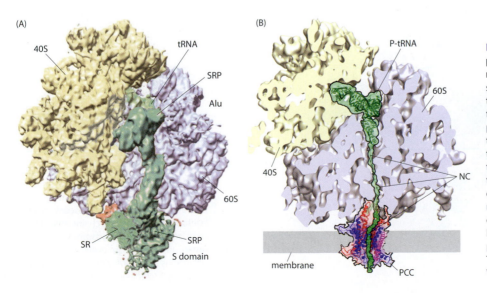

(A)

40S
tRNA
SRP
Alu
60S
SR
SRP
S domain

(B)

P-tRNA
60S
40S
NC
membrane
PCC

Figure 6.18 Direct translocation of nascent polypeptide chains into or through membranes. (A) When a protein is to be secreted across or inserted into a membrane, the signal recognition particle (SRP) binds to the nascent chain in the ribosome, slowing protein synthesis and facilitating coupling of the ribosome to the translocon by binding to the SRP receptor. (B) Binding of the ribosome to the translocon allows the nascent chain to move directly from the ribosomal tunnel exit across or into the membrane. SR, SRP receptor; NC, nascent chain; PCC, protein-conducting channel. (A, adapted from M. Halic et al., *Science* 312:745–747, 2006. With permission from AAAS. B, adapted from T. Becker et al., *Science* 326:1369–1373, 2009. With permission from AAAS)

Multiple ribosomes assembled into polysomes translate the same mRNA simultaneously

Typically, ribosomes initiate new rounds of translation on a given mRNA before the previous ribosome has finished translating. This results in an mRNA with multiple translating ribosomes, termed polyribosomes or polysomes. The density of ribosomes on a particular mRNA is largely determined by the rate of initiation. Polysomes were initially examined by sedimentation analysis and electron microscopy, but their polymorphism precluded in-depth structural characterization. Recently, cryo-electron tomography studies performed with *in vitro* systems (*E. coli* lysates) and *in vivo* mammalian systems have provided more detailed insight into their organization. The basic principles appear to be the same in bacterial and eukaryotic polysomes: the ribosomes are densely packed on the mRNA, and adjacent particles have preferred orientations. Although they can adopt different arrangements (pseudohelical, staggered in-plane, or mixed forms), the small subunits are always located toward the interior of the polysome, sequestering the mRNA but with the tRNA entrance sites remaining accessible. The exit sites for the nascent polypeptide chains face the cytosol. This arrangement ensures that the mRNA is read in a protected environment while the nascent proteins emerge sufficiently far apart to minimize the likelihood of unproductive interactions between them (**Figure 6.19**).

Many antibiotics target the ribosome and inhibit its function

Antibiotics are compounds that kill or inactivate microorganisms. Most are related to natural substances that are used by some organisms to neutralize attacks from others, but synthetic compounds have more recently come into use. Several antibiotics currently in clinical use inhibit ribosomal activity, targeting different steps in translation, principally elongation. Like most inhibitors, they bind at functional centers and do so via multiple mechanisms. These inhibitors are also valuable tools for the characterization of ribosomal function. Detailed coverage of antibiotic activity is beyond the scope of this section; instead, some representative examples are presented.

One group of inhibitors binds to the decoding site of the small subunit (**Figure 6.20A**). Many of these antibiotics are aminoglycosides. As described above, helix h44 of the small subunit is closely associated with decoding as well as with the ribosomal inter-subunit bridges. At the top of h44 are adenines 1492 and 1493, which participate in identifying correct Watson–Crick base pairs during decoding (see Figure 6.14A,B). Normally they stack in the helix without base-paired partners. To stabilize a cognate base pair between codon and anticodon, they have to flip out of the stacking conformation to interact in the decoding center. Some antibiotics, such as hygromycin B and paromomycin, bind to h44 and force the two adenines out of the helical stack. The ribosome then (mis-)identifies this conformation as correctly base-paired in the decoding site and induces EF-Tu to hydrolyze GTP. In

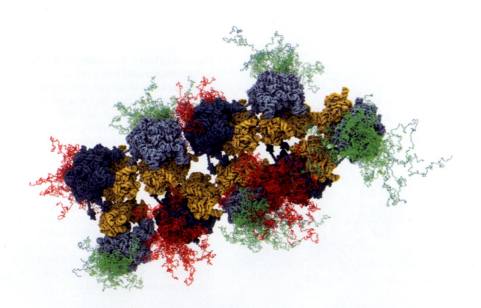

Figure 6.19 On a polysome, multiple ribosomes translate the same mRNA simultaneously. Shown here is a polysome model derived from electron tomography. Ribosomes seem to be arranged such that nascent chains from different ribosomes do not interact with each other. The figure shows 20 simulated nascent chain conformations. Nascent chains emerging from odd-numbered ribosomes are colored green, and those from even-numbered ribosomes are red (Adapted from F. Brandt et al., *Cell* 136:261–271, 2009.)

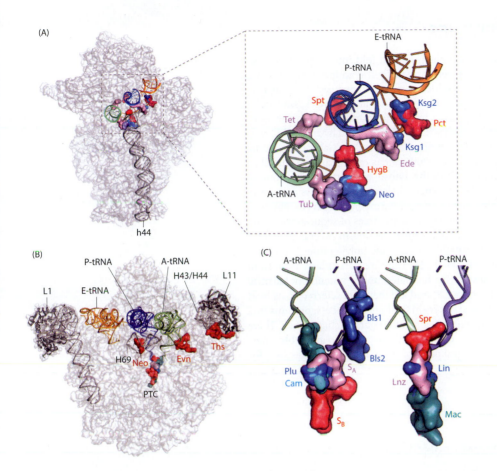

Figure 6.20 Binding sites of different antibiotics on the ribosome. (A) Overview with enlargement of antibiotic binding sites along the mRNA-binding channel of the 30S subunit: tetracycline (Tet), spectinomycin (Spt), kasugamycin (Ksg1, Ksg2), edeine (Ede), hygromycin (HygB), neomycin (Neo), procalcitonin (Pct), and tuberacyinomycins (Tub). The A-site tRNA (green), P-site tRNA (blue), E-site tRNA (orange), and h44 are highlighted as reference points. (B) Overview of the binding sites of Neo, evernimicin (Evn), and thiostrepton (Ths) on the 50S subunit. In addition to the A-site, P-site, and E-site tRNAs, H43/H44, H69, the PTC, and the L1 and L11 stalks are highlighted for reference. (C) Enlargement of the binding sites for blasticidin S (Bls1 and Bsl2), sparsomycin (Spr), lincomycin (Lin), linezolid (Lnz), macrolides (Mac), pleuromutilins (Plu), chloramphenical (Cam), and streptogramins A and B (S_A and S_B), relative to the A-site and P-site tRNAs. (Adapted from D.N. Wilson, *Nature Rev. Microbiol.* 12:35–48, 2014. With permission from Macmillan Publishers Ltd.)

this way, erroneous amino acids can be incorporated into proteins. Moreover, by stabilizing the A-site tRNA, these antibiotics strongly inhibit translocation. These effects on elongation have catastrophic consequences for the microorganism.

The peptidyl transfer site is also the target for numerous inhibitors (Figure 6.20B). Puromycin, one of the best known, mimics the terminal part of a tRNA, an adenine linked to a tyrosine. As such, it binds to the A site in the PTC, which leads to premature termination, because puromycin cannot be translocated. Instead, it simply falls off with its attached peptide. Numerous antibiotics of the macrolide family also bind in the same general area—for example, erythromycin, which binds further down the exit channel. They act by forming a plug that prevents the nascent peptide from traveling through the exit channel.

Antibiotics can also act by inhibiting translation factors. In the case of EF-Tu, the antibiotics kirromycin and aurodox bind between the G domain and domain III. In this way, kirromycin acts as a glue that locks the domains together. Thus, after GTP hydrolysis, the conformational change of the factor does not take place and EF-Tu remains stuck to the ribosome, where it blocks further translation by the affected ribosome and all others on the same mRNA.

Because the functional sites of the ribosome consist primarily of RNA, one might expect that mutations of these RNA molecules could lead to antibiotic resistance. However, there are several genes for ribosomal RNAs in most species. Thus, acquiring mutations in the ribosomal RNAs can be an inefficient means of avoiding inhibition by antibiotics. A more effective way to achieve resistance is through modifications of the bases in the RNA. Methylating enzymes with specificity for antibiotic binding sites are part of the resistance mechanisms. An alternative is to acquire mutations in the genes for ribosomal proteins, which are present in single copies in the genome. Ribosomal proteins can thus affect the structures of the RNA molecules in the functional sites, thereby creating resistance. Other ways in which to build resistance include actively pumping antibiotics out of the cell, chemically modifying the antibiotic itself—for example by acetylation—and transient binding of specialized factors to the ribosome that can remove bound antibiotics.

6.3   MOLECULAR CHAPERONES

After synthesis on the ribosome, the polypeptide chain should fold into a specific three-dimensional structure. In principle, this fold is determined by the amino acid sequence and, *in vitro*, most proteins are able to reach the folded state spontaneously (Section 1.3). However, the cellular milieu is more complex and poses additional challenges. Molecular crowding (Section 1.8) causes strong protein/protein interactions and may result in non-productive aggregation of unfolded or partly folded species with exposed hydrophobic residues. Nascent polypeptide chains on translating ribosomes cannot adopt stable folds before the minimum folding unit, usually a domain of 100–300 aa, has emerged from the ribosome tunnel. Thus, unfolded chains discharging into the cytosol are particularly prone to aggregation unless protected from doing so. This is where *molecular chaperones* come into play. These proteins bind to unfolded chains or to folding intermediates and thereby prevent aggregation, or they may promote folding by engaging in cycles of sequestration and release until the correct fold has been achieved. **Chaperones** belong to several structural classes (**Table 6.2**) and cooperate in networks capable of handling polypeptides from their initial synthesis throughout their life span (**Figure 6.21**). They are particularly important under conditions of stress. Some chaperones are up-regulated under potentially harmful conditions such as elevated temperature (Section 1.3) and are therefore referred to as **heat shock proteins**. In addition to proteins that assist others in the folding process, proteins that promote the assembly of multi-subunit complexes are also often named chaperones. There are many structurally unrelated assembly factors of this kind; they bind to assembly intermediates transiently and without becoming integral parts of the mature complexes (see, for example, Section 8.5).

Table 6.2 The major classes of heat shock proteins (Hsps) or chaperones in the three kingdoms of life.

Chaperone class		Kingdom	Organism	Name	Subunit mass (kDa)	Co-chaperones	Subcellular localization and activity
Ribosome-associated chaperones		Bacteria	*E. coli*	Trigger factor	48, monomer–dimer		Cytosol; folding of nascent chains; peptidyl-prolyl bond isomerization
		Archaea	*Thermoplasma acidophilum*	NAC	17, homodimer		Cytosol; protein folding/quality control (?)
		Eukarya	*H. sapiens*	NAC (α,β)	19, heterodimer	SRP	Cytosol; folding/quality control (?)
Chaperonins/Hsp60	Group I	Bacteria	*E. coli*	GroEL	57, 14-mer	GroES	Cytosol; folding of a subset of proteins; refolding of stress-denatured proteins
		Eukarya	*H. sapiens*	Hsp60	60, 14-mer	Hsp10	Mitochondrial matrix; folding of newly imported proteins
		Eukarya	*Arabidopsis thaliana*	Cpn60 (α,β)	57–58, hetero-14-mer	Cpn10, Cpn20	Chloroplasts; folding of Rubisco
	Group II	Archaea	*T. acidophilum*	Thermosome (α,β)	58, alternating 16-mer	Prefoldin	Cytosol; folding of a subset of the proteome; refolding of stress-denatured proteins
		Eukarya	*H. sapiens*	TRiC/CCT	57–60, 16-mer (8 different subunits)	Prefoldin, PhLP	Cytosol; folding of a subset of the proteome, including actin and tubulin; masking of toxic polyglutamine expansion protein aggregates

Table 6.2 *continued.*

Chaperone class		Kingdom	Organism	Name	Subunit mass (kDa)	Co-chaperones	Subcellular localization and activity
Hsp70		Bacteria	*E. coli*	DnaK	69, monomer	DnaJ, GrpE, ClpB	Cytosol; folding of nascent chains; facilitates protein aggregate disassembly and degradation of abnormal proteins; regulation of heat shock response
		Archaea	*T. acidophilum*	DnaK	66, monomer	DnaJ, GrpE	Cytosol; folding of nascent chains (?)
		Eukarya	*H. sapiens*	Hsp70, Hsc70, BiP (Endoplasmic Reticulum (ER) luminal variant)	71, monomer	Hsp40 (many specialized isoforms, such as auxilin), NEFs (Hsp110, HspBP1, BAG proteins), Hip, TPR-clamp proteins (HOP, CHIP)	Cytosol, Endoplasmic Reticulum (ER) lumen; folding of nascent chains; transport into ER; protein disaggregation and quality control; inhibition of polyglutamine fibril formation
Hsp90		Bacteria	*E. coli*	HtpG	71, dimer		Cytosol; protein refolding in stressed cells
		Eukarya	*H. sapiens*	Hsp90, Grp94 (ER luminal variant)	85, dimer	HOP, Cdc37, Aha1, TPR-clamp immunophilins, p23, CHIP	Cytosol; folding and conformational regulation of signaling proteins; reactivation or degradation of stress-denatured proteins; regulation of heat shock response
Hsp100		Bacteria	*E. coli*	ClpA, ClpB	83, 86, homohexamers	ClpP, SspB, ClpS, DnaK	Cytosol; protein remodeling and disaggregation for degradation
		Archaea	*T. acidophilum*	VAT	83, homohexamer		Cytosol; protein remodeling and disaggregation for degradation (?)
		Eukarya	*S. cerevisiae*	Hsp104, Cdc48	104, homohexamers	Hsp70, p47, Dsk2	Cytosol; protein disaggregation; cooperates with ERAD system
Small Hsps		Bacteria	*E. coli*	IbpA, IbpB	16, multiple oligomeric states	DnaK, DnaJ	Cytosol; prevents protein aggregation; cooperates with DnaK in refolding
		Archaea	*Methanocaldococcus jannaschii*	Hsp16.5	16, multiple oligomeric states		Cytosol; stabilizes unfolded proteins and prevents aggregation
		Eukarya	*H. sapiens*	α-Crystallin	16, multiple oligomeric states	Hsp70, Hsp40	Cytosol; prevents protein aggregation in the vertebrate eye lens

(Adapted from Y. Tang et al., *Cell* 128, 2007.)

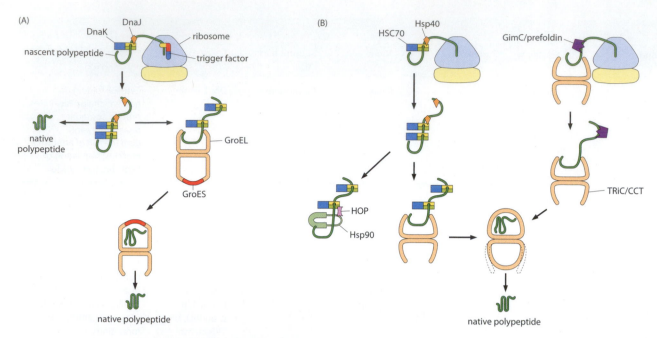

Figure 6.21 Simplified scheme for the *de novo* folding pathways in bacteria and eukaryotes. Chaperones such as trigger factor (A) or the Hsp70 system (B) interact with the polypeptide chain during translation and protect it from unproductive interactions. Proteins that cannot reach their native state by themselves are processed through a cascade of molecular chaperones. (Adapted from M.J. Young et al., *Nature Rev. Mol. Cell Biol.* 5:781–791, 2004. With permission from Macmillan Publishers Ltd.)

Nascent polypeptide chains are met by an array of chaperones

In bacteria, the first chaperone to become engaged with the nascent polypeptide chain is *trigger factor* (TF), a protein of 48 kDa in *E. coli* (**Figure 6.22**), which binds to ribosomes with a 1:1 stochiometry near the polypeptide exit site (see Figure 6.21A). TF is thought to monitor nascent chains and to recognize hydrophobic sequences. By sequestering them it keeps the newly synthesized polypeptide in a folding-competent state. There is no homolog of TF in archaea or in eukarya.

Members of the cytosolic 70 kDa heat shock protein (Hsp70) system (see Table 6.2), which exists in all three kingdoms of life (although not in all archaea), bind to nascent chains without being physically associated with ribosomes. The structural and mechanistic aspects of the Hsp70 system are best understood for the bacterial Hsp70 DnaK, its Hsp40 co-chaperone, DnaJ, and the nucleotide exchange factor GrpE (**Figure 6.23**). DnaK has two domains: a 44 kDa N-terminal ATPase domain and a 27 kDa C-terminal polypeptide-binding domain. The latter is subdivided into a β sandwich with a peptide-binding cleft and an α-helical lid-like segment. Hsp70 recognizes short (~7 aa) hydrophobic peptides in extended conformations, a structural feature that is quite common in nascent chains that have yet to fold. In the ATP-bound state, the α-helical lid over the peptide-binding cleft is in an open conformation and substrate proteins can exchange rapidly. When ATP is hydrolyzed and the lid closes, substrates become stably bound. The cycling between these two conformational states is regulated by two proteins, DnaJ and GrpE, the latter a homodimer of 20 kDa subunits. The J domain of DnaJ binds to DnaK and accelerates ATP hydrolysis, thereby promoting stable binding. GrpE induces the release of ADP from DnaK, after which the rebinding of ATP leads to release of the substrate protein and completion of the reaction cycle. The C-terminal domain of DnaJ, and similarly that of other Hsp40s, is a chaperone in its own right. It recognizes hydrophobic peptides and can thereby recruit DnaK to nascent chains.

DnaK is a very abundant cytosolic protein whose intracellular concentration (~50 μM) exceeds that of ribosomes (~30 μM). It is therefore likely that multiple Hsp70 molecules can act on the same nascent polypeptide chain (see Figure 6.21A). Release from DnaK gives the chain a chance to reach its native fold, while the rebinding of slow-folding intermediates to DnaK prevents unproductive interactions.

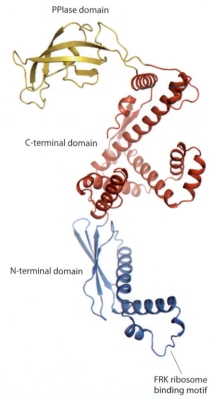

Figure 6.22 Structure of trigger factor from *E. coli*. Peptidyl-prolyl isomerase (PPIase) activity is not essential for trigger factor function, but this domain contributes to interactions with client proteins. The Phe-Arg-Lys (FRK) motif is critical for ribosome binding. (PDB 1W26)

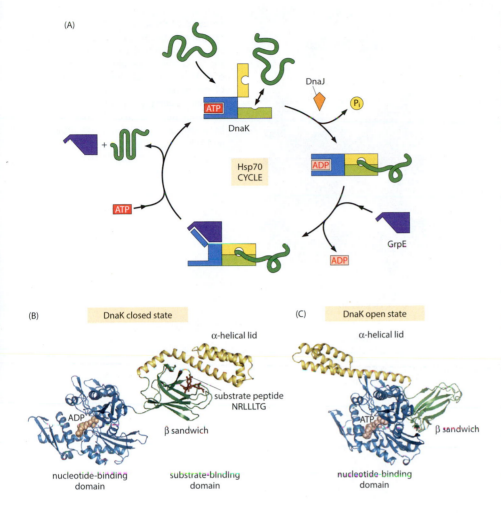

Figure 6.23 The Hsp70 system.
(A) Schematic representation of the DnaK folding cycle. In the ATP-bound state, DnaK interacts dynamically with partly folded substrate protein (open state). Binding of DnaJ and substrate protein synergistically triggers ATP hydrolysis by DnaK, resulting in stable association of DnaK with the substrate protein (closed state). Binding of the nucleotide exchange factor GrpE triggers dissociation of ADP. Rapid binding of ATP releases the substrate for a new folding attempt. (B) NMR-derived model of DnaK in the closed state. In this state, the N-terminal nucleotide-binding domain (NBD) and the bipartite substrate-binding domain (SBD) are flexibly tethered. ADP is bound at the center of the NBD. Hydrophobic peptides such as the model sequence NRLLLTG are bound in an extended conformation in a groove in the β-sandwich domain that is covered by the α-helical lid domain. Such binding motifs occur on average every 36 residues in nascent polypeptide chains. (PDB 2KHO) (C) Crystal structure of DnaK in the open state. Binding of ATP induces a twist in the NBD that enables tight binding of the linker region. In this conformation, DnaK is primed for ATP hydrolysis. The SBD subdomains rearrange and separately associate with the NBD. In the open state, the peptide-binding groove widens, explaining the lower substrate affinity of this state. Note that DnaK samples open and closed conformations all the time; only the frequency changes, depending on the nucleotide present. (PDB 4B9Q) (A, adapted from F. Hartl and M. Hayer-Hartl, *Science* 195:1852–1858, 2002.)

Eukaryotic Hsp70s, such as the mammalian 70 kDa heat shock protein HSC70, follow a similar ATP-dependent reaction cycle in which DnaJ homologs stimulate ATP hydrolysis. There are no orthologs of GrpE in the eukaryotic cytosol; they are, however, found in mitochondria and chloroplasts.

Some proteins require the assistance of chaperonins to reach their native states

Upon release from the ribosome, most newly synthesized proteins interact with the DnaK system, but a subset needs further assistance from the large multimeric **chaperonins**, as the Hsp60 subgroup of chaperones are called (see Table 6.2). Unlike the monomeric DnaK, which binds to nascent chains and maintains them in a folding-competent state, the chaperonins act further downstream and allow folding to be completed in the sequestered environment provided by the chaperonin complex. A proteome-wide analysis of selectively isolated GroEL and GroES complexes suggests that they interact with ~250 different proteins, of which ~85 seem to be strictly dependent on the action of chaperonins. Many of these substrates populate kinetically trapped intermediate states.

Chaperonins are evolutionarily conserved protein-folding machines

Chaperonins are found in all three kingdoms of life: archaea, bacteria, and eukarya. The ~55 kDa subunits form barrel-shaped assemblies with two rings face-to-face and seven to nine subunits per ring. Each subunit has three domains: equatorial, intermediate, and apical (see, for example, **Figure 6.24**). The domains interlock such that the apical domain is inserted into the intermediate domain, which, in turn, is inserted into the equatorial domain. Contacts between the equatorial domains, where both the N- and C-termini and

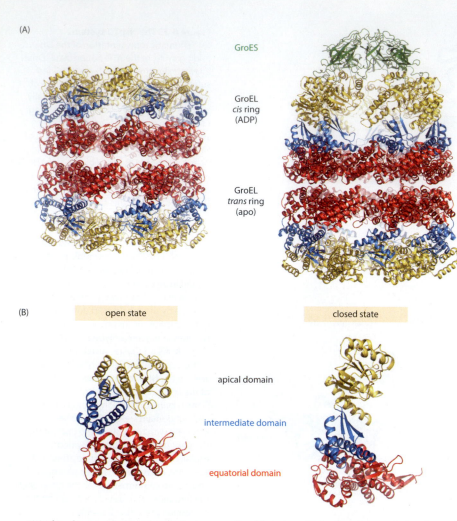

(A)

GroES

GroEL
cis ring
(ADP)

GroEL
trans ring
(apo)

(B)

open state   closed state

apical domain

intermediate domain

equatorial domain

Figure 6.24 Structure of the bacterial chaperonin GroEL. (A) Crystal structures of the apoGroEL tetradecamer (open state) and the GroEL–GroES complex in the presence of ADP (closed state). The domain structure of the chaperonin subunits is indicated by the color coding; see (B). (B) Conformational rearrangements in the GroEL subunit between open and closed states. (PDB 1SS8 and 1PF9)

an ATP-binding pocket are located, connect the subunits within the rings and also form the interface between the two rings. The apical domains are located at the upper and lower rims of the barrel, and the intermediate domains provide a link that transmits conformational changes triggered by ATP binding and hydrolysis to the apical domains.

Chaperonins fall into two groups: group I chaperonins are found in bacteria and in eukaryotic organelles of bacterial descent (Section 1.9), whereas group II chaperonins occur in archaea and in the cytosol of eukaryotic cells. In exceptional organisms (such as the archaeon *Methanosarcina mazeii*), group I and group II chaperonins coexist. The protein-folding function of group I chaperonins requires a partner protein or co-chaperonin, such as GroES in *E. coli*, which closes off the apical openings of the complex. Group II chaperonins possess a built-in closing mechanism; the apical domains with their helical protrusion can close the openings in an 'iris-like' manner. Group I chaperonins form seven-membered rings, whereas group II chaperonins have eight or nine members per ring. The arrangement of the rings across the equator is staggered in group I chaperonins, in contrast with the situation in the group II chaperonins, where each subunit makes contact with only one subunit of the opposite ring. This has consequences for allosteric effects related to ATP hydrolysis.

GroEL is an ATP-driven folding machine

The structure and mechanism of the bacterial chaperonin GroEL and its co-chaperonin GroES have been studied in great detail (see Figure 6.24). During its functional cycle, the heptameric GroEL–GroES rings undergo large-scale conformational changes. In the nucleotide-free state, the apical domain of GroEL exposes a hydrophobic cleft suitable for the binding of peptides in an extended or amphiphilic α-helical conformation. This state is characterized by a 'collapsed' ring conformation with only small internal cavities.

The binding of ATP primes the rings for a large conformational change: the apical domains undergo an upward and clockwise turning motion that is completed and stabilized upon the

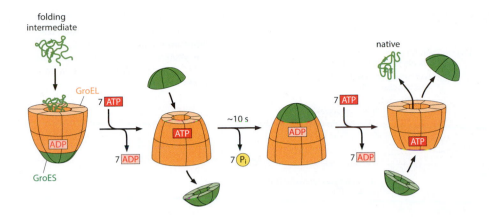

Figure 6.25 Protein folding cycle of GroEL. The substrate is bound to the apical domains in the *trans* ring of the GroEL–GroES complex. ATP binding subsequently primes the *trans* ring for the conformational transition to the closed state. During this stage, some substrate remodeling probably takes place. Binding of the GroES heptamer closes off the folding chamber and displaces the substrate into its cavity. After ATP hydrolysis in the *cis* ring, ATP binding to the *trans* ring triggers GroES and substrate release. (Adapted from Y.E. Kim et al., *Annu. Rev. Biochem.* 82:323–355, 2013.)

binding of a GroES heptamer (**Figure 6.25**). ATP hydrolysis is cooperative within rings; in other words, the hydrolysis of ATP in one subunit triggers ATP hydrolysis in neighboring subunits. Cooperativity between the two rings is negative: only one ring is active at a time. In the absence of bound substrate, GroES is an inhibitor of ATP hydrolysis, preventing non-productive conformational cycling. The binding sites of GroES on the apical domain of GroEL overlap with the substrate-binding sites. As a consequence, the binding of GroES induces the displacement of substrate into the now much expanded inner cavity, which can accommodate proteins of up to 60 kDa. This is sufficient for most genuine substrates of GroEL (see Figure 6.25).

Upon GroES binding, the ATP in the seven GroES-interacting subunits of GroEL (the 'cis ring') is hydrolyzed, and for a period of ~10 seconds the enclosed substrate has a chance to fold in the protected environment of the *cis* cavity, sometimes referred to as the **folding cage** or 'Anfinsen cage.' When ATP hydrolysis in the *cis* ring is completed, the opposing *trans* ring becomes primed for a new cycle of nucleotide and substrate binding. ATP binding to the *trans* ring induces a negative allosteric signal, causing the release of the ligands (ADP and GroES) from the *cis* ring and the opening up of the GroEL cavity, allowing the substrate to leave. If folding is still incomplete, it may rebind, either to the same or another GroEL, reiterating the folding cycles until folding is successful.

The crystal structure of the complex in its GroES-bound form shows that the inner walls of the folding cage are polar in character and that the hydrophobic (substrate-binding) patches of the apical GroEL domains become shielded upon GroES binding (**Figure 6.26**). Thus in the presence of enough water, which can enter the cage freely, the substrate protein is effectively isolated in solution. In fact, it has been shown that the model substrate rhodanese (~30 kDa) folds inside the cage with the same speed and efficiency as it does in dilute solution; that is, under artificial conditions that avoid aggregation.

Indeed, the physical environment of the cage may promote folding, and it has been shown for some proteins that folding inside GroEL–GroES can be faster than spontaneous folding under otherwise favorable conditions in bulk solution. Confinement in the cage may significantly reduce the number of accessible conformational states. Moreover, it has been suggested that the size of the cavities can be modulated to accommodate substrates of different sizes by means of flexible Gly-Gly-Met repeats at the protein C-terminal ends which extend into them.

Proteins in the 30–60 kDa range and with more complex domain topologies are especially prone to an accumulation of kinetically trapped intermediates. Local unfolding upon binding of substrates to multiple GroEL subunits and their motions during the ATP-driven cycle may destabilize misfolded intermediates and thereby provide the activation energy needed to proceed toward the native state.

The thermosome is the archetype of group II chaperonins

Thermosomes were first discovered in the hyperthermophilic archaeon *Pyrodictium occultum*, but they are constitutively expressed and very abundant in all archaea. Their expression levels are strongly increased after heat shock, and this increase correlates with the acquisition of thermotolerance. In *P. occultum* cells, which grow optimally at 105°C, thermosomes represent 6% of the total soluble proteins at 90°C, 11% at 100°C and 73% at 108°C as judged

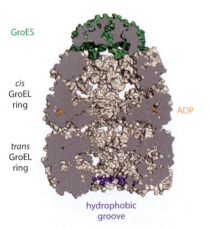

Figure 6.26 Cutaway view of the GroEL–GroES complex. Surface representation of the GroEL–GroES complex. The *trans* ring has a collapsed conformation, exposing a hydrophobic groove at the inner rim of the apical pore (purple). The *cis* ring and the GroES heptamer enclose a hydrophilic cavity for protein folding. The hydrophobic contact residues become buried at the interface between GroEL and GroES. (PDB 1PF9)

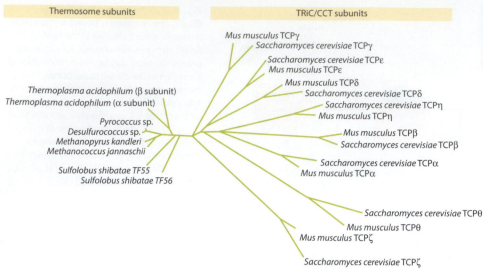

Figure 6.27 Dendrogram of group II chaperonins from archaea and eukarya. (From M. Nitsch et al., *J. Mol. Biol.* 267:142–149, 1997. With permission from Elsevier.)

from SDS-PAGE. Some thermosomes have a single type of subunit, most are built of two closely related subunits, and one thermosome has been reported to be composed of three types of subunit. Thermosomes with two kinds of subunit have 4-fold symmetry, indicating that the two subunits alternate around the ring. Occasionally, thermosomes with apparent 9-fold symmetry have been observed in preparations from one species, and it appears that some degree of polymorphism is characteristic for thermosomes even though the subset of particles deviating from the dominant symmetry is usually minor.

Sequence conservation among thermosome subunits is very high (50–70% identity), and also the subunits of the TRiC/CCT (tailless complex protein 1 ring complex/chaperonin-containing) chaperonin of eukaryotes can be aligned with them over essentially their entire length. In contrast, thermosome and TRiC/CCT subunits are similar to GroEL-like proteins only in their equatorial (N- and C-terminal) domains. Accordingly, the thermosome/TRiC/CCT family has been named 'group II chaperonins.' In a dendrogram of the group II chaperonins. the archaeal proteins form a single branch that is shorter than the eight branches into which the eukaryotic subunits have diverged (**Figure 6.27**).

Group II chaperonins undergo large-scale conformational changes

The conformational cycle of group II chaperonins has been studied in detail by X-ray crystallography and electron microscopy. Most high-resolution information was obtained with homomeric thermosomes (**Figure 6.28**), while structural studies of TRiC/CCT were complicated by their having eight distinct but very similar subunits (see below).

In thermosomes, the two rings seem to change their conformations independently, whereas TRiC/CCT displays negative inter-ring *cooperativity*, favoring asymmetric bullet-shaped complexes with one hemisphere open and the other closed. In the absence of nucleotide, the subunits assume an extended conformation in which the apical domains of neighboring subunits are not in contact and point outward (see Figure 6.28), giving each ring the appearance of an open bowl. Overall, this open state seems to be very dynamic, permitting substantial reorientations of the apical domains. Binding of ATP primes the complex for closure, inducing a ~45° counterclockwise rotation and a slight inward movement of the apical domains. Upon ATP hydrolysis, the equatorial domains perform a large inward rocking motion, and the cavity becomes closed by a further counterclockwise rotation of the apical domains, bringing the α-helical protrusions into close contact near the apex of the complex. In the closed state, each ring forms a chamber large enough to accommodate a protein of 60–70 kDa.

Information about the natural folding substrates of archaeal thermosomes is scarce, and so far the folding cycle has been studied with non-physiological model substrates such as rhodanese or green fluorescent protein. Tight binding of denatured protein occurs only in the open state and, as the cycle progresses, the substrate becomes transiently encapsulated. Thermosome mutants unable to attain the closed conformation have no folding activity. During closure and encapsulation, the high-affinity binding sites of the open state become

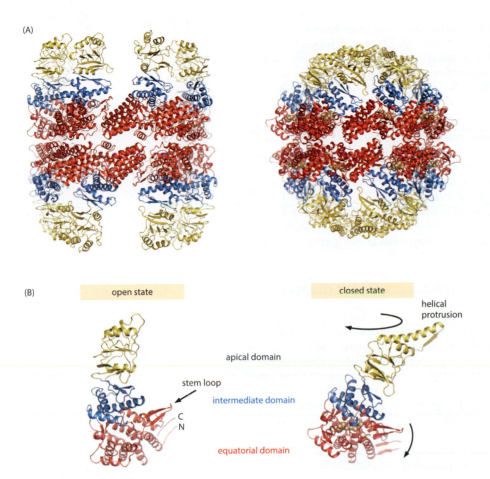

Figure 6.28 Structure of the archaeal thermosome. (A) Crystal structures of the open and closed conformations of the thermosome from *Methanococcus maripaludis*. (PDB 3KFK and 3KFB) (B) Structural rearrangements between the open and closed conformations within one subunit.

inaccessible to the substrate. The surface of the inner walls of the chamber is largely polar in character and unlikely to interact strongly with the substrate. A lidless mutant cannot retain the substrate in the cavity and fails to support folding. As with the group I chaperonins, a secluded volume shielding the substrate from unproductive interactions is required for folding activity.

The chaperonin TRiC/CCT of eukaryotes is built from eight distinct subunits

Given the sequence similarity of the TRiC/CCT subunits, it was a formidable challenge to determine the order of the subunits in the rings. The correct subunit positions were eventually established by chemical cross-linking in conjunction with mass spectrometry for yeast, bovine, and human TRiC/CCT (**Figure 6.29A**). Across the equator, there are two

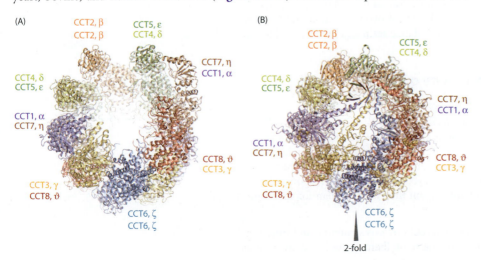

Figure 6.29 Structure of the eukaryotic chaperonin TRiC/CCT. (A) Structural model of the open conformation derived from a cryo-EM map of bovine TRiC/CCT in the absence of nucleotide. A tentative subunit assignment is shown. The apical domains are highly mobile and partly disordered in the structure. (PDB 4A0O) (B) Crystal structure of yeast (*Saccharomyces cerevisiae*) TRiC/CCT in the closed conformation. Note the 2-fold rotational axis of the otherwise asymmetrical arrangement. CCT6–CCT6 is a homotypic inter-ring contact. To fulfill its function, ATP hydrolysis activity is only required in the contiguous subunits CCT1, CCT4, CCT2, and CCT5. (PDB 4D8Q)

homotypic contacts (between subunits of the same kind) and three heterotypic contacts (between different subunits) (Figure 6.29B). The network of contacts between the different equatorial domains was revealed in the crystal structure of the closed conformation of the yeast complex. Only the N-terminal segment of subunit CCT4 traverses the wall of the complex between the equatorial domains, stiffening the complex at one end, whereas all others contribute to the β sheets at the bottom of the cavity. A stabilizing role has therefore been ascribed to CCT4.

The ordering of subunits in the heterooctamer has implications for the binding of substrates. Certain substrates could bind to two subunits across the ring. Indeed, the surface charges are largely segregated on the inner surface, resulting in two oppositely charged halves of the chamber. It is noteworthy that yeast TRiC/CCT is only sensitive to P-loop mutations that abolish ATP hydrolysis in four of the eight subunits, CCT1, CCT4, CCT2 and CCT5, and these are the four with the highest affinity for ATP. These four subunits are situated adjacent to each other in the ring and can close the chamber asymmetrically, as shown by crystallographic and EM studies. Whether this occurs in a sequential or in a 'power-stroke' mode is not yet clear.

Proteins with complex fold topologies are dominant clients of TRiC/CCT

For some time, TRiC/CCT was thought to be required primarily for the folding of the abundant cytoskeletal proteins actin and tubulin. However, proteomic studies have shown that ~10–15% of all newly made proteins visit the complex. Particularly abundant among them are proteins with a high propensity for β-sheet structures and complex fold topologies, which are liable to form long-lived folding intermediates. Many of the putative clients are components of large assemblies such as the *septin* rings involved in cell division, the APC/cyclosome (Section 12.5), and chromatin remodeling complexes (Section 2.4). Multidomain proteins constitute a large proportion of the clients, many of them too large to be accommodated in the cavity. It has been proposed that the asymmetry of the rings and the iris-like closing mechanism may allow partial closure and encapsulation of client proteins, leaving the remainder of the protein outside.

TRiC/CCT has been shown to bind and sequester amyloidogenic intermediates of polyglutamine protein aggregation *in vivo* (see Section 6.5). It has been proposed that they bind to the apical domains but, unlike GroEL, there are no obvious hydrophobic binding clefts. Some apical domains nevertheless seem to recognize hydrophobic peptide motifs, such as in von Hippel–Lindau protein, whereas others recognize folding intermediates, such as α helices of the amyloidogenic huntingtin protein (see Section 6.5). Electron density ascribed to bound tubulin was found close to the equator in the crystal structure of the open conformation of TRiC/CCT, and this was confirmed by chemical cross-linking. This region is accessible in group II chaperonins, in contrast to the respective region in the *trans* ring of GroEL.

Although TRiC/CCT has been shown to interact with nascent polypeptides, it is assumed that two other chaperone systems act upstream of the cytosolic chaperonin in the *de novo* folding of proteins, namely Hsp70 and prefoldin/GimC. Prefoldin is a starfish-shaped complex of six distinct subunits in eukaryotes or six subunits of two kinds in an α2β4 heterohexamer in archaea (**Figure 6.30**) that is thought to stabilize actin and tubulin until folding by TRiC/CCT commences.

Hsp90 chaperones regulate the activity of multifarious client proteins

Hsp90 is a very abundant chaperone in the cytosols of bacteria and eukaryotic cells, but it is absent from archaea. Unlike chaperones interacting promiscuously with nascent or unfolded polypeptide chains, Hsp90 interacts with a limited array of partly folded client proteins. Among these are many proteins involved in signal transduction (Chapter 12), including many kinases and E3 ligases, proteins involved in the immune response, or steroid receptors. In addition, several viral proteins rely on the Hsp90 system for maturation and activation.

Hsp90 clients belong to diverse protein families without obvious structural and functional similarities. Nevertheless, a feature common to many of them is the need to undergo

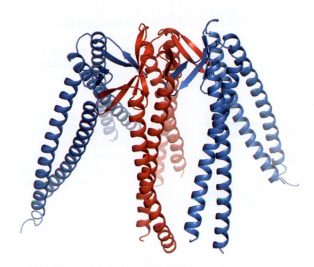

Figure 6.30 Structure of archaeal prefoldin from *Methanothermobacter thermautotrophicus*. The two α and four β subunits are indicated in red and blue, respectively. Both termini are at the tip of the tentacle. (PDB 1FXK)

conformational changes in order to reach their functional states. It is assumed that members of the bacterial Hsp90 system by themselves are able to fulfill their role in regulating the activity of their substrates, whereas eukaryotic Hsp90s rely on a multitude (more than 20) of co-chaperones for the recruitment of substrates and the remodeling of chaperone–substrate complexes during the functional cycle.

Hsp90 undergoes large-scale conformational changes during its reaction cycle with the help of its co-chaperones

Structurally, Hsp90 belongs to the gyrase, histidine kinase, and MutL superfamily of ATPases (Chapters 3 and 4). Crystal or EM structures are available for homodimeric Hsp90 from bacteria and eukarya in their open and closed conformations and in complex with various co-chaperones. Three structures reflect the high degree of flexibility that allows Hsp90 to interact with diverse client proteins. The Hsp90 monomer comprises three domains: the highly conserved N-terminal domain with the ATP-binding site, the middle domain, and the C-terminal domain (**Figure 6.31**).

Inhibition of ATP binding and hydrolysis by mutagenesis or inhibitors such as geldanamycin renders Hsp90 nonfunctional. The middle domain composed of α-β-α modules is required for substrate binding and the regulation of ATP hydrolysis. The peptide motif -Met-Glu-Glu-Val-Asp at the C-terminus is needed for tethering TPR-clamp domain co-chaperones such as the Hsp70–Hsp90 adaptor protein, HOP. HOP is needed for the transfer of nuclear receptors to Hsp90. For most kinase substrates, the specialized co-chaperone Cdc37, which binds to the N-terminal and middle domains of Hsp90, is required.

The Hsp90 dimer undergoes an ATP-regulated reaction cycle with large-scale conformational changes. In the open V-shaped form, it receives client proteins and then converts

(A)

(B)

nucleotide-binding domain   middle domain   C-terminal domain

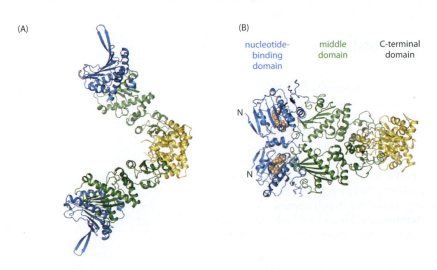

N

N

Figure 6.31 Structure of Hsp90. (A) Crystal structure of *E. coli* Hsp90 in the nucleotide-free open state. Two copies of Hsp90 interact via their C-terminal domains (yellow). In this, the apo state, the two arms of the Hsp90 dimer can sample a wide array of conformations. (PDB 2IOQ) (B) The closed state of Hsp90. In this crystal structure of yeast (*S. cerevisiae*) Hsp90 in complex with the ATP analog AMP-PNP and the co-chaperone p23 (omitted), the N-terminal nucleotide-binding domains bind tightly together, swapping their N-terminal strands between subunits. (PDB 2CG9)

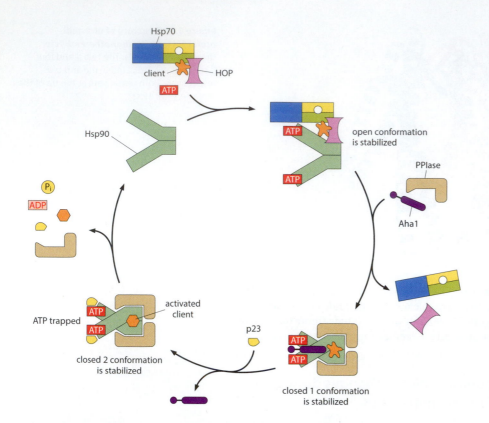

Figure 6.32 Folding cycle of eukaryotic Hsp90. Client (substrate) proteins are recruited from the upstream folding machinery via adaptor proteins such as the TPR-domain protein HOP, which links Hsp70 and Hsp90. Alternatively, Cdc37 mediates the transfer of numerous kinases to Hsp90. There are also several other specialized recruitment factors. Both HOP and Cdc37 inhibit ATP hydrolysis by Hsp90. In the next step, the recruitment factors are displaced by Aha1 and TPR-domain-containing peptidyl-prolyl isomerases (PPIase). Aha1 promotes association of the N-terminal nucleotide-binding domains of Hsp90. Specialized PPIases are required for the maturation of specific client proteins. Binding of p23 stabilizes the Hsp90–client complex. In the case of steroid receptors, additional binding of the steroid hormone is required for activation. Finally, ATP is hydrolyzed and the client is released. (Adapted from A. Röhl et al., *Trends Biochem. Sci.* 38:253–262, 2013. With permission from Elsevier.)

to the closed ('molecular clamp') form (**Figure 6.32**). The reaction cycle is driven by ATP binding, ATP hydrolysis, interactions with multiple co-chaperones, and post-translational modifications.

In the absence of co-chaperones, Hsp90 is enzymatically virtually inactive. For ATP hydrolysis, the N-terminal domains of Hsp90 have to pair up tightly, leading to a compact conformation in which the two subunits twist around each other. N-terminal pairing is promoted by the Aha1 co-chaperone. This 'tense' conformational state of Hsp90–substrate complex can be prolonged by binding of the inhibitory co-chaperone p23. As model client proteins, the nuclear receptors require their cognate hormone to be bound during this stage for activation. Finally, ATP hydrolysis triggers disassembly of the complex and release of the substrate.

If activation of a Hsp90 client fails, the protein is delivered to the 26S proteasome (Section 7.4) by the Hsp70–Hsp90 machinery for degradation. Together they constitute the major protein quality control system of the eukaryotic cell. The *proteostasis* capacity of cells is presumably regulated by the expression levels of co-chaperones with partly overlapping functions (such as ~40 Hsp40s and 11 Hsp70-NEFs in humans) and extensive post-translational modifications such as the phosphorylation and acetylation of Hsp70 and Hsp90.

Small heat shock proteins protect client proteins from aggregation

Small heat shock proteins (**sHsps**) were first discovered as proteins of 15–30 kDa accumulating in cells exposed to environmental or pathological stress. They are found in all three kingdoms of life. Many sHsps are expressed ubiquitously, but some are expressed only in specific tissues. They interact with partly unfolded or aggregation-prone proteins and thereby delay their aggregation. This 'holdase' function is of critical importance for cellular viability under stress conditions; impairment of the sHsp system is implicated in many diseases, including Alzheimer's and Alexander diseases. For an sHsp to become functional, it has to assemble into very large homomeric or heteromeric complexes of up to 50 subunits. This is a dynamic process with rapid subunit exchanges resulting in polydisperse structures, and this property poses a challenge to structure determination. Hybrid methods combining data from a variety of techniques are just beginning to provide more detailed insight into the structures of the functional complexes.

Figure 6.33 Structure of the αB crystallin ACD dimer. The molecule shown has the C-terminal IXI/V motif bound. The regions that define the dimer interface are in blue, and the β4–β8 grooves in inter-dimer interactions with the IXI/V motif are in red. (PDB 2KLR)

α-Crystallin domains form the core of all sHsps

All sHsps share a conserved central domain named the α-crystallin domain (ACD), flanked by a highly variable N-terminal region and a variable C-terminal region (**Figure 6.33**). αB crystallin is the predominant protein of the vertebrate eye lens and is a paradigm for sHSPs in general. The eye lens is composed mostly of post-mitotic cells lacking the machinery for protein synthesis and degradation and unable to turn over during their entire life span. And yet the eye lens must remain transparent, which requires an efficient system for preventing protein aggregation. ACDs are conserved from bacteria to humans. They all have six- to eight-stranded β-sandwich structures, forming homodimers via *domain swapping*. The divergent N-terminal regions seem to be responsible for oligomer assembly, because their removal results in the formation of much smaller complexes. The flexible C-terminal regions vary in length and sequence, with the exception of a three-residue motif of Ile or Val (the IXI/V motif). This motif has been shown to bind in the groove of the ACD and this interaction is intermolecular; that is, the IXI/V of one dimer binds to the ACD of another dimer.

6.4   NATIVELY UNFOLDED PROTEINS

In recent years, it has been recognized that many proteins have extended tracts of amino acids—in some cases, the entire polypeptide chain—that under physiological conditions (near-neutral pH, moderate ionic strength) have no specific structure. These tracts are said to be **natively unfolded** (or natively disordered, or intrinsically disordered, or intrinsically unstructured). Regions of this kind are invisible in crystal structures because their lack of order precludes coherent diffraction. In fact, in almost all known crystal structures of proteins, there are a few segments, usually only a few residues long and located in turns or at termini, that are not visualized. Natively unfolded regions differ from this trend primarily in their size, often exceeding 50 residues and in extreme cases extending to more than 1000 residues.

Natively unfolded regions may be recognized by multiple experimental approaches

How are regions of this kind to be recognized? NMR spectroscopy (see **Methods**) provides the most powerful technique because it can readily identify amino acids of high mobility, as is the case in disordered regions. Ultracentrifugation analyses tend to show unfolded proteins as physically larger, in terms of hydrodynamic radius, than would be expected for folded proteins of the same molecular weight. Similar outcomes of anomalous apparent size are recorded with other biophysical approaches such as dynamic light scattering or small-angle X-ray scattering (SAXS). Optical spectroscopy (circular dichroism) and EPR (electron paramagnetic resonance) spectroscopy can detect the absence of regular secondary structure (α helix and β sheet) and the predominance of random coil. If scanning calorimetry does not detect a phase transition at a characteristic temperature that would correspond to thermal denaturation, this serves as a strong indication that the protein is already unfolded. High sensitivity to proteolysis is another marker for a natively unfolded molecule: all of its peptide bonds are potentially exposed to proteases, not just those presented in surface loops.

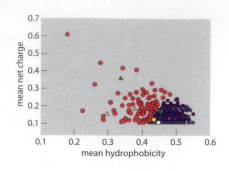

Figure 6.34 Folded and natively unfolded proteins segregate on a charge/hydrophobicity plot. The mean hydrophobicity of each protein was obtained as the sum of the hydrophobicities of all its residues divided by the total number of residues, using the numerical values for hydrophobicity assigned by Kyte J & Doolittle RF (*J. Mol. Biol.* 157:105–132, 1982), normalized to the range (0, 1). The mean net charge is the total number of charged residues (Asp, Glu, Lys, Arg) divided by the total number of residues. Totals of 275 folded proteins (blue squares) and 91 unfolded proteins (red disks) are plotted. Three proteins, α-synuclein (green disk), negative factor (yellow disk), and helix-destabilizing protein (white disk), map into the folded zone but when their natively unfolded portions are plotted separately (as triangles of the same colors), they map well into the unfolded zone. (From V. Uversky et al., *Proteins* 41:415–27, 2000. By permission of John Wiley and Sons.)

Natively unfolded proteins have distinctive amino acid compositions

In many cases, a propensity to be natively unfolded may be inferred from an appraisal of the amino acid sequence. Diagnostic sequence characteristics include:

- Low complexity (that is, sequences in which a few amino acids are represented in abnormally high levels, for example Pro-rich, Gln-rich, Gly-rich).

- Multiple repeats of a short peptide sequence, for example Pro-Glu-Val-Lys in titin (Section 14.4), although some such repeats assemble into well-defined folds, as with the canonical collagen tripeptide Gly-Pro-X (Section 11.7).

- A high content of uncharged hydrophilic amino acid residues (Asn, Gln, Ser, Thr); in some cases, also with many charged residues (Lys, Arg, Glu, Asp).

- Amino acids with large hydrophobic side chains (Phe, Trp, Tyr, Met, Ile, Leu, Val) are under-represented. As a rule of thumb, a sequence that is likely to fold into a globular domain has a minimum of ~30% of hydrophobic residues that pack together to form the water-excluding core (Section 1.3). If proteins are assigned two coordinates—net hydrophobicity and mean net charge—and plotted in two dimensions, the folded and unfolded classes partition into distinct regions (**Figure 6.34**).

Criteria based on these considerations as well as empirical ones relating to sequences known not to fold are employed in computer programs designed to identify natively unfolded regions. An example of an output from such an algorithm is shown in **Figure 6.35**. The best have been benchmarked as achieving ~75% accuracy.

These sequence-scanning algorithms have indicated that proteins with substantial unfolded tracts are common in eukaryotes (~25%) but much rarer in bacteria and archaea. In eukaryotes, they occur in many different kinds of protein but are particularly prevalent in the cytoplasmic domains of signaling proteins and in proteins involved in transcription. The estimated incidence of disordered structure in all proteins encoded by the human genome is given in **Figure 6.36**. Some examples described in other sections of this book are listed in **Table 6.3**.

Natively unfolded regions are often involved in regulation, folding when they engage interaction partners

Many natively unfolded polypeptides assume defined conformations, at least for part of their length, when they encounter an interaction partner, which can be another protein or a nucleic acid molecule. These conformations are often extended, as the folding chain segment becomes draped over the surface of the partner molecule. Such is the case in the system shown in **Figure 6.37**, in which a segment of polypeptide chain that is disordered when the protein is free in solution forms two orthogonal α helices upon binding to its partner. This interaction is induced by phosphorylation of a Ser in a helix-forming sequence.

Several motifs appropriate for different interaction partners may coexist and even overlap within a given unfolded region, making this an economical and versatile regulatory device. Post-translational modifications within such a region can steer the interactions toward particular partners, according to differing regulatory requirements at, for instance, different

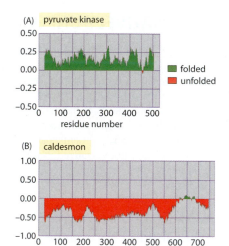

Figure 6.35 Prediction of natively unfolded regions in proteins. Shown are examples of profiles calculated for two proteins by the Fold-Index algorithm, using a 51-residue window; thus, the prediction for each amino acid also takes into consideration 25 on either side. Regions predicted to be folded have positive abscissae (green) and those predicted to be unfolded have negative abscissae (red). In these two examples, one protein (A) is predicted to be entirely folded, and the other (B) entirely unfolded. (From J. Prilusky et al., *Bioinformatics* 21:3435–3438, 2005.)

stages of the cell cycle. A well-known example is the modifications of the disordered N-terminal tails of histone proteins in the nucleosome particles of chromatin, giving rise to the so-called histone code (Chapter 2).

A large percentage of post-translational regulatory modifications (phosphorylation, acetylation, ubiquitylation, methylation) affect residues in natively unfolded regions, which are readily accessible to the modifying enzymes. This situation is well illustrated by the p53 tumor suppressor protein, which has a central folded DNA-binding domain (see Figure

Table 6.3 Some natively unfolded domains[a], their roles and structural transitions.

Natively unfolded region	Function/transition	Reference
Histone tails	NTDs of core histones; regulate chromatin structure, gene expression, etc.	Section 2.1
	CTDs of linker histones	
Factors UvrA, UvrB	Nucleotide excision-repair; unfolded → full tertiary structure	Section 4.3
RNA polymerase, Rbp1	C-terminal repeat region	Section 5.2
Transcription factor TFIIB	C-terminal 'box fold' domain	Section 5.3
Ribosomal proteins	Fold upon assembling onto the ribosomal particle	Section 6.2
Prion domains	Unfolded → cross-β (amyloid) in prion assembly	Section 6.6
Pyruvate dehydrogenase	Inter-domain linkers	Section 9.3
Clathrin light chain	Endocytic protein coat: unfolded → α helix on binding triskelion	Section 10.2
Nucleoporin FG regions	Steric filter; remain unfolded	Section 10.5
Intermediate filament end-domains	Interact with cytoplasmic proteins	Section 11.2
Gly/Ser-rich domains	In keratins and RNA-binding proteins	Section 11.2
Tight junction claudins and occludin cytoplasmic domains	Fold on binding cytoplasmic partners	Section 11.3
β-Catenin	Engages cadherin cytoplasmic domains in adherens junctions	Section 11.4
Gap junction connexin	Unfolded → partly folded cytoplasmic domains	Section 11.5
Focal adhesion adaptors	Couple focal adhesion to actin cytoskeleton	Section 11.6
Talin	In focal adhesions	Section 11.6
Elastin	Coating fibrillin microfibrils	Section 11.7
Pilin N-term extensions	Polymerization of type I pili; unfolded → β strand	Section 11.8
Protein kinase substrates	Signaling; unfolded regions are phosphorylated	Section 12.2
TGFβ receptor RII	NTD; complex assembly and signal transduction	Section 12.4
Tau	Regulates microtubule stability; unfolded → extended	Section 14.6
Flagellin D1	Unfolded → α helix; forms coiled coils in flagellar backbone	Section 14.11

[a]Most of these unfolded regions are parts of proteins that also have folded domains. When these regions fold, only part of them assumes a definite conformation.

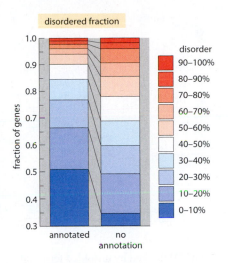

Figure 6.36 The distribution of proteins coded in the human genome according to increasing amounts of disorder. An amino acid residue is designated as disordered when there is a consensus between at least 75% of the predictors in the D2 P2 database for disordering at that position in the sequence. The predictions are shown for genes with and without annotation. (Adapted from R. Van der Lee et al., *Chem. Rev.* 114:6589–6631, 2014.)

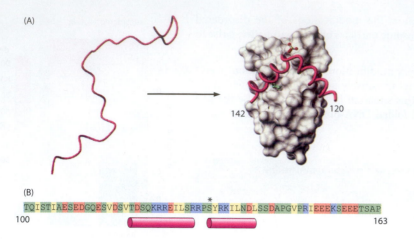

(A)

(B)

TQISTIAESEDGQESVDSVTDSQKRREILSRRPSYRKILNDLSSDAPGVPRIEEEKSEEETSAP

100   163

Figure 6.37 Binding-coupled folding. In this process, a natively unfolded polypeptide folds into an ordered structure as it binds to its partner. (A) In this example, the phosphorylated kinase-inducible domain (pKID) of the transcription factor cyclic AMP-response-element-binding protein (CREB) folds on forming a complex with the KID-binding (KIX) domain of CREB-binding protein. (B) The primary sequence of the KIX-binding region of CREB. Green, small amino acid residues, uncharged hydrophilic residues, and Pro; yellow, hydrophobic residues; red, acidic residues; blue, basic residues. Two α helices in pKID form on its binding to KIX in response to phosphorylation of Ser130. They are indexed in pink below the sequence. (From H.J. Dyson and P.E. Wright, *Nature Rev. Mol. Cell Biol.* 6:197–208, 2005. With permission from Macmillan Publishers Ltd.)

6.38C,D) flanked by unfolded terminal regions. The disorder of these regions was quite well detected by a predictive algorithm (**Figure 6.38A,B**). Almost all of the many modifications of p53 affect residues in the terminal regions. Moreover, a segment of the C-terminal region switches into an α helix that combines with the corresponding segments from three other molecules (Figure 6.38E). The binding partners are also unfolded before the interaction and thus do not present a pre-folded template.

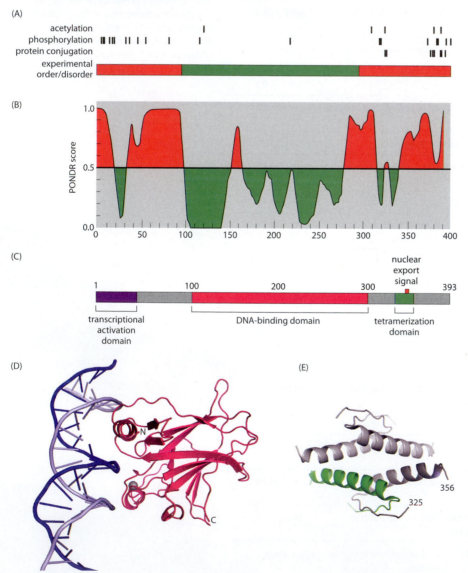

Figure 6.38 Natively unfolded regions in the tumor suppressor protein p53. (A) Below: map of unfolded (orange) and folded (green) regions of p53, determined experimentally. Above: map of post-translational modification sites. Almost all of them lie in the unfolded terminal regions. (B) Prediction profile of propensity to be unfolded (orange, scores > 0.5) or folded (green, scores < 0.5), according to algorithms in the PONDR package. The curve correlates quite well with the experimentally determined profile shown above. (C) Domain map as assigned by structural and functional criteria. (D) Structure of the DNA-binding domain (residues 102–292)–DNA complex, with α helices inserted in the major and minor grooves. The DNA-binding domain is stabilized by a Zn^{2+} (gray sphere). (PDB 1TSR) (E) A tetramer of the tetramerization domains (residues 1–44), which become ordered in the oligomer; one is colored green. The transcription activation domain remains unfolded. (PDB 1C26) (B, adapted from A.K. Dunker et al., *Curr. Opin. Struct. Biol.* 18:756–64, 2008. With permission from Elsevier.)

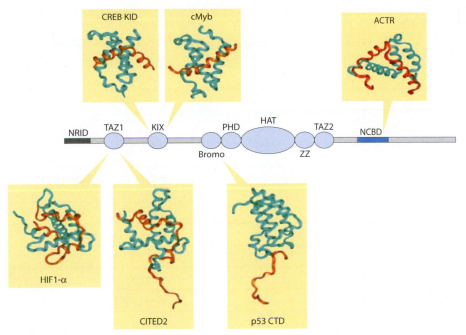

Figure 6.39 A protein with multiple globular domains connected by long disordered linkers. In this map of the transcriptional co-activator CREB-binding protein (p300), the domains are represented by bubbles of corresponding sizes and the linkers as bars of corresponding lengths. NRID, nuclear receptor-interacting domain; NCBD, nuclear receptor co-activator-binding domain. Structures have been determined for the globular domains of p300 (blue) with partly ordered transcription factors bound (red), whose folding was templated by binding to the corresponding p300 domain. TAZ1 domain with HIF1-α (PDB 1L8C) and CITED2 (PDB 1R8U); KIX domain with KID of CREB (PDB 1KDX) and cMyb (PDB 1SB0); Bromo domain with p53 CTD (PDB 1JSP). The NCBD domain, initially disordered, folds upon interacting with the ACTR domain of the p160 co-activator (PDB 1KBH). (From M. Fuxreiter et al., *Nature Chem. Biol.* 4:728–773, 2008. With permission from Macmillan Publishers Ltd.)

A common role for natively unfolded regions, particularly in eukaryotic regulatory proteins, is to serve as extended linkers between folded domains (**Figure 6.39**). This arrangement serves to keep functionally related domains in mutual proximity while affording degrees of freedom in their interaction with each other and with other factors. The timing of folding is also exploited in some systems for temporal regulation of assembly; that is, protein subunits being incorporated into larger structures may have domains that remain unfolded until a late stage of assembly. An example of this is flagellin, the principal subunit of bacterial flagella (Section 14.11).

6.5   PROTEIN MISFOLDING AND AMYLOID FIBRILS

The term **amyloid** was coined by the German physician Rudolph Virchow in the mid-nineteenth century to denote a class of abnormal deposits detected in human patients. These deposits are very stable, a property that facilitated their recovery from source tissues. Microscopy showed them to be fibrillar, and chemical analyses revealed that they are proteinaceous. Moreover, they are composed of cellular proteins—mostly fragments thereof—not parts of infectious agents such as bacteria or viruses. This crucial distinction was established when it proved possible to generate fibrils *in vitro* from proteolyzed IgG light chains that reproduced amyloid isolated from sera of patients suffering from myeloma, a cancer of the blood in which large amounts of a single abnormal antibody are produced.

It has emerged that each clinical amyloid is composed primarily of a single protein. Bearing in mind that human cells express thousands of proteins, the number that form clinically relevant amyloids is small, about 25. Some examples are listed in **Table 6.4**. However, many—perhaps most—other proteins may be converted into amyloids *in vitro*. The procedure typically involves unfolding by exposure to denaturing conditions such as low pH, followed by incubation under conditions that support fibrillation. In general, these conditions differ from those that prevail for *in vivo* folding, which involve near-neutral pH and physiological ionic strength (Section 6.3).

Table 6.4 Properties of some amyloid-forming proteins.

Protein	Disease	Precursor protein mass (kDa)	Amyloid fragment mass (kDa)	Precursor conformation
Aβ	Alzheimer's	110–135	4–5[a]	Unfolded
Tau	Alzheimer's	40–50	see [b]	Partly unfolded
α-Synuclein	Parkinson's	15	15	Unfolded[c]
Huntingtin with polyQ expansion	Huntington's	330	see [d]	Unfolded NTR (100 aa)[d]
Transthyretin (TTR)	Familial amyloidotic neuropathy, senile systemic amyloidosis	14	5–14	β-sheet fold (see Figure 6.44)
Ig light chain	Ig light-chain amyloidosis	25	10	β-sheet fold
β₂ Microglobulin	Hemodialysis-related amyloidosis	12	12	β-sheet fold, Ig-like
Amyloid A	Serum amyloid A, rheumatoid arthritis	12	5–10	α-helical fold
Amylin[e]	Type II diabetes	9	4	unfolded
Insulin	Injection-localized amyloidosis	3–4	3–4	α-helical fold
Lysozyme	Hereditary lysozyme amyloidosis	14	14	α/β fold
PrP (prion protein)	Spongiform encephalopathies[f]	30–35	27–30	NTR, unfolded. CTR, α-helical fold

[a] Two forms with 40 and 42 aa respectively.
[b] Tau in neurofibrillary tangles is often proteolytically degraded but not to a fragment of a unique size. It is susceptible to processing at multiple sites by the proteases calpain and caspase-3. See Section 14.6.
[c] Also has an α helical conformation when associated with membranes.
[d] PolyQ expansions take place within the first exon of huntingtin. A crystal structure of a fusion protein containing this exon revealed an N-terminal α helix, a poly17Q region, and a polyproline helix. The poly17Q region adopted multiple conformations, including α helix, random coil, and extended loop (MW Kim et al. [2009] *Structure* 17:1205–1212). There is immunochemical evidence that C-terminal parts of this large protein are removed from the molecules that enter intranuclear inclusions.
[e] Also called islet amyloid polypeptide (IAPP).
[f] Examples are scrapie (sheep), sporadic and 'variant' Creutzfeldt–Jakob disease (humans), kuru (humans), and BSE/'mad cow' disease (cattle).
NTR, N-terminal region (aa 1–120); CTR, C-terminal region (aa 121–253).

Amyloids share distinctive properties: fibril morphology, stability, dye-binding, and cross-β conformation

Although the precursor proteins vary, all amyloids share certain basic properties. They are fibrillar, insoluble, protease-resistant, and thermostable. Viewed by EM, the fibrils are straight, smooth-sided, and unbranched (**Figure 6.40**). Many have a tendency to bundle. Amyloid fibrils bind Congo Red and other dyes such as thioflavin T. After staining with Congo Red, a distinctive apple-green birefringence is observed (**Figure 6.41**).

(A) 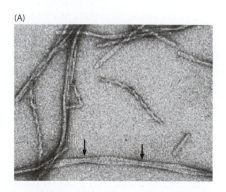 (B) 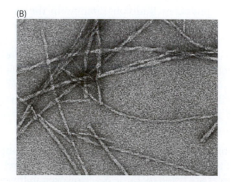 (C)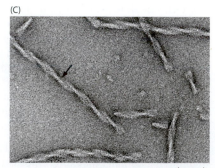

50 nm

Figure 6.40 Amyloid fibrils assembled *in vitro* from three different proteins. Electron micrographs are shown of negatively stained fibrils. (A) Human amylin, residues 8–37. (B) Aβ, residues 1–40. (C) A construct of the four repeats of human tau representing residues 244–372 (see also Figure 14.45B). These fibrils exemplify the 'paired helical filament' (PHF) morphology. (A, from C. Goldsbury et al., *J. Struct. Biol.* 130:352–362, 2000. With permission from Elsevier. B, from C. Goldsbury et al., *J. Struct. Biol.* 130:217–231, 2000. With permission from Elsevier. C, from S. Wegmann et al., *J. Biol. Chem.* 285:27302–27313, 2010. With permission from ASBMB.)

(A)

(B)

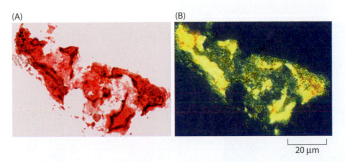

20 μm

Figure 6.41 Dye binding by amyloid fibrils. Light micrographs of aggregates of fibrils of the N-terminal 'prion domain' of the yeast protein Ure2p, stained with Congo Red. (A) Bright-field phase-contrast image. (B) Under polarized light, intense apple-green birefringence is seen, indicative of amyloid. (From K.L. Taylor et al., *Science* 283:1339–1343, 1999. With permission from AAAS.)

The most definitive evidence for an amyloid is **cross-β structure** detected by fiber diffraction with X-rays or electron diffraction (**Figure 6.42**). This term denotes a structure rich in β sheets whose strands run perpendicular to the fibril axis, with the H-bonds running parallel to the fibril axis. Thus oriented, they generate a strong meridional reflection at $(4.7 \text{ Å})^{-1}$, corresponding to the inter-strand spacing. Equatorial reflections are weaker and more variable, but one at $(9–10 \text{ Å})^{-1}$ is fairly common and relates to stacking of β sheets.

In globular proteins, the majority of β sheets are antiparallel (that is, adjacent strands run in opposite directions). However, the techniques of solid state NMR and EPR spectroscopy are capable of distinguishing between parallel and antiparallel β sheets, and most amyloids thus analyzed have been found to contain parallel β sheets (that is, with all strands running in the same direction).

Are amyloid fibrils polar structures, or are both ends the same? Time-resolved atomic force microscopy (AFM) (see **Methods**) has answered this question in several cases. For example, successive scans of amylin fibrils fixed to a substrate but still growing showed that the two ends of a given fibril grew at different rates in a manner that varied from fibril to fibril (**Figure 6.43**). These observations established that these fibrils are polar and that elongation is a stochastic process.

For most amyloidogenic proteins, their fibrillar conformation represents a misfolded state, markedly different from their native conformation (Section 1.3). However, there is a growing number of proteins for which amyloid is in fact the native state. These are called *functional amyloids*. Some examples are discussed at the end of this section.

Figure 6.42 Signature diffraction patterns for cross-β structure. (A) Schematic diagram of a fiber diffraction pattern typically obtained from bundles of aligned amyloid fibrils, and of an individual fibril with cross-β structure. (B) X-ray fiber diffraction patterns from amyloid fibrils of the yeast prion protein Sup35p-MN (see Section 6.6). Both the dried specimen (left) and the hydrated specimen (right) show the canonical meridional reflection at 4.7 Å (red arrows). In the dried sample, the fibrils are packed tightly together, giving coherent scattering that generates a broad equatorial reflection at 10 Å (blue arrow). In the wet specimen (right), the fibrils are less regularly packed and that reflection is not produced. (C) Left: electron diffraction pattern from a field of α-synuclein amyloid fibrils. The 4.7 Å reflection is marked with a red arrow. By suitably adjusting the EM lens currents, an image of the diffracting area is formed (right panel). The lateral packing of the fibrils in this field is not regular enough to diffract coherently; there is therefore no equatorial reflection. (A, adapted from R. Nelson and D. Eisenberg, *Adv. Prot. Chem.* 73:235–282, 2006. With permission from Elsevier. B, from A. Kishimoto et al., *Biochem. Biophys. Res. Comm.* 31:739–45, 2004. With permission from Elsevier. C, from L.C. Serpell et al., *Proc. Natl. Acad. Sci. USA* 97:4897–902, 2000. National Academy of Sciences, USA. Reprinted with permission.)

(A)

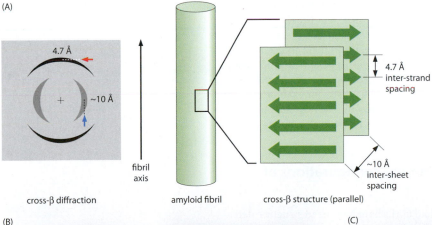

4.7 Å

~10 Å

fibril axis

4.7 Å inter-strand spacing

~10 Å inter-sheet spacing

cross-β diffraction   amyloid fibril   cross-β structure (parallel)

(B)

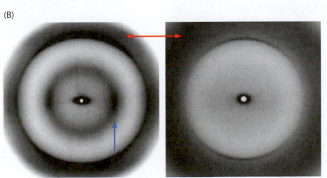

(C)

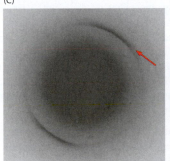

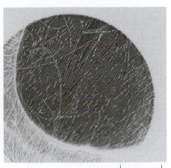

0.25 μm

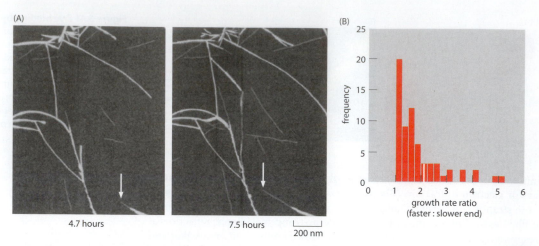

Figure 6.43 Asymmetric bidirectional growth of amylin protofibrils visualized by AFM. (A) Amyloid fibrils attached to a mica substrate and overlaid with a drop of amylin-containing buffer were scanned in an atomic force microscope. Pixel brightness corresponds to the height of the AFM probe tip above the substrate. Single protofibrils are present, recognizable by their 2.4 nm height (calibration not shown); they are the thinnest fibrils in this field. Thicker fibrils consisting of multiple protofibrils (the denser white structures) are also present. Scans of the same area at two time points are shown. (B) The protofibrils grow at both ends at different rates, indicating that they have a polar structure. (Adapted from C. Goldsbury et al., *J. Mol. Biol.* 285:33–39, 1999. With permission from Elsevier.)

Amyloid fibrils are polymorphic

The uniformity with which amyloids observe the properties listed above led to the idea that they all have essentially the same structure. This proposition, strikingly at odds with the diversity of folds exhibited by globular proteins, led to the idea that fibril structure might be dictated by interactions of the polypeptide chain backbone, which would not be sequence-specific. However, this scenario did not accord with the observation that proteins vary widely in their propensity for amyloid formation. It is now evident that although all amyloid fibrils share certain properties, they exhibit a variety of structures, not only among amyloids of different proteins but also among fibrils formed from a given protein.

At the basic level is the **protofibril**, which is a single strand, typically 4–8 nm in diameter. If a protein filament has an axial channel, even a very narrow one, it is usually detectable by negative staining EM: by this criterion, amyloid protofibrils tend to have solid cores. Protofibrils vary in the numbers of β sheets that they contain and in their connectivity. A fibril may consist of a single protofibril or of two or more protofibrils wound around a common axis or associating side by side in a ribbon. Viewed in projection by EM, many fibrils narrow and widen periodically with characteristic repeat distances (see, for example, Figure 6.40). These repeats relate to the twist of their β sheets. Two-protofibril arrangements, called **paired helical filaments**, exhibit the most extreme variations between the narrowest and the widest points (see, for example, Figure 6.40C).

Models of amyloid fibrils envisage differing configurations of β strands

The insolubility and unwieldy dimensions of amyloid fibrils have hampered structure determination by X-ray crystallography. An exception arises with proteins whose precursor conformations are rich in β sheets and which form amyloids in which their subunits retain much of their native conformation. However, they are perturbed in a way that allows interactions between subunits that give rise to polymerization; they are therefore called **gain-of-interaction amyloids**. Models may be built by adapting crystal structures of the precursor proteins. An example is shown in **Figure 6.44** for transthyretin, a transporter protein whose native structure is a homotetramer of 16 kDa subunits, but it accumulates as amyloid in at least two neuropathies (see Table 6.4).

For other amyloids, the fibrillar state is preceded by one in which the precursor protein is unfolded, either natively so or transiently as the protein converts into amyloid. For these fibrils, most current models are based on combining data from multiple sources. Specifically, X-ray fiber or electron diffraction can reveal the presence of cross-β structure (see

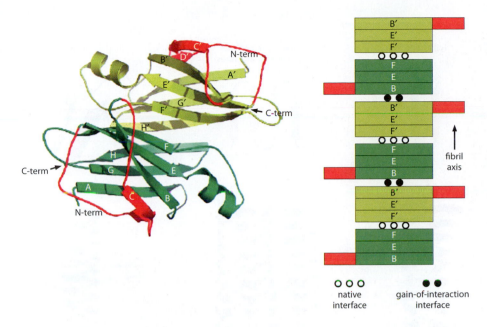

Figure 6.44 Gain-of-interaction model for transthyretin amyloid fibrils. Left: crystal structure of a dimer of human transthyretin. (PDB 2PAB) The red strands and loops are motifs proposed to change conformation so as to allow a fibril to form by direct stacking of these molecules. Right: diagram of the fibril model. Rectangles represent motifs colored as in the structure. Open circles represent the native dimer interface, maintained in the fibril model; filled circles represent the new interface, creating a continuous antiparallel β sheet in cross-β conformation. (From R. Nelson and D. Eisenberg, *Adv. Prot. Chem.* 73:235–282, 2006, and A.A. Serag et al., *Nat. Struct. Biol.* 9:734–739, 2002. With permission from Elsevier.)

Figure 6.42). EM can reveal the number of protofibrils per fibril, estimate their diameter, and measure crossover spacings. Mutational data on sites that affect fibrillogenesis and bioinformatic analysis can identify segments that are likely to form β strands. Solid-state NMR is now contributing increasingly detailed structural information. Scanning transmission EM (STEM) can provide measurements of mass-per-unit-length (MPL), which translate into the axial rise per subunit; in many cases, this matches the inter-strand spacing of 4.7 Å. This is illustrated in **Figure 6.45** for a set of proteins in which the amyloidogenic 'prion' domain of the yeast protein Ure2p was fused to globular domains of varying sizes. The filament MPLs are directly proportional to the subunit masses, consistent with a conserved fibrillar backbone of polymerized prion domains.

Three classes of protofibril model may be distinguished on the basis of the topology of their β sheets: **β arcades**, stacked **β serpentines** (also called **parallel superpleated β structures**),

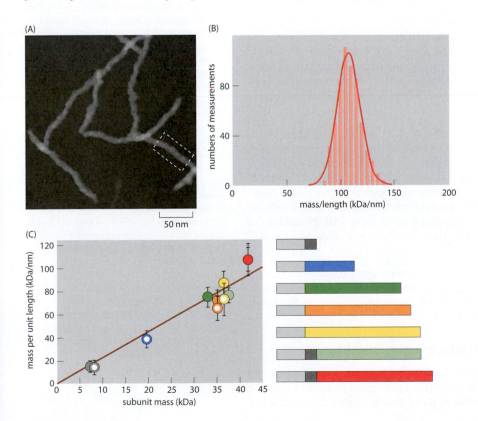

Figure 6.45 Mass-per-length measurements of amyloid filaments by STEM microscopy. (A) A STEM (scanning transmission electron microscope) dark-field image of unstained freeze-dried filaments of the Ure2p protein. To make a mass-per-length (MPL) measurement, the image density in a given segment (for example the dashed box) is integrated, and, after subtraction of the contribution from the support film, is proportional to the mass of the specimen in the box. (B) A unimodal distribution of measurements, fitted with a Gaussian curve. (C) Plot of mean MPL measurements from STEM analyses of a set of filaments assembled from fusion proteins in which globular domains of different sizes (colored blocks) were appended to the N-terminal 'prion domain' of Ure2p (gray). The filament MPLs are directly proportional to the subunit masses, with the slope of this line corresponding to one subunit per 4.7 Å axial step. (From U. Baxa et al., *Adv. Prot. Chem.* 73:126–180, 2006. With permission from Elsevier.)

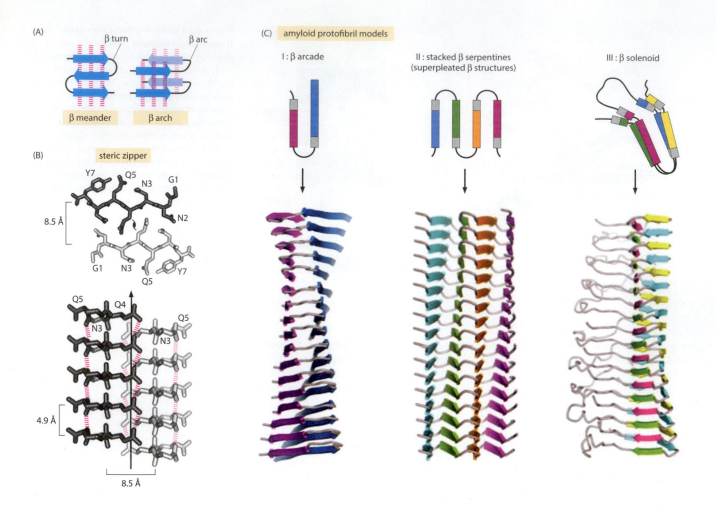

and **β solenoids** (see Figure 6.46C). The β arcade is a special case of the stacked β serpentine in which there are only two β strands per layer. All three models are explicitly cross-β structures but differ in how the β strands are arranged. β Solenoids are described below. In a β serpentine, the polypeptide chain zig-zags in a planar fold. This differs from the *β-meander* motif found in globular proteins (**Figure 6.46A**) in that the strands are rotated by 90° so that the backbone interactions that define a sheet are between planes—that is, along the fibril axis—while the interactions between sheets are perpendicular to the fibril axis and involve the side chains.

Crystal structures have been determined for a number of short peptides (typically, five to seven residues), whose sequences were identified as amyloidogenic by bioinformatic analysis. They show a **steric zipper** interaction between β sheets in which side chains are intercalated so closely that water is excluded, giving a 'dry' interface (Figure 6.46B). Interactions of this kind could stabilize the packing of β sheets in the models shown in Figure 6.46C. However, the packing of sheets may also involve other kinds of interaction, such as hydrophobic ones. Moreover, natural amyloid-forming polypeptides tend to be at least 30 residues in length, which is too long for an unbroken β strand; they should therefore form turns as well as strands.

Advances in solid state NMR spectroscopy have produced two experimentally determined protofibril structures. They depict, respectively, paired β arcades for the Aβ peptide associated with Alzheimer's disease, and a β solenoid with two turns per subunit for the fungal protein HET-s, as shown schematically in Figure 6.46C.

The native folds of β-solenoid proteins are amyloid-like

An extensive class of secreted bacterial proteins called adhesion factors have sequences containing tandem pseudo-repeats, ~19–30 aa in length. Each repeat forms one coil of a solenoid, consisting of short β strands connected by turns. Matching strands in successive

Figure 6.46 Packing of β strands in crystals and protofibril models. (A) In a β-meander fold, the main-chain bonds between β strands that stabilize an antiparallel β sheet (red dashes) lie in the plane of the sheet. In a β arch, the strands are rotated relatively about their axes by 90°, giving rise to two oppositely directed parallel β sheets. (B) Mating of β sheets in a 'cross-β spine' arrangement observed in a crystal structure of the heptapeptide GNNQQNY. (PDB 1YJP) Top: view along the proposed fibril axis, with the dyad axis marked. Bottom: side view. The Asn and Gln side chains in the space between the two sheets interdigitate to form a 'steric zipper.' The two sheets are offset axially by half of a 4.7 Å step. (C) Three classes of fibril models distinguished by the connectivity and arrangements of their parallel β sheets. Structures for fibrils of types I (Aβ) and III (HET-s) have been determined experimentally. (B, from R. Nelson and D. Eisenberg, *Adv. Prot. Chem.* 73:235–282, 2006. With permission from Elsevier. C, adapted from A.V. Kajava et al., *FASEB J.* 24:1311–1319, 2010. With permission from FASEB.)

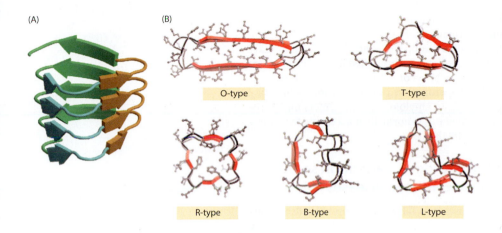

Figure 6.47 β-solenoid folds. (A) Archetypal β solenoid with three strands per coil, forming three parallel β sheets (blue, green, beige). (B) Cross sections of five classes of β solenoids with differing configurations of strands and turns, observed in crystal structures. In each case, two coils are viewed along the solenoid axis. The β strands are shown as red segments with the side chains in ball-and-stick representation and colored gray. O-type is from a protein stabilizer of an iron transporter (PDB 1VH4); T-type is from filamentous hemagglutinin, a 50 nm-long adhesion factor of *Bordetella pertussis* (PDB 1RWR); R-type is from the pentapeptide repeat protein of *Mycobacterium tuberculosis* (PDB 2BM5); B-type is from the α subunit of *Azospirillum brasilense* glutamate synthase (PDB 1EA0); L-type is from pertactin, an adhesion factor of *Bordetella pertussis* (PDB 1DAB). (From A.V. Kajava and A.C. Steven, *Adv. Prot. Chem.* 73:55–96, 2006. With permission from Elsevier.)

coils stack, giving parallel β sheets in cross-β conformation (**Figure 6.47A**). (The term **β sandwich** is also used for solenoids with two sheets per coil, and **β helix** for three sheets per coil). The repeat length is not necessarily conserved in a given molecule, because loops may be inserted at turn sites. Nor are the repeat sequences closely conserved except at key positions. Conserved residues include polar uncharged residues such as Asn recurring at the same position in successive coils, whose side chains stack into H-bonded ladders that stabilize the solenoid.

A variety of cross-sectional shapes have been observed in β-solenoid proteins (Figure 6.47B). Those with the shortest repeats are triangular in cross section and right-handed; longer repeats have L-shaped cross sections and a left-handed twist. Like amyloids, β solenoids are stable protease-resistant structures. Unlike amyloids, they are monomeric or trimeric, because capping (by another part of the polypeptide chain) prevents axial growth.

Fibril assembly proceeds in two phases: nucleation (slow) and elongation (a faster, templated process)

The kinetics of fibril assembly have been studied by light scattering: typically, a lengthy lag phase is followed by a gradual increase in turbidity, the growth phase. The lag phase is associated with nucleation, and the growth phase with fibril elongation (**Figure 6.48A,B**). The

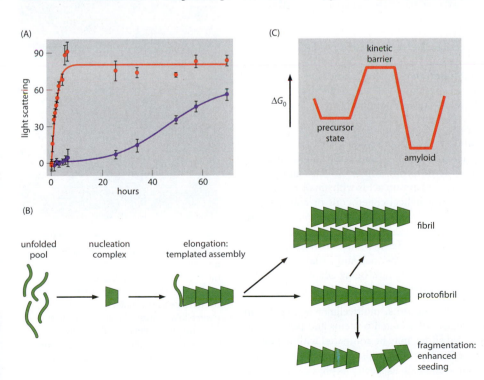

Figure 6.48 Templated assembly of amyloid fibrils. (A) Aggregation in a solution of polyQ-containing polypeptides (K2Q37K2) with seeding (red curve and data points) and without seeding (blue line and data points). The terminal dilysines were added to enhance solubility. (B) In this schematic pathway, there is a high energetic barrier to spontaneous nucleation but once nucleation has been achieved, fibrils elongate readily as monomers are deposited on one or both ends. The ends serve as templates that guide unfolded monomers into their amyloid conformation. Protofibrils may associate laterally to form fibrils. Fragmentation generates more free ends, accelerating growth (see (A)). (C) A hypothetical free energy diagram for this transition. (A, from S. Chen et al., *J. Mol. Biol.* 311:173–183, 2001. With permission from Elsevier.)

structures of the nucleation complexes have yet to be determined. In elongation, molecular surfaces presented on the growing fibril—usually at its ends—guide the recruitment and folding of additional subunits, a process called templated assembly. Elongation takes place, generally at different rates, at both ends. Fibrillation is promoted if some mechanism exists to augment the number of free ends. This may be accomplished by fragmenting existing fibrils by sonication. *In vivo*, there may be chaperone-like proteins capable of breaking down long fibrils into short segments. In forming an amyloid, a protein enters into a free energy state that is lower than that of its native fold but is normally inaccessible for kinetic and/or mechanistic reasons (Figures 1.12 and 6.48C).

Oligomeric assemblies may be the pathogenic agents in some neurodegenerative and other amyloid-related diseases

Amyloid deposits feature in several human diseases (see Table 6.4). Neuronal cells are particularly susceptible to disorders of this kind, but amyloids are also found in other tissues, as with amylin amyloid in the pancreas in type 2 diabetes, or serum amyloid A in multiple organs in rheumatoid arthritis. Some amyloids are intracellular but most are extracellular. Among the latter are the systemic amyloids, found in multiple body sites and derived from proteins secreted into the circulating plasma.

Several neuropathy-associated amyloidogenic proteins, including *huntingtin* (of Huntington's disease) have N-terminal tracts of polyglutamine (polyQ). In affected individuals, the length of these tracts is greater than the norm of 25–30 aa. Above a threshold of ~36, there is a correlation with disease: the longer the tracts, the earlier the age of onset and the more severe the symptoms. Consistent with this, the kinetics of aggregation of polyQ polypeptides *in vitro* becomes faster as the tract length increases.

Nevertheless, it has yet to be settled whether the fibrils are active pathogens or passive byproducts of the disease process. The possibility has been raised that the pathogenic factors are, in fact, small oligomers of the amyloidogenic protein. Observations in favor of this hypothesis include experiments in which oligomers were found to be more toxic to cultured cells than amyloid fibrils, and others in which oligomers were shown to interact with lipid vesicles, rendering them porous. In this context, the putative nucleation complexes are evident candidates. However, in some systems—for example α-synuclein—the ability to remodel membranes has been correlated with the formation of α helices, not β sheets.

For a growing number of proteins, amyloid represents the native functional state

In most studies to date, amyloid fibrils have been viewed as misfolded products, different in conformation from the native proteins and devoid of function, except perhaps in pathogenicity. Contrary to this view, however, several **functional amyloids** now have been documented. For these proteins, there is no evidence of alternative (non-amyloid) folds, and they have been assigned specific functional properties. They are found in bacteria, insects, and higher animals. We note here a few examples.

Silk proteins are synthesized in specialized glands of insects and spiders where they remain in a fluid state until secreted by spinning, whereupon they organize into highly ordered insoluble fibers. Many silks are based on the protein fibroin, whose size and amino acid composition vary but it generally has a high content of amino acids with small side chains (Gly, Ala, Ser). Although a few silks are based on α-helical coiled coils or collagen-like structures, many are rich in β sheets. Fiber diffraction patterns of the cross-β silks conform to the canonical amyloid pattern with a meridional reflection at $(4.7 \text{ Å})^{-1}$. Silk fibers can be extremely strong, with the tensile strength of steel, as in spider web silk. Silk from the moth *Bombyx mori* has been used as a textile prized for its tactile, thermal, and visual qualities for more than 3000 years. Motifs derived from silk proteins are now being adapted and combined to create novel biomaterials with defined properties.

There are many situations in which bacteria form **biofilms**, multicellular aggregates in which the individual cells are embedded in a matrix of secreted proteins and polysaccharides (**Figure 6.49**). This matrix, sometimes called slime, may be compared with the *extracellular matrix* of eukaryotic tissues (Section 11.7). Amyloid fibrils assembled along dedicated pathways are a major component of the slime and help to specify its physical and other

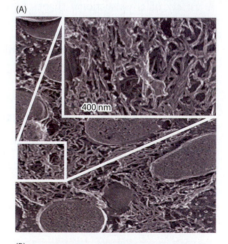
(A)

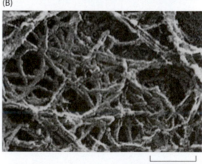
(B)

Figure 6.49 Curli amyloid fibrils in a biofilm. (A) A freeze-etch electron micrograph. The metal replica shows oval structures, which are cross-fractured bacterial cells. The material in the intercellular spaces is mainly curli fibrils. (B) Some intercellular fibrillar matrix at higher magnification. (From M.R. Chapman et al., *Science* 295:851–855, 2002. With permission from AAAS.)

properties. Extracellular microbial amyloids also participate in other cell/cell interactions such as adhesion and host invasion. One well-characterized biofilm-former is the **curli** protein family of *E. coli* and *Salmonella typhimurium*: it has seven members, each with a signal sequence that allows it to be transported through the inner membrane via the SecYEG translocon (see Section 10.7), whereupon the two fibril-forming curli subunits are conveyed through the outer membrane by the other five family members. The resulting fibrils meet the criteria for amyloid and are thought to be based on parallel β sheets.

A very different function is fulfilled by amyloid in mammals in the signaling events that initiate **programmed necrosis** (also called necroptosis), a process that leads to cell death by an alternative pathway to *apoptosis*, which is described in Section 13.6. This involves the activation of two kinases, RIP1 and RIP3 (RIP = receptor-interacting protein) in the assembly of a complex called the necrosome. Assembly depends on the amyloidogenic RHIM (RIP homotypic interaction motif) domains located toward their C termini (**Figure 6.50A**). RHIM-containing truncated versions of RIP1 and RIP3 co-assemble into amyloid fibrils *in vitro* (Figure 6.50B), and the necrosome is thought to be a filament with this backbone, decorated with kinase domains in a manner similar to the yeast prion proteins Ure2p and Sup35p described in the next section. In this way, the kinases are somehow activated. In support of this scenario, mutations in the RHIM domains have been shown to have adverse effects on amyloid assembly and kinase activation, and to inhibit necroptosis.

6.6 PRIONS

An infectious agent should have two basic properties: a mechanism for cell-to-cell transmission, and the ability to replicate in an infected cell. Viruses, which are nucleoprotein assemblies (Chapter 8), readily meet this prescription. According to the 'protein only' concept, proposed by John Griffith and subsequently developed by Stanley Prusiner and others, some proteins are capable of performing both functions in the absence of nucleic acids. They are called **prions**. Specifically, a prion is envisaged as a conformationally altered form of an endogenous protein that is transferrable into other cells, where it replicates by converting wild-type conformers into the same altered state. Nucleic acids enter the picture only indirectly in that the prion protein is a conventionally synthesized gene product.

The first discovered prion, involving the mammalian PrP (prion protein), is implicated in neurological diseases marked by progressive, irreversible, and ultimately fatal, damage to the brain. In its normal conformation, PrP (also called PrP^C, C for cellular) forms dimers of 30 kDa subunits that are protease-sensitive, dispersed, and attached to the outer surface of the plasma membrane by a GPI (glycosylphosphatidylinisotol) anchor. In the prion state, the protein, now called PrP^{Sc} (for **scrapie**, the sheep-borne form of the disease) is relatively resistant to proteases, aggregated, and also present in endocytic vesicles. In the 1990s, an epidemic erupted in the bovine population of the United Kingdom as a consequence of including material from sheep carcasses in cattle feed. It became known as 'mad cow' disease or BSE, bovine spongiform encephalopathy. The infection entered the human population as variant Creuzfeldt–Jakob disease (vCJD), caused by consuming prion-infected bovine products. The symptoms and pathology resemble those of sporadic **Creuzfeldt–Jakob disease (sCJD)**, a long-known syndrome that occurs spontaneously among elderly people at an annual rate of ~1 in 10^6; in contrast, the outbreak of vCJD that peaked over the period 1996–2004 affected much younger populations.

The existence of prions in the yeast *Saccharomyces cerevisiae* was inferred in 1994 on genetic grounds to explain the non-Mendelian transmission of certain traits. Compared with mammalian prions, for which infectivity assays in animals take many months to complete, yeast prions offer a faster, less expensive system with advanced genetic manipulability. Thus, work on yeast and other fungal systems has contributed greatly to what is now known about prions. Here we mainly address—in addition to PrP^{Sc}—two well-studied yeast prions, namely [URE3], associated with the protein Ure2p, and [*PSI*$^+$], associated with Sup35p; and one prion [Het-s] of the filamentous fungus, *Podospora anserina*, associated with the protein HET-s.

The emerging picture is that prions are amyloid forms of the proteins in question, whose replication is based on templated propagation of the corresponding filament (see Section 6.5). Their structures differ between prions and can vary for a given prion, giving rise to distinct **genetic strains**. The key property that distinguishes prions from non-infectious amyloids is their transmissivity, from cell to cell and from organism to organism.

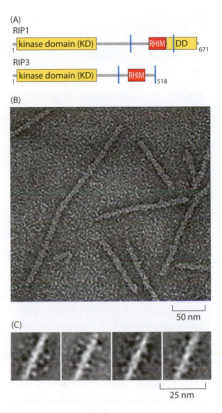

Figure 6.50 An amyloid involved in cell death-related signaling. (A) Domain maps of RIP1 and RIP3. DD, death domain (a six-helix bundle found in some proteins involved in inducing cell death). The blue bars mark the boundaries of the RHIM-containing constructs used to assemble the amyloid fibrils shown in the negatively stained electron micrograph in (B). In (C), class-average images illustrate protein decorating the fibril backbone. (From J. Li et al., *Cell* 150:339–350, 2012. With permission from Elsevier.)

Table 6.5 Properties of four prion proteins.

Protein	Activity	Prion	Loss or gain of function	Dispensable?
PrP	Not known	PrP^{Sc}	Gain	Yes
Ure2p	Regulate nitrogen catabolism	[URE3]	Loss	Yes
Sup35p	Translation termination	[*PSI*$^+$]	Loss (reduced)	No
HET-s	Not known	[Het-s]	Gain (heterokaryon incompatibility)	Not known

Prion domains are unfolded in the wild-type protein and amyloid in the prion

Some basic properties of these four proteins are summarized in **Table 6.5**, and their domain organizations are given in **Figure 6.51**. Each of the fungal proteins has an extended region that is *natively unfolded* in the wild-type protein but polymerizes into an amyloid fibril in the prion. This region is necessary and sufficient for inducing the prion phenotype and for *in vitro* assembly of the full-length proteins into filaments; accordingly, it is referred to as the **prion domain (PD)**. The position of this domain seems not to be critical: it is located at the C-terminus of HET-s and at the N terminus in Ure2p and Sup35p, and the Ure2p PD may be transposed to the C terminus without affecting its amyloidogenic properties.

The PDs of Ure2p and Sup35p have unusual amino acid compositions, being highly enriched in the uncharged polar residues Asn and Gln, and low in hydrophobic and charged residues. Remarkably, the sequences of these domains may be randomized without abrogating the proteins' ability to assemble into filaments and to induce the prion phenotype. This property reflects both their low sequence complexity and redundancy in their amyloidogenic capability. The correlation of elevated Asn and Gln content with a propensity for amyloid formation extends, in an extreme case, to fibrillation by the poly-Gln expansions of *huntingtin* and other neuropathy-associated proteins (see Table 6.4). In contrast, the PD of HET-s has a normal amino acid composition.

Complementing the PDs are globular domains that confer the activity (if any) of the wild-type protein. Structures have been determined in all four cases (**Figure 6.52**). That of Ure2p is α-helical: it is similar to the enzyme *glutathione S-transferase* (GST) in sequence (31% identity) and structure and is able to bind glutathione but not to hydrolyze it. The globular portion of Sup35p from *Schizosaccharomyces pombe*, which is 65% sequence-identical to the *S. cerevisiae* protein and therefore structurally similar, has three domains, one α/β and two

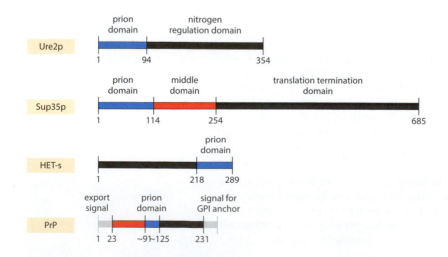

Figure 6.51 Domain organization of four prion proteins. Prion domains (PDs) are shown in blue, and globular domains in black. The M (middle) domain of Sup35p is an extended linker. The status of PrP, residues 23–91 (orange), is currently unclear. The PD assignment given for PrP is based on several observations that this region is important for the structural change in the transition from PrP^C to PrP^{Sc}. (Adapted from U. Baxa et al., *Adv. Prot. Chem.* 73:126–180, 2006. With permission from Elsevier.)

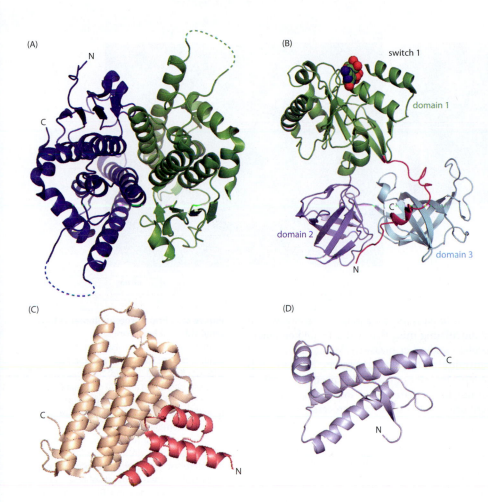

Figure 6.52 The globular domains of four prion proteins. (A) A dimer of the Ure2p C-terminal domain (CTD), residues 97–354. Each subunit has two domains and has the same fold as the enzyme glutathione *S*-transferase. (PDB 1HQO) (B) Sup35p CTD, as represented by residues 215–662 of its homolog, eukaryotic release factor 3 (eRF3). Domain 1 is a typical GTPase with a six-stranded β sheet and six α helices, shown with bound GDP; switch 1 is involved in nucleotide binding and Mg^{2+} coordination. The N-terminal extension (in magenta) is involved in interaction with another eukaryotic release factor, eRF1. Domains 2 and 3 are β barrels. (PDB 1R5N) (C) The HeLo domain of HET-s (aa 1–227) has nine α helices and two short β strands. In solution, it is a dimer with unfolded CTDs. The C-terminal region of the Helo domain (red part) is thought to unfold when incorporated into a growing filament. (PDB 2WVN) (A–C) are crystal structures. (D) NMR structure of the C-terminal domain of human Prp (aa 125–228). (PDB 1E1G)

with β folds. The globular domain of HET-s has homologs in other fungi and is structurally similar to the loss-of-pathogenicity (LOP-B) protein of *Leptosphaeria maculans*, a fungus that infects the rapeseed plant. The name HeLo domain (HET-s/LOP-B) has been assigned to this α-helical fold.

The situation is generally similar for PrP (see Figure 6.51). In this case, a signal sequence of 22 aa is removed by proteolysis, leaving an N-terminal region (res. 23–124) connected by a hydrophobic linker to a C-terminal domain with an α-helical fold (res. 125–231). The distal part of the N-terminal region appears to have a role comparable to that of the PDs of yeast prion proteins. It contains several octapeptide repeats and two positively charged clusters. On a related note, the Sup35-PD has several heptapeptide repeats.

The prion domains self-assemble to form amyloid fibril backbones of prion filaments

The PDs polymerize readily *in vitro*. PD fibrils are robust (those of Ure2p can withstand boiling) and exhibit all the hallmarks of amyloid (such as Congo red binding and cross-β structure; see Section 6.5). Moreover, the ability to polymerize is retained when the C-terminal domains of Ure2p and Sup35p are replaced by other globular domains (for example *green fluorescent protein*—see Figure 6.45). These and other observations have led to the unifying concept that the prion structures are based on amyloid fibril backbones surrounded by, or decorated with, the globular domains (**Figure 6.53A**).

To equate *in vitro* assembly products with the corresponding prions, they should be shown to be infectious. First demonstrated in the [HET-s] system, this has also been accomplished for [URE3] and [PSI⁺]. Fibrils assembled *in vitro* from purified, bacterially expressed protein were introduced into *spheroplasts* (derived from yeast cells by removal of the rigid cell wall) whose membranes had been permeabilized. The spheroplasts were then propagated and progeny cells were assayed for infection status, as discussed below.

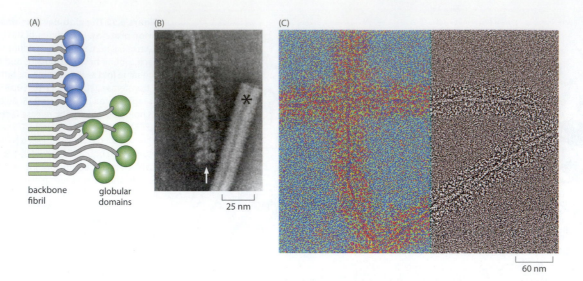

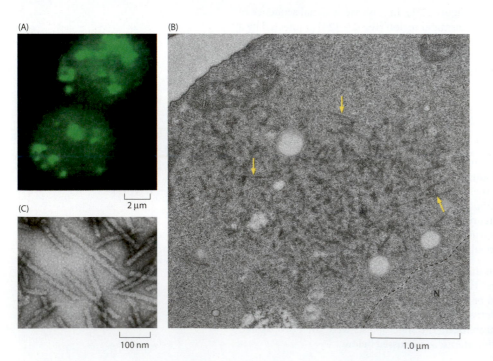

The infectivity of PrP^Sc-like aggregates assembled *in vitro* was for a long time controversial on account of low yields. More recently, by inoculating mice that had a 15-fold enhancement in their expression of a PrP-derived protein and other measures, significant levels of infectivity have been reported, as has its transmittal to the next generation. However, it is generally the case in all these systems that a large amount of material is needed to achieve modest levels of infectivity (orders of magnitude lower than in viral infections). It follows that the seeding of daughter fibrils in recipient cells is a low-efficiency process.

Further evidence that filaments assembled *in vitro* reproduce the prion has come from their morphological similarity to filaments in infected cells. In [URE3] cells expressing a Ure2p–GFP fusion, large round aggregates are visualized by fluorescence microscopy (**Figure 6.54A**). Thin-section EM of Ure2p-expressing cells detected aggregates containing filaments of about the same thickness (~20 nm) as filaments assembled *in vitro* (Figure 6.54B,C). Immuno-gold labeling EM confirmed that they contained Ure2p. Bundles of Sup35p-containing filaments have also been observed in [*PSI*^+] cells, and filaments have been seen in [Het-s] cells. The aggregation of PrP^Sc in prion-infected mouse brain has been demonstrated by immuno-gold EM.

Figure 6.53 Prion protein filaments have amyloid fibril backbones. (A) Schematic models of prion filaments of Ure2p (CTD, blue) and Sup35p (green); Sup35p has a longer linker (M domain) between the amyloid backbone and the globular CTD. (B) STEM dark-field micrograph of a negatively stained Ure2p filament shows globular domains surrounding a thin axial fibril (white arrow). Asterisk marks a tobacco mosaic virus particle, included for reference. (C) Electron micrographs of negatively stained filaments of full-length Sup35p, showing a central fibril surrounded by loosely ordered globular domains extending radially on either side to give a full width of ~60 nm; the left part is colorized to accentuate the region occupied by the linkers plus CTDs. (B, from U. Baxa et al., *J. Biol. Chem.* 278:43717–43727, 2003; C, adapted from U. Baxa et al., *Mol. Microbiol.* 79:523–532, 2011.)

Figure 6.54 Ure2p prion filaments assembled *in situ* and *in vitro*. (A) A fluorescent light micrograph of yeast cells expressing high levels of a Ure2p–GFP fusion shows concentrations of the protein in round inclusion bodies. (The normal level of Ure2p expression is low, making detection difficult.) (B) Thin-section electron micrograph of a cell over-expressing Ure2p shows an inclusion containing randomly oriented filaments ~20 nm in diameter. Examples are marked with yellow arrows. (C) Electron micrograph of negatively stained Ure2p filaments assembled *in vitro*. (A and C, from U. Baxa et al., *Adv. Prot. Chem.* 73:126–180, 2006. With permission from Elsevier. B, courtesy of Vladislav V. Speransky.)

Prion infection is accompanied by a loss or gain of function

Early versions of the prion hypothesis envisaged that the conformational switch to PrP^{Sc} would be accompanied by loss of the protein's original function, with pathological consequences. However, when mice in which the gene encoding PrP was knocked out were found to be viable, it followed that the function of PrP, if any, is not essential; instead, prion infection is associated primarily with a gain of function (neurotoxicity). The knockout mice also provided direct evidence that an animal's supply of PrP is central to disease progression, in that those animals were resistant to inoculation with PrP^{Sc} homogenates that induced the disease in wild-type animals.

[Het-s] is also a gain-of-function prion that operates in **heterokaryon incompatibility**, a proofing process for genetic compatibility when two *P. anserina* cells fuse. In the event of incompatibility, cell death (*apoptosis*) is induced and the mating of incompatible strains is thereby aborted. Although details differ, an outcome of this kind (cell death) also follows infection by PrP^{Sc}. However, infection by other prions is indeed accompanied by a loss or down-regulation of activities.

With [URE3] or [*PSI*], the prion phenotype is expressed simply as the ability of cells to grow on media containing certain nutrients, such that infection may be detected by a simple colorimetric assay. Ure2p is a negative regulator of the nitrogen catabolism regulation (NCR) pathway. In wild-type cells, it binds a transcription factor, Gln3p, preventing it from entering the nucleus, where it would otherwise promote the expression of certain genes involved in nitrogen catabolism. In [URE3] strains, this activity is lost. In [*PSI*$^{+}$] strains, the activity of Sup35p, which normally interacts with ribosomes in *translation termination* (see Section 6.2), is down-regulated, although some residual activity must be retained because Sup35p deletions are not viable.

How is the switch in functional status accomplished? In [URE3], it proceeds via a *steric blocking mechanism*: in the filament, the globular domain retains its native fold but is unable to bind Gln3p because its binding site is sequestered (**Figure 6.55A**). With [*PSI*$^{+}$], the globular domain of Sup35p also maintains its native conformation in filaments and appears to remain accessible to ribosomes, albeit with reduced activity. (Residual activity may also reflect low levels of soluble Sup35p.) Consistent with this scenario, Sup35p has a long (125 aa) M-domain linker that enables its C-terminal domains to be widely offset from the amyloid backbone (see Figure 6.53), allowing enough space for interaction.

As for [Het-s] and PrP^{Sc}, current data suggest that when the wild-type proteins are recruited into filaments, their globular domains become at least partly unfolded. The globular domain of HET-s is ordered in the crystal structure up to its α-helical C terminus (residue 227). Because the first β strand in the amyloid fibril starts at residue 225, it follows that some unfolding must take place. Moreover, as HeLo domain attachment points are only 9.4 Å apart and the domain is ~45 Å in diameter, a linker is needed. To provide it, as much as the last three α helices of the HeLo domain (see Figure 6.53C) may be unfolded (Figure 6.55B). Unfolding may be caused by physical stresses imposed as the polypeptide is drawn into the tightly compacted fibril backbone. Spectroscopic evidence suggests that the increase in β structure observed when PrP^{C} converts to PrP^{Sc} is large enough to require some restructuring of the α-helical C-terminal domain. An electron micrograph of PrP^{Sc} fibrils is shown in Figure 6.55C.

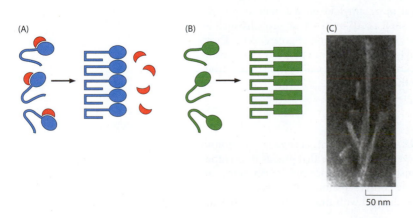

(A)   (B)   (C)

50 nm

Figure 6.55 Functional consequences of assembling prion proteins into filaments. Two scenarios are shown. (A) Inactivation by steric blocking. In its soluble form, Ure2p (blue) has unfolded PDs, and its CTDs bind the transcription factor Gln3p (orange). In the filament, the PDs polymerize into an amyloid fibril, sequestering the binding sites for Gln3p and leaving it free to enter the nucleus and promote the expression of *NCR* genes. (B) Conformational perturbation of the globular domain. Again, the PD is unfolded in the solution form of the protein and forms an amyloid fibril in the filament. On being incorporated into filaments, the globular domains are coerced to undergo a conformational change. This behavior is attributed to HET-s and, tentatively, to Prp. (C) Electron micrograph of negatively stained PrP^{Sc} fibrils (under-glycosylated). (From V.L. Sim and B. Caughey, *Neurobiol Aging* 30:2031–2042, 2009. With permission from Elsevier.)

In infection, fibrillation is nucleated by transmitted or spontaneously formed seeds

For fungal cells that undergo fusion during their replicative cycles, cytoplasmic mixing readily explains the mechanism of infection.

There are at least three ways in which an animal may become infected with PrP^{Sc}. The primary route involves the alimentary system, with infectious ingested material somehow traversing the blood–brain barrier (see Box 11.1). A second (in humans) is to receive a blood transfusion or an organ transplant from an infected person. The third source of infection involves spontaneous generation—in other words, somehow a viable 'prion seed' is formed, and infectious material propagates from this point of origin. This seems to be the origin of cases of sCJD. Unusually high frequency and early onset have been observed in a few families and correlated with mutations in the PrP gene. Spontaneous initiation also occurs in fungal prions at a rate of approximately 1 case per 10^6 cell cycles.

All mammals (but not fungi) have a PrP gene, and the proteins encoded by these genes are closely similar. Notwithstanding, the efficiency of infection by ingestion depends markedly on the species involved. Despite the sequence similarity between mammalian PrPs, the phenomenon of **species barriers** exists, concerning transmissivity. A plausible hypothesis is that its basis lies in some structural nuance(s) in the amyloid formed by a given PrP. Evidence for structural variability comes from experiments that showed differences in fragmentation patterns when material from different isolates was treated with proteinase K. Because the substrate PrP^{Sc} protein had the same sequence in each case, it must have been the conformation presented to the protease that differed.

For fungal prions, the property whereby strains are distinguished is cell color as determined by residual activity of the enzyme Ade2p. Ade2p is responsible for a step in the adenine synthesis pathway. In yeast cells lacking the enzyme, an intermediate product accumulates and is detected as a red pigment. If this block is due to an aberrant stop codon in the *ADE2* gene (giving premature termination), the [*PSI*$^+$] prion can suppress the mutation, and infected cells are white or a shade of pink, depending on the prion strain. With the [URE3] prion, *ADE2* expression is placed under the control of a promoter that allows no synthesis of ADE2p except in [URE3] strains grown in a nitrogen-rich medium. In this case, infected cells are strain-specific shades of pink, and uninfected cells are red. As a possible way to correlate prion structures with genetic strains, it is noteworthy that β arcades and super-pleated β structures (see Figure 6.46) should lend themselves to polymorphisms through slight adjustments in the positions of the turns and strands.

How widespread are prions?

So far, we have discussed just four prion proteins, two of which have PDs with high Asn and Gln contents. A search of the yeast genome found several hundred proteins with regions of elevated Asn and Gln contents, connecting to domains associated with many activities. A similar incidence was found in the genomes of *Caenorhabditis elegans* and *Drosophila*, but very few in bacteria or archaea. In principle, these domains should be capable of fibrillation. However, to qualify as a prion, a defined phenotype has to be identified. So far, this has been accomplished in only a few cases.

Hitherto, it has been the requirement for transmissibility that has distinguished prions from other amyloids. However, studies of several other amyloid-related neurodegenerative disorders, involving either secreted proteins such as Aβ or cytosolic proteins such as tau or α-synuclein, indicate that prion-like spreading to neighboring cells can occur. These considerations raise the possibility that these amyloids may be more properly classified as prions, or that the term prion should be reserved for amyloids that can be transmitted from animal to animal, not just to neighboring cells in the same animal.

6.7   SUMMARY

In protein synthesis, genetic information in DNA is translated into a sequence of amino acids. The DNA transcript, a mature mRNA, serves as a template that guides the synthesis of the polypeptide chain. The process of translation takes place on the ribosome, an RNA–protein complex whose properties are largely conserved among the three kingdoms of life, denoting ancient origins. tRNA molecules charged with their cognate amino acids are recognized by the mRNA and decode it via triplets of codon–anticodon base pairing. The synthesis involves, sequentially, initiation (the binding of the ribosome to the 5′ end of the mRNA), elongation (the addition of amino acids one by one to the 3′ end of the growing polypeptide chain), termination (release of the completed polypeptide chain), and recycling. Each of these steps involves the transient binding of ancillary protein factors.

For a newly synthesized polypeptide chain to become functional, it has to fold into its native conformation. Under ideal conditions, most proteins fold spontaneously either during synthesis (co-translational) or after its completion (post-translational). The amino acid sequence contains the information needed, and folding is governed by thermodynamics, a search for the lowest free energy form of kinetically accessible conformations. However, in the crowded environment of cells, folding can be compromised, resulting in misfolded proteins or spurious interactions that result in aggregation. To overcome this obstacle, folding machines called molecular chaperones safeguard the folding and assembly of client proteins without becoming an integral part of their mature structures.

Most polypeptide chains are destined to adopt a unique three-dimensional fold, but some— and sizable regions of others—lack an ordered structure under physiological conditions. They are referred to as *natively unfolded*. However, many of them undergo disorder–order transitions upon interacting with other proteins, and serve as adaptors, mediating protein/ protein interactions.

Despite elaborate protein quality control mechanisms in cells, involving both chaperones and proteolytic systems, protein misfolding is not a rare event. In many cases, it results in the formation of hyper-stable fibrillar aggregates called amyloids. Although many proteins with different folds can convert to amyloid, the fibrils all have cross-β structures, albeit with differing β-sheet topologies. Amyloid fibrils accumulate in several diseases, notably neuropathies. However, for the so-called functional amyloids, the fibrillar state is their native conformation.

Prions are a subset of amyloids that act as infectious agents. In an infected cell, the prion presents a template that recruits natively folded conformers of the same protein—via an unfolded intermediate—into the fibril. The key property distinguishing prions from other amyloids is their transmission from organism to organism (infectivity). In mammals, prions are responsible for several progressive, irreversible, ultimately fatal, neurodegenerative diseases, including 'mad cow' disease.

A section of this chapter is based on the following contribution

6.2   Elizabeth Villa, University of California, San Diego, USA and Anders Liljas, Lund University, Sweden.

REFERENCES

6.2   The ribosome

Becker T, Bhushan S, Jarasch A et al (2009) Structure of monomeric yeast and mammalian Sec61 complexes interacting with the translating ribosome. *Science* 326:1369–1373.

Brandt F, Etchells SA, Ortiz JO et al (2009) The native 3D organization of bacterial polysomes. *Cell* 136:261–271.

Frank J & Agrawal RK (2000) A ratchet-like inter-subunit reorganization of the ribosome during translocation. *Nature* 406:318–322.

Frank J & Gonzalez RL Jr (2010) Structure and dynamics of a processive Brownian motor: the translating ribosome. *Annu Rev Biochem* 79:381–412.

Liljas A (2004) Structural Aspects of Protein Synthesis. World Scientific.

Melnikov S, Ben-Shem A, Garreau de Loubresse N et al (2012) One core, two shells: bacterial and eukaryotic ribosomes. *Nature Struct Mol Biol* 19:560–567.

Ogle JM & Ramakrishnan V (2005) Structural insights into translational fidelity. *Annu Rev Biochem* 74:129–177.

Palade GE (1955) A small particulate component of the cytoplasm. *J Biophys Biochem Cytol* 1:59–68.

Petrov A, Chen J, O'Leary S et al (2012) Single-molecule analysis of translational dynamics. *Cold Spring Harbor Perspect Biol* 4:a011551.

Rodnina MV (2012) Quality control of mRNA decoding on the bacterial ribosome. *Adv Prot Chem Struct Biol* 86:95–128.

Simonović M & Steitz TA (2009) A structural view on the mechanism of the ribosome-catalyzed peptide bond formation. *Biochim Biophys Acta* 1789:612–623.

Voorhees RM & Ramakrishnan V (2013) Structural basis of the translational elongation cycle. *Annu Rev Biochem* 82:203–236.

Wilson DN (2014) Ribosome-targeting antibiotics and mechanisms of bacterial resistance. *Nature Rev Microbiol* 12:35–48.

6.3   Molecular chaperones

Agashe VR, Guha S, Chang HC et al (2004) Function of trigger factor and DnaK in multidomain protein folding: increase in yield at the expense of folding speed. *Cell* 117:199–209.

Delbecq SP & Klevit RE (2013) One size does not fit all: the oligomeric states of αB crystallin. *FEBS Lett* 587:1073–1080.

Ditzel L, Löwe J, Stock D et al (1998) Crystal structure of the thermosome, the archaeal chaperonin and homolog of CCT. *Cell* 93:125–138.

Hunt JF, Weaver AJ, Landry SJ et al (1996) The crystal structure of the GroES co-chaperonin at 2.8 Å resolution. *Nature* 379:37–45.

Kalisman N, Adams CM & Levitt M (2012) Subunit order of eukaryotic TRiC/CCT chaperonin by cross-linking, mass spectrometry, and combinatorial homology modeling. *Proc Natl Acad Sci USA* 109:2884–2889.

Kerner MJ, Naylor DJ, Ishihama Y et al (2005) Proteome-wide analysis of chaperonin-dependent protein folding in *Escherichia coli*. *Cell* 122:209–220.

Kim EY, Hipp M, Bracher A et al (2013) Molecular chaperone functions in protein folding and proteostasis. *Annu Rev Biochem* 82:323–355.

Klumpp M, Baumeister W & Essen LO (1997) Structure of the substrate binding domain of the thermosome, an archaeal group II chaperonin. *Cell* 91:263–270.

Krukenberg KA, Street TO, Lavery LA et al (2011) Conformational dynamics of the molecular chaperone Hsp90. *Q Rev Biophys* 44:229–255.

Langer T, Lu C, Echols H et al (1992) Successive action of DnaK, DnaJ and GroEL along the pathway of chaperone-mediated protein folding. *Nature* 356:683–689.

Leitner A, Joachimiak LA, Bracher A et al (2012) The molecular architecture of the eukaryotic chaperonin TRiC/CCT. *Structure* 20:814–825.

Ostermann J, Horwich AL, Neupert W & Hartl F-U (1989) Protein folding in mitochondria requires complex formation with Hsp60 and ATP hydrolysis. *Nature* 341:125–130.

Pearl LH & Prodromou C (2006) Structure and mechanism of the Hsp90 molecular chaperone machinery. *Annu Rev Biochem* 75:271–294.

Phipps BM, Hoffmann A, Stetter KO & Baumeister W (1991) A novel ATPase complex selectively accumulated upon heat shock is a major cellular component of thermophilic archaebacteria. *EMBO J* 10:1711–1722.

Röhl A, Rohrberg J & Buchner J (2013) The chaperone Hsp90: changing partners for demanding clients. *Trends Biochem Sci* 38:253–262.

Siegert R, Leroux MR, Scheufler C et al (2004) Structure of the molecular chaperone prefoldin: unique interaction of multiple coiled coil tentacles with unfolded proteins. *Cell* 103:621–632.

Xu Z, Horwich AL & Sigler PB (1997) The crystal structure of the asymmetric GroEL–GroES–(ADP)$_7$ chaperonin complex. *Nature* 388:741–750.

6.4   Natively unfolded proteins

Dunker AK, Silman I, Uversky VN & Sussmann JL (2008) Function and structure of inherently disordered proteins. *Curr Opin Struct Biol* 18:756–764.

Dyson HJ & Wright PE (2005) Intrinsically unstructured proteins and their functions. *Nature Rev Mol Cell Biol* 6:197–208.

Fuxreiter M, Tompa P, Simon I et al (2008) Malleable machines take shape in eukaryotic transcriptional regulation. *Nature Chem Biol* 4:728–737.

Namba K (2001) Roles of partly unfolded conformations in macromolecular self-assembly. *Genes Cells* 6:1–12.

Tompa P (2010) Structure and Function of Intrinsically Disordered Proteins. Chapman & Hall.

Uversky VN, Gillespie JR & Fink AL (2000) Why are 'natively unfolded' proteins unstructured under physiologic conditions? *Proteins* 41:415–427.

Van der Lee R, Buljan M, Lang B et al (2014) Classification of intrinsically disordered regions and proteins. *Chem Rev* 114:6589–6631.

6.5   Protein misfolding and amyloid fibrils

Craig CL & Riekel C (2002) Comparative architecture of silks, fibrous proteins and their encoding genes in insects and spiders. *Comp Biochem Physiol B* 133:493–507.

Goldsbury C, Baxa U, Simon MN et al (2011) Amyloid structure and assembly: insights from scanning transmission electron microscopy. *J Struct Biol* 173:1–13.

Kajava AV & Steven AC (2006) β-Rolls, β-helices, and other β-solenoid proteins. *Adv Protein Chem* 73:55–96.

Kajava AV, Baxa U, Wickner RB & Steven AC (2004) A model for Ure2p prion filaments and other amyloids: the parallel superpleated β-structure. *Proc Natl Acad Sci USA* 101:7885–7890.

Li J, McQuade T, Siemer AB et al (2012) The RIP1/RIP3 necrosome forms a functional amyloid signaling complex required for programmed necrosis. *Cell* 150:339–350.

Murphy RM (2007) Kinetics of amyloid formation and membrane interaction with amyloidogenic proteins. *Biochim Biophys Acta* 1768:1923–1934.

Nelson R & Eisenberg D (2006) Structural models of amyloid-like fibrils. *Adv Protein Chem* 73:235–282.

Omenetto FG & Kaplan DL (2010) New opportunities for an ancient material. *Science* 329:528–531.

Pham CLL, Kwan AH & Sunde M (2014) Functional amyloid: widespread in Nature, diverse in purpose. *Essays Biochem* 56:207–219.

Sawaya MR, Sambashivan S, Nelson R et al (2007) Atomic structures of amyloid cross-beta spines reveal varied steric zippers. *Nature* 447:453–457.

Sipe JD & Cohen AS (2000) Review: history of the amyloid fibril. *J Struct Biol* 130:88–98.

Tycko R (2011) Solid-state NMR studies of amyloid fibril structure. *Annu Rev Phys Chem* 62:279–299.

6.6   Prions

Aguzzi, A & Calella AM (2009) Prions: protein aggregation and infectious diseases. *Physiol Rev* 89:1105–1152.

Baxa U, Cassese T, Kajava AV & Steven AC (2006) Structure, function, and amyloidogenesis of fungal prions: filament polymorphism and prion variants. *Adv Protein Chem* 73:125–180.

Baxa U, Speransky V, Steven AC & Wickner RB (2002) Mechanism of inactivation on prion conversion of the *Saccharomyces cerevisiae* Ure2 protein. *Proc Natl Acad Sci USA* 99:5253–5260.

Bocharova OV, Breydo L, Parfenov AS et al (2005) *In vitro* conversion of full-length mammalian prion protein produces amyloid form with physical properties of PrP^{Sc}. *J Mol Biol* 346:645–659.

Caughey B, Baron, GS, Chesebro B & Jeffrey M (2009) Getting a grip on prions: oligomers, amyloids, and pathological membrane interactions. *Annu Rev Biochem* 78:177–204.

Greenwald J, Buhtz, C, Ritter C et al (2010) The mechanism of prion inhibition by HET-S. *Mol Cell* 38:889–899.

Michelitsch MD & Weissman JS (2000) A census of glutamine/asparagine-rich regions: implications for their conserved function and the prediction of novel prions. *Proc Natl Acad Sci USA* 97:11910–11915.

Saupe SJ (2011) The [Het-s] prion of *Podospora anserina* and its role in heterokaryon incompatibility. *Semin Cell Dev Biol.* 22:460–468.

Verges KJ, Smith MH, Toyama BH & Weissman JS (2011) Strain conformation, primary structure and the propagation of the yeast prion [*PSI*$^+$]. *Nature Struct Mol Biol* 18:493–499.

Westermark GT & Westermark P (2010) Prion-like aggregates: infectious agents in human disease. *Trends Mol Med* 16:501–507.

Intracellular Proteolysis: Protein Quality Control and Regulatory Turnover

7

7.1 INTRODUCTION

The population of proteins within a cell, its *proteome*, is regulated by both synthesis and degradation. The synthesis of proteins selected for expression takes place on ribosomes at rates that are governed by a variety of transcriptional and translational mechanisms. Proteins vary widely in their longevity: some regulatory proteins turn over in a few hours; at the other extreme, crystallin proteins in the eye lens last indefinitely. At the end of their terms, proteins are eliminated by proteolysis. Programmed degradation of regulatory proteins—including transcription factors, cyclins, and various inhibitors—has key roles in guiding the cell cycle (Section 13.2), pathways of differentiation, and apoptosis (Section 13.6). Misfolded, foreign, and otherwise aberrant proteins that might interfere with cellular functions are eliminated in a similar fashion in a process called **protein quality control**.

Whereas extracellular proteolysis is carried out by relatively small and simple secretory enzymes, also known as 'classic' proteases—for example, trypsin and chymotrypsin, which are familiar laboratory reagents—intracellular proteolysis must be tightly regulated if havoc is to be avoided. To meet this requirement, this activity is mainly carried out by large, multi-component, degradation machines.

Broadly speaking, there are two types of intracellular proteolysis: ATP-dependent and ATP-independent. Most proteolysis occurring in extracytosolic compartments is ATP-independent: the main site is the *lysosome*, but this activity also includes the processing of prohormones in the Golgi complex and secretory vesicles. Within the cytosol, both ATP-independent and ATP-dependent proteolysis take place. ATP-independent proteolysis is carried out by a number of enzymes, of which the most intensively studied belong to the *caspase* family and mediate the apoptotic program (Section 13.6). However, most cytosolic proteolysis is ATP-dependent and is carried out by large macromolecular machines that couple ATPase activity to proteolysis.

In bacteria and the mitochondria and chloroplasts of eukaryotes, these machines encompass a group of architecturally related proteases called Clp, Lon, and FtsH (Section 7.3). In the eukaryotic cytosol and nucleus, this activity is mediated by the **26S proteasome**, named for its sedimentation coefficient (Section 7.4). Archaea have a stripped-down version of the proteasome. These machines consist of two or three subcomplexes, with many subunits and masses in the 0.5–2.5 MDa range. In general, they consist of a core particle and one or two regulatory particles. The core particle houses the proteolytic activity and is capable of degrading oligopeptides or unfolded proteins on its own. To degrade folded proteins, a regulatory particle is required; it selects valid substrates and uses its ATPase activity to unfold them for presentation to the core particle. Collectively, we refer to these degradation machines as **unfoldase-assisted proteases**. This chapter focuses mainly on this form of proteolysis.

The peptides released from unfoldase-assisted proteases are processed further to single amino acids, to be recycled in a subsequent round of protein synthesis. In eukaryotes and archaea, this event proceeds stepwise, whereby the peptides are first reduced to dipeptides and tripeptides by the so-called **giant proteases** (Section 7.5) and then to amino acids by downstream aminopeptidases.

7.2   PRINCIPLES OF UNFOLDASE-ASSISTED PROTEOLYSIS

The key attribute of degradation machines is that their proteolytic active sites are sequestered in a cavity inside the core particle: thus, normal cellular proteins, which are folded, are never exposed to them. Substrates are selected by the regulatory particle. Once a substrate has entered the degradation chamber, proteolysis is conducted by enzymatic mechanisms similar to those of classical proteases.

Classical proteases selectively sever peptide bonds

Classical proteases are classified according to the cognate amino acid present at their active site: this can be a Ser (serine proteases), Cys (cysteine proteases), or Asp (**aspartyl proteases**) residue. A fourth class, the metalloproteinases, depends on metal ions for activity. The mechanisms of these enzymes are summarized in **Box 7.1**. Many proteases are initially synthesized as inactive precursors, known as *zymogens*, that have a peptide extension or **propeptide**—usually at the N terminus—that keeps the enzyme in an inactive conformation. In activation, the propeptide bond is cut, the propeptide is released, and the enzyme assumes an active conformation. This activating cleavage may be autocatalytic or catalyzed by another protease. **Exoproteases** detach residues from an end of the polypeptide chain, whereas **endoproteases** recognize and act on sites away from the termini. Classical proteases recognize and cleave at specific sites—'scissile sites'—wherever these are exposed on folded or unfolded proteins. These sites have particular amino acids in the positions before and after the bond to be broken—the scissile bond; in the sequence P_4'-P_3'-P_2'-P_1'-P_1-P_2-P_3-P_4, the scissile bond is between P_1' and P_1. P_1' and P_1 denote the first residue before and after the scissile bond, respectively. Different proteases vary in the stringency of their requirements for occupancy of these positions and, consequently, in their specificity.

Whereas classical proteases simply recognize exposed sites presenting scissile bonds, unfoldase-assisted proteases recognize specific motifs or tags present on valid substrates. Once a substrate has been selected, it is unfolded and degraded processively, starting from one end, and whittled down to peptides of 8–12 amino acids. An adapted form of the proteasome in which three core particle subunits are replaced—the **immunoproteasome**—has a role in the *adaptive immune system* in recognizing and processing foreign proteins, for example viral proteins, in the cytosol of an infected cell. The resulting peptides are exported to the endoplasmic reticulum, where they are bound by class I *major histocompatibility complex* (*MHC-I*) molecules and subsequently presented at the cell surface. Cells bearing them—*antigen-presenting cells*—activate the cytolytic response of T lymphocytes to selectively eliminate infected cells (see Section 17.4). More generally, the primary degradation products (peptides) are released from the proteasome and processed further to single amino acids, to be recycled in a subsequent round of protein synthesis.

In its various manifestations, the degradative process proceeds along pathways that may be called 'disassembly lines,' beginning with the recognition of a valid substrate, continuing with its unfolding and translocation into the degradation chamber where it is processively degraded, and completed with the dispersal and further processing of the resulting peptides (**Figure 7.1**).

Substrate specificity is conferred by the regulatory particles

Proteolytic activity inside cells has to be more selective than that of classical proteases because the cytoplasm and other compartments are densely populated with 'normal' proteins, most of which must be spared. At the same time, intracellular proteases have to be capable of recognizing and degrading a wide range of substrates. Core particle proteases cleave every 8–12 residues for most polypeptide chains, once degradation has commenced; in this sense, they are—paradoxically—less specific than classical proteases, whose scissile sites tend to be more sparsely distributed.

The conundrum of substrate selectivity is solved not by having extremely specific cleavage sites but by assigning responsibility for substrate selection to the regulatory particles, and this may involve a variety of mechanisms. The regulatory particle may recognize a particular sequence—a **degradation motif**, or **degron**—such as an unusual preponderance of exposed hydrophobic residues, indicative of a misfolded protein, or an appended non-native tag that

Box 7.1   Enzymatic mechanisms of proteolysis

As noted above, proteases fall into two broad classes: exo-proteases and endoproteases. The enzymatic mechanisms of endoproteases all proceed by nucleophilic attack on the carbonyl group of the peptide bond, followed by stabilization of the tetrahedral intermediate that forms. The way in which one such protease, chymotrypsin, acts by nucleophilic attack of an activated serine residue in its active site is described in detail in Figure 1.29.

Chymotrypsin is typical of a large group of homologous serine proteases, some involved in digestion, some in blood clotting, some in the immune response, among other contexts. All of them make use of the **catalytic triad** Asp/His/Ser to activate the hydroxyl group of the Ser side chain (**Figure 7.1.1A**), as described in Figure 129. However, the specificity of these enzymes is determined by their ability to recognize different amino acid side-chains in scissile sites, mainly the P_1' residue. Thus, chymotrypsin, which has a surface-exposed hydrophobic binding pocket, cleaves on the C-terminal side of bulky hydrophobic residues such as Phe, Tyr, and Leu. In contrast, trypsin, which has a negatively charged Asp replacing a neutral Ser toward the bottom of its binding pocket, cleaves on the C-terminal side of Lys and Arg residues. These enzymes have similar structures, indicating divergent evolution in specificity while a common bond-breaking mechanism has been retained. Another group of serine proteases, including subtilisin, which was originally isolated from the bacterium *Bacillus subtilis*, has the same type of catalytic triad in the active site but a different fold. Thus, this group appears to have arisen by convergent evolution toward the same mechanism (see Figure 7.1.1A). These enzymes are exploited as components of commercial products such as washing powder.

Nucleophilic attack on the >C=O group is the basis of all peptide bond cleavage by endoproteases, but the nucleophile is not always serine. In cysteine proteases, Cys replaces Ser in this role; it, too, is activated by an adjacent His in analogous fashion but without the participation of a neighboring Asp residue. The plant protease papain is the archetype of this mechanism (see Figure 7.1.1B), and many of the **cathepsins**, found predominantly in lysosomes (Section 7.1), are homologs. The much lower pK of the thiol group of Cys (~8.5) compared with Ser (>14) makes it easier for the activated nucleophile to form at the low pH (typically 4.5–5.0) found in lysosomes. The caspases, which are important in *apoptosis* (Section 13.6) and in cytokine maturation (interleukin-1-converting enzyme, ICE) in inflammation, share this mechanism, but they are structurally unrelated to papain.

Aspartyl proteases, unlike serine and cysteine proteases, do not form a covalent bond with the substrate as an intermediate; rather, an Asp in the active site activates a water molecule as the nucleophile, while a second Asp (this one in its protonated form) polarizes the peptide carbonyl group, facilitating the nucleophilic attack and formation of the tetrahedral intermediate (see Figure 7.1.1C). The archetypes here are pepsin, which operates at the acid pH (1.8–2.0 during food intake) of the stomach, and renin, which is important in the regulation of blood pressure. These enzymes have two similar domains, suggesting that they arose by gene duplication. This view is

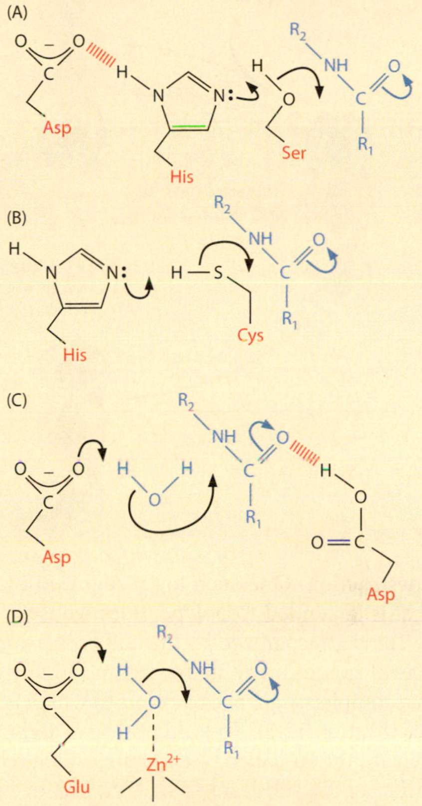

Figure 7.1.1 Four catalytic mechanisms employed by proteolytic enzymes. In all four, hydrolysis is initiated by nucleophilic attack on the >C=O group of the scissile peptide bond of the bound substrate (blue). (A) Serine proteases house a catalytic triad, Asp/His/Ser, to activate the hydroxyl group of a serine side chain. (B) In cysteine proteases, the mechanism is similar but the Ser is replaced by Cys, and there is no catalytically important Asp residue. (C) In aspartyl proteases, two Asp residues combine to activate a bound water molecule as a nucleophile and facilitate formation of the tetrahedral intermediate. (D) Metalloproteinases require a bound metal ion, usually Zn^{2+}, and a neighboring general base, usually an unprotonated Glu, to activate a bound water molecule as nucleophile.

reinforced by the discovery that the proteases of retroviruses such as HIV (Box 8.1), which promote virion maturation, are dimers of subunits similar to either half of the single chain of pepsin or renin.

Nucleophilic attack by an activated water molecule is also the mechanism employed by the metalloproteinases. In this case, an enzyme-bound metal ion, usually Zn^{2+}, activates the water molecule. A nearby base, often an unprotonated Glu side-chain, helps remove a proton from the bound water (see Figure 7.1.1D). Examples include the bacterial enzyme thermolysin and the matrix metalloproteinases, which are involved in remodeling the *extracellular matrix* (Section 11.7) and consequently in cell proliferation, migration, and so on. The exoprotease carboxypeptidase A shares the same mechanism but is structurally dissimilar.

Protease inhibitors are of major importance as drugs, as exemplified for proteasome inhibitors in Box 7.3.

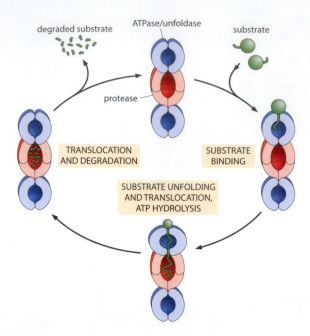

degraded substrate

Figure 7.1 Generic pathway for processive degradation of a protein substrate by an unfoldase-assisted protease. Processing starts with the binding of the substrate to the apical surface of an ATPase/unfoldase subcomplex, followed by its unfolding and translocation into the central proteolytic degradation chamber, powered by conformational changes induced by ATP hydrolysis. Processive degradation takes place and, finally, dispersal of the resulting peptide fragments.

marks it for destruction. One such tag recognized by Clp proteases is an 11-aa C-terminal degron that is appended to polypeptides whose synthesis has stalled on the ribosome (Section 7.3). The tag recognized by the 26S proteasome is *polyubiquitin*, which comprises multiple copies of ubiquitin, an abundant and highly conserved 76-aa protein. Tagging is carried out in a complex sequence of events in which specificity is conferred by *E3 ubiquitin ligase* enzymes (Section 12.5). The Clp proteases and the proteasome also degrade proteins in compliance with the so-called **N-end rule**, which requires having particular amino acids at the N terminus of the substrate—Phe, Tyr, Trp, or Leu for the Clps, and these plus a few more hydrophobic residues for the proteasome. There are many other degrons that also invite ubiquitylation.

Unfoldase-assisted proteases have stacked-ring architecture and modular organization

The architectural principle underlying these machines involves the stacking of rings of protein subunits. Core particles have two identical rings of protease subunits, stacked face-to-face, with the active sites lining the central cavity. Regulatory particles have rings of ATPase subunits that stack axially on one or both apical surfaces of the core particle (**Figure 7.2**). Both forms, with one or two regulators on board, are active. Additional factors—which are not ring-like in the 19S proteasome—may be appended to the regulatory particle. Alternatively, other proteasome activators lacking ATPase activity may bind at the apical sites (Figure 7.2B–F).

The rings generally have six or seven subunits and may be homomeric or heteromeric. The protease rings may be single or double rings. The proteasome core particle has two double rings, whose outer-ring subunits are proteolytically inactive but serve a scaffolding function in assembly (see Section 7.4). These rings are heteromeric in the eukaryotic proteasome,

Figure 7.2 Modular architectures of unfoldase-assisted proteases. (A) A complex consisting of two rings of protease subunits stacked face-to-face, creating a hollow compartment that is only accessible via axial channels; the ATPase/unfoldase mounts coaxially on one (as here) or both apical surfaces. Alternative architectures to this basic structure are shown in panels (B) to (G). (B) Regulators mounted at both ends of the protease: examples, ClpXP and ClpYQ. (C) The protease ring has two tiers: example, the archaeal proteasome. (D) Additional regulatory factor binds to the ATPase: a simple example is the binding of ClpS to ClpAP; a more complex one is the 26S proteasome. (E) Different regulators bind at the two ends: example, ClpXAP particles assembled *in vitro*. (F) A regulatory factor without ATPase activity binds directly to the protease: example, the 20S proteasome with activator PA28. (G) The ATPase has two tiers: example, ClpAP.

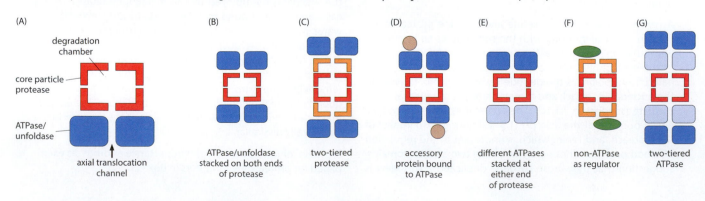

(A)

degradation chamber

core particle protease

ATPase/ unfoldase

axial translocation channel

(B) ATPase/unfoldase stacked on both ends of protease

(C) two-tiered protease

(D) accessory protein bound to ATPase

(E) different ATPases stacked at either end of protease

(F) non-ATPase as regulator

(G) two-tiered ATPase

whereby their heteromeric nature supports three distinct proteolytic activities. ClpPs from chloroplasts are also heteromeric. Docked onto the protease rings are homomeric or heteromeric unfoldases; the former occur in bacteria and archaea, and the latter in eukaryotes. These proteins belong to the extensive family of AAA+ ATPases, mechanoenzymes whose enzymatic mechanisms are described in Box 3.1, with other properties outlined in **Box 7.2**. In the context of intracellular proteolysis, the unfoldases recognize valid substrates, unfold them, and translocate them into the core particle. Most are single hexameric rings, but some, such as the bacterial unfoldase ClpA, have subunits with two ATPase domains that form two stacked tiers—in effect, a double ring (see Figure 7.2G and Section 7.3).

One key aspect of unfoldase-assisted proteases is the principle of **self-compartmentalization**. The proteolytic active sites are encapsidated in a nanocompartment which is accessible only via narrow axial channels that act as a steric filter. Small peptides may diffuse or unfolded proteins may wriggle into the degradation chamber to be processed there, but folded proteins are too large and are excluded; to enter the degradation chamber, they must be unfolded and fed into it. This elegant steric-blocking mechanism explains why normal intracellular proteins are spared.

7.3   UNFOLDASE-ASSISTED PROTEASES IN BACTERIA AND EUKARYOTIC ORGANELLES

In most bacteria and in mitochondria and chloroplasts—reflecting their common evolutionary origins (Section 1.9)—these machines are members of the Clp family (for caseinolytic protease) and the Lon and FtsH proteins. Here the term 'family' refers to similar quaternary structures, not tertiary structures, because the core particle subunits have several different folds. In *Escherichia coli*, there are two core particle proteases, ClpP and ClpQ (also called HslV, for Heat shock locus V), and four AAA+ ATPases—ClpX, ClpA, ClpB, and ClpY (also called HslU) (**Figure 7.3A**). ClpQ partners ClpY, and ClpP partners ClpX and ClpA, an arrangement that extends the range of substrates that may be processed by a single protease. In FtsH and Lon, both components are connected in the same polypeptide chain. FtsH also has a membrane-associating domain. Lon orthologs partition into two subfamilies: LonA, which has a substantial N-terminal gating domain; and LonB, whose N-terminal region includes a membrane anchor. LonB is found in archaea, where it and PAN protease

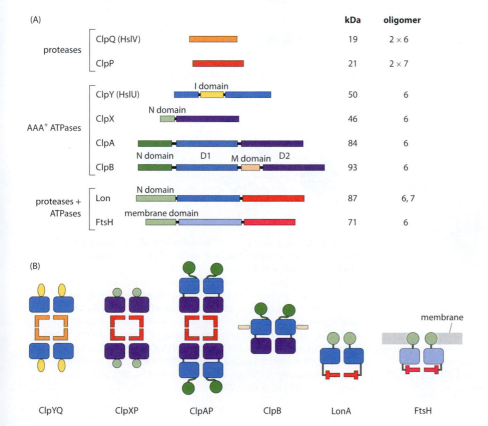

Figure 7.3 Organization of unfoldase-assisted proteases of *E. coli*. (A) Domain maps: proteases are in shades of orange and red; AAA+ domains are in shades of blue and mauve; accessory domains are in other colors. Bacterial Lons are hexamers, whereas yeast Lon is a heptamer. With ClpB, both hexamers and heptamers have been observed but the principal form is hexamers. (B) Schematic diagrams of the assembled complexes, color-coded as in (A). In Lon and FtsH, all three domains are connected in a single polypeptide chain.

Box 7.2   Mechanoenzymes of the AAA+ ATPase family

AAA+ ATPases carry out a strikingly diverse range of cellular functions—hence the catch-all name, ATPases associated with activities in cells. In the first instance, their relatedness was identified by their possessing Walker A and B motifs as well as an additional signature motif, the so-called second region of homology (SRH), a tract of ~15 amino acids (**Figure 7.2.1A**). The Walker A motif conforms to GXXXXGK(T/S), where X is any amino acid and either T or S occupies the last position. Also known as the 'P-loop,' it interacts closely with bound nucleotide. The Walker B motif, $\Phi\Phi\Phi\Phi$DE, where Φ is a hydrophobic amino acid, also interacts with bound nucleotide. Mutations in these motifs render the protein unable to hydrolyze and/or bind ATP. These proteins also share several other motifs: **sensor 1** is a polar residue, most often Asn, and **sensor 2** is usually Arg. The **arginine finger** is an SRH residue whose side-chain is directed into the nucleotide binding pocket. Another conserved feature is the 'pore loop' that protrudes into the axial channel of the oligomer.

Crystal structures have been determined for numerous AAA+ domains, confirming that they share a common fold. Most of these structures are for constructs that lack other domains (see below), which may reflect flexible inter-domain connections. The AAA+ domain is ~230 aa long and combines a larger N-terminal subdomain with an α/β fold (the Rossmann fold) with a smaller C-terminal subdomain consisting of four α helices. The ATP-binding site is positioned at the interface between subunits. Other conserved motifs are mapped in Figure 7.2.1B.

Comprehensive analysis of relevant amino acid sequences showed that the AAA family initially identified is part of a larger family, which became named the AAA+ ATPases. In turn, AAA+ ATPases are a subset of the superfamily of P-loop NTPases

and they constitute one of the six or seven clades of the alternatively defined superfamily of ASCE (for additional strand, catalytic E) nucleotidases. These motifs, the classification into clades, and the mechanism of ATP hydrolysis are described further in Box 3.1.

Genome-wide scans with consensus sequences have established that AAA+ proteins are found in all kingdoms of life and thus are of ancient origin, and that the family is extensive. *E. coli*, a typical bacterium, has 28 AAA+ proteins, the archeon *Methanococcus* has 15, and the AAA+ module has also been acquired by some viruses. Among eukaryotes, yeast (*Saccharomyces cerevisiae*) has ~50 AAA+ genes; the human genome codes for ~80; and plants such as *Arabidopsis thaliana* have even more (~140).

In their functional forms, most AAA+ ATPases assemble into rings or solenoids (single-start helices)—**Figure 7.2.2A**. In AAA+ proteins that remodel other proteins (such as the unfoldases in this chapter) or membranes, rings are favored; those that manipulate DNA (Chapter 3) often form helices. In some cases, assembly is nucleotide-dependent.

Despite the conservativity of their ATPase modules, AAA+ proteins vary in size from ~35 kDa to ~200 kDa. Two factors are involved. One is the number of AAA+ modules (each 25–30 kDa): there is usually one (class II proteins) or two (class I), but there are as many as six in the motor protein

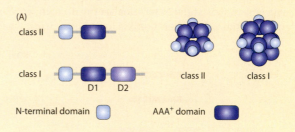

(A)

class II

class I

D1   D2

class II   class I

N-terminal domain ☐   AAA+ domain ◼

(B)

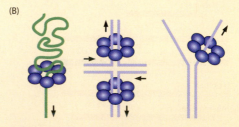

Figure 7.2.2 Arrangements of domains in AAA+ oligomers and interacting with substrates. (A) Domain maps of class I and class II AAA+ proteins. The N domains are sometimes located around the periphery, as shown here, but in other cases they extend from the apical surface where they are closer to the axial pore. (B) Diagram of inferred modes of action of several AAA+ ATPases involving the passage of substrate through their axial channels: left, a protein unfoldase; center, a Holliday junction motor protein (Chapter 3); right, a helicase unwinding double-stranded DNA (Chapter 3). (A, adapted from S.R. White and B. Lauring, *Traffic* 8:1657–1656, 2007; B, adapted from T. Ogura and A.J. Wilkinson, *Genes Cells* 6:575–597, 2001. Both with permission from John Wiley & Sons.)

(A)

GKT/S   DE   T RR

N ▬—☐—◼——│◼——◼—☐◻—────── C

N-linker   pore   SRH
Walker A   Walker B   sensor 1   sensor 2

(B)

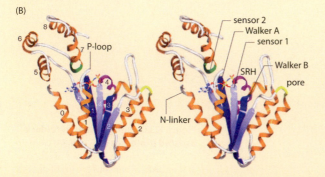

sensor 2
Walker A
sensor 1

P-loop

SRH

Walker B

pore

N-linker

Figure 7.2.1 Characteristic features of a AAA+ domain. (A) Map of motifs on the primary structure. The second region of homology (SRH) is exclusive to AAA+ proteins: Walker A and B occur more widely: (B) The AAA+ fold: on the left, the elements of secondary structure are labeled; on the right, the characteristic motifs. (Adapted from P.I. Hanson and S.W. Whiteheart, *Nature Rev. Cell. Mol. Biol.* 6:519–529, 2005. With permission from Macmillan Publishers Ltd.)

Box 7.2 *continued*

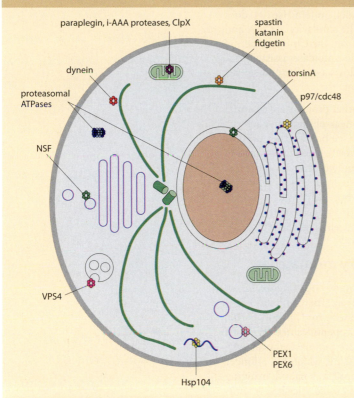

Protein	Function
NSF	Disassembly of SNARE complexes for vesicular transport (Section 10.4)
VPS4	Pinching off vesicles budding into the multivesicular body
PEX1 and PEX6	Dislocation of PTS receptor from the peroxisomal membrane for recycling into the cytosol
Clp/Hsp100 family	Unfoldase and disaggregase activity in protein quality control
Proteasomal ATPases	Unfoldase and translocase activity in protein degradation
TorsinA	Nuclear membrane remodeling
Paraplegin, ClpX, iAAA protease	Maturation of mitochondrial proteins through cleavage or membrane dislocation; protein quality control
p97/cdc48	Retrotranslocation of misfolded substrates during ERAD (Box 7.4); maintenance of ER & nuclear envelope; Golgi reassembly after mitosis
Cytoplasmic dynein	Minus-end-directed microtubule-associated motor protein (Section 14.7)
Katanin	Microtubule severing (Section 14.7)
Spastin	Microtubule severing (Section 14.7)
Fidgetin	Microtubule severing

Figure 7.2.3 Mapping of some AAA+ ATPases to particular compartments or cytoskeletal elements of a eukaryotic cell. AAA+ proteins involved in DNA replication are not shown. The green filaments are microtubules. TorsinA is associated with the nuclear membrane; NSF, with the Golgi; p97, with the endoplasmic reticulum; paraplegin, with mitochondria, etc. (Adapted from S. R. White and B. Lauring, *Traffic* 8:1657–1667, 2007. With permission from John Wiley & Sons.)

dynein (Section 14.7). Further variability enters in the accessory domains grafted onto the AAA+ platform. These domains vary widely in sequence and have several folds, including the ψ barrel, the double ψ barrel, and coiled coils. They are often at the N terminus of the AAA+ module but can also be at the C terminus, as in the clamp loader protein (Section 3.2), or inserted into a surface loop, as in the unfoldase ClpY (Section 7.3). These accessory domains and adaptor proteins are the key to the functional adaptability of AAA+ ATPases. They recognize appropriate substrates—which can be proteins or nucleic acids—and engage them with the force-generating machinery of the AAA+ domains. During their ATPase cycles, conformational changes are induced that bring forces to bear on the bound substrates—unfolding, translocating, remodeling, or disaggregating them, as the case may be. The axial channel through the center of a AAA+ ring is a functionally important feature, as many substrates are translocated through it (Figure 7.2.2B).

Despite the conserved AAA+ fold, the levels of ATPase activity possessed by these proteins vary widely, and some family members, for example some AAA+ modules of dynein (Section 14.7), do not hydrolyze ATP but instead have scaffolding roles. In some cases, the ATPase activity is stimulated by association of the AAA+ protein with substrates or other interacting proteins.

In sum, AAA+ ATPases represent an ancient lineage of mechanoenzymes in which the basic force-generating mechanism has been largely conserved and has been harnessed to a remarkable variety of cellular functions by the evolution of diverse sets of accessory domains and adaptor proteins. Some examples and the cellular locations where these proteins operate are summarized in **Figure 7.2.3**.

References

J.P. Erzberger and J.M. Berger (2006) Evolutionary relationships and structural mechanisms of AAA+ proteins. *Annu. Rev. Biophys. Biomol. Struct.* 35:93–114.

P.I. Hanson and S.W. Whiteheart (2005) AAA+ proteins: have engine, will work. *Nature Rev. Mol. Cell Biol.* 6:519–529.

L.M. Iyer, D.D. Leipe, E.V. Koonin et al. (2004) Evolutionary history and higher order classification of AAA+ ATPases. *J. Struct. Biol.* 146:11–31.

T. Ogura and A.J. Wilkinson (2001) AAA+ superfamily ATPases: common structure—diverse function. *Genes Cells* 6:575–597.

S.R. White and B. Lauring (2007) AAA+ ATPases: achieving diversity of function with conserved machinery. *Traffic* 8:1657–1667.

(Section 7.4) are the only machines of this kind. In *E. coli*, these components muster five degradation machines—ClpAP, ClpXP, ClpYQ, Lon, and FtsH (Figure 7.3B). ClpB, which is closely related to ClpA and is a homolog of the eukaryotic heat-shock protein Hsp104, does not partner a protease but instead disperses protein aggregates. The similarity of ClpA and ClpB is confirmed by the observation that transplanting a motif from the ClpP-interacting surface of ClpA onto ClpB renders it competent to partner ClpP in unfoldase-assisted proteolysis. For each of these machines, the degradation pathway follows a similar course (see Figure 7.1), but the specifics vary somewhat from system to system.

Only FtsH is essential in *E. coli*. However, some homologs of the other proteins are essential in other organisms, and their importance is underscored by their near-ubiquity in bacteria, as well as in mitochondria and chloroplasts. With some overlap (see below), they process different substrates, and their multiplicity represents another device whereby the range of substrates susceptible to intracellular proteolysis has been expanded. The Clp proteases serve as tractable model systems for investigating unfoldase-assisted proteolysis generally. Crystal structures have been determined for all of these proteins, but we focus here mainly on ClpAP, ClpXP, and ClpYQ, for which information on the fully assembled particles, their dynamic properties, and their interactions with substrates is quite well developed.

Assembly involves polymorphisms and symmetry mismatches

These proteins form hexameric or heptameric rings, and sometimes both (Figure 7.3A), depending on the assembly conditions or the organism in question. The ClpP and ClpQ proteases are bipolar double rings, heptameric for ClpP and hexameric for ClpQ. In *E. coli*, ClpP is initially assembled as an inactive zymogen with 15-aa propeptides occupying its active sites. Upon formation of the double ring, the propeptides are removed autocatalytically, activating the protease. In contrast, human mitochondrial ClpP, which has a longer propeptide (56 aa), remains as a stable but inactive single ring until it is paired, bringing about a conformational change that activates the now-sequestered proteolytic sites. A similar mechanism activates newly assembled proteasomes (Section 7.4). Lon protease rings do not pair to form a closed chamber: rather, the protease ring and the ATPase ring of a single hexamer enclose a gated chamber in which both unfolding and proteolysis take place. In FtsH, the active sites are also located in the compartment enclosed between the ATPase ring and the protease ring (see Figure 7.3B).

The ATPases ClpA and ClpX are hexamers. ClpB and ClpY have been found as both hexamers and heptamers, and both proteins have been solved crystallographically as hexamers. In the presence of ATP and Mg^{2+}, the rings form readily and stack axially on one or both ends of the protease. Because there are different orders of symmetry for ClpA and ClpX (6-fold) and for ClpP (7-fold), formation of the complex invokes a symmetry mismatch. Two possible functional implications of the mismatch have been suggested. First, it may serve as a 'quick release' device that allows the subcomplexes to disengage more easily than in a symmetry-matched system whose components are locked together in a cooperative interaction. Second, it could be conducive to mutual rotation of the rings during the processive processing of substrates (**Figure 7.4A**). Because only a small rotation is needed to transfer the key interaction to another pair of subunits (Figure 7.4B), the mismatch could facilitate rotation by making it easier for the two rings to remain in contact. Each rotary step might, in principle, be associated with a cycle of nucleotide hydrolysis.

A similar 7 : 6 symmetry mismatch is also observed in the proteasome—or, rather, a pseudo-symmetry mismatch in the case of heteromeric systems—but is not universal. The principal oligomeric form of ClpYQ, as visualized in crystallographic studies, is a symmetry-matched (6 : 6) complex. Lon and FtsH have collinear ATPases and proteases, ruling out a symmetry mismatch. There is, nevertheless, potential for symmetry-breaking in other ways; for instance, if the subunits in a given ring assume non-equivalent conformations or different nucleotide-binding states. Moreover, symmetry-matched systems may undergo other kinds of motions, such as screw displacements or axial contractions, to achieve similar dynamic effects.

Substrate proteins are marked for degradation by peptide signals

Among the first substrates to be identified for ClpXP were replication protein O of phage λ, and the repressor and transposase of phage μ. Subsequently, a systematic approach to the

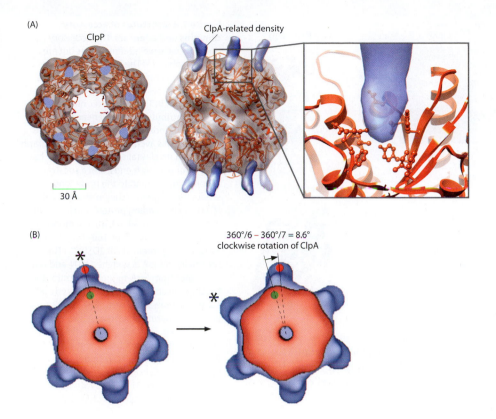

(A) ClpP · ClpA-related density

(B) 360°/6 − 360°/7 = 8.6°
clockwise rotation of ClpA

30 Å

Figure 7.4 The apical surface of ClpP heptamer presents interaction sites for ClpA hexamer. (A) The sites, marked by blue dots (left) or fingers (center), involve a crevice between ClpP subunits, where a surface loop of ClpA binds. This loop is relatively mobile; it was not seen in the ClpA monomer crystal structure but was detected by cryo-EM of the ClpAP complex. Because ClpA is a hexamer, only one of these sites on the ClpP heptamer can be occupied unless there are substantial departures from 6-fold symmetry. At right is shown the docking site between two ClpP subunits. (B) Diagram illustrating how relative rotation between coaxial rings could (hypothetically) be facilitated by a symmetry mismatch. The asterisk in the left-hand panel denotes the subunit pair on ClpA (red dot) and Clp (green dot) that are appropriately juxtaposed to interact. A small physical rotation (8.6°) shifts the juxtaposition to different subunits in the respective oligomers (asterisk, right hand panel). (A, from G. Effantin et al., *J. Biol. Chem.* 285:14834–14840, 2010. With permission from American Society for Biochemistry and Molecular Biology.)

identification of substrates was taken by engineering a proteolytically inactive ClpP with an appended affinity tag and expressing the gene for this protein in bacterial strains expressing *ClpX* or *ClpA*, or both. In each case, the inactivated ClpP with internalized substrates was harvested by affinity chromatography and the substrates were identified by mass spectrometry. About 60 proteins of diverse functionality were found to have been internalized by ClpXP, ~30 by ClpAP, and 10 by both proteases. Analysis of their amino acid sequences led to the conclusion that ClpX recognizes at least five degrons, each 5–6 aa long and located at one of the substrate's termini, usually the N terminus.

Most degrons are natural parts of the polypeptide chains in question. Some are not exposed—and so the protein survives—until some regulatory event occurs, such as the binding of a cofactor. One important degron is an exogenous peptide added at the C terminus of incomplete translation products whose synthesis has stalled on the ribosome. To rectify such situations, a system is mobilized that aborts the incomplete translation product, releasing the ribosome to resume operations (**Figure 7.5**). The key player is a remarkable RNA molecule, SsrA (for **s**mall **s**table **R**NA A), that functions as both a tRNA and an mRNA. SsrA recognizes a stalled ribosome and binds to its A site (see Section 6.2). Like an alanyl-tRNA, SsrA is charged with Ala, and this amino acid is first added to the C-terminal end of the stalled polypeptide chain. The SsrA then switches to the role of a surrogate mRNA, extending the nascent polypeptide chain by an additional 10 amino acids: ANDENYLYAA, in the single-letter code. The SsrA-appended polypeptide is then released, whereupon this tag is recognized by ClpXP, ClpAP, or FtsH, and degradation ensues.

Figure 7.5 Sequence of events in SsrA tagging. A translating ribosome stalls on a faulty mRNA. The distressed complex is recognized by an aminoacylated-SsrA RNA, which binds to the ribosomal A site and transfers the incomplete polypeptide chain (green) to the Ala-charged tRNA domain of SsrA (yellow). The SsrA (shown in orange) then replaces the faulty mRNA (red) and translation is resumed, adding more amino acids (red balls) until a stop codon is reached and the tagged protein is released. (Adapted from A.W. Karzai et al., *Nature Struct. Biol.* 7:449–455, 2000. With permission from Macmillan Publishers Ltd.)

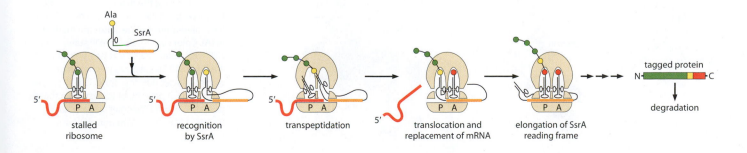

stalled ribosome · recognition by SsrA · transpeptidation · translocation and replacement of mRNA · elongation of SsrA reading frame · tagged protein · degradation

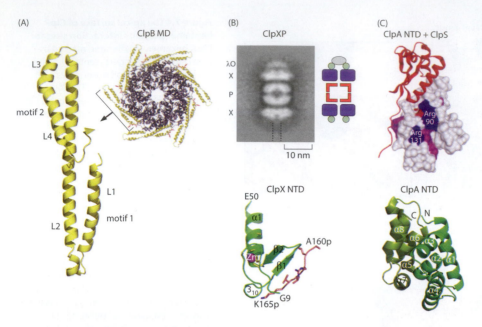

Figure 7.6 Structures of accessory domains and adaptors. The accessory domains have well-defined folds but are highly mobile. (A) The M domain (MD) of ClpB from *Thermus thermophilus* occupies a peripheral location on the D1 AAA+ ring. (PDB 1QVR) (B) The ClpX N-terminal domain (NTD) contributes the small near-axial protrusions indexed with short lines at the bottom of this averaged electron micrograph of the negatively stained complex. ClpP has ClpX bound at both ends, and a substrate, λO protein, is bound to the upper end. A schematic diagram is shown alongside. The NTD is a Zn⁺-binding protein and is shown here (below, from G9 to E50) as a monomer complexed with residues 160–165 of the Ssb adaptor protein. (PDB 2DS8) (C) The specificity of ClpA is switched to N-end rule substrates from SsrA-tagged substrates by binding the adaptor protein ClpS (red). (PDB 1LZW) (A, from S. Lee et al., *Cell* 115:229–240, 2003. With permission from Elsevier; B, top, from J. Ortega et al., *EMBO J.* 21:4938–4949, 2002; B, bottom, from E.Y. Park et al., *J. Mol. Biol.* 367:514–526, 2007. With permission from Elsevier; C, from K. Zeth et al., *Nature Struct. Biol.* 9:906–911, 2002. With permission from Macmillan Publishers Ltd.)

Identification of the SsrA peptide as a degron has been confirmed by making recombinant proteins in which it is appended to molecules such as GFP (green fluorescent protein) that would not normally be susceptible to a given protease, and observing the resulting degradation. Although its natural position is at the C terminus, the SsrA tag is also effective in most cases when attached at the N terminus of a recipient protein.

Accessory domains and adaptor proteins affect substrate selection

In addition to their AAA+ modules, ClpX, ClpA, and ClpB have other domains that, like their counterparts on other AAA+ ATPases, have been implicated in substrate binding. These are known as **accessory domains** (**Figure 7.6**). ClpX has an N-terminal domain (N domain) of ~60 aa, whereas those of ClpA and ClpB are longer at ~155 aa and consist of two similar subdomains, presumably reflecting a gene duplication event. On assembled hexamers, these domains surround the axial pore, which is the port of entry for substrates. ClpY has no N domain but has an insertion domain (I domain) that is positioned similarly on the apical surface of the ATPase ring (**Figure 7.7**). In contrast, the M domains of ClpB are distributed around the periphery of its D1 AAA+ tier (see Figure 7.6A): they are envisaged to act in conjunction with the *chaperone* DnaK in transferring substrate proteins to the axial pore, whereby they are pulled out of aggregates and allowed to refold.

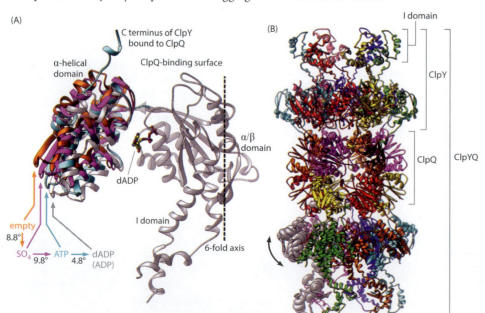

Figure 7.7 Conformations of ClpYQ (HslUV) protease in four different nucleotide states. (A) is an enlargement of the bottom left part of the ClpYQ complex, shown in (B), in which ClpY hexamers are bound at both ends of the bipolar ClpQ double hexamer. The I domains of ClpY protrude axially at both ends of the complex. The ClpY structure is a composite of PDB 1G3I and the I domain from PDB 1E94. The nucleotide-dependent conformational changes are confined to ClpY and affect the orientation of the 'small' domain (α-helical domain) of each ClpY subunit relative to the 'large' domain (α/β domain) of the neighboring subunit. Corresponding rotations of this domain for each of four nucleotide states are marked in (A). Color coding: silver, ADP state; cyan, ATP state; magenta, SO₄ state (which has 4 ATPs and 2 SO₄s bound per hexamer); orange, nucleotide-free state. The nucleotide, if present, binds between these domains. (From J. Wang et al., *Structure* 9:1107–1116, 2001. With permission from Elsevier.)

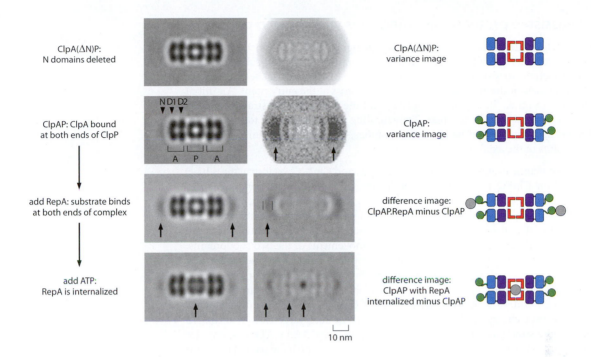

10 nm

These accessory domains are connected to the respective AAA+ rings by flexible linkers. Their mobility is affirmed by the fact that they tend to be poorly visible in cryo-EM (see **Methods**) reconstructions (**Figure 7.8**), indicative of large-scale random fluctuations, and that they are invisible in crystal structures unless captured by crystal contacts. Substrate selectivity may be modulated by regulatory proteins. For example, binding of ClpS to the N domain of ClpA (see Figure 7.6C) changes its specificity from SsrA-tagged substrates to N-end rule substrates (Section 7.2) whose N-termini lodge in a hydrophobic pocket on the surface of ClpS. Another class of adaptors regulates protease activity by binding to potential substrates so as to conceal their degrons.

Substrates are unfolded and translocated along an axial pathway

The binding of substrates to the apical surfaces of ClpA (Figure 7.8) and ClpX (Figure 7.6B) has been visualized by cryo-EM of complexes assembled with the ATP analog ATPγS, which supports assembly and substrate binding but not unfolding or translocation. The N domains are involved in the initial capture of substrates, which become bound to the apical surface of the AAA+ ring. However, the N domain of ClpA is, at least in part, dispensable: a mutant lacking it remains competent to process the model substrate bacteriophage protein RepA. Moreover, the importance of the apical surface of ClpX for substrate binding is confirmed by mutations in its conserved GYVG loop that alter its specificity.

To visualize the internalization of substrates, illustrated for ClpAP in Figure 7.8, they were first bound in the presence of ATPγS; excess ATP was then added to initiate translocation. To detect substrates that reached the degradation chamber, where they would normally be processed and the fragments released, protease inhibitors were used. When the same experiment was performed using an inactive ClpP whose degradation chamber was already half full with unprocessed propeptides, incoming substrates backed up at earlier points on the pathway. These observations established that translocation follows an axial route.

Complexes with ATPases mounted at both ends of the core particle could, in principle, be translocating substrates inward from both ends, simultaneously. It appears, however, that although substrates may bind at both ends, translocation proceeds only from one end at a time; in other words, there is *negative cooperativity*. Alternatively, the explanation may lie in the relative kinetics of the initiation and elongation stages of substrate translocation: the former slow and dependent on unfolding, and the latter fast. On the other hand, binding of the second ATPase may serve mainly to open the axial gate of the core particle to facilitate the release of degradation products (see below).

Figure 7.8 Substrate binding and internalization by ClpAP. These events are shown in averaged cryo-EM side views (left column), difference maps and variance maps (center column), and schematic diagrams (right column). In difference maps, one digital image is subtracted from another. Variance maps depict the pixel-by-pixel variability of the images used for averaging: regions of high variability can represent the locations of mobile components. The top two panels on the left compare wild-type ClpAP (ATPγS state) with ClpA bound at both ends of ClpP (row 2) and the complex in which the N domains of ClpA were deleted (row 1). Although their cumulative mass is fully ~60% of that of the D1 or D2 tier, the N domains are almost invisible in the averaged image, implying that they undergo large-scale fluctuations. The region in which they fluctuate is detected in the variance map, high variance appearing dark. On binding RepA substrate to wild-type ClpA (row 3), diffuse crescent-shaped densities become apparent (arrows) and are the strongest feature of the difference image (row 3). On the addition of ATP, substrate is translocated into the internal chamber of ClpP (arrow in row 4); the difference image shows that some density remains at the initial binding site and at earlier staging posts on the axial pathway (arrows). (Adapted from T. Ishikawa et al., *Proc. Natl Acad. Sci. USA* 98:4328–4333, 2001. With permission from National Academy of Sciences.)

Unfoldase-assisted proteases are machines with moving parts

The ATPase modules undergo major conformational changes during their functional cycles. The resulting movements apply forces that unfold and translocate bound substrates. For ClpA, this entails a trajectory of more than 100 Å from the initial binding sites (see Figure 7.8). The axial channels of the ATPases are only ~15 Å across at the entry point. An initial peptide enters the channel, 'threading the needle,' and the rest of the substrate protein is then unfolded and pulled into the complex. Translocating substrates appear to move along the axial channel in extended, non-native, conformations, which enables them to negotiate the narrowest points of the axial channel.

We discuss these dynamic events mainly for ClpX, ClpA, and ClpY. Crystal structures have been determined for ClpYQ in several nucleotide states (Figure 7.7). The orientation in which ClpQ binds to ClpY, with the I domains protruding from the distal surface, was determined both by small-angle X-ray scattering (SAXS) and by cryo-EM. The structural differences between the ADP and ATP states are small, but there is a more substantial change in the nucleotide-free state, corresponding to rigid-body movements of the two subdomains, suggesting that the power stroke may accompany nucleotide binding or release rather than hydrolysis. Further progress along the translocation pathway is assisted by interactions with loops that line the channel and move in response to global changes in the ring, ratcheting the substrate along. Models of the two AAA+ tiers of ClpA in the ADP and ATP states (**Figure 7.9**) show little change in the D1 ring, as in ClpYQ, but a large change in the D2 ring. The observed distinction supports the view that in two-tier AAA+ ATPases, one tier has low ATPase activity but provides a stable platform relative to which the second tier, with higher ATPase activity, may undergo more radical motions.

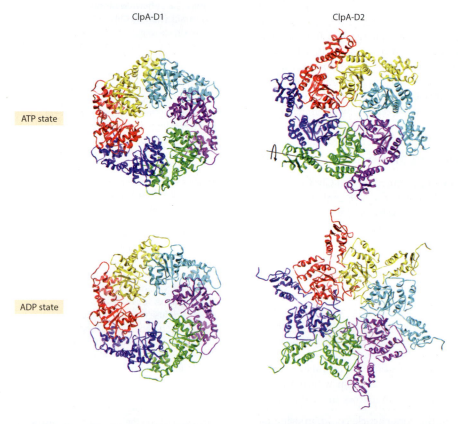

ClpA-D1   ClpA-D2

ATP state

ADP state

Figure 7.9 The D2 ATPase domains of ClpA switch conformation between their ATP and ADP states. The hexamer model of ClpA in the ADP state was derived from a crystal structure of the monomer. (PDB 1R6B) The ATP state was modeled by fitting the subdomains from the monomer/ADP structure into a cryo-EM map of the ATPγS-bound hexamer. The six subunits are in different colors. The D1 domain of a given subunit overlies the D2 domain of the next adjacent subunit round the ring. There is little structural change between the ADP and ATP states in the D1 tier but a major change in the D2 tier that can be described as a 20° rotation about the axis marked on the green subunit (top right). (From G. Effantin et al., *Structure* 18:553–562, 2010. With permission from Elsevier.)

Protein degradation is expensive in terms of its energetic requirements. The actual cost depends on the stability of the substrate and the resistance that it offers to unfolding. With ClpX, for example, it has been estimated that 30–80 molecules of ATP are hydrolyzed per 100 amino acids translocated, depending on the substrate.

The mechanics of substrate unfolding and translocation have been studied in the ClpXP system by 'single molecule' methods, using a dual optical trap. At one end, biotin-labeled ClpX was secured to a streptavidin-coated polystyrene bead; at the other end, the model substrate, SsrA-tagged GFP, was connected to a second, antibody-coated bead. With applied forces up to 13 pN, a translocation rate of 80 aa/sec was observed, falling off to zero translocation, or stalling, as the force was increased to 20 pN. The step-length observed was 1 nm, which was suggested to represent the axial stagger between different copies of a substrate-engaging pore loop in an asymmetric conformation of the ClpX hexamer.

Proteolytic active sites are sequestered inside gated chambers

Whereas the ATPase modules of all these proteolytic complexes share a common fold, their proteases vary in their folds. Those of ClpP, FtsH, and Lon are each unique, whereas ClpQ resembles the subunits of the 20S proteasome (see Figure 7.16, below). The soluble portion of FtsH has also been solved as a hexamer (**Figure 7.10**); its protease ring and ATPase ring are more closely associated than in ClpYQ or ClpAP. The protease is a Zn-dependent metalloproteinase whose active site includes a HEXXH motif set in a predominantly α-helical fold. The axial opening of the protease ring is small enough to exclude folded proteins, and substrates are unfolded and fed in from the ATPase side. The almost complete Lon protease of the archaeon *Thermococcus onnurineus*, comprising its gating (portal), ATPase, and protease domains, has been solved as a hexamer (**Figure 7.11B**). Its unfolding and degradation chambers merge to form a single compartment.

The ClpP subunit has two parts connected by a hinge: the 'N-loop,' in which residues 1–17 form a β-hairpin, and residues 18–183, which adopt a compact hatchet-shaped fold (Figure 7.11A). Two heptameric rings make up a barrel-like complex whose interior, the degradation chamber, is lined with the active sites. The exit route for the resulting peptides has not been established, but the axial channels offer an evident possibility. An NMR study has revealed that the inter-ring region of the ClpP barrel is dynamic, and fluctuations there may transiently open an alternative exit route.

Access to the degradation chamber is controlled by the N-loops. If they are deleted, the barrel still assembles and has a wide axial opening, ~30 Å across (**Figure 7.12E**). With full-length ClpP in the absence of ClpA, this opening is closed off by N-loops in 'down'

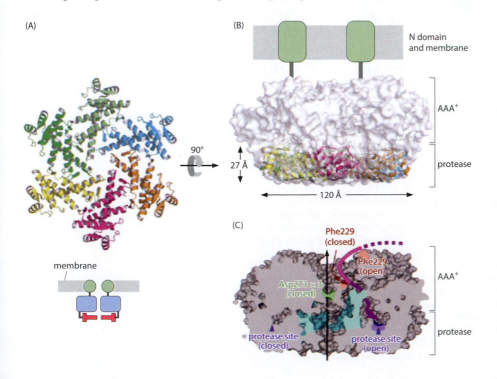

Figure 7.10 Structure of the soluble portion of FtsH from *Aquifex aeolicus*. This construct, lacking the membrane-inserting N-terminal domain (residues 1–125) was solved as a hexamer. (A) Top view of protease domains, residues 405–634. (B) Side view of the whole cytoplasmic portion shown as a transparent surface rendering, with the protease ring represented as a ribbon diagram. Inferred membrane association is shown schematically. (C) Cutaway view of the hexamer with active sites marked in two conformations—open and closed. These sites alternate in the structure solved. The putative pathway of incoming unfolded substrate is marked in magenta. (From R. Suno et al., *Mol. Cell* 22:575–585, 2006. With permission from Elsevier.)

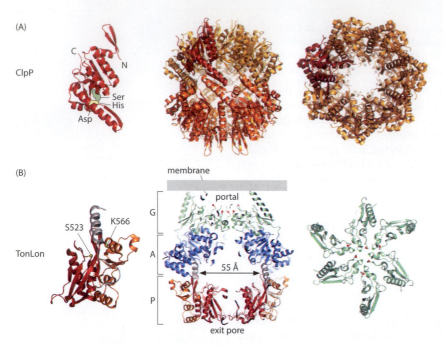

(A)

ClpP

(B)

TonLon

Figure 7.11 Structures of the proteolytic enzymes ClpP and LonB. (A) The *E. coli* ClpP subunit (left), and the side view (center) and axial view (right) of the tetradecamer. The two ClpP heptamers are in gold and orange, respectively, with one subunit in the gold ring in red. The catalytic triad residues are marked. (PDB 3MT6) (B) TonLon, a membrane-associated archaeal Lon protease from *T. onnurineus*, comprises a single hexameric ring with all three domains—gating (G, green), ATPase (A, blue) and protease (P, red)—as collinear parts of the same polypeptide chain. Left: the protease subunit with the active-site residues S523 and K566 are marked. Center: central slab of the complex in side view, showing the internal chamber and axial access points. Right: the G domain is formed from two inserts in the A domain. Two conformers of the A and G domains alternate around the ring, and represent states that differ with respect to ADP binding. The red and green balls mark side chains in the gating loops, in each case two sets of three. (Adapted from S.-S. Cha et al., *EMBO J.* 29:3520–3530, 2010. With permission from John Wiley & Sons.)

conformations (Figure 7.12C). The N-loops are not highly ordered; nor is the pore tightly sealed, because ClpP is able to cleave peptides and unfolded proteins in the absence of an ATPase. Nevertheless, they form a barrier that excludes folded proteins. When ClpP binds ClpA at a peripheral site relatively remote from the pore (see Figure 7.4), an allosteric signal is transmitted that causes the N-loops to switch into 'up' conformations, opening the gate (Figure 7.12A,D). ClpQ also has a gated axial pore that opens when ClpY binds (**Figure 7.13**).

A model of the fully assembled 1.3 MDa ClpAP complex is shown in **Figure 7.14**.

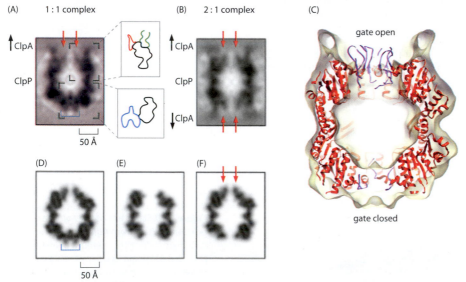

Figure 7.12 Binding of ClpA causes the axial gate of ClpP to open. (A, B) Central longitudinal sections through cryo-EM maps of ClpP with ClpA bound at the top end (A, 1 : 1 complex) or at both ends (B, 2 : 1 complex). Only the part of ClpA in immediate contact with ClpP is shown. In the upper insert in (A), contours delineate densities contributed by ClpP N-loops (red), the ClpP barrel (black), and part of ClpA (green). In the lower insert, they delineate the ClpP barrel (black) and pore-blocking N-loops (blue). N-loop densities in the 'up' (open) conformation are also marked by red arrows, and in 'down' (closed) conformation with blue bars, as visualized by cryo-EM (A) and as modeled and band-limited to the same resolution (D). (C) Crystal structure of ClpP with N-loops (magenta) fully resolved in the top tier (red) and only partly resolved in the bottom tier, fitted into the cryo-EM map of the 1 : 1 complex. (PDB 1YG6) (E) The same section from the crystal structure with the N-loop densities omitted, limited to the same resolution as the cryo-EM data; without the N-loops, the pore is ~30 Å wide. (F) Corresponding section from the crystal structure with N-loop-associated densities retained. (C, from G. Effantin et al., *J. Biol. Chem.* 285:14834–14840, 2010. With permission from American Society for Biochemistry and Molecular Biology.)

(A)

(B)

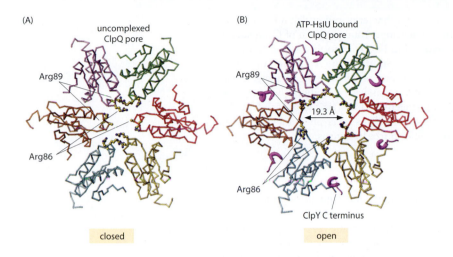

Figure 7.13 Binding of ClpY (HslV) causes the axial gate of ClpQ (HslU) to open. Insertion of the C termini of ClpY (HslU) subunits, shown in magenta in (B), into the clefts between adjacent ClpQ subunits elicits conformational changes in pore-lining loops in unbound ClpQ (A), causing the pore to open (B). This region is positively charged: two conserved arginines are marked. (From J. Wang et al., *Structure* 9:1107–1116, 2001. With permission from Elsevier.)

Proteases have regulatory roles in the replication cycles of bacteria and bacteriophages

Intracellular proteases effect the degradation of transcription factors and other regulatory proteins. The total number of such substrates is likely to be substantial; however, in bacteria only a few have been characterized in detail. These include two *sigma factors*, σ_s and σ_{32}, which bind to RNA polymerase, altering its specificity and promoting the expression of particular sets of genes (see Section 5.3). σ_s is a general stress-response factor; σ_{32} is a heat shock sigma factor. Under normal conditions, both proteins are expressed constitutively at low levels but are immediately intercepted and degraded by FtsH and ClpXP/RssB, respectively. In a stress condition, such as the accumulation of denatured cytoplasmic protein (for σ_{32}) or a variety of altered conditions such as elevated temperature (σ_s), these factors are up-regulated to the point of being able to elicit the appropriate transcriptional responses. In some cases, the protease does not act alone but in concert with another proteolytic activity that renders the substrate susceptible to degradation. One example of such a substrate is the UmuD' protein that is mobilized in the SOS response (Chapter 4) to repair double-strand breaks in DNA, and is eventually degraded by ClpXP.

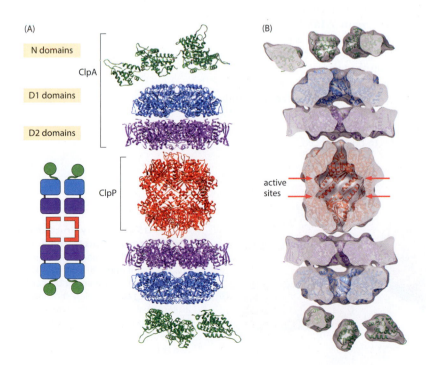

Figure 7.14 Model of the ClpAP holoenzyme in the ATP (ATPγS) state. The central tetradecamer of ClpP is PDB 1YG6. The model of ClpA was constructed by fitting the AAA+ domains from PDB 1K6K into a cryo-EM density map. The axial separation and relative azimuthal setting of ClpP and ClpA were determined by cryo-EM. The N-domain structure (PDB 1K6K) was placed in six random non-overlapping positions in the zone assigned to them (Figure 7.8). The flexible linkers connecting the NTD to D1 and D1 to D2 are not seen. Shown are (A, right) a ribbon diagram of the complex and (B) a cutaway view to convey the axial channel/pathway for substrates. (A, left) A schematic diagram. (Adapted from G. Effantin et al., *Structure* 18:553–562, 2010. With permission from Elsevier.)

Intracellular proteases also have regulatory roles in the bacterial cell cycle and in direct-ing the developmental behavior of bacteria. In the developmental program of *Caulobacter crescentus*, the intracellular levels of proteins involved in flagellar motility and chemotaxis (Section 14.11) are regulated by Clp and Lon proteases. Similarly, the replication of bacte-riophage λ in *E. coli* and, in particular, the crucial decision of whether the phage genome becomes incorporated into the host chromosome for passive replication (*lysogeny*) or embarks on DNA replication and the assembly of progeny virions (*lytic response*) depends on the processing of phage proteins by the host proteases. This latter phenomenon repre-sents a striking example of exploitation of a host cell resource by a virus.

7.4   THE PROTEASOME

In eukaryotic cells, proteins that are damaged or misfolded, and regulatory proteins that have reached the end of their lifespan, are marked for degradation by the covalent attach-ment of a small 76-aa protein, ubiquitin. Via a cascade of enzymatic reactions, an isopep-tide bond is formed between a Lys residue of the target protein and the C-terminal Gly of ubiquitin. Polyubiquitin (see Chapter 12), resulting from chain extensions, is a recognition signal for the 26S proteasome, a molecular machine of ~2.5 MDa that executes the destruc-tion of the target proteins. It comprises the barrel-shaped 20S core particle, of ~700 kDa, with one or two copies of the 19S regulatory particle, of ~900 kDa, bound to its end(s).

The core particle confines the proteolytic action to a nanocompartment sequestered from the cytosol, whereas the regulatory particle recruits substrates marked for degradation and prepares them for translocation into the core particle. The preparation includes several steps: the binding of a polyubiquitylated protein and its deubiquitylation; unfolding of the substrate protein; and opening of the gate that controls access to the interior of the core particle.

26S proteasomes exist only in eukaryotic cells, but 20S proteasomes are also found in prokar-yotes. The growing number of sequenced genomes has revealed a progressively clearer pic-ture of the proteasome distribution across the three kingdoms of life: 20S proteasomes are ubiquitous and essential in eukaryotes, ubiquitous but not essential in archaea, and rare and non-essential in bacteria, where other ATP-dependent proteases abound (Section 7.3).

The molecular architecture of 20S proteasomes is conserved from archaea to humans

In spite of differences in subunit complexity, the quaternary structure of 20S proteasomes is highly conserved across the three domains of life. Four seven-membered rings form a barrel-shaped complex 16 nm long and 12 nm in diameter (**Figure 7.15**). Most archaeal and bacterial 20S proteasomes are built from two subunits (α and β), whereas eukaryotic pro-teasomes typically contain 14 different but related subunits, which are classified as α-type or β-type on the basis of sequence similarity. Although 14 is the upper limit for the number of different subunits that can be accommodated in a single proteasome, vertebrates have nevertheless achieved an even higher degree of complexity: in addition to the constitutive subunits, they have several inducible β subunits that can replace their constitutive counter-parts. In *immunoproteasomes*, the incorporation of inducible β subunits allows specificity to be modulated and the diversity of immunogenic peptides to be enhanced.

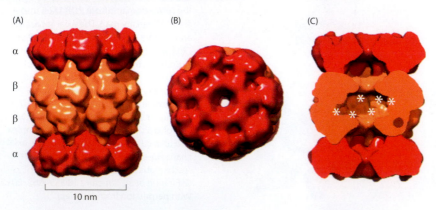

(A)   (B)   (C)

α

β

β

α

10 nm

Figure 7.15 Quaternary structure of the *Thermoplasma* 20S proteasome. (A) Side view showing the arrangement of α-subunit and β-subunit rings; (B) axial view showing an α ring with an opening at the center—the gate; and (C) cutaway view showing the arrangement of the inner cavities. Active sites, located in the central cavity, are marked with asterisks; the central cavity is flanked by two antechambers.

As determined by immuno-EM, the two outer rings of the core particle are formed by α subunits and the two inner rings by β subunits. In the simplest case, common in prokaryotes, the complex has 2-fold and 7-fold symmetry axes and an $\alpha_7\beta_7\beta_7\alpha_7$ stoichiometry. The eukaryotic complexes observe pseudo-7-fold symmetry and consist of two equal halves related by 2-fold symmetry; their stoichiometry can therefore be described as $\alpha_{1-7}\beta_{1-7}\beta_{1-7}\alpha_{1-7}$, with indices referring to the order in rings. Thus, each of the 14 subunits is present in two copies at precisely defined positions.

The two β rings enclose a central cavity with a diameter of ~5 nm, which harbors the catalytic sites (see below). It is connected via narrow constrictions to two slightly smaller outer cavities, the 'antechambers', each enclosed between one α and one β ring. The antechambers can serve as storage places for substrates before degradation, as shown by mass spectrometry and cryo-EM. Axial gates in the α rings lead into the antechambers (see Figure 7.15).

α subunits and β subunits have the same basic fold

Because of their relative simplicity, studies on prokaryotic proteasomes—in particular of the complex from the archaeon *Thermoplasma acidophilum*—have had a pivotal role in elucidating the structure and enzymatic mechanism of the proteasome and its assembly. As expected from their sequence similarity, the non-catalytic α-type and the catalytic β-type subunits have the same fold (**Figure 7.16**): a central five-stranded β sheet sandwich flanked on either side by α helices. Helices 1 and 2 mediate the interaction of the α and β rings.

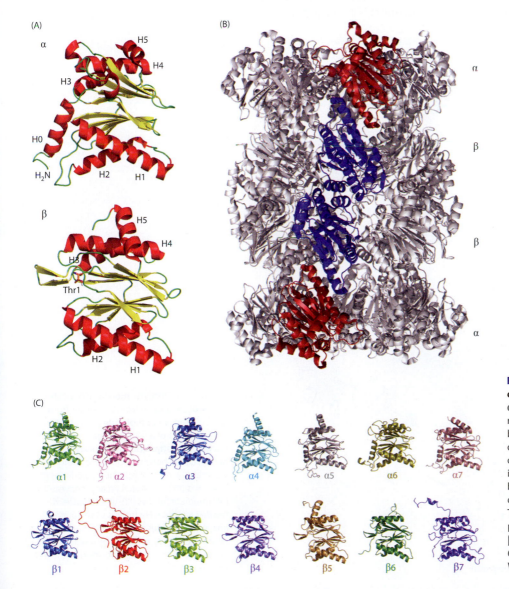

Figure 7.16 Structure of 20S proteasome core particle from *Thermoplasma*. (A) Top, α subunit; bottom, β subunit. The active-site residue Thr1 of the β subunit is shown in ball-and-stick representation. (B) Structure of the fully assembled tetradecameric core particle. One α subunit in each ring is highlighted in red and one β subunit in blue. (C) Structures of the α and β subunits of the yeast 20S proteasome core particle. The subtle differences determine the positioning of the subunits within the α and β rings. (From M. Groll et al., *ChemBioChem* 6:222–256, 2005. With permission from John Wiley & Sons.)

Helices 3 and 4, located on the opposite side, mediate the association of the β rings. In β subunits, the β-sheet sandwich is open on one side and closed on the other side by four hairpin loops. The open side forms the active-site cleft and is exposed to the central cavity. The α subunits have a highly conserved N-terminal extension; part of this extension forms an additional helix, helix O, which occupies the cleft in the β sandwich. Its location at the top of the α rings suggests a role in mediating the interaction with the 19S caps and other regulatory complexes. All 14 subunits of the eukaryotic core particle (see Figure 7.16) contain characteristic insertion and terminal elements that have key roles in specifying the order of these otherwise very similar subunits within the complex.

The N-terminal threonine functions as a single-residue active site

The 20S proteasome of eukaryotes was initially characterized as a 'multicatalytic' protease with chymotrypsin-like, trypsin-like, and peptidylglutamyl-peptide hydrolase activities, and it was thought to be an assembly of up to five different proteolytic components. However, the lack of sequence similarity to other proteases and the inconclusive nature of inhibitor studies indicated that it did not belong to one of the known protease families. Eventually, site-directed mutagenesis studies and the crystal structure of the *T. acidophilum* proteasome in complex with the competitive inhibitor *N*-acetyl-Leu-Leu-norLeu revealed that the active site is the N-terminal Thr of the β subunits. The hydroxyl group of Thr1 attacks the carbonyl carbon of the scissile bond, while its amino group serves as the primary proton acceptor, which enhances the nucleophilicity by stripping the proton from the side-chain hydroxyl; for steric reasons, a water molecule is likely to mediate the proton transfer (**Figure 7.17**). A single-residue active site like this is characteristic of N-terminal nucleophile

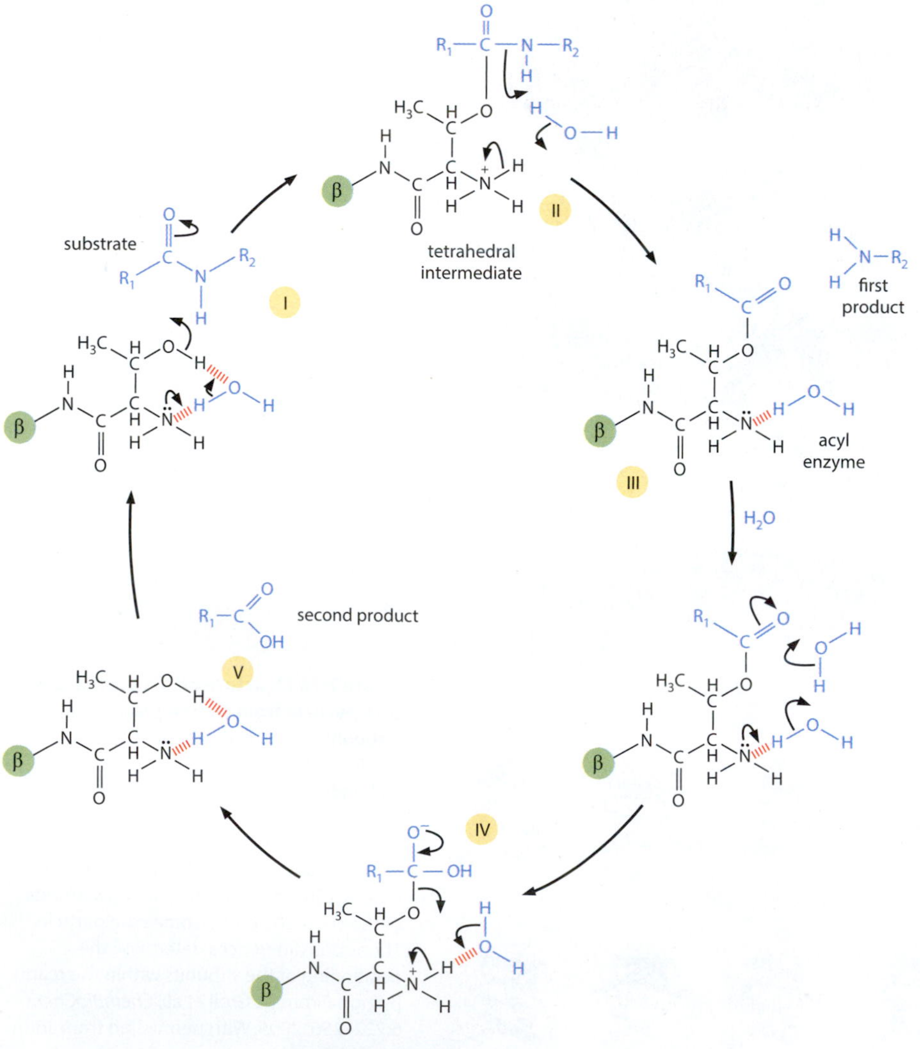

Figure 7.17 Catalytic mechanism of the 20S proteasome. A proton transfer from an activated water molecule at the hydroxyl group of Thr1 of a β subunit promotes nucleophilic attack on the carbonyl group of the scissile bond of a bound protein substrate (I). A tetrahedral intermediate is formed (II) from which an acyl enzyme intermediate follows (III), with release of the first product. Hydrolysis of the acyl enzyme generates another tetrahedral intermediate (IV), followed by release of the second product and regeneration of the free enzyme (V).

(Ntn) hydrolases, as is the autocatalytic removal of the propeptide, a process that is needed to expose the nucleophilic group (see below). Besides the N-terminal Thr, several other residues are required for proteolytic activity, but their exact roles await clarification. Proteasome inhibitors, such as peptide boronates that muzzle the N-terminal Thr have become valuable therapeutic agents for the treatment of several cancers (see **Box 7.3**). The barrel-shaped architecture of the 20S proteasome allows substrate proteins to be degraded in a processive manner, without the release of partly degraded protein. Whereas the 14 identical β subunits of the archaeal proteasome are proteolytically active, only six of those in the eukaryotic proteasomes are active, two copies each of β_1, β_2, and β_5. Nevertheless, the degradation products of both archaeal and eukaryotic proteasomes fall into the same size range of typically 8–12 amino acids. The mechanism underlying the generation of peptides of such a length is only poorly understood; possibly, peptides below this length can escape from the proteasome's interior rather easily.

Assembly of the complex precedes active site formation

The proteolytically active β subunits are synthesized as inactive precursors with N-terminal propeptides of varying lengths, which must be removed to allow the formation of active sites. Activation is delayed until assembly is complete and the active sites are sequestered from the cellular environment. Cleavage of the propeptides proceeds autocatalytically, relying on the active-site Thr and a conserved Gly at position -1 which appears to be the prime determinant of the cleavage site.

The pathway of 20S proteasome assembly and also the roles of the β-subunit propeptides differ somewhat between archaea, bacteria, and eukaryotes. The α subunits of archaeal *T. acidophilum* proteasomes form heptameric rings spontaneously, whereas β subunits remain monomeric, unprocessed, and hence inactive unless α rings are available to serve as a template for their assembly (**Figure 7.18**). In contrast, neither α nor β subunits of the bacterial *Rhodococcus* proteasome assemble by themselves, but complexes are formed as soon as both types of subunit are allowed to interact. Assembly proceeds via the formation of αβ heterodimers followed by half-proteasomes, built from one α ring and one β ring. The docking of two half-proteasomes triggers a conformational switch of a loop, S2–S3, that ensures the correct coordination of the active-site residues for propeptide cleavage (see Figure 7.18). The long propeptides (up to 65 aa) have dual functions: they support the initial folding of the β subunits and also promote assembly by increasing the contact area between subunits.

Assembly of the eukaryotic core particle is a more complicated process; 14 closely related but different subunits must be positioned correctly. First, α rings are formed, as in archaeal proteasomes, and these rings serve as templates for the sequential assembly of 'early' and

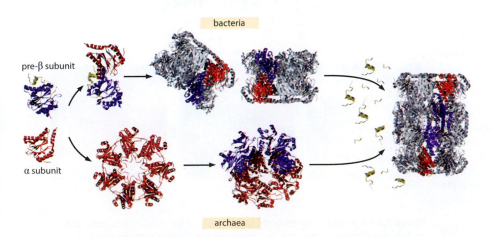

Figure 7.18 Assembly pathways of bacterial and archaeal 20S proteasomes. In bacteria, the first assembly intermediate is an αβ heterodimer, which then assembles into a half-proteasome. In archaea, the α subunits assemble to a heptameric ring that provides a platform for the assembly of the β subunits. In both cases, half-proteasomes pair to form the 20S complex; at this stage, the propeptides of the β subunits are removed autocatalytically, rendering the active sites functional.

Box 7.3   **Proteasome inhibitors as anti-cancer drugs**

Several early studies with peptide aldehydes as inhibitors of proteasomes hinted at a possible anti-cancer activity. Peptide aldehydes easily enter cells and reversibly inhibit the proteasome; however, the active aldehyde is unstable and rapidly oxidized to an inactive acid. Lactacystin is a natural product produced by bacteria of the genus *Streptomyces*. Although it cannot permeate cells and does not react with proteasomes, it converts spontaneously in culture medium to a reactive β-lactone that can traverse the plasma membrane. The β-lactone reacts with the hydroxyl group of the active-site Thr of the proteasome to form an acyl-enzyme conjugate. Lactacystin is more specific than peptide aldehydes, but the active β-lactone is also very unstable under intracellular conditions. Peptide aldehydes, and lactacystin and its derivatives, are useful reagents for dissecting the physiological roles of proteasomes; however, their lack of specificity and metabolic instability makes them unsuitable for use as therapeutic drugs.

These shortcomings stimulated the synthesis of other compounds. Boronic acid derivatives of tripeptide aldehydes turned out to be very potent inhibitors, 'capping' the N-terminal Thr of the β subunits. The boron atom interacts covalently with the nucleophilic oxygen ion pair of Thr1-O$^\gamma$, while one of the acidic boronate hydroxyl groups forms a hydrogen bond with Gly47, stabilizing the oxyanion hole. A second hydroxyl group bridges the N-terminal Thr amino group, functioning as a catalytic proton acceptor. The pyrazine moiety interacts with Asp114 of the β6 subunit. A defined water molecule is hydrogen-bonded to the inhibitor as well as to Asp114, Ala49, and Ala50 (**Figure 7.3.1**). A further improvement in specificity was obtained by truncation of the tripeptide to a dipeptide. One such dipeptide, bortezomib, fulfilled all the criteria for testing in animals and in human patients.

Bortezomib or Velcade® was found to be effective in the treatment of multiple myeloma (MM). MM is a differentiated clonal B-cell malignancy characterized by a high proliferation of plasma cells. The disease, which is typically contained in the bone marrow, suppresses all other bone marrow cellular elements, resulting in anemia, severe infections, and bleeding disorders. It is accompanied by osteoporosis, lytic skeletal lesions, pathological fractures, and hypercalcemia, which damages the kidneys. A hallmark of the disease is the massive production of circulating monoclonal immunoglobulin species that also contribute to the renal damage. MM accounts for 1% of all malignancies and 10% of hematological malignancies, with ~15,000 new cases per year in the USA alone. Velcade has also been approved for the treatment of mantle-cell lymphomas.

The cellular mechanisms underlying the anti-cancer potential of proteasome inhibitors are complex and little understood. Broadly, it appears that cancer cells and normal cells respond differently to proteasome inhibition. Normal cells often undergo cell-cycle arrest, whereas cancer cells tend to be driven into apoptosis. The inhibitors affect malignant cells via several mechanisms, one of which appears to be suppression of the transcriptional regulator, nuclear factor κB (NFκB). NFκB is constitutively activated in numerous malignancies, including MM, and appears to support several tumor-promoting activities. By inhibiting NFκB activation, proteasome inhibitors

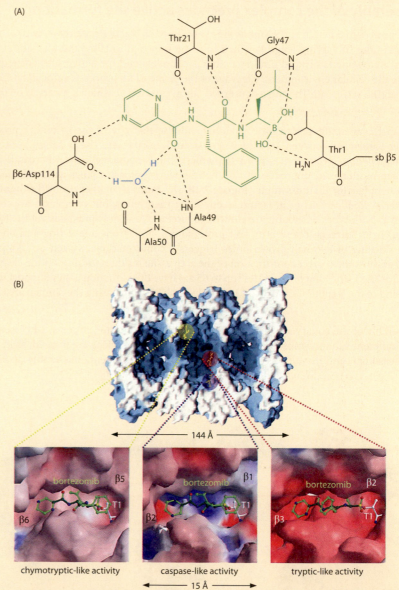

Figure 7.3.1 A proteasome inhibitor. (A) Structure of the dipeptidyl boronic acid proteasome inhibitor bortezomib (green) and its interactions with the active-site residues of subunit (sb) β5 ('chymotryptic' active site). (B) Crystal structure of bortezomib in complex with the yeast 20S proteasome. Top, cutaway overview, exposing internal structure; bottom, enlargements of bortezomib bound to distinct active sites. (B, adapted from M. Groll et al., *Structure* 14:4451–4456, 2006. With permission from Elsevier.)

sensitize malignant cells to the activity of other chemotherapeutic agents. Thus, the use of proteasome inhibitors in combination with other drugs, such as Hsp90 inhibitors (Section 6.3)
appears to be synergistic over the use of these drugs singly.

Another mechanism relates to the disruption of the *unfolded
protein response* (UPR), a cellular stress response related to the
endoplasmic reticulum (ER) that involves chaperone activity.
Cells that produce large amounts of immunoglobulins, such as
myeloma cells, are dependent on an intact ER-associated degradation (ERAD) system and the UPR to attenuate the ER stress
that these proteins induce (see Box 7.4). The combined effects
of an increased load of misfolded proteins along with suppression of the UPR result in an induction of apoptosis.

References

J. Adams (2002) Proteasome inhibition: a novel approach to
cancer therapy. *Trends Mol. Med.* 8:S49–S54.

A.F. Kisslev, W.A. van der Linden and H.S. Overkleeft (2002)
An expanding army attacking a unique target. *Chem. Biol.*
19:99–115.

A. McBride and P.Y. Ryan (2002) Proteasome inhibitors in the
treatment of multiple myelomas. *Expert Rev. Anticancer Ther.*
13:339–358.

'late' β subunits. Whereas prokaryotic core particles mature spontaneously, the assembly of
eukaryotic proteasomes is facilitated by maturation factors such as the proteasome assembly chaperones (PAC) and Ump1. These small proteins, 15–30 kDa in size, are found in half-
proteasome precursors but not in the mature core particle; they are degraded after fulfilling
their role in assembly.

Access to the proteolytic chamber is controlled by gated pores

The interior of the core particle is only accessible to unfolded polypeptides, which are able
to pass through the narrow opening at the center of the α-subunit rings. In archaeal proteasomes, this 13 Å-wide pore is delimited by turn-forming segments of a Tyr-Gly-Gly-Val
motif. In eukaryotic core particles, the N-termini of the α subunits project into this opening,
effectively closing it with several layers of interdigitating tails, whereas in archaeal proteasomes those residues are disordered (**Figure 7.19**). Thus, substrate access into the interior of
eukaryotic proteasomes requires substantial rearrangements of these N-terminal segments.
Those of the seven α subunits are divergent in sequence and different in structure, but highly
conserved across species. The binding of regulatory complexes to the α rings causes this gate
to open (see below), allowing substrates to enter. However, how degradation products leave
is still unclear: the dynamical structure of the complexes, as revealed by NMR spectroscopy
(see **Methods**), might create temporary openings that allow them to escape.

The 11S regulator acts as a gate opener

Several regulatory protein complexes have been described that bind to the ends of the
core particle, causing the α-ring gates to open. They are referred to as **proteasome activators** (PA) because they stimulate protein or peptide degradation. Best studied is the PA28

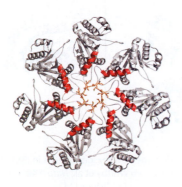
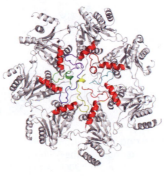

A. fulgidus CP   yeast CP wild type

**Figure 7.19 Open and closed states of
the axial pore of the core particle.** The
open α ring in the archaeal (*Archaeoglobus
fulgidus*) core particle (CP) and the closed
α ring in the yeast core particle. (From M.
Groll et al., *ChemBioChem* 6:222–256, 2005.
With permission from John Wiley & Sons.)

particle, which has 28 kDa subunits and is also called the 11S. PA28 is an ATP-independent activator that strongly stimulates the hydrolysis of small peptides but not of denatured or ubiquitylated proteins. It is a seven-membered ring built from one or two kinds of subunits, α and β—not to be confused with core particle subunits—that bind to the ends of the core particle. *In vitro* studies have shown that PA28 can enhance the generation of antigenic peptides; consistent with this, genes for PA28α and PA28β are found only in organisms with an *adaptive immune system* (see Chapter 17), and the expression of their genes is stimulated by γ-interferon. The PA28 γ-protein, also called Ki antigen, is a nuclear protein with sequence similarity to PA28α and PA28β but is not present in the cytosolic PA28 complex. It is considered an evolutionary precursor of PA28α and PA28β.

The mode of operation of PA28-like activators has been illuminated by the crystal structure of PA26, a relative from *Trypanosoma brucei* with low sequence similarity to PA28, in a hybrid complex with the yeast core particle (**Figure 7.20**). Subunits comprising four long α helices form a homoheptamer 90 Å in diameter and 70 Å in length. The heptamer has an axial pore 33 Å in diameter at the proteasome-binding end and 24 Å at the distal end. The loop between helices 2 and 3, referred to as the 'activation loop,' acts as a trigger for gate opening; upon binding into pockets on the α-ring surface of the core particle, a conformational change is induced (**Figure 7.21**).

PAN is an archetypal proteasome-activating nucleotidase

Genomic sequencing of archaea revealed the existence of genes with sequences similar to those of the ATPases found in eukaryotic 26S proteasomes (see below). The proteins encoded by these genes have N-terminal coiled-coil domains, which is a hallmark of proteasomal ATPases, and a C-terminal AAA+ domain (see Box 7.2). The proteins were found to have ATPase activity and to stimulate the degradation of substrate proteins by the core particle; hence the name proteasome-associated nucleotidase (PAN). Homologs are found in most but not all archaea. In some archaea and also in proteasome-possessing bacteria, the role of PAN is assumed by other members of the AAA+ family, such as homologs of the cell division control protein 48 (Cdc48).

Binding of the PAN hexamer to the core particle, which requires ATP binding but not hydrolysis, opens the α-ring gate. PAN and the other proteasomal ATPases all have a conserved C-terminal ΦYX motif (where Φ, sometimes given as Hb in the proteasome literature, is any hydrophobic residue) with a key role in gate opening. Similar to the PA28 activator, PAN binds to an inter-subunit pocket in the α ring, inducing a conformational change. As with the Clp proteases ClpAP and ClpXP (Section 7.3), the symmetry mismatch between the 7-fold α ring and the 6-fold ATPase ring means that its six C-termini cannot bind in an equivalent manner; this will favor a weak binding that may facilitate functional cycling. PAN has chaperone activity: it can prevent the aggregation of misfolded proteins and promote their refolding, acting in both cases in an ATP-dependent manner. An ATP-independent activity has been assigned to the N-terminal domain, which is assumed to

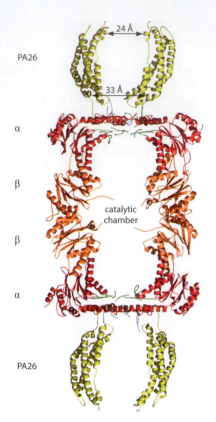

Figure 7.20 Structure of the PA26/20S proteasome complex. A ribbon representation is shown, cut away to reveal internal features. Residues of the gate in the α ring are in green. (From F.G. Whitby et al., *Nature* 408:115–120, 2000. With permission from Macmillan Publishers Ltd.)

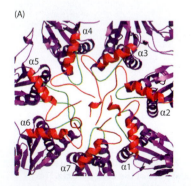
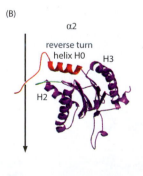
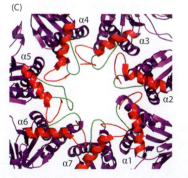
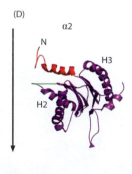

Figure 7.21 Conformational changes in the 20S proteasome α ring associated with PA26 binding and gate opening. (A) View of the α ring down the pseudo-7-fold axis in the closed unliganded state. (B) Structure of subunit α2 (unliganded) viewed ~90° around the horizontal from (A).

(C) As in (A), but for the PA26-liganded proteasome. (D) As in (B), but for the PA26-liganded proteasome. (From Whitby et al., *Nature* 408:115–120, 2000. With permission from Macmillan Publishers Ltd.)

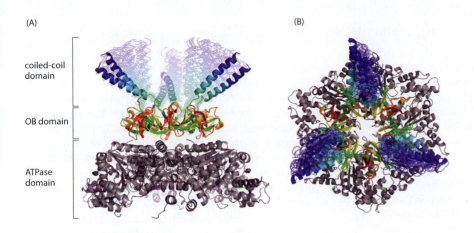

(A)

coiled-coil
domain

OB domain

ATPase
domain

(B)

Figure 7.22 Model of the PAN ATPase showing the conformational range of the coiled-coil domains. The coiled coils (dark blue) are shown in side view (A) and top view (B). The experimentally determined structure is shown in solid lines, and samplings of conformations accessible to the linker are added in translucent representation (rainbow colors). A model of the ATPase ring (gray) has been added to illustrate the relative domain sizes and overall shape of the complex. Given the lack of structural knowledge on the long connector between the OB and ATPase rings, no attempt was made to model this part of the structure. (From S. Djuranovic et al., *Mol. Cell* 34:580–590, 2009. With permission from Elsevier.)

have a key role in substrate recognition and binding. This activity can be enhanced and modulated by ATP hydrolysis in the nucleotidase domains, and it has been suggested that ATP hydrolysis fuels concerted motions in the N domains.

The crystal structure shows that the N-terminal domain comprises a domain with an oligonucleotide-oligosaccharide-binding fold (OB fold—see **Guide**) that forms a hexameric ring and, flexibly linked to it, a coiled-coil domain. The coiled coils pair adjacent subunits, bestowing 3-fold symmetry on the complex (**Figure 7.22**). The central pore in the OB ring ranges between 11 and 18 Å in diameter, implying that substrate proteins must be unfolded before they can pass through to the ring of nucleotidase domains.

The 19S regulatory particle links the ubiquitin system with the proteasome

In eukaryotic cells, the 20S core particle co-assembles with one or two 19S regulatory particles in an ATP-dependent manner to form the 26S proteasome, which is the executive arm of the ubiquitin pathway. 26S proteasomes are abundant in both the nucleus and the cytoplasm, and studies with proteasome inhibitors indicate that the cell makes use of this system for 80–90% of its degradation of cellular proteins. Not only is it used as machinery for the disposal of misfolded proteins, such as defective ribosomal products and proteins damaged by stress and aging, but it also controls the lifespan of many short-lived regulatory proteins. Defective or misfolded proteins in the endoplasmic reticulum must be translocated into the cytosol for degradation by the 26S proteasome, a process known as endoplasmic reticulum-associated degradation (ERAD; see **Box 7.4**).

Reports on the molecular composition of 19S regulatory particles have been confusing, because it is difficult to distinguish between constitutive components, transiently bound subunits, and interacting factors. However, it is now accepted that they comprise 18–19 different constitutive subunits, depending on species (**Figure 7.23**): their masses add up to ~900 kDa, in good agreement with the mass of the 19S particle as measured by scanning transmission EM. The regulatory particles can be considered as consisting of two subcomplexes that are referred to as the 'base' and the 'lid.' The complex formed by the base and the core particle is sufficient for the degradation of non-ubiquitylated protein substrates, but does not mediate the degradation of ubiquitylated substrates.

The base subcomplex recruits substrates and prepares them for degradation

Whereas the structure of the 20S core particle was determined by X-ray crystallography (see **Methods**) some time ago, the structure of the 26S holocomplex remained elusive because of its instability and compositional and conformational heterogeneity. Only recently was its subunit architecture established by cryo-EM integrated with other data, including high-resolution structures of several subunits (**Figure 7.24**). Six of the 19S subunits are AAA+ ATPases referred to as regulatory particle triple-A proteins (Rpt1–6); the remainder form

Box 7.4 Endoplasmic reticulum-associated degradation (ERAD)

The ER is the cellular compartment where membrane proteins and secretory proteins fold and are assembled and glycosylated. As its quality control system, it employs an elaborate protein machinery. This system, ERAD, identifies and degrades misfolded and incompletely folded proteins. The ER does not possess its own protease(s), relying instead on the cytosolic 26S proteasome. Thus, substrates destined for degradation are translocated from the ER to the cytosol, where they are ubiquitylated and degraded. Two E3 ubiquitin ligases mediate the ubiquitylation of ERAD substrates: defective proteins in the ER lumen (L-ERAD) or in the ER membrane (M-ERAD) are targeted by the E3 ligase Hrd1, whereas the E3 ligase Doa10 tags ER proteins with misfolded cytosolic domains (C-ERAD). To carry out the sequential ERAD steps, the E3 ligases associate with other proteins to form the larger **ligase core complexes**.

The steps of the ERAD pathway have been studied mostly for L-ERAD substrates such as the model substrate CPY*, a mutant of yeast carboxypeptidase Y (**Figure 7.4.1**). Step 1: E3 ubiquitin ligases and associated luminal recognition factors recognize misfolded proteins. Step 2: these proteins are translocated to the cytosol. Step 3: at the cytosolic face of the ER, substrates are polyubiquitinylated by the cascade of E1/E2/E3 and also E4 enzymes. Step 4: finally, the substrate is segregated from the ubiquitylation machinery by the AAA+ ATPase Cdc48 and degraded by the 26S proteasome. Some substrates do not require all of these steps and hence some subunits may be dispensable.

Most luminal substrates are recognized by their glycosylation patterns: after nascent proteins have been translocated into the ER, glycans are attached to particular Asn residues. These N-linked glycans are further trimmed to facilitate folding and chaperone recruitment; when folding is too slow, trimming by ER mannosidase 1 (ERMan1) and ER degradation-enhancing mannosidase-like proteins results in a terminal α-1,6-linked mannose that is recognized by the *lectins* ERMan1 and Htm1 and by the ubiquitin ligase complex subunit Yos9. In conjunction with Hrd3 and the chaperone Kar2, the substrate is recruited to the ligase complex. Next, the substrate is translocated to the cytosolic ER side, presumably via an as yet unidentified channel. Ubiquitylation is accomplished by a cascade of E1, E2, and E3 enzymes, specifically the E1 ubiquitin-activating enzyme in concert with the E2 ubiquitin-conjugating enzyme Ubc7, which is recruited to the Hrd1 complex via Cue1, and the cytosolic RING (for really interesting new gene) domain of Hrd1, which acts as an E3 ubiquitin ligase. The ATPase Cdc48 (in mammals, p97) is central to the final ERAD step and associates with the ligase core complex via subunit Ubx2. The Cdc48 hexamer, in complex with its cofactors Ufd1/Npl4 and Ufd2, assists in the translocation of substrates to the cytosol, promotes the polymerization of polyubiquitin chains, and finally segregates the substrate from the translocation/ubiquitylation machinery for degradation by the proteasome.

References

P. Carvalho, V. Goder and T.A. Rapaport (2006) Distinct ubiquitin-ligase complexes define convergent pathways for the degradation of ER proteins. *Cell* 126:361–373.

M.M. Hiller et al. (1996) ER degradation of a misfolded luminal protein by the cytosolic ubiquitin-proteasome pathway. *Science* 273:1725–1728.

C. Hirsch et al. (2009) The ubiquitylation machinery of the endoplasmic reticulum. *Nature* 458:453–460.

E.D. Werner, J.L. Brodsky and A.A. McCracken (1996) Proteasome-dependent endoplasmic reticulum-associated protein degradation: an unconventional route to a familiar fate. *Proc. Natl Acad. Sci. USA* 93:13797–13801.

Y. Ye, H.H. Meyer and T.A. Rapaport (2001) The AAA ATPase Cdc48/p97 and its partners transport proteins from the ER into the cytosol. *Nature* 414:652–656.

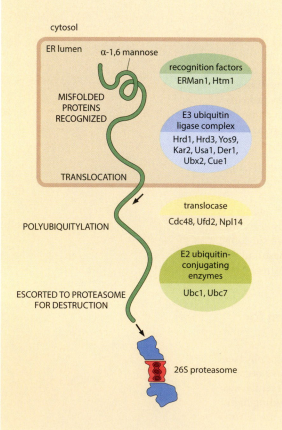

Figure 7.4.1 The ERAD pathway. A misfolded luminal protein (green) is recognized by protein factors on the basis of its glycosylation pattern. A large membrane-associated complex containing the E3 ligase Hrd1 is involved in retrograde transport of the substrate to the cytosol, where polyubiquitylation occurs. The E2 ubiquitin-conjugating enzymes for the polyubiquitylation reaction are Ubc1 and Ubc7. The AAA+ ATPase Cdc48 and its cofactors are required for the ATP-dependent retrotranslocation and for escorting of the substrate to the 26S proteasome.

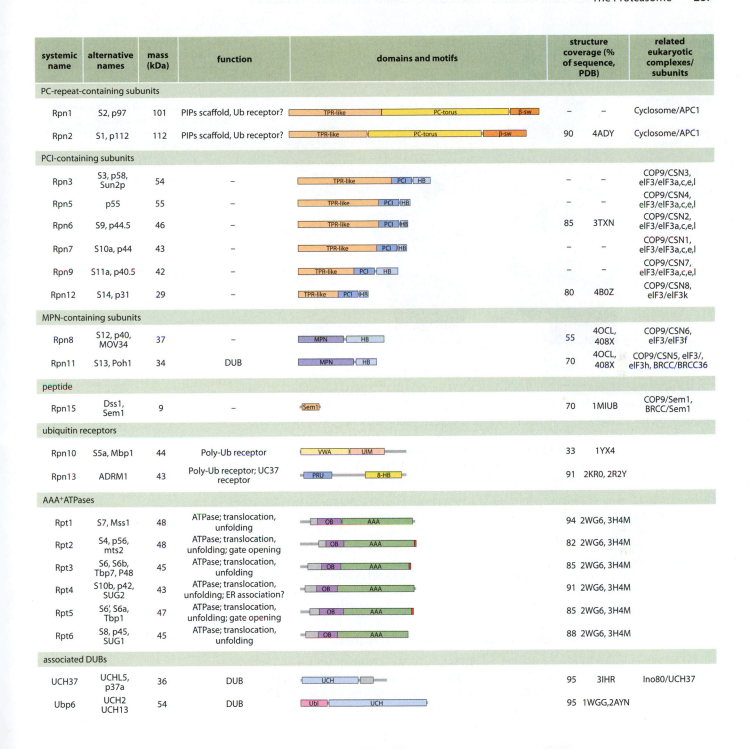

systemic name	alternative names	mass (kDa)	function	domains and motifs	structure coverage (% of sequence, PDB)		related eukaryotic complexes/ subunits
PC-repeat-containing subunits							
Rpn1	S2, p97	101	PIPs scaffold, Ub receptor?	TPR-like / PC-torus / β-sw	–	–	Cyclosome/APC1
Rpn2	S1, p112	112	PIPs scaffold, Ub receptor?	TPR-like / PC-torus / β-sw	90	4ADY	Cyclosome/APC1
PCI-containing subunits							
Rpn3	S3, p58, Sun2p	54	–	TPR-like / PCI / HB	–	–	COP9/CSN3, eIF3/eIF3a,c,e,l
Rpn5	p55	55	–	TPR-like / PCI / HB	–	–	COP9/CSN4, eIF3/eIF3a,c,e,l
Rpn6	S9, p44.5	46	–	TPR-like / PCI / HB	85	3TXN	COP9/CSN2, eIF3/eIF3a,c,e,l
Rpn7	S10a, p44	43	–	TPR-like / PCI / HB	–	–	COP9/CSN1, eIF3/eIF3a,c,e,l
Rpn9	S11a, p40.5	42	–	TPR-like / PCI / HB	–	–	COP9/CSN7, eIF3/eIF3a,c,e,l
Rpn12	S14, p31	29	–	TPR-like / PCI / HB	80	4B0Z	COP9/CSN8, eIF3/eIF3k
MPN-containing subunits							
Rpn8	S12, p40, MOV34	37	–	MPN / HB	55	4OCL, 4O8X	COP9/CSN6, eIF3/eIF3f
Rpn11	S13, Poh1	34	DUB	MPN / HB	70	4OCL, 4O8X	COP9/CSN5, eIF3/, eIF3h, BRCC/BRCC36
peptide							
Rpn15	Dss1, Sem1	9	–	Sem1	70	1MIUB	COP9/Sem1, BRCC/Sem1
ubiquitin receptors							
Rpn10	S5a, Mbp1	44	Poly-Ub receptor	VWA / UIM	33	1YX4	
Rpn13	ADRM1	43	Poly-Ub receptor; UC37 receptor	PRU / 8-HB	91	2KR0, 2R2Y	
AAA+ATPases							
Rpt1	S7, Mss1	48	ATPase; translocation, unfolding	OB / AAA	94	2WG6, 3H4M	
Rpt2	S4, p56, mts2	48	ATPase; translocation, unfolding; gate opening	OB / AAA	82	2WG6, 3H4M	
Rpt3	S6, S6b, Tbp7, P48	45	ATPase; translocation, unfolding	OB / AAA	85	2WG6, 3H4M	
Rpt4	S10b, p42, SUG2	43	ATPase; translocation, unfolding; ER association?	OB / AAA	91	2WG6, 3H4M	
Rpt5	S6′, S6a, Tbp1	47	ATPase; translocation, unfolding; gate opening	OB / AAA	85	2WG6, 3H4M	
Rpt6	S8, p45, SUG1	45	ATPase; translocation, unfolding	OB / AAA	88	2WG6, 3H4M	
associated DUBs							
UCH37	UCHL5, p37a	36	DUB	UCH	95	3IHR	Ino80/UCH37
Ubp6	UCH2, UCH13	54	DUB	Ubl / UCH	95	1WGG, 2AYN	

a heterogeneous group called the regulatory particle non-ATPases (Rpn1–3,5–13,15). Assembly of the base is mediated by several dedicated chaperones. Its role is to capture polyubiquitylated proteins, promote their unfolding, and facilitate translocation through the α rings of the core particle (**Figure 7.25**).

Although genetic data suggest some specialization of the six ATPase subunits, their structure and mode of interaction with the core particle is very similar to that of the PAN hexamer. In particular, the mechanism of gate opening appears to be conserved, even though only three of the six ATPase subunits contain the ΦΨX motif that triggers gate opening; Rpn13 serves as an integral polyubiquitin receptor and captures substrates via a pleckstrin-like (see **Guide**) ubiquitin receptor. The other known integral polyubiquitin receptor is Rpn10, which is part of neither the base nor the lid, suggesting that it integrates into the 26S proteasome late in assembly. It is positioned in close proximity to the N-terminal coiled

Figure 7.23 The subunits of the 19S regulatory particle. The subunits, their domain organization, nomenclature, and structures (where known) are listed. Molecular masses are given for the *Drosophila melanogaster* subunits. Ub, ubiquitin. PC, proteasome–cyclosome. PCI; proteasome-COP9-initiation factor; MPN, Mpr1-Pad1 N-terminal.

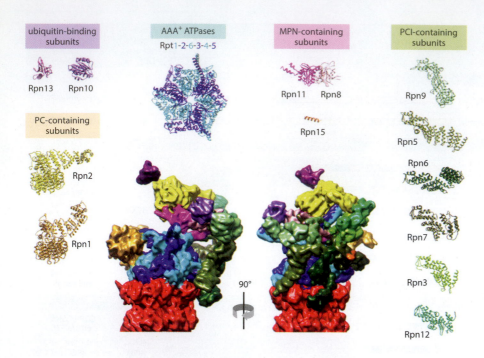

Figure 7.24 Subunit architecture of the 19S regulatory particle, and atomic structures of its canonical subunits. One half of a 26S proteasome (one 19S particle plus half of the 20S core particle (in red)), is seen from two different views (center). The color coding of the subunits corresponds to that of the atomic models. (PDB 4ADY, 2X5N, 3TXN, 4BOZ, 4CR2)

coils of Rpt4/Rpt5 and recognizes substrates through its C-terminal ubiquitin-interacting motif. Rpn1 and Rpn2 act as adaptors or scaffolds for substrate recruitment factors. The ubiquitin receptor Rpn13 binds to Rpn2, and the deubiquitylating enzyme (DUB) Uch37 is associated with Rpn13. The DUB Ubp6 binds transiently to Rpn1, as do the 'shuttling' ubiquitin receptors Rad23, Dsk2, and Ddi1. Polyubiquitylated (Ub) substrates (S) are recruited to the regulatory particle by the ubiquitin receptors Rpn10 and Rpn13; simultaneous binding to both receptors would increase affinity but remains to be proven.

Cycles of ATP binding and hydrolysis by the AAA+ ATPase heterohexamer drive the conformational changes in the regulatory particle required for substrate processing. Domain movements of the six ATPase subunits translocate the substrate through the central channel of the ATPase and unfold it. The active site of the DUB Rpn11 becomes positioned directly above the mouth of the ATPase channel, allowing Rpn11 to access the substrate; this results in the removal of its polyubiquitin chains. Furthermore, the AAA+ ATPase channel and the gate of α rings of the core particle become aligned, facilitating translocation of the substrate into the core particle.

The lid subcomplex serves to deubiquitylate substrates

The lid is built from nine RPN subunits: Rpn3,5–9,11–12 & 15. Six of them (Rpn3,5,6,7,9,12) have proteasome-COP9-initiation factor (PCI) domains (a stack of α-helices capped with a winged helix subdomain, see **Guide**). These subunits assemble in a horseshoe-like

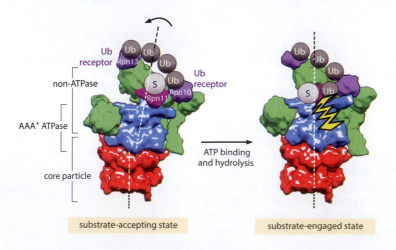

Figure 7.25 Mechanistic interpretation of regulatory particle functions. The gold flash indicates excision of ubiquitin. Dashed lines indicate the pseudo-7-fold axis of the core particle and the pseudo-6-fold axis of the ATPase. Some aspects of this model are still hypothetical and await experimental confirmation.

configuration, embracing the Rpn8/Rpn11 heterodimer and positioning it above the mouth of the Rpt1–6 heterohexamer. This involves large-scale conformational changes (shifts and rotations) of the lid subcomplex. All PCI subunits and Rpn8–11 contribute to the formation of a large helical bundle that governs the assembly of the lid and allows it to move as a unit in the functional cycle. Moreover, two of the PCI subunits, Rpn5 and Rpn6, make contact with the 20S core, stabilizing the otherwise weak interaction between the core and regulatory particles. Rpn15 is a small polypeptide that serves as a glue stabilizing the Rpn3/Rpn7 heterodimer and promoting its assembly. The main function of the lid is de-ubiquitylation of substrates prior to their unfolding and translocation into the 20S core. This function is performed by Rpn11, a metalloisopeptidase that cleaves polyubiquitin chains at the proximal site. Other deubiquitylation enzymes, such as Ubp6 and Uch37, interact with the 19S regulator transiently and are also implicated in the editing of polyubiquitin chains.

With the exception of Rpn15, all lid subunits have homologs in the signaling complex **COP9 signalosome (CSN)**, which cooperates with the ubiquitin–proteasome system. COP9 mediates the deneddylation—a process closely related to deubiquitylation—of Skp1-Cul1-Fbox (SCF) E3 ligases and, in concert with associated enzymes, it also mediates the phosphorylation of transcription factors such as c-Jun and p53, and the deubiquitylation of specific ligases and their substrates. The paralog of RPN11, CSN5, is the metalloproteinase that mediates deneddylation, but the functions of most of the other subunits are unknown. The subunit architecture of CSN appears to be closely similar to that of the lid.

7.5 GIANT PROTEASES

Giant proteases are homomultimeric enzymes of extraordinary size—up to 15 MDa. Discovered only in the 1990s, there are two classes of these machines—tricorn and tripeptidyl-peptidase II (TPPII)—which are unrelated in sequence and structure but have converged on common architectural principles. Their hallmark is the confinement of the proteolytic activity to internal cavities, which are formed as the subunits assemble into higher-order structures. This strategy of self-compartmentalization is reminiscent of the Clp proteases (Section 7.3) and the proteasome (Section 7.4) and provides the basis for substrate selectivity and processivity. Contrasting with their enormous size, giant proteases act mostly on relatively small substrates—oligopeptides released by the proteasome, whose catabolic functions they augment. However, when proteasomal functions are compromised—for example, in the presence of inhibitors—giant proteases are often up-regulated, which may reflect an ability to substitute for some of the cellular roles of proteasomes.

In archaea, tricorn protease is the archetypal giant protease

The search for regulatory factors of the archaeal proteasome led, serendipitously, to the discovery of tricorn protease in *T. acidophilum*. Tricorn is quite common in archaea and is also found in some bacterial genera. The 120 kDa monomers assemble into a toroidal structure formed as a trimer of dimers—hence the name tricorn. The hexamer can assemble further into an icosahedral capsid of 20 hexamers with a mass of 14.6 MDa, at normal cellular protein concentrations. Tricorn protease interacts functionally, and probably also physically, with an array of aminopeptidases. Together, the protease and aminopeptidases provide a pathway for the sequential and synergistic degradation of peptides released by the proteasome, that generates free amino acids (**Figure 7.26A**). The exact mode of interaction between tricorn and its associated aminopeptidases remains to be established. It is tempting to speculate that the capsid provides a scaffold for the aminopeptidases, positioning them such that substrates can be channeled readily from one active site to the next. Such an arrangement should increase the efficiency of degradation.

The tricorn monomer has a complex domain structure (**Figure 7.27**): the N-terminal part, a six-bladed β-propeller, β6, is followed by a seven-bladed β-propeller, β7. A PDZ-like domain (see **Guide**) flanked by two α-β domains, C1 and C2, forms the C-terminal half. Similar to prolyl iminopeptidase, the two β-propellers have an open 'Velcro-like' structure. The C-terminal domains containing the catalytic residues have an α-β hydrolase fold. Structural studies with chloromethyl ketone-based inhibitors and mutagenesis studies suggest that Ser965, His746, Ser745, and Glu1023 form a catalytic tetrad, and the spatial arrangement of Ser965, His746, and the oxyanion makes it likely that peptide bond hydrolysis follows the classical steps for trypsin-like serine proteases (see Chapter 1).

(A)   prokaryotic protein degradation pathway

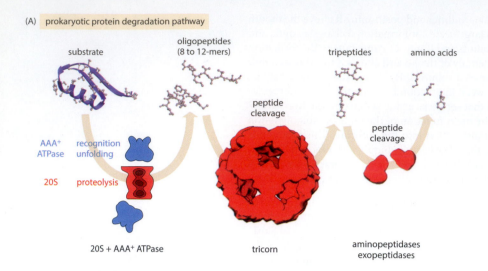

(B)   eukaryotic protein degradation pathway

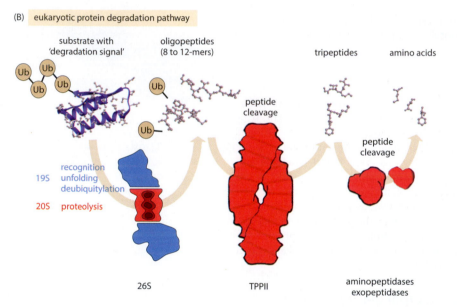

Figure 7.26 Protein degradation pathways. Pathways are shown for (A) archaea and some bacteria; and (B) eukaryotes. Substrates bearing degradation signals—ubiquitin (Ub) in eukaryotes—are targeted and degraded by the proteasome system. The degradation products are oligopeptides, which are cleaved further by the giant proteases tricorn or tripeptidylpeptidase (TPPII). An array of aminopeptidases completes the catabolic pathway.

Figure 7.27 Tricorn protease structure. (A) Tricorn monomer (PDB 1K32) as a ribbon diagram. Color coding: blue, 7-propeller domain; orange, 6-propeller domain; red, PDZ domain; purple and green, mixed domains C1 and C2. (B) Cutaway representation of a tricorn hexamer with one subunit inset. White asterisks mark the locations of active sites. (C) Cryo-EM density map of the icosahedral tricorn capsid with four tricorn hexamers inset. (Courtesy of Beate Rockel.)

Tricorn accepts substrates with a broad range of sequences upstream of the scissile bond but is restrictive with regard to the length of substrate downstream. These restrictions are imposed by a pair of basic residues, Arg131 and Arg132, delimiting the binding site of the substrate's C terminus; this gives tricorn the character of a carboxypeptidase releasing preferentially dipeptides and tripeptides. Although a role for the PDZ domain in substrate

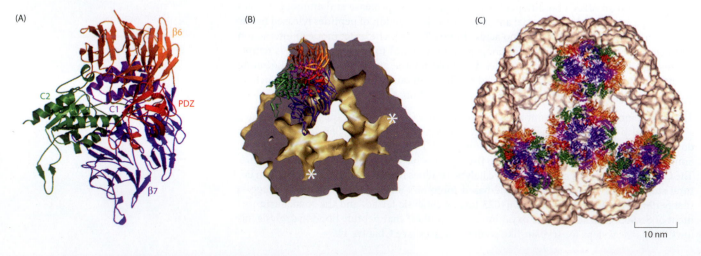

recognition has yet to be confirmed by structural data, the two β-propellers are likely to control substrate access to the active site and substrate release from it. In fact, both propeller axes are oriented toward the active site, where they almost intersect. These domains make tricorn an oligopeptidase by providing a size-exclusion mechanism that protects larger structured peptides and proteins in the cytosol from unwanted proteolysis.

Tripeptidyl peptidase II is an enzyme that counts in threes

TPPII is the functional equivalent of tricorn protease in the cytosol of eukaryotic cells: it cleaves off tripeptides successively from the N-termini of peptides such as those released by the proteasome. This exoprotease activity has been implicated in the trimming of proteasomal degradation products for antigen presentation to MHC-I, although its exact role in antigen processing is still under debate. More specifically, TPPII degrades cholecystokinin (CCK), a neuropeptide hormone found in the brain. CCK is known to be important in many physiological processes, including the regulation of satiety, pancreatic enzyme secretion, and neurotransmission. CCK, which is 115 aa long in humans, exists in various truncated forms from 58 to 4 aa. The highly conserved octapeptide CCK-8 is the predominant form among neuronal CCK peptides. TPPII has been identified as the CCK-8-inactivating peptidase in the brain, making it a potential target for the control of obesity. In addition to its exoprotease activity, TPPII also has a weak endoprotease activity; this appears to enable TPPII to compensate for a loss of functions normally performed by the proteasome. TPPII is up-regulated in diseases involving increased or uncontrolled proteolysis, and its inhibition causes radiation sensitivity in cancer cells. TPPII knockout mice are viable, but lack of TPPII results in the activation of cell death programs (Section 13.6) and a gradual decline in the effectiveness of the immune system, resulting in an immunosenescence-like phenotype.

TPPII has an unusual spindle-like architecture

TPPII monomers range from 138 kDa to 150 kDa in mass, depending on the species. The N-terminal region, approximately the first third of the polypeptide chain, contains the protease domain, which is homologous to the serine protease subtilisin (see Box 7.1). The remaining two-thirds are the building blocks of the unusual architecture of TPPII, which is a spindle-shaped assembly of two twisted strands with a total mass of 5–6 MDa. Each strand is a linear assembly of nine dimers in humans, or ten in fruit flies, in which the monomers are oriented head-to-head (**Figure 7.28**). Stacking of the dimers produces a linear array of cavities, harboring the active sites. Substrates can enter the rather complex system of cavities only via 'side windows' and have to pass through several narrow constrictions, ~10 Å across, to reach the active sites; therefore access is controlled by a size-exclusion mechanism (**Figure 7.29**).

Whereas TPPII spindles predominate at protein concentrations around 0.3 mg/ml *in vitro*, strands begin to grow longer at higher concentrations (> 1 mg/ml) and become polymorphic, and hence not conducive to crystallization. The structure was eventually solved by docking the crystal structure of the TPPII dimer into a cryo-EM-derived envelope of the holocomplex. The TPPII monomer can be divided into three parts: a subtilisin-like N-terminal domain with a long insertion between the catalytic Asp44 and His272 residues (see Figure 7.28A,B), a central domain mostly composed of β strands, and a C-terminal domain that is mainly α-helical (see Figure 7.28A,C). Together, the N-terminal and central domains form a ring structure with a central hole, with the active site on one side (see Figure 7.28D). The C-terminal domain is a curved protrusion, connected to the ring by the insertion. It strongly resembles a tetratricopeptide repeat and may mediate interactions with other proteins involved in the catabolic pathway. In the dimer, two monomers are oriented in a head-to-head fashion. Through this arrangement, the holes in the centers of the rings are covered reciprocally and the C-terminal protrusions assemble into a handle-like structure (see Figure 7.28D).

The arrangement of the catalytic triad residues (Asp44, His272, and Ser462) in the dimer is rather similar to that found in subtilisin. Notably different is the relative location of Ser462, which is displaced by a few Å from the position of the corresponding serine in subtilisin Carlsberg. Whereas the active-site Ser of subtilisin is located at the N-terminal end of a long α helix, that of TPPII is at the N terminus of a partly unwound α helix (see Figure 7.28B). This unwinding and the resulting displacement go together with the binding of a portion of loop L2 into the substrate binding-cleft (see below). The cleft is wider in TPPII, resulting

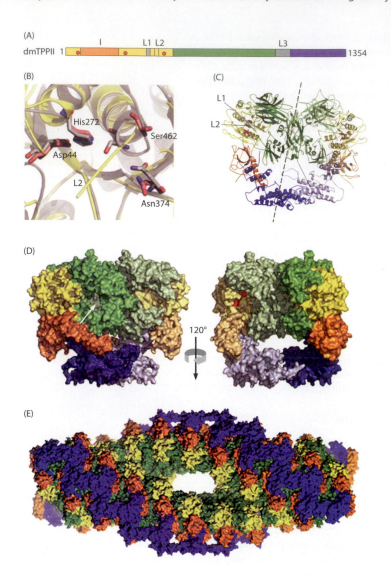

Figure 7.28 Tripeptidyl peptidase II structure. (A) Domain map of *Drosophila* TPPII. The subtilisin-like domain (residues 1–522, yellow; active-site residues Asp44, His272, and Ser462, marked by orange balls) contains loop L2 (residues 442–461; open box), loop L1 (residues 408–417; gray box), and a long insertion (residues 75–266; orange). The central domain (residues 523–1098, green) contains loop L3 (residues 1027–1098; gray). The C-terminal domain is residues 1099–1354 (blue). (B) Close-up view of the active-site region of TPPII. (PDB 3LXU) Subtilisin (gray) and TPPII (yellow) are superposed; active-site residues are shown in stick format for subtilisin (gray) and TPPII (red). (C) Ribbon diagram of a dimer; the 2-fold axis is indicated. Active-site residues are in stick format (red). The locations of unstructured loops L1 and L2, are marked on the monomer on the left. (D) Surface representation of a TPPII dimer shown in two orientations (color scheme as in (A), active-site residues in red). In the dimer, the open bottom of the bowl of one monomer is closed by the bowl of the other monomer (gray arrow). (E) Holocomplex as rendered by a hybrid X-ray crystallography–cryo-EM approach. Color coding of domains, as in (A). (Adapted from C.K.Chuang et al., *Nat. Struct. Mol. Biol.* 17:990–996, 2010.)

in a different mode of interaction with substrates, which form a two-stranded anti-parallel β sheet with one of the two strands of the binding cleft, namely Ser34–Gly342, instead of a three-stranded anti-parallel β sheet, as in subtilisin. Two highly conserved residues, Glu312 and Glu343, whose mutation abolishes exoprotease activity, have a critical role in recognizing the free N terminus of the substrate. They are positioned such that they effectively close off one end of the binding cleft, allowing only the three N-terminal residues of the substrate access to the cleft.

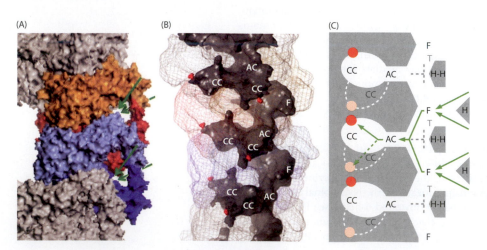

Figure 7.29 The cavity system of TPPII. (A) Surface rendering of a single strand of TPPII. Two dimers are highlighted in red and orange (dimer 1) and in light and dark blue (dimer 2). (B) Visualization of the cavity system as a gray surface inside a transparent strand. (C) Schematic representation of the layout of the cavity system. Routes that substrates can take to reach the active sites (red spheres) are shown by green arrows. CC, catalytic cavity; AC, antechamber; F, foyer; HH, helical handles; T, T-joint. (Adapted from C.K.Chuang et al., *Nat. Struct. Mol. Biol.* 17:990–996, 2010.)

(A)

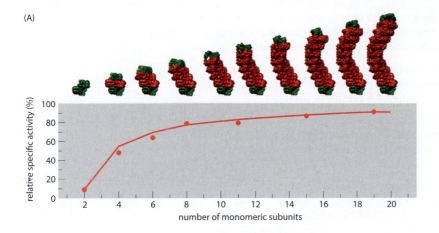

Figure 7.30 Mechanism of TPPII activation.
(A) Diagram showing the increase in specific activity as more dimers are added to the strands. Active dimers are shown in red, inactive ones in green. (B) Model for the assembly-dependent activation mechanism. In the inactive TPPII dimers (left), the L2 loops (green) are flexible, but a three-residue segment of L2 occupies the substrate binding cleft, and Ser462 at the active site (indicated by S and AS, respectively) is displaced. Upon association of the dimers (right), the catalytic chambers harboring the active sites are formed at the dimer–dimer interface. At the same time, the L2 loop residues (red) are removed from the substrate-binding cleft, and Ser462 is relocated to a catalytically active position. (Adapted from C.K.Chuang et al., *Nat. Struct. Mol. Biol.* 17:990–996, 2010.)

(B)

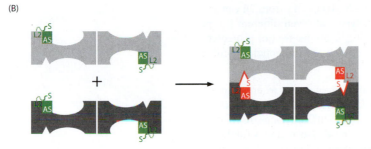

TPPII activity increases with assembly

Biochemical studies have shown that TPPII activation is boosted by assembly. Its oligomer size is dependent on concentration, and an increase in average oligomer size correlates with a hyperbolic increase in specific activity per subunit (**Figure 7.30A**). At the concentration prevailing in the cytosol, namely ~0.3 mg/ml TPPII, the fully assembled spindle is strongly favored. Tetramers have ~50% of the specific activity of the fully assembled complex, whereas dimers have only approximately 10% activity (see Figure 7.30A). This suggests that the formation of dimer–dimer contacts primes the active sites, which are located near the interface. In the dimer, a three-residue segment at the mobile loop, L2, carrying the active site serine is bound to the substrate-binding cleft. It precludes substrate binding and thereby provides an autoinhibitory mechanism. The spindle structure suggests that, as the dimers stack to form strands, the mobile loops and, concomitantly, Ser462 reposition themselves, assuming a proteolytically fully active conformation (Figure 7.30B). Thus, the priming of the active sites is coupled to their sequestration in a cavity formed by the assembly of dimers into spindles. This mechanism is fundamentally different from the activation of TPPI, a monomeric lysosomal enzyme of the subtilisin type. TPPI is synthesized as a zymogen.

7.6   SUMMARY

Within cells, proteolysis is needed both to remove foreign and misfolded proteins and as part of the regulatory processes that control development and the cell cycle, but it must be tightly regulated to avoid collateral damage. Accordingly, much intracellular proteolysis is performed by large ATP-dependent machines. These unfoldase-assisted proteases comprise core particles housing the proteolytic activity and regulatory particles that identify substrates by recognizing degradation motifs, also known as degrons. The regulatory particles, mechanoenzymes of the AAA+ family of ATPases, also unfold substrate proteins and translocate them into the core particles to be degraded.

The hallmark of unfoldase-assisted proteases is stacks of multimeric rings. The core particles are bipolar with six or seven subunits in each ring, and each half-complex can be a single or a double ring, but the catalytic active sites are always in the central cavity, the degradation chamber. Access to the degradation chamber is gated: binding of a regulatory particle opens the gate wide enough to allow the passage of substrate proteins unfolded and translocated by the regulatory particle.

The regulatory ATPases also form rings, which dock onto one or both ends of the core particles. Some of these machines have a 7 : 6 symmetry mismatch between the core particle and the regulatory ATPase; as a consequence, their interaction is weak, facilitating remodeling.

Each such protease recognizes multiple degrons: these can be peptides at internal sites that are only exposed when a protein is unfolded; alternatively, they can be covalently attached 'tags' such as the protein ubiquitin in eukaryotes, or the SsrA peptide in bacteria.

Bacteria have multiple unfoldase-assisted proteases, including three architecturally related proteases Clp, Lon, and FtsH. The protease can be a separate polypeptide, for example ClpXP, or fused with its companion ATPase, as with FtsH; and the ATPase may bind specificity-altering accessory proteins.

Protein degradation in archaea and in eukaryotes relies on the proteasome system. In archaea the core particle, or 20S proteasome, is built from two types of subunits, α and β, which form homoheptameric rings. It pairs with the PAN ATPase or other members of the AAA+ family, such as homologs of Cdc48. The eukaryotic 26S proteasome, which is found in the cytosol and the nucleus, is a molecular machine of 2.5 MDa, built from 34 canonical subunits. The 20S core is built from seven different α-type and seven different β-type subunits, forming heteroheptameric rings. The outer, α-ring, subunits are not proteolytically active but have a scaffolding role and control access to the inner proteolytic chamber formed by the β rings.

Activation of inner-ring subunits is deferred until assembly is complete, whereupon the propeptides that block the active sites are removed autocatalytically. Three of the seven inner-ring subunits of the eukaryotic 20S proteasome are active, with differing specificities. The 19S regulatory particle comprises two subcomplexes: a 'base' that includes the heterohexameric ATPase ring and subunits involved in the recognition and recruitment of ubiquitylated substrates; and a 'lid' that deubiquitylates substrates before transferring them to the base for unfolding and translocation into the core particle.

The peptides released from the unfoldase-related proteases are processed further to dipeptides and tripeptides by giant proteases. Tricorn protease, found in archaea and some bacteria, forms dimers of trimers that can assemble into porous icosahedral shells; tripeptidyl peptidase II, found in eukaryotic cells, assembles into a two-stranded 'spindle' with a total of 40 subunits.

REFERENCES

7.2 Principles of unfoldase-assisted proteolysis

Bukau B, Weissman J & Horwich A (2006) Molecular chaperones and protein quality control. *Cell* 125:443–451.

Dougan DA, Mogk A, Zeth K et al (2002) AAA+ proteins and substrate recognition, it all depends on their partner in crime. *FEBS Lett* 529:6–10.

Erdmann R (ed.) (2012) Special issue on AAA ATPases: structure and function. *Biochim Biophys Acta* 1823:1–198.

Mogk A, Haslberger T, Tessarz P et al (2008) Common and specific mechanisms of AAA+ proteins involved in protein quality control. *Biochem Soc Trans* 36:120–125.

Ogura T, Fujiki Y & Katayama T (eds) (2012) Special issue on AAA+ ATPases. *J Struct Biol* 179:77–250.

Sauer RT & Baker TA (2011) AAA+ proteases: ATP-fueled machines of protein destruction. *Annu Rev Biochem* 80:587–612.

Wickner S, Maurizi MR & Gottesman S (1999) Posttranslational quality control: folding, refolding, and degrading proteins. *Science* 286:1888–1893.

7.3 Unfoldase-assisted proteases in bacteria and eukaryotic organelles

Baker TA & Sauer RT (2012) ClpXP, an ATP-powered unfolding and protein-degradation machine. *Biochim Biophys Acta* 1823:15–28.

Cha S-S, An YJ, Lee CR et al (2010) Crystal structure of Lon protease: molecular architecture of gated entry to a sequestered degradation chamber. *EMBO J* 29:3520–3530.

Flynn JM, Neher SB, Kim YI et al (2003) Proteomic discovery of cellular substrates of the ClpXP protease reveals five classes of ClpX-recognition signals. *Mol Cell* 11:671–683.

Ishikawa T, Beuron F, Kessel M et al (2001) Translocation pathway of protein substrates in ClpAP protease. *Proc Natl Acad Sci USA* 98:4328–4333.

Keiler KC, Waller PR & Sauer RT (1996) Role of a peptide tagging system in degradation of proteins synthesized from damaged messenger RNA. *Science* 271:990–993.

Kress W, Maglica Z & Weber-Ban E (2009) Clp chaperone-proteases: structure and function. *Res Microbiol* 160:618–628.

Maillard RA, Chistol G, Sen M et al (2011) ClpX(P) generates mechanical force to unfold and translocate its protein substrates. *Cell* 145:459–469.

Sousa MC, Trame CB, Tsuruta H et al (2000) Crystal and solution structures of an HslUV protease-chaperone complex. *Cell* 103:633–643.

Wang J, Song JJ, Seong IS et al (2001) Nucleotide-dependent conformational changes in a protease-associated ATPase HsIU. *Structure* 9:1107–1116.

Yu AY & Houry WA (2007) ClpP: a distinctive family of cylindrical energy-dependent serine proteases. *FEBS Lett* 581:3749–3757.

7.4 The proteasome

Baumeister W, Walz J, Zühl F et al (1998) The proteasome: paradigm of a self-compartmentalizing protease. *Cell* 92:367–380.

Beck F, Unverdorben P, Bohn S et al (2012) Near-atomic resolution structural model of the yeast 26S proteasome. *Proc Natl Acad Sci USA* 109:14870–14875.

Djuranovic S, Hartmann MD, Habeck M et al (2009) Structure and activity of the N-terminal substrate recognition domains in proteasomal ATPases. *Mol Cell* 34:580–590.

Förster F, Unverdorben P, Śledź P et al (2013) Unveiling the long-held secrets of the 26S proteasome. *Structure* 21:1551–1559.

Glickman MH, Rubin DM, Coux O et al (1998) A subcomplex of the proteasome regulatory particle required for ubiquitin-conjugate degradation and related to the COP9-signalosome and eIF3. *Cell* 94:615–623.

Groll M, Bochtler M, Brandstetter H et al (2005) Molecular machines for protein degradation. *ChemBioChem* 6:222–256.

Lander GC, Estrin E, Matyskiela ME et al (2012) Complete subunit architecture of the proteasome regulatory particle. *Nature* 482:186–191.

Lasker K, Förster F, Bohn S et al (2012) Molecular architecture of the 26S proteasome holocomplex determined by an integrative approach. *Proc Natl Acad Sci USA* 109:1380–1387.

Löwe J, Stock D, Jap B et al (1995) Crystal structure of the 20S proteasome from the archaeon *T. acidophilum* at 3.4 Å resolution. *Science* 268:533–539.

Seemüller E, Lupas A, Stock D et al (1995) Proteasome from *Thermoplasma acidophilum*: a threonine protease. *Science* 268:579–582.

Tanaka K (2009) The proteasome: overview of structure and functions. *Proc Jpn Acad B* 85:12–36.

Whitby FG, Masters EI, Kramer L et al (2000) Structural basis for the activation of 20S proteasomes by 11S regulators. *Nature* 408:115–120.

Zhang F, Hu M, Tian G et al (2009) Structural insights into the regulatory particle of the proteasome from *Methanocaldococcus jannaschii*. *Mol Cell* 34:473–484.

Zwickl P, Seemüller E, Kapelari B et al (2002) The proteasome: a supramolecular assembly designed for controlled proteolysis. *Adv Protein Chem* 59:187–222.

7.5 Giant proteases

Brandstetter H, Kim JS, Groll M et al (2001) Crystal structure of the tricorn protease reveals a protein disassembly line. *Nature* 414:466–470.

Chuang CK, Rockel B, Seyit G et al (2010) Hybrid molecular structure of the giant protease tripeptidyl peptidase II. *Nature Struct Mol Biol* 17:990–996.

Geier E, Pfeifer G, Wilm M et al (1999) A giant protease with potential to substitute for some functions of the proteasome. *Science* 283:978–981.

Rockel B, Kopec KO, Lupas AN et al (2011) Structure and function of tripeptidyl peptidase II, a giant cytosolic protease. *Biochim Biophys Acta* 1824:237–245.

Schönegge A-M, Villa E, Förster F et al (2012) The structure of human tripeptidyl peptidase II as determined by a hybrid approach. *Structure* 20:593–603.

Tamura N, Lottspeich F, Baumeister W et al (1998) The role of tricorn protease and its aminopeptidase-interacting factors in cellular protein degradation. *Cell* 95:637–648.

Tamura T, Tamura N, Cejka Z et al (1996) Tricorn protease—the core of a modular proteolytic system. *Science* 274:1385–1389.

Yao T & Cohen RE (1999) Giant proteases: beyond the proteasome. *Curr Biol* 9:R551–R553.

Assembly of Viruses

8

8.1   INTRODUCTION

Assembly is the key final phase in the production of macromolecular machines. The protein building blocks are first synthesized and then folded. These components, sometimes together with nucleic acids and/or lipids, then have to be fitted together in appropriate numbers, at the correct positions, and in the right order. In some systems, folding and assembly are directly coupled. As a process, assembly is more similar to protein folding than to protein synthesis, in that it depends primarily on properties of the components themselves—assisted in more complex cases by ancillary factors—and does not require an external platform analogous to the ribosome.

There are two modes of self-assembly: simple and facilitated. In simple self-assembly, the components of the final complex contain all the information needed for assembly. They may be compared, albeit at the next level in the structural hierarchy, with polypeptide chains that fold without the assistance of other factors. Complexes of this kind may be cycled between the assembled and dissociated states by adjusting the environmental conditions, for example pH, temperature, ionic strength, the concentration of components, and the presence of divalent cations. Under any conditions, there is a dynamic equilibrium between the unassembled and assembled states or alternative assembly products. Plotting the predominant forms as a function of two condition parameters gives a **phase diagram** for the system. **Figure 8.1** shows an example for purified nucleosome core particles (see Section 2.2), which show a rich variety of phases.

In contrast, facilitated self-assembly involves irreversible changes in the assembling structure, resulting from, for example, regulatory proteolysis or the discarding of essential scaffolding proteins. Once such a complex is dissociated, its components are not capable of

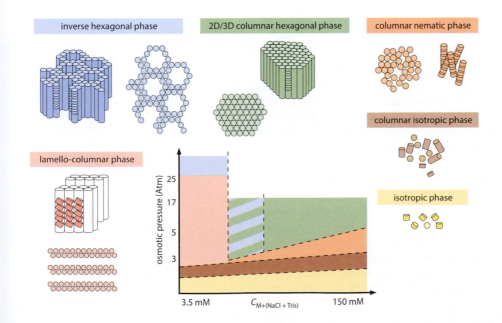

inverse hexagonal phase   2D/3D columnar hexagonal phase   columnar nematic phase

columnar isotropic phase

lamello-columnar phase

isotropic phase

osmotic pressure (Atm)

25

17

5

3

3.5 mM $C_{M+(NaCl + Tris)}$ 150 mM

Figure 8.1 Phase diagram for nucleosome core particles in the presence of monovalent ions. The core particle has a histone protein core around which ~168 bp of double-stranded DNA are wrapped (Section 2.2). The predominant states of aggregation are plotted as a function of two parameters: monovalent salt concentration (NaCl/Tris) and applied osmotic pressure. The states were determined by electron microscopy. In chromatin, core particles are concatenated by linker DNA and associated with additional proteins, leading to different forms of aggregation. (Adapted from F. Livolant et al., *Phil. Trans. R. Soc. A* 364:2615–2633, 2006. With permission from The Royal Society.)

recovering the assembled state. In many cases, the components are locked together so tightly by cooperative interactions that they tend not to disassemble spontaneously, even at very low concentrations. If their physiological role requires disassembly, it is effected by external factors such as mechanoenzymes, kinases, or proteases.

The products of assembly may be uniform (deterministic structures) or they may exhibit stochastic variability. Complexes that assume a discrete set of related structures are called **polymorphic** and those that form a continuously variable set are **pleiomorphic**.

The largest assemblies have thousands of constituents of dozens of different kinds and masses of many megadaltons. How is the construction of such elaborate entities orchestrated? How are their dimensions determined? And how are their components selected from all the macromolecules available inside a cell? Answering these questions is equivalent to solving a three-dimensional jigsaw puzzle with hundreds or thousands of pieces. Given this level of complexity, it is not surprising that there are not yet complete answers. For practical reasons, many aspects of assembly have been studied in greatest depth in viruses, whose relatively small gene sets and short replication cycles make them easier to study than cellular systems. These investigations have been further assisted by the discovery of conditionally lethal mutants, discussed below. This chapter addresses virus assembly as a paradigm for macromolecular assembly in general.

Virus assembly proceeds sequentially along pathways that resemble industrial assembly lines: each component is added only at a specific stage of the pathway, and the components are themselves the products of folding pathways. In systems of high complexity, subcomplexes are assembled on separate pathways that then converge as the subcomplexes are combined. Branching of pathways in this manner enhances efficiency in that it permits quality control at each branch point, increasing the probability of obtaining a correctly formed end product. Examples of a schematic branched pathway and an actual one (bacteriophage T4) are illustrated in **Figure 8.2**. The increase in speed to be gained from branching is similar to the performance enhancement in computation from parallelization, which distributes a calculation over multiple processors.

Although the focus of this chapter is on assembly mechanisms, we begin with an overview of general properties of viruses. We then consider five classes of viruses that exhibit varying degrees of complexity and have been well characterized. In addition, recognizing that viruses exhibit many machine-like behaviors, we outline two cases in which such behavior is particularly evident: the viral particles that serve as transcription and replication compartments, and virally mediated membrane fusion. Finally, we give a brief overview of prospects to exploit the self-assembly properties of viral proteins in nanotechnology.

8.2   PRINCIPLES OF VIRUS REPLICATION

A living entity can be defined, mechanistically, as a molecular system that is capable of self-replication in the presence of a suitable source of raw materials. Cells meet this definition, but viruses require a special environment—the host cell—to provide raw materials as well as needed functional devices like enzymes, ribosomes, and other machines.

A fundamental difference between the replication of cells and that of viruses is that, throughout the process, there is always at least one complete cell, whereas viruses are built from scratch—that is, *de novo*—in each generation. A growing cell synthesizes components; when it divides, it organizes their orderly partition between two daughter cells. With viruses, a single replication cycle may yield thousands of progeny. The infected cell becomes a factory for viral gene products that assemble into *virions* (mature, infectious, viral particles) and are then dispersed either by rupture of the host cell (**lysis**) or by secretion, leaving the cell intact and capable of supporting continued virus production.

An alternative strategy for the proliferation of viral genomes is **lysogeny**, in which the viral genome is transformed into double-stranded DNA (dsDNA), if not already in this form, and is either integrated into the host chromosome or established as an extrachromosomal genetic element. In either case, the lysogenic genome is replicated passively through successive cell divisions—in effect, a genetic hitchhiker—without producing progeny virions. After many cellular generations, some stimulus may induce the productive replication of the lysogen.

(A)

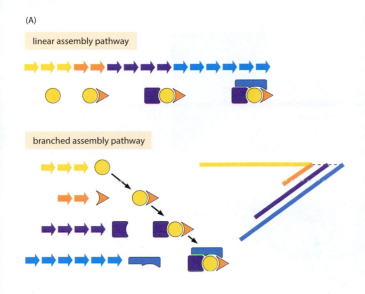

linear assembly pathway

branched assembly pathway

Figure 8.2 The efficiency and speed of assembly pathways are enhanced by branching. (A) The linear pathway has four stages (yellow, orange, purple, blue), involving three, two, four, and six steps respectively. If each step takes time t and has a probability p of being performed correctly, the total time for assembly is $15t$ and the probability of a perfect product is p^{15}. As the number of steps grows, the overall efficiency falls, even if individual steps have low error rates. For instance, if $p = 0.95$ (95% success), the efficiency is only 46%, and if $p = 0.9$ it falls to 21%. With the branched pathway, we make steps in adding subcomplexes at each branch-point and such a step takes time t' and has p' probability of success. Then the overall process takes a time of max $(3t + 3t', 6t+t')$ and has probability $p^3.(p')^3$ or $p^6.(p')$, respectively, of success. If, for simplicity, we take $t = t'$ and $p = p'$, the overall assembly time is reduced to $7t$ and the efficiency of assembly increases to p^7, which gives ~ 70% for $p = 0.95$ and ~ 48% for $p = 0.9$. (B) Branched assembly pathway of bacteriophage T4. The pathway has three main branches that respectively produce the nucleocapsid (~3000 protein subunits of 13 different kinds, ~80 MDa, plus the genome, one linear double-stranded DNA molecule ~100 MDa); the tail (~400 protein subunits of 20 different kinds, ~12 MDa); and the long tail-fibers (60 protein subunits of four different kinds, ~6 MDa). (Adapted from F.A. Eiserling and L.W. Black, in Molecular Biology of Bacteriophage T4 [J.D. Karam ed], pp 209–212. Washington DC: American Society for Microbiology, 1994.)

(B)

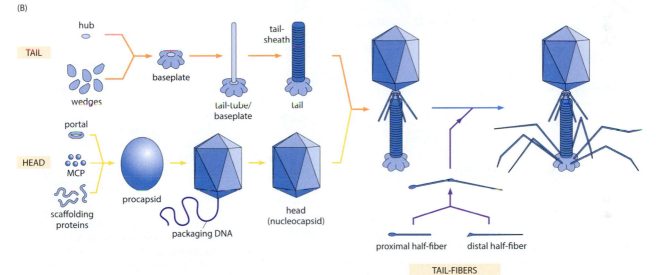

Viruses behave like machines and self-replicating automata

Self-propagating machines were defined by Norbert Wiener, a pioneer in the field of cybernetics: "*A machine is not only a form of matter but an agency for accomplishing certain definite purposes. And self-replication is not merely the creation of a tangible replica; it is the creation of a replica capable of the same functions.*" Weiner also defined living systems as having "*the power to learn and the power to reproduce themselves,*" two properties that are intimately related.

Viruses 'learn' through evolutionary probing by mutation and natural selection. Through their assembly properties, viruses are—quintessentially—self-propagating machines. A machine of this kind, considered in terms of its ability to replicate with high fidelity, is called an **automaton**. The theoretical basis of automata has been explored in considerable depth, initially by Alan Turing and John von Neumann, and subsequently by John Horton Conway and others, as in the online simulation, "Game of Life."

A virion is a 'smart' delivery vehicle for genetic information. A virion houses the viral genome and, in some cases, cargo proteins as well. Virions leave the cell in which they are produced and then identify and enter another susceptible cell for the next cycle of infection. The genome is delivered to the appropriate cellular compartment, usually the cytoplasm or the nucleus. As with any other delivery system, a key issue is raised by the following question: How is a host cell capable of supporting a productive infection to be recognized? The basis for virus–host compatibility (**tropism**) lies in complementarity between some

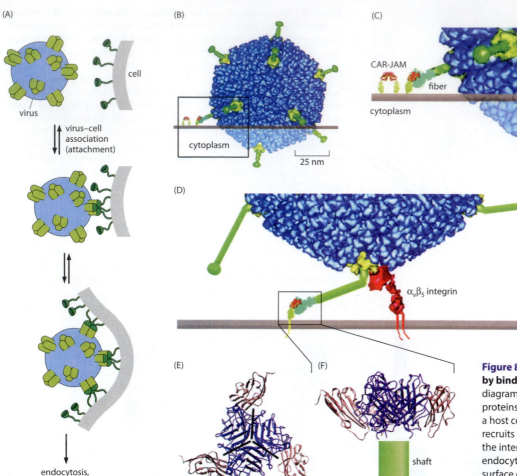

Figure 8.3 Viruses recognize host cells by binding to receptors. (A) Schematic diagram showing a virus with trimeric proteins distributed over its surface infecting a host cell. In the scenario shown, the virus recruits successively more receptors into the interaction zone as it is taken up by endocytosis. (B) Model showing the outer surface of adenovirus with fibers protruding from its vertices. The fiber has a shaft and a terminal knob. (C) A JAM junctional molecule called CAR, the coxsackie and adenovirus receptor, binds to the knob (coxsackie is a picornavirus, see Section 8.4). (D) An integrin receptor binds to the penton base (the vertex protein, in yellow). (E) Top view and (F) side view of a crystal structure of a knob/receptor complex showing three Ig-like CAR domains (red) binding around the knob (blue). Black lines in (E) mark the interfaces between knob subunits. (PDB 1KAC) (D, from G.R. Nemerow and P.L. Stewart, *Virology* 288:189–191, 2001; E and F, from G.R. Nemerow et al., *Virology* 384:380–388, 2009.)

external marker on the virion and receptor molecules displayed on the host cell's surface. Virus receptors also have other functions. Viral receptor-recognition proteins play similar roles to the localization signals that consign cellular proteins to the nucleus or other compartments. Infection by some viruses involves two receptors, for example by adenovirus, as outlined in **Figure 8.3**. Host recognition is followed by uptake of the virion in a sequence of endocytic reactions called cell entry. Many viruses have protein shells, called **capsids**, that may be compared with the coats of trafficking particles, such as *clathrin* (see Section 10.2), with the qualification that capsid size and content tend to be specifically defined for a given virus, whereas clathrin coats vary in size, depending on the cargo.

Helical and icosahedral symmetry are widely employed in virus architecture

Virions show great diversity in size, shape, and molecular composition (**Figure 8.4**). Unlike cells whose genomes are always encoded in dsDNA, those of viruses may be single-stranded or double-stranded, and they may consist of RNA or DNA. Genomes of single-stranded RNA occur in both positive-sense (mRNA-like) and negative-sense (the complementary strand) forms. Virions may be simple rods or spherical particles consisting only of a capsid protein and a genome, or they may be elaborate assemblies made up of many different kinds of molecules arranged in modules that may be rod-like, spherical, or have more complex shapes. They consist mainly of molecules encoded by the viral genome, but in some cases molecules from the host cell, such as lipids or *histone* proteins, are also used. Most virions have, as a major component, a regular polymer of one or more protein subunits complexed with the nucleic acid. This assembly, the **nucleocapsid**, may either be a linear filament or a closed shell. In the simplest systems, the nucleocapsid represents the mature virion; in more elaborate cases it is surrounded by a membrane, called the **viral envelope**, and/or additional protein shells.

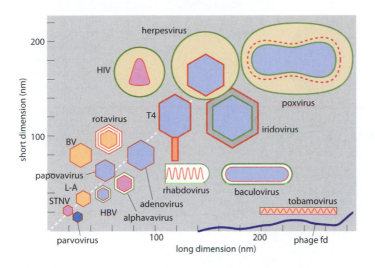

Figure 8.4 Viruses vary widely in size, shape, and architecture. A representative sampling of virions is mapped according to their longest and shortest dimensions. For spherical (mainly icosahedral) viruses, the shortest dimension applies to both dimensions. Some of these are mapped off the diagonal for space considerations; for example, HIV has a diameter of ~130 nm and herpesvirus one of ~230 nm. For the elongated virions—tobamovirus, fd, baculovirus, and rhabdovirus—the x-coordinates refer to their long dimensions but they are drawn approximately to scale. Capsids are red; membranes, including glycoproteins, are green. Genome types: double-stranded (ds)DNA, light blue; single-stranded (ss)DNA, dark blue; dsRNA, orange; ssRNA, violet. HIV, human immunodeficiency virus; BV, birnavirus; STNV, satellite tobacco necrosis virus; HBV, hepatitis B virus.

Unlike the majority of protein filaments, filamentous nucleocapsids have uniquely defined lengths. Genomic nucleic acids have specific lengths and, with rare exceptions, each protein subunit engages a fixed number of nucleotides, regardless of sequence. This number may be as low as three, as in tobacco mosaic virus (TMV; Section 8.3), or as high as nine for vesicular stomatitis virus (VSV; **Figure 8.5**). Thus, in filamentous nucleocapsids, genomic segments of single-stranded nucleic acid serve as **tape measures**.

Filamentous nucleocapsids also vary in their intrinsic flexibility: for instance, TMV forms rigid rods ~300 nm long and 18 nm in diameter, whereas the VSV nucleocapsid is highly flexible and ~8 nm across, with a contour length of ~3600 nm when fully extended. In the virion it is coiled into a regular superhelical solenoid with a diameter of 55 nm (see Figure 8.5). During VSV replication (synthesis of plus strands), the single-stranded RNA (ssRNA) is never uncoated; that is, it never leaves its groove in the protein sheath of the uncoiled nucleocapsid but is accessed by the viral polymerase.

Spherical capsids are protein shells with icosahedral symmetry

The **icosahedron** is a regular polyhedron with 5-3-2 point group symmetry: it has axes of 5-fold symmetry traversing its 12 vertices, axes of 3-fold symmetry at the centers of its 20 facets, and axes of 2-fold symmetry at the middle of its 30 edges (**Figure 8.6A**). Of the five

Figure 8.5 Coiling of a helical nucleocapsid. (A) Cryo-electron micrograph (cryo-EM) of the bullet-shaped rhabdovirus, vesicular stomatitis virus (VSV). Ectodomains of the trimeric G protein (glycoprotein) protrude as spikes. (B) Dark-field scanning transmission electron micrograph of negatively stained, uncoiled, highly flexible, nucleocapsids. On them, elongated N-protein subunits are stacked ~50 Å apart. (C) Inside the membrane is a helical lattice of M (matrix) protein in register with the helically coiled nucleocapsid, shown in a cryo-EM reconstruction. Purple = membrane. Pink = sheet of smeared G proteins. (D) Crystal structure of a complex consisting of a ring of 10 N-protein subunits with bound ssRNA. (E) The subunit has an N-terminal lobe (green) and a C-terminal lobe (yellow) and binds nine nucleotides in the groove between them. (A, courtesy of Naiqian Cheng; B, from D.J. Thomas et al., *J. Virol.* 54:598–607, 1985. With permission from American Society for Microbiology; D, from P. Ge et al., *Science* 327:389–393, 2010. With permission from AAAS; E, from T.J. Green et al., *Science* 313:357–360, 2006. With permission from AAAS.)

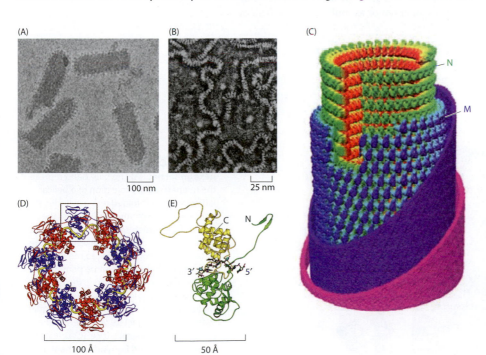

(A)

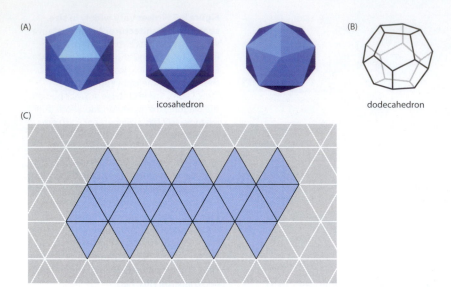

icosahedron

(B)

dodecahedron

(C)

Figure 8.6 The icosahedron is a favored capsid architecture. (A) An icosahedron, viewed along its axes of 2-fold, 3-fold, and 5-fold symmetry (left to right). (B) A dodecahedron viewed from an oblique angle. (C) Geometrical construction of an icosahedron considered as a folding of a portion cut out of a planar hexagonal net.

Platonic solids—tetrahedron, cube, octahedron, icosahedron, and dodecahedron (Section 1.4)—the last two enclose the largest volume for a given surface area. The relationship between the icosahedron and the dodecahedron (Figure 8.6B) goes further: they are 'dual' structures—that is, they may be mapped on to each other—and they are both recurring themes in viral architecture, although the icosahedron is more common. The arrangement of its 60 protein subunits (each an asymmetric object) determines whether a given capsid more closely resembles an orthodox (that is, flat-faceted) icosahedron or a dodecahedron. The packing of protein subunits in icosahedral or dodecahedral shells combines maximum capacity with simplicity of construction, namely repetitive interactions of identical or similar building blocks.

An icosahedron may be described as a folding of a hexagonal net out of which 60° wedges have been cut at strategic positions (Figure 8.6C). This construction may be generalized to realize other structures encountered in viruses: larger icosahedra; prolate icosahedra extended along a symmetry axis; irregular polyhedra, for example elongated forms with one end wider than the other; and open-ended tubes, called **polyheads**. The two fundamental symmetries—helical and icosahedral—converge in the latter structures, which may be described either as cylindrical foldings of the hexagonal net or as multi-stranded helices (**Figure 8.7**).

The protein oligomers that make up an icosahedral capsid are called **capsomers**. There may be 12 pentameric capsomers at the vertices, 20 trimers on the facets, or 30 dimers located on the edges, depending on which interactions predominate (**Figure 8.8**, top row). Regardless, the 60 subunits all have the same fold and occupy identical bonding environments; thus

(A)   (B)

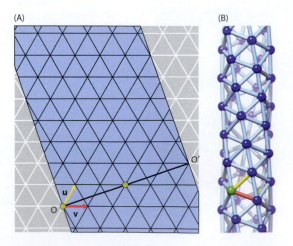

Figure 8.7 Cylindrical folding of a hexagonal net. (A) The blue band represents the (planar) radial projection of a helical lattice: when it is folded into a cylinder, the lattice lines run continuously across the seam. These structures may be described by the numbers of steps (h,k) taken along the two lattice vectors **u** and **v**, to move from the origin O back to the same point O′. The line O–O′ is the equator. Here, $(h,k) = (2,4)$ and the folding is left-handed. (The hand is that of the lattice lines closest to the equator; in this case, along **v**.) (B) Side view of the corresponding helical lattice. (Courtesy of James Conway and Bernard Heymann.)

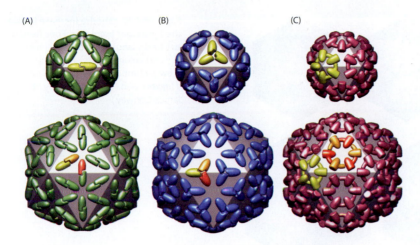

(A)   (B)   (C)

Figure 8.8 Icosahedral capsids may be built of dimers, trimers, or pentamers and hexamers. Top row: these alternative capsomer arrangements are illustrated for 60-subunit capsids. In each capsid, every subunit has the same (equivalent) structure and bonding pattern. Bottom row: in $T = 3$ capsids, the 180 subunits adopt three quasi-equivalent conformations, each present in 60 copies. The capsomers can be 90 dimers (A), 60 trimers (B), or 12 pentamers and 20 hexamers (C). One copy each of the three quasi-equivalent subunits is shown in yellow, ochre, and orange. (Courtesy of James Conway.)

they are equivalent. This design is employed by viruses with very small genomes, such as the plant virus satellite tobacco mosaic virus and animal viruses called parvoviruses, from Latin *parvus*, 'small.' A few cellular enzymes also have icosahedral symmetry (Section 9.3), whose primary role, in this case, is to enhance cooperativity in substrate processing.

The first high-resolution capsid structures were determined in the late 1970s by Stephen Harrison and Michael Rossmann for small plant viruses. They shared a subunit fold, a distinctive eight-stranded β barrel (the 'jelly roll'; see **Guide**), with each other and with the next several capsids that were solved, which included some animal viruses. Intriguingly, there is essentially no sequence similarity between these proteins and they belonged to viruses that infect widely divergent hosts. Technology for purifying the large quantities of virions needed for crystallographic analysis and for computational analysis of diffraction data from crystals with very large unit cells has gradually improved, and now there are at least eight folds that are exhibited by capsid-forming proteins. Several of these are discussed in the following sections. Most are unrelated to folds encountered in cellular proteins, but the capsid protein of Sindbis virus, an alphavirus, resembles the serum protease *chymotrypsin*.

Cryo-electron microscopy (cryo-EM) (see **Methods**) reconstructions of icosahedral viruses, first reported in the late 1980s, are now playing an increasingly important role and achieving resolutions that, in the best cases, equal those achieved by X-ray crystallography (see **Methods**). As crystals are not required, this approach is more widely applicable and it may readily be used to characterize virions in conformationally altered states and to study interactions of virions complexed with receptors or antibodies, which is rarely possible with crystallography.

Quasi-equivalence allows the assembly of larger capsids with more than 60 subunits

It seems likely that, as a general trend, viral complexity increased on an evolutionary timescale: as viruses acquired more functions, their genomes increased in size. (Arguments have been advanced that some of the smallest viruses may represent the end products of a progressive shedding of functions and that larger viruses may have been derived from degenerating cells. However, we shall focus here on the 'growth in complexity' scenario.) Genome expansion has been complemented by mutations affecting capsid assembly. Larger genomes may be accommodated by increasing the size of the capsid subunit, by segmenting the genome and distributing the segments among multiple particles, or by altering the architecture to employ more than 60 subunits. All three strategies have been adopted, but the most prevalent strategy has been the latter.

With more than 60 subunits, equivalence is no longer possible. To resolve this dilemma, the principle of **quasi-equivalence** was proposed by Donald Caspar and Aaron Klug in 1962. It states that larger capsids should invoke only minimal departures from equivalence; that is, small perturbations of the subunit fold and its interactions with neighboring subunits. Quasi-equivalence restricts foldings of the hexagonal net to those that maintain continuity of the lattice lines; in other words, they employ only bonds similar to those in the

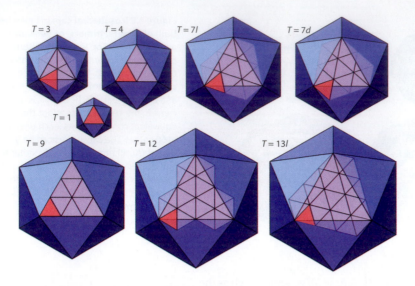

Figure 8.9 Icosahedra increase in size as the *T*-number increases. For a fixed unit cell size (red triangle), the diameter increases as $T^{1/2}$ and volume as $T^{3/2}$. $T = 7$ is the lowest skew *T*-number and has two enantiomorphs, *d* (*dextro*) and *l* (*levo*). (Courtesy of James Conway.)

corresponding 60-subunit structure. Such capsids have integral numbers of the basic triangle in each triangular facet (**Figure 8.9**). These **triangulation numbers (*T*)** conform to the selection rule $T = h^2 + hk + k^2$, where h and k are non-negative integers. *T*-numbers map on a hexagonal lattice called the **Goldberg diagram** (**Figure 8.10**). If h or k equals 0 or $h = k$, the lattice is said to be non-skew, but if h and k are both non-zero and $h \neq k$, the lattice is skew. Each skew lattice has two enantiomorphs, *levo* (*l*, left-handed) and *dextro* (*d*, right-handed). Considered as geometrical constructions, the enantiomorphs are related by simple reflection. However, proteins are not reflection-invariant: for capsids, enantiomorphs do not represent valid alternatives. For example, the capsid of bacteriophage HK97 is always $T = 7l$ and that of polyoma virus is always $T = 7d$.

The corresponding generalization of the basic dodecahedron with its 12 pentagonal facets is made by inserting additional hexagons between them, thereby generating **Fullerene** structures (**Figure 8.11A**), named after Buckminster Fuller, who developed the geodesic dome as an architectural form based on icosahedral/dodecahedral lattices.

The icosahedron is a geometric construction with flat facets and sharply defined vertices. Some capsids have this form, but others are almost perfectly spherical and still others have shapes that are intermediate between these extremes (Figure 8.11B). The deciding factor is how the subunits involved achieve a free-energy minimum in the assembled state.

The number of subunits per capsid prescribed by quasi-equivalence is 60*T*. The capsomers may still be all dimers or all trimers, but for pentamer-containing capsids their number remains fixed at 12 (one per vertex), complemented with increasing numbers of hexamers. For example, a $T = 3$ capsid could consist of 90 dimers or 60 trimers, or 12 pentamers plus 20 hexamers (see Figure 8.8, bottom row).

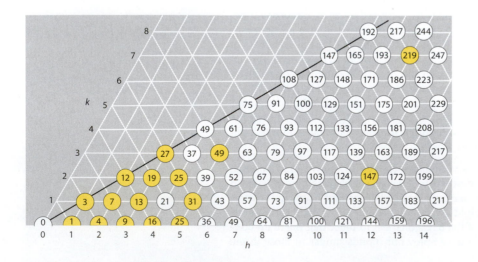

Figure 8.10 The Goldberg diagram specifies *T*-numbers compatible with quasi-equivalence. $T = h^2 + hk + k^2$. The yellow lattice points mark virus structures that have been observed. It seems certain that other high *T*-numbers will be discovered, especially among dsDNA bacteriophages and the adenovirus-like clade.

Figure 8.11 Capsid structures with varying tesselations and sphericity. (A) A $T = 16$ structure ($h,k = 4,0$) (upper, by triangulation; lower, as a Fullerene lattice), viewed along an axis of 3-fold symmetry. (B) Flat-faceted, intermediate, and spherical icosahedra, viewed along an axis of 2-fold symmetry. (A, courtesy of Bernard Heymann ; B, courtesy of James Conway.)

(A)

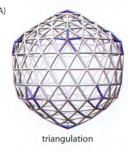

triangulation

fullerene

(B)

All of the lowest T-numbers—1, 3, 4, 7, 9, 12, 13, and 16—have been observed in different viruses, but only some of the higher T-numbers have been recorded. The largest one documented so far is $T = 219$ (see Section 8.5). If the subunit size is kept fixed, the enclosed volume increases rapidly with T, according to $T^{3/2}$. Thus, if a mutation were to switch a subunit from assembling $T = 1$ capsids to assembling $T = 3$ capsids, its enclosed volume would increase more than fivefold.

Larger viruses show progressively greater complexity in structure and composition

Early ideas of virus structure drew on the idea that protein folds are uniquely specified and the notion that viral genomes are too small to allow more than one gene to be assigned to the task of capsid construction. However, there are now many examples of proteins, including capsid subunits, that have multiple conformations. In addition, it is now apparent that some viruses have hundreds of genes, some of which may be dispensable under certain conditions. For example, in the phage T4 system with its 169-kilobase-pair (kbp) genome and roughly 300 gene products, at least 20 genes participate in capsid assembly. Quasi-equivalence invokes conformational variations among chemically identical subunits at different lattice sites. Variants may also be generated by modifying the subunit's amino acid sequence, such as via a *gene duplication* event followed by divergent evolution to yield two proteins with similar but not identical structures. For example, the T4 capsid has one protein forming the pentamers at 11 vertices (but not at the portal vertex; see below) and another related protein as the 155 hexamers that make up the rest of the lattice (see Figure 8.67B). In the gene duplication scenario, sibling subunits may be expressed separately or they may be connected in a **polyprotein**. The latter strategy is employed by **picornaviruses** (pico = very small; rna = RNA), whose members include poliovirus (Section 8.4). Their polyprotein has three jelly roll domains that are separated by proteolysis during assembly. As the capsid has 60 copies of each, it is formally a $T = 1$ particle, but the similarity of the three domains can be acknowledged by calling it a pseudo-$T = 3$ structure. The term pseudo-T-*number* refers to a mixed lattice built from chemically distinct but assembly-compatible subunits.

Quasi-equivalence allows small departures from uniformity in both the subunit fold and its interactions. Other viruses adopt a more radical form—**non-equivalence**—in which variations in the subunit fold remain small but there are multiple sets of quite different interactions between subunits. Two well-documented examples of non-equivalence are the 120-subunit inner capsids of double-stranded RNA (dsRNA) viruses with 60 copies each of two non-equivalent subunits (see Section 8.6); and the 'all-pentamer' capsids of papovaviruses, a family of tumorigenic dsDNA viruses. They have 72 capsomers in a $T = 7d$ lattice, but instead of 12 pentamers and 60 hexamers they have 72 pentamers (**Figure 8.12A,B**). Each subunit in a vertex-resident pentamer is equivalent to its four neighbors. Each of the other 60 pentamers has six neighbors with which they engage in different interactions via their flexible terminal regions. This non-equivalent packing arrangement is well described in terms of a tiling of quadrilateral 'rhombs' and 'kites' (Figure 8.12C).

Pairs of complementary interaction patches are the key to self-assembly

Is there a fundamental difference between enzymes and structural proteins? In the context of capsids, this distinction is somewhat superficial. The classical lock-and-key picture of enzyme–substrate interactions envisages complementarity between a patch on the enzyme surface (the lock) and the substrate (the key). With capsid subunits, the complementary patches are both on the surface of the same molecule (**Figure 8.13**); in other words, a capsid protein is its own substrate. For a lattice to be formed, at least two such pairs are needed,

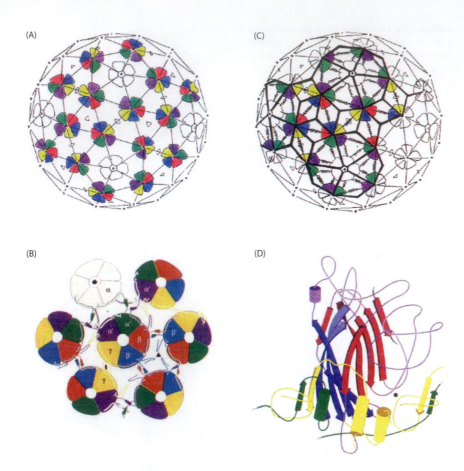

Figure 8.12 Non-equivalent capsid structure of simian virus 40 (SV40). Papovaviruses (for example SV40) are dsDNA viruses with circular supercoiled genomes packaged as viral chromatin with ~20 *nucleosomes* per virion. (A) The capsid has 72 pentamers arranged on a *T = 7d* surface lattice instead of the 12 pentamers + 60 hexamers expected for a quasi-equivalent structure. The six distinct (non-equivalent) monomers are shown in different colors. (B) Inter-capsomer interactions are mediated by C-terminal arms that assume several different extended structures and insert into neighboring capsomers. (C) The non-equivalent surface lattice represented as a regular tessellation of two quadrilateral forms, the kite and the diamond-shaped rhomb (marked by bold black lines). See Box 8.1. (D) The capsid subunit has a jelly roll core surrounded by short α helices and β strands that is conserved between the two kinds of pentamers—respectively, those with five and six neighbors. (A and C, from R. Twarock, *J. Theor. Biol.* 226:477–482, 2004. With permission from Elsevier; B and D, from R.C. Liddington et al., *Nature* 354:279–284, 1991. With permission from Macmillan Publishers Ltd.)

where the members of a given pair lie on opposite sides of the molecule and the two pairs are approximately coplanar. By adjusting their positions, the curvature of the shell may be altered, changing the size of the capsid. The number of pairs of *interaction patches* (Section 1.3) on a given molecule may exceed two, as in capsids whose subunits undergo large rotations during maturation, in which they switch to different sets of interaction patches.

In progressing from equivalence through quasi-equivalence to non-equivalence, only slight variations in the building block (capsid protein) have been invoked. The next step in increasing complexity invokes multiple subunits with different folds and interactions (namely, heterotypic assembly). In most cases, there is still one subunit, the **major capsid protein (MCP)**, that forms the principal building blocks. In some capsids, specialized proteins are inserted at individual sites. For example, in the capsids of tailed phages and herpesviruses, a *portal protein*, which is a 12-membered ring, is inserted at one 5-fold vertex, giving a striking symmetry mismatch with the surrounding MCP lattice (Section 8.5). The term 'portal' (gateway) refers to its role in providing a conduit for DNA to enter the capsid during assembly and to exit during infection. Some dsRNA viruses install at all 12 vertices, non-MCP subassemblies through which RNA molecules are imported and exported (Section 8.6).

Pathways are mapped by characterizing mutants for which assembly is blocked

There are essentially two stages to the characterization of an assembly pathway. The first involves identifying the molecular components and the order in which they interact. The second is to establish *in vitro* systems with purified molecules, allowing detailed investigation. The mapping of pathways was greatly advanced by work in the 1960s, by Robert Edgar, Richard Epstein and Eduard Kellenberger, on bacteriophage mutations conditionally lethal for assembly. Under permissive conditions—at a certain growth temperature or with a particular bacterial strain as host—the mutant replicates normally, allowing stocks of virus

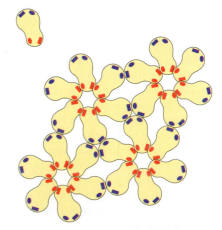

Figure 8.13 A protein lattice is generated by interactions between complementary surface patches. In this diagram, a hexagonal lattice is generated from an asymmetric subunit on which there are two pairs of complementary interaction patches (rectangular and oval; see Section 1.2 and Figure 1.10). Interactions are denoted by matching colors, for example, red pairing with red and blue with blue.

to be accumulated. Under non-permissive conditions—a different temperature or host strain—the mutated protein is defective, causing the assembly pathway to stall at the point when its function is needed. By examining the aberrant assembly products that accumulate under nonpermissive conditions, inferences could be made as to the order of assembly and the role of each gene product.

A major development that greatly facilitated assembly studies of animal viruses was the development of tissue culture methods that allow many viruses to be propagated and investigated with almost equal facility as bacteriophages. Moreover, the roles of particular gene products may now be investigated by using small regulatory RNAs to 'knock down' their levels of expression in infected cells.

Many capsids are initially assembled as precursor procapsids that subsequently mature

Assembly pathways fall into two broad classes. On one, genome packaging and capsid assembly are coupled processes whose end-product is a viable nucleocapsid. On the other pathway, a precursor particle—the **procapsid**—is first assembled and it then serves as the receptacle for genome packaging. Co-assembly is the route followed by all filamentous nucleocapsids and most small icosahedral nucleocapsids. Selectivity is conferred by a genomic motif, the **packaging signal**, which is recognized by the capsid protein.

The packaging of genomes into procapsids is effected by motor proteins driven by ATP hydrolysis. Pathways of this kind are followed by some viruses with genomes of dsDNA. Duplex DNA is a relatively stiff molecule, and its resistance to bending—and it must be bent to fit into the confines of a capsid—is overcome by the force exerted by the packaging motor and the mechanical strength of the capsid (Section 8.5). The genomes of dsRNA viruses, in contrast, are packaged as relatively pliable single strands, with synthesis of the complementary second strands taking place inside the capsid (Section 8.6).

The final step in assembly, which renders a virion infectious, is **maturation**. This process is often controlled by a protease that is incorporated into the procapsid and then activated. In this context, the role of proteolysis is not degradation but the triggering of structural changes. These changes stabilize the procapsid and/or actuate certain downstream reactions. For some enveloped viruses, maturation tends to be based on the processing of viral glycoproteins, rendering them capable of fusing with the host cell (**fusogenic**) and thus effecting viral entry (Section 8.7). For others, such as HIV (human immunodeficiency virus), maturation involves major changes in both the capsid and the envelope glycoproteins.

Capsids and crystals exhibit defects, symmetry-breaking, and dynamics

On account of its regularity and symmetry, a capsid seems at first sight to be just like a crystal except that it is hollow and of definite size. However, capsids are functional entities, whereas three-dimensional protein crystals appear in Nature only as inert storage forms. With this qualification, the analogy between capsids and crystals extends further. Crystal lattices can have defects of several kinds: vacancies (unoccupied lattice sites), inclusions (incorporated impurities), or grain boundaries (interfaces between two crystalline regions). Capsid surface lattices also accommodate inclusions in heterotypic assembly (see above). Grain boundaries may be mobile, moving through a crystal as it shifts into a single, more stable, phase. Such dynamic transformations have also been observed in viral capsids, moving like molecular tsunamis as the MCP lattice undergoes the transition from its precursor conformation into its stable mature conformation (**Figure 8.14**).

In general, virions are dynamic machines with moving parts. To withstand the challenges that it may encounter in the extracellular environment (such as extreme temperatures or corrosive chemical agents), a virion should be robust. However, structural changes must be allowed to take place at certain stages of the replication cycle. To reconcile these requirements, viral particles engage in dynamic behaviors. Such transitions, involving either entire capsids or individual components, will be described later in this chapter. These transitions may involve an entire capsid or individual components and they occur in several contexts, including virus-induced membrane fusion, genome packaging, procapsid stabilization, genome release, and cell entry.

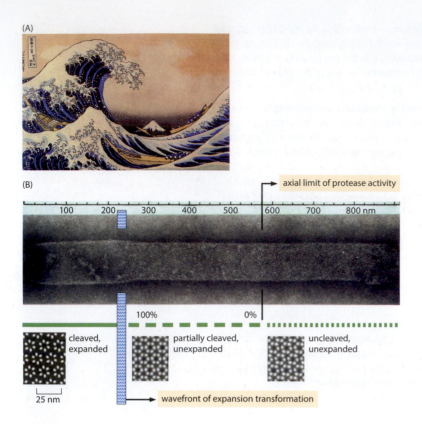

Figure 8.14 A molecular tsunami.
(A) Woodcut of a tsunami (tidal wave) by the Japanese artist, Katsushika Hokusai. (B) Two waves of structural transformation propagating along the axis of this giant T4 procapsid were captured in this electron micrograph of a negatively stained specimen. The wild-type T4 procapsid is ~100 nm long and 75 nm in diameter, with icosahedral $T = 13$ caps. Perturbing assembly produces some greatly elongated particles ('giants'), as here. Upon activation, the viral protease processes the scaffolding proteins and the capsid proteins, whereupon the surface shell enters a metastable state, preceding expansion. This particle flattened on the EM grid so as to present two overlaid hexagonal lattices that allowed MCP hexamers to be visualized by image-filtering of local patches. Three conformations of the surface lattice—cleaved/expanded, cleaved/unexpanded, and uncleaved/unexpanded—are shown underneath. A narrow (roughly 20 nm) transition zone or wavefront separates the expanded and unexpanded regions. In the unexpanded region, there is a broader transition zone between completely cleaved and uncleaved regions. It appears that proteolytic activity initiated in the left-hand cap and proceeded axially as far as the limit marked. Subsequently, expansion was triggered. Again, the transition started in the left-hand cap (see Section 8.5). (A, from *The Great Wave* by Katsushika Hokusai, 1830-32; B, adapted from A.C. Steven and J.L. Carrascosa, *J. Supramol. Struct.* 10:1–11, 1979. With permission from Elsevier.)

8.3   HELICAL VIRUSES

Here we address the assembly of helical viruses, relatively simple particles in which the nucleocapsid constitutes the whole virion. The capsid protein subunits associate via repetitive interactions to coat the genome, giving a filament with helical symmetry (see Section 1.4). We discuss two examples: tobacco mosaic virus (TMV), whose infection of tobacco plants causes a mottling (mosaic of spots) of the leaves; and filamentous bacteriophages, which infect Gram-negative bacteria and have entered widespread use as cloning vectors, as molecular scaffolds for peptide and protein display, and as the vehicle for DNA sequencing by the dideoxy method invented by Frederick Sanger. TMV was the first virus to be discovered—by Dmitri Ivanovsky in Russia and Martin Beijerinck in the Netherlands, in the 1890s. It was found to be an infectious agent smaller than a bacterium—one that could pass through filters that retained bacteria. It was also the first virus to be purified and investigated by means of X-ray diffraction and the first shown to be capable of self-assembly. By contrast, filamentous bacteriophages only assemble in infected cells while in transit across the cell membrane and require a membrane potential and the expenditure of cellular energy. Other, more complex viruses, for example vesicular stomatitis virus (see Figure 8.5), also have helical nucleocapsids but enclose them inside membranous envelopes.

TMV is a rigid rod containing a single-stranded RNA and only one type of capsid protein

TMV has an ssRNA genome of 6395 nucleotides (nt) (positive sense) that runs the length of the virion, encapsidated by ~2130 protein subunits (each of 158 aa), giving a total molecular mass of ~36 MDa. The virion is a stiff rod, 300 nm long and 18 nm in diameter (**Figure 8.15A**). The subunits pack in a right-handed helix that has a pitch (axial rise per turn) of 23 Å and $16^1/_3$ subunits per turn. There is an axial lumen of radius 20 Å, and the RNA is protected by being embedded at a radius of 40 Å between successive turns of the protein helix (Figure 8.15B). Two subunits in the helical array, one directly above the other, are illustrated in Figure 8.15C. The major part of a subunit is a right-handed, anti-parallel, four-helix bundle, pointing radially outward. Carboxyl–carboxylate pairs are found at an

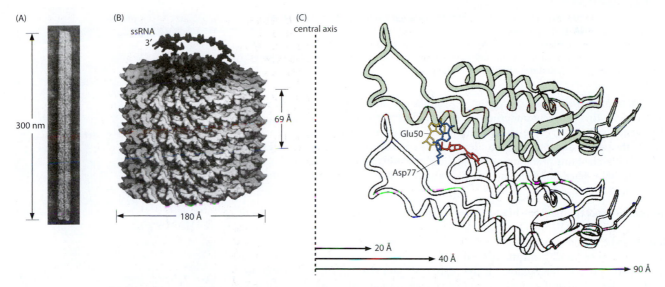

Figure 8.15 Structure of tobacco mosaic virus. (A) Electron micrograph of a negatively stained TMV virion. Note the cross-striations representing successive coils of the protein helix and the axial lumen penetrated by stain. (B) A portion (~5%) of the virion structure, as determined by X-ray fiber diffraction. The capsid is a one-start helix, pitch 23 Å, with 49 protein subunits in three turns. (See Figure 8.20B,C below for an explanation of one-start and multi-start helices.) Protein subunits embedding the RNA in the uppermost helical turn are omitted for clarity. (C) Ribbon diagram of two capsid protein subunits, one immediately above the other. Three nucleotides are shown in the groove between the subunits. (A, from J.T. Finch, *J. Mol. Biol.* 16:291–294, 1972. With permission from Elsevier; B and C, based on K. Namba et al., *J. Mol. Biol.* 208:307–325, 1989. With permission from Elsevier.)

inner radius (~25 Å) and between Glu50 of the lower subunit and Asp77 of the upper subunit at a radius of ~55 Å; these are potential Ca^{2+}-binding sites (the binding of Ca^{2+} ions stabilizes TMV and other plant virus capsids) and are important in controlling assembly. Three nucleotides interact with each protein subunit, their negative charges neutralized by three positively charged Arg side chains in a complementary groove in the protein. The protein–RNA interactions are mostly ionic and van der Waals (Section 1.3), consistent with the lack of RNA sequence specificity appropriate to accommodating a whole genome. However, the interaction between the phosphate of the second nucleotide and the carboxyl group of Asp116 generates a potential Ca^{2+}-binding site. If this Ca^{2+} dissociates, the resulting electrostatic repulsion would weaken the protein–RNA binding, thus facilitating disassembly during infection.

In the 1950s, Heinz Frankel-Conrat demonstrated that TMV RNA and protein can be separated by biochemical techniques and then reassembled under physiological conditions (neutral pH and moderate ionic strength) into infectious virus. In the absence of the RNA and at lower pH (around 5.0), the protein assembles into rodlike particles, essentially indistinguishable from virions except that they are of variable length (**Figure 8.16**). The viral protein does not self-assemble under physiological conditions. Several conclusions follow from these observations. First, it is the protein, not the RNA, that dictates the helical structure. Second, the RNA acts as a **molecular tape-measure** to specify the virion length. Third, there is a mechanism to prevent the production of empty (without RNA), and thus non-infectious, particles. To assemble, free protein subunits need to adopt a slightly altered conformation. In the absence of the viral RNA, this can be achieved by lowering the pH. Each of two pairs of juxtaposed carboxylate groups acquires a proton to form carboxyl–carboxylate pairs, thereby relieving the electrostatic repulsion between them in their double negatively charged state. It is this repulsion that helps maintain the protein in a conformation incapable of assembly. In contrast, in the presence of the RNA, the free energy of the RNA–protein interaction overcomes the repulsion and drives the protein subunits into the assembly-compatible conformation.

Polarity of the helical array is important for disassembly *in vitro* and *in vivo*

The helical TMV capsid is a polar structure; that is, one end is different from the other. Inside the capsid, the ssRNA is always oriented with its 5′ end (carrying the $m^7G^{5'}ppp^{5'}Gp$

with RNA

Figure 8.16 Helical rods assembled in the presence and absence of TMV RNA. In the presence of RNA (top), the capsid protein assembles at neutral pH and moderate ionic strength (physiological conditions); in the absence of RNA (bottom), protein-only rods of variable length are assembled but only at lower pH (~5). (Courtesy of Richard Perham.)

cap typical of translatable RNAs—see Section 5.2) at one end and its 3′ end at the other end (see Figure 8.15). This property underlies the observation that denaturing agents such as urea, detergents, or high pH, cause protein subunits to dissociate from the end that contains the 5′ end of the ssRNA and not, except under extreme conditions, from both ends (**Figure 8.17A**). The initial shedding of subunits coating the 5′ end probably reflects locally less tight binding to the RNA. Shedding continues from the same end, exposing more and more of the RNA from its 5′ end.

For infection to proceed, the RNA has to become exposed, and in electron micrographs of virions extracted from TMV-infected cells, ribosomes are seen to be clustered around one end, thought to be the one containing the 5′ end of the RNA (Figure 8.17B). It is likely that this dissociation of subunits from the naturally weaker end of the virion is caused by the effect on subunit–subunit interactions of the higher pH and the dissociation of stabilizing Ca^{2+} from the capsid at the much lower Ca^{2+} concentration inside the plant cell than outside. This exposes the capped end of the RNA. As ribosomes bind to it and translate the RNA, they push protein subunits off ahead of them in a process of co-translational disassembly. Indeed, biochemical experiments aimed at investigating disassembly in infected plant protoplasts indicate that ~70% of protein subunits are lost from the virion in the first few minutes, mainly from the 5′ end, followed by the release of subunits from the 3′ end.

without RNA

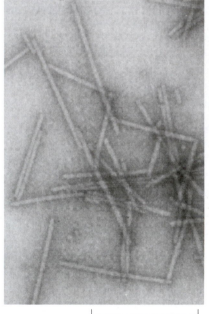

300 nm

TMV RNA is selected for encapsidation by recognition of an internal stem-loop structure

It would be reasonable to suppose that the assembly of TMV might be the reverse of disassembly, beginning at one end of the RNA and continuing toward the other. However, biochemical experiments revealed that the first part of the RNA to be covered with protein (identified by the protection conferred against nuclease activity) is an internal site, characterized by a stem-loop structure (**Figure 8.18A**). This sequence is located ~1000 nt from the 3′ end. Electron micrographs of particles at various stages of assembly similarly showed that the rod starts growing from a site ~1000 nt from one end of the RNA, and continues rapidly in the longer (5′) direction and much more slowly in the shorter (3′) direction (Figure 8.18B). The initial rate of encapsidation was found to depend on the square of the protein concentration, indicating that initiation is a second-order reaction; that is, one with two protein components. Under physiological conditions, the protein exists chiefly as a large assembly. This was proposed to be a two-layer disk seen in a crystal structure of the capsid protein, with each layer a ring of 17 subunits and with both layers pointing in the same direction. There are subtle conformational differences between the upper and lower layers that are thought to prevent further growth. The regions of polypeptide chain lying between radius 20 Å, lining the axial lumen, and 40 Å, where RNA is to bind, were disordered—unlike the same regions in the intact virion. These observations led to a proposal for the assembly pathway (Figure 8.18C,D) in which the RNA stem-loop acts as initiator (packaging signal), inserting into the central hole of a disk and being captured between the layers of protein subunits in it and a second disk. The disordered inner regions facilitate entry of the RNA and become ordered in the process. Further disks then add, one by one, from the top. The 5′ arm of the RNA is pulled through the lumen of the growing virion to create a traveling loop as each new disk is added. The RNA–protein interaction forces the subunits to switch into the conformation assumed in the helical capsid (see Figure 8.18). Assembly from the 3′ end is much slower and involves the addition of small oligomers (mostly trimers).

(A)   (B)

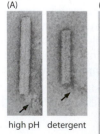

high pH   detergent

25 nm

ribosome-mediated disassembly

Figure 8.17 Disassembly of TMV virions *in vitro* and *in vivo* visualized by negative staining EM. (A) Virions perturbed by exposure to (left) high pH (pH 10) and (right) detergent (sodium dodecyl sulfate). Exposed RNA is seen at the bottom in both cases. (B) Ribosomes clustering around a partly disassembled virion extracted from TMV-infected tobacco cell protoplasts. (A, from T.M. Wilson et al., *FEBS Lett.* 64:285–289, 1976. With permission from Elsevier; B, from T.M. Wilson, *Virology* 137:255–265, 1984.)

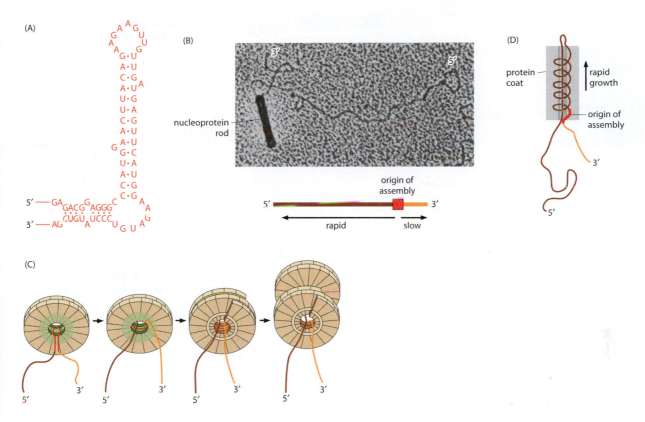

Figure 8.18 Model of TMV assembly *in vitro*. (A) The stem loop in the viral RNA ~1000 nucleotides from the 3′ end, which acts as the origin of assembly. (B) Electron micrograph of a partly assembled virion, contrasted by rotary shadowing with platinum. The long strand of RNA (5′ end, rapid growth) and the short strand (3′ end, slow growth) emerge from the same end of the rod. Below: map of the RNA with the stem-loop region marked in red. (C) In this model, assembly initiates with the stem-loop entering a protein disk axially and inserting between the two layers. Access is facilitated by disordering of the near-axial protein (green). The disk switches to a proto-helix in which the near-axial protein becomes ordered (brown) and an RNA loop emerges at the top of the rod to bind a second disk, which is merged into the growing helical rod. (D) Further growth proceeds with the 5′ end of the RNA being drawn up the central hole as a traveling loop and more disks being recruited, one after the other. In the process, the initiating stem-loop is melted into ssRNA (red). Encapsidation of the 3′ end is slower and follows a different mechanism. (A and D, adapted from P.J. Butler, *Phil. Trans. R. Soc. Lond. B* 354:537–550, 1999. With permission from The Royal Society; B, adapted from G. Lebeurier et al., *Proc. Natl Acad. Sci. USA* 74:149–153, 1977.)

Others have presented evidence that a short helical aggregate (~38 subunits) rather than the 34-subunit disk is the assembly-active protein species. Whichever is correct, it is agreed that the nucleoprotein rod grows rapidly in the 5′ direction. It is unclear how the much slower covering of the RNA toward its 3′ end takes place—it does not proceed very far until coverage of the 5′ end is almost complete—but it is likely that the protein adds in the form of monomers, dimers, and trimers. The RNA sequence AAGAAGUCG at the tip of the stem-loop (see Figure 8.18) is necessary and sufficient to nucleate assembly, after which the process, although bidirectional, is essentially sequence-independent.

Filamentous bacteriophages have long flexuous capsids enclosing circular ssDNA genomes

The long thin filamentous bacteriophages infect many Gram-negative bacteria, but the most studied systems are phages fd, M13, and f1 (whose genomes are more than 98% identical), which infect *Escherichia coli*, and Pf1 and Pf3, which infect *Pseudomonas aeruginosa*. While substantially slowing the growth of infected hosts, they reproduce without lysing or killing them and are secreted steadily from the bacterium, generating vast titers of progeny virions (up to 10^{13} per ml). The filaments are helical polymers of the major capsid protein, housing circular single-stranded genomes, with a few copies of four essential proteins at the ends. Bacteriophage fd is 890 nm long and ~7 nm in diameter and has a genome of 6407 nt, whose 9 genes yield 11 products (**Figure 8.19A**). Proteins pX and pXI arise from internal

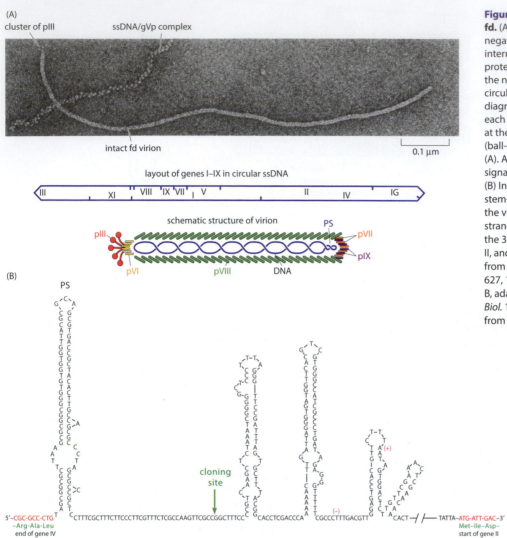

Figure 8.19 Structure of bacteriophage fd. (A) Top, electron micrograph of a negatively stained fd virion and an assembly intermediate of ssDNA coated with pV protein. Middle: layout of the nine genes and the noncoding, intergenic region (IG) in the circular ssDNA genome. Bottom, schematic diagram of a virion (not to scale). Five copies each of pIII/pVI and of pVII/pIX are located at the respective ends. Some pIII molecules (ball-and-stalks) are visible in the micrograph (A). A stem-loop in the IG (the packaging signal, PS) is at the pVII/pIX end of the virion. (B) In isolated DNA, the IG adopts a series of stem-loops. Only one, the PS, is retained in the virion. The origins of (+) strand and (−) strand DNA synthesis are indicated, as are the 3′ end of gene IV and the 5′ end of gene II, and a cloning site in the IG. (A, adapted from C.W. Gray et al., *J. Mol. Biol.* 146:621–627, 1981. With permission from Elsevier; B, adapted from R.E. Webster et al., *J. Mol. Biol.* 152:357–374, 1981. With permission from Elsevier.)

translational starts in genes II and I respectively. Some properties are listed in **Table 8.1**. The isolated DNA has a region of stem-loop structures (Figure 8.19B), the biggest of which (78 nt) is accommodated at one end of the virion. This feature, which serves as a packaging signal (PS), follows the 3′ end of gene IV and marks the beginning of a noncoding intergenic region (IG) of ~430 nt that runs until the 5′ start of gene II. The end of the virion with the packaging signal has five copies each of pVII and pIX, and the other end has five copies each of pIII and pVI. The protein sheath comprises ~2700 copies of pVIII (50 aa, a gently curving α helix with its long axis aligned close to that of the virion) in a shingled array, which overlap like fish scales. The protein subunits are arranged in stacked rings of five; that is, a five-start helix (**Figure 8.20A,B**). The number of nucleotides packaged per pVIII subunit is non-integral, at ~2.4.

Attachment of filamentous bacteriophages to pili initiates infection

Filamentous bacteriophages have a complex reproductive cycle that begins with the attachment of a virion to a bacterial pilus (Section 11.8), mediated by protein pIII (**Figure 8.21**). The target for fd, M13, and f1 is the *E. coli* sex pilus encoded by the conjugative F plasmid, and these bacteriophages are known as Ff phages. It seems that pili undergo spontaneous cycles of retraction and extension and that, during retraction, the virus is drawn toward the bacterial cell membrane where it is disassembled, beginning at the pIII/pVI end.

Table 8.1 Proteins encoded by the ssDNA of bacteriophage fd (M13, f1).

Protein	Number of aa residues in mature protein	Signal sequence	Inner (IM) or outer (OM) membrane	*Dependence on Sec	Location of N terminus	Function
pI	348	–	IM	No	Cytoplasm	Assembly
pII	410	–	Neither	n/a	n/a	DNA replication
pIII	406	18 aa	IM	Yes	Periplasm	Minor capsid protein
pIV	405	21 aa	OM	Yes	Periplasm	Exit channel
pV	87	–	Neither	n/a	n/a	Binding ssDNA
pVI	112	–	IM	No	Periplasm	Minor capsid protein
pVII	33	–	IM	No	Periplasm	Minor capsid protein
pVIII	50	23 aa	IM	No	Periplasm	Major capsid protein
pIX	32	–	IM	No	Periplasm	Minor capsid protein
pX	111 (C-terminal part of pII)	–	Neither	n/a	n/a	DNA replication
pXI	108 (C-terminal part of pI)	–	IM	No	Cytoplasm	Assembly

Pf1, although similar to fd, has some notable differences. Its capsid is longer (~1940 nm) and has ~7350 capsid subunits arranged as a one-start helix with 5.4 subunits per turn (Figure 8.20C). The genome has 7349 nt and is packaged at 1.0 nt per protein subunit. fd and Pf1 exemplify two structural classes: those with a five-start helix like fd are designated Class I, and those with a one-start helix like Pf1 are Class II. Pf3, like Pf1, has Class II symmetry, but with a DNA of 5833 nt and ~2500 pVIII subunits it resembles Class I in packaging ~2.4 nt per protein subunit. n/a, not applicable. * The Sec system is a general purpose bacterial secretion system that translocates a wide range of proteins across the inner membrane (Section 10.7).

Ff phages bind to the tip of the F pilus, as shown by electron microscopy (see **Methods**). pIII has several domains separated by Gly-rich linkers (**Figure 8.22B**). Its C domain remains tightly bound at the end of the virion but, on contacting a pilus, domains N1 and N2 detach, with N2 binding to the pilus and N1 released. When the pilus has retracted sufficiently, N1 encounters domain D3 of the bacterial protein TolA (Figure 8.22A). TolA is a component of the TolQRA complex and is anchored with its N-terminal region (D1) in the inner membrane, its central region (D2, largely α-helical) spanning the periplasm, and its C-terminal domain (D3) in the outer membrane. N1 binds tightly to D3 (Figure 8.22C), and the TolQRA complex mediates entry of the viral ssDNA into the cytoplasm. Pf1 is thought

Figure 8.20 Helical structures of bacteriophages fd and pf1.
(A) Arrangement of fd pVIII subunits. This 120 Å segment represents ~1.4% of a virion. The proteins are arranged on a five-start helix, as indexed on a radial projection in (B). In (A), the α-helical subunits are represented as columns of spheres. Subunits 0, 6, and 11 (the numbering is explained below) are shown in detail. The N termini are on the outside (the first few amino acid residues are disordered), and the C-terminal regions line the interior. Three N termini are marked with N. (B) Left, helical lattice; subunit positions are marked with crosses. (Here, the cylinder is opened out on to a plane.) This is a five-start helix with one of the helices drawn as a red line and the other four in white; that is, the structure is composed of stacked 5-fold rings, with a 16.2 Å axial repeat and a 36° rotation between rings. The subunits are numbered so that 0–4 are in the first ring, 5–9 in the second ring, and so on. Right, the helical lattice of Pf1, which follows a one-start helix with 5.4 subunits per turn and an axial rise (pitch) of 15.7 Å per turn. In both lattices, the dotted line indicates the angle between the capsid protein helix and the virion axis. (Adapted from D.A. Marvin et al., *J. Mol. Biol.* 355:249–309, 2006. With permission from Elsevier.)

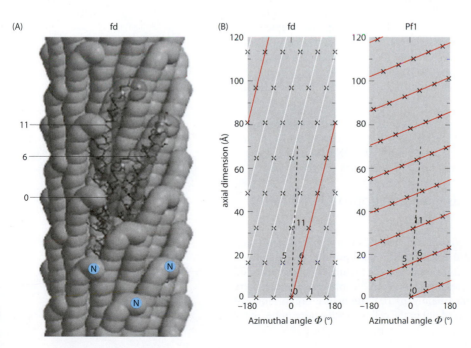

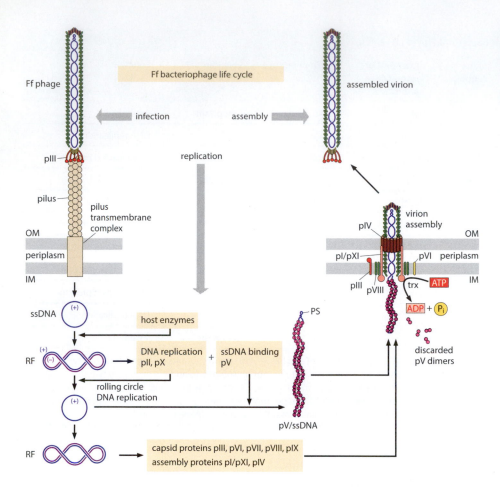

Figure 8.21 Replication cycle of Ff bacteriophage. Left, pIII subunits at one end of the virion bind to a pilus tip. Retraction of the pilus brings the virion into contact with the cell envelope and through the outer membrane (OM). Capsid proteins are shed and enter the inner membrane (IM), releasing the viral ssDNA into the cytoplasm where it is converted to the double-stranded replicative form (RF). Proteins pII, pX, and pV involved in DNA replication are synthesized and more (+) strands are synthesized. The RF directs the synthesis of capsid proteins and assembly factors, all of which, except pIV and pV, are inserted into the IM. Eventually, (+) strands are sequestered in pV/ssDNA complexes. Right, the exposed packaging signal (PS) directs the pV/ssDNA complex into virion assembly at the IM, whereby pVII and pIX are added and then pVIII progressively replaces pV. For this process, pI/pXI are required, together with the host protein thioredoxin (trx) and ATP hydrolysis. (Thioredoxin also serves as a processivity factor for the DNA polymerase of bacteriophage T7, as described in Section 3.3.) The growing virion exits from the cell through a ring of pIV subunits in the OM. Assembly is completed with the addition of pIII and pVI and release of the newly formed virion.

to follow essentially the same process, beginning with binding to a type IV pilus of *P. aeruginosa*, but Pf3 differs at least in binding to the side of the pilus.

Capsid protein subunits of an Ff phage are stripped from the ssDNA as it enters the *E. coli* cell, where it is replicated by a **rolling-circle** mechanism (**Figure 8.23A**). Host enzymes (an RNA polymerase to synthesize a primer, followed by DNA Pol III for DNA synthesis) use this (+) strand as template to generate the complementary (–) DNA strand. Once the double-stranded circle has been established (replicative form, RF), the old (+) strand is displaced by Pol III as it travels around the (–) strand. When the circle is complete, the covalently linked pII cleaves the DNA at the junction of the fresh and displaced DNA and ligates the ends of the displaced DNA. This releases a circular (+) strand and breaks the covalent bond between pII and the DNA. The new (+) strand can be used as a template for the synthesis of another RF. In this way, RF DNA builds up to ~50 copies per cell, thereby increasing viral protein synthesis.

The supercoiled RF dsDNA is now able to direct the synthesis of viral proteins (see Figure 8.21). Two of the earliest genes transcribed are II and X. PII selectively cleaves the (+) strand of the RF and becomes covalently attached to the new 5′ end of the DNA. The new 3′ end is then used as the starting point for the synthesis of a new (+) strand by Pol III.

Encapsidation of the DNA proceeds via a nucleoprotein filament intermediary

Filamentous phages are assembled and secreted from infected cells in a complex, multi-step process (see Figure 8.21). First, the genome is coated with a viral protein, pV, forming a filament in the cytoplasm, while capsid proteins destined for the mature virion are inserted into the inner membrane. The pV/DNA complex associates with the inner membrane; there pV is progressively replaced by other proteins (mainly, pVIII), the helical organization of the filament changes, and the nascent virion is fed out of the cell through a pore in the outer membrane.

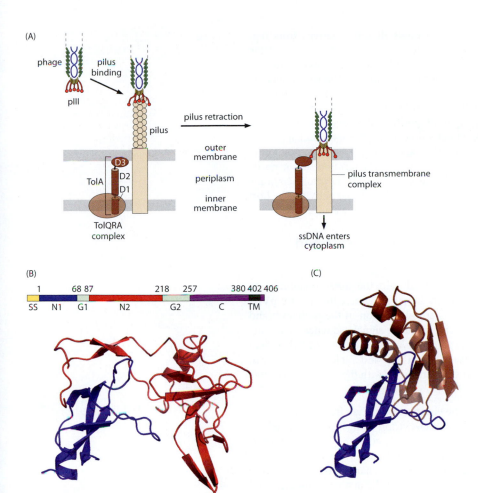

Figure 8.22 Cell entry pathway of Ff phage. (A) Domain N2 of pIII binds to an F pilus, releasing domain N1. The pilus retracts toward the outer membrane; N1 then binds to domain D3 of TolA. The TolQRA complex mediates entry of the viral DNA into the cytoplasm. (B) Domain map of pIII whose precursor form has an 18 aa signal sequence (SS, yellow). Gly-rich linkers G1 and G2 are in pale green; the C domain, which binds in the virion, is in purple; and the transmembrane segment (TM) is in black. The N-terminal 217 aa of pIII, encompassing N1 (blue) and N2 (red), form a horseshoe-shaped structure in which connecting linker G1 is not ordered. Mutagenesis studies indicate that pilus binding is affected by residues on the outer surface of N2. (PDB 1G3P) (C) Structure of N1 (blue) in complex with D3 (brown). The sets of N1 residues that mediate binding with D3 and N2 are very similar, despite the differences in the partner domains. (PDB 1TOL)

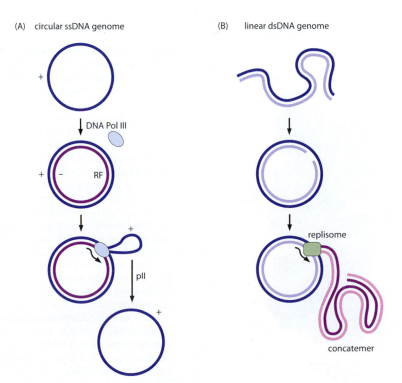

Figure 8.23 Rolling-circle mechanisms of DNA replication. (A) Mechanism for replication of circular ssDNA genomes of filamentous bacteriophages. The starting points for (+) and (−) strand synthesis are in the IG (see Figure 8.19A). (B) Proposed mechanism for some double-stranded DNA viruses with linear genomes, including late-stage replication of phage λ. The genome is circularized with a gap in one strand. Then a *replisome* of virally encoded proteins proceeds multiple times around the circle, spinning off multiple copies of the genome in a linear concatemer. This is the substrate for packaging, with single genomes being cut off by the packaging enzyme, terminase.

As pV accumulates, it binds to newly synthesized (+) strands, diverting activity from replication toward assembly. pV dimers bind cooperatively to the DNA (+) strand, generating a filament ~850 nm long and 8 nm in diameter (see Figure 8.19). It takes ~800 pV dimers to cover the DNA completely; that is, four nucleotides per monomer. The pV dimers are wound in a left-handed superhelix; the ssDNA in an extended conformation follows the inside of this superhelix, with the two halves of the circular DNA running in opposite directions (**Figure 8.24**). Crucially, the packaging signal is positioned at one end of the pV/DNA filament. Assembly proper begins with the interaction of the packaging signal with an inner membrane complex thought to consist of pI, pVII, and pIX. pI has an ATPase activity, which is likely to account for the need for ATP in assembly. Its N-terminal 253 aa reside in the cytoplasm and its C-terminal 75 aa are in the periplasm; its 20-aa transmembrane segment may help form a channel. pXI derives from pI but lacks ATPase activity. The host protein, thioredoxin, is an essential partner of pI in assembly but its role is unknown.

Ff virions assemble in the inner membrane and are secreted through the outer membrane

As they are synthesized, Ff capsid proteins are inserted into the inner membrane to await assembly (see Figure 8.21). After removal of its signal sequence, the mature pVIII (50 aa) spans the membrane once with its 20-aa N-terminal region in the periplasm and the C-terminal 11 aa in the cytoplasm. pIII, which also has a signal sequence, spans the membrane once, with only the C-terminal 5 aa in the cytoplasm. Insertion of pVIII is Sec-independent but that of pIII is Sec-dependent (see Section 10.7 and Table 8.1). pVI, pVII, and pIX lack signal sequences and seem to insert spontaneously, with their N-terminal ends in the periplasm.

Five copies each of pVII and pIX form the tip of the virion as it emerges from the cell. Then pV dimers are displaced by pVIII subunits as the nascent virion elongates. (Displaced pV dimers are recycled.) In this transition, there is an abrupt change in the DNA conformation and protein packing. Assembly continues until the distal end of the DNA has been reached. At this point, pIII and pVI are added and the virion is discharged from the cell. The nascent virion is secreted through a ring of 14 pIV subunits in the outer membrane. (pIV is secreted into the periplasm, where its C-terminal end becomes embedded in the outer membrane.) As seen in electron micrographs, this ring has an outer diameter of ~14 nm and an inner diameter of 6–8 nm, just wide enough for passage of the virion. pI/pXI interacts with N-terminal parts of pIV in the periplasm. The pIV exit port is homologous to the *secretins*, complexes involved in a variety of protein export events of Gram-negative bacteria (Section 10.7) and, like them, may be gated by a septum-like occlusion that closes the channel to compounds that would otherwise erroneously cross the outer membrane.

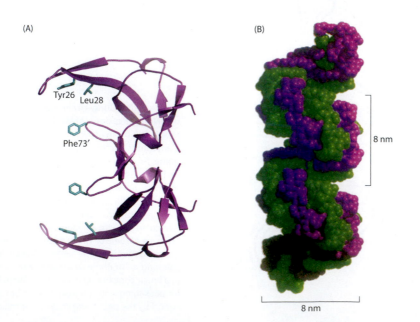

(A)

Tyr26 Leu28

Phe73′

(B)

8 nm

8 nm

Figure 8.24 Structure of the pV/ssDNA complex. (A) The pV dimer of phage f1. The two subunits are related by a 2-fold axis of symmetry (horizontal). Biochemical experiments indicate that Tyr26, Leu28, and Phe73 are involved in binding the ssDNA (Phe73′ comes from the other subunit). (PDB 1BGH) (B) Model of the arrangement of pV dimers in the left-handed superhelix. Alternating dimers are shown in magenta and green. Two and a half turns of the superhelix are shown. The model has eight dimers per turn in a helix of pitch 8 nm, with the DNA-binding residues pointing inward and the 2-fold axes at right angles to the helix axis. The two ssDNA strands would run from bottom left to upper right between adjacent dimers, in opposite directions. The distance between Tyr26 residues in adjacent dimers is ~15 Å, a distance easily spanned by four nucleotides in an extended conformation. (From M.M. Skinner et al., *Proc. Natl Acad. Sci. USA* 91:2071–2075, 1994 National Academy of Sciences, USA. With permission from PNAS.)

Packaging of ssDNA in Ff bacteriophages involves charge-matching

Apart from the 78-nt stem-loop at the pVII/pIX end of the virion, the ssDNA in an Ff phage has no regular base pairing between the two strands that run in opposite directions within the sheath-like capsid. If fully extended, a circular ssDNA arranged in this way (~3200 nt in each direction) would be approximately twice as long as the virion, so the DNA strands must be coiled in some way. The mature pVIII protein is arranged with its C-terminal region, possessing four positive charges, lining the sheath (**Figure 8.25A,B**), creating a complementary positively charged environment for the DNA.

Directed mutagenesis can be used to introduce DNA cloning sites into the IG (see Figure 8.19B). If extra DNA is cloned into this noncoding region, assembly can initiate and proceed as usual, provided that the packaging signal is undisturbed. It is this property that has made filamentous phages so valuable in biotechnological applications (Section 8.8). The resulting nucleocapsid is longer, in proportion to the amount of DNA inserted (Figure 8.25C). This is consistent with the genome's acting as a tape-measure.

In contrast, if the K48Q mutation is made in pVIII, decreasing the number of positive charges in its C-terminal region from four to three, the resulting virions are longer by one-third than in the wild type while still having a genome of normal length (see Figure 8.25C). No viable virions are produced with the K48E mutation, which lowers the net positive

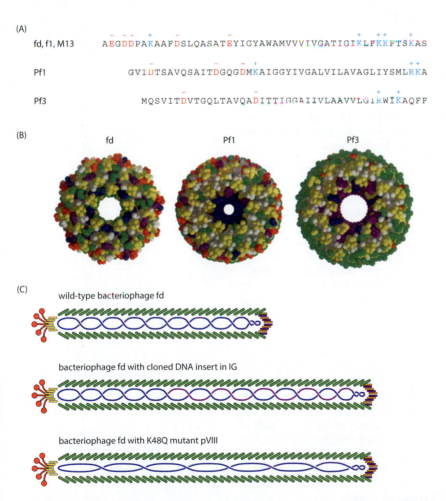

Figure 8.25 **Packing of DNA in filamentous bacteriophages.** (A) Amino acid sequences of the major capsid proteins, pVIII. The fd and Pf1 sequences are preceded by N-terminal signal sequences (not shown) that are cleaved off as the proteins are inserted into the inner membrane. (B) Cross sections of three capsids. Main-chain atoms are shown in yellow, acidic side chains in red, polar side chains in green, apolar side chains in white, aromatic side chains in purple, and basic side chains in blue. (C) Schematic arrangements of the genome (blue) in wild-type fd, a mutant with extra DNA (purple) cloned into its IG, and the K48Q mutant.

charge in the C-terminal region to two. However, hybrid capsids can be constructed by mixing K48E and wild-type subunits. Such particles have no fixed length but all are longer than wild-type virions, the average length increasing with the proportion of K48E subunits. These observations support a packaging mechanism in which the positive charge density lining the capsid dictates how the DNA is arranged. Lowering this charge density requires the DNA to adopt a more extended form with a lower negative charge density to match it. On its own, the K48E mutation is unacceptable, probably because it requires the DNA to be more extended than is physically possible.

The ssDNA in bacteriophages Pf1 and Pf3 may be inside-out

Pf1, with only two positively charged residues in the C-terminal region of its pVIII, packages DNA at 1.0 nt per subunit. For this, the DNA must be fully extended in the virion; indeed, the axial lumen has a smaller diameter (see Figure 8.25B). Pf3, however, also has Class II symmetry and the C-terminal region of its pVIII also has two positively charged residues. Nevertheless, its axial lumen is similar to that of fd and it packages 2.4 nt per subunit, like a Class I Ff phage. The positive charge density inside its capsid presumably has contributions from sources other than the pVIII Lys residues. A proposal for both Pf1 and Pf3 is that the DNA is unusually arranged with the phosphates facing inward and the bases interacting with the inside lining of pVIII subunits. This 'everted' DNA structure remains to be proved.

8.4   SMALL ICOSAHEDRAL VIRUSES

We now consider some of the smallest icosahedral viruses, with T-numbers of 1, 3, and 4, comprising 60, 180, and 240 subunits respectively, and diameters in the range 15–35 nm. In most cases, the MCP is capable of self-assembly and the genome is encoded in single-stranded nucleic acid, most commonly RNA. Only a few of these capsids are enveloped, so that responsibility for host cell recognition and entry falls to the MCP. In several systems, the binding of the virion to its cellular receptor has been described. Although these viruses share a common icosahedral geometry, they are richly diverse in the mechanisms that regulate their assembly and in how they function as genome delivery machines.

Viruses with small capsids are handicapped through their limited capacity for nucleic acid. Whereas some viruses persist successfully with a minimalist genome that encodes only an MCP and a polymerase, others have evolved strategies to overcome this limitation. One such strategy, followed by some RNA plant viruses, involves genome segmentation with different genes consigned to separate particles. This strategy, although evidently viable, has the disadvantage that a host cell must be infected by multiple virions if it is to receive all segments. Other small viruses depend on 'helper' viruses to provide them with some functions: in effect, such a virus—termed a satellite virus—is a parasite of a parasite. Examples include satellite tobacco necrosis virus (STNV), a $T = 1$ plant virus with its ssRNA genome of only 1.22 kilobases (kb), and adeno-associated virus, a parvovirus with a $T = 1$ capsid and a ssDNA genome of 4.68 kb.

Other strategies involve maximizing the use of limited coding capacity. A genomic tract may code for two proteins in different overlapping reading frames. This mechanism, employed both by MS2, a $T = 3$ phage with an RNA genome of 3.57 kb, and φX174, a $T = 1$ phage with a ssDNA genome of 5.39 kb, gives 'two genes for the price of one.' Hepatitis B virus (HBV) has a DNA genome of only 3.2 kb but generates a substantial set of gene products through a combination of alternative initiation sites, post-translational processing, and overlapping genes: every nucleotide in this genome is in a coding region, and more than half of the genome is expressed in more than one reading frame (**Figure 8.26**). S-antigen (surface antigen; the viral glycoprotein) is expressed in three forms with different sizes (S = small, M = middle, L = large) from three distinct in-frame initiation sites. A similar mechanism produces two forms of the MCP (core antigen and e-antigen) and the distinction between them is accentuated by post-translational proteolytic processing (see below).

Capsid proteins may self-assemble or co-assemble with the genome

As noted above, the first discovered capsid protein fold was the widely occurring jelly roll. Other folds found in small icosahedral viruses are those of phage MS2, hepatitis B virus,

(A)

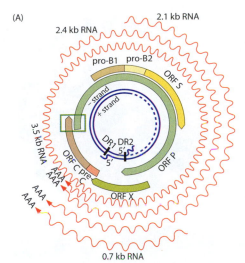

(B)

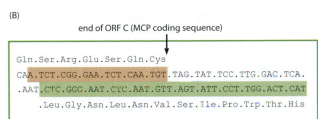

end of ORF C (MCP coding sequence)

```
Gln.Ser.Arg.Glu.Ser.Gln.Cys
CAA.TCT.CGG.GAA.TCT.CAA.TGT.TAG.TAT.TCC.TTG.GAC.TCA.
.AAT.CTC.GGG.AAT.CTC.AAT.GTT.AGT.ATT.CCT.TGG.ACT.CAT
    .Leu.Gly.Asn.Leu.Asn.Val.Ser.Ile.Pro.Trp.Thr.His
```

part of ORF P (polymerase coding sequence)

Figure 8.26 Dense packing of genetic information in a small viral genome. (A) Map showing the genes (ORF, open reading frame), the double-stranded and single-stranded regions of DNA in the HBV genome, and the transcripts. Several genes overlap in out-of-register reading frames. (B) Detail of the region boxed in (A), which codes for both the end of the core (MCP) gene and, in the relative –1 reading frame, part of the polymerase. (A, adapted from D. Ganem, in Fields Virology, vol 2, p 2706. Philadelphia: Lippincott-Raven, 1996.)

and the fungal virus L-A (all different; see below). These MCPs are programmed to self-assemble into capsids of defined sizes. Although the MCP alone can self-assemble, genomic nucleic acid will be incorporated if it is available. Some small viruses will encapsidate any available RNAs; others have a strong preference for their own genomes, which contain packaging signals in the form of stem-loop structures. These motifs interact with the MCP to initiate assembly; as it proceeds, further interactions between the growing protein shell and the genome lead to its compaction and, ultimately, enclosure (**Figure 8.27**). Once initiated, assembly must proceed rapidly and cooperatively, because partly assembled capsids are rarely observed. Packaging signals not only identify a nucleic acid molecule as being eligible for packaging, but also promote MCP assembly at concentrations lower than the critical concentration for self-assembly, thereby avoiding the fruitless production of empty capsids.

In some viruses, the genome binds directly to the capsid wall. In others, it binds to an MCP appendage rich in basic residues that extends into the capsid interior. Packaged genomes cannot, even in principle, conform to the same high order of symmetry as the capsid because they do not consist of 60 tandem repeats. Nevertheless, some virions have substantial contents of ordered nucleic acid. In others, the encapsidated genome is variable and/or asymmetric in structure so that it becomes smeared out in icosahedrally symmetrized density maps.

Tomato bushy stunt virus (TBSV) is an RNA plant virus with a $T = 3$ dimer-clustered capsid. In addition to a protruding projection domain and a shell domain with the canonical jelly roll fold (**Figure 8.28A**), its MCP has an N-terminal 'arm,' which was not seen in the crystal

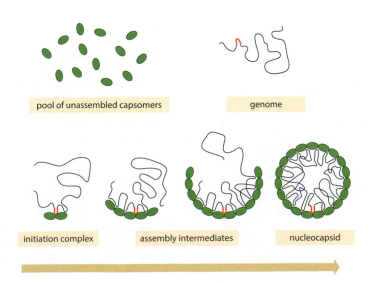

pool of unassembled capsomers

genome

initiation complex

assembly intermediates

nucleocapsid

Figure 8.27 Co-assembly of a viral genome and a capsid protein. Aggregation of MCP capsomers (green) is nucleated by the packaging signal motif on the genome (red), followed by the recruitment of progressively more capsomers into the growing shell, accompanied by additional interactions with the condensing genome.

Figure 8.28 Mapping of RNA and protein by neutron diffraction. (A) Outer surface of tomato bushy stunt virus (TBSV). (B, C) 'Contrast matching' was used to discriminate between its RNA and protein. Virion structures determined to ~16 Å resolution by neutron crystallography are illustrated in central sections viewed along a 2-fold axis. Densities are color-coded in descending order: red (highest), yellow, green, blue, dark blue (lowest). (B) In a 38:62 mixture of $D_2O:H_2O$, the capsid shell is not seen; the highest density is that of the RNA, distributed throughout the capsid interior. (C) In a 70:30 mixture, the highest densities are those of buffer and RNA; the protein is visible with negative contrast. This map reveals an inner protein shell, presumably formed by the N-terminal arms. (From P.A. Timmins et al., *Structure* 2:1191–1201, 1994. With permission from Elsevier.)

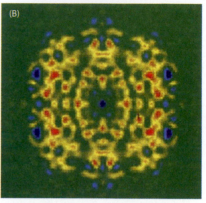

structure and was inferred to be flexible. It was, however, detected by neutron diffraction, a technique that allows selective visualization of protein and nucleic acid by a procedure called contrast matching. Either component may be camouflaged by suitably adjusting the mixture of heavy and light water (D_2O and H_2O) in the buffer. In a 38:62 mixture the protein is essentially invisible because its ability to scatter neutrons is no different from that of the surrounding buffer; consequently, only the nucleic acid is seen. The converse is true in a 70:30 mixture. When this technique was applied to TBSV crystals, RNA was found to be distributed throughout the capsid interior; however, there is also an inner shell of protein, offset by ~50 Å from the surface shell, which was assigned to (part of) the N-arm (Figure 8.28B,C).

Bacteriophage MS2 has a 3.57-kb genome of positive-sense ssRNA. The capsid is a *T* = 3 icosahedron formed from the MCP (14 kDa) plus a single copy of A-protein (44 kDa). The MCP has a β-sheet fold quite different from the jelly roll (**Figure 8.29**). Genomic RNA binds to sites on the inner surface of the capsid. In the original crystal structure of the virion, some faint density under the MCP dimer interface was tentatively assigned to RNA. Earlier studies had suggested a 21-nt stem-loop as likely to be the initiator of assembly. Its visualization was achieved by placing the same RNA motif at all symmetry-related sites. Crystals of

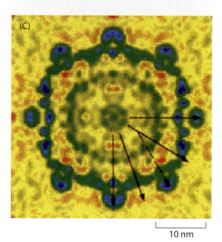

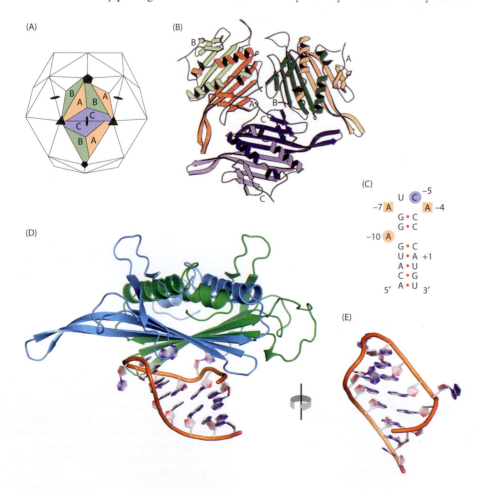

Figure 8.29 Bacteriophage MS2 capsid and its interaction with an assembly-nucleating RNA stem-loop. (A) The dimer-clustered *T* = 3 surface lattice. (B) Three MCP dimers from the virion structure. The quasi-equivalent subunits (A, B, and C) form two A–B dimers and one C–C dimer. (C) The 21-residue stem-loop portion of the MS2 ssRNA genome, which acts both as a regulator of gene expression and to initiate assembly. (D) Binding of the stem-loop under the interface of an A–B dimer. (E) Structure of the stem-loop. (Adapted from K. Vålegard et al., *Nature* 371:623–626, 1994. With permission from Macmillan Publishers Ltd; courtesy of Neil Ranson and Peter Stockley.)

empty capsids were perfused with the synthetic stem-loop, which penetrated into the capsids. The resulting density map revealed the structure of the stem-loop and its interaction with the MCP (see Figure 8.29C–E).

In MS2 assembly, binding of the stem-loop to an MCP dimer triggers assembly around that site, either because its interaction with the RNA switches the initiating dimer into a higher-affinity state for other MCP dimers or because their recruitment is enhanced by secondary interactions with the RNA, or both. As assembly proceeds, RNA attaches to other sites on the assembling shell.

Nodaviruses have ordered RNA and mature by autocatalytic proteolysis

Nodaviruses, which infect insects, mice, and fish, are notable in that a substantial fraction of their encapsidated ssRNA genomes are ordered. (The name refers to Nodamura, a village in Japan where the first virus of this kind was discovered.) The genome of flock house virus (FHV), an insect nodavirus, consists of two segments, 3.1 and 1.4 kb long, of positive-sense RNA. The $T = 3$ capsid packages one copy of each segment. Although the genome is single-stranded, many regions engage in intramolecular base pairing. The FHV crystal structure revealed a 10-nt, double-stranded segment underlying each dimer of C–C MCP subunits (**Figure 8.30**). Cryo-EM revealed that the amount of ordered RNA is even greater. In the reconstruction, the RNA forms a dodecahedral cage whose side length corresponds to ~25 bp of A-type RNA. As there are 30 such segments, they account for ~35% of the genomic RNA.

Proteolysis is widely employed in regulating virus assembly. Many viruses encode their own proteases; others enlist host enzymes. In FHV maturation, the proteolytic event is **autocatalytic**—that is, no enzyme is involved and the local chemical environment promotes the reaction. The FHV MCP is synthesized as a 407-aa precursor, the α subunit. After assembly, it is cleaved to the β subunit (363 aa) and γ peptide (44 aa). This reaction, required for infectivity, leaves the γ peptides inside the shell formed by the β subunits, which have the canonical jelly roll fold. Earlier, during assembly, the γ peptides are involved in selecting viral RNA for packaging. Asp75, which the capsid crystal structure showed to be closely apposed to the cleavage site, has a key role, because converting it to Glu, Asn, or Val abolishes cleavage.

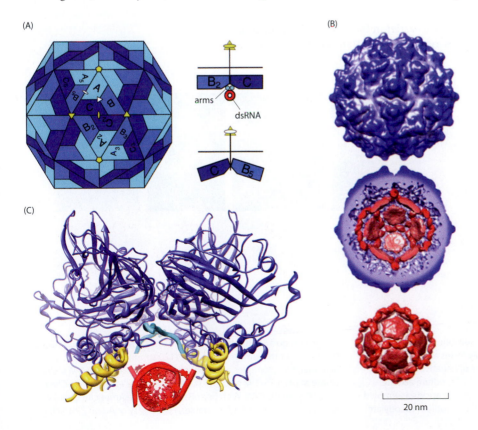

(A)

(B)

(C)

20 nm

Figure 8.30 Flock house virus (FHV) has a high content of ordered RNA. About 35% of the ssRNA genome is packaged as base-paired duplexes bound to the inner surface of the capsid. (A) The $T = 3$ surface lattice. Its three quasi-equivalent subunits are labeled A–C. The RNA-interacting sites lie between C–C subunits on the axes of 2-fold symmetry. A schematic duplex is shown at one site (red). (B) Cryo-EM reconstruction of FHV. Although the capsid is $T = 3$, the topography of the outer surface suggests a $T = 1$ structure; each of the 60 protrusions combines loops from three subunits (A, B, and C) grouped around a local 3-fold axis. The ordered RNA (red) appears as a dodecahedral shell with struts that show the helical twist. (C) Detail from the crystal structure of the FHV virion, showing a duplex under the 2-fold axis with B and C subunits at the left and B2 and C2 subunits at the right. (B, courtesy of Jeffrey Speir and Jack Johnson; C, from A.J. Fisher and J.E. Johnson, *Nature* 361:176–179, 1993. With permission from Macmillan Publishers Ltd.)

A likely mechanism is that Asp75 forms a H-bond with the carbonyl of the scissile peptide bond, rendering it susceptible to nucleophilic attack by water.

The capsid protein of a simple plant virus has polymorphic assembly products

Cowpea chlorotic mottle virus (CCMV) has a simple composition and was already available in large quantities in the pre-cloning era. These properties commended it as a model system for *in vitro* assembly studies. Under native conditions, the MCP, which has a jelly roll domain preceded by a basic (nucleic acid-binding) N-terminal domain and followed by a C-terminal 'arm,' assembles into a $T = 3$ capsid. CCMV has a positive-sense ssRNA genome whose four segments partition into three particles (the two smallest segments are packaged together). *In vitro*, the purified MCP polymerizes into a wide variety of structures depending on the pH, the ionic strength, and the presence or absence of RNA. Assembly products include icosahedral shells of other sizes, cylindrical tubes, and multilayered shells and tubes (**Figure 8.31**). Although the MCP shows little selectivity for its own genome, RNA molecules can affect the outcome of an assembly reaction; fragmented RNA favors smaller capsids of 60 non-equivalent dimers, a form otherwise seen only in the capsids of dsRNA viruses, whereas short RNA fragments favor 30-dimer $T = 1$ capsids and over-long RNA leads to strings of $T = 3$ particles.

Unassembled CCMV MCP is dimeric, and the virion crystal structure led to the proposal that the initiating complex may be a hexamer of dimers. However, during assembly, the predominant interactions switch to a hexamer/pentamer clustering of subunits such that the $T = 3$ capsid consists of 12 pentamers and 20 hexamers (**Figure 8.32**). Other assembly products have distinctive patterns of hexamers alone (in tubes), or pentamers alone ($T = 1$ capsids made of 60 non-equivalent dimers), or hexamers and pentamers (other quasi-equivalent icosahedral shells); presumably, assembly initiates differently in each case.

The assembly and, in particular, the curvature of a growing CCMV particle are influenced by the C-terminal arm of the MCP. In the $T = 3$ capsid, these arms form a 'β-hexamer' of short β strands underlying each hexamer of jelly roll domains (Figure 8.32B). There is no corresponding ring under the pentamers where the C-arms are disordered. The C-arm may be deleted without abrogating assembly. Virions with this deletion are less stable than the wild-type but remain infectious.

Although its capsid must provide a sturdy protective barrier in the extracellular milieu, an infecting virion must release its genome to allow replication and transcription to proceed.

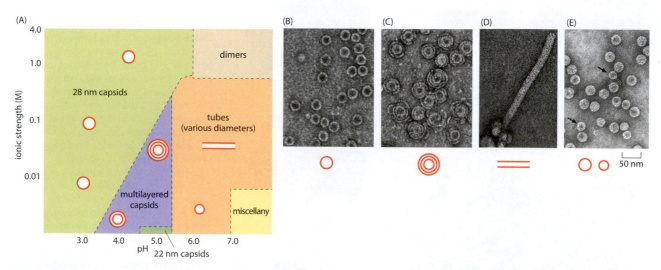

Figure 8.31 Phase diagram for polymorphic assemblies of CCMV capsid protein. (A) Map of predominant assembly forms as a function of pH and ionic strength. (B–E) Electron micrographs of negatively stained specimens. (B) Empty $T = 3$ capsids assembled in 0.1 M sodium acetate buffer at pH 4.67, 0.1 M ionic strength. (C) Multi-layered capsids assembled at pH 4.8, ionic strength 0.01 M. (D) Tube assembled in sodium cacodylate buffer with no added salt at pH 6.0, 0.01 M. (E) Mixture of $T = 3$ (180 subunits, 28 nm diameter) and $T = 1$ (120 subunits = 60 dimers, 22 nm diameter, two examples arrowed) capsids assembled in 0.1 M sodium citrate at pH 4.75, 1 M NaCl. (B–D, from L. Lavelle et al., *J. Phys. Chem. B* 113:3813–3819, 2009. With permission from The American Chemical Society, courtesy of Charles Knobler; E, courtesy of Adam Zlotnick.)

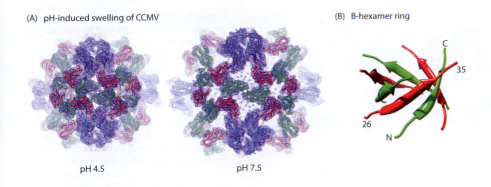

(A) pH-induced swelling of CCMV

pH 4.5   pH 7.5

(B) B-hexamer ring

C

35

26

N

Figure 8.32 CCMV capsid undergoes a pH-dependent structural transition.
(A) CCMV has a $T = 3$ capsid with hexamer/pentamer clustering of subunits. There are three quasi-equivalent versions (shown in blue, purple, and green) of a jelly roll subunit. Quasi-equivalence is expressed in terms of differing inter-subunit dihedral angles. Underlying the hexamers are 6-fold rings of parallel β strands (B). At acidic pH (here, ~4.5) and in the presence of metal ions, the capsid has a compact conformation. When shifted to neutral pH (here, ~7.5) and with the metal ions chelated, the capsid expands by ~10% and fissures open up between the capsomers. The hexamer/pentamer clustering of the MCP subunits is accentuated in the expanded state. (Courtesy of Jeffrey Speir and Jack Johnson.)

This requires partial or complete disassembly of the virion. Upon suitable adjustment of the ionic conditions, CCMV and other plant viruses undergo a concerted transition in which the MCP subunits undergo rigid-body rotations that result in expansion of the virion by ~10% and the opening up of fissures in the surface lattice (Figure 8.32A). CCMV is stable around pH 5 but if the pH is raised to 7 at moderate ionic strength, this expansion takes place. An additional level of control is imposed by divalent cations: bound Ca^{2+}, if present, blocks the expansion. It is thought that this transition may represent a prelude to genome release during a normal infection and that the fissures represent potential exit pathways for the genome.

Hepatitis B virus capsid is a porous compartment for retrotranscription

HBV is a human virus that causes liver disease on a massive scale, with an estimated 400 million persons chronically infected, mainly in Asia. Unusually for such a small virus, its capsid is enclosed within a lipoprotein envelope. Whereas, under native conditions, the MCPs of most icosahedral viruses assemble into capsids of defined size, HBV capsids are of two sizes: most observe $T = 4$ symmetry (120 dimers) but some are $T = 3$ (90 dimers) (**Figure 8.33B,C**). This dimorphism may indicate that HBV is at a transition point in its evolutionary history.

In nucleocapsid assembly, the genome is packaged in the form of ssRNA (the so-called **pregenome**), which has a stem-loop packaging signal. When the MCP is expressed recombinantly in *E. coli* it assembles into capsids and a similar amount of bacterial RNA is packaged instead of the pregenome, which is unavailable. In a normal infection, the viral reverse transcriptase (RT) is packaged along with the pregenome. This enzyme renders a DNA version of the genome while the pregenome is degraded. Thus the HBV capsid serves as a compartment in which genome replication takes place. As synthesis of the DNA genome in its mature form (which is linear and partly double-stranded) approaches completion, a structural transition is triggered in the capsid. Sites on the outer surface are altered, sending a signal that the nucleocapsid is ready for envelopment and to proceed on the cell exit pathway.

The MCP has an 'assembly' domain connected by a 9-residue linker to a basic (Arg-rich) C-terminal domain that binds the pregenome in assembly (Figure 8.33A). Dimers of the assembly domain (Figure 8.33D) drive capsid formation, as demonstrated by constructs that lack the CTD but nevertheless assemble efficiently. With these constructs, the proportion of larger ($T = 4$) capsids depends on the length of linker retained, increasing from ~20% for no linker to more than 90% for a full-length linker. The dimer interface involves a pairing of α-helical hairpins. The resulting four-helix bundle protrudes from the external surface as a spike, 25 Å long (see Figure 8.33B,C). The capsids are perforated with holes, ~20 Å across, large enough to allow the infusion of nucleotide triphosphates for retrotranscription but small enough to exclude nucleases. The capsid wall thus acts as a steric filter.

HBV also expresses an alternative form of the MCP known as e-antigen, which lacks the CTD and retains a 10-residue propeptide at the N terminus (see Figure 8.33A). This protein has essentially the same subunit fold as the assembly domain and it also dimerizes, but these dimers have a radically different mode of dimerization, one that precludes assembly into capsids (Figure 8.33E). To do so, the propeptide introduces steric hindrance at the dimer

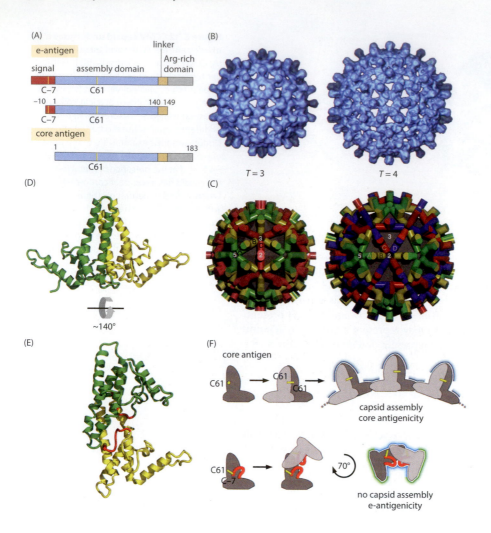

Figure 8.33 HBV capsid protein is expressed in two forms and its capsids are of two sizes. (A) Domain maps. Core antigen assembles into capsids, and e-antigen is secreted in unassembled form from infected cells. The precursor and mature (proteolytically processed) forms of e-antigen are shown. Two critical Cys residues are in yellow. (B) Cryo-EM reconstructions of $T = 3$ and $T = 4$ capsids assembled from a construct consisting of the assembly domain plus seven residues of the linker. (C) Models showing the packing of dimers and marking the quasi-equivalent subunits (A–C in $T = 3$ and A–D in $T = 4$). (D) The A-B assembly domain dimer, a building block of capsids. (PDB 1QGT) (E) The e-antigen dimer, in which the green subunit is rotated 140° relative to its setting in the core antigen dimer, is stabilized by two intramolecular disulfide bonds between Cys7 in the propeptide (magenta) and Cys61. (PDB 3V6F) The core antigen dimer has one intermolecular disulfide, Cys61–Cys61. (F) Diagram explaining how alternative dimerization affects the propensity to assemble and the antigenic make-up. Surfaces potentially presenting epitopes are traced in blue and green. (B, from J.F. Conway et al., *Nature* 386:91–94, 1997. With permission from Macmillan Publishers Ltd; C, courtesy of David Belnap; E, adapted from M.A. DiMattia et al., *Structure* 21:1–10, 2013. With permission from Elsevier; F, courtesy of Michael DiMattia.)

interface, where it is locked into place through the formation of intramolecular disulfide bridges involving a Cys residue in the propeptide. Most epitopes on the capsid are conformational, combining two or more loops from different subunits. As the inter-subunit interfaces are changed in the e-antigen dimer and additional surfaces are exposed, the antigenic character of the two dimers (core antigen and e-antigen) are very different (Figure 8.33F). E-antigen dimers are secreted from infected cells. Although the function of e-antigen has been unclear, a growing body of evidence suggests that it modulates the host immune system to repress its response against HBV capsids. It is plausible that the virus may have evolved this strategy to favor chronic infection over viral clearance after an acute infection.

Picornavirus assembly involves proteolytic processing of a polyprotein

Members of the extensive picornavirus family are associated with important diseases of humans (such as polio) and animals (such as foot-and-mouth disease). The poliovirus MCP is synthesized as a large polyprotein called P1, itself a fragment of an even longer translation product (**Figure 8.34A**). As assembly proceeds, P1 (97 kDa) is dissected into three subunits of ~25 kDa, each with a jelly roll fold (VP1, VP2, and VP3), and a fourth fragment, VP4. In the mature virion—also called the 160S particle, referring to its sedimentation coefficient— the three jelly rolls are arrayed on a pseudo-$T = 3$ lattice (**Figure 8.35A**), with VP4 having an extended conformation on the inner surface.

For viruses that lack an envelope, cell entry is orchestrated by interactions between the capsid and a receptor on the host cell's surface. Receptors are transmembrane proteins, many of which have *ectodomains* consisting of several concatenated Ig-like domains. The poliovirus receptor has three such domains: D1, D2, and D3. Its binding has been studied by decorating virions with soluble ectodomains and visualizing the complex by cryo-EM

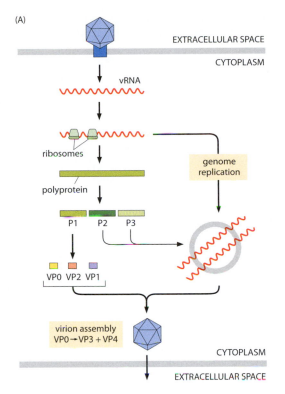

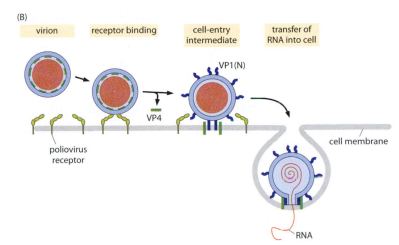

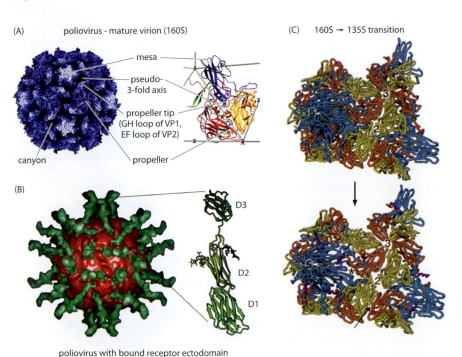

Figure 8.34 Cell entry pathway and replication cycle of poliovirus. (A) Replication cycle showing synthesis and processing of the polyprotein bearing the four capsid protein subunits. (B) Cell entry. VP4 (green) and the N termini of VP1 are on the inside of the mature virion. The receptor is monomeric and its ectodomain consists of three Ig domains. Receptors bind to the capsid, eventually with five receptors bound around a 5-fold symmetry axis. This interaction causes the virion to switch into the 135S state in which VP1(N) and VP4 are externalized, and receptor binding is relaxed. VP1(N) and VP4 are envisaged to form a pore that allows the RNA to be transferred into the cytoplasm of the host cell. The virion is thought to be taken up by endocytosis, with the viral genome delivered into the host cell cytoplasm as shown. (Courtesy of David Belnap.)

(Figure 8.35B). Ectodomain glycosylations are apparent as local protrusions. The receptor's binding site lies in the 'canyon,' a trench like depression that surrounds each 5-fold vertex of the capsid. Because the canyon is too narrow for antibodies to enter, sites on the canyon floor are spared from the requirement to be constantly mutating to evade immune surveillance. This simple *steric blocking* arrangement has been proposed to allow some (but not all) picornaviruses to maintain stable long-term relationships with their receptors.

Receptor binding to mature poliovirus elicits a transition to the 135S or A-particle state, whereby VP1, VP2, and VP3 undergo rigid-body movements, creating gaps between them (Figure 8.35C). These movements are reminiscent, on a microscopic scale, of the geophysical

Figure 8.35 Structure of poliovirus, receptor binding, and transitions during cell entry. (A) View of capsid ($T = 1$, pseudo-$T = 3$) along a 2-fold axis of symmetry, with enlargement showing ribbon diagrams and positions of the three jelly roll subunits VP1 (blue), VP2 (yellow), and VP3 (red). The three subunits are derived from a common polyprotein (PDB 1AL2). (B) Cryo-EM reconstruction of poliovirus (red) decorated with 60 copies of the receptor ectodomain (green), which consists of three Ig domains, D1, D2, and D3. Two carbohydrate moieties on D2 are shown. Binding of the receptor induces the virus to switch to the 135S state in which its affinity for the receptor is reduced. (C) This transition involves rigid-body movements (tectonic shifts) of VP1, VP2, and VP3 (shown here as modeled by fitting subunits from the virion crystal structure into a cryo-EM reconstruction), and externalization of VP4 and the N terminus of VP1 (not shown). (A, from D. Bubeck et al., *J. Virol.* 79:7745–7755, 2005. With permission from American Society for Microbiology; B, from W. Baumeister and A.C. Steven, *Trends Biochem. Sci.* 25:624–631, 2000. With permission from Elsevier; C, courtesy of David Belnap.)

poliovirus with bound receptor ectodomain

phenomenon of plate tectonics, in which rigid plates in the Earth's crust shift during an earthquake. In the 160S → 135S transition, the N-terminal region (~30 aa) of VP1 (the vertex-associated MCP) and VP4 are externalized. In a current model for cell entry, these components form transmembrane channels along the 5-fold axes, through which the genome passes into the host cell (Figure 8.34B).

Breathing capsids fluctuate in their ground-state conformations

Despite the impression given by crystal structures and cryo-EM reconstructions of capsids as static shells, they are in fact dynamic particles that, in solution, pulsate about a median conformation. Evidence for this **breathing** behavior comes from two kinds of experiments pointing to transient exposure of internal components. In one, nodavirions were incubated with proteases and the resulting cleavage products were identified by mass spectrometry. Referring to the crystal structure, cleavage sites were seen to be located *inside* the capsid wall. The most plausible explanation was that the capsids undergo fluctuations large enough for the scissile sites to be transiently exposed at the exterior surface. A second line of evidence involves poliovirus VP4 and the N-terminal region of VP1, which are deeply embedded inside the virion (see above) but are nevertheless accessible to neutralizing antibodies.

Breathing also provides an explanation for the mode of action of WIN compounds, antiviral drugs effective against rhinovirus—another picornavirus and the agent of the common cold. These are hydrophobic molecules of ~350 Da, related to the candidate antiviral agent arildone, which was developed at the Sterling Winthrop Research Institute. WIN compounds bind in a pocket at the base of the canyon and their binding reduces the propensity for proteolysis at internal sites. The implication is that breathing contributes to infectivity and that the drug has its antiviral effect through suppressing these motions.

8.5   LARGE ICOSAHEDRAL VIRUSES

Viruses whose genes are encoded in dsDNA have the largest viral genomes, greatly exceeding those of any ssRNA, ssDNA, or dsRNA virus. Herpesvirus genomes range up to ~250 kbp, bacteriophage G has 500 kbp, and the amoebal mimivirus exceeds 800 kbp. These overlap the genome size range of small cells, among which *Nanoarchaeum equitans* has 490 kbp and *Mycoplasma genitalium* 580 kbp. However, these cells are obligate symbionts and the minimum genetic capacity for a free-living organism has been estimated at ~1000 kbp. Almost all viral DNA codes for proteins. The largest genomes contain hundreds of open reading frames, capable of supporting a rich and diverse repertoire and contrasting vividly with those of the simplest ssRNA viruses, which code for only two components, a polymerase and a capsid protein.

Genetic complexity is complemented in the large size and elaborate architecture of these virions. For example, herpes simplex virus (HSV) is ~230 nm in diameter and consists of some 10,000 protein molecules of ~50 different kinds, in addition to membrane lipids and viral DNA, giving a cumulative molecular mass of ~800 MDa. Poxviruses and iridoviruses are even larger (see Figure 8.4). Herpesvirions have a large icosahedral capsid (*T* = 16) surrounded by a proteinaceous compartment called the tegument and the viral envelope (lipid bilayer plus viral glycoproteins). The tegument and envelope are pleiomorphic; nevertheless, three-dimensional structures may be determined for individual virions by an imaging technique called cryo-electron tomography (**Figure 8.36**).

The capsids of some dsDNA viruses may be the largest known entities in the biosphere whose structures are uniquely specified to the atomic level. Here, we focus on capsid assembly as practiced in two evolutionarily related groups (clades), each encompassing families

(A)

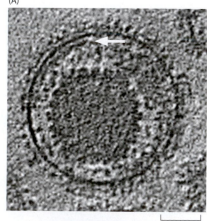

50 nm

(B)

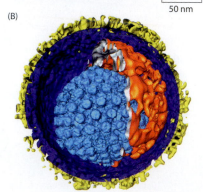

Figure 8.36 Cryo-electron tomogram of herpes simplex virus. (A) A central 1 nm-thick slice through the tomogram, showing that the nucleocapsid is positioned eccentrically. The white arrow points to a protein filament, possibly actin, in the tegument. (B) Segmented cutaway view showing internal components of the virion: light blue, capsid; orange, tegument; dark blue, inner surface of the envelope; yellow, ectodomain spikes of viral glycoproteins. (From K. Grünewald et al., *Science* 302:1396–1398, 2003. With permission from AAAS.)

of viruses that at first sight seem unlikely to be related: tailed phages, which replicate in bacteria; and herpesviruses, which infect a wide range of vertebrate hosts, including humans. Notwithstanding this, they share highly distinctive behaviors in capsid assembly that point to a common ancestry. Similarly, adenoviruses with their animal hosts, and phages such as PRD1, share common folds and virion architecture and, in all probability, common ancestry.

In evolution, one may expect complexity to develop progressively: as viruses acquire more functions, larger virions are needed to accommodate the additional genes responsible. Conversely, early viruses would be expected to be relatively small and simple. To the contrary, the inferred common origins of tailed phages and herpesviruses and of adenoviruses and PRD1-like phages imply that sophisticated assembly pathways evolved early, before the separation of prokaryotes and eukaryotes, and have been retained as these viruses co-evolved with their respective hosts.

Assembly proceeds in three stages: procapsid assembly, DNA packaging, and maturation

Tailed phages have three separate pathways for assembly of their capsids, tails, and tail-fibers (see Figure 8.2B). A generic pathway for capsid assembly is shown in **Figure 8.37**. First, a precursor procapsid is formed; then, the genome is packaged; concomitantly, maturation of the procapsid takes place. Maturation is the process that converts an innocuous ensemble of macromolecules into an infectious, replication-competent, virion. Different viruses have developed many variations on this theme, in some cases adding components; in others, substituting or bypassing certain steps. Nevertheless, much the same set of six proteins is used in all cases to assemble capsids of various sizes and shapes (see the key in Figure 8.37). Structures have been determined for representative examples of each of these proteins (**Figure 8.38**). In general, there is little sequence similarity among corresponding proteins but the core folds have been conserved and embellished, on a case-by-case basis, with additional domains.

These capsids conform to hexamer/pentamer clustering of their MCPs except that they have a different protein, the portal, at one vertex (see below). T-numbers of 7 and 13 are common, but other T-numbers are also found. Some phage capsids, for example T4, are prolate icosahedra extended along a 5-fold symmetry axis.

The first phage MCP structure to be determined was that of HK97 (Figure 8.38E and **Figure 8.39B**). It was solved as a component of the mature capsid. Subsequently, a crystal structure for the T4 vertex subunit and cryo-EM reconstructions of several other capsids have revealed similar folds. The emerging picture is that these proteins have a common core of ~30 kDa on to which additional domains may be grafted at insertion sites or at the termini (Figure 8.39A).

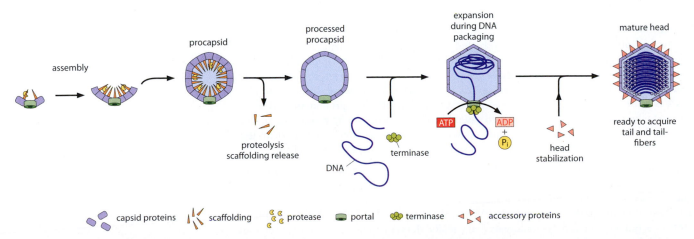

Figure 8.37 Capsid assembly pathway for a tailed phage capsid. The pathway begins with the assembly of a procapsid, which serves as a receptacle for DNA. DNA packaging is preceded, in many cases, by proteolytic processing. The procapsid scaffold is expelled before or during packaging. Driven by the motor protein terminase, packaging is accompanied by conformational changes that stabilize and increase the volume (capacity) of the maturing capsid. (Adapted from A.C. Steven et al., *Curr. Opin. Struct. Biol.* 15:227–236, 2005. With permission from Elsevier.)

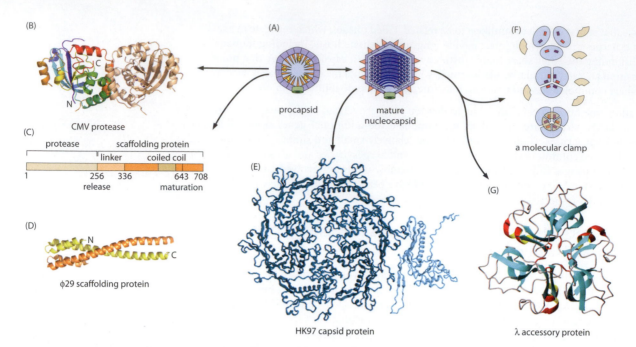

(B) CMV protease

(C)

protease   scaffolding protein

linker   coiled coil

1   256 336   643 708
release   maturation

(D)

φ29 scaffolding protein

(A) procapsid   mature nucleocapsid

(F) a molecular clamp

(E) HK97 capsid protein

(G) λ accessory protein

Figure 8.38 Folds of capsid proteins of tailed phages and herpesviruses. (A) Schematic models of the procapsid and mature nucleocapsid. (B) Dimer of cytomegalovirus (CMV) protease. It is likely to occupy the same location inside the procapsid as in HSV (see Figure 8.43 below). (PDB 1CMV) (C) Domain map of the CMV polyprotein for protease and scaffolding protein. (D) Fragment of the dimeric scaffolding protein of phage φ29. (PDB 1NO4) (E) Packing of subunits of phage HK97 MCP in the asymmetric unit of the mature capsid. The capsid is a $T = 7$ icosahedron whose asymmetric unit consists of a hexamer plus one subunit from the adjacent pentamer. The wedge-shaped A domains (see Figure 8.39B below) pack closely together around the symmetry axis. (PDB 1OHG) (F) The phage λ accessory protein gpD (116 aa) serves as a molecular clamp by binding as trimers around the 3-fold axes on the expanded capsid. These sites are created when the procapsid matures; that is, when the red and blue motifs on either side of the inter-subunit interface that make up the site become suitably juxtaposed. The capsid is thereby reinforced by a network of interactions involving strengthened MCP–MCP binding, MCP–gpD interactions, and gpD–gpD interactions. (G) Structure of gpD. This protein has little regular secondary structure and consists mainly of loops and turns. (PDB 1C5E)

Procapsid assembly involves portal, capsid protein, scaffolding protein, and protease

One of the few seemingly universal features of tailed phages is the **portal protein**, a ring of 12 subunits at the one vertex (**Figure 8.40**) to which the tail is later attached. The portal is also called the **connector** because it connects the head and tail. The φ29 portal subunit (35 kDa) is one of the smallest; others range up to more than 80 kDa. The architecture and distribution of α helices around the axial channel seem to be conserved, with larger subunits acquiring additional domains (see Figure 8.40E).

In addition to serving as a conduit for DNA, the portal probably initiates procapsid assembly by providing a nucleus from which the outgrowth of co-polymerizing MCP and scaffolding protein takes place. The portal-as-initiator hypothesis offers an explanation for how a portal ring comes to be incorporated at just one vertex. In some systems, the portal is not required for geometrically correct procapsids to be assembled; however, it is essential if DNA is to be packaged. If the critical concentrations for scaffolding protein and MCP are lowered when the portal is available, then assembly of portal-containing procapsids should prevent their subunit pools from building up to concentrations at which barren portal-less procapsids or alternative assembly products, such as tubular *polyheads*, are produced.

If a virus has a protease (many do, but some do not), it is incorporated into the procapsid. The protease is not needed for assembly but is required if the procapsid is to mature. The protease enters the assembling procapsid in an inactive form (a *zymogen*), to be auto-activated after the particle is complete. As yet, there are no crystal structures for phage proteases, but sequence analysis suggests that a major class resembles herpesvirus proteases for which four crystal structures have shown them to be serine proteases that share a unique fold (see Figure 8.38B). Activation releases the protease from its initial berth in the procapsid, allowing it to access cleavage sites on the scaffolding protein and/or the MCP. Cleavage primes the procapsid for DNA packaging and maturation.

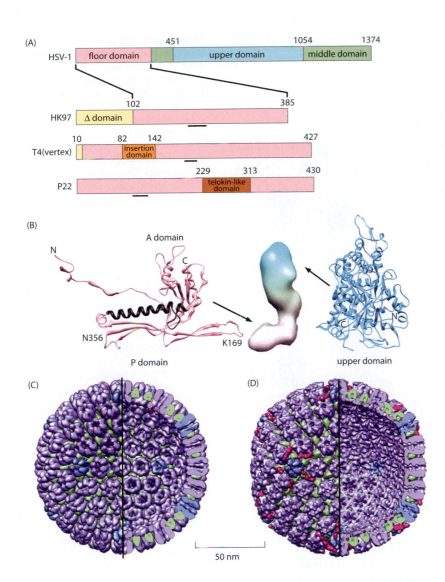

Figure 8.39 Major capsid proteins of tailed phages and herpesviruses. (A) The common core (floor) domain is shown in pink and additional domains are inserted at either end or at internal sites. (B) Left, the HK97 core domain in its mature conformation; its distinctive 'spine' α helix is in black. Its positions in the phage protein sequences are shown in (A) by black underlines. The wedge-like A domains pack around axes of 6-fold and 5-fold symmetry, with the P domains at the periphery of the resulting capsomers (see Figure 8.38E). N356 and K169 on neighboring subunits become cross-linked in the mature HK97 capsid. Center, an MCP monomer taken from a cryo-EM reconstruction of the HSV capsid shows the relative positions of the HK97-like floor domain and the upper domain (right). (PDB 1NO7) (C, D) Cryo-EM reconstructions of the HSV-1 procapsid (C) and mature capsid (D). In each case, the left half shows the outer surface and the right half shows a cutaway view of the inner surface. Light blue, MCP hexamers (capped with VP26 in D); mid-blue, MCP pentamers; green, triplexes; magenta (in D), UL25/UL17 heterodimers, bound when DNA packaging is complete or nearly so. This reconstruction is from a DNA-filled capsid from which DNA was removed computationally to expose the inner surface. (Courtesy of Bernard Heymann.)

Procapsid assembly is assisted by host chaperones and scaffolding proteins

Much of the protein folding in cells is conducted by *chaperone* proteins (Section 6.3). The first chaperone to be discovered, GroEL, was identified in the context of phage assembly. Its name refers to the Large subunit required for T4 to grow in *E. coli*. The small subunit GroES is the usual co-chaperone, but T4 substitutes its own protein gp31 (gene product) in this role, apparently because it increases the effective size of the chamber where folding takes place, making it large enough to accommodate the T4 MCP (56 kDa).

In addition to chaperones, scaffolding proteins are needed for the assembly of geometrically correct procapsids. Scaffolding proteins vary considerably in size and probably also in structure, but many are dimers, and α-helical coiled coils are common. Phage φ29 has a relatively small scaffolding protein of 25 kDa that functions as a dimer and forms a two-stranded coiled coil with the N termini of the two subunits back-folded to give a segment of four-helix bundle (see Figure 8.38D).

How do scaffolding proteins promote morphogenesis? Clues come from the structures produced when scaffolds are mutated or absent. A common phenotype is for the MCP to form tubes or misshapen particles that are aberrant foldings of the same hexagonal net that would otherwise be closed into an icosahedral shell. Scaffolding proteins act by guiding the growing MCP lattice into the correct curvature. In large procapsids such as those of T4 or HSV (see below), the scaffold forms a separate inner shell on which the MCP shell is overlaid,

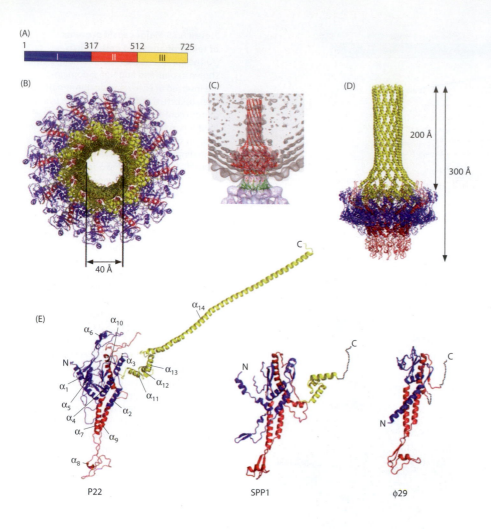

Figure 8.40 Portal proteins are dodecameric channels for the passage of DNA. (A) Domain map of the bacteriophage P22 portal protein. (B) Axial view of the P22 dodecamer, showing the axial channel for duplex DNA. (PDB 3LJ5) (C) Cryo-EM reconstruction in which the crystal structure of the portal (here shown in red) has been fitted, showing how the tubular extension could reach deeply in towards the center of the capsid. The green molecule is the fitted crystal structure of tail component, gp4. (D) Side view, showing how the long C-terminal α helices coil to form a tubular extension of the channel. (E) Structure of P22 subunit (83 kDa) with the same color coding as in (A). Other phages have portals with similar core structures but may have only a short C-terminal extension (for example SPP1, 55 kDa; PDB 2JES) or lack one altogether (for example φ29, 33 kDa; PDB 1FOU). (From A.S. Olia et al., *Nature Struct. Mol. Biol.* 18:597–602, 2011. With permission from Macmillan Publishers Ltd.)

thereby providing for the correct curvature—and hence, size—of the capsid. In other systems, scaffolding proteins attach to sites on the inner surface of the growing shell.

Once the procapsid is complete, the scaffolding protein is expelled, freeing up more space for DNA. Detachment of the scaffolding protein from the MCP shell is facilitated by conformational changes in the latter structure. Once detached, the scaffolding protein is expelled, in some cases after proteolytic breakdown. Capsids are not notably porous, and it is not clear how the scaffolding protein exits. One likely possibility is that holes open up transiently in the maturing shell.

Capsid maturation involves a massive conformational change

Conversion of the procapsid into the mature capsid is accompanied by a major transformation of the MCP lattice. This transformation is often called, simply, expansion because it is usually accompanied by a 10–20% increase in linear dimensions that greatly increases the internal volume (**Figure 8.41**). At the same time, the capsid wall becomes thinner and its shape changes from round to polyhedral. For HK97, this transition is based on rigid-body rotations of the MCP subunits, accompanied by some remodeling, that transfer the inter-subunit interactions to different molecular surfaces. Other maturation events appear to involve generally similar lattice transformations.

Capsids must be strong enough to resist the high outward pressure of DNA, which has been estimated at several tens of atmospheres. In contrast, procapsids are relatively labile. Three stabilization mechanisms have been distinguished. For some phages, such as T7 and P22, the interactions generated by expansion suffice. For others, such as HK97, expansion has little effect on stability but leads to the formation of covalent cross-links between neighboring subunits. The cross-link involves an isodipeptide bond between Asn356 and Lys169 on

Figure 8.41 Maturation pathway of phage HK97 capsid. Shown are cryo-EM reconstructions of five successive states. The precursor MCP, gp5 (385 aa), can self-assemble into a $T = 7l$ procapsid, Prohead I. Gp5 consists of a scaffolding protein, the Δ domain (102 aa; see Figure 8.39) fused with gp5*, the capsid building block. (A) Outer and inner surfaces of Prohead I. On the inner surface, six Δ domains form two masses (yellow) underlying each gp5* hexamer, and five Δ domains form a single mass under each pentamer. (B) Processed procapsid, Prohead II. The Δ domains have been degraded by the viral protease and the rest of the shell has undergone a subtle structural change. In (B)–(D), the left half shows the outer surface and the right half shows a cutaway view of the inner surface. (C) The Expansion intermediate is ~8% larger in diameter. Cross-linking starts in this state. The E-loops, at whose tip is Asn365, initially point radially outward so that Asn365 is ~30 Å from the Lys169 with which it will eventually cross-link. As the capsid matures, the E-loops rotate down into the plane of the shell, bringing more and more Asn365 and Lys169 pairs into mutual proximity. (D) The Balloon intermediate is almost fully expanded, and all cross-links have been formed except those involving penton subunits. (E) In the final step, the pentons move radially outward, the last cross-links are formed, and the mature, fully stabilized, capsid, Head II, is a flat-faceted icosahedron. (Courtesy of James Conway.)

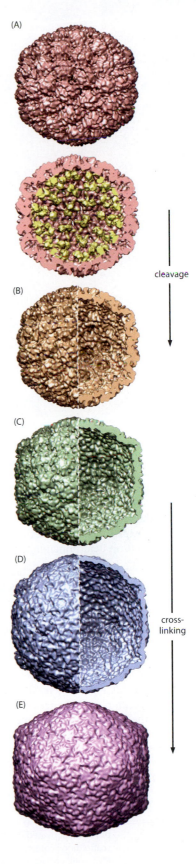

a subunit in a neighboring capsomer (Figure 8.39B). These bonds establish a network of 5-fold and 6-fold rings, interlinked like chain-mail or the Olympic rings. Alternatively, as with λ and T4, expansion creates binding sites on the outer surface for accessory proteins that serve as clamps (see Figure 8.38F).

How is maturation controlled? The mature capsid represents a lower free-energy state than the procapsid but one that is initially inaccessible. The protease converts the procapsid into a metastable particle primed for expansion. *In vivo* the expansion is triggered by DNA packaging, but *in vitro* it may be induced by exposure to chemical perturbants such as acidic pH or treatment with low concentrations of denaturants. The procapsids of different phages undergo similar expansions but vary in their stability and responsiveness to expansion-inducing agents. The inner surface of the HK97 capsid is strongly negatively charged, so that packaged DNA exerts outward pressure through electrostatic repulsion, thereby promoting expansion.

Herpesvirus capsids assemble in the cell nucleus along a phage-like pathway

The HSV replication cycle (**Figure 8.42**) involves extensive trafficking within the infected cell. After entry, the nucleocapsid is transported to and delivers the genome into the nucleus, where DNA replication and capsid assembly take place. Once the nucleocapsid is complete, it leaves the nucleus, embarking on the exit pathway on which it acquires its tegument and envelope. Phages replicate in the bacterial cytoplasm. Notwithstanding this, the capsid assembly pathway (**Figure 8.43**) of herpesviruses closely resembles that of tailed phages. First, a procapsid is assembled from the MCP, portal, scaffolding proteins, and protease. Then DNA is packaged from a concatemeric replicative form. The procapsid matures under control of the protease, whereby the scaffolding protein is cleaved and then expelled and the surface shell undergoes a stabilizing transformation. The remaining steps in assembly—the acquisition of a tegument and envelope, instead of a tail—are quite unphage-like, reflecting their co-evolution with quite different host cells.

However, there are some differences. First, herpesvirus capsids invariably have a T-number of 16, which is rare but not unknown for tailed phages, whose T-numbers vary widely. Herpesvirus MCPs (120–150 kDa) are much larger than those of tailed phages (30–50 kDa). This difference is accounted for by a large C-terminal domain with two subdomains—'upper' and 'middle' (see Figure 8.39A,B) appended to an N-terminal 'floor' domain that, according to cryo-EM data, shares the HK97 MCP fold. The CTDs protrude externally, giving a thick (12 nm) capsid wall. Second, in addition to a scaffolding protein that forms an inner shell, the HSV procapsid has an external scaffolding protein in the form of 'triplexes,' $\alpha_2\beta$ heterotrimers that occupy the 3-fold positions of the procapsid shell, where they coordinate

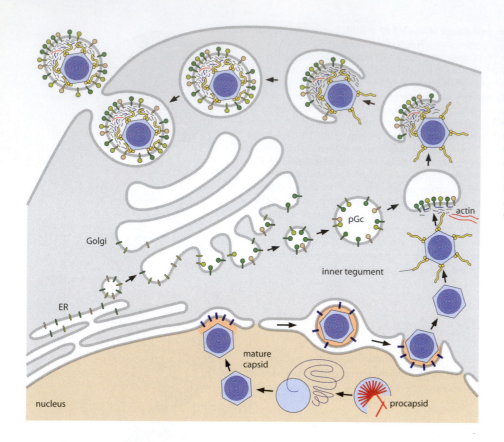

Figure 8.42 Replication cycle of herpes simplex virus. DNA (blue) is replicated and capsids are assembled in the nucleus of the infected cell (bottom right). After DNA packaging, the mature nucleocapsid buds through the inner nuclear membrane (primary envelopment) and is released into the cytosol by fusion of the membrane thus acquired with the outer nuclear membrane. It then migrates to a post-Golgi compartment, collecting the inner tegument on the way, and acquires more tegument as it buds into this compartment (secondary envelopment). Finally, the virion exits from the cell by exocytosis.

hexamers and pentamers of MCP (see Figure 8.39C,D). Thus, neighboring MCP capsomers interact only indirectly, via the triplexes. During maturation, the NTDs rotate, like phage MCPs, and thus establish direct contact between adjacent capsomers. The internal scaffolding protein is expelled during DNA packaging but the triplexes remain in place, switching their role from form determination to structural reinforcement.

Maturation creates binding sites on the outer surface for a small accessory protein [VP26 (12 kDa), in the case of HSV] that binds around the tips of the MCP hexamers. VP26 is not required for capsid assembly and does not enhance capsid stability; it is in fact dispensable, but if present it seems to act early in infection, helping to attach incoming nucleocapsids to the motor protein dynein for transport to the nucleus. (In some other herpesviruses, the protein corresponding to VP26 is indispensable.) VP26 does not bind to MCP pentamers, but other functionally important proteins do bind to these vertex sites. An elongated heterodimer of UL17 (75 kDa) and UL25 (63 kDa) binds to sites involving the vertex-proximal triplexes and the MCP pentamers. This protein has been implicated in nuclear exit, possibly by engaging receptors in the inner nuclear membrane. It also helps to load another protein, UL36, on to the tips of the MCP pentamers. UL36 (336 kDa) seems to provide a platform for recruiting proteins to the tegument (see Figure 8.36).

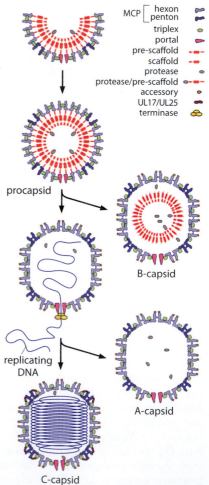

Figure 8.43 Capsid assembly pathway of herpes simplex virus. Procapsid assembly is thought to start on the portal, generating outward growth of the MCP co-assembling with external (triplex) and internal scaffolding proteins. The protease, activated upon release from its polyprotein, processes the internal scaffolding protein, which is expelled during DNA packaging. At this stage, the MCP shell undergoes a major conformational change, which creates binding sites for the accessory protein VP26 at the hexon tips. As packaging approaches completion, the capsid undergoes a second, subtler change that promotes binding of the UL17/UL25 heterodimer. The A-capsid and B-capsid are abortive particles resulting from failure at earlier stages in the pathway.

(A)
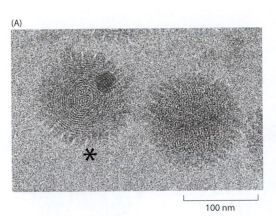

(B)
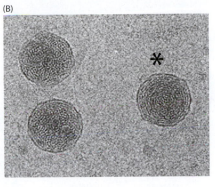

(C)
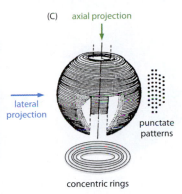

100 nm

Some viral DNAs are packed in coaxial spools by terminase, a motor protein

Condensation and decondensation of chromosomes is a widespread phenomenon through-out biology: the condensed state is quiescent and the decondensed state is transcriptionally active. With tailed phages and herpesviruses, the degree of compaction achieved in the con-densed state is extreme, reaching a near-crystalline state with a concentration of ~500 mg/ml. Cryo-EM of filled capsids viewed along their portal axes shows concentric ring patterns (so-called 'fingerprints'), whereas side views present punctate patterns corresponding to axial projections of bundled duplexes (**Figure 8.44**). Near the center of the capsid there is a less ordered region. This conformation minimizes the overall energy deficit incurred in electrostatic repulsion between DNA duplexes and from DNA bending.

How is this remarkable degree of compaction achieved? The replicating concatemer of DNA is the substrate for packaging, which is effected by a motor protein called **terminase**. An oligomer of large and small subunits, terminase has two main activities: it translocates DNA into the procapsid and it cleaves the packaged genome from the concatemer, terminating packaging. In some cases, there are specific DNA sequences called *pac* sites, where cutting takes place. Other phages perform **head-full packaging** whereby some physical mecha-nism (perhaps back-pressure reaching a critical threshold) senses that packaging should be terminated. Head-full packaging is less specific and is accompanied by terminal redun-dancy: for example, T4 packages ~102% of a genome, so that all virions have about the same amount of DNA encoding complete copies of all genes, but with ends that differ from virion to virion. The terminase large subunit is a *translocase* that hydrolyzes one molecule of ATP per two base pairs packaged. Terminases form ringlike oligomers that mount coaxially on the portal.

The capsid architecture of adenovirus is shared by some dsDNA bacteriophages

Adenoviruses, which cause respiratory and other infections in humans (adeno = adenoids), have attracted investigation since the earliest days of molecular virology. As noted above, adenovirus architecture has turned out to be strikingly similar to that of bacteriophage PRD1. Points of resemblance include the fold of the MCPs, which both have a double β-barrel core, embellished in the case of adenovirus by large external protrusions (**Figure 8.45C**). The same fold has since been detected in other viruses with diverse hosts, including plants (PBCV1, *Paramecium bursaria* chlorella virus type 1) and thermophilic archaea (STIV, *Sulfolobus* turreted icosahedral virus), suggesting descent from a distant common ancestor.

These MCP subunits form trimers, which are pseudo-hexamers when both β-barrel domains are considered. They are called **hexons**, and their shape is well suited to pack on a hexagonal lattice. At the vertices are pentamers of a different protein, penton base, which also has a β-barrel domain. Thus, the surface lattice of Ad (Figure 8.45A) has a pseudo-T-number of 25 (u,v coordinates of 5,0—see Section 8.2) and PRD1 a pseudo-T-number of 31 ($u,v = 5,1$), when all of the β barrels are counted separately.

There are four salient features of these virions. First, hexon packing is consolidated by small proteins at specific sites between hexons, where they serve as stabilizing clamps (also called

Figure 8.44 Genomic DNA is coaxially coiled in tailed phages and herpesviruses. Cryo-EMs of nucleocapsids of (A) HSV and (B) phage HK97. Asterisks mark particles oriented so as to project concentric ring motifs ('fingerprints'); other orientations typically project punctate motifs. (C) Diagram illustrating how these motifs are generated as projections of coiled spools. This arrangement is common to many tailed phages and herpesviruses but some, such as phage T5, diverge significantly from it. (Courtesy of Naiqian Cheng.)

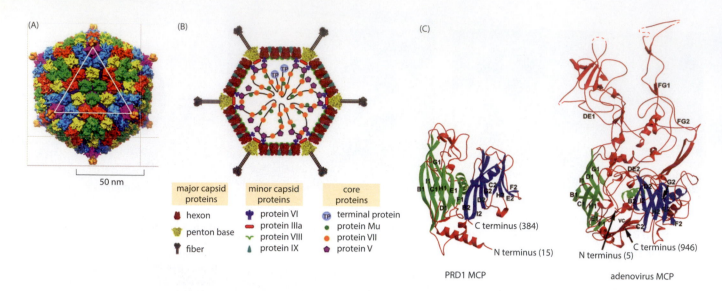

Figure 8.45 The complex molecular architecture of adenovirus. The virion has been solved to high resolution by both X-ray crystallography and cryo-EM, revealing the icosahedrally ordered components. (A) Outer surface viewed along a 3-fold axis; one of the 20 facets is marked by a white triangle. The asymmetric unit comprises one penton base subunit (magenta) and four hexon trimers (light blue, yellow, green, and red). In mid-blue is a stabilizing 'cement' protein. The vertex-mounted fiber of this strain (orange) is short and flexible. (B) Schematic section giving the placements of the minor capsid proteins and internal (core) proteins associated with the DNA. The length of these fibers is more typical than the strain in (A). (C) The MCP hexon subunit (PDB 1P30) compared with that of phage PRD1 (PDB 1CJD). Both have 'double-barrel' folds (the two β barrels are in green and blue), but adenovirus hexons also have elaborate, externally protruding 'towers.' (A, from V.S. Reddy et al., *Science* 329:1071–1075, 2010. With permission from AAAS; B, from G.R. Nemerow et al., *Virology* 384:380–388, 2009; C, from S.D. Benson et al., *Mol. Cell* 16:673–685, 2004. With permission from Elsevier.)

'cement' or 'glue' proteins—for example protein IX, Figure 8.45B). Second, at the vertices there is a symmetry mismatch between the fibers (3-fold) and penton base (5-fold). Third, high-resolution structures of PRD1 and adenovirus disclose filamentous networks connecting adjacent vertices that seem to specify the spacing between them and hence the number of hexons to be accommodated, and hence the pseudo-T-number (**Figure 8.46**). Finally, these viruses mostly contain an internal membrane between the inner surface of the capsid and the internal complex of DNA and associated proteins (see below, Figure 8.58); adenovirus, which is an exception, does not.

An increase in genome size, as a virus evolves, may be accommodated by an increase in the T-number. Adenovirus-like capsid proteins have proved conducive to adaptations of this kind. For instance, PBCV1 has a (pseudo) $T = 169$ capsid and a diameter of 185 nm, with a complement of 1680 hexons plus vertex components, and the enormous mimivirus is expected to have a pseudo-T-number of about 1000. Large capsids of this kind break down into sub-assemblies called **pentasymmetrons** and **trisymmetrons** (**Figure 8.47**), which consist respectively of pentagonal clusters of hexons arranged symmetrically around a vertex and triangular plates of hexons. PBCV1 pentasymmetrons contain 30 hexons and one penton, and the trisymmetrons have 66 hexons. It is not yet clear whether they also represent assembly intermediates.

8.6   DOUBLE-STRANDED RNA VIRUSES

Viruses whose genomes are encoded in dsRNA infect a broad range of hosts extending across the animal kingdom to plants, fungi, and bacteria. Despite this diversity, they share certain fundamental properties. Virion architecture is based on up to three nested protein shells with icosahedral symmetry (**Figure 8.48**). Most of the genomes are segmented: depending on the virus, they consist of up to 12 linear segments. Importantly, these virions are more than delivery vehicles for inactive genomes: their inner capsids provide protein-bound compartments where genome replication and transcription take place. As transcripts exit, their 5′ ends are capped with the 7-methyl-GMP (m^7.GMP) that is needed for

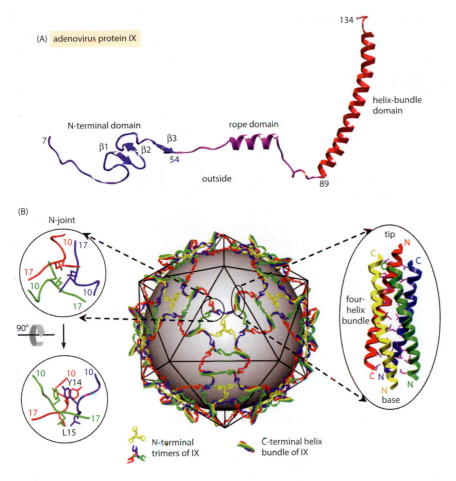

(A) adenovirus protein IX

134

helix-bundle domain

N-terminal domain

rope domain

7

β1 β2 β3

54

outside

89

(B)

N-joint

tip

N

C

C

N

four-helix bundle

90°

base

N

C

N

N

Y14

L15

N-terminal trimers of IX

C-terminal helix bundle of IX

Figure 8.46 A two-dimensional network of tape-measure proteins. The pseudo-$T = 25$ triangulation geometry of the Ad capsid appears to be guided by an underlying filamentous $T = 1$ template of protein IX, which was visualized by cryo-EM. (Courtesy of Hong Zhou; adapted from H. Liu et al., *Science* 329:1038–1043, 2010. With permission from AAAS.)

translation. This capping is performed by enzymatic machinery that is an integral part of the capsid. Thus, these viruses are macromolecular machines in the fullest sense of the term.

Double-stranded RNA is toxic to cells. Exposure of a cell cytoplasm to dsRNA may trigger a number of responses, including *apoptosis* (programmed cell death, see Section 13.6). From the perspective of an infected cell, apoptosis represents a self-sacrifice that limits the threat to the overall population. From the perspective of the virus, apoptosis curtails its potential for proliferation. To avoid eliciting such a response, replication proceeds without the dsRNA genome's coming into contact with the cytoplasm. RNA synthesis takes place within a viral particle: the replicated genome remains inside and transcripts are secreted. Thus, the viral polymerase must be present inside the viral particle. During virion assembly, transcripts are packaged into procapsids within which the second, complementary, strands are synthesized.

The virions share a distinctive architecture and the fold of the viral replicase, an RNA-dependent RNA polymerase. The triangulation geometry of the two inner capsids is conserved. There is greater variability in their outermost components, which interact with their respective hosts. Nevertheless, their shared core properties make it likely that many of these viruses evolved from a common ancestor. Figure 8.48 summarizes the essential features of several viruses that have been well characterized structurally and are illustrative for certain aspects of assembly. Rotavirus is an important human pathogen associated with infantile diarrhea; bluetongue virus (BTV) takes a toll on sheep in southern Europe; SA-11 reovirus is an innocuous murine virus; birnavirus is a pathogenic avian virus; L-A infects yeast; and φ6 is a bacteriophage.

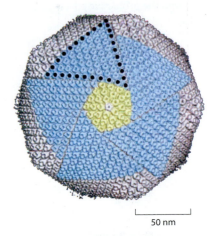

50 nm

Figure 8.47 Subassemblies in the capsid of a large icosahedral virus. PBCV1 has a pseudo-$T = 169d$ structure. Its MCP forms trimers of a subunit that is likely to have the 'double barrel' fold found in adenovirus and PRD1 hexons (see Figure 8.45C). The surface lattice subdivides into triangular and pentagonal blocks, called trisymmetrons (blue) and pentasymmetrons (yellow). They comprise 31 and 66 capsomers respectively, whereas the capsomers at the 5-fold vertices are likely to consist of a different protein. (From X. Yan et al., *Nature Struct. Biol.* 7:101–103, 2000. With permission from Macmillan Publishers Ltd.)

Figure 8.48 Double-stranded RNA viruses have up to three nested capsids. The innermost capsid (black), present in all except birnavirus, has a unique $T = 1$ 120-subunit architecture. The second layer (blue) is a $T = 13$ trimer-clustered shell whose lattice may either be complete (for example, in rotavirus) or have 'turrets' (yellow) at its 12 vertices (for example, in reovirus). Three viruses have an additional outer layer that is either a $T = 13$ shell (red; rotavirus and reovirus) or a membrane (gray; φ6).

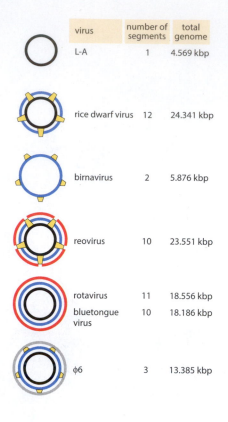

virus	number of segments	total genome
L-A	1	4.569 kbp
rice dwarf virus	12	24.341 kbp
birnavirus	2	5.876 kbp
reovirus	10	23.551 kbp
rotavirus	11	18.556 kbp
bluetongue virus	10	18.186 kbp
φ6	3	13.385 kbp

Double-stranded RNA viruses have one, two, or three nested protein shells

The innermost shell consists of 120 chemically identical subunits: it has $T = 1$ icosahedral symmetry with 60 protomers, each a homodimer of non-equivalent subunits. This architecture is found in almost all dsRNA viruses, but not in any other kind of virus. There appear to be (at least) three different subunit folds. As expected, the fold is conserved between family members, say, reovirus and BTV. The MCP fold of L-A is considered to be different, although it has a similar shape and packing arrangement (**Figure 8.49**).

The second layer is a trimer-clustered $T = 13l$ capsid, of which there are two variants. In rotavirus (**Figure 8.50**), this lattice is complete, consisting of 780 subunits (13×60), giving 260 trimers. In reovirus (**Figure 8.51**), the vertices are occupied by different proteins; it has a pseudo-$T = 13\,l$ lattice. These proteins are mounted on the vertices of the underlying inner capsid, where they form protrusions called turrets. To accommodate the turrets, five trimers are replaced at each vertex, leaving a gapped framework of 200 trimers ($260 - 12 \times 5$).

The third and outermost shell, found only in members of the reovirus family, including rotavirus, has the same $T = 13$ symmetry. Implanted in this shell are proteins that have roles in host cell recognition similar to those of glycoprotein spikes of enveloped viruses (see Section 8.7). φ6 has a lipoprotein envelope instead of a third capsid (see Figure 8.48).

Fungal viruses, the simplest dsRNA viruses, have only the inner shell. These viruses are transmitted from cell to cell by cytoplasmic mixing when an infected cell fuses with an uninfected cell; such mating is a frequent occurrence in fungi. Accordingly, these viruses can forgo the protective covering and cell entry apparatus afforded by the second and third shells.

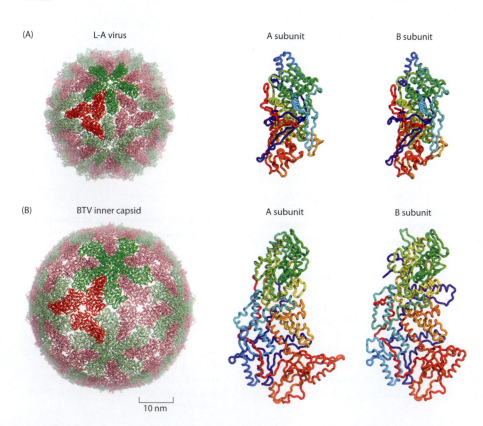

(A)   L-A virus   A subunit   B subunit

(B)   BTV inner capsid   A subunit   B subunit

10 nm

Figure 8.49 The 120-subunit capsids of L-A and bluetongue virus (BTV). Although similar in capsid architecture and MCP shape, they differ in the folds of these two large MCPs (70 kDa for L-A and 90 kDa for BTV). The protomers of these capsids consist of two chemically identical but structurally non-equivalent subunits, molecules A (green, clustered round the 5-fold axes) and B (red, clustered round the 3-fold axes). Their shapes are self-complementary, packing together so as to leave only very small gaps. (PDB 1M1C, 2BTV) (Courtesy of Jonathan Grimes.)

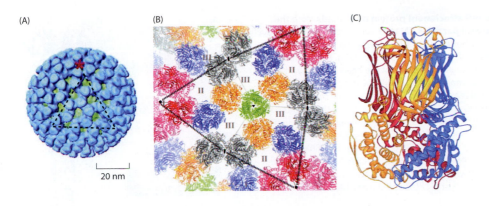

(A)   (B)   (C)

20 nm

Figure 8.50 The second capsid of rotavirus is a trimer-clustered $T = 13l$ particle. (A) Cryo-EM reconstruction of the two-capsid particle. Trimers are in blue, and the underlying inner capsid is in green. (B) One triangular facet of the surface lattice (dotted lines in (A)) with ribbon diagrams of viral protein VP6 (43 kDa). Quasi-equivalent trimers are in different colors: red, blue, gray, orange, and green (one-third). There are 13 quasi-equivalent subunits, forming $4^{1}/_{3}$ trimers. (C) VP6 trimer in side view, with the three subunits coded in blue, yellow, and red. Its outer surface is at the top. VP6 has a β-barrel upper domain and an α-helical base domain. (PDB 1QHD). (A, from J.A. Lawton et al., *Proc. Natl Acad. Sci. USA* 96:5428–5433, 1999 National Academy of Sciences, USA. With permission from PNAS; C, from M. Mathieu et al., *EMBO J.* 20:1485–1497, 2001 European Molecular Biology Organization. By permission of John Wiley & Sons.)

Birnaviruses (bi = two segments; rna = RNA) lack the 120-subunit inner shell. Nevertheless, their single capsid has the same *T*-number (13) as the second shell of other group members. Unusually for a capsid with the complexity represented by *T* = 13 geometry, birnavirus does not have a scaffolding protein. However, the C-terminal portion of its MCP has a key role in form determination: small truncations in it alter the capsid size.

Conformational changes and proteolytic processing promote infectivity

The cell-attachment proteins of reovirus and rotavirus are elongated molecules anchored on the second capsid that extend outward through holes in the third capsid to reach host cell receptors. The reovirus protein σ1 is a trimer resident at all 12 vertices, where there is a 3:5

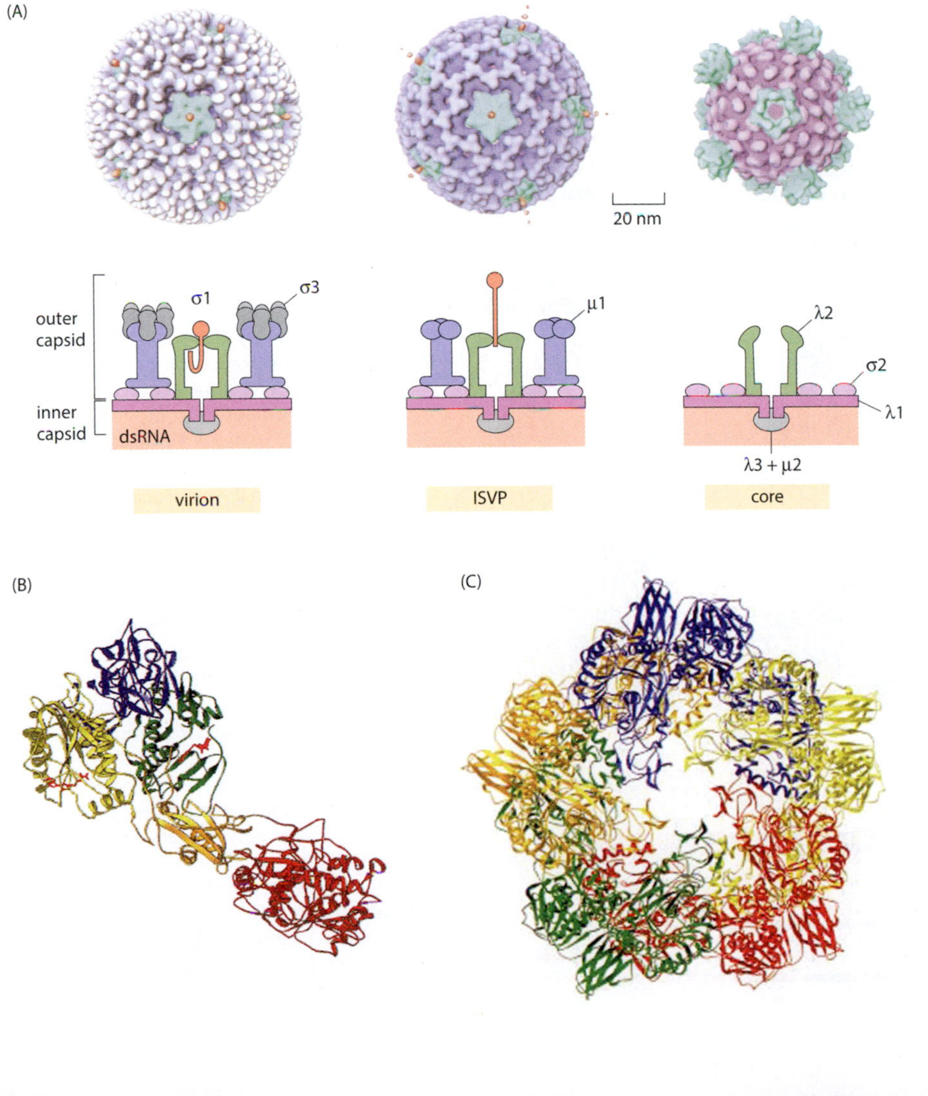

(A)

20 nm

outer capsid

inner capsid

dsRNA

σ1   σ3

μ1

λ2

σ2

λ1

λ3 + μ2

virion   ISVP   core

(B)   (C)

Figure 8.51 The reovirus inner capsid is overlaid with clamping proteins and vertex-mounted turrets. (A) Top, cryo-EM reconstructions of the virion, the ISVP (infectious subviral particle), and the core particle. Bottom, schematic diagrams showing the placement of proteins. The inner capsid is overlaid with 150 monomers of σ2 (47 kDa) and has, at its 5-fold vertices, pentameric 'turrets' of λ2 (147 kDa). In the ISVP and the virion, whole and partial hexameric rings of μ1 (75 kDa) form the remainder of a pseudo-*T* = 13 lattice. In the virion, μ1 is capped with σ3 (39 kDa). A trimer of σ1 (49 kDa) is associated with each turret. A single copy of the viral polymerase underlies each vertex (see Figure 8.54). (B) Ribbon diagram of a λ2 monomer, as viewed from the inside of the turret. Domains involved in capping reactions are shown in different colors: the GTPase domain is red, methylase-1 is yellow, methylase-2 is green, and the Ig-like domains are blue. Binding sites for *S*-adenosylmethionine are marked with small red patches. (C) At each 5-fold vertex is a pentamer of λ2 subunits (here, each is colored differently), along whose axial channel transcripts emerge to be capped. (A, from K. Chandran and M.L. Nibert, *Trends Microbiol.* 11:374–382, 2003. With permission from Elsevier; C, from K.M. Reinisch et al., *Nature* 404:960–967, 2000. With permission from Macmillan Publishers Ltd.)

Figure 8.52 The reovirus cell attachment protein σ1 extends from the surface of infectious subviral particles. (A) Electron micrograph of negatively stained ISVP with some σ1 subunits arrowed. (B) Domain organization of σ1 shown in an averaged electron micrograph (negative stain). The tip of the tail anchors σ1 in the ISVP. (C) Crystal structure of the head domain and the proximal portion of the tail domain (boxed in (B)). The head is a 3-fold ring of β barrels. The proximal tail is a succession of seven β-spiral motifs with a short coiled coil (cc). The β spiral is a ring of three copies of a three-stranded, antiparallel β sheet. This structure shows sialic acid-containing sugars in orange, bound to β3 (sialic acid is involved in the entry of reovirus into the cell). The remainder of the tail is a long three-stranded coiled coil. (PDB 3S6X) (A, from D.B. Furlong et al., *J. Virol.* 62:246–256, 1988; B, from R.D.B. Fraser et al., *J. Virol.* 64:2990–3000, 1990. A and B, with permission from American Society for Microbiology; C, from D.M. Reiter et al., *PLoS Pathogens* E1002166, 2011.)

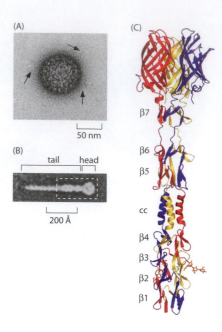

symmetry mismatch. It combines a string of motifs, including a three-stranded coiled coil and β spirals (**Figure 8.52B,C**), which are also found in its functional counterparts of other, quite different, viruses. For example, there are triple coiled coils in the glycoprotein spikes of enveloped viruses (Section 8.7), and β spirals in adenovirus fibers (Section 8.5). In reovirus, σ1 is in a retracted conformation from which it is released to extend from the cell-entry intermediate, the so-called infectious subviral particle (ISVP) (Figures 8.51A and 8.52A). Limited proteolysis of σ1, for example by trypsin, enhances infectivity.

The rotavirus spike protein VP4 is located at 60 of the 132 holes through its outer capsid; that is, those close to but not immediately at the vertices (**Figure 8.53A**). This protein has a fusion peptide, comparable to those of the fusogenic glycoproteins of enveloped viruses. VP4 has the unusual form of a non-equivalent trimer, in which the base is a regular triplet of three polypeptide chains, whereas the external portion (head) is dimeric (Figure 8.53C). Like σ1, VP4 undergoes limited proteolysis and concomitant structural transitions during entry into the host cell.

dsRNA virus capsids are protein-bound compartments for transcription and replication

The number of genomic segments ranges from 1 to as many as 12 (see Figure 8.48). Each virion contains one copy of each segment. No higher segment number has been observed,

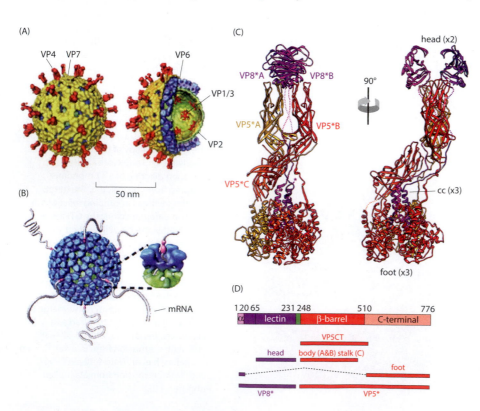

Figure 8.53 Mature (infectious) and transcriptionally active rotavirus particles. (A) Cryo-EM reconstruction showing the third (outermost) capsid, a *T* = 13 lattice of VP7 trimers (yellow) with 132 holes at the sites of 5-fold symmetry (12 sites) and local 6-fold symmetry (2 sets of 60 sites). The 6-fold holes closer to the vertices are occupied by the cell attachment protein VP4 (red). The cutaway view shows the second capsid of VP6 trimers (blue) and the *T* = 1 inner capsid of VP2 (green), with the polymerase VP1/3 (red) under each vertex. (B) Transcriptionally active particles lacking the outer layer secrete transcripts from their vertices. Only the segment close to the capsid (purple density) is sufficiently well ordered to be visualized; the rest is sketched in for illustrative purposes. (C) Crystal structure of VP4 in its mature cleaved form, VP5*/VP8*. The head (distal on the virion) is dimeric, and the foot is trimeric. 'cc' is a three-stranded coiled coil, from the region between the β barrel and the C domain. (D) Domain map of VP4 (VP5*/VP8*). The β-barrel domains of subunits A and B form the body and that of subunit C is in the stalk. The foot contains the C-terminal domains. The fragment generated from recombinant VP4 by successive cleavages with chymotrypsin and trypsin is called VP5CT. The segment between residues 231 and 248 is removed proteolytically. (B, adapted from H. Jayaram et al., *Virus Res.* 101:67–81, 2004. With permission from Elsevier; C and D, adapted from E.C. Settembre et al., *EMBO J.* 30:408–416, 2011 European Molecular Biology Organization. By permission of John Wiley & Sons.)

apparently because each segment has to associate with a vertex of the inner capsid, of which there are 12. This calls for a mechanism to ensure that the assembling particle specifically selects one copy each of its single-stranded transcripts. A model for how this is accomplished in the cystovirus system is described below. Segmentation enables a virus to recover from genomic damage by 'reassortment.' If a cell is multiply infected with viruses, each of which is mutated or damaged in one or more segments, a wild-type genome is regenerated in those progeny particles that receive a full complement of wild-type segments: this will happen in a certain percentage of cases, given random sampling of the pools for each segment.

In broad outline, the assembly pathway is quite well defined. First, the inner capsid is assembled and it contains the viral polymerase. Then single-stranded transcripts, initially produced by the infecting viral particle(s), are packaged—one for each segment. Next, the polymerase synthesizes the complementary negative-sense strands *in situ*. Finally, the outer capsid(s) are overlaid. Of particular note is the role of capsids in synthesizing and secreting transcripts. In rotavirus and other members of the reovirus family, the transcriptionally active agent is the double-layered particle (that is, the two innermost layers). With L-A and cystoviruses, it is only the inner capsid.

As with most icosahedral capsids with low T-numbers, the MCP of the inner capsid self-assembles. The inner capsid then serves as a scaffold for assembly of the outer capsid(s). Despite their mismatched T-numbers ($T = 1$ and $T = 13$), the vertices of the first and second capsids are in register and their curvatures are coordinated so that one fits snugly over the other. In turn, the second capsid serves as a template for the third shell. In the case of reovirus, the MCP for the inner capsid—the large $\lambda 1$ protein (101 kDa)—requires an additional protein, $\sigma 2$, in an external scaffolding role (see Figure 8.51). As this particle assembles, the $\lambda 3$ replicase (140 kDa) is bound under each vertex (**Figure 8.54**). The vertices are then decorated on their outer surface with turrets, which are pentamers of the $\lambda 2$ protein (144 kDa) (see Figure 8.51B,C). The enzymatic machinery for capping transcripts as they are secreted is housed in the turrets. $\lambda 2$ has three domains that catalyze sequential reactions in capping. They are arranged spatially accordingly and afford a good example of a macromolecular assembly line. In contrast, rotavirus and BTV lack turrets, and their capping machinery is located inside the capsid.

There are a few exceptions to this pathway. Birnavirus lacks an inner capsid and its $T = 13$ capsid assembles directly. With L-A, its RNA co-assembles directly with the MCP and its polymerase is incorporated as a 90-kDa C-terminal appendage to one or two of the 120 MCP subunits.

Conformational changes of the procapsid may regulate RNA selection by phage φ6

In this system, three proteins co-assemble with the MCP to form a dodecahedral procapsid with deeply indented vertices (**Figure 8.55A,B**). The MCP shell is decorated on its inner surface with three to five copies of the **RNA-dependent RNA polymerase** at sites close to the 3-fold axes, giving 15–20% occupancy of the twenty 3-fold sites. On the outside, hexamers of an NTPase are mounted over the vertices, creating turret-like features with yet another symmetry mismatch, 6:5.

The next step involves packaging three ssRNA segments in the order S, M, L (small, medium, large) by the NTPase, which has a helicase-like structure and activity (Figure 8.55D,E). Packaging is accompanied by major structural changes in the procapsid, which transforms from a deeply indented dodecahedron to an almost perfectly spherical shell

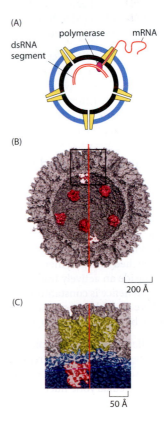

Figure 8.54 Secretion of mRNA by the vertex-mounted polymerase of reovirus core particle. (A) Diagram showing the two-layered capsid, a dsRNA segment binding to the polymerase, and an mRNA transcript leaving the virion. (B) Cutaway view of a cryo-EM density map of the virion, showing monomers of $\lambda 3$, the RNA polymerase (in magenta) under each vertex. The $\lambda 3$ structure was fitted computationally into the reconstruction. The red line connects two 5-fold axes. (PDB 1N35) (C) Enlargement of the region boxed in (B). $\lambda 2$ is in yellow, $\lambda 1$ in blue, and $\lambda 3$ in red. (From X. Zhang et al., *Nature Struct. Biol.* 10:1011–1018, 2003. With permission from Macmillan Publishers Ltd.)

Box 8.1   Assembly and maturation of human immunodeficiency virus (HIV)

AIDS (acquired immunodeficiency syndrome) broke out in the early 1980s as a lethal epidemic that rapidly spread to a global scale. Within a few years of diagnosis of this syndrome, its agent was identified as a retrovirus "of unprecedented complexity and virulence," now known as HIV (human immunodeficiency virus). Retroviruses are distinguished by the inclusion in their life cycles of a step in which the RNA of the infecting virion is retrotranscribed (retro = backward) into DNA by a viral enzyme, reverse transcriptase (RT). Retrotranscription was demonstrated by Howard Temin and David Baltimore in the early 1970s for Rous sarcoma virus, an oncogenic avian retrovirus, and is now known to be general among retroviruses, as well as in hepadnaviruses, such as hepatitis B. Its discovery represented a violation of the Central Dogma, then prevailing (see Section 1.2), that information invariably flows from DNA to RNA to proteins.

In response to the AIDS epidemic, a massive research effort was mobilized to investigate the molecular properties and vulnerabilities of HIV. As a result, despite its complexity, arguably more is now known about HIV than any other enveloped virus. The replication cycle of HIV is outlined in **Figure 8.1.1**. High-resolution structures have been determined for most of its gene products, in some cases as fragments (**Figure 8.1.2**).

The pursuit of an effective vaccine against HIV remains a work in progress, in large part because the mutability of HIV (RT is a relatively error-prone enzyme) promotes immune evasion. However, structural knowledge of the virus's key enzymes has contributed to the development of antiviral drugs. The enzymes in question are: the protease that controls maturation, the integrase that inserts the viral DNA into the host chromosome, and RT. Although resistance mutants against any single drug rapidly appear, the antiviral compounds are lastingly effective in suppressing viral replication (although not curing the infection), when used in combination in 'cocktail' therapy or HAART (highly active antiretroviral therapy).

A conical capsid is assembled inside the maturing HIV provirion

The principal building blocks for virion assembly are polyproteins called Gag and Gag-Pol. Gag has four components: matrix (MA), capsid (CA), nucleocapsid (NC), and a small terminal domain (p6). Gag-Pol, the minority species, is further extended by addition of the protease, RT, and integrase. As the nascent provirion buds out from the host cell membrane, Gag molecules associate laterally, forming spherical precursor particles (provirions) of variable size, each incorporating two copies of the viral genome (a single segment of negative-sense ssRNA) and a variety of host proteins. The diploidy of retroviral genomes is unusual, because viruses employ a variety of mechanisms to maximize the coding capacity of nucleic acid that they can accommodate and a redundant copy of the genome is counter to that trend. The diploidy of HIV may represent a measure to compensate for the low fidelity of its replicase, to ensure that a viable copy of each gene is delivered. The Gag shell is locally ordered with patches of a hexagonal honeycomb-like lattice. In budding, the provirion acquires some copies of the trimeric precursor glycoprotein gp160, although they are sparse (on average, perhaps 10–15 spikes per virion) compared with other enveloped viruses.

After the provirion separates from the host cell, it matures. This order of events is unusual in that maturation usually precedes release. In HIV maturation, the viral protease is activated and dissects the polyproteins into their constituent domains. MA remains associated with the envelope but CA polymerizes *de novo* into capsids containing the RNA and bound NC and RT.

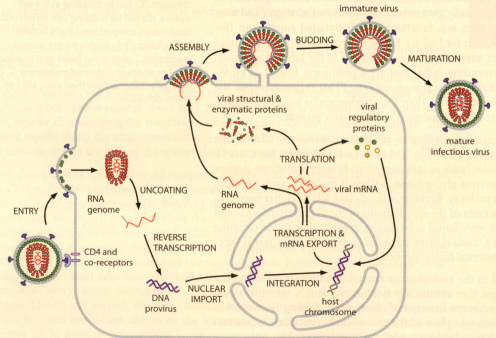

Figure 8.1.1 Replication cycle of human immunodeficiency virus (HIV). After the nucleocapsid has been released into the cytoplasm of the host cell, the genome is retrotranscribed. This takes place in a particle derived from the capsid, called the pre-integration complex. The DNA genome is transported into the nucleus and integrated into a host chromosome. Transcripts, both mRNAs and progeny genomes, are exported into the cytoplasm. In budding, two polyproteins, Gag and Gag-Pol, assemble to form a spherical shell lining a tract of cell membrane. The immature virion buds off the membrane. The viral protease is then activated, auto-releases from the Gag-Pol protein, and proceeds to process the polyproteins. Capsid protein CA reassembles *de novo* into a capsid, packaging NC with the viral RNA. (Adapted from B.K. Ganser-Pormillos et al., *Curr. Opin. Struct. Biol.* 18:203–217, 2008. With permission from Elsevier.)

Box 8.1 *continued*

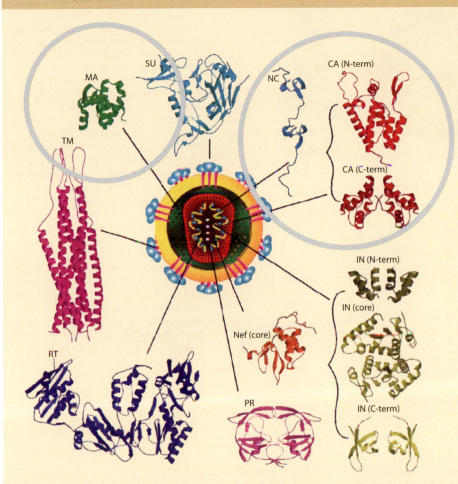

Figure 8.1.2 Structures of HIV proteins. Structures have been determined for most HIV proteins; their locations on or in the virion are marked. MA, matrix protein. The capsid protein CA has two domains (N-terminal and C-terminal) connected by a flexible linker; similarly, the integrase IN has three domains connected by flexible linkers. Protease PR is a homodimer. RT, reverse transcriptase; Nef, 'negative factor,' a non-structural protein. The parts of the trimeric Env glycoprotein shown are: TM, the transmembrane and adjacent portions in a six-helix bundle, representing the post-fusion conformation; and SU, a monomer of gp120, the ectodomain. (From B.G. Turner and M.F. Summers, *J. Mol. Biol.* 285:1–32, 1999. With permission from Elsevier.)

These capsids are polymorphic: in the case of HIV, many have distinctive bi-conical shapes (**Figure 8.1.3**), representing geometrical generalizations of regular fullerene lattices (see, for example, Figure 8.11A), which have a symmetrical distribution of 12 pentamers and a variable number of hexamers. Instead, bi-conical HIV capsids have five pentamers at the narrow end and seven pentamers at the wider end. Other retroviral capsids are different kinds of irregular polyhedrons but with six irregularly distributed pentamers at both ends. Typically, only about half of the CA subunits are used to form the capsid, the rest remaining unassembled. In some cases, a second capsid is formed within the same viral particle.

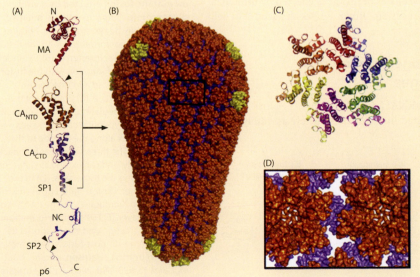

Figure 8.1.3 Bi-conical structure of the HIV capsid. (A) The structural proteins are initially synthesized as parts of the Gag polyprotein, which, in maturation, is cleaved into its component domains at the sites marked by black arrowheads. The domains are shown here in the radial order assumed in the immature virion (MA is outermost, next to the membrane). The CA protein reassembles inside the maturing virion. (B) The resulting capsids are polymorphic, many of them 'bi-conical'; that is, a closed fullerene lattice with seven CA pentamers (yellow) at one end and five at the other, together with 150–200 hexamers [N-terminal domains orange and the mostly hidden C-terminal domains (CTDs) blue]. (C) Structure of the CA hexamer. (D) Two hexamers interacting via their respective CTDs. (A, from B.K. Ganser-Pormillos et al., *Curr. Opin. Struct. Biol.* 18:203–217, 2008; B–D from J.A. Briggs and H.G. Kräusslich, *J. Mol. Biol.* 410:491–500, 2011. All panels with permission from Elsevier.)

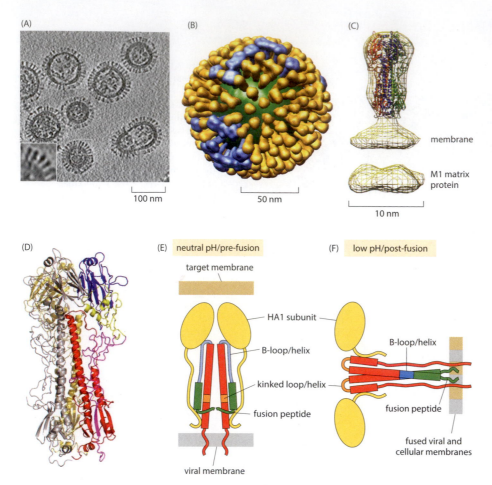

Figure 8.60 Influenza virus and the fusogenic transition of hemagglutinin (HA). (A) Slice from a cryo-electron tomogram of influenza viruses at neutral pH. A blow-up is shown at bottom left. (B) Model of the outer surface of an influenza virion from a cryo-electron tomogram with HA spikes (yellow) and neuraminidase spikes (NA, blue) implanted. The minority NA spikes tend to cluster. (C) Enlargement of an averaged HA spike showing the external part with ectodomain trimer fitted in, overlying the lipid bilayer and the underlying layer of M1 protein. (D) Structure of the ectodomain trimer (PDB 1ruz). One subunit is gray; the one behind is in pale shade. (E, F) Schematic diagrams indicating the changes in disposition of HA domains and motifs between its initial conformation (also called the neutral pH or pre-fusion state) and its final conformation (low pH or post-fusion state). The trimeric HA is shown as a dimer for clarity. (A and B, from A. Harris et al., *Proc. Natl Acad. Sci. USA* 103:19123–19127, 2006 National Academy of Sciences, USA. With permission from PNAS; C, courtesy of Giovanni Cardone; E and F, courtesy of Juan Fontana.)

viral glycoprotein(s)—enters this pathway with a fair degree of synchronization, a property that has facilitated the investigation of membrane protein synthesis and assembly in general.

The most dramatic event in the replication cycle of an enveloped virus is its accomplishment of membrane fusion. Fusion, enabling particles to transfer between compartments while maintaining the integrity of the cell, is effected by radical conformational changes in the glycoproteins. Because fusion, once triggered, is a rapid process, it has been difficult to capture the crucial intermediate states but there are now several crystal structures for ectodomains in the pre-fusion and post-fusion states. A generic model of a fusion pathway is shown in **Figure 8.61**. We summarize the main features of fusion as effected by influenza virus, a Class I system, and Dengue virus, a Class II system. Box 8.1 gives an account of cell entry by HIV, the agent of AIDS.

Influenza virus hemagglutinin, a fusogen, undergoes pH-dependent conformational changes

Influenza virus has a segmented genome of ssRNA (eight segments, negative sense). Unusually for an RNA virus, its genome is replicated in the nucleus, where assembly starts

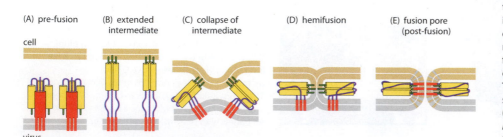

Figure 8.61 Generic pathway for membrane fusion effected by a viral glycoprotein. Some aspects are putative and there may be system-specific features. (A) Pre-fusion conformation, with the fusion peptides (green) sequestered. (B) 'Extended intermediate' conformation with the fusion peptides externalized and inserted into the target membrane. (C) Collapse of the extended intermediate: a segment of each of the three polypeptide chains in each spike folds back along the outside of the trimeric core. The three segments fold back independently, and can extend to different distances along the trimer axis. (D) In the metastable hemifusion state, apposed leaflets from the two membranes merge into a region of single bilayer. (E) Fusion pore formation. The core of the refolded trimer adopts its stable, symmetric, post-fusion conformation. (Adapted from S.C. Harrison, *Nature Struct. Mol. Biol.* 15:690–698, 2008. With permission from Macmillan Publishers Ltd.)

with the formation of ribonucleoprotein (RNP) particles in which an RNA segment is complexed with polymeric N-protein and the viral polymerase. The RNPs exit from the nucleus and are transported to the plasma membrane, either by motor proteins running along the actin cytoskeleton or by associating with the endodomains of the glycoproteins and being transported with them. Ultimately the RNPs are incorporated into virions budding from specialized regions of the plasma membrane (**Figure 8.62**). In most virions, the envelope is lined with a layer of matrix protein, but particles lacking this layer are also observed.

The envelope contains two glycoproteins, hemagglutinin (HA) and neuraminidase (NA) (see Figure 8.60A, B). The name HA derives from the virus's ability to aggregate red blood cells, its sialic acid receptor being abundant on their surface. Influenza virions are pleiomorphic, with both spherical and filamentous forms. Viral strains are classified according to the serotypes of their glycoproteins (16 for HA; 9 for NA); for example, an H1N1 virus caused the devastating 'Spanish flu' pandemic of 1919. The spikes are packed densely but not regularly (Figure 8.60B): a typical spherical virion, 90 nm in diameter, has ~350 HA trimers and 50 NA tetramers. HA is responsible for membrane fusion. NA has sialidase activity, which promotes the separation of nascent virions from the host cell and their passage through the respiratory tract; antiviral drugs inhibit this activity. NA also has a role in budding and virion morphogenesis.

Influenza virus enters host cells by receptor-mediated endocytosis. Its receptor-binding sites are situated on the outer surface of the HA trimer (Figure 8.60D). The virion is taken

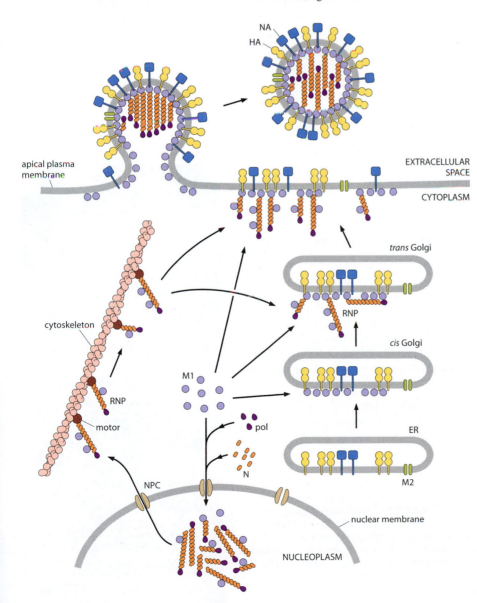

Figure 8.62 Assembly pathway of influenza virus. The glycoproteins HA (yellow) and NA (blue) are transported from the *trans* Golgi network to the budding site at a lipid raft. The ion channel protein M2 (lime green) follows the same route but does not require lipid rafts. Complexes of ribonucleoprotein particles (RNP; brown, with the polymerase in magenta) and the M1 matrix protein (mauve) are exported from the nucleus via nuclear pore complexes (NPC) and transported to the budding site. M1 connects RNPs to the glycoprotein endodomains. In budding, the plasma membrane bends outward to engulf the RNP complexes as it becomes lined with M1. Finally, the nascent virion pinches off from the host cell membrane. (Adapted from D.P. Nayak et al., *Virus Res.* 106:147–165, 2004. With permission from Elsevier.)

up into an endosome-like compartment. Once there, the action of a proton pump causes the internal pH to fall, triggering conformational changes in HA that lead to fusion of the viral envelope with the surrounding membrane (see Figure 8.59B). The HA precursor, HA0, is cleaved into fragments HA1 and HA2 by a host protease, thereby activating the molecule for fusion. Approximately six HA trimers are needed to form a fusion-competent complex.

An HA ectodomain trimer in its 'neutral pH' conformation is shown in Figure 8.60D. Structures have also been determined for the uncleaved precursor, HA0, and for a fragment of HA2 in its 'fusion pH' conformation. (Fusogenic activity peaks at about pH 5.) HA0 differs from the HA1/HA2 trimer only in the disposition of the loop containing the cleavage site, which is close to the surface of the viral membrane. Adjacent to the cleavage site—that is, at the N terminus of HA2—is the hydrophobic **fusion peptide**. After cleavage, the fusion peptide is tucked into a cleft in the stalk of HA2 ectodomain. Subsequently (upon acidification), it inserts into the target membrane. At this stage, however, the fusion peptides are still some 9 nm from the target membrane.

The 'fusion pH' structure results from a conformational major transition, involving the refolding of HA2. In this transition, an extended loop of HA2 switches to α helix and is merged with the two flanking helical segments to form a continuous helix 12 nm long, and the distal helical segment rotates through 180°, shifting the fusion peptide outward (Figure 8.60E,F). Concomitantly, the proximal segment of α helix is folded back. The three copies of the long central helix form a stable three-stranded coiled coil, with the fusion peptides at its tip.

As fusion proceeds, the coiled coil is thought to swivel around, becoming tangential to the apposed membranes instead of perpendicular to them, allowing fusion peptides from the same trimer to insert into both membranes (see Figure 8.61). The location of the HA1 moieties at this stage is unclear. From this point, the membranes are progressively deformed, leading first to the formation of a hemifusion pore and then to a fusion pore. The energy source driving this process is a release of conformational free energy stored in a metastable structure (the HA trimers). Once in the 'post-fusion' conformation, the molecule is no longer fusogenic. The conformation(s) critical for fusion are transient state(s) achieved when HA is appropriately juxtaposed with the apposed membranes.

The HIV glycoprotein, a heavily glycosylated molecule, is also a Class I trimer and also undergoes maturational cleavage as the gp160 (glycoprotein with an apparent molecular mass of 160 kDa) is separated into its gp120 and gp41 fragments.

Class II glycoproteins are arranged in icosahedral lattices

Alphaviruses are ssRNA (positive sense) viruses whose most studied members are Sindbis and Semliki Forest virus (SFV). Both are transmitted by mosquitoes to infect human, simian, and other species. Flaviviruses constitute another family of ssRNA viruses that includes the human pathogens Dengue virus, yellow fever virus, and West Nile virus. In both families, the virion envelopes contain icosahedrally ordered lattices of glycoprotein ectodomains. In alphaviruses, they form a quasi-equivalent $T = 4$ lattice with trimers of E1/E2 heterodimers (**Figure 8.63A**). The TM regions of E1 and E2 pair to form a coiled coil (Figure 8.63B). Flaviviruses, in contrast, have a $T = 1$ lattice featuring a non-equivalent packing of 90 dimers of a single glycoprotein (**Figure 8.64A**).

These glycoproteins are of Class II and have similar folds (Figure 8.64B), quite different from those of Class I (Figure 8.60D). The ectodomains (~50 kDa) are elongated molecules

(A)

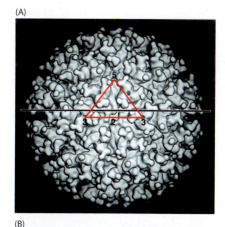

(B)

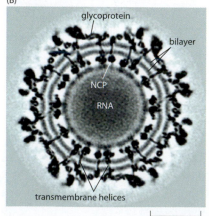

Figure 8.63 Structure of Sindbis virus, an alphavirus. (A) Surface rendering of a cryo-EM reconstruction shows the $T = 4$ icosahedral lattice of trimers of E1/E2 heterodimers viewed along an axis of 2-fold symmetry. The red triangle marks an asymmetric unit. The numbers 5, 3, and 2 mark the positions of symmetry axes. (B) A central section, taken at the level of the black line in (A), resolves details of the glycoprotein, including its transmembrane helices, the two leaflets of the bilayer, and the nucleocapsid protein (NCP), which is also arrayed on a $T = 4$ lattice. The RNA is not icosahedrally ordered and is smeared out in the reconstruction. (From W. Zhang et al., *J. Virol.* 76:11645–11658, 2002. With permission from American Society for Microbiology.)

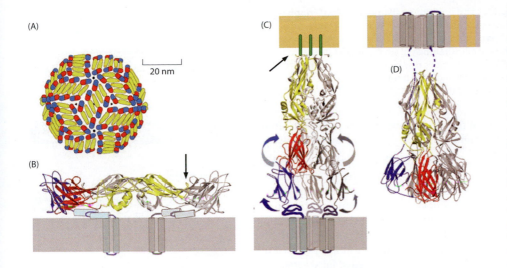

(A)

20 nm

(B)

(C)

(D)

Figure 8.64 The fusion pathway of flavivirus Class II glycoproteins. (A) Non-equivalent packing of 90 E protein dimers on the $T = 1$ icosahedral lattice. The red, yellow, and blue parts correspond to domains I, II, and III respectively, of the E ectodomain. (B) Side view of the pre-fusion dimer, based on the crystal structure of dengue E ectodomain, plus two short α helices and the transmembrane helical hairpin, represented as cylinders, from cryo-EM. The domains of one subunit are colored as in (A); the other subunit is in gray. The fusion loop is at the tip of domain II (arrow) buried at the interface with a neighboring subunit. (PDB 1OAN) (C) Extended intermediate conformation of the ectodomain trimer. The subunits have rotated through almost 90°. Domains I and II have associated into the core of the post-fusion conformation, but domain III has not yet flipped over (upper arrows) to dock against them. (D) Post-fusion ectodomain conformation. Domain III has reoriented, and the stem (shown schematically as a blue dotted line) connects it to the transmembrane anchor. The orientation of the post-fusion trimer relative to the membrane is not known. (PDB 1OK8) (Adapted from S.C. Harrison, *Nature Struct. Mol. Biol.* 15:690–698, 2008. With permission from Macmillan Publishers Ltd.)

consisting of three domains (I, II, and III), all rich in β sheets, with domain III having an Ig-like fold (see **Guide**). The fusion peptide—in this case, an internal loop rather than a terminal peptide—is located at the opposite end from the C-terminal domain III, which connects to the TM segment.

Infection by these viruses is also based on membrane fusion driven by conformational changes of the glycoproteins in response to acidification. The basic mechanism seems to be similar to that of Class I glycoproteins, although some specifics differ. The glycoprotein lattice forms a near-continuous canopy, and a sufficient area of viral membrane surface must be exposed to allow fusion to proceed. The process of flavivirus-induced fusion is represented by three ectodomain crystal structures in Figure 8.64B–D. On acidification, the glycoprotein subunits rotate so that their long axes are nearly perpendicular to the membrane instead of nearly tangential. They also regroup from dimers into trimers. At this time, the fusion peptide is exposed and becomes poised for insertion into the host cell membrane. Finally, in the transition between the 'extended intermediate' and 'post-fusion' conformations, domain III rotates relative to the rest of the molecule. Again, fusion requires the concerted action of multiple (about five) spikes.

8.8   VIRUS ENGINEERING AND NANOTECHNOLOGY

The ability of viral capsid proteins to self-assemble into defined higher-order structures suggests that suitably modified variants of these molecules might be used to produce particles with altered properties, designed for diverse purposes. There is great interest in using this approach to construct novel nanoparticles, nanomachines, and biomaterials. The prefix 'nano' means on the scale of nanometers, although virions are typically an order of magnitude bigger. Although still an emerging technology, this line of investigation has the potential for ramifications throughout biomedicine, the chemical industry, electronic devices, and other fields. Here we outline some of the initial developments.

Phage capsid proteins can be engineered to display peptides and proteins

In **phage display**, foreign peptides or proteins are inserted into structural proteins at exposed sites where they do not interfere with assembly or infection. The earliest generally successful method, based on filamentous bacteriophages (**Figure 8.65**), was introduced in 1985 by George Smith. Insertion is achieved by directed mutation of the gene encoding pIII (see Figure 8.19A). It can be arranged for displaying virions to have from one to all five of their copies of pIII in the modified form. Peptides can also be displayed in high abundance on pVIII, the major capsid protein (~2700 copies per virion). Although the length of displayed peptides is limited, in this case to six to eight amino acids, this is sufficient for many applications. Slightly longer peptides (up to 15 aa) can be accommodated by generating

Figure 8.65 Display of peptides and proteins on filamentous bacteriophages. (A) Wild-type virion. (B) Display of peptides or proteins (red stars) on pIII, either on all five copies or only some copies on virions produced by simultaneously expressing a second gene encoding wild-type pIII. (C) Display of small peptides (lateral squiggles) on all 2700 pVIII subunits. (D) Larger peptides may be displayed by co-expressing the fusion protein with wild-type pVIII. (Adapted from P. Malik and R.N. Perham, *Nucleic Acids Res.* 25:915–916, 1997. With permission from Oxford University Press.)

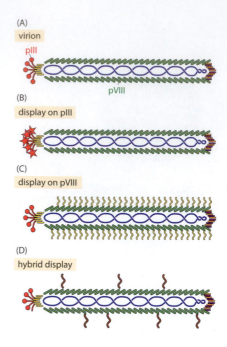

hybrid virions in which only a small fraction of the pVIII subunits carry the insert. Similar techniques have been developed to display proteins on spherical viruses (see below).

A technique known as **biopanning** can be used to identify displayed peptides that bind to a given target (small molecule, peptide, or macromolecule). With an insert of just six amino acids, a library of randomized inserts can contain up to 20^6 ($\sim 10^7$) different sequences. If a high-affinity interaction is sought, pIII display with its low copy number is the better approach. Virions capable of binding to the target are selected by their retention on an affinity column presenting the targeted ligand (**Figure 8.66**). Nonbinding virions are washed away; those that bind can then be eluted with a solution (for example at low pH) that disrupts the binding, to be propagated in bacteria. This procedure is repeated for several cycles to select for enhanced affinity. The sequences of the selected peptide inserts may then be determined by sequencing the phage DNA. Biopanning for short sequences is useful for identifying drug leads (the drug target is also the biopanning target) or linear epitopes on protein antigens (the antibody is the biopanning target). Alternatively, a domain or even a full-length protein may be displayed on pIII and its sequence changed by directed mutagenesis. Applications of this kind have been extensively exploited to develop new antibody specificities or generate higher affinities through display of libraries of antibody fragments (Fabs or single-chain variable fragments).

Display of peptides at high multiplicity on pVIII has somewhat different uses. Probably because of the repetitive display of peptides on this platform, filamentous phages are potent immunogens, invoking a powerful B- and T-cell-mediated immune response to displayed peptides. This display system is used to investigate immunological epitopes and to identify epitopes as a basis for designing new vaccines.

Full-length proteins can be displayed on icosahedral capsids

Proteins can be displayed in large numbers in regular arrays on the 'accessory' capsid proteins of dsDNA bacteriophages such as T4 and λ (**Figure 8.67**; see Section 8.5). Here, capsid assembly and display are decoupled: accessory proteins are not required for assembly but they bind avidly to mature capsids. The protein to be displayed is fused to the accessory protein, expressed separately, and used to decorate capsids produced by a mutant lacking the accessory protein. The example shown is for SOC-based display on T4. Proteins can also be displayed from either terminus of the gpD accessory protein of phage λ (see Figure 8.38). Different proteins may be displayed in close association on the same particle, facilitating studies of interactions between them.

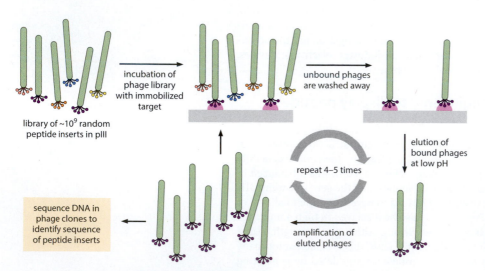

library of ~10^9 random peptide inserts in pIII

incubation of phage library with immobilized target

unbound phages are washed away

elution of bound phages at low pH

repeat 4–5 times

amplification of eluted phages

sequence DNA in phage clones to identify sequence of peptide inserts

Figure 8.66 Biopanning with filamentous bacteriophages. Starting with a library of phages with random inserts, those with affinity for the target of interest are selected by binding to the immobilized target. The inserts with the highest affinity are identified by repeating the cycle multiple times.

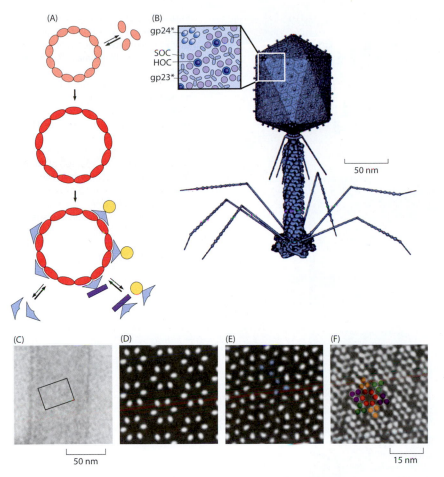

Figure 8.67 Display of proteins on accessory proteins of dsDNA bacteriophages. (A) A procapsid is first assembled (pink) that subsequently matures, creating binding sites for an accessory protein (light blue triangles). The mature capsid may be decorated with chimeric accessory proteins in which the added moieties (gold) may be domains or full-length proteins or proteins of different kinds (gold circle, blue rectangle). (B) Model of bacteriophage T4 with an enlargement mapping capsid proteins around a vertex. SOC (small outer capsid) and HOC (highly antigenic outer capsid) are accessory proteins that may be used for display. (C–F) Display of the V3 domain of HIV Env glycoprotein on phage T4 polyheads (tubular variants of the wild-type capsid). Polyheads were decorated with a SOC–V3 fusion. (C) Electron micrograph of a negatively stained polyhead flattened on the EM grid. Direct interpretation is difficult because of co-projection of the upper and lower sides of the flattened tube and a low signal-to-noise ratio. However, computer filtering separates the images of the two sides and eliminates noise. (D–F) Three filtered lattice patches corresponding in size to the box in (C). White features represent protein outcrops. (D) The mature surface lattice, consisting of hexamers of the MCP, gp23*; (E) as decorated with trimers of SOC (two examples are in blue) at the 3-fold positions; (F) displaying the SOC-V3 fusion, which contributes the larger densities (mauve, green, and orange) at the 3-fold positions. Monomer molecular weights: gp23*, 47 kDa; SOC 11 kDa; V3, 5.5 kDa. (B, adapted from P.G. Leiman et al., *Cell. Mol. Life Sci.* 60:2356–2370, 2003. With permission from Springer Science and Business Media; C–F, adapted from Z.J. Ren et al., *Protein Sci.* 5:1833–1843, 1996. With permission from John Wiley & Sons.)

The capsids of many small icosahedral viruses ($T = 1$, $T = 3$, or $T = 4$) have also been adapted as display platforms (**Figure 8.68A**). Knowledge of the capsid structure allows surface regions of the capsid protein to be identified (either loops or exposed N and C termini) to which polypeptides of interest may be inserted. Display of an Ig domain on the outer surface of the $T = 3$ capsid of flock house virus is illustrated in Figure 8.68B–D.

Another system that has been extensively studied is the hepatitis B virus capsid, where the main goal has been to display foreign epitopes. (Because the HBV virion is enveloped, its capsid is inaccessible to antibodies. For the anti-HBV vaccine, lipoprotein particles produced in yeast are used, see Figure 8.57.) HBV capsids have 90 or 120 protruding spikes,

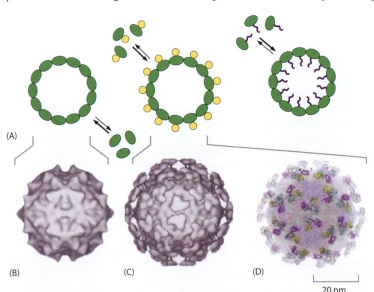

Figure 8.68 Display of an Ig domain on flock house virus. (A) Display on a self-assembling icosahedral capsid. A peptide or protein domain (gold) is inserted into an external surface site on the capsid protein (green). Alternatively, the inserted motif can be on the inner surface (purple squiggles) to promote the packaging of a designated cargo. (B–D) A viral Ig domain of 137 aa was inserted in the most externally exposed loop of the FHV capsid protein (see Figure 8.30). These cryo-EM reconstructions show the outer surfaces of (B) the capsid and (C) the capsid displaying triplets of the Ig domain. (D) An Ig structure (colored) has been fitted into the corresponding EM densities. (B–D, from G. Destito et al., *Curr. Top. Microbiol. Immunol.* 327:95–122, 2009. With permission from Springer Science and Business Media.)

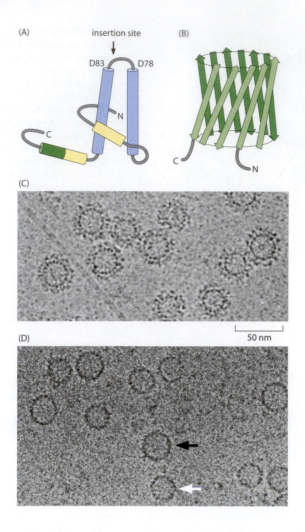

Figure 8.69 Display of green fluorescent protein at the tips of HBV capsid spikes. Diagrams are shown for tertiary structures of (A) the assembly domain of the capsid protein, and (B) green fluorescent protein (GFP). In the displayed protein, the GFP sequence flanked by two 10-aa linkers was substituted for two amino aids at the spike tip. (C) Cryo-electron micrograph of capsids displaying GFP. GFP molecules contribute the peripheral halos of punctate densities. (D) Cryo-electron micrograph of control (non-displaying) capsids (black arrow, $T = 4$ capsid; white arrow, $T = 3$ capsid). (C, from P.A. Kratz et al., *Proc. Natl Acad. Sci. USA* 96:1915–1920, 1999 National Academy of Sciences, USA. With permission from PNAS; D, from E. Kandiah et al., *J. Struct. Biol.* 177:145–151, 2012. With permission from Elsevier.)

spaced ~7 nm apart (see Figure 8.33). At the tip of each spike are two copies of a flexible 'immunodominant' loop into which peptides or even intact globular proteins may be inserted and displayed (**Figure 8.69**).

Another conceptually similar approach is to use the core complexes of multi-subunit enzymes such as the $T = 1$ icosahedral core particle of pyruvate dehydrogenase complex (Chapter 9) as a display platform.

Virus-like particles can be used to generate protective vaccines

The first vaccines entered medical practice long before viruses were discovered and the molecular basis of immunity was understood. In the early nineteenth century, Edward Jenner built on the observation that milkmaids did not contract the deadly disease of smallpox (~30% mortality), although they had the much milder disease of cowpox. Jenner demonstrated that inoculation with material from pustules on the skin of a milkmaid with an active cowpox infection conferred protection against smallpox. Both infections are caused by viruses in the same family as vaccinia virus (see Figure 8.57A), and it turned out that infection with cowpox virus raises antibodies that protect against smallpox. In the 1880s, Louis Pasteur produced a vaccine effective against rabies by infecting rabbits with the 'microbe' and 'attenuating' material taken from their spinal columns by drying. It is now known that rabies is caused by a neurovirulent virus in the same family as VSV (see Figure 8.5) and that the drying caused at least partial denaturation and hence loss of infectivity, while retaining sufficient antigenicity.

Over the years, vaccines have been developed against many viruses. Some were produced by cultivating a less pathogenic virus from the same family (as with Jenner's use of cowpox virus) or 'attenuating' the virus in question by propagating it in tissue culture. Another approach involves modifying the virus chemically or physically in such a way as to render it

Figure 8.70 Virus-like particles (VLPs) visualized by electron microscopy. (A) Electron micrograph of negatively stained VLPs produced by expressing the gene for the major capsid protein of human papillomavirus 16 in insect cells from a recombinant baculovirus. (B) Cryo-EM reconstruction of a VLP of Chikungunya virus produced by expressing three genes, for the capsid protein and two envelope glycoproteins, color-coded according to radius from the capsid center. (A, courtesy of John Schiller; B, from W. Akahata et al., *Nature Med.* 16:334–338, 2010. With permission from Macmillan Publishers Ltd.)

(A)
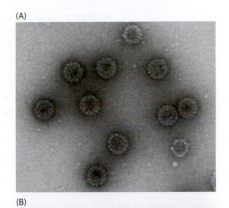

(B)
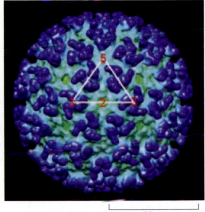

20 nm

non-infectious but still antigenically active. In contrast, much of the recent work on antiviral vaccines is based on exploitation of recombinant DNA technology.

Viral capsids and envelopes present epitopes in many closely clustered copies and thus make potent immunogens. This property can be exploited by using **virus-like particles (VLPs)** as the inoculum. VLPs resemble virions in that they are assembled from the same constituents and hence are antigenically similar, but they lack the genome and thus are incapable of replication. A notable success has been in the use of VLPs composed of L1, the capsid protein of human papillomavirus (**Figure 8.70A**) to confer protective immunity against cervical cancer. Another has been the use of multimeric glycoprotein particles in the absence of other viral components to make a vaccine against HBV (Figure 8.58). A related approach, involving enveloped VLPs produced by expressing the glycoproteins as well as the capsid protein of Chikungunya virus, is showing promise as a vaccine against a disease with severe arthritis-like symptoms (painful joint swelling) that affects millions, mainly in Africa (Figure 8.70B). Chikungunya virus is an alphavirus in the same family as Sindbis virus (see Figure 8.63).

Virus-like particles are also used in gene therapy and have applications in drug delivery and clinical imaging, and as nanotechnological devices

In creating a VLP, genetic engineering techniques may be used to modify a capsid's interaction with cells by altering structures on its outer surface, and/or to change its payload by altering structures on its inner surface to package materials other than the genome. The displayed moiety may be intended to bind the VLP to a targeted class of cells, as in display of peptides that have specificity for endothelial cells lining the vasculature. This has some potential for cancer detection and therapy. For applications in gene therapy, the capsid protein may be adapted to package a modified genome with a chosen insertion or a different nucleic acid molecule altogether and to deliver it to a particular kind of cell. (Because they contain genetic material, such particles are classified as engineered viruses, not VLPs.) Drug delivery is another area of potential application. The drug might be packaged by first disassembling the capsid and then reassembling it in the presence of a high concentration of the drug. Packaging efficiency may be increased by modifying the inner surface of the VLP, for instance to display a peptide with affinity for the drug.

Procedures have also been developed to modify the side chains of exposed amino acids for particular purposes. Attachment of fluorescent compounds to these residues is useful in imaging experiments because the modified virions exhibit several hundred copies of the fluorophore and thus offer a strong optical signal from a single nanoparticle (**Figure 8.71**).

The robustness of filamentous phages and the facility with which they can be aligned in arrays are attracting interest in the fabrication of electronic devices. For example, amorphous anhydrous iron phosphate has been deposited on an M13 virion with a Glu-Glu-Glu-Glu insert at the N termini of its pVIII proteins, and a binding tag for electrically conducting carbon nanotubes has been inserted in its pIII protein, potentially applicable as a cathodic material for lithium-ion batteries.

For VLPs to be used in humans, certain criteria must be met

If a VLP is to be approved for use as a vaccine in humans, the particles must be non-toxic and they must be biodegradable. This is normally not a problem, given the cell's resources for proteolytic degradation. However, they must be suitably long-lived *in vivo* to have the sought-after effect, and it must be technically and economically feasible to produce them in bulk. Here, plant viruses are advantageous in that some of them can be produced

Figure 8.71 Intravital imaging in the chick embryo with fluorescent cowpea mosaic virus (CPMV) particles. The term intravital microscopy refers to the observation of cells in which the protein has been rendered fluorescent or otherwise enhanced. Usually, these are tumor cells and the aim is to observe steps in the metastatic process. In this experiment, animals were injected with CPMV conjugated with fluorescein. The smallest punctae represent individual viral particles. (From M. Manchester and P. Singh, *Adv. Drug Deliv. Rev.* 58:1505–1522, 2006. With permission from Elsevier.)

inexpensively in gram quantities, and an effective dose of the modified capsid may require as little as a few micrograms. For applications of VLPs for drug delivery in humans, care must be taken that the display platform itself does not elicit an immune response that would compromise long-term use. In addition, there may be some risk in using genetically engineered human viruses for this purpose, because they may potentially revert to their wild type.

8.9   SUMMARY

There are two kinds of assembly processes: 'simple' self-assembly in which the components of the final product can assemble, disassemble, and reassemble upon appropriate changes of ionic conditions; and facilitated assembly, which is irreversible and requires morphogenic factors not present in the mature virion and/or processes such as limited proteolysis and conformational changes. In branched pathways, subcomplexes are first assembled, which then associate.

The simplest viruses are filamentous or icosahedral nucleocapsids built from a single protein subunit. Quasi-equivalence allows the construction of larger, more capacious, icosahedral capsids with more than 60 subunits. The subunits can be coordinated as dimers, trimers, or pentamers and hexamers. Genomes are selected for packaging by recognition of their stem-loop packaging signals. Larger viruses have a wider range of building blocks and progressively greater complexity.

As prototypic helical viruses, TMV and filamentous bacteriophages share certain features: helical arrays of the MCP, the genome acting as a tape-measure to specify virion length, and disassembly starting at a defined end. However, TMV initiates assembly at an internal site on its linear ssRNA and is capable of self-assembly. In contrast, filamentous bacteriophages have circular ssDNA and assembly proceeds from one end, a process requiring ATP hydrolysis and a host cell with an intact inner membrane.

Most MCPs of small icosahedral capsids are capable of self-assembly. Typically, the MCP has a rigid core domain with flexible extensions at one or both termini. Such extensions anchor the genome to the capsid, guide assembly of core domains, or facilitate cell entry. Among the mechanisms that regulate assembly is limited proteolysis, which can be enzymatic or autocatalytic. Many virions are dynamic particles that undergo structural changes in response to receptor binding or changes in ionic environment. These changes are based mainly on rigid-body rotations of core domains. The free-energy state represented by the virion is not a global minimum but rather a local minimum from which a suitable stimulus leads to a lower free-energy state and an altered structure.

Large icosahedral viruses form by facilitated assembly. Within a given clade, capsids vary in size and shape but are constructed from similar building blocks. Corresponding proteins have little sequence similarity but are conserved in structure, consisting of a common core on which additional domains may be grafted. First, a procapsid is assembled; then the genome is packaged; finally, maturation takes place. Procapsid assembly requires a nucleation complex, chaperones, scaffolding proteins, and the MCP. In maturation, which is often triggered by limited proteolysis, conformational changes stabilize the capsid, dismantle binding sites for scaffolding proteins, and create binding sites for accessory proteins.

Many dsRNA viruses share the following features. First, a non-equivalent capsid of 120 subunits is formed, containing the viral polymerase. Then ssRNA transcripts—one for each genomic segment—are packaged. Second-strand synthesis ensues, followed by the transcription and secretion of mRNAs. A second capsid containing the machinery for RNA capping is overlaid. In some cases, a third capsid with the infection machinery is added. Conformational changes promote RNA packaging and capping and activate cell attachment proteins.

Enveloped viruses enter and exit from their host cells by membrane fusion. The envelopes are appropriated from specialized patches of cellular membranes where viral glycoproteins accumulate. In entry, the glycoproteins have two functions: host cell recognition, and fusion. Three classes of fusogenic glycoproteins have been distinguished. During fusion, these glycoproteins undergo radical conformational changes. In pH-dependent cell entry, the signal for this transition is acidification in the endosome; in pH-independent entry, it is triggered by interaction with the receptor. Free energy released in these conformational changes provides the energy needed to deform the fusing membranes.

REFERENCES

8.1   Introduction

Caspar DL (1980) Movement and self-control in protein assemblies. Quasi-equivalence revisited. *Biophys J* 32:103–138.

Gardner M (1970) Mathematical games—the Game of Life. *Sci Am* 223(4):120–123.

Kellenberger E (1972) Assembly in biological systems and mechanisms of length determination in protein assemblies. In Polymerization in Biological Systems (Ciba Foundation Symposium 7) (Wolstenholme GEW & O'Connor M, eds), pp 189–206. Ciba Foundation.

Wiener N (1961) Cybernetics: or Control and Communication in the Animal and the Machine, 2nd ed. The M.I.T. Press and John Wiley & Sons, Inc.

8.2   Principles of virus replication

Caspar DLD & Klug A (1962) Physical principles in the construction of regular viruses. *Cold Spring Harbor Symp Quant Biol* 27:1–24.

Crick FH & Watson JD (1956) Structure of small viruses. *Nature* 177:473–475.

Epstein RH, Bolle A, Steinberg CM et al (1963) Physiological studies of conditional lethal mutations of bacteriophage T4D. *Cold Spring Harbor Symp Quant Biol* 28:375–394.

Flint SJ, Enquist LW, Racaniello VR & Skalka AM (2009) Principles of Virology: Molecular Biology, Pathogenesis, and Control of Animal Viruses, 3rd ed. ASM Press.

Harrison SC, Olson A, Schutt CE et al (1978) Tomato bushy stunt virus at 2.9 Å resolution. *Nature* 276:368–373.

Rao VB & Rossmann MG (eds) (2012) Viral Molecular Machines. Springer.

Rayment I, Baker TS, Caspar DL & Murakami WT (1982) Polyoma virus capsid structure at 22.5 Å resolution. *Nature* 295:110–115.

Roy P (ed.) (2005) Virus Structure and Assembly. Elsevier/Academic Press Inc.

8.3   Helical viruses

Butler PJG (1999) Self-assembly of tobacco mosaic virus: the role of an intermediate aggregate in generating both specificity and speed. *Phil Trans R Soc Lond B* 354:537–555.

Caspar DLD & Namba K (1990) Switching in the self-assembly of tobacco mosaic virus. *Adv Biophys* 26:157–185.

Day LA, Marzec CJ, Reisberg SA & Casadevall A (1988) DNA packing in filamentous bacteriophages. *Annu Rev Biophys Biophys Chem* 17:509–539.

Greenwood J, Hunter GJ & Perham RN (1991) Regulation of filamentous bacteriophage length by modification of electrostatic interactions between coat protein and DNA. *J Mol Biol* 217:223–227.

Russel M & Model P (2006) Filamentous phage. In The Bacteriophages, 2nd ed. (Calendar R ed.), pp 146–160. Oxford University Press.

Webster RE (2001) Filamentous phage biology. In Phage Display: A Laboratory Manual (Barbas III CF, Burton DR, Scott JK, Silverman GJ eds), pp 1.1–1.37. Cold Spring Harbor Laboratory Press.

Welsh LC, Symmons MF, Sturtevant JM et al (1998) Structure of the capsid of Pf3 filamentous phage determined from X-ray fibre diffraction data at 3.1 Å resolution. *J Mol Biol* 283:155–177.

Wu X & Shaw J (1996) Bidirectional uncoating of the genomic RNA of a helical virus. *Proc Natl Acad Sci USA* 93:2981–2984.

8.4   Small icosahedral viruses

Belnap DM, Filman DJ, Trus BL et al (2000) Molecular tectonic model of virus structural transitions: the putative cell entry states of poliovirus. *J Virol* 74:1342–1354.

Hogle JM (2002) Poliovirus cell entry: common structural themes in viral cell entry pathways. *Annu Rev Microbiol* 56:677–702.

Lewis JK, Bothner B, Smith TJ & Siuzdak G (1998) Antiviral agent blocks breathing of the common cold virus. *Proc Natl Acad Sci USA* 95:6774–6778.

Rossmann MG (1989) The canyon hypothesis. Hiding the host cell receptor attachment site on a viral surface from immune surveillance. *J Biol Chem* 264:14587–14590.

Steven AC, Conway JF, Cheng N et al (2005) Structure, assembly, and antigenicity of hepatitis B virus capsid proteins. *Adv Virus Res* 64:125–164.

Zlotnick A, Cheng N, Conway JF et al (1996) Dimorphism of hepatitis B virus capsids is strongly influenced by the C terminus of the capsid protein. *Biochemistry* 35:7412–7421.

Zlotnick A, Reddy VS, Dasgupta R et al (1994) Capsid assembly in a family of animal viruses primes an autoproteolytic maturation that depends on a single aspartic acid residue. *J Biol Chem* 269:13680–13684.

8.5   Large icosahedral viruses

Bamford DH, Grimes JM & Stuart DI (2005) What does structure tell us about virus evolution? *Curr Opin Struct Biol* 15:655–663.

Casjens SR (2005) Comparative genomics and evolution of the tailed-bacteriophages. *Curr Opin Microbiol* 8:451–458.

Dokland T (1999) Scaffolding proteins and their role in viral assembly. *Cell Mol Life Sci* 56:580–603.

Fane BA & Prevelige PE Jr (2003) Mechanism of scaffolding-assisted viral assembly. *Adv Protein Chem* 64:259–299.

Leiman PG, Kanamaru S, Mesyanzhinov VV et al (2003) Structure and morphogenesis of bacteriophage T4. *Cell Mol Life Sci* 60:2356–2370.

Steven AC, Heymann JB, Cheng N et al (2005) Virus maturation: dynamics and mechanism of a stabilizing structural transition that leads to infectivity. *Curr Opin Struct Biol* 15:227–236.

Tong L (2002) Viral proteases. *Chem Rev* 102:4609–4626.

Zlotnick A & Fane BA (2011) Mechanisms of icosahedral virus assembly. In Structural Virology (Agbandje-McKenna M, McKenna R eds), pp 180–202. RSC Publishing.

8.6   Double-stranded RNA viruses

Caston JR, Trus BL, Booy FP et al (1997) Structure of L-A virus: a specialized compartment for the transcription and replication of double-stranded RNA. *J Cell Biol* 138:975–985.

Clarke P, Richardson-Burns SM, DeBiasi RL & Tyler KL (2005) Mechanisms of apoptosis during reovirus infection. *Curr Top Microbiol Immunol* 289:1–24.

Coulibaly F, Chevalier C, Gutsche I et al (2005) The birnavirus crystal structure reveals structural relationships among icosahedral viruses. *Cell* 120:761–772.

Cremer T & Cremer C (2001) Chromosome territories, nuclear architecture and gene regulation in mammalian cells. *Nat Rev Genet* 2:292–301.

Dormitzer PR, Nason EB, Prasad BV & Harrison SC (2004) Structural rearrangements in the membrane penetration protein of a non-enveloped virus. *Nature* 430:1053–1058.

Jacobs BL & Langland JO (1996) When two strands are better than one: the mediators and modulators of the cellular responses to double-stranded RNA. *Virology* 219:339–349.

Norman KL & Lee PW (2005) Not all viruses are bad guys: the case for reovirus in cancer therapy. *Drug Discov Today* 10:847–855.

Patton JT (ed.) (2008) Segmented Double-Stranded RNA Viruses: Structure and Molecular Biology. Caister Academic Press.

Tao Y, Farsetta DL, Nibert ML & Harrison SC (2002) RNA synthesis in a cage—structural studies of reovirus polymerase lambda3. *Cell* 111:733–745.

8.7   Enveloped viruses

Ekiert DC & Wilson IA (2011). Attachment and entry: Receptor recognition in viral pathogenesis. In Structural Virology (Agbandje-McKenna M, McKenna R eds), pp 220–242. RSC Publishing.

Gibbons DL, Vaney MC, Roussel A et al (2004) Conformational change and protein-protein interactions of the fusion protein of Semliki Forest virus. *Nature* 427:320–325.

Harrison SC (2008) Viral membrane fusion. *Nature Struct Mol Biol* 15:690–698.

Kuhn RJ & Rossmann MG (2005) Structure and assembly of icosahedral enveloped RNA viruses. In Virus Structure and Assembly (Roy P ed.), pp 263–284. Elsevier Academic Press.

Lehmann MJ, Sherer NM, Marks CB et al (2005) Actin- and myosin-driven movement of viruses along filopodia precedes their entry into cells. *J Cell Biol* 170:317–325.

Mercer J, Schelhaas M & Helenius A (2010) Virus entry by endocytosis. *Annu Rev Biochem* 79:803–833.

Ono A & Freed EO (2005) Role of lipid rafts in virus replication. *Adv Virus Res* 64:311–358.

Rossman JS & Lamb RA (2011) Influenza virus assembly and budding. *Virology* 411:229–236.

Schowalter RM, Smith EC & Dutch RE (2011) Attachment and entry: viral cell fusion. In Structural Virology (Agbandje-McKenna M, McKenna R eds), pp 243–260. RSC Publishing.

White SH & von Heijne G (2005) Transmembrane helices before, during, and after insertion. *Curr Opin Struct Biol* 15:378–386.

8.8   Virus engineering and nanotechnology

Domingo GJ, Caivano A, Sartorius R et al (2003) Induction of specific T-helper and cytolytic responses to epitopes displayed on a virus-like protein scaffold derived from the pyruvate dehydrogenase multienzyme complex. *Vaccine* 21:1502–1509.

Fehr T, Bachmann MF, Bucher E et al (1997) Role of repetitive antigen patterns for induction of antibodies against antibodies. *J Exp Med* 185:1785–1792.

Manchester M & Steinmetz NF (eds) (2009) Viruses and Nanotechnology. Springer.

Schiller JT & Lowy DR (1996) Papillomavirus-like particles and HPV vaccine development. *Semin Cancer Biol* 7:373–382.

Smith GP & Petrenko VA (1997) Phage display. *Chem Rev* 97:391–410.

Sternberg N & Hoess RH (1995) Display of peptides and proteins on the surface of bacteriophage lambda. *Proc Natl Acad Sci USA* 92:1609–1613.

Ulrich R, Nassal M, Meisel H & Kruger DH (1998) Core particles of hepatitis B virus as carrier for foreign epitopes. *Adv Virus Res* 50:141–182.

ohesins

CipA

SdbA

SLH

bacterial
cell

CBM

Multienzyme Complexes:
Catalytic Nanomachines

9

9.1   INTRODUCTION

Most of the biochemical reactions on which life depends are too complicated to be accomplished in one step; much of metabolism, both *catabolic* (breaking down) and *anabolic* (building up), is found in the form of pathways. Metabolic pathways may have numerous steps and are made possible by successive enzymes acting on intermediates, transforming one or more starting compounds into one or more end-products. The possession of common intermediates and the interlocking of the pathways underpin the vast array of biochemical transformations that are found in a cell, a tissue, and an organism (Sections 1.6 and 1.7). An enzyme that catalyzes any one of the steps of a biochemical pathway must have access to the product of the preceding step. In eukaryotic cells, the restriction of pathways to particular membrane-delimited organelles, such as the citric acid cycle and fatty acid degradation found in mitochondria and thereby separated from glycolysis in the cytosol (Chapter 15), can serve to isolate the intermediates from competing reactions and raise their local concentrations. In some instances the enzymes that catalyze a particular sequence of reactions appear to operate as some sort of giant aggregate within a cell. For example, the enzymes that catalyze glycolysis have been reported to be physically associated (a *metabolon*), although any such aggregate must be one that readily falls apart, given that it has so far resisted attempts to purify and characterize it.

However, in prokaryotic cells and within a compartment or cytosol of a eukaryotic cell, a different type of enzyme segregation occurs, to facilitate the catalysis of multi-step reactions that are particularly difficult to bring about for one or more reasons. The participating enzymes form robust complexes, generally termed **multienzyme complexes** or catalytic nanomachines. They represent organized assemblies and come in many shapes and sizes. In the simplest case, two enzymes that catalyze successive reactions (more rarely, two nonsequential reactions) in a biosynthetic pathway may be found as a heterodimer. Much more elaborate complexes also exist, which catalyze overall reactions comprising several coordinated steps and contain multiple copies of various different enzymes. In these instances the molecular mass may reach 5–10 MDa or more, bigger than a ribosome.

It is often found that the genes encoding the components of a multienzyme complex in bacteria are part of an operon and are transcribed and regulated coordinately. This facilitates the synthesis of appropriate amounts of the constituent enzymes but does not necessarily correspond to their respective turnover numbers. These may vary widely, but disparity in turnover numbers is not necessarily a problem. In some instances, for example the pyruvate dehydrogenase complex (Section 9.4), a molecular architecture and assembly pathway that accommodates varying protein stoichiometry can allow major differences in turnover number. Particularly in eukaryotes, two or more enzymes that in bacteria are physically separate are often found together in a single multifunctional polypeptide chain with linker regions between them. These evolutionary products of gene fusion have the constituent enzyme activities co-represented, and this may account for some of the multifunctional enzymes found in Nature.

9.2   ACTIVE-SITE COUPLING AND SUBSTRATE
CHANNELING IN PROTEIN COMPLEXES

Multienzyme complexes have features that are not possible with a mixture of separate individual enzymes. These include restricting the need for intermediates to diffuse between distant active sites, the elimination of competition between different enzymes for use of the same substrate, the protection of labile intermediates, and the suppression of unwanted side-reactions.

Multienzyme complexes channel substrates and protect labile intermediates

Diffusion of small substrates between separate enzymes is not normally rate-limiting (Section 1.8), but proximity and the favorable orientation of multiple active sites can still offer advantages. In particular, if a substrate or intermediate is not a small, highly diffusible molecule but something much larger, say a protein or a nucleic acid, the juxtaposition and arrangement of the individual enzymes within a multienzyme complex can shorten the diffusion pathway between successive active sites and thereby speed up the overall reaction. In addition, within a cell, the product of one enzymatic reaction will sometimes serve as the substrate for two or more different successor enzymes, each leading in a different metabolic direction. The preferred outcome can be ensured if the intermediate is passed directly from one enzyme to another, a property called **substrate channeling**. For this reason, multienzyme complexes often occur at metabolic branch points, as shown below:

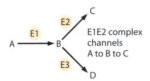

Moreover, an intermediate in a complicated reaction might be toxic or too reactive in a bulk aqueous environment to survive even a short diffusion from one enzyme active site to another. By passing the intermediate directly between active sites, multienzyme complexes can protect such labile intermediates (the 'hot potato' hypothesis of substrate channeling).

Another crucial, albeit rare, feature of multienzyme complexes is the ability to avert competing and wasteful side-reactions. Thus, if one of the enzymes in a complex can, in its free form, catalyze two variants of a reaction, the 'wrong' reaction can be suppressed by a conformational change induced by the assembly of the multienzyme complex. This ensures that the correct substrate is available to the succeeding enzyme. We shall meet examples of all these features in the multienzyme complexes described below. A further property of multienzyme complexes is the ability simultaneously to bring to bear a range of different enzymes on a substrate that is unusually refractory or otherwise difficult to act on. An example of this will be found in the degradation of cellulose described in Section 9.6.

Substrate channeling takes place through molecular tunnels or by the covalent attachment of intermediates to swinging arms

A catalytic intermediate can be passed directly between enzymes in a complex in two principal ways. In one, the intermediate passes through a **molecular tunnel** that links two active sites; the tunnel serves as a conduit from which bulk solvent is excluded and essentially turns the diffusion of intermediates from a three-dimensional into a one-dimensional problem. The concept is straightforward, along with the need for some form of gating mechanism. The other means of substrate channeling is to link the intermediate covalently to a prosthetic group that is in turn attached to an amino acid side chain by a specific post-translational modification of one of the enzymes. The elongated side chain constitutes a **swinging arm** that is free to move and, without letting go of it en route, transfer the intermediate to a second active site as far away as twice the length of the swinging arm. Three examples of swinging arms have been found so far, each based on a different prosthetic group: lipoic acid, biotin, and phosphopantetheine.

The vitamins **biotin** and **pantothenic acid** (the latter after conversion to **phosphopantetheine**) and the essential metabolite **lipoic acid** have different chemical structures (**Figure 9.1**). All three compounds exhibit stereoisomerism (see Box 1.1), and a preferred enantiomer (D in all three instances) is used *in vivo*. The structures and biosynthetic pathways for biotin and lipoic acid have some features in common and, as components of swinging arms, both are linked through their carboxyl groups to the N^6-amino group of specific lysine

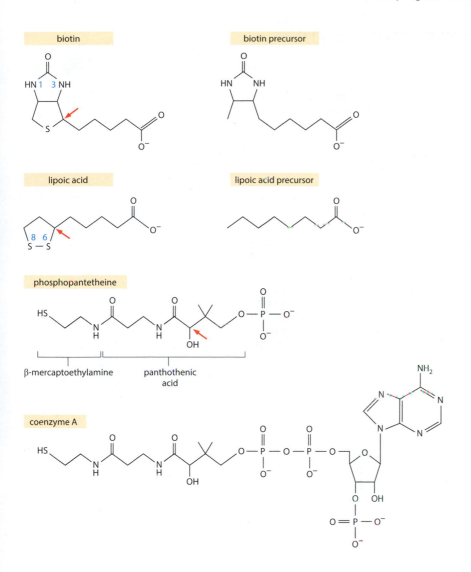

Figure 9.1 Structures of biotin, lipoic acid, and phosphopantetheine. On the right are shown the biosynthetic precursors of biotin and lipoic acid, into which sulfur atoms (one for biotin, two for lipoic acid) are inserted by additional enzyme systems. In phosphopantetheine, the –NH$_2$ group of β-mercaptoethylamine is in amide linkage with the –COOH group of panthothenic acid, and its –OH group has been phosphorylated. Chiral carbon atoms are indicated by red arrows and key atoms are numbered in blue. In coenzyme A, a molecule of phosphopantetheine is linked by a phosphoanhydride bond to the 5′ phosphate of a molecule of 3′-phospho-AMP.

residues. A dedicated enzyme (an ATP-dependent ligase—which interestingly also serves as a repressor of biotin synthesis in *Escherichia coli*) catalyzes the formation of the amide bond between biotin and the correct lysine residues in the relevant proteins. There is a similar ligase for attaching lipoic acid to its target proteins, but in *E. coli* and in mitochondria an additional biosynthetic route is possible. A specific enzyme can transfer an octanoyl group onto the lysine residue from its thioester link to an acyl carrier protein (ACP; Section 9.5), and the requisite two S atoms are then inserted into the octanoyl chain by the action of further enzymes. In contrast, certain other enzymes, phosphopantetheinyl transferases (PPTs), transfer the phosphopantetheinyl moiety from CoA to make a phosphodiester link with the hydroxyl group of specific serine side chains in designated proteins. In mammals, a single PPT operates on all proteins earmarked for this post-translational modification, whereas in yeast and bacteria there is a battery of PPTs, each operating on a specific target protein.

In all three instances, a flexible side chain ~15 Å in length is created, which enables the functional groups at the ends of the swinging arms (the heterocyclic rings in the case of lipoyl-lysine and biotinyl-lysine and the thiol group in the case of phosphopantetheinyl-serine) to connect different enzyme active sites up to ~30 Å apart in a complex. This was the basis of the original swinging arm hypothesis proposed separately by Lester Reed, David Green, and Fyodor Lynen as part of the mechanism of the lipoic acid-dependent 2-oxo acid dehydrogenase multienzyme complexes, the biotin-dependent carboxylases, and the phosphopantetheine-dependent fatty acid synthases. We describe each of them in turn after first looking at examples of substrate channeling through molecular tunnels.

9.3   MULTIENZYME COMPLEXES WITH TUNNELS

Some multienzyme complexes have a physical tunnel through which the product of one enzyme is passed directly to the active site of the next enzyme. Two very different enzyme complexes, tryptophan synthase and carbamoyl phosphate synthase, exemplify substrate channeling through tunnels. Although the intermediates being channeled are very different, the mechanisms have a number of features in common.

Tryptophan synthase has two active sites connected by a molecular tunnel

In bacteria, the enzyme tryptophan synthase, which catalyzes the last two steps in the biosynthesis of tryptophan, occurs as a stable $\alpha_2\beta_2$ heterotetramer (143 kDa). In fungi, the α and β subunits are fused into a single bifunctional polypeptide chain, generating an $(\alpha\beta)_2$ dimer. Animals do not possess the enzymes of aromatic biosynthesis. The bacterial α and β subunits are readily separated and each catalyzes a part-reaction: the α subunit (29 kDa) generates an indole intermediate, whereas a β_2-homodimer (subunit 43 kDa) catalyzes the condensation of indole with serine to yield tryptophan, for which it needs the help of a **pyridoxal phosphate** (**PLP**) prosthetic group (**Figure 9.2A**). The β_2 homodimer also catalyzes the conversion of serine to pyruvate and NH_3, but this is a side-reaction that is suppressed on assembly of the $\alpha_2\beta_2$ heterotetramer. Kinetic evidence, and the failure of radiolabeled free indole to become incorporated into the product tryptophan, indicate that the indole generated by the α subunit must pass directly to the active site of the β subunit in the $\alpha_2\beta_2$ heterotetramer.

Tryptophan synthase from *Salmonella typhimurium* is ~150 Å in length, with a roughly linear arrangement of the four subunits in the order $\alpha/\beta/\beta/\alpha$ (Figure 9.2B). The α subunit adopts the fold $(\beta/\alpha)_8$ of a TIM barrel. As observed so far for all enzymes based on this fold, the active site (revealed by the binding of a competitive inhibitor, indole propanol phosphate) is at the C-terminal end of the TIM barrel. The β subunit has two domains of roughly equal size: the N-terminal domain is built round four strands of parallel β sheet, whereas the C-terminal domain contains five strands of parallel and one strand of anti-parallel β sheet.

(A)

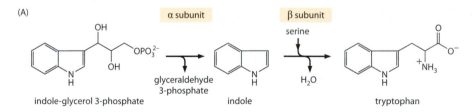

indole-glycerol 3-phosphate   indole   tryptophan

(B)

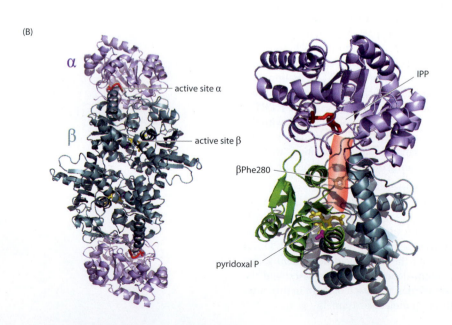

Figure 9.2 Structure and mechanism of tryptophan synthase. (A) Two-step reaction catalyzed by the α and β subunits in succession. The intermediate, indole, is generated by the action of the α subunit, and its subsequent conversion to tryptophan is catalyzed by the β subunit. (B) Structure of the enzyme from *S. typhimurium*. Left, the $\alpha_2\beta_2$ heterotetramer with positions of α and β active sites indicated. Right, close-up of an α/β pair. Indole propanol phosphate (IPP, a competitive inhibitor, shown in red) is bound in the active site of the α subunit; pyridoxal phosphate (yellow spheres) is in aldimine linkage with the α-amino group of the second substrate, serine, in the active site of the β subunit. In the resting enzyme, pyridoxal phosphate is in imine linkage with the N^6-amino group of βLys87 but this is a βK87T mutant (Thr87 in magenta). A 25 Å tunnel (sketched in pink) runs from the active site in the α subunit to that in the β subunit. Residues βGly93–βGly189 are shown in green; βPhe280 has been suggested as a possible gatekeeper in the tunnel. (PDB 2TRS)

The active site in the β subunit can be identified from the PLP attached by an imine bond (Schiff base) to the N^6-amino group of Lys87, at the interface of the two domains. Indole glycerol phosphate enters and D-glyceraldehyde 3-phosphate exits through the α subunit, and L-serine enters and L-tryptophan exits through the β subunit. The active sites of the α and β subunits in a functional α/β pair are ~25 Å apart but are connected by an intramolecular tunnel, large enough to accommodate up to four molecules of indole in transit from one active site to the next.

The tunnel in tryptophan synthase is gated

In the absence of ligands, the active sites of the α and β subunits are both exposed to solvent. However, when they are occupied, a disordered surface loop in the α subunit becomes ordered and moves to cover the active site of the α subunit. This is accompanied by a rigid-body rotation of the α subunit with respect to the β subunit and a movement of a β subunit subdomain between residues Gly93 and Gly189 (see Figure 9.2B). These conformational changes restrict solvent access to the active sites and to the tunnel, and prevent loss of the indole into the bulk solvent. Single-turnover experiments demonstrate that indole passes between the active sites very rapidly (>1000/s) and that the reaction of indole at the β subunit active site is essentially irreversible. Moreover, the reaction of L-serine with PLP at the β subunit active site, to form an amino-acrylate intermediate ready to react with an incoming indole molecule, increases the rate of cleavage of indole-3-glycerol phosphate approximately thirtyfold at the α subunit active site. This catalytic cycle is mediated by allosteric conformational changes triggered by substrate binding in the α subunit active site and by covalent bond formation between the second substrate, serine, and the PLP cofactor, together with the intermediate indole, in the active site of the β subunit. A monovalent cation (such as Na^+, K^+, or NH_4^+, although others will work) in the β subunit about 8 Å from its active site is essential for catalytic activity and for communication between α and β subunit active sites.

These features of the enzyme complex keep the two component reactions in phase and prevent accumulation of the indole intermediate, a prerequisite for efficient substrate channeling. Indole is far from unstable, but if it were to diffuse freely between the active sites of two separate enzymes, it could easily escape from the cell by passive diffusion across the cell membrane. Tryptophan synthase is thus a highly efficient device for the biosynthesis of tryptophan: it is capable of suppressing a side-reaction, sequestering and channeling an intermediate, and coordinating active sites.

Ammonia is channeled as a reaction intermediate in several different enzyme complexes

At least six different enzyme complexes that require NH_3 as a reactive intermediate have been found to have molecular tunnels to channel it between active sites: asparagine synthetase, carbamoyl phosphate synthase, glutamine phosphoribosylphosphate amidotransferase, glutamate synthase, imidazole glycerol phosphate synthase, and glucosamine 6-phosphate synthase. In all cases the NH_3 is generated by the hydrolysis of glutamine in a remote part of the relevant complex. The tunnels connecting the active sites differ in size and architecture and appear to have been evolved separately. Two examples, carbamoyl phosphate synthase (CPS) and glutamine phosphoribosylphosphate amidotransferase (GPAT), illustrate the key features, not least channeling a gas as a reactant.

CPS catalyzes the formation of carbamoyl phosphate, a key intermediate in the biosynthesis of arginine and pyrimidines in both prokaryotes and eukaryotes, and in the biosynthesis of urea in most terrestrial vertebrates. Four interlocking reactions make up the mechanism (**Figure 9.3A**). The enzyme from *E. coli* contains two different subunits, designated large (118 kDa) and small (42 kDa) (Figure 9.3B). The small subunit (SS) catalyzes the hydrolysis of the amide bond in glutamine to generate ammonia and glutamate, and is a member of the class I family of **amidotransferases**. These utilize a Cys/His/Glu *catalytic triad* in the active site (mechanistically similar to the Ser/His/Asp proteases described in Box 7.1). The large subunit (LS) comprises two homologous domains. The N-terminal one catalyzes the phosphorylation of bicarbonate (active site LS1), and the product, carboxyphosphate, reacts with NH_3 coming from the SS to form carbamate. This is then phosphorylated in the C-terminal domain (active site LS2) to yield the final product, carbamoyl phosphate. Three

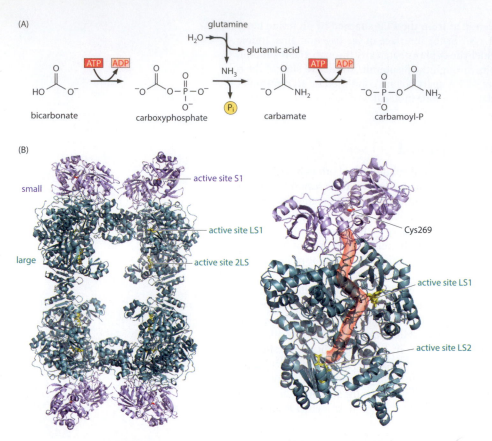

(A)

(B)

small

large

active site S1

active site LS1

active site 2LS

Cys269

active site LS1

active site LS2

Figure 9.3 Structure and mechanism of CPS. (A) The four-part reaction catalyzed by CPS. The hydrolysis of Gln to Glu and NH$_3$ is catalyzed by the SS (active site S1), the synthesis of carboxyphosphate and then carbamate takes place at the first active site (LS1) of the LS, and the phosphorylation of carbamate to form carbamoyl phosphate occurs at the second active site (LS2) of the LS. (B) Structure of the CPS from *E. coli*. Left, the hetero-octamer (LS$_4$SS$_4$) as crystallized. Right, close-up of the functional unit, an SS/LS heterodimer, in a different orientation. Cys269 (red) is part of the catalytic triad in the active site of the SS; an ADP molecule (yellow spheres) is bound in each of the two active sites of the LS. The intermolecular tunnel (>90 Å long) connecting all three active sites is sketched in pink. (PDB 1JDB)

of the intermediates (NH$_3$, carboxyphosphate, and carbamate) are reactive and unstable and require efficient channeling. The half-life of carbamate at neutral pH has been estimated to be only ~70 ms, and the enzyme-bound NH$_3$ must be sequestered from the bulk solvent because, if it becomes protonated, it will not react with carboxyphosphate.

The enzyme from *E. coli* forms an α$_4$β$_4$ hetero-octamer under certain conditions, but the functional unit is the SS/LS heterodimer. The active site in the SS is located ~45 Å from the active site LS1, which in turn is ~35 Å from the second active site (LS2) in the same LS (see Figure 9.3B). A tunnel more than 90 Å long and with a mean diameter of about 7 Å connects all three. Between the SS active site and active site LS1 the tunnel is mostly hydrophobic, whereas between the active sites LS1 and LS2 it is less so. As with tryptophan synthase, the active sites are coordinated: the rate of hydrolysis of glutamine in the SS is increased 500-fold when bicarbonate is phosphorylated by ATP in the LS. Ammonia is therefore dispatched into the tunnel only when its intended partner intermediate, carboxyphosphate, has been generated and is ready to receive it. The conformational changes in the protein that transmit this information from the LS to the SS are not yet precisely worked out, nor is the way in which the two active sites in the LS themselves might be coordinated. In eukaryotes, the CPS involved in pyrimidine production exists as part of a multifunctional complex (CAD) that contains <u>C</u>PS plus <u>a</u>spartate transcarbamoylase and <u>d</u>ihydroorotase activities.

GPAT catalyzes the first reaction in the *de novo* synthesis of purine nucleotides, the conversion of α-phosphoribosylpyrophosphate into 5-β-phosphoribosylamine. Two active sites are required; one, an amidotransferase to generate NH$_3$ by hydrolyzing Gln, and the other to catalyze the reaction of the NH$_3$ with the phosphoribosylpyrophosphate. The enzyme from *E. coli* (a homodimer of 55 kDa subunits) differs from CPS in that the amidotransferase is of class II, in which the thiol of an N-terminal Cys residue is the catalytic nucleophile (mechanistically similar to the proteasome described in Section 7.4) as opposed to a Cys/His/Glu triad. The second difference is that there is no preexisting tunnel between the two active sites in each subunit. A tunnel 20 Å long is formed by conformational changes within each subunit once both active sites are occupied with appropriate ligands. Overall, however, it is clear that a molecular tunnel has been an evolutionarily favored mechanism for transferring reactive and highly soluble NH$_3$ between enzyme active sites. Together with tryptophan synthase, the NH$_3$-utilizing enzymes underline the importance of this form of substrate channeling.

9.4 MULTIENZYME COMPLEXES WITH LIPOIC ACID OR BIOTIN IN THEIR SWINGING ARMS

Two of the three swinging arms commonly used to convey substrates or intermediates around in multienzyme complexes, namely lipoyl-lysine and biotinyl-lysine (see Figure 9.1), are found in two families of molecular machines that catalyze related but opposing types of reaction: exergonic oxidative decarboxylation for lipoyl-lysine and endergonic ATP-dependent carboxylation for biotinyl-lysine. The complexes in which they operate are structurally very different, as are the chemical mechanisms, but there are some important similarities; in particular, the constituent active sites are too far apart to be spanned by a single swinging arm, and elaborate movements of protein domains tethered by flexible linkers are required to ferry the swinging arms from one site to another.

Lipoic acid-dependent 2-oxo acid dehydrogenase multienzyme complexes are built round multisubunit cores

The lipoic acid-containing 2-oxo acid dehydrogenase complexes (2-OADH complexes) are essential components of the glycolytic pathway and the citric acid cycle (Chapter 15) and of the catabolism of branched-chain amino acids (leucine, valine, and isoleucine). In eukaryotes they are located in the mitochondria, and defects in them can lead to severe metabolic disorders, such as maple syrup urine disease and primary biliary cirrhosis. Each complex catalyzes the oxidative decarboxylation of a different metabolite: for pyruvate, the pyruvate dehydrogenase (PDH) complex; for 2-oxoglutarate, the 2-oxoglutarate dehydrogenase (OGDH) complex; and for branched chain 2-oxo acids, the branched-chain 2-oxo acid dehydrogenase (BCDH) complex. They are giant assemblies (~5–10 MDa) of multiple copies of three different enzymes: a 2-oxo acid decarboxylase (E1) dependent on thiamin diphosphate (ThDP) and Mg^{2+}, a dihydrolipoyl-acyltransferase (E2) containing a lipoyl-lysine residue, and an FAD-containing dihydrolipoyl dehydrogenase (E3). The overall reaction of *oxidative decarboxylation* takes place in four steps (**Figure 9.4A**). The first two, the decarboxylation of the 2-oxo acid (shown in the figure with its carboxyl group protonated; it would be unprotonated at physiological pH) and the reductive acylation of the lipoyl group on E2, are catalyzed by E1; the subsequent transfer of the acyl group from the dihydrolipoyl group to CoA is catalyzed by E2; and the reoxidation of the dihydrolipoyl group is catalyzed by E3 with NAD^+ as coenzyme.

In addition to its role in catalysis (**Box 9.1**), E2 is the physical core around which the rest of the structure is assembled. In the PDH complex from *Geobacillus stearothermophilus* (a Gram-positive moderately thermophilic bacterium), it is composed of 60 polypeptide chains (each 46 kDa) arranged with icosahedral (532) symmetry (see Section 1.4 for a discussion of symmetry). Other 2-OADH complexes have E2 cores (24-mers) with octahedral (432) symmetry. The E2 polypeptide chain has a classic domain-and-linker structure, one of the first to be elucidated (Figure 9.4B). The N-terminal domain (the **lipoyl domain**, ~9 kDa) is a flattened β barrel of two four-stranded β sheets with quasi-2-fold symmetry. The lipoyl-lysine side chain is near the tip of a protruding β turn between β strands 4 and 5, where it is essentially free to rotate. The lipoyl domain is followed by a long (~35 aa) segment of polypeptide chain that leads into a small (~4.5 kDa) domain to which can be bound, noncovalently but tightly (K_d ~10^{-9} M), either one E1 (an $\alpha_2\beta_2$ heterotetramer) or one E3 (an α_2 homodimer) component. Hence it has been named the peripheral <u>s</u>ubunit-<u>b</u>inding <u>d</u>omain (PSBD). The PSBD is one of the smallest known protein domains with a stable fold (Section 1.3), and it binds close to or across the 2-fold axis of its partner E1 or E3, in 1:1 stoichiometry. Another long linker (~35 aa) attaches the PSBD to the C-terminal domain (28 kDa). This domain both contains the acetyltransferase active site and assembles to form the icosahedral core. The linker between the lipoyl domain and PSBD is unusually rich in Ala/Pro residues and is known, from NMR spectroscopic (see **Methods**) studies of the intact complex, to be highly flexible, behaving very like a peptide of the same length free in solution (**Figure 9.5**).

Substrate channeling depends on mobility of the lipoyl domain

A curious feature of the mechanism of the PDH complex is the fact that free lipoic acid or lipoamide are very poor substrates for E1. There is no immediately obvious chemical reason for this, but a lipoyl domain serves vastly better: compared with free lipoamide, it has a

(A)

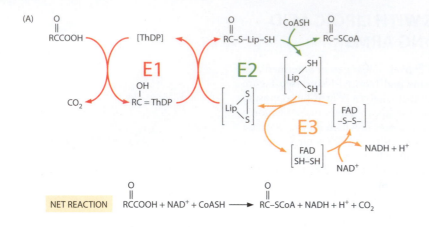

NET REACTION $\underset{\text{RCCOOH}}{\overset{O}{\|}}$ + NAD$^+$ + CoASH $\longrightarrow$ $\underset{\text{RC–SCoA}}{\overset{O}{\|}}$ + NADH + H$^+$ + CO$_2$

(B)

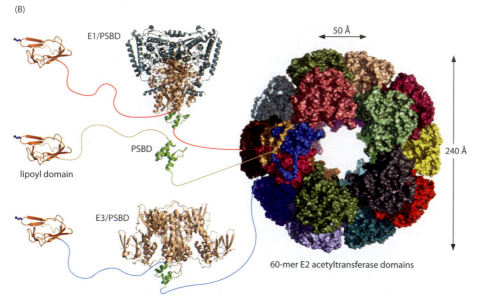

Figure 9.4 Reaction mechanism and schematic structure of a PDH complex. (A) Reactions 1 (decarboxylation of the 2-oxo acid) and 2 (reductive acylation of a lipoyl group) are catalyzed by E1, reaction 3 (acyl transfer from dihydrolipoyl group to CoA) is catalyzed by E2, and reaction 4 (reoxidation of dihydrolipoyl group) is catalyzed by E3. ThDP, thiamin diphosphate; LipS-S, lipoyl group covalently attached to a lysine residue in E2 in oxidized form; R-CO-S-Lip-SH, reductively acylated lipoyl group; Lip(SH)$_2$, dihydrolipoyl group. (B) Schematic model of the PDH complex of *G. stearothermophilus* (for clarity, the individual parts are not drawn to scale). The structural core (2.8 MDa) is a 60-mer of E2 chains, in which the C-terminal acetyltransferase domains (28 kDa) are arranged with icosahedral symmetry, viewed here across a 5-fold axis. For simplicity, only one of the 20 E2 trimers is shown in full, with each chain in a separate color (magenta, blue, and gold), except for the lipoyl domains (N-terminal, 9 kDa) in orange and the PSBDs (4.5 kDa) in green. Lys42, to which the lipoyl group is attached in the lipoyl domain, is depicted in blue. For the other 19 E2 trimers, only the acetyltransferase domains are shown and each trimer is depicted in a single color. An E1 ($\alpha_2\beta_2$) in beige (α, 38 kDa) and teal (β, 42 kDa) is shown bound to one PSBD, and an E3 (α_2) in beige (α, 55 kDa) is shown bound to another PSBD. For clarity, the third PSBD is shown without an E1 or E3 component bound. The wavy lines in magenta, blue, and gold between the lipoyl domain and the PSBD and between the PSBD and the acetyltransferase domain represent long (30–40 aa, ~100 Å) conformationally flexible linker regions. (PDB 1B5S, 1EBD, 1W85, 1W3D, 1LAB)

catalytic efficiency (k_{cat}/K_m—see Section 1.6) that is more than 10^4-fold higher. Moreover, the lipoyl domain must be the one from the E2 chain of the parent PDH complex, not even one from another 2-OADH complex. Decarboxylation of the 2-oxo acid is therefore dependent on recognition by E1 of the lipoyl domain presented to it by E2, and the true co-substrate is a lipoyl group on a 9 kDa protein. This mechanistic feature becomes more understandable in the light of the structure of E1 and the restricted access to its active sites (**Box 9.2**; see also Figure 9.12 below).

Box 9.1   Structure and mechanism of E2 in a 2-OADH complex

E2 catalyzes the transfer of the acyl group from the dihydrolipoyl domain to CoA. E2 is found in two closely related forms: 8 trimers of the acyltransferase domains arranged as a 24-mer with octahedral symmetry and 20 trimers of the acyltransferase domains making up a 60-mer with icosahedral symmetry (see the main text and Section 1.4). Two images of one of the eight acetyltransferase domain trimers from the octahedral E2 of the *Azotobacter vinelandii* PDH complex crystallized with dihydrolipoamide and coenzyme A are shown in **Figure 9.1.1A**. Each of the three active sites has a dihydrolipoamide close to the 'outer' surface and a coenzyme A close to the 'inside.' The right-hand panel gives a side view. The acyltransferase domain trimers in all 2-OADH complexes, whether octahedral or icosahedral, are very similar.

Each active site lies between two subunits. There is a highly conserved motif -Asp-His-Arg-X-X-Asp-Gly- in E2 enzymes; the underlined His and Asp are key residues in the active site

(represented by His610 and Asn614 in *A. vinelandii* E2, which is exceptional in having Asn instead of Asp in this position). A third key residue is a Ser from the apposing subunit (Ser558 in the *A. vinelandii* E2). In the proposed catalytic mechanism (Figure 9.1.1B), the Asp/His pair promotes nucleophilic attack of the thiol group of CoA on the thioester bond between the acetyl group and S8 of the dihydrolipoyl-lysine side chain in the lipoyl domain, and the tetrahedral intermediate is stabilized by H-bonding with the −OH of Ser558. The reductively acetylated lipoyl domain (LD) approaches the active site from the 'outside' of the octahedron/icosahedron, whereas CoA enters the active site from the large cavity on the 'inside.' After acyl group transfer, acetyl-CoA is discharged into the interior of the octahedron or icosahedron and the dihydrolipoyl domain is released on the outside. CoA and acetyl-CoA enter and exit from the octahedron through the large holes on the 4-fold faces (and equivalently on the 5-fold faces in icosahedral complexes).

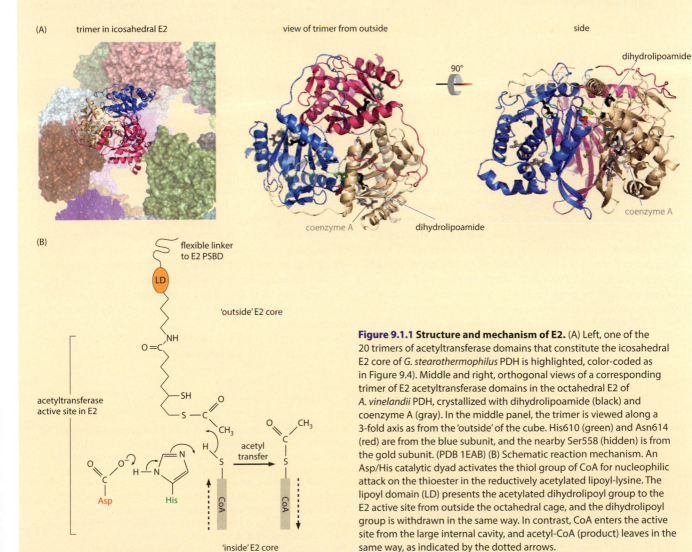

Figure 9.1.1 Structure and mechanism of E2. (A) Left, one of the 20 trimers of acetyltransferase domains that constitute the icosahedral E2 core of *G. stearothermophilus* PDH is highlighted, color-coded as in Figure 9.4). Middle and right, orthogonal views of a corresponding trimer of E2 acetyltransferase domains in the octahedral E2 of *A. vinelandii* PDH, crystallized with dihydrolipoamide (black) and coenzyme A (gray). In the middle panel, the trimer is viewed along a 3-fold axis as from the 'outside' of the cube. His610 (green) and Asn614 (red) are from the blue subunit, and the nearby Ser558 (hidden) is from the gold subunit. (PDB 1EAB) (B) Schematic reaction mechanism. An Asp/His catalytic dyad activates the thiol group of CoA for nucleophilic attack on the thioester in the reductively acetylated lipoyl-lysine. The lipoyl domain (LD) presents the acetylated dihydrolipoyl group to the E2 active site from outside the octahedral cage, and the dihydrolipoyl group is withdrawn in the same way. In contrast, CoA enters the active site from the large internal cavity, and acetyl-CoA (product) leaves in the same way, as indicated by the dotted arrows.

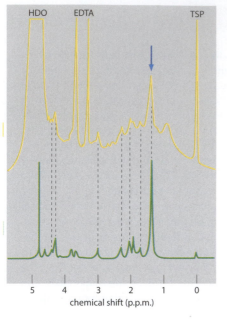

AAPAAAPAKQEAAAPAPAAKAEAPAAAPAAKA

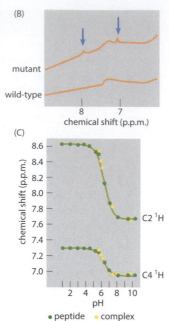

Figure 9.5 Flexible inter-domain linker regions in the PDH complex. (A) The ^{1}H NMR spectrum (in yellow) of the PDH complex of *E. coli*. The complex (~5 MDa) tumbles slowly with a rotational correlation time of ~10^{-5} s: a broad envelope of overlapping resonances is observed. A side-chain methylene proton would be expected to have a line width of ~8.5 kHz, but there is a sharp resonance with a line width of ~50 Hz (blue arrow), which is characteristic of methyl protons of Ala and Thr residues in small peptides. Other resonances are from EDTA in the buffer; HDO, deuterated water; and TSP. The lower spectrum (in green) is from a 32-aa peptide whose sequence (given below), rich in Ala and Pro, is that of an inter-domain linker in the complex. Its numerous sharp resonances, in particular one at 1.52 p.p.m. from the methyl protons of Ala residues, match counterparts in the spectrum of the intact complex. (B) Aromatic region of the ^{1}H NMR spectrum of a PDH complex in which a His replaces a Gln at position 9 of the linker. Two sharp resonances (arrows) present only in the spectrum of the mutant complex are characteristic of the C2 and C4 protons of His residues in small peptides. (C) By observing the pH-dependence of these resonances, the pKa of the His residue can be estimated. It is ~6.4 for both the complex and the peptide. (A and C, adapted from S.E. Radford et al., *J. Biol. Chem.* 264:767–775, 1989. With permission from American Society for Biochemistry and Molecular Biology; B, adapted from F.L. Texter et al., *Biochemistry* 27:289–296, 1988. With permission from American Chemical Society.)

Box 9.2   Structure and mechanism of E1 in a 2-OADH complex

E1 catalyzes the ThDP-dependent decarboxylation of the 2-oxo acid and the reductive acylation of the dithiolane ring at the end of a lipoyl-lysine side chain in a lipoyl domain that is part of E2. The structure of the $\alpha_2\beta_2$ E1 of *Geobacillus stearothermophilus* bound to its PSBD is shown in **Figure 9.2.1A**. The binding involves electrostatic interactions between a set of positive side chains from PSBD and negative side chains from the two β subunits of E1, together with some buried hydrophobic interactions. In addition, four bound water molecules form a network of H-bonds at the interface. The two active sites in the hetero-tetramer differ in local structure (Figure 9.2.1B), one having two surface loops from an α subunit partly obscuring the entrance to the active site ('closed' and inactive), the other being opened up as a result of the movement of the surface loops ('open' and active). This 'open/closed' structure can flip reversibly to 'closed/open,' thereby accounting for alternating active-site activity (half-sites reactivity or total negative cooperativity) exhibited by the enzyme.

A proton is removed from ThDP in the open active site (facilitated by Glu59α) to generate a carbanion at C2 on the thiazolium ring, thereby activating the ThDP to initiate nucleophilic attack on the 2-oxo acid (Figure 9.2.1C). The thiazolium ring is deeply buried at the bottom of a long funnel, down which the lipoyl-lysine side chain has to extend to become reductively acetylated (see Figure 9.12). The two ThDP molecules in the tetramer are at opposite ends of an interior tunnel lined with charged side chains and water molecules (Figure 9.2.1D). This is thought to be the channel of active-site communication, perhaps as a 'proton wire'; the proton that was removed to generate the active ThDP carbanion in what had been the inactive site enters the wire and then, by means of a **Grotthuss-like mechanism** (Figure 9.2.1E), a proton is rapidly discharged at the other end to protonate (and thus deactivate) the ThDP carbanion in the previously active site.

Figure 9.2.1 Structure and mechanism of E1. (A) The PSBD is in green. The two E1α and E1β chains are in complementary colors (E1α/E1α*, E1β/E1β*). The two ThDP are in yellow and the three Mg^{2+} in brown. Surface loops 203–212 and 267–290 in both α subunits are in cyan. One active site is 'closed'; the other is 'open.' (B) Close-up of the two active sites. E1α* and E1β* make the closed active site. E1α and E1β make the open active site. Color code as in (A). (C) Reaction mechanism of E1. Glu59α helps activate the ThDP cofactor by abstracting a proton from N1 and creating a carbanion at C2. His128β is the general acid/base 2. LD, E2 lipoyl domain. Only the dithiolane ring of the lipoyl-lysine residue is depicted. (D) Left, top view of the

E1/PSBD. Right, the same except that the protein has been omitted to reveal the two ThDP molecules (yellow stick representation), the three Mg^{2+} (brown spheres), and the collection of charged amino acid side chains that line the aqueous tunnel through the protein between the two ThDP molecules. (E) A Grotthuss mechanism for proton transfer (hopping) along a chain of water molecules. A proton enters on the left-hand side to create a hydronium ion (H$_3$O$^+$) and, by means of a succession of H-bond rearrangements, a hydronium ion on the right-hand side is created, from which a proton can be discharged. This accounts for the anomalously high rate of diffusion of protons through water. (PDB 1W85)

Box 9.2 *continued*

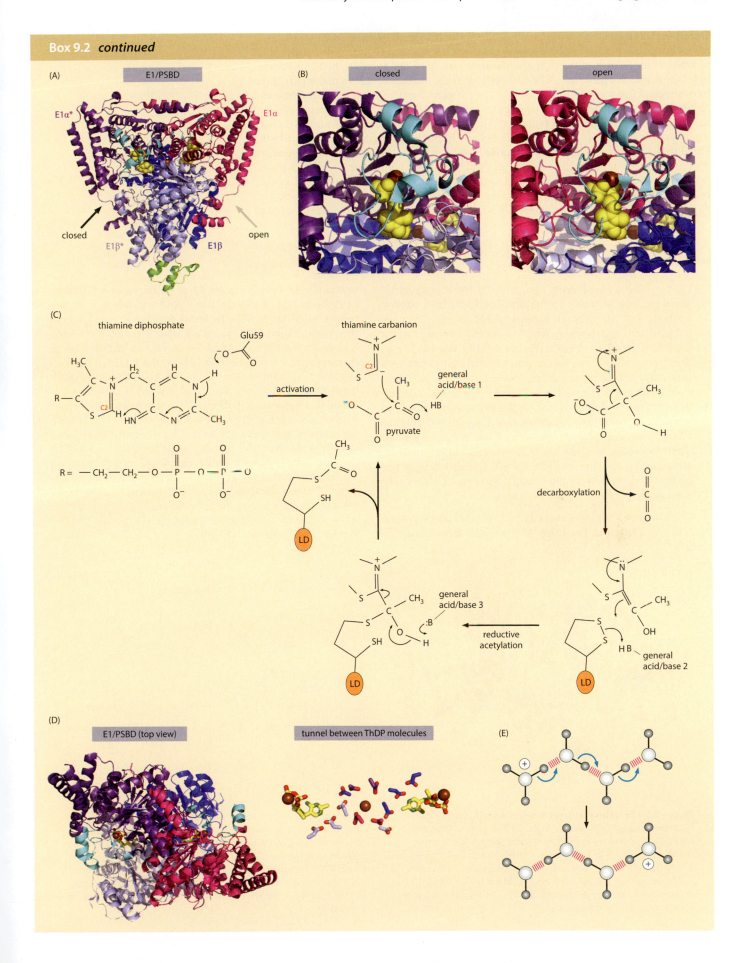

(A) E1/PSBD

E1α*   E1α

closed   open

E1β*   E1β

(B) closed   open

(C) thiamine diphosphate   thiamine carbanion

Glu59

activation   general acid/base 1

decarboxylation

pyruvate

reductive acetylation   general acid/base 2

general acid/base 3

(D) E1/PSBD (top view)   tunnel between ThDP molecules   (E)

Decarboxylation reactions are generally exergonic and difficult to reverse—hence the need for carboxylations catalyzed by biotin-dependent enzymes that require concomitant ATP hydrolysis (see below). In the oxidative decarboxylation of, say, pyruvate, by E1, the decarboxylation does not generate a free carboxylic acid but instead captures a major part of the free energy of the reaction in the form of the thioester bond between the acyl group and reduced lipoic acid, ready for acyl transfer to coenzyme A. This is crucial to energy-yielding reactions in the citric acid cycle (Chapter 15). A substrate for E1 the size of a lipoyl domain would diffuse much more slowly than free lipoic acid, but tethering the lipoyl domain to the PSBD in the E2 chain by means of a flexible linker keeps it within easy reach of the active sites of all three enzymes: E1 bound to its PSBD, E2 in the core assembly, and E3 bound to another PSBD. E2 and E3 (**Box 9.3**) catalyze readily reversible reactions and neither requires the lipoic acid to be attached to a lipoyl domain, but the vast increase in k_{cat}/K_m for E1 with the lipoyl group on the partner lipoyl domain greatly speeds up the overall reaction

---

Box 9.3 Structure and mechanism of E3 in a 2-OADH complex

E3 is a dimeric flavoprotein that catalyzes the reoxidation of the dihydrolipoyl group to the disulfide form. Each subunit contains four domains: an N-terminal FAD-binding domain, a Rossmann fold binding NAD⁺, a central domain, and a C-terminal interface domain (**Figure 9.3.1A**). The binding to the PSBD is largely due to electrostatic interactions between essentially the same set of positively charged side chains from PSBD as are involved in binding E1 (hence the competitive binding) and negatively charged side chains from both interface domains in the E3 dimer, together with hydrophobic interactions and H-bonds. However, there are no entrained water molecules as there are in the E1/PSBD interface. Thus, although the binding in E1/PSBD and E3/PSBD is tight (for both, $\Delta G \approx -50$ kJ/mol), the entropic and enthalpic contributions are very different in the two cases.

The E3 active sites lie at the subunit interfaces and therefore only the dimer can be active. A small redox-active disulfide

bridge (Cys42–Cys47) in each E2 chain becomes reduced by Lip(SH)₂; as part of this reaction Cys42 engages in a temporary mixed disulfide with the lipoyl group, and His446′ (from the opposite subunit) picks up a proton (Figure 9.3.1B). LipS₂ is released and the thiolate of Cys47 makes a charge-transfer complex with the neighboring FAD on the *si* face of its isoalloxazine ring. (The notation *si* and *re* is used to distinguish between the 'front' and 'back' faces when a substituent is added to a trigonal sp^2-hybridized atom in a planar molecule, creating a chiral center.) NAD⁺ can then enter its binding site on the *re* face of the isoalloxazine ring. The FAD becomes reduced to FADH₂, which in turn reduces NAD⁺ to NADH and H⁺. The disulfide bond between Cys42 and Cys47 is reformed but has to adopt a strained conformation with abnormal torsion angles. This may facilitate its opening, helping to explain how the enzyme can catalyze the reduction of NAD⁺ from Lip(SH)₂ when the redox potential for LipS₂/Lip(SH)₂ is −280 mV and that for NAD⁺/NADH is −310 mV.

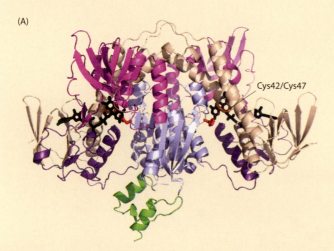

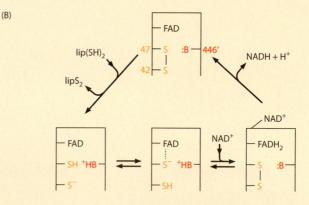

Figure 9.3.1 Structure and mechanism of E3. (A) Structure of *G. stearothermophilus* E3 bound to its E2 PSBD (green). Wheat, FAD-binding domain (residues 7–147); magenta, NAD⁺-binding domain (residues 148–276); purple, central domain (residues 277–345); light blue, interface domain (residues 346–461); green, PSBD; orange, Cys42/Cys47; red, His446′; black, FAD. (PDB 1EBD) (B) Schematic reaction mechanism. :B is His446′ from the opposite subunit (numbered in red) to Cys42/Cys47 (numbered in yellow).

Cys42 initiates the dithiol/disulfide interchange with lip(SH)₂. LipS2 is released and Cys47 makes a thiolate-flavin charge-transfer complex with the FAD. NAD⁺ is now able to bind, the FAD is reduced to FADH₂, and the Cys42-Cys47 disulfide is restored. Finally, the bound NAD⁺ is reduced at the expense of FADH₂ and released as NADH. (Adapted from N. Hopkins and C.H. Jr. Williams, *Biochemistry* 34:11757–11765, 1995. With permission from American Chemical Society.)

and provides a highly effective means of substrate channeling. It has also been suggested that free lipoic acid, like the indole intermediate in tryptophan synthase, could be lost from the cell by passive diffusion through the mitochondrial and/or cell membranes; covalently linking it to a protein domain prevents that, too.

Multiple active sites are coupled irrespective of their geometric arrangement

The lipoyl-lysine residue on a lipoyl domain is an essential intermediate in the overall reaction mechanism of a PDH complex (see Figure 9.4A). However, if lipoyl groups are excised by hydrolysis of the lipoyl-lysine amide bond (a reaction catalyzed by a separate enzyme, termed a lipoamidase), the catalytic activity of the complex does not decrease in direct proportion as expected; instead it remains high while many of the lipoyl groups are being lost. In the absence of CoA, the lipoyl groups on the E2 component in a complex can be reductively acetylated but are unable to undergo any further reaction. However, if in the absence of CoA the E1 components in the complex are progressively inhibited by exposure to thiamin thiothiazolone diphosphate (TTTDP), a catalytically inert molecule resembling thiamin diphosphate and which binds very tightly, the reductive acetylation of the 60 lipoyl groups on the E2 component remains close to maximal until almost all the E1 components have been inhibited (**Figure 9.6A**). How can this be?

The apparent surplus of lipoyl groups in the complex is attributable to the flexible linkers that tether the lipoyl domains to the E2 core. The likely length of the lipoyl domain plus its linker region (~120 Å, much longer than the 15 Å of the lipoyl-lysine side chain) should enable any given lipoyl group to reach multiple E1 active sites and, reciprocally, to permit many lipoyl domains to visit the active site of any one E1 component. However, to account for the reductive acetylation of all the lipoyl groups on the E2 core when only one or a very few of the bound E1 components remain active, a reductively acetylated lipoyl group must be able to pass on its substrate to a neighboring oxidized lipoyl group, which can then do the same with other nearby lipoyl groups in the E2 core (Figure 9.6B). Kinetic analysis shows that E1 catalyzes the slowest reaction in the PDH complex and that subsequent transacylation reactions are not rate-limiting. Thus, any lipoyl group reductively acetylated by any E1 bound to a PSBD on the E2 core is able to transfer its acyl group to CoA and cause a reduced lipoyl group to present itself for reoxidation by an E3 bound to another PSBD, however remote, on the E2 core. This remarkable system of active-site coupling is thought to operate in all 2-OADH complexes, a property only made possible by their highly aggregated structures.

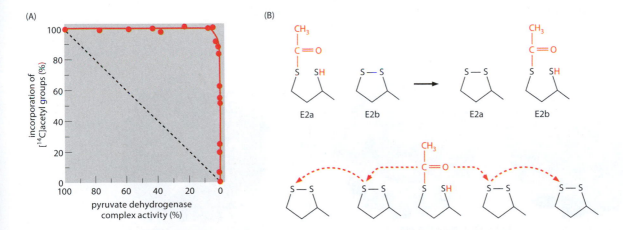

Figure 9.6 Active-site coupling by transacylation reactions in the E2 core. (A) The overall activity of the *G. stearothermophilus* PDH complex falls in direct proportion to the number of its E1 components inhibited by binding TTTDP. The reductive acetylation of the lipoyl groups on the E2 core measured in the absence of CoA would be expected to fall similarly (dashed line). However, close to the full 60 lipoyl groups per E2 core can be reductively acetylated until almost all the E1 components have been inactivated (red data points). (B) Top, a reductively acetylated lipoyl group (E2a, only the dithiolane ring is shown for clarity) can reductively acetylate a neighboring oxidized lipoyl group (E2b). It becomes reoxidized and is able to serve as a substrate for E1 again. Bottom, further transacetylation reactions, depicted as dotted arrows, can extend the process with other lipoyl groups around the E2 core. (Adapted from L.C. Packman, C.J. Stanley and R.N. Perham, *Biochem. J.* 213:331–338, 1983.)

The intact PDH complex is built of spherical protein shells

An early problem in understanding active-site coupling in PDH complexes was that FRET experiments (Section 1.5) between donor and acceptor fluorophores in the active sites of E1 and E3 indicated that they were further apart than the maximal ~30 Å spanned by a single lipoyl-lysine swinging arm in E2. The recognition that a lipoyl domain on the end of a long flexible linker region is the true substrate overcame this problem in part. The structural context comes from cryo-electron microscopy (see **Methods**) images, in which the E1 and E3 components of the *G. stearothermophilus* PDH complex form a spherical shell of protein subunits ~75–90 Å above the surface of the inner (acetyltransferase domain) E2 core (**Figure 9.7A,B**). The E1 or E3 components are each bound to a PSBD and must be held in their peripheral positions by the inner linker between the acetyltransferase domain and the PSBD. This inner linker (~35 aa) is long enough to cross the gap only once, leaving the lipoyl domain to dangle from the PSBD on the end of the highly flexible linker (35–40 aa) that tethers it to the PSBD. A lipoyl domain has to move backward and forward rapidly (the PDH complex turnover number is ~100/s) across the gap to visit the active sites of E1 and E3 in the outer protein shell and an E2 active site in the acetyltransferase domains that make up the inner core. This feature of the mechanism is presumably driven by the thermal energy of Brownian motion. The 2-oxo acid substrate and coenzyme NAD$^+$ are acted upon by the peripheral E1 and E3 subunits, and CoA and acetyl-CoA must pass through the gaps in the outer shell of E1 and E3 to reach E2. The orientation of the active sites in the E2 subunits (see Box 9.1) requires that CoA and acetyl-CoA pass through the gaps in the outer shell of E1 and E3 and respectively enter and leave the cavity in E2 through the big (50 Å) 'holes' on its 5-fold faces (see Figure 9.4 and Figure 9.7C). The icosahedral cores of the human and yeast PDH complexes have been reported to show some variation in particle diameter, and it has been suggested that they may undergo reversible expansion and contraction about a mean size, a kind of molecular 'breathing' (see Section 1.5), which may also contribute to the mechanism of those complexes by varying the distances between active sites.

The restricted space between the outer spherical shell of E1 and E3 subunits and the surface of the inner E2 acetyltransferase core means that the lipoyl domains, each flexibly attached to a PSBD to which an E1 or E3 subunit can be bound, are present in a local concentration of >1 mM. This is substantially greater than the K_m (<20 μM) measured for a lipoyl domain as substrate for E1 in the *E. coli* PDH complex. Thus E1, which has the lowest (rate-determining) turnover number of the three enzymes in the complex, is therefore working at its V_{max} at all times. In addition, because there is competition between E1 and E3 for binding to the 60 PSBDs of the E2 core, the stoichiometry of E1:E2:E3 is not fixed, nor is the assembly pathway uniquely ordered. However, as described above, the active sites of E1 and E3 bound to E2 will be coupled at all stages of assembly. Moreover, in principle, a complex will have many structural variants depending on how many, and where, copies of E1 and E3

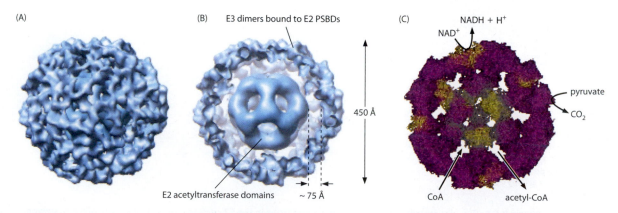

Figure 9.7 Model of the *G. stearothermophilus* PDH complex derived from cryo-EM images. (A) Outer surface of an E3E2 subcomplex assembled *in vitro* from a full E2 (60-mer) and 60 homodimers of E3, based on cryo-EM reconstructions. (B) A cross section of (A), showing the dodecahedral inner E2 core of 60 E2 acetyltransferase domains. A large (~75 Å) gap separates the outer shell of 60 E3 components from the surface of the inner E2 core. This gap is a little greater, ~90 Å, if the outer shell is composed of 60 E1 heterotetramers. The E3 dimers or E1 heterotetramers are bound to the PSBDs of the E2 component. (C) A model of the full PDH complex with ~50 E1 heterotetramers (purple) and ~10 E3 dimers (yellow), distributed over the surface of the E2 core (gray). This represents the typical composition of the native PDH complex. (From J.L.S. Milne et al., *J. Biol. Chem.* 281:4364–4370, 2006. With permission from American Society for Biochemistry and Molecular Biology.)

individually are bound; the total (E1 + E3) cannot exceed 60, although full occupancy may be difficult to achieve when both E1 and E3 are present. Maximal PDH complex activity is observed with ~42–48 E1 and ~6–12 E3 bound per E2 core, and the ability to bind more copies of the catalytically slow E1 than of the much faster E3, as often observed in purified complexes, may have been evolutionarily advantageous.

Eukaryotic PDH complexes have additional components and are subject to regulation by reversible phosphorylation

The PDH complexes of eukaryotes (found in the mitochondria) are similar to that of *G. stearothermophilus*, but with important differences. The E2 chain in their icosahedral cores contains two tandem lipoyl domains in its N-terminal region and the E3 component is bound not to the PSBD of the E2 chain but to a very similar E3-binding domain (E3BD) in a different E2-like polypeptide chain (E3BP) that has only one lipoyl domain and whose acetyltransferase domain is catalytically inactive. Exactly how these E3BPs insert into the 60-mer icosahedral E2 core is uncertain: arguments in favor of 12 E3BP:48 E2 and 20 E3BP:40 E2 have been advanced. In addition, the PDH complexes of eukaryotes are subject to control by phosphorylation and dephosphorylation reactions responsive to the metabolic state of the cell. A kinase bound to the complex phosphorylates specific target serine residues in the α subunits of the E1 heterotetramers, which inhibits the decarboxylation reaction and thus the activity of the whole complex (see Box 9.1); phosphorylation of E1 therefore regulates the formation of acetyl-CoA, the end product of glycolysis. The activity is restored by dephosphorylation of the serine residues by a dedicated phosphatase. The kinase and phosphatase are potentially valuable targets for therapeutic intervention in diabetes, heart ischemia, and cancer.

Only one or two copies of the kinase (a dimer) bound to the PDH complex are able to phosphorylate multiple copies of E1. They do not do so by dissociation and reassociation with the complex. Rather, they appear to migrate round the complex in search of E1 targets, probably mediated by a 'hand-over-hand' movement of the kinase from one mobile lipoyl domain to another, echoing the participation of lipoyl domains in the internal transacylation reactions described above (see Figure 9.6B). This is a further remarkable feat of dynamic protein/protein interaction within a catalytic machine.

Some 2-OADH complexes are based on octahedral rather than icosahedral symmetry

The PDH complexes of Gram-negative bacteria such as *E. coli*, and the OGDH complexes of all organisms, usually have E2 cores of octahedral rather than icosahedral symmetry. The acyltransferase domains of 24 E2 chains assemble as eight trimer clusters at the corners of a cube (see Section 1.4). They are not subject to control by phosphorylation, and in some instances the E2 chains have up to three lipoyl domains in tandem array in the N-terminal region (**Figure 9.8**). In the *E. coli* PDH complex, two of the three lipoyl domains can be deleted *in vitro* without serious detriment to the catalytic activity, and the 'extra' lipoyl domains may have evolved as a means of extending the reach of the outermost lipoyl domain in the complex. In octahedral complexes, the E1 components are usually homodimers (a fusion of the β and α components of the heterotetrameric versions) and bind predominantly to the acyltransferase domains that make up the E2 inner core, rather than competing with the E3 components for binding peripherally to the E2 PSBDs. However,

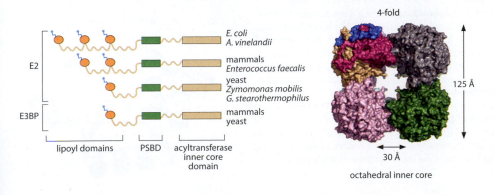

Figure 9.8 Structures of E2 chains of other PDH complexes. The E2 chains of some PDH complexes have one, two, or three lipoyl domains (orange) depending on the source. The lipoyl-lysine side chains are depicted in blue. In E2 cores of octahedral symmetry, 24 acetyltransferase domains aggregate as eight trimer clusters at the corners of a cube (Chapter 1.4). The structure shown is that of the octahedral inner acetyltransferase core (0.7 MDa) of the *A. vinelandii* PDH complex viewed across a 4-fold face; in one trimer cluster, the chains are colored magenta, gold, and blue as in Figure 9.4, whereas the other trimers (only three are visible in this view) are depicted in separate colors. E3BP is the E3-binding protein found as part of the E2 core of icosahedral eukaryotic PDH complexes. (PDB 1EAA)

despite their octahedral E2 cores, BCDH complexes have $\alpha_2\beta_2$ E1s and these are also subject to regulation by reversible phosphorylation. Taken overall, there is no simple correlation between E2 symmetry and the number of lipoyl domains or subunit structure of the E1 components.

There was no evidence of 2-OADH complexes in archaea until the recent discovery of the relevant genes in aerobic archaea and the isolation of a complex from the thermophilic archaeon *Thermoplasma acidophilum*. This complex displays activity against branched-chain 2-oxo acids and, to a lesser extent, against pyruvate, and is remarkable in that its E2 core has 42 polypeptide chains. Thus the assembly does not exhibit regular cubic point group symmetry (24 or 60 E2 chains, as described in Section 1.4). Instead, the E2 trimers form an oblate hollow spheroid with a maximum diameter of 220 Å and non-equivalent interactions. Other variations on the classic arrangements of E1, E2, and E3 components are emerging from the study of other organisms but it is likely that, in most critical respects, the octahedral complexes and these other variants share the key mechanistic features of the icosahedral *G. stearothermophilus* PDH complex.

A lipoylated protein is part of a glycine decarboxylase system in serine biosynthesis

An important metabolic conversion, particularly in plants, is the biosynthesis of serine from glycine catalyzed by a series of enzymes known as the glycine decarboxylase (or glycine cleavage) system. In animals, deficiencies lead to hypotonia and seizures shortly after birth (*non-ketotic glycinemia*). The reactions are analogous to those of the PDH complex (**Figure 9.9A**), but with critical differences. A decarboxylase (P), requiring pyridoxal phosphate (PLP) as cofactor, catalyzes glycine decarboxylation and the reductive aminomethylation of a lipoyl-lysine residue in the H-protein. This decarboxylase is mechanistically different from the E1 of the PDH complex. The aminomethylated H-protein interacts with a transferase (T) that uses 5,6,7,8-tetrahydrofolate as cofactor (acting analogously to E2 and CoA), releasing NH_3 and generating N^5,N^{10}-methylene-tetrahydrofolate. This is used by serine hydroxymethyl transferase (SHMT) in a C_1 transfer to make serine from another molecule of glycine. Finally, the dihydrolipoyl group on the H-protein is reoxidized by dihydrolipoyl dehydrogenase (L), equivalent to the E3 component of the PDH complex.

A major difference is that the enzymes in the glycine decarboxylase system do not form an overall complex, and they interact only transiently during this sequence of reactions. The H-protein (131 aa) is homologous to the lipoyl domains (~80 aa) of 2-OADH complexes but has some important changes, including an α helix between residues 29 and 35 and another helix at the C-terminal end. In the crystal structure of the oxidized form, the lipoyl-lysine side chain is found nestling against the nearby surface of the domain; it is not well ordered but is not invisible as a freely rotating classical swinging arm would be (Figure 9.9B). Moreover, in the reductively aminomethylated form, the swinging arm has moved in position: the aminomethylated dithiolane ring has become tightly sequestered in a surface cleft, of which the internal α helix (residues 29–35) forms one side. The relatively labile aminomethyl-sulfur bond is thereby protected from the surrounding solvent. Given that the H-protein is not part of a multienzyme complex and has to migrate through the cell until

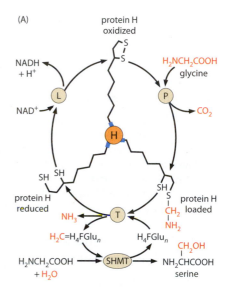

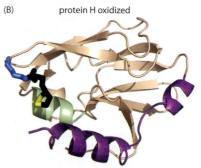

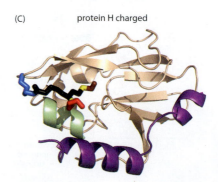

Figure 9.9 The glycine decarboxylase (cleavage) system. (A) H, lipoylated H-protein; P, PLP-dependent glycine decarboxylase; T, a tetrahydrofolate-dependent transferase; SHMT, a PLP-dependent serine hydroxymethyl transferase; H_4FGlu_n, 5,6,7,8-tetrahydrofolate, where *n* is the number (1–6) of Glu residues in the polyglutamate tail of the coenzyme. The lipoyl-lysine side chain of protein H is a substrate for proteins P, T, and L in turn. (B) Structure of the oxidized H-protein from pea leaf mitochondria. The side chain of Lys63 is in blue and the oxidized lipoyl group is in black (its disulfide in yellow). Two helical regions, residues 29–35 (green) and at the C-terminal end (purple), highlight some structural differences from the homologous lipoyl domain of 2-OADH complexes. (C) Structure of the aminomethylated H-protein. The S8 of the lipoyl group carries the aminomethyl group (brown), and the -SH of S6 is linked to a molecule of 2-mercaptoethanol (red) derived from the crystallization. After its reductive aminomethylation, the lipoyl-lysine arm is bound differently, having entered a cleft generated by a structural rearrangement in the H-protein. (PDB 1HPC, 1HTP)

it encounters the transferase T protein, the intermediate requires special protection. The aminomethyl-lipoyl group is thus a 'hot potato' intermediate (Section 9.2) and the transferase T must be able to trigger release of the lipoyl-lysine residue from its cleft and permit it to enter the transferase active site safely and deliver its cargo.

Biotin-dependent carboxylases also have a swinging arm mechanism

Carboxylation reactions are endergonic (energy-requiring) and difficult to catalyze owing to the stability and lack of reactivity of CO_2. In the cell, a family of multifunctional enzyme complexes makes use of a coupled hydrolysis of ATP to overcome the unfavorable free energy change. Like 2-OADH complexes, they use a swinging arm as part of the mechanism, in this instance a biotinyl-lysine side chain. Examples are acetyl-CoA carboxylase, which catalyzes the synthesis of malonyl-CoA from acetyl-CoA, and propionyl-CoA carboxylase, which converts propionyl-CoA into methylmalonyl-CoA. These are crucial substrates in fatty acid biosynthesis and degradation respectively, which are catalyzed by yet another family of multienzyme complex described in Section 9.5. A third biotin-dependent carboxylase, pyruvate carboxylase, catalyzes the conversion of pyruvate to oxaloacetate, an essential citric acid cycle intermediate and a starting point in gluconeogenesis.

Pyruvate carboxylase (PC) has many features in common with a 2-OADH complex. It usually functions as a tetramer and, typically of biotin-dependent carboxylations, the overall reaction takes place in two steps (**Figure 9.10A**): ATP-dependent carboxylation of a

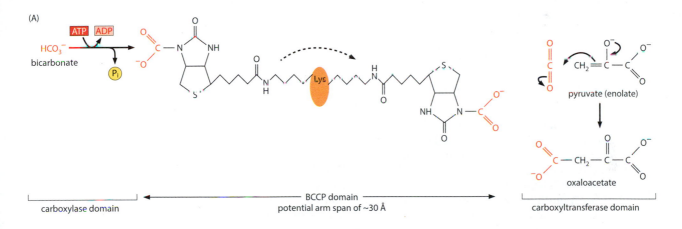

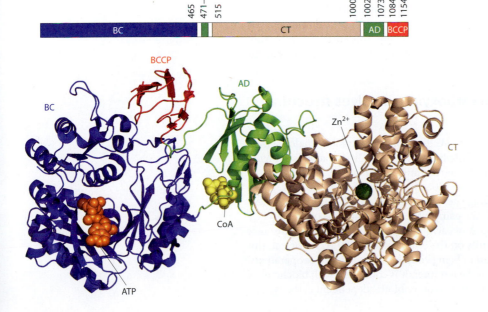

Figure 9.10 Biotinyl-lysine swinging arm in pyruvate carboxylase. (A) In an ATP-dependent reaction at the active site of a BC domain, bicarbonate is stripped of water, and the CO_2 is transiently captured as carboxyphosphate. This is the same reactive intermediate as in carbamoyl phosphate synthase (Figure 9.3) but here it carboxylates the N1 position of a biotin in a biotinyl-lysine swinging arm on the BCCP domain, with the release of P_i. The swinging arm then moves to the active site of a CT domain, where the CO_2 is released and reacts with the enolate form of pyruvate to generate oxaloacetate. (B) Structure of one of the subunits of the tetrameric PC of *R. etli*. BC, biotin carboxylase; BCCP, biotin carboxy carrier protein; AD, allosteric domain; CT, carboxyltransferase domain. CoA indicates where ethyl-CoA, an analog of acetyl-CoA, binds. The Zn^{2+} (green sphere) is at the CT active site. (PDB 2QF7)

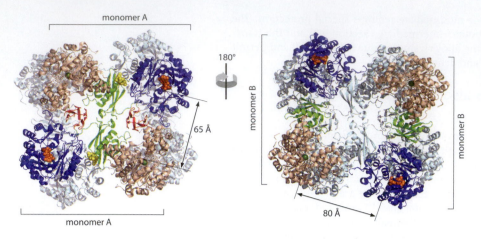

Figure 9.11 Movement of the BCCP domain between BC and CT active sites in PC. (A) Left, the front face of the *R. etli* tetramer showing two A monomers. The two B monomers on the back face are in pale blue. The shortest distance between ATP in a BC active site and Zn²⁺ in a CT active site on the top face is 65 Å, between BC and CT active sites in opposing monomers (arrow). Right, a 180° rotation of the tetramer about the vertical axis to show the back face of the tetramer. Color coding is as in (A) but reversed, with the A monomers at the back now in pale blue. The BCCP domain in B monomers is disordered and could not be modeled. There is a different orientation of the BC domain, and the distance between ATP in the BC active site and Zn²⁺ in the CT active site of the opposing monomer has increased to 80 Å. (PDB 2QF7)

biotinyl-lysine residue by a biotin carboxylase (BC) and subsequent transfer of the carboxy group to a second substrate by a carboxyltransferase (CT). The enzyme from *Rhizobium etli* is a dimer of dimers; each polypeptide chain (130 kDa) contains an N-terminal BC domain followed by a CT domain and a biotin carboxy carrier protein (BCCP) domain. The latter houses the biotinyl-lysine residue (Figure 9.10B). The structure of the BCCP domain (~80 aa) closely resembles that of the lipoyl domain, with its biotinyl-lysine residue protruding from a β turn between β strands 4 and 5. Moreover, the BCCP domain is attached to the C-terminal end of the CT domain by a long (~25 aa) flexible linker, which is unusually rich in Ala/Pro residues and resembles the linker that joins the lipoyl domain to the PSBD in the E2 chain of the PDH complex (see Figure 9.4). The CT domain is a TIM barrel, with a conserved Zn²⁺ in the active site. Between the CT and BC domains is an additional domain that comprises elements of the polypeptide chain between the BC and CT domains and the CT and BCCP domains. This domain has been termed the 'allosteric domain' (AD) because it is the binding site for ethyl-CoA, an analog of acetyl-CoA that is an allosteric activator of PC.

In the crystal structure, the dimers in the tetramer are asymmetric (**Figure 9.11**). Each monomer A is complete but each monomer B has a disordered BCCP domain and there are differences in the position and orientation of the BC domain. The shortest distance between the active site of a BC domain, where ATP is bound, and that of a CT domain, where the Zn²⁺ is located, is 65 Å. This is between BC and CT active sites in opposing A-type subunits in a dimer. The distance is much greater than the ~30 Å that can be spanned by a swinging arm of biotinyl-lysine alone, and the BCCP domain must therefore also move between the BC domain of its own monomer A and the CT domain of its opposing monomer A. This is made possible by conformational flexibility in the linker region that connects the BCCP domain to the CT domain, established from NMR experiments (see below). Free biotin is a poor substrate for the BC and CT components of the closely related acetyl-CoA carboxylase of *E. coli*, but attachment of the biotin to the biotinyl domain increases the catalytic efficiency (k_{cat}/K_m) greatly: 8000-fold and 2000-fold respectively. This ensures correct substrate channeling of carboxylated biotin, adding to the striking structural and mechanistic similarities to the lipoyl domain in 2-OADH complexes described above.

Allosteric regulation of pyruvate carboxylase involves structural rearrangements

PC is subject to allosteric activation by acetyl-CoA. The non-hydrolyzable analog, ethyl-CoA, binds in the central AD domain between the BC, CT, and BCCP domains (see Figure 9.10B) but only to the A subunit in each asymmetric dimer (see Figure 9.11). Compared with the B subunit, there is a 40° rotation and a translocation of nearly 40 Å in the BC active site, as measured from the ethyl-CoA binding site. This suggests that binding acetyl-CoA could activate PC by pushing the two active-site pairs in the A-type subunits on the top face of the tetramer closer together, to within 65 Å of each other, while increasing the distance between the active-site pairs in the B subunits on the bottom face to 80 Å. If correct, this would be a rare example of allosteric activation being coupled with negative cooperativity, only two active sites being fully functional in the tetramer. It is consistent with biochemical and kinetic evidence of half-sites reactivity in this and other biotin-dependent carboxylases.

(A)   PDH complex: AAPAAAPAKQEAAAPAPAAKAEAQAAAPAAKA
       acetyl-CoA carboxylase: AAPMMQQPAQSNAAAPATVPSMEAPAAA

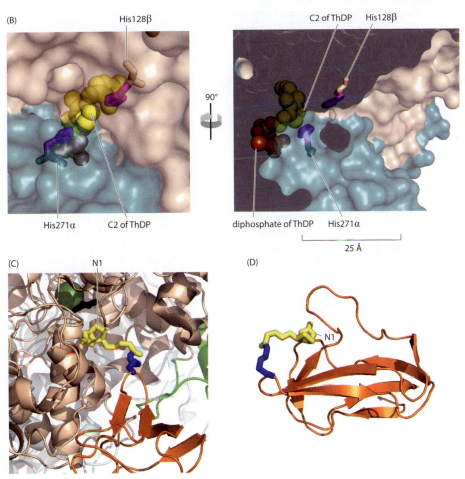

Figure 9.12 Flexibly tethered lipoyl and biotinyl domains enter enzyme active sites. (A) Amino acid sequences of flexible linker regions that tether a lipoyl domain in the *E. coli* PDH complex and the biotinyl domain in BCCP component of *E. coli* ACC. (B) Two views of an active site of the $\alpha_2\beta_2$ E1 component of the *G. stearothermophilus* PDH complex, in surface representation. Left, looking down a funnel-shaped entrance to the deeply buried ThDP (yellow) and Mg^{2+} (brown) between the α subunit (teal) and β subunit (wheat). Key residues, His128β and His271α, are indicated, as is the C2 atom (site of carbanion formation) in ThDP. (PDB 1W85) Right, side view showing how the entrance funnel widens toward the outside. At far right is a ribbon diagram of a lipoyl domain at the same scale. To reach the ThDP, the lipoyl-lysine side chain (yellow and blue) would clearly need be fully extended and the protruding β turn to which it is attached must also enter the funnel. Residues in the loop (green) between β strands 1 and 2 and in the lipoyl-lysine β turn make specific transient contact with E1 during reductive acetylation of the lipoyl group. (PDB 1LAB) (C) The biotinyl-lysine side chain (yellow and blue) in the protruding β turn of the biotinyl domain (orange) from monomer 4 of *S. aureus* PC inserted in extended form into the active site of the CT domain (wheat) of monomer 3. The N1 of the biotin is close to a substrate pyruvate (black) and metal ion (green) in the CT active site. (PDB 2QF7) (D) The biotinyl domain of the BCCP component of *E. coli* ACC. The biotinyl-lysine side chain (yellow and blue) is held tightly with the N1 of the biotin in contact with the additional large surface loop that extends out between β strands 2 and 3. (PDB 1BDO)

Some important differences have been reported for the PC from *Staphylococcus aureus* and humans. These include the relative positioning of the domains and an overall symmetrical structure for the tetramer apart from the disposition of the BCCP domains, but they may be related to effects of acetyl-CoA binding in the *R. etli* enzyme. In the *S. aureus* enzyme, one of the BCCP domains is found with its biotinyl group inserted into the active site of the CT domain of a neighboring subunit (**Figure 9.12C**), consistent with the need for two chains to partner in catalysis. Acetyl-CoA carboxylases (ACCs) show that biotin-dependent complexes can be constructed in multiple ways, again like the 2-OADH complexes. In eukaryotes, ACC is a dimer rather than a tetramer and each subunit contains the BC, CT, and BCCP activities fused together. In contrast, in *E. coli* the BC, CT, and BCCP activities reside in separate polypeptide chains and the ACC is a heterohexamer made up of two copies of each; however, it readily falls apart *in vitro* into CT and a BC–BCCP subcomplex. Despite these variations, the key features of the PC structure and mechanism are likely to be shared by all biotin-dependent carboxylases.

Swinging arms extend from flexibly tethered lipoyl and biotinyl domains to enter active sites

Lipoyl domains (see Figure 9.4) and biotinyl domains (see Figure 9.10) must migrate between active sites in their respective complexes as part of the catalytic activity. In the BCCP (156 aa) of *E. coli* ACC, the biotinyl domain is tethered by a long (~30 aa) linker. It is Ala/Pro-rich and resembles the flexible linkers in 2-OADH complexes (see Figure 9.12A). Indeed, BCCP mutants can be made in which the native linker is replaced by a corresponding linker from the E2 chain of the *E. coli* PDH complex, without loss of function. NMR studies of the PDH linkers (see Figure 9.5) indicate that all (>95%) of the Ala-Pro bonds are in the all-*trans* configuration; if this region were a random coil, the value would be 80–85%. This implies that the linkers are stiffened in places and, though flexible, are better regarded as extended and articulated, like lobster claws, rather than simply molecular string. Such a linker allows substantial relative movement of domains while preventing one from collapsing onto the other, and may also encourage the tethered domains to move in preferred trajectories and orientations between active sites.

The ThDP cofactor in the active site of E1 in a PDH complex is buried at the bottom of a long (~25 Å) funnel-shaped cavity (see Figure 9.12B). This is consistent with its activated form being a carbanion (see Box 9.2), which must be well protected from water. For the dithiolane ring at the distal end of the swinging arm to reach the cofactor, thereby to become reductively acylated, it must be inserted with the lipoyl-lysine side chain fully extended at the tip of the protruding β turn of the lipoyl domain. NMR studies of the transient interaction between the E1 component and the lipoyl domain for several different PDH complexes reveal a mosaic of interactions, among the most important sites on the domain being the β turn surrounding the lipoyl-lysine residue and the loop between β strands 1 and 2 that protrudes nearby (see Figure 9.12B). The specificity of these protein/protein interactions, allied perhaps with local conformational changes, is presumably what determines whether the lipoyl group is allowed entry to the active site of E1.

Similarly, in the crystal structure of *S. aureus* PC, the β hairpin and biotinyl-lysine extending from the biotinyl domain of one monomer have been captured inserted into the active site of the CT domain of a neighboring monomer (see Figure 9.12C). This is likely to be a physiologically relevant structure, because the N1 atom (the site of carboxylation) of the biotin is within 4.7 Å of the methyl group of a pyruvate (substrate) bound in the active site. Specificity in recognition of the biotinylated protein domain could account for the big rise in catalytic efficiency (k_{cat}/K_m) for biotin when it is attached to the biotinyl domain, as with the lipoyl domain in 2-OADH complexes. In some carboxylases, notably the *E. coli* ACC, the biotinyl domain differs from other biotinyl domains (and the homologous lipoyl domain) in that it has an additional (8 aa) loop region between β strands 2 and 3. The biotinyl-lysine nestles against this loop and does not swing freely (see Figure 9.12D). However, the N1 of the biotin ureido group abuts the protein and the swinging arm must be released for the N1 to be carboxylated at the BC active site and the subsequent carboxyl transfer at the CT site. Why the biotinyl-lysine arm in the BCCP protein of ACC should be immobilized in this way is unclear.

9.5   MULTIENZYME COMPLEXES WITH PHOSPHOPANTETHEINE SWINGING ARMS

Fatty acids are important to the cell in many guises, from components of membranes to major energy stores, and their biosynthesis in humans is an important target for therapeutic intervention in obesity, diabetes, and cardiovascular disorders. Fatty acid synthesis is based on the serial addition of C_2 units to an acyl chain in thioester linkage with a phosphopantetheine swinging arm, which is attached to a Ser residue in a small (~10 kDa) protein termed the **acyl carrier protein** (**ACP**). In the cells of animals and fungi, the enzymes responsible are found conscripted into giant complexes, fatty acid synthases (FASs), located in the cytosol and designated type I systems. In contrast, fatty acid synthesis in the smaller cells of bacteria (and in the mitochondria and chloroplasts of eukaryotes) is catalyzed by a set of separate monofunctional enzymes, designated type II. A special function in mitochondria is the generation of octanoyl chains that are the precursors of the lipoyl moieties of the mitochondrial 2-OADH complexes (Section 9.2). Fatty acid biosynthesis is a reductive endergonic process, requiring NADPH and, as substrates, acetyl-CoA and malonyl-CoA synthesized at the expense of ATP hydrolysis. This explains why in eukaryotic cells the type I synthases, responsible for the bulk of fatty acid synthesis, are located in the cytosol.

Fatty acid synthases come in two different forms

In fatty acid synthesis, the C_2 units are derived from malonyl-CoA and the process typically goes through up to eight or nine rounds of the reaction to produce the C_{16} or C_{18} fatty acids, palmitate or stearate. There are two kinds of type I FAS, one in animals (class II) and the other in yeasts and other fungi (class I). Their detailed reaction schemes are shown in **Figure 9.13**. In both classes, the process is initiated by the transfer of the acetyl group from

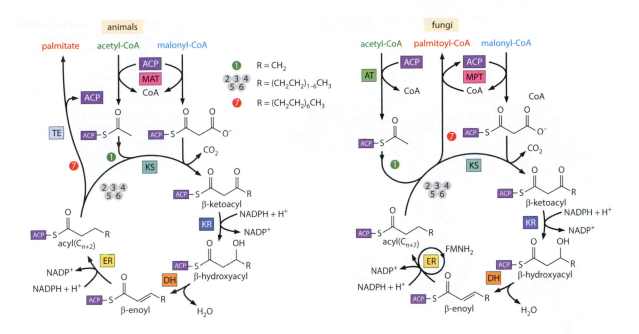

$$CH_3CH_2CH_2CH_2CH_2CH_2CH_2CH_2CH_2CH_2CH_2CH_2CH_2CH_2CH_2COO^-$$
palmitate

Figure 9.13 Reactions catalyzed by fatty acid synthases in animals and fungi. ACP, acyl carrier protein. The reactions catalyzed by β-ketoacyl-ACP synthase (KS), β-ketoacyl-ACP reductase (KR), β-hydroxyacyl-ACP dehydratase (DH), and β-enoyl-ACP reductase (ER) are common to both. In animal FAS, both loading reactions are catalyzed by malonyl/acetyl-CoA-ACP transacylase (MAT); release of product, predominantly palmitate, is catalyzed by thioesterase (TE). In the fungal FAS, acetyl group loading is catalyzed by acetyl-CoA-ACP transacetylase (AT), whereas malonyl group loading is catalyzed by malonyl/palmitoyl transferase (MPT), which also catalyzes release of product (less specifically palmitate) as acyl-CoA. Another difference is that the fungal ER requires FMN as cofactor. (Adapted from M. Leibundgut, T. Maier, S. Jenni and N. Ban, *Curr. Opin. Struct. Biol.* 18:714–725, 2008. With permission from Elsevier.)

acetyl-CoA to the thiol group of a phosphopantetheine swinging arm attached to the ACP domain in the multifunctional protein. Another acetyl-CoA is activated by carboxylating it to form malonyl-CoA, an ATP-dependent reaction catalyzed by a biotinylated ACC of the kind described above (Section 9.4), and the malonyl group is transferred to a second ACP. The system is now primed for fatty acid synthesis to begin.

A cycle of four reactions ensues, catalyzed by four enzymes in turn: a β-ketoacyl-ACP synthase (KS) condenses the acetyl-CoA with the malonyl-CoA to generate acetoacetyl-ACP with expulsion of the activating CO_2; β-ketoacyl-ACP reductase (KR) reduces the outermost >C=O group with NADPH as cofactor; β-hydroxyacyl-ACP dehydrase (DH) introduces a *trans* double bond by removing H_2O; and the double bond is reduced by β-enoyl-ACP reductase (ER) with NADPH again as cofactor. The alkyl chain has been elongated by a C_2 unit, and the product, butyryl-ACP, is then the substrate together with a fresh malonyl-ACP for a further round of the same four reactions. This process can be repeated up to seven times, at which point the C_{16} palmitoyl group is released from the ACP by an enzyme that hydrolyzes the thioester bond linking it to the phosphopantetheine swinging arm. Note that the fatty acid grows by the sequential addition of C_2 units derived from malonyl-CoA at its carboxyl end, the point of its thioester attachment to ACP. The strongly endergonic nature of the process is evident in that, to make palmitic acid, 14 NADPH and 7 malonyl-CoA are used, and each malonyl-CoA requires the hydrolysis of one ATP molecule in its synthesis from acetyl-CoA.

The animal and fungal FASs differ in the way in which the acetyl-CoA and malonyl-CoA are loaded onto ACP and in how the final product is released (see Figure 9.13). In animals, a bifunctional malonyl/acetyl-CoA-ACP transacylase (MAT) is responsible for loading both substrates, and a thioesterase (TE) hydrolyzes the fatty acyl-ACP bond on completion. In contrast, fungal FASs have a dedicated acetyl-CoA-ACP transacetylase (AT) to load the acetyl group, and a bifunctional malonyl/palmitoyl transferase (MPT) that both loads the malonyl-CoA and transfers the palmitoyl group from ACP to CoA for release as palmitoyl-CoA. The animal and fungal FASs also differ in the way in which the constituent enzyme activities are grouped in multifunctional polypeptide chains, and in their molecular architectures.

Animal FAS is a dimer of multifunctional polypeptide chains

The animal FAS (540 kDa) is a homodimer of giant (~2500 aa) polypeptide chains, each of which houses all the catalytic activities as individual domains (**Figure 9.14**). It forms a

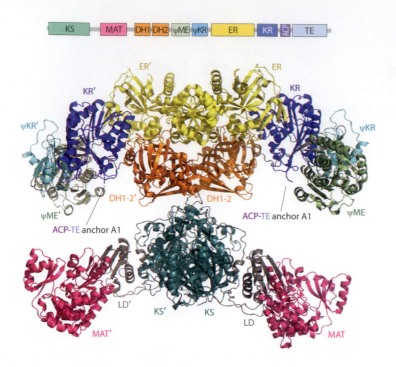

Figure 9.14 Structure of the porcine FAS. The abbreviations and color code are the same as in Figure 9.13. ψKR, a non-catalytic pseudo-KR domain; ψME, a non-catalytic domain resembling enzymes of the methyltransferase family. The ACP-TE domains are not visible but are attached to the C-terminal end of the KR domain (at the anchor point indicated) by flexible linkers. (PDB 2VZ9)

two-tier structure, the lower tier comprising the domains (MAT and KS) involved in loading the ACP and condensing the acyl-ACP and malonyl-ACP, whereas the upper tier contains those (KR, DH, and ER) that modify the β-carbon of the growing fatty acyl chain. The upper tier also contains two non-catalytic domains (ψKR and ψME, in which ψ stands for 'pseudo-'). The ψKR domain represents ~50% of the KR fold and has no NADP-binding site, but dimerizes with, and may help to stabilize, the catalytic KR domain. The ψME domain is homologous to the family of *S*-adenosyl-methionine-dependent methyltransferases but lacks a crucial active-site motif and may be an evolutionary relic of a common precursor of FAS and *polyketide synthases* (see below). The ACP and TE domains are not visible in the structure and are thought to be mobile, flexibly tethered by a linker region (12–14 aa, maximum length ~40 Å) at the C-terminal end of the KR domain and by another (23–26 aa, maximum length ~80 Å) between the ACP and TE domains.

The disposition of the two subunits in animal FAS creates two reaction compartments, left and right as depicted in Figure 9.14, each of which contains a full complement of catalytically active domains required for fatty acid synthesis. The arrangement of catalytic domains does not coincide with the order of their participation in the reaction cycle, and the active sites are found on the front and back faces of the complex. The length of its flexible retaining linker means that an ACP (tethered in the upper tier) is able to shuttle around and across the cleft between the tiers to deliver the substrate to each active site in its own compartment, and there is strong biochemical evidence that it can also interact with the KS and MAT domains (in the lower tier) of the neighboring compartment. For this to happen, there has to be a rapid swiveling motion of the upper tier relative to the lower tier around the 'waist' between them. This is perhaps reflected in the asymmetry in the two compartments visible in the crystal structure (see Figure 9.14) and is supported by EM images indicative of changed shapes. It has also been suggested that the two chambers may operate asynchronously, one being engaged in β-carbon modification while the other is working on chain elongation.

Fungal FAS is an α₆β₆ double-domed cage

The FAS of yeast and other fungi has a completely different architecture from that of animals. The complex (2.6 MDa) is composed of six α subunits (~210 kDa) and six β subunits (~230 kDa), with the constituent enzyme activities (see Figure 9.13) distributed between both types of subunit (**Figure 9.15A**). Unlike the animal FAS, there are numerous interdomain and intra-domain segments that contribute rigidity and define the arrangement of the catalytic domains. The six α subunits form a central disk, on either side of which is a dome formed by three β subunits (Figure 9.15B). The disk has D3 symmetry, three

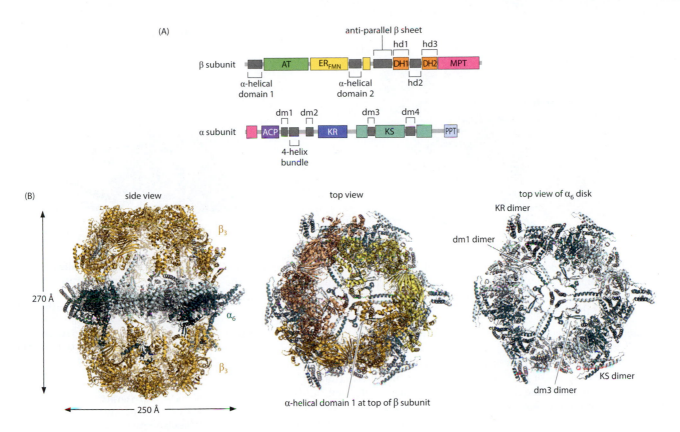

Figure 9.15 Structure of the fungal FAS. (A) Domain layout of the α and β subunits, using the same abbreviations and color code as in Figure 9.13. Non-catalytic domains unique to fungal FAS are depicted in gray; hd1, hd2, and hd3 represent 'hot-dog' domains within DH; dm1–4 are dimerization modules. PPT is phosphopantetheine transferase (Section 9.2; Figure 9.1). (B) The structure is that of the FAS from *Thermomyces lagunisosus*. Left, the α and β subunits are shown in single colors (teal and yellow respectively) for simplicity. Six α subunits constitute a central disk, three β subunits constitute an upper dome, and three a lower dome. Middle, in the top view, one of the three β subunits in the upper dome is in beige. The three copies of the N-terminal α-helical domain 1 generate the top of the dome. Right, the β subunits have been omitted to show the top view of the α₆ disk. The three α subunits interacting with the upper dome are in light teal; the three α subunits interacting with the lower dome are in darker teal. The positions of the KR and KS dimers and the dm1 and dm3 dimers that appear as spokes are indicated. (PDB 2UVB, 2UVC)

α subunits facing 'up' and three 'down,' whereas the three β subunits in each dome conform to C3 symmetry (for a description of symmetry, see Section 1.4). Thus, the disk divides the complex into two reaction chambers, each of which has three full sets of catalytic domains. There are numerous α/β subunit contacts where the β-subunit domes rest on the α-subunit disk, and each α subunit also projects its N-terminal 94 aa into the apposing dome where it makes contact with a β subunit, thereby creating an α/β functional unit. The α-subunit backbones are extensively intertwined, marked by homodimer interfaces mediated in places by four dimerization modules. These include what appear as three thin spokes between the disk hub and peripheral domains, which are made up of the dm1 module, and three thicker spokes formed from the dm3 module. This leaves six openings in the disk that connect the upper and lower chambers. The tip of each dome comprises three α-helical domain 1 segments located at the N-terminal ends of the three β subunits. The walls of the complex have open 'windows' in them, suitable for the entry and exit of small substrates and cofactors; in this they resemble the E2 cores of the 2-OADH complexes with their large openings on their 4-fold or 5-fold faces.

The structures of an α subunit from the disk and its partner β subunit in the upper dome are shown in **Figure 9.16**. The complementary parts of the MPT domain comprise the C-terminal end of the β subunit, which contains the catalytically important residues, and the N-terminal 94 aa of the α subunit. This is suggestive of a structural gene having been split at some point in evolution. There is no electron density for the α subunit between residues 95 and 324, a gap of ~60 Å between the end of the MPT fragment and the start of module dm1. This is where the ACP should be located, which makes it clear that the ACP must be flexibly

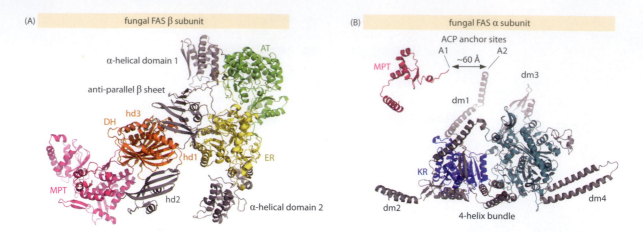

tethered on both sides. A second missing domain is the phosphopantetheine transferase (PPT), responsible for transferring the phosphopantetheinyl group from CoA to a specific serine residue in ACP to create the swinging arm. It, too, is thought to be flexibly tethered, in this instance at the C-terminal end of the α subunit.

Mycobacteria are unusual in having both a fungal-type (class I) FAS and a system of discrete enzymes like most bacteria. The class I FAS resembles the cagelike structure of the fungal FAS but is a 'slimmed-down' version: a hexamer of ~3100 aa subunits, each of which contains the enzyme domains of the β- and α-subunits of fungal FAS (totaling ~4000 aa) but lacking the C-terminal TE. The complex is able to produce C_{16}/C_{18} fatty acids plus C_{26} fatty acids for conversion to mycolic acids (C_{60} to C_{90} fatty acids), important components of the cell walls of mycobacteria such as *Mycobacterium tuberculosis*, the causative agent of tuberculosis. The openings in the walls of the mycobacterial FAS are larger than those of fungal FAS, in keeping with the larger products generated.

Flexible tethering of the ACP is an essential feature of the catalytic mechanism of FAS

To add one C_2 unit in fatty acid synthesis, the ACP has to visit a succession of active sites. At its N-terminal end in fungal FAS, it is tethered by a flexible linker (~45 aa) to the MPT fragment of the α subunit located at the side wall of a reaction chamber (attachment point A1). Its C-terminal end is tethered by a flexible linker (~25 aa) to the N-terminal end of the dm1 module ~60 Å away, at the hub of the central disk (attachment point A2). The distances between the various active sites within a reaction chamber suggest that one set, arranged in a circular path of ~55 ± 15 Å around the peripheral A1 attachment point (**Figure 9.17B**), would be the most efficacious. In any given α subunit, the KR active site faces into one

Figure 9.16 Structures of individual α and β subunits in the fungal FAS. (A) A β subunit of the upper dome. The abbreviations and color code are the same as in Figure 9.15. The MPT domain at the C-terminal end of the β subunit is incomplete without the part coming from an α subunit. (B) The partner α subunit from the disk. Residues 1–94 comprise the MPT fragment that complements the rest of the MPT domain in the β subunit. The ACP domain is not visible but is anchored by flexible linkers at residues 94 (C-terminal end of MPT fragment, site A1) and 325 (N-terminal end of dm1, site A2). The PPT domain is also not visible and is thought to be flexibly tethered at residue 1715 on the C-terminal side of the KS domain. (PDB 2UVB, 2UVC)

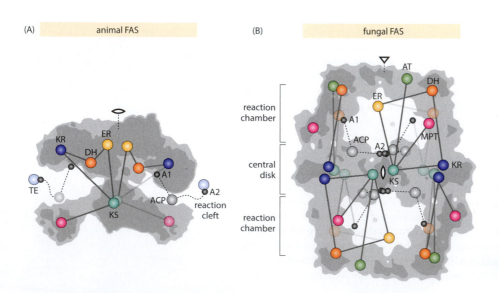

Figure 9.17 Movement of the ACP domain between active sites in FAS. Schematic view of the movement of an ACP domain between a set of active sites. (A) The two subunits in animal FAS are depicted in light gray and darker gray. The 2-fold axis is indicated by an open ellipse. Each ACP is flexibly tethered at its N-terminal end at site A1 (C-terminal end of the KR domain) but also has a TE domain flexibly attached to its C-terminal end (site A2). Each reaction cleft between the upper and lower decks has a full set of active sites, connected by solid lines, distributed between the front and back faces for visits by the ACP. Color code as in Figure 9.14. (PDB 2UV8) (B) The $α_6β_6$ fungal FAS. The 2-fold axis is indicated by an open ellipse (middle) and the 3-fold axis by an open arrowhead (top). Each of the six ACPs is flexibly tethered on both sides, at sites A1 (C-terminal end of MPT) and A2 (N-terminal end of dm1) in an α subunit. Each reaction chamber contains three full sets of active sites. Each set, connected by solid lines, is grouped around an A1 site. Color code as in Figure 9.15. (PDB 2UVB, 2UVC) (Adapted from M. Leibundgut, T. Maier, S. Jenni and N. Ban, *Curr. Opin. Struct. Biol.* 18:714–725, 2008. With permission from Elsevier.)

chamber and that of the KS faces into the other; the ACP anchors are located in the same chamber as that served by the KR. The proposed path of ACP circulation takes in active sites from two α subunits and two β subunits, consistent with biochemical experiments that demonstrated cross-linking between participating active sites from different subunits. Three such circular paths operate in each chamber.

The KS, KR, DH, and ER domains on a given path come in the order required in the catalytic mechanism, and the local concentration of the ACP domain is kept high. Both features are presumably advantageous. The N-terminal ACP linker resembles the linkers that allow motion to the lipoyl and biotinyl domains in their respective complexes, not least in being Ala/Pro-rich and thus likely to be flexible but also stiffened in places (Section 9.4). This may be helpful in directing the ACP into preferred trajectories and avoiding any tangling of the three ACPs in the same reaction chamber.

Animal FAS operates somewhat differently (Figure 9.17A). In each of the two reaction chambers, the ACP is tethered only at its N-terminal end (and with the TE domain dangling from its C-terminal end). In addition, the various active sites are found on the front and back faces of the dimer, and not so obviously arranged to correlate with the order in which they participate in the overall reaction (see Figure 9.14). Moreover, there is some flexibility in the animal FAS structure, allowing a swiveling motion around the 'waist', and this may contribute to active-site coupling.

The ACP can sequester the long acyl group on the phosphopantetheine arm and present it for reaction

Free ACPs and ACP domains display some interesting differences (**Figure 9.18A**). In the bacterial type II systems, as in plants, the separate ACP (a four-helix bundle of ~9 kDa) sequesters the fatty acyl chain attached to the phosphopantetheine arm in a long expandable cavity lined with nonpolar side chains as it migrates between successive enzymes. In contrast, the single-tethered ACP of the multifunctional animal FAS, though very similar in structure, does not sequester the acyl chain. The doubly-tethered ACP of the fungal FAS is different again in that it has the four-helix bundle but contains four additional helices at its C-terminal end, approximately doubling its size. There is no evidence yet as to whether it sequesters the growing acyl chain.

In a crystal structure of the yeast FAS, the three ACP domains in each chamber have been found stalled in the active sites of their respective KS domains, perhaps because there are no substrates present and this is the binding site of highest affinity under crystallization conditions (Figure 9.18B). The ACP is positioned at the entrance to the active site; one prominent contact is provided by the second α helix of the four-helix bundle, identified as a 'recognition helix' in the binding of ACPs by their cognate KSs, and another by the additional part of the yeast FAS domain. The latter abuts the dm3 spoke in the central disk (see Figure 9.15) projecting inward from the KS dimer. Electron density is visible only for the initial pantoic acid part of the swinging arm covalently linked to the Ser residue of the ACP, but an arm 18 Å long extended into the KS active site would place the phosphopantetheine thiol group next to the catalytic Cys, as required by the catalytic mechanism (see Figure 9.18B). It appears that, as with the lipoyl and biotinyl groups described above, the phosphopantetheinyl group has to be fully extended in reactions at the active site of a partner enzyme.

Acyl chain length is important in product release

In animal FAS, the action of the TE domain flexibly tethered at the C-terminal end of the ACP domain (see Figure 9.13) releases predominantly C_{18} palmitate as the FAS product. A crystal structure of the human TE domain in complex with the inhibitor Orlistat, a drug used in cancer therapy, points to the way this is achieved. TE has three hydrophobic pockets that are able to bind fatty acids with a maximum of 6, 12, and 16 or 18 carbon atoms respectively (**Figure 9.19A**). As the length of the fatty acyl group on ACP increases with successive rounds of FAS catalysis, eventually it enters the largest pocket in TE and the C_{16} or C_{18} alkyl chain is best able to position its thioester bond appropriately in the active site of the esterase. TE has a catalytic triad, Asp/His/Ser, in its active site, and hydrolysis proceeds by nucleophilic attack of the Ser on the thioester (Figure 9.19B), a mechanism similar to that of serine proteases (see Box 7.1). Products with shorter chain lengths can be released under conditions of limiting malonyl-CoA concentration, when the KS is compromised in

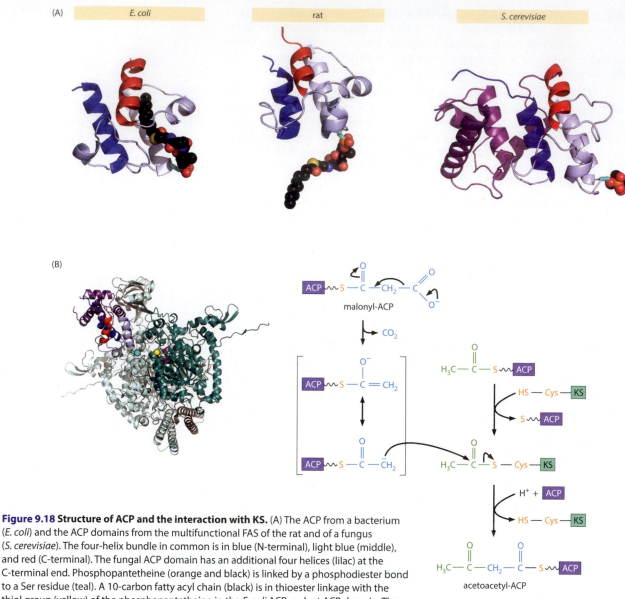

Figure 9.18 Structure of ACP and the interaction with KS. (A) The ACP from a bacterium (*E. coli*) and the ACP domains from the multifunctional FAS of the rat and of a fungus (*S. cerevisiae*). The four-helix bundle in common is in blue (N-terminal), light blue (middle), and red (C-terminal). The fungal ACP domain has an additional four helices (lilac) at the C-terminal end. Phosphopantetheine (orange and black) is linked by a phosphodiester bond to a Ser residue (teal). A 10-carbon fatty acyl chain (black) is in thioester linkage with the thiol group (yellow) of the phosphopantetheine in the *E. coli* ACP and rat ACP domain. The fatty acyl group is sequestered in a long expandable cavity in the *E. coli* protein but not in the rat ACP domain. (PDB 2FAE, 2PNG, 2UV8) (B) Left, the ACP domain of yeast FAS, colored as in (A), bound at the active site of the KS component (KS dimer in dark and light teal, as in Figure 9.16). Extension of the phosphopantetheine arm from the Ser residue (cyan sphere) in ACP would place the thiol group at the distal end of the arm next to the cluster of two His residues (magenta) and a Cys residue (yellow sphere) in the KS active site. Right, the catalytic mechanism of KS. The malonyl group (blue) on ACP is decarboxylated and the resulting carbanion (the incoming C_2 unit) attacks the thioester linking the acyl group (temporarily transferred from its ACP) to the active site Cys of KS. Only the first C_2 addition, to acetyl-CoA to make acetoacetyl-ACP, is depicted. Color code as in Figure 9.13.

its ability to initiate another round of C_2 unit addition and the low rate of release of shorter fatty acyl-CoA chains becomes appreciable.

In contrast, the MPT domain of fungal FAS uses a less precise measure of chain length (see Figure 9.19B). Its bifunctional active site transfers malonyl groups from CoA to ACP for synthesis, and fatty acyl chains from ACP to CoA for product release. The length of chain is determined mainly by the relative affinities and concentrations of substrates and potential products. Under normal physiological conditions, for example, the yeast FAS releases C_{16}

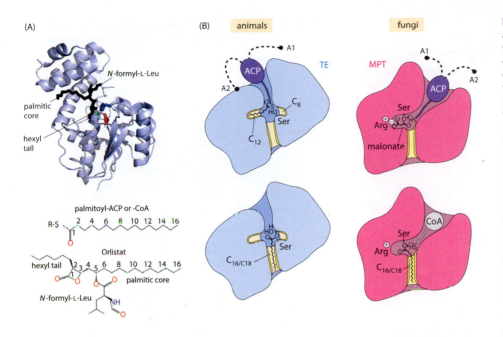

Figure 9.19 Determination of fatty acyl chain length. (A) Structure of the thioesterase domain of human FAS in complex with Orlistat, a drug used in cancer treatment. Orlistat (black) inhibits the enzyme by occupying three different hydrophobic clefts. Only the longest of these (the palmitic core) can efficiently harbor the C_{16}/C_{18} fatty acyl chain of fatty acyl-ACP and thereby present its thioester bond with ACP for hydrolysis in the TE active site. The catalytic triad is Ser2308 (cyan), His2481 (blue), and Asp2338 (red). In the structure shown, the inhibitor is already in ester linkage with Ser2308, representing the catalytic acyl-enzyme intermediate. (PDB 2PX6) (B) Schematic diagram of chain termination and release from animal and fungal FAS. Color code as in Figure 9.13. Left, in animal FAS, the ACP is flexibly anchored at one end (A1) to the KR component, and at the other end to the TE domain at A2 (Figure 9.17). Efficient release of the acyl group once it has reached C_{16}/C_{18} length is catalyzed by the TE domain, beginning with nucleophilic attack of the active-site Ser (upper sketch) and subsequent hydrolysis of the fatty acyl-Ser ester by a water molecule (lower sketch). Right, in the fungal FAS, the ACP is flexibly anchored at both ends (A1, A2). In the synthetic direction, the incoming malonyl group is transferred from CoA to the OH of an active-site Ser in MPT, where it is stabilized by making a salt bridge with a nearby Arg side chain. It can then be transferred to make a thioester bond with the phosphopantetheine of ACP (upper sketch). When it comes to fatty acid release, the fatty acyl group on ACP can displace the malonyl group and then, after nucleophilic attack on the thioester by the active-site Ser, it can be transferred to make a new thioester bond, this time with CoA. Finally, the fatty acyl-CoA is released. Longer (C_{16}/C_{18}) fatty acyl groups are favored, but shorter acyl groups can displace the malonyl group and generate fatty acyl-CoA products, depending on the conditions. (Adapted from M. Leibundgut, T. Maier, S. Jenni and N. Ban, *Curr. Opin. Struct. Biol.* 18:714–725, 2008. With permission from Elsevier.)

and C_{18} fatty acyl-CoAs in a ratio of ~2 : 3, but if malonyl-CoA levels are low, shorter acyl-CoAs can be released. This may be of evolutionary advantage to organisms living in more changeable environments.

Fatty acid degradation partly resembles fatty acid synthesis

Fatty acids are degraded by a process of **β-oxidation**, which looks somewhat like a reverse of fatty acid synthesis but has some distinct differences. First, it is an oxidative exergonic process (yielding acetyl-CoA and NADH) that in eukaryotes takes place in mitochondria. Second, no ACP is required; instead, the α-carboxyl group is activated by the formation of a thioester with CoA. This reaction takes place in the cytoplasm, catalyzed by one of at least three fatty acyl-CoA synthetases. They differ in acyl chain length specificity, but all proceed as follows:

$$R\text{-COOH} + ATP + CoA\text{-SH} \rightarrow R\text{-CO-S-CoA} + AMP + PP_i$$

The pyrophosphate (PP_i) will be hydrolyzed by a pyrophosphatase, and the similarity to the action of aminoacyl-tRNA synthetases (Section 6.2) is striking. The acyl group is then temporarily transferred from the thiol group of CoA to a hydroxyl group in a different small molecule, carnitine, and the acyl-carnitine ester is translocated across the mitochondrial inner membrane by a specific carrier protein.

Once inside the mitochondrial matrix, the acyl group is transferred back to the thiol group of CoA and a double bond is then introduced at the β-carbon by an FAD-dependent fatty acyl-CoA dehydrogenase (ACD). There are four ACDs that differ in their specificity for the length of the acyl group (**Figure 9.20A**). Inherited defects in the medium-chain-length ACD have been associated with ~5% of cases of *sudden infant death syndrome (SIDS)*. Acyl-CoAs up to ~ C_{12} are dealt with by sets of soluble enzymes that catalyze the remaining three reactions in the β-oxidation cycle, whereas the long-chain acyl-CoAs ($C_{12}–C_{18}$) are fed into a fatty acid oxidation (FAO) complex (Figure 9.20B).

In animals, the FAO complex is an $\alpha_4\beta_4$ hetero-octamer, whereas in bacteria it is an $\alpha_2\beta_2$ hetero-tetramer. The β subunit contains the enoyl-CoA hydratase (ECH), whereas the α subunit contains the NAD⁺-dependent hydroxyacyl-CoA dehydrogenase (HACD) and the ketoacyl-CoA thiolase (KACT) activities. Each completed set of reactions discharges one C_2 unit as acetyl-CoA for entry into the citric acid cycle and generates one NADH and one $FADH_2$ for exergonic oxidation in the respiratory chain (Chapter 15). The oxidation of the $FADH_2$ is unusual in that it feeds into the mitochondrial electron transport chain at the coenzyme Q level via an electron-transfer flavoprotein (ETF) and ETF–ubiquinone oxidoreductase. The existence of two forms of the FAO structure (see Figure 9.20B), one symmetric (form I) and the other asymmetric (form II) as a result of conformational rearrangement in

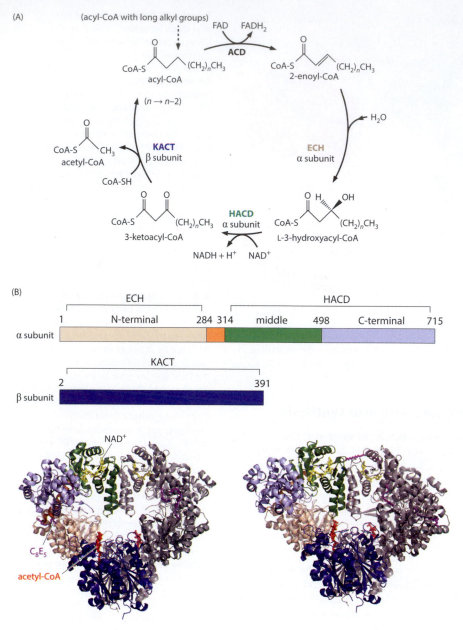

Figure 9.20 Degradation of fatty acids by β-oxidation. (A) Four enzymes carry out the reaction cycle of β-oxidation. ACD, FAD-dependent fatty acyl-CoA dehydrogenase; ECH, enoyl-CoA hydratase; HACD, NAD⁺-dependent hydroxyacyl-CoA dehydrogenase; KACT, ketoacyl-CoA thiolase. (B) Domain maps of the two subunits of *Pseudomonas fragi* FAO and the structure of the $\alpha_2\beta_2$ complex viewed along a twofold axis. The right-hand α/β pair is depicted in gray; in the left-hand pair, the domains are colored individually. Residues 284–309 in the α subunit form an α-helical linker between the ECH and HACD domains. C_8E_5, n-octylpentaoxyethylene, an analog of a fatty acid tail needed for stable crystals (orange); NAD⁺ is shown in yellow, and acetyl-CoA in red. Left, symmetrical dimer (form I), with two molecules of C_8E_5 in each HACD active site and two more with each ECH domain. Right, asymmetrical dimer (form II), which has C_8E_5 molecules bound to only one of the two α subunits. (PDB 1WDK, 1WDL) (A, adapted from M. Ishikawa et al., *EMBO J.* 23:2745–2754, 2004. With permission from John Wiley & Sons.)

the HACD domain and the α-helical linker between it and the ECH domain, suggests that reversible conformational change might be important in coordinating the separate catalytic activities.

Substrate channeling in the FAO complex is slightly leaky

The active sites of all three enzymes in the FAO complex face inward but are well separated and do not conform to the order in which they participate in the overall reaction. The ECH and HACD domains share a binding pocket for the 3′-phospho-ADP moiety of CoA, but KACT has a separate binding pocket. A plausible model of active-site coupling (**Figure 9.21**) is that the substrate is first noncovalently anchored through its 3′-phospho-ADP moiety in the binding pocket shared by the ECH and HACD domains, and the acyl group swings from the ECH active site into the HACD active site. From there it migrates to the KACT active site, and the 3′-phospho-ADP moiety detaches from the ECH/HACD-binding site and enters the CoA-binding site on the β subunit. This requires the form I/form II conformational change to permit the >C=O group in what is now the 3-position of the ketoacyl-CoA proper access to the KACT active site. Thus, unlike fatty acid synthesis, the intermediates in FAO move from site to site, but their binding to the enzyme complex is

Figure 9.21 Substrate channeling in FAO complex. Three views of successive binding modes of catalytic intermediates in the FAO cycle. Color code as in Figure 9.20. The enoyl-CoA (depicted at top) first binds with the 3′-phospho-ADP moiety of its CoA in a joint ECH/KACT binding site, thereby anchoring the acyl group in the active site of ECH (first diagram). The hydroxyacyl group swings across to the active site of HACD (second diagram), after which the 3-ketoacyl group migrates to the active site of KACT, and the 3′-phospho-ADP moiety enters the CoA-binding site on the β subunit (third diagram). This requires a conformational change between form I and form II for the correct presentation of the β >C=O group for attack by the SH of the catalytic Cys of KACT and the release of a molecule of acetyl-CoA.

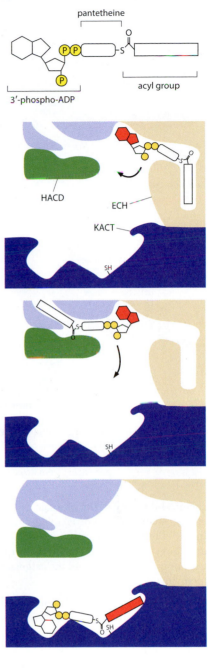

not covalent. Occasional dissociation and a small amount of leakage in substrate channeling (the 'leaky hosepipe' model) can therefore occur.

Successive repeats of the four-enzyme process degrade the fatty acyl-CoA, one C_2 unit at a time. If the acyl group has an odd number of C atoms, it cannot be degraded completely by β-oxidation but when it reaches a C_3 length (propionyl-CoA) a separate set of enzymes takes over, including a biotin-dependent propionyl-CoA carboxylase, which converts it into succinyl-CoA for entry into the citric acid cycle.

Polyketide synthases are related to animal FAS

Polyketides are synthesized by many bacterial species and constitute a structurally diverse family of natural products (more than 10,000 have been identified), many of which have found important uses as antibiotics, immunosuppressants, and anti-tumor agents (**Figure 9.22**). They are assembled by enzyme complexes, polyketide synthases (PKSs), which rely on ACPs with a phosphopantetheine swinging arm to ferry the growing product from one active site to the next. The simpler ones (type II) behave as successive monofunctional enzymes, resembling bacterial FAS. Others (type I) are giant enzyme complexes, similar to the animal FAS, but instead of the repetitive action of a single set of enzymes, extraordinary

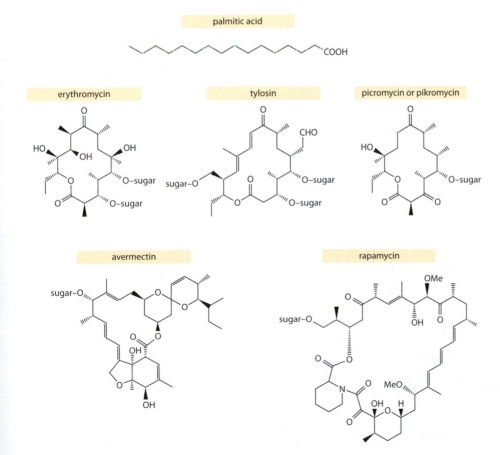

Figure 9.22 Some typical polyketide products of modular PKSs. Palmitic acid, the product of FAS, is shown for comparison. Many polyketides go on to further modification by the addition of sugar residues. (Adapted from S. Smith and S.-C. Tsai, *Nat. Prod. Rep.* 24:1041–1072, 2007. With permission from Royal Society of Chemistry.)

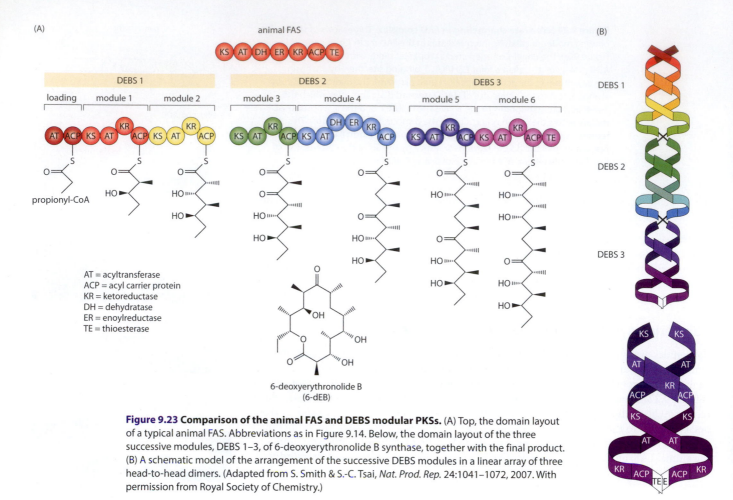

Figure 9.23 Comparison of the animal FAS and DEBS modular PKSs. (A) Top, the domain layout of a typical animal FAS. Abbreviations as in Figure 9.14. Below, the domain layout of the three successive modules, DEBS 1–3, of 6-deoxyerythronolide B synthase, together with the final product. (B) A schematic model of the arrangement of the successive DEBS modules in a linear array of three head-to-head dimers. (Adapted from S. Smith & S.-C. Tsai, *Nat. Prod. Rep.* 24:1041–1072, 2007. With permission from Royal Society of Chemistry.)

chemical diversity in the ultimate polyketide products is brought about by the transfer of an intermediate from one FAS-like complex to another after each round of chain extension. In FAS the starter compound is usually acetyl-CoA and the extension comes from the decarboxylation of a malonyl group; in contrast, in a PKS the starter is generally acetyl-CoA or propionyl-CoA and the extension derives from the decarboxylation of a malonyl or methylmalonyl group.

The 6-deoxyerythronolide B synthase (DEBS) of *Saccharopolyspora erythraea* illustrated in **Figure 9.23** synthesizes the aglycone precursor of the antibiotic erythromycin. Six different enzymatic modules, each of which comprises a set of FAS-like catalytic activities, are arranged as pairs in three huge (~350 kDa) multifunctional polypeptide chains. This makes a total of 28 active sites (the KR domain in module 3 is inactive). Each of DEBS 1, DEBS 2, and DEBS 3 form head-to-head dimers, which are arranged linearly in the overall complex (~6 MDa). (In PKSs the dimers are often held together by dimerization motifs at the ends of the modules and, in some cases, with dimerization elements (~55 aa) on the N-terminal side of the KR domain.) The order of functional domains in the modules is the same as that in animal FAS (which means it also differs from the order of the reactions catalyzed), but in several modules certain catalytic activities are missing, and the product of the module is correspondingly different. At the end of each reaction cycle, the product is transferred from the ACP of that module to the KS (and ACP) of the next module. Each module has its own AT domain for loading the ACP with a C_3 extender unit derived from methylmalonyl-CoA. Only the module at the end contains a TE-domain dimer to catalyze the release of the linear product followed by its cyclization. DEBS is one of the simpler PKSs. In other instances, many more modules are strung together: there are as many as 122 domains in nine proteins (4.7 MDa) in the assembly line for the polyketide ECO-02301 produced by *Streptomyces aizunensis*.

A complete structure for a PKS remains elusive, but the structures of individual domains and fragments have revealed their homology with corresponding domains of animal FAS.

The FAS domains are more closely related to their counterparts in PKS than to the mono-functional bacterial enzymes, suggesting a common evolutionary origin for PKS and FAS. This is further supported by the presence of an active ME domain in some PKSs compared with an inactive ψME domain in FAS. Unlike fatty acids, which are essential in energy metabolism and as components of other structures, notably membranes, polyketides are *secondary metabolites* and their origins are obscure.

Non-ribosomal peptide synthases also have phosphopantetheine swinging arms

Some relatively large peptides, such as the anti-tumor agent bleomycin, the current 'last-resort' antibiotic vancomycin, and the immunosuppressant cyclosporin, are synthesized without need of an mRNA or ribosome. Like a PKS, a non-ribosomal peptide synthase (NRPS) is a giant (MDa) structure (type I) in many fungi but, unlike the dimeric PKSs, the NRPS subunit is monomeric. In many bacterial species, the various NRPS enzymes are found as discrete interacting proteins (type II). A type I NRPS normally contains as many modules as there are amino acids in the final product (**Figure 9.24**). Each module contains an A (adenylation) domain, which activates a specific amino acid (which need not be a natural proteinogenic L-amino acid) by converting it into an aminoacyl adenylate in an ATP-dependent reaction (like an aminoacyl-tRNA synthetase). It then catalyzes nucleophilic attack on the aminoacyl adenylate by the thiol group of a phosphopantetheinyl swinging arm linked to a Ser residue in a peptidyl carrier protein domain (PCP, ~80–100 aa, with the same basic fold as the ACPs of FASs and PKSs), leaving the amino acid attached to the swinging arm in thioester linkage. The NH_2- group of the aminoacyl group on the swinging arm is now free for a nucleophilic attack on the thioester linkage holding the amino acid or peptide to the phosphopantetheinyl swinging arm of the PCP domain located in the preceding module. This reaction is catalyzed by the C (condensing) domain. A new peptide bond is thereby formed and the peptide chain is extended by one amino acid at the C-terminal end (the same direction of growth as in a ribosome). The system is now ready to repeat, with attack on the new peptidyl thioester by an incoming aminoacyl group that has been attached to its PCP in the next module.

In addition to its A, C, and PCP domains, a particular module may also contain a domain responsible for epimerization (L–D racemizaton of the Cα of the newly added amino acid) or other domains responsible for side-chain modification (such as cyclization or methylation). The completed product is finally released by the action of a thioesterase domain in the 'last' module, which may also in some cases generate cyclic peptides. Substantial individual tailoring (including glycosylation, halogenation, the addition of lipids, or chemical cross-linking) typically takes place after the initial synthesis. The similarities between PKS and NRPS are emphasized by the recognition of natural hybrid PKS–NRPS systems; in some instances NRPS C domains are capable of condensing polyketide intermediates with an aminoacyl group bound to a phosphopantetheine swinging arm in a NRPS PCP domain. An example is in the biosynthesis of the immunosuppressant rapamycin, which contains a pipecolate moiety (with a six-membered ring related to proline).

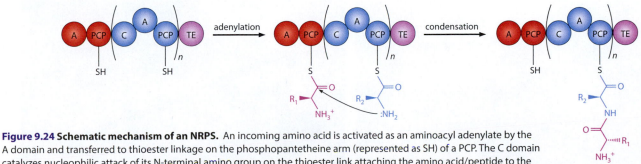

Figure 9.24 Schematic mechanism of an NRPS. An incoming amino acid is activated as an aminoacyl adenylate by the A domain and transferred to thioester linkage on the phosphopantetheine arm (represented as SH) of a PCP. The C domain catalyzes nucleophilic attack of its N-terminal amino group on the thioester link attaching the amino acid/peptide to the neighboring PCP, causing the polypeptide chain to elongate by one residue. The process is initiated with the A and PCP domains at the N-terminal end of the protein (shown in red) and continued with the repeats of A, PCP and C domains (shown in blue) until all the programmed amino acids have been incorporated. It often concludes with release of the peptide chain by the action of a thioesterase (TE).

9.6   THE CELLULOSOME

Another property of multienzyme complexes, different from the active-site coupling and substrate channeling made possible by tunnels and swinging arms, is the ability to orchestrate the collective action of a multiplicity of enzymes on a refractory substrate. This can enhance their individual activities by putting the right enzyme in the right place at the right time. A good example is the **cellulosomes**, large extracellular assemblies (>3 MDa) of hydrolytic enzymes generated by anaerobic microorganisms to break down plant cell walls.

In photosynthesis, plants capture carbon in the form of CO_2 as an essential part of the Earth's carbon cycle (Chapter 15). Much of this carbon is subsequently found in the cell walls of plants in the form of macromolecular polymers of sugars. Cellulose polymers, which account for ~15–40% of the cell wall, contain thousands of D-glucose units in 1→4β linkage, which makes them linear rather than coiled as in 1→4α linked starch and glycogen. They are laid down side by side in insoluble crystalline arrays (called 'microfibrils') held together by multiple inter-chain H-bonds. Hemicellulose polymers are shorter than cellulose and contain a variety of other sugars (pentoses and hexoses) in addition to glucose; they are branched instead of straight, and have a more amorphous structure. Pectins are different again, containing a variety of sugars but rich in galacturonic acid (many of the carboxyl groups are esterified with methanol); some are straight-chain polymers, some are branched. Hemicelluloses and pectin together make up ~30–40%. The remaining 20% is filled by lignin, a more hydrophobic polymer comprising various aromatic alcohols.

These polymers are refractory to conventional attack by enzymes, but organisms have evolved that can degrade them and thereby grant access to the carbon as an energy source. Ruminant animals are dependent on them for their digestive processes. Aerobic bacteria and fungi secrete a battery of free-standing hydrolases, endoglucanases, exoglucanases and ancillary enzymes, which synergistically attack the plant cell wall, exploiting imperfections and working to widen them. In anaerobic microorganisms, in contrast, the hydrolytic enzymes (cellulases, hemicellulases, and even pectinases) are assembled into a variety of cellulosomes.

The cellulosome is a multienzyme complex assembled on an inactive protein core

The cellulosome was first described as a discrete multifunctional protein complex from the anaerobic thermophilic bacterium *Clostridium thermocellum* (**Figure 9.25**). The various cellulolytic enzymes are recruited by a non-catalytic **scaffoldin** subunit. They are held together by tight interactions ($K_d < 10^{-9}$ M) between tandem arrays of **cohesin** modules (~150 aa) on the scaffoldin molecule and (usually) single copies of **dockerin** modules (~70 aa) at the C-terminal ends of the enzymes (**Figure 9.26A**). (Note that the similarly named proteins in cellulosomes are not the same as the cohesins in chromosomes, described in Section 13.4.) The number of cohesin modules in a scaffoldin varies between 1 and 11 but is usually more

Figure 9.25 A cellulose target acted on by cellulosomes of *C. thermocellum*. (A) An electron micrograph of a cellulosome protuberance gold-labeled with anticellulosome antibodies in the region between the bacterial cell surface and the cellulose. Much of the label is associated with the cellulose and is connected to the cell by extended fibrous material. (B) Cryo-TEM of the adsorption of cellulosomes on cellulose microcrystals from *Valonia ventricosa*. The arrows point to individual cellulosomes on the cellulose surface. (A, from E.A. Bayer, L.J. Shimon, Y. Shoham and R. Lamed, *J. Struct. Biol.* 124:221–234, 1998. With permission from Elsevier; B, courtesy of Claire Boisset, Edward Bayer, Bernard Henrissat and Henry Chanzy.)

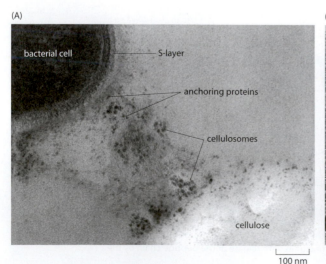

(A)

bacterial cell

S-layer

anchoring proteins

cellulosomes

cellulose

100 nm

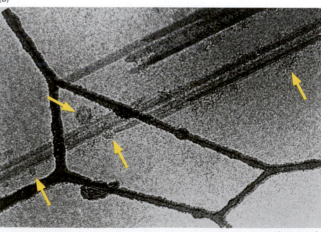

(B)

100 nm

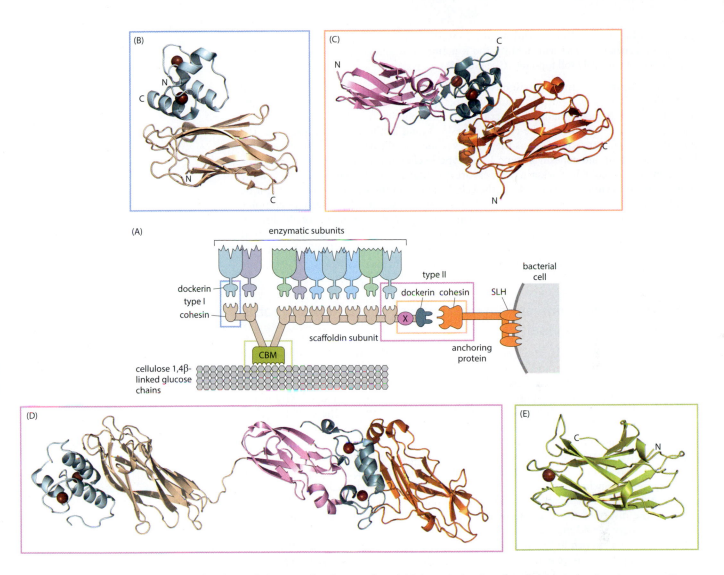

Figure 9.26 Structure of the cellulosome of _C. thermocellum_. (A) Schematic drawing of a cellulosome showing nine enzymatic subunits bound to a primary scaffoldin via type I cohesin/dockerin interactions. The primary scaffoldin attaches to the cellulose substrate via its CBM domain. The primary scaffoldin also has an X-module and is bound by a type II cohesin/dockerin interaction to an anchoring protein which, in turn, binds to the bacterial S-layer. The different modules of the various cellulosomal subunits are separated by distinctive linker segments. The color code is the same throughout the figure. (B) The type I cohesin/dockerin heterodimer, with two Ca^{2+} (brown spheres) in the dockerin. (PDB 1OHZ) (C) The type II cohesin/dockerin heterodimer with an X-module (pink) covalently linked to the N-terminal end of the dockerin and two Ca^{2+} (brown spheres) in the dockerin. (PDB 2B59) (D) A ternary complex comprising (from the left) a type I dockerin module from a cellulose enzyme (Cel9D glycoside hydrolase); a type I cohesin module; an X-module and a type II dockerin module from the C-terminal end of the CipA primary scaffoldin; and a type II cohesin module from the SdbA anchoring protein. (PDB 4FL4) (E) The CBM with bound Ca^{2+} (brown sphere). (PDB 1NBC)

than 4, and the scaffoldin also contains a cellulose-specific **carbohydrate-binding module** (**CBM**) for substrate targeting. The various modules are all joined together by linker regions.

In some cellulosomes, including that of _C. thermocellum_ (see Figure 9.26A), the primary, enzyme-binding, scaffoldin (with type I cohesin) is supplemented by another type of cohesin-containing scaffoldin (with type II cohesin). This is an anchoring scaffoldin that serves to bind the primary scaffoldin and its complement of enzymes onto the bacterial cell surface, usually via a type II dockerin module on the primary scaffoldin and an S-layer homology (SLH) module. Type I cohesin domains cannot distinguish one enzyme-dockerin from another, which generates heterogeneity in the complement of cellulosomal enzymes that are bound; the ensemble may also vary according to the plant cell wall target. A type I cohesin/dockerin interaction is shown in Figure 9.26B. The dockerin contains two duplicated segments of ~22 aa. The first 12 aa of each segment resemble the Ca^{2+}-binding loop of the EF-hand motif (see **Guide**), with highly conserved Ca^{2+}-binding Asp or Asn residues. Calcium is crucial for dockerin stability and function. The cohesin is a nine-stranded flattened β barrel with jelly roll topology (see **Guide**).

A type II cohesin/dockerin interaction is depicted in Figure 9.26C. On the N-terminal end of the dockerin module is an X-module; its function is unknown, but it may stabilize the dockerin structure and enhance interaction with the cohesin. This dockerin also has two Ca^+ bound and is specific for the type II cohesins found in the anchoring proteins. The type II cohesin has the same fold as the type I version, but the orientations and interfaces between dockerin and cohesin in type I and type II complexes are clearly different. The type I and type II cohesin/dockerin partners are not interchangeable, which means that the interactions governing the assembly of the cellulosome and the way in which it might be attached to the bacterial cell surface are kept distinct. The structure of a ternary complex (Figure 9.26D) shows how the C-terminal trimodular part of the primary scaffoldin can bind the type I dockerin of a cellulolytic enzyme and can itself be bound through a type II dockerin to a type II cohesin of an anchoring scaffoldin. From small-angle X-ray scattering data, the 13-aa linker between the X-module and the type I cohesin module appears to be highly flexible.

Different bacteria generate different sorts of cellulosomes capable of extensive heterogeneity

Cellulosomes have been classified into two major types. The 'simple' cellulosomes of some bacteria (for example, mesophilic *Clostridium cellulovorans*, *C. cellulolyticum*, and *C. acetobutylicum*) possess a single scaffoldin that contains an N-terminal CBM, between five and nine cohesins (depending on the species), and between one and six X-modules (**Figure 9.27**). The inter-module linkers are relatively short (5–19 aa). The scaffoldin gene occurs as part of a cluster on the chromosome, and numerous genes encoding cellulosomal enzymes are located downstream of it. Other cellulosomal genes are dispersed throughout the genome, frequently in small clusters. In contrast, other bacteria such as *C. thermocellum*, *Acetivibrio cellulolyticus*, and *Ruminococcus flavefaciens* produce a 'complex' type of cellulosome with anchoring scaffoldins (see Figure 9.26A). In these instances, several genes encoding scaffoldins are clustered on the chromosome but the genes for the cellulosomal enzymes are dispersed, often clustering in different locations. The inter-module linkers in their primary scaffoldins are longer (13–49 aa in *C. thermocellulum*) than in simple cellulosomes and are rich in Pro and Thr residues. In the anchoring scaffoldins, the linkers are longer again, often more than 100 aa, and are *O*-glycosylated on Ser and Thr residues.

The additional scaffoldins in complex cellulosomes can provide assembly points for more than one copy of the primary scaffoldin. In *C. thermocellum*, for example, three different anchoring scaffoldins make it possible for 9, 18, or 63 cellulosomal catalytic subunits to be incorporated into cellulosomes (**Figure 9.28**). The most elaborate system yet found comes from *R. flavefaciens*, a major symbiont in cattle, in which two scaffoldins (ScaA with two cohesins and ScaC with just one) have cohesin/dockerin interactions with ScaB (nine cohesins). These interacting scaffoldins bring with them their dockerin-bearing enzymes, and a new (type IIIe) dockerin at the C-terminal end of ScaB binds to a single type IIIe cohesin at the N-terminal end of a fourth scaffoldin, ScaE, which is covalently anchored to the cell envelope (see below). The potential for heterogeneity in binding dockerin-bearing enzymes is extraordinary.

In some instances a simple cellulosome is found attached to the bacterial cell surface. How it is bound is unclear, but an X-module may be involved. The complex cellulosomes are bound to the cell surface, normally through SLH modules in their anchoring scaffoldins (see Figure 9.28). S layers (surface layers) are crystalline protein arrays in the outermost layer of the cell envelopes of many bacteria and archaea, where they have a protective role

Figure 9.27 Scaffoldins of some 'simple' cellulosomes. CbpA, CipC and CipA are scaffoldins in the cellulosomes of three different mesophilic *Clostridia*. All have an N-terminal CBM but they differ in the number and arrangement of their dockerins and X-modules. Color code as in Figure 9.26.

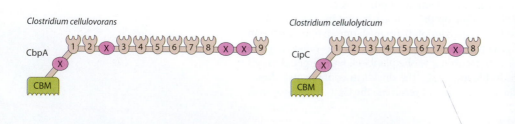

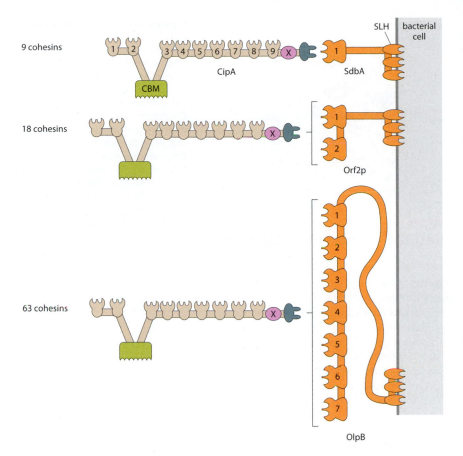

Figure 9.28 Extra scaffoldins in complex cellulosomes. Complex cellulosomes contain additional anchoring scaffoldins, as depicted here for *C. thermocellum*. The dockerin module on the primary scaffoldin (CipA) can bind to cohesins displayed on three different types of secondary scaffoldin encoded by genes SdbA, Orf2p, and OlpB. These allow 9, 18, and 63 type I cohesin modules respectively to be deployed and to bind catalytic enzymes in these increasingly complex cellulosomes.

(Section 1.9). The SLH modules of scaffoldins consist of two or three repeating units of ~60 aa and are likely to be structurally related to the three tandem SLH domains of the surface S-layer protein, Sap, of *Bacillus anthracis*, which form a three-pronged spindle that engages with cell wall polysaccharides. The scaffoldins bind selectively either to the secondary cell-wall polymer or directly to the peptidoglycan layer of the cell envelope. The elaborate cellulosome of *R. flavefaciens* exhibits yet another way of binding to the cell surface: its anchoring scaffoldin ScaE has no SLH modules but, in addition to its N-terminal type III cohesin, it has a sortase motif (an exposed -Leu-Pro-X-Thr-Gly- sequence) in its C-terminal part. This is cleaved enzymatically on the C-terminal side of the Thr residue, and the new α-carboxyl group makes an amide bond with the N-terminus of a penta-Gly moiety provided by a cell wall precursor.

The type I cohesin/dockerin interaction is plastic and not confined to cellulosomes

The dockerin module contains an internal repeat of a Ca^{2+}-binding motif (see Figure 9.26B), and the first structure of a type I cohesin/dockerin complex from *C. thermocellum* revealed a dual mode of binding in which the dockerin module has rotated through 180°. This is illustrated by the cohesin/dockerin complex from *C. cellulolyticum* (**Figure 9.29**). Despite the sequence similarities between the corresponding modules from the different bacteria, there is no cross-binding interaction between them. Significantly, the key residues Ala16/Leu17 and Ala47/Phe48 in the *C. cellulolyticum* dockerin are replaced by Ser11/Thr12 and Ser45/Thr46 in the *C. thermocellum* dockerin. In contrast, the type II cohesin/dockerin complex (see Figure 9.26C) has only a single mode of binding. The existence of a dual mode of binding in the type I complex may confer additional flexibility in cellulosome assembly and in the disposition of catalytic subunits.

There is increasing evidence from genome analysis that dockerins and cohesins are not confined to cellulosomes but occur widely in all three kingdoms of life. Their biological roles are as yet unknown, but only a few are found in proteins involved in carbohydrate metabolism.

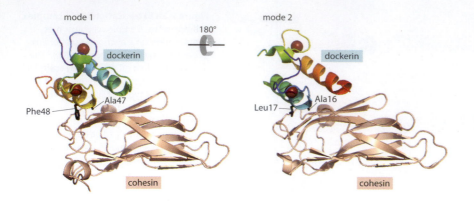

Figure 9.29 Dual mode of interaction between type I dockerins and cohesins. The type I dockerin from *C. cellulolyticum* can bind to its partner type I cohesin in two ways, involving a 180° rotation of the dockerin. The dockerin molecules are color-coded through the spectrum from N-terminus (blue) to C-terminus (red). The two Ca²⁺ are shown as brown spheres. Mode 1 is for the A16S/L17T mutant, with the hydrophobic interaction dominated by Ala47 and Phe48 in the N-terminal helix 1 of the dockerin. Mode 2 is for the A47S/F48T mutant, in which the hydrophobic interaction is dominated by Ala16 and Leu17 in the C-terminal helix 3 of the dockerin. (PDB 2VN5, 2VN6)

The carbohydrate-binding module (CBM) anchors the cellulosome to the polymeric substrate

The CBMs in scaffoldins fall into three categories, displaying three different specificities: type A interacts with crystalline polysaccharides, mostly cellulose; type B binds to internal regions of single glycan chains; and type C targets the ends of polymers. CBMs range in size from ~35 aa to more than 170 aa, and those of bacteria are generally much larger than their fungal counterparts. The family 3 modules are ~150 aa long; as an example, the type A CBM3 (155 aa) module from the primary scaffoldin of *C. thermocellum* is depicted in Figure 9.26E. It has a nine-stranded β-sandwich fold with jelly-roll topology, somewhat similar to the cohesin structure, and a bound Ca²⁺ probably confers stability. Conserved, surface-exposed residues are found in two locations. One is dominated by a linear strip of aromatic and polar residues, resembling those found in other bacterial CBMs and the much smaller fungal modules, and this is likely to be the site of the very tight interaction with crystalline cellulose chains. The side chains of three aromatic residues are predicted to stack against the glucose rings, and the polar amino acids could make contact with the oxygen atoms and hydroxyl groups of glucose moieties. The other conserved residues are contained in a shallow groove on the opposite side of the molecule, but their function is still uncertain.

The cellulosome enzymes resemble the free cellulolytic enzymes secreted by other microorganisms. Many of the latter have a CBM appended to their catalytic domains, as, for example, do the enzymes from the cellulosomes of *C. thermocellum* and *R. flavefaciens*. These enzyme-borne CBMS are in addition to any CBMs that are part of the cellulosome scaffoldins. CBMs vary widely and help to direct the catalytic domains of the enzymes to, and prolong contact with, their polysaccharide targets. A scaffoldin-borne CBM delivers the entire cellulosome complex to the cellulosic substrate, and the other enzyme-associated CBMs possess ancillary binding functions that may help target the catalytic domains.

The *R. flavefaciens* cellulosome has no CBMs, but the cohesin module of the ScaE scaffoldin, which is covalently bound to the peptidoglycan of the *R. flavefaciens* cell, is able to bind the dockerin module of another protein, CttA. CttA has two CBMs and thus can act as a bridge to anchor the bacterial cell close to the plant cell wall, ready for cellulosomes bound to other ScaE molecules to act on it.

Cellulosomes possess a wide range of cellulolytic activities

The degradative enzymes carried by cellulosomes attack their target in many different ways. They include numerous glycoside hydrolases (mainly cellulases and hemicellulases), carbohydrate esterases (for example, acetyl and feruolyl esterases), and polysaccharide lyases (for the degradation of plant cell wall pectins). Endoglucanases hydrolyze the bond between any two glucose residues along the cellulose chain. Exoglucanases hydrolyze a cellobiose unit (two glucose residues) from the end of the chain and then continue to hydrolyze successive cellobiose units processively. β-Glucosidases hydrolyze the cellobiose units to two glucose molecules. Hemicellulases help expose the cellulose fibrils by degrading the xylans, mannans, xyloglucans, arabinoxylans, and so on. The wealth of choice among enzymes to be recruited into cellulosomes is illustrated by the fact that the *C. thermocellum* genome encodes 72 polypeptides with type I dockerin modules, and in *R. flavefaciens* this number may exceed 200.

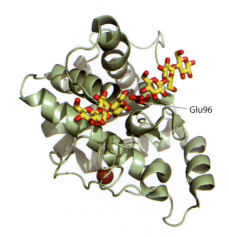

Figure 9.30 Structure of the catalytic domain of glycoside hydrolase *Ct*Cel124. The C-terminal catalytic module (220 aa) of this cellulolytic enzyme from *C. thermocellum* is shown bound to two cellotriose molecules (carbon in yellow stick format; OH groups in red) representing a cleaved substrate. A catalytically important residue, Glu96, is depicted in red stick format, and a bound Ca²⁺ is shown as a brown sphere. (PDB 2XQO)

The catalytic domain of one example, a glycoside hydrolase called *Ct*Cel124 from the *C. thermocellum* cellulosome, is shown in **Figure 9.30**. It is an endoglucanase that may act at the interface between crystalline and amorphous regions of cellulose, synergistically exposing potential substrates for exo-acting cellulases. Its α_8 superhelical fold differs from the six folds observed in other cellulase families but resembles that of some other hydrolases (for example, black swan lysozyme G) of very different specificity.

The cellulosome is thus a means of assembling an array of enzymes that could enhance the synergy between them in catalyzing a set of related reactions. Together with the structural flexibility conferred by the intermodular linkers and possibly the dual mode of type I cohesin/dockerin binding, the complex can bring together a wide range of catalytic activities in an assembly that can adapt and respond to a substrate as difficult as a plant cell wall.

The modular construction of multienzyme complexes opens the way to their redesign for selected purposes

The multienzyme complexes described in this chapter are modular machines, each put together from a variety of constituent enzymes, and many of these enzymes themselves are permutations and combinations of catalytically active domains. Their individual steps underlie the orderly construction of a succession of intermediates on what are essentially assembly lines. Intervention at various stages to interrupt the assembly or introduce different specificities would cause them to synthesize different end-products. For example, it is possible to change the coenzyme specificity of a PDH complex from NAD^+ to $NADP^+$ by redesigning the E3 component. Different modules catalyzing variants of their part-reactions have been introduced in PKSs, and different polyketides have emerged. Similarly, the products of NRPSs can be varied by manipulation of the modules that add one amino acid at a time to the growing peptide chains. The potential ability to generate new pharmaceutical compounds by using protein engineering to permute and recombine modified domains is intensifying structural and mechanistic interest in PKSs and NRPSs. This could be of major importance in the search for new pharmaceutical compounds, of value for example in tumor therapy or to combat the potentially disastrous emergence of bacterial resistance to known antibiotics.

Because of their modular construction around the non-catalytic scaffoldins, cellulosomes are also obvious targets for the creation of designer multienzyme complexes. In native cellulosomes, a single type of cohesin/dockerin interaction, means that assembly of the complex is a quasi-random process. The complexes produced by a bacterium are therefore heterogeneous in their enzyme content and disposition. However, by generating cohesin/dockerin pairs of differing specificity, complexes of any desired stoichiometry and catalytic specificity can be assembled on modified scaffoldins (**Figure 9.31**). The assemblies need not be restricted to cellulolytic activities, and they hold promise for a variety of biotechnological applications. Designer cellulosomes are of particular interest for their potential for converting plant cellulosic biomass to useful products. It has been estimated that, each year, microorganisms degrade ~10^{11} tons of plant biomass, releasing energy equivalent to 640 billion barrels of crude oil. Major benefits would accrue if, for example, some of the degradation could be directed to producing fermentable sugars for fuel.

9.7   SUMMARY

Enzymes that catalyze multi-step biochemical transformations are often found as multienzyme complexes. These catalytic nanomachines limit the diffusion pathway of reaction intermediates, channel them between successive active sites, and if necessary, protect them against potential loss from the cell, adventitious degradation, and unwanted side-reactions. Multienzyme complexes can also gather together a collection of enzymes to act synergistically on refractory substrates.

Substrates can be channeled through molecular tunnels. Tryptophan synthase has a tunnel ~25 Å long, through which an indole intermediate generated by the α subunit is passed to the active site in the β subunit. In other multienzyme complexes, ammonia is generated by the hydrolysis of glutamine to glutamic acid in one subunit, and passed through a molecular tunnel as long as ~90 Å to participate in a reaction in an active site of another subunit.

(A) wild-type cellulosomes

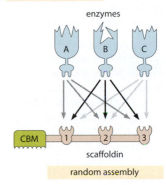

(B) designer cellulosomes

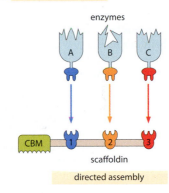

Figure 9.31 An approach to designer cellulosomes. In wild-type scaffoldins, the type I cohesin/dockerin interactions are unspecific and generate no particular order in the structure. To create specific cohesin/dockerin interactions, scaffoldins can be engineered to bind enzymes in desired orders and with chosen specificities.

Some tunnels are not preformed but are generated when the two active sites are occupied by ligands. Tunnels need to be gated, usually by allosteric conformational changes that control access to their entrances and exits.

Another mode of substrate channeling is the covalent attachment of intermediates to a prosthetic group in a swinging arm. Two examples are lipoic acid and biotin, both of which are linked to the N^6-amino groups of specific lysine residues in small protein domains (9 kDa) flexibly tethered by long linker regions to enzymes in their respective complexes. In the giant (5–10 MDa) 2-oxo acid dehydrogenase (2-OADH) complexes, the lipoyl domains of the E2 component couple reactions in multiple E1 (decarboxylase) and E3 (dehydrogenase) subunits assembled round an icosahedral (60-mer) or octahedral (24-mer) E2 transacylase core. Biotin-dependent carboxylases acting on acetyl-CoA, pyruvate, and propionyl-CoA perform the carboxylation of biotin in an ATP-dependent carboxylase subunit, and ferry it on a biotinyl domain to the active site in a carboxyltransferase subunit, where the CO_2 is reacted with the substrate. The lipoyl and biotinyl domains bring about large increases in k_{cat}/K_m compared with free lipoic acid or biotin, and ensure multienzyme complex specificity.

Fatty acid synthases (FASs) have a phosphopantetheine swinging arm attached to a serine side chain in a mobile acyl carrier protein (ACP) flexibly linked to the complex. A fixed set of enzymes act in cycles on substrates and intermediates linked as thioesters to the swinging arm. The iterative addition of C_2 units (from acetyl-CoA) generates fatty acids, which are released from the swinging arm when they reach the required length. Two types of FASs are known: one (from animal cells) is a giant dimer of multifunctional polypeptide chains; the other (from fungi) consists of $\alpha_6\beta_6$ dodecamers in the shape of double–domed cages. Fatty acid degradation is carried out by a similar set of enzymes in small complexes, but acting on fatty acyl-CoA intermediates. Polyketide and non-ribosomal peptide synthases resemble FASs in the successive reactions with intermediates carried on phosphopantetheine swinging arms in ACPs, but they differ in that other enzyme modules work successively on them as they grow in complexity and are passed from module to module.

The cellulosomes of anaerobic bacteria exemplify another kind of nanomachine. These extracellular multienzyme complexes bring together multiple cellulolytic enzymes for the synergistic degradation of refractory cellulose polymers (plant cell walls). The enzymes are assembled on a non-catalytic scaffold protein by the interaction of their dockerin modules with cohesin modules on the scaffold. Carbohydrate-binding modules on the scaffoldin and/or enzymes anchor the assembly to the cellulose substrate.

A section of this chapter is based on the following contribution

9.6   Edward Bayer, Weizmann Institute of Science, Rehovot, Israel and Raphael Lamed, Tel Aviv University, Israel.

REFERENCES

9.1   Introduction

Williamson MR (2012) How Proteins Work. Garland Science.

9.2   Active-site coupling and substrate channeling in protein complexes

Perham RN (2000) Swinging arms and swinging domains in multifunctional enzymes: catalytic machines for multistep reactions. *Annu Rev Biochem* 69:961–1004.

Reed LJ (1974) Multienzyme complexes. *Acc Chem Res* 7:40–46.

9.3   Multienzyme complexes with tunnels

Barends TRM, Dunn MF & Schlichting I (2008) Tryptophan synthase, an allosteric molecular factory. *Curr Opin Chem Biol* 12:593–600.

Dunn MF (2012) Allosteric regulation of substrate channeling and catalysis in the tryptophan synthase bienzyme complex. *Arch Biochem Biophys* 519:154–166.

Miles EW, Rhee S & Davies DR (1999) The molecular basis of substrate channeling. *J Biol Chem* 274:2193–2196.

Niks D, Hilario E, Dierkers A et al (2013) Allostery and substrate channeling in the tryptophan synthase bienzyme complex: evidence for two subunit conformations and four quaternary states. *Biochemistry* 52:6396–6411.

Raushel FM, Thoden JB & Holden HM (1999) The amidotransferase family of enzymes: molecular machines for the production and delivery of ammonia. *Biochemistry* 38:7891–7899.

Raushel FM, Thoden JB & Holden HM (2003) Enzymes with molecular tunnels. *Acc Chem Res* 36:539–548.

9.4   Multienzyme complexes with lipoic acid or biotin in their swinging arms

Blanchard CZ, Chapman-Smith A, Wallace JC & Waldrop GL (1999) The biotin domain peptide from the biotin carboxyl carrier protein of *Escherichia coli* acetyl-CoA carboxylase causes a marked increase in the catalytic efficiency of biotin carboxylase and carboxyltransferase relative to free biotin. *J Biol Chem* 274:31767–31769.

Bocanegra JA, Scrutton NS & Perham RN (1993) Creation of an NADP-dependent pyruvate dehydrogenase multienzyme complex by protein engineering. *Biochemistry* 32:2737–2740.

Cronan JE Jr (2002) Interchangeable enzyme modules. Functional replacement of the essential linker of the biotinylated subunit of acetyl-CoA carboxylase with a linker from the lipoylated subunit of pyruvate dehydrogenase. *J Biol Chem* 277:22520–22527.

Cronan JE Jr & Waldrop GL (2002) Multisubunit acetyl-coA carboxylases. *Prog Lipid Res* 41:407–435.

Domingo G, Chauhan HJ, Lessard IAD et al (1999) Self-assembly and catalytic activity of the pyruvate dehydrogenase multienzyme complex from *Bacillus stearothermophilus. Eur J Biochem* 266:1136–1146.

Frank RAW, Pratap JV, Pei XY et al (2005) The molecular origins of specificity in the assembly of a multienzyme complex. *Structure* 13:1119–1130.

Frank RAW, Titman CM, Pratap JV et al (2004) A molecular switch and proton wire synchronize the active sites in thiamine enzymes. *Science* 306:872–876.

Guest JR (1987) Functional implications of structural homologies between chloramphenicol acetyltransferse and dihydrolipoamide acetyltransferase. *FEMS Microbiol Lett* 44:417–422.

Hiromasa Y & Roche TE (2003) Facilitated interaction between the pyruvate dehydrogenase kinase isoform 2 and the dihydrolipoyl acetyltransferase. *J Biol Chem* 278:33681–33693.

Jones DD, Stott KM, Reche PA & Perham RN (2001) Recognition of the lipoyl domain is the ultimate determinant of substrate channeling in the pyruvate dehydrogenase multienzyme complex. *J Mol Biol* 305:49–60.

Jordan SW & Cronan JE Jr (2003) The *Escherichia coli lipB* gene encodes lipoyl (octanoyl)-acyl carrier protein:protein transferase. *J Bacteriol* 185:1582–1589.

Lengyel JS, Stott KM, Wu X et al (2008) Extended polypeptide linkers establish the spatial architecture of a pyruvate dehydrogenase multienzyme complex. *Structure* 16:93–103.

Knowles JR (1989) The mechanism of biotin-dependent enzymes. *Annu Rev Biochem* 58:195–221.

Marrot NL, Marshall JJT, Svergun DI et al (2012) The catalytic core of an archaeal 2-oxoacid dehydrogenase multienzyme complex is a 42-mer protein assembly. *FEBS J* 279:713–723.

Menefee AL & Zeczycki TN (2014) Nearly 50 years in the making; defining the catalytic mechanism of the multifunctional enzyme, pyruvate carboxylase. *FEBS J* 281:1333–1354.

Nemeria NS, Arjunan P, Chandrasekhar K et al (2010) Communication between thiamin cofactors in the *Escherichia coli* pyruvate dehydrogenase complex E1 component active centers: evidence for a 'direct pathway' between the 4′-aminopyrimidine N1′ atoms. *J Biol Chem* 285:11197–11209.

Perham RN (1991) Domains, motifs and linkers in 2-oxo acid dehydrogenase multienzyme complexes: a paradigm in the design of a multifunctional enzyme. *Biochemistry* 30:8501–8512.

Reed LJ & Hackert ML (1990) Structure–function relationships in dihydrolipoamide acyltransferases. *J Biol Chem* 265:8971–8974.

Roberts EL, Shu N, Howard MJ et al (1999) Solution structures of apo and holo biotinyl domains of acetyl CoA carboxylase of *Escherichia coli* determined by triple-resonance nuclear magnetic resonance spectroscopy. *Biochemistry* 38:5045–5053.

Roche TE & Hiromasa Y (2007) Pyruvate dehydrogenase kinase regulatory mechanisms and inhibition in treating diabetes, heart ischemia, and cancer. *Cell Mol Life Sci* 64:830–849.

Vijayakrishnan S, Callow P, Nutley MA et al (2011) Variation in the organization and subunit composition of the mammalian pyruvate dehydrogenase complex E2/E3BP core assembly. *Biochem J* 437:565–574.

Waldrop GL, Holden HM & St Maurice M (2012) The enzymes of biotin dependent CO_2 metabolism: what structures reveal about their reaction mechanisms. *Prot Sci* 21:1597–1619.

Weaver LH, Kwon K, Beckett D & Matthews BW (2001) Corepressor-induced organization and assembly of the biotin repressor: a model for allosteric activation of a transcriptional regulator. *Proc Natl Acad Sci USA* 98:6045–6050.

Yu X, Hiromasa Y, Tsen H et al (2008) Structures of the human pyruvate dehydrogenase complex cores: a highly conserved catalytic center with flexible N-terminal domains. *Structure* 16:104–114.

9.5 Multienzyme complexes with phosphopantetheine swinging arms

Boehringer D, Ban N & Leibundgut M (2013) 7.5 Å cryo-EM structure of the mycobacterial fatty acid synthase. *J Mol Biol* 425:841–849.

Brignole EJ, Smith S & Asturias FJ (2009) Conformational flexibility of metazoan fatty acid synthase enables catalysis. *Nat Struct Mol Biol* 16:190–197.

Hur GH, Vickery CR & Burkart MD (2012) Explorations of catalytic domains in non-ribosomal peptide synthetase enzymology. *Nat Prod Rep* 29:1074–1098.

Jenni S, Leibundgut M, Boehringer D et al (2007) Structure of fungal fatty acid synthase and implications for iterative substrate shuttling. *Science* 316:254–261.

Joshi AK, Zhang L, Rangan VS & Smith S (2003) Cloning, expression, and characterization of a human 4′-phosphopantetheinyl transferase with broad substrate specificity. *J Biol Chem* 278:33142–33149.

Keatinge-Clay AT (2012) The structures of type I polyketide synthases. *Nat Prod Rep* 29:1050–1073.

Khosla C, Tang Y, Chen AY et al (2007) Structure and mechanism of the 6-deoxyerythronolide B synthase. *Annu Rev Biochem* 76:195–221.

Lomakin IB, Xiong Y & Steitz TA (2007) The crystal structure of yeast fatty acid synthase, a cellular machine with eight active sites working together. *Cell* 129:319–332.

Lowry B, Tobbins T, Weng C-H et al (2013) *In vitro* reconstitution and analysis of the 6-deoxyerythronolide B synthase. *J Am Chem Soc* 135:16809–16812.

Maier T, Leibundgut M & Ban N (2008) The crystal structure of a mammalian fatty acid synthase. *Science* 321:1315–1322.

Maier T, Leibundgut M, Boehringer D & Ban N (2010) Structure and function of eukaryotic fatty acid synthases. *Q Rev Biophys* 43:373–422.

Nguyen C, Haushalter RW, Lee DJ et al (2014) Trapping the dynamic acyl carrier protein in fatty acid biosynthesis. *Nature* 505:427–431.

Walsh CT, O'Brien RV & Khosla C (2013) Nonproteinogenic amino acid building blocks for nonribosomal peptide and hybrid polyketide scaffolds. *Angew Chem Int Ed* 52:7098–7124.

Wong FT & Khosla C (2012) Combinatorial biosynthesis of polyketides—a perspective. *Curr Opin Chem Biol* 16:117–123.

Zheng J, Gay DC, Demeler B et al (2012) Divergence of multimodular polyketide synthases revealed by a didomain structure. *Nat Chem Biol* 7:615–621.

Zheng J, Fage CD, Demeler B et al (2013) The missing linker: a dimerization motif located within polyketide synthase modules. *ACS Chem Biol* 8:1263–1270.

9.6 The cellulosome

Bayer EA, Belaich JP, Shoham Y & Lamed R (2004) The cellulosomes: multienzyme machines for degradation of plant cell wall polysaccharides. *Annu Rev Microbiol* 58:521–554.

Brás JLA, Cartmell A, Carvalho ALM et al (2011) Structural insights into a unique cellulose fold and mechanism of cellulose hydrolysis. *Proc Natl Acad Sci USA* 108:5237–5242.

Currie MA, Adams JJ, Faucher F et al (2012) Scaffoldin conformation and dynamics revealed by a ternary complex from the *Clostridium thermocellum* cellulosome. *J Biol Chem* 287:26953–26961.

Cuskin F, Flint JE, Gloster TM et al (2012) How nature can exploit nonspecific catalytic and carbohydrate binding modules to create enzymatic specificity. *Proc Natl Acad Sci USA* 109:20889–20894.

Doi RH & Kosugi A (2004) Cellulosomes: plant-cell-wall-degrading enzyme complexes. *Nat Rev Microbiol* 2:541–551.

Fontes CMGA & Gilbert HJ (2010) Cellulosomes: highly efficient nanomachines designed to deconstruct plant cell wall complex carbohydrates. *Annu Rev Biochem* 79:655–681.

Kern J, Wilton R, Zhang R et al (2011) Structure of surface layer homology (SLH) domains from *Bacillus anthracis* surface array protein. *J Biol Chem* 286:26042–26049.

Levy-Assaraf M, Voronov-Goldman M, Rozman Grinberg M et al (2013) Crystal structure of an uncommon cellulosome-related protein module from *Ruminococcus flavefaciens* that resembles papain-like cysteine peptidases. *PLoS ONE* 8:e56138 (doi:10.1371/journal.pone.0056138).

Ragaukas AJ, Williams CK, Davison BH et al (2006) The path forward for biofuels and biomaterials. *Science* 311:484–489.

Rincon MT, Čepeljnik T, Martin JC et al (2005) Unconventional mode of attachment of the *Ruminococcus flavefaciens* cellulosome to the cell surface. *J Bacteriol* 187:7569–7578.

Salama-Alber O, Jobby MK, Chitayat S et al (2013) Atypical cohesin–dockerin complex responsible for cell-surface attachment of cellulosomal components: binding fidelity, promiscuity, and structural buttresses. *J Biol Chem* 288:16827–16838.

adaptor
protein

Transport

10

10.1 INTRODUCTION

Eukaryotic cells are generally much larger than prokaryotic cells. To avoid havoc, eukaryotic cells are subdivided into functionally distinct membrane-enclosed compartments or organelles. Most of the genetic material of eukaryotes is stored in a single organelle, the nucleus. The nucleus is suspended in the cytosol, together with all other organelles such as the endoplasmic reticulum, the Golgi apparatus, lysosomes, mitochondria, and (in plants) chloroplasts. This compartmentation brings about the problem of how molecules are exchanged between organelles. For example, most nucleic acids are synthesized in the nucleus, but proteins are synthesized (using RNA templates) in the cytoplasm. Several fundamentally different mechanisms allow the exchange of molecules between compartments.

Some compartments are topologically equivalent (**Figure 10.1**) and so materials can be exchanged by budding off small membrane-enclosed vesicles from one compartment and their fusion with another compartment. Thus, cargo within the vesicle is delivered from one compartment to another. This chapter will describe how these vesicles are formed (Section 10.2), the protein machines (dynamins) responsible for their budding (Section 10.3), and the machinery responsible for membrane fusion (Section 10.4).

Vesicle transport between topologically equivalent compartments does not require the crossing of membranes but does require their fission and fusion. However, cells do need to import or export molecules from the interior of the cell to the exterior, or between discrete organelles such as mitochondria, which all involve the crossing of membranes. The details of transmembrane transport are covered in Chapter 16.

Large amounts of materials are exchanged between the cytosol and the nucleus. The nuclear membrane is punctuated by pore complexes, which are gated to control what enters and leaves the nucleus. The structure of these pores is described in Section 10.5 and the mechanism of nuclear import and export in Section 10.6.

Bacteria are not physically compartmentalized like eukaryotic cells but still need mechanisms to import and export materials across their bounding membranes. Pathogenic bacteria also have machines that allow them to insert materials into host cells. These will be described in Section 10.7.

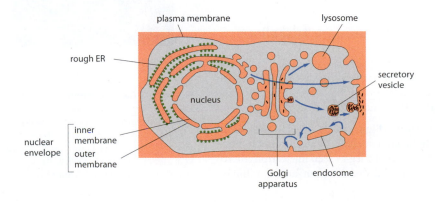

Figure 10.1 Topological relationship between cellular compartments. Cycles of membrane budding, vesicle formation, and fusion allow the exchange of material between topological compartments (shown in red) and the exterior of the cell. Arrows indicate main directions of outbound and inbound traffic. Some organelles like mitochondria or chloroplasts are not involved in vesicular membrane traffic. (Adapted from B. Alberts et al. *Mol. Biol. Cell,* 6th Edition, 2015)

10.2 CLATHRIN-MEDIATED ENDOCYTOSIS

The uptake of material into a cell by invagination of the plasma membrane and its internalization in a membrane-enclosed vesicle is referred to as endocytosis. When the selection of cargo is mediated by membrane-spanning receptors, we speak of receptor-mediated endocytosis. In the secretory pathway, molecules destined to be moved out of the cell become enclosed in vesicles that fuse with the plasma membrane to release their cargo to the exterior. Both processes involve large membrane fluxes.

It has been estimated that complete turnover of the plasma membrane occurs in less than 1 hour and that a protein bound for the plasma membrane takes less than 40 minutes to journey from its site of synthesis in the endoplasmic reticulum through the Golgi apparatus to its final destination. Vesicular traffic must not only allow for the exchange of components between membrane-enclosed compartments; it must, at the same time, maintain the differences in composition that exist between different membranes. Therefore, membrane traffic must be selective. Only a subset of proteins and lipids of a donor membrane is allowed to enter the carrier vesicle, effectively preventing the 'homogenization' of membranes.

The budding of vesicles from a donor membrane and their fusion with a target membrane can be broken down into several steps. Budding starts with the coordinated deformation and invagination of patches of the donor membrane, and the selective recruitment of cargo for inclusion into the carrier vesicle. Next there is scission of the membrane bridge between donor and carrier membrane, and intracellular movement of the carrier. Finally, there is targeted delivery of the cargo by fusion of the carrier vesicle with that of the acceptor organelle. These steps involve the concerted action of dozens of different proteins. Most studies have focused on carrier vesicles coated by three types of coat protein: **clathrin** mediates traffic between the plasma membrane and endosomes, and COPI and COPII mediate movements between the Golgi apparatus and the endoplasmic reticulum.

Clathrin-coated pits and vesicles are transient molecular assemblies that transport a wide range of different cargoes

Clathrin-coated structures were the first carrier vesicles to be discovered and analyzed in detail because of the distinctive morphology of budding coated pits in electron micrographs (**Figure 10.2**), the ease with which coated vesicles could be purified, and the importance of clathrin-coated structures for receptor-mediated endocytosis. The 'life cycle' of typical endocytic clathrin-coated vesicles, from coat assembly and cargo recruitment to coat disassembly is on a timescale ranging from seconds to a few minutes. Several structural proteins and a larger number of regulatory proteins, some present only in small numbers, participate in a tightly coordinated sequence of events. Clathrin-coated vesicles are believed to transport approximately 50% of all endocytic cargo from the plasma membrane to endosomes. Clathrin-mediated endocytosis is also the major route for the rapid recycling of membrane components at synapses between nerve cells (see Section 10.4). Most of these coated vesicles range in size between 70 and 150 nm and accommodate a variety of cargoes ranging from small receptor-bound ligands to larger objects such as viruses. In most cases the clathrin coat fully surrounds the vesicle containing the cargo, except where the vesicle is too large for complete closure of the clathrin coat. Clathrin-coated pits also allow entry of some invasive bacteria such as *Listeria*.

The building blocks of clathrin-coated structures are 'triskelions'

Crystal structures of parts of clathrin, adaptors, and other clathrin-associated molecules together with cryo-EM (see **Methods**) studies of fully or partially (re-)assembled clathrin coats, have led to a fairly detailed picture of the framework of the coat, suggesting how clathrin assembly is coupled to cargo incorporation and a mechanism of uncoating.

The building block of a clathrin coat is a spidery trimer (**Figure 10.3**) called a **'triskelion,'** which has three 'heavy chains' (each 190 kDa, 1675 aa) associated at a central hub. Each triskelion leg is about 475 Å long. The characteristic curl of a leg as seen in electron micrographs allows distinguishing of a 'proximal leg,' a 'knee,' a 'distal leg,' an 'ankle,' a 'linker,'

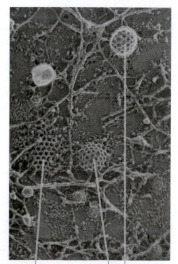

flat clathrin coat   deeply curved clathrin coat

Figure 10.2 Morphogenesis of clathrin coats. Electron micrograph of metal-shadowed plasma membrane showing clathrin-coated pits at different stages of invagination. (Adapted from L.M. Fujimoto et al., *Traffic* 1:161–171, 2000. With permission of John Wiley & Sons.)

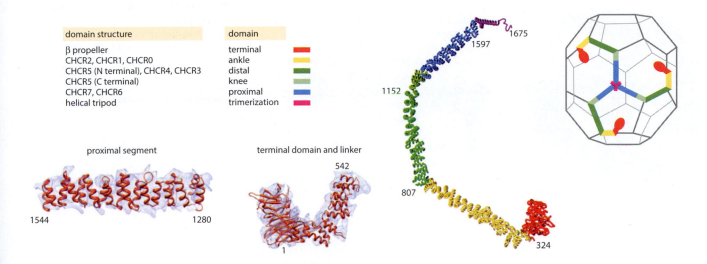

domain structure

β propeller
CHCR2, CHCR1, CHCR0
CHCR5 (N terminal), CHCR4, CHCR3
CHCR5 (C terminal)
CHCR7, CHCR6
helical tripod

domain

terminal
ankle
distal
knee
proximal
trimerization

proximal segment

terminal domain and linker

Figure 10.3 Molecular architecture of clathrin coats. Schematic representation of a clathrin coat with a single triskelion highlighted (right). The color coding of the domains is explained in the table (left). Below are crystal structures of two domains (proximal segment and terminal domain with linker). They were fitted into the EM density of a leg of a triskelion extracted from a reconstructed clathrin coat. (Middle, adapted from M. Edeling et al., *Nature* 7:32–44, 2006. With permission from Macmillan Publishers Ltd.)

and a 'terminal domain'. The thickness of a leg is rather uniform, except for the terminal domain, a globular element at the N terminus of the heavy chain. A single light chain (~25 kDa) associates with each heavy-chain proximal segment, through a central segment of 71 residues with a clearly recognizable α-helical heptad repeat (see below). This description is based on a conventional clathrin heavy chain that is expressed in all metazoans, including humans. A second heavy-chain homolog in humans has a similar sequence but appears not to interact with the light chains or with conventional clathrin; it is supposed to have a specialized role in the regulated secretory pathway in muscle and adipocytes. Vertebrate genomes encode two light chains, designated LCa and LCb, each of which can have differentially spliced variants.

The main motif of the clathrin heavy chains is an extended α-helical zigzag

The central region of the 1675 aa vertebrate heavy chain (residues 331–1630), forms a continuous set of 40 transversely directed α-helical zigzags (Figure 10.3). A crystal structure showed that the N-terminal domain (residues 1–330) is a seven-bladed β propeller, connected to α-helical zigzags (residues 331–494). A second crystal structure showed that a substantial part of the proximal segment is likewise an α-helical zigzag. The latter structure also revealed a super-repeat of about 145 residues or five helical α-zigzags—the clathrin heavy-chain repeat (CHCR). Residues 1280–1576 contain eight complete repeats, designated CHCR0-7. A cryo-EM reconstruction combined with the two crystal structures has yielded a model of the triskelion (Figure 10.3). Homology-modeled structures for the remaining CHCRs were docked into the cryo-EM density. The resulting model covers almost the entire length of a leg, without gaps or overlaps. The EM reconstruction also revealed a threefold symmetric feature projecting inwards from the hub: a tripod of 50 Å long α helices (residues 1598–1630). The remaining C-terminal residues (1631–1675) do not have clearly resolved secondary structure. This C-terminal segment hangs under the hub and contains the sequence QLMLT, to which the Hsc70 ATPase which drives the uncoating reaction (see below) can attach.

Clathrin light chains are mostly unstructured in isolation but adopt an extended organization (see Section 6.4) when bound to the heavy chain. The only well-ordered part of the light chain is an extended 71-residue α helix, formed by the central, heptad-repeat segment, which interacts with a portion of the heavy chain relatively close to the triskelion hub. The C-terminal part of the light chain projects toward the hub and hence lies near a vertex in a coat. The 100 Å long light-chain helix interacts with a surface formed by interhelical loops in the heavy-chain proximal leg (**Figure 10.4**). The axis of the light chain, which tilts relative to the axis of the heavy chain, roughly follows the right-hand twist of the heavy-chain α-helical zigzags. The N- and C-terminal portions of the light chain may adopt locally ordered structures in association with other proteins.

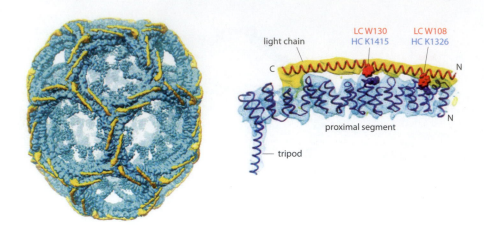

Figure 10.4 Clathrin light chains associate with the heavy chains. Localization of clathrin light chains as obtained by EM difference imaging. Light chains, shown in yellow, align with the proximal segments of the triskelions. (From Fotin et al., *Nature* 432:573–579, 2004. With permission from Macmillan Publishers Ltd.)

Assembly of clathrin-coated structures and recruitment of cargo requires helper proteins

The most common clathrin-coated structures begin as coated pits, invaginations in which assembly of a curved clathrin lattice deforms the underlying membrane and eventually yields, by membrane constriction and scission, a coated vesicle (**Figure 10.5**). In some instances constriction depends solely on completion of the clathrin lattice; in others constriction is aided by local actin polymerization propelling the coated membrane into the interior of the cell. Membrane scission of endocytic coated pits depends on the action of dynamin (Section 10.3). Other still unidentified molecules carry out this step in endosomal clathrin coats. COPI- and COPII-mediated pathways have a similar sequence of events (**Box 10.1**).

Cargo sorting, the process of selective incorporation of membrane-bound proteins into the coated vesicles, is controlled by adaptors. Located in the space between the outer clathrin scaffold and the cargo-bearing lipid vesicle, they bind simultaneously to clathrin and membrane-spanning proteins and/or phospholipids. The most prominent adaptors are the so-called APs (adaptor proteins)—the heterotetrameric AP2 complex (see below) and its homologs. As their name implies, they have a number of binding partners and are essential

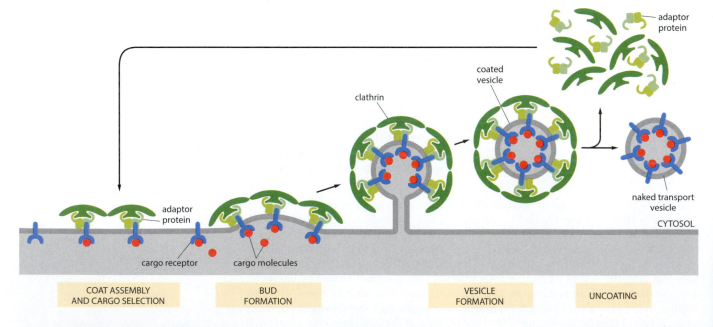

Figure 10.5 Morphogenesis of a clathrin-coated vesicle. Sequence of events in the formation of clathrin-coated vesicles. (Adapted from T. Kirchhausen, *Nat. Rev. Mol. Cell Biol.* 1:187–198, 2000. With permission from Macmillan Publishers Ltd.)

Box 10.1 The COPI and COPII membrane traffic pathways

COPI and COPII vesicles commute between the endoplasmic reticulum (ER) and the Golgi complex; COPI primarily from the Golgi to the ER and between Golgi cisternae, and COPII from the ER to the Golgi. COPII-coated vesicles were originally discovered in *Saccharomyces cerevisiae* using genetic approaches in conjunction with cell-free assays monitoring the transfer of marker proteins. Most components of the yeast COPII system have counterparts in mammalian cells (**Table 10.1**).

COPII vesicular transport can be reconstituted with a minimal system comprising the following cytosolic components: the Sec13p–Sec31p heterodimer, the Sec23p–Sec24p complex, and the small GTPase Sar1p. Sar1p has a particularly important role; its activation initiates coat formation. The GDP-bound form of Sar1p is normally cytosolic and is recruited to the ER membrane upon interaction with Sec12p, an ER membrane protein serving as a guanine exchange factor (GEF) for Sar1p. Sar1p-GTP then promotes the association of the Sec23p–Sec24p adaptor complex with cargo proteins. Members of the p24 family of transmembrane proteins are thought to serve as

cargo adaptors. In addition to recruiting the Sec23p–Sec24p complex, Sar1p-GTP activates Sec23p to bind SNARE proteins (Section 10.4) required for targeting and fusion to the target membranes. ER membrane patches populated with Sec23–Sec24p and Sar1p then recruit Sec13–Sec31p, the complex required for coat formation, the enforcement of membrane curvature, and eventually budding. Sec13p is a WD40 β propeller. Sec31p also has an N-terminal propeller followed by an α solenoid reminiscent of clathrin. Unlike clathrin, however, the β propellers of Sec13p–Sec31p form the vesicles of the assembled COPII coats. Finally, after GTP hydrolysis, induced by Sec23p, Sar1p-GDP is released, which results in vesicle uncoating and fusion to the target membrane.

Whereas COPII-dependent traffic is unidirectional (ER to Golgi), COPI appears to be more versatile (Golgi to ER and between Golgi cisternae) and this is reflected by a higher degree of complexity. The COPI coatomer is a complex of seven proteins (α, β, β′, γ, δ, ε, and ζ) (see Table 10.1). Of these, three (α, β′, and ε) assemble into a heterotrimer forming the coat. The remaining

(Box continues on next page.)

Table 10.1 Coatomer proteins

Protein	Yeast	Mammalian	Structural features and binding partners
COPII			
Sar1p	Sar1p	hSar1p	Small GTPase; Ras family
Sec13p–Sec31p	Sec13p	hSec13p	WD40 repeats (β propeller)
	Sec31p	hSec31p	WD40 repeats (β propeller)
Sec23p–Sec24p	Sec23p	hSec23p	Homology with Sec24p; GTPase activating protein for Sar1p
	Sec24p/Iss1p/Lst1p	hSec24p	Homology with Sec23p
	Sec12p	?	GEF, guanine exchange factor for Sar1p
	Sec16p	?	Membrane protein; forms ternary complex with Sec23p–Sec24p
COPI			
ARF1	yARF1/2/3	ARF1	Small GTPase; Ras family
Coatomer	Ret1p	α-COP	WD40 repeats (β propeller)
	Sec26p	β-COP	Binds ARF, ADP-ribosylation factor
	Sec27p	β′-COP	WD40 repeats (β propeller)
	Sec21p	γ-COP	Binds members of p24 family
	Ret2p	δ-COP	Weak sequence identity to μ-AP (adaptor protein)
	Sec28p	ε-COP	
	Ret3p	ζ-COP	Weak sequence identity to σ-AP (adaptor protein)
ARFGAP	Glo3p	ARFGAPs	GTPase activating protein for ARF
ARFGEF	Gea1p/Gea2p	ARFGEFs	GEF, guanine exchange factor for ARF

Box 10.1 *continued*

four form a complex similar to the heterotetrameric clathrin adaptors. A crystal structure is available for the αβ' subcomplex. The β' subunit forms two tandem WD40 β propellers followed by an α solenoid and the central part of the α subunit contributes another α-solenoidal segment. Jointly, they assemble into a triskelion-like trimer, with the dual β propellers of β' at its center and the overlapping anti-parallel α solenoids of the α and β' chains radiating outwards (**Figure 10.1.1**). The C-terminal part of α-COPI interacts with ε-COP. This subcomplex is flexibly linked to the αβ' subcomplex and probably projects inwards from the αβ' lattice, enabling it to recruit the βγδζ-adaptor complex. COPI vesicles capture and deliver cargo carrying C-terminal sorting signals with KKXX or KXKXX motifs (X is any amino acid). The γ subunit is the coatomer subunit recognizing these motifs.

The initial event in the COPI pathway is the association of the GTPase ARF1 (ADP-ribosylation factor 1) in its active form to the membrane (ARF1-GTP). The ARF protein family has many members and recruitment of ARF1 to its target membrane involves a specific association with its cognate GEF (guanine exchange factor). ARF 1 is myristoylated, which anchors it to the membrane. In the GTP-bound state the myristoyl group is exposed; upon GTP hydrolysis it becomes sequestered, resulting in the release of ARF1. Similar to Sar1p in the COPII system, ARF1 is believed to act as a timer, triggering the release of the coatomers and thus preparing the vesicles for fusion (**Figure 10.1.2**).

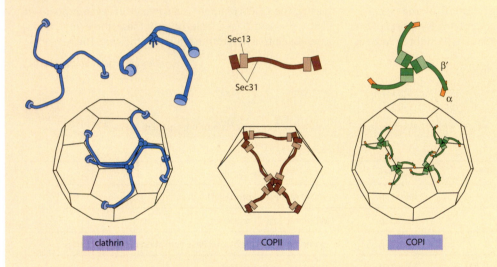

clathrin   COPII   COPI

Figure 10.1.1 Schematic representation of three coat proteins and their modes of assembly. The coat proteins of vesicles involved in membrane traffic have common architectural principles. This includes the structural motifs that form the open lattice structures. The basic elements are β propellers and α-helical zigzags, known as α solenoids. The latter form stiff but still compliant structures, which can accommodate vesicles of variable size and shape. In clathrin the basic unit is the triskelion with the β propeller domains at the tips of the legs. In the assembled coat, there is a triskelion at each vertex. In COPII, the heterodimeric Sec31p/Sec13p complex forms the coat. The N-terminal β propeller domain of Sec31p is at the tip of the rodlike complex. In the cuboctahedral lattice the propeller domains pack around the vertices. In COPI, the N-terminal half of the β' subunit and a central segment of the α subunit form the assembly unit. The triskelion structure shown schematically corresponds to a crystal structure but the arrangement of the triskelions in the lattice is speculative. (Adapted from Harrison & Kirchhausen *Nature* 466:1048-1049, 2010. With permission from Macmillan Publishers Ltd.)

for the formation of clathrin-coated pits at the plasma membrane. Other, more specialized adaptors, often referred to as sorting proteins, such as β-arrestins or epsin, are recruited to the clathrin coat by interacting with both clathrin and APs. The protein and/or lipid signals that initiate clathrin coat assembly are not yet known. Cargo recruitment and early association of the coat with dynamin (see Section 10.3) appear to stabilize the assembling coats; in their absence, coated pits fail to complete their assembly and the coat dissociates. In some cell types, these 'abortive coats' can outnumber the fully assembled ones.

Box 10.1 *continued*

Figure 10.1.2 Sequence of events in the formation of COPI-coated vesicles. (Adapted from Popoff et al. *Cold Spring Harb Persp Biol*, 3: 11, 2011.)

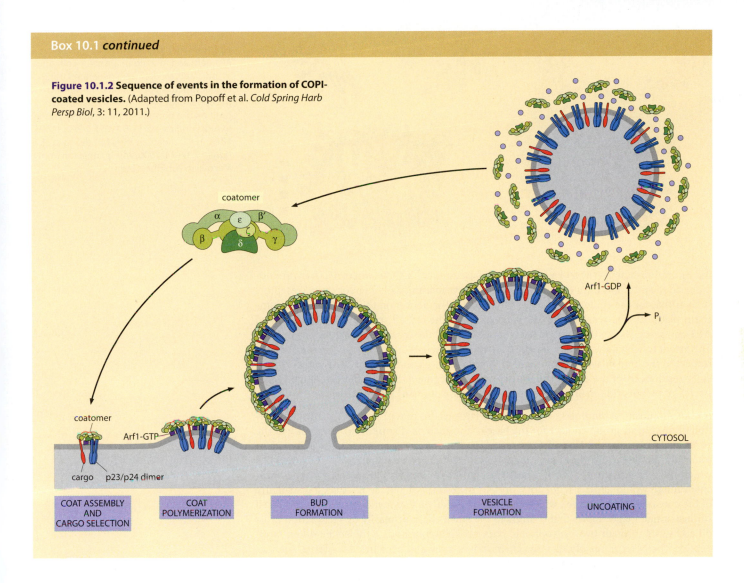

Heterotetrameric adaptors are the most abundant non-clathrin components of coated vesicles and mediate interaction with endocytic signals

Vertebrate genomes encode four types of adaptors (AP1–AP4). AP2 is the predominant adaptor for the plasma membrane, whereas AP1, AP3, and AP4 localize to endosomes. The adaptors comprise two large chains (γ and β_1 in AP1 and α and β_2 in AP2), a medium chain (μ_1 in AP1 and μ_2 in AP2), and a small chain (σ_1 in AP1 and σ_2 in AP2). The β, μ, and σ subunits are homologous (~40–80% sequence identity), whereas α and γ are more diverse (~25% identity). Each large chain has a compactly folded N-terminal half composed of a set of helical zigzag repeats (closely related to the so-called Armadillo repeats which form helical solenoids, see **Guide**), an extended (~100 residue) hinge segment, and a C-terminal 'appendage domain' or 'ear' (**Figure 10.6**). The heterotetrameric complex has a pseudo-two-fold symmetric 'brick-shaped' base, formed by the N-terminal regions of the large chains and the homologous regions of the medium and small chains. The unpaired C-terminal domain of the medium chain rests against this base, with the large-chain appendages loosely connected. The AP base (the heterotetrameric complex without the large-chain appendages and their linkers) faces the membrane, and its interactions with various endocytic signals (see below) direct incorporation of cargo into an assembling coated pit.

The medium chain has an apparently flexible hinge between its N-terminal domain (the small-chain homolog) and its C-terminal domain. Structures of this C-terminal domain with a bound peptide identify clearly the site on the medium chain for one important cargo

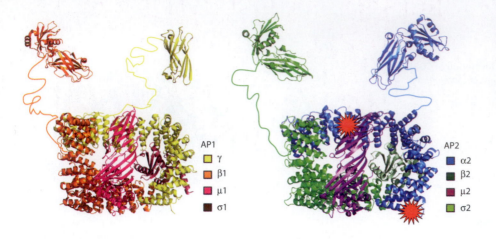

Figure 10.6 Crystal structures of the adaptors AP1 and AP2. The subunits of AP1 (left) and AP2 (right) are color-coded as indicated. Models were generated using crystal structures of subunits. The linkers to the appendages (top) were modeled. The position of a phosphoinositide mimic bound to AP2 is indicated. (PDB 1W63, 1GYU, 1GW5, 1B9K, 1E42) (Adapted from M. Edeling et al., *Nature* 7:32–44, 2006. With permission from Macmillan Publishers Ltd.)

AP1

- ■ γ
- ■ β1
- ■ μ1
- ■ σ1

AP2

- ■ α2
- ■ β2
- ■ μ2
- ■ σ2

incorporation signal—the so-called 'tyrosine-based motif', YXXΦ (where Φ is a hydrophobic residue and X any amino acid), found on the cytoplasmic tail of endocytosed receptors such as that for the epidermal grown factor (EGF). The pocket on the side of the medium-chain platform that receives the Tyr is occluded in the crystal structures of the AP2 and AP1 cores, suggesting that both crystals have captured the core in a 'closed' conformation. The extent to which the adaptor needs to 'open' may depend on the position of the Tyr with respect to the lipid bilayer. In some endocytosed receptors, the relevant Tyr is only seven to eight residues from the end of the transmembrane segment, but in others, a much longer stretch of polypeptide chain (or even an entire kinase domain) may intervene.

A phosphoinositide-binding site on one of the large chains of AP2 probably marks the position of lipid head-group contact for the base of the adaptor (Figure 10.6). Flexibility of the connection between the N- and C-terminal medium-chain domains may help the adaptor accommodate to the range of potential distances between the lipid head-group layer to which it docks and the Tyr motif it must recognize. The exposed location of the medium-chain hinge would allow a regulatory protein to restrain or release it. A crystal structure of the core of AP2 in complex with a peptide with a diLeu endocytic motif from the T-cell surface antigen CD4 shows that it binds to the small chain primarily by recognition of its diLeu motifs through two hydrophobic pockets. To accommodate the acidic diLeu motif peptide, the N terminus of one large chain moves away and becomes disordered. Interactions between the adaptors and their target membranes are weak, allowing rapid equilibration between cytosol and membrane surface, but linkage to other contacts can reinforce the association. For example, co-association of two adaptors with the same clathrin triskelion will greatly lengthen their joint residence time. This sort of ternary interaction may be one of the early events in coated vesicle formation.

Reconstituted coated vesicles provide insights into interactions in a clathrin lattice

Clathrin coats are highly polymorphic. However, preparations of clathrin 'cages' assembled *in vitro* from purified clathrin triskelions and AP2 adaptors contain enough particles with the same coat geometry to support cryo-EM reconstructions (**Figure 10.7**). On average, these cages contain about one AP2 heterotetramer for every one or two triskelions, but the AP2s are not visible in the EM image reconstructions, because they do not occupy unique positions. The reconstructions show that when triskelions assemble into a coat, the legs interdigitate to create a lattice of open hexagonal and pentagonal faces. A triskelion hub lies at each vertex, and each leg extends along one of the three edges. The proximal parts of the legs radiate toward the three neighboring vertices and project gently inward towards the lumen. The legs bend smoothly at the knees, and the distal parts of the legs extend along a second edge toward a second vertex, in such a way that the distal legs are situated beneath the proximal leg of another triskelion. The ankles of three converging chains cross each other about 75 Å beneath the apex of the triskelion centered on the second vertex. The linkers place the terminal domains at an internal position along a third edge, so that they face toward the third vertex. A space created by the three knees as they bend around a vertex allows the helical tripod at the hub of the triskelion centered on that vertex to

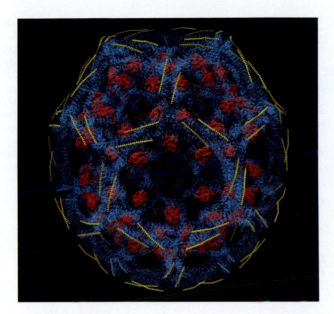

Figure 10.7 Model of a reconstituted clathrin coat. Structures of the clathrin heavy (blue) and light (yellow) homology models were fitted into a cryo-EM reconstruction. Auxilin, a protein required for uncoating, is in red. (From Fotin et al., *Nature* 432:573–579, 2004. With permission from Macmillan Publishers Ltd.)

project inward, terminating in the only disordered segment (residues 1630–1675) in the entire heavy chain. The direct distance between two N-terminal domains of a clathrin triskelion in a coat is about 45 nm—similar to the corresponding span measured by EM when free clathrin is sprayed onto a mica substrate. This suggests that a clathrin triskelion has a relatively fixed three-dimensional conformation.

Clathrin coats isolated from cells vary greatly in size and shape

Adaptability of lattice curvature is critical for function, as coated pits must engulf cargoes of differing sizes and shapes. No two of about one hundred coats isolated from calf brain and analyzed in detail by cryo-electron tomography had exactly the same lattice structure (**Figure 10.8**). Any closed, polyhedral lattice with three edges meeting at each vertex and with only hexagonal or pentagonal faces has a geometrical requirement for exactly 12 pentagons. Inclusion of a heptagonal face, as sometimes observed, requires an additional pentagon. None of the lattices analyzed had a highly symmetric design. Three of the smallest coats made by assembly with APs *in vitro* are built from 28, 36, and 60 triskelions and referred to as the 'mini-coat,' 'D6-hexagonal barrel,' and 'soccer ball,' respectively. The hexagonal barrel (along with less symmetrical structures of the same size) is probably the smallest polyhedron that can accommodate a transport vesicle.

The molecular structure of clathrin and its contacts within coats of different diameter provide a relatively simple molecular mechanism for achieving variable curvature with a single assembly unit. The transverse helical zigzags of a triskelion leg generate a stiff but compliant structural element that can bend smoothly as required, whereas the trimer contact and the pucker it generates at the hub of a triskelion are invariant. The only extended contact within an assembled lattice is between a proximal leg and a distal leg from the triskelion centered on a neighboring lattice point. This pair of segments (which also bears a light chain) defines one side of an edge in the polyhedral lattice; it has only tenuous contacts with the proximal-distal pair running anti-parallel to it along the same edge, so that the crossing angle between the two pairs (in the cross-sectional plane containing the edge) can vary. Crossing-angle variation, together with compliant bending of the knees and somewhat variable ankle crossings, allows a growing lattice to accommodate cargoes of various sizes, while still imposing curvature on the membrane being engulfed. The flat arrays seen by EM on ventral membranes (Figure 10.2) have numerous defects and dislocations, with only a limited number of uninterrupted hexagonal facets. Thus, the reverse bend at the knee required to form a flat lattice probably brings with it considerable strain.

Vesicles enclosed in clathrin coats are substantially smaller in diameter than the coats themselves (Figure 10.8). About 150–200 Å separate the outer head-group layers from the inner margins of the clathrin lattices, allowing plenty of space for APs and for entry and exit of auxilin, Hsc70, and other factors (see below). Cryo-electron tomography analysis of small

(A)

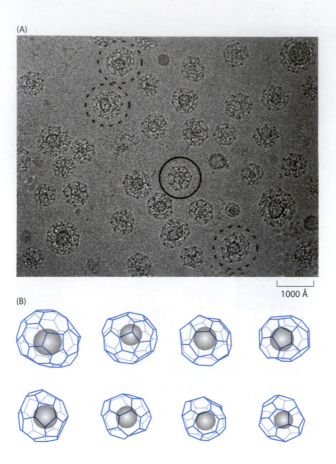

1000 Å

(B)

Figure 10.8 Cryo-EM of clathrin-coated vesicles isolated from bovine brain. (A) The micrograph shows the polymorphism of the clathrin-coated vesicles isolated from cells. The particle marked by a continuous circle shows a symmetrical (DC) coat, while the particles marked by a dashed circle are larger and have the vesicle located eccentrically. (B) Gallery of structures as derived from an electron tomographic study. (Adapted from Y. Cheng et al., *J Mol Biol* 365:892–899, 2007. With permission from Elsevier.)

coated vesicles from brain shows the lipoprotein vesicles to be offset from the center of the clathrin shell, so that is in closer contact with one pole of the coat than with the opposite pole. A likely explanation for this eccentricity is that APs (which bind lipid head-groups) participate intimately in initiating coat assembly and that at early stages of invagination, the membrane is tightly apposed to the growing lattice. Vesicle pinching, which depends on dynamin, does not require APs—indeed they might get in the way. Thus, the terminating pole of the clathrin coat is likely to be nearly devoid of APs and to have much sparser contacts with the enclosed membrane.

Clathrin boxes mediate interactions with heavy chains

The elaborate pattern of interactions between clathrin heavy chains in a lattice and their binding to light chains have been described above. Clathrin heavy chains also have a number of other binding partners. Members of one important group are proteins with short peptide sequences (called 'clathrin boxes') that contact various sites on the clathrin terminal domain. The hinge regions of large chains in the AP1, AP2, and AP3 adaptors contain a so-called 'type I' clathrin box (LΦXΦD/E, where Φ represents a hydrophobic residue and X is any residue), which binds in the groove between blades 1 and 2 of the terminal β propeller domain of clathrin. Likewise, an unstructured loop in β-arrestin2, an adaptor that helps to couple GPCRs to clathrin-dependent endocytosis, contains a clathrin box. The BAR domain protein amphiphysin I (see below) contains both a type I clathrin box and a type III clathrin box (PWXXW; also called a W-box), 21–25 residues apart in the unstructured loop between the BAR domain and the C-terminal SH3 domain. The type III box binds in the deep pocket on the top surface of the clathrin β propeller. β-Arrestin is a dimer, and its two clathrin boxes can probably enhance avidity by associating with two independent terminal domains. These interactions allow a wide range of relative orientations and positions for the adaptors or other regulatory proteins with respect to the clathrin terminal domain that recruits it. The position of the clathrin box in largely unstructured hinge regions of the APs makes very few demands on the placement of the adaptor with respect to membrane or cargo, because the hinge is over 100 residues long.

Auxilin and Hsc70 are required for uncoating

Clathrin vesicles uncoat as soon as they have budded. Recycling of clathrin triskelions for further rounds of coated-vesicle formation requires that a fully formed clathrin coat recruit the agents of its own disassembly—the Hsc70 'uncoating ATPase' and the J-domain protein, auxilin. Hsc70 is a generic Hsp70-type chaperone (see Section 6.3), an ATP-driven molecular clamp that binds and releases a short, hydrophobic peptide. J-domain proteins provide target selectivity for Hsp70 family members. Auxilin has a J-domain of roughly 80 residues at its C terminus, preceded by its clathrin-binding region. Vertebrates have two auxilin isoforms: both have a domain that is structurally similar to the PTEN lipid phosphatase, but with no catalytic activity. One isoform has an additional N-terminal protein kinase module (GAK). The clathrin-binding region (of as yet unknown structure) folds against the J-domain and contacts three different surfaces on three triskelions, as presented within a coat lattice. This leaves no detectable affinity for a free triskelion. Its binding to a coated vesicle is further enhanced by interaction of the PTEN-like domain with lipid.

Auxilin directs Hsc70 to the unstructured, C-terminal segment of the clathrin heavy chain which contains the consensus motif QLMLT for binding to the Hsc70 clamp. The mode of action of Hsc70 in dismantling a coat has been investigated by cryo-EM difference mapping between the D6 barrel with and without bound Hsc70, although interpretation has been hampered by substoichiometric occupancy and hence weak density. In one account (**Figure 10.9**), Hsc70 is assigned to positions directly under the vertices where it is proposed to bind to one of the three disordered C-terminal regions of clathrin at this site. In another, Hsc70 appears to have two sites, one on the side of the spars that run between vertices as well as a vertex-adjacent site. In both cases, the uncoating is envisaged to involve capturing clathrin legs exposed by thermal fluctuation and gradually extricating triskelions from the coat lattice by many such interactions. EM of isolated Hsc70-bound triskelions shows densities bound near the vertices.

Membrane invagination and budding have substantial energy costs

To divide a spherical phospholipid bilayer into two smaller vesicles requires substantial energy, with details depending on such factors as the intrinsic curvature imparted to the bilayer by its lipid composition. In a cell, rapid lateral diffusion of lipids means that some of the energy can be supplied by other processes, even at some distance—for example, processes that use or remove lipids for which the relevant curvature of the invagination is particularly unfavorable. But even 1.5–2 kcal/mole of net favorable free energy per clathrin heavy chain from lattice interactions would accumulate to 250 kcal/mole for a coat of 60 triskelions—for example, the 'soccer-ball' lattice, roughly 50 nm in outer diameter.

Although clathrin assembly can readily provide the net free energy needed to generate a separated vesicle, there are substantial kinetic barriers. The most obvious is at the fission stage, where it is clear that the dynamin GTPase activity (see Section 10.3) must in some way enable the neck of a coated pit to reorganize into an unstable transition state that leads to pinching. Inhibition of the dynamin GTPase with the compound **dynasore** leads to accumulation of coated pits at two stages—pits with a fully constricted neck, as might be expected from the properties of dynamin collars, and pits just at the point at which a neck is beginning to form. Thus in addition to the well-known kinetic barrier between a narrow neck and a budded vesicle, there also appears to be a barrier between an unconstricted 'dome' and the initial formation of a re-entrant ring at its base.

Other contributions to curvature stabilization have been proposed to come from the binding of BAR domain proteins endophilin and/or amphiphysin or from insertion into the outer leaflet of the bilayer of the amphipathic helix of epsin, a membrane curvature-driving protein. BAR domain (named after Bin, amphiphysin, and Rvs—the initial set of proteins in which the conserved sequence signature of this structure was recognized, see **Guide**) forms elongated bundles of three bent α helices. Dimerization generates a symmetrical, arc-like molecule, the curvature of which varies from molecule to molecule (**Figure 10.10**). The amphiphysin BAR domain dimer can subtend about 60–75° of a circle about 300 Å in diameter; that of formin-binding protein 17 (FBP17), about 45° of a 600 Å diameter circle. A number of family members can remodel spherical liposomes into coated tubes

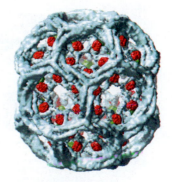

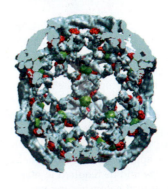

Figure 10.9 Localization of the uncoating molecules Hsc70-auxilin. EM difference imaging of Hsc70-auxilin-clathrin coats in an outside (top) and a cutaway (bottom) view. Clathrin is in blue, auxilin in red and Hsc70 in green. (Adapted from Fotin et al., *Nature* 432:64–653, 2004. With permission from Macmillan Publishers Ltd.)

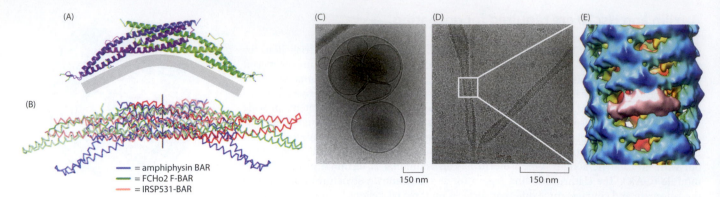

(A)

(B)

= amphiphysin BAR
= FCHo2 F-BAR
= IRSP531-BAR

(C)

150 nm

(D)

150 nm

(E)

when present in large excess. Conserved positive charges face the concave side of the BAR domain arc, and these presumably interact with negatively charged lipid head-groups in the cytosolic face of a suitably curved membrane bilayer. Mammalian amphiphysins and endophilins comprise sets of closely related proteins in which an N-terminal, membrane-inserting amphipathic helix precedes the BAR domain (hence the designation N-BAR for this subgroup of BAR proteins). None of the BAR domain proteins have been found in isolated coated vesicles; they appear to participate as regulators, rather than as structural elements. BAR domains and epsins are unlikely to be major contributors to stabilizing global curvature of a typical clathrin-enclosed vesicle, both because their preferred curvature appears in many cases to be too strong and because, in the case of epsin, too many copies would be required to make a substantial dent in the energy balance. As a sensor of curvature at a constricting neck, however, BAR domain association could indeed participate in bilayer fission. Amphiphysin and endophilin associate through their SH3 domains with the proline-rich, C-terminal region of dynamin, and either protein could help direct dynamin to the vesicle neck.

10.3 DYNAMINS ARE VERSATILE MOLECULAR MACHINES

GTPases of the dynamin superfamily have multiple roles in membrane and organelle remodeling and also in other cellular processes (**Figure 10.11**). Examples are cytokinesis and centrosome cohesion or the remodeling of F-actin-rich structures such as phagocytic cups or actin comet tails which allow bacterial pathogens to move inside their eukaryotic

Figure 10.10 Structures of BAR domain proteins. (A) In the amphiphysin BAR structure two BAR domains form a crescent-shaped homodimer with positively charged residues along the membrane interacting concave face. (B) Different BAR domain proteins assume different curvatures. The N-BAR domain of endophilin remodels vesicles (C) into membrane tubes (D) at a protein-to-lipid ratio of 1:20. This process is reminiscent of the neck formation of clathrin-coated vesicles. (E) A helical 3D reconstruction of a membrane tube with endophilins forming a periodic array. Colors (warm to cold) indicate the distance from the center of the tube. One asymmetric unit (that is, one N-BAR dimer) is highlighted in dark pink. (Courtesy of Naoko Mizuno.)

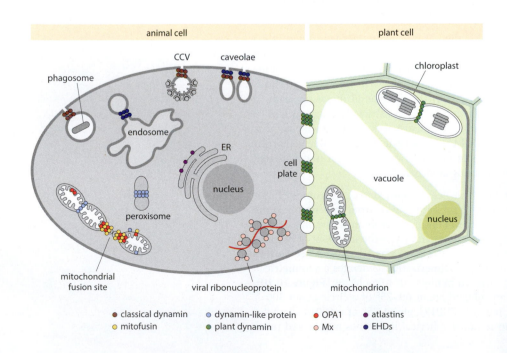

animal cell

plant cell

phagosome

CCV   caveolae

endosome

ER

nucleus

peroxisome

mitochondrial
fusion site

viral ribonucleoprotein

chloroplast

cell
plate

vacuole

nucleus

mitochondrion

● classical dynamin   ● dynamin-like protein   ● OPA1   ● atlastins
○ mitofusin   ● plant dynamin   ○ Mx   ● EHDs

Figure 10.11 Cellular functions of dynamin and dynamin-related protein in animals and plant cells. Classical dynamins (maroon) function in the budding of clathrin-coated vesicles at the plasma membrane, cleavage furrow, Golgi, and endosome, but also in clathrin-independent budding events at caveolae and phagosomes. Dynamin-1-like protein (light blue) is involved in division of organelles such as mitochondria and peroxisomes. The OPA1 (optic atrophy 1, red) and mitofusin (yellow) families are involved in mitochondrial fusion, and therefore antagonize the function of dynamin-1-like protein. Myxovirus resistance (Mx) proteins (pink) are induced by interferons and confer resistance against RNA viruses. Eps15 homology domains (EHDs) mediate membrane trafficking pathways from the plasma membrane and at intracellular compartments. Atlastins are mediators of membrane fusion at the endoplasmic reticulum (ER). Plants contain many different dynamin proteins (light green). Some of them have similar functions to those in animals, but others have functions that are unique to plants, such as formation of the cell plate or chloroplast division. (Adapted from Praefcke & McMahon, *Nat. Rev. Mol. Cell Biol.* 5:133–147, 2004. With permission from Macmillan Publishers Ltd.)

hosts. In membrane remodeling events, dynamins polymerize into helical lattices on specific membrane patches, enforcing drastic changes in curvature as a prelude to fission or fusion. Dynamins are viewed as mechano-chemical enzymes; that is, the energy from GTP binding and hydrolysis is converted into a mechanical force that deforms the membrane. While most studies of dynamins have focused on their role in clathrin-mediated endocytosis, recent structural studies of bacterial **dynamin-like proteins** (BDLPs) have furthered our understanding of the mechanisms underlying membrane restructuring.

Dynamin and dynamin-like proteins (DLPs) share structural and mechanistic features

The classical fission dynamins are large (~100 kDa) proteins composed of five conserved modules: an N-terminal GTPase domain (G domain), a middle domain (MD), a pleckstrin homology (PH) domain (see **Guide**), a GTPase effector domain (GED), and a C-terminal proline/arginine-rich domain (PRD) (**Figure 10.12A, B**). The G domains share the canonical fold of small *Ras-like GTPase domains* (see Chapter 12) but with an additional dynamin-specific sequence inserted between the G2 and G3 motifs.

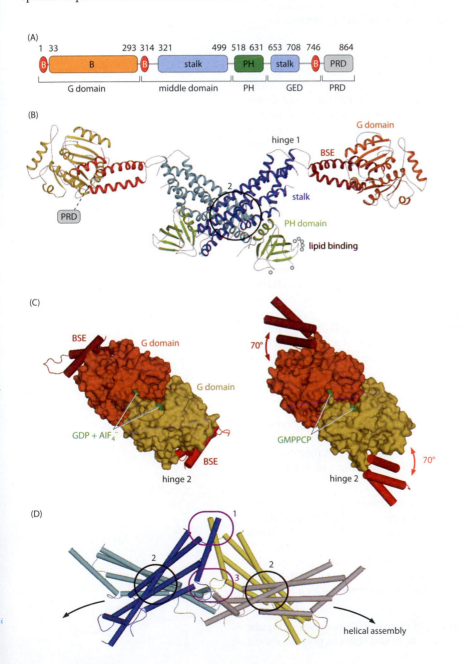

Figure 10.12 Structure of dynamin.
(A) Domain organization of dynamin, as derived from the crystal structures. The sequence-derived domain boundaries are shown below. Bundle signaling element (B). (B) Crystal structure of the dynamin 1 dimer, which assembles via a central stalk interface. (PDB 3SNH) (C) Surface representation of GTPase domain dimers of dynamin in the presence of GDP-AlF₄ (PDB 2X2E, left) and the non-hydrolyzable GTP analogue CMPPCP (PDB 3ZYC, right). The 70° rotation of the BSE (shown in red helices) between the two nucleotide loading states was suggested to act as a power stroke in dynamin. (D Assembly of stalks via two additional interfaces mediates polymerization of dynamin. Adjustments of the interfaces 1 and 3 during assembly of stalk dimers have been proposed to induce the formation of helical dynamin filaments. (Adapted from K. Faelber et al. *Structure* 20:1621–1628, 2012. With permission from Elsevier.)

Compared with other GTPases, dynamins have a low (micromolar) affinity for nucleotides, the affinity for GDP being even lower than for GTP. Dynamin's basal GTPase activity increases approximately 100-fold upon polymerization. GTPase stimulation is mediated by dimerization of the G domains, which assemble head-to-head in a GTP-dependent interaction (Figure 10.12C). Assembly leads to a rearrangement of catalytic residues *in cis* and the positioning of the catalytic water molecule, followed by GTP hydrolysis. In the GDP-bound form, the G domains then dissociate. The mechanism of dimerization-induced GTPase stimulation appears to be a common feature of the dynamin superfamily. Adjacent to the G domain is a three-helix bundle called the bundle signaling element (BSE) (Figure 10.12A–C). Its first helix is derived from the N terminus of dynamin, whereas the second helix follows at the C-terminal end of the G domain. The third helix is derived from a C-terminal region of the molecule. The BSE senses the nucleotide-loading state of the adjacent G domain: around a hinge region, it undergoes a 70° rotation relative to the G domain in response to nucleotide hydrolysis. This movement was suggested to act as a power stroke in dynamin and DLPs.

The dynamin stalk is an anti-parallel four-helix bundle whose boundaries do not coincide with the sequence-derived domain boundaries (Figure 10.12A). Thus, three of the stalk helices are provided by the MD with the fourth helix from the N-terminal part of the GED. The stalks assemble in a criss-cross fashion via three distinct interfaces to mediate formation of a helical dynamin filament (Figure 10.12D).

The PH domain, inserted between the MD and the GED, serves to anchor dynamin to lipid bilayers. By inserting residues into the lipid bilayer, it contributes to the generation of membrane curvature. Furthermore, the PH domains regulate the assembly of dynamin by forming autoinhibitory interactions with the stalk. At the C terminus, the PRD has been shown to interact with several endocytic proteins. It is therefore believed to target dynamin to membrane patches destined for remodeling. Several of the PRD interaction partners contain a BAR or F-BAR domain, suggesting that they recognize membrane regions with high curvature or help in maintaining curvature (see Section 10.2). For example, the F-BAR proteins, syndapin and FBP17, co-localize to the necks of clathrin-coated pits and are believed to act early in pit formation before the neck becomes constricted.

GTP binding and hydrolysis drive the constriction of the dynamin polymer

EM studies of dynamin bound to lipid tubules led to a tentative assignment of domains in the dynamin polymer and provided closer insights into the mechanism of polymerization and constriction. A three-layered structure was revealed: the G domains and BSE constitute the outer layer; the stalks mediate assembly of the dynamin filament in the middle layer; and the PH domains facilitate membrane binding in the inner layer. The G domains dimerize across neighboring filaments of a dynamin helix. In the presence of a nonhydrolyzable GTP analog, the polymer was observed to constrict via large-scale rearrangements in the stalk regions.

Different models for dynamin's fission activity can be reconciled with the structural observations. The observed GTP-dependent constriction of the dynamin polymer suggested that dynamin induces a narrowing of the vesicle neck causing membrane fission (constrictase model, **Figure 10.13**). In a microscopy-based live assay of dynamin fission, a rotary movement of the dynamin oligomer was observed following GTP hydrolysis. This rotation might be caused by a sliding of neighboring filaments; accordingly, it was proposed that twisting of the vesicle neck contributes to vesicle scission.

In the presence of a more rigid lipid template, the GTP-bound dynamin polymer formed a compact structure whereas an open helix was formed in the presence of GDP. This relaxation might be induced by dissociation of the G domains in the GDP-bound state. These observations led to the 'poppase' model, where GTP hydrolysis induces a break in the vesicle neck by extending it. Other (passive) models suggest that GTP binding induces polymerization and its hydrolysis leads to disassembly and release of the dynamin coat, leaving behind a highly curved and unstable naked lipid tube ready to relieve stress by a fission or fusion event (Figure 10.13).

A structural view of dynamin's action including aspects of these models is shown in **Figure 10.14**. In this model, dynamin is recruited to the vesicle neck and starts to polymerize via

'active models' of dynamin-induced fission

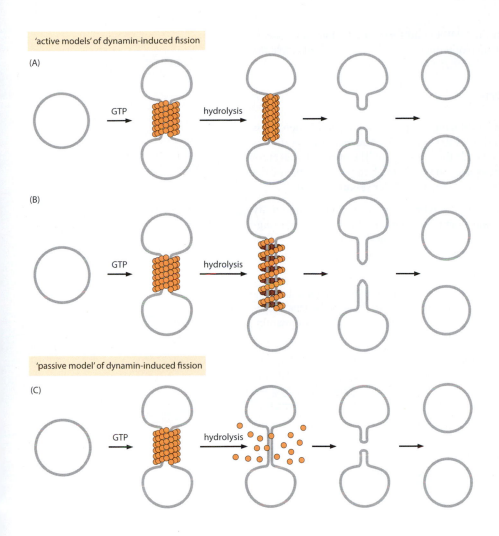

'passive model' of dynamin-induced fission

Figure 10.13 Models for dynamin membrane scission activity. Schematic drawing of the sequence of events according to active and passive fission models. In the 'constrictase' model, GTP hydrolysis leads to constriction of the vesicle neck by sliding of adjacent filaments (A). In the context of a long helix, this can be observed as a rotary movement. (B) In the 'poppase' model, GTP hydrolysis induces an extension of the vesicle neck inducing it to break. (C) According to the passive model, GTP hydrolysis results in a release of the dynamin coat from the lipid tube, leaving it in an energetically unfavorable state resulting in fission. (Adapted from H.H. Low et al. *Cell* 139:1342–1352, 2009. With permission from Elsevier.)

the stalks into a helical filament. When the growing dynamin filament has embraced the bud neck once, GTP-loaded G domains from neighboring helical turns dimerize thereby closing the dynamin helix. G domain dimerization induces nucleotide hydrolysis, followed by the power stroke via the BSE. It was proposed that this power stroke pulls dynamin filaments against each other. This might be visible in the light microscopy-based assay as a rotary movement ('twistase') and leads to a narrowing of the vesicle neck, for example, the dynamin polymer acts as a contracting sling ('constrictase'). Increased membrane

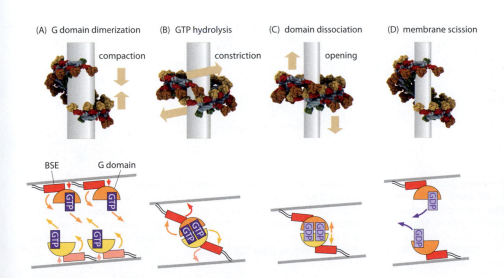

Figure 10.14 Structural model for dynamin action. (A) Dynamin oligomerizes via the stalks around the neck of a clathrin-coated vesicle. (B) GTP-loaded G domains from opposing filaments dimerize, leading to GTP hydrolysis and the power stroke of the bundle signaling element (BSE). The power stroke pulls the filaments against each other leading to constriction of the underlying membrane tubule (C). In the GDP-bound form, G domains dissociate (D). (Adapted from K.Faelber et al., *Structure* 20:1621–1628, 2012. With permission from Elsevier.)

curvature destabilizes the vesicle neck. In the GDP-bound form, the G domains dissociate according to the 'poppase' model. It is currently unclear at which step of this cycle the membrane may break.

Structure and mechanism of other DLPs

Dynamin-like proteins (DLPs) have a similar domain organization, but the PH domain is replaced by other lipid-binding motifs, and the PRD is absent. Sequence conservation is high in the GTPase domain but drops off towards the C terminus. The overall architecture of the stalk is conserved in all DLP structures solved so far, although the dimensions and the relative orientation of the G domain with respect to the stalk differ (**Figure 10.15**).

Myxovirus-resistant proteins (Mx proteins) are antiviral effectors in innate immunity (Figure 10.15A). Like dynamin, they can bind to lipid surfaces and polymerize into ring-like structures. However, they were also suggested to oligomerize around the helical ribonucleoprotein complexes of negative-strand RNA viruses such as influenza virus (see Section 8.2), therefore interfering with viral replication.

The first structure solved for a full-length DLP was that of the human guanylate-binding protein 1 (GBP1), which is implicated in the defense against intracellular pathogens. Compared with dynamin, the stalk is more elongated (Figure 10.15B). Similar to dynamin, GTPase activation also involves dimerization of G domains.

Eps15 homology domain (EHD)-containing proteins are eukaryotic DLPs involved in several membrane-trafficking pathways (Figure 10.15C). Despite having a dynamin-related G domain, they bind ATP rather than GTP. Like dynamin, EHD2 polymerizes in ringlike structures around tubulated liposomes leading to ATPase stimulation. Both dimerization and further polymerization are mediated by assembly interfaces in the G domain. The lipid-binding sites are located at the tips of the helical domains.

Closely related to the GBPs are the atlastins, GTPases which mediate homotypic membrane fusion at the ER where they are anchored by two transmembrane helices (Figure 10.15D).

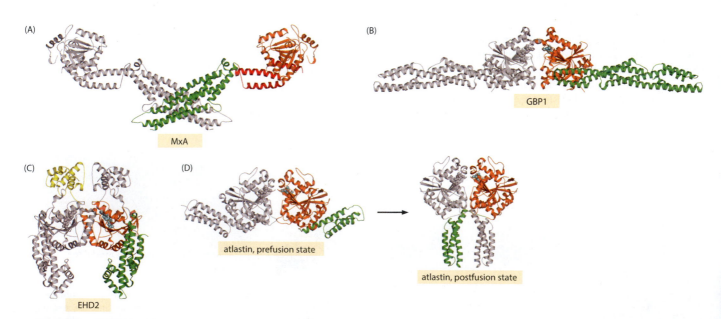

Figure 10.15 DLP structures. All dynamin-like proteins (DLPs) have an N-terminal G domain (orange) and a helical stalk. Some members additionally have a bundle signaling element (BSE) between G domain and stalk. The architecture of the basic dimeric building block differs between the different subfamilies. (A) MxA has a related architecture to dynamin but lacks the PH domain and the PRD. (PDB 3SZR) (B) GBP1 has an elongated shape with a highly extended stalk. Dimerization is mediated via GTP-dependent interactions of the G domain. (PDB 1F5N) (C) EHD2 dimerizes via the G domains. It has only one helical domain. (PDB 2QPT) (D) Atlastin was crystallized in two conformations containing different nucleotides. They differ in the relative orientation of G domain and stalk and may represent two snapshots during endoplasmic reticulum (ER) fusion events. (PDB 3Q5D, 3Q5E)

The structure of an atlastin dimer in two conformations suggested that a large-scale domain movement of the stalk versus the G domain mediates tethering and fusion of opposing membranes.

Bacterial dynamin-like protein undergoes alternative conformational changes when bound to lipid membranes

Structural characterization of a bacterial DLP (BDLP) from *Nostoc punctiforme* provided another example of the conformational changes that dynamins can undergo upon polymerizing on membranes. Like eukaryotic dynamins, BDLP is capable of tubulating lipid vesicles. However, BDLP was suggested to mediate membrane fusion rather than membrane fission similar to the eukaryotic mitofusins involved in mitochondrial fusion (see Figure 10.11).

The monomer of *N. punctiforme* BDLP has a compact shape, comprising an extended G domain embraced by the BSE, and the stalk (**Figure 10.16A**). Despite very low sequence similarity, the BSE and stalk have a strikingly similar architecture to that of dynamin. Dimerization is mediated by the conserved interface in the G domain; however, in BDLP, this does not result in GTPase activation. The 'paddle' is an insertion at a position equivalent to the PH domain in eukaryotic dynamins. It serves as a membrane-anchoring motif.

The crystal structure of the full-length BDLP monomer was fitted into a cryo-EM reconstruction of BDLP-coated lipid tubes. According to this model, the compact monomers undergo a large-scale conformational change to an extended conformation in which the G domains homodimerize across their nucleotide-binding pockets (Figure 10.16D). This allows for a mechanism by which the nucleotide-loaded state can control protein–protein interactions in the assembly. The stalk mediates self-assembly and the paddle region interacts with the lipid bilayer. Assembly on liposomes results in a highly curved membrane tubule; the diameter of the inner leaflet is only 10 nm and the outer leaflet appears to be disordered.

A 'passive' polymerization/depolymerization model has been suggested for BDLP (Figure 10.13). In it, GTP binding to a BDLP dimer triggers a series of conformational changes that allow it to bind to lipid bilayers. The concomitant change from a compact to an extended conformation exposes polymerization surfaces (Figure 10.16D). With the assembly of more and more molecules, the BDLP coat enforces high membrane curvature. When GTP is turned over to GDP, the polymer becomes destabilized and the BDLP coat is disassembled. The lipid bilayer is left behind in a strained energetically unfavorable state that is highly

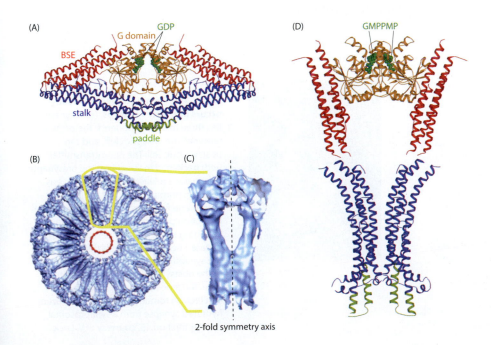

Figure 10.16 EM-based helical reconstruction of a bacterial dynamin-like protein (BDLP)-coated lipid tube. (A) Crystal structure of BDLP dimer with GDP bound. (PDB 2J68) (B) EM-based helical reconstruction of a BDLP-GMPPNP coated lipid tube. The tube is shown in cross section to the helical axis. Lipid is in red, protein in blue. The yellow box shows two BDLPs forming the dimeric unit. (C) Close-up image of the area boxed in (B). (D) BDLP-GMPPNP model indicating the large conformational changes the dimer undergoes upon association and polymerization on a lipid tube. (PDB 2W6D) (B and C, from Low et al., *Cell* 139:1342–1352, 2009. With permission from Elsevier.)

(A) BSE · G domain · GDP · stalk · paddle

(B) (C) 2-fold symmetry axis

(D) GMPPMP

fusogenic and might allow membrane fusion to occur. Despite their structural similarity, the polymerization modes of BDLP and dynamin and their mechanisms of membrane remodeling appear to be fundamentally different.

10.4 THE MACHINERY OF SYNAPTIC VESICLE FUSION

The human brain contains approximately 100 billion cells, known as neurons. Neurons are highly branched, which allows them to come into close apposition with each other at sites called **synapses** where they communicate. Indeed, most neurons form thousands of synapses with other neurons. The brain is thus an extremely complex cellular network, and its ability to perform an enormous variety of functions depends critically on the ability of neurons to send signals to each other at synapses, in spatially defined and precisely timed patterns.

Neurotransmitter release is an exquisitely regulated form of membrane fusion

Chemical synaptic transmission provides the most common means for interneuronal communication (**Figure 10.17**). Transmission involves small molecules called **neurotransmitters** (for example, acetylcholine; see Chapter 16) that are packaged into **synaptic vesicles** in presynaptic terminals. When an electrical signal called an action potential reaches the terminal, voltage-gated Ca^{2+} channels open and the resulting influx of Ca^{2+} into the terminal induces synaptic vesicle **exocytosis**, a process that involves fusion between the vesicle and plasma membranes. The neurotransmitters then diffuse across the narrow space between the two cells, the synaptic cleft, and bind to receptors in the postsynaptic cell, inducing there a signal transmitting the nerve impulse.

Neurotransmitter release is very fast, occurring less than 0.5 ms after Ca^{2+} influx. Such high speeds are possible because, after docking onto the plasma membrane, synaptic vesicles undergo a priming reaction that leaves them ready for release. Although in the temporal order of events that leads to neurotransmitter release docking occurs first, then priming, and then membrane fusion, below we discuss first the mechanism of membrane fusion because this is the step that has been studied most extensively and because vesicle priming can be better understood on the basis of what is known about fusion. Vesicle docking is perhaps the least understood of these steps, but is believed to depend on interactions between proteins from the vesicle and the plasma membranes, some of which are mentioned below. It is also important to note that neurotransmitter release is not just a means to send signals between neurons. The probability that a synaptic vesicle undergoes exocytosis is modulated through diverse processes that depend on the usage patterns of the synapse. This plastic

(A)

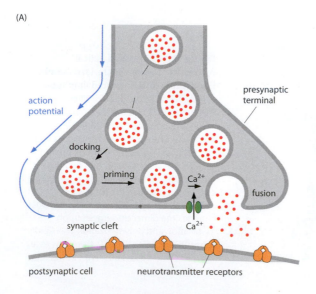

(B)

Figure 10.17 Synaptic vesicle fusion occurs in a series of steps. (A) A synapse has three main components: the presynaptic terminal, the synaptic cleft, and the postsynaptic cell. The neurotransmitters (red) packaged in vesicles at the presynaptic terminal are released in a series of steps: docking, priming, and Ca^{2+} triggered membrane fusion when an electrical signal (an action potential) induces opening of voltage-gated Ca^{2+} channels (green). The released neurotransmitters diffuse through the synaptic cleft and bind to receptors in the postsynaptic plasma membrane, inducing a postsynaptic response. (B) Electron tomographic slice of a frozen-hydrated synapse from a synaptosomal cellular fraction. (B, courtesy of V. Lucic.)

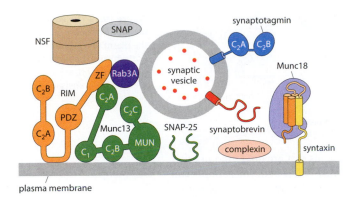

Figure 10.18 Schematic view of the neurotransmitter release machinery. Selected components of the synaptic vesicle fusion machinery, indicating their localization and illustrating some of their domain structures as well as a few of the interactions between them. Each protein is described in more detail in the ensuing figures.

nature of synaptic exocytosis, which is generally referred to as presynaptic plasticity, is thought to underlie some forms of information processing in the brain.

The release machinery includes a conserved core and components specialized for its tight regulation

In all eukaryotes, members of several protein families are involved in most types of intracellular membrane fusion, suggesting that basic core machinery underlies a universal mechanism of membrane fusion. These families include **N-ethylmaleimide sensitive factor (NSF)**, soluble NSF attachment proteins (**SNAPs**), SNAP receptors (**SNAREs**), Sec1/Munc18 (SM) proteins, small GTPases from the **Rab** family, and some tethering factors. Here we focus on the members of these families that are involved in synaptic vesicle fusion (**Figure 10.18; Table 10.2**), but some key findings made from other systems are also discussed. Proteins within each family share properties that underlie their primary functions, but the members involved in neurotransmitter release have some unique properties that evolved to meet the tight regulatory requirements of this process.

Table 10.2 Selected proteins involved in synaptic vesicle fusion

Protein family	Main isoforms*	Length[†]	Function and comments
Syntaxin	1A, 1B	288, 288	SNARE from the plasma membrane. Central role in membrane fusion
SNAP-25	A, B	206, 206	SNARE from the plasma membrane. Central role in membrane fusion
Synaptobrevin	2	116	SNARE from synaptic vesicles. Central role in membrane fusion
NSF		744	Hexameric ATPase. Together with SNAPs, disassembles the SNARE complex
SNAP	α, β	295, 297	Binds to the SNARE complexes, assisting NSF in SNARE complex disassembly
Munc18	1	594	SM protein crucial for neurotransmitter release. Controls SNARE complex assembly and might function directly in membrane fusion
RAB3	A	220	Small GTPase from the Rab family. Likely functions in docking and perhaps in priming
Munc13	1, 2	1735, 1555	Active zone protein. Crucial for priming and some forms of presynaptic plasticity. Orchestrates SNARE complex assembly together with Munc18-1, and might function directly in membrane fusion
RIM	1α, 2α	1615, 1555	Rab3 effector. Key role in priming and some forms of presynaptic plasticity. Likely provides a scaffold to organize the active zone
Synaptotagmin	1, 2	421, 422	Ca^{2+} sensor for synchronous neurotransmitter release
Complexin	1. 2	134, 134	Dual active and inhibitory roles in release, controlling the Ca^{2+}-triggering step in tight coupling with the SNAREs and synaptotagmin

*Main neuronal isoforms involved in synaptic vesicle fusion. [†]Number of residues in rat for main isoforms.

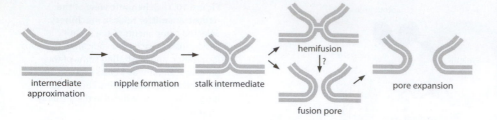

intermediate approximation   nipple formation   stalk intermediate   hemifusion   ?   fusion pore   pore expansion

Figure 10.19 The stages of the stalk mechanism of membrane fusion. Fusion of the plasma membrane (planar membrane, bottom) and the synaptic membrane (curved membrane, top) are shown here. The same mechanism is envisaged for fusion of two planar membranes.

This tight regulation depends on dozens of additional proteins that are not found in other membrane traffic systems and have specialized roles in release (a few of the most important among them are listed in Table 10.2 and are discussed below). The mechanisms of release and its regulation are complex, and at least some of the models described below are likely to be oversimplifications. An effort is made to clearly distinguish between concepts that are well established and ideas that remain to be demonstrated.

Membrane fusion is believed to occur through a stalk mechanism

The neurotransmitter release machinery is required to fuse the membranes of a vesicle and a cell. Fusion must occur without membrane rupture, as this would alter the ionic gradients across the plasma membrane that are essential for synaptic transmission itself. Theoretical calculations, supported by diverse experimental data, suggest that fusion of biological membranes occurs by a so-called 'stalk mechanism' (**Figure 10.19**) in which the two membranes are first brought into close proximity and then bent to form structures called 'nipples.' Merger of the proximal leaflets of the two bilayers leads to the formation of a 'stalk intermediate,' which under certain conditions, may arrest into a hemifusion intermediate. Subsequent merger of the distal leaflets creates a fusion pore, which can then expand. Each one of these steps requires substantial amounts of energy.

SNAREs are central components of the membrane fusion apparatus

SNARE proteins play crucial roles in all steps of the secretory and endocytic pathways. The SNAREs that mediate synaptic exocytosis are the synaptic vesicle protein synaptobrevin (also called VAMP) and the plasma membrane proteins syntaxin and SNAP-25 (synaptosomal associated protein of 25 kDa); note that SNAP-25 is unrelated to the acronym SNAPs defined above for soluble NSF attachment proteins (see Figure 10.18). These proteins are the targets for neurotoxins, extremely toxic proteases produced by bacteria from the *Clostridium* family that strongly inhibit neurotransmitter release. These findings, together with genetic data, demonstrated the key importance of these three SNARE proteins for neurotransmitter release. SNAREs are characterized by sequences called SNARE motifs that contain 60–70 aa and have a high propensity to form coiled coils. SNAP-25 contains two SNARE motifs, whereas syntaxin and synaptobrevin have one each that precedes a C-terminal transmembrane (TM) region (**Figure 10.20A**). In addition, syntaxin contains a long N-terminal region including an autonomously folded domain (H_{abc}) that forms a three-helix bundle (Figure 10.20A, B).

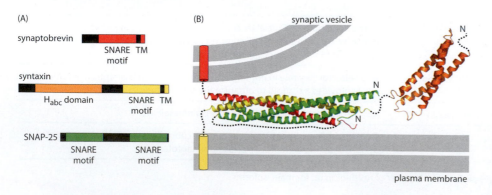

(A)

synaptobrevin
SNARE motif   TM

syntaxin
H_{abc} domain   SNARE motif   TM

SNAP-25
SNARE motif   SNARE motif

(B)   synaptic vesicle

N   N   N

plasma membrane

Figure 10.20 The neuronal SNARE complex. (A) Domain diagrams of the neuronal SNAREs synaptobrevin, syntaxin, and SNAP-25. (B) Ribbon diagrams of the solution NMR structure of the syntaxin H_{abc} domain (1BR0) and of the neuronal SNARE complex formed by the SNARE motifs of synaptobrevin, syntaxin, and SNAP-25 (1SFC). For illustration purposes, the synaptobrevin and syntaxin TM regions are represented as red and yellow cylinders, respectively, and linker sequences of the three proteins are represented by dashed black lines. (B, from I. Fernandez et al., *Cell* 94:841–849, 1998. With permission from Elsevier; and R.B. Sutton et al., *Nature* 395:347–353, 1998. With permission from Macmillan Publishers Ltd.)

(A)

(B)

(C)

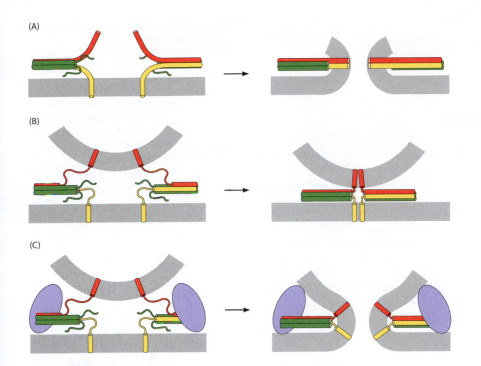

Figure 10.21 Models of SNARE-mediated membrane fusion. (A) Minimal model whereby the SNAREs alone can induce membrane fusion. The model assumes that that SNARE motifs and TM regions of synaptobrevin and syntaxin form continuous helices. The color coding is the same as in Figure 10.20. (B) Model illustrating how flexibility in the linkers between the SNARE motifs and TM regions of synaptobrevin and syntaxin would uncouple SNARE complex assembly from the membranes and allow assembly of the complex without membrane fusion. (C) Model illustrating how assembly of the SNARE complex while bound to an unidentified bulky protein(s) (purple) that may bind to the membranes or push them apart would allow the SNARE complex to apply force on the membranes even if there is flexibility in the linkers between the SNARE motifs and TM regions of synaptobrevin and syntaxin.

Synaptobrevin, syntaxin, and SNAP-25 form the highly stable SNARE complex in which the four SNARE motifs form a bundle of four parallel α helices (Figure 10.20B). The parallel arrangement of the helices implies that assembly of the SNARE complex brings the vesicle and plasma membranes into close proximity, suggesting a universal model of intracellular membrane traffic whereby the free energy of SNARE complex formation drives membrane fusion, much like some viral proteins induce membrane fusion through coiled-coil interactions. Although this overall notion is widely accepted, it remains to be established how the energy of SNARE complex formation is transduced into membrane fusion.

A popular model postulates that the SNAREs alone constitute a minimal membrane fusion machinery (**Figure 10.21A**), with the SNARE motifs and TM regions forming continuous helices to exert mechanical force on the membranes. However, there is evidence for some flexibility in the linkers between the SNARE motifs and TM regions, which would severely disrupt any such application of force and might allow SNARE complex assembly without fusion (Figure 10.21B). A modified model whereby the SNAREs can exert force on the membranes much more efficiently, even if there is flexibility in the linkers, predicts that the assembling SNARE complexes are bound to a bulky factor or factors. This factor may interact in some way with both membranes, providing support points to help the SNAREs bend the membranes (Figure 10.21C).

Reconstitution experiments with purified components are fully consistent with the crucial importance of SNARE complex formation for membrane fusion, and indicate that the SNARE transmembrane regions may play a role by destabilizing the bilayer structure of the membrane, thus facilitating formation of the stalk intermediate. However, reconstitution experiments also suggest that membrane fusion without membrane rupture requires the involvement of other proteins such as Munc18 and Munc13 for neurotransmitter release, or homologs of these proteins for other membrane traffic systems.

Multiple proteins with key roles in release bind to the SNARE complex and influence its assembly or disassembly. These observations emphasize that the SNARE complex is at the heart of the membrane fusion machinery and underline the importance of regulating SNARE complex assembly to control neurotransmitter release. Such regulation is performed in part by syntaxin itself, since its N-terminal H_{abc} domain binds intramolecularly to the SNARE motif to form a so-called closed conformation that is incompatible with formation of the SNARE complex (**Figure 10.22**). Hence, syntaxin must undergo a large structural change during exocytosis to adopt the open conformation required for SNARE complex assembly. This transition from closed to open syntaxin is likely a key event during priming of synaptic vesicles to a release-ready state.

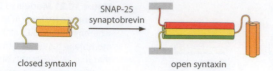

closed syntaxin   open syntaxin

SNAREs are recycled by NSF and SNAPs

After synaptic vesicle fusion, the resulting SNARE complexes reside on the plasma membrane. These complexes are called *cis*-SNARE complexes to distinguish them from the *trans*-SNARE complexes that form between two membranes before fusion. *cis*-SNARE complexes need to be disassembled in order to sort synaptobrevin back to synaptic vesicles and thus recycle the SNAREs for another round of fusion. Because SNARE complexes are so stable, their disassembly requires the energy of ATP hydrolysis, which thus indirectly provides the energy for membrane fusion. The disassembly reaction is brought about by NSF, a specialized ATPase that binds to SNARE complexes through adaptor proteins called SNAPs (no relation to SNAP-25) (**Figure 10.23A**). NSF and SNAPs perform this function not only at the synapse but also at all membrane compartments where SNARE-dependent fusion occurs, and are therefore universal SNARE complex disassembly factors.

Figure 10.22 Model of conformational switch of syntaxin during neurotransmitter release. The diagrams illustrate how isolated syntaxin adopts a closed conformation where the H$_{abc}$ domain (orange) binds intramolecularly to the SNARE motif (yellow), whereas in the SNARE complex syntaxin adopts an open conformation. The color coding is the same as in Figure 10.20. The gray bars depict the membranes.

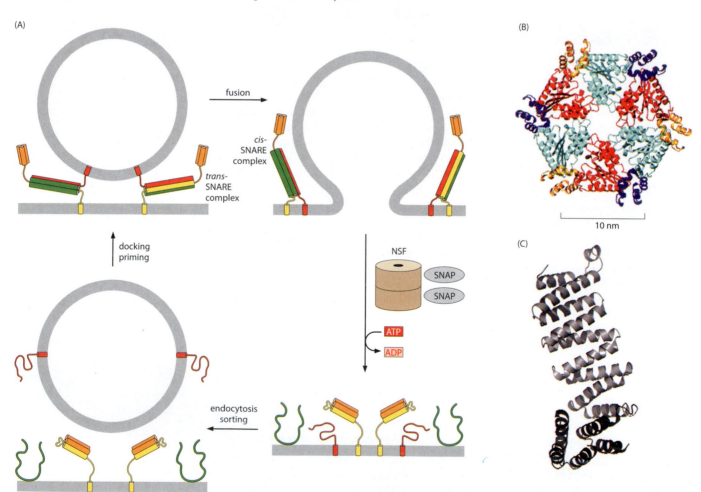

Figure 10.23 Assembly–disassembly cycle of the SNARE complex.
(A) After formation of *trans*-SNARE complexes between the vesicle and plasma membranes, the membranes fuse and the resulting SNARE complexes are on the same membrane (and hence are called *cis*-SNARE complexes). The *cis*-SNARE complexes bind to NSF through SNAPs, and are dissociated by the ATPase activity of NSF. Sorting of synaptobrevin to the vesicles and perhaps formation of the syntaxin closed conformation prevent reassembly of the *cis*-SNARE complexes. (B) Ribbon diagram of the hexameric structure formed by the D2 domain of NSF. The protomers have been colored alternately in cyan/blue and red/orange. (PDB 1D2N) (C) Ribbon diagram of the structure of Sec17p, the yeast SNAP homolog. The helical hairpins are colored in light gray, and the helical bundle in dark gray. (PDB 1QQE) (B, from C.U. Lenzen et al., *Cell* 94:525–536, 1998. With permission from Elsevier. C, from L.M. Rice et al., *Mol. Cell* 4:85–95, 1999. With permission from Elsevier.)

NSF contains three domains, including two ATP-binding domains (D1 and D2) from the *AAA-ATPase superfamily* (see Box 3.1), and an N-terminal substrate-binding domain. The D2 domain forms a hexameric structure (Figure 10.23B) that gives the complex a cylindrical shape. The structure of the yeast SNAP homolog Sec17p showed that proteins in this family form a sheet of helical hairpins and helical bundles (Figure 10.23C). Although the mechanism of SNARE complex disassembly is still unclear, EM (see **Methods**) and X-ray studies suggest that SNAP molecules cover the surface of the SNARE complex and may be used as levers to untwist the SNARE four-helix bundle when, after binding to NSF, ATP hydrolysis by the NSF D1 domain drives large overall conformational changes in the entire complex.

NSF-SNAP function is not limited to disassembling the ternary *cis*-SNARE complexes formed between synaptobrevin, syntaxin, and SNAP-25, as NSF-SNAP also disassemble additional complexes resulting from the promiscuity of the syntaxin and SNAP-25 SNARE motifs. For instance, syntaxin forms oligomers and diverse complexes with SNAP-25, some of which represent kinetic traps that hinder binding to synaptobrevin. NSF and SNAPs disassemble at least some of these unproductive complexes. What prevents binary or ternary *cis*-SNARE complexes from forming again after disassembly by NSF and SNAPs, which would lead to a futile expenditure of ATP? Clearly, sorting of synaptobrevin to synaptic vesicles contributes to preventing re-assembly of the ternary *cis*-SNARE complexes. Moreover, formation of the closed conformation in syntaxin after the disassembly reaction by NSF and SNAPs prevents or at least slows down its re-association with itself and SNAP-25 (Figure 10.23A). As described next, Munc18 also contributes to preventing such re-association.

Munc18 orchestrates SNARE complex assembly together with Munc13

Like the SNAREs, SM proteins play critical roles in most types of intracellular membrane traffic, which is particularly well emphasized by the total abrogation of neurotransmitter release observed in central synapses of mice lacking the neuronal SM protein Munc18. This strong phenotype is believed to arise because Munc18 plays multiple functions in neurotransmitter release, including a central role in organizing SNARE complex assembly.

Munc18 is a soluble, arch-shaped protein that binds to the SNAREs in at least two different modes. Munc18 forms a tight binary complex with the closed conformation of syntaxin via a large cavity that wraps around syntaxin (**Figure 10.24A, C**, top panel). This interaction is not generally conserved in other types of intracellular membrane traffic and may have arisen to meet the tight regulatory requirements of neurotransmitter release. Munc18 binding stabilizes the closed conformation of syntaxin and hinders SNARE complex formation. Therefore, this interaction is, in principle, inhibitory. However, the binding of Munc18 to syntaxin likely also has a positive role because it stabilizes both proteins.

Another type of Munc18–SNARE interaction involves assembled SNARE complexes, where syntaxin is open (Figure 10.24C, middle and bottom panels). This interaction appears to be conserved in most types of intracellular membrane traffic and, although

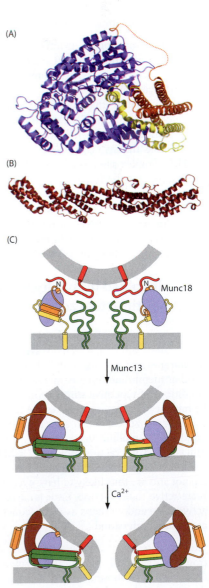

(A)

(B)

(C)

Munc18

Munc13

Ca²⁺

Figure 10.24 Orchestration of SNARE complex assembly by Munc18 and Munc13. (A) Ribbon diagram of the crystal structure of Munc18 (blue) bound to the closed conformation of syntaxin (H_{abc} domain in orange, remaining sequences in yellow). The dashed curve depicts the sequence connecting the H_{abc} domain to the syntaxin N terminus. (PDB 3C98) (B) Ribbon diagram of Exo70, which is homologous to the Munc13 MUN domain. (PDB 3SWH) (C) Model of how Munc18 orchestrates SNARE complex assembly together with Munc13. In the top panel, Munc18 (blue) is bound to the closed conformation of syntaxin, which also involves interactions with the syntaxin N terminus as shown in the crystal structure of panel (A). The Munc13 MUN domain (brown) helps to open syntaxin, leading to partially assembled SNARE complexes with bound Munc18 and Munc13 MUN domain (middle panel). This pathway of SNARE complex assembly is resistant to NSF-SNAP. Membrane fusion upon Ca²⁺ influx (bottom panel) involves additional proteins such as synaptotagmin and complex (not shown), but Munc18 and/or Munc13 may also cooperate in fusion by acting as the bulky factors that help bending the membranes as in the model of Figure 10.21C. (A, from K.M. Misura et al., *Nature* 404:355–362, 2000. With permission from Macmillan Publishers Ltd.)

no three-dimensional structure is available, is believed to involve binding of Munc18 to the N terminus of syntaxin and to the four-helix bundle formed by the SNARE motifs (Figure 10.24C, middle and bottom panels). The interaction of Munc18 with the syntaxin N terminus also participates in the binary Munc18/syntaxin complex (Figure 10.24C, top panel), thus providing a link between the two types of Munc18/SNARE complexes.

The transition from the binary Munc18/syntaxin complex to the quaternary Munc18/ SNARE complex is mediated by Munc13 (unc13 in *Caenorhabditis elegans*) and involves opening of syntaxin (Figure 10.24C). Munc13 is a large multidomain protein that plays an essential role in release through a large C-terminal module called the MUN domain and also has multiple regulatory roles associated with its other domains (see below). The MUN domain is homologous to factors that mediate tethering in other membrane compartments and hence is expected to have an elongated structure built of helical bundles similar to those that have been determined for some of these tethering factors (shown for the tethering factor Exo70 in Figure 10.24B). The MUN domain is responsible for the activity of Munc13 in opening syntaxin, forming a template for SNARE complex formation together with Munc18. Thus, when the quaternary Munc18/SNARE complex forms, it probably remains bound to the Munc13 MUN domain (Figure 10.24C, middle and bottom panels). The resulting state is believed to be metastable and to contain partially assembled SNARE complexes that are poised to induce membrane fusion upon Ca^{2+} influx (Figure 10.24C, bottom panel). Hence, formation of this metastable state most likely underlies synaptic vesicle priming. Because Munc18 and the Munc13 MUN domain are large (ca. 70 kDa), they could play the role of the bulky factor(s) involved in the model of Figure 10.21C to cooperate with the SNAREs in bending the membranes and inducing membrane fusion.

While the nature of the primed state and thus the direct participation of Munc18 and Munc13 in membrane fusion remain unclear, there is now little doubt that Munc18 and Munc13 orchestrate SNARE complex assembly through a pathway that starts with the Munc18/syntaxin complex. This mechanism contrasts with the previous widely accepted model postulating that syntaxin/SNAP-25 complexes in the plasma membrane bind to synaptobrevin to form SNARE complexes and induce membrane fusion (Figure 10.21A). Reconstitution experiments have shown that this pathway is inhibited by NSF-SNAP, which disassemble the syntaxin/SNAP-25 complexes, and that Munc18 and Munc13 mediate SNARE complex assembly in an NSF-SNAP resistant manner. These results correlate with reconstitution experiments of yeast vacuolar fusion and strongly suggest that orchestration of SNARE complex assembly in an environment that strongly favors SNARE complex disassembly by NSF-SNAP constitutes the function that makes Munc18 and Munc13 so crucial for neurotransmitter release. Docking interactions involving Munc18 and/or Munc13 may also favor the Munc18-Munc13-dependent pathway leading to membrane fusion over mechanisms involving only SNARE proteins.

Rab is involved in the docking of synaptic vesicles at some presynaptic active zones

A key event for any type of intracellular membrane traffic is the initial apposition of one membrane to another, which involves interactions between proteins residing on the apposed membranes and is commonly referred to as **docking** or **tethering**. Compared to docking at other membrane compartments, synaptic vesicle docking needs to be rapid since, under conditions of repetitive stimulation, the supply of synaptic vesicles at release sites needs to be replenished. The release sites are specialized areas of the presynaptic plasma membrane called active zones, which are populated with large proteins that are believed to facilitate rapid vesicle mobilization, docking, priming, and release. These proteins include, among others, Rab3-interacting molecule (RIM), Munc13, bassoon, piccolo/aczonin, liprin, RIM-binding protein, and ELKS. Some of these proteins are conserved in invertebrates but others are not and are thus likely to have evolved to meet particularly tight requirements of vertebrate active zone function. Little is currently known about the structural organization of the active zone and, perhaps as a consequence, the mechanism(s) of synaptic vesicle docking are poorly understood.

Studies of traffic at diverse membrane compartments have revealed that small GTPases from the Rab family play a critical role in docking. Typically, these proteins cycle between GTP- and GDP-bound states, assisted by GAP and GEF proteins (see Box 12.1). The

GTP-bound form is anchored on a transport vesicle through a prenyl group covalently attached to a C-terminal cysteine (see Chapter 1), and binds to an effector on the target membrane (**Figure 10.25A**). In this way, Rab provides a bridge between the two membranes and ensures vesicle-targeting specificity. The Rab effectors can be very diverse, and in some cases are highly elongated proteins, which help in capturing transport vesicles as they pass near the target membrane. After GTP hydrolysis, Rab proteins dissociate from the vesicles. Hence, factors that control GTP hydrolysis or GDP/GTP exchange by Rab proteins can indirectly regulate docking.

The most abundant Rab protein on synaptic vesicles is Rab3, but it is unclear whether it has the same function that Rab proteins commonly perform in other types of intracellular membrane traffic. While strong blocking of traffic in other membrane compartments is commonly observed in the absence of the corresponding Rab protein, genetic deletion of Rab3 leads to only a mild impairment of neurotransmitter release in mice. This suggests that there is some functional redundancy in mammalian brain synapses.

RIM (unc in *C. elegans*) is a Rab3 effector localized at the presynaptic active zone, and some evidence indicates that the docking defects resulting from deletion of RIM in *C. elegans* result from disruption of Rab3/RIM binding. Hence, it seems likely that this interaction, which follows the paradigm of other Rab proteins, contributes to docking. The interaction involves a zinc finger (ZF) domain (see **Guide**) and two adjacent α helices at the RIM N terminus, and can be modeled based on the homology between RIMs and rabphilin, another Rab3 effector, together with the crystal structure of a Rab3/rabphilin complex (Figure 10.25B). It is unclear why docking depends also on syntaxin, Munc18, and Munc13, and whether this dependence is related to Rab3. SNARE complex formation does not appear to mediate docking, since docking is not altered by deletion of the synaptobrevin gene in neuroendocrine cells.

Munc13 and RIM govern synaptic vesicle priming and presynaptic plasticity

Among the components of presynaptic active zones, Munc13 and RIM are particularly important because of their central functions in diverse forms of presynaptic plasticity, in addition to the roles discussed above in controlling SNARE complex assembly (for Munc13) and Rab3-dependent docking (for RIM). The multiple functions of Munc13 and RIM arise because of their multidomain nature, which is illustrated for their major isoforms in F**igure 10.26A**. Both proteins contain multiple C_2 domains, which are widespread protein modules that often bind Ca^{2+} and phospholipids, although most of the Munc13 and RIM C_2 domains are Ca^{2+}-independent.

The functions of Munc13 and RIM are clearly related, as they bind tightly to each other and the phenotypes resulting from genetic deletion of these proteins have some common features, including severe impairments in vesicle priming. The finding that the phenotype observed in the absence of RIM is more severe that that observed in the absence of Rab3 suggests that RIM has a function beyond acting as a Rab3 effector. One such function is to provide a scaffold to organize the active zone, as RIM binds to several other active zone proteins. In addition, RIM plays a key role in vesicle priming through its tight interaction with Munc13. This interaction involves the RIM ZF domain and the Munc13 C_2A domain (Figure 10.26B). The latter forms a homodimer, which strongly inhibits the essential function of Munc13 in vesicle priming. Binding of the RIM ZF domain to the Munc13 C_2A domain dissociates the homodimer and forms the RIM/Munc13 homodimer, activating Munc13 (Figure 10.24C). RIM/Munc13 binding can occur simultaneously with the RIM/Rab3 interaction, albeit with some rearrangement in the binding mode. Since RIM and Rab3A control diverse forms of synaptic plasticity, these findings have suggested a model whereby a Munc13 homodimer to Munc13/RIM heterodimer switch, perhaps modulated by Rab3 (Figure 10.26C), provides a link between synaptic vesicle priming and presynaptic plasticity.

In addition to its association with RIM-dependent plasticity through the C_2A domain, Munc13 is involved in presynaptic plasticity processes that depend on diacylglycerol (through the C_1 domain) or on Ca^{2+} (through a calmodulin-binding sequence and the C_2B domain). All these results have led to the hypothesis that the crucial priming activity of

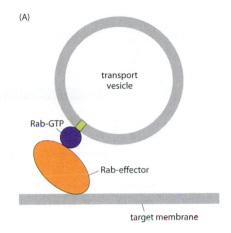

(A)

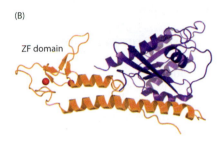

(B)

Figure 10.25 Rab–Rab effector interactions. (A) General model of Rab-mediated docking in intracellular membrane traffic whereby the GTP-bound form of a Rab protein is associated with a transport vesicle through a lipidic anchor (green) and binds to a Rab effector on the target membrane. (B) Ribbon diagram of the crystal structure of Rab3A (purple) bound to an N-terminal fragment of rabphilin (orange, with zinc ions in red). Note that Rab3A binding primarily involves two rabphilin α helices that are adjacent to a zinc finger (ZF) domain. NMR data have suggested that the Rab3A/RIM binding mode is analogous. (PDB 1ZBD) (B, from C. Ostermeier & A.T. Brunger, *Cell* 96:363–374, 1999. With permission from Elsevier.)

(A)
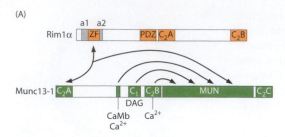

(B)
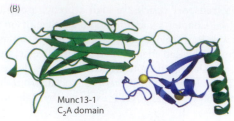

Munc13-1
C₂A domain

(C)
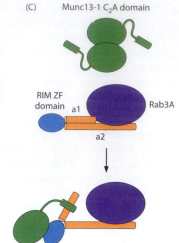

the MUN domain is controlled by intramolecular interactions of MUN with these various Munc13 domains and that modulation of these intramolecular interactions by diverse agents constitutes the basis for the corresponding forms of presynaptic plasticity (Figure 10.26A). While the mechanisms underlying the diverse forms of RIM- and Munc13-dependent plasticity remain to be elucidated, the ideas outlined above illustrate the extreme complexity of this system, a complexity that is fundamental for the ability of the brain to perform an amazingly diverse number of functions.

Synaptotagmin triggers Ca²⁺-dependent neurotransmitter release

A hallmark of synaptic vesicle fusion is its acute dependence on Ca^{2+} ions, which indicated early on that at least one Ca^{2+} receptor must be involved in neurotransmitter release. Actually, release in mammalian central synapses exhibits two phases: a major synchronous component that emerges in less than 0.5 ms after Ca^{2+} influx, and a slower, asynchronous component. While the Ca^{2+} sensor(s) for the latter is unknown, it is well established that the synaptic vesicle protein **synaptotagmin** (46 kDa) acts as a Ca^{2+} sensor for synchronous release.

Most of the cytoplasmic region of synaptotagmin is formed by two C_2 domains (Figure 10.18) which form similar β sandwich structures and bind multiple Ca^{2+} ions through loops at the top of the sandwich (**Figure 10.27**). These loops also mediate Ca^{2+}-dependent phospholipid binding, and mutations in these loops that alter the apparent Ca^{2+} affinity of synaptotagmin cause parallel changes in the Ca^{2+} dependence of neurotransmitter release. Moreover, synchronous release (but not the asynchronous component) is abolished upon deletion of the synaptotagmin gene in mice. In addition to demonstrating the role of synaptotagmin as a Ca^{2+} sensor in synchronous release, these results showed that Ca^{2+}-dependent phospholipid binding is critical for synaptotagmin function. Intriguingly, mutations that disrupt Ca^{2+} binding to the synaptotagmin C_2B domain abolish neurotransmitter release, whereas analogous mutations in the C_2A domain lead to milder impairments. This suggests that an additional activity of the C_2B domain is crucial for release.

There are two attributes of the C_2B domain that could explain its preponderant role in release. One is its ability to bind simultaneously to two apposed membranes through its Ca^{2+}-binding loops at the top face and through basic residues at the bottom face of the β sandwich. This suggests that the synaptotagmin C_2B domain can bring two membranes together to help induce fusion, much like the SNAREs do, but in a Ca^{2+}-dependent manner that could confer the tight Ca^{2+} dependence of synaptic vesicle fusion (Figure 10.27B). The top loops are negatively charged before Ca^{2+} binding and could thus repel the target membrane, which is also negatively charged, and thereby could hinder membrane fusion before Ca^{2+} influx. Upon Ca^{2+} binding, the top loops become positively charged, now attracting the target membrane. In this way synaptotagmin could bring the two membranes together to initiate fusion. In addition, the highly positively charged surface of the C_2B domain could help in bending the membranes to accelerate fusion. In this model, the C_2A domain plays an auxiliary role by assisting in the overall Ca^{2+}-dependent binding of synaptotagmin to membranes.

The second important attribute of the C_2B domain is its role in binding to the SNARE complex. This interaction involves the side of the C_2B domain β sandwich, thus allowing simultaneous binding to the SNARE complex and the membranes (Figure 10.27B), and helping the SNAREs and synaptotagmin to cooperate in membrane fusion. The mechanism of action of synaptotagmin is still unclear and alternative models have been proposed. For instance, it has been suggested that Ca^{2+}-dependent phospholipid binding of synaptotagmin and the insertion of hydrophobic residues in the membrane bilayer helps trigger fusion by altering membrane tension or by inducing a positive membrane curvature. In addition,

Figure 10.26 Coupling between Munc13 and RIM. (A) Domain diagrams of the two major Munc13 (Munc13-1) and RIM (RIM1α) isoforms, illustrating the notion that the Munc13 MUN may be the focal point for regulation of diverse presynaptic plasticity processes. In this model, diverse domains of Munc13 interact intramolecularly with the MUN domain, and diverse agents that govern different types of presynaptic plasticity (RIM, Rab3, Ca²⁺, and DAG) modulate the crucial activity of the MUN domain in synaptic vesicle priming by altering these intramolecular interactions. (B) Ribbon diagram of the crystal structure of the RIM zinc finger (ZF) domain (blue, with zinc ions in yellow) bound to the Munc13 C₂A domain and a C-terminal helical extension (green). (PDB 2CJS) (C) Model of the switch from a Munc13 homodimer to a Munc13/RIM heterodimer, which allows formation of a tripartite Munc13/RIM/Rab3A complex and may connect vesicle priming to presynaptic plasticity. The RIM/Rab3A interaction is modeled on the crystal structure of a rabphilin/Rab3A complex (Figure 10.25B); NMR data suggest that this interaction is slightly altered upon Munc13 binding to RIM. (Adapted from J. Lu et al., *PLoS. Biol.* 4:e192, 2006.)

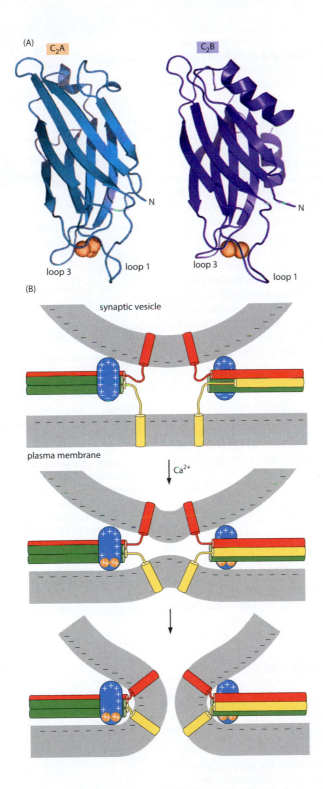

Figure 10.27 Ca²⁺-triggering of neurotransmitter release by synaptotagmin. (A) Ribbon diagrams of the Ca²⁺-bound structures of the synaptotagmin C_2 domains, with the Ca²⁺ ions shown in orange. (PDB 1BYN, 1KSW) (B) Model of how synaptotagmin might cooperate with the SNAREs to induce Ca²⁺-triggered membrane fusion and neurotransmitter release. Like the SNAREs, the synaptotagmin C_2B domain (blue) binds to the two apposed membranes simultaneously and brings them together, but in a Ca²⁺-dependent manner that confers the Ca²⁺ dependence of membrane fusion. The plus and minus signs indicate the electrostatic charges of the C_2B domain and the membrane. The Ca²⁺ ions that bind to one face of the C_2B domain are represented by orange circles with two plus signs each. Before Ca²⁺ influx, the Ca²⁺-binding region of the C_2B domain is negatively charged and may repel the target membrane. Ca²⁺ binding would switch the electrostatic potential of the C_2B domain Ca²⁺-binding region, inducing binding to the target membrane. The highly positive electrostatic potential of the C_2B domain might help to bend the membranes accelerating membrane fusion. (A, from Shao et al., *Biochemistry* 37:16106 16115, 1998. With permission from ACS. B, from I. Fernandez et al., *Neuron* 32:1057–1069, 2001. With permission from Elsevier.)

the interaction of the C_2B domain with the SNARE complex likely helps to release the inhibitory function of complexin in fusion, as discussed in the next paragraph.

Complexin plays both active and inhibitory roles

Complexin, a small (15 kDa) soluble protein, binds tightly to the SNARE complex through a central α-helical region (**Figure 10.28A, B**). It epitomizes the need of balancing activating and inhibitory interactions in an exquisitely regulated process such as neurotransmitter release. A selective impairment of the Ca²⁺-triggered step of release is observed in the absence of complexin in mice. However, an opposing inhibitory function is also suggested by other observations, including reconstitution and cell–cell fusion assays. These indicate that, upon Ca²⁺ binding, synaptotagmin releases the complexin-induced inhibition of fusion, in keeping with biophysical data suggesting that synaptotagmin competes with complexin for binding to membrane-anchored SNARE complexes, through the C_2B domain.

While there is no doubt that a tight interplay between complexin, synaptotagmin, and the SNAREs controls Ca^{2+}-triggering of neurotransmitter release, the underlying mechanisms are still to be elucidated. Rescue experiments of neurotransmitter release in complexin knockout mice show that a helical region called the accessory helix, which precedes the central SNARE-binding helix (Figure 10.28A), is partly responsible for the inhibitory function of complexin. Based on the crystal structure of the complexin/SNARE complex (Figure 10.28B), a simple model postulates that the accessory helix inhibits membrane fusion because it points to the vesicle membrane (Figure 10.28C). Since both the membrane and the accessory helix are negatively charged, they would experience strong electrostatic and steric repulsion upon full SNARE complex assembly. Hence, the accessory helix may contribute to keeping the SNARE complex partially assembled before Ca^{2+} influx (Figure 10.28C), as depicted also in Figure 10.24C, middle panel. In this model, full SNARE complex assembly and membrane fusion requires melting of the accessory helix or dissociation of complexin, which may be mediated by synaptotagmin by an as yet unknown mechanism. The basis for an active role of complexin in release is also unclear. Some evidence suggests that complexin binding stabilizes the SNARE complex and hence could help to exert more force on the membranes after the inhibitory action of the accessory helix is released, but this and other existing models of complexin function remain to be validated.

How did synaptic vesicle fusion arise?

A fascinating aspect of neurotransmitter release is to unravel how a primitive machinery that initially emerged to fuse membranes in eukaryotic cells evolved to control synaptic vesicle fusion in such a beautifully regulated manner. Some fundamental concepts are well established (for example, the central role of the SNAREs in fusion and the Ca^{2+}-sensing function of synaptotagmin), but many mechanistic ideas remain to be proven. The identification of crucial components of the release machinery and of critical interactions between them has provided some clues. However, many of the interactions have been identified in the absence of membranes, and increasing evidence indicates that membranes can strongly influence some of the interactions within the release machinery. This is hardly surprising given the very nature of the biological process controlled by this machinery. Hence, biochemical and structural experiments in the presence of one or even better two membranes will be essential to obtain a complete and coherent picture of the mechanism of neurotransmitter release that may integrate the various models presented in the figures in this chapter.

10.5   NUCLEAR PORE COMPLEXES

The nucleus is the defining feature of eukaryotic cells

The nucleus is a compartment for the storage and processing of the cell's genetic material. It is the largest organelle in the eukaryotic cell, typically occupying 6–10% of the cellular volume. Its size varies from species to species, ranging from 1 μm in yeast to up to 10 μm in mammals.

The nucleus is surrounded by the nuclear envelope (NE) separating the nucleoplasm from the cytoplasm (**Figure 10.29A**). The NE comprises two parallel membranes, the outer (ONM) and the inner (INM) nuclear membranes. The two membranes merge at sites where they are perforated by nuclear pores (Figure 10.29B) but a barrier function at their junction preserves the difference in their protein compositions. Furthermore, the ONM is continuous with the endoplasmic reticulum (ER) and, like the ER membrane, is studded with ribosomes. Between the ONM and INM is the perinuclear space, which connects to the lumen of the rough ER. The INM has a protein lining, the nuclear lamina, whose meshwork (see Section 11.2) has a role in maintaining the shape of the nucleus and in providing anchoring sites for chromatin (see Section 2.2) and other nuclear components.

The confinement of the genetic material to the nucleus necessitates mechanisms for the bidirectional exchange of a large number of macromolecules between the nucleoplasm and the

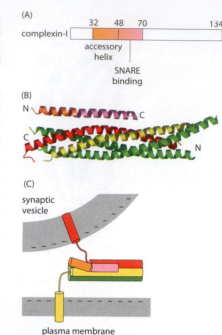

Figure 10.28 Complexin function in neurotransmitter release. (A) Domain diagram of complexin. (B) Crystal structure of the complexin/SNARE complex with complexin color-coded as in (A), and the SNAREs colored as in Figure 10.20. (PDB 1KIL) (C) Model of how the complexin accessory helix might inhibit neurotransmitter release due to electrostatic and steric hindrance with the vesicle membrane. The minus signs represent negative charges in the membranes and the complexin accessory helix. The model postulates that, because of the repulsion with the vesicle membrane, the accessory helix prevents C-terminal assembly of the SNARE complex. (B, from X. Chen et al., *Neuron* 33:397–409, 2002. With permission from Elsevier.)

(A)
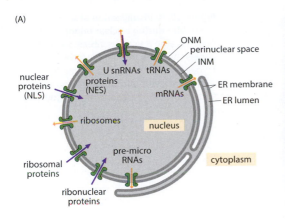

(B)
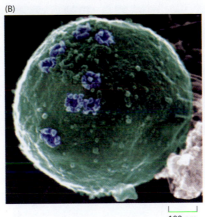
100 nm

(C)
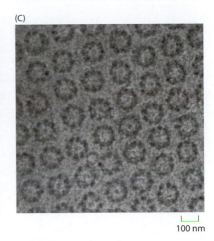
100 nm

cytosol. All nuclear proteins (histones, individual ribosomal proteins, polymerases, transcription factors, RNA processing factors, etc.) need to be imported into the nucleus after their synthesis on cytosolic ribosomes. Conversely, ribosomal subunits, messenger RNAs, transfer RNAs, and many other RNAs need to be exported to the cytosol after they have been transcribed and processed. Certain proteins conduct their function through shuttling between both compartments and are thus subject to both nuclear import and export. It is estimated that in a growing eukaryotic cell more than a million macromolecules per minute traverse the NE.

Nuclear pore complexes are the gateways for nucleocytoplasmic transport

The bidirectional exchange of macromolecules across the nuclear envelope passes through the **nuclear pore complexes** (NPCs). Unlike other aqueous channels, the NPCs do not span a single bilayer but perforate the NE at sites where the INM and ONM fuse through sharply bent membrane junctions. In yeast and in typical mammalian cells, there are around 5–15 pores per μm^2 (Figure 10.29B) but the pore density can be significantly varied depending on the cellular differentiation state.

NPCs were first observed in 1950 in electron micrographs of NEs isolated from amphibian oocytes, but whether they made the NE permeable to macromolecules remained controversial for quite some time. Compelling evidence for the existence of pores and their role in nucleocytoplasmic transport came from a classic experiment performed by C.M. Feldherr in 1962. He microinjected colloidal gold particles, 2.5–5.5 nm in diameter, and coated with polyvinyl-pyrrolidone, into the cytosol of the ameba *Chaos chaos*. After a few minutes, he observed the gold particles both in the cytosol and in the nucleus. Some particles were found centrally within the 'annuli' of the NE, which he concluded demonstrated that "passage through these structures may be restricted to a central channel" (**Figure 10.30**).

The NPC is a large and elaborate transport machine

Ions, metabolites, and many small proteins (<40 kDa) diffuse passively through NPCs, while larger molecules need to be actively transported. Active nuclear transport is a highly regulated, rapid, and efficient process. *In vitro* studies and single-molecule imaging analysis suggest an import rate of 1000 molecules per second per NPC; because of the enormous size of the central channel, several cargoes might traverse an individual NPC simultaneously.

The orchestration of multiple concurrent transport events requires an elaborate structure. The NPC is indeed one of the largest protein assemblies in eukaryotic cells: estimates of its molecular mass based on scanning transmission EM mass measurements range from 60 MDa for yeast to up to 125 MDa for vertebrate cells. The overall dimensions of the NPC are indicated in **Figure 10.31**. The outer diameter of the NPC is similar in all species, at 90–120 nm, whereas the height appears to be more variable (60–85 nm without the nuclear basket). However, despite some variability in size, the overall structural organization of NPCs has been conserved throughout evolution.

Figure 10.29 Nuclear transport and nuclear pore complexes. (A) The double membrane envelope surrounding the nucleus consists of an inner nuclear membrane (INM), which is bound to the nuclear lamina, and an outer nuclear membrane (ONM), which is continuous with the endoplasmic reticulum. Nuclear pores (green) cross the envelope providing aqueous channels for nuclear transport. The most important nucleocytoplasmic transport pathways are shown. The arrows indicate the direction of transport into and out of the nucleus. (B) Nuclear pore complexes in the nuclear envelope of *S. cerevisiae* visualized by scanning electron microscopy. The image was pseudocolored to highlight and differentiate the nuclear envelope (green) and nuclear pore complexes (blue). (C) Electron micrograph of frozen-hydrated nuclear envelope from *Necturus* showing nuclear pore complexes. (B, image courtesy of Elena Kiseleva et al. *Nat. Protoc*, 2007, 2, 1943. The work was supported by the Wellcome Trust (UK) & RFBR (Russia). C, image courtesy of Chris Akey.)

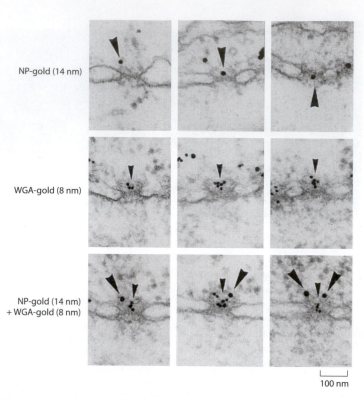

NP-gold (14 nm)

WGA-gold (8 nm)

NP-gold (14 nm)
+ WGA-gold (8 nm)

100 nm

Figure 10.30 Visualization of single molecules during nuclear import. Gold-labeled cargo is used for visualization. The panels show electron micrographs from cryo-sections of nuclear envelopes from *Xenopus* oocytes transporting cargo labeled with 8 nm and 14 nm gold beads. (Courtesy of Nelly Pante.)

The NPC forms an octagonal multilayered cylinder from which distinct substructures emanate on the nuclear and cytosolic faces. The scaffold is formed of three rings. The spoke ring occupies the space between the INM and ONM showing C2 symmetry in EM-based three-dimensional reconstructions. The cytoplasmic ring and the nuclear ring are coaxial with the spoke ring. Eight seemingly flexible filaments protrude from the cytoplasmic ring, which are free at their distal ends. On the nucleoplasmic side, an elaborate structure, called the nuclear basket, is suspended from eight filaments joining it to the nucleoplasmic ring (Figure 10.31). The overall arrangement displays some 'plasticity,' which is indicative of dynamic changes as they might occur during active transport.

The central channel of the NPC is believed to be filled with a meshwork of natively unfolded domains (see Section 6.4), which are tethered to its walls. They are thought to constitute a semi-permeable barrier which prevents the diffusion of larger molecules through the NPC unless they are bound to specific transporter molecules (see below). A structure of the NPC *in situ* obtained by cryo-electron tomography revealed the existence of another system of channels near the periphery (that is in close proximity to the NE). It has been proposed that

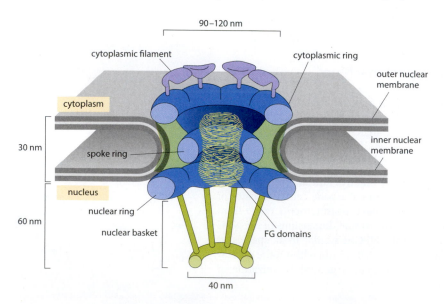

90–120 nm

cytoplasmic filament
cytoplasmic ring
outer nuclear membrane
cytoplasm
30 nm
spoke ring
inner nuclear membrane
nucleus
60 nm
nuclear ring
nuclear basket
FG domains
40 nm

Figure 10.31 The main structural components of the nuclear pore complex. These include the central framework embedded in the nuclear envelope membrane, the cytoplasmic ring and cytoplasmic filaments, the nuclear ring, and the nuclear basket.

these channels serve as a passageway for smaller (<40 kDa) macromolecules or for membrane proteins destined for the INM.

The many Nups are built up from domains with only a few folds

The past decade has seen progress in revealing the structures of individual Nups by X-ray crystallography (see **Methods**). Structure predictions grouped **nucleoporins** into a small set of fold classes that account for 76% of the mass of the yeast NPC (**Figure 10.32**). These five classes include the phenylalanine-glycine (FG) domains, coiled-coil domains, scaffold proteins composed of β propellers, α-helical domains, or a tandem combination of both. Meanwhile, after overcoming problems in expression of full-length proteins or of truncated versions thereof, crystal structures are available for representatives of each fold class (**Figure 10.33**).

Figure 10.32 Nucleoporins that make up the nuclear pore complex (NPC). Domain architecture of nucleoporins from *S. cerevisiae* as determined by X-ray crystallography or prediction (where experimental structures are still lacking). Abundance and derived mass calculations are based on available Nup/NPC stoichiometries. Nucleoporins specific to metazoa are italicized. (Adapted from Brohawn et al. *Structure* 17:1156-1168, 2009. With permission from Elsevier.)

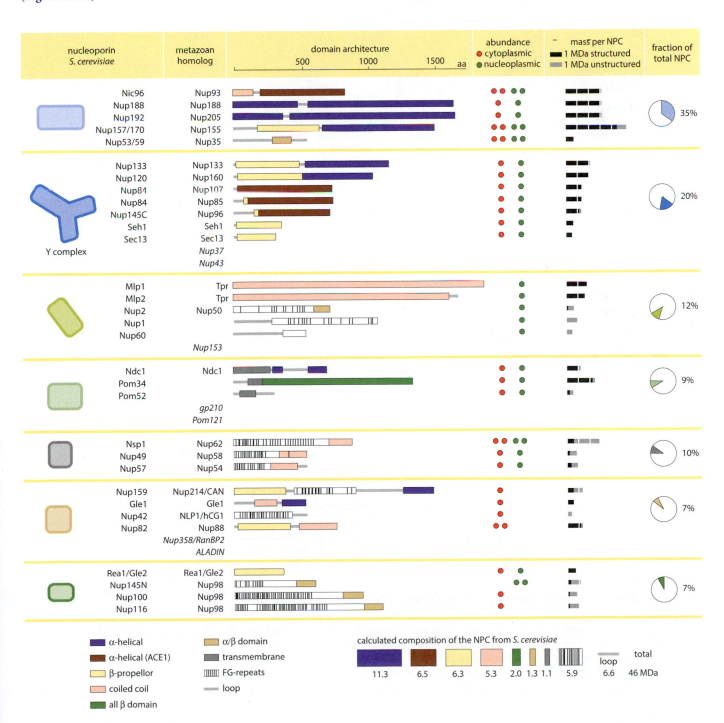

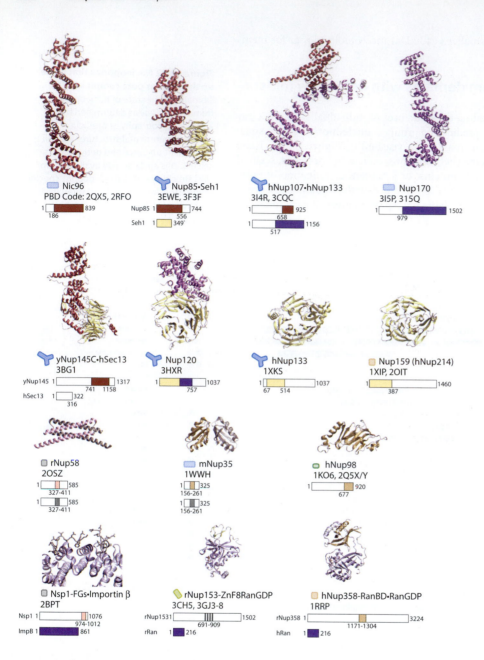

Figure 10.33 Structures of nucleoporins.
All representative nucleoporin structures (Nups) published to date; PDB accession codes are indicated. Structures are colored according to structural motifs as in Figure 10.38. Residue information for each crystallized fragment is given below the structure. Structures are shown in the assembly state that is supported by crystallographic and biochemical evidence. Structures are from *S. cerevisiae* unless noted otherwise (h, human; m, mouse; r, rat). (Adapted from Brohawn et al. *Structure* 17:1156–1168, 2009. With permission from Elsevier.)

The β-propeller fold (see **Guide**) is very abundant in eukaryotic proteins, having various functions. Although many proteins in this class present a signature WD 40 repeat motif (see **Guide**), only two (Sec13 and Seh1) of the eight conserved Nups that fold into β propellers contain this motif. β Propellers in the NPC have an architectural role, serving as protein–protein interaction sites: β propellers found near the periphery recruit accessory factors, whereas those inside the NPC are likely used to connect subcomplexes. Most of the eight β propeller-containing Nups have been structurally characterized (Figure 10.33). Of these, NuP133, NuP120, and NuP159 have seven-bladed β propellers at the N terminus, while Sec13 and Seh1 have open six-bladed β propellers with partner proteins that insert the seventh blade into the propeller.

More than half of the mass of the NPC scaffold is made up of α-helical domains. The structures of helical-rich Nups have thus far revealed three different α-helical folds. The first of these, the so-called ancestral coatomer element 1 (ACE1) domain is formed by 30 helices arranged in a J-like topology, plus an N-terminal coiled-coil domain. This fold is the critical element for establishing the scaffold and is postulated to be analogous to that found in vesicle coat proteins inducing membrane curvature (see Section 10.2), suggesting a common evolutionary origin for these proteins. Four Nups contain ACE1 domains: Nic96, NuP84, NuP85, and NuP145C. The second α-helical fold is present in NuP133 and NuP170, which form an extended and stretched helical stack. The third fold was observed in NuP120, fully integrating a β propeller and an α-helical domain.

A key structural element of NPCs comprises the phenylalanine-glycine (FG) repeat domains occupying much of the 50 nm wide central channel (see Figure 10.31). Some 13% of the NPC mass is made up of FG repeat domains. They can be further classified as FXFG domains (where X stands for a variable aa) and GLFG domains. They are components of about 10 Nups, tethering them to the walls lining the central channel. The intrinsic disorder of the FG domains is well documented but little is known about the intervening non-FG sequences, which are poorly conserved, though rich in polar and charged residues. The FG domains are naturally unfolded domains and act as a barrier to repel nonspecific cargo. For translocation through the NPC to occur, cargo must be attached to receptors (the karyopherins or importins, see Section 10.6) that are capable of binding to individual FG repeats. The structure of karyopherins and their role in the transport of cargo through NPCs is described in Section 10.6.

Nups form stable subcomplexes

The Nups are organized in discrete subcomplexes or modules; multiple copies of each module are arranged around the central symmetry axis (**Figure 10.34**). It has been established

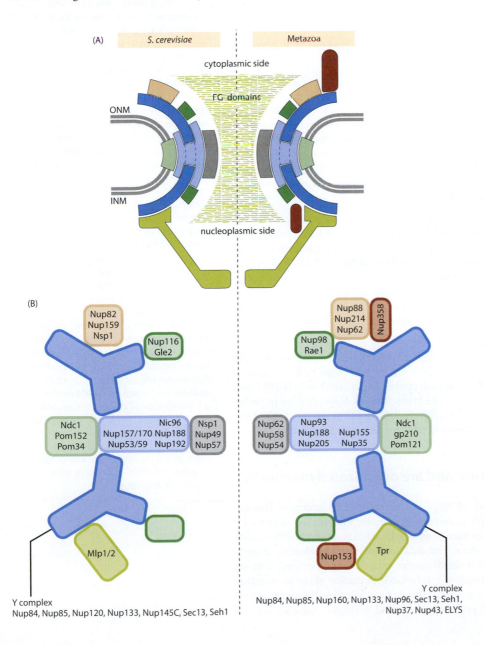

Figure 10.34 Modular organization of Nups and subcomplexes. Schematic representation of (A) the nuclear pore complex (NPC) and (B) the modular assembly of the Nups. The NPC is built from ~30 Nups, which are organized in a small set of subcomplexes. The cartoons show the major subcomplexes that make up the latticelike scaffold (blue colors), the membrane attachment (light green), the phenylalanine-glycine (FG) network (gray), the cytoplasmic filaments (light-brown), and the nuclear basket (lime) of the NPC. *S. cerevisiae* modules are shown on the left, and metazoan modules, with specific additions, are shown on the right. A few peripheral Nups are left out for clarity. This is a simplified representation, connections are not to be taken literally, and box sizes are not proportional to molecular masses. (Adapted from Brohawn et al. *Structure* 17:1156–1168, 2009. With permission from Elsevier.)

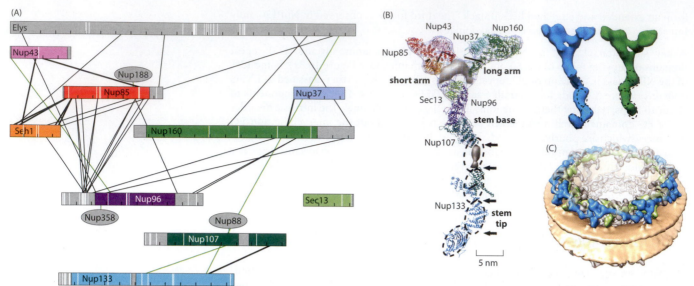

by biochemical means that identifiable subcomplexes remain assembled during mitosis and the concomitant disintegration of the NE.

There are two major scaffolding complexes: while the cytoplasmic and nuclear rings are made up by the so-called Y-complex, the Nic96 complex is thought to reside in the spoke ring (Figure 10.34). The multimeric Y-complex, known as NuP84-complex in yeast, is very stable and is essential for NPC assembly. It is composed of seven universally conserved Nups. In higher eukaryotes, three additional Nups, NuP37, NuP43, and ELYS/MEL-28, are members of this complex (**Figure 10.35**). Electron microscopic maps provide a framework for fitting into them crystal structures of the subunits. Often additional information is required and obtained, for example, through labeling or cross-linking experiments. To date, only the Y-complex has been characterized by this hybrid approach.

In the context of the NPC, the Y-complex is thought to localize symmetrically at the cytoplasmic and nucleoplasmic faces of the NPC. In between copies of the Y-complex, at the spoke ring, lies the Nic96 complex. This complex is not as well characterized as the Y-complex, perhaps reflecting lower stability. It is formed by Nic96, NuP53/59, NuP188, and NuP192, and it tethers another complex (Nsp1) in the center of the NPC. The transmembrane protein Ndc1 interacts with NuP157/150 and NuP53/59, thus connecting this complex to the fused INM and ONM. Other transmembrane Nups, for example, Pom34 and Pom152 in yeasts, appear not to be conserved throughout the tree of life.

In the NPC, coiled-coil domains play structural roles in the more peripheral subcomplexes. Six proteins that line the channel are held together by coiled-coil interactions: the Nsp1 complex is composed of Nsp1, NuP57, and NuP49, and forms the bulk of the central transport barrier. Another copy of Nsp1 in each spoke is associated with NuP82 and NuP159, localized at the cytoplasmic side. Also the yeast proteins MIp-1/2 (Tpr in vertebrates) that form the nuclear basket contain coiled coils. They are thought to attach to the nuclear ring through NuP60 and serve as docking sites for the recruitment of accessory factors to the NPC.

NPCs have a modular architecture and are organized dynamically

To date, most individual Nups or modules of Nups can be assigned to structural features of the NPC as revealed by cryo-EM in only a coarse and tentative manner. To address the assignment problem, modeling approaches have been developed that integrate all available biochemical, biophysical, and structural data. The data include immunoblotting experiments estimating the Nup stoichiometry; hydrodynamics experiments providing data on volumes and shapes; immuno-EM providing low-resolution information about localization of Nups; affinity purification determining the composition of modules; overlay experiments probing direct interactions; and cryo-EM providing shape and symmetry information. All

Figure 10.35 Integrative structure determination of the human Nup107 subcomplex. (A) The spatial restraints identified by cross-linking-mass spectrometry are shown as connections between the primary structures of the Nup107 subcomplex proteins. Spatial restraints that map unto available crystal structures are shown in black. Cross-links accounting for verified inter-subcomplex interactions with the Nup107 subcomplex as well as inter-subcomplex interactions with other nuclear pore complex (NPC) components are shown in green; cross-links not covered by structural models are shown in gray. Identified phosphorylation sites are indicated by white lines within the primary structures. (B) The structural model of the isolated Nup107 subcomplex is shown side-by-side with the inner and outer Nup107 subcomplex segmented from the EM tomography map shown in MB 10.5.04/10.37. Available crystal structures or homolog models were fitted into the EM map based on their shapes and the spatial restraints shown in (A). Arrows and dashed ellipses indicate the positions of flexible hinges and static connectors, respectively. Regions for which no crystal structures are available are shown isosurface rendered. (C) Visualization of the orientation of the Nup107 subcomplex within the cytoplasmic ring. The segmented outer and inner Y-complexes are shown in blue and green, respectively. (Adapted from Khanh Huy Bui et al. *Cell* 155:1233–1243, 2013. With permission from Elsevier.)

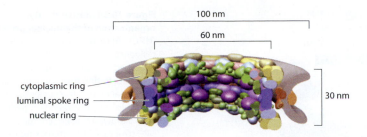

Figure 10.36 Molecular architecture of the nuclear pore complex (NPC). Computationally generated model of the NPC from *S. cerevisiae*. (From Alber et al., *Nature* 450:695–701, 2007. With permission from Macmillan Publishers Ltd.)

these data were translated into sets of spatial restraints and the relative positions and proximities of individual Nups or modules of Nups were calculated by an optimization process satisfying all available restraints. The model emerging from this analysis (**Figure 10.36**) can be interpreted as a Nup interaction map, which provides useful information that cannot be retrieved from any single experimental source. However, although it can localize individual proteins in the core of the NPC where multiple interactions occur, it cannot map interactions in the filamentous structures emanating from the core.

The most detailed pictures of complete NPCs have been obtained by cryo-electron tomography with subsequent subtomogram averaging (**Figure 10.37**). Although, at the resolution of ~3 nm the tomographic reconstructions do not permit the direct fitting of high resolution structures, taken together with complementary information they provide the structural framework that can be annotated with high-resolution structures. The combination of cryo-electron tomography with cross-linking and quantitative mass spectrometry has revealed the position and orientation of a Y-complex within the fully assembled NPC (Figure 10.35).

While many structural Nups are stably integrated into the NPC at all times, some of the natively unfolded Nups are very dynamic. Nup dynamics have been probed systematically by fluorescence recovery after photobleaching (FRAP) of green fluorescent protein (GFP)-tagged Nups in living mammalian cells. The FRAP measurements revealed a wide range of residence times, from seconds to days (**Figure 10.38**). Mass spectrometric experiments have revealed that some structural Nups are extremely long lived in nondividing cells and can literally last for life in rat brain. The loss of these Nups has been implied to go along with a partially compromised nuclear permeability barrier during aging. Taken together with stoichiometric measurements of Nups in different cell types, a picture emerges in which the central components of the NPC show low turnover consistent with their role as a structural scaffold, whereas the peripheral components showed a high turnover. This is indicative of targeting functions or regulatory roles in the translocation, recycling or release of transport complexes.

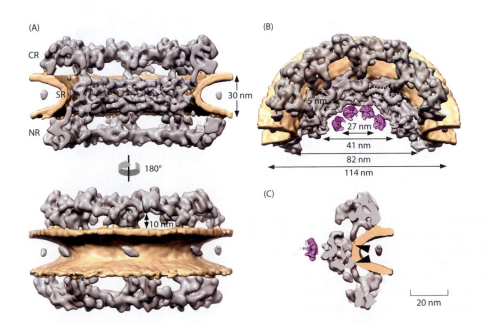

Figure 10.37 Structure of the human nuclear pore complex (NPC) as determined by cryo-EM. (A) Isosurface-rendered view of the NPC (cut in half) in two different orientations. Membranes are rendered pale orange, the cytoplasmic (CR), spoke (SR), and nuclear rings (NR) are marked. (B) Same as (A) but seen as top view. An additional inner density (purple) is visible only at a lower isosurface threshold. Crescent-shaped features reminiscent of the structures of some members of the Nup93/Nic96 subcomplex are highlighted with dots in (B); openings that might be relevant for the nuclear membrane (NM) protein import are indicated in (A) and (B). (C) The spoke ring complex connects to the membrane at two distinct sites per asymmetric unit (arrowheads). (Adapted from Khanh Huy Bui et al. *Cell* 155:1233–1243, 2013. With permission from Elsevier.)

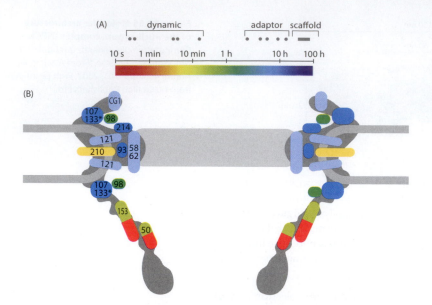

Figure 10.38 Map of the dynamic organization of the nuclear pore complex (NPC). (A) Distribution of the average residence times of 19 different nucleoporins displayed on a logarithmic pseudocolor scale, illustrating three dynamic classes. (B) Spatial map of the nucleoporins based on published localizations superimposed onto a drawn-to-scale scheme representing NPC structure. The asterisks represent other members of the Nup107–160 complex: Seh1, Sec13, Nup37, Nup43, and Nup85. The average residence time for each nucleoporin is represented quantitatively by the pseudocolor scale in (A). (Adapted from Rabut et al. *Curr Opin Cell Bio* 16:314–321, 2004. With permission from Elsevier.)

10.6 NUCLEAR IMPORT AND EXPORT

As described in Section 10.5, macromolecules cross the nuclear envelope via nuclear pore complexes (NPCs). The central channel of NPCs is, in principle, large enough to allow the passage of ions and small macromolecules (<40 kDa) by diffusion. However, even macromolecules smaller than this diffusion limit (such as tRNAs or histones) are unable to traverse the NPCs on their own and need to associate with dedicated transport factors.

Nucleocytoplasmic transport is mediated by transport factors

Nucleocytoplasmic transport involves several steps. First, cargo is recognized and bound by transport factors either in the nucleus or in the cytoplasm. Next, the transport complexes interact with the FG repeats lining the NPC central channel, mediating their translocation. Once on the opposite side of the nuclear envelope, the cargo is released. Finally, the transport factor is recycled back to its original compartment, allowing it to carry out another round of transport. With the notable exception of mRNA export factors, the majority of nucleocytoplasmic transport factors belong to the karyopherin family.

Despite the large number and the diversity of cargoes, their recognition by **karyopherins** is very specific: only proteins destined for the nucleus are imported and only properly processed RNA and ribonucleoprotein particles (such as ribosomal subunits) are exported. In addition, specific proteins are also exported (such as eIF1A, a small translation initiation factor that can passively diffuse into the nucleus and has to be actively brought back to the cytoplasm). The karyopherin β family includes 14 members in yeast and more than 20 in humans. Karyopherins are generally classified as importins or exportins. **Importins** mediate transport into the nucleus, **exportins** from the nucleus to the cytoplasm. However, certain karyopherins are known to function as bidirectional transport factors, importing certain cargoes and exporting others. In this context, the nomenclature in the literature is somewhat misleading: importin 13, for example, is a bidirectional transport factor that can also export cargoes while transportin (also known as karyopherin β2) is strictly an import factor.

A Ran GTP gradient acts as a cargo pump

Cargo uptake and release are regulated by **Ran**, a Ras-related GTPase that is present in the nucleus in the GTP-bound state and in the cytoplasm in the GDP-bound state. This difference in Ran states is maintained with the help of regulators: the Ran <u>G</u>TPase <u>a</u>ctivating <u>p</u>rotein (Ran GAP) is localized on the cytoplasmic side of the NPCs and therefore promotes hydrolysis of RanGTP to RanGDP in the cytoplasm. The Ran guanine exchange factor (RCC1) is localized on chromatin and thus restricts the loading of Ran with GTP to the nucleus (**Figure 10.39**). Karyopherins respond differently to the presence of RanGTP

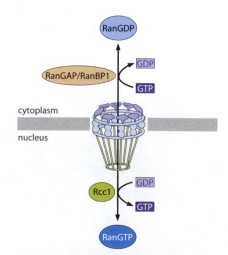

Figure 10.39 The RanGTP-GDP cycle. When RanGTP enters the cytoplasm, the bound nucleotide is hydrolyzed to GDP by RanGAP (which is cytosolic). When RanGDP is imported into the nucleus (by a dedicated transport factor, NTF2, not shown in this figure), the nucleotide is exchanged by Rcc1 (which is nuclear) and GTP is loaded.

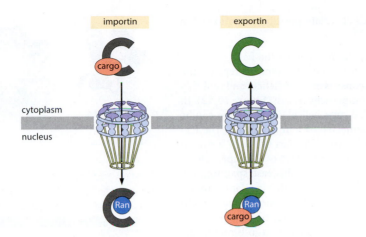

Figure 10.40 Directionality of import and export pathways. The karyopherin transport factors (importins and exportins) respond to the presence of RanGTP in the nucleus by unloading and loading cargoes, respectively.

and RanGDP. Importins bind their cargo in the absence of RanGTP (in the cytoplasm) and release it in the presence of RanGTP (in the nucleus). Conversely, exportins bind their cargo in the presence of RanGTP (in the nucleus) and release it upon RanGTP hydrolysis (in the cytoplasm) (**Figure 10.40**). The presence of RanGTP signals a nuclear location to the karyopherins and the chemical potential of the RanGTP gradient across the nuclear envelope acts as a unidirectional cargo pump.

Importins interact with cargo directly or via adaptor molecules

Crystallographic studies have provided detailed insights into the modes of interactions between karyopherins and their cargo and the role of RanGTP in the binding and release of cargo during the transport cycle. A widespread mechanism of nuclear import relies on the presence of short linear motifs embedded in intrinsically unstructured regions of the cargo, the so-called nuclear localization signals (NLSs). The first NLS was discovered in the 1980s as a short sequence motif characterized by a high proportion of positively charged residues (the so-called classical or c-NLS). The positively charged residues are grouped in either a single cluster (monopartite c-NLSs such as in the SV40 T antigen, ~8 amino acids in length) or in two clusters separated by a linker (bipartite c-NLSs such as in nucleoplasmin, ~16 amino acids in length). The c-NLS sequences are recognized by an adaptor molecule, importin α, which in turn binds the transport factor, importin β (also known as karyopherin β1). Another well-studied import motif is the so-called PY-NLS. PY-NLSs are also positively charged, unstructured sequences, but are longer than c-NLSs (about 20–30 residues). The different length reflects the different size of the molecule they bind to, namely importin α and transportin. Although linear NLSs for other importins have yet to be characterized or discovered, it is clear that cargoes can also be recognized via globular domains. In contrast to the linear c-NLSs and PY-NLSs, in this case the 'signal' is a complex one in that it involves different portions of the sequence in the context of a folded polypeptide chain.

Importin α has two domains: an N-terminal importin β-binding (IBB) and a C-terminal c-NLS-binding domain. The C-terminal domain comprises 10 tandem repeats known as Armadillo (ARM) motifs, each of them built of three α helices. These motifs pack side-by-side forming an elongated curved molecule (**Figure 10.41**). The c-NLS binds stretched out onto the concave surface, which provides multiple binding pockets that are exploited by the side chains of different NLS peptides in different ways. Thus, importin α presents a binding

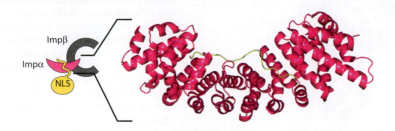

Figure 10.41 Structure of the ARM repeat domain of importin bound to the nuclear localization signal (NLS) of nucleoplasmin. In the schematic (left) the part of the ternary import complex visualized in the crystal structure (right) is indicated. Importin is red, the NLS is yellow. (PDB 1EE5)

platform that is both specific (only certain amino acids at certain positions are allowed in the binding pockets) and versatile.

PY-NLSs are recognized directly by transportin, a typical member of the karyopherin β family. Karyopherin β proteins are all α-helical molecules made up of tandem repeats, the so-called HEAT motifs (see **Guide**). Each HEAT motif consists of two α helices and typically 18–20 of them pack in an almost parallel fashion with a superhelical twist (**Figure 10.42**). In the case of transportin, the molecule adopts an extended superhelical structure that can be described as formed by two crescent-shaped arches connected by a hinge region. The PY-NLS stretches onto the concave surface of the superhelix, covering part of the N-terminal and C-terminal arches as well as the hinge region. The concave surface of the superhelix is also used by importin β to bind the IBB domain of importin α, in this case spanning the C-terminal arch and the hinge region. Importin β binds cargo not only via adaptor molecules such as importin α, but also directly, as in the case of the hormone-related protein PTHrP or the transcription factor SREBP2. PTHrP binds to the N-terminal arch, while SREB2 binds to the C-terminal arch. Thus, importin β provides a large cargo-binding surface on the concave side of its two arches and different cargoes can bind to different regions. A similar versatility is observed in the case of importin 13. This karyopherin binds two globular proteins using different portions of the superhelix: Ubc9 at the concave surface formed by the N-terminal arch and the hinge region and the Mago-Y14 complex at the concave surface formed by the hinge region and C-terminal arch.

Importins adopt shapes complementary to their cargoes

While transportin does not undergo large conformational changes upon cargo binding, importin β shows remarkable flexibility. The overall conformation of IBB-bound importin is a closed 'snail-like' structure, with the C-terminal arch wrapped tightly around the IBB (**Figure 10.43,** left). In the SREBP2-bound structure, the overall conformation is more open compared with the snail-like IBB structure. Similarly, in the RanGTP-bound and free importin β structures, the pitch of the superhelix is more expanded than it is in the IBB importin (Figure 10.43, right). The importin 13 superhelix also adopts different conformations upon binding different cargoes. The flexibility in assuming different conformations can be attributed to the segmented architecture of the multiple repeating units. Changes in the relative orientation of consecutive HEAT repeats can be accommodated and can be propagated along the entire molecule. In addition to the conformational rearrangements of HEAT repeats, local changes also play key roles, for example at a loop in the hinge region that in importin β and transportin is critically involved in the RanGTP-mediated cargo dissociation.

The binding of cargo and RanGTP to importins is mutually exclusive

Karyopherin β proteins bind RanGTP at the concave surface spanning the N-terminal arch and the hinge region. The hinge region of importin β and transportin features a conserved acidic loop that is capable of interacting with either cargo or Ran. In the case of transportin,

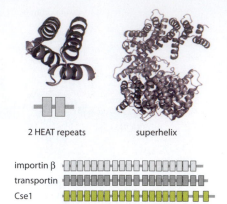

2 HEAT repeats   superhelix

importin β
transportin
Cse1
Crm1

Figure 10.42 Architecture of HEAT repeat karyopherins. The sequence of importin (the founding member of the karyopherin family) contains tandem helical repeats (in light gray) that form a superhelical structure (upper panel, right). Other karyopherins are also made up of similar building blocks and display a regular architecture (in dark gray and green, lower panel).

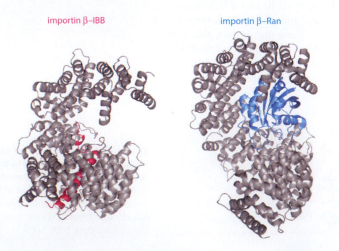

importin β–IBB   importin β–Ran

Figure 10.43 Structure of importin in two different states of its transport cycle. Importin is shown in gray. On the left importin is bound to the importin β-binding (IBB) domain of importin (red) and on the right to RanGTP (blue). The structures are viewed with their N-terminal arch in the same orientation, and show different conformations at their C-terminal arch. (PDB 1QGK, 2BKU)

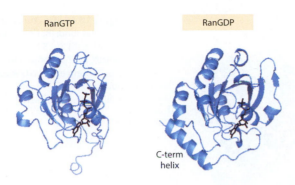

Figure 10.44 Conformational changes of the RanGTPase upon nucleotide binding. Ran is shown in blue with GTP/GDP in black. In the GDP-bound structure (right), the C-terminal helix is bound to the core domain of Ran, while it is displaced in GTP-bound Ran (not shown in the structure because exposed to solvent and disordered). (PDB 1QBK, 3GJ0)

RanGTP-mediated cargo release involves an allosteric rather than a direct mechanism: RanGTP binding is accompanied by a conformational change in a long acidic loop present in the hinge region of transportin, forcing it to bind to the PY-NLS-binding site and thereby dissociating it from karyopherin. In the case of importin β, binding of cargo and Ran is mutually exclusive, as they would clash near the central hinge region of importin, and a similar mechanism is used by importin 13.

Why is RanGTP but not RanGDP able to bind to importin β? Ran is a small globular protein with a core domain (known as the G domain) that contains the nucleotide-binding pocket (**Figure 10.44**). Similar to other Ras-related GTPases, it features two loops (referred to as switch I and II) that alter their conformation depending on the state of the bound nucleotide. In the GTP-bound Ran, the switch I and II regions are close to the γ-phosphate of the nucleotide, while they move apart in GDP-bound Ran (where the γ-phosphate is absent). In addition, Ran has a third region that switches conformation: the C-terminal portion of the molecule forms an α helix that packs against the G domain in the GDP-bound form but moves away from the G domain in the GTP-bound state (Figure 10.44). In the GTP-bound conformation, the C-terminal α helix of Ran would sterically clash with the karyopherin.

The mode of RanGTP binding is similar amongst karyopherins

The binding mode of RanGTP to importins such as importin β and transportin is similar to that observed with the exportins such as Cse1, Xpo5, and Xpot. RanGTP always binds to the concave side of the N-terminal arch. The N-terminal half of these karyopherins is the most conserved part of the molecules, suggesting that different members of the family share a common mechanism of RanGTP recognition. Shape complementarity appears to be a key aspect of the recognition. Although the interactions involve different residues in the karyopherins studied to date, in all cases the two N-terminal HEAT repeats bind the switch I and II regions of Ran, while the hinge region between the two arches binds the core GTPase that is unmasked by the unfolding of the third switch region.

Despite the similarities in RanGTP recognition, importin β and transportin use RanGTP binding to dissociate their cargo, whereas Cse1, Xpo5, and Xpot use RanGTP to associate with their cargo. Importin 13 is a 'dual duty' karyopherin that can use RanGTP to either bind or dissociate the cargo. The export cargo eIF4A binds the concave surface of importin 13 at the hinge region. Counterintuitively, this is a surface that is also used by the import cargo Mago-Y14. Subtle changes in the shape and surface charge of the two cargoes determine whether they bind cooperatively with RanGTP or antagonistically.

Cse1 is required for the export of importin α

In the c-NLS transport pathway, at the end of a round of import both importin α and importin β need to be recycled to the cytoplasm. While importin β shuttles back in complex with RanGTP, importin α needs a dedicated transport factor, Cse1. In the absence of cargo, Cse1 has a closed ringlike conformation with the two arches touching each other. Upon binding RanGTP and importin α, Cse1 opens up, enclosing the GTPase between the two arches and providing a platform for importin α to bind (**Figure 10.45A,B**). Importin α and RanGTP both bind near an insertion situated on the eighth HEAT repeat. While importin β can recognize its adaptor, importin α, only when an NLS-containing protein is bound, Cse1 only exports the adaptor without bound cargo, ensuring that NLS-carrying proteins are

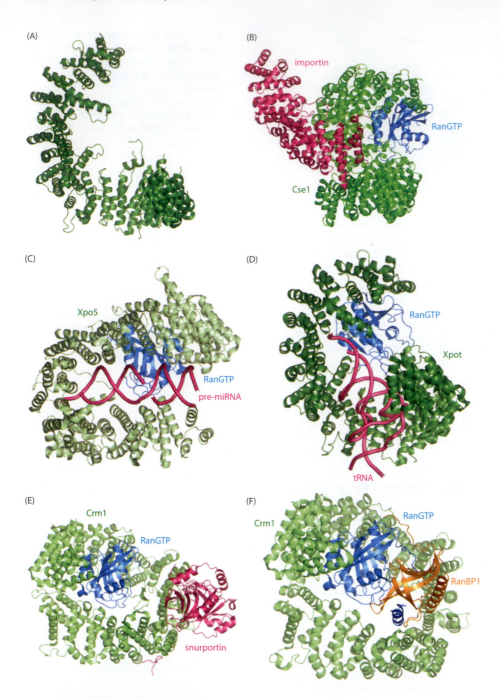

(A)

(B) importin

RanGTP

Cse1

(C) Xpo5

RanGTP
pre-miRNA

(D) RanGTP

Xpot

tRNA

(E) Crm1

RanGTP

snurportin

(F) Crm1

RanGTP

RanBP1

Figure 10.45 Structures of exportins. (A) Cse1 unbound; (B) Cse1 bound to RanGTP-importin; (C) Xpo5 bound to RanGTP-pre-miRNA; (D) Xpot bound to RanGTP-tRNA; (E) Crm1 bound to RanGTP-snurportin; (F) Crm1 bound to RanGTP-RanBP1. (PDB 1Z3H, 1WA5, 3A6P, 3ICQ, 3GJX, 3M1I)

not inadvertently returned to the cytoplasm. This raises the question as to how two karyopherins, importin β and Cse1, identify two different states of the same molecule. Importin β binds the N-terminal IBB domain of the adapter importin α, a region not involved in the binding of the NLS, but which autoinhibits importin α by occupying the NLS recognition groove in the absence of NLS cargoes. The structure of the ternary complex Cse1-RanGTP-importin α shows that, in contrast to importin β, Cse1 binds a large portion of the globular ARM repeat domain of importin β. Thus, the autoinhibitory sequence between the NLS-binding domain and the N-terminal arch of Cse1 is trapped, locking importin α in its auto-inhibited state.

Xpot and Xpo5 export tRNAs and pre-micro RNAs

Following transcription and processing in the nucleus, small structured RNAs, such as tRNAs and pre-micro RNAs, must be exported to the cytoplasm where the translation machinery is localized (see Section 6.2). The exportins Xpot and Xpo5 recognize these

RNAs specifically but in a sequence-independent manner. Moreover, they are capable of selecting correctly processed RNAs and rejecting immature precursors.

In the pre-miRNA export complex, the two arches of Xpo5 assume a U-shaped conformation (Figure 10.45C). The double-stranded stem of 23 bp RNA is sandwiched by the inner surface of the U-shaped structure and is positioned such that the two nucleotide overhang at the 3′ end fits snugly into a short tunnel situated in the crook of the curve connecting the two arches. RNAs with an extended 5′ end would not fit into the structure without causing clashes with the 3′ end, explaining the preference for correctly processed overhangs.

In the tRNA export complex, the C-terminal arch of Xpot wraps around the tRNA TϕC loop while the N-terminal arch makes contact with the acceptor stem; the 3′-CCA sequence and the 5′-phosphate are the characteristics of mature tRNAs (Figure 10.45D). This mode of recognition explains why correct folding of tRNAs and correct 5′ and 3′ end processing are critical for tRNA export. Xpot can, however, export unspliced tRNAs. The anticodon arm of tRNA, where intervening sequences are inserted, is exposed to the solvent and is not involved in the binding to Xpot. Thus, Xpot rejects tRNAs that are not correctly folded or processed, but cannot discriminate between spliced and unspliced tRNAs because it does not probe this part of the structure. RNA binding by Xpot and Xpo5 is achieved not only via shape complementarity, but also via electrostatic complementarity: the inner concave surface of both exportins is lined with conserved, positively charged residues that contact the sugar–phosphate backbone of the nucleic acids, explaining why recognition is not sequence-dependent.

Crm1 exports nuclear proteins that carry an export signal

Crm1 is an exportin for nuclear proteins carrying a short linear motif with a distinct pattern of leucine residues, which is referred to as a nuclear export signal (NES). Cmr1 also recognizes a folded protein in the case of snurportin, an adaptor molecule required for the nuclear import of some small ribonuclear proteins (Usn RNPs) after their processing in the cytoplasm. Snurportin, however, also contains an NES. Crm1 has a ringlike architecture with an intermolecular interaction formed between the N-terminal and the C-terminal HEAT repeats (Figure 10.45E). Currently, there is no high-resolution structure of Crm1 in the unbound state, but EM and SAXS studies suggest that there will be no major conformational change in the ringlike structure of Crm1.

In contrast with all other karyopherin–cargo complexes known to date, snurportin binds to the outer convex surface of the Crm1 ring. It does so in a bipartite manner via a globular domain and an unstructured region that has an NES sequence (Figure 10.45E). The structures of Crm1-RanGTP-snurportin and of Crm1-snurportin are rather similar. Given the invariance of snurportin-bound Crm1 structure in the presence and in the absence of RanGTP, how is cargo dissociated in this export pathway? NES dissociation requires a complex of RanGTP with its effector RanBP1. In the structure of the ternary Crm1-RanGTP-RanBP1 complex, RanGTP positions RanBP1 such that it shifts the internal loop at the hinge region, thereby causing a conformational change in the NES-binding pockets and preventing NES binding (Figure 10.45F). The NES is thus dissociated from Crm1 via an allosteric mechanism, with the loop at the hinge region sensing RanGTP binding and relaying this information to the cargo-binding site. Interestingly, the related protein RanBP3 enhances NES binding to Cmr1 rather than destabilizing it as in the case of RanBP1.

The karyopherins bind to the FG repeats of the nuclear pore complex

Transport of karyopherin-loaded cargoes is mediated by the binding of these complexes to the FG repeats of the NPGs (see Section 10.5). Different models of how the FG repeats of nucleporins mediate translocation are discussed in Section 10.5. In terms of the interaction with karyopherins, FG repeats have been shown to bind to the convex side of the importin β superhelix. The FG repeat binds with relatively few interactions: the primary interaction is via the Phe side chain binding into a hydrophobic pocket, while the Gly appears to provide the conformational flexibility required to access this pocket. The binding site is located as the convex side of the karyopherins, opposite the (concave) side used to bind cargo and RanGTP, which explains how the transport complexes can interact with the NPC while carrying their cargo.

10.7 BACTERIAL EXPORT AND SECRETION COMPLEXES

The envelope of a Gram-negative cell is a multilayered structure (see **Figure 10.46**). There are two membranes separated by a periplasmic space. Within the periplasm is a stress-bearing peptidoglycan layer that protects the cytoplasmic inner membrane (IM) against rupture, which could result from changes in osmotic pressure. The outer membrane (OM) is an atypical lipid bilayer with phospholipids in the inner leaflet and a glycolipid, the lipopolysaccharide, in the outer leaflet. Many of the proteins embedded in the two membranes facilitate the import and export of substrates as diverse as nutrients, toxins and drugs, virulence factors or components of adhesion systems.

As part of the machinery by which they infect other cells, many bacterial pathogens have developed highly specialized and elaborate molecular machines, termed **injectisomes** or **nanoinjectors**. These enable them to deliver bacterial virulence factors directly into a eukaryotic host cell. They essentially connect the bacterial cytoplasm with that of the host cell and form a conduit for the translocation of bacterial effector proteins. Once inside, the effector molecules subvert eukaryotic cellular processes in a variety of ways, making the target cell susceptible to invasion and infection. To prevent effectors from causing havoc inside the bacterium itself, chaperones sequester them in a nonfunctional state. Collectively the injectisomes, the effectors, and their chaperones are referred to as **secretion systems** (SS). A synopsis of several bacterial export and secretion systems is shown in **Figure 10.47**.

The Sec system is a general purpose secretion system in bacteria

The secYEG translocon is required for the translocation of a wide range of soluble and secreted proteins across and for the insertion of membrane proteins into the IM. The bacterial **Sec system** is structurally and functionally related to the translocon of the endoplasmic reticulum of eukaryotes (see Section 6.2).

Translocation via the Sec system begins with a targeting phase, during which proteins are directed to the membrane by hydrophobic sequences, which are either cleavable signal

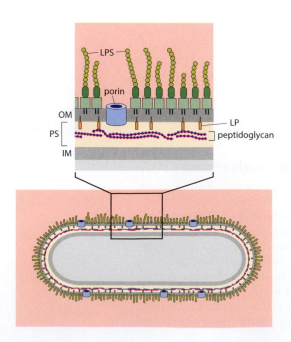

Figure 10.46 The cell envelope of Gram-negative bacteria. There is a layered architecture, with the inner membrane (IM) and the outer membrane (OM) enclosing the periplasmic space (PS). The peptidoglycan layer is found in the PS and is anchored to the OM via a lipoprotein (LP). The outer leaflet of the OM is mostly composed of lipopolysaccharides (LPS). Several species of porins make the OM permeable for nutrients.

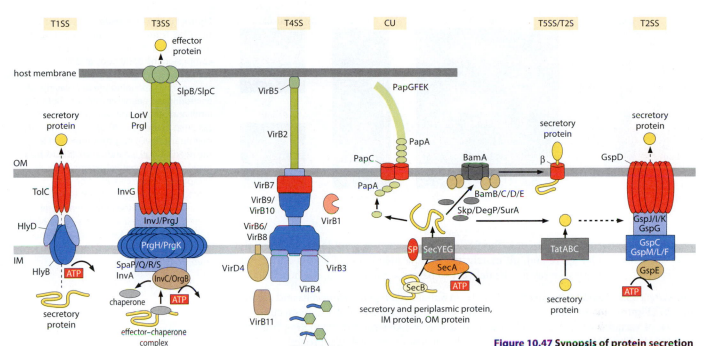

Figure 10.47 Synopsis of protein secretion systems in Gram-negative bacteria. The type 1 system (T1SS) is represented by the hemolysin A secretion pathway of pathogenic *E. coli*; the type 3 system (T3SS) by the *Salmonella enterica* system; the type 4 system (T4SS) by the Vir BD proteins of *Agrobacterium tumefaciens*; the chaperone-usher system (CU) is represented by the P pilus secretion system of uropathogenic *E. coli* (see Section 11.8); the type 2 system (T2SS) by the *E. coli* pathway for the secretion of heat-labile enterotoxin. The general secretion pathways Sec and Tat across the inner membrane are also included. Proteins secreted via the Sec system can be targeted to the T2SS or to the CU but also many periplasmic as well as inner and outer membrane proteins use this pathway. Most secretory proteins are folded in the periplasm with the help of chaperones (not shown). The β barrel outer membrane proteins are targeted to an outer membrane insertion complex (BamABCD). β represents the translocator domain of the autotransporter (T5SS) and the two-partner system (TPS).

sequences or transmembrane segments in the case of IM proteins. The SecYEG translocon must associate with other cellular components that provide the driving force necessary for polypeptide translocation or insertion. In bacteria, the SecY channel can either bind to the ribosome to translocate polypeptides during their synthesis (*co-translational translocation*), or associate with the cytoplasmic SecA ATPase to transport secretory proteins after completion of their synthesis (*post-translational translocation*).

The twin-arginine system translocates folded proteins across bacterial inner membranes

Bacteria utilize a second general transport system for protein translocation across the IM, the **twin-arginine translocation (Tat) system**. In contrast with Sec-dependent export, the Tat pathway transports folded proteins. These are targeted to the Tat pathway by N-terminal signal peptides that contain a characteristic twin-arginine sequence motif (-SRRX-FLK-), where X is any aa. The Tat system is composed of two or three membrane-bound components, TatA, TatB, and TatC. TatA and TatB each contain one membrane-spanning domain, whereas TatC contains six membrane-spanning domains. The Tat system differs from the Sec system in that it is driven by the proton motive force at the bacterial cytoplasmic membrane. Tat homologs were first identified in chloroplasts as a protein translocation system relying on the pH gradient across the thylakoid membrane (see Chapter 15). TatB and TatC oligomerize into large ~360–700 kDa complexes, without the participation of TatA. TatBC heterodimers represent functional units and can assemble into oligomeric structures in which the N-terminal region of TatC functions as the primary recognition site for twin-arginine signal sequences. TatA is the most abundant of the Tat components and mediates the actual translocation event upon association with TatBC. TatA assembles into large homo-oligomeric complexes varying in size from ~100 kDa to 700 kDa. Single-particle EM of different purified *E. coli* TatA complexes has shown a pore-forming structure with a lid-like element.

The type 3 secretion system (T3SS) is a bacterial nanoinjector

The T3SS or injectisome is found in many Gram-negative pathogens including *Pseudomonas*, *Yersinia*, *Salmonella* or enteropathogenic *Escherichia coli* strains (EPECs) and is essential for their virulence. The T3SS is a multi-subunit apparatus built from more than 20 different

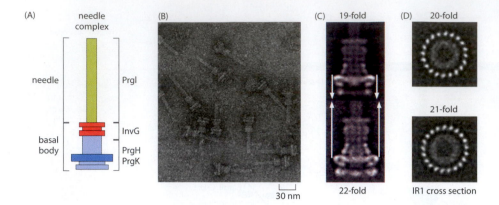

Figure 10.48 The needle complex of the T3SS from *Salmonella typhimurium*. (A) Schematic of the needle complex comprising the basal body and the extracellular needle. The basal body is a stack of rings spanning the periplasm and anchoring the needle complex to both the inner and outer membrane. The proteins assigned to the inner (PrgH, Prgk) and outer membrane (InvG) ring structures and to the needle (Prgl) are indicated. (B) Electron micrograph of negatively stained needle complexes. Image classification and averaging revealed significant variations in the diameter of the inner membrane rings (C) and in their symmetry (19–22-fold) (D). (Adapted from Marlovits et al. *Science* 306:1040–1042, 2004. With permission from AAAS.)

proteins spanning the bacterial IM, the periplasmic space, the peptidoglycan layer, the OM, the extracellular space and, ultimately, the plasma membrane of a host cell. It thus forms a continuous passageway for the delivery of bacterial proteins into the host cells.

The core T3SS apparatus has been isolated from *Salmonella typhimurium* cells and studied by EM (**Figure 10.48**). This 'needle complex' (NC) consists of a cylindrical basal complex, a stack of variably sized rings, and an extended needle-like structure emanating from the basal body. The EM data suggest that the needle encloses a channel of approximately 25 Å in diameter, which would be sufficient for the passage of (partially) unfolded proteins. The basal complex provides a platform for the recruitment and assembly of several essential IM proteins and an associated ATPase (EscN in EPEC). These proteins are highly conserved and closely related to proteins of the bacterial flagellar export apparatus (see Section 14.11). The basal complex is thought to provide a regulated pore for the selection of substrates and it must be able to switch from the export of needle/tip components during NC assembly to secretion of effector proteins after connecting to the eukaryotic host cell.

The T3SS-associated ATPase is needed for the recruitment of secretion substrates

The T3SS ATPase plays a key role in the initial stages of secretion. Its substrates (the effectors) are bound in a partially unfolded form in the bacterial cytoplasm by a specific chaperone. The ATPase provides a docking site for the chaperone/effector complex and it provides the energy required for releasing the effector from the chaperone and delivering it to the needle complex. It is not clear, however, whether it also powers the movement of the effectors from the needle base to the tip.

The crystal structure of the EPEC ATPase, EscN, is very similar to the flagellar ATPase, FliI (see Section 14.11). Surprisingly, both EscN and FliI have a fold reminiscent of the F_1 ATPase (see Section 15.4) rather than the AAA-ATPases otherwise involved in disassembly functions (see Section 10.4). The monomers of EscN are composed of three domains: an N-terminal six-stranded β barrel required for the assembly of the hexameric complex; an ATPase domain with the typical nucleotide-binding α/β Rossmann fold, and a C-terminal domain which is more divergent from the F_1 ATPase. Mutagenesis experiments suggest that the C-terminal domain provides a docking site for chaperone/effector complexes. Helices that are truncated or missing when compared with the F_1 ATPase might be complemented by helices within the chaperone.

The basal complex of T3SS is a stack of variably sized rings with a modular architecture

The basal complex of T3SS is built of only three proteins (**Figure 10.49A**). In *S. typhimurium* these are PrgH and Prgk (IM ring) and InvG (OM ring). The IM ring is larger in diameter (~170 Å) than the OM ring. Prgk belongs to a highly conserved (YscJ) family of periplasmic lipoproteins anchored in the outer leaflet of the IM by N-terminal lipidation and, in many cases, additionally by a C-terminal transmembrane helix. The crystal

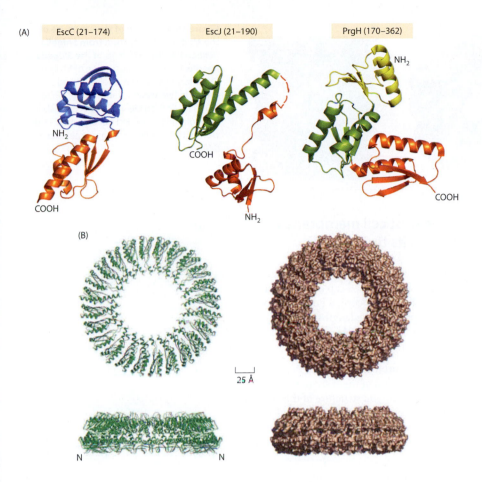

(A) EscC (21–174)   EscJ (21–190)   PrgH (170–362)

(B)

25 Å

N   N

Figure 10.49 Structures of the T3SS basal body proteins EscC (InvG), EscJ (Prgk), and PrgH (EscD). (A) These three proteins have similar folds which led to the hypothesis that this conserved fold might provide a ring-building motif allowing for the assembly of variably sized basal body rings characteristic of the T3SS. (PDB 3GR5,1YJ7,3GR0) (B) Model of the 24-subunit EscJ ring in ribbon and surface representation. The side views show the two-layered structure. The N termini of all subunits are located at the wide face of the ring. (Adapted from Spreter et al., *Nat. Struct. Mol. Biol.* 16:468–476, 2009 and Yip et al., *Nature* 435:702–707, 2005. With permission from Macmillan Publishers Ltd.)

structure of EscJ, the YscJ homolog of EPEC, shows the monomers packed into a superhelical assembly comprising 24 subunits (Figure 10.49B). This suggests that in the basal body EscJ might form a 24-subunit symmetrical ring structure. Such a ring would have similar dimensions to the IM rings as visualized by EM and can be docked into the EM map of the *S. typhimurium* basal body. In this model all the C-terminal transmembrane anchors and the N-terminal lipidation sites point into the same direction, as required for anchoring of the complete ring to the outer leaflet of the IM (**Figure 10.50**). The Prgk/EscJ ring is thought to function as a specialized membrane patch for the recruitment of other IM components of T3SS and an initiator for the NC assembly.

In *S. typhimurium*, the second protein of the IM ring is PrgH. The exact stoichiometry of PrgH:PrgK is unknown and EM single-particle analysis suggests a variable (19–21-fold) symmetry for the inner ring of the NC, with a majority showing 20-fold symmetry (Figure 10.48). The mode of interaction between the PrgH and PrgK rings remains to be established.

The OM component of the T3SS basal body is formed by a member of the secretin family of proteins. These are integral OM proteins that assemble into homomultimeric ring structures (12–14 mers) with large central pores and are involved in a variety of different export systems. All secretins consist of two domains: a C-terminal 'homology domain' that is embedded in the OM and a less conserved N-terminal periplasmic domain. For efficient assembly and OM targeting, the secretins rely on the assistance of a specific lipoprotein called pilotin. The crystal structure of MxiM, the pilotin of *Shigella*, suggests that the pilotin binds to the C-terminal helix of the secretin, displacing lipid from a pocket and thereby allowing the pilotin to direct the secretin to the OM. It appears that the secretins use an assembly and membrane insertion pathway distinct from β barrel OMPs such as most porins and members of the general porin family of OM diffusion channels characterized by β barrel structures. They may be more similar in structure to the *E. coli* Wza protein involved in capsular polysaccharide secretion (see below).

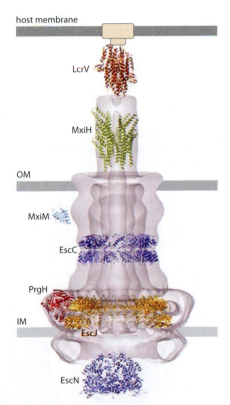

host membrane

LcrV

MxiH

OM

MxiM

EscC

PrgH

IM

EscJ

EscN

Figure 10.50 Hybrid model of the T3SS needle complex. The available structures of components were fitted into the EM density of the assembly. (Adapted from Moraes et al., *Curr Opin Struct Biol* 18:258–266, 2008. With permission from Elsevier.)

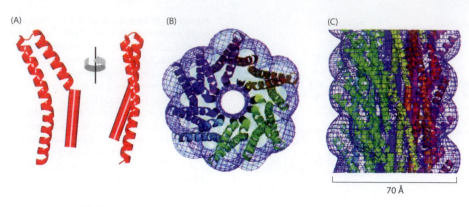

(A) (B) (C)

70 Å

Figure 10.51 Fitting of the structure of the MxiH needle protein from *Shigella flexneri* **into the EM map of the** *Shigella* **needle.** (A) Ribbon illustration of MxiH with the N-terminal helix modeled (cylinder) shown in two views rotated by 90°. (PDB 2CA5) (B) End-on view of a 40 Å thick slice through the assembled needle and (C) side view of the needle. Each monomer is colored differently.

Insertion of the T3SS translocon into the host cell membrane is mediated by the needle and an adaptor at its tip

The extracellular components of the T3SS NC consist of three components: the needle, a helical polymer built of subunits of the EscF/PrgI family; a bell-shaped cap or needle extension, depending on the species; and the translocation pore (translocon), a hetero-oligomer forming a channel upon insertion into the host cell membrane. The lengths of the needles is species- or even strain-specific; in *Yersinia* the length is approximately 60 nm and is controlled by the YscP protein, which is supposed to act as a *molecular tape measure*.

The structures of a number of truncated monomeric needle proteins have been determined crystallographically or by NMR spectroscopy (see **Methods**). All of them share a two-helix bundle coiled-coil motif linked by a conserved turn. The coiled-coil structure of the *Shigella flexneri* MxiH subunit has been docked into medium-resolution EM structures of intact needles. The resulting hybrid model (**Figure 10.51**) provides some insights into the mode of interactions between the needle subunits and of conformational changes that may relay signals along the needle.

The tip proteins are adaptors linking the needle to the translocon. Several high-resolution structures of tip proteins are available; they all show a conserved central coiled-coil domain, but the N- and C-terminal domains vary, conferring specificity for the needle and translocon with which they interact. By docking the *Yersinia* tip protein LcrV to the tip of the *Shigella* needle, five monomers could be placed there, providing a model for the needle-tip complex (**Figure 10.52**). For EspA, the tip protein of EPEC, it has been shown that for its export through the needle complex it must associate with a chaperone keeping it in a monomeric form.

The adaptor complex is required for the insertion of the translocon into the host cell membrane (**Figure 10.53**). The two hydrophobic translocon proteins are not present at the tip of the needle before contact with the host cell. They are secreted only when contact is established and need the adaptor as an assembly platform. Currently, no high-resolution structures are available for any component of the translocon nor are the oligomerization state and

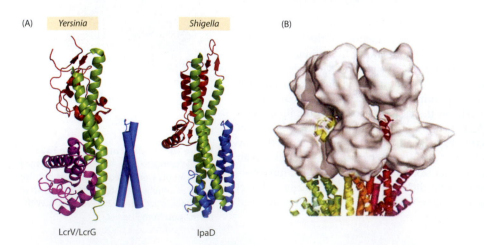

(A) *Yersinia* *Shigella* (B)

LcrV/LcrG IpaD

Figure 10.52 Crystal structures of various adaptor proteins. (A) The structures of the adaptor proteins from *Yersinia* (LcrV) and *Shigella* (IpaD) reveal a common architecture. They all have a central coiled-coil domain (green), a needle-distal domain (red), and a needle-proximal domain (purple/blue). (PDB 1R6F, 2J0O) (B) Model of a pentameric LcrV adaptor complex fitted onto the tip of a T3SS needle.

stoichiometry of the pore complex known. Low-resolution EM and AFM (see **Methods**) data of the EPEC translocon embedded in red blood cell membranes indicate a nonsymmetrical 6–8-subunit hetero-oligomeric complex formed by two proteins, EspB and EspD.

The type 4 secretion system transports DNA and/or proteins across the cell envelope

Type 4 secretion systems (T4SSs) are multi-subunit complexes that are found in many Gram-negative bacteria and are involved in a wide variety of processes ranging from the delivery of virulence factors into eukaryotic host cells to the conjugative transfer of genetic material. In *Helicobacter pylori* or in *Legionella pneumophila*, T4SSs mediate the injection of virulence proteins into host cells, causing gastric ulcers or legionnaire's disease, respectively. The *Agrobacterium tumefaciens* T4SS delivers oncogenic DNA and proteins into plant cells, whereas the *Bordetella pertussis* T4SS secretes a toxin into the environment.

The archetypal *A. tumefaciens* T4SS comprises 12 proteins named VirB1–11 and VirD4. Other species may have additional components or homologs of only some of these proteins. The T4SS has an extracellular pilus associated with it, built from a major (VirB2) and a minor (VirB5) subunit. Three ATPases, VirB4, VirB11, and VirD4, each of them forming hexamers with distinct non-redundant functions, are essential for substrate secretion and pilus biogenesis. Structural studies by EM and X-ray crystallography have provided a detailed picture of the architecture of the T4SS. The structure showed the T4SS to be formed of an ~1 MDa outer membrane core complex and a bipartite ~2.5 MDa inner membrane complex (IMC), connected by a stalk (Figure 10.47).

The outer membrane core complex of T4SSs spans both membranes of Gram-negative bacteria but primarily locates in and near the outer membrane

Fourteen copies each of three proteins, VirB7, VirB9, and VirB10 (or TraN, TraO, and TraF in the homologous *E. coli* system encoded by the conjugative plasmid pKM101) form a 1.05 MDa complex. The complex appears to be divided into two cup-shaped halves oriented face-to-face and only weakly connected near the equator (**Figure 10.54**). They are referred to as the inner and outer layer to indicate their location with respect to the IM and OM. The double-layered walls enclose large internal chambers accessible only via axial openings that are 1–2 nm wide in the outer layer and 5 nm in the inner layer. Relatively large-scale conformational changes that result in a widening of the pore in the outer layer are necessary for substrate secretion and/or pilus biogenesis, and it has been suggested that the ATPases associated with the inner membrane control these conformational changes.

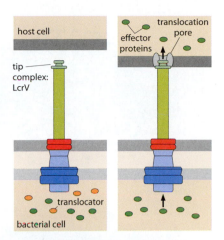

Figure 10.53 Functional model of the LcrV tip-adaptor complex. (A) When not in contact with host cell, LcrV forms an adaptor complex at the tip of the needle. (B) Once in contact with a host cell membrane, the adaptor assists with the assembly of the translocation pore, which is inserted into the host cell membrane. Thus, by providing an assembly platform, a passageway is created for effector proteins. (Adapted from G.R. Cornelis, *Nat. Rev. Microbiol.* 4:811–825, 2006. With permission from Macmillan Publishers Ltd.)

Figure10.54 Cryo-EM structure of the T4SS core complex. The complex comprises two subcomplexes, a 1 MDa outer membrane core and a 2.5 MDa inner membrane complex. (Courtesy of Gabriel Waksman.)

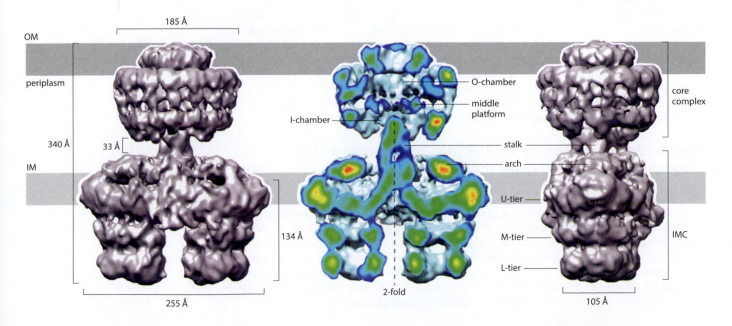

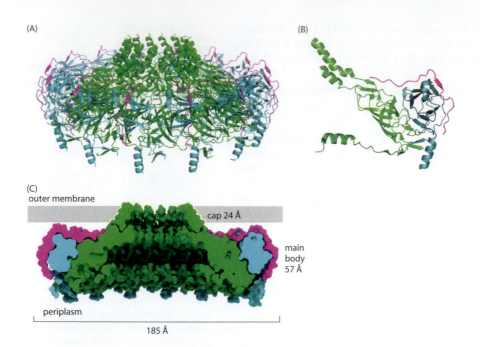

(A)

(B)

(C)
outer membrane
cap 24 Å
main body 57 Å
periplasm
185 Å

Figure 10.55 Crystal structure of the T4SS outer membrane complex. (A) Complex structure and (B) the heterotrimeric unit from which it is assembled in ribbon diagrams. (PDB 3JQO) (C) Tetradecameric complex in a space-filling cutaway. The three subunits contributing to the complex are TraN (magenta), TraF$_{ct}$ (green), and TraO$_{ct}$ (cyan). The 'cap' inserts into the outer membrane.

For the outer layer (~0.6 MDa) of the complex a crystal structure has been obtained after removing proteolytically the inner layer from the core complex (**Figure 10.55**). This subcomplex contains 14 copies each of the C-terminal domain of TraF (VirB10), the C-terminal domain of TraO (VirB9), and the full-length TraN (VirB7). This structure revealed that the OM-inserting region in the O-layer is formed by an α-helical barrel, where each of the 14 VirB10 subunits contributes two α helices to form a channel.

The OM core complex retained its overall shape whether purified in isolation or within the fully assembled T4SS, implying that the VirB10 N termini (known to be inserted in the IM) in the T4SS must adopt a much more extended conformation in comparison with their compact fold in the isolated OM core complex, in order to extend to the IM. This extended region of VirB10, potentially along with other components, likely forms the stalk, the thin flexible region connecting the core/OM complex with the IMC.

The inner membrane complex of T4SS is composed of VirB3, VirB4, VirB6, VirB8, and the VirB10 N termini

Two VirB4 hexamers, one on each side of the complex, are associated with the inner membrane at the cytoplasmic side. VirB3 is a small IM protein that binds to VirB4 and assists its membrane localization. VirB6 proteins are polytopic IM proteins. VirB8 subunits are bitopic proteins with a short cytoplasmic N-terminal domain, a TM region, and a large C-terminal periplasmic domain. VirB8 interacts with many other VirB proteins, including VirB4 and VirB10, and is likely central for the assembly of the IMC. Each VirB IMC components is present in multiples of 12, with 12 subunits of VirB3, VirB4, and VirB8, and 24 subunits of VirB6. Also present in the T4SS structure are 12 subunits of VirB5, likely participating in the formation of the stalk between the OM core complex and the IMC.

The translocation pathway of T4SS switches between a pilus biogenesis and a DNA transfer mode

The substrates are first recruited by VirD4 (also termed the 'coupling protein'), which brings substrates to the T4SS. ATPase activity of all three ATPases is necessary for substrate transfer to the IMC components VirB6 and VirB8. The substrate is then passed onto VirB9, VirB10 and VirB2 (pilin) as it enters the OM core complex and exits through the OM and the pilus. In DNA transport systems such as conjugation systems, the T4SS switches between two different modes, a pilus biogenesis mode, succeeded by a DNA transfer mode. A pilus is required for conjugative transfer and thus has to be formed first. Once the pilus is formed and contact with a recipient cell is established, a conformational change occurs in the system leading to the T4SS complex switching to its DNA transfer mode. VirB11 appears to be the switch 'mastermind' in that, depending on whether VirB11 is associated with VirB4 or VirD4, the T4SS will operate in its pilus biogenesis or DNA transfer mode, respectively.

The type 5 secretion system (T5SS) is a simple but diverse family of transporters

The most widespread of the specialized protein secretion pathways is the **autotransporter** or type 5 secretion system (T5SS). Both the autotransporter and the related **two-partner system (TPS)** rely on the Sec system for transport across the IM. Classic autotransporters are composed of a cleavable signal sequence, an N-terminal passenger domain ranging in molecular mass from <20 to >400 kDa, and a C-terminal translocator domain of ~30 kDa. Although the 'classic' autotransporters are monomeric, there are a few trimeric examples.

The signal peptides target the proteins to the Sec complex and initiate translocation across the IM. Subsequently, the translocator domain is inserted into the OM and the passenger domain is channeled through it into the extracellular space where it can unfold its toxic or virulence function (**Figure 10.56A**). Many passenger domains are cleaved from the cell surface and/or post-translationally modified. More than 1000 autotransporters have thus far been identified in bacterial genomes by sequence profile analysis.

Crystal structures of several passenger domains are available, either of the full-length proteins or of fragments; they were all found to contain β solenoid motifs, as shown for pertactin and Hbp (Figure 10.56B). The transporter domain structures of Na1P, an autotransporter from *Neisseria meningitidis*, and Hia, a trimeric autotransporter from *Haemophilus influenzae*, have also been solved (Figure 10.56C). The Na1P transporter domain crystallized as a monomeric 12-stranded β barrel with a 30 residue segment of the passenger domain traversing the pore in an α-helical conformation. The Hia transporter domain is also a single 12-stranded β barrel, but it is assembled from three subunits that

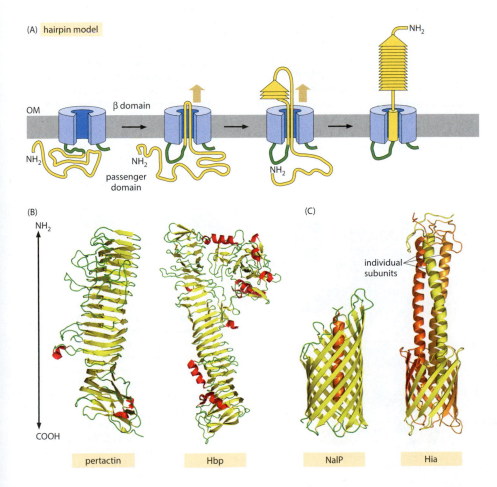

Figure 10.56 The T5SS or autotransporter system. (A) Schematic of the hairpin model of action of a generic autotransporter. (B) Crystal structures of autotransporter domain structures of the passenger domains of the autotransporter's pertactin (PDB 1DAB) and Hbp (PDB 3AK5) showing the characteristic β solenoid motifs. (C) Structures of C-terminal fragments of the autotransporters NalP (1YUN) and the trimeric Hia (3EMO). In NalP β strands are colored yellow and the α helix is orange; in the Hia structure the individual subunits are shown in different shades of yellow/orange. (A, adapted from N. Dautin & H. D. Bernstein, *Annu. Rev. Microbiol.* 61:89–112, 2007.)

each contribute four β strands. A short passenger domain derived from each of the three subunits is found inside the β barrel.

Multi-subunit complexes couple the biosynthesis and the export of capsular polysaccharides

Capsular polysaccharides (CPSs) are protective structures on the surface of many bacteria. They are well-established virulence factors, often protecting the cell from phagocytosis initiated by an opsonin antibody in blood serum and complement-mediated killing (see Section 17.3). Biosynthesis and assembly of CPSs is a complex process. Activated precursors (nucleotide monophospho- and diphosphosugars) are incorporated into nascent polysaccharide chains by enzymes that are associated with the IM. They are subsequently moved across the periplasm and OM by a dedicated translocation pathway (see **Figure 10.57**). Translocation across the OM requires Wza, a member of the outer-membrane auxiliary (OMA) family. Wza forms a homo-octameric complex and is linked to the IM biosynthesis machinery by Wzc, which forms a tetrameric complex. Cyclic phosphorylation and dephosphorylation of Wzc are essential for CPS assembly.

Wza is the only CPS transporter for which a high-resolution structure is available. The octomeric ring complex of Wza (14 nm high) appears as a stack of four rings; each monomer contributes one domain to each of the rings. The uppermost ring inserts into the OM, and the lower three rings protrude into the periplasmic space (**Figure 10.58**). Wza was the first OM protein to be found not forming a β barrel but an α-helical membrane-spanning domain. No high-resolution structure is available for the Wza-Wzc heterocomplex. A hybrid structure in which the Wza crystal structure was docked to the structure of Wzc, obtained by negative stain EM, suggests that the periplasmic domains of Wza may undergo major conformational changes upon complex formation.

10.8 SUMMARY

Eukaryotic cells are divided into membrane-enclosed compartments and transport mechanisms are needed for the exchange of macromolecules between these compartments. Transport between topologically equivalent compartments relies on fission and fusion of vesicles. These are often coated with a protein such as clathrin, which is a spidery trimer that forms a coat around a budding membrane vesicle. Clathrin coated vesicles isolated from cells can vary greatly in size and shape and can accommodate a wide range of cargoes. Adaptor proteins recruit cargo and also help the assembly of the clathrin coat and vesicle formation. Scission of the vesicle relies on dynamin and dynamin-like proteins, GTPase motor proteins that drive conformation changes that lead to a constriction of the vesicle neck.

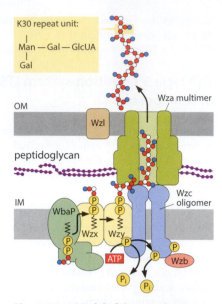

Figure 10.57 Model of the complex carrying out the synthesis and export of capsular polysaccharide in *E. coli*. Repeats of the polysaccharide are assembled on a lipid acceptor at the interface between the cytoplasm and the inner membrane in a reaction initiated by the UbaP enzyme. The newly synthesized and -PP-linked repeats are flipped across the inner membrane in a process involving Wzx. This provides the substrates for Wzy-dependent polymerization; the polymer grows by transfer of the growing chain to the incoming and -PP-linked repeat unit. Continued polymerization requires transphosphorylation of C-terminal tyrosine residues in the Wzc oligomer and dephosphorylation by the Wzb phosphatase. Wza together with Wzc provides a channel for the translocation of the polysaccharide to the cell surface. The role of Wzl, an integral outer membrane protein, remains unclear. (Adapted from C. Whitfield, *Annu. Rev. Biochem.* 75:39–68, 2006.)

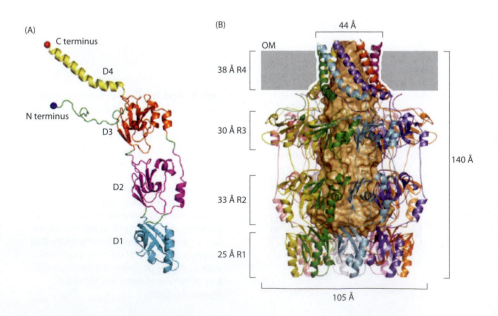

Figure 10.58 Structure of the Wza translocation complex. (A) The Wza monomer is composed of four domains (D1, D2, D3, D4) that form the 8-fold symmetric stack of rings shown in (B) (R1, R2, R3, R4). The outer membrane-spanning part of Wza is all α-helical. The internal void of 15.000 Å³ is represented by the space-filling volume. The internal cavity has a polar surface facilitating the passage of carbohydrate. (PDB 2J58) (Adapted from C. Whitfield & J.H. Naismith, *Curr Opin Struct Biol* 18:466–474, 2008. With permission from Elsevier.)

Vesicle fusion also relies on GTPases to provide the energy for the necessary membrane rearrangements. The fusion of synaptic vesicles is one of the best-studied systems. The transmembrane SNAREs are central components of the membrane fusion apparatus. Conformational changes in SNARE proteins are brought about by interaction with various binding partners, including Sec1/Munc18 (SM) proteins. GTPases of the Rab family (for example RIM, Munc13) are involved in docking of synaptic vesicles. The binding of Ca^{2+} to synaptotagmin alters charge and attracts or repels membranes, leading to cargo (neurotransmitter) release.

Transport into and out of the double-membrane enclosed nucleus is through the nuclear pore. The outer and inner nuclear membranes are joined to leave gaps in which nuclear pore complexes (NPCs) are formed. One of the largest complexes in eukaryotic cells, NPCs control the passage of macromolecules into and out of the nucleus. Nucleoporins form a octagonal, multilayered cylinder with three rings, decorated by a basket on the nucleoplasmic side and filaments on the cytoplasmic side. The central pore is filled by a meshwork of phenylalanine-glycine (FG) repeat domain proteins tethered to the walls. Macromolecule cargo is bound by transporters (karyopherins); importins move cargo in and exportins move cargo out. A RanGTP gradient acts as a cargo pump; GTP-bound Ran is found inside nucleus and GDP-bound Ran in the cytoplasm. Importins bind to cargo in the absence of RanGTP and release it in the nucleus in the presence of RanGTP. Exportins bind to their cargo in the presence of RanGTP and release it upon RanGTP hydrolysis in the cytoplasm. Importins adopt complementary shapes to their cargoes. The exportins Xpot and Xpo5 bind tRNAs and miRNAs, respectively, in a sequence-independent manner, but the RNAs must have been processed with the correct 3′ and 5′ ends. Karyopherins interact with FG domain proteins in the nuclear pore, causing transient rearrangements allowing cargo to pass through.

Gram-negative bacteria are bounded by a cell envelope comprising inner and outer membranes separated by a periplasmic place. Secretion systems (SS) are elaborate molecular machines allowing for the transfer of pathogenic bacterial molecules into the environment or directly into eukaryotic host cells. The type 3 SS nanoinjector is a stack of variably sized rings, plus a needle with an adaptor at its tip to penetrate the cell membrane of the host cell. The type 4 SS is a large bipartite assembly spanning the entire cell envelope. The type 5 SS use the standard secretion (Sec) system to traverse IM, then insert a translocator domain into the OM through which a passenger domain is channeled into the extracellular space. Following cleavage and/or post-translational modification, it can unfold its toxic function.

Sections of this chapter are based on the following contributions

10.2 Tomas Kirchhausen, Harvard Medical School, Boston, USA.
10.3 Jenny Hinshaw, National Institutes of Health, Bethesda, Maryland, USA and Oliver Daumke, Max-Delbrück-Centrum for Molecular Medicine, Berlin, Germany.
10.4 Jose Rizo-Rey, UT Southwestern Medical Center, Dallas, USA.
10.5 Elizabeth Villa, University of California, San Diego, USA and Martin Beck, European Molecular Biology Laboratory, Heidelberg, Germany.
10.6 Elena Conti, Max-Planck Institute of Biochemistry, Martinsried, Germany.
10.7 Natalie Strynadka and Thomas Spreter, University of British Columbia, Vancouver, Canada.

REFERENCES

10.2 Clathrin-mediated endocytosis

Cheng Y, Boll W, Kirchhausen T et al (2007) Cryo-electron tomography of clathrin coated vesicles – structural implications for coat assembly. *J Mol Biol* 365:892–899.

Edeling MA, Smith C & Owen D (2006) Life of a clathrin coat: insights from clathrin and AP structures. *Nature Rev Mol Cell Biol* 7:32–44.

Evans PR & Owen D (2002) Endocytosis and vesicle trafficking. *Curr Opin Struct Biol* 12:814–821.

Fotin A, Cheng Y, Grigorieff N et al (2004) Structure of an auxilin-bound clathrin coat and its implications for the mechanism of uncoating. *Nature* 432:649–653.

Fotin A, Cheng Y, Sliz P et al (2004) Molecular model for a complete clathrin lattice from electron cryomicroscopy. *Nature* 432:573–579.

Heldwein EE, Macia E, Wang J et al (2004) Crystal structure of the clathrin adaptor protein 1 core. *Proc Natl Acad Sci USA* 101:14108–14113.

Kirchhausen T (2000) Three ways to make a vesicle. *Nat Rev Mol Cell Biol* 1:187–198.

Stagg SM, LaPointe P & Balch WE (2007) Structural design of cage and coat scaffolds that direct membrane traffic. *Curr Opin Struct Biol* 17:221–228.

10.3 Dynamins are versatile molecular machines

Byrnes LJ & Sondermann H (2011) Structural basis for the nucleotide-dependent dimerization of the large G protein atlastin-1/SPG3A. *Proc Natl Acad Sci USA* 108:2216–2221.

Chappie JS, Acharya S, Leonard M et al (2010) G domain dimerization controls dynamin's assembly-stimulated GTPase activity. *Nature* 465:435–440.

Chappie JS, Mears JA, Fang S et al (2011) A pseudoatomic model of the dynamin polymer identifies a hydrolysis-dependent powerstroke. *Cell* 147:209–222.

Chen Y-J, Zhang P, Egelman EH & Hinshaw JE (2004) The stalk region of dynamin drives the constriction of dynamin tubes. *Nat Struct Mol Biol* 11:574–575.

Daumke O, Lundmark R, Vallis Y et al (2007) Architectural and mechanistic insights into an EHD ATPase involved in membrane remodelling. *Nature* 449:923–927.

Faelber K, Posor Y, Gao S et al (2011) Crystal structure of nucleotide-free dynamin. *Nature* 477:556–560.

Ford MGJ, Jenni S & Nunnari J (2011) The crystal structure of dynamin. *Nature* 477:561–566.

Low HH & Löwe J (2010) Dynamin architecture - from monomer to polymer. *Curr Opin Struct Biol* 20:791–798.

Praefcke GJ & McMahon HT (2004) The dynamin superfamily: universal membrane tubulation and fission molecules? *Nat Rev Mol Cell Biol* 5:133–147.

Roux A, Uyhazi K, Frost A & De Camilli P (2006) GTP-dependent twisting of dynamin implicates constriction and tension in membrane fission. *Nature* 441:528–531.

10.4 The machinery of synaptic vesicle fusion

Arac D, Chen X, Khant HA et al (2006) Close membrane-membrane proximity induced by Ca^{2+}-dependent multivalent binding of synaptotagmin-1 to phospholipids. *Nat Struct Mol Biol* 13:209–217.

Brunger AT (2005) Structure and function of SNARE and SNARE-interacting proteins. *Q Rev Biophys* 38:1–47.

Chen X, Tomchick DR, Kovrigin E et al (2002) Three-dimensional structure of the complexin/SNARE complex. *Neuron* 33:397–409.

Fernandez-Chacon R, Konigstorfer A, Gerber SH et al (2001) Synaptotagmin I functions as a calcium regulator of release probability. *Nature* 410:41–49.

Grosshans BL, Ortiz D & Novick P (2006) Rabs and their effectors: achieving specificity in membrane traffic. *Proc Natl Acad Sci USA* 103:11821–11827.

Hanson, PI, Roth R, Morisaki H et al (1997). Structure and conformational changes in NSF and its membrane receptor complexes visualized by quick-freeze/deep-etch electron microscopy. *Cell* 90:523–535.

Jahn R & Scheller RH (2006) SNAREs–engines for membrane fusion. *Nat Rev Mol Cell Biol* 7:631–643.

Ma C, Su L, Seven AB et al (2013) Reconstitution of the vital functions of Munc18 and Munc13 in neurotransmitter release. *Science* 339:421–425.

Mima J, Hickey CM, Xu H et al (2008) Reconstituted membrane fusion requires regulatory lipids, SNAREs and synergistic SNARE chaperones. *EMBO J* 27:2031–2042.

Misura KM, Scheller RH & Weis WI (2000) Three-dimensional structure of the neuronal-Sec1-syntaxin 1a complex. *Nature* 404:355–362.

Reim K, Mansour M, Varoqueaux F et al (2001) Complexins regulate a late step in Ca^{2+}-dependent neurotransmitter release. *Cell* 104:71–81.

Rizo J & Xu J (2015) The Synaptic Vesicle Release Machinery. *Annu Rev Biophys* 44:339–367.

Schoch S, Castillo PE, Jo T et al (2002). RIM1alpha forms a protein scaffold for regulating neurotransmitter release at the active zone. *Nature* 415:321–326.

Sollner T, Whiteheart S, Brunner M et al (1993) SNAP receptors implicated in vesicle targeting and fusion. *Nature* 362:318–324.

Sudhof TC (2004) The synaptic vesicle cycle. *Annu Rev Neurosci* 27:509–547.

Sudhof TC & Rothman JE (2009) Membrane fusion: grappling with SNARE and SM proteins. *Science* 323:474–477.

Sutton RB, Fasshauer D, Jahn R & Brunger AT (1998) Crystal structure of a SNARE complex involved in synaptic exocytosis at 2.4 Å resolution. *Nature* 395:347–353.

10.5 Nuclear pore complexes

Alber F, Dokudovskaya S, Veenhoff LM et al (2007) The molecular architecture of the nuclear pore complex. *Nature* 450:695–701.

Beck M, Lucic V, Förster F et al (2007) Snapshots of nuclear pore complexes in action captured by cryoelectron tomography. *Nature* 449:611–615.

Brohawn SG, Partridge JR, Whittle JRR & Schwartz TU (2009) The nuclear pore complex has entered the atomic age. *Structure* 17:1156–1168.

Bui K H, v. Appen A, DiGiulio A et al (2013) Integrated structural analysis of the nuclear pore complex scaffold. *Cell* 155, 1233–1243.

D'Angelo MA & Hetzer MW (2008) Structure, dynamics and function of nuclear pore complexes. *Trends Cell Biol* 18:456–466.

Fahrenkrog B & Aebi U (2003) The nuclear pore complex: nucleocytoplasmic transport and beyond. *Nat Rev Mol Cell Biol* 4:757–766.

Feldherr CM (1962) The nuclear annuli as pathways for nucleocytoplasmic exchanges. *J Cell Biol* 14:65–72.

Lim RYH, Fahrenkrog B, Köser J et al (2007) Nanomechanical basis of selective gating by the nuclear pore complex. *Science* 318:640–643.

Tran EJ & Wente SR (2006) Dynamic nuclear pore complexes: life on the edge. *Cell* 125:1041–1053.

10.6 Nuclear import and export

Cook A & Conti E (2009) Nuclear export complexes in the frame. *Curr Opin Struct Biol* 20:247–252.

Cook A, Bono F, Jinek M & Conti E (2007) Structural biology of nucleocytoplasmic transport. *Annu Rev Biochem* 76:647–671.

Fried H & Kutay U (2003) Nucleocytoplasmic transport: taking an inventory. *Cell Mol Life Sci* 60:1659–1688.

Stewart M (2007) Molecular mechanism of the nuclear protein import cycle. *Nat Rev Mol Cell Biol* 8:195–208.

10.7 Bacterial export and secretion complexes

Cascales E & Christie PJ (2004) Definition of a bacterial type IV secretion pathway for a DNA substrate. *Science* 304:1170–1173.

Chandran V, Fronzes R, Duquerroy S et al (2009) Structure of the outer membrane complex of a type IV secretion system. *Nature* 462:1011–1016.

Cornelis GR (2006) The type III secretion injectisome. *Nat Rev Microbiol* 4:811–825.

Dautin N & Bernstein HD (2007) Protein secretion in gram-negative bacteria via the autotransporter pathway. *Annu Rev Microbiol* 61:89–112.

Fronzes R, Schäfer E, Wang L et al (2009) Structure of a type IV secretion system core complex. *Science* 323:266–268.

Low HH, Gubellini F, Rivera-Calzada A et al (2014) Structure of a type IV secretion system. *Nature* 508:550–553.

Marlovits TC, Kubori T, Sukhan A et al (2004) Structural insights into the assembly of the type III secretion needle complex. *Science* 306:1040–1042.

Moraes TF, Spreter T & Strynadka NCJ (2008) Piecing together the type III injectisome of bacterial pathogens. *Curr Opin Struct Biol* 18:258–266.

Spreter T, Yip CK, Sanowar S et al (2009) A conserved structural motif mediates formation of the periplasmic rings in the type III secretion system. *Nat Struct Mol Biol* 16:468–476.

Whitfield C & Naismith JH (2008) Periplasmic export machines for outer membrane assembly. *Curr Opin Struct Biol* 18:466–474.

Whitfield C (2006) Biosynthesis and assembly of capsular polysaccharides in *Escherichia coli*. *Annu Rev Biochem* 75:39–68.

Yip CK, Kimbrough TG, Felise HB et al (2005) Structural characterization of the molecular platform for type III secretion system assembly. *Nature* 435:702–707.

Connectivity and Communication

11.1 INTRODUCTION

The individual cells that make up tissues—and multicellular organisms generally—must be capable of maintaining their structural integrity and functional morphology when subjected to physical stresses. Their resilience depends both on the response of cells to applied forces and on the strength and flexibility of intercellular contacts. Moreover, cells must communicate with neighboring cells to carry out many processes, for example *homeostasis*, or to follow pre-programmed pathways of differentiation. This chapter focuses on macromolecular assemblies that effect physical coupling between neighboring cells, and those responsible for cell-to-cell communication or a cell's response to contact with its surroundings. These assemblies are notably diverse in their building blocks and connective mechanisms, but one striking common feature is the near-universality of protein modules with immunoglobulin (Ig)-like folds (see **Guide**). Mostly, these proteins are secreted outward from the cytoplasm, but in one case, nuclear lamins (Section 11.2), they are transported inward into the nucleus. Another recurring theme is regulatory phosphorylation.

The size and shape of a eukaryotic cell are determined primarily by its cytoskeleton

The cytoskeleton is a network of protein filaments that governs a cell's internal organization and its physical resilience. The filaments are of three kinds: *actin filaments* (F-actin), *microtubules* (MTs), and *intermediate filaments* (IFs), so called because they appear to be intermediate in diameter between F-actin (8 nm) and MTs (25 nm) (**Figure 11.1**). In addition to their structural roles, the latter two systems engage in motility and are described in Sections 14.2 and 14.6 respectively. IFs, in contrast, are nonmotile but are important contributors to cell architecture and connectivity (Section 11.2). IF proteins differ from actin and tubulin in forming a large superfamily, whereas actin and tubulin have only a few isoforms and the diversity of their filaments stems mainly from sets of associated proteins. Moreover, IF architecture is based on the bundling of coiled coils, whereas F-actin and MTs are polymers of globular subunits. IFs permeate the cytoplasms of most eukaryotic cells, where they have biomechanical roles, tailored to the specific requirements of each cell type.

(A)　　　　　(B)　　　　　(C)

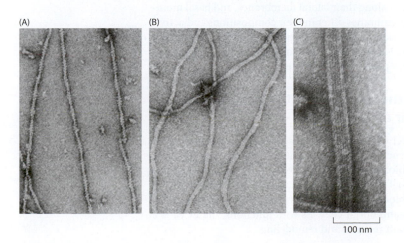

100 nm

Figure 11.1 Cytoskeletal protein filaments imaged by negative staining electron microscopy. (A) F-actin; (B) intermediate filaments (vimentin); (C) a microtubule. (Courtesy of Ueli Aebi.)

Recently, it has transpired that prokaryotes also express cytoskeletal proteins that are distant relatives of actin and tubulin; there is also at least one case of a bacterial IF-like protein, *crescentin*. Their roles as putative cytoskeletal elements are under active investigation. However, unlike eukaryotic cells, the sizes and shapes of most prokaryotes are dictated primarily by the *sacculus*, a rigid layer of cross-linked polysaccharide strands (*peptidoglycan*) in the cell envelope (Section 1.8); it is, in effect, an exoskeleton.

Eukaryotic cells communicate with each other at specialized contact regions

The regions where cells meet and their plasma membranes become closely apposed are known as **junctions**. The term **synapse** is also used, mainly for junctions between neural cells. The molecular compositions and structures of junctions differ from those of the rest of the plasma membrane. Typically, they have an outer layer of transmembrane receptor proteins whose *ectodomains* (extracellular portions) bind to the cell on the other side of the junction, while their *endodomains* (intracellular portions) connect to cytoplasmic proteins and, in many cases, to the cytoskeleton.

In addition to their adhesive role in coordinating multicellular bodies, junctions also inform a cell about its neighbors. Intercellular communication can take place in several ways. The binding of extracellular ligands by a receptor may induce conformational changes in its cytoplasmic domain that elicit a signaling response. Small water-soluble molecules may be exchanged between coupled cells through the aqueous channels of a gap junction (Section 11.5). Cytoplasmic vesicles may discharge their contents into the extracellular milieu, to be detected by receptors on the other side of the synapse (Section 16.3). Mechanical forces may also transmit signals between cells via the junctions.

The cells that form the outer surface of an organism must be integrated cohesively to prevent the loss of material to the outside and the ingress of foreign materials, including microbial pathogens. Accordingly, peripheral cells are well endowed with junctions, of which there are four main kinds: **tight junctions**, **adherens junctions**, **desmosomes**, and **gap junctions**. There are also two kinds of junctions—**focal adhesions** and **hemidesmosomes**—that connect cells to the **extracellular matrix** (ECM), a network of secreted proteins and polysaccharides. A schema of the various kinds of junctions and their connections to the cytoskeleton is given in **Figure 11.2**. Originally distinguished by morphological criteria, junctions are also classified by physiological properties: they are dynamic entities and must be assembled, disassembled, and redistributed during development, tissue remodeling, or wound repair, among other processes.

Junctions are particularly abundant in *epithelia*, which consist of one or more layers of closely packed cells covering an animal or lining its internal cavities. For example, the epidermis, the outer layer of skin, is a stratified epithelium whose layers represent successive stages of differentiation. The outermost layers are dead cells that gradually *desquamate*; that is, they are shed, with replacements moving up from beneath. *Endothelia* are single layers of flattened cells that line blood vessels or surround internal organs. An epithelial cell has several distinct surfaces: its apical membrane faces the outside (for example, the lumen of the gut), adjacent cells attach to one another along their lateral membranes, and basal membranes connect to the ECM. These three membranes differ in their compositions and activities from each other and from tissue to tissue. For example, in the gut, the apical membrane contains transporters for the uptake of digestive products, and in the kidney it contains the Na^+/K^+ ATPase (Section 16.6) that ensures proper ionic balance.

Cells secrete macromolecules that allow them to communicate with other cells and sense their environment

One way in which eukaryotic cells interact with each other and with their surroundings is via the ECM which varies in thickness, structure, and composition, depending on the tissue (Section 11.7). In connective tissue, the biomass of ECM can exceed that of the cells from which it is secreted, and it is the ECM that confers the essential mechanical properties of connective tissues such as cartilage and blood vessels. An important additional role of the ECM is its involvement in signaling into and out of cells. As with junctions, the ECM is a dynamic structure and is subject to constant renewal and remodeling.

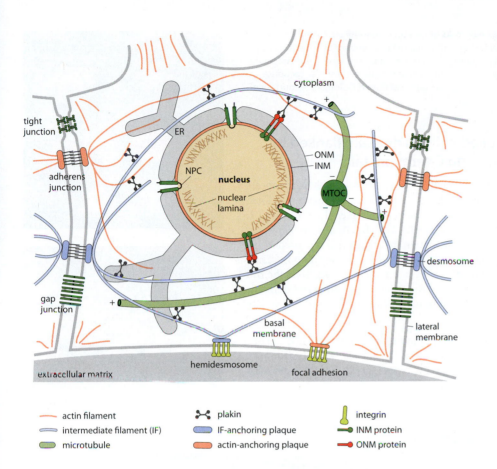

cytoplasm

tight
junction

ER

ONM
INM

NPC

nucleus

MTOC

nuclear
lamina

adherens
junction

desmosome

gap
junction

basal
membrane

lateral
membrane

hemidesmosome

focal adhesion

extracellular matrix

— actin filament

▬ intermediate filament (IF)

▬ microtubule

⋈ plakin

▬ IF-anchoring plaque

▬ actin-anchoring plaque

▮ integrin

▬ INM protein

▬ ONM protein

Figure 11.2 Junctional complexes and cytoskeletal connections in an epithelial cell. IFs are connected to other cytoskeletal filaments (F-actin and MTs) and cellular structures, including the outer nuclear membrane (ONM), by plakin-type cytolinkers. On the interior side of the inner nuclear membrane (INM) is the nuclear lamina, containing type V IF proteins (lamins). ER, endoplasmic reticulum; MTOC, microtubule-organizing center; NPC, nuclear pore complex. (Adapted from H. Herrmann et al., *Nature Rev. Mol. Cell Biol.* 8:562–573, 2007. With permission from Macmillan Publishers Ltd.)

Prokaryotes also interact with other cells and with their environments via macromolecules that are secreted and presented on their surfaces; such proteins are known as **adhesins**. A prominent class of bacterial adhesins is the *pili* family of protein filaments, described in Section 11.8. In addition, some bacteria have outer coatings (**capsules**) of complex polysaccharides that are anchored in their envelopes (Section 1.8). There is little if any relationship between bacterial secretions and eukaryotic ECM components, reflecting that the respective systems evolved after their ancestral divergence.

11.2 INTERMEDIATE FILAMENTS

IFs are found in most eukaryotic cells, often in impressive abundance, coursing through the cytosol and connecting the plasma membrane to the nucleus (**Figure 11.3A**). In some cells, such as terminally differentiated *keratinocytes* in the epidermis, they bundle into aggregates (Figure 11.3B) that fill much of the cytosol. In general, IFs have structural roles, as major determinants of cell shape and in enabling cells to withstand physical stress. Another role assigned to them is *mechanotransduction*, whereby IFs physically transmit signals across a cell or through a tissue.

There are ~65 IF genes in humans that code for proteins of 50–200 kDa. At the other extreme, the worm *Caenorhabditis elegans* has only four, and yeast and plants have none, although both have actin and tubulin. The overall picture, then, is that IFs are primarily an elaboration of the cells of multicellular animals and they have diversified as evolution has produced 'higher' (that is, more complex) organisms.

Immunofluorescent light microscopy has revealed that only a few IF genes are expressed in any given cell type. Some neoplastic cells express distinctive combinations of IF proteins, and this property assists in the differential diagnosis of tumors. However, a detailed rationale for patterns of IF gene expression is likely to emerge only when a fuller understanding is achieved of the functional requirements that different cells make of their IF cytoskeletons.

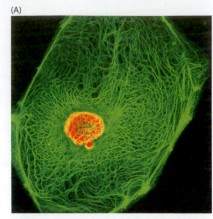

Figure 11.3 Intracellular networks of keratin intermediate filaments.
(A) A PtK2 cell labeled for keratin and visualized by immunofluorescent light microscopy. (B) Electron micrograph of a thin section through bundles of keratin filaments in newborn mouse epidermis: left and central, transverse sections; upper right, longitudinal section. (A, courtesy of Robert Goldman; B, courtesy of Peter Steinert.)

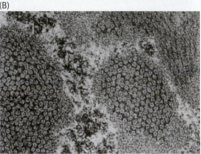

50 nm

IF proteins have a conserved coiled-coil rod domain, flanked by highly variable N-terminal heads and C-terminal tails

All IF proteins have a central α-helical 'rod' domain of ~315 aa, made up of heptad repeats (**Figure 11.4B**). These domains dimerize, with the helices parallel and in register, to form coiled coils. The run of heptads is interrupted in three places with linkers L1, L12, and L2, and the rod has been viewed as consisting of four segments: 1A, 1B, 2A, and 2B. However, crystal structures of IF protein fragments have now shown that the coiled coil is continuous across L2 and that the α helices are continuous across L1 although they do not form a coiled coil (Figure 11.4A). No crystallographic data are yet available for L12. Thus the rod domain is discussed here in terms of three segments: 1A, 1B, and 2. Whereas rod domains are largely conserved, the N-terminal head and C-terminal tail domains vary widely in size and sequence.

IF proteins are classified into types I–VI on the basis of their rod domain sequences. *Keratins*, the IFs of epithelial cells, account for more than half of the human IF genes. They are distinguished by the net charges of their rod domains into types I (acidic) and II (neutral or

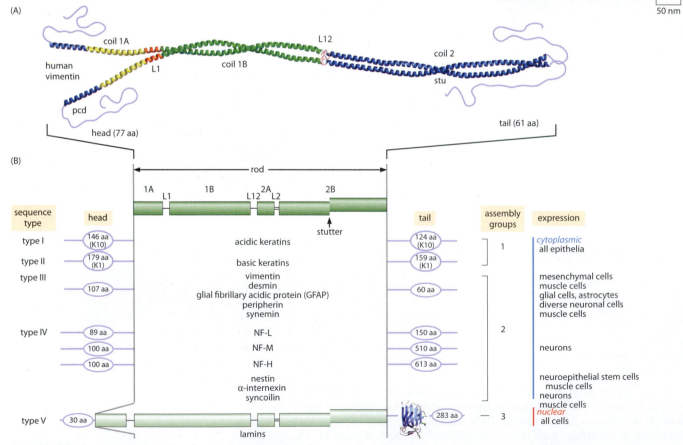

Figure 11.4 IF rod domain structure and the classification of IF proteins. (A) Molecular model of the vimentin dimer. The colors denote segments solved in crystal structures of the respective fragments. The pcd (pre-coil domain) is thought to be α-helical but not to form a coiled coil; stu marks a 'stutter,' a three-residue deletion in the succession of heptads that leads to an underwinding of the coiled coil, as in the downstream segment shown here. (B) IF proteins fall into five types based on their rod domain sequences, which correlate with the kinds of cells in which they are expressed. Type V nuclear lamins have a four-heptad insertion in helix 1B. The sizes of some representative end domains (head, tail) are listed. Two proteins specific to the eye lens, filensin and phakanin may constitute a sixth type. (A, from A.A. Chernyatina et al., *Proc. Natl Acad. Sci. USA* 109:13620–13625, 2012 National Academy of Sciences, USA. With permission from PNAS.)

basic). When a type I keratin is expressed, so is a nearly equal amount of a type II partner, which has ~8 kDa of additional sequences flanking the rod domain. The requirement for co-expression reflects the formation of assembly-competent heterodimers; some keratins can form homodimers in assembly experiments *in vitro* but they are unable to progress into filaments. A second subcategorization distinguishes between 'hard' and 'soft' keratins. The former are expressed in outgrowths of the epidermis such as hair, horn, and nails, and the latter in softer epithelia such as skin and the mucosa.

Type III IFs are found in cells of mesenchymal lineage (that is, deriving from the mesoderm, one of the three primary germ layers in the very early embryo): **vimentin** is expressed in fibroblasts and **desmin** in muscle cells. Neural cells are rich in IF proteins, most prominently the type IV triplet of the **neurofilament** proteins NF-L, NF-M, and NF-H (L stands for light, M for medium, and H for heavy). They are distinguished by having exceptionally long tail domains. These IF proteins form assembly-competent homodimers (for example vimentin) or facultative heterodimers (for example NF-L/NF-M).

Type V IFs are the **lamins**, of which four isoforms, A–D, are encoded in the human genome. Their rod domain is longer than the norm by a four-heptad insertion (see Figure 11.4B). Unlike other IF proteins, which are all found in the cytoplasm, lamins reside in the nucleus. Moreover, their expression is constitutive, not confined to a few cell types.

The model of the vimentin homodimer shown in Figure 11.4A was pieced together from crystal structures of several rod domain fragments. It is striking that the pitch of the coiled coil varies markedly between segments; for instance, the two helices are almost parallel—that is, they have a very long pitch—in the segment immediately after L12. The head and tail domains, which probably lack defined structures (see below) are conveyed schematically. Given the high level of conservation in rod domain sequences, other homomeric IF proteins are likely to have similar structures. However, a crystal structure of helix 2B of the heterodimeric K5/K14 keratin pair differs in revealing a more regular coiled coil, in which the two α helices interact via asymmmetric hydrogen bonds and salt bridges.

Most IF proteins, except lamins, assemble into nonpolar filaments

The next level in the structural hierarchy involves the pairing of dimeric molecules into tetramers. Analysis of proteolytic fragments derived from cross-linked IFs led to the conclusion that there are three kinds of tetramers in which the dimers are respectively: anti-parallel, half-staggered; anti-parallel, in register; and parallel, in register. Studies on the assembly of vimentin defined conditions in which anti-parallel tetramers 65 nm long (**Figure 11.5A,B**) were the only building blocks, implying that the resulting filaments are nonpolar; that is, they lack directionality—a rare property in protein polymers.

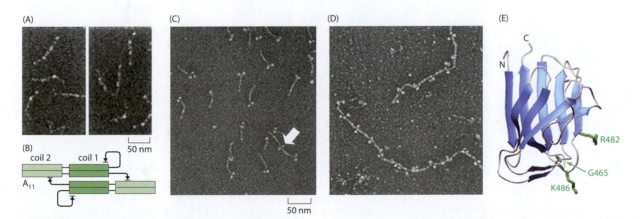

Figure 11.5 Subfilamentous assemblies of IF proteins. (A) Electron micrograph of metal-shadowed 'rodlets' 65 nm long, which are tetramers of a headless vimentin construct. Two dimers pair in anti-parallel orientation, with overlap of their coil-1 domains, are called A11 tetramers, as shown in (B). (C) Electron micrograph of dimers of rat liver lamin A/C, prepared by freeze-drying and rotary shadowing. The globules at one end are C-terminal domains whose positions reflect the parallel in-register mode of dimerization (the arrow points to an overlap region). (D) Lamin dimers assembled head-to-tail into linear strands by dialysis against physiological buffer. (E) Structure of the Ig domain in the human lamin A/C tail. (PDB 1IVT) (A, from N. Muecke et al., *J. Mol. Biol.* 340:97–114, 2004; D, from L. Kreplak et al., *Exp. Cell Res.* 301:77–83, 2004; E, from I. Krimm et al., *Structure* 10:811–823, 2002. All with permission from Elsevier.)

In an alternative classification scheme based on their modes of assembly, IF proteins have been assigned to three groups. Groups 1 and 2 lead, via different intermediates, to similar filaments. In group 2, tetramers associate laterally into 'unit-length filaments' (ULFs) that then associate end-to-end and anneal to yield mature nonpolar IFs. Observations by light microscopy of IF protein complexes large enough to be ULFs being transported intracellularly, apparently along MTs, suggest that a similar assembly pathway may be operative *in vivo*. In group 3 (lamins), dimers associate head to tail to form polar strands that subsequently associate laterally. Although it is well established that lamins are major components of the *nuclear lamina*, a meshwork underlying the nuclear envelope (see Figure 11.2), and tetragonal meshworks of filaments have been visualized on isolated nuclei of *Xenopus laevis* oocytes, the substructure of nuclear laminae in general remains unclear.

IFs have backbones of packed rod domains surrounded by protruding end domains

Structure determination has been difficult because IFs appear smooth-surfaced when observed by negative staining or cryo-EM (see **Methods**) (**Figure 11.6C–E**). Nevertheless, the rod domain is predicted to be ~47 nm long (315 Å × 1.5 Å), and a 47 nm axial repeat has been observed in X-ray fiber diffraction from porcupine quill, which contains well-ordered keratin IFs. It has been discovered, using rotary-shadowing EM, that several kinds of IFs all have a 22 nm axial repeat (Figure 11.6B) and that the long, flexible, C-terminal tails of neurofilament subunits NF-M and NF-L extend widely from the filament backbone as 'side-arms' (Figure 11.6A). These side-arms are thought to function as spacers and are subject to regulatory phosphorylation.

Measurements of mass per unit length by scanning transmission EM (STEM; Figure 11.6E,F) have shown that, for vimentin and several keratin IFs, this quantity is proportional to subunit mass; in other words, the number of molecules per unit length is constant. This is consistent with the conservation of the 22 nm axial repeat. Taking the molecular length to be that of the rod domain (47 nm), the observed mass per unit length corresponds to 16 dimers in cross-section. Variants with 12 and 8 dimers in cross section have also been observed, representing immature IFs or polymorphic variants. This quantization (8–12–16), together with EM observations of unraveling IFs (see Figure 11.6D), suggests that IFs consist of multiple protofibrils, each with four dimers in cross section, coiling around a central axis.

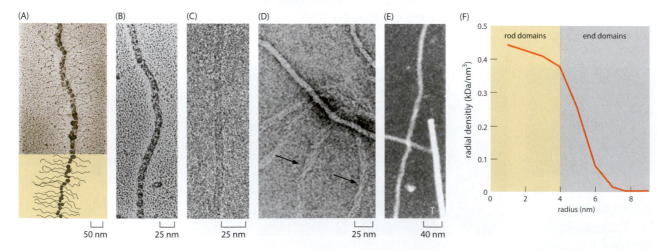

Figure 11.6 Electron micrographs of IFs. (A) Rotary-shadowed neurofilaments. (B) Rotary-shadowed bovine epidermal keratin IFs show a regular (22 nm) axial beading. (C) Cryo-electron micrograph of a keratin IF isolated from rat whisker follicle shows a narrow axial channel. (D) Negatively stained bovine epidermal keratin IF, showing one compact smooth-sided filament and several IFs unraveling to reveal subfilaments (arrows). (E) Dark-field STEM micrograph of unstained desmin/vimentin copolymer IFs prepared by freeze-drying. T, tobacco mosaic virus included for mass calibration. (F) Radial density profile calculated from STEM dark-field micrographs of keratin IFs from rat whisker follicle. (A, from S. Hisanaga and N. Hirokawa, *J. Mol. Biol.* 202:297–303, 1988. With permission from Elsevier; B, courtesy of Margaret Bisher, Peter Steinert, and Alasdair Steven; C and F, adapted from N.R. Watts et al., *J. Struct. Biol.* 137:109–118, 2002. With permission from Elsevier; D and E, courtesy of Alasdair Steven.)

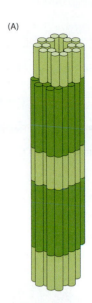

Figure 11.7 Models of a generic IF backbone. These models embody some well-established properties of rod domains, but the exact packing arrangement is still uncertain. (A) Model of annealed unit-length filaments, with 16 dimeric coiled coils in cross section, organized as eight anti-parallel A11 tetramers (see Figure 11.5B). The overlap regions (helix 1) are darker. The top segment (light) shows only eight coiled coils, pending the addition of the next ULF at that end. For clarity, the width of the filament is exaggerated and no twist is shown. (B) Hierarchical model of α helices: coiled coils; protofilaments; protofibrils; IFs, showing alternation of handedness at successive steps. More detailed models have been proposed, combining chemical cross-linking data with other constraints, by R.D.B. Fraser and D.L.D. Parry, *J. Struct. Biol.* 151:171–182 (2005). (A, adapted from H. Herrmann et al., *Nature Rev. Mol. Cell Biol.* 8:562–572, 2007. With permission from Macmillan Publishers Ltd.; B, adapted from N.R. Watts et al., *J. Struct. Biol.* 137:109–118, 2002. With permission from Elsevier.)

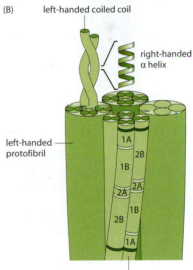

The high contrast of STEM dark-field images of freeze-dried specimens is advantageous for detecting diffuse and/or disordered peripheral material. The radial density profile shown in Figure 11.6F for keratin IFs reveals a dense inner zone (to a radius of ~4.5 nm) and a diffuse outer zone (radii between 4.5 and 7.5 nm), whose masses correlate with those of the rod domains and the end domains respectively. The same keratin IFs have a small axial channel that is detected in frozen-hydrated specimens (Figure 11.6C).

IF architecture is based on backbones of bundled rod domains from which the head and tail domains protrude laterally. The backbones have similar structures, reflecting sequence conservation in the rod domains; some known features are illustrated in **Figure 11.7**. In contrast, the heads and tails are highly variable and their peripheral locations present them to binding partners. So far, only one high-resolution structure has been determined for (part of) an end domain, namely an Ig domain in a lamin tail (see Figure 11.5E). It appears that most other IF end domains are, partly or wholly, *natively unfolded* (see Section 6.4). This is consistent with their low sequence complexity; those of epidermal keratins have unusually high contents (~60%) of Gly and Ser, whereas those of 'hard' keratins are rich in Cys, which enables them to cross-link with Cys-rich 'matrix' proteins. The same architectural principle of a filament backbone of packed coiled coils surrounded by functional domains recurs in myosin filaments (Section 14.5).

IFs organize into higher-order structures that integrate cells and tissues

In many contexts, IFs are gathered together in bundles called **tonofilaments** that couple at one end to the plasma membrane at *desmosome* junctions (Section 11.4), and at the other end to the nuclear envelope. IFs connect to desmosomes, each other, and other cytoskeletal elements via cytolinker proteins of the *plakin* family, which are elongated dumbbell-shaped dimers (**Figure 11.8**). All except epiplakin have a long coiled-coil rod domain. Their C-terminal domains, containing multiple homology domains A, B, and C, bind to IFs, and their N-terminal domains (NTDs, shown in yellow in Figure 11.8) connect to other structures. For example, the NTDs of plectin and BPAG1 have an actin-binding domain, consisting of two calponin homology domains (see **Guide**).

IFs also serve as scaffolds that define certain intracellular compartments; examples include the vimentin IF cage that surrounds lipid globules in *adipocytes*, and IFs involved in the formation of the *aggresome*, a perinuclear compartment for protein degradation. Thus, IFs serve to integrate cellular space.

The dead cells of the outer layers of the epidermis and of hair or horn consist almost exclusively of bundles of matrix-embedded keratin IFs. These structures may be viewed as composite biomaterials, comparable to man-made composites such as fiberglass. The mechanical properties of a composite depend on the interactions between its two components, fibrils and matrix, and on the ratio in which they are mixed.

Mutations in IF genes underlie numerous human diseases

Keratin IFs are major constituents of skin, and a number of dermatological conditions with symptoms typically involving flaking or blistering, have been correlated with point

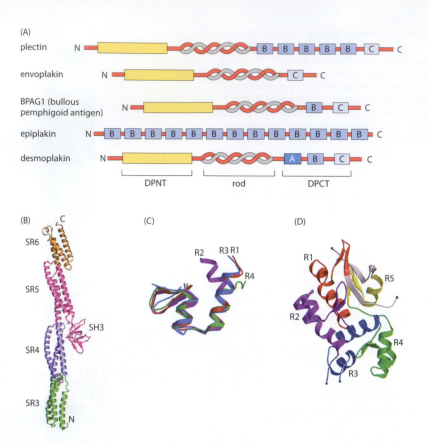

Figure 11.8 Domain structures of plakin cytolinker proteins. (A) Domain maps of some plakins. The rod domains are coiled coils (this segment of the paired molecule is added in gray to convey the coiled coil). (B–D) Structures of some plakin domains. (B) Fragment of desmoplakin N-terminal domain (DPNT) with its spectrin-like repeats (SR3–SR6) and an inserted SH3 domain. (PDB 3R6N) (C) Plakin repeats from DPCT are superimposed to show their similarity. Each repeat has two short anti-parallel β strands followed by two α helices. (D) DPCT subdomain C. Unlike other structures composed of short repeat motifs that form extended structures, the 38-residue plakin repeats (R1–R5) coalesce to form a globular domain. The five plakin repeat motifs are color-coded, and non-repeat sequences are in gray. (PDB 1LM7) (B, from H.J. Choi and W.I. Weis, *J. Mol. Biol.* 409:800–812, 2011. With permission from Elsevier; C and D, from H.J. Choi et al., *Nature Struct. Mol. Biol.* 9:612–620, 2002. With permission from Macmillan Publishers Ltd.)

mutations in keratin genes. For example, in epidermolysis bullosa simplex (EBS), a severe blistering disease, mutations map to sites in the rod domain. Desmin is expressed in muscle, where it is associated with the I-band, which connects adjacent sarcomeres (Section 14.5). Desminopathies include muscular dystrophy and heart disease, in both cases reflecting impairment of heart muscle function. Laminopathies encompass a wide range of pathological effects, including muscular dystrophy and heart disease (as with desmin), but also lipodystrophy and *progeria*, a premature aging disease. Residues in the lamin Ig domain where mutations correlate with Dunnigan-type familial lipodystrophy, a rare disease involving the loss of subcutaneous fat from certain regions of the body and insulin resistance, are marked in green in Figure 11.5E.

11.3   TIGHT JUNCTIONS

As their name implies, **tight junctions** are the regions where the plasma membranes of two interacting cells come closest together (**Figure 11.9A**). Their other name, occluding junctions, reflects their dual functions: first, selectively blocking the passage of materials as small as ions or as large as metastatic cancer cells along the paracellular route between adjacent cells; and second, dividing the plasma membrane of an epithelial cell (or other 'polarized' cells) into two zones, apical and basolateral (see Figure 11.2). According to the *fluid mosaic* concept, intrinsic membrane proteins are free to diffuse laterally and rotationally unless immobilized by being anchored to higher-order structures. The tight junction poses a barrier to such movements. As a result, the *apical membrane* and the *basolateral membrane* differ in composition and in their functions.

Tight junctions consist of networks of paired intramembrane strands

In electron micrographs of transverse thin sections, tight junctions show local sites of particularly close apposition called 'kissing points' (see Figure 11.9A). Freeze-fractured EM specimens (Figure 11.9C) reveal that kissing points represent the cross-cutting of strands of intrinsic membrane proteins at sites where the two membranes come together. Thus a

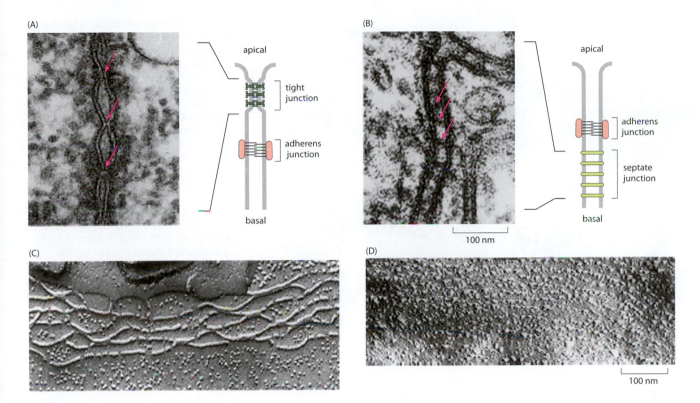

(A)

apical

tight
junction

adherens
junction

basal

(B)

apical

adherens
junction

septate
junction

basal

100 nm

(C)

(D)

100 nm

tight junction consists of several interconnected strands running approximately parallel to each other and perpendicular to the paracellular pathway. The number of strands in a junction is thought to correlate with its efficiency of occlusion. This hypothesis is based on an observed correlation between measurements of transepithelial electric resistance (TER) in various epithelia and the numbers of strands in their tight junctions. However, TER and other measures of occlusion efficiency are likely to depend also on the proteins present in the junction.

It has been established by immuno-EM that the main components of tight junction strands are **claudins**, a family of intrinsic membrane proteins with subunit masses of ~25 kDa; moreover, exogenous expression of claudins in cultured cells that otherwise lack them is accompanied by the appearance of strands in their membranes. Humans and mice have 24 claudins that are differentially expressed in various tissues, each cell type expressing a specific subset. A second major component is **occludin** (60 kDa); although it is not clear that occludin can form strands, its incorporation into claudin-based strands has been demonstrated.

As deduced from their amino acid sequences, claudins and occludin each have four transmembrane segments, two extracellular loops, one cytoplasmic loop, and both termini in the cytoplasm (**Figure 11.10A**). The TM portion of a claudin is shown in Figure 11.10C. The two extracellular loops of claudins, ECL1 (50 aa, including two conserved cysteine residues) and ECL2 (25 aa), are responsible for *trans* interactions between the paired membranes and also regulate the paracellular movement of ions (**Figure 11.11**). Claudins vary in their propensity to allow passage to various ions. On the cytoplasmic side, their C-terminal regions possess a four-residue motif, Glu-Ser/Thr-Asp-Val, which can recruit PDZ proteins (see **Guide**) to the junction. These binding partners include the large multidomain proteins, ZO-1, ZO-2, and ZO-3 (ZO for *zonula occludens*, a Latinized version of occluding junction), two of which are capable of binding F-actin and thus linking it to the junction.

Occludin has a long C-terminal region that appears to be involved in signaling and interacts with numerous cytoplasmic proteins, including the ZO scaffolding proteins and F-actin. The crystal structure of a 106-aa segment reveals a kinked anti-parallel coiled coil (Figure 11.10B) that has a large positively charged surface where the ZO-1 binding site resides; this surface also targets occludin to the junction. The remaining 34 aa of this construct were disordered in the crystal and it may be that other cytoplasmic portions of both claudins and occludin are unfolded until they encounter binding partners.

Figure 11.9 Tight junctions and septate junctions. Thin-section electron micrographs of of (A) a tight junction, and (B) septate junctions. Pink arrows mark 'kissing points' and 'septa' respectively. Alongside are diagrams showing the positioning of these junctions relative to adherens junctions in vertebrates and invertebrates (arthropods) respectively. Only the tight junction (A) and the septate junction (B) are represented in the micrographs. (C, D) Freeze-fracture electron micrographs show filamentous strands of intramembrane particles in tight junctions (C), and strands of less closely connected particles in septate junctions (D). (Adapted from M. Furuse and S. Tsukita, *Trends Cell Biol.* 16:181–188, 2006. With permission from Elsevier.)

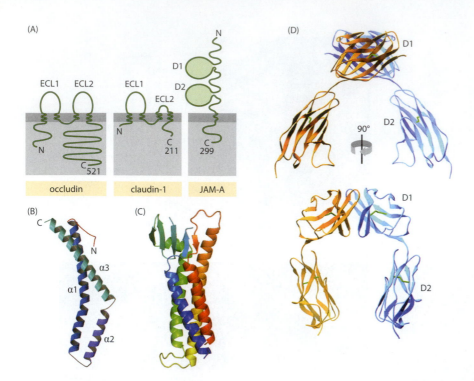

Figure 11.10 Transmembrane topology and domain structures of tight junction proteins. (A) Organization of occludin, claudin-1, and JAM-A. ECL, extracellular loops. JAM-A has two Ig ectodomains, D1 and D2. The C-terminal amino acids are numbered. (B) The C-terminal segment (residues 416–522) of occludin (PDB 1WPA). (C) The murine claudin mCld15 has four α-helices in its transmembrane portion and a β-sheet plus two short α-helices in its extracellular portion (PDB 4P79). Linear strands of monomers seen in the crystal may relate to junctional strands *in situ*. (D) A dimer of the human JAM-1 ectodomain. The subunits associate via their D1 domains. (PDB 1NBQ) (A, adapted from S. Tsukita et al., *Nature Rev. Mol. Cell Biol.* 2:285–293, 2001. With permission from Macmillan Publishers Ltd; C, from Suzuki et al., *Science* 244:304–306, 2014. With permission from AAAS.)

The architecture of junctional strands is not yet clear, although it is known that claudins are their main constituents and immuno-EM evidence has shown that different claudins can coexist in the same strand. Their oligomeric status is not known. *Connexins*, which have a similar transmembrane topology, form hexamers in gap junctions (Section 11.5), but if 6-fold symmetry were to apply to claudin oligomers, they would be expected to form two-dimensional arrays rather than linear strands.

When genes coding for claudins are knocked out, the mice exhibit an embryonic-lethal phenotype. In contrast, occludin-deleted mice appear normal at birth but subsequently develop a variety of aberrant conditions including growth retardation, chronic inflammation, and mineral deposition in the brain. The latter phenotype may reflect that the blood–brain barrier (**Box 11.1**), of which tight junctions are an important component, is compromised. In humans, mutations in claudin genes have been associated with several diseases, including hereditary hypomagnesemia, in which total body magnesium homeostasis is perturbed, and hereditary deafness, in which the organization of the inner ear is developmentally impaired.

Junctional adhesion molecules (JAMs) serve as virus receptors

A third major class of tight junction components is the JAMs, of which there are three isoforms in humans: JAM-A, JAM-B, and JAM-C, plus four related proteins. JAMs have ectodomains consisting of two Ig-like domains, a single-pass transmembrane segment, and a short cytoplasmic domain (see Figure 11.10A). Crystal structures for ectodomains show that the two Ig domains are not collinear but have a 60° bend between them, and

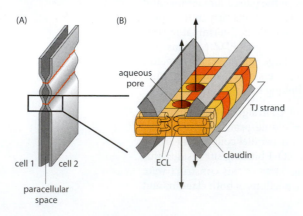

Figure 11.11 Pores allow the selective diffusion of ions along the paracellular pathway. (A) Passage through the paracellular space between these two cells is restricted by two tight junction filaments (brown). (B) Enlargement in which the extracellular loops (ECLs) of apposing claudins create a barrier that is traversed only through small aqueous pores. The arrangement of claudin molecules in a strand is not known; here, subunits are shown illustratively in light, mid, and dark brown. TJ, tight junction (Adapted from M. Furuse and S. Tsukita, *Trends Cell Biol.* 16:181–188, 2006. With permission from Elsevier.)

Box 11.1 The blood–brain barrier

The brain cells that make up the central nervous systems (CNS) of vertebrates are partitioned off from the rest of the body, as first observed by Paul Ehrlich in the 1880s in terms of a failure of dyes introduced into the bloodstream of experimental animals to enter the CNS. Nevertheless, a pathway is required for the delivery of nutrients into the brain and for the egress of potentially toxic products. The agent principally responsible for these selective sieving functions is the **blood–brain barrier** (BBB). The brain is densely pervaded with the microvasculature, a system of narrow tubes (the microcapillaries), each a cylindrical monolayer of endothelial cells. The lumens of these tubes are part of the bloodstream. They are surrounded by pericytes, astrocytes, and microglia in associations called *neurovascular units* (NVU) (**Figure 11.1.1**). Within the endothelial monolayer, the interfaces between the cells are sealed with tight junctions (Section 11.3), which block the passage of materials along the intercellular spaces (the *paracellular* pathway). Accordingly, materials to be transferred into the brain—mostly small molecules—are conveyed by transcytosis; that is, they are imported through the apical membrane into the cytoplasm of the endothelial cell, migrate across it, and are exported on the other side through the basolateral membrane. Most invertebrates have no BBB; those that do form their tight junctions between modified glial cells.

The vertebrate BBB has other components. Surrounding the tube of endothelial cells is a layer of *basement membrane*, a specialized form of extracellular matrix (Section 11.7), and much of its surface is covered with a class of cells called pericytes or by the end-feet of *astrocytes* (see Figure 11.1.1). Pericytes are contractile cells that contribute to the regulation of blood flow by changing the microcapillary diameter, and they are thought to contribute to the regulation of gene expression in the underlying endothelial cells. Astrocytes provide the cellular link to the neurons and are thought to guide the differentiation of BBB endothelial cells that distinguishes them from other endothelia: they also contribute to management of intracerebral water. Also nearby are microglia, immune cells whose role is to intercept infectious agents that somehow traverse the BBB and enter the CNS. The development, maintenance, and functioning of the BBB are complex processes involving many macromolecular assemblies and are still at a relatively early stage of characterization. Here, we briefly outline the locations and some properties of macromolecular machines that are covered in greater detail elsewhere in this book. Often, they involve BBB-specific members of protein superfamilies.

BBB tight junctions are built from claudin-3, in addition to containing the endothelia-specific claudin-5 (Section 11.3). Associated with the tight junctions and perhaps regulating their assembly are adherens junctions with VE-cadherin (Section 11.4). Gap junctions containing connexin Cx43 (Section 11.5) form communication channels between astrocytic processes but not with the endothelial cells. The inward-rectifying K-channel Kir4.1 (Section 16.2) and the water channel aquaporin AQP4 (Section 16.4) are present in the astrocytic feet, where they respectively contribute to homeostatic regulation of potassium ions and the expulsion of metabolically produced water, which leaves the brain by other pathways (in other words, not by transcytosis). The basement membrane of

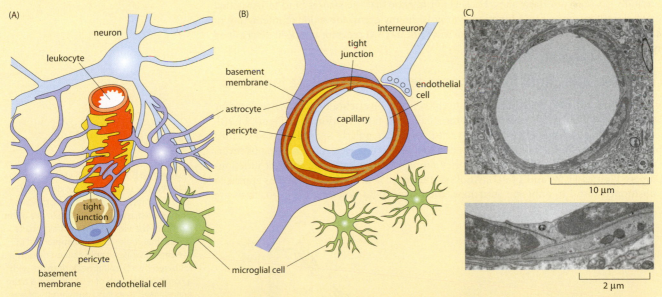

Figure 11.1.1 Organization of the blood–brain barrier (BBB).
(A) Schematic drawing of a neurovascular unit (NVU), indicating the physiologic interactions between endothelial cells and adjacent pericytes, astrocytes, microglia, and neurons in the central nervous system. (B) Cross-sectional diagram of the BBB showing a microcapillary endothelial cell, surrounded by basal lamina and astrocytic endfeet. (C) Electron micrograph showing a cross section through a murine microcapillary and surrounding tissue with (at the bottom) an enlargement of a junctional region. (A, adapted from N.J. Abbott and A. Friedman, *Epilepsia* 53(Supp. 6):1–6, 2012. With permission from John Wiley & Sons; B, adapted from N.J. Abbott, L. Rönnbäck and E. Hansson, *Nature Rev. Neurosci.* 7:41–53, 2006; C, from A. Armulik et al., *Nature* 468:557–561, 2010. B and C, with permission from Macmillan Publishers Ltd.)

Box 11.1 *continued*

BBB endothelial cells contains laminins α4 and α5 and collagen IV, but not fibrillar collagens (Section 11.7).

Maintenance of an intact and properly functioning BBB is essential; failure to do so can have dire consequences. The BBB can be compromised in a variety of ways, of which a few examples are: stroke (cerebral hemorrhage); infection of the brain by microorganisms (for example by bacteria in meningitis) or viruses (for example by the neurotropic herpes simplex virus; Chapter 8.5); multiple sclerosis, in which an impairment of the BBB is accompanied or caused by deficiencies in the assembly of basement membrane and/or tight junctions; and trauma to the brain, inducing the production of bradykinin, a 9-aa peptide that acts on endothelial cells, causing blood vessels to expand and blood pressure to drop. In turn, bradykinin causes the release of the cytokine interleukin-6, leading to opening of the BBB.

References

Abbott NJ, Rönnbäck L & Hansson E (2006) Astrocyte–endothelial interactions at the blood–brain barrier. *Nature Rev Neurosci* 7:41–53.

Engelhardt B & Sorokin L (2009) The blood–brain and blood–cerebrospinal fluid barriers: function and structure. *Semin Immunopathol* 31:497–511.

Schoknecht K & Shalev H (2012) Blood–brain barrier dysfunction in brain diseases: clinical experience. *Epilepsia* 53(6):7–13.

that subunits dimerize through their N-terminal D1 (membrane-distal) domains (Figure 11.10C). This mode of engagement may be employed as a *cis* interaction in junctions; however, the exact location of JAMs in junctions has yet to be determined.

Several viruses have been identified as using tight-junction proteins as their receptors. They include coxsackie virus (a picornavirus; Section 8.3), adenovirus (a double-stranded DNA virus; Section 8.5), and reovirus (a double-stranded RNA virus; Section 8.6). All three of these viruses are non-enveloped, implying that their mode of cell entry does not involve the fusion of a viral membrane with a host cell membrane (Section 8.5). The receptor is in each case a JAM. At first sight, the close apposition of tight-junction-coupled cells would appear to exclude viruses from these sites. However, it may be that the viruses bind to JAMs that are not closely integrated into junctional strands. In any event, the mode of interaction of the trimeric *host attachment protein* of reovirus with its JAM receptor has been determined (**Figure 11.12**).

In invertebrates, tight junctions are replaced by septate junctions

Tight junctions are ubiquitous in vertebrate epithelia. In invertebrates, they are replaced by related but morphologically distinct structures called **septate junctions** (see Figure 11.9B). Septate junctions do not have kissing points; instead, regularly spaced densities called septa connect the two interacting membranes. Freeze-fracture electron micrographs reveal strands of intramembrane particles but they are less closely knit than in tight junctions (compare Figures 11.9C and 11.9D). Another difference is that septate junctions appear to be positioned distal to adherens junctions, whereas in tight junctions their positions are

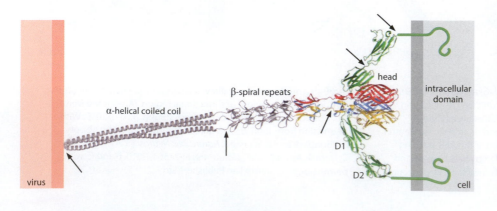

Figure 11.12 JAMs act as virus receptors. JAM-A is the receptor for reovirus (Section 8.6). The viral host attachment protein σ1, a homotrimer, is initially packed in the outer capsid of the virion. During infection, its N-terminal region remains implanted in the capsid while a triple coiled-coil domain followed by a β-spiral 'neck' domain, with a combined length of ~30 nm, and a head are released to extend outward to interact with JAM-A. A crystal structure for the head and a neck fragment of σ1 complexed with three JAM-A ectodomains (colored regions; only two ectodomains are shown, for clarity) shows that the D1 domain of JAM-A binds to the head–neck junction. Arrows mark hinge regions. (PDB 3EOY) (From E. Kirchner et al., *PLoS Pathogens* 1:1–12, 2008.)

reversed (see Figure 11.9A,B). The inferred relationship between the two kinds of structures has been confirmed by the identification of two claudin-like proteins in *Drosophila* septate junctions.

11.4   ADHERENS JUNCTIONS AND DESMOSOMES

Adherens junctions and **desmosomes** have key roles in coordinating cell–cell adhesion in epithelial tissues (see Figure 11.2). Their molecular architectures are quite similar and approximately mirror-symmetric across the junction (**Figure 11.13**). At the surfaces of both interacting cells are arrays of **cadherin** proteins, whose ectodomains form the intercellular contacts, while their endodomains connect, via adaptor proteins, to the underlying cytoskeletons. The principal difference between them is that adherens junctions are linked to the actin cytoskeleton, whereas desmosomes are attached to the IF system.

These junctions have multilayered organizations. *In situ* structures have been studied in greater detail for desmosomes which are more highly ordered, whereas more information on the molecular components and their interactions has been obtained for adherens junctions. Current mappings of the respective sets of cadherins, adaptors, and associated cytoskeletal components are given in **Figure 11.14**. On the cytoplasmic side of an adherens junction membrane (shown on only one side of the junction), the membrane-proximal portion of the cadherin cytoplasmic tail binds to p120 and the distal portion of the tail to β-catenin or to its close relative plakoglobin. In turn, β-catenin binds α-catenin, which is required for force transduction between cadherins and F-actin. It is thought that when the actin cytoskeleton is anchored to the cadherins, as during cytoskeletal contraction in cell morphogenesis, the catenins and other proteins (blue ovals in Figure 11.14A) provide the linkage. These connections are discussed further below.

Through their interactions with the cytoskeleton, junctions confer shape and mechanical strength on cells and tissues. Adherens junctions have a vital role during tissue morphogenesis, when coordinated movements of adherent cells generate distinct organs. Desmosomes in concert with IFs enable tissues to resist physical stress. Similar junctions are also found in endothelia, cardiac muscle, and neuronal synapses. It is likely that they have specializations needed for the functioning of these particular tissues, but their core components are similar to those in the better characterized epithelial junctions. Hemidesmosomes (see Figure 11.2) are junctions that appear to consist of one half of a desmosome interfaced with the *extracellular matrix*, except that their receptor molecules are integrins, as in focal adhesions (Section 11.6), rather than cadherins.

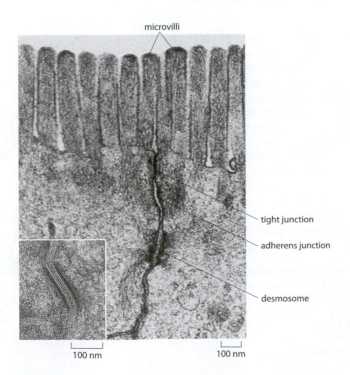

microvilli

tight junction

adherens junction

desmosome

100 nm   100 nm

Figure 11.13 Electron micrograph showing junctions coupling two epithelial cells. Tonofilaments (bundles of keratin IFs) are seen connecting to the desmosome, while actin filaments emanating from microvilli are linked to the adherens junction and the tight junction. The inset shows a thin section of a well-ordered desmosome in an intercalated disc, the junction between cardiac myocytes. Desmosomes in the intercalated disc connect to desmin IFs and have a distinctive layered structure, containing outer and inner 'dense plaques.' (Main figure, from D.H. Cormack in Ham's Histology, 9th ed. Philadelphia: J.B. Lippincott Co., 1987; inset, courtesy of Pauline Bennett.)

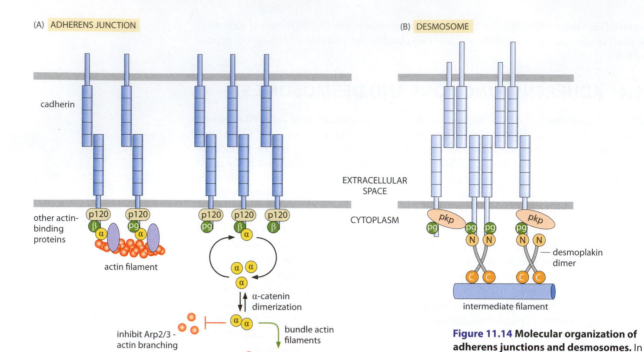

(A) ADHERENS JUNCTION

cadherin

EXTRACELLULAR SPACE

CYTOPLASM

other actin-binding proteins

actin filament

α-catenin dimerization

inhibit Arp2/3 - actin branching

bundle actin filaments

(B) DESMOSOME

desmoplakin dimer

intermediate filament

Figure 11.14 Molecular organization of adherens junctions and desmosomes. In both junctions, apposing cadherins interact *in trans* through their EC1 domains. For their packing in a native desmosome, see Figure 11.16B. Desmosomal cadherins have longer cytoplasmic tails than do classical cadherins. They bind plakoglobin (pg), plakophilin (pkp), and desmoplakin connecting to IFs.

Mutations in junctional proteins—particularly in desmosomal proteins—give rise to a variety of diseases. Some skin-blistering diseases, such as epidermolysis bullosa simplex (EBS), arise from mutations in proteins that link desmosomes to keratin IFs, as well as mutations in the keratins themselves (noted in Section 11.2). Other blistering diseases such as pemphigus vulgaris are caused by autoantibodies against desmosomal cadherins. The symptoms of these conditions reflect impairment of the tissue integrity through defects in cell–cell adhesion.

Classical cadherins and desmosomal cadherins have similar multi-domain structures

The best characterized cadherins are the so-called 'classical' cadherins found in vertebrate adherens junctions, and the desmosomal cadherins, **desmogleins** and **desmocollins**. All are single-pass transmembrane proteins whose N-terminal ectodomains comprise five tandem extracellular (EC) domains. The EC domain has an Ig-like fold and some are glycosylated (**Figure 11.15A**). Membership of the cadherin superfamily can be defined by the presence of one or more EC domains, but it is not known whether all such proteins function in adhesion. Classical cadherins fall into two subfamilies, with 6 type I and 13 type II cadherins in humans. Type I cadherins are broadly distributed but segregate according to embryonic germ layer or a particular tissue, whereas type II cadherins often have overlapping patterns of expression, particularly in the nervous system. Desmosomal cadherins are less diverse: there are four desmogleins and three desmocollins. Cadherins found in invertebrate adherens junctions, e.g. *Drosophila* DE- and DN-cadherins, play roles in tissue patterning, as do their vertebrate counterparts. However, these proteins contain more EC domains (up to 34 of them!), and their adhesive mechanisms may differ from those of vertebrate cadherins.

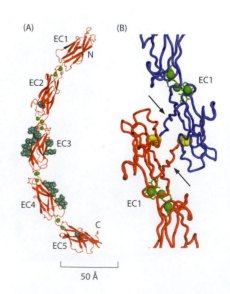

Figure 11.15 Cadherin ectodomain interactions and strand swapping. (A) Structure of the C-cadherin ectodomain. Turquoise spheres represent ordered carbohydrates. (B) Strand exchange in the *trans* EC1/EC1 interaction of C-cadherin. (PDB IL3W). Calcium ions binding between EC1 and EC2 are shown as green spheres. The side-chain of each Trp2 (yellow) inserts into the partner molecule. (Adapted from S. Pokutta and W.I. Weis, *Annu. Rev. Cell Dev. Biol.* 23:237–261, 2007. With permission from Annual Reviews.)

50 Å

Cadherin-mediated cell–cell adhesion is homophilic and Ca²⁺-dependent

Cadherin expression often correlates with the formation of specific tissues, and different cadherins mark successive cell layers of mature tissues, such as the epidermis. When cells are mixed *in vitro*, they sort according to cadherin type: cells expressing the same cadherin stick together; in other words, adhesion is homophilic. However, the outcome of a cell-sorting experiment depends not only on the cadherins expressed but also on their density at the cell surface, the shear force used in mixing the cells, and other parameters. Some cell layers expressing different cadherins adhere to one another, and experiments with purified proteins have demonstrated binding between closely related cadherins, suggesting that, under some circumstances, cellular cadherins also interact heterophilically.

Cadherins effect adhesion only in the presence of Ca^{2+} (hence their name). Ca^{2+} binds between successive EC domains, imparting rigidity and a characteristic curvature to the molecule (see Figure 11.15A). Electron tomograms of desmosomes depict curved rods of a thickness and length that correspond to cadherins projecting from the two apposed membranes (**Figure 11.16**).

The outermost domain, EC1, confers adhesive specificity, as established by multiple approaches. Experiments in which EC domains of different classical cadherins were exchanged showed that cell-sorting specificity resides in EC1. Cross-linking of engineered disulfides also demonstrated EC1/EC1 interactions between cells. The same EC1–EC1 interface has been observed consistently in several crystal structures. Electron tomography has demonstrated that only the outermost domains interact in desmosomes (see Figure 11.16B). Finally, fluorescence resonance energy transfer (FRET) spectroscopy has also shown cadherins pairing via EC1/EC1 interactions.

In the classical cadherin EC1/EC1 interaction, there is *domain swapping* involving the tips of the N-terminal 'A' β strands (Figures 11.15B). A conserved tryptophan residue binds in a hydrophobic pocket in the partner molecule, and additional contacts are made between the A strand and the partner molecule. In type II cadherins, a second conserved tryptophan residue also inserts into the hydrophobic pocket on the partner, and there is a more extensive interface between EC1 domains. Residues in these interfaces are conserved within types I and II but differ between them, thus explaining the potential for intra-type heterophilic interactions and the observed rarity of inter-type heterophilic binding.

Cadherins in junctions have both *trans* and *cis* interactions

A junction contains many cadherin molecules and their multivalency has important implications for the forming, maintaining, and remodeling of cell-to-cell contacts. The primary interaction takes place in *trans* between apposing monomers (see Figure 11.16C), with

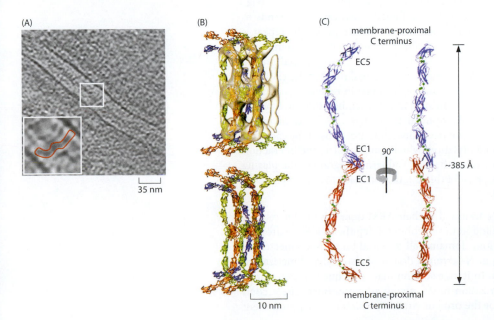

(A)

35 nm

(B)

10 nm

(C)

membrane-proximal C terminus

EC5

EC1   90°

EC1

~385 Å

EC5

membrane-proximal C terminus

Figure 11.16 Interactions of cadherin ectodomains in desmosomes. (A) Thin (2.4 nm) slice through a tomographic reconstruction of a frozen-hydrated thin section of human epidermis, containing a desmosome. The inset shows an enlargement of the boxed area. (B) Local packing of cadherin ectodomains on both sides of the desmosome. Top: a slab of density from a cryo-tomogram, with the C-cadherin ectodomain crystal structure fitted in. Bottom: the tomographic density has been removed to convey the cadherins better. (C) Side views of a *trans* dimer. Each curved ectodomain has five tandem Ig domains, EC1–EC5. (B, from A. Al-Amoudi et al., *Nature* 450:832–837, 2007. With permission from Macmillan Publishers Ltd; C, from S. Pokutta and W.I. Weis, *Annu. Rev. Cell Dev. Biol.* 23:237–261, 2007. With permission from Annual Reviews.)

rather low affinity (K_d = 3–100 μM). It has been proposed that the weakness of these interactions is crucial to cadherin-specific cell sorting. The initial contact occurs via cadherins presented on dynamic *lamellipodia* (Section 14.9). If the local cadherin concentration is lower than the dissociation constant, few *trans* dimers will form and small differences in affinity will produce large differences in the number of *trans* dimers in the contact zone. As long as the membranes remain in contact, more cadherins can diffuse into this zone, strengthening the nascent junction. Conversely, a contact zone with few interacting cadherins may disassemble before enough cadherins have been recruited to stabilize it. If *trans* interactions were strong, paired membranes would remain in contact long enough for even heterotypic contact zones to be stabilized. Thus, low affinity seems to ensure that stable contacts form only between cells expressing the same or closely related cadherins.

The clustering of cadherins in junctions raises the question of how the cadherins are recruited into the contact zone and, in particular, of *cis* interactions on the same cell surface. A conserved *cis* interface between EC1 of one molecule and EC2 of a neighbor has been observed in multiple crystal structures of full-length ectodomains of type 1 classical cadherins. They all form a similar lattice that has been proposed to mimic quasi-crystalline packing at the adhesive interface. A similar lattice has also been seen in EM reconstructions of artificial adherens junctions formed by binding E-cadherin ectodomains to liposomes. Mutation of conserved residues at this interface disrupts the lattice and, importantly, disrupts the formation of adherens junctions between cells.

Electron tomography of desmosomes has revealed a fairly regular organization of their cadherins, with an average spacing of ~7.5 nm between molecules in the same membrane (see Figure 11.16B). The tomograms also revealed a quasi-periodic arrangement, with *trans* dimers alternating with *cis* dimers (see Figure 11.14B); however, it is not clear whether there is long-range order or a more liquid-like arrangement. Nor has it been possible to define the *cis* interfaces with certainty, although the tomograms suggest that the membrane-proximal domains EC4 and EC5 are involved.

Cadherins interact with the actin cytoskeleton via catenins and other proteins

Actin filaments associate with the cytoplasmic side of adherens junctions, and in many epithelial cells they form a belt encircling the cytoplasm. During tissue development, cell-to-cell contacts are maintained as the cells undergo shape changes caused by constriction of the actin cytoskeleton. Adherens junctions provide anchorage points that allow forces to be transduced between cells. Here, the conserved intracellular 'tails' of classical cadherins (~150 aa) have a crucial role. Their association with the actin cytoskeleton involves intermediary proteins, primarily the *catenins* (Figure 11.14A and **Figure 11.17**).

The membrane-distal two-thirds of the cadherin tail binds β-catenin or its close relative, **plakoglobin**. This portion of cadherin, which is unfolded when unbound, binds to the central domains of the latter proteins, which comprise 12 ARM repeats (the repeat is an α-helical subdomain named after armadillo, the *Drosophila* ortholog of β-catenin; see **Guide**)—Figure 11.17B. These repeats stack in a solenoid that has a pronounced groove where the cadherin interaction takes place. The membrane-proximal third of the cadherin tail binds the ARM domain of p120, a β-catenin homolog involved in regulating cadherin turnover and coordination with the cytoskeleton. In structure it resembles the corresponding domain of the desmosomal p120 relative, **plakophilin-1**, which is an ARM domain like the central domain of β-catenin but with a pronounced kink. Both portions of the cadherin tail are subject to phosphorylation and ubiquitylation, which can modulate the interaction with β-catenin and regulate turnover. Newly synthesized cadherins migrate to the plasma membrane bound to β-catenin, and disruption of this interaction leads to proteolytic degradation by the proteasome (Section 7.4).

The parts of β-catenin and plakoglobin N-terminal to their ARM domains bind α-catenin (Figure 11.14A) of which three forms are found in vertebrates, E (epithelial), N (neuronal), and T (testis and heart). αE-catenin has four domains, all α-helical bundles, connected by flexible loops (Figure 11.17A,C). Pairing of N-terminal domains, also called dimerization domains, produces an asymmetric dimer. In it, one subunit may subsequently be replaced by β-catenin, which has a similar dimerization motif, giving a heterodimer. The second domain of α-catenin has a binding site for the protein vinculin, which is also a prominent

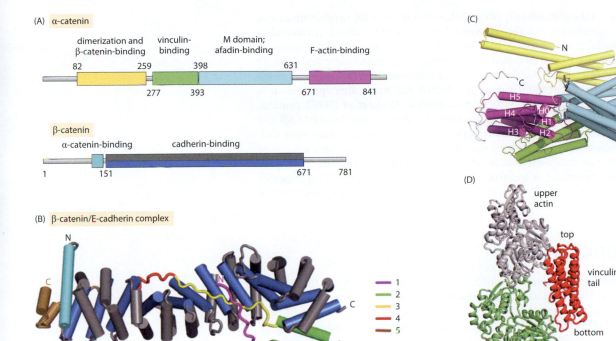

Figure 11.17 Domains of α-catenin and β-catenin, and their interactions.
(A) Domain maps of both proteins.
(B) Complex of the ARM domain of β-catenin with the cytoplasmic portion of E-cadherin bound to it. The α helices of β-catenin are blue and gray rods, with the blue helices lining the superhelical groove. The β-catenin-binding portion of E-cadherin has an extended conformation and, as shown, is divided into five segments, which are color-coded. Segment 3 (yellow) contains conserved residues required for this interaction. (C) Structure of α-catenin, showing its four helix-bundle domains. (D) The C-terminal domain is similar to a vinculin domain that binds to F-actin, as shown here. (B, from S. Pokutta and W.I. Weis, *Annu. Rev. Cell Dev. Biol.* 23:237–261, 2007. With permission from Annual Reviews; C, from E.S. Rajaraman and T. Izard, *Nature Struct. Mol. Biol.* 20:188–193, 2013. With permission from Macmillan Publishers Ltd; D, from M.E. Janssen et al., *Mol. Cell* 21:271–281, 2006. With permission from Elsevier.)

component of focal adhesions (see Section 11.6 and Figure 11.28B). The M domain (middle), comprising two four-helix bundles (see Figure 11.17C), binds afadin, another protein with affinity for F-actin. The C-terminal domain of α-catenin binds to F-actin. This domain is similar in sequence and structure to the vinculin C-terminal domain, which forms a five-helix bundle and has been shown by cryo-EM to contact two successive actin subunits along the filament (Figure 11.17D). It is plausible that α-catenin binds to F-actin in a similar fashion. Thus, αE-catenin binds to both β-catenin and F-actin, as well as to other actin-binders such as vinculin and afadin (see above). Binding to β-catenin weakens its affinity for F-actin by sterically impeding that interaction. The binding of αE-catenin to vinculin is mechanosensitive; in other words, it is promoted by forces that relieve autoinhibitory intramolecular interactions such that the helical bundle of the vinculin-binding domain unfolds, exposing one or more binding sites for vinculin. Thus it appears that α-catenin can adopt a variety of conformations that modulate its affinities for different junctional components.

In addition to participating in cadherin–actin linkage, α-catenin has been implicated in regulation of the actin cytoskeleton. Initial cell-to-cell contacts involve interactions between cadherins presented on *lamellipodia*; as a contact matures, membrane movement abates. Lamellipodia are based on branched actin networks whose formation requires the Arp2/3 complex (Section 14.9; see Figures 14.67 and 14.68). It was found that αE-catenin can inhibit Arp2/3 activity *in vitro*. To explain the change in membrane dynamics upon cell contact formation, it has been proposed that cadherin accumulation leads to the recruitment of α-catenin; then a high local concentration of α-catenin inhibits actin polymerization, leading to cessation of lamellipodial motion.

Desmosomes are coupled to the intermediate filament network

Desmosomal and classical cadherins are generally similar except that the cytoplasmic tails of desmogleins, at 360–484 aa, are much longer. They include a sequence homologous to the β-catenin-binding region of classical cadherins that interacts with plakoglobin (see Figure 11.14B). Plakoglobin is the only protein common to desmosomes and adherens junctions, and its presence in adherens junctions is thought to lead to the later establishment of desmosomes. Desmocollin tails are shorter, ~180 aa, but also contain a sequence homologous to the catenin-binding region. Desmosomes also contain *plakophilins*, which are p120 homologs whose ARM-repeat domain structure is known but whose binding partners and functions are not yet clear.

Desmoplakin belongs to the plakin family of *cytolinkers* that connect IFs to other structures (see Figure 11.2); specifically, it is thought to connect the desmosomal cadherin–plakoglobin plaque to IFs. Desmoplakin is a very large homodimeric protein with a 1056-aa N-terminal domain (DPNT) followed by a 900-aa coiled-coil-forming region and a C-terminal (DPCT) region of 950 aa (Figure 11.8A). DPCT contains three subdomains, each composed of 4.5 copies of a 38-aa **plakin repeat** motif (see Figure 11.8C) that binds directly to IFs. It is reported to bind to plakoglobin and, possibly, plakophilins. The core of DPNT contains six spectrin-like repeats, one of which has an inserted SH3 domain (see Figure 11.8B and **Guide**). *Spectrin* was first discovered in erythrocyte envelopes, where it co-assembles with actin to form a cytoskeletal web underlying the plasma membrane. It contains a series of three-helix bundle repeats that can unfold locally in response to stress, imparting elasticity to the web. Given the importance of spectrin for the viscoelastic properties of erythrocytes (Box 1.3), it may be that the DPNT domain confers similar mechanical properties to the cadherin–IF linkage.

11.5 GAP JUNCTIONS

Gap junctions are double-layer arrays of transmembrane channels that connect the cytosols of two interacting cells, allowing them to exchange water-soluble molecules, up to ~1 kDa. In electron micrographs of transverse thin sections, gap junctions have a characteristic hep-talaminar appearance (**Figure 11.18B**). Each membrane is a trilaminar component with its two hydrophilic surfaces represented by dark-staining layers separated by a hydrophobic light-staining region. The two triplets are separated by a seventh light-staining layer, the 'gap.' Gap junction channels cluster in hexagonal arrays called **plaques** (Figure 11.18A,C).

Gap junctions have several essential roles in multicellular organisms. They have been impli-cated in embryonic development (mice knocked out for the gene encoding the major gap junction protein in the heart, Cx43, die soon after birth), in tissue homeostasis, and in controlled cell growth and differentiation. Mutations in genes coding for gap junction pro-teins have been linked to diseases that include oculodentodigital dysplasia, a developmental disease whose symptoms can include small eyes, underdeveloped teeth, and arrhythmias; X-linked Charcot–Marie–Tooth syndrome, a neuropathy; and several skin diseases. In tis-sues where gap junction channels have a critical homeostatic role, such as the uptake of K^+ ions after an action potential has impinged, impairment of intercellular communication leads to tissue atrophy, resulting in macroscopic symptoms such as deafness and cataracts.

Gap junction channels are composed of paired hexameric rings (connexons)

Gap junctions are built from one or more members of the **connexin** protein family, which are designated by Cx followed by the subunit mass in kilodaltons, for example Cx26. A different nomenclature is used for their genes, based on the classification of the protein sequences into three groups (α, β, and γ) and numbered according to their order of dis-covery; for example, the gene *GJβ2* codes for the second β-connexin discovered (which codes for human Cx26). The human genome encodes 21 connexins, and the mouse genome

Figure 11.18 Gap junction channels pack in hexagonal arrays. (A) Freeze-fracture electron micrograph of a rat liver gap junction. The specimen was immuno-gold-labeled with anti-Cx32 antibodies before the surfaces exposed by fracturing were shadowed with heavy metal. The asterisk marks the area enlarged × 3 in the inset, which illustrates the close packing of particles (channels) and intervening areas of lipid bilayer. (B) Electron micrograph of a transverse thin section through a liver gap junction, showing the canonical heptalaminar appearance. (C) Electron micrograph of a negatively stained gap junction plaque isolated from HeLa cells exogenously expressing Cx26. The inset area, enlarged ×3, shows its regular hexagonal lattice. The more ordered packing of channels compared with gap junctions *in situ* (A) results from lipid extraction by the detergents used in the isolation procedure. (A, courtesy of John Rash; B, courtesy of Daniel Goodenough; C, from A. Hand et al., *J. Mol. Biol.* 315:587–600, 2002. With permission from Elsevier.)

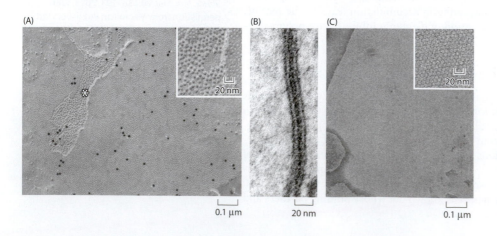

(A)

(B)

(C)

20 nm

20 nm

0.1 μm

20 nm

0.1 μm

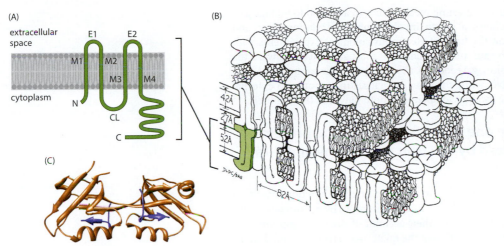

Figure 11.19 Organization of connexins in gap junctions.
(A) Transmembrane topology of a connexin subunit with a long C-terminal cytoplasmic region. (B) Model of a gap junction. (C) Crystal structure showing dimers of the second PDZ domain of ZO-1 complexed with the C-terminal nine residues of Cx43 in which the latter peptide (blue) adopts a β-strand conformation in an intermolecular interaction that involves *domain swapping* (cf. Figure 11.15B). (PDB 3CCY). (B, original drawing by D.L.D. Caspar and D. Goodenough, published in *J. Cell Biol.* 74:605–628, 1977; C, adapted from J. Chen et al., *EMBO J.* 27:2113–2123, 2008. With permission from John Wiley & Sons.)

encodes 20. Each connexin has four transmembrane helices, two extracellular loops, a cytoplasmic loop (CL), and cytoplasmic terminal regions (**Figure 11.19A**).

A gap junction channel consists of two connexin hexamers, stacked back-to-back (Figure 11.19B). The hexamers are called **connexons** or hemichannels. Connexons pair via their extracellular loops, which form a tight seal such that a continuous aqueous channel runs along their common axis. The extracellular loops are important in cell–cell recognition and the formation of channels. Because the sequences of the extracellular loops are conserved, connexons containing different connexins are able to dock together to form heterotypic channels. The cytoplasmic regions influence the channel's gating properties. The C-terminal regions, which account for most of the variation in connexin subunit masses, contain motifs that interact with cytoplasmic proteins (for example Figure 11.19C), but are mostly unfolded in the absence of such an interaction.

Whether unpaired connexons have a definite function remains controversial. Newly synthesized connexins are transported from internal membranes (Golgi or ER) to the plasma membrane in the form of connexons, which are probably closed. When they enter the plasma membrane, connexons are still dispersed and are consequently not detectable by currently available light or electron microscopic methods. The number of open connexons in a plasma membrane must be low to minimize the leakage of extracellular molecules or ions into the cell or the loss of needed molecules or ions from the cell. However, some connexins form nonjunctional connexons that open to allow the passage of measurable currents when mechanical stress is applied to the plasma membrane. Furthermore, stimuli such as external ATP release or Ca^{2+} wave propagation can trigger the opening of closed connexons during signaling cascades.

Gap junctions allow a variety of molecules to pass between communicating cells

The channels exercise some selectivity according to size, charge, or shape, a property known as **permselectivity**. Each connexin has a characteristic profile; for example, Cx32 is anion-selective whereas Cx43 is cation-selective. Permselectivity is reflected in differing diffusion rates measured for similar-sized molecules, such as fluorescent Alexa dyes or nucleotides. The differences in selectivity between connexins have been attributed primarily to sequence variations in their C-terminal regions (see below).

Connexons are usually homomeric, although in some tissues, such as eye lens, heteromers have been demonstrated. Channels are called **heterotypic** if they comprise two different connexons, and **homotypic** otherwise. In heterotypic channels, connexons usually pair with members of the same class (for example Cx43/Cx40, which are α–α channels). The ability of connexins to form heterotypic channels has led to the hypothesis that cells exploit this property to fine-tune their physiological properties.

Gap junction channels fulfill different functions, depending on the tissue; for instance, they can form ion channels (for example Cx43 in heart) or metabolic channels (for example

Cx32 in liver). They also employ a variety of gating mechanisms: in addition to voltage gating, channel closure is influenced by chemical gating, involving phosphorylation, Ca^{2+}, and pH. It has been inferred, primarily on the basis of electrophysiological experiments with mutated connexins, that there are two gates per hemichannel: a fast gate located toward the cytoplasmic surface (V_j gate) and a slow gate located near the extracellular surface ('loop gate'). The fast gate is voltage-sensitive, while the slow gate closes in response to external stimuli such as Ca^{2+}. Similar gates are thought to operate in both halves of the channel.

Gap junction channels were one of the first membrane protein complexes to be characterized by electron microscopy and X-ray diffraction

In 1977, Don Caspar and colleagues published their rendition of a gap junction, based on an analysis of X-ray diffraction from stacked plaques isolated from rat liver, correlated with EM images of the same plaques (Figure 11.19B). Their account of the Cx32 gap junction remains remarkably accurate today in its depiction of the quaternary structure and dimensions of the connexons and lipid bilayers. In 1999, electron crystallography of an exogenously expressed Cx43 construct that formed large well-ordered plaques confirmed the presence of four transmembrane helices per connexin subunit. Recently, electron crystallography of a Cx26 mutant revealed a protein density within the pore vestibule (arrow in **Figure 11.20B**) and identified it as part of the N terminus (NT).

In 2009, an X-ray crystal structure was reported for the detergent-solubilized Cx26 channel (Figure 11.20A,C–F). It showed the following features: first, how the four transmembrane helices are connected, with M1 and M2 forming the major and minor pore-lining helices; second, that the cytoplasmic entrance is positively charged, while the transmembrane pore is negatively charged; and third, that the NT resides within the pore, making specific contacts with M1. It should be noted that almost all modifications to the NT—additions, deletions, and substitutions—do not affect the assembly and trafficking of channels but instead render them nonfunctional; that is, not supporting dye transfer or electrical activity. These observations, combined with electrophysiological measurements of site-directed mutants, suggested that the NT may act as a voltage sensor. The current hypothesis that the NT acts to keep the channel open by specific interactions within the pore explains why modifications

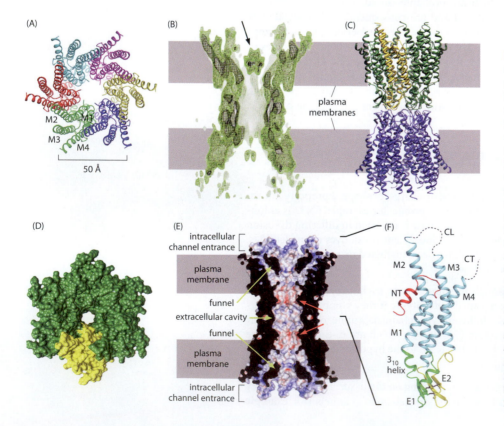

Figure 11.20 Structure of gap junction channels. (A, C–F) X-ray crystal structure of the Cx26 channel (PDB 2ZW3). (A) Top view, showing the arrangement of transmembrane helices of one connexon, M1–M4 labeled in the green subunit. (D) Top view showing the cytoplasmic surface of a connexon in space-filling representation. (B) Side-view of the Cx26 channel from an electron crystallographic study, showing a density (arrow) in the axial pore of the upper connexon. (C) Side-view of the two apposed connexons; one subunit is in yellow. (E) Cutaway side-view of the pore with surface patches colored for negative (red) and positive (blue) charges. (F) Fold of Cx26 with various motifs marked (see Figure 11.19A). M1 and M2 are the pore-lining helices (see (A)); NT is the N-terminal helix. Segments not visualized and presumably disordered are shown as dashes (A, C–F, adapted from S. Maeda et al., *Nature* 458:597–602, 2009. With permission from Macmillan Publishers Ltd; courtesy of Shoji Maeda and Tomitake Tsukihara; B, courtesy of Atsunori Oshima, Gina Sosinsky, and Yoshi Fujiyoshi.)

to the NT render the channel nonfunctional. This structure also revealed that the extracellular loops contain three short β strands; the conserved cysteine residues that form three disulfide bonds all lie in these strands.

As the cytoplasmic loop (CL) and C-terminal region (CT) are invisible in the Cx26 crystal structure, they were inferred to be disordered. NMR (see **Methods**) studies of the corresponding peptides found the CL and CT of Cx43 (60 and 151 aa respectively) and the CT of Cx40 (131 aa) to be mostly disordered. Resonances associated with the disordered regions shifted in response to phosphorylation, low pH, or interactions with other proteins, indicative of conformational changes. The last (C-terminal) nine residues of Cx43, disordered according to NMR, have a β-strand-containing conformation in a crystal structure of this peptide complexed with a domain of ZO-1 (see Figure 11.19C). Thus it appears that this and possibly also other parts of the CT become ordered when they engage binding partners, a recurring theme with junctional proteins.

Connexon conformations are influenced by changes in [Ca²⁺], pH, phosphorylation state, and binding partner

In topographic imaging by atomic force microscopy (AFM) (see **Methods**), repeated measurements may be made on the same sample. If a sample changes conformation in response to a treatment, the change can be observed directly, as in the opening of Cx26 hemichannels in the absence of Ca^{2+} and their closure in its presence (5 mM). Hemichannels register a similar response to changes in pH (open at acidic pH, closed at neutral pH) (**Figure 11.21**). Acid-induced closure of gap junction channels has been observed by activity assays in tissue culture systems and attributed a functional role in tissues; however, it is not known whether acidification affects the connexin itself or other proteins or compounds that act to close the channel. For Cx26, taurine (2-aminoethane sulfonic acid, an organic acid found in bile and in the cytoplasms of many cells) has a role in the acid-induced closing of hemichannels.

Cells modulate gap junctional communication by regulating the synthesis, transport, gating, and turnover of channels. Cx43, the most widely expressed connexin, is synthesized and trafficked through the ER and Golgi (**Figure 11.22**), assembling into connexons in the *trans* Golgi network. The connexons are then transported in vesicles to the plasma membrane, where they pair with apposing connexons and are recruited into plaques. Plaques undergo constant turnover whereby newly synthesized channels are added in at the periphery, as shown by correlative microscopy in which the same region of a specimen was imaged by light microscopy and subsequently by EM (**Figure 11.23**). Old channels, in contrast, are degraded from the center of the plaque (see Figure 11.22A). Cx43 has a half-life of 3–6 hours, such a short turnover time being typical of connexins. The sole exception is gap junctions in eye lens, where cells do not turn over once the lens is mature and connexin half-life is several days.

The binding of cytosolic proteins can affect both plaque size and channel properties, including gating and trafficking. Cx43 can interact with the tight-junction-associated proteins, ZO-1 and ZO-2 (Section 11.3), and with N-cadherin and β-catenins (Section 11.4). The fusion protein Cx43-GFP (GFP, green fluorescent protein) does not bind ZO-1 and forms abnormally large gap junctions, but it can be rescued from this state by co-expressed wild-type Cx43. It has been proposed that ZO-1 controls the rate of channel accretion at the periphery of Cx43 gap junctions by serving as a scaffold that tethers connexons and other

Figure 11.21 Gap junction dynamics visualized by atomic force microscopy.
(A) Topogram of the cytoplasmic surface of a Cx26 plaque. The labile cytoplasmic domains are displaced by the AFM cantilever with very small applied forces as in regions that appear void (such as in the dashed circle). The inset is an averaged topogram, enlarged ×2.8. Protein peaks (bright yellow) extend 1.7 nm above the lipid bilayer (dark). (B, C) Topograms of the extracellular surface at low pH (B) and neutral pH (C). Cx26 connexons close in response to low pH by a rotation of the connexin subunits that constricts the pore at the extracellular surface. To obtain these images, single-layer arrays of connexons were produced by AFM force dissection of plaques and the exposed extracellular surfaces were imaged. (A, from D.J. Müller et al., *EMBO J.* 21:3598–607, 2002. With permission from John Wiley & Sons; B and C, from J. Yu et al., *J. Biol. Chem.* 282:8895–8904, 2007. With permission from the American Society for Biochemistry and Molecular Biology.)

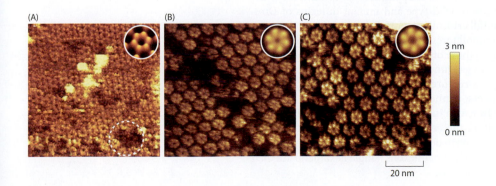

(A)   (B)   (C)

3 nm

0 nm

20 nm

(A)

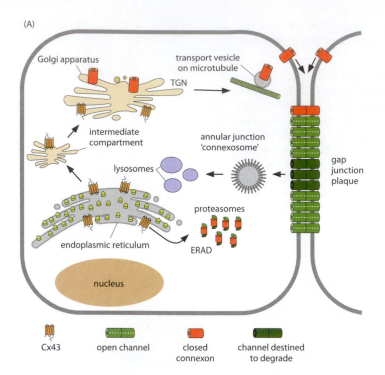

(B)

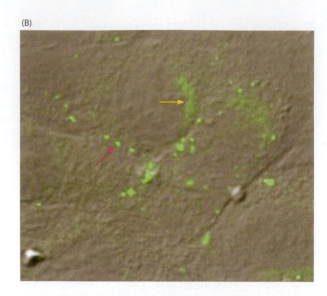

Figure 11.22 Synthesis and assembly of gap junctions. (A) Diagram illustrating the life cycle of connexin Cx43. The connexosome is an aggregate of senescent channels en route to degradation in lysosomes. TGN, *trans* Golgi network; ERAD, endoplasmic reticulum-associated protein degradation (see Box 7.4). (B) Normal rat kidney cells shown by differential interference contrast light microscopy, overlaid with green immunofluorescence labeling of Cx43. The cells have punctate-appearing gap junctions (an example is indicated by the pink arrow) at their periphery and a Golgi-resident pool of connexons (an example is indicated by the yellow arrow), en route to the plasma membrane. (A, adapted from a figure by D. Laird; B, courtesy of Daniela Boassa and Gina Sosinsky.)

proteins, thus regulating plaque size. The second of the three PDZ domains of ZO-1 binds to the CT of Cx43 (see Figure 11.19C). In fact, most proteins with PDZ domains are found at specialized membrane regions such as synapses, junctions, or apical–basolateral interfaces.

Most connexins contain multiple sites for phosphorylation, and their phosphorylation influences important biological processes, such as cell migration and proliferation. Cx43, for example, has a highly regulated life cycle, in which phosphorylation occurs at specific sites in its CT. These events take place during all stages of the cell cycle and they can affect any of the following: the proteins that interact with Cx43; the kinetics and/or localization of Cx43 trafficking; and channel assembly, gating, and turnover. The CT of Cx43 has 21 serine residues, and 12 of them, as well as two tyrosine residues, are phosphorylated by various kinases; for example, MAP kinase phosphorylates Ser255, Ser279, and Ser282, altering the gating properties of the channel. Key to understanding gap junction regulation are the questions of how phosphorylation events turn channels on and off, and how they may act as signals, flags, or gatekeepers for channel activity.

The connexin-like superfamily includes innexins, pannexins, and vinnexins

Sequence analysis has led to the designation of three additional protein families as 'connexin-like': **innexins** (invertebrate connexin analogs); **pannexins** (vertebrate analogs of innexins, pan from the ancient Greek word for 'all'); and **vinnexins** (viral connexin analogs, identified in genomes of an insect virus family). Little is known about vinnexins, except that they form functional channels in standard gap junction coupling assays. Innexins are widespread among invertebrates. Characterization of wild-type and mutant innexins of *Drosophila* and *C. elegans* has shown that they fulfill critical functions in gap junctions. Innexins have been localized to many tissues by immunofluorescence and they display developmental spatio-temporal relationships analogous to those of connexins. However, although there is an extensive literature of EM observations of gap junction-like structures in invertebrates, few innexins have yet been localized to these structures by immunogold labeling. In mammals, the pannexin family has three members: pannexin-1, -2, and -3 (Panx1–3), which share 50–60% sequence identity but only ~16% with innexins or connexins. Panx1 has wide tissue expression, Panx2 is expressed solely in the central nervous system, and Panx3 is found primarily in skin and spleen. Pannexins were originally envisaged to form gap junction-like channels, but recent evidence indicates that pannexons (single hexamers) are their only functional entity, acting as ATP release channels.

(A)

(B)

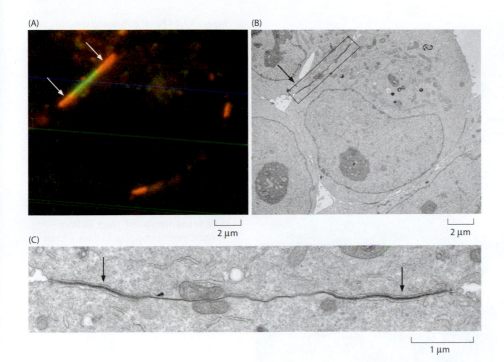

2 µm

2 µm

(C)

1 µm

Figure 11.23 Newly synthesized Cx43 gap junction channels are incorporated at the plaque periphery. A Cx43 construct with a tetra-Cys motif at its C terminus was transfected into HeLa cells and expressed. The cells were then perfused with a derivative of the dye fluorescein, which has a high affinity for the tetra-Cys tag and becomes brightly fluorescent upon binding it (green in (A), which shows a confocal image of a labeled gap junction plaque). After the unbound dye was flushed out, another tetra-Cys-binding dye that generates red fluorescence was introduced. (B) For EM visualization, electron-dense reaction products were formed by the photoconversion of diaminobenzidine and they contrast labeled regions, as shown here in electron micrographs of thin sections. (C) Higher magnification of the boxed region in (B), showing staining of the peripheral portions of the junction (arrows). (From G. Gaietta et al., *Science* 296:503–507, 2002. With permission from AAAS.)

11.6 FOCAL ADHESIONS

Focal adhesions (FAs) were first observed in about 1970 in electron micrographs of cultured mammalian cells, as localized contact regions between adherent cells and flat inorganic surfaces ('substrata') (**Figure 11.24A,B**). These contacts were found to involve elongated structures 2–10 µm long and 0.25–0.5 µm in thickness, with a separation of less than 15 nm to the substratum as measured by interference-reflection microscopy (IRM)—much closer than the typical cell–substratum spacing of 100 nm. At these sites, cytoplasmic plaques appeared to connect to actin filaments. Recent observations by correlative fluorescence microscopy with cryo-electron tomography have confirmed that FAs engage a dense meshwork of actin filaments (Figure 11.24C,D).

Do these structures found in cultured cells have counterparts *in vivo*? Typically, a *metazoan* cell interacts with other cells both by direct contacts and via the extracellular matrix (ECM; Section 11.7). Unlike substrata, ECMs are three-dimensional and have complex physical and chemical properties. Nevertheless, the contacts that cells make with the ECM *in vivo* are thought to closely resemble the FAs formed by cultured cells, which are easier to manipulate. Many studies have employed substrata coated with defined matrix components and recent research has turned to three-dimensional matrices that mimic physiological environments more closely. Nevertheless, there are likely to be subtle differences in composition, architecture, and signaling responses between FAs assembled *in vitro* and those assembled *in vivo*, which probably possess additional tissue-dependent components.

FAs have two distinct functions: an adhesive function that provides a robust physical link between the ECM or substratum and the actin cytoskeleton, and a signaling function that transmits information about the cell's surroundings to its interior. In this way, FAs contribute to several essential activities, including migration, tissue development, *homeostasis*, and even survival.

Focal adhesions are large assemblies with complex protein compositions

Attempts to define the molecular composition of FAs by conventional biochemical approaches met with limited success, for two reasons: first, because FAs partition into the insoluble phase in cell fractionation studies; and second, because their compositions vary dynamically, as demonstrated by time-resolved microscopy. In fact, microscopy has been

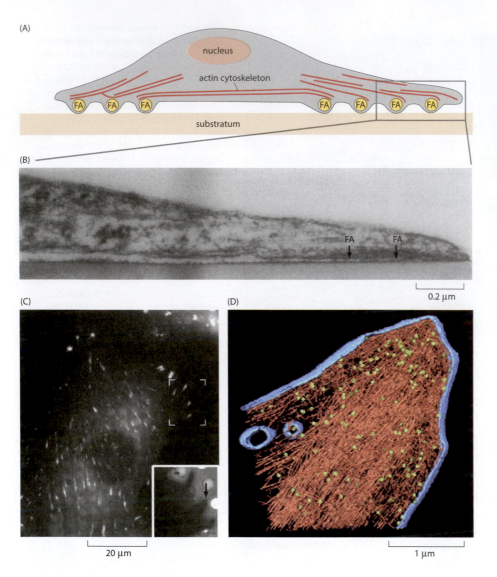

Figure 11.24 Focal adhesions (FAs) mediate close contact between cells and substrata. (A) Diagram of an adherent mammalian cell. (B) Electron micrograph of a transverse thin section of the leading edge of a migrating chick heart fibroblast. (C, D) Correlative microscopy of adherent rat embryo fibroblasts using light microscopy with fluorescently labeled paxillin to locate FAs (C), followed by cryo-electron tomography (D) of an FA-containing area (arrowed in panel (C) inset). The tomogram shows that FAs are rich in actin filaments (tan) and that there are many donut-shaped particles (green), as yet unidentified, close to the membrane (blue). (B, from M. Abercrombie et al., *Exp. Cell Res.* 67:359–367, 1971. With permission from Elsevier; C and D, from I. Patla et al., *Nature Cell Biol.* 12:909–915, 2010. With permission from Macmillan Publishers Ltd.)

central to their characterization. The first FA components were identified by combining the ability to visualize adhesion sites by IRM with immunofluorescent light microscopy. This approach showed that FAs contain, in addition to actin, the cytoskeletal protein α-actinin (Section 14.5), vinculin (see below), and the Src kinase. In turn, these markers allowed other FA proteins to be identified by immunofluorescent co-localization. So far more than 150 FA proteins have been identified, of which ~90 are stably associated intrinsic components and ~60 are transiently associated peripheral components. Some 690 pairwise interactions have been identified or inferred among these components, giving rise to a complex interaction network known as the **adhesome**. However, FAs vary not only between cell types but also within a given cell. In early stages of adhesion, small dynamic **focal complexes** form at the cell edge and develop over time into mature FAs, which are larger and more stable.

In view of the great diversity of FA proteins, it is helpful to group them according to function (**Figure 11.25**). At the core of FAs are adhesion receptors that interact with the ECM or substratum and transmit signals across the plasma membrane. The cytoplasmic domains of activated receptors then recruit actin binders/regulators and signaling molecules to adhesion sites either directly or via adaptor proteins.

Focal adhesion assembly starts with integrin activation and is orchestrated by adaptor proteins

The first step in FA formation is the engagement of adhesion receptors. The primary receptors are **integrins**, transmembrane proteins comprising one α-subunit and one β-subunit. Both subunits contribute to divalent cation-dependent binding to extracellular ligands

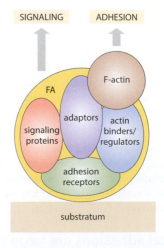

Figure 11.25 Focal adhesions contain several functional classes of proteins. Members of these classes cooperate in a hierarchical manner to mediate adhesion and to transmit adhesion-dependent signals into the cell.

such as the ECM proteins, *fibronectin* and *collagen*. Bound integrins transmit signals into the cell in a process called **outside-in signaling**. In a reverse process called **inside-out signaling**, integrin-mediated adhesion can be modulated by changes in the interactions of their tails with intracellular activators. Integrins undergo large conformational changes in response to signals. Evidence from EM (**Figure 11.26A**) and X-ray crystallography (see **Methods**) (Section 11.7) suggests that the ectodomains can adopt three conformational states: bent, extended, and extended with open headpiece. These are believed to correspond to three functional states: 'low affinity' (inactive), 'high affinity' (active), and 'ligand bound.' FRET measurements have shown that whereas the bent inactive state has closely associated cytoplasmic tails, both inside-out and outside-in signaling induce their separation (Figure 11.26B). Tail separation could promote signaling by exposing binding sites for effector molecules, such as *talin* (Section 11.7). Clustering of integrins at adhesion sites in response to a high density of binding sites on the ECM promotes FA assembly.

By monitoring FA proteins fused to fluorescent proteins, assembly has been followed in real time in living cells; these observations imply that FA assembly is a complex process involving the sequential recruitment of its components. All steps depend on specific protein/protein interactions. Most FA proteins have several domains. While some domains have enzymatic functions, others serve solely to provide binding sites; many FA proteins consist entirely of protein-interaction domains and have been named **adaptors** or scaffolding proteins. These domains are interspersed with *natively disordered* linker regions, as illustrated for the adaptors, Grb2 and paxillin in **Figure 11.27A**. Paxillin has an N-terminal region of more than 300 aa, containing five LD (Leu-Asp) motifs, a polyproline (PP) motif, and two phosphotyrosine (pY) motifs; the N-terminal region precedes four LIM domains (see **Guide**). The structures of three interaction domains with bound motifs are shown in Figure 11.27B. SH2 domains recognize pY-containing motifs; SH3 domains bind to PP peptides that adopt a polyproline-II helix conformation when bound. The α-helical LD motifs of paxillin interact with distinct domains of different proteins. One important binding partner is the C-terminal CH (Calponin Homology) domain of the adaptor α-parvin shown in Figure 11.27B. Although the linkers are not long (10–70 aa), their disordered nature allows them to extend into the cell and present short motifs to which cytoplasmic proteins may bind. These and other such interactions afford a basis for adhesome networks.

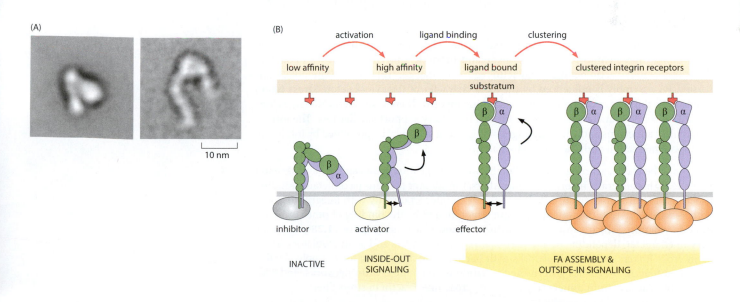

Figure 11.26 Mechanisms of integrin signaling. (A) Bent and extended conformations observed in averaged EM images of negatively stained integrin $\alpha_V\beta_3$ ectodomains. (B) Inactive integrins in the 'low-affinity' bent conformation are activated by binding activator proteins such as talin (see Figure 11.28), causing them to switch into a 'high-affinity' conformation that promotes ligand recognition ('inside-out signaling'). Conversely, engagement with ligands in the ECM leads to extension of the ectodomains and separation of cytoplasmic tails in the 'ligand-bound' conformation, and to integrin clustering ('outside-in signaling'). Conformational changes and integrin clustering are thought to contribute to outside-in signaling through cytoplasmic effectors (namely adaptors, actin binders/regulators, and signaling proteins). (A, from J. Takagi et al., *Cell* 110:599–611, 2002. With permission from Elsevier; B, diagram by M. Hoellerer, loosely based on J.A. Askari et al. *J. Cell Sci.* 122:165–170, 2009.)

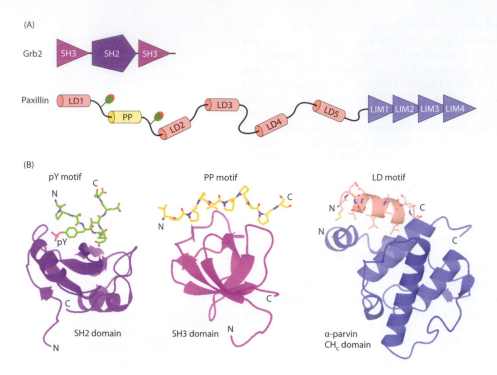

Figure 11.27 Conserved protein modules mediate interactions in FAs. (A) Domain maps of the adaptor proteins Grb2 and paxillin. LD motifs are leucine-rich, beginning with Leu-Asp (LD in the one-letter code). The phosphotyrosine (pY) motifs are in green/orange. PP, polyproline. (B) Crystal structures are shown of the Grb2 SH2 domain complexed with a pY-containing peptide (PSpYVNV) (PDB 1JYR), the second SH3 domain of ponsin with the Pro-rich peptide (VPPPVPPPPS) of paxillin (PDB 2O9V), and the LD1 motif of paxillin bound to the C-terminal CH domain of α-parvin (PDB 2VZD).

The adaptors also participate in signaling by recruiting and juxtaposing key signaling molecules; in this role, they have been compared to molecular switchboards that connect and disconnect signaling pathways as required. Consistent with their importance in the adhesome, some adaptors (such as paxillin and Grb2) have been found to be essential in mice, in that knocking out their genes is lethal.

Binding of FAs to the actin cytoskeleton and actin polymerization are coupled processes

Signals from integrin receptors, transmitted through FAs, affect cytoskeletal architecture; in so doing, they help to regulate cell shape and motility. This is achieved with adaptors that connect to the actin network, and actin regulators that control filament assembly. Conversely, FA assembly is influenced by the local character of the cytoskeleton: **stress fibers** (contractile bundles of actin filaments containing myosin II) are associated with large elongated FAs, while branched networks at the cell periphery support smaller FAs. The formation of new adhesions at the leading edges of migrating cells may be prevented by inhibiting actin polymerization with the drug cytochalasin D.

For FAs to be stable, a strong physical link between integrin receptors and the cytoskeleton is needed. This generally involves, among others, the adaptor proteins paxilin, talin, kindlin, α-actinin, filamin, and tensin, which are all able to interact directly with integrin tails and with F-actin. The resulting interactions are strengthened by the binding of other proteins, notably vinculin, which can associate with talin and α-actinin (**Figure 11.28**), and they vary in character, depending on the participating actin binders and actin regulators. Thus, actin cross-linkers can create networks with particular architectures: for example, flexible V-shaped filamin molecules govern the formation of dynamic F-actin networks, and rod-shaped bipolar α-actinin dimers (Figure 14.23) cross-link F-actin in stress fibers.

Initiation of actin polymerization is tightly controlled by nucleating factors such as the formin mDia and Arp2/3 (Figure 14.67). Formins catalyze the *de novo* formation of unbranched filaments in stress fibers and *filopodia*, whereas Arp2/3 nucleates the outgrowth of branches from preexisting filaments in *lamellipodia*. Conversely, FAs control Arp2/3 activity by regulating the 'nucleation-promoting factors' cortactin, WASP, and WAVE. Growth of actin filaments is stimulated by FA-associated elongation factors, including formins and Ena/Vasp proteins, and local actin dynamics are regulated by the filament-severing protein cofilin and the G-actin-binding protein profilin (Section 14.2).

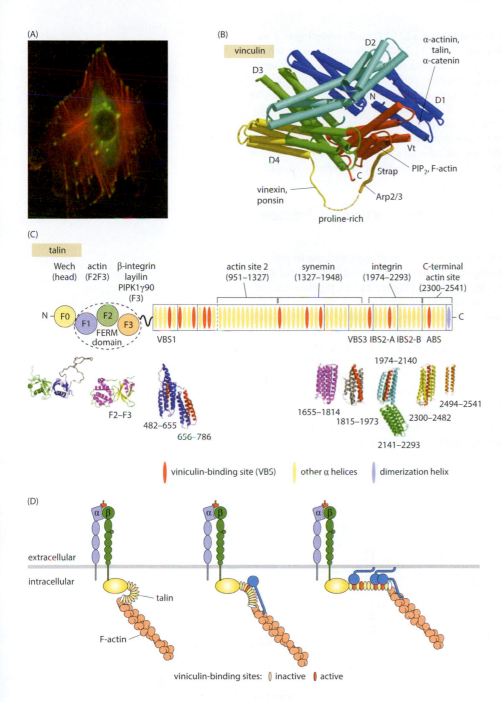

(A)

(B)

vinculin

D2

α-actinin, talin, α-catenin

D3

N

D1

Vt

D4

Strap   PIP₂, F-actin

vinexin, ponsin

C

Arp2/3

proline-rich

(C)

talin

Wech (head)   actin (F2F3)   β-integrin layilin PIPK1γ90 (F3)   actin site 2 (951–1327)   synemin (1327–1948)   integrin (1974–2293)   C-terminal actin site (2300–2541)

N — F0   F1 F2 F3   FERM domain   VBS1   VBS3 IBS2-A IBS2-B ABS   C

F2–F3

482–655

656–786

1655–1814   1974–2140   1815–1973   2300–2482   2141–2293   2494–2541

vinculin-binding site (VBS)   other α helices   dimerization helix

(D)

extracellular

intracellular

talin

F-actin

vinculin-binding sites: inactive active

Figure 11.28 FAs are coupled to the actin cytoskeleton. (A) Immunofluorescent image of a mouse fibroblast cell stained with antibodies against actin (red) and vinculin (green), showing coordinated localization (yellow). (B) Crystal structure of chicken vinculin in a closed conformation (PDB 1ST6). Binding sites for some FA components are marked. The 'strap' hinders access to a basic region involved in lipid PIP₂ binding and restrains the Pro-rich region, inhibiting the binding of vinexin, ponsin, and Arp2/3. (C) Domain map of talin (2541 aa), with binding sites for other proteins marked. VBS, vinculin-binding site; IBS, integrin-binding site; ABS, actin-binding site. Crystal structures have been determined for F2/F3 and for several helical bundles in the 'tail,' which comprises 62 α helices, represented by ovals. Vinculin-binding helices are red and the C-terminal dimerization helix is gray. (D) Schematic diagram showing talin connecting an integrin to F-actin, and its opening up as more vinculin molecules bind. This model invokes the C-terminal actin-binding site of talin. Talin is a dimer but is shown as a monomer for clarity. (A, courtesy of Tina Izard; B, from C. Bakolitsa et al., *Nature* 430:583–586, 2004. With permission from Macmillan Publishers Ltd.)

The binding sites of FA adaptors are exposed in 'active' conformations and sequestered in 'inactive' conformations

This property has been demonstrated for vinculin and talin and is probably widespread among FA proteins. **Vinculin** is a 116 kDa monomeric protein that associates with actin through its binding to talin. A crystal structure of full-length vinculin (see Figure 11.28B) revealed an auto-inhibited 'closed' conformation in which the C-terminal tail domain (Vt, red) is held pincer-like by the head (Vh, domains D1–D4). These domains are all α-helical bundles. Interactions between the head and tail regulate the binding of partner proteins (some sites are marked). For example, the interaction of Vt with Vh blocks its interaction with F-actin (a case of steric inhibition), and contacts between Vt and D1 prevent the conformational changes in D1 (a reorganization of the helices) that are required for talin and α-actinin binding (a case of allosteric inhibition).

Talin has a 50 kDa FERM domain head (see **Guide**) and a 225 kDa tail consisting of a succession of five-helix and four-helix bundles (see Figure 11.28C). It terminates in an α helix that forms a coiled coil with the same motif in a second subunit. Talin has many binding sites for other FA components, including integrins, actin, and vinculin. The vinculin sites map to particular α helices present in many of the bundles, and these sites must be exposed to be active. In FAs, vinculin does not bind integrins directly but via talin (see Figure 11.28D) or α-actinin, which binds to the cytoplasmic tail of the integrin β-subunit. As more vinculin molecules bind, talin opens up. Among the factors that influence talin's conformation is mechanical stress. In a plausible model, talin attaches by its head to an integrin cytoplasmic domain and by the tip of its tail to F-actin, and opens up in response to applied stress, exposing progressively more vinculin-binding sites.

Small GTPases coordinate the assembly of the actin cytoskeleton and FAs

The machinery responsible for actin assembly must be modulated appropriately to produce such diverse actin-based structures as lamellipodia, filopodia, and stress fibers. Three small GTPases of the Rho family, namely Cdc42, Rac, and Rho (**Figure 11.29**), regulate the activation of proteins upstream of F-actin nucleation and thus coordinate actin assembly. The GTPases cycle between inactive GDP-bound conformations and active GTP-bound conformations, in which they bind and thereby activate downstream effectors. The equilibrium between these two forms is governed by activating guanine nucleotide-exchange factors (GEFs) and inactivating GTPase-activating proteins (GAPs). Additional regulation comes from their subcellular localization; Rho GTPases are prenylated at their C termini and cycle between membranes, where they are active, and the cytoplasm, where they are sequestered in an inactive state by the inhibitor RhoGDI.

FAs signal to the GTPases through the GEFs and GAPs. Proteins with GEF or GAP activity are recruited to FAs through adaptors such as paxillin and are regulated by FA-associated kinases. Local activation of GTPases leads to the activation of signaling molecules,

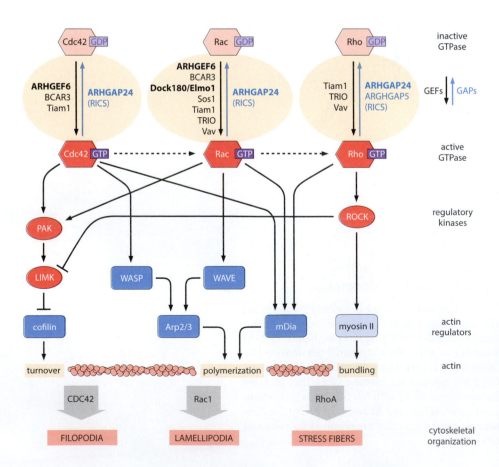

Figure 11.29 Small GTPases regulate actin organization. Activity of the Rho family GTPases Cdc42, Rac1, and RhoA is locally regulated by GEFs and GAPs that are stably (names in bold) or transiently (other names) associated with FAs. Active GTPases use signaling pathways involving protein and lipid kinases to coordinate the activity of regulators that control the formation of various actin-based structures.

including protein Ser/Thr kinases, the actin nucleator mDia, and lipid kinases that give rise to signaling lipids such as PIP_2. Many of the signaling pathways target actin regulators, thereby controlling its polymerization. The three GTPases have differing effects on actin regulators: thus Cdc42 promotes filopodia formation, Rac1 stimulates membrane ruffling and lamellipodia protrusion, and RhoA activity leads to the assembly of stress fibers. The antagonistic effects of Cdc42/Rac1, which stimulate protrusion, and RhoA, which enhances contractility, help to coordinate the actin cytoskeleton during cell migration.

Protein phosphorylation has a central role in FA regulation and signaling

A striking characteristic of FAs is their accumulation of phosphorylated proteins in response to integrin stimulation (Figure 11.30). Several Tyr kinases phosphorylate FA proteins, a process that can be reversed by FA-associated phosphatases. Regulated cycles of phosphorylation and dephosphorylation accompany the assembly and disassembly of FAs, and inhibition of Tyr kinases blocks the formation of FAs and the assembly of stress fibers. Two Tyr kinases, FAK and Src, are at the heart of integrin-mediated signaling pathways. Both are recruited to the cytoplasmic tails of ligand-bound integrins whose clustering brings the kinases close together, enabling them to activate through *trans*-autophosphorylation. Once activated, Src and FAK phosphorylate a plethora of FA proteins including adhesion receptors, adaptors, actin binders/regulators, GTPase regulators, protein kinases, and lipid kinases.

One of the commonest interaction domains in FA proteins is SH2, which, as noted above, binds to short pY-containing motifs (see Figure 11.27B). Tyrosine phosphorylation promotes FA assembly by switching on SH2-mediated interactions, and dephosphorylation switches them off, leading to disassembly. Controlling protein associations also influences downstream signaling through certain effector molecules. Thus, phosphorylation controls the recruitment and activation of GAPs and GEFs, which regulate Rho family GTPases. FAK and Src govern the organization of the actin cytoskeleton, both indirectly by controlling actin regulators via the GTPases, and directly through the phosphorylation of actin-binding proteins and actin regulators.

Focal adhesions are signaling centers

FAs also serve as sensors that detect and transmit information about a cell's surroundings. First, through their adhesion receptors, they register which parts of the cell are in contact with the ECM. Second, they sense the chemical composition of the ECM. Cell type-specific expression of adhesion receptors dictates the particular ECM components to which a given cell may bind, restricting it to certain environments. Third, FAs can sense if the cell is under mechanical stress.

FAs integrate this information and muster a cellular response by modulating their rates of assembly and turnover, and by remodeling the actin cytoskeleton. However, their signaling function extends much further. Integrin signaling through FAs is essential for the survival of many cell types in which loss of adhesion to the ECM triggers a special form of *apoptosis* (Section 13.6) called **anoikis**. Most cells proliferate only when they can form appropriate ECM contacts, as a result of cross-talk from integrin signaling to pathways that regulate cell cycle progression. Recently it has been shown that integrin-mediated tension helps to define the plane across which a cell divides. Thus, cell/matrix interactions are communicated to the mechanical process of cell division, with important consequences for tissue architecture. Loss of the adhesion dependence of cell survival, proliferation, and deregulation of cell adhesion and motility can have devastating consequences by enabling the growth of tumor cells and promoting metastasis.

Figure 11.30 FA signaling is mediated by phosphorylation. Tyr-phosphorylated proteins accumulate in FAs. Immunofluorescent microscopic localization of pY in an attached human foreskin fibroblast highlights FAs around the cell periphery. (From B. Geiger et al., *Nature Rev. Mol. Biol.* 2:793–805, 2001. With permission from Macmillan Publishers Ltd.)

11.7   THE EXTRACELLULAR MATRIX

The ECM is a complex assembly of glycoproteins and polysaccharides that mediates cell-to-cell interactions and confers distinctive physical properties on tissues. Its components are secreted and assembled into networks by interactions that include covalent cross-linking. The ECM evolved in conjunction with multicellular animals, and its basic features have been conserved although the number of proteins and the complexity of their assemblies have grown. Most ECM proteins have elaborate domain organizations and are multifunctional. Many are oligomeric. Most, if not all, are incorporated into networks with varying mechanical properties; some are elastic, whereas others are rigid with high tensile strength. These networks can be highly ordered (collagen fibrils), flexible (elastic or fibronectin fibrils), complex (basement membranes), and dynamic (all). Tissues with prominent ECMs include tendon, cartilage, skin, bone, and blood vessels. ECM networks are macromolecular machines in the sense that they embody defined sets of macromolecules coordinated in specific (albeit stochastically variable) spatial relationships and acting in concert.

ECM provides a load-bearing scaffold that supports and anchors cells and segregates tissues but also has more sophisticated functions. The ECM is linked to cell surfaces by receptors that connect it to intracellular cytoskeletal structures, mediating the bidirectional transduction of mechanical forces and modulating signaling pathways. Interactions of a cell with its ECM provide spatial cues for migration and modulate differentiation status and stem cell properties, among other functions. *Growth factors* can be activated or sequestered by binding to ECM components and in this way can be presented to cells in a prescribed temporal or spatial order. Throughout life, the ECM undergoes constant remodeling by extracellular proteases.

The ECM is not compartmentalized by membranes and so ions can migrate by diffusion. However, most ECM components carry permanent charges. This leads to nonuniform distributions of ions by *Donnan effects*, resulting in osmotic pressures. Osmotic swelling is important for the elasticity of cartilage, in which the sulfated and carboxylated glycosaminoglycan (GAG) chains of proteoglycans and hyaluronan provide an excess of negative charges and bind water molecules.

Calcium and other ions have important roles in stabilizing the ECM and in promoting its assembly. Many ECM proteins bind Ca^{2+} ions and are even denatured in their absence. ECM proteins also nucleate the deposition of calcium phosphate and calcium carbonate in teeth and bone. The concentration of free Ca^{2+} ions in the ECM is about six orders of magnitude higher than inside cells, and almost equal to the plasma level of 2 mM.

ECM is built from a diverse assortment of fibrous proteins, glycoproteins, and proteoglycans

Seven of the many hundred ECM components of vertebrates are diagrammed in **Figure 11.31A**. Tenascin C has six subunits, linked by coiled-coil domains and disulfide bonds. Fibronectin has two disulfide-linked subunits. Fibrillin is a monomer. In laminin, three different subunits are connected via the triple coiled coil in its long arm. Nidogen, α-dystroglycan and integrin $\alpha_5\beta_1$ are prominent binding partners. Thrombospondin 1 is a trimer unified by a triple coiled coil. Perlecan and aggregan are **proteoglycans**, which are ECM proteins with one or more O-linked GAG side-chains. Aggregan interacts with hyaluronan a non-proteinaceous GAG produced at the cell surface.

Except for fibronectin, each of these proteins represents a large multi-gene family: for example, in humans, there are 16 isoforms of laminin, and *splice variants* (see Figure 1.3) generate additional complexity. Most ECM proteins are highly elongated, consisting of multiple copies of the same or different domains, which allows them to engage in multivalent interactions. The domain organizations of these and other ECM components may be viewed in the SMART or CDART databases in which they are also linked to three-dimensional structures. A CDART map for fibronectin is shown in Figure 11.31B: it has 12 fn1, 2 fn2, and 15–17 fn3 domains. Some of them expose the Arg-Gly-Asp (RGD) motif (for example, Figure 11.31E), which is bound by integrin receptors at the cell surface. Many ECM proteins are glycosylated.

Interactions between domains in the same polypeptide can alter its overall conformation and functional status. These interactions can be between adjacent domains, but can also

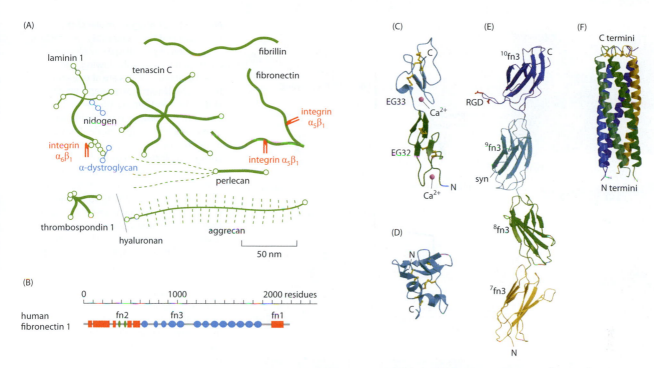

Figure 11.31 Filamentous organization of ECM glycoproteins and proteoglycans. (A) Drawings of a few major ECM components, approximately to scale. ECM proteins typically have many domains arranged like beads on a string and represented here by thick curved lines (except for the long arm of laminin, which is a three-stranded coiled coil), larger globular domains by spheres, and GAGs by dashed lines. (B) CDART map for fibronectin: type 1 domains (fn1) are in red, fn2 domains in green, and fn3 domains in blue. (C–F) Some ECM domain structures. (C) A pair of Ca^{2+}-binding EGF-like domains from fibrillin 1. (D) The sixth TB domain of fibrillin 1. (E) Four tandem fn3 domains from fibronectin 1. The RGD binding site for integrin $\alpha_5\beta_1$ resides in an exposed loop and its binding is enhanced by a synergy site (syn). (F) The five-chain coiled coil of cartilage oligomeric matrix protein (COMP), also called thrombospondin 5 (TSP-5). (From E. Hohenester and J. Engel, *Matrix Biol.* 21:115–128, 2002. With permission from Elsevier.)

involve domains that are widely separated in the primary sequence. Crystal structures have been determined for many ECM domains; a few examples are shown in Figure 11.31C–F.

Collagens all have triple-helical domains but assemble into diverse higher-order structures

Collagens are the quintessential ECM proteins. The 28 collagens identified in mammals are denoted by roman numerals, in order of discovery. They are the most abundant proteins of the ECM—and, indeed, of entire organisms: collagens I, II, and III account for ~80% of total protein in many mammalian tissues. A collagen molecule has three α chains (here meaning polypeptide chains), which can be homomeric or heteromeric. Their hallmark is the presence of a 'triple-helical' domain in which all three α chains have multiple repeats of the tripeptide Gly-X-Y, where X is frequently Pro and Y is often 4(*R*)-hydroxyPro. The Gly-X-Y repeats of each chain are folded into a left-handed polyproline-II helix with three residues per turn. Three α chains coil into a right-handed triple helix with 10 or 10.5 residues per turn (**Figure 11.32C,D**). The chains are staggered by one residue per chain, giving rise to two isoforms with arrangements 1–2–3 and 1–3–2. The small side-chain of Gly fits into the narrow space between α chains, and the X and Y side-chains point radially outward. Local clusters of these side-chains provide sites for interactions with receptors, matrix proteases, and other protein.

Triple-helical domains can be as long as 300 nm (in collagen I; Figure 11.32A) or as short as 14 nm (in the minicollagen of *Hydra*). Non-helical domains have important roles in oligomerization and in initiating fibril assembly. For example, the globular C-terminal domains of collagen I (visible at one end of the molecules in Figure 11.32A) serve to align three α chains in parallel and in register to nucleate formation of the triple helix; and similarly for collagens II, III, and V.

The complex processes of collagen biosynthesis and fibrillation are mediated by a large retinue of chaperones and other proteins, including glycosyltransferases, isomerases, and

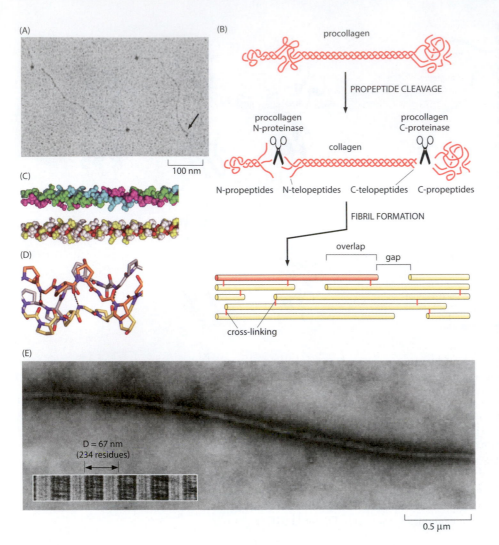

(A)

100 nm

(B)

procollagen

PROPEPTIDE CLEAVAGE

procollagen
N-proteinase

procollagen
C-proteinase

collagen

N-propeptides   N-telopeptides   C-telopeptides   C-propeptides

FIBRIL FORMATION

overlap

gap

cross-linking

(C)

(D)

(E)

D = 67 nm
(234 residues)

0.5 μm

Figure 11.32 Collagen: molecular structure and fibril assembly. (A) Electron micrograph of procollagen I molecules contrasted by rotary shadowing. The C-terminal propeptides are at one end and the smaller N-terminal propeptides diverge at the other end (arrow). (B) The pathway on which procollagen is processed and assembled into collagen fibrils. The fibril diagram shows the staggered packing of collagen molecules. In some collagens, the telopeptides engage in intermolecular cross-linking. (C) Crystal structure of human collagen III G991–G1032. (PDB 3DMW) It is colored (top) by chain (green, turquoise, magenta) and (bottom) by amino acid residue (red, Gly; yellow, X; white, Y) to show the differing exposures of their side-chains. (D) A triple-helical segment with the sequence (Gly-X-Y), where X and Y are both Pro, stabilized by H-bonds between the NH groups of Gly and the CO group of the X(Pro) in an adjacent chain. The chains are offset in one-residue steps: thus, Gly in the yellow chain is at the same level as X(Pro) of the gray chain and Y(Pro) of the red chain. (E) Electron micrograph of an *in vitro*-assembled collagen I fibril, negatively stained. Inset: enlargement showing the banding pattern of D-repeats. (B, adapted from T. Starborg et al., *Methods Cell Biol.* 88:319–345, 2008. With permission from Elsevier; C, courtesy of Jürgen Engel, adapted from S.P. Boudko et al., *J. Biol. Chem.* 283:32580–32589, 2008.)

propeptidases. The assembly pathway of fibrillar collagen is outlined in Figure 11.32B. The terminal domains of collagens I, II, III, and V are removed by propeptidases after assembly of the triple helix. The cleavage sites depend on sequences in all three chains; misalignment caused by deletions in certain genetic diseases impairs their processing, resulting in severe disturbance of the collagen fibrils. In other collagens, the non-helical domains are retained in the assembled state. Disruption of the assembly of the triple helix by mutations in either the helical or non-helical domains leads to osteogenesis imperfecta, in which patients have brittle bones that deform or fracture easily.

Different collagens assemble into higher-order structures with distinct morphological and mechanical properties. In fibrils of collagen I (see Figure 11.32B,E), the triple-helical domains pack laterally with an axial stagger of 67 nm between neighboring molecules (the so-called D-repeat). Each molecule contributes to four overlap regions (D-repeats), followed by a 40 nm gap region. The gap is often occupied by small proteoglycans such as decorin. In EM of fibrils stained with heavy metals, the D-repeat shows a banding pattern that corresponds to positive staining of charged amino acids. Quite differently, collagen VI forms thin beaded filaments (**Figure 11.33**). The beads (so-called large and small segments) consist mainly of multiple von Willebrand factor A (vWA) domains and the connecting linkers consist of triple-helical domains arranged as two anti-parallel dimers.

The high tensile strength of collagen fibrils (comparable to that of steel) stems from weak but very numerous electrostatic and hydrophobic interactions in the long overlap regions. In some tissues, fibrils are further stabilized by covalent cross-links initiated by lysyl oxidase. In tendon, there are micrometer-thick bundles of fibrils of collagens I, III, and V; in bone, the fibrils are mainly collagen I; blood vessel walls contain collagen III; and articular cartilage has very thin fibrils of collagen II. Collagen V is distributed throughout the body.

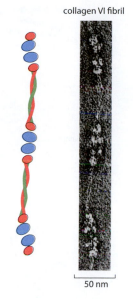

collagen VI fibril

50 nm

Figure 11.33 Beaded filaments of collagen VI. Right: thin slab from an electron tomogram of a negatively stained filament of collagen VI, depicting large and small beads mediating the interaction between tetramers. In the interpretive diagram (left), the small segments (red) are composed of 24 C-terminal vWA domains and the large segments (blue) of 40 N-terminal vWA domains; the packings of the vWA domains have yet to be determined. (From C. Baldock et al., *J. Mol. Biol.* 330:297–307, 2003. With permission from Elsevier.)

Fibril architecture depends not only on the collagen isoform but also on binding partners within the ECM. For example, decorin and biglycan are proteoglycans that bind to the surface of fibrils and regulate assembly. Tenascin X regulates the packing of fibrillar collagen in the dermis by binding to decorin.

Fibrillins constitute a major fibrillar system in many connective tissues

Fibrillins are large (~350 kDa; 140 nm long) flexible glycoproteins made up of 43 EGF-like domains interspersed with other domains, including seven TB domains, homologous to TGFβ-binding domains (**Figure 11.34A**; EGF (TGF) is epidermal (transforming) growth factor). Fibrillin is secreted from cells as profibrillin and cleaved by furin proteases into molecules competent to fibrillize. Multimerization proceeds through interactions of the C-terminal domains, which coalesce into a globule, and engage the N-terminal regions of other multimers, which remain extended, to form beaded fibrils 10–12 nm wide (Figure 11.34B). Mature microfibrils contain additional components accreted onto the fibrillin scaffold, some of which are indicated schematically in Figure 11.34A.

In addition to their structural role, fibrillin microfibrils regulate the activity of the growth factors TGFβ-1 and TGFβ-2 and members of the bone morphogenetic protein (BMP) family. These proteins are secreted from cells as latent propeptides; the N-terminal region of fibrillin provides a high-affinity binding site that holds them in the ECM in an inactive state. Fragments of fibrillin compete to release the growth factors from microfibrils, leading to their activation. Active TGFβ or BMP bind to cell surface receptors, thereby activating signaling pathways that regulate gene expression. In this way they control cell proliferation, the secretion of ECM components, and tissue patterning during development. In the absence of fibrillin, TGFβ activation and signaling increase, with pathological consequences that include tissue fibrosis, which involves the excessive deposition of ECM.

Fibrillin microfibrils also serve as the template for the deposition of tropoelastin, the soluble form of elastin—a major component of connective tissues (**Figure 11.35**). **Elastin**, a 65 kDa protein of low sequence complexity, rich in Gly, Ala, and Pro, was one of the first *natively unfolded* molecules (Section 6.4) to be recognized as such: it assembles into fibrils through

(A)

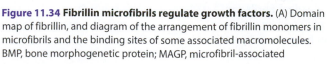

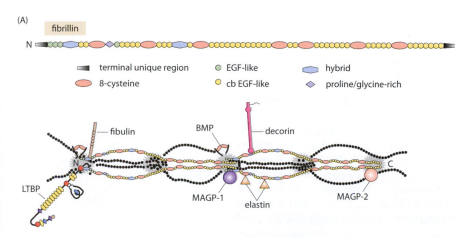

(B)

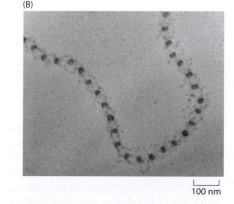

100 nm

Figure 11.34 Fibrillin microfibrils regulate growth factors. (A) Domain map of fibrillin, and diagram of the arrangement of fibrillin monomers in microfibrils and the binding sites of some associated macromolecules. BMP, bone morphogenetic protein; MAGP, microfibril-associated glycoprotein; LTBP, latent TGF-binding protein; cb, calcium-binding. (B) Electron micrograph of a rotary-shadowed microfibril. (Adapted from F. Ramirez and L.Y. Sakai, *Cell Tissue Res.* 339:71–82, 2010. With kind permission from Springer Science and Business Media.)

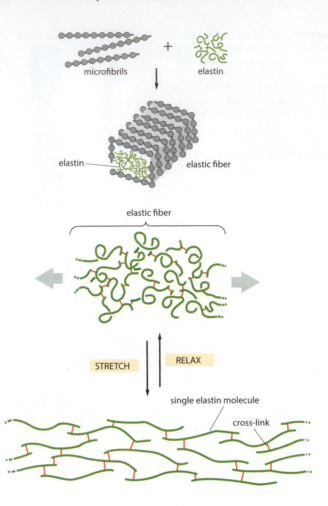

Figure 11.35 Fibrillin-templated assembly of elastic fibrils. Covalent cross-links (marked in red) within this 'elastic fiber' confer its characteristic recoil properties. In this model, elastin molecules in random coil conformations extend and contract when the fibril is subjected to and relieved from applied stress. (Adapted from B. Alberts et al., Molecular Biology of the Cell, 6th ed. New York: Garland Science, 2015.)

its association with fibrillin microfibrils. The resulting elastin matrix, cross-linked by lysyl oxidase, forms a dense, basket-like, meshwork of elastic microfibrils. Its cross-linking, coupled with the reversible unfolding properties of elastin molecules, allows the ensemble to behave like a rubber band. These fibrils are prominent in smooth muscle layers, including blood vessel walls where their elastic recoil properties allow a more uniform flow of blood and decrease the load on the heart. Elastin is found only in vertebrates and is thought to have been important for the evolution of their closed, high-pressure, blood circulation systems. Genetic defects in fibrillin result in Marfan's syndrome, a condition in which patients have shortened lifespans as a result of weakening and rupture of blood vessels.

Fibronectin evolved later than other ECM proteins and is specifically a component of vertebrate ECM

Fibronectin (FN) is a 440 kDa dimeric glycoprotein with domains that interact with integrins, collagens, tenascin C, and proteoglycans distributed along both subunits (see Figure 11.31A,B). When the dimer is secreted from cells, the two subunits are folded compactly, apparently through intramolecular interactions between fn3 repeats 2–3 and fn3 repeats 12–14 and other binding sites (colored red in **Figure 11.36A**). The primary cell surface interaction is through an RGD site that binds integrin $\alpha_5\beta_1$. Tensile forces generated within the cell by the contractile actin cytoskeleton and transmitted via the active integrin cause the FN dimer to open and extend on the cell surface (Figure 11.36B). In this way, sites within the N-terminal region become exposed, leading to the lateral assembly of dimers and fibrillogenesis (Figure 11.36C). Cross-linking mediated by the enzyme transglutaminase then stabilizes the assembly, creating a flexible meshwork in which individual fibrils can extend and retract. These transitions may reflect either the force-induced unfolding of fn3-type domains (see Figure 11.31E) or the switching of whole molecules between compact and extended forms.

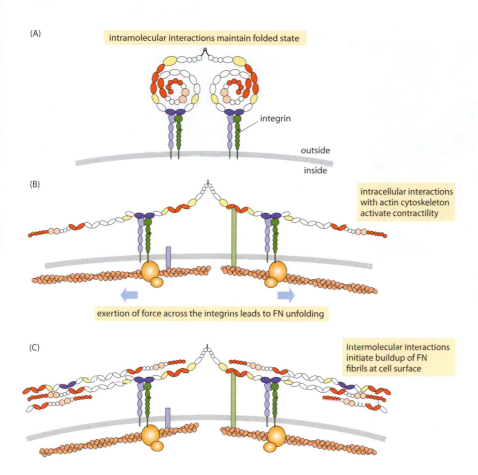

(A) intramolecular interactions maintain folded state

integrin

outside

inside

(B) intracellular interactions with actin cytoskeleton activate contractility

exertion of force across the integrins leads to FN unfolding

(C) intermolecular interactions initiate buildup of FN fibrils at cell surface

Figure 11.36 Assembly of fibronectin fibrils. (A) Soluble FN dimer binds to integrins via the RGD sites in their heads. FN subunits are dimerized via C-terminal disulfide bonds. (B) Binding of FN to integrins and other receptors (light blue and green bars), including syndecan-4, induces reorganization of the actin cytoskeleton and activates signaling complexes (gold balls). Cell contractility induces conformational changes in integrin-bound FN, exposing previously sequestered binding domains. (C) Fibrils form through FN/FN interactions and are stabilized by transglutaminase-mediated cross-linking. (Adapted from Y. Mao and J.E. Schwarzbauer, *Matrix Biol.* 24:389–399, 2005. With permission from Elsevier.)

FN is essential to mammals, and the FN fibril meshwork is a ubiquitous component of connective tissues. Soluble FN, which circulates in plasma at a concentration of ~300 μg/ml, is one of the primary components of the wound repair ECM that is laid down at sites of vascular injury.

Basement membranes, a specialized ECM, support epithelial and endothelial cell layers

Basement membranes are not *lipid bilayer* structures like true membranes; rather, they are 100 nm- thick layers of ECM associated with the basal surfaces of epithelial and endothelial cells (**Figure 11.37A**). They contain **laminins**, collagen IV, nidogen/entactin, and perlecan, among other components (Figure 11.37B). Extracellular deposition of laminin is essential for basement membrane assembly, beginning at the two-cell stage of development. The three chains (α, β, and γ) of this trimeric protein are linked by a triple coiled coil of ~60 nm in the 'long arm' (bold green in Figure 11.37C); it ends with five LG domains of the α chain, which provide the binding sites for $\alpha_6\beta_1$ integrin and α-dystroglycan. These interactions bind laminin directly to the cell surface. The short arms consist of many EGF-like domains, three of which provide the binding site for nidogen/entactin (see Figure 11.37C). At different stages of development and in different organs, basement membranes contain laminin isoforms with distinct integrin-binding capabilities. Sulfated lipids (sulfatides), which bind to laminin, are also essential for basement membrane formation. Numerous interactions between basement membrane components, some of which have been characterized *in vitro* with purified proteins (denoted by dashes in Figure 11.37B), produce the final complex network.

ECM assemblies are remodeled through the actions of extracellular proteases

Several groups of ECM proteases belong to the superfamily of zinc-dependent metalloproteinases (Box 7.1): the matrix metalloproteinases (MMPs); members of the a disintegrin

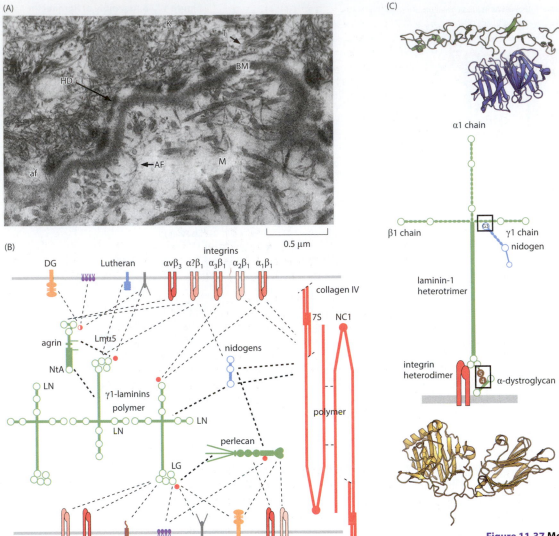

(A)

(C)

(B)

Figure 11.37 Molecular organization of basement membranes. (A) Electron micrograph of a thin section of a dermal–epidermal junction in human skin. BM, basement membrane; K, keratinocyte; HD, hemidesmosome; M, ECM with many sectioned collagen fibrils; AF, anchoring fibrils. (B) Diagram summarizing interactions (as dashes) between basement membrane components. It includes three molecules of laminin (LN), a key component. The gray horizontal lines indicate membranes. DG, dystroglycan; HNK1, human natural killer 1 (a carbohydrate); Lmα5, laminin chain α5; NC1, non-collagenous domain 1; NtA, N-terminal agrin. (C) Domain organization of a laminin. The crystal structure of a pair of LG domains (PDB 1H30) is shown below, and that of the nidogen/entactin complex (PDB 1NPE) is shown above (see boxes). (A, from R. Burgeson, in Molecular and Cellular Aspects of Basement Membranes. New York: Academic Press, 1993; B, from P.D. Yurchenco et al., *Matrix Biol.* 22:521–538, 2004. A and B, with permission from Elsevier; C, from T. Sasaki et al., *J. Cell Biol.* 164:959–963, 2004. With permission from Rockefeller University Press.)

and metalloproteinase domain (ADAM) family; and the ADAM domain with thrombospondin type 1 repeats (ADAMTS) subfamily. These enzymes activate extracellular signaling to cells (ADAM proteins) and process procollagens for fibril assembly (ADAMTS proteins). The proteases ADAMTS and MMP also release proteins or proteoglycans from ECM networks, allowing for changes in chemical and mechanical signaling to cells (see below), and alterations of the ECM that facilitate cell migration. These processes are important in normal development and tissue repair and also contribute to diseases; for example, ADAMTS-4 activity is implicated in osteoarthritis, and MMP activity may assist in tissue invasion by tumor cells.

MMPs include both membrane-anchored and secreted forms. Thus, ADAM proteins are transmembrane glycoproteins with proteolytic and adhesive activities at the cell surface, whereas ADAMTS proteins are secreted. All have multiple domains. Crystallographic studies of the catalytic domains of several ADAM and ADAMTS family members have revealed near-identical structures. An example of a membrane-anchored ADAM protein is shown in **Figure 11.38**; it is a 'sheddase' that processes membrane-bound substrates. Substrate specificity is thought to be conferred by ancillary domains, which may also localize the proteases within the ECM or at cell surfaces. For example, the Cys-rich domain of ADAM12 interacts with syndecans, whereas the disintegrin and Cys-rich domains of ADAM13 bind fibronectin and laminin. Under normal conditions, metalloproteinase activity is tightly regulated by the binding of extracellular inhibitors.

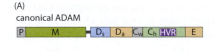

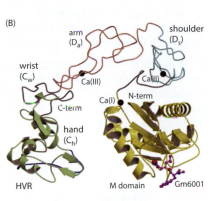

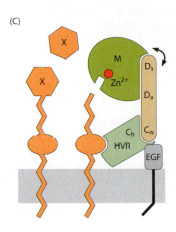

Figure 11.38 ADAM proteins are membrane-anchored metalloproteinases. (A) Domain map of a canonical ADAM protein. P, propeptide; M, metalloproteinase domain; D, disintegrin-like domain; HVR, hypervariable region; C, cysteine-rich; E, EGF-like domain. (B) The ectodomain of the catrocollastatin/VAP2B proteinase (PDB 2DW1). The same color-coding as in (A) is used. Bound Ca^{2+} is represented as black spheres, and Zn^{2+} (in the M domain) is in red. The hydroxamic inhibitor Gm6001 is bound to the M domain. (C) Model for 'shedding' by membrane-bound ADAM family proteins, in which a membrane-bound substrate molecule is recognized by the HVR and presented to the M domain for cleavage, which releases domain X. (From S. Takeda, *Semin. Cell Dev. Biol.* 20:146–152, 2009.)

Integrins are bidirectional signaling machines

The primary cellular receptors for ECM interactions are integrins, an extensive family of $\alpha\beta$ heterodimers with large extracellular heads and small cytoplasmic tails. The predominant integrin-binding sites on ECM proteins are RGD-containing motifs that lodge in a cleft between the α and β heads. However, some integrins have other specificities: for example, $\alpha_1\beta_1$ binds through the I domain of the α subunit (**Figure 11.39A**) to a site on the triple helix of collagen I. Different cell types express sets of integrins that enable them to bind particular combinations of ECM components.

Richard Hynes has proposed that integrins are signaling machines that switch between conformational states with differing affinities for ligands (see Section 11.6 and Figure 11.26B). In so doing, they communicate between the ECM and the cytoskeleton. In *inside-out signaling*, the integrin, initially in its bent inactive conformation, is activated by the binding of the FERM domain in the talin head (see Figure 11.28) to a phosphorylation site in the β-subunit tail (see Figure 11.39A). The cytoplasmic tails of the two subunits separate and the extracellular heads switch into an elongated state with the ligand-binding site open (Figure 11.39B). In *outside-in signaling*, ligand binding results in a 'swing-out' motion of the β_A, hybrid, and PSI domains (Figure 11.39C). The binding of multivalent ECM ligands and interactions between integrin tails and can both lead to clustering, which promotes the assembly of *focal adhesions* (Section 11.6).

Although integrins are the most important class of ECM receptors, they are not the only one. The transmembrane proteoglycan syndecan-4 acts as a co-receptor with integrins, through the interaction of its GAG chains with the heparin-binding domain of fibronectin. Collagens also bind to discoidin domain receptors, glycoprotein VI on platelets, leukocyte-associated Ig-like receptors, and members of the mannose receptor family.

In mechanotransduction, physical stimuli elicit biological responses such as gene expression

In the foregoing account we have noted several instances in which the application of mechanical forces cause conformational changes in proteins that alter their interactions with other cellular constituents, for instance in the remodeling of talin as it links the actin cytoskeleton to integrin receptors (see Figure 11.28D) and in the fibrillation of fibronectin (see Figure 11.36). These events are examples of **mechanotransduction** and they occur widely in ECM

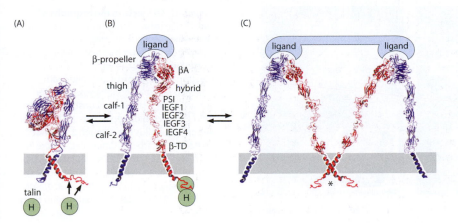

Figure 11.39 Mechanism of inside-out and outside-in signaling. This model is based on the crystal structures of integrins $\alpha_V\beta_3$ and $\alpha_{II}\beta_3$. The α and β subunits are in blue and red respectively. The integrin subdomains are denoted by name. (A) Inside-out signaling is initiated by binding of the FERM head domain (H) of talin (shown) or kindlin (not shown) to the cytoplasmic tail of the β subunit. This causes the two integrin tails to separate (B), accompanied by conformational changes in the ectodomains that render the integrin active, that is ligand-binding. IEGF, integrin epidermal growth factor domain; TD, terminal domain; PSI, plant-specific insert domain. (C) Integrin oligomerization leads to receptor clustering via the transmembrane domains (asterisk). In the absence of an extracellular ligand or tensile force, the integrin will revert to its bent form. (From B.S. Coller and S.J. Shattil, *Blood* 112:3011–3025, 2008. With permission from American Society for Hematology.)

systems. Among the earliest processes determined to be guided by mechanical forces were the formation and remodeling of bone. A dependence on mechanical load was first recognized in the healing of bone fractures. The structures under repair orientate to the gradient of the load, indicating that the bone-forming osteoblast cells sense the mechanical gradient. It is now recognized that essentially all ECM-producing cells are mechanosensitive and that the synthesis and spatial deposition of most ECM proteins depend on forces acting on the tissue, including gravity, tension, compression, and hydrostatic pressure.

ECM networks transmit vectorial mechanical signals between cells. It has been calculated that their transmission is orders of magnitude faster than any known chemical signaling mechanism. Mechanical coupling of the ECM with the cell nucleus (via IFs) is also envisaged. The identity of the putative mechanotransducing receptors is still under discussion, but integrins and stretchable ion channels are plausible candidates.

11.8 BACTERIAL PILI

Bacteria have extracellular protein filaments called **pili** (pilus = Latin for 'hair') or **fimbriae** (fimbria = Latin for 'thread'), **Figure 11.40A**. Pili participate in a wide range of adhesive functions, including the colonization of animal tissues, the formation of biofilms, congregation (microcolony formation), the intercellular transfer of DNA (**conjugation**), the uptake of DNA (**transformation**), and a form of locomotion called **twitching motility**. Pili also serve as the receptors for some bacteriophages (Figure 11.40B). There are many kinds of pili and they differ from the other extracellular filaments expressed by bacteria, the motility-generating flagella (Section 14.11), in composition and structure as well as function.

Pili were originally thought to be found exclusively on Gram-negative bacteria but it is now clear that some Gram-positive organisms have pili as well. (The differing cell envelope architectures of Gram-negative and Gram-positive cells, which have implications for how pili may be assembled and anchored, are described in Section 1.8 and Section 10.6). There is as yet no comprehensive classification system. Six kinds, called types 1–5 and F-pili (F for fertility) were distinguished by Charles Brinton in 1965, based on morphology and other properties, but use of this system is now confined to type 1 pili of *E. coli*, type 4 pili, and F-pili. Some names relate to functional properties; for example, P-pili are involved in pyelonephritic infections that result in inflammation of the kidney. Other names refer to bacterial strains. Here, we focus on a few pili that have been well characterized structurally: P-pili of *E. coli* and the related type 1 pili, which attach the bacteria to the bladder in urinary tract infections (**Figure 11.41**); and type 4 pili, an extensive class that includes the pili of *Neisseria gonorrhoeae* (often abbreviated to *Ngo*, or GC for *gonococcus*), the agent of a sexually transmitted disease.

The five types of pili are distinct from the seven types of bacterial secretion systems (I–VII) described in Section 10.6. Thus, for example, type 4 pili assemble via the type II secretion system; they are also called T4p or type IV pili. However, confusingly, type IV secretion systems also produce pili named T4SS pili, which are distinct from type 4 pili.

(A)

(B)

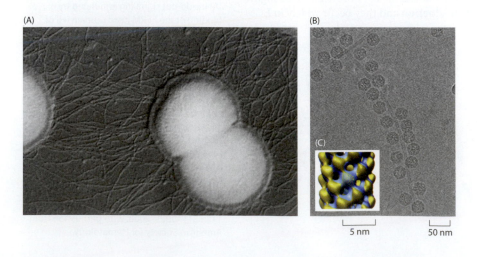

5 nm   50 nm

Figure 11.40 Electron micrographs of pili. (A) Electron micrograph of a *Neisseria gonorrhoeae* cell expressing type 4 pili, contrasted by heavy metal-shadowing. The bacterium is ~1 μm in width. (B) Cryo-electron micrograph of an F-pilus with attached MS2 bacteriophages. (C) Outer surface of an F-pilus, as reconstructed from cryo-EM. Each blob represents a 7 kDa pilin subunit. The pilus has a lumen 2 nm wide (not seen). (A, from L. Craig et al., *Nature Rev. Microbiol.* 2:363–378, 2004. With permission from Macmillan Publishers Ltd; B, courtesy of Neil Ranson; C, from Y.A. Wang et al., *J. Mol. Biol.* 385:22–29, 2009. With permission from Elsevier.)

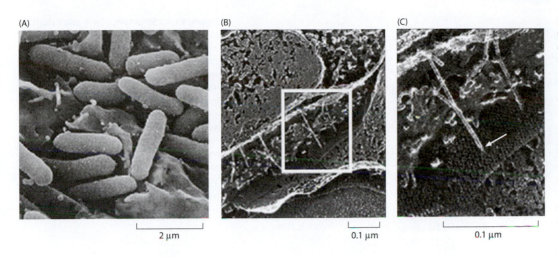

(A)

(B)

(C)

2 µm

0.1 µm

0.1 µm

Figure 11.41 *E. coli* adheres via type 1 pili to epithelial cells lining the bladder. (A) Scanning electron micrograph showing the bacilliform cells colonizing this tissue in mouse. (B) Electron micrographs of specimens prepared by freeze-fracture and contrasted by rotary shadowing. (C) Enlarged area of the area boxed in (B), showing pili extending from the bacterial surface and contacting (white arrow) an array of *uroplakins* on the host cell. (From M.A. Mulvey et al., *Science* 282:1494–1496, 1998. With permission from AAAS.)

The binding specificity of pili is conferred by adhesin proteins

The protein subunits that make up a pilus are called **pilins**. The filament shaft is generally a homopolymer of a pilin of 15–20 kDa, and is 5–10 nm in diameter and typically ~1 µm long. The **adhesin** is the component that determines binding specificity. In most cases it is located at the tip and may consist of a single protein or, as with type 1 pili and P-pili, an extended structure called a **fibrillum**, composed of several different proteins (**Figure 11.42A–C**). In other cases—for instance type 4 pili and F-pili—there are receptor-binding regions all along the shaft.

The type 1 adhesin binds to mannose sugars and that of P-pili to a tetrasaccharide containing galactose sugars. Insight into their attachment processes has been aided by crystal structures for these adhesins, which have two domains with Ig-like folds. The type 1 adhesin has a pocket that accommodates mono-mannose residues presented on **uroplakins**, membrane proteins that form hexagonal arrays on the surfaces of epithelial cells lining the urinary tract (see Figure 11.41C). The affinity of this adhesin for mannose exhibits sequence-dependent variations that correlate with the degree of uropathogenicity.

The fibrilla of these pili are elongated, segmented, structures (see Figure 11.42A–C). That of the P-pilus consists of the actual adhesin PapG, connected by an adaptor protein (PapF) to a strand of PapE subunits, then a second adaptor (PapK) that connects the fibrillum to the pilus shaft, composed of PapA subunits. The genes coding for Pap pilins are clustered in the *Pap* operon. Their type 1 homologs are expressed from the *Fim* operon and are called FimH, and so on.

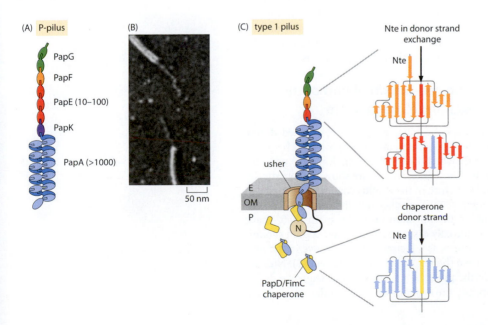

(A) P-pilus

PapG
PapF
PapE (10–100)
PapK
PapA (>1000)

(B)

50 nm

(C) type 1 pilus

Nte in donor strand exchange

Nte

usher

E
OM
P

N

chaperone donor strand

Nte

PapD/FimC chaperone

Figure 11.42 The fibrilla of P-pili and type 1 pili assemble via the chaperone/usher pathway. The pilin subunits are mapped in (A) and (C). P-pilins are denoted PapG, and so on; their homologs in type 1 pili, denoted FimH, and so on (Fim for fimbria), are shown in the same colors. The type 1 fibrillum lacks a counterpart to PapK. E, extracellular space; OM, outer membrane; P, periplasm. (B) Electron micrograph showing fibrilla extending at the tips of type 1 pili. The specimen was contrasted by rotary shadowing. (C) The donor strand exchange mechanism. At left, schematic model: N is the N-terminal domain of the usher. At right, topology maps of (lower) a chaperone-bound pilin subunit and (upper) two assembled pilin subunits. In the pilin/chaperone complex, the Nte of the chaperone (yellow) completes the fold of the pilin (blue) which is said to be donor strand-complemented (dsc). When the pilin is added into the growing pilus, the chaperone donor strand is replaced by the Nte of the preceding pilin subunit (donor strand exchange). Nte = N-terminal extension. (B, from M.J. Kuehn et al., *Nature* 356:252–255, 1992. With permission from Macmillan Publishers Ltd; C, adapted from H. Remaut et al., *Cell* 133:640–652, 2008. With permission from Elsevier.)

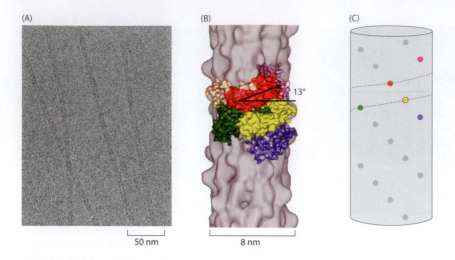

(A)

(B)

13°

(C)

50 nm

8 nm

Figure 11.43 Structure of a P-pilus shaft. (A) Cryo-electron micrograph of P-pili. The axial channel is faintly visible. (B) Three-dimensional reconstruction, with the crystal structure of the elongated pilin subunit (PapA, PDB 1PDK; see Figure 11.44A) fitted in at five locations, showing their packing and the orientations relative to the pilus axis. (C) The helical lattice: each point marks the position of a protein subunit in the front half. (A, courtesy of Esther Bullitt; B, from X.Q. Mu and E. Bullitt, *Proc. Natl Acad. Sci. USA* 103:9861–9866, 2006. With permission from National Academy of Sciences.)

The P-pilus shaft is a straight hollow rod 8 nm in diameter with a 2 nm axial channel. It is based on a single-start right-handed helix with 3.3 subunits per turn and a pitch of 2.3 nm (**Figure 11.43**).

Assembly of P-pili and type 1 pili follows the 'chaperone/usher' secretory pathway

The factors that govern the rate of pilin synthesis and the number and length of pili that a bacterium may deploy are not well understood. There are, however, indications that the degree of piliation reflects a response to the cell's environment: natural isolates tend to be more heavily piliated than laboratory strains passaged in liquid culture. The assembly and disassembly of pili are complex, highly regulated, processes.

P-pili and type 1 pili assemble via the **chaperone/usher pathway** (see Figure 11.42C). Pilin subunits are secreted through the inner membrane, then bound by a periplasmic chaperone until incorporated into growing pili. The pilus traverses the outer membrane via the 'usher,' whose translocation domain is a ring-like pore whose architecture resembles that of outer membrane porins. (**Porins** constitute an extensive family of bacterial outer membrane channels with transmembrane β-barrel folds: many are trimers.) The usher pore has a 24-stranded β barrel with a kidney-shaped cross section (appropriately, for P-pili!) (**Figure 11.44B**). In a crystal structure of the detergent-solubilized pore, its axial channel was blocked by an insertion domain called the 'plug' and a β hairpin capped with an α helix. In a subsequent analysis of the ternary usher/chaperone/pilin complex for the type 1 system, these elements were displaced and the pilin subunit (FimH) was visualized traversing the channel, which was considerably altered in shape relative to the plug-inserted conformation. The N-terminal and C-terminal domains of the usher protein (FimD) both reside on the periplasmic side of the channel, where they can interact with secretory proteins in transit.

Pilin subunits are incorporated at the cell-proximal end of the growing pilus by a 'donor strand exchange' mechanism

Assembly of type 1 pili and P-pili involves *domain swapping* in a process called **donor strand exchange**. Their pilins have incomplete Ig-like folds, lacking the seventh β strand. In the chaperone/pilin complex, the chaperone (which itself has two Ig domains) provides the missing strand, completing the pilin's fold. These interactions are shown schematically in Figure 11.43C, and in terms of the crystal structure of the P-pilus chaperone complexed with two pilin subunits in Figure 11.44A. Apart from the adhesin at the outer tip, all subunits in a given pilus—be they components of the fibrillum or of the main shaft—have N-terminal extensions (Ntes). These Ntes are initially disordered, but when a pilin subunit is incorporated into a growing pilus, its Nte inserts as a β strand into a groove in the preceding subunit, displacing that of the chaperone (hence 'donor strand exchange'). Ntes are conserved between species but not between the pilins of a given species. Sequence differences in Ntes and in the docking grooves specify the order in which the different pilins are incorporated.

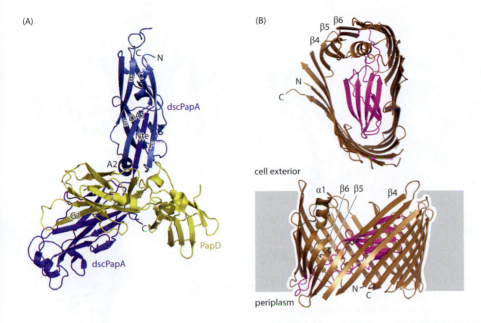

(A)

(B)

Figure 11.44 Complexes of the chaperone/usher pilus secretion pathway.
(A) Structure of the P-pilin/chaperone complex, illustrating donor strand exchange. The complex consists of two PapA subunits in differing conformational states (light blue and mid-blue; dsc stands for 'donor strand complemented') plus one copy of the PapD chaperone (yellow). The Ig fold of the lower PapA subunit is completed by insertion of the G1 strand of PapD, and its Nte is inserted into the corresponding groove on the upper PapA subunit. The Nte of the upper subunit is not seen and is presumably disordered. (PDB 2UY7) (B) Translocation domain of the PapC usher. The full-length protein is 809 aa; this construct, residues 135–640, encodes the β barrel that embeds in the outer membrane of *E. coli*; it was solubilized with detergent for crystallographic analysis. Top: axial view; bottom: side-view. Some β strands are labeled. (PDB 2VQI) (A, adapted from D. Verger et al., *PLoS Pathogens* 3:674–682, 2007; courtesy of Gabriel Waksman; B, from H. Remaut et al., *Cell* 133:640–652, 2008. With permission from Elsevier.)

The network of interactions generated by the donated strands enables the pilus to resist shear forces that might otherwise break it, detaching the bacterium from a colonized surface. Different reinforcement mechanisms are employed in other pili, as in a network of covalent cross-links (isopeptide bonds) between adjacent subunits in pili of the Gram-positive bacterium *Streptococcus pyogenes*.

The employment of proteins with related structures to form segments of a mixed filament recurs in the bacterial flagellum, but with different building blocks (Section 14.11). Another major difference between pili and flagella is that they grow at opposite ends: pilins are inserted at the proximal end, whereas flagellin subunits are secreted along the axial channel of the filament to be added at its distal tip.

Type 4 pilins have α-helical N-terminal extensions that pack to form the pilus backbone

Crystal structures have been determined for several type 4 pilins, including those of *Ngo* pili and the PAK pili of *Pseudomonas aeruginosa*. They resemble P-pilins in having an Nte protruding from a globular domain; however, in this case it is an α helix instead of a β strand (**Figure 11.45C**). The N-terminal half of the *Ngo* Nte is hydrophobic, and these motifs pack together to form the backbone of the pilus shaft (Figure 11.45D,E). Its C-terminal half forms part of the globular domain, which also has a conserved four-stranded β sheet and two variable elements, the αβ motif and the C-terminal D-region. The same architectural principle of bundling single α helices in a filament backbone recurs in bacterial flagella (Section 14.11), filamentous bacteriophages (Section 8.3), and type II bacterial secretion (Section 10.6).

X-ray fiber diffraction patterns of type 4 pili show a strong meridional reflection at a spacing of 0.51 nm, indicating that the α helices are aligned with the filament axis (Figure 11.45B,E). EM studies have revealed that there are four subunits per turn with a helical pitch of 4.1 nm in a filament 6 nm in diameter with an axial channel 1 nm wide. A model in which subunits with the crystal structure are packed according to this conserved helical geometry places the variable elements (the αβ motif and the D-region) strategically in externally exposed positions (see Figure 11.45D).

Phase variation allows the antigenic character of pili to change without altering their basic architecture

The exposed position of pili on the outside of the cell and the repetitive nature of their structures suggest that pili should be prominent antigens. This is indeed the case, and pili have been included in some acellular vaccines, for example against *Bordetella pertussis*, the agent of whooping cough. However, bacteria have evolved strategies for immune evasion;

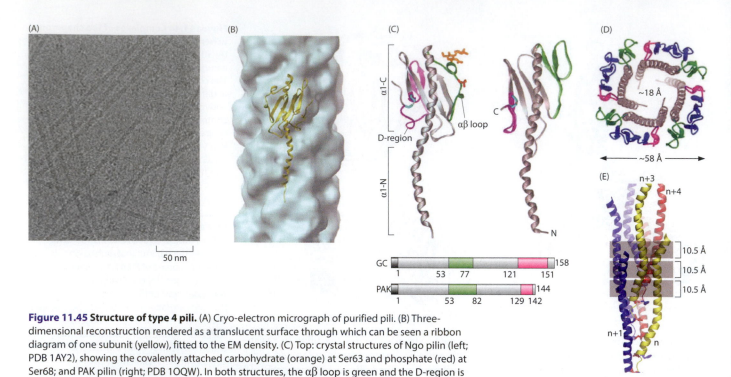

Figure 11.45 Structure of type 4 pili. (A) Cryo-electron micrograph of purified pili. (B) Three-dimensional reconstruction rendered as a translucent surface through which can be seen a ribbon diagram of one subunit (yellow), fitted to the EM density. (C) Top: crystal structures of Ngo pilin (left; PDB 1AY2), showing the covalently attached carbohydrate (orange) at Ser63 and phosphate (red) at Ser68; and PAK pilin (right; PDB 1OQW). In both structures, the αβ loop is green and the D-region is magenta. Bottom: domain maps marking the leader sequences (dark gray), the αβ loops (green) and D-regions (magenta). (D, E) The packing of Ntes in the pilus backbone: (D) axial view; (E) side view. (A, B, E, from L. Craig et al. *Mol. Cell* 23:651–662, 2006. With permission from Elsevier; C and D, from L. Craig et al., *Nature Rev. Microbiol.* 2:363–378, 2004. With permission from Macmillan Publishers Ltd.)

one, practised by *Ngo* and called **phase variation**, is based on horizontal gene transfer. The genome of this bacterium contains ~15 silent pilin genes (*PilS* loci), all lacking the sequence for the conserved Nte that is required for pilus assembly but coding for different versions of the variable regions that are exposed on the outer surface of the pilus. These cassettes may be inserted into the currently expressed full-length pilin gene by recombination; the resulting Nte-containing variant is selected when the bacterium comes under pressure from the adaptive immune system. The parts of the pilin that are affected are primarily the αβ motif and the D-region, including its 'immunodominant' loop (see Figure 11.45C). Combinatorial calculations based on the mutable sites estimate the total number of variants to be ~10^7. Meanwhile, the Nte that effects pilus assembly remains unchanged; in this way, the pilus retains its basic architecture and adhesive properties while switching its antigenic character.

Pilus retraction by motor-driven depolymerization is the mechanism for twitching motility

After removal of its signal sequence, the type 4 pilin is anchored in the inner membrane via its hydrophobic Nte. Thereafter, regulatory proteins, including a mechano-enzyme called PilT, feed them into the growing pilus. PilT belongs to the **traffic ATPase** family, part of the secretion superfamily of NTPases. Crystal structures have been determined for PilT from the hyperthermophile *Aquifex aeolicus*, which is likely to be similar to *Ngo* PilT (~50% identical in sequence). The monomer has an N-terminal PAS domain (see **Guide**), a RecA-like ATPase domain (see Box 3.1), and a C-terminal α-helical domain (**Figure 11.46A**). The protein has been visualized both as a symmetric hexamer and as a highly asymmetric hexamer (Figure 11.46B,C), on which basis it was proposed that ATPase-driven rigid-body subunit rotations between these states might provide the force-generating mechanism.

Thus, these cells have two populations of pilin subunits. One comprises pilins assembled into filaments. Subunits in the second pool are anchored in the inner membrane, where they await incorporation into filaments or to which they are returned from disassembling pili. In certain circumstances, pili embark on rapid shortening (disassembly), whereby whatever is attached to the pilus is hauled in toward the cell body as pilin subunits detach from the proximal end. In *twitching motility*, a bacterium attaches to a substrate by its pili tips, and

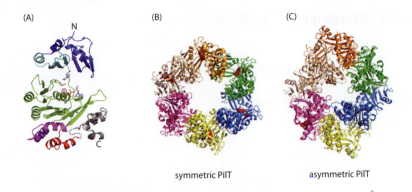

symmetric PilT   asymmetric PilT

Figure 11.46 Alternative conformations of a PilT traffic ATPase. (A) A single subunit whose N-terminal domain (NTD; blue) is connected by a linker to the core ATPase domain (green). ADP is bound to the core domain at the NTD-facing surface. At the C-terminal end, on the opposite side from the NTD, are several α helices, including the conserved helix J (red). Mutations in J abolish the ability of PilT to retract pili. (B, C) Symmetric and highly asymmetric hexamers. Bound ATP molecules are in red. The J helices lie on the underside of the hexamer, as viewed. It has not yet been possible to correlate these conformations with particular states of the hexamer in solution (symmetric hexamers in both ATP-bound and ADP-bound states have been observed). (PDB 2EWW & 2GSZ). (A, from K.A. Satyshur et al., *Structure* 15:363–376, 2007. With permission from Elsevier.)

then drags itself across that surface by pilus retraction. Similar mechanisms operate in bacterial conjugation to bring mating cells close together.

The mechanics of retraction have been studied for type 4 *Ngo* pili by *optical trapping* (see **Methods**) and light microscopy. Pilus retraction is rapid and processive, occurring at a rate of more than 1 μm/s (that is, ~1000 pilin subunits per second). Disassembly is effected by the PilT ATPase, operating in reverse mode relative to its action in pilus assembly. The forces that must be exerted by the optical tweezers to resist retraction have been measured at ~100 pN; if this force is generated by a single retracting pilus, it makes PilT one of the most powerful molecular motors on record.

11.9 SUMMARY

The ability of cells to interact with each other and their surroundings depends on a diverse group of externally exposed proteins that are either embedded in the plasma membrane or secreted. Many of them are large multidomain molecules. There is also is a widespread incidence of proteins with immunoglobulin-like folds. Within cells, intermediate filament (IF) proteins serve as important determinants of cellular architecture.

Direct cell-to-cell communication occurs at specialized regions of the plasma membrane called junctions, which contain receptor proteins that bind externally to neighboring cells and internally to cytoplasmic proteins. In many cases, the cytoplasmic domains are natively disordered until they engage a binding partner, and many are subject to regulatory phosphorylation. Types of junctions include:

- Tight junctions, which block the passage of materials through the space between adjacent cells and divide the membranes of polarized cells into distinct zones

- Adherens junctions and desmosomes, which are linked to the actin cytoskeleton and the IFs system respectively, and have key roles in coordinating cell–cell adhesion in epithelial tissues

- Gap junctions, which are gated transmembrane channels that connect the cytoplasms of two interacting cells and have roles in embryonic development, tissue homeostasis, and the control of cell growth and differentiation

- Focal adhesions, initially identified as localized contact regions between adherent cells in culture and flat inorganic surfaces (substrata), contain adaptors that connect to the actin network, and actin regulators that control filament assembly; the adaptors also recruit and juxtapose key signaling molecules.

The extracellular matrix (ECM) is an elaborate extracellular assembly of glycoproteins and polysaccharides that mediate cell/cell interactions via cell surface receptors that connect the ECM to intracellular cytoskeletal structures. In this way they modulate signaling pathways. The ECM also provides a load-bearing scaffold that supports and anchors cells and segregates tissues, and allows the bidirectional transduction of mechanical forces.

Bacteria have extracellular protein filaments called pili (or fimbriae) that participate in a wide range of adhesive functions, including the colonization of animal tissues, biofilm formation, microcolony formation, the intercellular transfer of DNA, DNA uptake, and a form of locomotion called twitching motility.

Sections of this chapter are based on the following contributions

11.4   William I. Weis, Stanford University School of Medicine, USA.

11.5   Gina E. Sosinsky, University of California, San Diego, USA.

11.6   Maria K. Hoellerer, University of Oxford, UK and Martin E.M. Noble, Newcastle University, UK.

11.7   Jürgen Engel, University of Basel, Switzerland and Josephine C. Adams, University of Bristol, UK.

REFERENCES

11.2 Intermediate filaments

Herrmann H & Aebi U (2004) Intermediate filaments: molecular structure, assembly mechanism, and integration into functionally distinct intracellular scaffolds. *Annu Rev Biochem* 73:749–789.

Herrmann H, Bar H, Kreplak L et al (2007) Intermediate filaments: from cell architecture to nanomechanics. *Nature Rev Mol Cell Biol* 8:562–573.

Herrmann H, Strelkov SV, Burkhard P et al (2009) Intermediate filaments: primary determinants of cell architecture and plasticity. *J Clin Invest* 119:1772–1783.

Lee C-H, Kim M-S, Chung BM et al (2012) Structural basis for heteromeric assembly and perinuclear organization of keratin filaments. *Nature Struct Mol Biol* 19:707–715.

Omary MB (2009) 'IF-pathies': a broad spectrum of intermediate filament-associated diseases. *J Clin Invest* 119:1756–1762.

Other aspects of IF biology and pathologies are reviewed in several articles in *J Clin Invest* 119:1763–1848.

Parry DAD & Steinert PM (1995) Intermediate Filament Structure. RG Landes Co.

Wang E, Fischman D, Liem RKH et al (eds) (1985) Intermediate filaments. *Ann NY Acad Sci* 455:1–826.

11.3 Tight junctions

Bazzoni G (2003) The JAM family of junctional adhesion molecules. *Curr Opin Cell Biol* 15:525–530.

Kirchner E, Guglielmi KM, Strauss HM et al (2008) Structure of reovirus σ1 in complex with its receptor junctional adhesion molecule-A. *PLoS Pathogen* 4:e1000235.

Kostrewa D, Brockhaus M, D'Arcy A et al (2001) X-ray structure of junctional adhesion molecule: structural basis for homophilic adhesion via a novel dimerization motif. *EMBO J* 20:4391–4398.

Krause G, Winkler L, Piehl C et al (2009) Structure and function of extracellular claudin domains. *Ann NY Acad Sci* 1165:34–43.

Li Y, Fanning AS, Anderson JM & Lavie A (2005) Structure of the conserved cytoplasmic C terminal domain of occludin: identification of the ZO-1 binding surface. *J Mol Biol* 352:151–164.

Tsukita S, Furuse M & Itoh M (2001) Multifunctional strands in tight junctions. *Nature Rev Mol Cell Biol* 2:285–293.

11.4 Adherens junctions and desmosomes

Boggon T, Murray J, Chappuis-Flament S et al (2002) C-cadherin ectodomain structure and implications for cell adhesion mechanisms. *Science* 296:1308–1313.

Drees F, Pokutta S, Yamada S et al (2005) α-Catenin is a molecular switch that binds E-cadherin/β-catenin and regulates actin filament assembly. *Cell* 123:903–915.

Kowalcyk AP & Green KJ (2013) Structure, function and regulation of desmosomes. *Prog. Mol Biol Transl Sci* 116:95–118.

Leung CL, Green KJ & Liem RK (2002) Plakins: a family of versatile cytolinker proteins. *Trends Cell Biol* 1:37–45.

Nelson WJ, Dickinson DJ & Weis WI (2013) Roles of cadherins and catenins in cell-cell adhesion and epithelial cell polarity. *Prog Mol Biol Transl Sci* 116:3023.

Niessen, CM & Gottardi CJ (2008) Molecular components of the adherens junction. *Biochim Biophys Acta* 1778:562–571.

North AJ, Bardsley WG, Hyam J et al (1999) Molecular map of the desmosomal plaque. *J Cell Sci* 112:4325–4336.

Pokutta S & Weis WI (2007) Structure and mechanism of cadherins and catenins in cell–cell contacts. *Annu Rev Cell Dev Biol* 23:237–261.

Rangarajan ES & Izard T (2013) Dimer asymmetry defines α-catenin interactions. *Nature Struct Mol Biol* 20:188–193.

11.5 Gap junctions

Dobrowolski R & Willecke K (2009) Connexin-caused genetic diseases and corresponding mouse models. *Antioxid Redox Signal* 11:283–295.

Harris A & Locke D (2008) Connexins: A Guide. Springer.

Laird DW (2006) Life cycle of connexins in health and disease. *Biochem J* 394:527–543.

Maeda S, Nakagawa S, Suga M et al (2009) Structure of the connexin 26 gap junction channel at 3.5 Å resolution. *Nature* 458:597–602.

Oshima A, Tani K, Hiroaki Y et al (2007) Three-dimensional structure of a human connexin26 gap junction channel reveals a plug in the vestibule. *Proc Natl Acad Sci USA* 104:10034–10039.

Solan JL & Lampe PD (2009) Connexin43 phosphorylation: structural changes and biological effects. *Biochem J* 419:261–272.

Sosinsky GE & Nicholson BJ (2005) Structural organization of gap junction channels. *Biochim Biophys Acta* 1711:99–125.

Willecke K, Eiberger J, Degen J et al (2002) Structural and functional diversity of connexin genes in the mouse and human genome. *Biol Chem* 383:725–737.

11.6 Focal adhesions

Brunton VG, MacPherson IR & Frame MC (2004) Cell adhesion receptors, tyrosine kinases and actin modulators: a complex three-way circuitry. *Biochim Biophys Acta* 1692:121–144.

Huveneers S & Danen EHJ (2009) Adhesion signaling—crosstalk between integrins, Src and Rho. *J Cell Sci* 122:1059–1069.

Nobes CD & Hall A (1995) Rho, Rac and Cdc42 GTPases regulate the assembly of multinuclear focal complexes associated with actin stress fibers, lamellipodia and filopodia. *Cell* 81:53–62.

Streuli CH & Akhtar N (2009) Signal co-operation between integrins and other receptor systems. *Biochem J* 418:491–506.

Vicente-Manzanares M, Choi CK & Horwitz AR (2009) Integrins in cell migration—the actin connection. *J Cell Sci* 122:199–206.

Wegener KL, Partridge AW, Han J et al (2007) Structural basis of integrin activation by talin. *Cell* 128:171–182.

Zaidel-Bar R, Itzkovitz S, Ma'ayan A et al (2007) Functional atlas of the integrin adhesome. *Nature Cell Biol* 9:858–867.

11.7 The extracellular matrix

Bork P, Downing AK, Kieffer B & Campbell ID (1996) Structure and distribution of modules in extracellular proteins. *Q Rev Biophys* 29:119–167.

Bosman FT & Stamenkovic I (2003) Functional structure and composition of the extracellular matrix. *J Pathol* 200:423–428.

Engel J (2007) Visions for novel biophysical elucidations of extracellular matrix networks. *Int J Biochem Cell Biol* 39:311–318.

Hohenester E & Engel J (2002) Domain structure and organisation in extracellular matrix proteins. *Matrix Biol* 21:115–128.

Hynes RO (2002) Integrins: bidirectional, allosteric signaling machines. *Cell* 110:673–687.

Mao Y & Schwarzbauer JE (2005) Fibronectin fibrillogenesis, a cell-mediated matrix assembly process. *Matrix Biol* 24:389–399.

Porter S, Clark IM, Kevorkian L & Edwards DR (2005) The ADAMTS metalloproteinases. *Biochem J* 386:15–27.

Ramirez F & Dietz HC (2009) Extracellular microfibrils in vertebrate development and disease processes. *J Biol Chem* 284:14677–14681.

Sasaki T, Fässler R & Hohenester E (2004) Laminin: the crux of basement membrane assembly. *J Cell Biol* 164:959–963.

Van der Rest M & Garrone R (1991) Collagen family of proteins. *Faseb J* 5:2814–2823.

Wang N, Tytell JD & Ingber DE (2009) Mechanotransduction at a distance: mechanically coupling the extracellular matrix with the nucleus. *Nature Rev Mol Cell Biol* 10:75–82.

11.8 Bacterial pili

Allen WJ, Phan G & Waksman G (2012) Pilus biogenesis at the outer membrane of Gram-negative bacterial pathogens. *Curr Opin Struct Biol* 22:500–506.

Craig L & Li J (2008) Type IV pili: paradoxes in form and function. *Curr Opin Struct Biol* 18:267–277.

Craig L, Pique ME & Tainer JA (2004) Type IV pilus structure and bacterial pathogenicity. *Nature Rev Microbiol* 2:363–378.

Hung CS, Bouckaert J, Hung D et al (2002) Structural basis of tropism of *Escherichia coli* to the bladder during urinary tract infection. *Mol Microbiol* 44:903–915.

Lillington J, Geibel S & Waksman G (2015) Biogenesis and adhesion of type 1 and P pili. *Biochim Biophys Acta* 1850:554–564.

Mattick JS (2002) Type IV pili and twitching motility. *Annu Rev Microbiol* 56:289–314.

Merz AJ, So M & Sheetz MP (2000) Pilus retraction powers bacterial twitching motility. *Nature* 407:98–102.

Paranchych W & Frost LS (1988) The physiology and biochemistry of pili. *Adv Microb Physiol* 29:53–114.

Phan G, Remaut H, Allen WJ, Pirker KF et al (2011) Crystal structure of the FimD usher bound to its cognate FimC–FimH substrate. *Nature* 474:49–52.

Proft T & Baker EN (2009) Pili in Gram-negative and Gram-positive bacteria: structure, assembly, and their role in disease. *Cell Mol Life Sci* 66:613–635.

Sauer FG, Mulvey MA, Schilling JD et al (2000) Bacterial pili: molecular mechanisms of pathogenesis. *Curr Opin Microbiol* 3:65–72.

CELLULAR
SPACE

signaling
molecules

CYTOPLASM

transmembrane
receptor

transducers
and amplifiers

effectors

second
messengers

sensors and
effectors

cellular
responses

Signaling

12.1 INTRODUCTION

Cells need to communicate with each other to control and coordinate their growth, movement, differentiation, and other forms of behavior. Signaling can be achieved by direct cell/cell contact, as in integrin signaling (Chapter 11) and the immune response (Chapter 17), or by secretion of a molecular messenger by one cell and response to the messenger by another. Some small signaling molecules (such as membrane-soluble steroid hormones or ions from the extracellular space) can enter the target cell in various ways (Chapter 16). Others (such as growth factors and protein hormones) operate by binding to transmembrane protein receptors at the cell surface. The binding brings about changes in the cytoplasmic part of the receptor that trigger the start of a specific intracellular response (**signal transduction**). Transduction involves the recruitment of other proteins as effectors; second messengers (specific small molecules) may be generated and enzyme cascades may be initiated. Enzyme cascades commonly involve the post-translational modification of target proteins, notable among which are kinase-catalyzed phosphorylation and phosphatase-catalyzed dephosphorylation (Section 1.7). The outcomes are changes in gene expression, cell division, metabolism, and motility (**Figure 12.1**).

Cells are often exposed to more than one signal, and a major strategy in ensuring specific responses is localization of the signaling complexes at discrete subcellular locations or on scaffolds—protein arrays that can bring together and organize several proteins that signal sequentially. Scaffold proteins tether and orient substrates, they can mediate pathway branching, and they may also promote allosteric regulation. Co-localization can be a potent

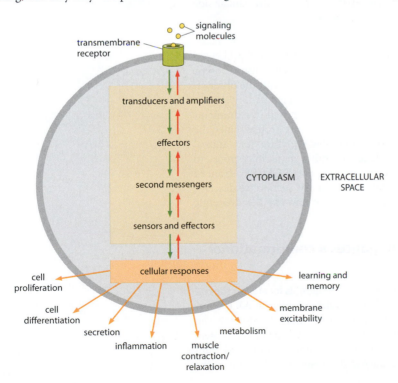

Figure 12.1 Signaling receptors and cellular responses. Extracellular signaling molecules act on transmembrane receptors, and changes in the intracellular portion of the receptors relay information to signaling pathways in the cytoplasm. Signaling is a dynamic process, consisting of ON mechanisms (green arrows), by which information flows down a pathway, opposed by OFF mechanisms (red arrows) that turn it off again. Signaling is not just linear; pathways may intersect and interact, giving cross-talk by which several pathways are coordinated. (Adapted from M.J. Berridge, *Cell Signaling Biology*, http://www.cellsignalingbiology.org.)

force for regulation of networks, as in signaling through protein kinase A and the tyrosine kinases (discussed below).

Signaling molecules can be secreted by a cell's immediate neighbors (**paracrine signaling**), by the same cell (**autocrine signaling**), or at a distant site in the organism and then transported to a particular cellular target (**endocrine signaling**). For example, the hormone insulin is synthesized in the beta cells of the pancreas and transported in the blood to liver and muscle where it exerts its action. In this chapter, we focus on three endocrine mechanisms: signaling through G-protein-coupled receptors (for example in response to epinephrine); signaling through tyrosine kinase receptors (for example in response to epidermal growth factor or insulin); and signaling through cytokine receptors. We also describe how signaling complexes are targeted for degradation when their activity needs to be switched off, a process mediated by the ubiquitylation system discussed in Chapter 7.

12.2 SIGNALING THROUGH G-PROTEIN-COUPLED RECEPTORS

The human genome encodes a large family (802 proteins identified) of **G-protein-coupled receptors** (**GPCRs**), with members involved in vision, the sense of smell, and responses to hormones and neurotransmitters. Signals that activate GPCRs include epinephrine, histamine, opioids, and serotonin. GPCRs initiate signaling through transducer G protein complexes that bind and exchange the nucleotides GTP and GDP. In turn, these **G proteins** (the name derives from the essential GTPase activity discovered by Martin Rodbell and Alfred Gilman) activate effector enzymes, such as adenylyl or guanylyl cyclases or phospholipases, which generate the **second messengers** cAMP or cGMP, or diacyl glycerol and inositol trisphosphate respectively. The second messengers then stimulate other sensors or effectors, which leads ultimately to responses in gene expression, in metabolic activity, and in cell movement. The GPCRs are regulated by a feedback desensitization mechanism mediated by the phosphorylation of protein kinases. An analysis in 2006 found that of the 324 targets of 1357 unique US Food and Drug Administration (FDA)-approved drugs, 27% were GPCRs; there is therefore keen interest in understanding GPCR signaling mechanisms.

All GPCRs are membrane proteins with seven transmembrane helices (each 25–30 aa); the N-terminal region is found on the extracellular face and the C-terminal region in the cytoplasm. The terminal regions, and the loops connecting the helices, range from a few to hundreds of amino acid residues in length. Activating molecules (**agonists**) such as hormones and neurotransmitters bind to GPCRs at sites that are accessible from the extracellular side and are found either in the N-terminal or transmembrane regions.

Bacteriorhodopsin is a light-driven proton pump from the halophilic archaeon *Halobacterium salinarium* (previously, *H. halobium*). Its structure was visualized by Nigel Unwin and Richard Henderson who used electron crystallography to resolve seven transmembrane rods of density which were interpreted as α-helices. This was consistent with the observation that seven long runs (~25–30 aa) of hydrophobic residues were separated by shorter segments (~ 8 aa) susceptible to limited proteolysis and therefore exposed in extramembrane regions. These data led to a model of a seven-transmembrane-helix protein. Later, the related protein rhodopsin provided the first X-ray crystallographic structure of an animal GPCR, helping to show how the visual receptor is able to transmit a signal in response to light. Similarly, the structures of epinephrine- and adenosine-regulated GPCRs advanced the understanding of how these GPCRs respond to hormones. In this section we describe two now classic examples of GCPR signaling, namely rhodopsin and the β_2-adrenergic receptor.

Light absorption by its retinal cofactor induces a conformational change in rhodopsin

Rhodopsin, the protein responsible for the response of the eye to light, is located in the rod outer segments of the retina. A cofactor, 11-*cis*-retinal (a derivative of vitamin A), is covalently linked by an imine bond (Schiff base) generated by the elimination of water between the –CH=O group of the retinal and the N^6-amino group of a Lys residue of the protein opsin (**Figure 12.2**). Absorption of light by retinal causes it to isomerize to the all-*trans* conformation, thereby inducing a conformational change in the opsin and activation of the

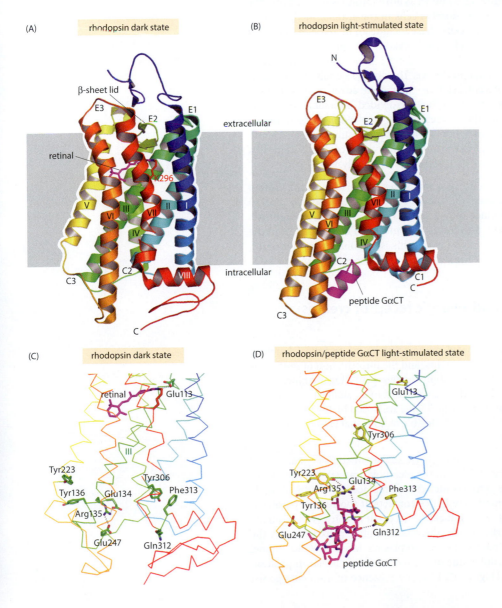

(A) β-ionone ring

11-*cis*-retinal

Lys296

Glu113

(B)

11-*cis*-retinal

light

all-*trans*-retinal

Figure 12.2 The structure of retinal and its response to light. (A) In bovine rhodopsin, retinal is attached in Schiff-base (imine) linkage to the N^6-NH$_2$ group of Lys296. In the inactive state, the imine is protonated and forms an ionic contact with Glu113 from helix III. (B) The 11-*cis* and all-*trans* conformations of retinal. The double bond between C11 and C12 is in red. The absorption maximum of free retinal is at 370 nm; that of the protonated Schiff base is at 440 nm. The retinal in rhodopsin has an absorption maximum at 500 nm, the wavelength of green light, at which the eye is most sensitive.

signaling pathway through a G protein called **transducin**. The bond between retinal and opsin is broken, and retinal rapidly dissociates from the protein. Newly synthesized 11-*cis*-retinal then attaches covalently to regenerate rhodopsin. The system is remarkably sensitive: it can detect a pulse of light with as few as five photons. It is also almost noise-free; in the absence of light, the half-life for the *cis/trans* isomerization of retinal is more than 400 years.

The structure of rhodopsin from bovine rod outer segments revealed a fold with seven transmembrane helices (**Figure 12.3A**). The X-ray diffraction was performed under dim red light so that the rhodopsin was in its inactive dark state. Spectroscopic measurements

(A) rhodopsin dark state

β-sheet lid
E3   E2   E1
retinal
K296
extracellular
V   VI III VII II I
IV
C2
C3   VIII   intracellular
C

(B) rhodopsin light-stimulated state

N
E3   E2   E1
V   VII II I
VI III
IV
C2
C3   C1   C
peptide GαCT

(C) rhodopsin dark state

retinal   Glu113
III
Tyr223   Tyr306
Tyr136   Glu134   Phe313
Arg135
Glu247   Gln312

(D) rhodopsin/peptide GαCT light-stimulated state

Glu113
Tyr306
Tyr223   Glu134   Phe313
Arg135
Tyr136
Glu247   Gln312
peptide GαCT

Figure 12.3 The structure of bovine rhodopsin. (A) Rhodopsin (348 aa) in the dark inactive state, colored from the N terminus in blue to the C terminus in red. There are seven transmembrane helices (I–VII) and an additional helix (VIII) at the C-terminal end, three extracellular loops (E1–E3), and three cytoplasmic loops (C1–C3). (C1 is obscured in this view). Retinal is bonded to Lys296. Sugar, lipid, and water molecules have been omitted for simplicity. (PDB 1U19) (B) Light-stimulated opsin (color-coded as in A) in complex with the GαCT peptide from transducin. To obtain the complex, rhodopsin was illuminated with light to produce active metarhodopsin II and then co-crystallized with the 11-mer peptide; meanwhile the all-*trans* retinal was lost. The last 22 residues of opsin are missing, assumed to be disordered. There are substantial shifts in helices V and VI compared with dark state rhodopsin. (PDB 3DQB) (C) Side-chain positions (Cα atoms in green) in dark-state rhodopsin, viewed as in (A). Arg135 H-bonds to Glu134 (helix III) and Glu247 (loop C2). (D) Side-chain positions (Cα atoms in yellow) in the opsin/GαCT peptide complex, viewed as in (B). Arg135 now H-bonds to a main-chain carbonyl oxygen of the GαCT peptide (magenta) and to Tyr223 (helix V). Glu134 and Glu247 are turned away. Gln312 from the C-terminal helix H8 H-bonds to a main-chain carbonyl oxygen of the peptide. Tyr306 and Phe313 have also shifted positions.

revealed that within milliseconds rhodopsin passes through at least three transient states to a fourth metastable state. Formation of this state, called metarhodopsin II, is coupled to the transfer of a proton from the Schiff base to Glu113 and then indirectly to Glu134, which is part of a short sequence (-Glu-Arg-Tyr-) that is conserved in many GPCRs (Figure 12.3C). In rhodopsin, protonation of Glu134 appears to be the trigger for the signal transduction, in which metarhodopsin II binds to transducin and initiates guanine nucleotide exchange in this protein (see below).

The first glimpse of how a GPCR communicates with its G protein partner came from a structure of activated opsin (without retinal) bound to an 11-aa peptide (GαCT) derived from the C-terminal region of the α subunit (Gα_t) of the heterotrimeric G protein (G$\alpha\beta\gamma$), transducin. The major structural changes in rhodopsin involve helices V and VI, with helix VI tilting outward by 6–7 Å and creating a crevice on the intracellular side of the protein within which the GαCT peptide binds in an α-helical conformation (Figure 12.3B). Significantly, the GαCT peptide cannot bind to the dark-adapted state of rhodopsin and, in the active conformation, the so-called ionic lock is broken (Figure 12.3D). Changes in H-bonds and ionic contacts help delineate a route of communication from the retinal-binding site to the Gα peptide-binding site some 28 Å away.

The G-protein-coupled adrenergic receptors are sensitive to agonists and antagonists

At times of stress, the hormone **epinephrine** (also known as **adrenaline**) is released into the blood from the chromaffin granules of neuroendocrine cells in the adrenal medulla. Epinephrine elicits a variety of responses in target cells; these include an increase in heartbeat in preparation for an emergency response, accelerated breakdown of glycogen in muscle cells to provide energy, inhibition of lipid breakdown in adipose tissues, inhibition of insulin secretion in the pancreas, and pupil dilation in the eye. The diverse signals are mediated through the binding of epinephrine to different, cell type-dependent, **adrenergic receptors**. The receptors fall into two main classes, which are further divided into subtypes on the basis of their response to synthetic ligands. The α-receptors are involved in smooth muscle contraction of blood vessels (α_1), decrease in action of the gastrointestinal tract (α1), and inhibition of insulin release (α_2). The β-receptors are mostly involved in cardiac output (β_1), smooth muscle relaxation (β_2), glycogenolysis (β_2), and enhanced lipolysis (β_3).

GPCRs are able to switch between several conformational states. Agonists push the protein to adopt the conformation of the active state and increase the receptor response; partial agonists also favor the active state but induce only a sub-maximal response; **antagonists** drive the receptor toward the inactive state, preventing the binding of other ligands. In the absence of ligands, equilibrium favors the inactive state. This is in contrast to rhodopsin, which is completely inactive in the dark state. The structures and properties of some adrenergic receptor ligands are listed in **Table 12.1**.

Binding of an agonist to the β_2-adrenergic receptor creates a G protein-binding site

Successful crystallization of GPCRs has involved overcoming many technical barriers, including low expression levels of recombinant proteins, and the introduction of mutations to combat conformational flexibility and improve stability in detergents. Ligand binding has been used to cause the protein to adopt a defined state and assist in crystallization, and also to throw light on the receptor mechanism. Further problems are posed by the hydrophobic surfaces of the GPCRs and the presence of inherently flexible regions that are not conducive to the formation of lattice contacts in crystals.

A more stable structure of the β_2-adrenergic receptor β_2AR was obtained by crystallizing it in the presence of the inverse agonist carazolol (K_d in the picomolar range), also known as a β-blocker (see Table 12.1), and with the cytoplasmic C3 loop replaced with a small stable protein, T4 lysozyme, to facilitate the formation of lattice contacts. It showed the predicted inactive conformation (**Figure 12.4A**). Proline-induced kinks occur at conserved positions in helices II, V, VI, and VII and, as in rhodopsin, there is an eighth helix (VIII) in the intracellular C-terminal region that runs parallel to the cytoplasmic face of the membrane. Carazolol occupies a site near the extracellular side of the receptor and similar in position to the retinal-binding site in rhodopsin (Figure 12.4A). The presence of a potent agonist

Table 12.1 Chemical structures of some agonists, antagonists, and inverse agonists of adrenergic receptors.

Agonists	Antagonists or inverse agonists	

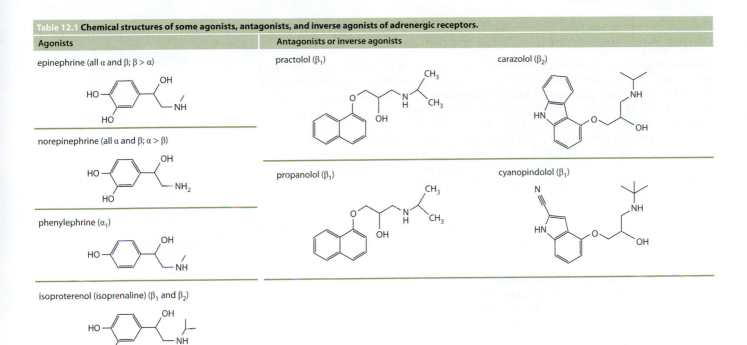

epinephrine (all α and β; β > α)

norepinephrine (all α and β; α > β)

phenylephrine (α_1)

isoproterenol (isoprenaline) (β_1 and β_2)

terbutaline (β_2)

practolol (β_1)

propanolol (β_1)

carazolol (β_2)

cyanopindolol (β_1)

and an antibody fragment (a nanobody) directed against the active agonist-bound state of β_2AR was critical in its crystallization (see Table 12.1); on binding, the nanobody inserts a peptide loop into the cavity that in the corresponding opsin binds the C-terminal region of the Gα subunit of transducin (Figure 12.4B). The inactive-to-active transition involves the repacking of helices III, V, VI, and VII, which is accompanied by small rotations about the Pro residues in helices V and VI that cause the G protein recognition site to undergo large changes and to open. In contrast, the changes at the agonist site are small, but the numerous interactions with the agonist help explain its high affinity and specificity.

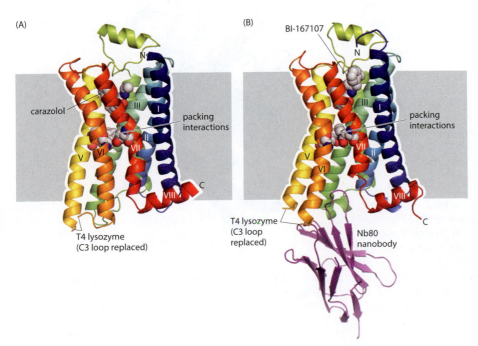

Figure 12.4 Structure of the β_2-adrenergic receptor. (A) The receptor in its inactive R state with the inverse agonist carazolol bound. Carazolol (see Table 12.1) is bound toward the extracellular side. (PDB 2RH1) (B) The receptor in its active R* state co-crystallized with the potent agonist BI-167107 and the nanobody Nb80. (PDB 3P0G)

Heterotrimeric G proteins act as molecular transducers that couple the activation of GPCRs to intracellular responses

Heterotrimeric G proteins are composed of three subunits: α, β, and γ (45, 37, and 9 kDa respectively). Gα subunits are members of the Ras GTPase family and bind GDP and GTP (**Box 12.1**). Gαβγ is tethered at the inner surface of the plasma membrane by a long-chain fatty acyl (C_{14} myristoyl) attached to the N-terminal amino acid or by a C_{16} palmitoyl group attached to a Cys thiol group of the Gα subunit, and by isoprenyl (C_{15} farnesyl or C_{20} geranylgeranyl) groups linked to a Cys residue at the C terminus of Gγ (**Figure 12.5**). In response to activation of a GPCR, the G protein heterotrimer binds to the GPCR cytoplasmic loops; conformational changes in Gα make the bound GDP dissociate, and the nucleotide-free Gα forms a high-affinity complex with the activated receptor. Gα can then bind GTP, and this causes it to dissociate from Gβγ. Gα alone has a slow intrinsic GTPase activity that returns the protein to its GDP-bound state—a form in which it can reassociate with Gβγ to complete the cycle (**Figure 12.6**). The time taken for the GTP to be hydrolyzed dictates how long the stimulus lasts. Throughout the process, Gα and Gβγ remain tethered to the membrane by their lipid modifications.

Gα and Gβγ both function as signal transducers to effector proteins. Effectors of Gα include adenylyl cyclases and cyclic GMP phosphodiesterase (discussed below) and phospholipase Cγ; those of Gβγ include various kinases, the inward-rectifying potassium channels, and voltage-gated calcium channels (Section 16.2).

In the human genome, there are 17 genes that, with splice variants, code for 21 Gα subunits (350–400 aa). They share > 40% sequence identity and can be divided into four main classes: $Gα_s$, $Gα_i$, $Gα_q$, and $Gα_{12}$. The Gα subunits, though homologous, vary in their regulatory

---

Box 12.1   G proteins

G proteins cycle between an inactive GDP-bound state and an active GTP-bound state. In its active state, a G protein interacts with effector proteins and induces downstream signaling events. Signaling terminates when GTP is hydrolyzed to GDP by the G protein. The best-known G protein, Ras, is a member of a superfamily with more than 150 members. Other families are named Rho, Rab, Ran, and Arf. These small G proteins are typically between 20 and 25 kDa and share a common structure. The G protein fold is also found in a domain of larger proteins that include the heterotrimeric G protein subunit Gα and the factors EF-Tu and EF-G involved in protein synthesis (see Section 6.2).

Ras (166 aa) comprises a six-stranded β sheet surrounded by six α helices (**Figure 12.1.1**). The guanine nucleotide binds at the C-terminal end of the β strands and close to the N-terminal region of the first α helix. The major regions of contact include the following: the N/TKXD^{119} motif and Ala146 (from the SAK motif), which contact the guanine base and provide specificity for guanine over adenine; the P loop, with the sequence G^{10}XXXXGKS/T^{17}, and the DXXG^{60} motif, which contact the β and γ phosphates of GTP. The conformational differences between the GDP- and GTP-bound states are confined to two regions: switch region I (residues 32–38 in Ras) and switch region II (residues 59–67 in Ras), which come together to create a binding site for the γ phosphate of GTP. In the GTP-bound state, the main-chain NH groups of Thr35 (switch I) and Gly60 (switch II) make H-bonds with the γ phosphate, and the Thr35 side chain binds Mg^{2+}, which is associated with the β and

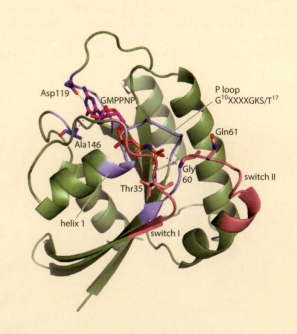

Figure 12.1.1 The G protein fold. p21 H-Ras in complex with the GTP analog GMPPNP (guanosine-5′-(γ-imino)triphosphate). (PDB 5P21) Switch regions I and II are shown in salmon.

Box 12.1 *continued*

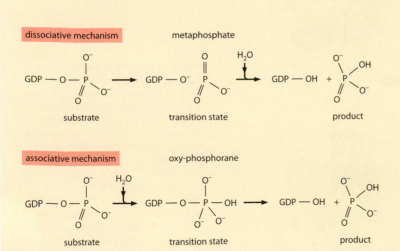

Figure 12.1.2 GTP hydrolysis in the Ras/Ras-GAP complex. The guanidinium group of the Arg side chain from GAP helps stabilize the negative charge on the GTP phosphates during catalysis, and the Arg main-chain carbonyl oxygen H-bonds to the side chain of Gln61 from Ras (in blue) and assists in correct alignment. A Gln residue is too weak a base to abstract a proton from the water, but Gln61 probably aligns the water for an in-line attack and increases its nucleophilicity.

factor signaling is SOS (Son of Sevenless), which, in combination with Grb2 (growth factor receptor binding protein 2), activates GDP/GTP exchange for Ras (Section 12.3). The affinities of nucleotide-free G proteins for GDP and GTP are very similar, but the intracellular concentration of GTP is ~10 times that of GDP, thus favoring GTP binding. There are also effectors that promote the GTP-bound state.

The GTPase activity of a G protein is low, and a **GTPase-activating protein** (**GAP**) usually assists catalysis. GAPs bind the G proteins and either contribute an essential group to the catalytic site or bring about conformational changes to promote catalysis. Ras-GAP provides an Arg residue to Ras that helps stabilize the transition state of the reaction and align the Gln that promotes attack by water (**Figure 12.1.2**). In the heterotrimeric G proteins, the Gα subunit already has an Arg residue with this function, and the role of the GAP is to promote the correct orientation of catalytic groups. The GEFs and GAPs for different G proteins vary in sequence and structure and are usually multidomain proteins.

The GTPase reaction proceeds through an in-line displacement reaction with a dissociative or associative transition state and inversion of stereochemistry at the γ phosphate (**Figure 12.1.3**). The nature of the transition state has been the subject of much controversy and it may be that different G proteins use differing proportions of the two mechanisms. In the associative mechanism, bond-making takes place before bond-breaking, leading to an oxy-phosphorane intermediate. In the dissociative mechanism, the leaving-group bond is broken before the attacking-group bond is made, leading to a metaphosphate-like intermediate. Such an intermediate is supported by kinetic isotope effect (KIE) measurements for Ras-GTP hydrolysis and is generally observed for the hydrolysis of phosphate triesters in solution. KIEs refer to the change in reaction rate when an atom in a substrate is replaced with a heavier isotope, and is defined as the ratio of the reaction rate with the unlabeled substrate to that with the labeled substrate (see also Chapter 15, Box 15.3).

γ phosphate groups. Gln61 promotes attack by water on the γ phosphate of the bound GTP. After the hydrolysis, the switch regions relax into a GDP-specific conformation.

Interaction of the G protein with **guanine exchange factors** (**GEFs**) induces the release of GDP so that it can be replaced with GDP. For the heterotrimeric G proteins, the GPCR serves as the GEF (Section 12.2). One of the best-known GEFs in growth

Figure 12.1.3 Mechanisms for the hydrolysis of nucleotide phosphotriesters.

(A)

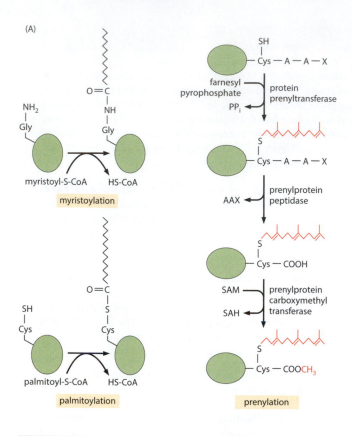

myristoylation

palmitoylation

prenylation

(B)

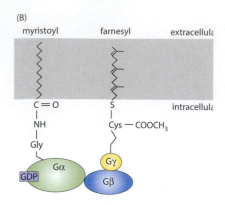

Figure 12.5 Lipidation of proteins as membrane anchors. (A) Left, enzyme-catalyzed myristoylation of an N-terminal Gly residue and palmitoylation of the thiol group of a Cys residue (often C-terminal) of a target protein. Myristoylation is irreversible; thio-palmitoylation is reversible *in vivo*. Right, prenylation of a CAAX box (where A is an aliphatic amino acid and X is Ser, Thr, Met, Leu, Ala, or Gln) at the C terminus of a target protein is catalyzed by the sequential action of three enzymes. The farnesyl group (red) is in thioether linkage with the Cys thiol group. The methyl group (red) that esterifies the new C-terminal COOH group derives from *S*-adenosylmethionine (SAM), leaving *S*-adenosylhomocysteine (SAH). (B) Model of a heterotrimeric G protein anchored to the plasma membrane by myristoyl and farnesyl groups.

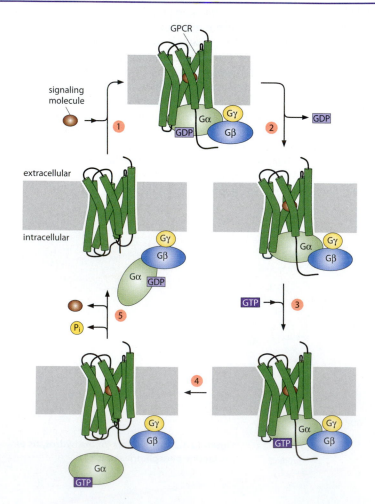

Figure 12.6 GDP/GTP exchange in a heterotrimeric G protein. The Gα subunit has GDP bound in the resting state. Step 1, on binding the signaling molecule, for example adrenaline, there is a change in conformation of the GPCR that signals to the G protein and stabilizes the nucleotide-free state. Step 2, GDP is released. Step 3, Gα binds GTP. Step 4, Gα/GTP and Gβγ dissociate and signal to downstream effectors. Step 5, Gα hydrolyses the GTP to GDP, the GPCR releases the adrenaline and reverts to the inactive conformation, and Gα (with GDP now bound) re-associates with Gβγ. For simplicity, the membrane anchors of the G protein subunits are omitted.

properties. For example, Gα$_s$ activates all nine isoforms of adenylyl cyclase, whereas Gα$_i$ inhibits only two isoforms. There are six types of Gβ subunit (~340 aa) encoded by five genes, all of which are closely related, and 12 subtypes of the Gγ subunits (~75 aa) that are more diverse. With a few exceptions (for example transducin), the βγ heterodimers appear to be interchangeable in their association with α subunits.

The Gα subunit adopts different conformations in the GDP- and GTP-bound complexes, whereas Gβγ is essentially unchanged

The G protein Gα$_{i1}$β$_1$γ$_2$ is activated by the α$_2$-adrenergic receptor in response to the hormone epinephrine. The Gα subunit has a projecting N-terminal helix which interacts with Gβ; an α-helical domain; and a Ras-like GTPase domain consisting of a six-stranded β sheet surrounded by six helices (**Figure 12.7A**). The Gβ subunit has an N-terminal helix followed by a module of seven β sheets, each with four antiparallel strands, which form the blades of a β-propeller structure. The outside strand of each sheet together with the inner three strands of the following sheet form a WD40 motif (see **Guide**). The Gγ subunit consists of two helices; one forms a coiled-coil with the N-terminal helix of Gβ, and the other interacts with the β-propeller domain (see **Guide**).

Gα and Gβγ come together at two interfaces. The most extensive interactions involve the switch II region (see Box 12.1) of Gα, which interacts with loops and turns from the β propeller of Gβ (see Figure 12.7A). These interactions distort the γ-phosphate recognition

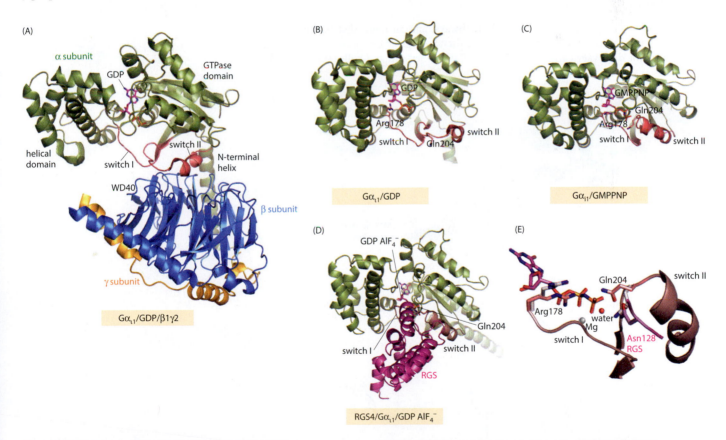

Figure 12.7 Structure and activation of a G protein. (A) The heterotrimeric complex Gα$_{i1}$/GDPβ$_1$γ$_2$ from *Rattus norvegicus*. The switch regions I (residues 178–186) and II (residues 199–216) in Gα are shown in salmon in all five panels. (B) The Gα$_{i1}$/GDP subunit from the complex. (C) Structure of the free Gα$_{i1}$ subunit complexed with the GTP analog GMPPNP. The switch II region has greatly changed, bringing Gln204 into proximity with the γ phosphate of the nucleotide. (D) The Gα$_{i1}$ subunit in complex with the regulator RGS4 and GDP AlF$_4^-$. AlF$_4^-$ binds to the site that is usually occupied by the γ phosphate of GTP and mimics a hydrolytic intermediate. RGS4 contacts the switch I and switch II regions. (E) Close-up view of the GTP-binding site (the GTP has been modeled on the GDP/AlF$_4^-$ structure), showing how association with RGS4 changes the conformation of catalytic residues, Arg178 and Gln204. Arg178 and Gln204 are shown with Cα atoms in dark salmon in the RGS complex, and in gray in the Gα$_{i1}$/GMPPNP complex. Asn128 from the RGS (magenta) conflicts with the position of Gln204 in the Gα$_{i1}$/GMPPNP complex. The enforced shift of Gln204 enables it to accept a H-bond from a water molecule and donate a H-bond to the γ phosphate. Gln204 is thus able to promote GTP hydrolysis by aligning the water molecule and increasing its nucleophilicity for in-line attack on the γ phosphate bond. (PDB: Gα$_{i1}$(GDP)β$_1$γ$_2$, 1GP2; Gα$_{i1}$(GMPPNP), 1CIP; RGs4/Gα$_{i1}$(GDP/AlF$_4^-$), 1AGR)

regions of the G protein, favoring the GDP-bound (over the GTP-bound) form of Gα. The second interface involves the N-terminal helix of Gα, which contacts the side of the β-propeller domain. Most of the contact residues are conserved in Gα and Gβ subunits, suggesting that they are likely to be common to other Gα and Gβγ interactions.

In comparing the Gα subunit in the heterotrimer with free Gα(GDP) and Gα(GTP), the major differences are in the switch I and II regions (Figure 12.7B,C). In Gαβγ, the switch II loop of Gα adopts a different conformation from that observed for the free Gα(GTP). In Gα(GTP), rearrangement of switch II allows the catalytic Gln residue to approach the γ phosphate of the GTP analog in the active site. There is no significant change in the structure of Gβγ on associating with Gα.

In Gα, two residues, an Arg and a Gln (Arg178 and Gln204 in Gα$_{i1}$), have key roles in catalysis. Both are involved in binding the GTP γ phosphate, and, by analogy with the Ras protein (see Box 12.1), Gln204 is likely to promote the attack by a water molecule, leading to hydrolysis of GTP. Typical rates for Gα-catalyzed hydrolysis of GTP are 2–4/min; this is too slow for deactivation of Gα(GTP), but the reaction is enhanced ~2000-fold by the RGS4 protein (regulator of G-protein signaling). RGS4 does not participate directly in the hydrolysis of GTP; rather, it binds Gα and induces switch I and switch II to adopt their active conformations and position the catalytic residues (Figure 12.7D).

Signaling from G proteins to adenylyl cyclase produces the second messenger, cyclic AMP

GPCRs and G proteins signal to multiple pathways, among them that leading from the β-adrenergic receptor to adenylyl cyclase and thence to glycogen phosphorylase (**Figure 12.8**). A major discovery came in 1957, when Earl Sutherland and colleagues showed that

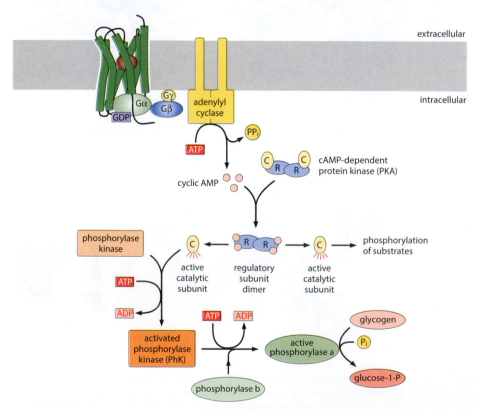

Figure 12.8 Epinephrine signaling through the β-adrenergic receptor to stimulate glycogen breakdown. Activation of the receptor by epinephrine signals to the heterotrimeric G-protein complex. The activated Gα/GTP subunit stimulates the enzyme adenylyl cyclase to produce cyclic AMP (cAMP) from ATP. cAMP activates protein kinase A (PKA), also known as cAMP-dependent kinase, by binding to the regulatory subunits (R) and releasing the catalytic subunits (C). The latter phosphorylate and activate phosphorylase kinase (PhK). PhK phosphorylates glycogen phosphorylase b (GPb), generating active phosphorylase a (GPa). GPa catalyzes the phosphorolysis of glycogen to yield glucose 1-phosphate.

(A)

ATP

3′,5′-cyclic AMP

adenylyl cyclase

PP_i

(B)

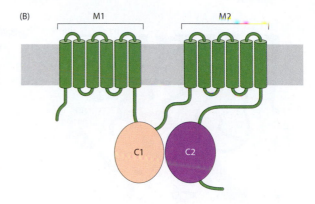

M1   M2

C1   C2

Figure 12.9 Adenylyl cyclase catalyzes the formation of cyclic AMP from ATP.
(A) The precursor and products. PP_i, pyrophosphate. (B) A topology diagram of adenylyl cyclase. The catalytic site is formed where the domains C1 and C2 come together.

the breakdown of glycogen in liver in response to epinephrine is dependent on 3′,5′-cyclic adenosine monophosphate. Cyclic AMP (cAMP) is synthesized from ATP by the action of adenylyl cyclase (**Figure 12.9A**), a membrane-bound enzyme that is activated by the G protein subunit $G\alpha_s$(GTP) after the GPCR has been activated by epinephrine. It can also be activated by other factors including forskolin (a diterpene plant product; Figure 12.9A), and is inhibited by a class of compounds called P-site inhibitors.

Adenylyl cyclase has 10 isoforms that vary in size from 1077 to 1610 aa. It comprises a transmembrane domain M1 with six α helices, a catalytic domain C1, a second transmembrane domain M2 also with six α helices, and another catalytic domain C2 (Figure 12.9B). The two C domains (each ~ 40 kDa) are brought together to form the active enzyme and associate tightly with activated $G_s\alpha$ (GTP) (**Figure 12.10A**). The primary structures and folds of C1 and C2 are similar, but only the heterodimer C1/C2 is active when stimulated by $G_s\alpha$ (homodimers are inactive). The myristoylated N-terminal Gly and S-palmitoylated Cys2 of $G_s\alpha$ must face toward the membrane, as must the N termini of the C1 and C2 domains to connect with their transmembrane domains (see Figure 12.10A). Thus, intracellular substrates should have direct access to the catalytic site. Importantly, the switch II region of $G_s\alpha$

(A) plasma membrane

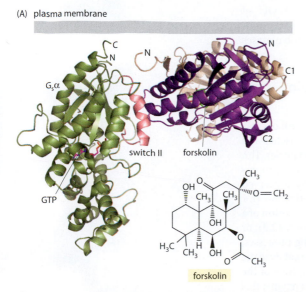

$G_s\alpha$

GTP

switch II   forskolin

C
N
N
C1
C2

forskolin

(B)

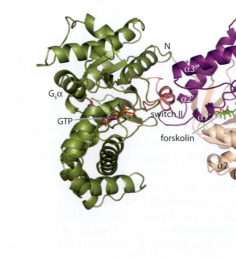

$G_s\alpha$

GTP

forskolin

switch II

N
C2
α3″
α2″
α1″
α5′
ATP analog
α2
β1
C1
catalytic site
α7
β7
C

Figure 12.10 Structure of adenylyl cyclase in complex with $G_s\alpha$.
(A) A view with the top surface oriented toward the membrane. A GTP analog (GDPγS) is bound to the $G_s\alpha$ subunit (green). Forskolin (inset) is bound between the C1 (beige) and C2 (mauve) domains of adenylyl cyclase. (PDB 1AZS) (B) The complex rotated ~70° about the horizontal axis in (A) to show the forskolin and the catalytic site with a substrate ATP analog (2′,3′-dideoxy-ATP) bound. Secondary structure elements from the C2 domain have a superscript prime.

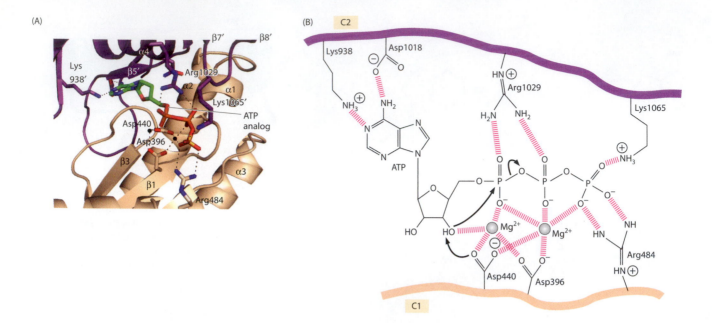

Figure 12.11 Catalytic mechanism of adenylyl cyclase. (A) The catalytic site lies between the C1 and C2 domains. Color code as in Figure 12.10. Two active-site residues from C1, Asp440 and Asp396, and one from C2, Arg1029, are conserved in all adenylyl cyclases. Two bound metal ions (small black spheres, Mg²⁺ in the cell) contact the Asp carboxyl groups and the triphosphate (orange) of the ATP analog with octahedral geometry. (PDB 1CJT) (B) Asp396(C1), Arg484(C1), Lys938(C2), Asp1018(C2) and Lys1065(C2) promote the correct orientation of the ATP.

Asp440 acts as a general base to abstract a proton from the 3′-OH of the ribose ring, allowing nucleophilic in-line attack of the 3′-oxyanion on the α phosphate group. Arg1029 from the C2 domain and the two Mg²⁺ ions orient the triphosphate of ATP and neutralize the negative charge at the α phosphate so as to stabilize the transition state formed at the α phosphate as the attacking group comes in before the leaving group has left. The reaction thus proceeds with inversion of configuration, and the products are 3′,5′-cyclic AMP and pyrophosphate.

inserts into the C2 domain at a groove formed by its α1 and α2 helices and the loop at the end of α3 (Figure 12.10B). As described above, only the activated GTP-bound form of Gα has the switch II region in the correct conformation for binding to adenylyl cyclase. This interaction is not possible with Gαβγ.

The Gₛα subunit binds some 30 Å from the catalytic site in C1/C2 but induces conformational changes at the interface between C1 and C2 that promote catalysis. The proposed catalytic mechanism (**Figure 12.11B**) involves in-line attack of the 3′-OH of the ATP ribose on the α phosphate of the 5′-triphosphate, with the elimination of pyrophosphate. In this, with its use of two bound metal ions, it resembles DNA polymerases (Chapter 3).

Protein kinase A is activated by cAMP and triggers a protein kinase cascade

Protein phosphorylation, a ubiquitous mechanism for signaling and for the control of catalytic activity in eukaryotes (Section 1.7), was discovered in 1955 by Edmond Fischer and Edwin Krebs, working on glycogen phosphorylase. **Protein kinases** phosphorylate Ser, Thr, or Tyr residues in their target proteins (**Box 12.2**). Kinases vary in size, regulatory properties, and subcellular localization, but all contain a catalytic core recognizable by conserved sequence motifs. Phosphorylation of a target can have several outcomes: activation (for example glycogen phosphorylase); inhibition (for example glycogen synthase); creation of a recognition site for other signaling molecules (for example SH2 domains; Section 12.3); or the transition between structural disorder and order (Section 6.4). In signaling processes, kinases often act in cascades, the phosphorylation of one enzyme activating it to catalyze the phosphorylation of the enzyme that follows. At each step in the cascade, every copy of the activated kinase is able to phosphorylate and activate many copies of its target enzyme; this can lead to a massive amplification of the starting signal.

cAMP activates protein kinase A (PKA). In the absence of cAMP, PKA is an inactive heterotetramer with two regulatory subunits (R, 42 kDa) and two catalytic subunits (C, 38 kDa).

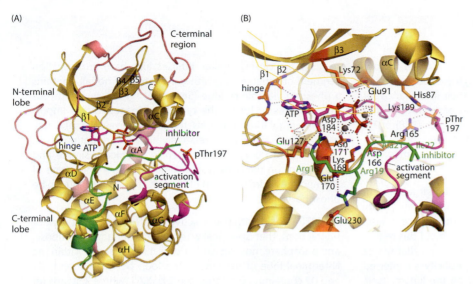

Figure 12.12 The C subunit of protein kinase A in complex with ATP and an inhibitor peptide.
(A) The C subunit of mouse PKA is shown with the kinase core in gold and the terminal extensions in salmon. In this isoform, Ser10, Thr197 (labeled), and Ser338 are all phosphorylated (phosphates in red). The activation segment is in magenta and an inhibitor peptide (PKI) is in green. (PDB 1ATP) (B) Details of the polar contacts at the ATP- and PKI-binding site in PKA, including water molecules. H-bonds are shown as gray dotted lines. The Gly-rich loop from the β1–β2 turn is shown as a thin line so as not to obscure the ATP. pThr197 binds to three basic groups (His87, Arg165, and Lys189). The catalytically important residues Asp166 and Lys168 are marked. Asp184, part of the Asp-Phe-Gly motif, contacts the metal ions (gray spheres). In PKI, the Ser to be phosphorylated in a true substrate peptide is replaced by Ala (Ala21). As part of the PKA recognition motif (RRXS*/T*Φ), Arg18 and Arg19 in PKI are recognized by Glu127, Glu170, and Glu230; the large nonpolar residue (Ile22) docks into a hydrophobic pocket formed by residues 198–205, part of the activation segment.

Binding of cAMP to the R subunits causes dissociation of the heterotetramer and the release of catalytically active C subunits:

$$C_2R_2 \text{ (inactive)} + 2cAMP \rightarrow 2C \text{ (active)} + (R/cAMP)_2$$

The C subunit fold has two lobes connected by a hinge region (**Figure 12.12A**). The smaller N-terminal lobe is mostly β sheet, but with one α helix (αC) whose orientation is crucial for ATP binding. The larger C-terminal lobe is mostly α helix but also contains an 'activation segment' of variable length that connects two conserved motifs, -Asp-Phe-Gly- and -Ala-Pro-Glu- (starting at residue Asp184 and finishing at Glu208 in PKA). ATP binds between the two lobes, with its adenine in a hydrophobic pocket and contacting the hinge region. These interactions are conserved in almost all protein kinases, and the hydrophobic pocket is the target site in other kinases for many clinically important inhibitors (Section 12.3). The activation segment in many, but not all, protein kinases contains a phosphorylatable residue (Thr197 in PKA) that is important for regulation. This pThr interacts with three basic groups (Figure 12.12B), one that immediately precedes the catalytic Asp, a second near the start of the activation segment, and a third from the αC helix (residues Arg165, Lys189, and His87 respectively in PKA). It acts as an organizing center, linking the two lobes and the catalytic site. Myristoylation of the N-terminal Gly (see Figure 12.5) may also be important in increasing stability and as a site for allosteric regulation.

With phosphorylation of the Thr residue, the activation segment becomes ordered and helps create the binding site that recognizes preferred amino acid residues around the phosphorylation site in a target protein (RRXS*/T*Φ in PKA—see Box 12.2). This was first defined by the binding of a natural heat-stable inhibitor, PKI, to PKA. PKI binds tightly (K_d = 2 nM) through an N-terminal helix and a pseudosubstrate-binding sequence in which the Ser residue that would be phosphorylated in a true substrate has been replaced with Ala (see Figure 12.12B).

The regulatory R subunit (four isoforms: RIα, RIβ, RIIα, RIIβ, with non-redundant properties and subcellular localizations) inhibits the C subunit by acting as a pseudosubstrate,

Box 12.2 Protein kinases

The human genome encodes 518 protein kinases (http://kinase.com), ~2% of the genome, making them the third most populous protein family and reflecting their importance for signal transduction and metabolism. Most of them phosphorylate the hydroxyl groups of Ser or Thr residues (388) or Tyr residues (90) in proteins; the remaining 40 have other substrates. They all catalyze the transfer of the γ phosphate of ATP to the substrate.

For most kinases, amino acid side chains in the target protein close to the phosphorylation site define specificity (**Table 12.2.1**). These side chains are exposed on the surface in less-well-ordered regions, allowing the kinase to mold this region of a target protein to an extended conformation that fits its catalytic site. For many kinases, further specificity is conferred by remote docking sites located either on the kinase itself, as in mitogen-activated protein kinase (MAPK) or on separate domains or subunits, as in the Cdk2/cyclin complexes (Section 13.2).

Protein kinases have a common bilobal catalytic core of ~290 aa with a conserved topology and certain conserved residues. Inactive kinases adopt several different conformations, but in the active state the conformations of the conserved residues are closely similar. Kinases also have an activation segment of variable length that in many, but not all, contains a phosphorylatable residue (Ser or Thr for the **serine/threonine kinases** or Tyr for the **tyrosine kinases**) that is important for regulation. Catalysis requires a divalent metal ion, usually Mg^{2+}, which has a relatively high concentration in the cell. The phosphoryl transfer step is chemically simple and dependent on the correct orientation of the γ phosphate of ATP and the hydroxyl group of the Ser, Thr, or Tyr residue to be phosphorylated.

The ATP is held within the active site through a network of interactions that specifically recognizes its adenine, ribose, and phosphate moieties and orients it for catalysis. Within the N-terminal lobe of the kinase, the loop between strands β1 and β2 containing the sequence GXGXXG (where X stands for any amino acid) is known as the 'glycine-rich loop.' It contacts the β phosphate of ATP through main-chain H-bonds from the NH moieties of the glycine residues. The α and β phosphates of the ATP are oriented by ion-pairing with the N^6-amino group of a Lys residue (Lys72 in protein kinase A [PKA], whose residue numbering is used here). This in turn is aligned by contact with a Glu (Glu91) from the αC helix (**Figure 12.2.1A**).

The triphosphate group is also chelated by two Mg^{2+} ions. One (Mg_1) chelates the β and γ phosphates and is bound to Asp184 of the -DFG- motif at the start of the activation segment; the second (Mg_2) is bound to the α and γ phosphates, to an Asp and an Asn (Asp184 and Asn171 respectively). The substrate hydroxyl is H-bonded to a catalytically important Asp (Asp166) that occurs in the sequence Arg-Asp in most protein kinases. In Ser/Thr kinases, a nearby Lys residue (Lys168) contacts the γ phosphate and is poised to stabilize the developing charge on the transition state during catalysis. In Tyr kinases, the stabilizing residue is an Arg four residues away to allow for the bulkier Tyr residue.

Structures of kinases bound to short peptide substrates show that the peptide binds to the catalytic cleft in an extended conformation and can be selectively recognized by three or four amino acids flanking the residue to be phosphorylated (see Table 12.2.1). The cleft is composed of residues from both lobes of the kinase and in particular from the activation segment. In some kinases, this sequence is structured, but in most it undergoes a phosphorylation-dependent restructuring in response to upstream protein kinase signaling.

Kinetic studies indicate that binding of one substrate does not exclude the other, although at the high ATP concentrations found in cells there is a preference for ATP binding first. There is no covalent phosphoryl intermediate and the reaction proceeds with inversion of configuration at the phosphorus; that is, the attacking group comes in from the opposite direction to that taken by the leaving group. The transition state for the intermediate could be either dissociative or associative, as for GTP hydrolysis. Evidence from several kinases indicates that the dissociative mechanism is preferred, which is similar to that of phosphoryl transfer mechanisms in solution, but a few kinase reactions may use the associative mechanism.

Table 12.2.1 Selected protein kinases and their preferred substrate specificities.

Name	Full name	Consensus sequence
Ser/Thr kinases		
PKA	Protein kinase A	-R-R-X-**S/T**-Φ
PhK	Phosphorylase kinase	-R-X-X-**S/T**-Φ-R
Cdk2	Cyclin-dependent protein kinase 2	**S/T**-P-X-K/R
ERK2	Extracellular signal-regulated kinase 2	-P-X-**S/T**-P
Plk1	Polo-like kinase 1	-D/E/N-X-**S/T**-Φ/not P
Aurora B	Aurora B	-R-R/K-**S/T**-(not P)
Tyrosine kinases		
Irk	Insulin receptor kinase	-D-**Y**-M-M
c-Src	Cellular form of the Rous sarcoma virus transforming agent	-E-E-I-**Y**X-X-F
Csk	C-terminal Src kinase	-I-**Y**-M-F-F
EGFR	Epidermal growth factor kinase	-E-E-E-**Y**-F

The Ser, Thr, or Tyr residues phosphorylated are indicated in bold. Φ is a hydrophobic residue. Some kinases (such as Plk1 or Aurora B) discriminate against Pro in the P + 1 site.

Box 12.2 *continued*

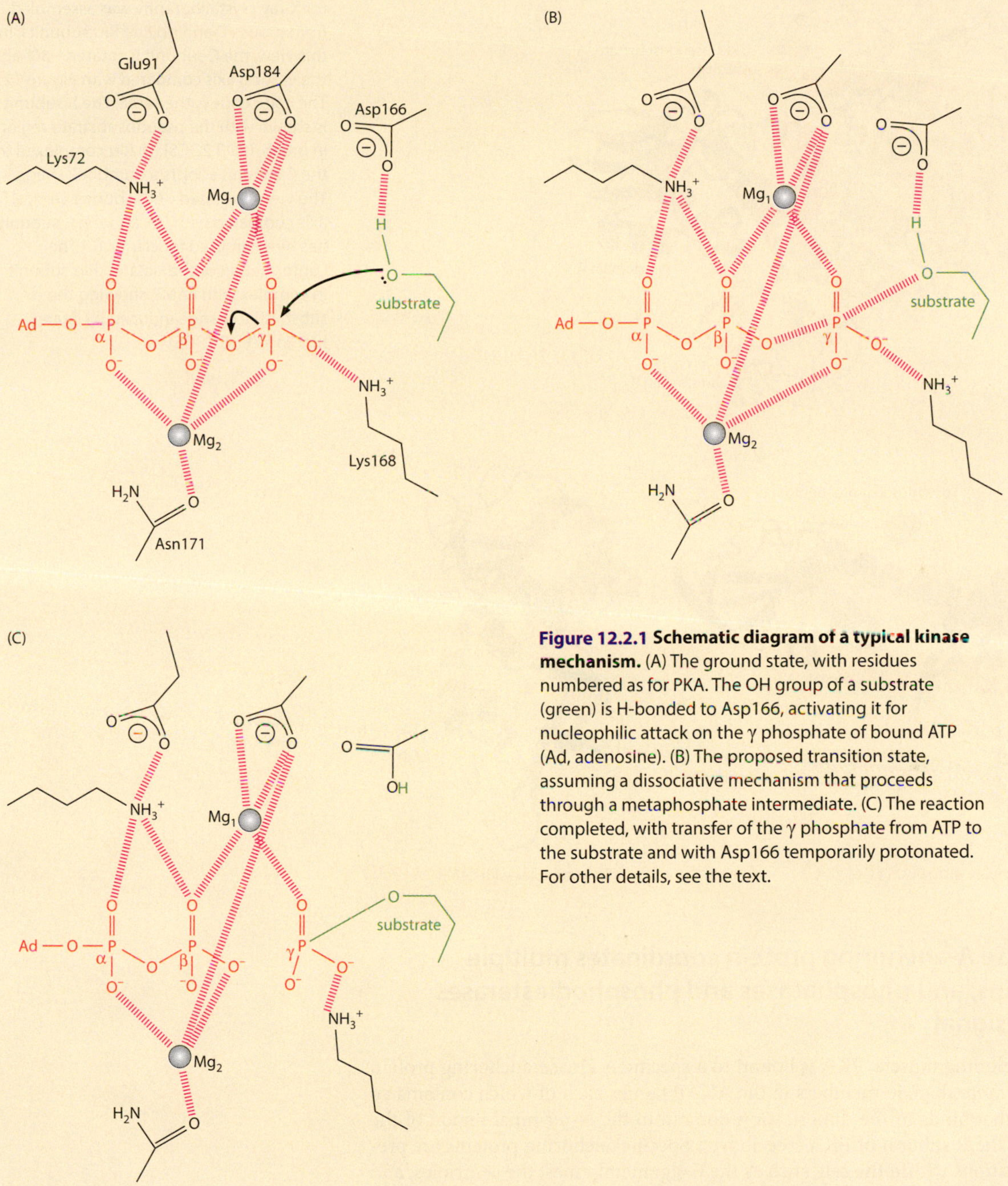

Figure 12.2.1 Schematic diagram of a typical kinase mechanism. (A) The ground state, with residues numbered as for PKA. The OH group of a substrate (green) is H-bonded to Asp166, activating it for nucleophilic attack on the γ phosphate of bound ATP (Ad, adenosine). (B) The proposed transition state, assuming a dissociative mechanism that proceeds through a metaphosphate intermediate. (C) The reaction completed, with transfer of the γ phosphate from ATP to the substrate and with Asp166 temporarily protonated. For other details, see the text.

resembling PKI (**Figure 12.13A**). As with many other signaling proteins, R subunits are modular. A helical dimerization/docking domain (~45 aa) at the N-terminal end is followed by a flexible region (~80 aa) containing a pseudosubstrate sequence that docks into the active site cleft of the C subunit, and then two cAMP-binding domains, A and B (each ~120 aa).

The R subunit wraps around the C-terminal lobe of the C subunit; in doing so, it undergoes substantial changes in conformation, in contrast with negligible changes in the C subunit (Figure 12.13B). In the presence of cAMP, an ordered disentangling occurs, beginning with cAMP binding to the more exposed B domain (the so-called 'gatekeeper' domain). Then, with cAMP binding to the A domain, the R subunit finally adopts a closed structure that can no longer bind to the C subunit (Figure 12.13C) and the latter, now catalytically active, is released.

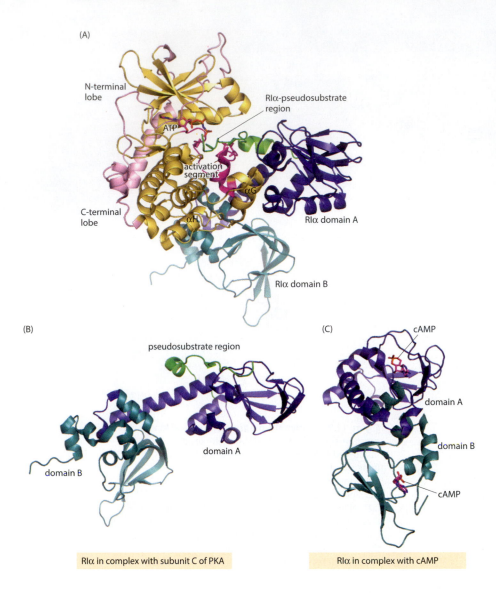

(A)

N-terminal lobe

ATP

RIα-pseudosubstrate region

activation segment

αG

C-terminal lobe

αH

RIα domain A

RIα domain B

(B)

pseudosubstrate region

domain A

domain B

RIα in complex with subunit C of PKA

(C)

cAMP

domain A

domain B

cAMP

RIα in complex with cAMP

Figure 12.13 The C subunit of PKA in complex with the R subunit in the C₂R₂ heterotetramer. (A) A C_2R_2 heterotetramer for X-ray crystallography was assembled from mouse C and bovine RIα subunits. In this view, the C subunit is rotated ~30° about the vertical axis compared with Figure 12.12. The color code is the same. The R subunit is shown with the pseudosubstrate region in green. (PDB 2QCS) (B) RIα complexed to the C subunit adopts a dumbbell shape. The view is rotated ~60° about a vertical axis compared with (A), and the C subunit has been omitted for clarity. (C) The conformation of the isolated RIα subunit in complex with cAMP, showing the substantial change induced by ligand binding. (PDB 1RGS)

A protein kinase A-anchoring protein coordinates multiple effector proteins, and phosphatases and phosphodiesterases terminate the signal

In many tissues, including muscle, PKA is bound to a specific A kinase-anchoring protein (AKAP). There are more than 50 members in the AKAP family, each of which contains an amphipathic helix that binds to the dimerization domain in the N-terminal region of the relevant isoform of the R subunit of PKA (see above). Specific anchoring proteins are present in different locations within the cell, such as the Golgi membranes, the centrioles, and the cytoskeleton, and each AKAP contains a unique sequence motif that targets it to the correct subcellular address. Through simultaneous interaction with signaling enzymes and their substrates, AKAPs form multivalent signal transduction complexes, integrating and regulating signaling pathways. For example, the 300 kDa muscle-specific AKAP, mAKAP, associates with PKA, with guanine nucleotide exchange factors that signal to other kinases, with the ryanodine receptor that regulates calcium release from the endoplasmic reticulum (Section 14.4), and with the phosphatase PP1.

Termination of a signal is as important as its initiation. PP1 is one of a family of phosphatases that dephosphorylate phosphoproteins, thus returning them to their starting state. A relatively small number of phosphatases are directed against a wide variety of phosphoproteins by virtue of different regulatory proteins. mAKAP also interacts with PP2A, a phosphodiesterase that hydrolyzes cAMP to AMP and thus terminates the signal. In mammals, no fewer than 50 genes encode phosphodiesterases; these enzymes, like the kinases, are necessarily tightly regulated. Also like the kinases, in the absence of a signal input, they are inactive because an autoinhibitory peptide sequence blocks the catalytic site. Signals such as Ca^{2+} ions or phosphorylation (for example PKB/Akt activating PDE3—see Section 12.3) induce conformational changes that displace the autoinhibitory peptide so that the phosphodiesterase becomes active. The phosphodiesterase (PDE6) that degrades cGMP in the retina is unique in that it is controlled by a trimeric G protein, transducin. Note that in the liver it is the generation of cAMP that turns on glycogenolysis, whereas in the retina it is the degradation of cGMP that leads to the closure of ion channels, membrane hyperpolarization, and transduction of the visual signal (see above).

PKA activates phosphorylase kinase, which then activates glycogen phosphorylase

PKA phosphorylates many targets. For example, phosphorylation of the transcription factor CREB (cAMP-response-element-binding protein) signifies its translocation from the cytoplasm to the nucleus, where it interacts with DNA at the cAMP-response element. Similarly, phosphorylation of the β_2-adrenergic receptor changes the receptor specificity from G_s complexes to G_i complexes and leads to the activation of Ras and the ERK (extracellular signal-regulated protein kinase) pathway. The classical and best understood action of PKA is the regulation of glycogen metabolism: PKA phosphorylates phosphorylase kinase to activate glycogen breakdown, and phosphorylates glycogen synthase to inhibit glycogen synthesis.

Phosphorylase kinase (PhK) integrates hormonal (epinephrine-mediated), neuronal (Ca^{2+}-mediated), and metabolic (ADP-mediated) signals to control glycogen breakdown in muscle and liver. In response to external stimuli, PhK catalyzes the conversion of inactive glycogen phosphorylase b (GPb) to active glycogen phosphorylase a (GPa) through phosphorylation of Ser14 in GPb. The amino acid sequence surrounding the phosphorylation site on GPb is important in conferring specificity, but other regions act as substrate docking sites. PhK (1.3 MDa) is a tetramer of heterotetramers, thus $(\alpha\beta\gamma\delta)_4$. Electron microscopy (see **Methods**) shows the four protomers in a butterfly-like bilobal arrangement, and phosphorylase molecules appear to bind to each lobe (**Figure 12.14**). The α and β subunits are regulatory and the target of phosphorylation by PKA; ADP binds to the β subunits and stimulates activity; and neuronal stimulation is effected by Ca^{2+} ions that bind to the δ subunit (called calmodulin) (see **Guide**) of PhK, thus coupling the stimulus for muscle contraction (Section 14.4) to the production of energy.

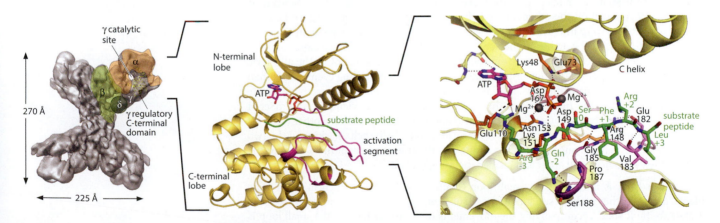

Figure 12.14 Structure of phosphorylase kinase. Left, the holoenzyme $(\alpha\beta\gamma\delta)_4$ as determined by cryo-EM, showing the likely locations of the α, β, γ, and δ subunits inferred from biochemical data and antibody labeling of the α and β subunits. Center, the catalytic subunit (γ) in complex with an inactive analog (AMPPNP) of ATP and a substrate peptide. The kinase activation segment is in magenta. Right, details of the interactions of the peptide substrate in the active site. The peptide (Arg-Gln-Met-Ser-Phe-Arg-Leu-) is numbered from −3 at the N terminus, with 0 at the phosphorylatable Ser. Selected H-bonds are drawn with dotted lines. Arg(−3) of the peptide contacts Glu110 and the ATP ribose, as in PKA.

Gln(−2) contacts Ser188 main chain. The side chain of Met(−1) has been hidden for clarity. The side chain of Ser(0) is directed toward what would be the γ phosphate of AMPPNP and is poised for an in-line nucleophilic attack promoted by Asp149, with Lys151 in a position to stabilize the negatively charged pentacoordinate transition state. The peptide from sites (+1) to (+3) makes an antiparallel β sheet with residues 183–185 in the activation segment. Phe(+1) docks into a nonpolar pocket shielded by Val183 and Pro187. Arg(+2) contacts Glu182, the Glu that replaces the pThr residue present in many other kinases. (PDB 2PHK) (A, from C. Venien et al., *Structure* 17:117–127, 2009. With permission from Elsevier.)

The γ subunit is the catalytic subunit. It has a kinase core similar to that of PKA, and also a calmodulin-binding domain. The regulatory subunits appear to be inhibitory, directly or indirectly restraining the kinase activity. Phosphorylation of the β subunits (and to a lesser extent, the α subunits) gives rise to activation. PhK does not require phosphorylation on its activation segment for activity; it has a Glu residue in place of the pThr of PKA, and the side-chain carboxylate has a similar role to that of the pThr phosphate in organizing the activation segment for recognition of the substrate peptide.

Glycogen phosphorylase undergoes conformational changes in response to phosphorylation and allosteric effectors

Glycogen phosphorylase initiates the breakdown of glycogen by catalyzing the repetitive phosphorolysis of the 1,4-α-glycosidic bond that links the terminal glucose to the non-reducing end polysaccharide chain, and releasing glucose 1-phosphate (**Figure 12.15A**). In muscle cells, phosphoglucomutase converts glucose 1-phosphate to glucose 6-phosphate, which is oxidized by the glycolytic pathway to generate ATP to meet the energy needs of the cell (Chapter 15), whereas in liver a phosphatase can hydrolyze glucose 1-phosphate to glucose, which is then transported in the blood for use in other tissues. Glycogen phosphorylase is controlled allosterically (Section 1.7) by reversible phosphorylation catalyzed by PhK and by noncovalent binding of ligands. The low-activity form (GPb) found in resting muscle is activated by AMP (which stabilizes the R state) and inhibited by glucose 6-phosphate and ATP (which stabilize the T state). When the AMP concentration rises as ATP is broken down by muscle activity, GPb is activated; conversely, if an increase in glycolysis causes glucose 6-phosphate levels to rise, GPb is inhibited. However, metabolite concentrations take time to change and, for a 'fight-or-flight' response, the hormone epinephrine turns on the PKA/PhK signaling pathway. This can bring about rapid conversion of GPb to the highly active GPa in the R state (within 1 second); GPa does not depend on metabolite concentrations for regulation and is fully active in the absence of AMP.

GP is a dimer in which the catalytic sites are at the center of each 97 kDa subunit, well away from the subunit interface. In GPb, the N-terminal region of each subunit curls back on an acidic surface region of the same subunit, which attracts the cluster of positively charged residues close to Ser14 (Figure 12.15B). Phosphorylation of Ser14 introduces two new negative charges carried by the phosphate group, causing the less well-ordered N-terminal region to rotate and Ser14 to shift by ~34 Å (Figure 12.15C). The pSer is then close to, and can interact specifically with, two basic residues, Arg69 from one subunit and Arg43' from the other, at the subunit interface (**Figure 12.16A**). In GPb (T state), AMP also binds near this site (~12 Å away) on the other side of helix α2, which carries Arg69 (Figure 12.16B), and its phosphate group interacts with two Arg residues, Arg309 and Arg310, in the same subunit. As the enzyme switches to the active R state, the subunit interface closes and the adenosyl moiety of AMP makes more substantial contacts (Figure 12.16C). The allosteric mechanism of phosphorylase thus rests on phosphorylation and binding events more than 45 Å distant from the catalytic sites. As the enzyme switches from the low-activity T state to the active R state, the tightening of the subunit interface in the vicinity of the pSer14 sites is accompanied by a rotation of ~10° of one subunit with respect to the other (see Figure 12.15C). In the T state, the interface is characterized by helix/helix packing of the two so-called 'tower helices,' one from each subunit. On the transition from the T state to the R state, the two helices pull apart and change their angle of tilt.

Each tower helix is followed by a small loop (containing Asp283), which in the T state blocks access to the catalytic site for the large substrate, glycogen. On transition from the T state to the R state, accompanying the movements of the tower helices, the small loop is displaced and the catalytic site becomes accessible. Glycogen phosphorylase also contains the cofactor pyridoxal phosphate (PLP), in Schiff-base (imine) linkage with the N^6-amino group of Lys680. The phosphate group of PLP (Figure 12.16D) acts as a general acid/base to promote attack by inorganic phosphate on the glycosidic bond retaining the terminal glucose at the non-reducing end of the glycogen polysaccharide chain. In the T state, the negatively charged Asp283 in the small loop is directed toward and repels the inorganic phosphate required for phosphorolysis of the glycosidic bond (Figure 12.16E). Conversely, in the R state, the loop is displaced and Asp283 is replaced by the positively charged group of Arg569. This favors catalysis by creating a substrate-binding site for the inorganic phosphate (Figure 12.16F).

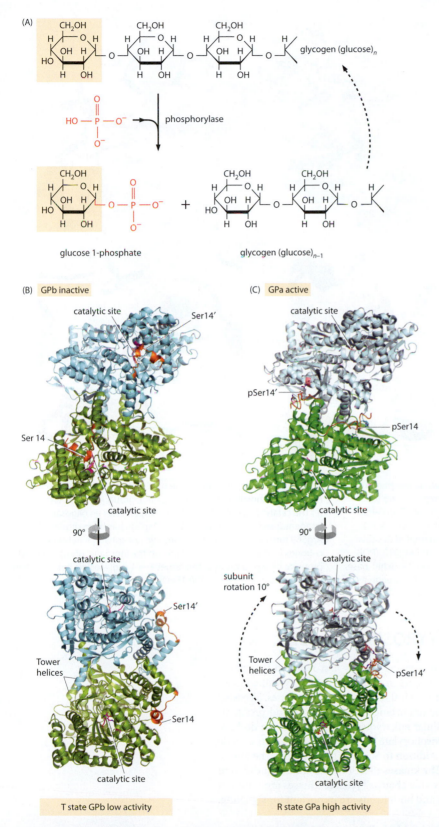

Figure 12.15 Glycogen phosphorylase. (A) Repetitive phosphorolytic cleavage of the terminal 1,4α-glycosidic bond in glycogen, releasing glucose 1-phosphate. (B) Structure of the inactive GPb dimer viewed down the 2-fold axis of symmetry and, below, rotated by 90°. The N-terminal region (residues 10–21) is in orange. (C) The high-activity phosphorylated GPa dimer viewed down the 2-fold axis of symmetry and, below, rotated by 90°. Color code as in (B). Ser14 shifts 40 Å on phosphorylation and tightens the subunit/subunit interface on its side; and one subunit (cyan) rotates through 10° with respect to the other subunit (green), weakening the subunit interface on the left as viewed in the lower image. (PDB 1A81 and 1GPA)

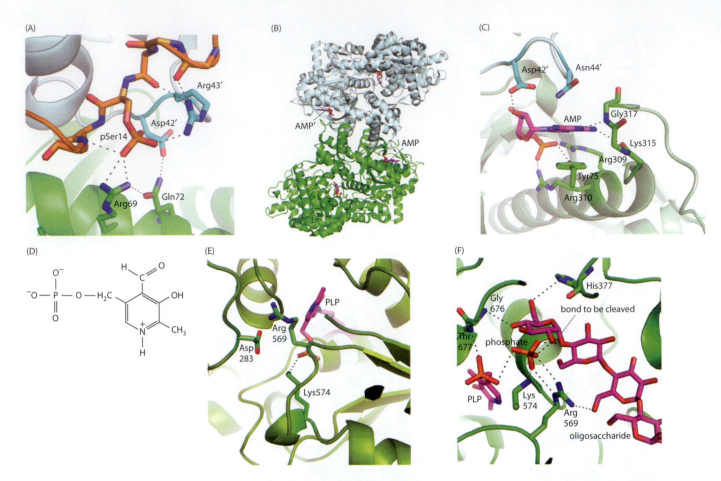

Figure 12.16 Allosteric structural changes and catalytic mechanism of phosphorylase. (A) In GPa, the major contacts from pSer14 are to Arg69 from one subunit and Arg43' from the other subunit. (B) In GPb, AMP binds ~12 Å from Ser14 and changes the subunit interactions. (C) Details of the AMP-binding site. The molecule has been rotated to show the contacts from the 5'-phosphate of AMP to Arg309 and Arg310. (PDB 1PYG) (D) The cofactor, pyridoxal 5'-phosphate (PLP). (E) PLP in Schiff-base linkage with Lys680 interacts with positively charged Lys574 through its 5'-phosphate group. The negatively charged carboxyl group of Asp283 points into the site, preventing the binding of inorganic phosphate required for phosphorolysis, and the basic group Arg569 is turned away. (F) The catalytic site of active GPa in complex with an oligosaccharide (magenta) and inorganic phosphate. Asp283 has been displaced and Arg569 turns in to contact the inorganic phosphate. A network of H-bonds (dotted lines) from PLP line up the inorganic phosphate and promote phosphorolysis of the target $1\rightarrow4$-α-glycosidic bond holding the terminal glucose residue. (PDB 1E4O)

12.3   SIGNALING THROUGH TYROSINE KINASES

Studies of the transforming properties of tumor-causing viruses first revealed that phosphorylation of the –OH groups of Tyr residues occurs *in vivo* and that the tyrosine kinase activity of a viral enzyme, v-Src, correlates with the viral transforming properties. Tyrosine phosphorylation has turned out to be of major importance in cell growth and proliferation. Tyrosine kinases are exclusive to multicellular eukaryotes, although bacteria do have enzymes related to ATPases that are able to phosphorylate Tyr residues, and yeasts encode dual-specificity kinases that can modify Tyr in addition to Ser and Thr residues. The fold of tyrosine kinases closely resembles that of Ser/Thr kinases (Section 12.2) but with different activation segments that serve to recognize Tyr side chains. Tyrosine kinases are found in the cytoplasm and associated with membranes, and both forms are important for signaling.

Receptor tyrosine kinases are activated by dimerization

Receptor tyrosine kinases (**RTKs**) are regulators of metabolism, cell growth and proliferation, responding to external stimuli of growth factors and hormones. Humans have 58 RTKs, which fall into 20 subfamilies. They have a common architecture: a ligand-binding domain in the extracellular region, a single helix spanning the membrane, and a kinase domain plus C-terminal region in the cytoplasm (**Figure 12.17**). In general, ligand binding

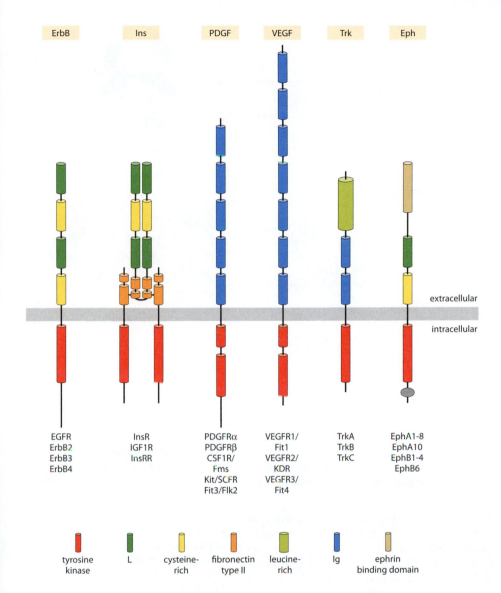

Figure 12.17 Receptor tyrosine kinase families. Selected human receptor RTKs are grouped into families and illustrated schematically, with family names above and family members listed below. Domains are colored as shown at the bottom. ErbB, erythroblastoma B; Ins, insulin; Eph, ephrin (Adapted from M.A. Lemmon and J. Schlessinger, *Cell* 141:1117–1134, 2010. With permission from Elsevier.)

causes the receptor to dimerize and the kinase domain of one subunit to phosphorylate its partner. The pTyr residues act as docking sites for downstream signaling molecules. The region between the transmembrane helix and the kinase domain (the 'juxtamembrane' region) is also important for regulation.

Ligand binding can bring about dimerization in various ways. In some, the ligand is itself a dimer. For example, vascular endothelial growth factor, VEGF) cross-links two VEGF receptor (VEGFR) molecules (**Figure 12.18A**). In others, such as the epidermal growth factor receptor (EGFR), the EGF ligand binds to the receptor and, without contributing to a dimer interface, induces large conformational changes that result in dimerization through receptor/receptor interactions (Figure 12.18B). In addition, some RTKs are intrinsically dimeric; an example is the insulin receptor, in which disulfide bonds link the two chains. The binding of insulin to the outside of the cell induces conformational changes in the kinase domain and phosphorylation in the activation segment.

The insulin receptor tyrosine kinase domain (IRK) is a model for tyrosine kinase activation

The **insulin receptor** regulates lipid, protein, and carbohydrate metabolism, including glucose homeostasis, in response to the hormone insulin. The receptor is a heterotetramer with two β chains that span the cell membrane and two extracellular α chains (see Figure 12.17). Insulin binds to the α chains, inducing a conformational change in the cytoplasmic portion of the β chain that promotes autophosphorylation of Tyr residues: two in the

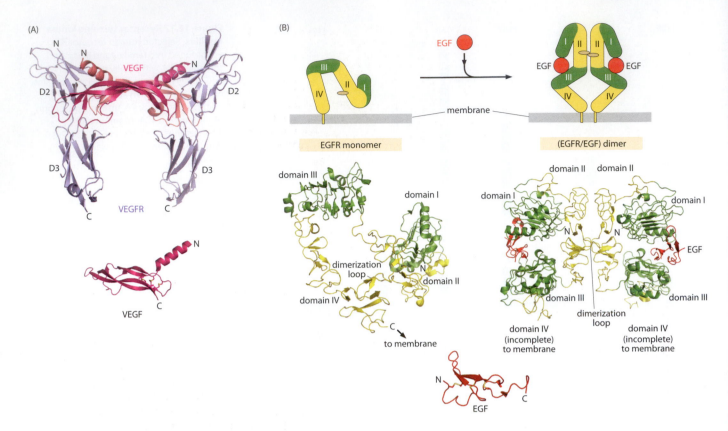

juxtamembrane region, three in the activation segment of the kinase domain (IRK), and two in the C-terminal tail. Several other RTKs are also activated by phosphorylation of Tyr residues in their kinase activation segment, including insulin-like growth factor receptor (IGFR) and nerve growth factor receptor (TrkA).

The unphosphorylated inactive form of IRK is subject to an autoinhibitory mechanism. One Tyr residue in the activation segment, Tyr1162, is located in the catalytic site and competitively inhibits access to a peptide substrate, while other residues at the start of the activation segment restrict access to the ATP binding site (**Figure 12.19A**). In the activated triphosphorylated IRK (pTyr1158, pTyr1162, and pTyr1163), a major conformational change in the activation segment opens up access for ATP and creates a platform for recognition of a peptide substrate (Figure 12.19B). Phosphorylation of Tyr1163 (corresponding to pThr197 of PKA [Section 12.2] and pThr160 of Cdk2 [Section 13.2]) is the crucial modification that stabilizes this new conformation. The substrate peptide derived from insulin receptor substrate 1 (IRS-1) binds in a short antiparallel β-sheet conformation with the substrate Tyr directed toward the γ phosphate of ATP (Figure 12.19C). In comparison with Ser/Thr kinases such as PKA, the substrate Tyr is positioned further out of the catalytic cleft by the activation segment to allow for the longer side chain.

The EGFR kinase is activated through asymmetric dimerization

The EGFR family has four members that share 40–45% sequence identity: EGFR (HER1 or ErbB1), HER2 (ErbB2), HER3 (ErbB3), and HER4 (ErbB4). (The name HER comes from human EGF receptor and ErbB from the erythroblastoma viral gene.) They are activated by members of the EGF family, which includes EGF, **transforming growth factor α (TGFα)**, and 10 other potential ligands. Family members can form both homodimers and heterodimers. HER2 is unusual in that it does not bind any EGF family ligand, but forms heterodimers with other ligand-bound EGFR family members. Up-regulation of EGFR and HER2 occurs in many human cancers, and these receptors have been intensively pursued as therapeutic targets, leading to several compounds being approved for clinical use (**Box 12.3**).

Unlike the insulin receptor, EGFR family members do not require phosphorylation on the activation segment in order to be active: noncovalent interactions of the activation segment are sufficient to assist alignment (**Figure 12.20A,B**). In the inactive conformation of the

Figure 12.18 Ligand-induced dimerization of receptor tyrosine kinases. (A) Dimeric VEGF binds two VEGFR monomers. Residues 112–215 of a human VEGF monomer are shown separately below. The two subunits in the VEGF dimer are linked by two disulfide bridges. Domains D2 and D3 of VEGFR (residues 120–326) have Ig-like folds. (PDB 2X1W) (B) Upper, EGF binds to the extracellular region of the EGFR monomer (whose cytoplasmic region is not shown); this induces dimerization in the cell membrane. Lower left, the EGFR extracellular region (~ 620 aa) has four domains: two leucine-rich repeat units (domains I and III), each arranged as a right-handed α-helical barrel structure, and two cysteine-rich repeat domains (domains II and IV). Lower right, the binding of EGF (red) to domains I and III induces a substantial rearrangement to a more elongated conformation, and dimerization is induced by interactions between the two domains II. Bottom, EGF (53 aa) consists of three loops linked by disulfide bonds and two short antiparallel β sheets. VGFR and EGFR are glycosylated, but for simplicity this is not shown here. (PDB: inactive EGFR, 1NQL; active EGFR, 1IVO). (Adapted from A.W. Burgess et al., *Mol. Cell* 12:541–552, 2003. With permission from Elsevier.)

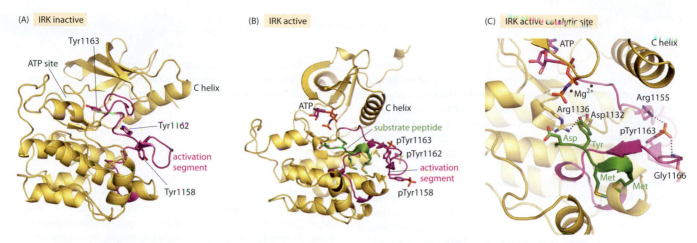

Figure 12.19 The insulin receptor tyrosine kinase domain. (A) In the inactive conformation, the activation segment and Tyr1162 block access to the substrate-binding site. The C helix is swung out. (B) In the active conformation, Tyr1158, Tyr1162, and Tyr1163 are phosphorylated, and the C helix has swung in. The activation segment forms a platform for substrate recognition; in the 18-residue peptide substrate, only residues Gly-Asp-**Tyr**-Met-Asn-Met- are ordered (the **Tyr** is the phosphorylation target). The N-terminal lobe has rotated ~21° with respect to the C-terminal lobe to close the interface between them. (C) pTyr1163 makes contacts with Arg1155 and with the Gly1166 main chain from the activation segment, which helps align the activation segment for substrate recognition (side chains in green). (PDB: inactive IRK, 1IRK; active IRK, 1IR3)

kinase domain, the C helix is rotated outward so that contributions to the ATP-binding site are distorted and the activation segment blocks access to the catalytic site. Activation occurs allosterically when the kinase domains associate as an asymmetric dimer: the C-terminal lobe of one kinase domain (the activator) makes contacts with the C helix of the other

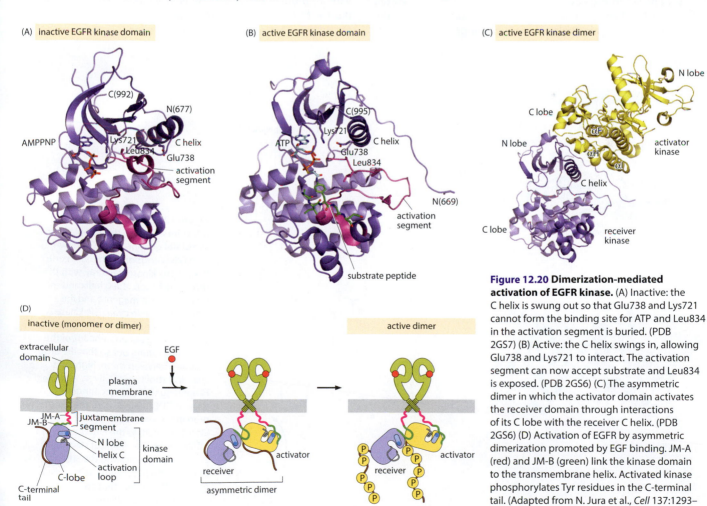

Figure 12.20 Dimerization-mediated activation of EGFR kinase. (A) Inactive: the C helix is swung out so that Glu738 and Lys721 cannot form the binding site for ATP and Leu834 in the activation segment is buried. (PDB 2GS7) (B) Active: the C helix swings in, allowing Glu738 and Lys721 to interact. The activation segment can now accept substrate and Leu834 is exposed. (PDB 2GS6) (C) The asymmetric dimer in which the activator domain activates the receiver domain through interactions of its C lobe with the receiver C helix. (PDB 2GS6) (D) Activation of EGFR by asymmetric dimerization promoted by EGF binding. JM-A (red) and JM-B (green) link the kinase domain to the transmembrane helix. Activated kinase phosphorylates Tyr residues in the C-terminal tail. (Adapted from N. Jura et al., *Cell* 137:1293–1307, 2009. With permission from Elsevier.)

Box 12.3 Protein kinase inhibitors for the treatment of cancer

Kinases are important targets for anti-cancer drug development. These enzymes are mutated or up-regulated in many different cancers and there is hope that if their signaling function can be inhibited, the proliferation of tumor cells will be arrested. At least 13 inhibitors have been approved for clinical use; many more are in clinical trials. All except three of the approved inhibitors target the ATP-binding site on the kinase. Despite the conserved nature of the adenine-binding pocket, there are enough adjacent pockets that can bind both polar and nonpolar groups to enable a potent inhibitor to be designed with sufficient specificity to avoid the inhibition of other kinases. Inhibitors for both active and inactive kinase conformations have been developed. Targeting the inactive conformation can give high specificity; targeting the active conformation is favored where the disease state arises from activating mutations, but such inhibitors are generally less specific. Mutations that impart drug resistance are a potential risk with both approaches.

Around 25% of patients with breast cancer have an over-expressed form of the gene for the EGF receptor HER2, and there is a significant correlation between over-expression of the *Her2* gene and progression of malignancy. These observations led to the development of Herceptin®, a humanized monoclonal antibody against the extracellular region of the receptor. Herceptin (trastuzumab) was approved for clinical

use in 1998 and, in combination with standard chemotherapy, has led to favorable outcomes for many (but not all) patients. Herceptin was the first example of a gene-based therapy for cancer.

Glivec® (imatinib), an inhibitor of Abelson tyrosine kinase (Abl), was approved for clinical use in chronic myeloid leukemia (CML), a hematological stem cell disorder characterized by excessive myeloid (non-lymphocytic leukocyte) proliferation. In CML, the N terminus of Abl is fused to another protein, Bcr, as a consequence of a chromosomal translocation event. Bcr–Abl is constitutively active and this leads to uncontrolled signaling. If the disease is recognized in the early stages, treatment with Glivec can be > 95% successful. Glivec is highly specific, binding preferentially to an inactive conformation that is unique to Abl (**Figure 12.3.1A**). Its only other major target is the tyrosine kinase receptor c-Kit, which is up-regulated in gastrointestinal cancers. In 2001, Glivec was also approved for treatment of these cancers. Further compounds have been developed to increase potency and to combat drug-resistant mutants. These include dasatinib (which targets Abl in its active state) and nilotinib (targeting the inactive state).

Various protein kinase inhibitors have been approved for the treatment of renal cancer. These target receptor tyrosine kinases that are essential for angiogenesis, the process

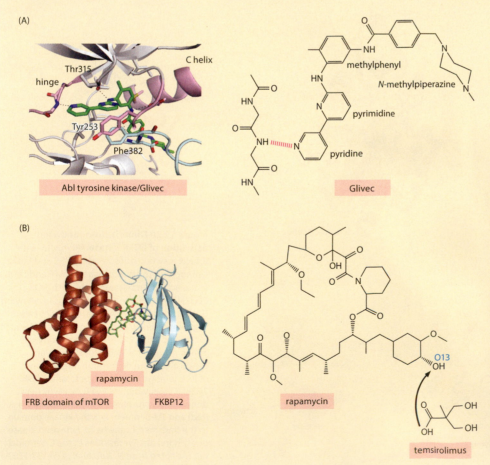

Figure 12.3.1 Protein kinase inhibitors. (A) Left, the ATP-binding site of Abl tyrosine kinase is shown with Glivec (carbon atoms in green) bound. The kinase is in gray, with the glycine-rich loop, the C helix and the hinge region in magenta and the activation loop in cyan. The kinase is in the inactive conformation with the C helix swung out and Phe382 of the DFG motif swung out so that the drug is bound between them. Right, the chemical structure of Glivec, with the H-bonding to the kinase hinge region shown. (PDB 1IEP) (B) Left, the FKBP12 protein presents rapamycin (carbon atoms in green) to the FRB domain of mTOR. Right, the chemical structure of rapamycin, indicating the site of esterification (O13) with the carboxyl group of the adduct that converts it to temsirolimus. (PDB 2FAP)

Box 12.3 *continued*

of new blood vessel formation necessary for the tumor to obtain nourishment and to spread. The compounds Sutent® (sunitinib), Nexavar® (sorafenib), and Votrient® (pazopanib) all target vascular endothelial growth factor receptor (VEGFR), platelet-derived growth factor receptor (PDGFR), and c-Kit.

Three derivatives of the immunosuppressant rapamycin (a macrolide natural product—see Section 9.5) have been shown to inhibit the Ser/Thr kinase mTOR, which is a key regulator of cell growth, cell proliferation, and protein synthesis. These compounds, the first of which was Torisel® (temsirolimus), licensed in 2007 for the treatment of renal cell carcinoma, do not target the ATP-binding site on the kinase. Rather, the rapamycin analog binds to the 12 kDa protein FKBP12, a peptidyl-prolyl *cis–trans* isomerase. The complex is a potent inhibitor of mTOR, binding tightly to its substrate recognition domain, FRB (Figure 12.3.1B). This action of rapamycin is different from its action in immunosuppression, thought to be due to the inhibition of calcineurin, a Ca^{2+}-dependent protein phosphatase that is an essential mediator of T cells in the immune response (Section 17.4).

With the impressive results for Glivec and Herceptin, the stage has been set for further elaboration of gene-based therapies, in combination with standard chemotherapy, in which drug combinations can be based on the patient's individual genetic fingerprint.

domain (the receiver) to promote an active conformation of the receiver (Figure 12.20C). In this, the C-terminal lobe of the activator acts like the cyclin that activates Cdk2 (Section 13.2). Mutational studies have shown that the activator kinase need not be catalytically active to activate the receiver kinase, which permits the activation of both homodimers (such as EGFR/EGFR) and heterodimers (such as HER2/HER3).

In many RTKs (for example the insulin receptor), the juxtamembrane region has an auto-inhibitory role that is relieved by phosphorylation. However, analysis of deletion mutants suggests that the juxtamembrane region in EGFR is unusual in that it assists in activation. NMR (see **Methods**) studies indicate that dimerization of the extracellular region of EGFR leads to pairing of the transmembrane helices. Changes in the transmembrane helix are transmitted by the juxtamembrane region to the kinase domain, ensuring that activation does not occur unless EGF is bound to the receptor. Combining all these results, a mechanism has been proposed for signaling through the transmembrane helices to promote both kinase dimerization and activation (Figure 12.20D).

Src is regulated by domain interactions and phosphorylation

Around one-third of human tyrosine kinases are not themselves receptors; instead, they control cell proliferation and differentiation in response to the activation of other cell surface receptors. This group of 32 non-receptor kinases includes its founder member, **Src** (there are nine Src kinases in the human genome), and Lck, which is involved in B-cell and T-cell signaling (Section 17.4). Src has six domains: a myristoylated N-terminal domain involved in membrane binding and subcellular localization; a unique region; two peptide-binding domains (an SH3 domain and an SH2 domain); a kinase domain (SH1); and a C-terminal tail (**Figure 12.21A**). The SH2 and SH3 (Src homology) domains (see **Guide**) dock to substrates that contain pTyr residues and polyproline regions respectively (see the structure of the SH2-pY domain in Figure 11.27B). Down-regulation of Src activity occurs when another kinase, Csk, phosphorylates Tyr527 in the C-terminal domain. In v-Src (of the cancer-causing avian retrovirus Rous sarcoma virus), this Tyr is missing and the Src activity is unregulated, driving pathways for oncogenic growth and differentiation.

The inactive form of Src has a compact structure in which the SH2 and SH3 domains are located on the opposite surface of the kinase from the catalytic site (Figure 12.21B), and pTyr527 interacts with the SH2 domain. Inactivity is ensured both by displacement of the kinase C helix from its conformation in the active enzyme through interactions of the SH2-to-SH3 linker, and by the engagement of the SH2 and SH3 domains at each end of the two kinase lobes, preventing the lobes from arranging themselves in the position required for catalysis.

The SH2–SH3 linker adopts a conformation similar to that of a polyproline helix (although in Src it contains only one Pro), allowing it to dock into the SH3 domain (see Figure 11.27B). The sequence surrounding pTyr527 in Src is different from the optimal sequence

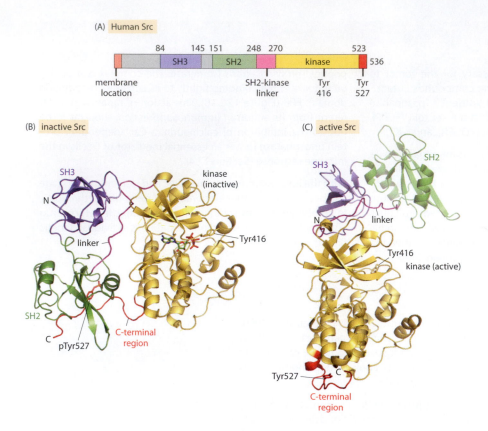

(A) Human Src

(B) inactive Src

(C) active Src

Figure 12.21 Inactive and active forms of Src. (A) Domain map for human Src. (B) Inactive Src with Tyr527 phosphorylated. An ATP analog (in stick representation) is bound at the catalytic site in the kinase domain. Color code as in (A). (PDB 2SRC) (C) Active Src with Tyr527 unphosphorylated. The SH3 and SH2 domains are rotated ~130° relative to their positions in inactive Src. In both structures, the construct lacked the N-terminal region depicted in gray in (A). (PDB 1Y57)

for binding to Src SH2 domains; the C-terminal tails of the Src family kinases are therefore not among the tightest-binding ligands for SH2 domains. Hence, proteins containing high-affinity pTyr peptides are able to displace the bound pTyr527. In inactive Src, Tyr416 in the activation segment is directed into the catalytic site; for activation, pTyr527 must be dephosphorylated, Tyr416 must be phosphorylated, and the constraints of the SH2 and SH3 domains relieved by their binding to ligands (pTyr and Pro-rich regions of proteins, respectively), which breaks their inhibitory grip. Thus, Src kinase has several levels of control.

With Tyr527 not phosphorylated, the SH2 and SH3 domains rearrange to form an elongated structure (Figure 12.21C). The kinase domain is in the active conformation, but the absence of a phosphoryl group on Tyr416 means that the activation segment is not properly ordered. The C-terminal region, with no phosphoryl group on Tyr527, binds to a hydrophobic pocket within the C-terminal lobe.

Growth factor receptor tyrosine kinases signal to the small G protein Ras

Growth factor receptors signal via a series of recognition and phosphorylation events that lead to the activation of transcription factors. The EGF signaling pathway is depicted in **Figure 12.22**, omitting for simplicity other pathways and cross-talk between them. In fact, the signaling system is better perceived as a network rather than a pathway. The pTyr residues in the C-terminal region of the activated receptor act as docking sites for the adaptor protein Grb2 (growth factor receptor binding protein 2). Grb2 (217 aa) consists of one SH2 domain sandwiched between two SH3 domains and has no catalytic function (see also Figure 11.27). Instead, the N-terminal SH3 domain binds to a Pro-rich region of the protein SOS (Son of Sevenless), thereby recruiting SOS to the membrane, whereas the C-terminal SH3 domain of Grb2 binds to Gab (Grb2-associated binder). Gab belongs to a family of large multi-site docking proteins that also have a role in signaling, for example through phosphatidylinositol 3-kinase (see below). Association of the Grb2/SOS complex with the tyrosine kinase receptor at the plasma membrane leads to its binding to **Ras**, a small G protein, which is tethered to the membrane by lipid (both palmitoyl and farnesyl) modification (see Figure 12.5A). The guanine exchange factor (GEF) activity of SOS then causes Ras to exchange its GDP for GTP.

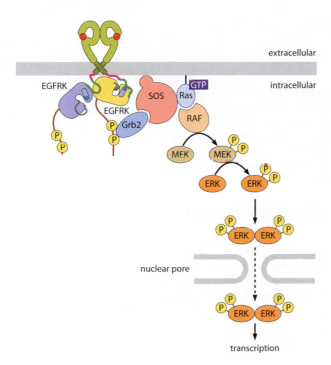

Figure 12.22 Regulation of the Ras/MAP kinase pathway by receptor tyrosine kinases. The pathway is exemplified by EGF signaling (see Figure 12.20D). The participation of Gab is omitted for simplicity. EGFRK, EGFR kinase. For other details, see the main text. (Adapted from B.D. Gomperts, I.M. Kramer and P.E.R. Tatham, Signal Transduction, 2004.)

SOS (1330 aa) is a multidomain protein (**Figure 12.23A**). The N-terminal domain contains two tandem histone folds that modulate interactions with the membrane and are part of the autoinhibitory apparatus. These are followed by a Dbl homology (DH) domain and a pleckstrin homology (PH) domain (see **Guide**). The next two domains, the Ras exchanger motif (Rem) domain and the Cdc25 domain, are both required for the Ras-specific GDP/GTP exchange activity. The C-terminal region contains the docking site, around -PPPVPPRRR- (residues 1150–1158, single-letter code), for Grb2. The binding of Ras to the catalytic site of SOS disrupts the Ras nucleotide-binding site, and any GDP or GTP is expelled. The loading of GTP to Ras in preference to GDP is due solely to their intracellular concentrations, that of GTP being about 10-fold higher; if Ras is empty, it will preferentially bind GTP. Ras-GTP binds at a second site on SOS located between the Rem and Cdc25 domains on the opposite side of the molecule to the catalytic site (Figure 12.23B); binding here allosterically enhances the GDP–GTP exchange activity of SOS toward Ras. Ras-GDP can also bind at the allosteric site but with approximately 10-fold lower affinity, and when bound it sets an intermediate level of activity. The distortion of Ras caused by binding to SOS^{cat} is shown in Figure 12.23C: the two key regions of Ras—switches 1 and 2—that mediate Ras-GTP

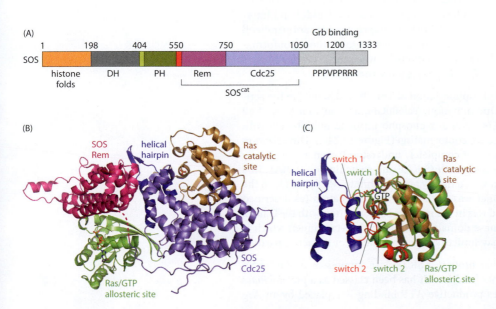

Figure 12.23 SOS recognition of Ras. (A) A domain map of SOS. (B) Structure of human SOS^{cat} (Rem–Cdc25 domains) in complex with Ras. Color code as in (A). Empty Ras (brown) binds at the SOS catalytic site with its nucleotide-binding site (partly obscured in this view) distorted by the helical hairpin of the Cdc25 domain. The binding of Ras/GTP (green) at a second site between the Rem and Cdc25 domains allosterically enhances the GDP–GTP exchange activity of SOS. (C) Ras (brown) at the catalytic site with Ras/GTP (green) from the allosteric site superimposed, showing how the helical hairpin of the Cdc25 domain (blue) displaces the switch 1 and switch 2 regions (in red) of Ras bound at the catalytic site. There is also some displacement of the phosphate-binding loop and the loop that contacts the nucleotide base. (PDB 1NVV)

recognition are displaced by the helical hairpin loop of the Cdc25 domain. In intact SOS, the DH and PH domains block the allosteric binding site and diminish exchange activity, suggesting that there are further levels of regulation of SOS that are likely to involve the targeting of SOS to the membrane.

The activation of Ras has profound consequences; it is therefore no surprise that SOS is subject to complex regulatory control. The presence of the allosteric binding site for Ras in SOS means that Ras helps tether SOS to the membrane. This increases the specific activity of SOS by up to 500-fold compared with SOS and Ras free in solution. The influence of membrane surface environment on the association and clustering of membrane-localized proteins is still poorly understood but is crucial for the next stage in the signaling: the activation of another kinase, RAF, so called because it was first discovered in rat fibrosarcoma.

The Ras/RAF/MEK1/ERK2 signaling pathway leads to activation of transcription

Ras-GTP initiates a three-tiered kinase cascade consisting of RAF, MEK1, and ERK2/ MAPK (see Figure 12.22). RAF, a Ser/Thr kinase, is activated on binding to Ras-GTP, and then phosphorylates the **MAP kinase kinase**, MEK1, on its activation segment. MEK1, in turn, activates another kinase, **MAP kinase** (mitogen-activated protein kinase), by phosphorylating two residues, a Tyr and a Thr, on its activation segment. Once phosphorylated, MAP kinase (also known as **ERK2**, extracellular signal-regulated protein kinase 2) migrates from the cytoplasm to the nucleus, where it phosphorylates and activates several transcription factors that control the expression of genes required for cell growth. This signaling process, operating solely through protein phosphorylation, is able to activate gene transcription within an hour of receptor stimulation.

C-RAF (RAF-1), the founding member of the RAF family, was identified as the human homolog of a viral oncoprotein. Two RAF paralogs, A-RAF and B-RAF, were later identified; B-RAF mutations are associated with 60% of malignant mutations for melanoma and also with other cancers. B-RAF comprises an N-terminal region that includes the Ras-binding domain (RBD); a cysteine-rich domain (CRD), which is involved in membrane association and also contributes to Ras binding; a Ser/Thr-rich region containing a conserved Ser (Ser365) with a regulatory function; a Ser/Thr kinase domain; and a C-terminal region, which also has a Ser (Ser729) and serves a regulatory function (**Figure 12.24A**). In its inactive state, the regulatory domains of B-RAF are phosphorylated (probably by PKA) and fold over the kinase domain. The pSer residues in these domains can also interact with 14-3-3 proteins (see **Guide**), which stabilize the inactive state and prevent recruitment to the plasma membrane.

When RAF binds to Ras-GTP, it adopts an open conformation that relaxes the interactions between the regulatory domains and the kinase domain. Recruitment to the membrane is accompanied by dephosphorylation of Ser365 and dissociation of 14-3-3 protein. All Ras-binding proteins have a common RBD, which has a ubiquitin fold (see **Guide**) and interacts preferentially with GTP-bound Ras (Figure 12.24B). An intermolecular antiparallel β sheet is formed, including part of Ras switch 1, one of the two regions whose conformation is altered when GTP is bound. The difference in affinity between the GTP-bound state ($K_d = 0.13$ μM) and the GDP-bound state ($K_d = 46$ μM) is more than 100-fold.

For maximal activation, B-RAF requires phosphorylation of Thr599 and Ser602 in the activation segment of the kinase domain. One mutation, Val600Glu, accounts for 90% of all B-RAF mutations in human cancers. Glu600 acts as a phospho-mimetic and interacts with Lys507 from the C helix to stabilize the active conformation (Figure 12.24C). Dimerization also has a role in activation but one that is not completely understood; even a catalytically compromised B-RAF is capable of dimerizing with C-RAF and switching on C-RAF. In a crystal structure of the Val600Glu mutant with a tightly bound inhibitor PLX4032 (K_i in the nanomolar range), one kinase in the dimer was in the active conformation (as described above) and the other with the drug bound was in the inactive conformation with the C helix swung out (Figure 12.24D). The two kinase domains form a head-to-tail dimer, which is different from that of EGFR kinase but may similarly align the C helix for the active partner.

RAF activity is also dependent on another protein, KSR (kinase suppressor of Ras). KSR is structurally similar to RAF but does not bind Ras. It has been classed as a *pseudokinase* because the conserved Lys that mediates productive ATP binding is replaced by an Arg

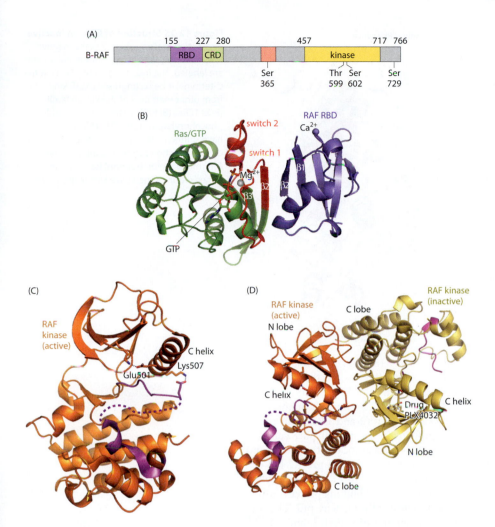

Figure 12.24 Structure of B-RAF and its activation. (A) Domain map of B-RAF, with sites of phosphorylation (Ser, Thr) indicated. (B) Structure of the RBD domain in complex with Ras/GTP. The Ras used in this structure was Rap, whose sequence is 57% identical to that of Ras. (PDB 1GUA) (C) B-RAF kinase domain in the active conformation owing to the activatory mutation Val600Glu in the activation segment (magenta) mimicking a phosphorylated residue. Glu600 interacts with Lys507 from the C helix. The contact with Glu501 holds Lys483 in the correct position to bind ATP. A break in the electron density for the activation segment is shown as a dotted line. (PDB 3OG7) (D) In complex with the potent inhibitor PLX4032, a RAF dimer is found in which one subunit is active and the other subunit, to which the drug is bound, is inactive. The association between the N and C lobes is head-to-tail. (PDB 3OG7)

residue. This decreases but does not eliminate kinase activity. KSR is constitutively bound to MEK and can form dimers with RAF and bind to ERK. It is thought to act as a scaffold protein that facilitates signaling by helping to coordinate the interaction of two or more components of the signaling pathway. Scaffold proteins can also prevent inappropriate cross-talk by insulating different cascades from each other.

Activation of MEK and ERK2 leads to activation of transcription

MEKs (~400 aa) exist as two isoforms, MEK1 and MEK2. They have an N-terminal regulatory domain, a catalytic kinase domain, and a C-terminal region. Near their C termini, MEKs have a segment of ~24 aa termed a versatile docking domain (VDD), which they use to associate with RAF. MEKs are activated by RAF-catalyzed phosphorylation of two Ser residues in their activation segment. Docking domains are also important for the subsequent recognition of ERK2 by MEK1. In resting cells, inactive MEK1 forms a stable complex with inactive ERK2. This interaction facilitates phosphorylation of ERK2 by MEK1; on activation, MEKs act as dual-specificity kinases and phosphorylate Tyr and Thr residues in the activation segment of ERKs, their only physiological substrates.

There are four main MAP kinases: ERK1 and ERK2, the prototype MAPKs, which respond to growth factor signaling; the JNK (Jun N-terminal kinase) family, which are stress-activated; the p38 family, also stress-activated; and the ERK5 family, which appear to respond to both mitogens and stress. ERK2 (42 kDa) achieves maximum activity when two residues, Thr183 and Tyr185, in the activation segment are phosphorylated by MEK. ERK2 undergoes lobe closure, allowing the catalytic and ATP-binding sites to take up their correct conformations for catalysis, and the activation segment is reorganized to create the substrate recognition surface (**Figure 12.25**). pThr183 and pTyr185 nestle in their new positions in

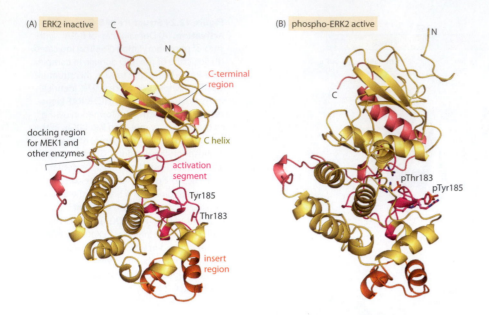

(A) ERK2 inactive

C-terminal region

docking region for MEK1 and other enzymes

C helix

activation segment

Tyr185

Thr183

insert region

(B) phospho-ERK2 active

pThr183

pTyr185

Figure 12.25 Structure of ERK2 in inactive and active states. (A) ERK2 in the inactive (unphosphorylated) state. Functional regions are labeled. The insert region of ~50 aa in the C-terminal lobe distinguishes MAP kinases from other members of the kinase family. (PDB 1ERK) (B) ERK2 in the active (doubly phosphorylated) state. pThr183 and pTyr185 have shifted positions, and the conformation of the activation segment is altered. The C-terminal region (salmon) has a role in dimerization of the active kinase. (PDB 2ERK)

a bed of positively charged side chains (Arg68, 146, 170, 187, 189, and 192). Single phosphorylation of ERK2 is not sufficient to promote these changes (it gives less than 1% of the catalytic activity of the doubly phosphorylated form). Activation of ERK2 also promotes dimerization; the dimer dissociation constant changes from 20 μM for non-phosphorylated ERK2 to 7.5 nM for phosphorylated ERK2 (pERK2). Both pERK2 and ERK2 can translocate into and out of the nucleus, interacting with proteins in the nuclear pore complex (Section 10.5), but phosphorylation and dimerization favor nuclear localization. A second import mechanism that requires energy and cytosolic factors occurs with pERK2 but not with ERK2. These and other findings suggest that the distribution of ERK2 is controlled not only by import and export processes but also by the presence of binding proteins with a range of affinities in both the cytoplasm and the nucleus. Once in the nucleus, pERK2 phosphorylates and activates several transcription factors, including p53, Ets1/2, and p62[TCF] (Elk1). ERK also stimulates protein synthesis by regulating the initiation factor 4E (eIF-4E). In total, more than 160 ERK substrates have been identified.

Growth factor receptor responses also activate at least two other pathways

On activation by EGF, the pTyr residues that are generated in the cytoplasmic part of the EGFR also recruit and activate phospholipase Cγ (PLCγ), a Ca^{2+}-dependent enzyme that breaks down certain membrane phospholipids and initiates additional signaling pathways (**Figure 12.26A**). PLCγ, unlike other PLC isoforms, contains two SH2 domains that recognize the phosphorylated receptor plus a PH domain (see **Guide**) that binds to membrane 3-phosphorylated phosphatidylinositol phosphates (PIPs). The action of PLCγ on phosphatidylinositol 4,5-bisphosphate (PIP2), a minor component of the inner leaflet of the plasma membrane, leads to the formation of diacylglycerol (DAG) and the water-soluble inositol 1,4,5-trisphosphate (IP3) (Figure 12.26B). DAG remains in the membrane where, among other effects (including prostaglandin formation), it activates protein kinase C (PKC). PKC phosphorylates several targets essential for growth, and also provides negative feedback by phosphorylating growth factor receptors, leading to their inactivation. IP3, a second messenger, interacts with an ion channel in the membrane of the endoplasmic reticulum, giving rise to an increase in the concentration of cytosolic Ca^{2+}, from ~0.1 μM to as high as 10 mM. In association with the Ca^{2+}-binding protein calmodulin, this triggers events such as muscle contraction (Section 14.4) and glucose mobilization. The third pathway also involves recognition of pTyr by the SH2 domain of another enzyme, phosphatidylinositol 3-kinase (PI-3K). PI-3K phosphorylates PIP2 to generate phosphatidylinositol 3,4,5-trisphosphate (PIP3), another signaling phospholipid, which activates other isoforms of PKC and protein kinase B (PKB). PKB, sometimes called Akt, plays an important part in encouraging cell survival by suppressing apoptosis (Section 13.7) and promoting anabolic activity such as glycogen synthesis (turning off GSK3; see above).

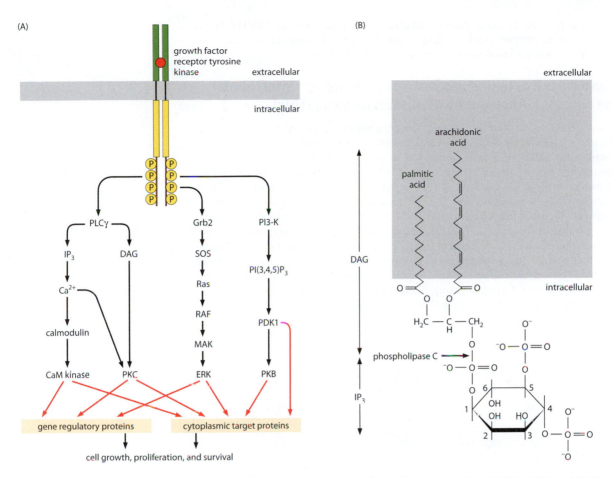

Figure 12.26 Signaling through phosphatidylinositol phosphates. (A) Outline of signaling pathways beginning with PLCγ, Grb2, and PI-3K. (B) Action of PLCγ on phosphatidylinositol 4,5-bisphosphate (PIP₂) to generate diacylglycerol (DAG), which remains in the membrane, and inositol 1,4,5-trisphosphate (IP₃), which is soluble and is released into the cytoplasm.

12.4 CYTOKINE SIGNALING

Cytokines are small (< 200 aa) soluble proteins that signal directly to the nucleus via a pathway that involves receptor binding, kinase activation, and the phosphorylation and activation of transcription factors. They are important in cell growth and development, the immune response (innate and adaptive), the inflammatory response, cell differentiation, the inhibition of cell proliferation, and response to viral infection. Cytokines fall into various structural classes, including the following: the helical cytokines such as human growth hormone (hGH), interleukin 2 (IL-2), and interferon γ (IFNγ), where the defining structural feature is a four-helix bundle; the cystine-knot growth factors of the **transforming growth factor β (TGFβ) family**; the trimeric tumor necrosis factor (TNF) family, for example TNFα, which is produced by macrophages during acute inflammation and has a role in apoptosis (Section 13.7); and the β-trefoil growth factors such as interleukin 1α (IL-1α) and interleukin 1β (IL-1β) which are involved in the inflammatory response. The interleukins— the name derives from the observation that many of these molecules act between white blood cells (leukocytes), where many of them are also synthesized—were numbered in the order in which they were discovered; the number does not correspond with any particular function. Chemokines are cytokines that bring about local inflammation by recruiting leukocytes through chemotaxis and subsequently activating them. They are small proteins (< 100 aa) characterized by four Cys residues that form two pairs of disulfide bridges and are classified according to the spacing between the first two Cys residues.

Cytokine receptors are grouped into families on the basis of structural features common to their ligands: class I cytokine receptors for helical cytokines; class II cytokine receptors for interferons and IL-10, somewhat like class I receptors; TGFβ receptors; TNF receptors; and immunoglobulin superfamily receptors (such as that for IL-1). In this section, we focus on the signaling mechanisms for class I (hGH) and class II (IFNγ) receptors and for the TGFβ receptors. Chemokine receptors are G-protein-coupled receptors (Section 12.2) and so have signaling pathways that are less direct than those of other cytokines.

Class I and class II cytokine receptors signal through intracellular tyrosine kinases (JAKs) and transcription factors (STATs)

IFNγ is released by *T cells* and *natural killer cells* and is a potent immunoregulator. It induces the production of MHC class II molecules and promotes the expression of CD4 in T cells, thus increasing the ability of T cells to respond to viral protein fragments displayed on *antigen-presenting cells* (Section 17.3). The human IFNγ receptor (class II) consists of two copies of two subunits, α (472 aa) and β (316 aa) (**Figure 12.27A**), but in unstimulated cells the two subunits are not associated. Their N-terminal extracellular regions are variably glycosylated and comprise two fibronectin type III (FnIII) domains (see **Guide**); these are followed by a transmembrane helical region (~23 aa) and a cytoplasmic region that has no enzyme activity but binds inactive forms of intracellular tyrosine kinases of the **JAK** family. JAK originally stood for 'just another kinase', but when its function was better understood it was renamed 'Janus kinase' after the Roman god of beginnings and transitions, usually depicted facing both ways (hence January). There are four JAKs (JAK1, JAK2, JAK3, and TYK2). JAK1 binds to the α subunit of the IFN receptor and JAK2 to the β subunit, close to

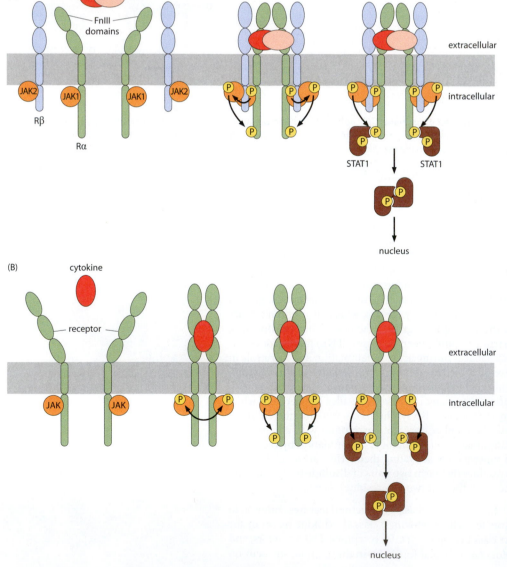

Figure 12.27 Cytokine signaling.
(A) Schematic view of the IFNγ receptor. The IFNγ homodimer binds to Rα, promoting receptor dimerization and recruitment of Rβ. JAK1 (bound to Rα) and JAK2 (bound to Rβ) are brought into juxtaposition. They phosphorylate and activate each other (JAK2 first, then JAK1) and then phosphorylate Tyr440 toward the C-terminal end of the intracellular region of both Rα subunits. Each pTyr recruits a STAT1 through the latter's SH2 domain, leading to STAT phosphorylation, dissociation, dimerization, and entry to the nucleus to promote transcription. (B) A single-subunit receptor with three FnIII extracellular domains, a transmembrane helix, and a JAK-binding intracellular domain. Binding of a cytokine promotes receptor dimerization and the two JAKs phosphorylate and activate each other. This is followed by STAT recruitment, phosphorylation, dimerization, and entry to the nucleus, as in (A).

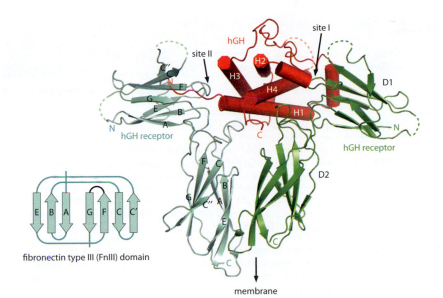

fibronectin type III (FnIII) domain

Figure 12.28 hGH in complex with the extracellular domains D1 and D2 of the hGH receptor. In the hGH structure (red), helices are shown as cylinders for simplicity. The receptor domains (green, α; light blue, β) have fibronectin III-type folds, with the topology shown in the inset. Disordered regions are shown with dotted lines. The C-terminal ends of the D2 domains (marked C) connect to the transmembrane regions of the receptor subunits. (PDB 3HHR)

the cell membrane in both instances. Binding of IFNγ, a dimer, to the extracellular domains of the α subunits induces receptor dimerization, which brings the intracellular domains of the α and β subunits into juxtaposition. Juxtaposition allows activation of their associated JAK kinases, which first phosphorylate themselves and then Tyr440 toward the C-terminal end of the intracellular region of the α subunit. The pTyr residues in turn provide docking sites for SH2 domains in a set of transcription factors termed **STATs** (signal transducer and activator of transcription).

There are seven members of the mammalian STAT family (STAT1–6, including the closely related STAT5A and STAT5B), and alternative splicing generates several additional isoforms. Each family member has a non-redundant function: in general, STAT1 and STAT2 signal in *innate immunity*, STAT4 and STAT6 in *acquired immunity*, and STAT3 and STAT5s in tissue injury and inflammation. In the human IFNγ receptor, a STAT1 molecule (750 aa) is recruited to each α subunit and, once bound through its SH2 domain, is phosphorylated on Tyr701 by the JAK kinases. The phosphorylation induces dissociation from the receptor, and the phosphorylated STAT subunits dimerize. The dimers migrate to the cell nucleus, bind to IFNγ-activated sequences (GAS) in the promoter regions of target genes, and drive specific gene expression commensurate with an antiviral state for the cell.

In other instances, class II and class I cytokine receptors may be homodimers and, for class I receptors, may have more than two FnIII domains per subunit and a conserved sequence motif, -Trp-Ser-X-Trp-Ser-, toward the end of their extracellular regions that is necessary for protein folding and cell surface binding. However, the overall mechanism remains the same (Figure 12.27B). The hGH receptor (class I) is a case in point: the extracellular regions of the two identical subunits comprise two FnIII domains connected by a four-residue Pro-rich linker, and the four-helix pituitary GH, although itself an asymmetric monomer, binds across the symmetric receptor dimer (**Figure 12.28**). There is evidence that the dimer is preformed in the membrane and that GH binding induces relative rotation of the intracellular regions, thereby bringing their associated JAK2 kinase molecules into juxtaposition. Transphosphorylation activates the kinases, which then phosphorylate and activate STATs 5A and 5B, but STATs 1 and 3 have also been implicated. Thus the transcription of many genes associated with GH activity will be initiated. It has also been reported that Src tyrosine kinase (see above) may be activated by GH independently of JAK2.

Overall, great diversity in cytokine signaling can be generated by the combination of cytokine receptors, their associated JAKs, and the STATs they activate (**Table 12.2**).

The gp130 and γ_c subunits are common to many class I cytokine receptor assemblies

The homodimeric nature of the hGH receptor is unusual; most class I cytokine receptors form heterodimers or heterotrimers or even higher oligomers, which feature a common receptor subunit as a signal transducer together with a non-signaling cytokine-specific

Table 12.2 Cytokine/JAK/STAT combinations and responses.

Cytokine receptor type	Cytokine	JAKs activated	STATs activated	Physiological response
Homodimeric receptors	hGH	JAK2	STAT1, STAT5A	Stimulates growth by inducing the production of insulin-like growth factor 1 (IGF1)
	Erythropoietin	JAK2	STAT5A	Stimulates the production of erythrocytes
Class I: common receptor gp130	IL-6	JAK1, (JAK2, TYK2)	STAT1, STAT3	Cytokine with a wide variety of biological functions; a potent inducer of the acute phase response of inflammation. It plays an essential role in the final differentiation of B cells into Ig-secreting cells
	IL-11	JAK1, (JAK2, TYK2)	STAT3	Directly stimulates the proliferation of hematopoietic stem cells and megakaryocyte progenitor cells located in bone marrow and induces megakaryocyte maturation resulting in increased platelet production
	LIF	JAK1, (JAK2, TYK2)	STAT3	Can induce terminal differentiation in leukemic cells; induces hematopoietic differentiation in normal and myeloid leukemia cells, neuronal cell differentiation, and the stimulation of acute-phase protein synthesis in hepatocytes
	CNTF	JAK1, (JAK2, TYK2)	STAT3	A survival factor for various neuronal cell types
	OSM	JAK1, (JAK2, TYK2)	STAT3	A growth regulator; inhibits the proliferation of a number of tumor cell lines
Class I common receptor β_c	GM-CSF	JAK2, (JAK1)	STAT3, STAT5	Stimulates the production of granulocytes and macrophages
	IL-3	JAK2, (JAK1)	STAT3, STAT5, STAT6	Induces granulocytes, macrophages, mast cells, stem cells, erythroid cells, eosinophils, and megakaryocytes
	IL-5	JAK2, (JAK1)	STAT3, STAT5, STAT6	Induces terminal differentiation of late-developing B cells to immunoglobulin secreting cells
Class 1 common receptor γ_c	IL-2	JAK1, JAK3	STAT3, STAT5	Stimulates the proliferation and differentiation of activated T cells
	IL-4	JAK1, JAK3	STAT6	Stimulates B-cell proliferation and differentiation; regulates T_H1 and T_H2 development
	IL-7	JAK1, JAK3	STAT3, STAT5	Promotes memory T-cell survival
	IL-15	JAK1, JAK3	STAT3, STAT5	Promotes memory T-cell survival
Class II receptors	IFNα/β	JAK1, TYK2	STAT1, STAT2	Increases cell resistance to viral infection
	IFNγ	JAK1, JAK2	STAT1	Activates macrophages; stimulates MHC expression
	IL-10	JAK1, TYK2	STAT1, STAT3 STAT5	Inhibits the synthesis of several cytokines, including IFNγ, IL-2, IL-3, TNF, and GM-CSF produced by activated macrophages and by helper T-cells

CNTF, ciliary neurotrophic factor; GM-CSF, granulocyte macrophage-colony stimulating factor; hGH, human growth hormone; IFN, interferon; IL, interleukin; LIF, leukemia inhibitory factor; OSM, oncostatin; T_H1, T_H2, helper T cell types 1 and 2.
(Adapted from Cell Signaling Technology (2010) http://www.cellsignal.com/reference/pathway/jakstat_utilization.html, and from UniProtKB http://www.ebi.ac.uk/uniprot/.)

subunit (the α subunit). There are three major shared receptor subunits, namely gp130, the common γ subunit (γ_c), and the common β subunit (β_c). They share a structural blueprint for the assembly of cytokine receptor signaling complexes but employ substantially different signaling topologies.

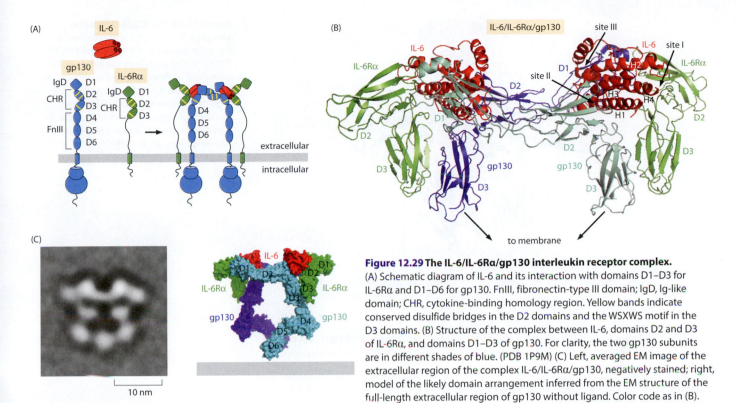

Figure 12.29 The IL-6/IL-6Rα/gp130 interleukin receptor complex.
(A) Schematic diagram of IL-6 and its interaction with domains D1–D3 for IL-6Rα and D1–D6 for gp130. FnIII, fibronectin-type III domain; IgD, Ig-like domain; CHR, cytokine-binding homology region. Yellow bands indicate conserved disulfide bridges in the D2 domains and the WSXWS motif in the D3 domains. (B) Structure of the complex between IL-6, domains D2 and D3 of IL-6Rα, and domains D1–D3 of gp130. For clarity, the two gp130 subunits are in different shades of blue. (PDB 1P9M) (C) Left, averaged EM image of the extracellular region of the complex IL-6/IL-6Rα/gp130, negatively stained; right, model of the likely domain arrangement inferred from the EM structure of the full-length extracellular region of gp130 without ligand. Color code as in (B). (PDB 3L5H and 1P9M) (A, adapted from P.J. Lupardus et al., *Structure* 19:45–55, 2011. With permission from Elsevier. C, courtesy of Christopher Garcia.)

The cytokine-binding site of the gp130 subunit recognizes at least nine cytokines, including Il-6, IL-11, IL-27, leukemia inhibitory factor (LIF), and oncostatin-M (OSM). IL-6 is secreted by T cells and macrophages and stimulates the immune response, notably during infection and after trauma, especially tissue damage leading to inflammation. The cytokine (212 aa in humans) first forms a low-affinity ($K_d \approx 10$ nM) complex with the IL-6Rα subunit (51 kDa), whose intracellular region (~277 aa) does not bind JAK. The IL-6/IL-6Rα heterodimer then binds to a preformed dimer of gp130 (2 × 101 kDa) in the membrane to form a competent signaling complex (cytokine $K_d \approx 10$ pM) composed of six subunits, two each of IL-6, IL-6Rα, and gp130 (**Figure 12.29A**). The major cytokine recognition site of IL-6Rα is at the elbow region between its D2 and D3 domains, and the two main contacts from IL-6 with gp130 are in the bend between domains D2 and D3 of gp130 (sites I and II). A third site (site III) is created between the N-terminal end of IL-6 and domain D1 of gp130, crossing over from the other subunit in the dimer (Figure 12.29B). Thus, neither IL-6 nor IL-6Rα on their own can bind gp130. A bend between the D4 and D5 domains in gp130 brings the D6 domains together (Figure 12.29C). As with the hGH receptor, the membrane-proximal interactions stabilize the complex and are likely to ensure the correct orientation of the transmembrane and cytoplasmic domains for JAK activation.

The γ_c subunit is common to a wide range of class I cytokine receptors, among them those for IL-2, IL-4, IL-7, IL-9, IL-15, and IL-21. Certain mutations in γ_c abolish the activity of all γ_c-dependent cytokine receptors and result in X-linked severe combined immunodeficiency diseases (X-SCIDs), in which the T and NK (natural killer) cells are profoundly reduced in number. Without effective treatment, patients die of opportunistic infections before 1 year of age, but in most cases they can be cured by bone marrow transplantations. IL-2 is the primary cytokine responsible for the rapid expansion, differentiation, and survival of antigen-selected T-cell clones during an immune response. It is also important for B-cell, NK cell, and regulatory T-cell function. IL-2 has therapeutic applications as an immune adjuvant in certain types of lymphoproliferative diseases and cancers, and IL-2 antagonists can prevent organ transplant rejection. IL-2 (153 aa) binds tightly ($K_d \approx 10$ pM) to its receptor, IL-2R, which is composed of three subunits: IL-2Rα (30 kDa), IL-2Rβ (58 kDa), and γ_c (41 kDa). All three are important for recognition and signaling, and a critical number of IL-2Rs must be triggered before an individual T cell will make the irrevocable decision to pass through

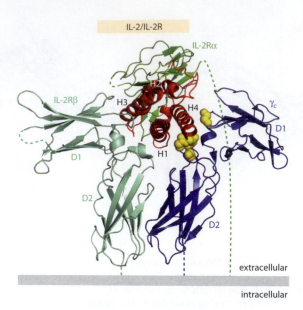

IL-2/IL-2R

IL-2Rα

IL-2Rβ

γc

H2
H3   H4
H1

D1

D1

D2

D2

extracellular

intracellular

Figure 12.30 The IL-2/IL-2 receptor complex. IL-2 (four-helix bundle, red) is shown in complex with its receptor, IL-2Rα/IL-2Rβ/γc. The path from the extracellular domain of each receptor protein (green, Rα; cyan, Rβ; blue, γc) to the transmembrane regions is shown by a dashed line. The positions of five missense mutations in γc that give rise to X-SCID are marked with yellow spheres. (PDB 2B5I)

the G1 restriction point in the cell cycle to undergo DNA replication and subsequent mitosis (Chapter 13). However, exactly how the cell senses this threshold is not understood.

The structure of the quaternary complex of IL-2 with the extracellular regions of the IL-2 receptor IL-2Rα/IL-2Rβ/γc has features similar to those of IL-6 and hGH bound to their receptors. The four-helix bundle IL-2 is clamped between the elbow regions of two different receptor subunits (D1 of IL-2Rβ, and D1 of γc) that are related by a pseudo-two-fold axis of symmetry (**Figure 12.30**). Neither IL-2 nor IL-2Rβ alone can bind to γc, but a combination of weak interactions creates a high-affinity complex for IL-2/IL-2Rβ/γc. The 'flat' surface of γc in this region forms a degenerate recognition surface that allows it to recognize several different cytokines. Signaling occurs through the intracellular domains of IL-2Rβ and γc, which bind JAK1 and JAK3 respectively. The interface formed between IL-2 and γc is not extensive, but five missense mutations in this region of γc have been reported in patients with X-SCID and are mapped in Figure 12.30.

The β$_c$ subunit is common to another set of cytokine receptor assemblies

Granulocyte–macrophage colony-stimulating factor (GM-CSF, 127 aa in humans) controls the production of blood cells and also regulates dendritic and T cells. Abnormalities in GM-CSF receptor signaling have been implicated in several diseases, including rheumatoid arthritis. Its receptor is a heterodimer of a ligand-specific α subunit (GMRα, 42 kDa in humans) which has no JAK-binding properties in its 54 aa intracellular region, and a β$_c$ subunit (97 kDa in humans), which is shared with IL-3 and IL-5 receptors and binds JAK2 in its ~437 aa intracellular region. The receptors are expressed at low levels (~100–1000 per cell) on the surface of hematopoietic cells. The GMRα subunit has three extracellular domains, D1–D3, of which (D1) is Ig-like and D2–D3 are FnIII domains (**Figure 12.31A**). It binds the cytokine with modest affinity ($K_d \approx 1$ nM), and GM-CSF/GMRα then associates with β$_c$ to form the high-affinity ($K_d \approx 100$ pM) receptor, leading to dimerization and receptor activation. The receptor takes the form of a hexamer with 2:2:2 stoichiometry of (GM-CSF/GMRα/β$_c$)$_2$. The extracellular region of β$_c$ is composed of four FnIII domains, and GM-CSF binds to domain D2 of GMRα and to β$_c$ at a site in the elbow region between domain D1 from one β$_c$ subunit and domain D4 of the second β$_c$ subunit. The contribution from two different symmetrically related molecules is unusual and underscores the versatility of the FnIII domain in cytokine recognition.

In the dimer, the C-terminal ends of the extracellular regions of the two β$_c$ subunits are ~120 Å apart, making it hard to understand how this mode of dimerization could lead to the two intracellular regions coming together to activate the JAK2 kinases bound to them. However, in the crystal lattice two hexameric dimers come together head-to-head around a two-fold axis of symmetry to form a dodecamer. This brings the C-terminal tails

(A)

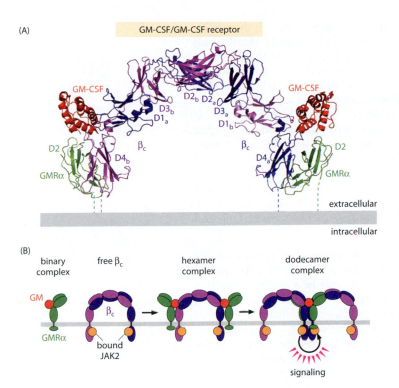

GM-CSF/GM-CSF receptor

GM-CSF

D3ᵦ
D1ᵦ
D2ᵦ D2ₐ
D3ₐ
D1ᵦ

βc

βc

D2
D4ᵦ
GMRα

GM-CSF

D4ₐ
D2
GMRα

extracellular

intracellular

(B)

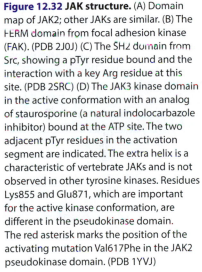

binary complex

free βc

hexamer complex

dodecamer complex

GM

βc

GMRα

bound JAK2

signaling

Figure 12.31 The GM-CSF/GM-CSF receptor complex. (A) The receptor is composed of two subunits, GMRα and βc (one blue and one magenta), which bind two copies of the cytokine GM-CSF in 2:2:2 stoichiometry. All four extracellular domains (D1–D4) of each βc subunit are shown, but only the D2 domains of the two GMRα subunits. Paths from the C-terminal ends of the extracellular domains to the membrane are shown by dotted lines. (PDB 3CXE) (B) Two head-to-head hexamers make up a dodecamer, bringing the two βc receptors together and promoting dimerization of JAK2 and transphosphorylation. (Adapted from G. Hansen et al., *Cell* 134:496–507, 2008. With permission from Elsevier.)

of neighboring βc domains D4 and GMRα domains D2 into close proximity (~10 Å), suggesting that this may be physiologically relevant (Figure 12.31B). The importance of the dodecamer is supported by studies of directed mutations that show that disrupting the sites of interaction in the dodecamer assembly leads to loss of function.

Signaling through JAKs and STATs induces transcription of target genes

JAK kinases (the first to be isolated was TYK2 in 1990) have a unique domain organization: the N-terminal region contains a FERM domain (see **Guide**) and an SH2 domain, and these are followed by two tyrosine kinase domains, only one of which is active (**Figure 12.32A**). The FERM domain comprises three subdomains (F1, F2, and F3), each

(A)

35		370	480	545	806	845	1132
	FERM		SH2		pseudokinase	Tyr-kinase	

Val617Phe
JAK2

(B) FERM (FAK)

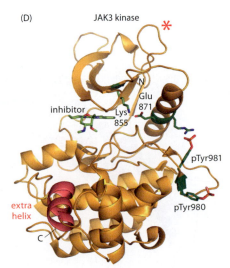

F1

F3

F2

(C) SH2 Src

Arg

pTyr

(D) JAK3 kinase ✱

N

Glu 871

inhibitor

Lys 855

pTyr981

extra helix

pTyr980

C

Figure 12.32 JAK structure. (A) Domain map of JAK2; other JAKs are similar. (B) The FERM domain from focal adhesion kinase (FAK). (PDB 2J0J) (C) The SH2 domain from Src, showing a pTyr residue bound and the interaction with a key Arg residue at this site. (PDB 2SRC) (D) The JAK3 kinase domain in the active conformation with an analog of staurosporine (a natural indolocarbazole inhibitor) bound at the ATP site. The two adjacent pTyr residues in the activation segment are indicated. The extra helix is a characteristic of vertebrate JAKs and is not observed in other tyrosine kinases. Residues Lys855 and Glu871, which are important for the active kinase conformation, are different in the pseudokinase domain. The red asterisk marks the position of the activating mutation Val617Phe in the JAK2 pseudokinase domain. (PDB 1YVJ)

of which is similar in topology to previously recognized domains, although there is little sequence similarity (Figure 12.32B). F1 contains a five-stranded mixed β sheet packed against an α helix with a short 3_{10} helix similar to that in ubiquitin (Section 12.5); F2 is composed of five α helices and has a topology similar to that of an acyl-CoA-binding domain (Chapter 9); and F3 has the topology of a pTyr-binding (PTB) (see **Guide**) or pleckstrin homology (PH) domain. The FERM domain binds to the intracellular region of the receptor, close to the membrane for the gp130 in the IL-6R (see Figure 12.29A). The FERM/receptor interface serves to position the JAK kinase domain so that ligand-induced receptor dimerization and reorientation can lead to mutual JAK activation. The sequence and structure of the JAK SH2 domain is similar to those of canonical SH2 domains, but the crucial Arg that contacts the pTyr in SH2 recognition proteins such as Src (Figure 12.32C) is either missing (JAK2, TYK2) or can be mutated (JAK1, JAK3) without interfering with cytokine signaling. It has been proposed the SH2 domain does not act as a pTyr recognition domain in signaling but may act by interacting with other binding partners such as SOCS (suppressor of cytokine signaling; Section 12.5).

The kinase domain of JAK3 shows a typical kinase fold composed of N-terminal and C-terminal lobes linked by a hinge region (Figure 12.32D). A short extra helix inserted into the C-terminal region conserved in all JAK structures may be a site for interaction with the FERM domain, and two pTyr residues (980 and 981) on the activation segment are similar to two of the three Tyr residues that become phosphorylated in the insulin receptor tyrosine kinase (Section 12.3), indicative of a similar activation mechanism. The pseudokinase domain lacks catalytic activity because key residues that recognize the triphosphate moiety of ATP are different. However, patients with myeloproliferative neoplasms (a group of blood cancers) express a mutant form of JAK2 with a Val617Phe mutation in the JAK2 pseudokinase domain, leading to unregulated kinase activity and suggesting some regulatory role of the normal pseudokinase domain. JAKs are important targets for drug therapy against inflammatory and immunological diseases.

Their partner STATs were first identified as transcription factors that are activated by interferons α and γ. They have a modular structure with seven domains (**Figure 12.33A**). The N-terminal domain regulates the dimerization of STATs in their inactive state and is involved in cooperative DNA binding. The C-terminal transactivation domain varies among family members and modulates the transcriptional activation of target genes. The STAT SH2 domains act as docking sites for one or more cytokine-stimulated pTyr residues

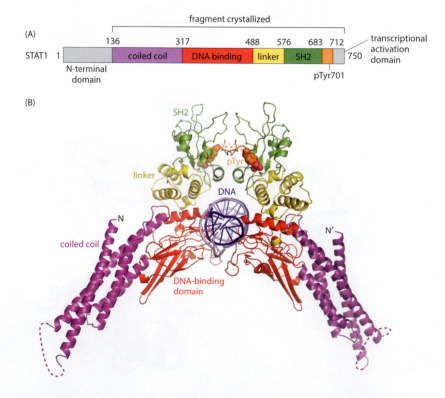

Figure 12.33 The STAT1/DNA complex.
(A) Domain map of human STAT1.
(B) Structure of STAT1 co-crystallized with an 18-bp DNA duplex containing the core sequence TTCCGTAA, which corresponds to part of the c-Fos promoter. The view is normal to the 2-fold axis of symmetry of the dimer and down the axis of the DNA molecule. The side chain of pTyr701 in each subunit (orange spheres) crosses over to contact the SH2 domain of the opposing subunit. Breaks in electron density are indicated by dotted lines. (PDB 1BF5)

on the intracellular region of the receptor and, once localized, the STATs are phosphorylated by JAKs at a conserved Tyr in their C-terminal regions (see Figure 12.27). STATs then form dimers in which each SH2 domain binds the pTyr on the opposing subunit. STATs may also be activated by other tyrosine kinases such as Src and Abl, and it is possible that JAKs can be activated by other routes.

As transcription factors, the two pTyr/SH2 interactions provide the only protein/protein contacts between the STAT monomers arranged with two-fold symmetry in the dimer. The DNA-binding domains have an immunoglobulin fold (see **Guide**), which is unusual for DNA-binding proteins but is also observed for the DNA-binding regions of the tumor suppressor protein p53 and the p50 subunit of the transcription factor NFκB (Figure 12.33B). They grip the DNA like a pair of pliers, contacting the DNA through loops at the ends of the β sheet.

TGFβ signals through a combination of RI and RII Ser/Thr kinase receptors

During development, TGFβs (there are three isoforms that share ~75% sequence identity) influence pattern formation, cell proliferation, differentiation, extracellular matrix production, and apoptosis. In adults, they are involved in tissue repair, immune regulation, and signaling inhibition of proliferation for most cells. TGFβ is also one of a number of growth factors that regulate the self-renewal and differentiation of stem cells and are used to maintain stem cell populations in culture. Transcription factors activated in response to TGFβ signaling often act in combination with other transcription factors, diversifying the genes transcribed. One such gene encodes the INK4 inhibitor p15, which controls cell cycle progression from G0 to G1 (Section 13.2). Mutations in TGFβ receptors or receptor-activated transcription factors appear in up to 50% of some human cancers, whereas loss of receptors produces tumors that are not responsive to growth inhibition by TGFβ.

TGFβ3 is cleaved proteolytically from a precursor of 412 aa and forms a homodimer of subunits (112 aa) linked by a disulfide bond between Cys77 and Cys77′. TGFβ, together with nerve growth factor (NGF) and platelet-derived growth factor (PDGF), belongs to the class of cystine-knot proteins. The core of the subunit contains three disulfide bridges; Cys44–Cys109 and Cys48–Cys111 form a ring wide enough for the Cys15–Cys78 disulfide bridge to pass through (**Figure 12.34A**). This creates a rigid knot-like structure that replaces the hydrophobic core commonly found in globular proteins.

Figure 12.34 Interaction of TGFβ3 with its receptor. (A) The TGFβ3 homodimer. The two subunits (one in red, one in pink) are cross-linked by a disulfide bridge between Cys77 and Cys77′. The cystine knot in each subunit is sketched below, in which the three disulfides are shown in stick representation. (B) The TGFβ3/Tβ-RI/Tβ-RII complex. Upper, viewed down the 2-fold axis of symmetry from above toward the membrane; lower, the extracellular domains viewed parallel to the plane of the membrane. (PDB 2PJY)

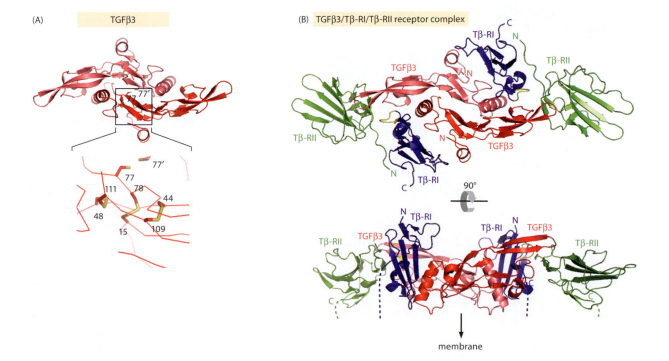

TGFβ receptors have two subunits, RI and RII, each with an extracellular binding domain, a transmembrane region, and a cytoplasmic Ser/Thr kinase domain. Another type of subunit, RIII, lacks kinase activity but may act to concentrate ligands near the RII subunit. Despite their Ser/Thr kinase activity, TGFβ receptors have sequences that are more similar to those of tyrosine kinases, suggesting a possible evolutionary link. Members of the TGFβ super-family bind to the extracellular domains of a specific RI/RII dimer, a combination selected from the seven RI and five RII receptors. The dimeric TGFβ binds to its RI receptor with high affinity and is responsible for cooperative recruitment of the low-affinity RII subunit. This brings the intracellular kinase domains together so that the RII receptor kinase, which is constitutively active (master), can phosphorylate and activate the RI receptor (slave). The structure of TGFβ3 in complex with the extracellular regions of its RI receptor (Tβ-RI) and RII receptor (Tβ-RII) reveals the two-fold symmetric TGFβ3 cytokine at the center of the heterotetramer (Figure 12.34B). How this places the intracellular kinase domain of RI where it can be phosphorylated by RII remains to be established.

Phosphorylation of SMADs by RI induces transcription of target genes

In its intracellular region, the RI receptor has a conserved run of ~30 aa immediately pre-ceding the kinase domain that includes the sequence TTSGSGSGLP (one-letter code), hence its name GS region. Many receptor kinases are activated by phosphorylation on the activation segment (Section 12.3), but activation of the Tβ-RI kinase (with its typical kinase fold) is achieved by multiple phosphorylation of the GS region by the RII Ser/Thr kinase (**Figure 12.35**). Phosphorylation displaces the GS region from its inhibitory binding site against the N-terminal lobe of the kinase. The next steps in TGFβ signaling are somewhat similar to those of the JAK/STAT pathway. **SMAD** proteins (~50 kDa) are transcription fac-tors linked to the receptor by a membrane-associated adaptor protein (SARA, for SMAD anchor of receptor activation). SMADs are named after the first two such proteins identified: Sma in *Caenorhabditis elegans* and MAD in *Drosophila melanogaster*). Eight isoforms are found in humans: regulatory R-SMADs (isoforms 1–3, 5, and 8), co-regulatory Co-SMADs (isoform 4), and inhibitory I-SMADs (isoforms 6 and 7). Only R-SMADs are phosphoryl-ated by receptor RI kinases on two Ser residues at their C termini: SMAD2 and SMAD3

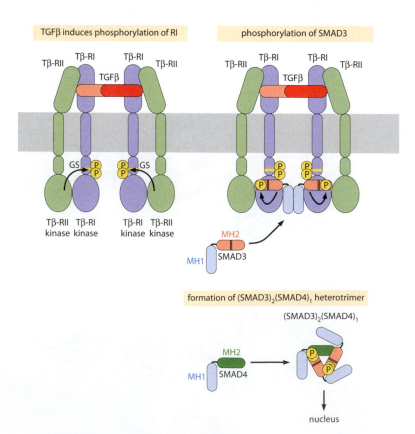

Figure 12.35 TGFβ signaling. Left, the TGFβ dimer binds to the Tβ-RI receptor subunits, promoting receptor association and recruiting the Tβ-RII receptor subunits. The intracellular kinase domain of Tβ-RII is constitutively active and is now able to phosphorylate the kinase domain of Tβ-RI in the GS region, activating the Tβ-RI kinase. Right, receptor SMADs (SMAD3 shown here), composed of MH1 and MH2 domains, are recruited to the Tβ-RI subunits and phosphorylated at two Ser residues in the C-terminal region. As depicted below, phosphorylated R-SMADs then dissociate and assemble with SMAD4 to form heterotrimers that translocate to the nucleus to initiate transcription. The bar on the MH2 domain of SMAD3 represents the L3 loop.

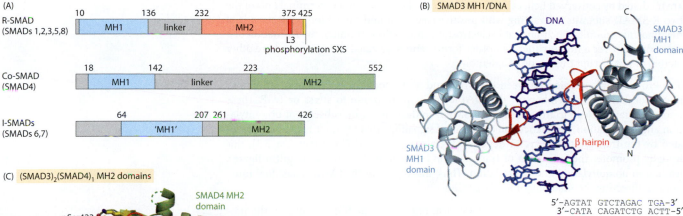

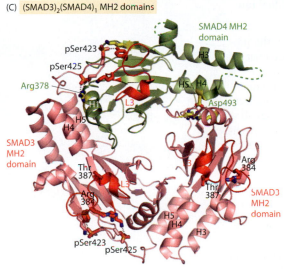

Figure 12.36 Structure of SMAD domains. (A) Domain maps of SMAD subunits. The residue numbering for R-SMADS is that of SMAD3, and that for I-SMADS is that of SMAD7, in which the MH1 domain is distantly related and does not bind DNA. (B) The MH1 DNA-binding domain of SMAD3 in complex with DNA, illustrating the β-hairpin interaction. The sequence of the double-stranded DNA-binding element used is shown in the inset. (PDB 1MHD) (C) The heterotrimer of MH2 domains (SMAD3)$_2$(SMAD4)$_1$. The two pSer residues (423, 425) at the C-terminal end of each SMAD3 make contact with side chains from a neighboring subunit (side chains in stick representation). The important contacts between Asp493 of SMAD4 and a SMAD3, and between pSer425 of a SMAD3 and Arg378 of SMAD4, are shown with side chains in yellow stick representation. The L3 loops are in red. Residues Arg384 and Thr387 (labeled) in each SMAD3, are important for specific receptor interactions. (PDB 1U7F)

phosphorylated by TGFβ or activin receptors, and SMAD1, SMAD5, or SMAD8 by bone morphogenetic protein receptors. SMAD3 is illustrated in Figure 12.35. Phosphorylated R-SMADs dissociate from the receptors and, in the presence of Co-SMAD4, form signaling heterotrimers, (R-SMAD)$_2$(SMAD4)$_1$, which then enter the nucleus and activate or inhibit the transcription of specific genes.

SMAD family proteins have three domains: an N-terminal MAD homology domain 1 (MH1) that can interact with other proteins, carries a nuclear localization signal, and binds to specific DNA sequence motifs; a Pro-rich linker region that contains two phosphorylation sites (Ser and Thr); and a C-terminal MH2 domain that can bind to receptors, interact with other proteins, and mediate homo- and hetero-oligomerization (**Figure 12.36A**). SMAD2 does not bind to DNA because of inserts in its MH1 domain, which interfere with the DNA-binding region. In SMADs 1, 2, 3, 5, and 8, at the C-terminal end of the MH2 domain is an SXS motif (X is any amino acid) housing the Ser residues that become phosphorylated when the SMAD binds to activated receptors. Among the target genes activated by TGF signaling are those for the I-SMADs: SMAD6 and SMAD7. These bind to activated receptors and inhibit signaling by competing for R-SMAD binding sites and by recruiting a ubiquitin ligase that ubiquitylates the receptor, thus leading to degradation. The I-SMADs also recruit protein phosphatases that dephosphorylate the receptor and thereby terminate signaling.

DNA recognition by SMADs is highly unusual

The crystal structure of the MH1 domain of SMAD3 in complex with a palindromic DNA-SMAD-binding element (SBE) shows two MH1 molecules bound identically on either side in the major groove of the DNA, where they make specific H-bonds with both the bases and the phosphodiester backbone (Figure 12.36B). To make contact with the DNA, the SMAD domain employs a β hairpin that is highly conserved across all SMADs. This is highly unusual—most DNA-binding proteins interact with DNA through an α-helical motif that docks into the major groove. The SMAD trimeric structure is also unusual, exemplified in the (SMAD3)$_2$(SMAD4)$_1$ complex of MH2 domains (Figure 12.36C). Inter-subunit contacts are mediated by conserved helices and loops and the phosphorylated C-terminal tails of the two R-SMAD subunits interacting with positive charges round the L3 loops/β8 strands of neighboring subunits. Such favorable electrostatic interactions promote the formation of a heterotrimer over a homotrimer, but other forms cannot be ruled out *in vivo*, where other factors might disrupt this mode of assembly.

The Pro-rich linker region between the MH1 and MH2 domains is the site of inhibitory phosphorylations by members of the MAPK family in response to stress or EGF. These phosphorylations lead to the retention of SMAD in the cytoplasm, subsequent ubiquitylation and degradation, and hence the cessation of signaling. This region is also phosphorylated by the transcriptional cyclin-dependent kinases Cdk8 and Cdk9 (Section 5.3); this initially promotes the activation of transcription but eventually attracts ubiquitin ligases, leading to ubiquitylation and degradation. Thus, this region of the SMADs has dual functions in directing the final outcome of signaling.

TGFβ cytokines mediate a diversity of biological responses. The SBE is found in the promoter regions of many TGFβ-responsive elements, often in multiple copies and/or adjacent to binding sequences for other transcription factors. The trimeric association of R-SMADs and SMAD4, together with additional DNA-binding factors, could facilitate their combinatorial binding to multiple gene sequences, which, together with the binding of adjacent transcription factors, could provide a mechanism for diversity of response. TGFβ signaling can also involve non-SMAD pathways; different, and not yet completely elucidated, mechanisms are involved in TGFβ activation of the different MAPK pathways leading to activation of the MAPK family members ERK, JNK, and p38.

12.5   UBIQUITYLATION

Ubiquitylation is a process for tagging proteins with **ubiquitin** or polyubiquitin for downstream signaling and processing, first discovered in the late 1970s by Avram Hershko, Aaron Ciechanover, and Irwin Rose in studies of protein degradation in crude extracts of reticulocytes. Ubiquitin is a small (76 aa) conserved protein present in all eukaryotic cells. Ubiquitylation is part of a key mechanism in the selective control of many biological processes. It is involved in the degradation of normal regulatory proteins or abnormal proteins by the proteasome (Section 7.4) and in endocytosis and the down-regulation of receptors and transporters (Chapter 10). The programmed degradation of regulatory proteins, such as cyclins, inhibitors of cyclin-dependent kinases, and anaphase inhibitors, is essential for cell cycle progression (Chapter 13). Cell growth and proliferation are further controlled by the ubiquitin-initiated degradation of tumor suppressors, proto-oncogenes, numerous transcriptional regulators, and components of signal transduction systems.

There are four main types of ubiquitylation (**Figure 12.37A**). In monoubiquitylation, a single ubiquitin molecule is attached through an isopeptide bond between the carboxyl group of its C-terminal Gly and the N^6-amino group of a Lys residue in the protein substrate. In multiubiquitylation, single ubiquitin molecules are attached to amino groups of multiple Lys residues (or N-terminal residue) of the protein substrate. In polyubiquitylation, a ubiquitin polymer, sometimes of mixed linkage, is assembled on a single ubiquitylated site. Finally, in linear ubiquitylation, ubiquitin chains are assembled in a head-to-tail fashion by linking the N terminus of a bound ubiquitin to the C-terminal Gly of an additional ubiquitin. Monoubiquitylation of either histones or PCNA (proliferating cell nuclear antigen) is involved in DNA replication (Chapter 3) and the DNA-damage response pathway (Chapter 4). Multiubiquitylation is associated with protein localization, protein interactions, and modulation of protein activity. Polyubiquitylation signals are responsible for recognition of target substrates for degradation by the proteasome and for coordinating a wide variety of cellular processes.

Ubiquitin has an N-terminal globular domain with a β sheet that presents a hydrophobic surface capable of interacting with many other proteins, followed by a flexible C-terminal tail (Figure 12.37B). Ubiquitin monomers contain seven Lys residues and can form various polyubiquitin and linear ubiquitin chains by means of different isopeptide linkages. Lys48-linked ubiquitin chains are the main signal for targeting to the proteasome; Lys29- and Lys11-linked ubiquitin polymers are also used, but less frequently. Lys63-linked polyubiquitin chains are associated with DNA repair pathways and cell surface receptor-mediated activation of signaling pathways.

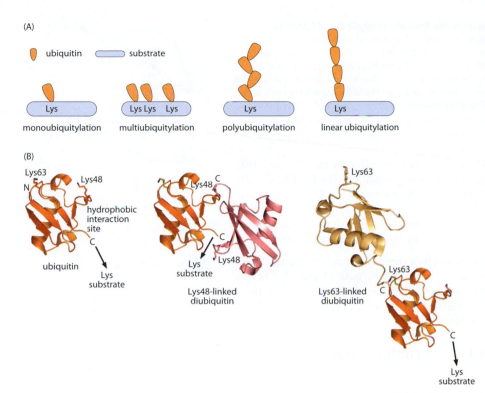

(A)

ubiquitin   substrate

| monoubiquitylation | multiubiquitylation | polyubiquitylation | linear ubiquitylation |

(B)

Lys63   Lys48
N
hydrophobic interaction site
C
ubiquitin   Lys substrate

Lys48   C
C
Lys substrate   Lys48
Lys48-linked diubiquitin

Lys63
Lys63
C
Lys63-linked diubiquitin
C
Lys substrate

Figure 12.37 Ubiquitin and ubiquitin oligomers. (A) Schematic of the four types of ubiquitylation of protein substrates. (B) Left, the structure of ubiquitin; the C-terminal Gly is making an isopeptide link with a substrate Lys residue. (PDB 1YIW) Middle, the structure of Lys48-linked diubiquitin. The second ubiquitin is in pink. There is an isopeptide bond between Lys48 of the first ubiquitin and the C-terminal Gly of the second. (PDB 1AAR) Right, the structure of Lys63-linked diubiquitin. The first ubiquitin is in yellow, the second in beige. There is an isopeptide bond between Lys63 of the first ubiquitin and the C-terminal Gly of the second. (PDB 1JF5) (A, adapted from V. Nagy and I. Dikic, *Biol. Chem.* 391:163–169, 2010.)

Post-translational modification of proteins also occurs with the **ubi**quitin-like proteins (UBLs) **SUMO** (**small ubiquitin-like modifier**) and **Nedd8** (**neural precursor cell expressed, developmentally down-regulated 8**). SUMO modification has many functions including regulating protein stability, nuclear/cytoplasmic transport, and transcription factors. The only known substrates of Nedd8 modification are the cullin subunits of SCF (Skip1–Cullin–F-box) ubiquitin E3 ligases (described below), and Nedd8 is implicated in cell cycle progression and cytoskeletal regulation. Unlike ubiquitylation, 'sumoylation' and 'neddylation' do not trigger protein degradation.

Protein ubiquitylation is catalyzed by a series of three enzymes

The first enzyme (E1) in the series activates the C-terminal carboxyl group of ubiquitin by adenylating it (**Figure 12.38**). The ubiquitin~adenylate then undergoes nucleophilic attack by a Cys in the catalytic site of E1 to generate an E1~ubiquitin thioester (chemically analogous to the formation of aminoacyl-tRNA described in Section 6.2). The ubiquitin is then transferred from E1 to the catalytic Cys of the E2 enzyme, generating an E2-ubiquitin thioester (E2~Ub) but, before this transfer, the E1 must bind a second ubiquitin at the adenylation site and undergo a major conformational change. In the third and final step, the activated ubiquitin is transferred from E2~Ub to the nucleophilic acceptor Lys of the protein substrate, a reaction catalyzed by E3, a protein-ubiquitin ligase. The RING and U-box domains of the E3 ligase (described below) promote the transfer of the activated ubiquitin by aligning the active E2~Ub thioester bond optimally for nucleophilic attack by the deprotonated N^6-NH$_2$ of the substrate Lys. (**Figure 12.39**). In crystal structures of E2~Ub

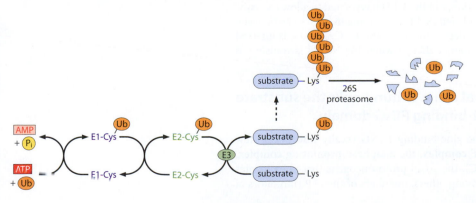

Figure 12.38 The polyubiquitylation reaction. E1, ubiquitin-activating enzyme; E2, ubiquitin-conjugating enzyme; E3, ubiquitin ligase. Further ubiquitin molecules can be added (dotted arrow) to the growing polyubiquitin chain on the target Lys residue of the substrate protein. The ubiquitylated substrate protein is designated for degradation by the 26S proteasome. (Adapted from L.A. Passmore and D. Barford, *Biochem. J.* 379:513–525, 2004.)

Figure 12.39 Transfer of ubiquitin from the catalytic Cys of E2 to the acceptor Lys of a substrate protein. (A) The RING domain of the E3 ligase (not shown) positions the thioester bond between the catalytic Cys of the E2 and the C-terminal Gly of ubiquitin optimally for nucleophilic attack. The catalytic machinery is provided by the E2. (B) The tetrahedral intermediate is stabilized by the formation of an oxyanion hole in the E2. (C) The tetrahedral intermediate has collapsed, leaving the ubiquitin now with an isopeptide link of its C-terminal Gly to the target Lys of the substrate. (Adapted from L.A. Passmore and D. Barford, *Biochem. J.* 379: 513–525, 2004.)

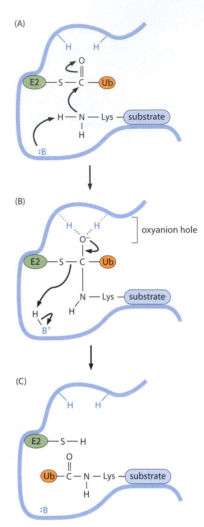

conjugates bound to RING E3 ligases, the catalytic site is provided by the E2. The RING domain of E3 interacts with both the E2 and ubiquitin moieties of the E2~Ub, thereby locking the C-terminal tail of the Ub into an active-site groove on the E2. This mediates structural changes at the E2 catalytic site, and a repositioned Asp residue promotes catalysis by deprotonating the $^+NH_3$ group of the target Lys of an incoming substrate. The tetrahedral intermediate that forms is stabilized by an oxyanion hole (Section 1.6) generated in E2.

E2s are classified into 17 subfamilies, characterized by a core ubiquitin-conjugating (UBC) domain that includes the catalytic Cys residue and interacts with both E1 and E3. E3s specifically couple a designated protein substrate to one or more cognate E2s. In humans, there are two E1s, ~30 E2s, and ~600 E3s, reflecting the importance and specificity of E3s for selective ubiquitylation as a biological control process.

Activation of UBLs depends on a major conformational change in E1

The E1s for ubiquitin, SUMO and Nedd8, have a common molecular framework. The E1 for ubiquitin is a monomer, whereas those for SUMO and Nedd8 are heterodimers, SAE1/UBA2 and NAE1/UBA3 respectively, in which the second subunit (UBA2 or UBA3) acts as a scaffold.

The various activities of E1 depend on the modularity and malleability of its structure, which consists of an adenylation domain, a catalytic Cys-containing domain, and a ubiquitin fold domain (UFD) that resembles ubiquitin and binds E2. The UBL initially binds in a large groove mainly in the adenylation domain, with its C terminus fastened in the adenylation active site by a 'crossover loop' that links the adenylation domain and the catalytic Cys-containing domain. In this conformation, the catalytic Cys is more than 30 Å from the C terminus of the UBL.

Studies of chemically trapped intermediate analogs of the SUMO E1 revealed that after the initial adenylation reaction, a major structural transformation occurs; some structures associated with the adenylation reaction are dismantled and replaced by new ones associated with thioester bond formation. The domain containing the catalytic Cys rotates, bringing the Cys adjacent to the C terminus of the UBL. In this conformation, the catalytic Cys can attack the C-terminal UBL~adenylate, the E1~UBL thioester is formed, and AMP is released.

Insight into UBL transfer from E1 to E2 came from the structure of a stable E1/E2/ATP. Mg^{2+}/Nedd8(T)/Nedd8(A) complex, in which one Nedd8(T) is in thioester linkage at the E1 catalytic site and a second Nedd8(A) has bound at the adenylation site (**Figure 12.40A**). In this structure, the catalytic Cys-containing domain has essentially returned to the position it occupied during the adenylation reaction, but the UFD has rotated to allow the associated E2 to face the E1 catalytic Cys and bind Nedd8(T). There remains a 20 Å gap between the E2 Cys (replaced here with an Ala to render E2 inactive) and E1 Cys, and it is inferred that there must be an additional conformational change when the Nedd8 is transferred from E1 to E2 in an activated enzyme (Figure 12.40B).

In E3 cullin RING ligases, a central cullin subunit links the substrate recognition complex and the E2-binding RING domain

The largest class of E3 ligases contains the zinc-binding **RING (really interesting new gene)** domain. This class includes the **SCF complex**, the **anaphase-promoting complex**, also known as the **cyclosome** (thus APC/C), the c-Cbl proto-oncogene that ubiquitylates activated tyrosine kinase receptors, and, among others, members of the IAP (inhibitors of

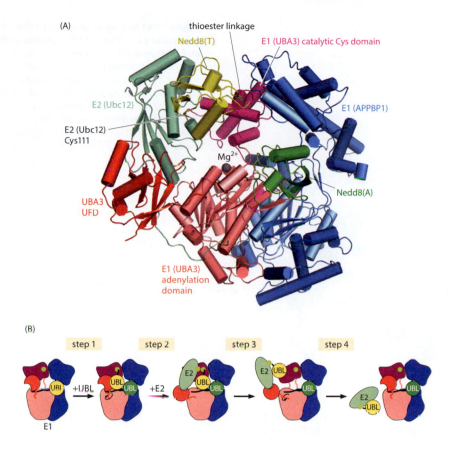

(A)

E2 (Ubc12)

E2 (Ubc12)
Cys111

thioester linkage

Nedd8(T)

E1 (UBA3) catalytic Cys domain

E1 (APPBP1)

Mg²⁺

UBA3
UFD

Nedd8(A)

E1 (UBA3)
adenylation
domain

(B)

step 1   step 2   step 3   step 4

+UBL   +E2

E1

Figure 12.40 Transfer of the thioester-linked UBL from E1 to E2. (A) The Nedd8 E1 enzyme (UBA3/APPBP1) in complex with E2 (Ubc12). The domains of E1 UBA3 are labeled; APPBP1 is the scaffold protein. The C-terminal–COOH group of Nedd8(T) is in thioester linkage with the active-site Cys in the E1 UBA3 catalytic domain. The second UBL Nedd8(A) has bound at the adenylation site with its C-terminal region reaching into the ATP site, which is marked by Mg²⁺ (blue sphere). Nedd8 E2 (Ubc12) is located between Nedd8(T) and the UBA3 UFD, with its catalytic Cys (Cys111) ~20 Å from the catalytic Cys of E1. (PDB 2NVU) (B) A model for successive steps in the transfer of the thioester-linked UBL from E1 to E2. The color scheme is as in (A). The C-terminal tail of the UBL is in black. Catalytic Cys residues in E1 and E2 are marked by a small yellow-green circle. Step 1, the first UBL to bind (Nedd8(T)) forms a thioester link with the Cys in the E1 catalytic domain and a second UBL (Nedd8(A)) is bound. Step 2, Nedd8(T) clashes with the UFD, displacing it, and E2 can then bind. Step 3, the catalytic Cys of E2 attacks the thioester bond holding the UBL, Nedd8(T), to E1. Step 4, the E2~UBL dissociates and the system reverts to its starting state for another round. (Adapted from D.T. Huang et al., *Nature* 445:394–398, 2007. With permission from Macmillan Publishers Ltd.)

apoptosis) protein family (Section 13.7). The U-box domain ligases adopt the same overall fold as RING proteins but lack Zn-binding sites, whereas the HECT (homology to E6 C terminus) domain family of E3 ligases differs mechanistically and structurally from both RING and U-box domain ligases in that they form an enzyme–ubiquitin intermediate (making a thioester bond with the ubiquitin C terminus) before substrate ligation. HECT E3 enzymes (not discussed further here) regulate the trafficking of many receptors, channels, transporters, and viral proteins.

Two important cullin RING E3 ligases (CRLs) are SCF and the APC/C. The **SCF complex** (the prototype member) consists of three invariant subunits, Skp1 (an adaptor, 163 aa), Cul1 (cullin, 776 aa), and Rbx1 (the RING domain-containing protein, 108 aa), plus a variable subunit, the **F-box** subunit (the substrate receptor protein). Members of the F-box family (~40 encoded in humans, ~400 to >1000 aa) share an N-terminal F-box motif, a three-helix bundle of ~40 aa that binds Skp1. Their specificity in binding substrate proteins is defined by the protein/protein interaction modules, such as leucine-rich repeats (LRRs) or WD40 domains, that follow (**Figure 12.41A**). At this stage the E3 ligase is catalytically inactive until, as described below, it undergoes neddylation to bring the E2~Ub into efficacious proximity to the substrate protein. Different F-box proteins with widely different substrate specificities promote the ubiquitylation of proteins such as cyclins, cyclin-dependent protein kinase inhibitors, IκBα (the inhibitory protein of transcription factor NFκB), and β-catenin (a cytoskeletal adaptor protein and transcription factor), targeting them for subsequent degradation by the proteasome. Thus they are important in the selective regulation of a range of biological activities: the cell cycle, the immune response, cell–cell adhesion, and signaling networks.

The human genome encodes seven **cullins** (Cul1, 2, 3, 4A, 4B, 5, and 7), which organize specific multi-subunit ubiquitin E3 ligases. Three distinct adaptor-binding domains have been defined: the Cul1-binding Skp1 (S-phase kinase association protein 1) binds F-box proteins; Cul2 and Cul5-binding elongin BC binds SOCS (suppressor of cytokine signaling) boxes; and Cul4-binding DDB1 (damaged DNA binding protein 1) binds DCAFs (DDB1 and Cul4A associated factors). For Cul3-based ligases, the adaptor and substrate receptor functions are combined into a single protein that binds the substrate through a C-terminal domain and Cul3 directly through an N-terminal BTB (broad-complex, tramtrack, and bric

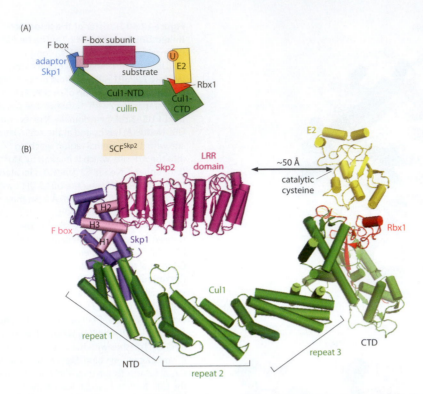

Figure 12.41 A composite model of the catalytically inactive SCF^{Skp2} complex with an E2 enzyme. (A) A diagram of the complex. (B) A composite structural model (color scheme as in part A) assembled from Skp1/Cul1/Rbx1/F-box(Skp2) (PDB 1LDK and 1LDJ), Skp1/Skp2 (PDB 1FQV), and c-Cbl/UbcH7 (PDB 1FBV). The α helices of the F-box motif of Skp2 (labeled) are connected by a linker region to the substrate recognition module, composed of 10 LRRs

à brac) domain, also known as POZ (pox virus and zinc finger) domain (see **Guide**). CRL substrate recognition diversity is conferred by a variety of protein recognition domains. These domains include WD40, LRR, Kelch, and ankyrin and armadillo repeat motifs (ARM) (see **Guide**).

The E3 ligase substrate recognition motif is termed a *degron*, a generic term for a degradation signal (Section 7.2). For many SCFs, substrate recognition is dependent on phosphorylation of the substrate degron. Thus, a primary form of SCF regulation relies on the activities of kinases and phosphatases. For example, GSK-3 (glycogen synthase kinase 3) generates the phospho-degron on β-catenin, the transcription factor involved in the Wnt signaling pathway, which is important for cell adhesion and gene regulation. This attracts the action of SCF$^{β-TrCP1}$ (SCF with the F-box protein β-TrCP1). Similarly, the action of SCF$^{β-TrCP1}$ on IκB is initiated by the IκB kinase (IKK), which phosphorylates IκB and targets it for destruction. This allows NFκB to be liberated and to enter the nucleus. Other CRLs can recognize degrons that have been modified by glycosylation or prolyl hydroxylation.

CRLs are assembled on a rigid cullin subunit and can be prevented from assembling by an inhibitor protein

In a model of the fully assembled SCF^{Skp2} complex (SCF with the F-box subunit Skp2), Rbx1 and Skp1 subunits are bound at opposite ends of a rigid Cul1 subunit (Figure 12.41B). Cul1 is curved, with an N-terminal domain (NTD, ~415 aa) comprising three repeats of an α-helix bundle (cullin repeat). The three cullin repeats create an arc-like structure spanning 110 Å, and the NTD binds the Skp1 adaptor through helix/helix interactions. The C-terminal domain (CTD, ~360 aa) of Cul1 is composed of a four-helix bundle, an α/β domain with a five-stranded β sheet that contains as its second strand the N-terminal segment of Rbx1, and two copies (WH-A and WH-B) of a winged-helix motif (see **Guide**). This is shown in more detail in **Figure 12.42A**. The CTD binds the RING domain of Rbx1 mostly through the intermolecular β sheet and a V-shaped groove formed by the α/β domain on one side and the WH-B domain on the other. The Skp1-binding site on the NTD is ~100 Å from the Rbx1-binding site on the CTD.

The Rbx1 RING domain comprises a three-stranded β sheet (60 aa) and an α helix whose interactions with two large loops are stabilized by two tetrahedrally coordinated Zn^{2+} ions (see Figure 12.42A). These loops and the α helix form a shallow hydrophobic groove that interacts with the E2. In a complex of an E2 (UbcH7) with the RING domain of the proto-oncogene c-Cbl (an E3 of the RING family), two loops of the E2 pack in and around the

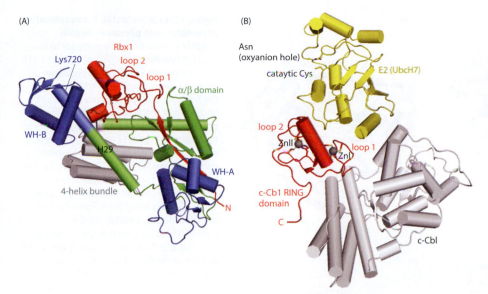

Figure 12.42 Rbx1 RING-domain interactions. (A) Interaction of Rbx1 with cullin CTD in the SCF^{Skp2} complex. The N-terminal region of Rbx1 (red) inserts a β strand into the sheet of the Cul1 α/β domain. Lys720 marks the site for neddylation. (PDB 1LDJ). (B) E2/RING domain interactions in a c-Cbl/UbcH7 complex. The RING domain (red) is in the C-terminal region of c-Cbl. The N-terminal region, which differs from the cullin ligases, is shown in gray. In the RING domain, three copies of a Cys-X-X-X-Cys motif plus a fourth pair of ligands provided by two His residues constitute two tetrahedrally coordinated binding sites for ZnI and ZnII (gray spheres). (PDB 1FBV)

shallow hydrophobic groove on the RING domain (Figure 12.42B). The conformation of UbcH7 in the complex is very similar to that of an isolated E2, indicating that binding to the RING domain does not affect its conformation. The closest distance between the E2 catalytic Cys and any RING domain residue is 15 Å, and the main role of the RING domain thus appears to be to position the substrate protein and the E2~Ub optimally for ubiquitylation of the target lysine residues. The distance between the tip of the C-terminal Skp2 LRR domain and the catalytic Cys of E2 is ~50 Å for the SCF^{Skp2} complex (see Figure 12.41B). This is sufficient to accommodate the substrate, p27^{Kip1} (a CDK inhibitor), to be ubiquitylated by SCF^{Skp2}. Depending on the F-box protein, the same basic SCF structure can accommodate protein substrates of many different shapes and sizes.

An inhibitor protein can prevent Cul/Rbx complexes from assembling with adaptor proteins and from displaying any catalytic activity. Cullin-associated and neddylation-dissociated 1 inhibitor (CAND1) has 27 tandem HEAT repeats (see **Guide**) that wrap around Cul1/Rbx1 like a solenoid. The C-terminal region of CAND1 binds to Cul1 on a surface that overlaps the Skp1-binding site and prevents Skp1 from binding, whereas the CAND1 N-terminal region binds at the junction between the Cul1 WH-B domain and the Rbx1 RING domain and thus locks the Rbx1 RING domain in place against Cul1.

Some F-box subunits use WD40 domains to recognize phosphoprotein substrates

The basic modular architecture first visualized in the SCF^{Skp2} structure is conserved in other CRLs, including those with different F-box subunits whose binding of substrate proteins is based on WD40 domains. These include SCF$^{β-TrCP1}$—a CRL that mediates the degradation of transcription factors such as β-catenin—and the phosphatase Cdc25A; the orthologous human SCF^{Fbw7} and yeast SCF^{Cdc4} CRLs regulate G1/S and G2/M transitions in the cell cycle. In such F-box subunits, an N-terminal F-box motif is connected through a linker to the C-terminal WD40 domain (**Figure 12.43**). The linker regions of β-TrCP1 (β-transducin-repeat-containing protein 1), Fbw7 (F-box/WD-repeat-containing protein 7) and Cdc4 are conserved, with (at least) two rigid α helices forming a stalk that joins the F-box to the β propeller of WD40 repeats. In β-TrCP1 the β propeller is seven-bladed, whereas Cdc4 and Fbw7 have eight-bladed β propellers. In all instances, binding to the F-box subunit is optimized by multi-site phosphorylation of the substrate.

All known substrates of the SCF$^{β-TrCP1}$ complex contain the degron DSGΦXS (where Φ is a hydrophobic amino acid and X is any amino acid). Phosphorylation of both Ser residues in the degron is required for substrate recognition and ubiquitylation. In the SCF$^{β-TrCP1}$ complex bound to a β-catenin substrate, the β-catenin peptide is at the narrow top face of the WD40 β propeller, with the six residues of the phospho-degron at the center of the channel (see Figure 12.43B). The negatively charged phosphates of both pSer residues (pSer33 and pSer37) make electrostatic interactions with Arg residues and additional H-bonds with polar groups located at topologically similar sites on WD40 repeats 2 and 5 respectively.

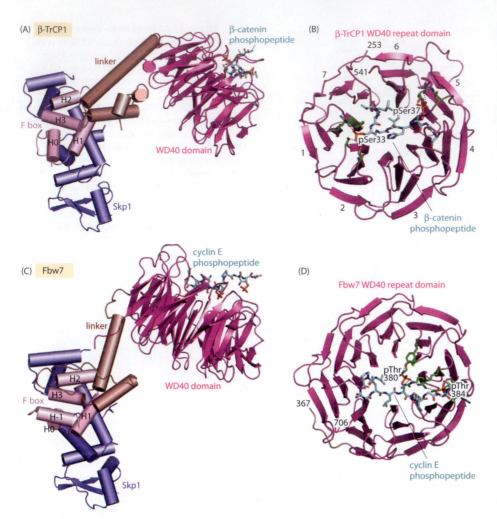

Figure 12.43 Skp1/WD40 F-box subunits in complex with phospho-degron peptides. The orientation is similar to that of SCF^Skp2 in the upper part of Figure 12.41B. (A) Complex of Skp1/β-TrCP1/β-catenin phospho-degron peptide. The helices of the F-box motif of β-TrCP1 are labeled. (B) End-on view of the seven blades of the β propeller, residues 253–541. The blades are numbered 1–7. pSer33 and pSer37 in the degron peptide are indicated; phosphate-contacting residues (Tyr271, Arg285, Ser309, Ser325, Arg431, and Ser448) are shown, but labels are omitted for simplicity. (PDB 1P22) (C) Complex of Skp1/Fbw7/cyclin E phospho-degron peptide. (D) End-on view of the eight blades of the β propeller, residues 367–706. pThr380 and pThr384 in the degron peptide are indicated; phosphate-contacting residues (Arg441, Ser462, Arg465, Arg465, Arg479, and Tyr519) are shown, but labels are omitted for simplicity. (PDB 2OVQ)

The invariant Asp (Asp32) of the degron makes specific electrostatic and polar interactions, the hydrophobic residue (Ile35) contacts a nonpolar pocket, and the conserved Gly (Gly34) packs in a hydrophobic environment with little space to spare for a residue other than Gly. The interaction of Skp1/Fbw7 with the phospho-degron peptide of the target protein cyclin E is similar, although the WD40 domain forms an eight-bladed β-propeller and two pThr residues replace the two pSer residues (see Figure 12.43C,D).

SCF^β-TrCP1 ubiquitylates Lys19 of β-catenin, which is in a position relative to the β-catenin phospho-degron similar to that of the target Lys residues (Lys21 and Lys22) of IκBα with respect to its phospho-degron. If constructs of β-catenin are prepared in which Lys19 is moved relative to the phospho-degron, the rate of ubiquitylation is substantially reduced. Both lines of evidence support the view that SCF and other RING E3 ligases promote catalysis by orienting the protein substrate optimally to present the target Lys to the catalytic Cys in the E2 active site (see above).

Neddylation activates CRLs by releasing the RING domain of Rbx1

The SCF structures pose the intriguing question of how to bridge the gap between the catalytic Cys of the E2 and the target Lys residue(s) of the substrate. The E3 ligase activity is stimulated by neddylation of the cullin. The site of attachment of Nedd8 on Cul1 was found to be Lys720 on the WH-B domain in its CTD, at the rim of the cleft formed by conserved residues from WH-B and the Rbx1 RING domain (see Figure 12.42A). The transfer of Nedd8 to the cullin is autocatalyzed by the E2~Nedd8 conjugate (Ubc12~Nedd8) bound to the Rbx1 subunit. The catalytic Cys on the E2 is ~35 Å from Lys720 on the cullin and they must be brought closer for auto-neddylation to occur.

The structure of Rbx1 in complex with the CTD of the cullin Cul5 to which Nedd8 is attached in isopeptide linkage (Nedd8/Cul5CTD/Rbx1) reveals the large underlying conformational changes. Neddylation of Lys724 of Cul5 (equivalent to Lys720 of Cul1) causes

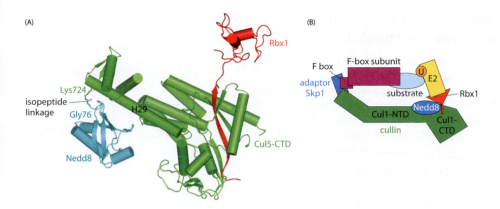

Figure 12.44 Neddylation of Cul5 liberates the Rbx1 RING domain. (A) Overview of the CTD of Cul5 with an isopeptide linkage between Lys724 and the C-terminal Gly76 of Nedd8, in complex with Rbx1. (PDB 3DQV) (B) Schematic structure of the assembled E3 ligase, similar to Figure 12.41A and using same color code, showing how neddylation of Cul1 can liberate Rbx1 and allow the E2~Ub to approach the substrate bound to the F-box subunit for ubiquitylation to occur.

the RING domain of Rbx1 to detach from its binding site on the cullin CTD. It remains linked to the Cul5 scaffold by its N-terminal region entering the intermolecular β sheet with the Cul5 CTD (**Figure 12.44A**). The Rbx1 RING domain should then be free to move about relative to Cul5 Lys724 and to substrates bound to the F-box subunit.

In small-angle X-ray scattering studies, Cul1/Rbx1 was found to adopt an open conformation on neddylation, and the flexible linker attaching Rbx1 to Cul1 was found to be important; making the linker shorter abolished neddylation, whereas making it longer and more rigid reduced the rate of substrate modification. Overall, these findings suggest that CRLs exist in equilibrium between a closed (inactive) conformation and an open (active) conformation in which the RING domain is released from its binding site on the Cul CTD. In the open conformation, Rbx1/E2~Ub can assume multiple positions relative to individual protein substrates and their target Lys residues. Neddylation of the cullin stimulates CRL activity by shifting the equilibrium toward the open conformation (Figure 12.44B).

The APC/C is a large multi-subunit CRL regulated by phosphorylation and co-activators

Progression through the mitotic phase of the cell cycle and G1 is governed by the APC/C. Activated early in mitosis, it ubiquitylates cyclin A and the kinase Nek2A at prometaphase (the cell cycle is described in Sections 13.2 and 13.3). At metaphase, it ubiquitylates securin and cyclin B, two inhibitors of the transition to anaphase. At the exit from mitosis, the key APC/C substrates are Cdc20 (its own co-activator) and B-type cyclins and mitotic kinases. During G1, the APC/C ensures low levels of mitotic cyclins.

Early in mitosis, Cdc20 activates APC/C when the activity of Cdh1 (a structurally related co-activator) is low. APC/C^{Cdc20} inactivates cyclin-dependent kinase (CDK) activity by stimulating the ubiquitylation of cyclins, leading to their degradation. CDK phosphorylation inactivates Cdh1; the destruction of cyclins therefore raises Cdh1 activity. APC/C^{Cdc20} and APC/C^{Cdh1} have overlapping but distinct substrate specificities, and switching between them means that different cell cycle regulators are degraded during separate phases of the cycle. APC/C^{Cdc20} activity is inhibited by the mitotic checkpoint complex (MCC), a multi-protein complex activated by the spindle assembly checkpoint (SAC) (Section 13.5).

The classical APC/C degron is the D-box (or destruction box), a nine-residue motif (-RXXLXX[I/V]XN-, in one-letter code) first characterized in B-type cyclins. Another APC/C degron, the KEN motif (in one-letter code), is commonly present in APC/C substrates in addition to the D-box. Efficient ubiquitylation by either APC/C^{Cdc20} or APC/C^{Cdh1} of substrates that harbor both D and KEN boxes is usually dependent on both degrons.

Without co-activators, the core APC/C is inactive as an E3 ubiquitin ligase and unable to bind most APC/C substrates. Co-activators act as substrate-recognition subunits, analogous to the F-box subunits of the SCF, and they also stimulate APC/C E3 ligase activity. They promote the formation of a ternary complex of APC/C, co-activator, and substrate. The co-activators Cdc20 and Cdch1 contain a C-terminal WD40 domain (similar to many F-box subunits) that interacts with D-box and KEN-box degrons; Apc10, one of the APC/C subunits, has also been shown to be necessary for D-box-mediated substrate recognition, dependent on co-activator. Thus, substrates bind to the APC/C at a bipartite site created from both co-activator and Apc10 subunits.

The APC/C is assembled from subunits containing multiple sequence repeats and undergoes major structural change

In vertebrates, 14 genes encode APC/C subunits, most of which are conserved in other eukaryotes (Table 12.3). The molecular masses and stoichiometries of subunits have been established by mass spectrometry and X-ray crystallography (see **Methods**) of recombinant assemblies. The catalytic core of the APC/C is composed of the cullin subunit Apc2 and the RING-domain subunit Apc11, analogous respectively to the cullin and Rbx1 subunits of SCF. Apc2 and Apc11 form a tight complex and together are able to catalyze ubiquitylation, but without substrate specificity.

The largest APC/C subunit, Apc1, has a C-terminal region composed of 11 tandem repeats of 35–40 aa termed the PC motif and shared with the Rpn1 and Rpn2 subunits of the proteasome regulatory particle (Section 7.4). The small subunits Apc12, Apc13, and Apc16

Table 12.3 Core and regulatory subunits of the APC/C.

S. cerevisiae	H. sapiens	Mass (kDa)[a]	Stoichiometry[a]	Structural motif	Function
Core subunits					
Apc1	Apc1	196	1	PC repeats	Scaffolding subunit
Apc2	Apc2	100	1	Cullin homology	Catalytic center
Cdc16	Apc6	95	2	TPR	Scaffolding subunit
Cdc27	Apc3	85	2	TPR	Cdh1/Doc1 binding
–	Apc7			TPR	Cdh1/Doc1 binding
Apc4	Apc4	73	1	Unknown	Scaffolding subunit
Apc5	Apc5	77	1	TPR	Scaffolding subunit
Cdc23	Apc8	70	2	TPR	Scaffolding subunit
Mnd2	–	43	1	Unknown	Ama1 inhibitor
Apc9	–	30	ND	Unstructured	Cdc27 stabilizing
Doc1	Apc10	26	1	Doc homology/IR motif	Substrate recognition
Apc11	Apc11	19	1	RING H2	Catalytic center
Cdc26	Cdc26	14	2	Unstructured	Cdc16 stabilizing
Swm1	Apc13	19	1	Unstructured	Cdc23 stabilizing
–	Apc16	–		Unknown	Unknown
Total mass (kDa)		1127–1158			
Co-activators					
Cdc20	Cdc20	55		WD40/IR motif	Substrate recognition
Cdh1	Cdh1	63		WD40/IR motif	Substrate recognition
Ama1	–	67		WD40/IR motif	Substrate recognition
Inhibitors					
Mad2	Mad2			Mad2 fold	APC/C inhibitor
Mad3/BubR1	Mad3/BubR1			TPR/pseudosubstrate	APC/C inhibitor
Bub3	Bub3			WD40	APC/C inhibitor
–	Emi1/Emi2			Pseudosubstrate	APC/C inhibitor
Acm1	–			Pseudosubstrate	APC/C inhibitor

[a] Molecular masses and stoichiometry are for the *S. cerevisiae* APC/C.
ND, not determined.

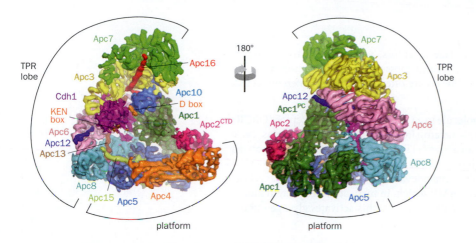

Figure 12.45 Model of the human APC/C in complex with Cdh1 substrate. Left, the molecular envelope from the cryo-EM map segmented and color-coded according to subunit assignments. The degron-recognition sites for D-box and KEN box are indicated, as is the catalytic module of Apc2^{CTD}/Apc11. Right, as (A) but rotated 180° about a vertical axis. For other details, see the text. (Adapted from Chang et al., *Nature* 513:388–393, 2014. With permission from Macmillan Publishers Ltd.)

stabilize the overall organization of the APC/C. Approximately 50% of the APC/C is contributed by subunits (Apc3, Apc5, Apc6, Apc7, and Apc8) that contain multiple tetratricopeptide repeats (TPRs). TPRs contain repeats of a pair of antiparallel α helices denoted A and B; in tandem array, they create a right-handed super-helix.

The overall architecture and subunit organization of human APC/C have been defined by cryo-EM (see **Methods**) reconstruction. Development of multi-protein expression systems that allowed APC/C complexes to be reconstituted from their constituent subunits to form functional E3 ligases made it possible to identify individual APC/C subunits in the density map. Then, fitting atomic models of APC/C subunits into the map yielded a pseudo-atomic model of APC/C in complex with co-activator Cdh1 and substrate. The APC/C measures ~250 Å in its longest dimension and has a triangular shape with an inner cavity (**Figure 12.45**). The back and top of the complex are formed from the bowl-shaped TPR lobe—an assembly of the four canonical TPR proteins (Apc3, Apc6, Apc7, Apc8) and their accessory subunits (Apc12, Apc13, Apc16). The base of the APC/C comprises the platform subunits (Apc1, Apc4, Apc5). Apc1 extends from the platform to contact the TPR lobe. The degron-recognition module of Apc10 and Cdh1 is located at the top of the cavity, with Apc10 interacting extensively with Apc1. The catalytic subunits of APC/C, Apc2 (cullin), and Apc11 (RING) are positioned at the periphery of the platform such that the CTD of Apc2 and its associated Apc11 are at the front of the cavity below and in close proximity to the degron-recognition module.

Cdh1 binding promotes a profound allosteric transition of the APC/C involving displacement of the cullin-RING catalytic subunits relative to the degron recognition module of Apc10 and Cdh1 (**Figure 12.46**). This stimulates the E3 ligase activity of the APC/C by enhancing its affinity for E2~ubiquitin and by increasing access of the substrate Lys to the E2~ubiquitin thioester bond. Thus, both SCF and APC/C are E3 ligases that set up a scaffold to recognize their substrate proteins and to present them appropriately for ubiquitylation by E2, the RING domain playing a prominent part.

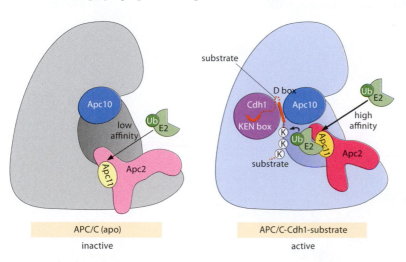

Figure 12.46 Allosteric structural change in APC/C on binding Cdh1. Schematic representation of a profound allosteric transition of the APC/C (left panel) on binding Cdh1 (right panel), which brings E2~Ub into efficacious proximity with the protein substrate and the target Lys residues (marked K). For other details, see the text.

12.6 SUMMARY

Extracellular signals are detected by cell surface receptors, which then pass the signal through a chain of intracellular transducers, second messengers, and effectors to bring about a change in cell behavior. Signaling pathways are often branched or interconnected, forming a network of signaling responses.

The binding of an extracellular signaling molecule to a G-protein-coupled receptor (GPCR) creates a binding site for the transducer, a heterotrimeric G protein. G proteins cycle between an inactive GDP-bound state and an active GTP-bound state. Activated G proteins separate into subcomplexes α and $\beta\gamma$, each of which is able to activate downstream effectors. In response to the hormone epinephrine, the relevant Gα signals to adenylyl cyclase, which produces the second messenger, cyclic AMP, which, in turn, activates protein kinase A (PKA). Phosphorylase kinase is activated by PKA; in turn, this activates glycogen phosphorylase to initiate glycogen breakdown.

Kinases that phosphorylate Ser/Thr or, in other instances, Tyr residues are found in many signaling pathways initiated by hormones or cytokines. Dimerization of the receptor is often a key feature of the initiation process. Phosphorylation can result in enzyme activation, enzyme inhibition, the creation of recognition sites for recruitment of other proteins, and conformational transitions from order to disorder or *vice versa*. Kinases are often linked in series, providing signaling cascades that greatly amplify the signal at each step, as in the initiation of glycogen breakdown by epinephrine. Kinase inhibitors are important in the treatment of cancer.

Class I cytokine receptors are assembled from more than one subunit, with different receptors often having a common subunit. The intracellular regions of activated class I and II cytokine receptors associate with multidomain tyrosine kinases (JAKs), which, in turn, create binding sites for other proteins, transcription factors called STATs. Once phosphorylated, STATs translocate to the nucleus to bring about changes in gene expression. TGFβ cytokines signal to the nucleus via Ser/Thr kinase receptors, through phosphorylation of transcription factors called SMADs.

Signaling can be terminated by the action of specific phosphatases. In addition, covalent attachment of the protein ubiquitin to proteins marks them for inactivation by degradation. Ubiquitylation is catalyzed by three enzymes acting serially: E1 catalyzes the C-terminal adenylation of ubiquitin and then transfers it from a Cys residue in its active site to a Cys residue in the active site of E2. E3 ligases catalyze the transfer of ubiquitin from E2 to a Lys residue of a designated substrate. The largest class of E3s is that of the cullin RING ligases, which have separate domains for binding E2~ubiquitin and the protein substrate. The substrate bears a recognition motif termed a degron, which is often phosphorylated, and the E3 ligase must undergo a major conformational change to catalyze ubiquityl transfer from E2 to the target protein.

Sections of this chapter are based on the following contribution

12.5   David Barford, MRC Laboratory of Molecular Biology, Cambridge, UK.

REFERENCES

12.1 Introduction

Berridge MJ (2014) Module 1: introduction. In Cell Signalling Biology, pp 1.1–1.69. Portland Press (doi:10.1042/csb0001001; http://www.biochemj.org/csb/001/001.pdf).

Hunter T & Pawson T (eds) (2012) The Evolution of Protein Phosphorylation. Royal Society. (Also *Phil Trans R Soc B* 367:2512–2668.)

Hynes N, Ingham PW, Lim WA et al (2013) Signalling change: signal transduction through the decades. *Nat Rev Mol Cell Biol* 14:393–398.

Lim W, Mayer B & Pawson T (2015) Cell Signaling: Principles and Mechanisms. Garland Science.

Marks F, Klingmüller U & Müller-Decker K (2009) Cellular Signal Processing: An Introduction to the Molecular Mechanisms of Signal Transduction. Garland Science.

12.2 Signaling through G-protein-coupled receptors

Brautigan DL (2013) Protein Ser/Thr phosphatases—the ugly ducklings of cell signaling. *FEBS J* 280:324–345.

Endicott JA, Noble MEM & Johnson LN (2012) The structural basis for control of eukaryotic protein kinases. *Annu Rev Biochem* 81:587–613.

Good M, Zalatan JG & Lim WA (2011) Scaffold proteins: hubs for controlling the flow of cellular information. *Science* 332:680–686.

Groves JT & Kuriyan J (2010) Molecular mechanisms in signal transduction at the membrane. *Nat Struct Mol Biol* 17:659–665.

Henderson R & Unwin PN (1975) Three-dimensional model of purple membrane obtained by electron microscopy. *Nature* 257:28–32.

Kobilka BK (2011) Structural insights into adrenergic receptor function and pharmacology. *Trends Pharmacol Sci* 32:213–218.

Lassila JK, Zalatan JG & Herschlag D (2011) Biological phosphoryl-transfer reactions: understanding mechanism and catalysis. *Annu Rev Biochem* 80:669–702.

Lebon G, Warne T & Tate C (2012) Agonist-bound structures of G protein-coupled receptors *Curr Opin Struct Biol.* 22:482–490.

Rasmussen SG, DeVree BT, Zou Y et al (2011) Crystal structure of the β_2 adrenergic receptor–G_s protein complex. *Nature* 477:549–555.

Scott JD, Dessauer CM & Taskén K (2013) Creating order from chaos: cellular regulation by kinase anchoring. *Annu Rev Pharmacol Toxicol* 53:187–210.

Taylor SS, Keshwani MM, Steichen JM & Kornev AP (2012) Evolution of the eukaryotic protein kinases as dynamic molecular switches. *Phil Trans R Soc B* 367:2517–2528.

Wittinghofer A & Vetter IR (2011) Structure–function relationships of the G domain, a canonical switch motif. *Annu Rev Biochem* 80:943–971.

12.3 Signaling through tyrosine kinases

Brognard J & Hunter T (2011) Protein kinase signalling networks in cancer. *Curr Opin Genet Dev* 21:4–11.

Endres NF, Barros T, Cantor AJ & Kuriyan J (2014) Emerging concepts in the regulation of the EGF receptor and other receptor tyrosine kinases. *Cell* 39:437–446.

Grangeasse C, Nessler S & Mijakovic I (2012) Bacterial tyrosine kinases: evolution, biological function and structural insights. *Phil Trans R Soc B* 367:2640–2655.

Hunter T (2014) The genesis of tyrosine phosphorylation. *Cold Spring Harb Perspect Biol* 6:a020644 (doi:10.1101/cshperspect.a020644).

Johnson LN (2009) Protein kinase inhibitors: contributions from structure to clinical compounds. *Quart Rev Biophysics* 42:1–40.

Juna N, Zhang X, Endres NF et al (2011) Catalytic control in the EGF receptor and its connection to general kinase regulatory mechanisms. *Mol Cell* 42:9–22.

Lemmon MA, Schlessinger J & Ferguson KM (2014) The EGFR family: not so prototypical receptor tyrosine kinases. *Cold Spring Harb Perspect Biol* 6:a020768 (doi:10.1101/cshperspect.a020768).

Levitzki A (2013) Tyrosine kinase inhibitors: views of selectivity, sensitivity, and clinical performance. *Annu Rev Pharmacol Toxicol* 53:161–185.

Pawson T (2013) Molecular mechanisms of SH2- and PRB-domain-containing proteins in receptor tyrosine kinase signaling. *Cold Spring Harb Perspect Biol* 5:a008987 (doi:10.1101/cshperspect.a008987).

Toettcher JE, Weiner OD & Lim WA (2013) Using optogenetics to interrogate the dynamic control of signal transmission by the Ras/Erk module. *Cell* 155:1422–1434.

Yasui N, Findlay GM, Gish GD et al (2014) Directed network wiring identifies a key protein interaction embryonic stem cell differentiation. *Mol Cell* 54:1034–1041.

12.4 Cytokine signaling

Baker SJ, Rane SG & Reddy EP (2007) Hematopoietic cytokine receptor signaling. *Oncogene* 26:6724–6737.

Brooks AJ, Dai W, O'Mara ML et al (2014) Mechanism of activation of protein kinase JAK2 by the growth hormone receptor. *Science* 344:1249783 (doi:10.1126/science.1249783).

Broughton SE, Hercus TR, Lopez AF & Parker MW (2012) Cytokine receptor activation at the cell surface. *Curr Opin Struct Biol* 22:350–359.

Bruce DL & Sapkota GP (2012) Phosphatases in SMAD regulation. *FEBS Lett* 586:1897–1905.

Dahl M, Maturi V, Lönn P et al (2014) Fine-tuning of Smad protein function by poly(ADP-ribose) polymerases and poly(ADP-ribose) glycohydrolase during transforming growth factor β signaling. *PLoS ONE* 9:e103651 (doi:10.1371/journal.pone.0103651).

Massagué J (2012) TGFβ signalling in context. *Nat Rev Mol Cell Biol* 13:616–630.

O'Shea JJ, Holland SM & Staudt LM (2013) JAKs and STATs in immunity, immunodeficiency, and cancer. *N Engl J Med* 368:161–170.

Wang X, Lupardus P, Laporte SL & Garcia KC (2009) Structural biology of shared cytokine receptors. *Annu Rev Immunol* 27:29–60.

Zhang L, Zhou F & ten Dijke P (2013) Signaling interplay between transforming growth factor-β receptor and PI3K/AKT pathways in cancer. *Trends Biochem Sci* 38:612–620.

12.5 Ubiquitylation

Capili AD & Lima CD (2007) Taking it step by step: mechanistic insights from structural studies of ubiquitin/ubiquitin-like protein modification pathways. *Curr Opin Struct Biol* 17:726–735.

Chang, L & Barford, D (2014) Insights into the anaphase-promoting complex: a molecular machine that regulates mitosis. *Curr Opin Struct Biol* 29:1–9 (doi:10.1016/j.sbi.2014.08.003).

Deshaies RJ & Joazeiro CA (2009) RING domain E3 ubiquitin ligases. *Annu Rev Biochem* 78:399–434.

Duda DM, Scott DC, Calabrese MF et al (2011) Structural regulation of cullin-RING ubiquitin ligase complexes. *Curr Opin Struct Biol* 21:257–264.

Herzog F, Primorac I, Dube P et al (2009) Structure of the anaphase-promoting complex/cyclosome interacting with a mitotic checkpoint complex. *Science* 323:1477–1481.

Komander D (2009) The emerging complexity of protein ubiquitination. *Biochem Soc Trans* 37:937–953.

Petroski MD & Deshaies RJ (2005) Function and regulation of cullin-RING ubiquitin ligases. *Nat Rev Mol Cell Biol* 6:9–20.

Rotin D & Kumar S (2009) Physiological functions of the HECT family of ubiquitin ligases. *Nat Rev Mol Cell Biol* 10:398–409.

Schulman BA & Harper JW (2009) Ubiquitin-like protein activation by E1 enzymes: the apex for downstream signalling pathways. *Nat Rev Mol Cell Biol* 10:319–331.

Ulrich HD & Walden H (2010) Ubiquitin signalling in DNA replication and repair. *Nat Rev Mol Cell Biol* 11:479–489.

Williams RL & Urbe S (2007) The emerging shape of the ESCRT machinery. *Nat Rev Mol Cell Biol* 8:355–368.

cyt c

dimerization

cleavage

executioners

casp-7

casp-3

substrates

APOPTOSIS

casp-9

initiator

apoptosome

The Cell Cycle and Programmed Cell Death

13

13.1 INTRODUCTION

Cells grow, divide, and die as part of life. For unicellular organisms, each cell division creates two new organisms. For multicellular organisms, life starts from a single cell, the fertilized egg, followed by numerous cell divisions in the developing embryo. A process of differentiation produces the cells that make up the different tissues of an animal. In the adult, some cells (such as neurons) cease dividing when they have differentiated to their final state. Others may continue to grow and divide, for example epithelial cells, those of the epidermis, and those that line the walls of the gastrointestinal tract; others such as cardiomyocytes, liver and bone cells are renewed in response to injury.

Similarly, cells die. For unicellular organisms, death of a single cell will have no major impact on the population. For multicellular organisms, however, the selective orderly deletion of cells, termed **apoptosis** or programmed cell death, is a key process governing normal development and maintenance. For example, in the worm *Caenorhabditis elegans* 959 cells are present in the adult but a further 131 cells were generated and then selectively eliminated. Apoptosis is also a quality control system. It ensures that cells whose DNA is damaged beyond repair and are potentially cancerous are eliminated, as are lymphocytes of the adaptive immune system after they have served their purpose in overcoming infection. In addition, it can act as a defense mechanism after viral infection.

Cell division and cell death are of crucial importance and subject to rigorous control. The two processes must be kept in balance: the adult human has a total of some 10^{14} cells, of which $\sim 10^{11}$ are replaced each day. The eukaryotic cell cycle functions as a series of steps, each of which is checked and must be successfully completed before the next can be initiated. These include DNA replication and the separation of the replicated chromosomes, one set for each daughter cell. An unchecked failure at any point could have catastrophic consequences. Similarly, once apoptosis is initiated it cannot be undone; no half-dead cells can be left undisposed of. In cancer, cells have become unresponsive to the control signals that normally regulate their growth and division, and they may fail to die when they should because of abnormal survival factors. Understanding the exquisite control mechanisms of cell division and cell death has given insight into some of the processes that go wrong and is contributing to advances in cancer therapy.

In this chapter, we focus first on the mammalian cell cycle. We describe the macromolecular assemblies that constitute and regulate the cell cycle machinery, many of them based on the sequential activation and inactivation of protein kinases and phosphatases. We then turn to the assemblies involved in apoptosis.

13.2 THE CELL CYCLE

An overview of the eukaryotic **cell cycle** is shown in **Figure 13.1A**. In response to external signals, cells enter the first gap phase, **G1**, and are readied for DNA replication. DNA replication takes place in **S phase** and the replicated DNA is packaged immediately into nucleosomes (Chapter 3). The duplicated chromosomes (**sister chromatids**) are held together in pairs by the protein *cohesin*. During S phase and the second gap phase, **G2**, the *centrosome*, the microtubule-organizing center, is also duplicated (Section 13.5) and the cells then enter **M phase** (mitosis).

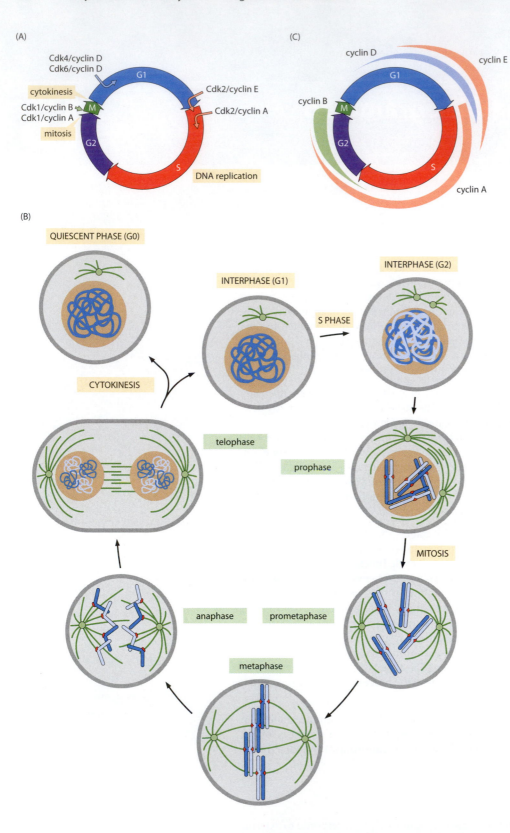

Figure 13.1 The cell cycle. (A) The four phases of the cell cycle (G1, S, G2, and M phase). The lengths of the curved arrows correspond to typical times for the 24-hour division of a mammalian cell. In vertebrate somatic cells, different Cdk/cyclin complexes become active at different stages in the cell cycle. CDK activity is dependent on association with cyclins and phosphorylation. (B) Cellular events occurring at various stages of the cell cycle. For simplicity, four chromosomes are represented. The DNA is blue and the duplicated copy is light gray, centrosomes are rings with green microtubules emanating from them, the kinetochores at the centromeres are red dots, the nucleus is depicted in beige, and the plasma membrane is shown as a thick gray line. (C) Concentrations of the cyclins fluctuate during the cell cycle, controlled by gene transcription and proteasomal degradation. (A, adapted from T.D. Pollard and W.C. Earnshaw, Cell Biology. Elsevier Science, 2002; B & C, adapted from I.M. Cheeseman and A. Desai, *Nat. Rev. Mol. Cell Biol.* 9:33–46, 2008. With permission from Macmillan Publishers Ltd.)

In **mitosis**, one copy of the duplicated genome is delivered to each daughter cell (Figure 13.1B). Mitosis has several phases recognizable by light microscopy. As cells progress from G2 to **prophase**, the sister chromatids condense into compact rods, and large macromolecular complexes—**kinetochores**—assemble on the **centromere** regions of chromosomes (Section 13.4). During prophase, the duplicated centrosomes separate toward opposite regions of the cell nucleus. There follows breakdown of the **nuclear envelope** as cells enter

prometaphase. *Microtubules* emanating from the centrosomes attach to the kinetochores to form the **mitotic spindle**. By **metaphase**, all chromosomes are attached via their kinetochores to microtubules.

For the cell to enter **anaphase**, each chromatid pair must be bi-oriented; that is, each kinetochore on a chromosome is bound to its proximal centrosome, thus building up tension across the kinetochores. The cohesin linking the sister chromatids is then cleaved and the sister chromatids are moved apart to opposite spindle poles. Subsequently, during **telophase**, the chromatid masses de-condense and the nuclear envelope re-forms to generate two daughter nuclei.

Mitosis is followed by **cytokinesis** (see Figure 13.1B). The cell membrane grows inward at the site of division, and a contractile ring of actin and myosin II filaments (Sections 14.2 and 14.3) pinches the mother cell to form two daughter cells. A single set of chromosomes and one centrosome are delivered to each daughter cell. Other more plentiful materials, such as ribosomes and mitochondria, are divided between the daughter cells.

Interphase is the interval between one round of mitosis and the next. The gap phases, G1 and G2, are important to allow growth. They also include the **checkpoints** where, if the preceding step is imperfectly executed, progression to the next phase can be blocked. (The term checkpoint can refer to the entire regulatory system that monitors a condition, for example DNA damage [Chapter 4], and can send a negative signal to block progression. Here we use the term to denote a transition point in the cell cycle.) There are three checkpoints: the first is at mid-to-late G1 (also called Start, G1/S, or the restriction point) and ensures that environmental conditions are appropriate for the cell to enter the cell cycle; the second, the G2/M checkpoint, ensures that the chromosomal DNA is fully replicated and undamaged; and the third, the spindle assembly checkpoint (SAC) at metaphase-to-anaphase, ensures that all sister chromatids are aligned on the mitotic spindle.

The cell cycle may be repeated to realize multiple rounds of cell division or, in response to external signals, cell proliferation can be arrested and the cells enter a nondividing quiescent stage called **G0** (see Figure 13.1B). In a rapidly dividing mammalian cell, the cell cycle takes ~24 hours, of which ~9 hours are spent in G1, ~10 hours in S phase, ~4.5 hours in G2 and ~0.5 hours in M phase (see Figure 13.1A). These times vary between organisms. Cell growth usually accompanies cell division, but these processes may not be coupled. For example, muscle cells grow considerably without dividing, whereas the early cell divisions of the fertilized egg take place in the absence of growth.

13.3 TRANSIENT ASSEMBLIES OF KINASES, CYCLINS, AND PHOSPHATASES

In the late 1970s, experiments with temperature-sensitive mutants of the yeast *Schizosaccharomyces pombe* established that a single gene, *cdc2* (cell division cycle 2), was required in G1 to commit the cell to the onset of S phase, and then in G2 to the onset of mitosis. A homologous gene, *cdc28*, was identified in *Saccharomyces cerevisiae*. In the early 1980s, in studies on protein synthesis in sea urchin eggs, it was observed that levels of an intracellular protein, now called **cyclin**, oscillated in synchrony with the cell cycle. The products of the *cdc2* and *cdc28* genes were found to be Ser/Thr protein kinases that require a cyclin for activity, implying that the rise and fall in cyclin concentration controls the cell cycle through kinase activation. These **cyclin-dependent kinases** (**CDKs**) are conserved throughout all eukaryotes.

The numbers of CDKs and cyclins vary between organisms. *S. cerevisiae* and *S. pombe* have a single CDK, Cdk1, whereas in higher organisms different CDKs bind different cyclins and initiate successive stages of the cell cycle (see Figure 13.1A, C). Activity is also regulated by the combined action of several protein kinases and phosphatases (**Table 13.1**). In multicellular eukaryotes, the concentrations of CDKs remain constant during the cell cycle, but those of cyclins vary. CDK concentrations are also much higher than those of the cyclins, so that essentially all the cyclins are complexed with their respective CDKs. After they have completed their function, cyclins are ubiquitylated and degraded (Section 12.5). Thus it is changes in the cyclin levels that regulate activity.

Table 13.1 **Protein kinases and phosphatases of the mammalian cell cycle.**

Kinases	Function	Structure (PDB)
Cdk4/cyclin D	Initiates G1 phase	3G33, 2W96
Cdk6/cyclin D	Initiates G1 phase	1JOW, 1G3N (with viral cyclin); 1BLX, 1BI7, 1BI8 (with INK4 inhibitor)
CDK2/cyclin E	Active during G1/S phase	1W98
CDK2/cyclin A	Active during S/G2 phase	1HCK (inactive monomeric Cdk2); 1FIN (inactive Cdk2/cyclin A); 1JST (active phospho-Cdk2/cyclin A); 1QMZ (active phospho-Cdk2/cyclin A with substrate peptide); 2CJM (pY15-phospho-Cdk2/cyclin A)
Cdk1/cyclin B	Essential cell cycle kinase; required for entry into mitosis	Structure of Cdk1 not available; 2JGZ (Cdk2/cyclin B)
Cdk7/cyclin H/Mat1 (CAK)	Cdk activating kinase for Cdk1, Cdk2, Cdk4, and Cdk6	Individual structures of 1UA2 (Cdk7), 1KXU (cyclin H), and 1G25 (Mat1)
Wee1	Tyrosine kinase responsible for inhibition of Cdk1 and Cdk2 at G1/S and G2/M checkpoints	1X8B
Myt1	Dual-specificity kinase responsible for inhibition of Cdk1 at G2/M checkpoint	3P1A
Nek2	Phosphorylates C-Nap1 in maturation of centrosomes	2W5A
Aurora A	Associates with centrosomes and spindle microtubules during mitosis; phosphorylates Plk1	1OL5 (Aurora A in complex with TPX2)
Aurora B	Important for sister chromatid cohesion and segregation; signals tension between kinetochores; associates with chromosomal passenger proteins INCENP, borealin and survivin	2BFX (Aurora B in complex with IN-box of the activator INCENP)
Polo-like kinase (Plk1)	Numerous functions in centrosome maturation, entry to mitosis, and cytokinesis	2OU7, 3D5W (Plk1 catalytic domain); 1Q4O, 1UMW (polo-box domain)
Bub1	Part of the mitotic checkpoint complex (MCC); phosphorylates H2A	2E7E (kinase domain), 3ESL (N-terminal domain tetratricopeptide motif)
Mps1	Dual-specificity kinase; part of mitotic checkpoint complex (MCC)	2ZMC (kinase domain)
Cdc7/Dfb4 (DDK)	Important for assembly of origins of replication and in meiosis	4F9B
Phosphatases	**Function**	**Structure (PDB)**
Cdc25	Part of checkpoint at G1/S and G2/M; dephosphorylates pTyr15 on Cdk2 and Cdk1 and thereby activates the CDKs	1YMK (Cdc25B phosphatase)
Kinase-associated phosphatase (KAP)	Dephosphorylates pThr160-Cdk2, thereby inactivating kinase for exit from mitosis	1FQ1 (KAP in association with pCdk2)
Cdc14	Dual-specificity phosphatase; dephosphorylates CDK targets	1OHE (Cdc14 with a peptide ligand)
PP2A	Checkpoint phosphatase regulating entry into and exit from mitosis; regulates cohesin protection	2IAE, 2NYM (holoenzyme); 3FGA (PP2A with Sgo1 [shugoshin])
PP1	Ser/Thr phosphatase with several different roles regulated by targeting subunits; PP1 is implicated in the tension-sensitive kinetochore dephosphorylation	1JK7 (PP1 in complex with okadaic acid); 1FJM (PP1 in complex with microcystin)

Human cells contain hundreds of CDK substrates. Despite advances in mass spectrometry and proteomics, identification of the full range of phosphorylated substrates is incomplete, but progress is being made. For example, a Cdk1 engineered to accept an ATP analog that no other kinase in the cell could use revealed 72 Cdk1 phosphorylation sites in 68 proteins in HeLa cells. The downstream consequences of phosphorylation of these substrates are understood in some cases.

Mitogenic signals initiate the cell cycle

In mammalian cells, the cell cycle is initiated in response to mitogenic signals from outside the cell, such as epidermal growth factor (EGF) or platelet-derived growth factor (PDGF), or from signals derived from monitoring growth inside the cell. Mitogenic signals trigger the Ras/Erk2 pathway (Section 12.3) that culminates in expression of the transcription factors Myc and AP1 (also known as Fos/Jun). These transcription factors promote the expression of three isoforms of cyclin D, of which cyclin D1 is the most important. Cdk4 and Cdk6 bind the D cyclins, and their action promotes entry into G1 (see Figure 13.1A). A major substrate of Cdk4/cyclin D is the retinoblastoma protein, Rb (named for its role in an inherited eye cancer in children). Rb is a tumor suppressor and it is inactivated in almost all human cancers. It binds the transcription factor E2F/DP1 and blocks transcription. On becoming phosphorylated, Rb is inactivated and dissociates from E2F/DP1, allowing transcription to proceed. The gene for cyclin E is one of the major genes transcribed, and cyclin E binds to Cdk2. Cdk2/cyclin E further phosphorylates Rb and also promotes the assembly of the DNA pre-replication complex (Chapter 3), preparing the cell for the transition from G1 to S phase. In addition, E2F/DP1 promotes transcription of the gene for cyclin A, and Cdk2/cyclin A phosphorylates substrates that promote progression through S phase.

During G2, cyclin A associates with Cdk1, and Cdk1/cyclin A promotes chromosome condensation in the early stages of mitosis. Cyclin B concentrations also rise in G2 (Figure 13.1C). Cdk1/cyclin B participates in the breakdown of the nuclear envelope through phosphorylation of nuclear pore complexes (Section 10.2) and nuclear lamins (Section 11.2) and arrests protein synthesis by phosphorylation of the ribosomal elongation factor elF2α (Section 6.2). Cdk1/cyclin B has a critical role in initiating mitosis through activation of the anaphase-promoting complex/cyclosome (APC/C) (Section 12.5).

CDKs activated by cyclins are stabilized by phosphorylation of a Thr residue in their activation loop, catalyzed by the nuclear CDK-activating kinase (CAK), a complex of Cdk7, cyclin H, and a third protein, Mat1. Unlike other CDKs, CAK is active throughout the cell cycle and, provided that Mat1 is present, does not require phosphorylation for activity.

Cyclin destruction inactivates CDKs

Cyclins and cell cycle-inhibitory proteins are targeted for destruction by ubiquitylation and degradation by the proteasome (Section 7.4), events that are crucial for the timing of the cell cycle transitions. Phosphorylation of the G1 cyclins D and E marks them for ubiquitylation by the Skp1–Cullin–F-box (SCF) complex. On entry into S phase, cyclin E is autophosphorylated, which targets it to the SCF^{Cdc4} complex. Mutations in the F-box protein Cdc4 of SCF result in failure to destroy cyclin E at the right time, leading to chromosomal instability and cancer.

Cyclins A and B are ubiquitylated by the APC/C, which recognizes their 'destruction box' sequence motifs. The APC/C requires the activator Cdc20 or Cdh1. Cdk1/cyclin B initiates a reaction that leads to its own destruction: it activates the APC/C by phosphorylating APC/C subunits, a reaction that promotes Cdc20 binding to form APC/C^{Cdc20}. On entry into anaphase, cyclin B is ubiquitylated by APC/C^{Cdc20} and immediately destroyed (Section 12.5). At metaphase, cyclin A is ubiquitylated by APC/C^{Cdc20}. Cyclin A can be degraded before cyclin B because of the association of the Cdk2/cyclin A complex with a third protein, Cks, that recruits the complex to the phosphorylated APC/C. This allows cyclin A to be degraded regardless of whether the spindle assembly checkpoint (described below) is active or not. Concentrations of cyclins A and B are kept low during G1 by the APC/C activated by association with Cdh1 (APC/C^{Cdh1}). At the transition from G1 to S phase, Cdk2/cyclin E phosphorylates Cdh1, causing it to dissociate, with a loss of APC/C activity and a consequent rise in cyclin A and cyclin B levels.

At the cell cycle checkpoints, the CDKs respond to a multitude of signals and feedback loops, which ensure that later events are dependent on the completion of earlier events. Many of the substrates of Cdk2/cyclin A are transcriptional regulators such as pRb (see above; the prefix p denotes the phosphorylated form of a protein). This allows the production of a variety of proteins that promote progression through S phase. Other substrates for Cdk2/cyclin A include Cdc6, a protein important in the formation of the pre-replication complex for DNA synthesis; p27, an inhibitor of Cdk2/cyclin A and Cdk2/cyclin E; and proteins important for histone expression (NPAT) and centrosome duplication. The same substrates are generally phosphorylated also by Cdk2/cyclin E but with some differences related to their different levels during G1 and S phase. Cyclin E levels rise in mid G1 (see Figure 13.1C) and fall abruptly in S phase, whereas cyclin A levels rise in S phase and remain high until entry into mitosis. During G1, the complex APC/C^{Cdh1} is active and triggers the ubiquitylation and degradation of cyclin A but not of cyclin E; after phosphorylation, the latter is ubiquitylated by the SCF complex.

Cdk2/cyclin A is a model for the structural basis of CDK regulation

Cdk2 (298 aa) has a minimal protein kinase fold comprising two lobes linked by a short hinge region. The N-terminal lobe is mostly β sheet with one α helix, the C helix. The C-terminal lobe is mostly α helices and a few short stretches of β sheet (**Figure 13.2A**). ATP binds between the two lobes. In the absence of cyclin, Cdk2 is inactive. ATP is able to bind, but the C helix is swung out and a Glu (E in a P^{45}STAIRE^{51} motif, where superscript numbers refer to the residue number in human Cdk2) is not able to contact a crucial Lys residue that aligns the triphosphate moiety of ATP for phosphoryl transfer. The activation segment, which comprises the region from a conserved D^{145}FG motif to the APE^{172} motif, packs against the C helix in the 'out' position and partly blocks the ATP site. (The fundamentals of protein kinase activity are described in Section 12.2 for PKA.)

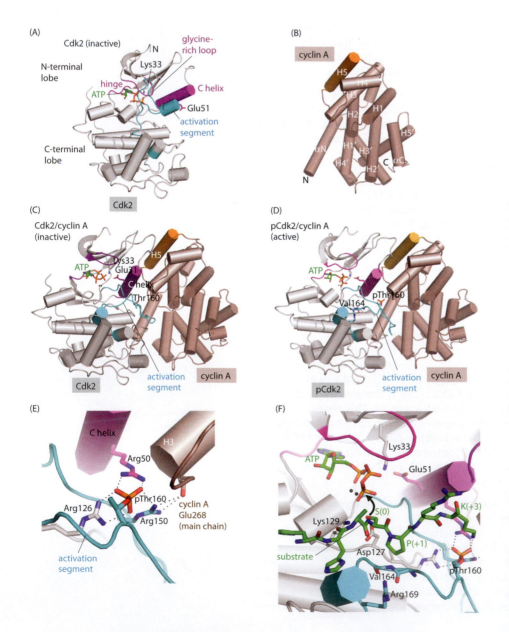

Figure 13.2 Activation of Cdk2 by cyclin A and phosphorylation. (A) Inactive Cdk2 (gray) with functionally important regions highlighted in color. Helices are depicted as cylinders. (PDB 1HCK) (B) Cyclin A. Helices are labeled αN, H1, and so on, from the N terminus. (PDB 1QMZ) (C) Inactive Cdk2/cyclin A. On binding cyclin A, Cdk2 undergoes substantial conformational changes. The H5 helix (orange) of cyclin A docks against the C helix of Cdk2. (PDB 1FIN) (D) Active Cdk2/cyclin A after phosphorylation of Thr160. The activation segment undergoes further conformational change. (PDB 1JST) (E) Details of the contacts made by pThr160. (PDB 1QMZ) (F) Details of the catalytic site of pCdk2/cyclin A in the vicinity of the Ser-Pro region of the substrate. ATP is bound in the active conformation with the triphosphate contacting Lys33 and the γ-phosphate contacting Mg^{2+}. The substrate peptide is aligned so that the Pro side chain docks into a hydrophobic pocket created by the activation segment in which the main-chain carbonyl oxygens of Val163 and Val164 contact Arg169. The Ser side chain is directed toward the γ phosphate of ATP. Phosphoryl transfer is promoted by attack of the Ser OH group on the γ phosphate (curved arrow) and is assisted by residues Asp127 and Lys129. (PDB 1QMZ)

Cyclin A (432 aa) comprises an N-terminal region (residues 1–174), which contains the 'destruction box' for recognition by the APC/C, and a C-terminal region (residues 175–432), which is responsible for activating Cdk2. The C-terminal region, hereafter referred to as cyclin A*, is composed of two cyclin box folds, each of which has five α helices in which the central helix H3 is surrounded by four other helices (Figure 13.2B). In addition, there is an N-terminal helix (αN) that is important for Cdk activation, and a short C-terminal helix (αC) that has no defined role. The cyclin box fold is common to all cyclins (despite sequence identities of <20% in some cases). On interaction with cyclin A* (tight, $k_d \approx 10$ nM), the conformation of Cdk2 changes considerably but there is no change in cyclin A* (Figure 13.2C). The H5 helix of cyclin A* contacts the C helix of Cdk2 and the Ile residue of the PSTAIRE motif is buried. This promotes a rotation and translation of the C helix so that Glu51 of the PSTAIRE contacts the Lys residue that helps align the triphosphate of ATP for phosphoryl transfer. The Cdk2 activation segment swings away from the ATP-binding site, adopting a more open conformation.

Despite these changes, the activation segment is not in the correct conformation to recognize the protein substrate. Activation requires phosphorylation by CAK of Thr160. In uncomplexed inactive Cdk2, Thr160 in the activation segment is buried close to the ATP site; binding of Cdk2 to cyclin A* exposes it for phosphorylation by CAK, leading to further conformational changes in the activation segment (Figure 13.2D). pThr160 forms an organizing center, in which the phosphate makes contacts with Arg50 from the C helix and Arg126 (which precedes the catalytic Asp127) and Arg150 from the activation segment (Figure 13.2E). This also strengthens the pCdk2/cyclin A* interaction. These changes align the activation segment for its role in presenting the target protein to the ATP site for phosphorylation.

CDKs phosphorylate substrates on Ser or Thr residues (P0 position) that are followed by a Pro (P +1 position). Cdk2 has an optimal sequence preference Ser/Thr-Pro-X-Arg/Lys, where X is any amino acid and which, when bound, directs its Ser-OH group toward the γ-phosphate of ATP (Figure 13.2F). Phosphoryl transfer is assisted by the carboxylate group of Asp127 and the positive charge on Lys129, which stabilizes the transition state. The Pro in the P +1 position helps align the Ser residue by docking into a hydrophobic pocket created by Val163 and Val164, whose carbonyl oxygen groups are directed away from the peptide substrate toward Arg169. Hence only Pro, whose side chain is covalently linked to the amide nitrogen, can be accommodated at this site; any other amino acid at this position in the peptide would require an H-bond acceptor for its NH group. The side chain of Lys in the P +3 position contacts pThr160, thus providing an additional role for the phosphoryl group in substrate recognition.

To ensure the complete deactivation of Cdk2 at the end of the cell cycle after the destruction of cyclins, pThr in the activation segment is dephosphorylated by the Ser/Thr phosphatase, kinase-associated phosphatase (KAP).

Cyclins can bind kinase substrates at a remote hydrophobic patch

Several important **S-phase** substrates of pCdk2/cyclin A, including Rb and the related protein p107, lack the basic Lys or Arg residue in the P +3 position. After the discovery that p107 could pull down Cdk2/cyclin A in cell extracts, it was found that cyclin A recognizes the motif Arg/Lys-X-Leu-Z-Φ (abbreviated to RXL), where Z denotes a spacer residue that may or may not be present and Φ is a nonpolar residue. This makes a poor substrate with a sub-optimal phosphorylation motif into an effective one. The RXL motif binds to a hydrophobic patch on the surface of the cyclin, ~40 Å from the catalytic site of Cdk2 (**Figure 13.3A**).

The Ser/Thr-Pro-X and RXL motifs are separate and there are no restrictions on the conformation of the substrate polypeptide chain between them, other than a minimum number (~15 aa) required to span the 40 Å distance between the two binding sites on the Cdk/cyclin. The RXL-binding site is conserved in cyclins A, E and D but not in cyclin B. Although the structure of cyclin B is similar to that of cyclin A, the hydrophobic patch has a Gln residue in place of Glu220 and there is a Lys residue in place of Glu224 in cyclin A (Figure 13.3B). These changes in charge weaken the affinity of cyclin B for the RXL motif, leaving Cdk1/cyclin B with a more relaxed specificity, requiring only the Ser/Thr-Pro motif for substrates.

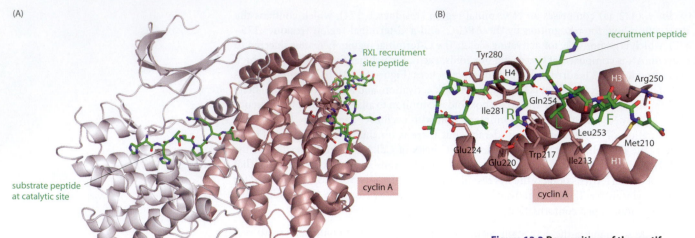

(A)

RXL recruitment
site peptide

substrate peptide
at catalytic site

cyclin A

pCdk2

(B)

Tyr280

H4

Ile281   R

Glu224

Glu220   Trp217

recruitment peptide

X

Gln254   H3   Arg250

L   F

Leu253

Met210

Ile213   H1

cyclin A

Figure 13.3 Recognition of the motif RXL(X)Φ by cyclins A, D, and E. (A) Structure of active pCdk2/cyclin A with a substrate peptide (green) bound at the catalytic site of pCdk2 and a recruitment peptide (green) from the pre-replication complex protein, Cdc6, bound at the RXL recruitment site of cyclin A. (B) Details of the interactions of cyclin A with the recruitment peptide HTLKGRRLVFDN (green, RXL(X)Φ recognition motif in bold). The peptide is bound in an extended conformation, with an ion pair between its Arg residue and Glu220 of cyclin A, and the Leu and Phe side chains docking into a hydrophobic pocket lined by cyclin A residues Met210, Leu213, and Trp217 from the H1 helix and Leu253 from H3 helix. Hydrogen bonds are shown as dashed red lines. (PDB 2CCI)

Cell cycle inhibitors regulate CDK activity

A variety of anti-proliferative signals arrest the cell cycle. The CDK inhibitors (CKIs) induce arrest in G1 and S phase and fall into two families: the Kip and Cip family, which includes p27^{Kip1}, p21^{Cip1}, and p57^{Kip2}, inhibits all the G1 and S-phase Cdks but not Cdk1/cyclin B; and the INK4 family, which includes p15, p16, p18, and p19 CKIs and is specific for Cdk4 and Cdk6. In many diverse cancers, p27^{Kip1} levels are low, owing to up-regulation of the SCF complex that ubiquitylates it. Alterations to p16^{INK4a} in cancer are particularly common, occurring in 80% of certain tumors.

p27^{Kip1} (198 aa in humans) comprises an N-terminal region (residues 22–106) that is competent to inhibit Cdk2/cyclinA and Cdk2/cyclin E, and a C-terminal region that contains sites for phosphorylation. In the complex pCdk2/cyclin A/p27^{Kip1} (residues 22–106), p27^{Kip1} makes contacts with both cyclin A and Cdk2 (**Figure 13.4A**). It exploits an RXL

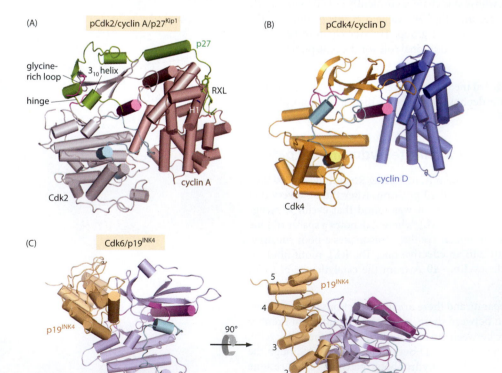

(A)   pCdk2/cyclin A/p27^{Kip1}

glycine-
rich loop   3$_{10}$ helix   p27

hinge   RXL

H1

Cdk2   cyclin A

(B)   pCdk4/cyclin D

cyclin D

Cdk4

(C)   Cdk6/p19^{INK4}

p19^{INK4}

p19^{INK4}

90°

5

4

3

2

1   Cdk6

Cdk6

Figure 13.4 The cyclin-dependent kinase inhibitors (CKIs). (A) pCdk2/cyclin A in complex with the inhibitor p27^{Kip1} (green), which extends from the RXL site on cyclin A to the catalytic site of Cdk2. The glycine-rich loop (distorted), the C helix and the hinge region are in magenta and the activation segment is in cyan. (PDB 1JSU) (B) The Cdk4/cyclin D complex has a more open conformation than that of Cdk2/cyclin A. (PDB 3G33) (C) Cdk6 bound to the inhibitor p19^{INK4}. The view on the right shows the arrangement of the five ankyrin repeats (numbered). (PDB 1BLX)

motif (residues R^{30}XLFG^{34}) to bind to the hydrophobic patch on cyclin A but inhibits Cdk2 by extending over the N-terminal lobe of Cdk2 with a short helical stretch and inserting a 3$_{10}$ helix into the ATP site. This explains why p27^{Kip1} does not inhibit Cdk1/cyclin B, because cyclin B lacks an RXL-binding site (see above).

Somewhat surprisingly, the Cip or Kip inhibitors are activators of Cdk4/cyclin D and Cdk6/cyclin D complexes. The association of pCdk4 with cyclin D is much less intimate than that of pCdk2 with cyclin A (Figure 13.4B) and the pCdk4/cyclin D complex is inactive. Binding of p27^{Kip1}, through recognition of the RXL motif by the cyclin D hydrophobic site and through other interactions, may enhance Cdk4/cyclin D contacts. Cdk4/cyclin D must assemble with p27^{Kip1} to translocate to the nucleus. The majority of p27^{Kip1} in proliferating cells is associated with Cdk4/cyclin D, and the complex allows cells to respond to mitogenic or anti-mitogenic signals by scavenging or releasing p27^{Kip1}.

As an activator, p27^{Kip1} is phosphorylated on Tyr88 and Tyr89. In the complex with pCdk2/cyclin A, these two residues are located in the 3$_{10}$ helix where p27^{Kip1} contacts the ATP-binding site of Cdk2 (see Figure 13.4A). The kinase for these phosphorylations *in vivo* has not been definitively identified, but the tyrosine kinases Abl, Src, or Csk1 are all competent *in vitro* and are up-regulated in proliferating cells. Tyr-phosphorylated p27^{Kip1} is no longer an inhibitor of Cdk2/cyclin E and Cdk2/cyclin A. It seems likely that the pTyr-p27 may promote Cdk4/cyclin D assembly by binding to the cyclin D RXL-recognition but is then unable to dock into the Cdk4 catalytic site.

Active Cdk2/cyclin E or Cdk2/cyclin A phosphorylates p27^{Kip1} on Thr187 in the C-terminal region (not present in the construct used in the crystal studies). The presence of pThr187 targets p27^{Kip1} for ubiquitylation by the SCF complex SCF^{Skp2}. Thus, the CDK target for inhibition by p27^{Kip1} is also able to promote destruction of the inhibitor—an example of negative feedback.

INK4 inhibitors bind to Cdk4 or Cdk6, interfering with the binding of cyclin D and thereby preventing activation. p16^{INK4} and p19^{INK4} bind on the opposite side of the kinase to the cyclin; in doing so, the inhibitor contacts both the N-terminal and C-terminal lobes of the kinase and promotes conformational changes that distort the catalytic site and prevent cyclin binding (Figure 13.4C). p16 and p19 consist of four and five ankyrin repeats (see **Guide**) respectively. This 30 aa feature has an L-shaped structure with a pair of anti-parallel helices forming the stems and a β-hairpin/loop region projecting out at 90°. The L-like repeats stack to give an extended concave surface.

Phosphorylation and dephosphorylation regulate CDK activity

At the checkpoints in G1 (Start) and G2 (G2/M), the cell cycle is paused by holding Cdk2/cyclin A and Cdk1/cyclin B in an inactive state. This is achieved by phosphorylations on Thr14 and Tyr15 (human Cdk2 numbering), catalyzed by the kinases Myt1 and Wee1 respectively. Myt1 is located in the membranes of the endoplasmic reticulum and the Golgi apparatus; it phosphorylates cytoplasmic Cdk1/cyclin B. Wee1 is found predominantly in the nucleus. Both Myt1 and Wee1 are active during most of the cell cycle but their activity decreases abruptly during mitosis. Although Wee1 phosphorylates Tyr residues, its structure is closer to that of a Ser/Thr kinase. It requires no phosphorylation on the activation segment for activity.

Phosphorylation of Tyr15 in pCdk2/cyclin A causes no conformational change, and pCdk2 can still bind ATP. However, the pTyr15 phosphate group moves close to the protein substrate-binding site and blocks binding of the target protein (**Figure 13.5A**). Catalytic activity is not completely abolished but diminished nearly 200-fold. Dephosphorylation and activation of the pThr14 and pTyr15 pCDKs are catalyzed by the Cdc25 dual-specificity phosphatases. There are three isoforms (A, B, and C) of Cdc25. Cdc25A is active at both G1/S and G2/M, whereas Cdc25B and Cdc25C are active only at G2/M. Cdc25 activity in the nucleus rises during prophase to prometaphase. As with most phosphatases, substrate recognition is dependent on a binding site remote from the Cdc25 active site, identified by mutagenesis studies.

For activity, Cdc25 requires phosphorylation by Plk1 (see below). Plk1 is composed of a kinase domain followed by a polo-box domain (PBD) that recognizes a specific phosphopeptide motif Ser-pThr/pSer-Pro. Phosphorylation of Cdc25 by the low-activity Cdk complex (pTyr^{15}/pThr^{160}Cdk2/cyclin A or the similarly phosphorylated Cdk1/cyclin B)

(A)

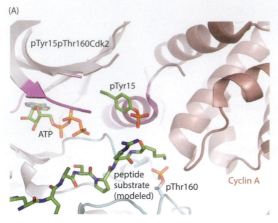

Figure 13.5 Inhibition of CDKs by phosphorylation. (A) Close-up of pTyr15pThr160Cdk2/cyclin A in complex with the ATP analog AMPPNP, plus a modeled peptide substrate. Phosphorylation of Tyr15 blocks peptide substrate binding. (PDB 2CJM) (B) Feedback loops govern entry to mitosis. Orange denotes proteins in an inhibited state; green denotes molecules in the active state. Cdk1/cyclin B is activated by phosphorylation of Thr161. The kinase domains (K; blue dot) of kinases Wee1 and Myt1 inactivate pCdk1/cyclin B by phosphorylation of Thr14 and Tyr15 on Cdk1. The inhibition is relieved by Cdc25 phosphatases, here Cdc25C. Extreme left, inactive Cdc25C is phosphorylated at Thr130 on its regulatory (R) domain (red dot) by Cdk1/cyclin B, which creates a recognition site for the polo-box domain (PBD) of Plk1. Plk1 is recruited and its kinase domain activates Cdc25C by phosphorylating its R domain at Ser198. Active Cdc25C phosphatase (PP domain; yellow dot) dephosphorylates pThr14 and pTyr15 of Cdk1 to give active pCdk1/cyclin B. Active pCdk1/cyclin B also phosphorylates Wee1 at Ser123 and Myt1 at Thr455 on their R domains. These sites recruit Plk1 through its PBD, and Plk1 phosphorylates Wee1 at Ser53 and Myt1 at Thr495. Phosphorylated Wee1 and Myt1 are ubiquitylated by the SCF and degraded by the proteasome.

(B)

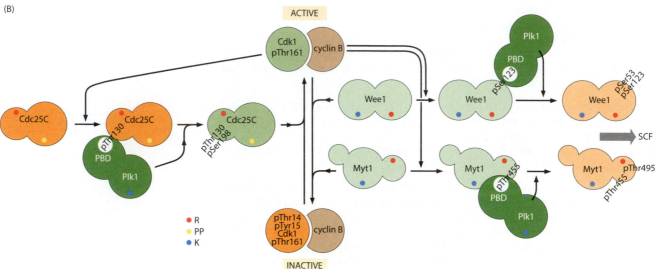

creates a recognition site for the PDB of Plk1. Plk1 binds to this site and phosphorylates Cdc25, thus activating the phosphatase. In turn, the phosphatase activates Cdk1/cyclin B or Cdk2/cyclin A, allowing the cycle to be amplified by a feedback activation loop (Figure 13.5B). Cdk1/cyclin B also phosphorylates Myt1 and Wee1, in a region outside their kinase domains, and this phosphorylation creates a recognition site on each kinase for the PDB of Plk1. Further phosphorylation by Plk1 targets Myt1 and Wee1 to the SCF for ubiquitylation and destruction, thereby allowing the cell cycle to progress. Although this description does not include all the subtleties of the feedback activation loop, especially what initiates it, the intricate interplay of phosphorylation and dephosphorylation in cell cycle progression is clear.

Cdk1/cyclin B is the master kinase in mitosis but other kinases play critical roles at other stages

As shown in gene knock-out experiments, Cdk1 can fulfill the functions of other cell cycle CDKs and is essential for cell cycle progression. Cdk1/cyclin B has roles in chromosome condensation, breakdown of the nuclear envelope, and spindle formation, in addition to activation of the APC/C and segregation of chromosomes. Cdk1/cyclin B activity first appears at the centrosome in prophase but later in mitosis is also present on the spindle microtubules; it is distributed throughout the cell after nuclear envelope breakdown. Some of the well-established substrates have been described above, but proteomics has identified 68 proteins as substrates for Cdk1. For some, the downstream consequences of phosphorylation are understood, such as phosphorylation of phosphatase PP1a that inhibits the enzyme DNA ligase I, and also of nuclear lamins for nuclear envelope breakdown.

Polo-like kinase (Plk1) is named after the phenotype of its mutation in *Drosophila mela-nogaster*, which results in abnormal spindle poles. Higher eukaryotes have four family

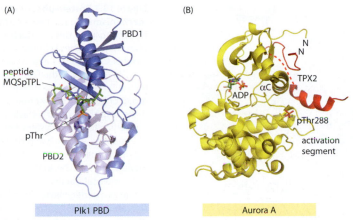

Figure 13.6 Structures of polo-like kinase (Plk1) and Aurora A kinase. (A) The polo box domain (PBD1) of Plk1 in complex with a phosphopeptide (its carbon atoms in green). Also highlighted but not labeled are some of the important side chains (His538, Lys540) that recognize the pThr and those that create a nonpolar site (Trp414, Phe535). (PDB 1Q4K) (B) Catalytic domain of Aurora A (yellow, residues 122–403) in complex with residues 1–43 of TPX2 (red). The kinase is phosphorylated on Thr288 and is in its active conformation. Part of the TPX2 peptide docks against the β sheet of the N-terminal lobe, followed by a disordered region (red dashes), and the remainder packs against the αC helix and the activation segment, promoting the active conformation. (PDB 1OL5)

members, but *D. melanogaster* and *S. cerevisiae* have just one, Plk1. Plk1 has roles in construction of the mitotic spindle, in maturation of the centrosomes, and in cytokinesis. Its localization is regulated in time and space through binding of its PBD to previously phosphorylated PBD recognition sites on its target substrates. The PBD is composed of two repeats, PBD1 and PBD2, each comprising a six-stranded β sheet and an α helix. The phosphopeptide-binding site is situated at the shallow cleft between the two β sheets (**Figure 13.6A**). Plk1 activity also requires phosphorylation of a Thr residue on its activation segment, catalyzed in human cells by Aurora A kinase several hours before entry into mitosis. The combined effects of phosphorylation of Plk1 and binding of its PBD to a phosphorylated peptide raise its catalytic efficiency ~1500-fold (as measured by k_{cat}/K_m).

The Aurora A and B kinases both become active during mitosis but differ in their subcellular locations. Aurora A, in complex with a spindle assembly protein called TPX2 (targeting protein for Xklp2), contributes to spindle assembly at the centrosomes. After breakdown of the nuclear envelope, the presence of free TPX2 in the vicinity of the chromosomes is thought to help nucleate microtubules that are subsequently incorporated into the mitotic spindle. TPX2 binds to the N-terminal lobe of the Aurora A kinase, making interactions with the β sheet, the αC helix, and the start of the activation segment (Figure 13.6B), thereby stabilizing the kinase in an active conformation (compare with the N-terminal and C-terminal extensions of protein kinase A [PKA]; Section 12.2). Aurora A functions in centrosome maturation (Section 13.4), mitotic checkpoint control, and the phosphorylation and activation of Plk1, a reaction that is enhanced by the binding of Aurora A to another protein, bora (539 aa).

Aurora B is associated with kinetochores and interacts with three chromosomal passenger proteins, INCENP (inner centromere protein), borealin, and survivin. It helps control the structure and segregation of sister chromatids (Section 13.4).

Phosphatases dephosphorylate substrates to terminate the cell cycle

As cells exit from mitosis, protein phosphatases dephosphorylate substrates to switch off the signal and allow the cycle to restart (see Table 13.1). There are 518 protein kinases encoded in the human genome (see Box 12.2), but many fewer phosphatases. The phosphatases fall into three categories: Tyr phosphatases, dual-specificity phosphatases, and Ser/Thr phosphatases. The number (~100) of Tyr and dual-specificity phosphatases approximately matches that (90) of Tyr kinases. They have a Cys residue in their catalytic site to promote hydrolysis of the phosphate ester (**Figure 13.7A**). Examples in the cell cycle include Cdc25 and KAP. The Ser/Thr phosphatases, such as PP1 and PP2A, use a two-metal mechanism, in which the metals bind and activate a water molecule to make a nucleophilic attack on the phosphate group (Figure 13.7B). Only ~30 Ser/Thr phosphatases are encoded in the human genome, many fewer than the Ser/Thr kinases (~428), but their versatility is enhanced by combining a common catalytic subunit with different regulatory subunits.

In *S. cerevisiae*, the phosphatase Cdc14 dephosphorylates key substrates at the onset of anaphase. The human homologs of Cdc14 are less well understood, but they have the potential,

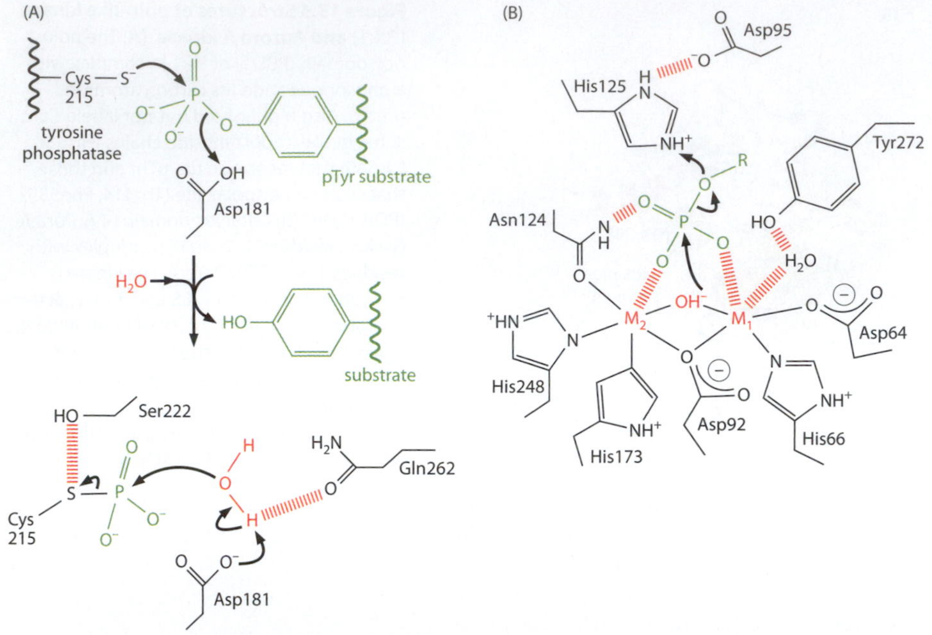

(A)

(B)

Figure 13.7 Protein phosphatase mechanisms. (A) The two-step mechanism for Tyr phosphatases and dual-specificity phosphatases. These contain a conserved motif of CXAGXGRS/T and conserved Asp and Gln residues. In the first step (numbering for the Tyr phosphatase PTP1B), the thiol group of Cys215 (the first residue, in bold, in the conserved motif) attacks the tyrosine phosphate to form a cysteinyl phosphate intermediate, and an Asp residue (Asp181) acts as a general acid to protonate the leaving phenolic OH. In the second step, a water molecule, activated by the Asp residue acting as a general base and oriented by a Gln residue (Gln262), attacks the cysteinyl phosphate, leading to phosphate release. (B) The Ser/Thr phosphatase two-metal-ion mechanism. The numbering corresponds to the phosphatase PP1. The metals are Fe and Mn, but the assignment was not definitive in the crystal structure. This is a single-step mechanism in which the metals activate a water molecule to attack the phosphate ester. His125 acts as a general acid to protonate the leaving ester oxygen. Each metal is six-coordinated: M1 by two His residues, an Asp residue and an Asn residue, and M2 by a His residue, two Asp residues and a water, with both metals also coordinated by a phosphorus oxygen and the activated water molecule.

through the recognition of pSer/pThr-Pro motifs, to dephosphorylate CDK targets. Cdc14 is a dual-specificity phosphatase, similar in mechanism to Cdc25 and KAP. It uses a Cys thiol group as a nucleophile and an Asp carboxylate as a general acid to promote hydrolysis of the phosphate ester. Specificity for Pro in the P +1 position is mediated by a hydrophobic site without an H-bond acceptor, which makes it unfavorable to bind a residue with a peptide NH group (compare Figure 13.2F above).

Protein phosphatase 1 (PP1) is a ubiquitous Ser/Thr phosphatase that is important in control of the cell cycle and in many other processes. It consists of a catalytic subunit with highly conserved metal-binding residues and a regulatory subunit that confers substrate specificity. At least 100 putative versions of the regulatory subunit have been identified. The PP2A phosphatase is a regulator of entry into and exit from mitosis but also has other important roles in, for example, signaling, tau dephosphorylation, and DNA damage. It accounts for up to 1% of proteins in some tissues. PP2A exists as a heterodimeric core enzyme consisting of a scaffold subunit (also known as the A subunit or PR65) and a catalytic subunit (C subunit). Both subunits have two isoforms, α and β. The core enzyme binds a regulatory B subunit which targets the trimeric holoenzyme to different substrates. The B subunits come from one of four families (B, B′, B″ and B‴), each of which has from two to five isoforms.

The A subunit in PP2A has 15 tandem HEAT repeats, each comprising a pair of anti-parallel helices (see **Guide**). The C subunit is a typical metal-dependent phosphatase and is inhibited by several naturally occurring tumor-inducing toxins such as okadaic acid (from shellfish) and microcystin-LR (from a blue-green alga). The PP2A enzyme that is important in chromosome segregation (Section 13.3) has a B′ subunit (B56) with a structure (eight HEAT-like repeats) similar to that of the A subunit, and it contacts both the A and C subunits (**Figure 13.8B**). In contrast, the PP2A holoenzyme with a B subunit is important for dephosphorylation of the microtubule-binding protein tau (implicated in forming neurofibrillary tangles in neurodegenerative diseases (Section 14.6 and Box 5.4). The B subunit has seven WD40 repeats (see **Guide**). Regulatory B and B′ subunits resemble the substrate presentation subunits of the SCF complex (Section 12.5) in that both use multiple repeats of structural motifs to provide a binding surface.

13.4 SISTER CHROMATIDS IN MITOSIS

After DNA replication in S phase, each pair of sister chromatids is held together to prevent jumbling and to ensure that one copy is delivered to each daughter cell (see Figure 13.1). As a cell enters mitosis (S/M), the chromatids are condensed from a malleable tangled mass to compact rods, which are then aligned by bipolar attachment to the mitotic spindle. Holding

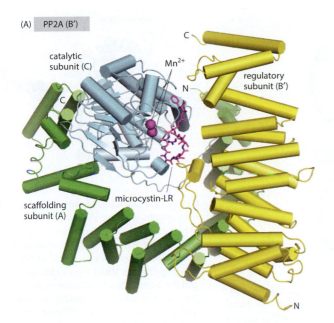

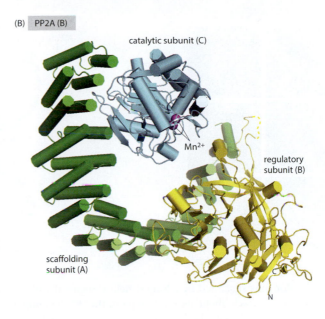

Figure 13.8 Protein phosphatase structures. (A) The human PP2A, showing scaffolding subunit A composed of 15 HEAT repeats (green), the regulatory subunit B'-γ1 composed of 8 HEAT repeats (yellow), and the catalytic subunit C (cyan) with the cyclic heptapeptide inhibitor mycrocystin-LR (magenta) bound. The catalytic site is marked by two Mn²⁺ atoms. Helices are shown as cylinders. (PDB 2NPP) (B) PP2A in complex with the regulatory subunit B. The color coding is as in (A); the molecule is tilted slightly to show the WD40 conformation of the regulatory subunit. (PDB 3DW8)

the sister chromatids together enables them to resist the forces exerted by the spindle when it attempts to pull them apart, until such time as they are ready to be separated. As robust rods, they are easily moved around during spindle assembly and more readily separated in anaphase.

Cohesin holds sister chromatids together until anaphase

Cohesin is a complex of four proteins: Smc1, Smc3, Scc1, and Scc3 (**Table 13.2**). It is very different from the 'cohesin' found in cellulosomes (Section 9.6). The Smc (structural maintenance of chromosomes) proteins Smc1 and Smc3 contain N-terminal and C-terminal

Table 13.2 Cohesin subunits and associated proteins.

Generic name (alternative name in humans)	Number of amino acids	Comment
Smc1	1233	Contains N- and C-terminal ATP-binding motifs (residues 32–39 and 1128–1163); two coiled-coil regions (residues 163–503 and 660–935) linked by a hinge (residues 504–659)
Smc3	1217	Contains N- and C-terminal ATP binding motifs (residues 32–39 and 1115–1150); two coiled-coil regions (residues 179–503 and 669–916) linked by a hinge (residues 505–667)
Scc1 (Rad21)	631	Scc1 contains the cleavage sites by separase
Scc3 (SA1 and SA2)	1258	Scc3 contains phosphorylation sites. In isoform SA1 these are Ser1062, Ser1065, Ser1067, Ser1068, and Ser1070. There are similar sites in isoform SA2
Associated proteins		
Pds5 (Pds5A and Pds5B)	1337	Pds5 is involved in maintenance of the cohesin complex; contains HEAT repeats
Wapl	1190	Wapl is involved in cohesin dissociation
Sororin	252	Sororin is involved in the establishment or maintenance of cohesin

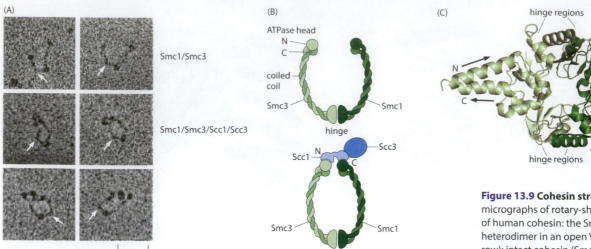

globular domains separated by a long α-helical segment with a central hinge region. Each protein doubles back about its hinge region, thereby forming a long (50 nm) anti-parallel coiled coil that brings the N and C domains together to form an *ATP-binding cassette* (*ABC*). ABCs are characterized by their Walker A and Walker B motifs (Box 3.1). In an Smc protein, Walker A is in the N-terminal domain and Walker B is in the C-terminal globular domain, ~1000 aa apart. Smc1 and Smc3 associate via their hinge domains to form a heterodimer, which adopts a V-shaped structure with the globular ABC domains at the ends of the flexible arms (**Figure 13.9**).

Scc1 (sister chromatid cohesion protein 1) belongs to the kleisin superfamily (kleisin is based on the Greek word for 'closure'). It has three parts: an N-terminal domain that interacts with the ATPase head of Smc3, a C-terminal domain that interacts with the ATPase head of Smc1, and a linker region. Scc1 forms a bridge connecting the two heads of the Smc1/Smc3 heterodimer to form a tripartite cohesin ring (see Figure 13.9A,B) that is essential for sister chromatid cohesion. The fourth protein, Scc3, binds to a site in the C-terminal half of the Scc1 linker (see Figure 13.9B).

The Smc1 coiled-coil α helices emerge on the surface opposite to where Scc1 binds (**Figure 13.10A**). The two Smc1 ATPase heads come together in a symmetrical dimer, the Walker A motif from the N-terminal domain of one head complemented by the Walker B motif from the C-terminal domain of the same head. The ATPase site is completed by the conserved

Figure 13.9 Cohesin structures. (A) Electron micrographs of rotary-shadowed complexes of human cohesin: the Smc1/Smc3 heterodimer in an open V conformation (top row); intact cohesin (Smc1/Smc3/Scc1/Scc3) showing ring-like structures (middle row); and more open structures (bottom row). The non-Smc subunits appear as separate globules between the two ATPase heads. One of the coiled-coil arms is often kinked (white arrows). (B) Models of the Smc1/Smc3 and Smc1/Smc3/Scc1/Scc3 complexes. (C) Structure of the hinge regions of *Thermotoga maritima* Smc homodimer, mimicking the Smc1/Smc3 hinge regions, viewed down the pseudo-2-fold axis of symmetry. One Smc is in pale green and the other is in dark green. The coiled coils are anti-parallel and intramolecular. (PDB 1GXL) (A, from D.E. Anderson et al., *J. Cell. Biol.* 156:419–424, 2002; B, adapted from B. Alberts et al., Molecular Biology of the Cell, 6th ed. New York: Garland Science, 2015.)

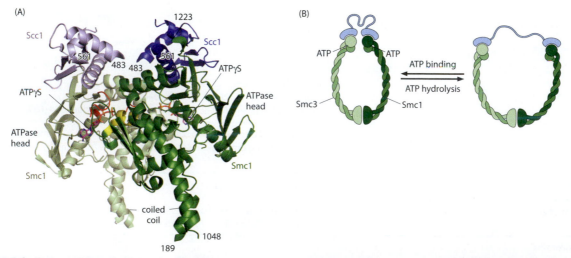

Figure 13.10 Cohesin Smc1 ATPase heads in complex with Scc1 C-terminal domains. (A) Two *S. cerevisiae* Smc1 ATPase heads (light and dark green) form a symmetric dimer. Each head associates with an Scc1 C-terminal domain (light and dark blue). The Smc1 molecule is labeled with the N- and C-terminal residues of the construct used for crystallization: 2–189 and 1048–1223, representing the N-terminal and C-terminal domains of the ATPase head respectively. ATPγS (red), a non-hydrolyzable analog of ATP, binds between the two domains, the one binding to the light green Smc1 is closer to the viewer and the one binding to the dark green Smc1 is further from the viewer. (PDB 1W1W) (B) Diagram of the Smc1/Smc3/Scc1 heterotrimer, showing how ATP hydrolysis promotes the separation of the Smc heads. (Adapted from K. Nasmyth and C.H. Haering, *Annu. Rev. Biochem.* 74:595–648, 2005.)

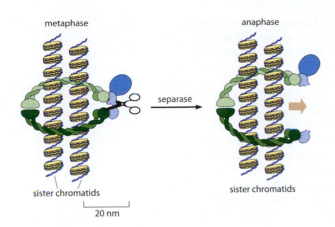

metaphase

anaphase

separase

sister chromatids

sister chromatids

20 nm

Figure 13.11 A model of the cohesin/ sister chromatid complex. From S phase to metaphase, the Smc1/Smc3/Scc1/ Scc3 cohesin complex encircles the sister chromatids (shown here with DNA in blue wrapped around nucleosomes) holding the sister chromatids together. At anaphase, Scc1 is cleaved by separase, the cohesin ring opens, and the sister chromatids can separate. (Adapted from B. Alberts et al., *Molecular Biology of the Cell*, 6th ed. New York: Garland Science, 2015.)

signature motif of ATPases (LSGGE/Q/KS/T) that contacts the γ-phosphate. This is supplied by the opposing Smc1 ATPase head. In the native cohesin heterodimer of Smc1/ Smc3, the two ATP heads, one from Smc1 and the other from Smc3, come together. The C-terminal domain of Scc1 has a winged helix domain (WHD) structure (see **Guide**) composed of three α helices followed by two β strands. Many WHDs are involved in DNA binding, but Scm1 blocks the potential DNA-binding sites, making it unlikely that Scc1 in complex with Smc1 can bind DNA. It has been proposed that hydrolysis of the bound ATP transiently drives the heads apart, thus providing a mechanism for widening the rings while maintaining their integrity (Figure 13.10B).

Its 35 nm diameter suggests that cohesin is able to trap sister chromatids inside its ring (**Figure 13.11**). The initial loading of cohesin occurs before S phase by means of a low-affinity interaction with DNA. During DNA replication in S phase, the replication fork transforms the loaded cohesin into a high-affinity form that surrounds the sister chromatids. This is assisted by ATP hydrolysis and involves the enzyme Eco1 (establishment of cohesin 1), an acetyl transferase which acetylates two Lys residues in the ABC globular head region of Scm3. This action of Eco1 counteracts the action of two proteins, Pds5 and Wapl (wings apart-like), that somehow favor the opening of the cohesin ring (mechanism unknown). When chromosome duplication is complete, the cohesin complexes are arrayed along the arms of the sister chromatids at intervals of 10–15 kbp, with a higher concentration at the centromeric regions.

Interphase (G2) to metaphase is characterized by chromosome condensation and then proteolytic cleavage of cohesin

As the cell enters mitosis, the sister chromatids are compacted and organized into discrete units, still held together by cohesin, that are capable of being pulled apart in anaphase. These are the structures commonly seen in optical micrographs of mitotic chromosomes. Condensation results in a 10,000-fold decrease in chromosome length; it begins early in prophase and is complete by the end of metaphase. It is promoted by the enzyme DNA topoisomerase IIα (TopoIIα), which decatenates sister chromatids (Chapter 3). (DNA catenation occurs when two replication forks meet and the DNA strands intertwine.) The five-subunit complex **condensin**, related both structurally and functionally to cohesin, stabilizes the condensed state of the chromosomes. Its two core subunits, Smc2 and Smc4, interact with three non-Smc subunits.

It is proposed that condensin forms a ring that encircles the DNA like cohesin does, but the process is not well understood. Cohesin is lost from the arms of the chromatids, a process mediated by Wapl and the kinases Plk1 and Aurora B. However, loss of cohesin from the centromeres is prevented by the protein shugoshin (Japanese for 'guardian spirit'), Sgo1, which also has a role in monitoring the attachment of kinetochore microtubules and delaying anaphase until secure attachments have been achieved. The retention of cohesin at the centromeres is needed for the sister chromatids to resist the pulling force in the mitotic spindle until the cell is ready to progress to anaphase.

The cell is now preparing to enter anaphase. A thiol protease, separase, cleaves Scc1 at two sites between its N-terminal and C-terminal domains, thereby rupturing the cohesin ring that holds the sister chromatids together (see Figure 13.11). Human separase (2120 aa) has 26 HEAT repeats, an unstructured region of 280 aa, and two caspase-like domains (see Section 13.6), only one of which is active. Until the metaphase/anaphase transition, separase is inhibited by binding to the protein securin, which has been phosphorylated by Cdk1. Human securin (202 aa) has an intrinsically disordered structure in solution. Early in mitosis, the APC/C is phosphorylated by Cdk1/cyclin B and then activated by assembly with Cdc20, whose availability is controlled by the Mad1 and Mad2 proteins (Section 13.4). Once the spindle assembly checkpoint has been satisfactorily met (see Section 13.5), phosphorylated securin is dephosphorylated by Cdc14, making it a target for ubiquitylation by APC/C^{Cdc20} and destruction by the proteasome. Separase is then able to cleave Scc1 and release the sister chromatids. APC/C^{Cdc20} also ubiquitylates cyclin B and targets it for destruction. In human cells, the destruction of securin and cyclin B begins in metaphase and takes about 20 minutes, suggesting a gradual release of the sister chromatids from cohesion. However, the process takes place abruptly, and it is not clear how this is achieved.

13.5   KINETOCHORES IN ANAPHASE

A large bipolar array of microtubules, the mitotic spindle, pulls sister chromatids apart in anaphase, thereby segregating the two sets of chromosomes to opposite ends of the cell (see Figure 13.1B). Microtubules are polar filaments built of heterodimers of α- and β-tubulin; their minus ends expose α-tubulin subunits, and the plus ends expose β-tubulin subunits (Section 14.6). In the spindle of animal cells, the minus ends are embedded in the centrosome (see below) and the plus ends point toward the center of the cell. Microtubules emanating from two oppositely oriented centrosomes overlap at their plus ends in an anti-parallel array at the midzone of the spindle. The plus ends of other microtubules are anchored in kinetochores, large protein complexes assembled on the centromeres (a special region of heterochromatin) of both sister chromatids. Multiple microtubules are bundled together to form kinetochore fibers that link the centrosomes to the chromosomes.

Microtubules exhibit dynamic instability (see Section 14.6). Kinetochore microtubules depolymerize at their minus ends and new tubulin is added to the plus ends. Surprisingly, the plus ends can grow and shrink while remaining attached to the kinetochore. It is these dynamic properties of microtubules that are mainly responsible for the forces that move the chromosomes. Depolymerization at the plus ends generates a force that pulls the kinetochore toward the pole (the mechanism is not yet fully understood), while dismantling of microtubules at their minus ends also contributes to pulling the chromatids toward the poles.

Spindle assembly begins early in mitosis and there is a tension-sensing mechanism to ensure that anaphase does not begin until each sister chromatid has been securely connected through its kinetochore to the opposite pole and bi-orientation has been accomplished for all sister chromatids. Only then does release from this **spindle assembly checkpoint** (SAC) permit the breakdown of cohesin and the irreversible segregation of sister chromatids toward opposite spindle poles of the cell.

Kinetochores are assembled on centromeric DNA associated with a novel histone, CENP-A

In electron micrographs, the kinetochore appears as a tri-laminar disk, with an inner kinetochore (sometimes called the inner plate) connected to the dense centromeric chromatin, an electron-translucent middle layer, and an outer kinetochore (also called the outer plate) to which the ends of microtubules are attached (**Figure 13.12**). The outer and inner kinetochores are connected by fibrous links. Electron tomography has indicated that individual kinetochore microtubules are attached to the outer plate by several fibers that appear either to embed their plus ends in a mesh or to extend out to bind to the microtubule walls. In the unbound state in the absence of microtubules, the fibrous network appears to rearrange. Structural analysis of kinetochores has been hampered by difficulties in obtaining homogeneous preparations. The inner kinetochore is embedded in a 'mat' of centromeric chromatin, which makes biochemical extraction difficult.

Centromeres are almost always located at the same site on a chromosome, generation after generation. The centromeric DNA sequence varies between species but the function is conserved. In *S. cerevisiae*, it is only ~125 bp long, with a defined sequence (**Figure 13.13A**). Because of its small size, it is referred to as a point centromere. In other eukaryotes,

(A)

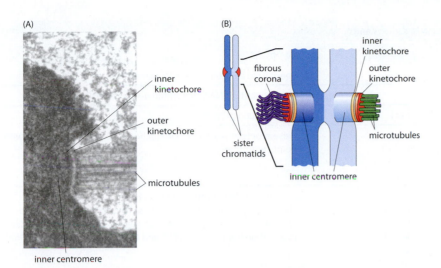

inner
kinetochore

outer
kinetochore

microtubules

inner centromere

(B)

inner
kinetochore

fibrous
corona

outer
kinetochore

microtubules

sister
chromatids

inner centromere

Figure 13.12 The kinetochore. (A) A thin-section electron micrograph of a metaphase kinetochore. (B) Schematic diagram of a condensed mitotic chromosome with paired sister chromatids (dark and light blue). Right, the inner centromere of one chromatid contacts the inner kinetochore; microtubules are attached to the outer kinetochore. Left, the dark blue sister chromatid is unattached and a fibrous corona is observed in the absence of bound microtubules. The fibrous corona contains spindle checkpoint proteins and other proteins that are removed from kinetochores upon microtubule attachment. (A, from B.F. McEwen et al., *Chromosoma* 107:366–375, 1998. With permission from Springer Science and Business Media. B, from I.M. Cheeseman and A. Desai, *Nat. Rev. Mol. Cell Biol.* 9:33–46, 2008. With permission from Macmillan Publishers Ltd.)

centromeres are usually dispersed over long arrays of repetitive DNA (regional centromeres). Human centromeres contain a 171 bp sequence called **α-satellite DNA**, which is repeated in tandem arrays ranging in size from 200 to 4000 kbp (Figure 13.13B).

In all eukaryotes, centromere specification is mediated by the protein CENP-A (centromeric protein A). CENP-A (140 aa) shares 62% sequence identity with histone H3 (138 aa; Chapter 2); it is found exclusively at centromeres and resides there throughout the cell cycle. The exact composition of CENP-A-containing nucleosomes may vary in different organisms (**Figure 13.14A**). Part of CENP-A (residues 76–115), when substituted into H3, selectively targets the protein to centromeres, where it can functionally replace CENP-A. This region, termed the centromeric A-targeting domain (CATD), consists of loop 1 (L1) and the α2 helix of the H3 histone fold (Figure 13.14B). In the nucleosome, the CENP-A/CENP-A interface is rotated relative to the normal H3/H3 interface; there are stronger nonpolar interactions at the CENP-A/H4 interface that make the CENP-A complexes more stable; and loop L1 has a different conformation. These structural changes are presumably associated with functional changes.

Kinetochores contain many proteins that form stable complexes

Differences in centromere size and organization are reflected in differences in the number of microtubules attached. The *S. cerevisiae* kinetochore attaches to a single microtubule; those of *S. pombe* to 2–5 microtubules; and those of vertebrates to bundles of 15–30 microtubules, the kinetochore fibers (K-fibers). From these observations, it has been proposed that kinetochores attaching to multiple microtubules comprise modular arrays of a unit that provides end-on attachment to a single microtubule. About two-thirds of kinetochore proteins form stable complexes (**Table 13.3**) and some of these are organized in higher-order assemblies.

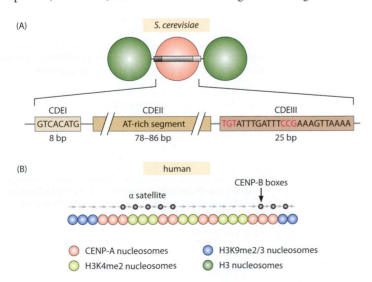

(A) *S. cerevisiae*

CDEI   CDEII   CDEIII
GTCACATG   AT-rich segment   TGTATTTGATTTCCGAAAGTTAAAA
8 bp   78–86 bp   25 bp

(B) human

α satellite   CENP-B boxes

- CENP-A nucleosomes
- H3K4me2 nucleosomes
- H3K9me2/3 nucleosomes
- H3 nucleosomes

Figure 13.13 Organization of centromeric DNA in yeast and human chromosomes. (A) The point centromere of *S. cerevisiae*. The CDEI, CDEII, and CDEIII regions of the 125-bp centromeric DNA assemble into a nucleosome-like structure containing the histone H3 variant Cse4 (pink disk). This nucleosome is flanked by H3-containing nucleosomes (green disks). The CDE DNA sequences shown are those for chromosome 3. (B) In human chromosomes, the centromere forms on a small subdomain of the α-satellite DNA repeats (gray circles). The α-satellite DNA (200-4000 kbp) contains a sequence known as the CENP-B box (CTTCGTTGGAAACGGGA), which binds in a sequence-specific manner to the CENP-B protein and facilitates kinetochore formation. Human centromeres contain blocks of CENP-A nucleosomes (pink disks) interspersed with histone H3-containing nucleosomes (green disks). H3 can be post-translationally modified by methylation on Lys4 (H3K4me2) or Lys9 (H3K9me2/3). (Adapted from R.C. Allshire and G.H. Karpen, *Nat. Rev. Genet.* 9:923–937, 2008. With permission from Macmillan Publishers Ltd.)

Complex	H. sapiens	S. cerevisiae	Comment
Table 13.3 Centromeric and kinetochore proteins in *H. sapiens* and *S. cerevisiae*.			
	CENP-A	Cse4	Specialized centromeric nucleosome, a variant of histone H3
	CENP-B		CENP-B binds CENP-B box within α-satellite DNA
		Cbf1	CDEI-binding protein
Cbf3 complex		Cep3 Ndc10 Ctf13 Skp1	Present only in yeast. Core kinetochore complex in organisms with point centromeres
CCAN (constitutive centromere associated network) (human); Ctf19 complex (*S. cerevisiae*)	CENP-C CENP-H CENP-I CENP-K CENP-L CENP-M CENP-N CENP-O CENP-P CENP-Q CENP-R CENP-S CENP-T CENP-U CENP-W	Mif2 Mcm16 Ctf3 Mcm19 Ch14 Mcm21 Ctf19	Inner kinetochore protein network required for accurate chromosome segregation but whose function is not clear. Because of weak sequence similarity across species, it is sometimes difficult to ascertain homology. CENP-T and CENP-W have histone-like folds
Knl1 complex	KNL1 Zwint	Spc105	Part of the KMN network (Knl1, Mis12, Ndc80)
Mis12 complex	Mis12 DSN1 NNF1 NSL1	Mtw1 Dsn1 Nnf1 Nsl1	Part of the KMN network
Ndc80 complex	NDC80 NUF2 SPC24 SPC25	Ndc80 Nuf2 Spc24 Spc25	Part of the KMN network
Dam1 complex		Dam1 Duo Spc34 Dad1 Spc19 Ask1 Dad2 Dad3 Dad4 Hsk3	Ring-forming microtubule-binding protein; found only in yeast
Ska1 complex	Ska1 Ska2 Ska3Rama		Microtubule-binding complex recruited to kinetochores via the KMN network. Stimulates microtubule oligomerization
Mitotic checkpoint complex (MCC)	Mad1 Mad2 Bub1 Bub3 BubR1 Cdc20	Mad1 Mad2 Bub1 Bub3 BubR1 Cdc20	The critical spindle checkpoint effector
Chromosome passenger complex (human); Ipl1 complex (*S. cerevisiae*)	Aurora B INCENP Survivin Borealin	Ipl1 Sli15 Bir1	Aurora B is a kinase that controls several aspects of mitosis, including chromosome condensation, centromere organization, kinetochore/microtubule attachment, spindle checkpoint, and cytokinesis

(Adapted from full table in I.M. Cheeseman and A. Desai, *Nat. Rev. Mol. Cell Biol.* 9:33–46, 2008.)

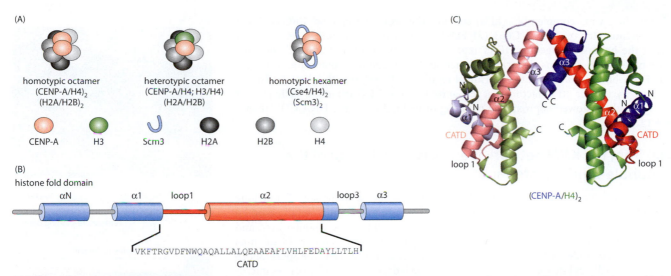

(A)

homotypic octamer
(CENP-A/H4)₂
(H2A/H2B)₂

heterotypic octamer
(CENP-A/H4; H3/H4)
(H2A/H2B)

homotypic hexamer
(Cse4/H4)₂
(Scm3)₂

CENP-A   H3   Scm3   H2A   H2B   H4

(B)
histone fold domain

αN   α1   loop1   α2   loop3   α3

VKFTRGVDFNWQAQALLALQEAAEAFLVHLFEDAYLLTLH
CATD

(C)

α3   α3
α2
α1   α2   α1
CATD   CATD
loop 1   loop 1

(CENP-A/H4)₂

Figure 13.14 Nucleosome composition and CENP-A structure.
(A) In non-centromeric regions, histone H3 assembles into octameric nucleosomes containing two H2A, H2B, H3, and H4 subunits. In centromeric chromatin, histone H3-like CENP-A can assemble into homotypic octamers, in which it replaces both H3 subunits, or into heterotypic octamers containing one H3 and one CENP-A subunit. In *S. cerevisiae*, Scm3 replaces the H2A/H2B dimers, forming an unusual hexamer comprising pairs of Scm3, Cse4 (CENP-A), and H4 subunits.

(B) Histone H3 has four α helices. The CENP-A targeting domain (CATD) (red) is sufficient to localize histone H3 to centromeres when substituted into human H3. (C) Structure of the human (CENP-A/H4)₂ heterodimer complex. CENP-A (residues 60–140) is in blue, and the CATD domain is in red. H4 (residues 21–103) is in green. (PDB 3NQJ) (A, from R.C. Allshire and G.H. Karpen, *Nat. Rev. Genet.* 9:923–937, 2008. With permission from Macmillan Publishers Ltd. B, adapted from B.E. Black et al., *Nature* 430:578–582, 2004. With permission from Macmillan Publishers Ltd.)

For instance, the KMN network contains ten proteins in three distinct complexes (Knl1, Mis12, and Ndc80) and its assembly is implicated in microtubule attachment.

The kinetochore of *S. cerevisiae* contains ~60 proteins and exceeds 5 MDa in mass. The point centromere of *S. cerevisiae* subdivides into three DNA elements: CDEI, CDEII, and CDEIII (see Figure 13.13). The 8 bp CDEI motif binds to the helix–turn–helix protein (see **Guide**) Cbf1; the 76–84 bp AT-rich sequence of CDEII binds to the homotypic hexameric nucleosome (Cse4/H4/Scm3)₂ (see Figure 13.14A); and CDEIII, a 56 bp DNA region, has an imperfect palindrome sequence with a ~25 bp core and binds to the four-subunit Cbf3 complex. Cbf3 also associates with the CDEII-distal sequence of 50–60 bp; in its absence, the recruitment of all outer layers of the kinetochore is lost. Together, the centromeric DNA elements and associated proteins assemble into a single Cse4-containing centromeric nucleosome that recruits the next level of proteins to the kinetochore (**Figure 13.15A**). The Cse4 nucleosome and associated proteins form a platform for the hierarchical assembly of complexes that connect the kinetochore to the microtubule interface. However, the identity of the proteins that form the physical linkages between the inner and outer kinetochores is still uncertain. Mif2 is the most likely candidate. The Mtw1 complex (Mis12 in human), which contacts Mif2 (CENP-C in humans), is part of an intermediate layer that also includes the Spc105 (Knl1) and Ndc80 complexes in the KMN network.

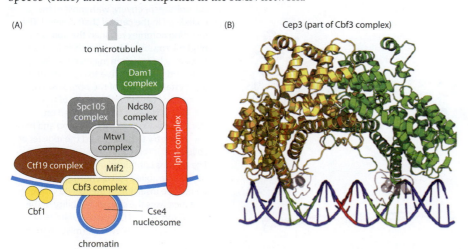

(A)
to microtubule
Dam1 complex
Spc105 complex   Ndc80 complex
Mtw1 complex
Ctf19 complex   Mif2   Ipl1 complex
Cbf3 complex
Cbf1   Cse4 nucleosome
chromatin

(B) Cep3 (part of Cbf3 complex)

Figure 13.15 A model of kinetochore proteins bound to the centromere of *S. cerevisiae*. (A) The 125-bp centromeric DNA is depicted in blue. A Cbf1 dimer binds the CDEI region. The Cbf3 complex binds the CDEIII region. The Ctf19 complex is part of the inner kinetochore. Mif2 (homologous to human CENP-C) acts as a linker protein to connect with intermediate protein complexes such as Mtw1 (Mis12), Spc105 (Knl-1), and Ndc80. The Ndc80 complex extends into the microtubule-binding region and, together with the Dam1 complex, contacts the microtubule directly. The Ipl1 complex spans from the inner to the outer kinetochore. The assignment of proteins to different layers in the kinetochore is tentative and based largely on the presumed structural similarity between *S. cerevisiae* kinetochores and those of higher eukaryotes. (B) Structure of dimeric Cep3 (residues 47–608), part of the Cbf3 complex. One subunit is in green and the other, in yellow. The N-terminal Zn₂Cys₆ DNA binding fragment (gray) has been modeled. The structure has been docked onto the major groove of CDEIII centromeric DNA with binding sites of TGT (left) and CCG (right) (bases in green). The two bases in red represent the putative Ctf13-binding site. (PDB 2QUQ). (A, adapted from I.M. Cheeseman et al., *J. Cell Biol.* 157:199–203, 2002; B, from J.J. Bellizi et al., *Structure* 15:1422–1430, 2007. With permission from Elsevier.)

The *S. cerevisiae* Cbf3 complex (450 kDa) contains the proteins Cep3/Ndc10/Ctf13/Skp1 with a stoichiometry of 2:2:1:1. Cep3 (71 kDa) has two domains, both α-helical, with several of the helices arranged in HEAT repeats, a fold that is likely to provide docking sites for other Cbf3 proteins. It forms a dimer, and each subunit contains a sequence-specific DNA-binding motif in its N-terminal region (1–48 aa), one that has similarities to the Zn_2Cys_6 transcription factors that recognize CGG or CCG triplets, usually present as a pair. This has allowed approximate mapping of its binding to the DNA CDEIII sites (Figure 13.15B).

Ndc80 from the KMN network is a crucial constituent of the microtubule-binding interface

The four-subunit Ndc80 complex (~170 kDa) is shaped like a dumbbell 57 nm long (**Figure 13.16**), and extends from the intermediate layer toward the microtubule-binding interface. A heterodimer of Ndc80 and Nuf2 subunits contributes one globular head and most of the 45 nm coiled-coil central shaft. A heterodimer of Spc25 and Spc24 subunits constitutes the remainder of the central shaft and the other globular head. The Spc24/Spc25 globular head mediates kinetochore association, and the Ndc80/Nuf2 head mediates microtubule binding, thereby creating a key link between the kinetochore and microtubules.

To facilitate crystallization a truncated ('bonsai') version of the human complex was created by fusing N-terminal segments of Ndc80 and Nuf2 to C-terminal segments of Spc25 and Spc24 respectively, thereby shortening the long coiled-coil domain between the globular heads (**Figure 13.17A**). The structure revealed that the globular head of the Ndc80/Nuf2 fusion product is composed of two tightly packed calponin homology (CH) domains (see **Guide**). The Ndc80 subunit also has an unstructured, positively charged region (80 aa) at its N-terminal end, which proved to be required for high-affinity microtubule binding. This region also contains a consensus motif (R/KR/KXS/T) for Ser/Thr phosphorylation by Aurora B kinase. On phosphorylation, the addition of a double negative charge to a Ser/Thr side chain is likely to disrupt electrostatic interactions in Ndc80 complexes. The kinetochore/microtubule interactions are enhanced by the Mis12 complex (Mtw1 in yeast) and Knl1 complex (Spc105 in yeast), part of the KMN network. These two complexes are also targets of Aurora B, and phosphorylation of them at spatially distinct sites contributes to reducing the affinity of the kinetochore/KMN network for microtubules.

Cryo-EM (see **Methods**) has revealed that the human Ndc80 (bonsai) tetrameric complex docks into a site between the α and β subunits of the microtubule and adopts an angle of ~45° away from the outer face of the microtubule lattice. It also exhibits a fixed polarity in which the Spc24/Spc25 dimer points away from the microtubule lattice in the direction of the plus end.

(A)

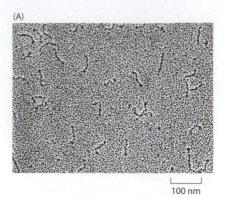

100 nm

(B)

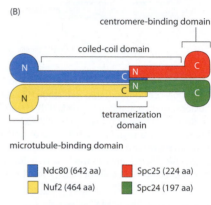

Figure 13.16 Organization of the Ndc80 complex. (A) Electron micrograph of rotary-shadowed Ndc80 complexes. (B) The molecular layout of the heterotetrameric complex. (A, from R.R. Wei et al., *Proc. Natl Acad. Sci. USA* 102:5363–5367, 2005; B, from C. Ciferri et al., *Cell* 133:427–439, 2008. With permission from Elsevier.)

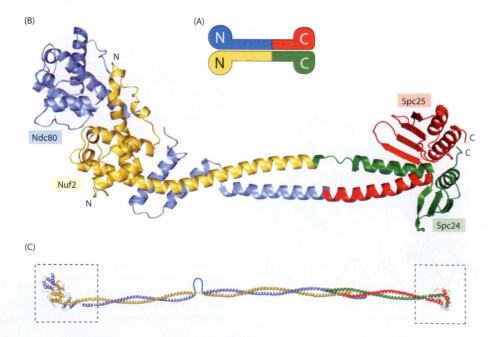

Figure 13.17 Structure of the Ndc80 complex. (A) A 'bonsai' version of the human Ndc80 complex based on the topology shown in Figure 13.16. The C termini of the Ndc80 and Nuf2 subunits were fused to the N termini of the Spc25 and Spc24 subunits respectively, with most of the coiled coil in the central shaft deleted. The resulting complex retained the ability to bind microtubules *in vitro* and to localize to kinetochores when injected into living cells, and was amenable to crystallization. (B) Crystal structure of the bonsai Ndc80 complex. (PDB 2VE7) (C) Model of the full-length Ndc80 complex, based on electron microscopy, cross-linking and mass spectrometry that identified the register of coiled-coil interaction in the central shaft. The regions contained in bonsai Ndc80 are boxed. The coiled coil is interrupted by a 50-residue insertion in the Ndc80 sequence that increases the overall flexibility of the Ndc80 rod. (B and C, from C. Ciferri et al., *Cell* 133:427–439, 2008. With permission from Elsevier.)

Vertical and horizontal arrangements for the microtubule-binding interface have been proposed

The abundance of the various protein complexes in a single kinetochore/microtubule attachment site has been quantified using fluorescence microscopy of GFP-tagged *S. cerevisiae* kinetochore proteins. Measured relative to the two Cse4 molecules in the centromeric nucleosome, the ratios range from 1–2 to 16 for different complexes. Differential labeling fluorescence microscopy and single-molecule studies have yielded a map of average protein positions for the human kinetochore. There is no overall detailed EM or X-ray (see **Methods**) model but, combining this evidence with knowledge about the Ndc80 complex, at least two layouts appear possible.

In a 'vertical' layout (shown schematically in **Figure 13.18A**), five to eight KMN complexes, comprising Spc105, Mtw1, and Ndc80 complexes in *S. cerevisiae*, transmit the forces exerted by a bound microtubule 25 nm in diameter, to a single Cse4/CENP-A nucleosome 10 nm in diameter. In a 'horizontal' layout (Figure 13.18B) the complexes do not bind directly to the single Cse4/CENP-A nucleosome but instead link with neighboring H3 nucleosomes and distribute the pulling forces over several contact points.

A role for H3 nucleosomes in kinetochore assembly in vertebrates is gaining acceptance after the discovery that inner kinetochore components such as CENP-C, CENP-T, and CENP-W (see Table 13.3) are recruited to H3 nucleosomes rather than to CENP-A nucleosomes. Apart from the Cbf3 and Dam1 complexes (see below), all yeast kinetochore complexes have counterparts in higher eukaryotes that occupy analogous positions in the kinetochore. The similarity in composition, abundance ratios, and hierarchical patterning of kinetochore complexes from yeast to humans supports the concept of a modular design for kinetochores that bind multiple microtubules.

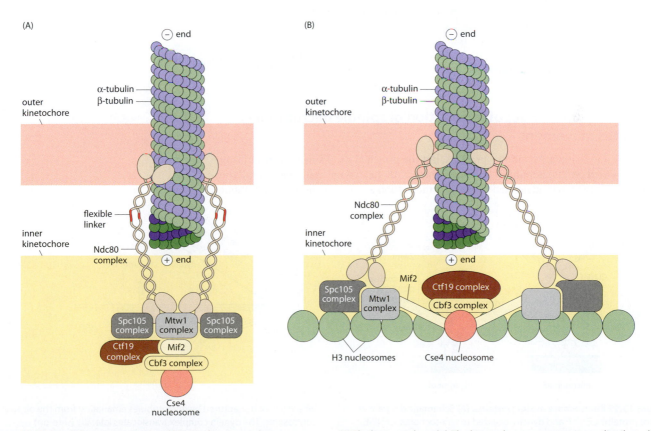

Figure 13.18 Possible arrangements of kinetochore complexes.
(A) The vertical model. All kinetochore proteins are ultimately connected to the single Cse4 (CENP-A) nucleosome of *S. cerevisiae*. Only two KMN (Spc105 [Knl1 in human], Mtw1 [Mis12 in human], Ncd80) complexes are shown here. One Mtw1 complex obscures the other one behind.

(B) The horizontal model. The kinetochore components are distributed horizontally. Specifically, the KMN network components are linked to the kinetochore core by Mif2 (CENP-C in humans), and also establish contacts with H3 nucleosomes.

Motor proteins speed up the processes of kinetochore/ microtubule attachment

For the initial capture of kinetochores, a search-and-capture mechanism is favored, whereby microtubules radiating out from centrosomes explore space until they encounter and are selectively stabilized by interaction with kinetochores. In addition, motor proteins have an accessory function in chromosome alignment. In higher eukaryotes, the only motors known to be associated with kinetochores are the minus end-directed *dynein* (Section 14.7) and the plus end-directed CENP-E.

The interaction of dynein with kinetochores involves several proteins, including the adaptor proteins dynactin, the Rod–ZW10–Zwilch (RZZ) complex and its binding partner spindly, lissencephaly 1 (LIS1), Zwint (part of the Knl1 complex), and the NDE (nuclear distribution protein E) and NDEL (NDE-like) proteins. Some of these proteins are shown schematically in **Figure 13.19A**. Several of the interactions occur simultaneously but others appear to be specific to a particular mitotic stage. Dynein is involved in the initial capture of spindle microtubules and in the ensuing rapid movement of chromosomes toward the centrosomes (Figure 13.19B). Dynein-powered movement along the microtubule facilitates end-on attachment of the kinetochore by moving the chromosome into the vicinity of the centrosomes where microtubule density is high.

Alignment of sister chromatids at the center of the spindle in metaphase is termed **congression**. The association of CENP-E with kinetochores is aided by CENP-F, a coiled-coil protein (see Figure 13.19A). CENP-E transports the mono-oriented sister chromatids along the kinetochore fiber to the spindle center, where they are captured by microtubules from the opposite pole (Figure 13.19B). This process speeds up congression and bi-orientation. Human CENP-E, a member of the kinesin-7 family (Section 14.7), is a dimer. Each giant subunit (2663 aa) has three main domains: an N-terminal motor domain (residues 1–327) that includes the ATP-binding and microtubule-binding sites; a 230 nm coiled-coil domain (residues 336–2471); and a C-terminal domain (residues 2472–2663). The kinetochore-binding region (residues 2126–2476) is located just before the C-terminal domain. The motor domain with the neck or linker region has a characteristic kinesin fold. The linker (neck) region that connects to the rest of the kinesin motor is in a docked conformation, which is typical for plus end-directed motors (see Section 14.7). At the end of mitosis, CENP-E is degraded.

Polymerization and depolymerization of spindle microtubules generates movement

Congression is facilitated by motor proteins, but the principal mechanism of chromosome motion is based on interactions of the kinetochore with the dynamic plus ends of microtubules, notably the forces generated by the polymerization and depolymerization of tubulin

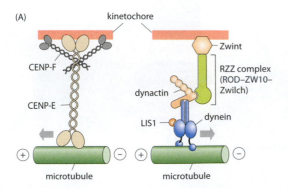

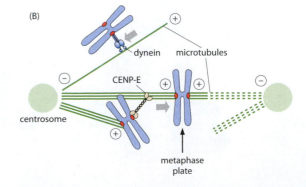

Figure 13.19 Kinetochore motor proteins. (A) Schematic diagram of motor proteins CENP-E and dynein localized to kinetochores. CENP-E contacts microtubules and the kinetochores in association with CENP-F. Dynein and associated proteins contact the kinetochore through the ROD–ZW10–Zwilch (RZZ) complex and Zwint. (B) CENP-E translocates along the kinetochore fiber of an already bi-oriented chromosome to move a mono-oriented chromosome toward the metaphase plate, where it may be captured by microtubules emanating from the opposite centrosome. The dynein complex translocates laterally (side-on) associated kinetochores to the vicinity of the minus end of spindle poles, where they may be captured end-on by microtubules. The diagram is not drawn to scale. (A and B, from I.M. Cheeseman and A. Desai, *Nat. Rev. Mol. Cell Biol.* 9:33–46, 2008. With permission from Macmillan Publishers Ltd.)

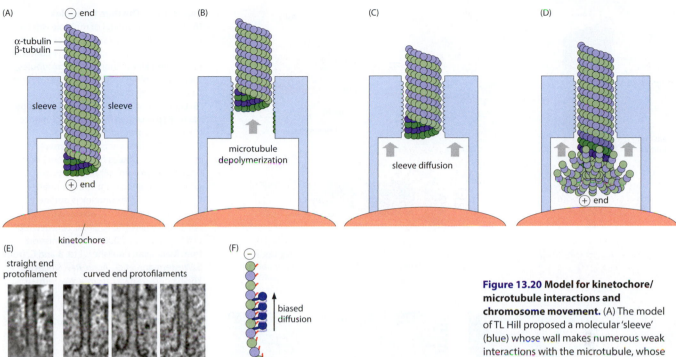

Figure 13.20 Model for kinetochore/ microtubule interactions and chromosome movement. (A) The model of TL Hill proposed a molecular 'sleeve' (blue) whose wall makes numerous weak interactions with the microtubule, whose plus end is free to grow and shrink. (B) Upon microtubule shortening and in the absence of applied forces, the number of interactions between the sleeve and the microtubule is reduced, so that several unoccupied sites are generated. (C) Diffusion of the sleeve in the direction of microtubule depolymerization maximizes occupancy of available binding sites. (D) During depolymerization, force is generated by curling protofilaments, which exert pressure against the base of the sleeve. The sliding of the microtubule inside the sleeve will allow a kinetochore-bound microtubule to 'treadmill' (Section 14.6). (E) Slices from electron tomograms of stained thin sections of cells showing kinetochore microtubules with a range of straight and curved plus ends. (F) A model for biased diffusion of the Ndc80 complex along microtubules in response to microtubule curling and depolymerization. The Ndc80/Nuf2 head is in dark blue, the N-terminal region of Ndc80 is in pale blue, and the negatively charged C-terminal region of tubulin is in red. As the geometry of the microtubule protofilament changes from straight to curved, the spacing of the monomers is no longer correct for Ndc80 binding and the complex dissociates at this site. It then makes contacts further along, thereby shifting it toward the minus end of the microtubule. (E, from K.J. Van den Beldt et al., *Curr. Biol.* 16:1217–1223, 2006. With permission from Macmillan Publishers Ltd. F, from G.M. Alushin et al., *Nature* 467:805–810, 2010.)

(Section 14.6). Microtubules alternate between phases of growth and shrinkage, exhibiting dynamic instability.

End-on attachment of kinetochores to dynamically unstable microtubules must be maintained throughout multiple cycles of assembly and disassembly at the attachment site. In a long-standing model, kinetochores provide a sleeve for microtubule binding. The microtubule and the sleeve establish a large number of equivalent interactions (**Figure 13.20A–C**) and, as the activation energy for the transition between binding sites is small relative to the overall binding energy of a microtubule–sleeve interaction, diffusion on the microtubule surface is energetically low-cost. In the absence of load, sleeve diffusion on a shortening microtubule (from depolymerization at the minus end) results from the attempt of the sleeve to maximize its overlap with the microtubule. However, in the presence of load, the amount of overlap between the sleeve and the microtubule is reduced, leading to possible detachment upon the application of tension.

In electron micrographs, the ends of protofilaments in shortening microtubules at kinetochores are flared at their plus ends (Figure 13.20D,E). 'Curling' of protofilaments is made possible by the fact that the energy from hydrolysis of the GTP bound to the β-tubulin subunits is stored in the form of mechanical strain. When disassembly begins, protofilaments undergo a straight-to-curved transition that can perform mechanical work. This could couple chromosome movement to microtubule depolymerization through a 'conformational wave' in which the depolymerizing microtubules form curved protofilaments that push against the sleeve. A biased diffusion model has been proposed on the basis of the structure of the Ncd80 (bonsai) complex bound to microtubules (see above), in which chromosome movement is coupled to microtubule depolymerization via the Ncd80 complex (Figure 13.20F); the complex can only bind to straight microtubules and hence shifts are favored that will accommodate this interaction.

The Dam1 complex of *S. cerevisiae* forms rings around microtubules (**Figure 13.21A–D**), and two-color fluorescence microscopy assays have demonstrated the processive movement of Dam1 rings for several micrometers at the ends of depolymerizing microtubules. Detached Dam1 rings have 16-fold symmetry (Figure 13.21E), in contrast with the 13-fold protofilament lattice of a microtubule (Section 14.6). This symmetry mismatch may serve to 'lubricate' sliding of the ring over the microtubule surface. Interaction of Dam1 with other protein complexes of the kinetochore such as Ndc80 may facilitate poleward movement of the chromosomes.

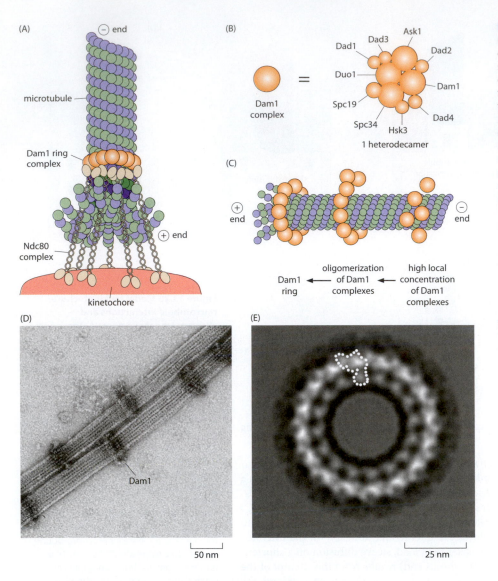

(A)

− end

microtubule

Dam1 ring complex

Ndc80 complex

kinetochore

+ end

(B)

Dam1 complex =

Dad1 Dad3 Ask1
Duo1 Dad2
Spc19 Dam1
Spc34 Dad4
Hsk3

1 heterodecamer

(C)

+ end   − end

oligomerization   high local
of Dam1   concentration
Dam1 ← complexes ← of Dam1
ring   complexes

(D)

Dam1

50 nm

(E)

25 nm

Figure 13.21 The Dam1 complex.
(A) A schematic model of multiple Ndc80 complexes in *S. cerevisiae* interacting with a single microtubule and the Dam1 complex connecting kinetochores to microtubules, thereby coupling chromosome movement to microtubule depolymerization.
(B) Dam1 complexes, containing one copy each of 10 essential proteins, oligomerize (C) into rings around microtubules.
(D) Electron micrograph of negatively stained microtubules bearing Dam1 rings (assembled *in vitro*). (E) Axial view of negatively stained Dam 1 ring, enhanced by averaging. One complex is outlined by a dotted contour. (A, adapted from I.M. Cheeseman and A. Desai, *Nat. Rev. Mol. Cell Biol.* 9:33–46, 2008. With permission from Macmillan Publishers Ltd. B and C, from S. Westermann et al., *Annu. Rev. Biochem.* 76:563–591, 2007; D, from S. Westerman et al., *Mol. Cell* 17:277–290, 2005. With permission from Elsevier. E, from Wang et al., *Nat. Struct. Mol. Biol.* 14:721–726, 2008. With permission from Elsevier.)

No homolog of the Dam1 complex has been identified in higher eukaryotes, but the three-subunit Ska1 complex containing the protein Rama1 (also called Ska3; see Table 13.3) localizes to kinetochores and microtubules during mitosis and is able to couple a cargo to the depolymerization of a microtubule. Thus Ska1 might have a role similar to that of Dam1. However, microtubule-coupling devices, other than the Ndc80 complex, may not be needed in organisms in which (unlike *S. cerevisiae*) there are multiple kinetochore/microtubule connections; in these instances, the processivity of chromosomal movement might be ensured by redundancy among the microtubule-binding sites.

The spindle assembly checkpoint monitors the state of kinetochore/microtubule attachment before the metaphase-to-anaphase transition is permitted

At the spindle assembly checkpoint (SAC), the progression of chromosome alignment is coordinated with the activation of the anaphase-promoting complex (APC/C) described in Section 12.5. The presence of a single kinetochore that has not been captured by microtubules and which is not under tension is enough for the SAC to keep the APC/C inactive. To achieve this, components of the SAC are recruited to unattached or tensionless kinetochores (that is, kinetochores that have not attained bi-orientation), and the **mitotic checkpoint complex** (**MCC**), a four-subunit anaphase inhibitor composed of Bub3, BubR1, and Cdc20 bound to C-Mad2 (see Table 13.3), is assembled. (The KMN network is also required for the recruitment of SAC proteins.) The MCC holds the APC/C in check by binding to it like a pseudo-substrate inhibitor and preventing activation by Cdc20 (**Figure 13.22A**).

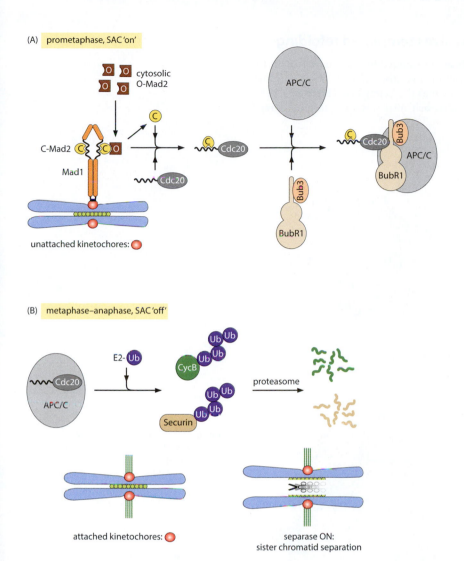

(A) prometaphase, SAC 'on'

cytosolic
O-Mad2

APC/C

C-Mad2

Mad1

Cdc20

Cdc20

Cdc20

Bub3

Bub3

BubR1

BubR1

APC/C

unattached kinetochores:

(B) metaphase–anaphase, SAC 'off'

Cdc20

E2-Ub

CycB

Ub Ub
Ub Ub

Ub Ub
Ub Ub

proteasome

APC/C

Securin

attached kinetochores:

separase ON:
sister chromatid separation

Figure 13.22 The spindle assembly checkpoint (SAC). (A) In prometaphase, the sister chromatids are in the process of attaching to spindle microtubules. Kinetochores devoid of microtubules recruit the Mad1/C-Mad2 complex, which binds cytosolic O-Mad2. A conformational heterodimer of C-Mad2 (yellow circle) and O-Mad2 (brown square) is formed, and the O-Mad2 molecule changes its conformation to that of C-Mad2, 'priming' it to bind Cdc20. The C-Mad2/Cdc20 complex associates with BubR1/Bub3 to form the MCC, which binds to the APC/C and inactivates it. (B) Sister chromatid separation at anaphase is triggered when the MCC disassembles from the APC/C and the Cdc20-bound APC/C is activated, leading to the ubiquitylation and proteasomal degradation of cyclin B and securin. The destruction of securin releases active separase, which cleaves cohesin and allows sister chromatid separation; the destruction of cyclin B results in a loss of Cdk1 activity, which is essential for mitotic exit. (From M. Mapelli et al., *Curr. Opin. Struct. Biol.* 17:716–725, 2007. With permission from Elsevier.)

The first step in the assembly of the MCC is the conversion of Mad2 (<u>m</u>itotic <u>a</u>rrest <u>d</u>eficient 2) to a conformation that can bind and sequester Cdc20. Mad2 alternates between two conformational states, known as open-Mad2 (O-Mad2) and closed-Mad2 (C-Mad2). To bind to Cdc20, Mad2 needs to be in the C-Mad2 state. In prometaphase, a Mad1/Mad2 complex is recruited to kinetochores and provides a local source of C-Mad2 (see Figure 13.22A). The C-Mad2 bound to Mad1 acts as a receptor for O-Mad2 and by transiently dimerizing with it promotes its conversion to C-Mad2. A single Mad1/Mad2 complex is able to convert multiple copies of O-Mad2 to C-Mad2, reminiscent of the template-based conformational conversion of prion proteins (Section 6.6). The new C-Mad2 molecule binds Cdc20 (Mad1 and Cdc20 bind in the same pocket on Mad2). The C-Mad2/Cdc20 complex then associates with the BubR1/Bub3 complex to form the MCC, which binds to the APC/C and prevents its activation. One unattached kinetochore appears to be sufficient to generate enough C-Mad2 to sequester all the Cdc20 in the cell—or at least all the Cdc20 required for anaphase. The process is very efficient, given that even 10% of the normal level of Cdc20 suffices to initiate the degradation of securin and cyclin B.

When all chromosomes have become securely attached and bi-oriented, the MCC can no longer bind and is disassembled. Cdc20 is then able to activate the APC/C and target its two major substrates in metaphase, securin and cyclin B. The ensuing degradation of securin relieves its inhibition of the protease separase, which selectively cleaves cohesin, thereby allowing sister chromatid separation to proceed (Figure 13.22B).

Conversion of O-Mad2 to C-Mad2 involves templated refolding

Human Mad2 (25 kDa) comprises a central β sheet surrounded by three α helices. In the O-Mad2 conformation, the N-terminal region forms the first strand, β1, and the C-terminal region forms the outer two strands, β7 and β8, known as the 'safety belt' (**Figure 13.23A**). On conversion to the C-Mad2 conformation, the 'safety belt' displaces the N-terminal strand

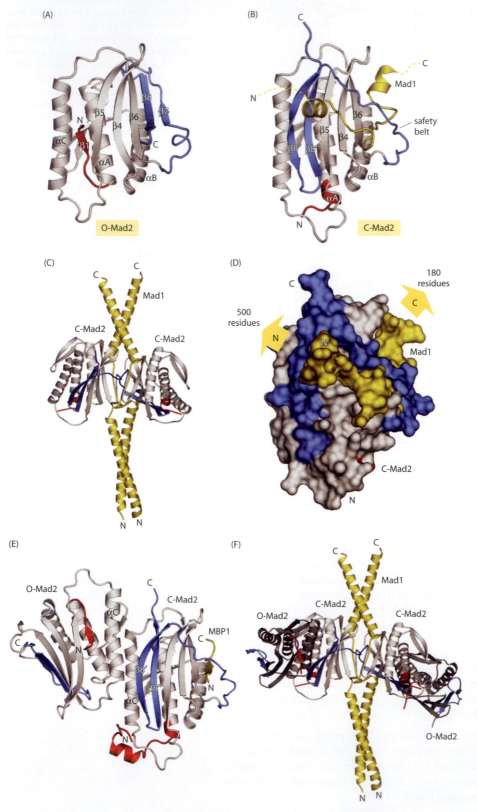

Figure 13.23 Structures of Mad2 complexes. (A) O-Mad2 with the structural elements that are important for the Mad2 conversion highlighted: the C-terminal tail, known as the 'safety belt' (in blue), and the N-terminal region (in red). (B) C-Mad2 and part of Mad1 (in gold), taken from the Mad1/C-Mad2 structure. The C-terminal region (blue) has displaced the N-terminal (red) region of O-Mad2, while the rest of the structure (gray) does not change. (PDB 1GO4) (C) The Mad1/C-Mad2 complex is a 2:2 tetramer. Mad1 is in gold and C-Mad2 in gray, red, and blue. The Mad1/C-Mad2 core complex has two identical binding sites for O-Mad2. (PDB 1GO4) (D) Surface view of the Mad1/C-Mad2 complex in the same orientation as in (B), illustrating the safety-belt binding mechanism. When a ligand (in this case Mad1, colored gold) is trapped under the safety belt (blue), the C-Mad2 conformation is stabilized. Only a short segment of Mad1 is shown here. (E) A view of the O-Mad2/C-Mad2 conformational dimer. Its asymmetry is demonstrated by the position of the safety belt (blue) in the O-Mad2 and C-Mad2 moieties. The C-Mad2 conformation was stabilized by binding a 12-residue peptide (MBP1) from Mad1. (PDB 2V64) (F) A model of the Mad1/C-Mad2 complex bound to two molecules of O-Mad2. The model is based on the crystal structures of the Mad1/C-Mad2 core complex and of the O-Mad2/C-Mad2 conformational dimer shown in (C) and (E). (From M. Mapelli et al., *Curr. Opin. Struct. Biol.* 17:716–725, 2007. With permission from Elsevier.)

which then assumes a helical conformation, the rest of the structure remaining unchanged (Figure 13.23B). In the Mad1/C-Mad2 complex (Figure 13.23C,D), the 'safety belt' wraps around part of Mad1 and stabilizes the C-Mad2 conformation. The mechanism underlying O-Mad2 to C-Mad2 conversion is not yet clear (Figure 13.23E,F), but it has been proposed that part of the binding energy from dimerization is used to destabilize the hydrophobic core of O-Mad2, facilitating its conversion to C-Mad2. C-Mad2 is so tightly bound to Mad1 that it never dissociates within the timescale of mitotic activation (~30 minutes in human cells). The cellular concentration of Mad2 is estimated to be ~100 nM, which is more than tenfold that of Mad1: hence there is plenty of free O-Mad2 ready to be converted to C-Mad2 and bind to Cdc20, triggering activation of the APC/C.

The APC/C is activated almost immediately after the last sister chromatid pair becomes bi-oriented. Many SAC proteins are transported by dynein along microtubules toward the centrosome. The protein p31[comet] resembles C-Mad2, despite having only 11% sequence identity, and promotes the rapid dismantling of checkpoint complexes; it binds to C-Mad2 in C-Mad2/Mad1, thereby preventing the conversion of more O-Mad2 to C-Mad2. Without C-Mad2 to bind to, Cdc20 is able to activate the APC/C.

Kinetochores are able to distinguish between correct and erroneous microtubule attachments

A distinguishing feature of bi-orientation is that it leads to tension between (inter) kinetochores and also within (intra) sister kinetochores. Improper attachments, for example both sisters in a pair connected to the same pole (syntelic attachment) or one sister attached to both poles (merotelic attachment), fail to generate inter-kinetochore tension. The lack of tension activates a correction pathway that breaks the incorrect attachment and allows the sister chromatids to re-attempt bi-orientation.

The key components of the kinetochore/microtubule interface are the KNL1/Mis12/Ndc80 complexes (KMN network). The kinase Aurora B is an important factor in the tension-sensing process. It is part of the chromosome passenger complex (CPC; see Table 13.3) and is targeted to the inner centromere by its interaction with a three-helix bundle of three regulatory proteins: INCENP (inner centromere protein), survivin (human, 142 aa), and borealin (human, 280 aa- **Figure 13.24A**). The C-terminal region of INCENP (the IN-box, residues 790–856 of the 863 aa protein) binds and activates Aurora B (Figure 13.24B), whereas the N-terminal region (residues 1–56) binds to survivin (142 aa) and the N-terminal region (residues 10–109) of borealin. In the presence of full-length INCENP, this core is sufficient to target Aurora B to the central spindle, but targeting to the centromere requires the C-terminal region of borealin. Survivin has a Zn-binding BIR domain (see **Guide**), characteristic of inhibitors of apoptosis (Section 13.7), but it is not clear whether survivin is involved in regulating apoptosis. As described above, Aurora B controls the binding of the Ndc80 complex; phosphorylation of other subunits of the KMN network by Aurora B, including the subunit Dsn1 of the Mis12 complex and an N-terminal microtubule-binding domain of KNL1, are also needed to inactivate the microtubule-binding properties of the KMN network.

Figure 13.24 Aurora B kinase and regulatory proteins of the chromosomal passenger complex (CPC). (A) The core of the CPC formed by survivin (green), the N-terminal region of borealin (magenta, residues 10–109) and the N-terminal region of INCENP (beige, residues 1–56; full length 863 aa, with residues 500–730 forming a coiled coil). Survivin (green) has a long C-terminal helix. Borealin and INCENP interact with the survivin helix to form a triple helix bundle. (PDB 2QFA)(B) Aurora B kinase (gray) in complex with the IN-box (residues 790–847, orange) of INCENP. The kinase is in an intermediate state for activation. The activation segment (cyan) has Thr248 phosphorylated, and the C helix (magenta) is swung in, but the transition to the fully activated state requires a further movement to form an ion pair at the ATP site. The IN-box forms a crown on the N-terminal lobe of the kinase and appears to activate the kinase by an allosteric mechanism. (PDB 2BFX)

(A)

(B)

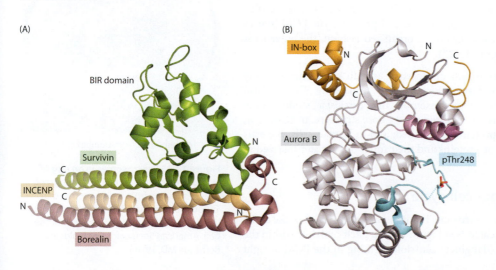

If kinetochore/microtubule interactions do not generate sufficient tension, Aurora B phosphorylates proteins in the outer kinetochore and thereby reduces the strength of the interaction with the microtubules. One obvious way in which this might be achieved is by somehow using mechanical stress to modulate the kinase activity of Aurora B. However, experiments using sensors based on fluorescence resonance energy transfer (FRET) targeted to the centromere by using a domain of CENP-B (centromere protein B) or to the kinetochore by using protein Mis12, demonstrated that the centromere-targeted sensor was phosphorylated by Aurora B under all conditions, but the kinetochore-targeted sensor was phosphorylated only when tension was low (for example, when microtubules were disassembled by treating the cells with the anti-neoplastic agent nocodazole). This outcome implied that phosphorylation of a substrate depends on its spatial separation from Aurora B. Thus when tension is maintained, some substrates are pulled away from the inner centromere and are safe from phosphorylation; their grip on microtubules is therefore not diminished and the mitotic spindle is permitted to pull the sister chromatids apart. This is an elegant mechanism for converting mechanical stress into a biochemical event.

Kinetochores are dynamic assemblies

Whereas CENP-A, CENP-B, and the components of the constitutive centromere associated network (CCAN) are localized at kinetochores throughout the cell cycle, most other components, including the Knl-1, Ndc80, Ska1, and RZZ complexes, CENP-E, CENP-F, dynein, kinetochore-based *microtubule-associated proteins* (MAPs), and the components of the SAC are recruited only in mitosis. The regulatory steps that allow the assembly of kinetochores at the beginning of mitosis and their disassembly at the end of mitosis are not well established but are likely to involve phosphorylation. In this sense, there is not a single kinetochore structure; rather, a time-dependent ensemble of structures.

13.6   CENTRIOLES AND CENTROSOMES

In animal cells, the microtubules of the mitotic spindle are nucleated in organelles called **centrosomes**, which serve as **microtubule-organizing centers** (**MTOCs**). They were first observed in light micrographs of cells in the 1880s (they range in size from 1 μm to 5 μm) and are found only in metazoans (**Figure 13.25**); other, but related, kinds of MTOCs are observed in fungi and plants, such as the *spindle pole body* (*SPB*) of *S. cerevisiae*. Inside a centrosome, there are two orthogonally attached **centrioles**, large protein aggregates based on tubular bundles of microtubules. Each centriole is a cylinder ~500 nm in length and ~200 nm in diameter and, typically, at its proximal end has a cartwheel structure with a central hub 20–25 nm in diameter, from which nine spokes emanate to contact triplet microtubules arranged with 9-fold symmetry (**Figure 13.26**). Toward its distal end, the nine triplet microtubules are replaced by nine doublets. The cartwheel comprises about six layers, spaced 15–20 nm apart. Although this canonical structure of centrioles is observed across a wide range of eukaryotes, there are variations: sometimes the microtubules are found as doublets or singlets, as in *Drosophila* or *C. elegans* respectively, and the structure and assembly of the *C. elegans* centriole differs somewhat (see below).

The centrioles are surrounded by **pericentriolar material** (**PCM**), in which super-resolution light microscopy has revealed ordering among coiled-coil proteins. The PCM houses sites that nucleate the outgrowth of microtubules in an organized fashion, the outer ends being the plus ends (Section 14.6 and **Figure 13.27**).

Defining the components of centrosomes is difficult because, as with kinetochores, many components serve signaling functions and are associated only transiently. Thus we still lack a good molecular picture of how centrosomes assemble. In *S. cerevisiae*, studies of the SPB have indicated that about 17 proteins are required for the core structure; in *C. elegans*, similar studies suggested that about 20 proteins are needed; and in human cells, the estimates are closer to 100 proteins. The major proteins and their functions are summarized in **Table 13.4**.

Centrosomes duplicate once per cell cycle

In interphase (G1) cells, there is only one centrosome but it duplicates during S phase at the same time as the chromosomes are duplicated (see Figure 13.1B). Each centriole within the centrosome is duplicated, with the new (daughter) centriole forming in the PCM at right

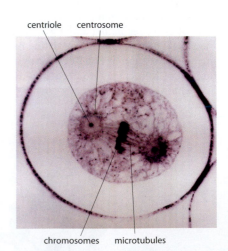

centriole   centrosome

chromosomes   microtubules

Figure 13.25 Centrosomes in an egg of the roundworm *Ascaris*. This photograph of a thin section from a slide prepared by Theodor Boveri (*ca.* 1880) was taken much later by J. Gall. (From J.G. Gall, Views of the Cell. American Society for Cell Biology: Bethesda MD, 1996.)

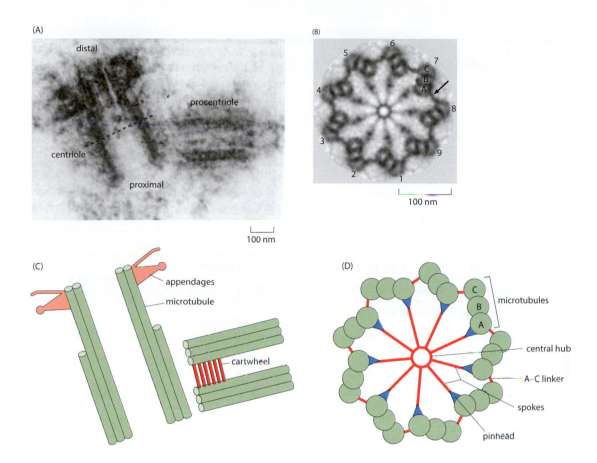

angles to the existing (mother) centriole (**Figure 13.28**). In *C. elegans*, there is no central hub to a cartwheel structure; rather, it appears as a central tube, ~150 nm long and ~60 nm in diameter, whose initial growth is followed by the recruitment of singlet microtubules (**Figure 13.29**). The microtubules form at random positions along the central tube but with the conventional 9-fold symmetry, suggesting that the central tube acts as a template for the symmetry.

During late S phase and G2, the daughter centrioles elongate, only reaching their full length during mitosis, and a model has been proposed in which attached or 'engaged' centrioles are unable to duplicate until they have 'disengaged', akin to the 'licensing' model of semi-conservative DNA replication (Chapter 3). However, one essential difference is that the mother centriole is not copied but simply aids in recruiting the material required for the assembly of the daughter centriole. Each duplicated centriole then recruits PCM to become a centrosome, but the two centrosomes stay physically attached until late G2. (Separately from the cell cycle, the mother centriole also has a major role in generating cilia and flagella—see Section 14.8.)

Figure 13.26 Centriole and cartwheel architecture. (A) Electron micrograph showing a thin section of a resin-embedded centriole and daughter centriole (procentriole) from a human cell. The dashed line separates regions with distal and subdistal appendages on the sides of the distal part of the centriole. (B) Cross section of the proximal part of a centriole from *Chlamydomonas reinhardtii*, highlighting the central hub and the nine spokes radiating toward the triplet microtubules, enhanced by 9-fold rotational symmetrization. Neighboring A and C microtubules are connected by a linker (arrowed). (C) Schematic representation of a human mother centriole/daughter centriole pair, illustrating the cartwheel not visible in (A). The cartwheel is shown with six layers at a spacing of ~15–20 nm. (D) Schematic representation of the cartwheel viewed from the proximal end, as in (B). Each spoke contacts an A microtubule through a 'pinhead'. (Adapted from P. Gönczy, *Nat. Rev. Mol. Cell Biol.* 13:425–435, 2012. With permission from Macmillan Publishers Ltd.)

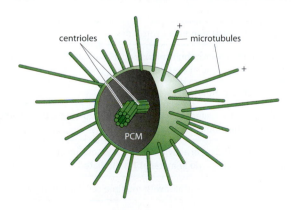

Figure 13.27 Schematic diagram of the centrosome, showing centrioles (green), pericentriolar material (PCM, gray), and microtubules (dark green).

Table 13.4 Centrosome proteins in *Homo sapiens*, *D. melanogaster*, and *C. elegans*.

Protein	Summary of localization and function
SAS-4	A centriole protein
SAS-6	A centriole protein required for centriole duplication
Cep192	Localizes to both centrioles and the PCM
Plk4 (SAK)	Localizes to the centriole throughout the cell cycle
Cep135	Important for centriole biogenesis
CP110	Important for centriole biogenesis. It localizes to the distal tips of both parental and nascent centrioles, where it assembles into a cap-like structure
Sfi1	Interacts with centrin 2, co-localizing to the centrosome
Centrin	Centrin localizes to the distal lumen of centrioles and is important for centriole duplication
Centrobin	A substrate of Nek2 kinase *in vitro* and *in vivo*. Localizes specifically to the daughter centriole
ε-Tubulin	Localizes to the subdistal appendages of mature centrioles
C-Nap1 (Cep250)	Localizes to the proximal ends of both mother and daughter centrioles and associates with rootletin. Substrate of Nek2
Nek2	Kinase activity of Nek2 is required for centrosome cohesion and centriole separation. Nek2 localizes to the centrosome
PP1	Type I protein phosphatase; complexes with Nek2
Rootletin	Localizes to the centrosome, forming centriole-associated fibers during interphase; functions together with C-Nap1
Plk1	Essential for multiple events in mitosis
Aurora A	Localizes to the PCM of the centrosome and the spindle microtubule
Asp	A PCM protein
γ-Tubulin	An indispensible component of nucleation complexes for the outgrowth of microtubules from the MTOC
GCP2	A component of γTuRC and γTuSC. Co-localizes and physically interacts with γ-tubulin
GCP3	A component of γTuRC and γTuSC
GCP4	A component of γTuRC
GCP5	A component of γTuRC. GCP5 and γ-tubulin co-localize to the centrosome and interact directly
GCP6	A component of γTuRC. GCP6 and γ-tubulin co-localize to the centrosome and interact directly
NEDD1	Binds and recruits γTuRC to the centrosome
Pericentrin (kendrin)	Localizes to the PCM. Anchors γTuRCs at the centrosome through interaction with GCP2 and GCP3
AKAP450	Associates with γ-tubulin through binding with GCP2 and GCP3
Cep215	Localizes to the PCM and the centriolar cylinders. Associates with γTuRC and anchors γTuRC at the centrosome. Also required for centrosome cohesion
ASPM	Localizes to the centrosome and the minus ends of central spindle microtubules
Ninein	Localizes to the mother centriolar appendages and the microtubule minus end. Has a role in anchoring microtubules and in promoting microtubule nucleation by docking with γTuRC at the centrosome
Centriolin (CEP110)	Localizes to the subdistal appendages of the maternal centriole
EB1, EB2, EB3	EB1 and its relatives EB2 and EB3 are microtubule tip-associated proteins that localize to the mother centriole and the spindle microtubule. EB1 forms a cap at the end of the mother centriole
Dynactin p150glued	A multi-subunit complex required for most cytoplasmic dynein activities in eukaryotes. Transport of centrosomal proteins including γ-tubulin, centrin, pericentrin, ninein, and pericentriolar material 1 is mediated by the dynein/dynactin complex
Dynein	A cytoplasmic, multi-subunit minus end-directed microtubule motor with multiple cellular functions

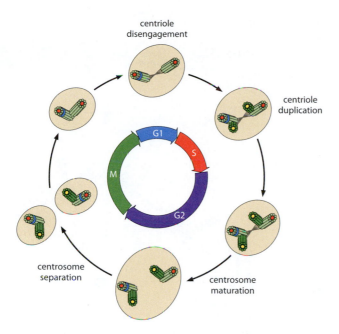

centriole
disengagement

centriole
duplication

centrosome
separation

centrosome
maturation

Figure 13.28 The centriole cycle. Centrioles are in green; the light blue bands represent the distal appendages that are present on the mother centriole. The daughter centriole is in darker green. The centrioles are surrounded by the PCM (beige).

The stage at which centrosomes separate varies according to cell type but is usually at G2/M under control of the cell cycle machinery. Key players are Cdk2 and separase (Section 13.3), suggesting that disengagement begins with separase cleaving one or more centriolar proteins. During prophase, the centrosomes associate with the nuclear envelope. When it breaks down, heralding cell division, the two centrosomes migrate to opposite poles of the cell, and microtubules anchored in them can build the mitotic spindle and interact with the chromosomes (see Figure 13.1B). After cell division, each daughter cell has inherited one centrosome and one copy of the genetic material. Thus, the parent cell must have two centrioles and two spindle poles so that each daughter can inherit the same genetic material. Any variation in the number of centrosomes can lead to aneuploidy (the loss or gain of chromosomes), an abnormality frequently observed in cancer cells.

Laser ablation experiments that removed the daughter centriole showed that the PCM maintains the capability to make a new daughter centriole, suggesting that formation of the first daughter centriole normally inhibits the formation of subsequent daughters. Moreover, centrioles are not essential for mitosis; if they are ablated by laser irradiation, a morphologically normal spindle can form, whose microtubules are organized by motor proteins. However, centrosomes are essential for the survival of the organism; in their absence, cells do not position their spindle poles effectively and lack fidelity in cell division.

Centriole duplication is controlled by a conserved set of proteins

Centriole duplication proceeds in two stages: (1) a 'trigger' or signal that initiates daughter centriole assembly; followed by (2) step-wise assembly of proteins to build daughter centrioles. A master regulator is the polo-like kinase Plk4: in *D. melanogaster* and human cells, overexpression of this kinase induces the near-simultaneous formation of multiple daughter centrioles. Cdk2 (Section 13.2), in combination with cyclin E (at G1/S) or cyclin A (in S phase), also participates in coordinating the cell cycle with the centriole cycle.

The development of genome-wide RNA-mediated interference (RNAi) screens in *C. elegans*, *D. melanogaster*, and cultured human cells has defined a conserved set of proteins required for centriole duplication (see Table 13.4). In *C. elegans* embryos, SPD-2 (equivalent to human Cep192) is required to recruit the Ser/Thr kinase ZYG-1 (the likely functional homolog of human Plk4) to the site of daughter centriole assembly. ZYG-1 phosphorylates at Ser123, retaining it at the site of daughter centriole assembly. This is followed by the initiation and growth of a central tube (see Figure 13.29) composed of two coiled-coil proteins, SAS-6 and SAS-5. The central tube elongates until it reaches a critical length, whereupon another coiled-coil protein, SAS-4, is required for the assembly of singlet microtubules around the tube. It is the mother centriole that recruits proteins such as ZYG-1, SAS-4, and SAS-6, seeding and regulating self-assembly of the daughter centriole. SAS-6 is important

Figure 13.29 Centriole duplication in *C. elegans*. The mother and daughter central tubes are red and yellow respectively, and microtubules are green.

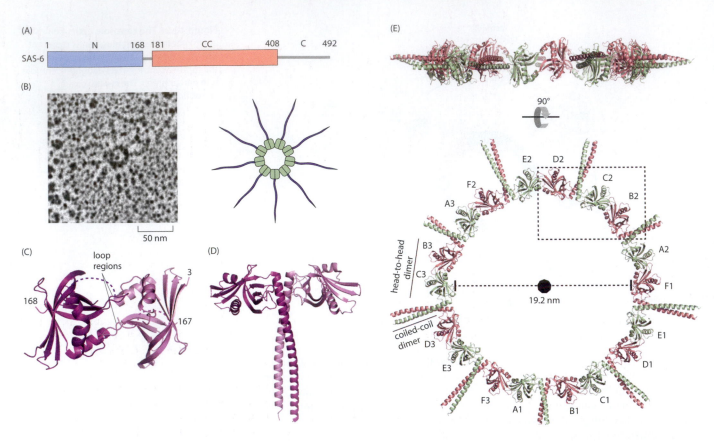

for formation of the tube-like structure but appears not to have a permanent structural role. Relatives of these five proteins, initially identified in *C. elegans*, are found in many other species, including humans.

In human and other cells, SAS-6 and Cep135 are important for defining the cartwheel structure (see Figure 13.26). CP110, which is a target for phosphorylation by Cdk2/cyclin E, together with SAS-4 controls centriole length. The proteins Sfi1, centrin, centrobin, γ-tubulin, and ε-tubulin also localize to centrioles and are required for their duplication. Electron tomography has indicated that there is no cartwheel in centrioles in the G1 phase of the cell cycle; it is removed from the mother centriole at mitotic exit. In contrast, at the beginning of the assembly process, the cartwheel is observed before centriolar microtubules.

The role of SAS-6 in generating a scaffold with 9-fold symmetry has emerged from studies of *C. elegans* and of Bld12, an ortholog from the green alga *Chlamydomonas reinhardtii*. SAS-6 has an N-terminal globular domain, a coiled-coil domain, and a disordered C-terminal region (**Figure 13.30A**). In electron micrographs, SAS-6 has a globular head (N-domain) attached to a 35 nm rod, consistent with ~220 aa (CC) in a coiled-coil conformation. Bld12 assembles into a ~22 nm ring from which nine flexible tails emanate radially (Figure 13.30B). The N-domain (168 aa) forms a dimer, in which the globular heads interact through a conserved hydrophobic patch at the dimer interface (Figure 13.30C); and a construct of the N-CC region of Bld12 has a similar structure for the N-domains, with a projecting dimeric coiled-coil (Figure 13.30D). These results support a model in which nine dimers associate into a ring through their N-domains with their coiled-coil regions projecting outward. This has been confirmed in a crystal structure of a construct (residues 97–320) of the homologous SAS-6 protein of the protozoan *Leishmania major* (Figure 13.30E), a ring with dimensions (inner diameter 19 nm, thickness 3.5–5 nm) consistent with the EM image. The existence of a cartwheel structure in *C. elegans* is still debated, given the requirement for SAS-6 to initiate the growth of the central tube (see above).

Duplicated centriole pairs remain associated until G2/M

After duplication of the mother centrioles, the two centriole pairs function as a single centrosome until their separation at the G2/M transition (see Figure 13.28). Cytoskeletal dynamics and post-translational regulation through kinases and phosphatases have been

Figure 13.30 Structure of centriole protein SAS-6. (A) Domain map of *C. elegans* SAS-6, showing the N-terminal domain (N), the coiled-coil region (CC), and the C-terminal region (C). (B) Left, electron micrograph of rotary metal-shadowed N–CC portion of *C. reinhardtii* SAS-6 homolog Bld12, which has assembled into a ring; right, a schematic interpretation of a central ring with nine projecting coiled-coil tails. (C) Crystal structure of the N domain from *C. elegans*, showing two heads associating through loop regions to form dimers. Directed mutagenesis has confirmed the importance of the loop region. (PDB 3PYI) (D) Crystal structure of a dimeric N–CC molecule from *C. reinhardtii*, indicating association through the coiled-coil regions. (PDB 3Q0X) (E) Crystal structure of an N–CC construct (residues 97–320) of SAS-6 from *Leishmania major*, demonstrating a ring of nine dimers with dimensions consistent with those found in electron micrographs. Subunits are colored alternately in magenta and green to aid identification. Coiled-coil dimers are labeled A1/B1, A2/B2, and so on; head-to-head dimers are B1/C1, D1/E1, and so on. (B, adapted from D. Kitagawa et al., *Cell* 144:364–375, 2011. With permission from Elsevier. E, from M. van Breugel et al., eLife 2014;3:e01812. DOI: 10.7554/eLife.01812, 2014.)

implicated in the mechanisms that govern centriole cohesion. In EM studies, fibers can be observed connecting the duplicated centriole pairs. C-Nap-1, a large protein (2442 aa) almost all of which is predicted to be in a coiled-coil conformation, interacts with Cep135 (1140 aa), another predicted coiled-coil protein (see Table 13.4); this may provide a docking site at the proximal end (the end nearest to the daughter centriole) of the mother centriole for these fibers. At the G2/M transition, C-Nap-1 is phosphorylated by the kinase Nek2 and this, together with the inhibition of type I phosphatases, is thought to promote the dissociation of C-Nap-1 from the centrosome. In addition, rootletin (2017 aa, the N-terminal 1700 aa of which are predicted to form yet another coiled-coil), a protein distantly related to C-Nap-1, interacts with C-Nap-1 and is also phosphorylated by Nek2. Rootletin is so named because it was found to be the major structural protein of ciliary rootlets, a cytoskeletal-like structure in ciliated cells. It contributes to the linker between mother centrioles of duplicated centrosomes.

Centrioles are sites of assembly of PCM

Functional centrosomes need PCM proteins as well as centrioles, but the links are poorly characterized. Centrosomes contain relatively little PCM during interphase but accumulate three-to-fivefold more in preparation for mitosis. This increases their size and capacity for microtubule nucleation. One key PCM protein is γ-tubulin (Section 14.6). The onset of centrosome maturation coincides with the activation of Cdk1 and is also regulated by Aurora-A and Plk1 kinases (see Table 13.4). NEDD1 mediates the recruitment of γ-tubulin ring complexes to centrosomes. (NEDD1 = neural cell expressed developmentally regulated Down-regulated protein, also known as WD-GCP; see Table 13.4.) NEDD1 contains WD40 repeats, and its phosphorylation by Plk1 indirectly assists the localization of γ-tubulin to the centrosomes, possibly by recruiting other proteins such as Cep192 and Cep215. Among other protein components of the centrosome, pericentrin (also known as kendrin), AKAP450, and Cep215 (see Table 13.4) have roles in γ-tubulin recruitment and anchoring. These are coiled-coil proteins that are thought to form a lattice-like structure on to which other components dock.

Microtubules are nucleated by the γ-tubulin ring complex

During the cell cycle, the number of microtubules emanating from centrosomes increases, with a concomitant increase in γ-tubulin ring complexes. γ-Tubulin is related to other tubulins but forms oligomers from monomeric, not dimeric, building blocks. These complexes, which act as templates for the assembly of αβ-tubulin (see Figure 14.39), exist in two forms: a ~220 kDa complex called the γ-tubulin small complex (γTuSC) and a larger ~2.2 MDa complex called the *γ-tubulin ring complex* (γTuRC).

In humans, γTuSCs are composed of two γ-tubulin subunits and one subunit each of GCP2 and GPC3 (Spc97 and Spc98 in *S. cerevisiae*; see Table 13.4). In animal cells, multiple γTuSCs assemble with subunits GCP4, GCP5, and GCP6 into the γTuRC, which forms a microtubule-nucleating array of γ-tubulin. The γTuRC is localized to the centrosome by one or more attachment factors. A model for how the γTuSC might function in the SPB of *S. cerevisiae* (see below) can be extended to γTuRCs. The microtubules are organized such that the minus ends are proximal to the centrosome and their more dynamic plus ends extend outwards. In addition to initiating microtubule assembly, the γTuRC stabilizes microtubules by capping their minus ends.

Microtubules anchor on the subdistal appendages of the mother centrioles (that is, on the side away from the contact to the daughter centrosome) and also in the PCM; this depends on several proteins, including ninein and the motor protein dynein (see Table 13.4). Ninein anchors the γTuRC to the centriole: its C-terminal region interacts with the centriole and its N-terminal region interacts with the γTuRC.

The kinases Aurora A and Plk1 promote microtubule nucleation, although their substrates are not well characterized and their phosphorylations are counteracted by protein phosphatase 1 (PP1). In addition, Plk1 phosphorylates the protein Asp (abnormal spindle protein), a conserved PCM-associated protein that helps anchor γTuRCs to the centrosome. Ubiquitylation may also have a role in centrosome maturation; the ubiquitin ligase activity of the tumor suppressor protein BRAC1 modifies γ-tubulin, which leads to the loss of microtubule-nucleating activity.

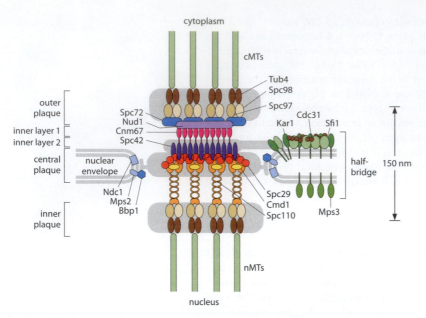

Figure 13.31 The yeast spindle pole body (SPB). The arrangement of the SPB proteins (Table 13.5) within the layered structure, as deduced from immuno-electron microscopy, FRET, and other measurements, is shown as a schematic diagram. Tub4 is γ-tubulin. (From S.L. Jaspersen and M. Winey, *Annu. Rev. Cell Dev. Biol.* 20:1–28, 2004. With permission from Annual Reviews.)

The yeast SPB suggests a structure for the centrosome and microtubule initiation

The most intensively studied example of assembly of a centrosome-like structure is the *S. cerevisiae* SPB, whose major structural components have been identified by a combination of genetics and mass spectrometry and their locations determined by immuno-electron microscopy. The mass of the SPB, including microtubules and microtubule-associated proteins, amounts to ~1–1.5 GDa, about twentyfold greater than that of the yeast nuclear pore complex, itself a very large complex (Section 10.2). It is composed of a stack of protein layers called plaques that is embedded in the nuclear envelope, but remains intact during mitosis (**Figure 13.31**). The outer plaque faces the cytoplasm and nucleates cytoplasmic microtubules, the central plaque spans the nuclear envelope, and the inner plaque anchors the nuclear microtubules. The cytoplasmic microtubules help orient the nucleus during cell division, while the nuclear microtubules form the mitotic spindle. In contrast with the centrosome, where microtubules emanate in all directions, the SPB microtubules are directed.

Although the morphology of the SPB is quite different from that of the centrosome, some proteins share similar functions in the two assemblies (**Table 13.5**). Many SPB proteins have a coiled-coil component, and their likely arrangement in the nuclear envelope is shown schematically in Figure 13.31. During G1/S, the SPB is duplicated. The half-bridge, a dense region within the nuclear envelope on one side of the central plaque, elongates during G1 and amorphous satellite material is deposited on its distal cytoplasmic side. The satellite next expands into a duplication plaque, a layered structure similar to the cytoplasmic half of a mature SPB. Finally, the duplication plaque is inserted into the nuclear envelope and the nuclear SPB components are assembled. During late S phase and G2, the two SPBs remain side-by-side, connected by the bridge. Nuclear microtubules are initiated by both SPBs. The bridge must be severed to allow the SPBs to separate and form the poles of the mitotic spindle. As cells enter mitosis, the SPBs separate and a short bipolar spindle is formed, from which microtubules seek out the kinetochores of the chromosomes to form the mitotic spindle.

The organization of the SPB proteins is focused on Spc42

The five core proteins of the central plaque and the adjacent inner layer 2 are Spc42, Spc29, Spc110, Cmd1 (calmodulin) (see **Guide**), and Cnm67 (see Figure 13.31). Spc42 is a coiled-coil protein and appears to make contact with all proteins of the central plaque.

Table 13.5 *S. cerevisiae* spindle pole body proteins.				
Protein	**Human homolog**	**Localization within SPB**	**Role**	**Likely structure**
Tub4	γ-Tubulin	γ-Tubulin complex	Microtubule nucleation	α/β topology
Spc97	GCP2	γ-Tubulin complex	Microtubule nucleation	
Spc98	GCP3	γ-Tubulin complex	Microtubule nucleation	Coiled coil
Spc42		Central plaque, inner layer 2	SPB core	Coiled coil
Spc29		Central plaque	SPB core	Coiled coil
Spc110		Central plaque to inner plaque	Spacer, γ-tubulin binding protein	Coiled coil
Cmd1	Calmodulin	Central plaque	Spc110 binding protein	
Cnm67[a]		Inner layer 1, outer plaque	Spacer	
Cdc31	Centrin	Half-bridge	SPB duplication	α helical EF hand, like calmodulin
Sfi1	Sfi1	Half-bridge	SPB duplication	α helix
Kar1		Half-bridge	Membrane protein, SPB duplication	
Mps3		Half-bridge	Membrane protein, SPB duplication	Coiled coil
Ndc1		SPB periphery	Membrane protein	
Msp2		SPB periphery	Membrane protein	Coiled coil
Bbp1		SPB periphery	Half-bridge linker to membrane	Coiled coil

[a]Although not essential, Cnm67 deletion mutants exhibit slow growth and microtubule abnormalities.
(Adapted from S.L. Jaspersen and M. Winey, *Annu. Rev. Cell Dev. Biol.* 20:1–28, 2004.)

Overexpression of Spc42 produces a hexagonal array (**Figure 13.32**) that is continuous with the central plaque and encircles the SPB. It has been suggested that in native SPBs, Spc42 molecules form a hexagonal lattice in the cytoplasmic part of the central plaque that acts as an internal scaffold for assembly. Spc110 is predicted to be a filamentous coiled-coil protein that acts as a spacer separating nuclear microtubules from the central plaque; its coiled-coil region extends ~60–80 nm from the central plaque to the inner plaque, and its N-terminal region binds the γ-tubulin complex.

Tub4 is the yeast homolog of γ-tubulin. It forms a complex with Spc97 and Spc98 in the outer plaques and inner plaques from which microtubules grow out into the cytoplasm and the nuclear compartment, respectively. The Tub4/Spc97/Spc98 complex is a TuSC complex, equivalent to that of animal cells, and it initiates the microtubules binding to the SPB through Spc110 on the nuclear face and through Spc72 on the cytoplasmic face (see Figure 13.31). Isolated γTuSCs form tilted rings (that is single turns of a low-pitch helix) and, in the presence of Spc110, these generate solenoidal filaments with 6.5 γTuSCs per turn (**Figure 13.33A,B**). Because there are two γ-tubulin subunits per γTuSC, this equates to 13 γ-tubulins per turn and matches the 13-fold symmetry of microtubules in axial projection. A single γTuSC/Spc110[1–220] complex computationally extracted from the reconstructed filament is shown in Figure 13.33C, and a model for how several of these complexes might attach through Spc110 to the MTOC and, at the distal end, how their γ-tubulins might nucleate microtubule assembly, is shown in Figure 13.33D. These ideas can be extended to the nucleation of microtubules by the TuRCs in metazoans.

Cdc31 (centrin) and Sfi1 have essential functions in SPB duplication. Cdc31 binds to three proteins in the half-bridge: Kar1, Mps3, and Sfi1 (see Figure 13.31). Kar1 and Mps3 have transmembrane domains that probably traverse the lipid bilayers of the half-bridge. Sfi1 (946 aa) contains ~20 tandem repeats of ~33 aa with the consensus sequence AX_7LLX_3F/LX_2W (single-letter code; X is any amino acid). In a complex of three repeats of the Sfi1 consensus sequence with centrin, the Sfi1 molecule forms a long α helix, with a centrin molecule contacting each repeat. The centrins are rotated about 65° clockwise successively

⊢——⊣
50 Å

Figure 13.32 A crystalline lattice of Spc42 protein. When Spc42p is overexpressed in *S. cerevisae,* it forms a two-dimensional crystalline lattice surrounding the SPB. This image is a computationally averaged electron micrograph of that lattice, isolated and negatively stained. Protein is white and stain is black. A similar image, at a lower resolution, was obtained from micrographs of isolated SPBs, suggesting that the same structure was present in both specimens. (From E. Bullitt et al., *Cell* 89:1077–1086, 1997. With permission from Elsevier.)

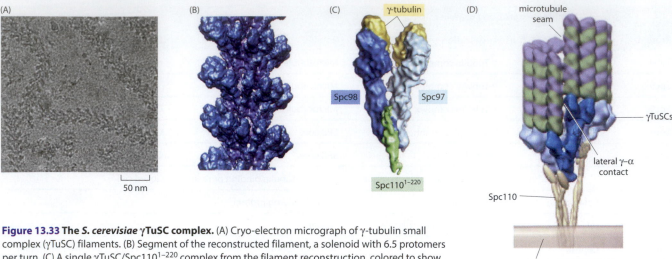

Figure 13.33 The *S. cerevisiae* γTuSC complex. (A) Cryo-electron micrograph of γ-tubulin small complex (γTuSC) filaments. (B) Segment of the reconstructed filament, a solenoid with 6.5 protomers per turn. (C) A single γTuSC/Spc110$^{1–220}$ complex from the filament reconstruction, colored to show the proposed segmentation of electron density into five subunits. (D) Model of a nucleation complex with a segment of a microtubule growing out from it. The γTuSC is attached directly to the MTOC by Spc110. It is envisaged that a conformational change in Spc98 promotes nucleation by rearranging γ-tubulin into an exact microtubule template. (From J.M. Kollman et al., *Nature* 466:879–882, 2010. With permission from Macmillan Publishers Ltd.)

around the Sfi1 α helix with local distortions of the helix to allow the centrin/centrin interactions (**Figure 13.34**). Centrin, a calmodulin-like molecule with four EF-hands, retains the same conformation when bound to Sfi1, regardless of the presence or absence of calcium.

Centrosomes have other roles beyond cell division

In animal cells, centrioles can also be tethered at the cytoplasmic membrane, where they are known as *basal bodies*. Here they act as a template for the formation of an *axoneme*, the microtubule-based structure that provides rigidity and motility to cilia and flagella (Section 14.8). Cilia protrude from the cell surface and participate in a multitude of developmental processes and signaling events. Motile and sensory cilia have roles in cell motility, in hearing, and in expelling pathogens from the respiratory tract. It is unclear how a centriole structure is turned into a basal body and *vice versa*.

Centrosomes also have a role in the immune system, at the immunological synapse between an infected target cell displaying an MHC/peptide complex in contact with the T-cell receptor (TCR) from a cytotoxic T lymphocyte (CTL) (Section 17.4, especially Figure 17.27). Upon TCR engagement, a centrosome relocates to the point of TCR signaling. Secretory lytic granules associated with microtubules migrate toward the centrosome, where they dock and release their contents at the plasma membrane.

There is growing awareness of a possible role for the centrosome as a signaling hub, fully equipped with kinases and phosphatases. For example, at the G2/M transition of the cell cycle, the key regulatory step of Cdk1 activation is facilitated by centrosomal co-localization with Plk1. At the end of mitosis, association of inactivating phosphatases with centrosomes promotes the completion of cytokinesis and exit from the cell cycle.

Figure 13.34 Structure of an Sfi1/centrin complex. The region of Sfi1 (salmon, residues 218–306) containing three repeats of the consensus sequence AX$_7$LLX$_3$FLX$_2$W is bound to three centrin molecules (I, II, and III) depicted in blue, green, and yellow respectively. Three Ca^{2+} ions (red spheres) are bound in each of centrins I and II. Centrin III is partly disordered (dashed region). The eight helices of the EF-hand (calmodulin-like) structure of the centrin molecules are numbered I–VIII for centrin I. Some of the side chains of key residues in the Sfi1 repeats that contact the centrin molecules are labeled. (PDB 2DOQ) (Adapted from S. Li et al., *J. Cell Biol.* 173:867–877, 2006.)

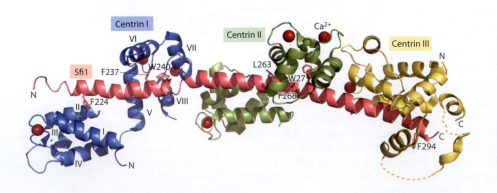

13.7 APOPTOSIS (PROGRAMMED CELL DEATH)

Apoptosis, or programmed cell death, is a key process governing development and homeostasis in multicellular organisms. The selective deletion of cells had long been noted by embryologists (for example, removal of the webbing between the digits of hands and feet during development), before the term apoptosis (derived from the Greek for 'falling off', as leaves from a tree) was coined by John Kerr, Andrew Wyllie, and Alastair Currie in 1972 to describe an orderly program of cell death as a basic biological process. However, it took another 20 years to gain widespread acceptance, and the first understanding of its molecular mechanisms only began to emerge with detailed genetic studies of the worm *C. elegans* by Sydney Brenner, Robert Horvitz, and John Sulston. In addition to signals that control cells during development, many stress conditions such as hypoxia, DNA damage (Chapter 4), mitochondrial dysfunction, or viral infection can initiate apoptosis, as can growth-related stresses such as amino acid availability and cell cycle checkpoints, if not immediately corrected. Apoptosis must be distinguished from the death of cells caused by trauma, injury, or lack of blood supply. In such deaths, *necrosis*, the cell swells and bursts, spilling out its contents. Moreover, the discharge of cytokines, previously confined to the intact cell, will often induce an inflammatory response. In contrast, a cell undergoing apoptosis shrinks, the cytoskeleton collapses, the nuclear envelope disappears, the chromatin condenses and the DNA becomes fragmented, and the surface chemistry of the cell changes. The cell membrane may also form irregular bulges or blebs, caused by disruption of the underlying cytoskeleton; indeed, the cell may break up into membrane-delimited entities called apoptotic bodies. Either way, the dying cell becomes a target for a macrophage (a specialized phagocytic cell) that can engulf it and any apoptotic bodies, ensuring that any debris is rapidly cleared. In this way, the cell contents are never released, no inflammation ensues, and the orderly process is elegantly completed.

The recognition by a macrophage of a cell undergoing apoptosis relies on the appearance of the negatively charged phospholipid, phosphatidylserine, in the outer leaflet of the apoptotic cell membrane. In normal healthy cells, this phospholipid is confined to the inner leaflet: apoptosis must somehow cause it to flip across. It is also likely that there are other signaling molecules, as yet unidentified, expressed on the surface of healthy cells to inhibit phagocytosis, and these are presumably no longer present during apoptosis.

Given the finality and irreversibility of apoptosis, the process must be tightly controlled. Thus, it is not surprising that a multitude of structural, protein-synthetic, and degradative mechanisms, and their corresponding assemblies, are involved. This includes both membrane-dependent and soluble signaling systems, membrane permeabilization, protease activation, and inhibitor and enhancer systems mediated by protein complexes. Many processes involve ATP, and the dynamics of ATP production are described in Section 15.4. This section addresses the molecular basis of the elaborate cellular activity involved in apoptosis, much of it only recently unraveled and some aspects still under intensive investigation.

Apoptosis proceeds by intrinsic and extrinsic pathways

Two main routes of apoptosis are illustrated in **Figure 13.35**, both of which culminate in the activation of a cascade of *cysteine proteases* (see Box 7.1) termed **caspases**. The *intrinsic pathway* requires the participation of mitochondria; in a cell required to self-destruct, for example one in which severe radiation damage to DNA may cause uncontrolled proliferation, activation of the intrinsic pathway causes a group of proteins belonging to the Bcl-2 protein family to release cytochrome *c*, among other factors, from the mitochondrial intermembrane space. (Bcl was the first example of this group, involved in B-cell lymphoma.) Cytochrome *c* then binds to a protein called Apaf-1 (apoptosis-protease-activating factor 1) and this initiates assembly of the *apoptosome*. This large (1.1 MDa) protein complex is a signaling platform that triggers the activation of the initiator protease, caspase-9.

In contrast, in the *extrinsic pathway*, a cell is targeted for apoptosis. A so-called death ligand (FasL), a protein projecting from a cell that is inducing the death of a target cell, binds to a *death receptor* (Fas) projecting from the surface of the target cell. This leads to the formation of so-called **death-inducing signaling complexes (DISCs)**. The DISC, acting through adaptor molecules such as FADD (Fas-associating death-domain-containing protein) is the counterpart of the apoptosome in the extrinsic pathway and a platform for the activation of the initiator (or apical) caspases-8 and -10.

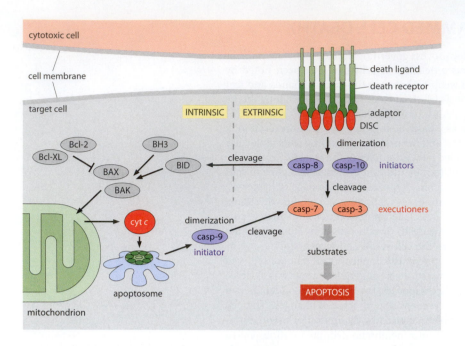

Figure 13.35 Initiation of apoptosis. There are two main apoptopic pathways. Left, in the intrinsic pathway, intracellular signals control proteins of the Bcl-2 family (BAX, BAK, Bcl-XL, BH3, BID), which cause cytochrome *c* (cyt *c*) to be released from mitochondria. Cytochrome *c* in turn promotes assembly of the multimeric apoptosome (in *C. elegans*, cytochrome *c* is not involved and the assembly is controlled differently; see Figure 13.42). The apoptosome activates initiator protease caspase-9 (casp-9) by causing it to dimerize, whereupon casp-9 irreversibly activates executioner caspases-7 and -3. The action of the executioner caspases on other protein substrates leads inevitably to cell death. Right, in the extrinsic pathway, death ligands clustered in the membrane of a cytotoxic cell bind and cluster death receptors in the membrane of the target cell, leading to the formation of a DISC in the affected cell. The initiator caspases-8 and -10 bind to the DISC and become activated by dimerization. They, in turn, activate the executioner caspases-7 and -3 as in the intrinsic pathway.

The activation of caspase-9 or caspases-8 and -10 (*initiator caspases*) is brought about by what is referred to as proximity-induced dimerization of the monomeric caspases, which are inactive, on their respective activation platforms. The dimeric caspases, which are active, then cleave and activate certain downstream caspases, the *executioner* (or effector) caspases-3 and -7, which in turn act on a host of substrates, ultimately leading to the demise of the affected cell. Several different mechanisms ensure that these various processes are tightly controlled.

Caspases mediate an intracellular proteolytic cascade

A cascade of caspases (<u>c</u>ysteinyl <u>a</u>spartate-<u>s</u>pecific <u>p</u>rot<u>eases</u>, which cleave after an Asp residue) lies at the heart of the mechanism and regulation of apoptosis. Although the initiator and executioner caspases share a common overall fold, they are activated by distinct mechanisms: proximity-induced dimerization for initiator caspases, and proteolytic cleavage for executioner caspases.

The key to initiating the caspase cascade lies in bringing inactive monomers of initiator caspases into close proximity to create active dimers. This task is performed by the two most prominent molecular assemblies in cell death: the soluble apoptosome (intrinsic pathway) and the membrane-dependent DISC (extrinsic pathway). Both assemblies recruit and dimerize initiator caspases via adaptor domains of the death domain superfamily.

Death domains (see **Guide**) are globular domains composed of six helices, which are used in a variety of cellular pathways including apoptosis. The activation of the initiator caspase, caspase-9, at the start of the intrinsic pathway is illustrated in **Figure 13.36A**. Caspase-9 uses its <u>c</u>aspase <u>r</u>ecruitment <u>d</u>omain (CARD; a death domain) to interact with the CARD domains displayed by the apoptosome (Figure 13.36B); two inactive monomers of caspase-9 are brought into juxtaposition and an active dimer is generated. Initiator caspase activation can thus be traced to the assembly of the apoptosome or DISC.

Active executioner caspases are $\alpha_2\beta_2$ tetramers, composed of large (α, 33 kDa) and small (β, 11 kDa) subunits. They are generated from inactive monomeric **pro-caspases** by proteolytic removal of the N-terminal prodomains and further cleavage to create the α and β subunits. A healthy cell is continuously making pro-caspases and it is crucial that they not be activated prematurely. In the caspase fold, an α-helix/β-sheet type of scaffold, essentially all the key catalytic residues are located in loop regions that connect elements of regular secondary structure. The executioner caspases are constitutive dimers and have a disordered arrangement of several loops in the uncleaved zymogen. These loops become ordered in the activated caspase, and this sets up the protease active site (Figure 13.36C). The disorder

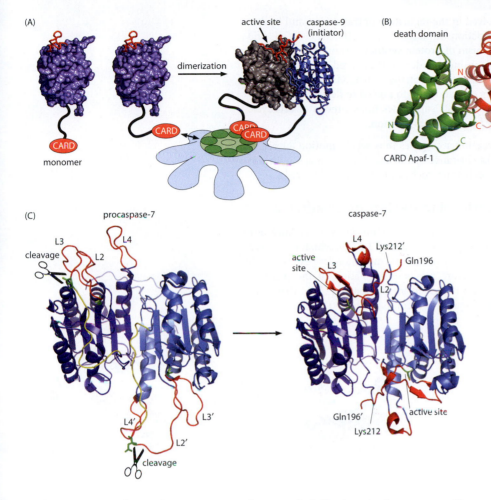

Figure 13.36 Different mechanisms of caspase activation. (A) Inactive initiator caspase-9 monomers display a CARD domain. The rendering of caspase-9 is putative because no structure is available for the monomeric state. Binding to CARD domains (green) of the death domain superfamily displayed on the apoptosome induces dimerization. (PDB 1JXQ) (B) Interaction of the CARD domain of caspase-9 (red) and the Apaf-1 CARD of the apoptosome (green). (PDB 3YGS) (C) The executioner caspase-7 zymogen, a dimer, is activated by proteolytic cleavage. Left, the procaspase-7 zymogen; right, the caspase-7 active enzyme. One monomer is in dark blue and the other in light blue. In the zymogen, all loops are highly flexible and no defined active site is present. Loop regions regulating activity are displayed in red. Further flexible loops are in yellow. Cleavage occurs at an Asp198 (scissors; green), which occurs in loops of different conformations in the respective subunits because of disorder in loop 2 (L2). After cleavage, loops L2, L3, and L4 adopt ordered conformations, allowing the formation of the catalytic site. Cleavage after Asp198 results in the two ends shifting apart as represented by residues Gln196 and Lys212. (PDBs: procaspase-7 1GQF; caspase-7 1F1J)

in the zymogen explains why it is inactive; cleavage of a linker loop in the zymogen allows the loops to adopt the ordered structure that constitutes the substrate-binding cleft, poising the active-site Cys residue for nucleophilic attack on the peptide bonds of target substrates. Once the zymogens have been cleaved, executioner caspases are irreversibly active, and completion of the cell death program is guaranteed. This all-or-nothing activation mechanism is perfectly suited to the executioner caspases.

Caspases bring about a multitude of changes in the cell that lead ultimately to its death

Once activated, the executioner caspases cleave particular substrates, leading to the apoptotic cell phenotype. These cleavages occur mainly in loop regions of target proteins and are dependent on the presentation of the executioner caspase consensus sequence, DXXD↓G/S/T (the downward arrow represents cleavage on the C-terminal side of the Asp), and they can initiate both gains and losses of function in a diversity of pathways.

Poly(ADP-ribose) polymerase (PARP) is a nuclear protein recognizing DNA strand breaks and is important in DNA repair (Chapter 4). It is inactivated by cleavage by caspase-3 or -7. This forms part of a hallmark feature of apoptosis, namely the fragmentation of chromosomal DNA. The process is mediated by specific endonucleases, most importantly DFF40/CAD (DNA fragmentation factor 40/caspase activated DNase). This endonuclease is normally bound to the inhibitor DFF45/ICAD, cleavage of which by caspase-3 releases DFF40/CAD. Monomeric DFF40/CAD can then dimerize and the active nuclease degrades the chromosomal DNA. Cell membrane blebbing or apoptotic body formation is caused by cleavage of proteins such as gelsolin and p21-activated kinase, which are directly involved in cytoskeleton and membrane remodeling. Caspase cleavage of protein kinase Cδ and subsequent stimulation of so-called *scramblases* facilitates the movement of phospholipids across the membrane, including the transfer of phosphatidylserine into the outer leaflet, where it attracts macrophages to engulf the dying cell.

Furthermore, caspase-3 substrates are involved in the regulation of translation and lead to altered protein synthesis. Cleavage of translation initiation factors eIF4GI, eIF4GII, and p97/DAP5 by caspase-3 changes the mechanism of protein synthesis from dependence on mRNA capped at the 5′ end (Section 5.2) to cap-independence. The altered enzymatic activity sequentially up-regulates the synthesis of apoptotic proteins such as XIAP, Bcl-2, Apaf-1, and p97 to attempt, first, to rescue the dying cell, and then to push it to destruction. At the same time, loss of cap-dependent protein synthesis suppresses the synthesis of cell cycle proteins such as cyclins and CDKs, thereby inhibiting cell division.

The highly controlled activation of the executioner caspases thus sets in motion a train of events that ensures death, degradation, and clearance of the affected cell in a way that is vastly different from other forms of death, such as necrosis, with their attendant risks.

Assembly of the apoptosome is critical to the intrinsic pathway

The **apoptosome** consists of a heptamer of Apaf-1 to which seven molecules of cytochrome c are bound. Apaf-1 (142 kDa) is a multidomain protein (**Figure 13.37A**), which exists as a monomer until it combines with cytochrome c released from mitochondria as the result of an apoptotic stimulus. Its domain architecture mediates regulation of the intrinsic pathway. As found in the crystal structure of a truncated form (Apaf-1 1–591) that lacks the WD40 repeat domain, monomeric Apaf-1 adopts a compact shape (Figure 13.37B).

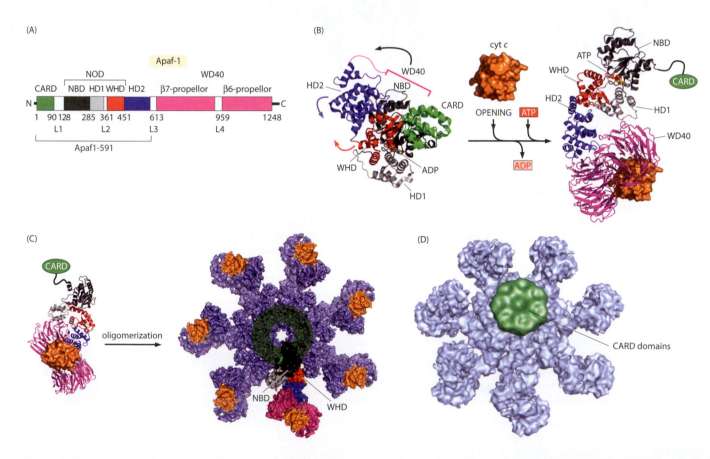

Figure 13.37 Apoptosome formation. (A) Domain map of Apaf-1: NOD, nucleotide oligomerization domain; NBD, nucleotide-binding domain; HD1, helical domain 1; WHD, winged helix domain; HD2, helical domain 2. (B) Apaf-1 (residues 1–591) adopts a compact shape with a central ADP molecule bound. The WD40 region was not present in the structure but is represented schematically by the magenta bar in the locking position. Binding of cytochrome c to the WD40 domain releases the lock, ADP/ATP exchange occurs, accompanied by a conformational opening of Apaf-1 (inferred from the EM apoptosome structure) through the WHD and HD2 domains. The CARD domain (green) is now flexibly attached. (PDB 1Z6T) (C) Assembly of the apoptosome. Open Apaf-1 molecules with cytochrome c bound (orange), (the model is rotated by ~90° with respect to the right-hand side of (B)), assemble into the heptameric apoptosome as inferred from cryo-EM. The NBD domains form an inner ring, and alternating WHD/HD1 domains form an outer ring. Only one Apaf-1 molecule is colored (as in A) for clarity. The positions of cytochrome c molecules have been tentatively modeled. The likely sites for docking the CARD domains are in green. (PDB 3IZA) (D) The apoptosome. In this cryo-EM structure, CARD complexes are found as a ring-like arrangement overlying the apoptosome. The model is tilted slightly with respect to (C). (Cytochrome c-bound model and density map for CARD-bound apoptosome kindly provided by Drs S. Yuan and C.W. Akey.)

In this state, the NBD and HD1 domains of the NOD domain plus the WHD domain nestle around a bound molecule of ADP. The HD2 domain interacts further with the NBD and WHD domains, adding to the stability of this closed form. The CARD domain makes extensive contacts with the NOD and WHD domains. Contact surfaces that would be used for binding the CARD domain of caspase-9 are buried and inaccessible in this structure. From electron micrographs, it appears that the WD40 repeats, which were not present in Apaf-1 1–591, interact with the compact structure and lock Apaf-1 in the closed form. Thus, in the absence of an apoptotic signal, Apaf-1 remains safely monomeric and cannot interact with its target caspase.

Upon receipt of an apoptotic stimulus, cytochrome *c* is released from the mitochondria and binds to the WD40 domains of monomeric Apaf-1, releasing them from their locking function. Although Apaf-1 retains its compact structure, it is then possible for the bound ADP to exchange with ATP (or dATP), and this is accompanied by an extensive opening of Apaf-1 as detected by electron microscopy (see **Methods**) (Figure 13.37D). Whereas the NBD and HD1 of the NOD domain largely retain their relative orientation, the WHD flips outward and rotates ~120° around its axes, with the HD2 following this motion (Figure 13.37B). The WD40 domains bound to cytochrome *c* have at this point moved away from the NOD. At the same time, the CARD domain is released from contact and left attached to the rest of the protein by a flexible linker.

Apaf-1 is now able to assemble into the heptameric apoptosome (Figure 13.37C). The NOD domain is a member of the AAA+ ATPase superfamily of molecular machines that use ATP hydrolysis to perform a wide range of mechanical functions (see Boxes 3.1 and 7.2). However, the NOD domain of Apaf-1 serves no such purpose; instead, its assembly plays a key if passive role in transmitting the apoptosis signal. The NBDs of seven Apaf-1 monomers come together to form the central ring of an oligomer, as with most AAA+ oligomers, although it is a heptameric ring rather than the more usual hexamer. (Some examples of other heptameric AAA+ proteins exist, and the potential for variability is further illustrated in the observation that the apoptosomes of *D. melanogaster* and *C. elegans* [see below] have octameric rings.)

Another difference from classical AAA+ ATPase assembly is that the WHD and HD1 domains of Apaf-1 participate in the oligomerization to form an outer ring (Figure 13.37C). The WHD undergoes the most extreme structural change during the opening and oligomerization of Apaf-1, which are dependent on cytochrome *c* and ADP/ATP exchange. At the end of the assembly process, seven CARD domains are exposed and in mutual proximity. Binding to the CARD domains of caspase-9 brings caspase monomers together and promotes their activation as dimers (see Figure 13.37D).

Proteins of the Bcl-2 family are required for cytochrome *c* release from mitochondria

The release of cytochrome *c* from mitochondria (Chapter 15) is an essential starting point in the assembly of the apoptosome. It occurs through a process termed mitochondrial outer-membrane permeabilization (MOMP), which also causes the release of other apoptotic factors, such as the protein SMAC described below. It is therefore of great importance, not only for the intrinsic pathway but also for apoptosis in general. Disruption of the outer mitochondrial membrane is achieved by a set of signaling assemblies formed by members of the **Bcl-2** protein family. There is as yet only a partial understanding of the intricate regulation of these proteins and how they bring about membrane permeabilization.

Generally, there are three groups of Bcl-2-like proteins: the anti-apoptotic members, most prominently Bcl-2 and Bcl-XL, which block cytochrome *c* release; and the pro-apoptotic members of this family, which promote cytochrome *c* release and are divided into the so-called BH3-only proteins and the proteins BAX and BAK (**Table 13.6**). The latter are often referred to as effector proteins because they have the main role in membrane permeabilization; mutant mouse cells lacking them cannot carry out the intrinsic pathway of apoptosis. The pro-apoptotic effectors, the anti-apoptotic Bcl-2 proteins, and some of the BH3-only members all share a common fold (**Figure 13.38A**). A feature of this fold is the presence of two predominantly hydrophobic α helices (helices 5 and 6), which are surrounded by six amphipathic α helices. In addition to this fold, several Bcl-2 members also have a C-terminal transmembrane (TM) helix.

Table 13.6 BCL-2 protein family.

Anti-apoptotic	Pro-apoptotic	
	BCL-2 effector proteins	BH3-only proteins
BCL-2	BAX	BID[a]
BCL-XL	BAK	BIM[a]
BCL-w		BAD
MCL-1		BIK
A1		BMF
		HRK
		Noxa
		PUMA

[a]Reflecting their possible direct involvement in MOMP, these are also referred to as 'direct activator BH3-only proteins.'

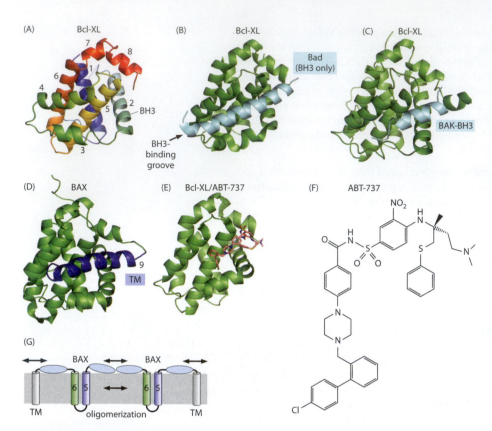

Figure 13.38 Structures of Bcl-2 proteins.
(A) Structure of Bcl-XL showing the typical helical fold of Bcl-2 family proteins in rainbow colors from N terminus to C terminus (no helix 9 TM). (PDB 1MAZ) (B–E) Interactions involving the BH3-binding groove as a focal point of regulation. (B) BH3-only protein Bad (cyan) interacting with Bcl-XL (green). (PDB 2BZW) (C) BH3 domain of effector BAK (cyan) interacting with Bcl-XL. (PDB 1BXL) (D) TM segment of BAX (dark blue, helix 9) interacting with its BH3 groove. (PDB 1F16) (E) The anti-cancer drug ABT-737 interacting with Bcl-XL. (PDB 2YXJ) (F) ABT-737. (G) A schematic model for BAX pore formation. Hydrophobic helices 5 and 6 are inserted into the membrane, where, together with helix TM, they assemble into oligomers that form pores through contacts mediated by the remainder of the protein, which resides peripherally on the membrane.

On the basis of biochemical analysis and sequence homologies, four common regions (sometimes referred to as domains) have been characterized. Structurally, these Bcl-2 homology (BH) regions reside around helix 1 (BH4), helix 2 (BH3), helices 4 and 5 (BH1), and helices 7 and 8 (BH2) (see Figure 13.38A). However, although for example the entire structure of the BH3-only protein BID shows a close resemblance to the typical Bcl-2 fold, other BH3-only proteins may be structurally similar to Bcl-2 only in the BH3 (helix 2) region (Figure 13.38B).

The underlying proposal is that the pro-apoptotic members BAX and BAK are bound to the anti-apoptotic proteins Bcl-2 and Bcl-XL; thus sequestered, they are prevented from forming pores in the mitochondrial outer membrane. This interaction is mediated by the binding of the BH3 region of BAX and BAK to the BH3-binding groove of the anti-apoptotic proteins (Figure 13.38C). In a structure of the protein BAX containing the hydrophobic helical segment TM, this groove is intriguingly occupied by the TM helix binding back onto the protein, making the protein soluble (Figure 13.38D). Other Bcl-2 proteins such as the anti-apoptotic members Bcl-2 and Bcl-XL contain a TM segment in their sequence, and this is also believed to occupy the central groove. The sequestering of pro-apoptotic Bcl-2 members via the BH3-binding region would be expected to displace the TM segment, forcing it outward and enabling it to initiate interaction with the mitochondrial membrane.

A simple possibility is that MOMP can be turned on by overexpression of genes encoding BH3-only proteins, for example as a result of radiation damage. The BH3-only proteins then bind to the BH3-binding groove of anti-apoptotic proteins, and by doing so they release the previously bound BAX or BAK proteins. Thus freed, BAX and BAK are able to insert into the outer mitochondrial membrane, oligomerize, and form pores. More intricate models have been developed, but this simple view illustrates the central role of BH3 helices and their interaction with the BH3-binding domain of anti-apoptotic Bcl-2 proteins in the regulation of MOMP. The role of the BH3-binding region of the anti-apoptotic Bcl-2 proteins has been identified as a prime target for cancer treatment in which small molecules mimicking the BH3 helical region of BH3-only proteins are used to bind to anti-apoptotic Bcl-2 proteins (Figure 13.38E, F), displacing BAX and BAK and initiating MOMP and apoptosis in malignant cells.

The mechanism of oligomerization of pro-apoptotic members and their auxiliary proteins and of pore formation remains unclear. One of several possible models is illustrated in

Figure 13.38G, in which a pair of hydrophobic helices inserts directly into the membrane. This is accompanied by significant changes in the tertiary structure of the protein, followed by interactions between the inserted proteins and their oligomerization. Other models depend on processes such as BAX or BAK largely retaining their tertiary structure and forming dimers, which are then inserted into the membrane and oligomerize to form the pore. In addition, a direct involvement of certain BH3 proteins, namely the so-called direct activator BH3-only proteins BID and BIM, in pore formation has been proposed. In any event, the inserted effector proteins lead to extensive pore formation and disruption of the membrane, the release of cytochrome *c* and other proteins, and a loss of membrane potential—all fostering the initiation of apoptosis.

The extrinsic pathway of apoptosis is mediated by membrane-associated DISCs

The second main means of triggering a caspase cascade, the extrinsic pathway, is brought about by DISCs (see above). Like the apoptosome, the DISC induces the dimerization of initiator caspases, in this case caspases-8 and -10. However, DISCs do not possess ATPase activity and operate by means of **death receptors**, death domains (DDs), and **death effector domains (DEDs)**.

At the heart of DISC formation are death receptors, in particular the Fas and TRAIL receptors. These receptors possess extracellular Cys-rich repeat domains, which can interact with a death ligand. This extracellular region is anchored by a single transmembrane helix, which leads into an intracellular death domain (**Figure 13.39**, left panel). Contrary to initial models, death receptors most likely exist as preformed trimers in *lipid rafts*, specialized microdomains of the plasma membrane, more ordered and tightly packed, that can serve as organizing centers for signaling molecules. Here they are poised for activation, but their intracellular death domains are unable to recruit the key adaptor protein FADD (Fas-associated death domain) and form the DISC. Structural work on the complex of the DDs of Fas and FADD revealed that the Fas DD must first undergo a conformational change that opens it up to make it capable of binding the FADD DD; in the absence of an extracellular trigger, the Fas DD adopts the canonical six-helix bundle of death domains. To bind the DD of FADD, helix 6 of the Fas receptor DD must rotate away to expose the binding site (Figure 13.39, middle panel). This change occurs upon activation of the receptor or, more precisely, when the receptors become highly clustered upon binding to a cluster of ligands. These death ligands are anchored in the plasma membrane of, and are presented by, another cell, for example a cytotoxic T cell. Inside the target cell, the DDs of the death receptor, in this case Fas, are brought close together, and the open forms of the DDs can stabilize one another, albeit by an as yet unknown mechanism (**Figure 13.40**). The DDs of multiple FADD molecules can now bind to the open DDs of Fas, thus creating an assembly of FADD proteins. The DEDs of the clustered FADDs attract the DEDs of initiator caspases-8 and

Figure 13.39 Opening of death domains in death receptors of DISC. Death domains (DDs) are central element of DISCs. Left: in the cytoplasm, the DD of the death receptor Fas adopts a six-helix globular fold. (PDB 1DDF) In the absence of an external stimulus, the extracellular regions (from a TRAIL receptor 2, PDB 1D0G), the transmembrane helix (pink) and the intracellular domains are present in lipid rafts as preformed trimers. The DDs are predominantly in the more stable closed conformation that is incapable of interacting with FADD-DD. Middle: on association with an extracellular death ligand, the Fas-DD undergoes an opening whereby helix 6 (residues 290–302) and residues 302–320 become continuous with helix 5. The C-terminal residues 321–327, which were disordered in the isolated DD, form another helix and the Fas-DD can now bind the FADD-DD. (PDB 3EZQ) Right: a stable FAS-DD/FADD-DD complex is formed.

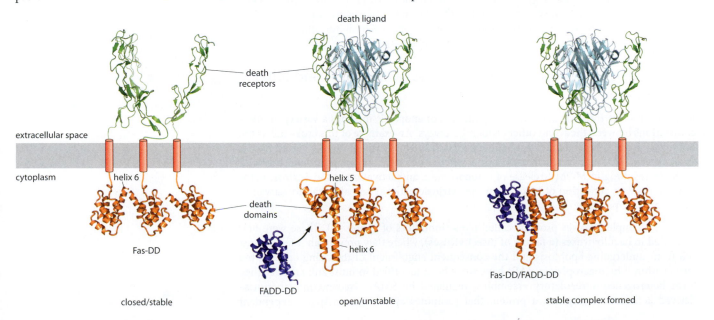

death ligand

death receptors

extracellular space

cytoplasm

helix 6

death domains

Fas-DD

helix 5

helix 6

FADD-DD

Fas-DD/FADD-DD

closed/stable

open/unstable

stable complex formed

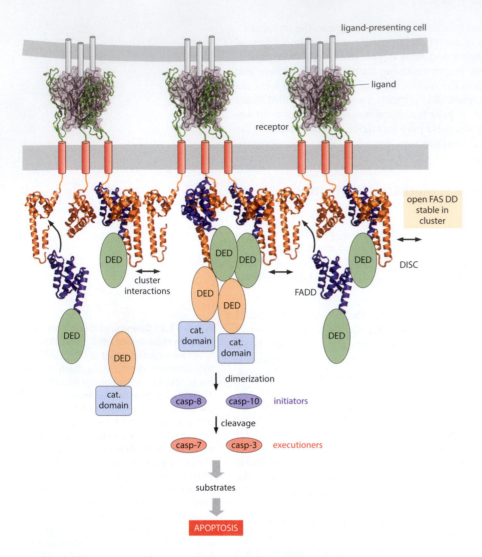

Figure 13.40 Activation of death receptor and DISC formation. Highly clustered membrane-anchored trimeric death ligands (TRAIL, in gray) are presented by a cytotoxic (ligand-presenting) cell. Binding to death receptors (TRAIL receptor 2, in green) on the target cell forces them into clusters, and the DDs (from Fas receptor, orange) are brought into close proximity. Open forms of the DDs are now stabilizing each other and permit the binding of FADD DDs (blue) and the clustering of FADD. FADDs also possess DEDs (green), which they now use to recruit the tandem DEDs (light orange) of caspases-8 and -10 to the DISC, which leads to the dimerization and activation of the caspases. (Based on PDBs: TRAIL receptor 2/death ligand 1D0G; Fas-DD 1DDF; Fas-DD/FADD 3EZQ)

-10, completing the DISC. The monomeric caspases are thus brought into close proximity, dimerize, and become activated. In this way, apoptosis is initiated.

Often, caspase-8 forms heterodimers with a homolog termed FLIP (FADD-like IL-1a-converting enzyme inhibitory protein), which contains a DED. FLIP is itself catalytically inactive but is able to activate caspase-8 efficiently by forming the heterodimer. It can then become cleaved by the active caspase-8 and turn into a caspase-8 inhibitor, allowing the cell to evade apoptosis and promoting drug resistance in chemotherapy.

There is cross-talk between the pathways of apoptosis and other cellular pathways

Regulation of the intrinsic and extrinsic pathways of apoptosis includes a variety of interconnections between them and other cellular pathways. An example of this cross-talk is the BH3-only protein BID, which is a substrate for caspase-8 and caspase-10 (see Figure 13.35). Once cleaved, BID migrates to mitochondria, where it activates pro-apoptotic effector Bcl-2 proteins, resulting in MOMP, release of cytochrome *c*, and activation of the intrinsic pathway. In this way, activation of caspases in the extrinsic pathway can also lead to activation of the intrinsic pathway.

Another example involves proteins called IAPs (inhibitors of apoptosis). These were first identified in baculoviruses (a family of insect viruses), where they prevent an infected host cell from undergoing apoptosis and the consequent engulfment of the dying cell and the virus within it by macrophages. IAPs have since been identified in most animal cells; they form heterogeneous regulatory assemblies regulated by SMAC (second mitochondria-derived activator of caspase), a protein that promotes cytochrome *c*/Apaf-1-dependent

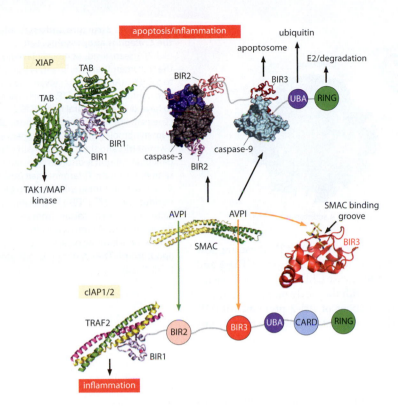

Figure 13.41 Other signaling assemblies regulate apoptosis. Top, multifaceted interaction networks are mediated by IAP proteins and regulated by SMAC. XIAP undergoes multiple interactions through its BIR domains. Bottom, structure of the XIAP relative cIAP-1/2 BIR1 domain with TRAF2. The other domains can form dimeric or trimeric complexes and add further linkages to the signaling network. SMAC (center) can affect contacts and influence degradation by binding through its N-terminal residues (sequence Ala-Val-Pro-Ile [AVPI]) to the SMAC-binding groove on BIR2/3 (center-right), illustrated for XIAP BIR3/SMAC. (PDBs: SMAC 1FEW; BIR3/SMAC 1G73; caspase-3/BIR2 1I3O; caspase-9/BIR3 1NW9; TAB1/BIR1 2POP; TRAF2/cIAP2 3M0A)

caspase activation and which is released upon MOMP (**Figure 13.41**). A prominent member of the IAP family is the protein XIAP, which possesses multiple BIR (baculovirus IAP repeat) domains. The second BIR domain (BIR2) of XIAP is an inhibitor of caspases -3 and -7, and the BIR3 domain is an inhibitor of caspase-9, thus very effectively suppressing apoptosis. Moreover, the first BIR domain (BIR1) has been found to dimerize with the BIR1 domain of another XIAP and together they bind the activator protein TAB, leading in turn to the activation of TAK1 and the MAP kinase cascade. XIAP also possesses a C-terminal UBA domain, which can interact with monoubiquitin and diubiquitin, and a RING domain (see **Guide**), which can function as an E3 ligase and bind E2 proteins (see Section 12.5), initiating the degradation of XIAP or bound proteins (Figure 13.41). This multitude of interactions illustrates the complexity of signaling networks, which are also regulated through MOMP and SMAC.

SMAC also targets the BIR2 and BIR3 domains of the cIAP (cellular IAP) proteins, which have a domain architecture very similar to that of XIAP (see Figure 13.41). In this case, it leads mostly to their degradation. The BIR1 domain of cIAP was found to interact with proteins of the TRAF (TNF-receptor-associated factor) family. These trimeric multidomain proteins further interact with a variety of target proteins, regulating inflammation and SMAC-mediated degradation of the cIAPs, which was found mainly to propagate inflammation. Thus, MOMP and the release of SMAC affect a spectrum of assemblies regulating a variety of processes in addition to cell death.

The apoptosome of *C. elegans* presents a simple mode of caspase activation

C. elegans has been a key model organism in understanding fundamental aspects of apoptosis. In terms of caspase activation, it possesses a rather simple pathway that consists solely of the CED-4 apoptosome, which activates the CARD-containing caspase CED-3. Its apoptosome is similar to that of humans but is octameric and built from four dimers of CED-4 (**Figure 13.42**). The CED-4 protein (571 aa) resembles Apaf-1 but lacks the WD40 repeats whereby that protein senses cytochrome *c*. Instead, CED-4 is regulated by a Bcl-2 family protein called CED-9 (280 aa), which binds to CED-4 dimers and prevents them from assembling into an apoptosome.

The stimulus for apoptosome formation in *C. elegans* lies in another Bcl-2 family protein, the BH3-only protein EGL-1 (106 aa) which binds tightly to CED-9 in the heterotrimer

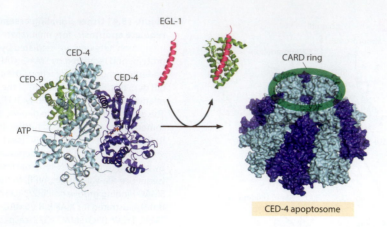

Figure 13.42 Structure and assembly of the *C. elegans* apoptosome. Left, CED-9 (Bcl-2) (green, residues 48–251 lacking the transmembrane segment) is bound to a CED-4 (Apaf-1) dimer (cyan and blue) that contains a bound ATP, which seems to have only a structural role. (PDB 2A5Y) (The names in parentheses are those of the human homologs.) The CED-4 dimers are asymmetric, and the CARD domain of the dark blue subunit is disordered. Middle, a segment from the C-terminal half (residues 45–87) of EGL-1 (magenta) binds to CED-9 (residues 68–237). (PDB 1TY4) Right, this causes CED-9 to dissociate from the CED-4 dimer; the CED-4 dimers, now free, are able to assemble into the octameric *C. elegans* apoptosome. The CARD ring is indicated. (PDB 3LQQ)

$(CED-9)_1(CED-4)_2$, thereby releasing the CED-4 dimers (see Figure 13.42). These are now able to assemble into the octameric apoptosome, which then activates caspase CED-3 and propagates apoptosis. This use of a Bcl-2 family protein to release CED-4 dimers for assembly into an apoptosome makes a fascinating comparison with the vastly more complicated intrinsic pathway in humans, which requires a large number of Bcl-2 proteins, MOMP, cytochrome *c*, and ADP/ATP exchange. It is remarkable how adaptable the structural motifs and assemblies of cell death turn out to be.

13.8 SUMMARY

The phases of the cell division cycle are regulated by the combined action of protein kinases and phosphatases. The key regulators are cyclin-dependent kinases (CDKs), which remain inactive until bound by their partner cyclins. CDK concentrations remain constant, but the concentrations of cyclins vary systematically through the cell cycle. Binding of a specific cyclin alters the conformation of the CDK, thereby allowing access to the ATP-binding site and substrate recognition. Checkpoints hold back the cell cycle at three crucial stages: at the start; until DNA has been fully replicated and checked (end of S phase); and until sister chromatids have been confirmed as properly aligned for segregation (end of mitosis).

After DNA replication, the chromatids are condensed to compact rods (a process called cohesion) and sister chromatids are held together by cohesin, a 35 nm diameter ring complex of three proteins. Chromatid separation is brought about by the mitotic spindle, a large bipolar array of microtubules nucleated in centrosomes; these organelles are 1–5 μm in diameter. In a centrosome, nine triplet microtubules emanate from a centriole, a cylindrical structure 0.5 μm by 0.2 μm, whose proximal end has a cartwheel structure with ninefold symmetry. The plus ends of the microtubules anchor in kinetochores, large protein complexes assembled on the centromere, a specific region of chromosomal DNA. The single centrosome in a cell embarking on the cell cycle is duplicated at the same time as DNA is replicated.

Chromatid movement toward the centrosomes at the poles of the cell is driven by the polymerization and depolymerization of microtubules, but not before their correct attachment to the kinetochores has been verified. A mitotic checkpoint complex (MCC) binds to the anaphase-promoting complex (APC/C), inhibiting it. When all chromatids have become securely attached to microtubules and bi-oriented, the APC/C inhibition is released. An inhibitor of the protease separase is degraded, allowing it to cleave cohesin, thereby permitting the sister chromatids to be pulled apart and daughter cells to separate (cytokinesis).

Apoptosis affords a means of disposing of cells that have become seriously damaged or need to be removed as part of normal development. Two pathways exist, both dependent on the initiation of a cascade of cysteine proteases (caspases). The intrinsic pathway begins with the action of Bcl-2 proteins to release cytochrome *c* from mitochondria of a damaged cell, which initiates assembly of the apoptosome, a 1.1 MDa protein complex with 7-fold symmetry (unusually, 8-fold in *C. elegans*). Binding of monomeric initiator caspases to the apoptosome triggers their activation by proximity-induced dimerization. In the extrinsic pathway, a death ligand displayed on the surface of one cell binds to a death receptor on the surface of the target cell, inducing apoptosis. A death-inducing signaling complex (DISC) is created in the target cell that is the counterpart of the apoptosome in the intrinsic pathway and which similarly can dimerize and activate an initiator caspase. The initiator caspases then activate downstream executioner caspases (which are constitutive dimers) by limited proteolysis; in turn, the executioner caspases act on a host of substrates to bring about the demise of the affected cell. Among these substrates are inhibitors of certain endonucleases that are cleaved by executioner caspases, enabling the endonucleases to degrade chromosomal DNA; proteins involved in cytoskeleton and membrane remodeling, thereby causing the characteristic blebbing of apoptotic cells; and scramblases that transfer phosphatidylserine into the outer leaflet of the cell membrane, where it attracts macrophages to engulf the dying cell. The intrinsic and extrinsic pathways are capable of cross-talk between themselves and with other cellular pathways, for example through inhibitor-of-apoptosis proteins (IAPs), first discovered in baculoviruses.

Sections of this chapter are based on the following contributions

13.5   Andrea Musacchio, Max Planck Institute of Molecular Physiology, Dortmund, Germany.

13.6   Anthony A. Hyman, Max Planck Institute of Molecular Cell Biology and Genetics, Dresden, Germany and Laurence Pelletier, Mount Sinai Hospital, Toronto, Canada.

13.7   Stefan Riedl, Sanford-Burnham Medical Research Institute, La Jolla, USA.

REFERENCES

13.2 The cell cycle

Morgan DO (2007) The Cell Cycle: Principles of Control. New Science Press.

Novák B, Hunt T & Nasmyth K (eds) (2011) Theme issue: 'The Cell Cycle'. *Phil Trans R Soc B* 366:3493–3652.

13.3 Transient assemblies of kinases, cyclins, and phosphatases

Barford D, Das AK & Egloff MP (1998) The structure and mechanism of protein phosphatases: insights into catalysis and regulation. *Annu Rev Biophys Biomol Struct* 27:133–164.

Bayliss R, Sardon T, Vernos I & Conti E (2003) Structural basis of Aurora-A activation by TPX2 at the mitotic spindle. *Mol Cell* 12:851–862.

Brotherton DH, Dhanaraj V, Wick S et al (1998) Crystal structure of the complex of the cyclin D-dependent kinase Cdk6 bound to the cell-cycle inhibitor p19INK4d. *Nature* 395:244–250.

Brown NR, Noble ME, Endicott JA & Johnson LN (1999) The structural basis for specificity of substrate and recruitment peptides for cyclin-dependent kinases. *Nat Cell Biol* 1:438–443.

Cheng KY, Noble ME, Skamnaki V et al (2006) The role of the phospho-CDK2/cyclin A recruitment site in substrate recognition. *J Biol Chem* 281:23167–23179.

Gray CH, Good VM, Tonks NK & Barford D (2003) The structure of the cell cycle protein Cdc14 reveals a proline-directed protein phosphatase. *EMBO J* 22:3524–3535.

Lowery DM, Lim D & Yaffe MB (2005) Structure and function of Polo-like kinases. *Oncogene* 24:248–259.

Pavletich NP (1999) Mechanisms of cyclin-dependent kinase regulation: structures of Cdks, their cyclin activators, and Cip and INK4 inhibitors. *J Mol Biol* 287:821–828.

Ray A, James MK, Larochelle S et al (2009) p27Kip1 inhibits cyclin D-cyclin-dependent kinase 4 by two independent modes. *Mol Cell Biol* 29:986–999.

Reinhardt HC & Yaffe MB (2013) Phospho-Ser/Thr-binding domains: navigating the cell cycle and DNA damage response. *Nat Rev Mol Cell Biol* 14:563–580.

Rudolph J (2007) Cdc25 phosphatases: structure, specificity, and mechanism. *Biochemistry* 46:3595–3604.

Shi Y (2009) Serine/threonine phosphatases: mechanism through structure. *Cell* 139:468–484.

Takaki T, Echalier A, Brown NR et al (2009) The structure of CDK4/cyclin D3 has implications for models of CDK activation. *Proc Natl Acad Sci USA* 106:4171–4176.

Tonks NK (2013) Protein tyrosine phosphatases—from housekeeping enzymes to master regulators of signal transduction. *FEBS J* 280:346–378.

Welburn JP, Tucker JA, Johnson T et al (2007) How tyrosine 15 phosphorylation inhibits the activity of cyclin-dependent kinase 2/cyclin A. *J Biol Chem* 282:3173–3181.

Xu Y, Xing Y, Chen Y et al (2006) Structure of the protein phosphatase 2A holoenzyme. *Cell* 127:1239–1251.

13.4 Sister chromatids in mitosis

Anderson DE, Losada A, Erickson HP & Hirano T (2002) Condensin and cohesin display different arm conformations with characteristic hinge angles. *J Cell Biol* 156:419–424.

Haering CH, Farcas AM, Arumugam P et al (2008) The cohesin ring concatenates sister DNA molecules. *Nature* 454:297–301.

Nasmyth K & Haering CH (2009) Cohesin: its roles and mechanisms. *Annu Rev Genet* 43:525–558.

Nasmyth K & Haering CH (2005) The structure and function of SMC and kleisin complexes. *Annu Rev Biochem* 74:595–648.

Peters JM, Tedeschi A & Schmitz J (2008) The cohesin complex and its roles in chromosome biology. *Genes Dev* 22:3089–3114.

Viadiu H, Stemmann O, Kirschner MW & Walz T (2005) Domain structure of separase and its binding to securin as determined by EM. *Nat Struct Mol Biol* 12:552–553.

13.5 Kinetochores in anaphase

Alushin GM, Ramey VH, Pasqualato S et al (2010) The Ndc80 kinetochore complex forms oligomeric arrays along microtubules. *Nature* 467:805–810.

Chang L & Barford D (2014) Insights into the anaphase-promoting complex: a molecular machine that regulates mitosis. *Curr Opin Struct Biol* 29C:1–9.

Chang L, Zhang Z, Yang J et al (2014) Molecular architecture and mechanism of the anaphase-promoting complex. *Nature* 513:388–393.

Cheeseman IM (2014) The kinetochore. *Cold Spring Harb Perspect Biol.* 6:a015826.

Cheeseman IM, Chappie JS, Wilson-Kubalek EM & Desai A (2006) The conserved KMN network constitutes the core microtubule-binding site of the kinetochore. *Cell* 127:983–997.

Falk SJ & Black BE (2013) Centromeric chromatin and the pathway that drives its propagation. *Biochim Biophys Acta* 1819:313–321.

Hill TL (1985) Theoretical problems related to the attachment of microtubules to kinetochores. *Proc Natl Acad Sci USA* 82:4404–4408.

Liu, D, Vader, G, Vromans MJ et al (2009) Sensing chromosome bi-orientation by spatial separation of Aurora B kinase from kinetochore substrates. *Science* 323:1350–1353.

London N & Biggins S (2014) Signalling dynamics in the spindle checkpoint response. *Nat Rev Mol Cell Biol.* 15:736–747.

Mapelli M & Musacchio A (2007) MAD contortions: conformational dimerization boosts spindle checkpoint signaling. *Curr Opin Struct Biol* 17:716–725.

McIntosh JR, Grishchuk EL, Morphew MK et al (2008) Fibrils connect microtubule tips with kinetochores: a mechanism to couple tubulin dynamics to chromosome motion. *Cell* 135:322–333.

Musacchio A (2011) Spindle assembly checkpoint: the third decade. *Phil Trans R Soc B* 366:3595–3604.

Wan X, O'Quinn RP, Pierce HL et al (2009) Protein architecture of the human kinetochore attachment site. *Cell* 137:672–684.

Welburn JP, Vieugel M, Liu D et al (2011) Aurora B phosphorylates spatially distinct target to differentially regulate the kinetochore-microtubule interface. *Mol Cell* 38:383–392.

Westermann S, Drubin DG & Barnes G (2007) Structures and functions of yeast kinetochore complexes. *Annu Rev Biochem* 76:563–591.

13.6 Centrioles and centrosomes

Azimzadeh J & Marshall WF (2010) Building the centriole. *Curr Biol* 20:R816–R825.

Badano JL, Teslovich TM & Katsanis N (2005) The centrosome in human genetic disease. *Nat Rev Genet* 6:194–205.

Bettencourt-Dias M & Glover DM (2007) Centrosome biogenesis and function: centrosomics brings new understanding. *Nat Rev Mol Cell Biol* 8:451–463.

Bornens M & Azimzadeh J (2007) Origin and evolution of the centrosome. *Adv Exp Med Biol* 607:119–129.

Jaspersen SL & Winey M (2004) The budding yeast spindle pole body: structure, duplication and function. *Annu Rev Cell Dev Biol* 20:1–28.

Kollman JM, Polka JK, Zelter A et al (2010) Microtubule nucleating γ-TuSC assembles structures with 13-fold microtubule like symmetry. *Nature* 466:879–882.

Lüders J & Stearns T (2007) Microtubule-organizing centres: a re-evaluation. *Nat Rev Mol Cell Biol* 8:161–167.

Marshall WF (2009) Centriole evolution. *Curr Opin Cell Biol* 21:14–19.

Nigg EA (ed.) (2004) Centrosomes in Development and Disease. Wiley-VCH.

Pelletier L, O'Toole E, Schwager A et al (2006) Centriole assembly in *Caenorhabditis elegans*. *Nature* 444:619–623.

Winey M & O'Toole E (2014) Centriole structure. *Phil Trans R Soc B* 369:20130457 (doi:10.1098/rstb.2013.0457).

13.7 Apoptosis (programmed cell death)

Boatright KM, Renatus M, Scott FL et al (2003) A unified model for apical caspase activation. *Mol Cell* 11:529–541.

Holcik M & Sonenberg S (2005) Translational control in stress and apoptosis. *Nat Rev Mol Cell Biol* 6:319–327.

Leber B, Lin J & Andrews DW (2007) Embedded together: the life and death consequences of interaction of the Bcl-2 family with membranes. *Apoptosis* 12:897–911.

Lu M, Lin SC, Huang Y et al (2007) XIAP induces NF-κB activation via the BIR1/TAB1 interaction and BIR1 dimerization. *Mol Cell* 26:689–702.

Qi S, Pang Y, Hu Q et al (2010) Crystal structure of the *Caenorhabditis elegans* apoptosome reveals an octameric assembly of CED-4. *Cell* 141:446–457.

Riedl SJ & Salvesen GS (2007) The apoptosome: signalling platform of cell death. *Nat Rev Mol Cell Biol* 8:405–413.

Sattler M, Liang H, Nettesheim D et al (1997) Structure of Bcl-xL-Bak peptide complex: recognition between regulators of apoptosis. *Science* 275:983–986.

Scott FL, Stec B, Pop C et al (2009) The Fas–FADD death domain complex structure unravels signalling by receptor clustering. *Nature* 457:1019–1022.

Shiozaki EN, Chai J, Rigotti DJ et al (2003) Mechanism of XIAP-mediated inhibition of caspase-9. *Mol Cell* 11:519–527.

Wilson NS, Dixit V & Ashkenazi A (2009) Death receptor signal transducers: nodes of coordination in immune signaling networks. *Nat Immunol* 10:348–355.

Yuan S & Akey CW (2013) Apoptosome structure, assembly, and procaspase activation. *Structure* 21:501–515.

Yuan S, Yu X, Topf M et al (2010) Structure of an apoptosome-procaspase-9 CARD complex. *Structure* 18:571–583.

Motility

14

14.1 INTRODUCTION

Movement is essential for life and takes place on many levels. Multicellular organisms, such as a raptor in pursuit of prey, can move rapidly, as can single-celled microorganisms, such as chemotactic bacteria homing in on sources of nutrients, if speed is considered as the time needed to move one body-length. Movement is also pervasive at the subcellular level. The interiors of eukaryotic cells are in constant flux through such dynamic processes as the transport of cargoes between compartments. In prokaryotic cells, motility drives cell division. Even viruses have specialized forms of motile apparatus, in addition to commandeering the motile systems of host cells.

Biological motions exhibit great diversity in terms of scale, speed, and form—which may include swimming, stepping, gliding, pushing, and pulling. However, at the chemical level, only a few distinct mechanisms have been identified. The predominant energy source for force generation involves the hydrolysis of a nucleotide triphosphate—usually ATP, sometimes GTP—with an accompanying release of free energy. For ATP, the amount of energy released is 30.5 kJ/mol under standard conditions but can vary substantially, up to ~50 kJ/mol, depending on the physiological context. The diversity and complexity of the motions ultimately generated by this simple reaction have their basis in the structures of the molecular machines that transduce the chemical free energy released into mechanical work.

In eukaryotes, the protein filaments of the **cytoskeleton** play a central role in intracellular motility. **Actin filaments (filamentous actin (F-actin)**; Section 14.2) and **microtubules (MTs**; Section 14.6) provide linear tracks on which **motor proteins** run, moving their cargoes (**Figure 14.1**). A motor protein typically has an ATPase domain, a cargo-binding domain, a track-binding motif, and regulatory domains or subunits. The ATPases are conserved in structure, whereas the other components are highly diverse. Successive steps in

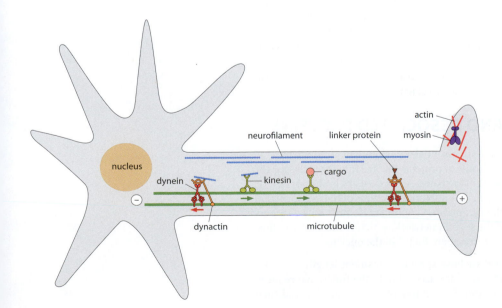

Figure 14.1 Schematic diagram of a neuronal axon, with various cytoskeletal components. Neurofilaments are strictly structural components; actin filaments and associated myosin motors are at the axon tip; and bidirectional motorized transport takes place on microtubules, whose cargoes can include neurofilament segments, vesicles, organelles, RNA granules, and receptors. (Adapted from E. Chevalier-Larsen and E.L. Holzbaur, *Biochim. Biophys. Acta* 1762:1094–1108, 2006. With permission from Elsevier.)

the ATPase cycle are accompanied by conformational changes that are transmitted to other parts of the motile complex. The motor alternates between states in which it respectively binds to and detaches from the track—an essential requirement for movement. The motors that run along F-actin belong to the myosin superfamily (Sections 14.3 and 14.5). MTs support two kinds of motors (Section 14.7): the *kinesin* motor domain is similar to that of *myosin,* whereas the *dynein* motor domains belong to another superfamily of *mechanoenzymes,* the AAA+ ATPases (Box 3.1 and Box 7.2).

The same proteins that afford tracks for motorized transport also generate force actively by another mechanism. The subunits of F-actin and MTs are constantly attaching to and dissociating from filament ends. When a high-affinity end abuts another structure (such as a membrane), local thermal fluctuations transiently expose that end, creating opportunities for subunits to bind. Once a subunit has been added, it restricts the ability of the membrane to re-enter space previously accessible to it; in this way, a force is exerted on the membrane (Section 14.9). The reverse reaction—depolymerization—can also generate motility, as in the vital process of *mitosis* (see Chapter 13).

For a long time, it was thought that the cytoskeleton is a hallmark distinguishing eukaryotes from prokaryotes. It is now known that prokaryotes have proteins with similar folds to those of actin and **tubulin**, despite little similarity in their amino acid sequences, and that these proteins are also involved in motility. However, no evidence has yet surfaced for prokaryotic counterparts of myosin, kinesin, or dynein. The explanation for this discrepancy may lie in cell dimensions: most eukaryotic cells are so large that diffusion, as retarded by molecular crowding effects (Box 1.4 and Section 1.8) is simply too slow for many transportation tasks. Because prokaryotes are much smaller, diffusion often suffices, but a force-generating apparatus is needed for various aspects of cell division: the underlying mechanism appears to be (as in eukaryotes) protein polymerization. It is plausible that, during evolution, polymerization was the first motile mechanism to appear and subsequently the same proteins came to assume roles in motorized transport.

Protein filaments also underlie the movement of whole cells through liquid media. They are called **flagella** (singular *flagellum*, Latin for whip). Bacterial flagella and those of eukaryotic cells are similar in function but quite different in structure and motile mechanism. The bacterial flagellum (Section 14.11) is a long extracellular filament of the protein *flagellin*, which is rotated by a motor embedded in the cell envelope, thus serving as a propeller. The direction of rotation—and hence of movement—can switch in response to signals from *chemoreceptor* proteins, allowing the bacterium to swim toward a source of nutrients, a behavior called *chemotaxis*. The eukaryotic flagellum (Section 14.8), 10-fold thicker, consists of an ordered bundle of MTs enclosed within a protrusion of the plasma membrane. Its undulations ('beating') are generated by dynein motors attached to the MTs. This mechanism is more akin to muscle contraction than it is to bacterial swimming.

In addition to cyclic mechanisms that repeat as long as fuel supplies last, there are macromolecular complexes that perform single convulsive movements at precisely programmed times and locations (Section 14.10). In motile systems of this kind, the energy source is conformational free energy. The initial state of the system represents a kinetically accessible local minimum of free energy, and the final state, a lower minimum that is reached via a pathway of conformational changes triggered in response to a specific signal. In this transition, conformational free energy is released and used to generate force.

14.2   ACTIN FILAMENTS AND ASSOCIATED PROTEINS

Actin, a 43 kDa motility-associated protein, is an essential and abundant component of all eukaryotic cells. It forms a highly conserved family of proteins that fall into three classes: the α, β, and γ isoforms, with distant relatives in prokaryotes. Actin exists as both monomeric **G-actin** (G for globular) and polymeric F-actin. The subunit structures of the two forms are similar but not identical (see below). In the cytoplasm, G-actin serves primarily as a pool that feeds the assembly and remodeling of the F-actin network, which is constantly in flux. Smaller amounts of actin, in non-filamentous form, are found in the nucleus.

F-actin permeates the cytoplasm in networks whose specifics—filament lengths, orientations, cross-linking, and intracellular locations—vary according to the functional requirements of a given cell. *In vivo*, the assembly and disassembly of actin filaments, and their organization into networks, are regulated by a large number of **actin-binding proteins (ABPs)**. The activities of these proteins are, in turn, under the control of signaling pathways (Section 11.6).

The actin-based cytoskeleton has many roles, including control of cell shape, motor-based transport of organelles and other cargoes, *cytokinesis,* regulation of ion transport, and receptor-mediated responses. In muscle, it is engaged in force generation as the principal constituent of the thin filaments (Section 14.5).

Actin filaments participate in motility in two ways. In one, they provide tracks for *myosin* motor proteins. Movement along these tracks is powered by cyclic conformational changes in the myosins, coupled to ATP hydrolysis. In the second mechanism, motile force is generated by actin polymerization.

F-actin is a polar two-stranded filament

Electron micrographs portray F-actin as a filament, 8–9 nm in diameter (**Figure 14.2A**). It may be described as having two strands that turn slowly round each other to form right-handed 'long-pitch' helices; or, alternatively, as a single left-handed helix—the so-called 'genetic' helix—whose axial repeat contains 13 subunits in six turns. The two descriptions may be reconciled in a 'radial projection' (Figure 14.2B). The axial rise per subunit is 2.75 nm along the genetic helix or 5.5 nm along a long-pitch helix. The spacing between 'crossovers' (narrowest points in side views) is ~36 nm.

The polarity of F-actin is not immediately apparent in electron micrographs but is enhanced when filaments are 'decorated' with myosin. In the absence of ATP (the rigor state), myosin heads bind strongly to F-actin, giving a distinctive arrowhead pattern (Figure 14.2C). The two ends are denoted as the 'pointed end' and the 'barbed end,' accordingly. When undecorated F-actin is visualized at higher resolution, its polarity also becomes apparent (Figure 14.2D).

Another naming convention designates the barbed end as the 'plus end' because this is the end of more rapid growth as filaments assemble; conversely, the pointed end is the 'minus end.' The actin subunit at the pointed (minus) end exposes its ATP-binding site to solvent (see below). Here we use the 'pointed/barbed' terminology. All actin filaments shown with their axes vertical in figures have their barbed ends at the bottom. Subunit structures are also oriented consistently with this convention.

Actin filaments are dynamic: they grow and shorten, coupled to ATP hydrolysis

Actin polymerization has been studied extensively *in vitro* with purified protein. Polymerization is energetically disfavored until conditions support the formation of nuclei, consisting of three subunits. This initial stage of polymerization, the *lag* phase, is followed by the rapid addition of subunits in the *elongation* phase. Solvent conditions that promote

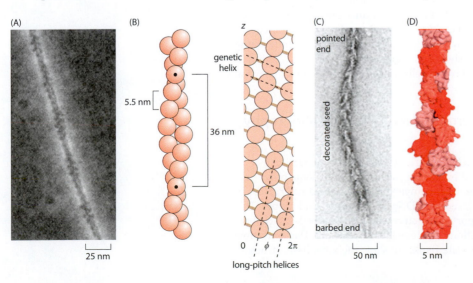

(A)   (B)   (C)   (D)

genetic helix

5.5 nm

36 nm

pointed end

decorated seed

barbed end

z

0   ϕ 2π

long-pitch helices

25 nm   50 nm   5 nm

Figure 14.2 Helical structure of F-actin. (A) Dark-field scanning transmission electron micrograph of a negatively stained actin filament. (B) Helical geometry of the filament; at the right is shown the radial projection of its lattice. (C) Electron micrograph of negatively stained F-actin, decorated with myosin heads except in the bottom segment. (D) Structure of F-actin derived from X-ray fiber diffraction and the crystal structure of G-actin. Subunits are shown in shades of red. (A, from M.O. Steinmetz et al., *J. Cell Biol.* 138:559–574, 1997. With permission from Rockefeller University Press; C, courtesy of Roger Craig.)

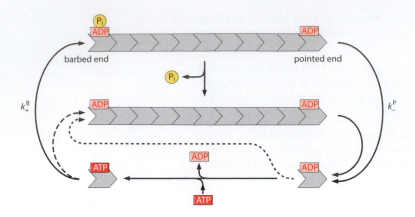

Figure 14.3 Actin subunits detach from the pointed end and add at the barbed end. The affinity of ATP-G-actin for the barbed end depends on whether it is occupied by ADP-P_i (that is, P_i has not yet detached) or ADP. k_+^B is the association rate constant of ATP-G-actin to barbed ends, and k_-^P is the dissociation rate constant of ADP-G-actin from pointed ends. (Adapted from B. Bugyi and M.-F. Carlier, *Annu. Rev. Biophys.* 39:449–70, 2010. With permission from Annual Reviews.)

polymerization include high ionic strength (KCl concentrations >50 mM), neutral or slightly acidic pH, high Mg^{2+} concentrations, and elevated temperature.

In the absence of filament-stabilizing ABPs, actin assembles and disassembles continuously. In the steady state, a pool of monomers coexists with the filament population in a *dynamic equilibrium*, whereby individual filaments lengthen or shorten as they exchange monomers with the unassembled pool, while the overall length distribution is maintained. The concentration of the monomer pool is the *critical concentration*, and it is the minimum concentration at which polymers can form and persist; its value depends on solvent conditions.

Subunits may be added or lost at either end of the filament, but association occurs mostly at the barbed end and dissociation at the pointed end. *In vivo*, G-actin almost always contains ATP complexed with Mg^{2+}, whereas most F-actin subunits contain ADP. As a filament matures, the ATP is hydrolyzed, phosphate is released, and eventually ADP-actin monomers are shed. They then undergo nucleotide exchange, yielding ATP-actin monomers that are available for polymerization. (Unlike F-actin, G-actin is a poor ATPase.) Whereas ATP bound to G-actin is readily exchangeable, the ADP of F-actin is non-exchangeable.

The critical concentration for subunit addition at the barbed end is ~12-fold lower than for the pointed end. This difference results in unidirectional growth of the filament at the barbed end through a continual flux of subunits from the pointed end. This cycling reaction is called **treadmilling** (**Figure 14.3**). Treadmilling is also performed by MTs (see Figure 14.42).

In general, filament length depends on ease of nucleation: if conditions favor nucleation, many short filaments are produced; conversely, if nucleation is disfavored, a smaller number of longer filaments results. In either case, filament length (*l*) is not uniquely defined but observes an approximately exponentially decaying distribution $(1 - \exp(-\kappa l))$ whose coefficient κ depends on the buffer conditions and which is well described by a simple thermodynamic model. However, actin filaments with uniquely specified lengths are produced *in vivo* when certain ABPs are present: examples include the ~1 μm-long thin filaments of **striated muscle** whose lengths are specified by giant proteins of the *nebulin* family acting as tape-measures (see Figure 14.20), and the short actin filaments (~40 nm long) found in the membrane-associated cytoskeleton of red blood cells, whose size is specified by several ABPs acting in concert.

The structure of G-actin was determined by X-ray crystallography of co-crystals

Because G-actin polymerizes readily, it has to be modified to inhibit polymerization if crystals are to be obtained. This may be achieved by forming a complex with some other protein that blocks polymerization by **capping**. This complex may then yield *co-crystals*. In the first actin crystal structure, it was complexed with DNase I, a pointed-end capper. This interaction is not physiological, but most reported structures have used barbed-end cappers such as gelsolin or profilin, whose interactions are biologically relevant. There are now dozens of G-actin structures in the Protein Data Bank. All show the same basic structure (**Figure 14.4**).

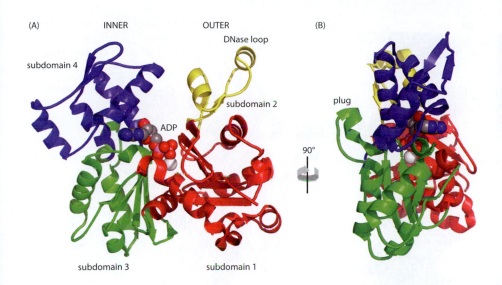

(A) INNER   OUTER

DNase loop

subdomain 4

subdomain 2

ADP

plug

90°

subdomain 3   subdomain 1

(B)

Figure 14.4 Structure of G-actin. (A) The names of the outer and inner domains refer to their positions in F-actin, the inner domain being closer to the axis. (PDB 1J6Z) Many capping proteins bind in the cleft between subdomains 1 and 3. (B) At right-angles to the view in (A). Note the twist between the inner and the outer domains. The plug is an insertion that makes an important contact between the two long-pitch helices in F-actin.

G-actin is a flat molecule built of two domains ('outer' and 'inner,' each comprising two subdomains), with ATP (or ADP) bound between them. Subdomains 1 and 3 are both α–β domains with 5-stranded β sheets and the same topology. They may have originated by gene duplication. The β-phosphate oxygen sits in a hairpin loop in subdomain 1 and the γ-phosphate in a similar loop in subdomain 3. The polypeptide chain crosses over twice between subdomains 1 and 3. Subdomains 2 and 4 appear to be insertions in subdomains 1 and 3 respectively. Subdomain 2 shows variations among known crystal structures. In particular, the 'DNase I-binding loop' can assume several different conformations and in many cases is disordered. In the structure shown in Figure 14.4A, it has a short α helix. This loop is also involved in inter-subunit interactions in F-actin (see below). Its conformation may be responsive to nucleotide-binding status, and the α-helical state may be favored in the ADP-bound form.

F-actin structure has been determined both by X-ray fiber diffraction and by reconstruction of cryo-electron micrographs

The most detailed structural data on F-actin have come from two approaches: X-ray diffraction of oriented gels, and cryo-EM (see **Methods**). Centrifuge pellets of F-actin are birefringent, indicating that they contain domains of filaments packed in parallel bundles. After further manipulation to optimize alignment, such gels yield X-ray fiber diffraction patterns in which *layer-line* reflections extend to resolutions of 7 Å or higher. These reflections are positioned according to the helical geometry of the filament, and the distribution of intensity over them is specified by the distribution of density in the filament. To interpret these data, subunits with the G-actin crystal structure were placed on a helix with the correct geometry, and a diffraction pattern was calculated. This procedure was repeated for all possible orientations of the subunit; the orientation chosen was the one that best matched the observed pattern. This solution was then refined by adjusting the subunit structure in F-actin to further improve the fit (**Figure 14.5A**). For the finest details, the outcome depends on the refinement method, but there is a consensus about the major changes in G-actin when it polymerizes to F-actin (Figure 14.5C):

- A 20° rotation of the inner domain relative to the outer domain. This twist makes the F-actin subunit flatter than G-actin.
- The DNase I-binding loop, an α helix in this G-actin structure, appears as an open loop in F-actin.

F-actin structures have also been determined by cryo-EM (Figure 14.5A,B). Many three-dimensional reconstructions have been performed for filaments in various states and with different proteins bound. In the structure shown in Figure 14.5B, secondary structure elements (α helices) are directly visualized. Again, interpretation was assisted by inserting the G-actin structure into the density map and adjusting it to optimize the fit. Many of the changes between G-actin and F-actin, such as the flattening out of the subunit (see above), were reproduced, but there were also new insights. For example, the domain rotations were found to open the nucleotide-binding pocket, causing Gln137 to move closer to

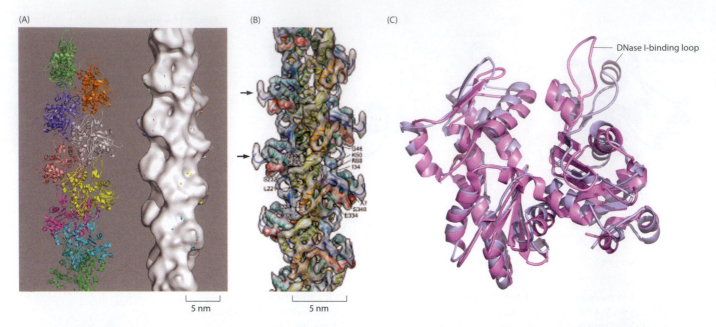

(A)   (B)   (C)

DNase I-binding loop

5 nm   5 nm

the γ-phosphate. This shift offered an explanation for the ATPase activation mechanism of F-actin, as Gln137 was known to be crucial for ATP hydrolysis. In addition, the negatively charged N-terminal residues (Asp-Glu-Asp-Glu)—not seen and apparently disordered in the crystal structure—which are involved in the interactions of F-actin with many ABPs, were visualized as a peripheral motif (arrows in Figure 14.5B).

In F-actin, the strong inter-subunit bonds are in the axial direction

These bonds connect neighboring subunits along the long-pitch helices. Strong bonding is needed because actin filaments are essentially inextensible, and bonding across this interface must be able to withstand applied stresses. The motifs involved in this interaction are identified in **Figure 14.6A**. The lateral bonding across the axis between the two long-pitch helices is much weaker and involves the 'plug' motif (Figure 14.6B), which forms a bridge between the two helices. When G-actin is incorporated into F-actin, the plug retains its native conformation. Another bond between the long-pitch helices is made higher up.

In vivo, the structure and stability of filaments are controlled by actin-binding proteins and can be affected by drugs

As outlined above, a propensity to self-assemble is an inherent property of actin; *in vivo*, however, this propensity is supervised by ABPs, which regulate the lengths, stability, and interactions of the filaments. There are a myriad of ABPs, as actin is involved in many cellular activities: a few examples are summarized in **Table 14.1** and **Figure 14.7**. Some ABPs promote the rapid nucleation of filaments. This function may be performed by scaffolding proteins such as formin homology (FH) (see **Guide**) proteins or nucleating centers such as the *Arp2/3* complex. The FH2 domain dimer apparently ratchets along at the barbed end, recruiting the addition of actin monomers. Alternatively, the number of seed filaments may be increased by severing existing filaments, as by cofilin. The *in vivo* processes are ~100 times faster than *in vitro* treadmilling.

Figure 14.6 Contacts between subunits in F-actin. The helix axis is vertical. (A) Longitudinal contacts. Contact residues in the blue subunit are colored yellow. From left to right these are residues (in rabbit actin) 241, 244–245, 200, 204–205, 62–64, and 40–48 (the DNase I-binding loop). The interacting residues in the green subunit are colored red and are 324, 290, 291, 286–288, 139, 143, and 166–171. (B) Lateral contacts. Parts of two long-pitch helices are shown: yellow and green subunits on the left, and red and blue subunits on the right. The 'plug' of the blue subunit is arrowed. The yellow residues in it (267–270) interact with residues 39–40 (magenta, part of the DNase I-binding loop) and 171–173 (red) in subunits in the neighboring long helix. Further up, residues 194–196 in the blue subunit (yellow) contact residues 110 and 113 (red) in the subunit.

Figure 14.5 Structure of F-actin. (A) Left: Ribbon diagram for nine subunits in different colors. Right: cryo-EM envelope of F-actin at 16 Å resolution with the atomic structure of the subunit enclosed within it. (B) Cryo-EM structure at 8 Å resolution with the atomic structure of subunits, as obtained by molecular dynamics refinement of the G-actin subunit, inset. (C) Comparison of G-actin and the F-actin subunit. G-actin (PDB 1J6Z) is in gray and the F-actin subunit in pale magenta. The orientation is as in Figure 14.4A. (B, from T. Fujii et al., *Nature* 467:724–729, 2010. With permission from Macmillan Publishers Ltd.)

(A)

(B)

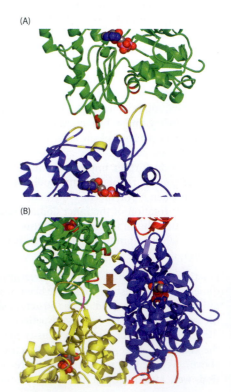

TABLE 14.1 PROPERTIES OF SOME ACTIN-BINDING PROTEINS.

ABP	Molecular mass (kDa)	Function	Structure	Reference
Formin	50–127	Barbed-end capper; nucleates filament growth	Large family of multidomain proteins having in common a 400-aa formin homology 2 domain	
Profilin	14–16	Binds to and sequesters G-actin	7-stranded β sheet + α helices	Section 14.9
Talin	225	Anchoring actin cytoskeleton to focal adhesions	FERM domain plus multiple four- and five-helix bundles	Section 11.6
Vinculin	116	Anchoring actin cytoskeleton to focal adhesions	Figure 11.28	Section 11.6
Arp2/3	15–40	Nucleates outgrowth of branched filaments in lamellipodia	Complex of seven subunits, totalling 225 kDa. Subunits Arp2 and Arp3 are actin mimics	Section 14.6
Calponin	29–34	Regulation of contractility and cytoskeleton organization in smooth muscle	Figure 14.55C (a calponin-homology domain)	
Fimbrin	63	Cross-linking filaments; member of calponin-homology family	Figure 14.7B	Section 14.2
α-Actinin	103	Cross-linking actin filaments in Z-line of muscle and other contexts	Elongated antiparallel dimer. Figure 14.23	Section 14.5
Tropomyosin	~40	Filament stabilization; Ca^{2+}-dependent regulation of muscle	Dimeric coiled coil; spans 7 actin subunits along long-pitch helices. Figure 14.28	Section 14.5
Nebulin	600–900[a]	Tape-measure; specifies length of thin filaments in muscle	185 35-residue repeats, predominantly α-helical	Section 14.7
α-Catenin	95	Connects F-actin to adherens junctions	Forms dimers; three main domains, C-terminal domain is an α-helical bundle similar to vinculin ABD	Section 11.4
Gelsolin	80	Filament severing and barbed-end capper	6 tandem homotypic domains	
Scruin	110	Cross-linking filaments in acrosomal process	Two domains, possibly β-propeller-like. Figure 14.72	Section 14.10

[a]Molecular mass and number of 35-residue repeats vary in different muscles. Human nebulin has 185 repeats.

Other ABPs include capping proteins that bind specifically to filament ends (for example, gelsolin), and proteins that sequester G-actin and thus augment the monomer pool (for example, profilin). The tropomyosins are a conserved family of coiled-coil ABPs that bind

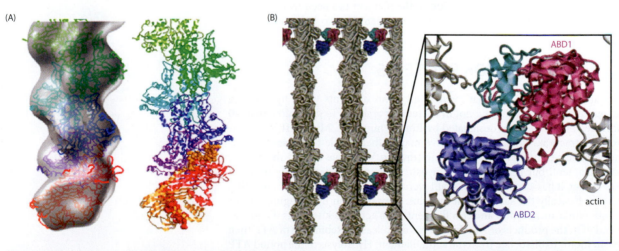

Figure 14.7 Interactions of ABPs with F-actin. (A) Binding of a 30 kDa capping protein (CP) to the barbed end. CP is shown in orange and red. Actin subunits are other colors. Left, cryo-EM-derived molecular envelope, with atomic structures fitted in. Right, the resulting model. (B) Cross-linking of F-actin by fimbrin. The high-resolution structure of fimbrin was fitted into an EM density map. ABD, actin-binding domain. (A, from A. Narita et al., *EMBO J.* 25:5626–5633, 2006. With permission from John Wiley & Sons; B, from N. Volkmann et al., *J. Cell Biol.* 153:947–956, 2001. With permission from Rockefeller University Press.)

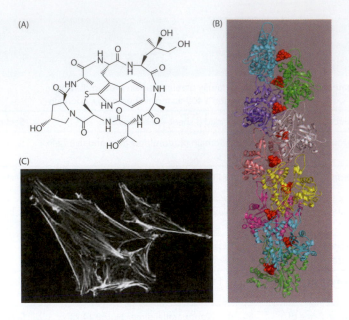

Figure 14.8 Actin filaments with phalloidin bound. (A) Chemical structure of phalloidin. (B) Phalloidin (shown in red) binds in the cleft between the two long-pitch helices, next to the hydrophobic plug, stabilizing F-actin. (C) Fluorescent light micrograph of actin filaments labeled with rhodamine-coupled phalloidin. (C, from H. Nakagawa et al., *J. Cell Science* 114:1555–1565, 2001. With permission from Company of Biologists.)

all along the filament, stabilizing it against spontaneous or ABP-mediated depolymerization. Tropomyosin also helps to regulate the interaction of myosin with actin in striated muscle (Section 14.5).

Some ABPs change the helical periodicity of F-actin. The largest change so far observed is that induced by cofilin, which binds F-actin cooperatively by bridging two longitudinally associated subunits. This interaction alters the twist of the actin helix from –166° per subunit to –162° so that cofilin-decorated filaments have shorter crossovers (27 nm versus the usual 36 nm). Cofilin binding destabilizes the F-actin helix primarily by weakening the longitudinal bonds.

Other ligands than proteins can affect actin filament assembly. *Latrunculin*, a sponge-derived toxin that blocks polymerization, has been used in several crystallographic studies of G-actin. *Cytochalasin A* is a fungal metabolite that caps the barbed end, promoting disassembly. It readily permeates into cells and is widely used in cell biology studies. Another ligand is *phalloidin*, a toxic bicyclic heptapeptide derived from a poisonous mushroom (**Figure 14.8**). Phalloidin prevents filaments from depolymerizing by trapping the subunits in the F-actin conformation. It is also an important reagent in cell biology because a rhodamine dye can be attached to phalloidin; this compound renders actin filaments fluorescent and hence visible by light microscopy. Its position in the filament has been worked out by analyzing fiber diffraction patterns: it lies in the cleft between the two long-pitch helical strands.

14.3   THE MYOSIN MOTOR PROTEINS

Myosin motors are *mechanoenzymes* that move along actin filaments while hydrolyzing ATP. They are found in all eukaryotes but not in bacteria or archaea. To date, more than 20 subfamilies have been identified and designated by Roman numerals. Myosin II is responsible for muscle contraction. Other myosins drive many other motile processes, such as *cytokinesis* and vesicular transport. All except one (myosin VI) move toward the barbed end of the actin filament, but it appears that myosin VI is not fundamentally different from the others: rather, it has a special insert that reverses its polarity. The force-generating cycle appears to be basically the same for all myosins: they function by swinging a **lever arm** through ~60° while anchored to an actin filament. This action is known as the **power stroke**; at the end of it, the products of ATP hydrolysis are released. Rebinding of ATP then causes the myosin motor to dissociate from the actin filament. Hydrolysis of the bound ATP enables the lever arm to switch back to its original position, allowing the myosin to bind once again to actin, starting a new cycle.

Myosin II was the first member of the superfamily to be discovered. In 1864, Wilhelm Kühne reported obtaining an 'Eiweiß' (in German, 'egg white,' now a generic term for protein),

(A)

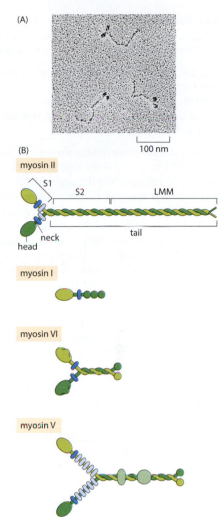

Figure 14.9 Myosin molecules. (A) Electron micrograph of rotary-shadowed myosin II, showing the tail and two heads. (B) Diagrams of myosins II, I, VI, and V. All except myosin I are dimers. The two ovals on the neck of myosin II represent the essential light chain (blue) and the regulatory light chain (gray). Light chains on the other myosins are also represented by ovals. Myosin II may be cleaved into subfragments S1 and S2 and light meromyosin (LMM). S2 and most of LMM are coiled coils. A flexible region at the S2–LMM junction gives rise to the kinks seen in (A). Polymerized LMMs form the myosin filament backbone. (Adapted from R. Craig and J.L. Woodhead, *Curr. Opin. Struct. Biol.* 16:204–212, 2006. With permission from Elsevier.)

which he called myosin, by extracting minced beef with salt. It was one of the first proteins to be studied. Nearly a century later, myosin was found to have ATPase activity. Shortly afterward, Albert Szent-Györgyi and Bruno Straub showed that Kühne's myosin preparation also contained another protein, which they called actin. They further found that purified actin and myosin can form tiny threads that contract when ATP is added. This experiment posed a serious challenge to the vitalistic theory that saw animal locomotion as a defining attribute of life, thought to transcend the inanimate laws of physics and chemistry.

In the latter part of the twentieth century, researchers began to discover myosin variants in other tissues. Some of these 'unconventional' myosins were dimers with two heads, like myosin II; others were monomers with a single head. All have tails (**Figure 14.9**). Appraisal of their amino acid sequences showed that the heads are highly conserved, whereas the tails vary widely in size and sequence. New members added to the superfamily were classified according to their structural properties, primarily sequence similarity in the head domain. The tail attaches to the cargo, be it a myosin filament as with myosin II, or a membrane vesicle as with myosin I. Each myosin interacts with a specific set of cargoes, while retaining the same force-generating machinery in its head(s). In some cases, such as myosin II, paired heads function independently; in other cases, such as myosin V, the activities of two heads are coordinated and their transport is *processive.*

Myosin heavy chains have three domains—head, neck, and tail

All myosins have one or two heavy chains of 150 kDa, each associated with one or more light chains (15–20 kDa) that are calmodulin-like (calmodulins are a family of regulatory Ca^{2+}-binding proteins). Myosin II is a dimer with two heavy chains and four light chains (Figure 14.9B). Proceeding from the N terminus of the heavy chain, we encounter the following components:

- The head, containing the ATPase, also known as the *motor domain*, the term mainly used in this account; or *cross-bridge*, in the context of muscle. Crystal structures of the motor domain have been determined for myosins I, II, V, and VI. All consist of a 7-stranded β sheet surrounded by α helices (see below).
- The neck or *lever arm*. This consists of an α helix whose length varies in different myosins. It contains from one to six IQ motifs, defined by the consensus sequence IQX_3RGX_3R where X can be any amino acid. Each IQ motif binds a light chain.
- The tail. In dimeric myosins, the subunits pair by forming a coiled coil in the tails; that of myosin II is exceptionally long (160 nm).

To illustrate the structural diversity of myosins, the domain organizations of myosins I, V, and VI are compared in Figure 14.9B. All three of these proteins engage in vesicular transport. Myosin I is a monomeric myosin whose tail consists of several globular subdomains. Myosin V is a dimer, with an exceptionally long neck to which six light chains bind, that moves processively by a 'hand-over-hand' mechanism. Myosin VI is a dimer with a short coiled coil in its tail and one light chain per neck.

The first crystal structure of a myosin motor domain depicted subfragment 1 from chicken myosin II without bound nucleotide

Motor domains may be isolated from rabbit muscle as a proteolytic product called subfragment 1 (S1). S1 is an actin-activated ATPase whose competence to generate motility is demonstrated by its ability to transport actin filaments in *in vitro* assays (see below). In the absence of nucleotide, S1 binds tightly to F-actin, yielding 'decorated' actin (see Figure 14.2C), also called **rigor complexes** (from the post-mortem condition, *rigor mortis*). Many biochemical and structural studies have addressed S1 and its interactions with actin.

Further proteolysis at exposed loops splits S1 into three fragments, named for their apparent molecular masses: 25 K (N-terminal), 50 K (middle), and 20 K (C-terminal) (**Figure 14.10**). These fragments remain associated under non-denaturing conditions, and cleaved S1 retains its ATPase activity. The 50K domain consists of two subdomains, named upper and lower.

The crystal structure of S1 is shown in **Figure 14.11**. It depicts an elongated molecule composed of a central β sheet together with numerous α helices that form a deep cleft at one end. This cleft separates the upper and lower 50K domains; both are involved in actin binding, which is accompanied by closure of the cleft. The ATP-binding site, lying close to the apex of the cleft, consists of a *P-loop* motif (see Box 3.1), flanked by the so-called Switch 1 (SW1) and Switch 2 (SW2) segments. Distal to the ATPase is the 'converter' domain that rotates through ~60° during the force-generating power stroke. This domain is the anchor for the lever arm that amplifies structural changes induced in the motor domain by ATP binding and product release. It is held in place by two interacting helices: the outer end of the relay helix and the SH1 helix.

Flexible loops (not seen) connect site A with site B (loop 1) and site C with site D (loop 2). The N-terminal part of loop 2, which is highly charged and partly ordered, forms part of the actin-binding site. The length and composition of loop 1 has considerable influence over the binding of nucleotides.

Myosin II has an N-terminal extension of ~80 residues containing an SH3-like subdomain (see **Guide**). It appears to be needed for communication between the functional regions within the motor domain.

A four-state cycle describes ATP hydrolysis, actin binding, and force generation

Myosin alone is not a good ATPase; rather, the motor is a product-inhibited ATPase whose activity is stimulated by binding to actin, which serves as a nucleotide exchange factor. In 1971, Richard Lymn and Edwin Taylor proposed a four-state cycle to explain the kinetic properties of actin/myosin complexes. (Here, 'state' refers both to nucleotide binding status and the corresponding structure; bold numbers refer to **Figure 14.12**.) Originally proposed for muscle myosin II, this cycle appears to apply to all myosins.

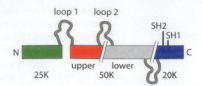

Figure 14.10 The three domains of myosin S1. SH1 and SH2 are conserved reactive thiol groups. SH1 is often used as a site for anchoring fluorescent markers.

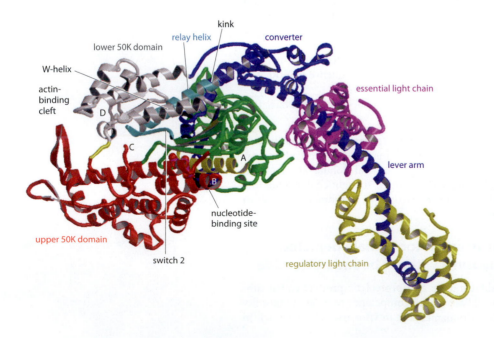

Figure 14.11 The myosin motor domain. The molecule is shown in the orientation it would take on binding to an actin filament, as viewed from the pointed end. The N-terminal region is green, and the P-loop and adjoining helix are dull yellow. The lower 50K domain (gray) is the primary actin-binding site. The disordered loop 1 (unseen, between points A and B) marks the N-terminal boundary of upper 50K. Upper and lower 50K are also connected by another disordered loop (loop 2) between points C and D. The long C-terminal helix (dark blue) carries two light chains. In this conformation (post-rigor state; Figure 14.12), the lever arm is in the 'down' position, as in rigor. The coloring follows subdomain boundaries. For clarity, the proximal end of the relay helix is shown light blue although it is part of the lower 50K domain. The distal end (beyond the kink) is firmly attached to the converter domain. In the post-rigor structure, the relay helix is straight (no kink). (PDB 2MYS) (From M.A. Geeves and K.C. Holmes, *Adv. Prot. Chem.* 71:161–193, 2005. With permission from Elsevier.)

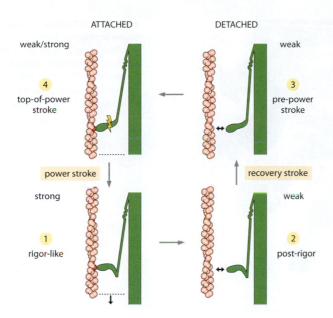

Figure 14.12 The four-state cycle of Lymn and Taylor, with swinging lever arm. Binding of ATP to actin-bound myosin (state **1**) leads to rapid dissociation of the motor domain but without hydrolysis of ATP, giving state **2**. The motor domain then performs the recovery stroke, putting the lever arm in its 'up' conformation in state **3**. This form is the ATPase. Subsequent rebinding to actin (state **4**) leads to product release and the moving 'down' of the lever arm as the motor switches to (**1**) in the power stroke. (Adapted from M.A. Geeves and K.C. Holmes, *Adv. Prot. Chem.* 71:161–193, 2005. With permission from Elsevier.)

(**1**) In the absence of nucleotide, motor domains bind tightly to actin, forming rigor complexes. *In vivo*, ATP is abundant, so this state—which is reached at the end of the power stroke—is short-lived. Because the nucleotide-free state reached *in vitro* may not be identical to that obtained *in vivo*, it is referred to as 'rigor-like.'
(**2**) On binding ATP, the motor rapidly dissociates from the actin filament.
(**3**) It then hydrolyzes ATP, retaining the products, ADP.P_i. This reaction primes the motor domain to change shape (**2**→**3**) in the **recovery stroke** that enables subsequent reattachment to actin.
(**4**) Binding to actin causes the motor to change its shape back again, thereby translocating the cargo (here, a myosin filament). This crucial transition (**4**→**1**) is the power stroke; at the end of it, ADP is released and the motor switches to state **1**. Another ATP molecule can now bind, causing rapid release of the motor from the actin filament, and the cycle starts over again.

In some contexts, it is convenient to distinguish between the 'weak' binding states of myosin (**2**, ATP bound, and **3**, ADP.P_i bound) and its 'strong' binding states (**4**, ADP bound, and **1**, no nucleotide).

Throughout the cycle, myosin has two roles: changing its shape; and, concomitantly, changing its affinity for actin. Both processes are controlled by ATP binding and hydrolysis. However, each myosin isotype proceeds through the cycle with its own kinetics. Myosins whose strongly bound states are relatively long-lived are said to have long *duty cycles*. Moreover, myosins vary in the force that they exert on the actin filament and in the step-length through which they translocate in the power stroke.

Myosin motors undergo nucleotide-dependent conformational changes

To date, several dozen structures of myosin motors (or fragments thereof) have been deposited in the Protein Data Bank. They represent several subfamilies and the binding of various nucleotide analogs. The Lymn–Taylor cycle affords a framework for assigning the observed conformations. Most fall into three classes that can be correlated with three of the four states in the cycle (namely **1**, **2**, and **3**). However, the crystal structures are all of myosin heads without actin, whereas two of the states are actin-associated. To describe them, actin/myosin complexes have been visualized by reconstructing cryo-electron micrographs and then fitting in crystal structures of G-actin and S1. A reconstruction of the rigor complex is shown in **Figure 14.13**.

The major changes undergone by the motor domain as it changes state affect (i) the position of the converter domain; (ii) the kink in the relay helix; (iii) the positions of SW1 and SW2; (iv) the actin-binding cleft—open or shut; (v) the position of the P-loop; and (vi) the twist of

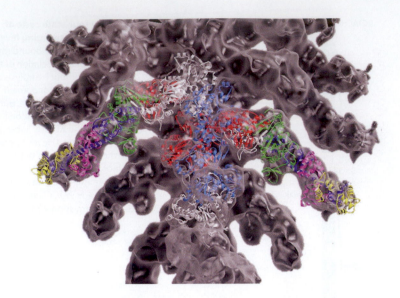

Figure 14.13 Decorated actin filament (rigor complex). A cryo-EM reconstruction of rabbit skeletal actin decorated with chicken S1 represented as a transparent surface into which two actin subunits (light blue) and two motor domains, color-coded as in Figure 14.11, have been fitted.

the central β sheet. These properties are not independent. SW1-open and actin cleft-closed are strongly coupled. The position of the P-loop and β sheet twist are also coupled, as are the kink in the relay helix and the position of the converter domain. The transitions at successive steps in the cycle may be summarized as follows.

Most crystal structures fall into two broad categories, depending on the position of the converter domain and lever arm, which rotate through 60° relative to the body of the motor. These categories may be equated with the detached motor in states **2** and **3**. The transition between these states is referred to as the *recovery stroke* or repriming of the cross-bridge. In state **2**, SW2 is 5 Å from the γ-phosphate of ATP; in state **3**, they make contact, allowing a H-bond to form between the amide group of a glycine residue and the γ-phosphate. Formation of this bond is an essential part of the enzymatic mechanism. In going from state **2** to state **3**, the lower 50K domain of which SW2 is part rotates through 5° about the W-helix, forcing the relay helix up against the β sheet. The relay helix responds by forming a kink near its midpoint that rotates the converter domain through 60° into its state **3** 'up' conformation (**Figure 14.14**).

In state **3**, the β sheet is held in the same conformation as in state **2** by the nucleotide bound in the active site, which constrains the P-loop and stops the twisting of the β sheet that occurs in the rigor-like state **1** (see below). The binding of ATP places the γ-phosphate into the pocket formed by SW1 and the P-loop, which promotes the inward movement of SW2. The effects of this movement are two-fold: the relay helix is kinked and thereby 'reprimed,' and the active site is formed. In state **3**, the nucleotide is completely enclosed by SW1 and SW2. This conformation seems to be the enzymatically active form of myosin.

(A) state 2   (B) state 3

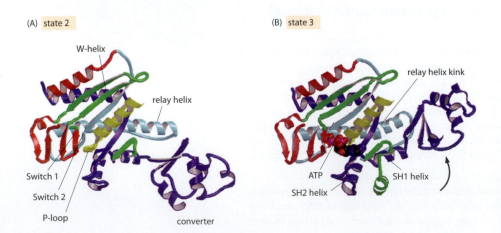

Figure 14.14 Conformational changes in the recovery stroke. This is the transition from state **2** (A, post-rigor) to state **3** (B, pre-power stroke). Movement of SW2 causes a kink to form in the relay helix (light blue) that rotates the converter domain (dark blue) through 60°. (From M.A. Geeves and K.C. Holmes, *Adv. Prot. Chem.* 71:161–193, 2005. With permission from Elsevier.)

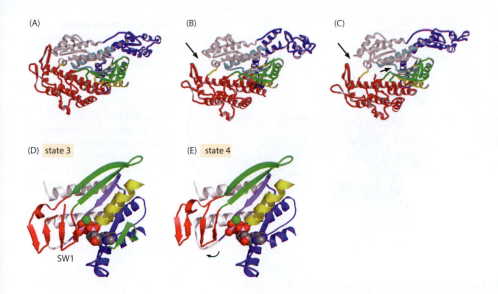

Figure 14.15 Closure and opening of the actin-binding cleft. (A–C) Here the motor is viewed along the actin helix, as in Figure 14.11A. The actin-binding cleft is between the upper (gray) and lower (red) 50K domains. In (A) (strongly bound; state **1**) the cleft is shut. In (B) (weakly bound; state **3**) its outer end is fully open (arrow) but the inner end, next to the nucleotide-binding pocket, is closed. Closure is brought about by SW2 being in its 'closed' conformation. In (C) (weakly bound; state **2**), both the outer and the inner end are open (arrows). (D, E) Strong binding to actin opens the nucleotide-binding pocket. In so doing, SW1 is moved away from the γ-phosphate. (D) State **3**; (E) strongly bound state **4**. As shown, state **4** was modeled from crystal data on state **1**. This view is at right-angles to that shown in (A). (From M.A. Geeves and K.C. Holmes, *Adv. Prot. Chem.* 71:161–193, 2005. With permission from Elsevier.)

The binding of myosin to actin and its binding of nucleotides are mutually antagonistic

This is one of the most distinctive properties of myosin. ADP release from myosin is accelerated 10^4-fold by its binding to actin. Conversely, ATP binding to rigor complexes causes rapid dissociation of the myosin. Structural studies and fluorescence experiments have indicated that communication between the actin-binding and nucleotide-binding sites involves the opening and closing of the cleft between the upper and lower 50K domains. Cryo-EM reconstructions have demonstrated that the 50K cleft shuts on strong binding to actin. Its shutting involves a local rotation of the upper 50K domain. This movement results in the opening of the ATP-binding site (**Figure 14.15**), thus explaining the reciprocity of ATP binding and actin binding.

As nucleotide-free myosin V is constitutively in the strong-binding form, a crystal structure for the myosin V motor appears to be a good model for the rigor-like state **1** reached at the end of the power stroke.

Between states **1** and **2**, the β sheet twists through ~8°. One effect of this twist is to displace the P-loop into an open environment where nucleotide exchange would be rapid. As **1** is an actin-bound state, the implication is that actin facilitates the release of ADP and phosphate. Conversely, rebinding of ATP would force the P-loop back into its former position in state **2** and bring about untwisting of the β sheet. When state **2** is compared with state **1**, SW1 moves so as to enclose the polyphosphate moiety of the ATP. As SW1 is embedded in the upper 50K domain, this movement opens the actin-binding cleft, producing a weak binding state (see Figure 14.15A,B). The recovery stroke, states **2** to **3**, then follows. Then, on rebinding to actin (state **4**), the actin-binding cleft should close and SW1 should open, exposing the nucleotide to solvation (see Figure 14.15D,E), leading to loss of the γ-phosphate.

The ephemeral nature of this top-of-power-stroke state **4** complicates structural analysis, and the following account was obtained by modeling. The geometry of the motor's attachment to F-actin is assumed to be the same as in rigor, as determined by cryo-EM, although the lever arm is differently oriented. The resulting structure is compared with the rigor-like conformation (state **1**) in **Figure 14.16**.

The power stroke is the transition from top-of-power stroke state **4** to rigor-like state **1**. At present, it is best modeled by looking at the recovery stroke and imagining it running backward. The real power stroke may differ because it takes place while myosin is bound to actin. The position of the converter domain depends on whether or not the relay helix is kinked. Removing the kink causes the lever arm to rotate by 60° from the 'up' conformation to the 'down' conformation; this is the pivotal structural event in the power stroke. It is brought about by binding to actin, which may cause the β sheet to twist, favoring the unkinked form of the relay helix. The position of the converter domain is stabilized by the interaction of the outer end of the relay helix with the SH1 helix. These two helices rotate together as the converter domain rotates.

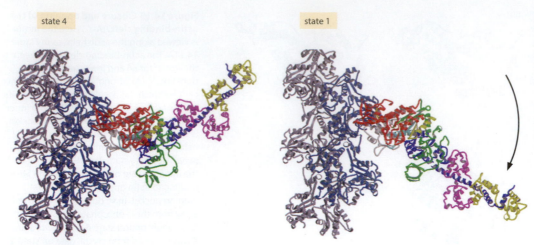

state 4   state 1

Figure 14.16 The power stroke. Models are shown for the strongly bound states **4** and **1**. In each case, the actin filament is shown in blue and gray on the left. The myosin subdomains are colored as in Figure 14.11. Because the crystallography was performed on truncated motors lacking most of the lever arm, the lever arm was taken from the chicken myosin II structure and assumed to be in the same position as in state **3**. (From M.A. Geeves and K.C. Holmes, *Adv. Prot. Chem.* 71:161–193, 2005. With permission from Elsevier.)

In vitro motility assays support the swinging lever arm model of contraction

Initial evidence for conformational changes in actin-bound myosin came from X-ray diffraction studies of insect flight muscle and led to a model of contraction that envisaged the rotation of whole motor domains. A firmer basis for mechanistic proposals was established when a detailed model of S1-decorated actin was obtained. It showed that the neck region, carrying the light chains, was offset from the actin filament in a position that would make it an excellent lever arm. Moreover, spectroscopic studies indicated that most of the motor does not move during the contraction cycle. Together, these observations led to the **swinging lever arm hypothesis**, which envisages that the actin–myosin interface remains unchanged and movement is limited to a swinging of the lever arm that 'rows' the actin filament past the myosin tail or, otherwise viewed, shifts the myosin tail and cargo relative to the actin filament.

This idea was tested with *in vitro* motility assays, using genetically engineered myosins. In such assays, the movement of individual actin filaments across a lawn of myosin molecules attached to a substrate is monitored by fluorescence microscopy (**Figure 14.17**). In

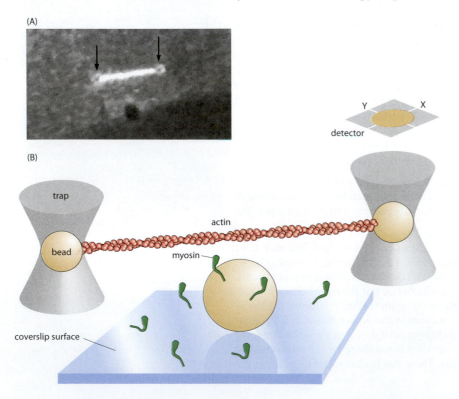

(A)

(B)

trap
actin
bead
myosin
coverslip surface
detector
Y   X

Figure 14.17 Single-molecule mechanics. (A) Micrograph of an actin filament rendered fluorescent by binding rhodamine-labeled phalloidin (see Figure 14.8). (B) Schematic of an actin filament attached to two polystyrene beads that are held in the focal points of two intense infrared beams. (Individual protein molecules are too small to be captured in this way.) A third polystyrene bead serves as an anchor for myosin heads. The myosin is brought up to the tethered actin. Contact between actin and the myosin head reduces the Brownian motion of the filament dramatically. The force and displacement produced by binding a myosin head can then be measured. (From M.J. Tyska and D.M. Warshaw, *Cell Motil. Cytoskel.* 51:1–15, 2002. With permission from John Wiley & Sons.)

the presence of ATP, actin filaments are transported with a speed that is characteristic of the myosin in question. The speed of transport was found to be proportional to the length of the lever arm, consistent with the lever arm hypothesis. Furthermore, this experiment allowed a determination of the distance from actin of the lever arm's axis of rotation. The value obtained was consistent with position of the axis of rotation as inferred from the pseudo-atomic model; that is, far from the surface of actin. On the basis of these and other observations, the lever arm mechanism is now widely accepted.

The polarity of a myosin motor can be reversed by re-orienting the converter domain through genetic engineering

Most myosins move toward the barbed end of F-actin. However, myosin VI was found to move toward the pointed end. As this unorthodox polarity correlated with an unusual insertion immediately after the converter domain, it was suggested that the insertion might reposition the lever arm so as to point at 180° to the normal direction. This idea was tested, successfully, by reversing the polarity of myosin I (which normally moves toward the barbed end) with the insertion of a putative direction-inverter. This inverter consisted of a four-helix bundle from human guanylate-binding protein-1 and a lever arm of two α-actinin repeats. In motility assays *in vitro*, this construct did indeed move toward the pointed end, demonstrating that myosin movement can be reversed by simply inverting the lever arm without any modification to the basic mechanisms that cause the converter domain to swing (**Figure 14.18A**).

When the crystal structure of the myosin VI motor domain was determined, it showed that its lever arm is indeed re-oriented, as predicted. An insertion at the end of the converter domain binds a light chain to a different surface of the converter domain from that used in normal myosins. The effect of this altered interaction is to rotate the lever arm through ~180°.

The obverse polarity-switching experiment is to produce an artificial forward-moving motor from myosin VI. This has been done by removing the inserted sequence that binds the unusual light chain (namely the dark purple helix in Figure 14.18B) and substituting the lever arm of myosin V. This construct moves forward along actin filaments in much the same way as myosin V itself.

The forces generated by individual myosin motors and their step-lengths have been measured by optical trapping

For many years, physiological studies correlating force generation in muscle with structural events were conducted by applying force transducers to the ends of muscle fibers that were monitored simultaneously by X-ray diffraction and light microscopy. The fibers could be stimulated to generate force or subjected to other treatments. Inevitably, such measurements reported averages over millions of unsynchronized myosin molecules—some attached, some not. More recently, the invention of *optical traps* in which actin filaments are tethered to polystyrene beads have made it possible to measure the force and displacement

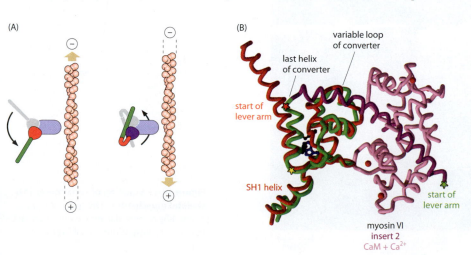

Figure 14.18 Inversion of the power stroke in myosin VI. (A) The motor domain is shown as a short blue cylinder, the converter domain and lever arm as a ball and stick. Left, normal movement of the lever arm; right, movement of the modified lever arm, which has an insertion that turns it through 180°. Now the swing of the converter toward the pointed (minus) end of F-actin causes the end of the lever arm to move toward the barbed (plus) end. (B) The converter regions of myosins VI and V are overlaid (myosin V is in red). The converter domain follows immediately after the SH1 helix and before the lever arm. The myosin VI converter (green) is extended by insert 2 (purple) plus a light chain (pink; CaM (calmodulin), with calcium ions as red balls), which redirect the lever arm helix. To alllow comparison, the SH1 helix and converter of myosin V (red) are overlaid. (A, from G. Tsiavaliaris et al., *Nature* 427:558–561, 2004. With permission from Macmillan Publishers Ltd; B, from H. Park et al., *Proc. Natl Acad. Sci. USA* 104:778–783, 2007 National Academy of Sciences. With permission from PNAS.)

(step-length) arising from the interactions of a filament with individual myosin molecules (see Figure 14.17). These 'single-molecule' experiments have determined that the average force exerted by a tethered cross-bridge varies among different myosins, from ~2 pN to ~5 pN. In another kind of single-molecule experiment, individual actin filaments rendered fluorescent by a dye have been visualized while being transported over a 'lawn' of immobilized myosin molecules.

The step-lengths are also dependent on myosin type and, as described above, are roughly proportional to the length of the lever arm. The myosin II motor has a step-length of ~10 nm, which agrees well with the axial shift of the tip of the lever arm between its two extreme positions in the ATPase cycle (see above and Figure 14.16). Myosin V, in contrast, has a step-length of 35 nm, which matches an axial repeat (13 subunits) of the actin filament, so that this motor—which is processive and capable of taking more than 100 steps per run—can move axially along the actin filament without needing to spiral around it. More detailed analysis has shown that this 35 nm step actually has two components—one, a displacement of 20–25 nm generated by the power stroke and another, of 10–15 nm resulting from a biased diffusional search by the detached cross-bridge for its next actin-binding site.

14.4 FORCE GENERATION IN MUSCLE

All myosin molecules operate as mechanoenzymes, employing the energy released by ATP hydrolysis to generate nanoscale forces whose application accomplishes a wide variety of tasks. However, it is only in muscle that the activities of millions of myosin motors are coordinated—geometrically and in terms of regulation—to generate forces on a macroscopic scale. A **muscle fiber** is a bundle of many **myofibrils**, each containing a bundle of actin and myosin filaments within a single cell. In **striated muscle**, the myofibrils are aligned in register in units called **sarcomeres**. Viewed by light microscopy (**Figure 14.19**), the sarcomeres present striations running perpendicular to the fiber axis.

They are spaced at intervals of ~2.5 μm (the actual value depends on the state of contraction of the muscle). Thus, a typical human fiber 25 cm long contains ~10^5 serially arranged sarcomeres. Within a sarcomere, the filaments are oriented perpendicular to the visible striations; that is, parallel to the axis of the fiber. Thus a muscle is a sophisticated and highly coordinated machine for generating force, whose structure is organized on multiple hierarchical levels.

The nature and significance of the cross-striations, which were first observed some 300 years ago (**Box 14.1**), remained enigmatic until 1954 when studies by Andrew Huxley and Rolf Niedegerke and by Jean Hanson and Hugh Huxley showed that sarcomeres are built of two kinds of filaments, thick and thin. The thick filaments are composed of myosin and a supporting cast of binding proteins (see below and Section 14.5), together with the thin filaments of actin and a set of actin-binding proteins (Section 14.2 and below). Longitudinal

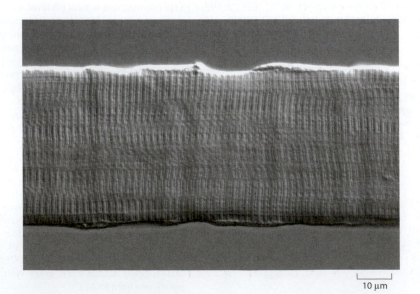

10 μm

Figure 14.19 Stacking of sarcomeres in a skeletal muscle fiber. The specimen was imaged by differential interference contrast light microscopy. (Courtesy of Evelyn Ralston.)

Box 14.1   Research on muscle contraction has deep roots

Theorizing about the mechanism(s) that generate force in muscles goes back more than 2000 years. One of the philosophical questions posed in classical times was the origin of animal locomotion, the *spiritus animalis* that was thought to be an intrinsic property of life. Erasistratus (third century BC) of the Alexandrian school associated the *spiritus animalis* with the muscles. The *spiritus* or *pneuma* was thought to course along the nerves and make the muscles swell and shorten. At the beginning of the second century AD, Galen, the last classical physiologist, expanded these ideas, introducing a primitive metabolism involving the four humors (blood, yellow bile, black bile, and phlegm). Galen also made a detailed anatomical examination of muscles and understood that they worked in antagonistic pairs, and that the heart was a muscle for pushing blood into the arteries. In the ensuing millennium, nothing much happened and even the classical physiological insight that muscles pull rather than push seems to have been forgotten because at the beginning of the sixteenth century, on the basis of his own anatomical examinations, Leonardo da Vinci felt compelled to write "*perchè l'ufizio del muscolo è di tirare e non di spingere*" ("because the action of muscle is to pull and not to push").

Fifty years later, Vesalius coined the name *Machina Carnis* (the machine of flesh), recognizing that the production of force resided in the flesh (muscle) itself. In the early seventeenth century, Descartes proposed a neuromuscular machine not unlike that of Erasistratus: the nerves were to carry a fluid from the pineal gland (which was also thought to be the seat of the soul) to the muscles, which made them swell and shorten. Soon afterward, Swammerdam showed that muscles contract at constant volume, which invalidated this whole class of pneumatic theories. However, other mechanical models were soon proposed. On an observational level, an epochal discovery was made by the microscopist van Leeuwenhoek, who was able with his simple instrument to detect the myofibrils and cross-striations in muscle fibers. His drawings (**Figure 14.1.1**) clearly showed 'globules' that we would now identify as sarcomeres (see Figures 14.19 and 14.20). Considering that sarcomeres are only ~2.5 μm long, this was a remarkable accomplishment. On the basis of van Leeuwenhoek's observations, Croone suggested that the 'globules' may serve as units of contraction.

Alongside such mechanical thinking, vitalism, maintaining that animal movement was an intrinsic property of life that transcends the laws of physics and chemistry, survived into the 20th century. Finally, given the advent of electron microscopy (see **Methods**) and X-ray diffraction to establish a clear structural basis, the whole fabric of metabolic biochemistry and thermodynamics combined to establish the concept that muscle is a chemical machine driven by the breakdown of ATP. However, the sliding filament hypothesis (see Figure 14.22), once formulated, did not find immediate acceptance. The view then current was that myosin was a long negatively charged polypeptide without much structure (an early and, in this case, erroneous appearance of the 'natively unfolded' concept (Section 6.4)) that shortened on the addition of Ca^{2+}. The fact that there was no evidence to support this model did not detract from its widespread acceptance. It was argued that the cross-striations of skeletal muscle could not be of great significance because smooth muscle contracted without even having them. However, electron microscopy removed any lingering doubts by demonstrating that, when cross-striated muscle contracts, sets of interdigitating filaments slide past each other without either altering their lengths significantly.

Reference

Holmes KC (2004) Introduction to a Discussion Meeting on myosin, muscle, and motility. *Phil Trans R Soc B* 359:1813–1818 (doi:10.1098/rstb.2004.1581).

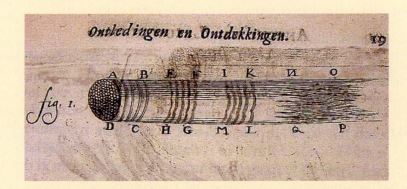

Figure 14.1.1 Substructure of skeletal muscle fibers as visualized by 17th-century microscopy. Drawing by A. van Leeuwenhoek in 1682. The transverse striations demarcate the sarcomeres. The lateral packing of fibrils was also observed. The legend reads "Dissections and Discoveries" in Dutch.

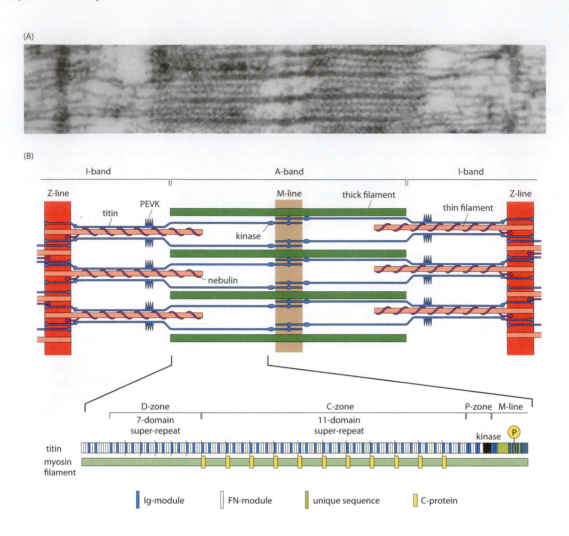

sections through myofibrils showed 'cross-bridges' connecting the thick and thin filaments (**Figure 14.20A**). These are now known to represent myosin motor domains.

Thick and thin filaments are packed in hexagonal arrays

Low-angle X-ray fiber diffraction patterns showed that the sarcomere contains hexagonally packed filaments, as equatorial reflections were found at spacings in ratios of 1:√3:2, and so on, as in a hexagonal lattice. This was confirmed by electron microscopy of transverse sections that revealed hexagonal arrays of thin and thick filaments. The thick filaments, composed mainly of myosin, are positioned at the hexagonal lattice points. The thin actin-containing filaments occupy its three-fold axes in mammals, birds, and fish or its two-fold axes in insects and other arthropods (**Figure 14.21A,B**). Transverse thin sections of skeletal muscle preserved in vitreous ice are shown in Figure 14.21C,D from regions of overlap (containing both thick and thin filaments) and non-overlap (containing only thick filaments).

Muscles contract and force is generated by a sliding filament mechanism

Force is generated when a muscle contracts. The investigators who first detected the thick and thin filaments also proposed that, to generate a contraction, the two kinds of filaments move past each other without altering their lengths. This is the now well-validated **sliding filament mechanism** (**Figure 14.22**).

The sarcomere is the contractile unit of a myofibril. Sarcomeres are demarcated at both ends by **Z-lines** (also called Z-disks; see Figure 14.20B). Actin filaments emanate with opposite polarities from both sides of a given Z-line. The bipolar thick filaments are held in register at the M-line. Thus filament polarity alternates every half-sarcomere, but neighboring

Figure 14.20 Molecular architecture of the sarcomere. (A) Electron micrograph showing cross-bridges as densities connecting the thick and thin filaments in rigor muscle. In this sectioning plane (see Figure 14.21B), two thin filaments are seen between each pair of thick filaments. (B) Diagram showing the thick-filament-containing A-band and the thin-filament-containing I-band. Thin filaments are anchored with their barbed ends in the Z-line, and thick filaments with their bare zones (overlap regions) in the M-line. Titin runs as a continuous strand from the M-line to the Z-line. There are thought to be six titin molecules per half-sarcomere. PEVK is a repeating tetrapeptide in an elastically extensible region of titin. (A, from H.E. Huxley, *J. Biophys. Biochem. Cytol.* 3:631–648, 1957; B, adapted from R. Craig and J.L. Woodhead, *Curr. Opin. Struct. Biol.* 16, 204–212, 2006 [top] and C.C. Gregorio et al., *Curr. Opin. Cell Biol.* 11:18–25, 1999 [bottom]. With permission from Elsevier.)

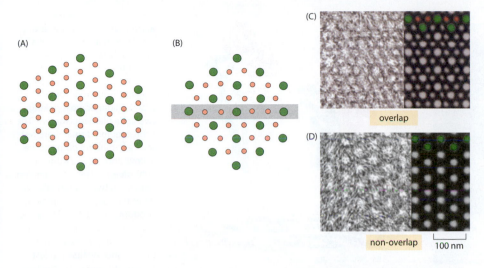

(A)   (B)   (C) overlap   (D) non-overlap   100 nm

Figure 14.21 Hexagonal arrays of thick and thin filaments. (A) Arthropod muscle; (B) mammalian skeletal muscle. The gray slab in (B) indicates the approximate thickness of the longitudinal section shown in Figure 14.20A. (C, D) Electron micrographs of transverse cryo-sections (frozen, hydrated) of rabbit skeletal muscle in overlap (C) and non-overlap (D) regions. Protein density is light. The specimen is in the relaxed state; in other words, ATP is plentiful and in (C) the myosin heads are not tightly bound to actin. The right-hand panels show images enhanced by averaging. Individual heads are not seen in (C), only averaged density distributions in which the heads are largely smeared out. (From B.L. Trus et al., *Biophys J.* 55:713–724, 1989. With permission from Elsevier.)

myosin and actin filaments always have the same relative orientation. Two giant proteins, **titin** and **nebulin**, serve as tape-measures that respectively specify the lengths of the myosin and actin filaments. The part of titin that does not overlap with myosin contains a series of elastic elements that are repeats of the PEVK motif. Titin also serves as a stress sensor via its C-terminal kinase domain, positioned close to the M-line. The thick filaments also contain regularly spaced C-proteins. Their function in skeletal muscle is not clear, but in heart muscle, C-proteins reduce the cross-bridge cycling rate and have a role in controlling contraction at low Ca^{2+} levels. Nebulin is described further in Section 14.2 and Table 14.1.

The Z-line serves as a scaffold that runs transverse to the myofibril axis. Actin filaments coming from both sides are anchored in this scaffold, as are the ends of titin molecules. Initially, the Z-line was viewed as a passive constituent of the sarcomere, serving only to coordinate the actin filaments and to transmit the force generated by the myosin filaments. However, the discovery of many additional components has implicated the Z-line in other functions—notably signaling (Chapter 12). The thickness, composition, and interactions of the Z-line vary considerably between different muscles. Nevertheless, one component generally present is desmin intermediate filaments (Section 11.2) that appear to play a coordinating role.

Another feature common to all Z-lines is the pairwise cross-linking of thin filaments by a rod-like α-helical protein called **α-actinin** (**Figure 14.23**). It is an antiparallel dimer of 103 kDa subunits made up of three domains: an N-terminal actin-binding domain; a central rod domain consisting of four 120-residue repeats; and a C-terminal calmodulin-like

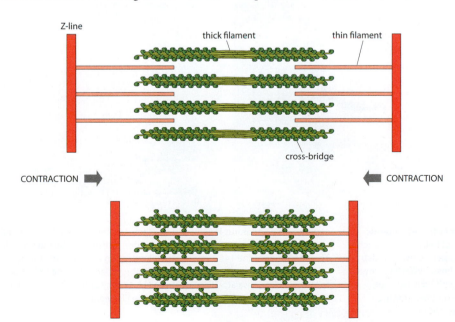

Z-line   thick filament   thin filament   cross-bridge   CONTRACTION   CONTRACTION

Figure 14.22 The sliding filament mechanism. In a contracting muscle, two sets of inextensible protein filaments move further into overlap, shortening the sarcomere.

(A)

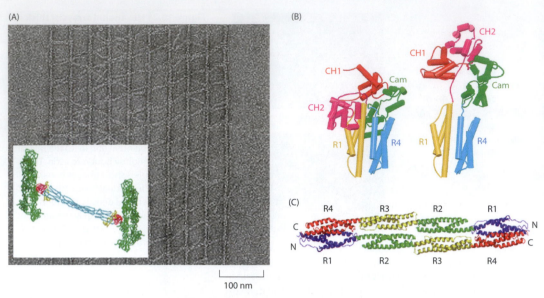

100 nm

(B)

(C)

Figure 14.23 Cross-linking of actin filaments by α-actinin. (A) Electron micrograph of a negatively stained array of actin filaments connected by α-actinin struts at various angles. Inset, model of an α-actinin dimer cross-linking two actin filaments (green). The actin-binding domains are red, the Ca^{2+}-binding domains are yellow, and the rod domains, light blue. (B) The actin-binding domain in two conformations. It has two calponin-homology subdomains (CH1 and CH2, red and magenta respectively), the Ca^{2+}-binding domain (Cam, green), and the first and last repeats (R1 and R4) of the two rod domains (yellow and blue). (C) The two antiparallel rod domains, consisting of four subdomains, each a bundle of three α helices (the spectrin repeat motif). (B, from J. Liu et al., *J. Mol. Biol.* 338:115–125, 2004 and C.M. Hampton et al., *J. Mol. Biol.* 368:92–100, 2007. With permission from Elsevier; C, from B. Sjöblom et al., *Cell. Mol. Life Sci.* 65:2688–2701, 2008. With permission from Springer Science and Business Media.)

Ca^{2+}-binding domain. Crystal structures have been determined for all three domains (or subdomains or homologs thereof) and these elements have been assembled into a model, guided by EM reconstructions of the complete molecule. Each rod domain repeat forms a three-stranded coiled coil, a motif that recurs in other members of the same superfamily, such as *spectrin* and dystrophin, mutations in which cause muscular dystrophy. In the 20 nm-long α-actinin dimer, there is almost complete overlap between the two subunits. The actin-binding domain is connected to the rigid rod domain by a hinge that allows considerable latitude in the cross-linking geometry. Its juxtaposition with a Ca^{2+}-binding domain suggests that cross-linking patterns in the Z-line are responsive to Ca^{2+} concentration.

Force generation is accompanied by structural changes in the myofibrils

Frog muscles yield detailed low-angle X-ray diffraction patterns with a prominent meridional reflection at a spacing of $(14.5 \text{ nm})^{-1}$, corresponding to the principal axial repeat of myosin molecules in the thick filament (**Figure 14.24A–C**). The distribution of intensity in the pattern changes when the muscle fiber is activated and, with intense synchrotron

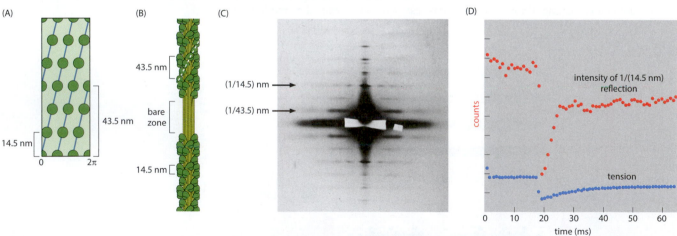

Figure 14.24 Helical organization of myosin filaments and dynamic studies of muscle activation. (A) Radial projection of helical lattice: each lattice point represents the position of a myosin II molecule (two heads). (B) Model of vertebrate thick filament, with pairs of heads following three coaxial right-handed helices (one is marked with a white dashed line). (C) X-ray diffraction pattern from relaxed vertebrate muscle, showing myosin layer-lines. (D) When an excised muscle mounted in an

experimental apparatus is activated by electrical stimulation, the resulting changes in the fiber tension and the diffraction pattern can be monitored. The intensities of the 14.5 nm meridional reflection (red) and the tension (blue) are plotted as a function of time. An abrupt fading of the reflection follows quick release at ~18 ms. (C, from H.E. Huxley and W. Brown, *J. Mol. Biol.* 30:383–434, 1967. With permission from Elsevier.)

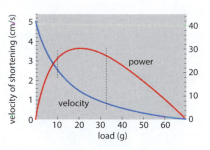

Figure 14.25 Relationship between load (P) and speed of shortening (v) for a frog muscle. Measurements were made for isotonic contraction at 0°C. The velocity (v, blue curve) depends on the (variable) load P according to an empirically derived relationship, the Hill equation $(P + 14.35)(v + 1.03) = 87.6$. The power developed (P.v) is given by the red curve (right-hand ordinate, in units of dynes). The region in which maximum power is developed is marked by the black dashed lines. (Based on the Hill Equation in A.V. Hill, *Proc. R. Soc. Lond. B* 126:136–195, 1938.)

radiation, these changes can be monitored with time resolution in the millisecond range (Figure 14.24D). Because unattached motor domains in relaxed muscle are not well ordered, they do not contribute coherently to the 14.5 nm reflection. However, its intensity increases in contracting muscle as a result of the motor domains (cross-bridges) attaching to actin and consequently becoming better ordered. The intensity of this reflection also depends on the cross-bridge tilt-angle: if they are at right angles to the filament axis, the 14.5 nm reflection is strong; as they depart from this orientation, it weakens. Thus time-resolved fiber diffraction allows cross-bridge dynamics to be monitored during a contraction. Computational modeling in three dimensions is then used to match the full diffraction pattern and thereby follow the molecular processes involved.

Muscle achieves ~50% efficiency as a force-generating machine

Muscle is a machine for turning chemical energy from the hydrolysis of ATP into mechanical work. At its best, muscle is nearly 50% efficient, with the remaining 50% of the energy being dissipated as heat. The force that a muscle develops is inversely related to the velocity of shortening (**Figure 14.25**), a relationship determined by Archibald Hill in 1938. This curve shows that the power output is maximized at intermediate speeds of shortening, namely 1–3 cm/s.

It was long assumed that the Hill equation comes about because faster-moving motor domains do not exert so much force. However, this appears not to be the case. The force decreases at high velocity because the fraction of interacting motors—determined by measuring the high-frequency stiffness of contracting muscle—goes down. By combining X-ray diffraction data on the intensity of the 14.5 nm reflection and on the calculated displacement of the centers of gravity of the motor domains, it has been possible to estimate the rotation of the lever arm, and in this way to conclude that myosin II motors execute a 6–7 nm power stroke, regardless of shortening velocity.

Analyses of this kind have led to a simple and self-consistent picture of the performance and efficiency of myosin motor domains as elemental contraction units. In each cycle, a motor hydrolyzes one molecule of ATP. The free energy of ATP hydrolysis under physiological conditions is 85 zJ/molecule (zJ = zeptoJ = 10^{-21} J). The efficiency of muscle is ~45% at high load, giving an available free energy of 38 zJ. These numbers are quite consistent with the work done by a myosin molecule, as determined by single-molecule force measurements (Section 14.3), on the basis of a constant 5 pN force applied throughout a stroke of 7 nm (work done, ~35 zJ).

Muscles are switched on by excitation–contraction coupling

Muscle activity is regulated by controlling the availability of Ca^{2+}. The arrival of an action potential at the axon terminal causes acetylcholine to be released into the synaptic cleft, whereupon it diffuses across the cleft and binds to nicotinic acetylcholine receptors in the motor endplate (Section 16.3). The receptor channels then open, resulting in a local depolarization and initiating an action potential that spreads across the surface of the muscle fiber into the transverse tubules (**Figure 14.26**). There it elicits the release of Ca^{2+} by opening voltage-sensitive Ca^{2+} channels, also called dihydropyridine receptors (DHPRs). This releases a pulse of Ca^{2+} that in turn causes the release of substantially more Ca^{2+} by binding to the calcium release channels of the sarcoplasmic reticulum—the ryanodine receptors (RyRs). Ryanodine is a poisonous plant-derived alkaloid that binds to the RyR with high (nanomolar) affinity, locking it in an open-like conformation. In humans, this interaction causes massive muscle contractions.

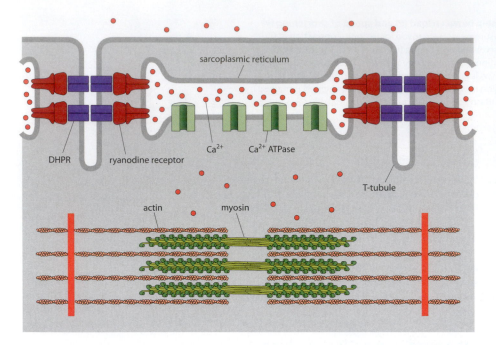

Figure 14.26 Excitation–contraction coupling in a skeletal muscle cell. When an action potential propagates into the transverse tubules (T-tubules), it induces a conformational change in the voltage-gated DHPR Ca^{2+} channels that causes them to open. In turn, Ca^{2+} (small red spheres) opens the RyRs so as to release more Ca^{2+} from the sarcoplasmic reticulum (SR) into the sarcoplasm (the cytoplasm of a muscle cell), where it initiates a contraction. The Ca^{2+}-ATPase (Section 16.6) pumps Ca^{2+} back into the SR. In skeletal muscle, the DHPR channels and the RyRs associate as tetrads (top left and right). (Adapted courtesy of Noriaki Ikemoto.)

RyRs are very large membrane proteins organized as tetramers of 560 kDa subunits. Their structure has been studied by cryo-EM of detergent-solubilized molecules in both 'open' and 'closed' conformations (**Figure 14.27A**). The tetramer has a mushroom-like shape whose stalk traverses the membrane and is thought to enclose a single Ca^{2+} channel. The cytoplasmic domains have an elaborate architecture and possess binding sites for a number of regulatory proteins, including calmodulin. DHPRs, in contrast, are funnel-shaped heteropentamers with α_1, α_2, β, γ, and δ subunits, totaling 430 kDa.

In skeletal muscle, the DHPRs are functionally coupled to the RyRs in structures called **calcium release units**. Their coupling requires the juxtaposition of four DHPRs with the four subunits of a single RyR. The four DHPRs linked to binding sites on RyR cytoplasmic domains define the corners of a square, known as a tetrad, which is revealed by freeze-fracture electron microscopy (Figure 14.27B). A model for a tetrad array is shown in Figure 14.27C. An increase in the Ca^{2+} level in the cytoplasm, mediated by RyRs releasing the ion from the sarcoplasmic reticulum, initiates a contraction by the binding of Ca^{2+} to troponin

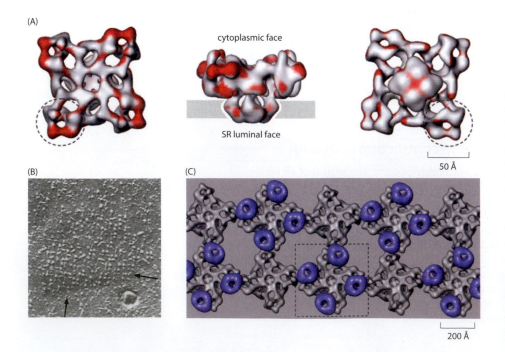

Figure 14.27 Complexes involved in calcium release into the cytoplasm from the sarcolemma. (A) Three views of a cryo-EM reconstruction of detergent-solubilized RyR. The regions that showed the greatest differences between the 'open' (100 mM Ca^{2+} and 100 nM ryanodine) and 'closed' (Ca^{2+} chelated with EGTA) states are shaded in red. (B) Freeze-fracture electron micrograph in which the fracture plane delineates tetrads of knob-like features that have been identified as individual DHPR receptors. (C) Model of a tetrad array of DHPRs (blue) bound to RyRs. (A and C, from I. Serysheva, *Biochemistry (Moscow)* 69:1226–1232, 2004; B, from C. Franzini-Armstrong and J.W. Kish, *J. Muscle Res. Cell Motil.* 16:319–324, 1995. All with permission from Springer Science and Business Media.)

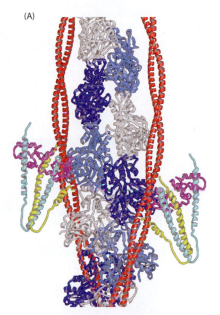

Figure 14.28 Access to the myosin-binding sites on actin is controlled by troponin/tropomyosin. (A) Model of a thin filament. Shown are actin subunits (blue, light blue, and white), tropomyosin (red) in the blocked position (Ca^{2+}-free EGTA form), troponin C (magenta), troponin I (TnI) (sky-blue), and troponin T (TnT) (yellow). The troponin structures are incomplete: most of TnC and ~80% of TnI, but only ~35% of TnT, are shown here. (B) The three positions that tropomyosin may occupy when binding to actin, whose domains (1–4) are marked on one subunit: blocked (red), closed (yellow), and open (green). (A, from A. Pirani et al., *J. Mol. Biol.* 357:707–717, 2006; B, from J.H. Brown and C. Cohen, *Adv. Prot. Chem.* 71:121–160, 2005. Both with permission from Elsevier.)

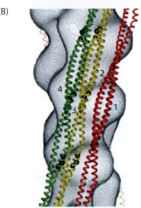

on the thin filaments, as described below. To switch off a contraction, Ca^{2+} pumps driven by ATP hydrolysis rapidly deplete the cytoplasmic Ca^{2+} (Section 16.6).

Contraction of striated muscle is regulated by changes in the interactions of troponin and tropomyosin with actin filaments

These two proteins control muscle activity via a Ca^{2+}-regulated *steric blocking mechanism.* Tropomyosin is a two-chain coiled coil that is supercoiled so as to follow the same geometry as the long-pitch actin helix (**Figure 14.28A**). It lies along the surface of actin, to which it is weakly bound. Tropomyosin molecules associate end-to-end along the actin helix, with each tropomyosin spanning seven actin subunits. Troponin, which has three subunits (C, T, and I), binds to actin at the inter-tropomyosin junctions. In resting muscle, the concentration of cytoplasmic Ca^{2+} is low and tropomyosin occludes the myosin-binding site on actin, thereby inhibiting contraction. In activation, the Ca^{2+} concentration rises, and binding of Ca^{2+} to troponin C induces a conformational change that is transmitted to tropomyosin, causing it to change its position. The myosin motor domain can then bind, but with some difficulty. However, the first motor to bind pushes the tropomyosin clear of the neighboring binding sites, allowing the next motors to bind more easily; that is, the binding of motors in response to elevated [Ca^{2+}] is highly *cooperative.* The three positions of tropomyosin on actin have been distinguished: they are called 'blocked,' 'closed,' and 'open' (Figure 14.28B). Evidence for this mechanism has come from X-ray fiber diffraction and EM data.

14.5 MYOSIN FILAMENTS

Myosin II is the only member of the myosin superfamily that assembles into filaments. In the sarcomeres of striated muscle, these 'thick' filaments are organized in precisely aligned arrays (Section 14.4). In smooth muscle, filament organization is less regular and does not give rise to visible striations. In non-muscle cells, myosin II transiently forms filamentous aggregates that interact with actin filaments to generate various types of motility, such as constriction of the contractile ring during *cytokinesis.*

Myosin II is a dimer with two heads and a long α-helical coiled-coil tail. At physiological ionic strength, the coiled coils pack together to form the filament backbone, with the heads exposed at its surface where they can interact with actin (**Figure 14.29**). This architecture, based on lateral packing of coiled coils, is reminiscent of *intermediate filaments* (Section 11.2). The C-terminal part of the tail ('light meromyosin'; see Figure 14.9) is responsible for filament formation. The N-terminal part (S2) connects the head to the filament backbone, allowing it free movement during the force-generating cycle.

Myosin filaments are bipolar (see Figure 14.29), a property that enables them to pull oppositely directed thin filaments in the two halves of the sarcomere toward each other. A filament is built by parallel tail/tail interactions in the two outer parts and from both antiparallel and parallel interactions in the central 'bare zone.'

Myosin filaments have helical arrangements of molecules

In both halves of a bipolar filament, the tails are arranged in multi-start helices. In filaments of different muscles and species, the number of helical starts (which equals the order of rotational symmetry) varies from three to seven, and their *helical pitch* ranges from 30 to 48 nm. A three-start model for the vertebrate thick filament is shown in Figure 14.24A,B. However, the axial spacing between 'crowns' of molecules is always 14.5 nm. This spacing is

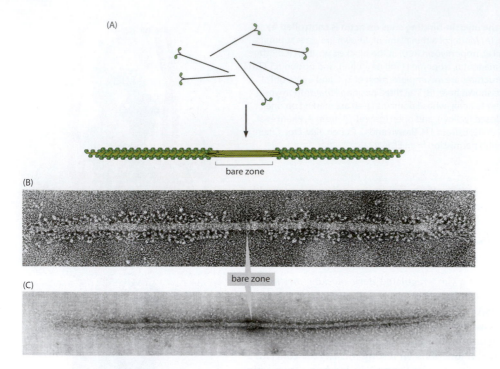

Figure 14.29 Assembly of myosin II into bipolar filaments. (A) Diagram of self-assembly. (B, C) Electron micrographs of native filaments extracted from muscle and visualized by rotary shadowing (B) and negative staining (C). Flattening and shrinkage effects cause the filaments in (B) to look larger than those in (C). The disordering of the heads is exaggerated by the preparative treatment. (B, from J. Trinick and A. Elliott, *J. Mol. Biol.* 131:133–136, 1979. With permission from Elsevier; C, courtesy of John Trinick and Peter Knight.)

based on a 28-residue repeat in the tail, in which positively and negatively charged patches alternate. Electrostatic attractions are maximized when adjacent tails are staggered by odd multiples of 14 residues, with an especially strong interaction at a stagger of 98 residues, corresponding to the 14.5 nm repeat. This helical symmetry is supported by the strong meridional reflection at a spacing of $(14.5 \text{ nm})^{-1}$ in the X-ray diffraction pattern from relaxed vertebrate muscle (see Figure 14.24C), taken together with the strong layer-line at one-third of this spacing, namely $(43.5 \text{ nm})^{-1}$.

However, the structures of myosin filament backbones are not yet understood in detail. The great length (160 nm), small diameter (2 nm), and tight packing of tails make analysis difficult. Theoretical considerations suggest that different filament symmetries arise as variations on a small set of interactions, with minor adjustments generating the observed diversity. In one model, individual tails are the assembly units (**Figure 14.30A**). In another model, they assemble into subfilaments, of which several associate to form the filament backbone (Figure 14.30B–E). In this scheme, packing of molecules is similar in subfilaments of different muscles, while variations in subfilament number and tilt produce the different symmetries. The myosin-related reflections in X-ray diffraction patterns from crustacean muscles are well described by a subfilament-based model. Direct support for the existence of subfilaments comes from cryo-EM reconstructions of tarantula filaments (**Figure 14.31B**). Subfilaments may also occur in vertebrate muscle. There are also data suggesting that vertebrate myosin filaments have three larger helical components, each made of three subfilaments.

In relaxed thick filaments, there is regulatory cross-talk between the two heads of a myosin molecule

In the absence of nucleotide, myosin heads bind tightly to actin filaments (the 'rigor' state), but when ATP is bound, they detach from actin (the 'relaxed' state). Their conformations in relaxed filaments vary between species, but a common theme is emerging. The most detailed information comes from tarantula filaments, which are especially well ordered. Fitting of atomic models of S1 into cryo-EM reconstructions of isolated filaments reveals that the two heads of a given myosin molecule interact with each other asymmetrically (Figure 14.31). As a result, there is physical interference between the actin-binding site of one head and the converter domain (Section 14.3) of the other head that suggests a simple model for switching off myosin activity, namely steric inhibition. Similar interactions are seen in other muscles whose activity is regulated via myosin (as in most invertebrates; in some other muscles, actin is also involved). Switching on depends on Ca^{2+}-dependent phosphorylation of

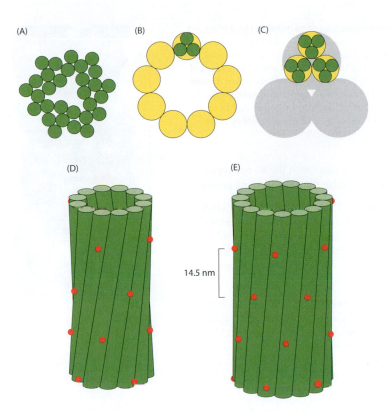

Figure 14.30 Models for packing myosin tails in filaments. (A) Individual tails are the units for assembly (each green disk represents a tail in cross section). (B) Tails assemble in groups of three, forming subfilaments that assemble into filaments. (C) Subfilaments assemble into higher-order subfilaments and then into filaments. (D, E) Lateral views of type B models with a ring of 12 tails with 4-fold rotational symmetry (D) and 15 tails with 5-fold rotational symmetry (E). Red spots mark head attachment sites. Models (B), (D) and (E) predict a hollow core. With relatively thick myosin filaments (such as those from mollusks), the central region is partly or completely filled with paramyosin, a headless variant of myosin. (A–C, from M.W. Chew and J.M. Squire, *J. Struct. Biol.* 115:233–249, 1995. With permission from Elsevier; D and E, from J.S. Wray, *Nature* 277:37, 1979. With permission from Macmillan Publishers Ltd.)

the regulatory light chain or Ca^{2+} binding to the heads. This disrupts the head/head interaction, freeing the heads to hydrolyze ATP and interact with actin.

Thick filaments contain accessory proteins with structural and regulatory roles

Although actin subunits are able to self-assemble into filaments, their state of aggregation in cellular contexts is defined by binding proteins (Section 14.2). The same holds true for myosin II. Just as the length of actin filaments in muscle cells is marked out by the protein nebulin, that of myosin filaments is specified by titin. The largest polypeptide known (~3 MDa), titin runs for more than 1 μm from the middle of the thick filament to the Z-line (see Figure 14.20B). The part of titin that interacts with the myosin filament consists of a long series of fibronectin Type III-like (Fn) domains (132 in all; see Figure 11.20B) and Ig domains (up to 166 in all; see **Guide**) that run along the surface of the thick filament. There are

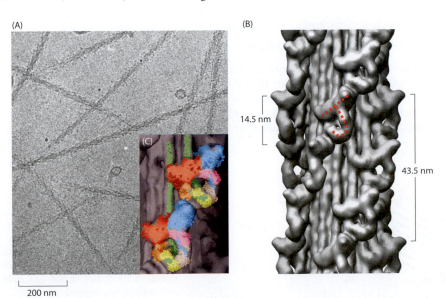

Figure 14.31 Thick filaments from tarantula striated muscle. (A) Cryo-electron micrograph. (B) Three-dimensional reconstruction. Each pair of heads appears as a tilted J. Subfilaments and S2 appear as parallel strands in the filament backbone. (C) Enlargement showing two heads fitted into each J-motif. In each J-motif, one motor domain is blue, the essential light chain is pink, and the regulatory light chain is beige. In the other head, these domains are respectively red, orange, and yellow. Two S2 tail domains in the filament backbone are green. (PDB 1I84) (From J.L. Woodhead et al., *Nature* 436:1195–1199, 2005. With permission from Macmillan Publishers Ltd.)

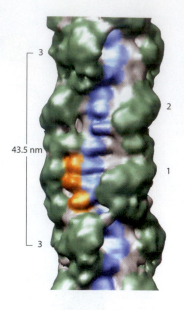

Figure 14.32 Three-dimensional reconstruction of vertebrate thick filament in the MyBP-C region. Every 43.5 nm axial repeat contains three crowns of paired myosin heads (labeled 1, 2, and 3). These crowns are non-equivalent. The pairs of heads (green) on crowns 1 and 2 fit well with the interacting-head motif (Figure 14.31B) but crown 3 does not. The blue strand of titin-associated density consists of eleven 4 nm beads in one 43 nm repeat. The orange beads probably represent three Ig or Fn domains of MyBP-C. (From M.E. Zoghbi et al., *Proc. Natl Acad. Sci. USA* 105:2386–2390, 2008 National Academy of Sciences. With permission from PNAS.)

11 such domains in each 'super-repeat' of titin which is in register with the 43 nm repeat of the myosin filament, and they specify the positions of other thick filament-binding proteins.

In vertebrate muscle, myosin-binding protein C (MyBP-C) binds at every third level of heads, at 11 sites, 43 nm apart, in each half-filament (see Figure 14.20B). Like titin, MyBP-C is constructed from Fn and Ig domains, and it may have a role in sarcomere assembly. In EM reconstructions, three of its domains appear to run alongside titin, but the rest have not been visualized (**Figure 14.32**). These proteins add to the complexity of vertebrate thick filaments.

Because MyBP-C binds to every third 14.5 nm level, myosin molecules are not all equivalent, and differ again from myosin in the distal regions of the filament that lack MyBP-C. Two of the three levels have the same head/head interaction as in smooth muscle and invertebrate striated muscle, whereas the third appears to be more mobile. The finding of head/head interactions in vertebrate thick filaments is surprising, because these filaments are not regulated via myosin. Possibly, the interactions in vertebrate myosin are weaker and insufficient to prevent activity.

Invertebrate striated muscle myosin filaments also have accessory proteins, including a smaller version of titin, that contribute to their mechanical properties. Invertebrate filaments also have the coiled-coil protein paramyosin, which resembles the myosin tail and is thought to occupy the filament core, where it may strengthen these filaments against the high loads they must bear.

Most vertebrate and invertebrate thick filaments are connected at their bare zones by protein bridges constituting the M-line (see Figure 14.20B). These bridges hold the thick filaments in longitudinal register and provide an anchor point for one end of titin. In vertebrates, the principal M-line proteins, such as MyBP-C and titin, are built from Fn and Ig domains. One component, creatine kinase, functions as an enzyme.

In summary, myosin filament assembly is based on self-associative properties of the tail but is fine-tuned by other factors. Depending on the system, they can include non-myosin components (for example paramyosin, titin, and MyBP-C), chaperones, and the phosphorylation state of the heavy or light chains (see below).

In smooth muscle, myosin filaments have a side-polar structure and their state of assembly is regulated by phosphorylation

In vertebrate smooth muscle, myosin filaments have a non-helical structure in which the orientation of the molecules reverses on opposite sides of the filament, rather than at opposite ends (**Figure 14.33**). This arrangement is called 'side-polar.' Antiparallel inter-tail interactions take place all along the filament, and the bare zones are at the ends. Accordingly, side-polar filaments can pull oppositely directed actin filaments toward each other along opposite sides.

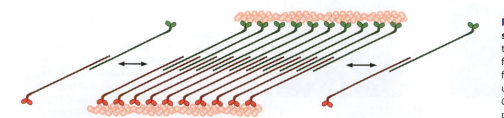

Figure 14.33 Side-polar filament structure. Antiparallel tail overlaps occur throughout the filament. Myosin II isoforms from non-muscle cells form side-polar filaments *in vitro*, but the structure *in vivo* is uncertain. (From R. Craig and J.L. Woodhead, *Curr. Opin. Struct. Biol.* 16:204–212, 2006. With permission from Elsevier.)

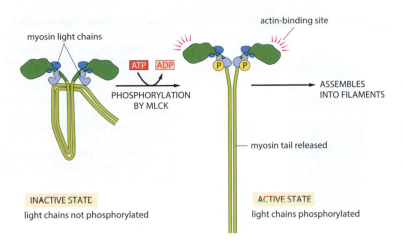

myosin light chains

actin-binding site

ATP ADP

PHOSPHORYLATION
BY MLCK

ASSEMBLES
INTO FILAMENTS

myosin tail released

INACTIVE STATE
light chains not phosphorylated

ACTIVE STATE
light chains phosphorylated

Figure 14.34 Phosphorylation-dependent assembly of myosin from non-muscle and smooth muscle cells. In relaxing conditions, filaments disassemble into molecules that fold their tails into a compact form and are inactive. Head/head interactions occur in the inactive state but are not shown here. On phosphorylation of the regulatory light chain, the tails extend and the molecules assemble into filaments. MLCK, myosin light chain kinase. (From B. Alberts et al., Molecular Biology of the Cell, 6th ed. New York: Garland Science, 2015.)

Myosin filament populations in smooth muscle are subject to phosphorylation-dependent remodeling. Under *in vitro* relaxing conditions (ATP, low [Ca^{2+}], regulatory light chain dephosphorylated), filaments disassemble into molecules with folded tails and low ATPase activity (**Figure 14.34**). On phosphorylation, the tails unfold and the myosin assembles into filaments with high ATPase activity, capable of interacting with actin. It is unclear whether they are side-polar or bipolar. This reaction provides a mechanism for storing inactive molecules and for their transient assembly into filaments when needed. In smooth muscle cells, myosin is present as filaments (probably stabilized by binding proteins), which are ready for immediate interaction with actin when contraction is switched on. In some smooth muscles, however, filament number or length can increase, presumably by assembly from a pool of the tail-folded molecules. This recruitment may enhance contraction by remodeling the contractile apparatus for optimal performance under the extremes of length that occur during smooth muscle contraction.

14.6   MICROTUBULES AND ASSOCIATED PROTEINS

Microtubules (MTs) are hollow filamentous polymers of the protein tubulin. They are essential components of all eukaryotic cells, where they support a multitude of functions in motility and cytoarchitecture (for example, **Figure 14.35**). MTs generate movement by two basic mechanisms: by polymerizing, whereby their growing ends exert force on other structures on which they impinge; and by providing tracks along which motor proteins pull their cargoes. They serve as one of the two major systems for intracellular transport, the other being F-actin. Although MTs and F-actin have different sub-unit structures and filament geometries, they share a number of basic behaviors: both are dynamic systems that undergo constant remodeling, according to the evolving needs of a given cell; and both bind 'associated proteins' that regulate their states of assembly and confer functional specificity.

The building block of MTs is a heterodimer of α-tubulin and β-tubulin. The two 50 kDa subunits share 30–40% sequence identity, probably reflecting origins in an ancient gene duplication event. They assemble into a two-dimensional lattice with inbuilt curvature. As it grows, this lattice rolls up into a tube ~25 nm in diameter (**Figure 14.36**). The 8 nm-long heterodimers polymerize head-to-tail in linear strands called **protofilaments**, which associate side-by-side in the hollow tubes. Because all protofilaments in a given MT point in the same direction, they form a polar structure: the end with β-tubulin subunits is called the 'plus end,' and the end with α-tubulin subunits, the 'minus end.' Their polarity is crucial for the roles of MTs in directed movement inside cells.

There is a discontinuity or 'seam' in the helical lattices of most microtubules

When MTs are assembled *in vitro* from purified tubulin, the number of protofilaments varies from 9 to 16. This polymorphism reveals some flexibility in the bonds between adjacent protofilaments, although MTs are quite stiff; that is, resistant to bending. In cells, MTs grow out from a nucleating complex called the microtubule-organizing center (MTOC), which

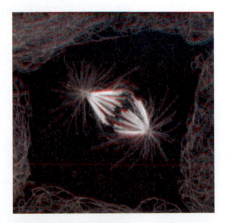

Figure 14.35 Microtubules in a mitotic spindle. Fluorescent light micrograph. (Courtesy of Julie Canman and Edward (Ted) Salmon.)

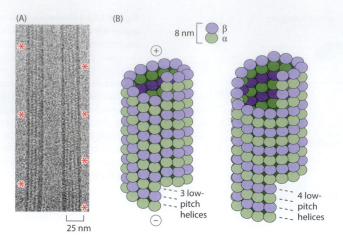

(A)

(B)

8 nm [β / α]

3 low-pitch helices

4 low-pitch helices

25 nm

Figure 14.36 Microtubule structure.
(A) Cryo-EM of MTs assembled *in vitro*. In these images, the near and far sides of the MTs are superposed, so that striped (asterisks) and fuzzy patches alternate. Their axial spacing reflects the helical twist of the protofilaments and the number of protofilaments: the MT on the left has 14 protofilaments, and the one on the right has 15. (B) Models of MTs; left, a 13-protofilament MT with three low-pitch helices and a seam; right, a 15-protofilament MT with four low-pitch helices and no seam.

presents a template (see below). Under these conditions of controlled nucleation, most MTs have 13 protofilaments. In such MTs, the protofilaments run straight. In MTs with more or fewer protofilaments, they wind slowly around the axis, producing periodic Moiré patterns in EM images (see, for example, Figure 14.36A).

Neighboring protofilaments are offset by 9 Å, so that laterally adjacent subunits lie along 'low-pitch' helical strands (see Figure 14.36B). MTs with 13 or 14 protofilaments have three such helices. Narrower or wider MTs have two or four low-pitch helices. If adjacent subunits along the low-pitch helices are of the same type, the pattern is called a B-lattice. If they alternate (α–β–α, and so on), the result is an A-lattice. To determine whether A-lattices or B-lattices apply, α and β subunits must be mapped on individual MTs. However, they are so similar that they cannot be distinguished in electron micrographs; to do so, one must decorate MTs with a protein that recognizes the distinction between them and visually amplifies it (**Figure 14.37**). According to such analyses, B-lattices are the norm. However, the B-lattice for a 13-protofilament MT always has (at least) one A-like discontinuity or seam (see Figure 14.36A). In fact, all MT lattices have such a seam, except for the 15-protofilament (see Figure 14.36B) and the rarer 12-protofilament and 16-protofilament variants. The seamless nature of 15-protofilament MTs has allowed their three-dimensional structures to be reconstructed from electron micrographs by Fourier–Bessel methods that exploit helical symmetry. Other methods allow the reconstruction of seamed MTs as well.

The MT seam is an unusual property in protein filaments. However, there may be advantages to such an arrangement: the seam expands the range of distinct binding sites that

microtubule outer surface (motor-decorated)

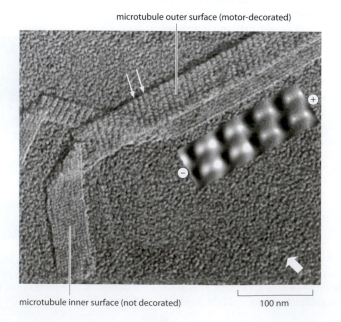

microtubule inner surface (not decorated)

100 nm

Figure 14.37 Electron micrograph of microtubules decorated with a motor protein. The specimen was prepared by freeze-drying and metal shadowing (the thick white arrow at bottom right indicates the shadowing direction). The MTs have flattened into double ribbons. The insert shows the outer surface with bound motors, enhanced by averaging. There is one motor per tubulin heterodimer, and their packing follows the low-pitch helices of both MTs (arrows) continuously, indicating that they have B-lattices. (From A. Hoenger and H. Gross, *Methods Cell Biol.* 84:425–444, 2008. With permission from Elsevier.)

(A)

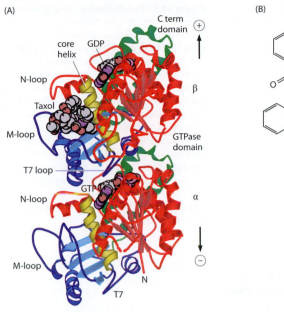

(B)

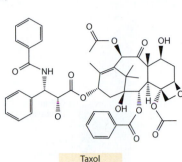

Taxol

Figure 14.38 Structure of tubulin. (A) In this heterodimer, each subunit has two globular domains separated by the core helix (H7, yellow), and each GTP domain (red) has a nucleotide-binding site on its plus end surface. GDP is bound to β-tubulin and GTP to α-tubulin. The T7 loop completes the nucleotide-binding site. The M-loop makes lateral contacts with neighboring GTP domains. The surface that is on the outside of assembled MTs is dominated by the C-terminal domains (green); each has two long α helices, H11 and H12, connected by a U-turn, which are more clearly seen in the view shown in Figure 14.43. 13 residues at the C terminus of α-tubulin and 9 of β-tubulin are very negatively charged; they do not show up as ordered density but are assumed to project outward by charge repulsion. (PDB 1JFF) (B) Chemical structure of Taxol.

may be recognized by associated proteins, thereby expanding the scope for regulation. For instance, proteins in the EB1 family (EB = end-binding; Section 14.7) have tubulin-binding domains that fit into the grooves between protofilaments; Mal3, the EB1 homolog of *Schizosaccharomyces pombe*, binds preferentially along the seam.

As with other proteins that have a propensity to polymerize, this property has thwarted attempts to obtain well ordered three-dimensional crystals of tubulin. However, it was found that Zn^{2+} causes protofilaments to line up in sheets, several micrometers across and only one molecule thick, instead of tubes. These sheets enabled Eva Nogales and Ken Downing to determine the protein fold by electron crystallography (**Figure 14.38A**). Subsequently, this fold was confirmed by X-ray crystallography (see **Methods**) of a tubulin complex (see below). The α and β subunits have similar folds, as expected from amino acid sequence similarity, but differ in several functionally significant respects. For example, the activation domains (blue) provide the T7 loops that complete the nucleotide-binding pockets. In a protofilament, the T7 loop of α-tubulin—but not that of β-tubulin—is able to promote the hydrolysis of GTP. The activation domains also carry the M and N loops that effect lateral contacts. The activation domain of β-tubulin has a binding site for *Taxol*, a MT-stabilizing drug whose structure is shown in Figure 14.38B. Its binding site on β-tubulin is shown in Figure 14.38A. α-Tubulin has an extended loop in place of the Taxol pocket.

Assembly, disassembly, and stability are influenced by GTP hydrolysis

Free in solution, the tubulin heterodimer binds two GTP molecules. The GTP in the α subunit is trapped by the overlying β subunit and is not exchangeable, but the GTP on the β subunit is surface-exposed and is exchangeable. After assembly, it is hydrolyzed to GDP, to be exchanged for GTP when the dimer becomes soluble again. The GTP analog GMP-CPP is hydrolyzed only very slowly by tubulin, and MTs with this nucleotide bound are very stable, implying that the GTP state of β-tubulin also has a stabilizing effect. When the exchangeable GTP is hydrolyzed, the bonds that the heterodimer makes with its neighbors are weakened. However, each heterodimer added to a growing MT has GTP in the β site, adding stability. It is thought that the assembled lattice is in a strained state as a result of storing energy from conformational shifts induced by GTP hydrolysis, and releases it during depolymerization.

If there are enough GTP-bound β subunits capping the plus end, the MT remains assembled, but if the cap is lost, disassembly may suddenly start from this end; this event is known as a **catastrophe**. For reasons that are still unclear, disassembly may stop before the MT

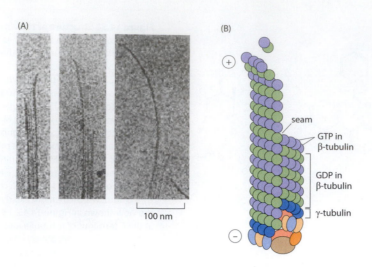

(A)

(B)

seam

GTP in
β-tubulin

GDP in
β-tubulin

γ-tubulin

100 nm

Figure 14.39 The distinct ends of growing microtubules. (A) Cryo-EM images of MTs *in vitro* growing at their splayed-out plus ends. (B) Model of a MT growing at its plus end. A narrow sheet of protofilaments often grows ahead of the rest of the tube, as here. Heterodimers with GTP-bound β subunits attach at the plus end, followed by hydrolysis of their nucleotide. Proteins such as EB1 stabilize plus ends by binding to the advancing sheet (not shown here; see Figure 14.55). The minus end of this 13-protofilament MT is capped with a γTurc complex that initiated assembly. (A, from D. Chrétien et al., *J. Cell Biol*. 129:1311–1328, 1995. With permission from Rockefeller University Press.)

disappears completely; this is known as a **pause**. GTP-bound subunits may attach to a paused end, rescuing assembly for a while. This stop-and-start behavior, called **dynamic instability**, makes it possible for one MT to be depolymerizing while another nearby is assembling.

Intracellular assembly of microtubules initiates in specific nucleation complexes

As noted above, MTs in cells grow out of organizing centers (MTOCs), which results in a more tightly defined assembly product with 13 protofilaments. The minus end is in contact with the MTOC and usually remains anchored there. MTOCs are ring-like complexes, ~25 nm in diameter, assembled from a third kind of tubulin, γ-tubulin, and associated proteins (**Figure 14.39**). These **γ-tubulin ring complexes (γTurcs)** are found in *centrosomes* (Section 13.5). As the first visible step in cell division, antiparallel sets of MTs push the two halves of the centrosome apart to form the poles of a spindle (**Figure 14.40B**). Smaller complexes, γTuscs, are found elsewhere in the cytoplasm and are thought also to initiate MT assembly with the help of accessory proteins. *In vitro*, when both ends are free, assembly and disassembly can occur at either end, although at different rates. The plus end is more dynamic than the minus end, where there are α-tubulin subunits with unhydrolyzed GTP. This property of nucleated assembly, together with the key attribute of structural polarity, makes it possible for the population of MTs in a cell to be controlled with a system of hubs.

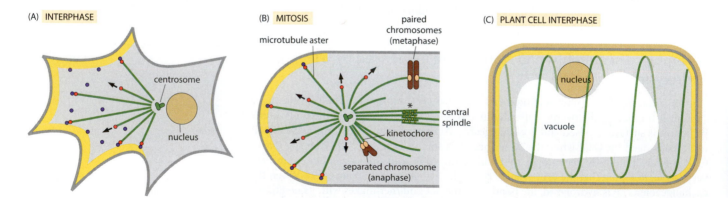

(A) INTERPHASE

centrosome

nucleus

(B) MITOSIS

microtubule aster

paired chromosomes (metaphase)

central spindle

kinetochore

separated chromosome (anaphase)

(C) PLANT CELL INTERPHASE

nucleus

vacuole

Figure 14.40 Cellular sites for capturing microtubule plus ends. (A) In an interphase animal cell, MTs (green) nucleate near the nucleus in complexes where their minus ends remain anchored. Their plus ends (red dots) grow and shrink until they encounter target complexes (blue dots). (B) During cell division, MTs grow out from two spindle poles (one is shown here), and some of them polymerize further to push the poles apart at the start of the division process. The two antiparallel sets of MTs (asterisk) growing from opposite poles are cross-bridged by accessory proteins. Other MTs grow until they capture a kinetochore at the center of a chromosome (yellow dots) or other sites at chromosome ends. MTs growing in other directions find targets in the cell cortex (yellow), making an 'aster'. (C) In interphase plant cells, helical MT arrays contacting the cell membrane help to determine the direction of growth.

Uncapped microtubules grow and shorten at both ends, asymmetrically

MTs continue to grow as long as the free tubulin concentration remains above the *critical concentration*. In populations of MTs in which both ends are free, minus ends stop growing first because their critical concentration is higher. During rapid assembly at the plus end, there is often a narrow sheet of protofilaments that takes the lead (see Figure 14.39A). Its slight outward curvature is thought to indicate that the initial conformation after assembly differs both from the straight conformation assumed in closed lattices and from the more curved disassembled state (see below). Growing plus end sheets appear to be rare *in vivo*, where MT growth is more controlled.

Disassembly at either end appears to be a cooperative process, as protofilaments at the ends of disassembling MTs splay apart and bend outward (**Figure 14.41A,B**). In some circumstances, segments of protofilaments are shed as spirals that can close into single or double rings, 30–40 nm across (Figure 14.41C).

It has been proposed that catastrophes, pauses, and rescues at the plus end are caused by random loss or restoration of the GTP-tubulin cap. However, MTs assembled *in vitro* also display some dynamic instability at their minus ends, where the GTP content is thought not to vary. Thus, the events responsible for stochastic changes in MT behavior may be spontaneous conformational changes propagated along individual protofilaments or small bundles.

Dynamic instability allows microtubules to search for targets

Because of dynamic instability, MTs that happen to assemble in the 'wrong' place in a cell usually disappear quickly. MTs initiated from an organizing center have stable minus ends but their plus ends grow and shrink unless stabilized in some way. A growing MT may probe a large volume of cytoplasm until it captures a suitable target. This is the process by which MTs growing out from the poles of a meiotic or mitotic *spindle* become attached to kinetochores (Section 13.7; see Figure 14.40B). *Kinetochores* contain kinesin motors (Section 14.7), and if one of them makes contact with any part of a MT, it can travel to the plus end. The MT may then continue to grow at the plus end, pushing the kinetochore with it.

When the free tubulin concentration is low enough to favor disassembly at the minus end but high enough for assembly to continue at the plus end, *treadmilling* may occur (**Figure 14.42**). (Actin also exhibits this behavior.) *In vitro*, treadmilling is partly suppressed by dynamic instability, which depletes plus ends and replenishes the pool of heterodimers, promoting bidirectional growth. *In vivo*, however, there are accessory proteins that can inhibit instability while allowing assembly to continue. During mitotic *metaphase*, tubulin may continue to assemble on MT plus ends embedded in kinetochores. At the same time, the minus ends of the same MTs may shed dimers if they have been severed from their

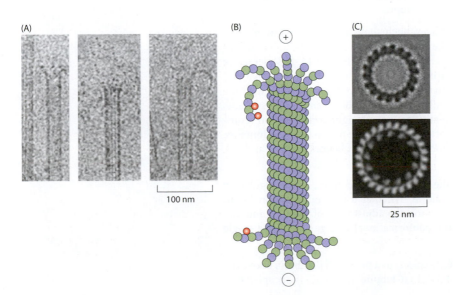

(A)

(B)

(+)

(C)

100 nm

25 nm

Figure 14.41 Microtubule disassembly. (A) Cryo-EM images of MTs spontaneously shortening *in vitro*. (B) Diagram of a MT shrinking at both ends. Accessory proteins (red), such as some members of the kinesin family (Section 14.7), can control depolymerization. (C) Protofilaments shed from depolymerizing MTs curl into spirals and rings. Top, averaged cryo-EM of a 9-heterodimer tubulin ring formed by treating MTs with the drug cryptophycin-1, which accentuates bending by binding between heterodimers. Bottom, averaged negatively stained EM of a tubulin ring complexed with kinesin subunits (one per tubulin heterodimer), on the inside of the ring, which corresponds to the outer surface of the MT. (A, from D. Chrétien et al., *J. Cell Biol.* 129:1311–1328, 1995. With permission from Rockefeller University Press; C, from E. Nogales et al., *Curr. Opin. Struct. Biol.* 13:256–261, 2003 [top]. With permission from Elsevier; C.A. Moores et al., *Mol. Cell* 9:903–990, 2002 [bottom].)

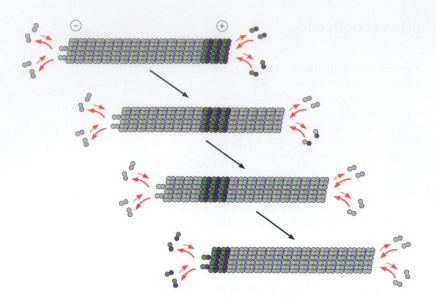

Figure 14.42 Microtubule treadmilling. This reaction can occur when the concentration of free heterodimers is higher than the critical concentration for the plus end and lower for the minus end. (Adapted from R.L. Margolis and L. Wilson, *BioEssays* 20:830–836, 1998. With permission from John Wiley & Sons.)

initiation sites in the centrosome. In this case, there is a flow of tubulin from centrosome to kinetochore. Plus end-directed motor proteins associated with kinetochores are thought to push these treadmilling MTs toward their minus ends, so there may be a flux of tubulin from kinetochore to centrosome. The helical bands of MTs that line the walls of plant cells (see Figure 14.40C) also undergo treadmilling rather than dynamic instability. Their presence allows growth to occur while the shape of the cell is maintained.

The state of tubulin assembly can be altered by drug binding

In addition to GTP/GDP, other factors may influence tubulin polymerization. Of particular importance to regulating MT networks in cells are binding proteins, of which a few well-characterized examples are discussed below. Certain drugs exert similar effects. For instance, **colchicine** promotes disassembly by binding to and sequestering tubulin heterodimers. This compound, extracted from crocus plants, is widely used in studies to ascertain how a phenomenon of interest is affected by eliminating the population of MTs in a cell. Although highly toxic, colchicine is used to treat gout, and its potential for cancer therapy is under assessment.

Other drugs, of which Taxol (also known as paclitaxel) is the best known example, stabilize MTs against disassembly. Taxol is found in the Pacific yew tree. Binding of Taxol to its site on the β subunits (see Figure 14.38) locks them in place, effectively paralyzing the MT. Cells need to be able to remodel their cytoskeletons as the cell cycle proceeds, so this reaction is cytotoxic. Because tumor cells have particularly rapid rates of cell division, they are especially vulnerable to Taxol, which is used in cancer therapy.

Some drugs perturb intra-protofilament interfaces, such that they cannot straighten to associate into MTs. Colchicine binds between the subunits in a heterodimer; others, such as vinblastine, interfere with the interaction between heterodimers (**Figure 14.43B**). Here, we distinguish between 'straight' (assembly-compatible) and 'curved' (assembly-incompatible) conformations of tubulin. A curved state is seen in crystals of tubulin complexed with the tubulin-sequestering protein **stathmin**, and involves ~12° bends between tubulin subunits (Figure 14.43B,D). Changes in the subunits relative to their straight conformation (compare Figure 14.43A) include a small rotation between the GTPase and activation domains. Contacts between subunits are maintained by local movements of helices H6, H7, and H8, and loop T5.

The changes in tubulin structure induced by the binding of nucleotides, drugs, regulatory proteins, or motor proteins tend to be quite subtle. Whether a heterodimer remains in the polymer or dissociates depends stochastically on the adjustment of multiple interactions, rather than being driven by a substantial conformational change affecting a single interaction.

MT-stabilizing drugs bind to the inner surface. Taxol sits in a pocket in β-tubulin above the β sheet of the activation domain (see Figure 14.38). In α-tubulin, this space is occupied by a

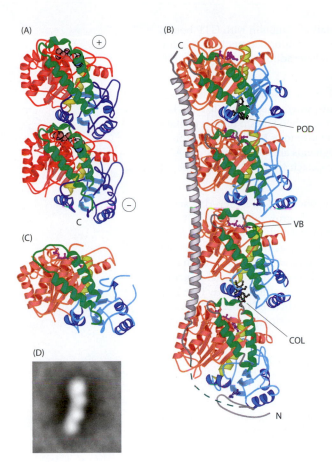

(A)

(B)

(C)

(D)

Figure 14.43 Tubulin in straight and curved conformations. (A) Side view of the straight structure shown in another view in Figure 14.38. (B) The curved structure determined from co-crystals of stathmin (silver) complexed with two tubulin heterodimers. The binding sites of depolymerizing drugs are marked: COL, colchicine; POD, podophylotoxin; VB, vinblastine. (PDB 1SA0) (C) Structure of γ-tubulin. (PDB 1Z5V) (D) Averaged scanning transmission electron micrograph of stathmin-induced depolymerization products, negatively stained. (From M.O. Steinmetz, *J. Struct. Biol.* 158:137–147, 2007. With permission from Elsevier.)

loop that is thought to open up into a T-shaped structure, exposing a hydrophobic surface that mates with a hydrophobic patch on β-tubulin. Several other compounds with diverse structures, including epothilones, discodermolide, and eleutherobin, compete with Taxol for MT binding, apparently because they bind to the same pocket on β-tubulin. Although Taxol and other MT-stabilizing drugs are found in only a few organisms, they bind to a site that is conserved throughout all eukaryotes. It may therefore be that these poisons are mimicking stabilizing agents, such as tau (see below), that are present in most eukaryotic cells.

Destabilizing proteins cause microtubules to disassemble

Some proteins inhibit assembly by sequestering tubulin. One such is **stathmin** (17 kDa), which has a central role in regulating MT assembly in cells, especially in spindle formation (Chapter 13). Stathmin cannot bind to assembled MTs but readily binds one or two heterodimers (see Figure 14.43B), reducing the concentration of free tubulin. When it falls below the critical concentration, pre-existing MTs disassemble. Tubulin complexed with stathmin is disabled both by being held in a curved conformation and by having its ends capped. Stathmin has an N-terminal domain that caps one tubulin heterodimer, and its C-terminal domain is a long α helix that spans four tubulin subunits. Evidence from NMR spectroscopy (see **Methods**) indicates that stathmin free in solution is natively unfolded, particularly the N-terminal domain, and assumes the conformation observed in the crystal structure only when bound to tubulin. During mitosis, when many MTs are assembled, the disassembling activity of stathmin is lowered by phosphorylation of four serine residues, three of which are in the capping domain. Dephosphorylation of these residues is needed for a cell to exit from mitosis.

Because GTP promotes the assembly of MTs in which tubulin has its straight conformation, whereas rings and coils cast off from disassembling MTs have GDP-containing tubulin in the curved conformation, it might be supposed that tubulin conformation is directly controlled by the nucleotide. However, when a protofilament closes into a ring, it bends at all inter-subunit interfaces, both between and within heterodimers, even though there is still GTP in the intradimer interface. Thus tubulin conformation is influenced but not fully controlled by the bound nucleotide. In this context, crystals of γ-tubulin with GTP bound showed a structure (see Figure 14.43C) similar to the curved conformation of αβ-tubulin. However, its subunits may straighten out when αβ-tubulin heterodimers bind to them. It is not known whether γ-tubulin ever hydrolyzes GTP.

More directly active depolymerizing proteins are found in the kinesin superfamily (Section 14.7). They bind to MTs and migrate to the plus ends, where they destabilize the lateral bonding between protofilaments (see Figure 14.41B,D). When kinesin motors contain ATP, they bind strongly to protofilaments. This interaction promotes bending where this is physically possible, as at a free end, and protofilament segments are able to break off. ATP hydrolysis weakens the binding and allows the motor to detach from tubulin, whereby it may diffuse back to the MT to remove another segment.

Other proteins destabilize MTs by severing them, exposing uncapped ends. Severing activity is especially important during mitosis. Spindle MTs initially grow out from MTOCs in the centrosomes, but during *metaphase* and *anaphase* the minus ends of kinetochore MTs, exposed through the action of severing proteins, shed subunits as the chromosomes are moved toward the poles (see Figure 14.40B). Severing is also essential for the growth and maintenance of nerve cell axons and dendrites, where a steady flow of cytoskeletal proteins, including segments of cytoplasmically assembled MTs, issues from the cell body (see Figure 14.1). Although these MTs are stabilized by **microtubule-associated proteins (MAPs)**, their ends remain dynamic, so that tubulin can be exchanged during transport; however, MT assembly is never initiated within a neural process.

Katanin and **spastin** are proteins with documented severing activity. Katanin is a heterodimer of 60 kDa and 80 kDa subunits that forms larger oligomers during its active cycle. The 60 kDa subunit is a member of the AAA+ family of ATPases (Boxes 3.1 and 7.2). The N-terminal region of the 80 kDa subunit contains six WD40 repeats (see **Guide**) and is responsible for targeting the protein to the centrosome. The severing activity of katanin yields MT segments short enough to be readily transportable. The severing mechanism involves the initial creation of a defect in the side of a MT that allows subunits to escape.

Spastin (~80 kDa) is a homolog of the 60 kDa katanin subunit. It appears to disrupt MTs by pulling on the negatively charged C termini of tubulin subunits so as to thread them in an at least partly unfolded state through the axial channel of a hexameric ring of AAA+ domains (**Figure 14.44**). Mutations in human spastin, mapping in its AAA+ domain, occur in ~40% of hereditary spastic paraplegias, which are neurodegenerative diseases characterized by degeneration of the terminal axons of spinal neurons, resulting in progressive weakening of the lower limbs. Presumably, impairment of the ability of spastin to remodel MTs underlies the symptoms of this disease.

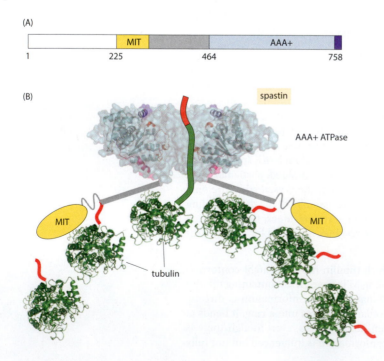

(A)

	MIT		AAA+	
1	225	464		758

(B)

spastin

AAA+ ATPase

MIT

MIT

tubulin

Figure 14.44 Microtubule severing by spastin. (A) Spastin has an N-terminal domain, a MT interacting and trafficking (MIT) domain, an extended linker (gray), and a C-terminal AAA+ module. A crystal structure for the AAA+ module as a nucleotide-free monomer was used to construct a hexamer on the basis of SAXS (small-angle X-ray scattering) data. Shown is a hypothetical model for the hexamer extracting tubulin subunits from the MT wall. The red tags are the C-terminal tails of tubulin subunits. (From A. Roll-Mecak and R.D. Vale, *Nature* 451:363–367, 2008. With permission from Macmillan Publishers Ltd.)

Microtubule-associated proteins stabilize the polymers

MAPs have been studied for almost as long as tubulin—intentionally or otherwise—because of their tendency to co-purify. Proteins such as MAP1A/B, MAP2, MAP4, and MAPT (tau) are all elongated molecules that can extend over the tubulin lattice and cross-link hetero-dimers. They therefore have a stiffening effect on MTs. MAPs vary widely in their amino acid sequences but many are positively charged, which is consistent with their binding in extended form over the negatively charged outer surface of a MT.

Tau has attracted particular interest because it has been implicated in several neurodegen-erative diseases ('tau-opathies'), including Alzheimer's disease. One of the two abnormal amyloid-containing deposits that accumulate in Alzheimer's brains—*paired helical fila-ments*—is composed mainly of tau (Section 6.5). Human tau occurs in six isoforms with molecular masses of 38–48 kDa, which represent alternatively spliced expression products of the *tau* gene. Spectroscopic data indicate that, in solution, tau is natively unfolded, and remains in an extended conformation when bound to MTs. A domain map of tau is shown in **Figure 14.45B**. Its N-terminal projection domain is followed by a proline-rich 'assembly' domain and then three or four copies of a 31–32-residue repeat, and finally a short end-domain. The repeat motifs bind tau to MTs; although the binding of individual repeats is weak, together they achieve a firm attachment.

The negatively charged projection domain is anchored on the MT surface by the assembly domain and is repelled outward by electrostatic interactions, whereby it determines the spacing between MTs in axons (sufficient separation is needed to avoid impeding motor-ized transport). There is less of a consensus regarding the location of assembly domain and the repeat motifs. According to one model, based on gold-labeling EM experiments (Figure 14.45A,C), the repeats bind to the inside of the MT, linking neighboring protofilaments, with the Pro-rich segment threading through a small gap in the surface lattice. According to another model, also based on gold-labeling EM data, the assembly domain is entirely on the outer surface, with the repeats running along the ridges of individual protofilaments. These models may differ because the experiments on which they are based used, in one case, MTs co-assembled with Taxol, and in the other case, Taxol added to pre-assembled MTs. Dynamically active MTs in an axon may include both kinds of interaction with tau.

The MT-binding repeats recur in many MAPs, whose sequences are otherwise quite dif-ferent. Other MAPs also have projection domains which, for MAP2 and MAP4, are sub-stantially longer than that of tau; larger projection domains correlate with greater inter-MT spacings. The domains may also be rearranged: MAP1 has a C-terminal projection domain and a long N-terminal domain with many positively charged residues that binds to the outer surface of MTs. MAP1, despite its overall dissimilarity to tau, can apparently substitute for it when tau is missing; mice that have either gene knocked out appear normal, but a double knock-out is lethal.

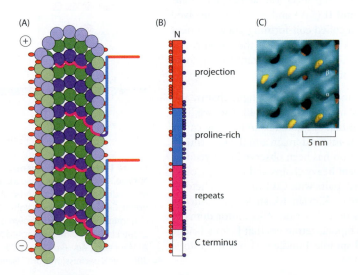

(A) (B) (C)

N

projection

proline-rich

repeats

C terminus

5 nm

Figure 14.45 Binding of a MAP (tau) to a microtubule. (A) Cutaway diagram showing the proposed interaction. (B) Domain map of human four-repeat tau. The negatively charged C termini (red external tags) of tubulin interact with positive charges in the Pro-rich domain of tau. The tau repeat motifs (purple) appear to bind to tubulin close to the Taxol site (Figure 14.38). The negatively charged N-terminal domains of tau (orange) are envisaged to project outward, repelled electrostatically from the MT surface. (C) Inner surface of an EM reconstruction of a MT assembled with tau. Yellow flecks represent nano-gold particles, which have a core of 40–50 gold atoms, ~1 nm in diameter, labeling the repeat region. (From S. Kar et al., *EMBO J.* 22:70–77, 2003. With permission from John Wiley & Sons.)

MAP activity is controlled by phosphorylation

Although MAPs mostly have net positive charge, they have many sites that are susceptible to charge alteration by phosphorylation, and this form of regulation is key to their activity. Tau, for instance, has a high incidence (~20%) of susceptible residues (serine, threonine, and tyrosine) and an extended, therefore kinase-accessible, conformation. The overall picture is complex, because many kinases have been implicated in phosphorylating tau (and other MAPs). Nevertheless, some general statements can be made. Phosphorylation—notably, of the repeat regions—disfavors MAP binding to MTs, thereby decreasing MT stability, a property that promotes MT dynamics, as in mitosis. At another level, phosphorylation appears to promote tau/tau interactions, and this property impacts some neurological diseases, in which excessive phosphorylation of neuronal tau leads to less stable axonal MTs and the accumulation of paired helical filaments.

14.7   MOTORIZED TRANSPORT ALONG MICROTUBULES

The motor proteins that use MTs as their tracks fall into two families: **kinesins** and **dyneins**. Most kinesins move toward the plus end, although an important subset is minus end-directed. All dyneins move toward the minus end. Processive kinesins move linearly along individual protofilaments, making multiple steps without losing contact with their tracks. Processive dynein motors, in contrast, can drift sideways, a property that may allow them to side-step obstacles, including kinesins coming from the opposite direction.

Kinesin families have distinctive domain arrangements

There is a large and varied superfamily of kinesin-like proteins (KLPs); currently, 14 families are distinguished and designated by arabic numerals, for example kinesin-1. They have in common a motor domain of ~320 residues with ATPase activity (**Figure 14.46**). Most KLPs have this domain at their N terminus. These so-called N-type kinesins move toward the plus ends of MTs.

A smaller group that includes the *Drosophila* protein Ncd (non-claret, disjunctional) have their motor domain at the C terminus (hence, C-type), and they move toward the minus end. These motors, or 'heads,' are complemented with 'tail' domains, which bind the cargoes to be transported. Usually, heads and tails are connected by a coiled-coil stalk. However, KLPs belonging to the mitotic centromere-associated kinesin (MCAK) family have domains on both sides of the motor domain and hence are termed M-type (M, for middle). It is not yet clear whether they work as motors, but some bind preferentially to MT ends, where they apparently trigger a *catastrophe* by inducing a destabilizing conformational change (see Figure 14.41).

Many motor domains have been visualized by X-ray crystallography, showing only small variations (**Figure 14.47**). Intriguingly, their fold turned out to be similar to that of the myosin motor domain (see Figure 14.11), based on a 7-stranded β sheet with associated α helices. As in the myosin motor, loops called Switches I and II (SWI and SWII) are involved in the control of ATP hydrolysis. Many kinesins have a coiled coil-forming 'stalk' segment via which subunits dimerize. In the prototypic N-type kinesin, kinesin-1, the coiled coil is connected to the motor domain by the 'neck linker,' which has an important role in force generation, and is followed by its cargo-binding C-terminal domain.

Kinesin molecules vary in their oligomeric character, in their content of light chains, and particularly in their tail domains. Some examples are shown in **Figure 14.48**. The two light chains of the kinesin-1 dimer (KLCs) assist the C-terminal tail domains in binding cargo. When kinesin-1 is folded as shown, contact between the motor domain and the C-terminal domain inhibits the ATPase activity. A similar interaction has been observed in myosin II (see Figure 14.34). The kinesin-2 dimer has two different heavy chains and one copy of an associated protein (KAP). Kar3, a kinesin-14 from yeast, pairs with Cik1 or Vik1 in heterodimers that bind to tubulin but have no ATPase activity. Kinesin-13, an M-type kinesin, dimerizes via two short coiled-coil-forming segments on either side of its motor domain. Kinesin-5 (for example XKLP2 from *Xenopus*) forms bipolar tetramers that form a bridge between antiparallel arrays of MTs, as in the 'central spindle' bundle that pushes mitotic poles apart in *anaphase* (see Figure 14.10B; Section 13.5).

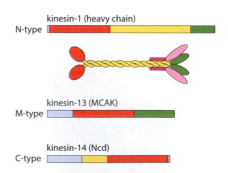

Figure 14.46 Domain organization of N-type, C-type, and M-type kinesins. Red, motor domain; yellow, coiled-coil domain; green, C-terminal domain; light blue, N-terminal domain. Also shown is a diagram of the kinesin-1 dimer. (Adapted from A. Marx et al., *Adv. Prot. Chem.* 71:299–345, 2005. With permission from Elsevier.)

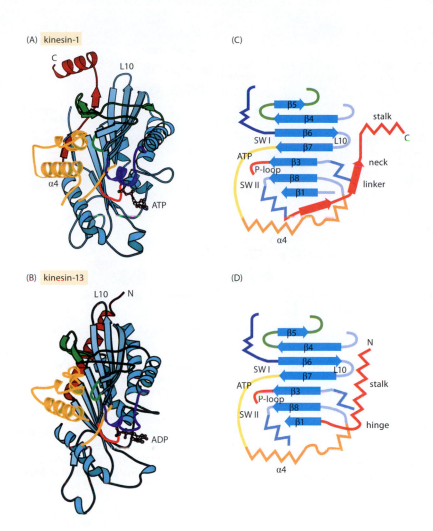

(A) kinesin-1

(B) kinesin-13

(C)

(D)

Figure 14.47 Kinesin structures.
(A) Kinesin-1 motor domain plus neck linker and part of the stalk. (PDB 2KIN)
(B) Kinesin-13 motor domain plus part of its stalk. (PDB 1CZ7) (C, D) Their topology maps. The neck linker of plus end-directed kinesins (for example kinesin-1) is fixed to the side of the motor domain in what is thought to represent the ATP-bound conformation but is disordered in the ADP-bound state. A minus end-directed motor domain (for example kinesin-13) binds its stalk in the ADP-bound state and releases it in the ATP-bound state.

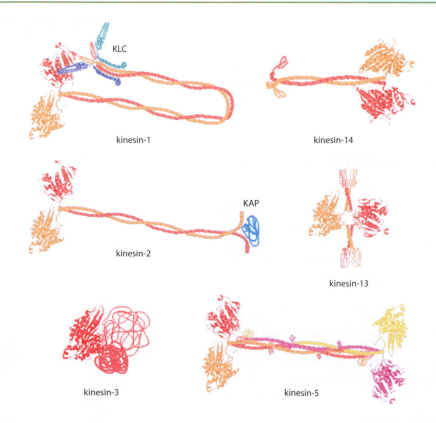

kinesin-1

kinesin-14

kinesin-2

kinesin-13

kinesin-3

kinesin-5

Figure 14.48 Models of some kinesins.
Kinesins 1, 2, 13, and 14 are dimers; the two heavy chains are colored red and yellow. Kinesin-3 is a monomer, and kinesin-5 is a bipolar tetramer. The kinesin light chains (KLC) are shown in cyan and blue. KAP, kinesin-associated protein. As shown, the tail domains are based mainly on coiled-coil predictions. (Courtesy of Linda A. Amos.)

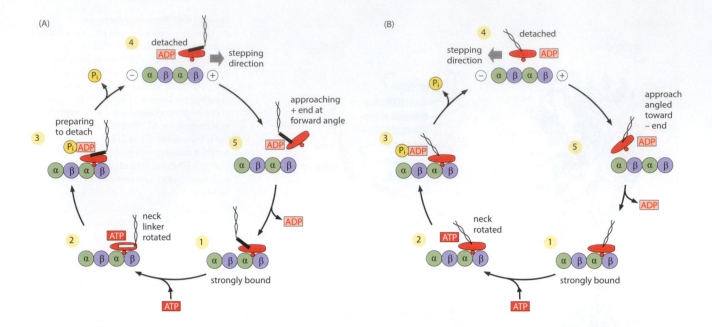

Interaction of kinesin with microtubules is controlled by the ATPase cycle of the motor domain

The force generated by a kinesin motor has been measured by 'single-molecule' techniques to be ~5 pN, and the average step-length is 8 nm (see Figure 1.23), the spacing of hetero-dimers along a protofilament. Considering kinesin as an engine transducing energy from ATP hydrolysis, this represents ~50% efficiency.

The functional cycles for two members of the kinesin family are shown schematically in **Figure 14.49**. The basic requirement is for a motor domain to bind tightly to tubulin during part of the ATPase cycle (stages 1 and 2 in Figure 14.49A) and weakly or not at all during the other stages.

To ensure progress along the track, there should be a regulatory conformational change. Unlike myosins and dyneins, kinesins bind only weakly to their tracks when they contain ADP (stage 3); their re-binding to tubulin releases the ADP (stages 4–5), and the nucleo-tide-free motor domain then binds strongly to the MT (stage 1). (The nucleotide-free 'rigor state' is similar in this respect for all three motor proteins.) Binding of ATP to kinesin motor domains is thought to bring about a crucial conformational change. Nevertheless, the motor remains strongly bound to the MT while ATP binds and its hydrolysis is catalyzed. When phosphate has been released, the motor is able to detach and revert to its original ADP-bound conformation, ready to find a new binding site.

Most kinesin motors cannot operate singly but must collaborate in groups

Some kinesins are monomers and others are dimers in which only one head is active at any time. Multi-step movement of a cargo along a MT may be accomplished by non-processive motors if more than one molecule is bound to the same cargo. The cooperation of two motors to effect directed movement is illustrated in **Figure 14.50**. The direction of move-ment is determined by nucleotide-dependent changes in the way in which the stalk (which connects the motor domain to the tail) associates with the motor domain. For a plus end-directed kinesin, the angle of the stalk in the ADP-bound state is such that an unattached motor domain can bind most easily to a site closer to the plus end (Figure 14.50A, stage 5). While the motor domain is bound to the MT, the neck linker attaches to the side of the motor domain (Figure 14.50A, stage 2), biasing the diffusion of the cargo toward the plus end. If another motor domain attached to the same cargo then binds to a new site, forward movement is established and this is the point in the cycle when force is exerted on the cargo. Now another step can be initiated. The more motors there are working together, the more likely it is that a step made by one motor will pull another one close to a binding site.

Figure 14.49 Conformational changes in kinesin ATPase cycles. Cycles are shown for a plus end-directed (A) and a minus end-directed kinesin (B). In (A) the neck linker is free (black bar) at most stages but closely associated with the motor domain when ATP is bound (white bar, stage 2). In contrast, in (B) the stalk associates closely with the ADP-bound motor domain after detachment from the MT (stage 5). At each stage in the cycle, the likely state is shown (although detachment/reattachment to the MT is possible at any stage). Simply attaching to and detaching from the MT does not lead to movement along it; this requires at least two motors acting out of step (Figure 14.50).

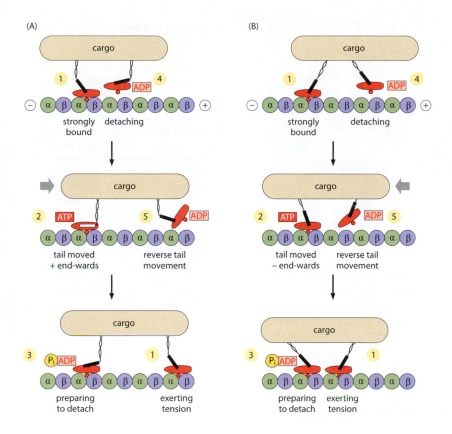

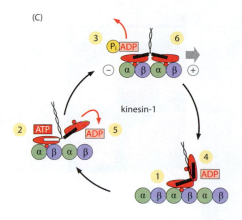

Figure 14.50 Cooperation between kinesin motors produces directed movement. Motors are shown working in pairs in plus end-directed (A) or minus end-directed (B) transport. Swinging of the stalk of an attached motor (stages 1 to 2) allows the other one to find a new binding site (stages 4 to 5). In (A), this site is closer to the plus end; in (B), to the minus end. The direction of movement is determined by the angles of tilt between a motor and its stalk, before attachment (stage 5) and then before detachment (stage 3). After the cargo has settled in the new position (stages 3 and 6, both motors bound), the left motor can continue through stages 4–6 of its cycle, while the other goes through the bound conformational change. Stage 6 differs from stage 1 only in that the cooperating motor is shown still attached. (C) Plus end-directed processive movement by a two-headed kinesin (kinesin-1).

Minus end-directed kinesins work in a similar way except that the stalk associates with the side of the motor domain when this is not bound to tubulin (Figure 14.50B, stage 5). The stalk starts off angled toward the plus end, and when freed by binding of the motor domain to a MT, the stalk can only rotate toward the minus end (Figure 14.50B, stage 2).

In contrast, a processive dimer such as kinesin-1 can move for micrometers along a MT without falling off. The two motor domains alternate in taking 8 nm steps along a protofilament. The dimer geometry requires that they must move past each other on opposite sides of the shared stalk. This motion is like two-legged walking, except that the two legs are basically identical, not mirror images.

Kinesin movement is best explained by a Brownian ratchet mechanism

For myosin motors, the *power stroke* has provided a very useful concept (Section 14.3). In this clockwork-like mechanism, energy derived from ATP hydrolysis drives a sequence of conformational changes in the motor domain that produces the cyclic motion of a lever arm. In the largest change, the power stroke, a step is taken along the track. Instead, kinesin appears to employ a *Brownian ratchet* mechanism (Section 14.9), whereby the motor domain shifts to its next binding site along the track by random (Brownian) motion driven by thermal energy. Instead of directly generating force, the binding and hydrolysis of ATP provide signals to control the interactions of the motor with the track. Thus, movement to a new site and the exertion of force after arrival may be considered as separate processes. A longer lever arm would simply allow a wider search for a new binding site, rather than amplifying a switch-like movement in the motor domain.

The binding of motors can be visualized by cryo-EM of decorated microtubules

In vivo, motor proteins are sparsely distributed over any given MT. *In vitro*, however, a large excess of motors may be used to saturate the MT lattice, with one kinesin attached to every tubulin heterodimer (**Figure 14.51A**). Cryo-electron micrographs of MTs decorated in this way may be reconstructed to visualize the interaction of the motor domain with the

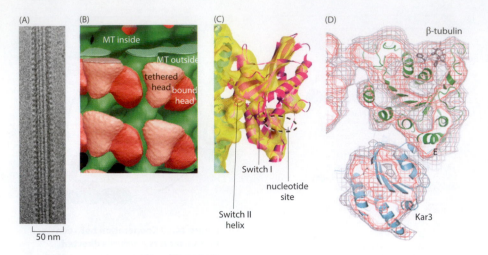

Figure 14.51 **Interaction of kinesin-1 with a microtubule.** (A) Cryo-EM of a MT decorated with kinesin-1 monomers. (B) Cryo-EM reconstruction of dimeric kinesin-1 motor domains in the nucleotide-free state. The MT is shown in green. (C, D) Cryo-EM density maps with kinesins and tubulin fitted into them. (C) A monomeric kinesin-1 motor domain in side view. (D) Axial view from the plus end of a slab of density from a reconstruction of the monomeric minus end-directed kinesin Kar3, binding to a MT. ADP state, stabilized by Taxol (bound at top right). (A, courtesy of Andreas Hoenger; B, adapted from M.C. Alonso et al., *Science* 316:120–123, 2007; C, adapted from C.V. Sindelar and K.H. Downing, *J. Cell Biol.* 177:377–385, 2007. With permission from Rockefeller University Press; D, adapted from K. Hirose et al., *Mol. Cell* 23:913–923, 2006. With permission from Elsevier.)

MT. Examples are shown in Figure 14.51. In a dimeric kinesin-1 construct in the rigor state (Figure 14.51B), the two motors are in distinct conformations. In the reconstruction with a monomeric kinesin-1 motor (Figure 14.51C), the SWII motif positioned at one end of helix α4 (see also Figure 14.47) is seen to make a primary contact with tubulin, thus explaining the MT-dependent stimulation of ATP hydrolysis. Figure 14.51D shows the ADP-bound state of the minus end-directed kinesin-14, Kar. By comparing this structure with other nucleotide states, it was possible to demonstrate that the duty cycle of this motor domain involves a twist of its central β sheet, as in myosin.

Dynein is a ring-shaped molecule with six AAA+ ATPase domains

The functional unit for dynein is a monomer, dimer, or trimer of heavy chains, depending on the isoform (**Figure 14.52C**), together with several light chains (LCs), intermediate chains (ICs), and/or light-intermediate chains (LICs) associated with the stem (also called the tail). The heavy chain, a massive polypeptide of ~500 kDa, is an unusual member of the AAA+ ATPase family (Boxes 3.1 and 7.2). Instead of forming rings of six subunits, dynein has six AAA+ domains in tandem, called P1–P6 (Figure 14.52A). P1 binds and hydrolyzes

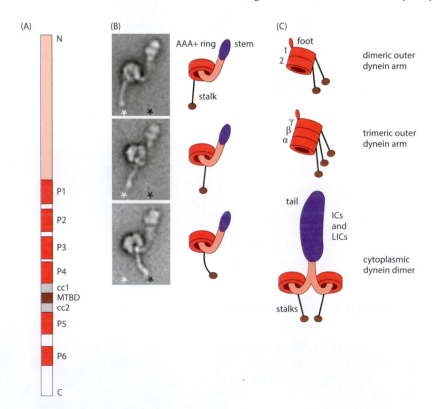

Figure 14.52 **Structures of dynein molecules.** (A) Domain map of a heavy chain. (B) Negatively stained EMs, enhanced by averaging, of a monomeric dynein in different conformations, with graphic interpretations. The molecule in the top panel has ADP and vanadate bound, considered equivalent to having ADP + P_i bound. The lower images show the nucleotide-free (apo) state in which the AAA+ ring (red disk) has rotated, bringing the stalk and stem closer together. Asterisks mark a reference point. (C) Axonemal outer-arm dyneins (Section 14.8) are heavy-chain heterodimers in most species but heterotrimers in the flagella of some protists. The AAA+ rings stack so that the stalks point in the same direction. Cytoplasmic dynein is a homodimer. IC, intermediate chain; LIC, light/intermediate chain. (B, from S.A. Burgess et al., *J. Struct. Biol.* 146:205–216, 2004. With permission from Elsevier.)

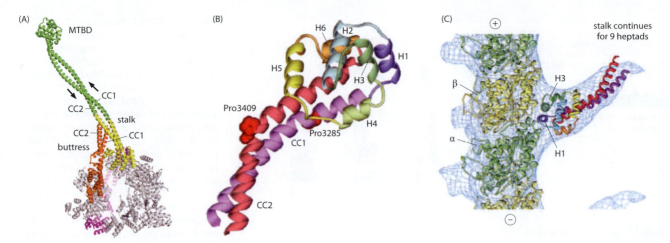

Figure 14.53 Structure of dynein and its mode of binding to microtubules. (A) The AAA+ ring plus the stalk. The buttress is a second coiled coil, emanating from the P4 AAA+ domain, that appears to interact with and stabilize the basal region of the stalk coiled coil. The purple region is part of the 'linker' N-terminal domain. MTBD, MT-binding domain; CC, coiled coil. (B) Structure of the MTBD plus ~30% of the stalk coiled coil. Two conserved prolines are in red. (C) Side view of a MT protofilament with the MTBD bound, depicted in a cryo-EM density map with crystal structure fitted in. (A, from A.P. Carter et al., *Science* 331:1159–1165, 2011; B and C, from A.P. Carter et al., *Science* 322:1691–1695, 2008. All with permission from AAAS.)

ATP. Measurements of ATP binding and sequence considerations indicate that P2, P3, and P4 together bind between one and three molecules of ATP, but P5 and P6 do not bind the nucleotide. The substructure of dynein has been visualized by EM of a monomeric axonemal dynein (Figure 14.52B). The six AAA+ modules form a ring from which the stem/tail and stalk protrude. EM studies of mutant dyneins with green fluorescent protein inserted separately into each of the six AAA+ modules have established that their positions around the ring follow their order in the primary sequence. A crystal structure of the motor domain (with stalk deleted) plus ~25% of the N-terminal domain (**Figure 14.53A**) has shown that the AAA+ ring is markedly asymmetric and that the six modules vary in having different insertions of secondary structure elements into the standard two-domain AAA+ fold.

Of the two appendages on dynein, the stalk interacts with MTs, and the stem binds associated proteins and engages cargoes. (Note that the kinesin 'stalk' connects to the cargo-binding tail.) The dynein stalk is contributed by the region between AAA+ modules P4 and P5; this region contains two heptad-rich tracts, each of ~100 residues, which pair to form an antiparallel coiled coil with a globular MT-binding domain (MTBD) at its tip (Figure 14.53B). A crystal structure of a fusion protein containing the MTBD and the proximal portion of the coiled coil (Figure 14.53B), conveying a weak binding state, shows a compact packing of six α helices. Three of these (H1, H3, and H6) bind to the interface between α-tubulin and β-tubulin subunits (Figure 14.53C).

The ATPase cycle of dynein powers minus end-directed transport

The ATPase cycle of dynein differs from that of the kinesins and is more like that of myosin; it is ATP binding, rather than ATP hydrolysis and loss of P_i, that causes dynein to detach from the track. A stepping cycle is given in **Figure 14.54** for transport by two monomeric dyneins attached to the same cargo.

Cytoplasmic dynein appears to operate differently, acting as a processive dimer whose two heads take turns in attaching their stalks to the MT (a detailed understanding of this process is still lacking).

Two key questions in dynein-based motility have been: first, how are conformational changes associated with ATP binding and hydrolysis in the AAA+ ring communicated to MTBD, some 15 nm distant at the tip of the stalk? and second, how do they affect its binding to MTs (and *vice versa*)? One proposal is that two-way communication is effected by relative sliding of the two strands of the coiled coil in the stalk (Figure 14.54B). The presence of kinks in the coiled coil at positions occupied by two conserved proline residues (see Figure 14.53B) is generally consistent with this idea.

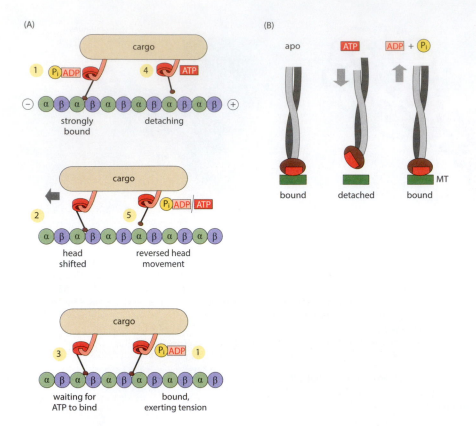

Figure 14.54 The ATPase cycle of dynein. (A) The dynein stalk is directed toward the minus end when first attaching (state 1, top left). The hydrolysis products, ADP and P_i, are released, causing the head (motor domain) to shift so that the attached cargo is moved toward the minus end (state 2). This shift allows a second dynein to bind at a new site (head at right, state 5). After a waiting period (state 3), the first head binds ATP and detaches (state 4). While detached, its motor hydrolyzes ATP and reverses its conformation (state 5; this part of the diagram is now referring to the first head). It then rebinds and is ready for another cycle (state 1, bottom right). Molecules cooperating in this way could be the two halves of a processive dimer or separate motors bound to the same cargo. Through their cooperation, the cargo/dynein complex stays continuously in contact with the MT. (B) The duty cycle of dynein's ATPase may involve a relative sliding of the two coiled-coil strands in the stalk (dark and light gray). In this scheme, the MT-binding site (red) is maintained while the products of ATP hydrolysis are lost and the AAA+ ring shifts (states 1–3 in (A)), but detaches when ATP binds (state 4).

Cytoplasmic dynein has important roles at both ends of microtubules

Whereas cells generally contain many kinesins, each with specific functions, usually only a single dynein heavy chain is expressed. The large size of dynein may equip it for multiple roles, not only in minus end-directed transport but also in the regulation of MT plus ends at various sites in the cell. One example of its role as a minus end-directed transporter is in keeping the many small compartments of the Golgi complex near the cell center during *interphase*. The minus ends of interphase MTs are anchored in and around the *centrosome*, near the nucleus (see Figure 14.40A).

Another example is in **fast retrograde axonal transport**. Axonal transport (see Figure 14.1) is vital for the growth and maintenance of neurons, and disruption of this process underlies many neurodegenerative diseases. Some of these diseases have been correlated with mutations in motor proteins, notably kinesin, and another with mutation in the dynein-associated protein p150, as outlined below. In fast axonal transport, vesicles are carried away from the cell body toward the nerve terminal by kinesins moving along MTs; others being returned to the cell body for membrane recycling are moved by cytoplasmic dynein. Both motor proteins are also involved in the transport of cytoskeletal proteins. Neurofilaments, for example, appear to be moved by kinesins (see Figure 14.1). Dynein is probably responsible for conveying short segments of MTs on which the transport of everything else depends.

Where growing MT plus ends reach the cell membrane, they are held in place by dynamic links consisting of cytoplasmic dynein bound to the membrane by the **dynactin** complex and tethered to an MT plus end by dedicated components. A schematic model is shown in **Figure 14.55A** of such a link as may occur in the axonal cell cortex; in it, dynactin connects to a membrane-bound actin filament and dynein to the MT plus end. This elaborate link is formed from several individually large subcomplexes, diagrammed in Figure 14.55B. Dynactin consists of a 37 nm-long filament of actin-like subunits capped at both ends, from which the p150 rod extends laterally. At its tip is a CAP-Gly domain (cytoskeletal-associated protein, glycine-rich), which binds to the MT. Its structure (Figure 14.55D) is quite different from that of the dynein MTBD (Figure 14.55E). Also present around the plus end are EB1 and CLIP170, both dimeric proteins. They appear to recognize, and may

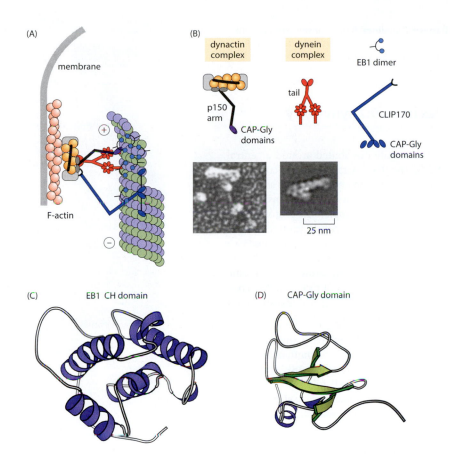

Figure 14.55 Cytoplasmic dynein as part of a plus end complex. (A) Dynein/dynactin and CLIP170 are shown making a mobile link from a MT end to F-actin. CLIP170 and EB1 target the complex to the plus end. The subcomplexes are diagrammed in (B). This dynein has two heavy chains and several associated chains. Dynactin (~1.25 MDa) is based on a doubly capped filament of actin-related protein, ARP1. EMs of purified dynactin are shown in (E). The averaged negatively stained image (right) brings out the subunits in the ARP1 filament but the flexible p150 side-arm is smeared out and lost. The metal-shadowed image (left) visualizes the p150 arm. The MT-binding affinity of EB1 resides in its CH domain (in (C): PDB 1PA7). Both dynactin and CLIP170 bind to tubulin via CAP-Gly domains (in (D): PDB 1LPL); the dynactin p150 subunit has one such domain at its N terminus. Each CLIP170 subunit has two CH domains in tandem, one of which may be used to hold on to the C terminus of EB1. (B, from H. Imai et al., *J. Mol. Biol.* 359:833–839, 2006 [left] and D.A. Schafer et al., *J. Cell Biol.* 126:403–412, 1994 [right]. With permission from Elsevier and Rockefeller University Press.)

maintain, specific tubulin conformations in the out-curving endpiece. CLIP170, like p150, terminates in a CAP-Gly domain. EB1 has a <u>c</u>alponin-<u>h</u>omology (CH) domain (see **Guide**) (Figure 14.55C).

The overall complexity of this link assembly hints at sophisticated control mechanisms operating on multiple levels. The motor domains of dynein generate movement toward the minus end of the attached MT. Because dynein is anchored on the membrane, these forces have the effect of driving the MT in the direction of its plus end.

14.8 MOTILE ORGANELLES BUILT FROM MICROTUBULES

The **cilia** and **flagella** of eukaryotic cells are elongated cellular protrusions that generate propulsion by performing a rhythmic motion called *beating* (**Figure 14.56**). They have similar architectures, consisting of a bundle of MTs, known as an **axoneme**, surrounded by a tubular extension of the plasma membrane. Their motility is generated by clusters of MT-bound dynein motors whose ATPase activity causes neighboring MTs to slide axially relative to each other. These local movements are coordinated throughout the axoneme to produce the beating motion of the flagellum or cilium.

Many unicellular organisms have a single flagellum, typically 50 μm long by 2 μm in diameter, protruding from one pole of the cell. Its beating is translated into locomotion by shear forces in the surrounding medium, whereby the cell is propelled in the direction away from the pole with the flagellum. Motility is an essential attribute of flagellated cells; for instance, nonmotile mammalian sperm result in male sterility. Cilia tend to be shorter than flagella—about 15 μm long by 2 μm wide—and are found in arrays on exposed surfaces of multicellular tissues, including many epithelia, where their beating generates turbulence that serves to repel invading microbes. Nevertheless, cilia are points of entry for several viruses, and their immobilization by viral infection is followed by bacterial infestation of the respiratory mucosa, giving rise to the characteristic symptom of nasal discharge.

Figure 14.56 Beating of a bull sperm flagellum. A time-lapsed series of light micrographs; each bend starts at the head and propagates toward the tip. (From I.H. Riedel-Kruse et al., *HFSP J.* 1:192–208, 2007. With permission from Taylor & Francis.)

The axonemal bundle of microtubules has a 9:2 symmetry mismatch

An axoneme has a ring of nine doublet MTs symmetrically distributed around a central complex containing two singlet MTs (**Figure 14.57**). The 9 + 2 arrangement is conserved across eukaryotes, although in some species the center is occupied by a different structure, and many species have a surrounding cylinder of additional MTs or other proteins that are thought to confer extra strength. Each doublet consists of a one complete 13-protofilament MT—the A-tubule—on to which is grafted an incomplete B-tubule with 10 protofilaments. Axonemal MTs are interconnected by an elaborate network of associated proteins. Two columns of dynein molecules—the outer and inner arms—are fixed to each A-tubule and interact dynamically with the neighboring B-tubule. A cross-linking protein with elastic properties, called **nexin**, also connects adjacent doublets. A column of radial spokes connects each outer doublet MT to the central complex. There are also over 100 less abundant protein components whose roles and locations are largely unknown.

An axoneme grows out from a template known as a **basal body**, which is the structural equivalent of a centriole (Section 13.5). The basal body resembles a short length of axoneme except that it contains nine triplet MTs instead of nine doublets (**Figure 14.58**). An incomplete C-tubule is attached to each B-tubule in the same manner as the B-tubule is attached to an A-tubule. The center is filled with an amorphous matrix except near the basal end, where there is a cartwheel that is the first structure to appear when a new basal body or centriole is being assembled and therefore may serve as a starting template for the 9-fold symmetry.

Intra-flagellar transport (IFT) is effected by cytoplasmic motor proteins, which are directed toward the plus ends of the MTs. Assemblies of substantial size are transported to and from the tips by kinesin-2 and cytoplasmic dynein, apparently running along the outer surface of outer doublet MTs.

Dynein heads are stacked in the outer and inner arms

In most species, the outer arms are two-headed—that is, their dyneins are dimeric—but they have three heads in several well-studied protozoan systems, including *Chlamydomonas*, whose axonemes have been visualized by cryo-electron tomography (see Figure 14.57A and **Figure 14.59A**). The triplets of outer-arm dynein heads attach to the A-tubules at axial

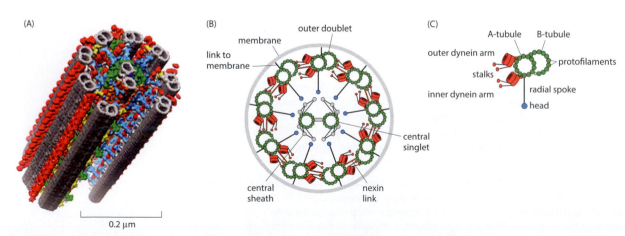

Figure 14.57 Organization of a eukaryotic flagellum. (A) Cryo-electron tomogram of a *Chlamydomonas* axoneme, with one MT doublet removed to expose internal structure: MTs gray; dyneins (three per arm) red; radial spokes blue. (B) Diagram of cross section through a mammalian axoneme, viewed from its tip; that is, from the MT plus ends. (C) Enlarged cross section of a doublet MT, showing the complete A-tubule and incomplete B-tubule. The first protofilament of the B-tubule at the outer junction makes direct contact with tubulin in the A-tubule, but a column of additional proteins (not shown) makes the link at the inner junction. The shared wall is strengthened by longitudinal filaments composed of tektins, which are related to intermediate filament proteins. (A, courtesy of Takashi Ishikawa.)

Figure 14.58 **The axoneme and the basal body both observe 9-fold symmetry.** Electron micrographs of thin sections through (A) the axoneme and (B) the basal body of a sea-urchin sperm tail. (From R.W. Linck and R.E. Stephens, *Cell Motil. Cytoskel.* 64:489–495, 2007. With permission from John Wiley & Sons.)

(A) axoneme

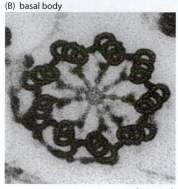

(B) basal body

50 nm

intervals of 24 nm, corresponding to three 8 nm steps of tubulin heterodimers. The column of inner arm dyneins has a more complex arrangement with three distinct isotypes in each axial repeat; they are spaced unequally but on average are 32 nm apart. The radial spokes also have unequal axial spacings, matching those of the inner dynein arms. Axonemal beating is apparently controlled by a regulatory complex that connects the bases of the inner arms and the radial spokes. Activation is initiated at the base of the flagellum. Dynein arms further along the doublets are thought to be turned on by a bending wave that passes along the axoneme and is detected via the radial spokes.

Beating results from dynein-driven sliding of adjacent microtubules

In its basic beating mode, a sperm flagellum undulates rapidly in a planar quasi-sinusoidal waveform (see Figure 14.56), with a frequency of 30–40 Hz. The plane in which beating takes place is perpendicular to that of the central pair of singlet MTs (this is one functional correlate of the 9:2 symmetry mismatch). Under particular conditions, for example close proximity to a fixed surface—which affects hydrodynamic drag—or physiological impairment, the waveform and frequency may differ. Nevertheless, the geometrical simplicity of the basic waveform has allowed quantitative analysis of the force-generating mechanism in terms of classical hydrodynamic theory in which the viscous flow fields around the flagellum are calculated. Other flagella, such as the pair on a *Chlamydomonas* cell, undergo more complicated motions with non-symmetric forward and backward strokes in their beating cycle. In this case, the central complex appears to rotate so as to drive rotation in the direction of bending.

Cilia also beat asymmetrically, but they are shorter than flagella and packed close together; accordingly, the motions of neighboring cilia must be coupled and they are brought into concerted motion by hydrodynamic forces. With both flagella and cilia, the energy source for motility resides ultimately in ATP hydrolysis by MT-bound dynein motors, interacting with adjacent MTs. The dynein arms on one doublet MT slide toward the minus end of the neighboring doublet, at the base of the axoneme. This movement has been demonstrated by measuring the sliding of doublet MTs in arrays in which the cylinder has been opened up by proteolytic treatment.

Sliding in opened cylinders and beating of closed tubules can be observed, even when the outer dynein arms have been extracted with buffer or inactivated by mutation. Their role appears to be enhancement of the motile force generated by the inner arms. Longitudinal sliding within the axoneme produced by the ATPase activity of dynein is converted to bending by other cross-bridging proteins such as nexin and the radial spokes. Globular domains at the inner ends of the radial spokes are thought to slide in a ratchet-like manner against a sheath-like structure surrounding the central complex (**Figure 14.60**).

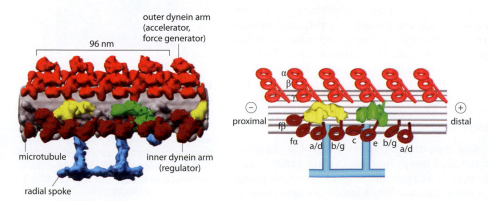

outer dynein arm (accelerator, force generator)

96 nm

α β γ

proximal   fβ   fα   a/d b/g   c   e b/g a/d   distal

(−)   (+)

microtubule   inner dynein arm (regulator)

radial spoke

Figure 14.59 ***Chlamydomonas* axoneme imaged by cryo-electron tomography.** A, segmented rendering; B, component map. The dynein regulatory complex (green) and dynein-associated intermediate chain/light chain (yellow) connect to the inner dynein arms. Of the eight inner-arm dynein heads, f is heterodimeric and a, b, c, d, e, and g are monomeric. These dyneins have distinct mechanical roles and are selectively omitted in certain mutant strains. (Adapted from K.H. Bui et al., *J. Cell Biol.* 183:923–932, 2008. With permission from Rockefeller University Press.)

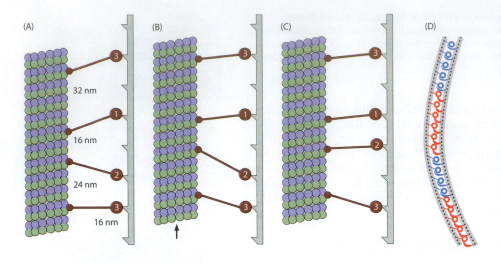

Figure 14.60 Controlling sliding and bending in axonemes. (A) The radial spokes project from the A-tubule of a doublet MT at unequal spacings (but each is a multiple of 8 nm), whereas the projections from the central complex to which they bind follow a regular 16 nm repeat. (B) Sliding of the outer doublets causes attached radial spokes to bend or pivot. When the bending angle exceeds 30° or so, the head detaches and then reattaches (for example molecule 2 in (B)→(C)). (D) Model of bending induced by the dyneins along a given A-tubule protofilament being clustered according to nucleotide-binding state—red, ADP; blue, apo (no nucleotide)—with differing intermolecular spacings according to that state. (From T. Movassagh et al., *Nature Struct. Mol. Biol.* 17:761–767, 2010. With permission from Macmillan Publishers Ltd.)

Nexin links limit the distance that the dynein arms can move before being forced to detach, whereupon they are carried back to their original positions by sliding movements on the opposite side of the axonemal cylinder. The combined interactions of all the cross-bridging elements convert relative sliding between pairs of doublet MTs into rhythmic bending of the whole axoneme. Much of the evident complexity seems to be needed for precise control of the beating pattern.

14.9 POLYMERIZATION/DEPOLYMERIZATION MACHINES

Forces generated by the polymerization of protein subunits into filaments and their depolymerization from filaments underlie a number of fundamental cellular processes. Substantial forces can be generated by an actin polymerization machine, as illustrated by the striking (albeit pathological) example given by the deformation of sickled erythrocytes by filaments of the mutant hemoglobin, HbS (see Box 1.3). The polymerization forces thus exerted must overcome the resistance presented by the viscoelastic membrane-lining cytoskeletons of these cells. Similar deformations may be induced in giant lipid vesicles formed in the presence of actin or tubulin subunits by switching conditions to favor polymerization of protein trapped inside the vesicle (**Figure 14.61**).

The absence of any strong evidence for motor proteins in prokaryotes suggests that the forces applied during their septation and cell division are based on polymerization reactions.

Brownian ratchets generate force by channeling random motions in particular directions

Protein filaments are in a state of dynamic exchange with the unassembled pool, with subunits randomly attaching to and detaching from filament ends. If the pool concentration exceeds the *critical concentration*, filaments tend to lengthen; below it, they shorten. Force generation by harnessing random fluctuations is also the principle underlying the *Brownian ratchet* devised by Richard Feynman (**Figure 14.62**) and viewed as a perpetual motion machine. In this case, the fluctuations are of gas molecules colliding with vanes mounted on a frictionless axle whose turning is biased by a ratchet. This virtual machine fails as a perpetual motion machine because thermal fluctuations in the mechanism may also allow rotational steps in the opposite direction from that favored by the ratchet. However, the biasing of random motions does indeed occur in macromolecular Brownian ratchets, but in the context of external energy sources—often, the hydrolysis of nucleotide triphosphates—so that the Second Law of Thermodynamics remains inviolate.

Another way to envisage the biasing of random motions invokes populations of particles that are partitioned by a periodic asymmetric potential energy function (**Figure 14.63**). When the potential is switched off, the particles in each well can diffuse away from their points of origin. When it is switched on again, a proportionately larger fraction of newcomers is

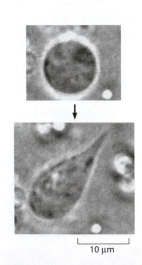

10 μm

Figure 14.61 Actin polymerization can deform a membrane vesicle. Phase-contrast micrographs of an actin-containing liposome before (top) and after (bottom) electroporation to allow ions to enter and promote polymerization. (From H. Miyata et al., *Proc. Natl Acad. Sci. USA* 96:2048–2053, 1999 National Academy of Sciences. With permission from PNAS..)

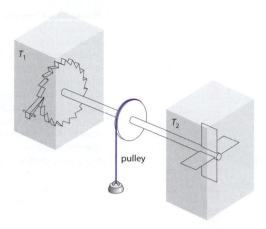

Figure 14.62 Feynman's Brownian ratchet. Fluctuations in the impinging of gas molecules on the vanes in the right-hand compartment will cause the axle to turn. The ratchet in the left-hand compartment allows rotation to take place in one direction only. In principle, this biased rotation could do work, raising a load on the pulley. If the two compartments are at the same temperature ($T_1 = T_2$), this is an apparent violation of the Second Law of Thermodynamics. (Adapted from a figure in The Feynman Lectures on Physics, vol. 1. Addison Wesley Longman, 1970.)

trapped in the shallow-sloped side of each well. In this way, cycling on and off leads to a net flux of particles in one direction. Here the asymmetry of the potential function is essential for a net flux to be achieved.

This model is applicable (as a simplification) to the case of kinesin, whereby switching on of the potential corresponds to the motor switching into a tight-binding state and the asymmetry arises both from structural polarity of the MT and the conformation of kinesin. The notion of order coming from disorder is an unsettling one but, as Dean Astumian has observed, "any microscopic machine must either work with Brownian motion or fight against it, and the former seems to be the preferred choice."

To give a comprehensive account of the *in vivo* operation of a polymerization machine is complicated by a number of factors, including uncertainty as to the local concentrations of polymerizable subunits, the availability of free ends, the interactions of regulatory proteins, and molecular crowding effects. Nevertheless, the basic interaction is straightforward (**Figure 14.64A**). At equilibrium, the critical concentration of free subunits (C_c) is related to the kinetic rate constants for the 'on' and 'off' reactions by $C_c = k_{off}/k_{on}$. If the filament is under a load and already abutting an object (Figure 14.64B,C), C_c is increased because it is harder for subunits to intercalate than if the end were freely exposed. Thus, $C_c' = C_c \exp(Fd/kT)$, where F is the applied force, d is the axial step per subunit in the filament, k is Boltzmann's constant, and T is the temperature. For actin monomers and tubulin heterodimers at concentrations in the mg/ml range, this relationship yields theoretical values for forces in the piconewton (pN) range, which is comparable with the forces that have been measured for myosin and kinesin motors. It is important to note that structural polarity is essential if a filament is to serve as a force-generating polymerization machine (this is equivalent to the asymmetry of the potential function in Figure 14.63). For this reason, intermediate filaments, which are nonpolar (Section 11.2), are ineligible, just as they are for motorized transport.

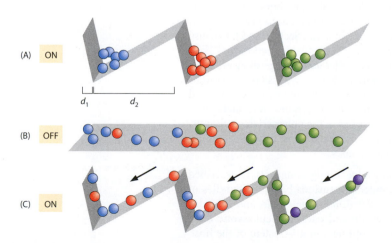

(A) ON

d_1 d_2

(B) OFF

(C) ON

Figure 14.63 Cycling of an asymmetric potential generates a net flux of particles. (A) The populations of identical non-interacting particles in each potential well are color-coded accordingly. (B) When the potential is switched off, the particles may diffuse away from their original positions. (C) When it is switched on again, they are again confined to particular wells. The rate of the net flux of particles to the left depends on the asymmetry factor d_1/d_2, the diffusion constant D, and the cycle time T. (Adapted from a figure courtesy of H. Linke.)

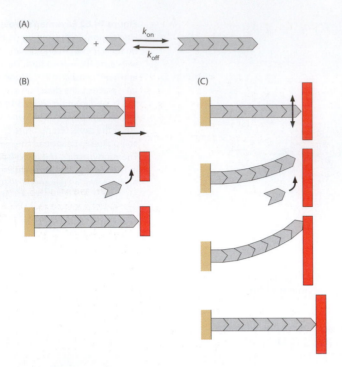

Figure 14.64 Force generation by a growing polymer. (A) Addition and loss of subunits from the end of a filament. More generally, subunits can be bound or lost from both ends. (B) The left end of the growing filament is anchored and the right end abuts the object (red) to be pushed. Fluctuations transiently open spaces in which additional subunits may bind to the end of the filament, preventing the object from re-entering this space, thereby pushing it. In (C), random bending motions transiently expose the filament end allowing recruitment of additional subunits, whereupon straightening pushes the object to the right. (Adapted from J.A. Theriot, *Traffic* 1:1–28, 2000. With permission from John Wiley & Sons.)

It has proved more difficult to measure the forces generated by individual growing actin filaments or MTs by 'single-molecule' methods than to measure those of motor proteins, but such data as have been acquired suggest that the force does not exceed ~1 pN and that filaments in local bundles act independently.

Tubulin, actin, and other proteins function as polymerization engines

A number of proteins are capable of operating as **polymerization engines**. The MSP (<u>m</u>ajor <u>s</u>perm <u>p</u>rotein) of the nematode generates motility by polymerizing into filaments at the leading edge: subsequently, as new filaments form and the *lamellipodium* pushes forward, preexisting ones are moved to the rear and dismantled there. This activity has been reconstituted *in vitro*.

In bacteria, several processes are powered by polymerization. One of them involves the separation of daughter molecules after replication of low-copy plasmids by growing filaments of ParM, an actin homolog. This reaction resembles, in greatly simplified form, the separation of eukaryotic chromosomes after replication (Chapter 13). The proteins that form and re-form the circumferential ring complexes that govern how bacterial cells grow and divide also depend on polymerization reactions. The best characterized of these have turned out to be distant relatives of tubulin.

Actin and tubulin are indeed the most versatile molecules of this kind: both proteins generate force actively through their polymerization reactions. These reactions are coupled to the hydrolysis of GTP (for tubulin; Section 14.6) and ATP (for actin; Section 14.2). Because the GTP-bound and ATP-bound subunits have high affinities for filament ends, they have a strong tendency to polymerize, and the growing ends of filaments can push against cellular structures. Nucleotide hydrolysis, promoted by polymerization, yields GDP-bound or ADP-bound subunits that have low affinity for the polymer; this pushes the system toward depolymerization, and the shrinking ends can pull on attached cargoes. Although polymerization engines depend ultimately on intrinsic properties of the polymerizing subunits, associated proteins are essential for regulating their polymerization and depolymerization cycles and for interfacing the filaments with loads and cargoes.

Actin polymerization drives protrusion of the leading edge of migrating cells, as described below. It also has key roles in the formation of cellular protrusions based on bundles of actin filaments such as *filopodia, microvilli*, and the *stereocilia* of hair cells, and in endocytosis (Section 10.3). Some pathogenic bacteria invade and parasitize eukaryotic cells. Once they have entered a cell, they move about by co-opting a motile system of the host.

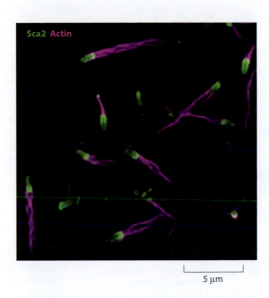

5 µm

Figure 14.65 *In vitro* formation of actin comets. The Sca2 protein of the Gram-negative bacterium *Rickettsia* (shown in green in this immunofluorescent light micrograph) nucleates the assembly of bundles of long unbranched actin filaments (purple, stained with Alexa dye-coupled phalloidin; see Figure 14.8). Sca2 shares many properties with the actin-regulating formin proteins of eukaryotes. (From C.M. Haglund et al., *Nature Cell Biol.* 12:1057–1063, 2010. With permission from Macmillan Publishers Ltd.)

Surface-exposed bacterial proteins at one end of the cell serve as nucleation centers that recruit polar bundles of actin filaments (**Figure 14.65**) whose polymerization pushes the bacterium around in the host cell cytoplasm. By visual analogy, these networks are called 'actin comet tails.'

Polymerization of tubulin underlies many aspects of the management of eukaryotic chromosomes, for example the formation and positioning of the mitotic spindle (see Figure 14.40B). Conversely, tubulin depolymerization drives chromosome segregation during anaphase, spindle positioning in metazoan cells and spindle length control, and the coupling of MTs to the kinetochore (Section 13.4). The addition or removal of subunits from the ends of dynamic MTs tends to be regulated not simply by the nucleotide-binding status of tubulin subunits but also through the activity of protein polymerases and depolymerases that can be elaborate, structurally as well as operationally (see Section 14.7).

Cell migration is powered by polymerization engines

Many cultured cells can crawl across solid substrates. Their rates of progression are slow compared with swimmers, for example several micrometers (a fraction of a cellular body-length) per minute, but they look livelier when accelerated in time-lapse movies. Such movements are termed **cell migration**. They are thought to recapitulate the behavior of cells in developing tissues or wound healing, as shown in the zebrafish system. Migratory cells have a marked structural polarity with respect to the direction of motion. Their movement takes place in three parts (**Figure 14.66**): first, the leading edge is pushed forward by outgrowth of actin filaments in the cross-linked network of the lamellipodium; second, forward progress is consolidated and reversal is blocked by the establishment of firm contacts with the substrate in *focal adhesions* (Section 11.6); third, older focal adhesions that are now at the trailing edge disengage and the rear of the cell is pulled forward. Each of these steps involves a complex interplay of the polymerizing and depolymerizing filaments with regulatory proteins. We focus here on lamellipodial protrusion and the machines that regulate this process.

Although based on the propensity of actin to polymerize, this process relies heavily on an extensive cast of associated proteins. The sequence of events is outlined in **Figure 14.67**. The signals that migration should proceed in a particular direction originate outside the cell and are transmitted into the cell by integrin receptors (steps 1–3 in Figure 14.67; see also Section 11.6 and Figure 11.26). They involve a cascade of phosphorylation events. The actin filaments in the lamellipodial network are oriented with their barbed ends pointing toward the cell membrane (or approximately so), so that polymerization forces are directed outward. The entire network serves as a platform that ensures that the polymerization force pushes the membrane outward instead of pushing the growing filament back into the cytoplasm.

New filaments nucleate on the sides of existing filaments and grow out as branches—an unusual form of higher-order actin assembly, where cross-linking is common but branching

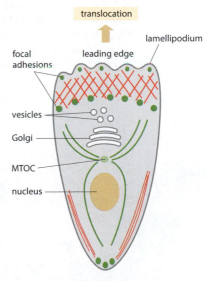

translocation

focal adhesions

leading edge

lamellipodium

vesicles

Golgi

MTOC

nucleus

Figure 14.66 A polarized migrating cell. In this diagram the cell is viewed from above. Red lines represent actin filaments and green lines represent MTs. MTOC, microtubule-organizing center. (Adapted from A.J. Ridley et al., *Science* 302:1704–1709, 2003. With permission from AAAS.)

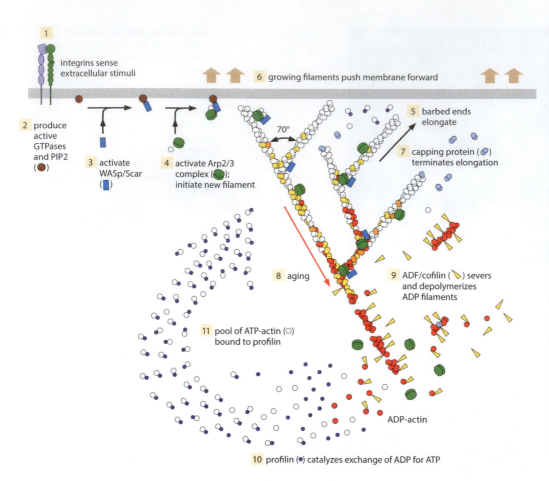

Figure 14.67 Model of an actin-based polymerization engine pushing out the leading edge of a migrating cell. (1) Extracellular signals activate receptors and, through them, (2) the Rho-family GTPases and PIP₂, which in turn (3) activate the actin-regulators, WASp/Scar (see Section 11.6). (4) Branch points with the Arp2/3 complex are nucleated on existing actin filaments. (5) Outgrowth at the barbed ends of new branches push the membrane forward (6). (7) Filament growth is terminated by a capping protein. (8) Filaments mature by hydrolysis of bound ATP (white subunits turn yellow) followed by dissociation of Pᵢ (subunits turn red). (9) ADF/cofilin promotes severing of ADP-actin filaments and dissociation of ADP-actin from pointed ends. (10) Profilin catalyzes nucleotide exchange (turning the subunits white), expanding the pool of polymerization-competent subunits (11). (Adapted from T.D. Pollard and G.G. Borisy, *Cell* 112:453–465, 2003. With permission from Elsevier.)

is rare. At the branch-point is a 7-subunit assembly called the *Arp2/3 complex* (Arp = actin-resembling protein; **Figure 14.68**) that nucleates outgrowth of the daughter filament. Two of its subunits are actin homologs, suggesting that the nucleation surface may resemble that of the actin trimer that nucleates F-actin assembly in a purified *in vitro* system (Section 14.2).

The ability of a polymerization engine to run depends on continuous availability of a high concentration of polymerizable subunits and on keeping the concentration of free filament ends relatively low. In dynamic lamellipodia, the natural propensity of actin-ADP subunits

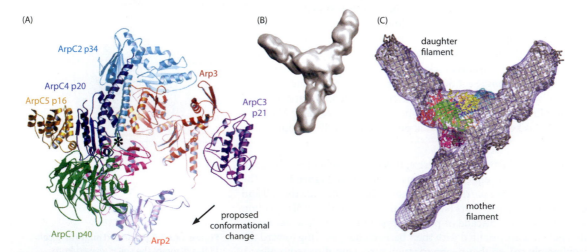

Figure 14.68 Structure of an actin filament branch point. At the branch point, the Arp2/3 complex grafts on to the side of the mother filament. (A) The crystal structure of the complex is thought to represent an inactive form because the two actin-like subunits, Arp2 and Arp3, are too far apart to act synergistically as a nucleation center; thus a major conformational change was suggested to accompany activation. (PDB 1K8K) (B) Electron tomographic (ET) reconstruction of a branch point. (C) The complex fitted into the ET density after modification consistent with a substantial conformational change. (From I. Rouiller et al., *J. Cell Biol.* 180:887–895, 2008. With permission from Rockefeller University Press.)

to shed from the pointed ends of filaments is boosted by the activities of a severing protein (ADF-cofilin) and a chaperone-like protein (profilin) that suppress any tendency for spontaneous nucleation. In this way, filament growth is restricted to existing barbed ends, where formin proteins further enhance the natural tendency of free subunits to add. If filaments at the growing edge were to become too long, they would bend, diverting the growth force away from the leading edge. To prevent this, they are capped after a certain point to prevent further growth.

14.10 MOTILITY POWERED BY SUPRAMOLECULAR SPRINGS

In other sections of this chapter, we consider a variety of motility mechanisms based on the hydrolysis of nucleotide triphosphates and conversion of the chemical energy thus released into physical work. Here we discuss several motile systems that operate on an entirely different principle—controlled release of conformational free energy from a macromolecular assembly in a state that represents a local, not a global, free energy minimum (**Figure 14.69**); as such, it is a store of potential energy. This energy is released in response to the appropriate trigger via appropriate conformational changes. All known systems of this kind, which resemble coiled springs under tension, are protein assemblies. Unlike motors and other cyclically repetitive machines, each such assembly (except for *resilin*—see below) generates a single powerful convulsive movement. To repeat its action, the system would have to be disassembled and reassembled *de novo*—an unlikely event, as the components of the spent system are, typically, tightly bonded together.

Contractile bacteriophage tails are dynamic gene-delivery systems

In terms of numbers, tailed bacteriophages are the most abundant biological entity on Earth, with a total population estimated at $\sim 10^{31}$. Their 'tails' are multiprotein assemblies that act as their infection organelles: they must recognize a susceptible host cell and deliver the phage genome into its cytoplasm. Because bacterial cell envelopes are rigid multilamellar structures (Section 10.6), infection must proceed by other mechanisms than the membrane fusion pathways employed by animal viruses (Section 8.7). Two classes of tails may be distinguished: contractile (for example, that of phage T4) and non-contractile (for example, that of phage λ). Both incorporate enzymes that degrade peptidoglycan in the cell wall, creating an access pathway. Beyond that, contractile tails are powerful motile devices that thrust through the entire cell envelope to deliver their genomes.

The assembly pathway of the archetype, T4, is outlined in Figure 8.2B. Its tail has four main components: the long tail-fibers; the baseplate; the rigid tail-tube through which DNA will pass; and, surrounding it, the contractile **tail-sheath**. The tail-sheath consists of 23 stacked hexameric rings of a 68 kDa protein called gp18 (gp = gene product), giving a length of 96 nm. In infection, after the long tail-fibers have bound to lipopolysaccharide receptors on the outer surface of the bacterium, a signal is transmitted to the baseplate, which undergoes an elaborate structural change that has two main consequences: first, the short tail-fibers unfold and extend out of the baseplate, attaching via their tips to secondary receptors on the cell surface; second, a wave of transformation is initiated in the baseplate-proximal part of the tail-sheath and propagates along its length (**Figure 14.70**). As it contracts, thrusting

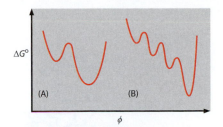

Figure 14.69 Free-energy cascades pass through progressively lower local minima. A dynamic system of this kind initially enters a metastable, kinetically accessible, precursor state (left-most minimum) and subsequently switches to a lower free energy ('mature') state. (A) A single precursor state precedes the mature state. (B) A cascade of three local minima precedes the final state. ϕ, conformational coordinates (for a macromolecular assembly, a high-dimensional space).

Figure 14.70 T4 tail-sheath contraction and host cell penetration. (A) The tips of the long tail-fibers are bound to their receptors, and the baseplate engages its receptor. (B) A conformational change in the baseplate induces the tail-sheath to contract in a transition that propagates wave-like toward the capsid. The short tail-fibers unfold and bind by their tips to the cell surface. The tail-tube is pushed into and through the cell envelope. (C) Tail contraction is complete. The tip of the tail-tube has penetrated into the host cell cytoplasm. (Adapted from F. Arisaka, *Chaos* 15:047502, 2005. With permission from American Institute of Physics.)

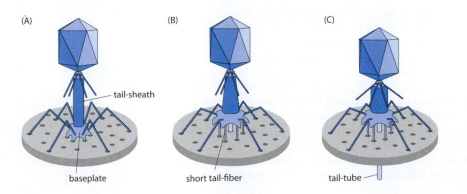

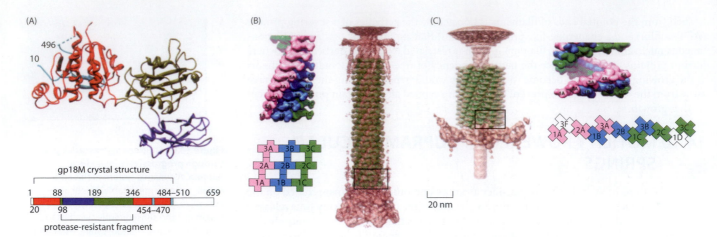

Figure 14.71 Contraction of the bacteriophage T4 tail-sheath.
(A) Structure of the tail-sheath subunit, gp18, residues 20–510 of this 659-aa protein. The domain map (below) is color-coded consistently with the ribbon diagram. (PDB 1FOA) (B, C) Cryo-EM reconstructions of the tail before contraction (B) and after contraction (C). The tail-sheath is a six-start helix in both conformations; that is, it consists of stacked hexameric rings (subunits in a ring labeled A–F; rings labeled 1, 2, 3,

and so on). The angular offset between adjacent rings generates helical strands, of which parts of three are shown (pink, blue, and green). The accompanying diagrams show their nearest-neighbor interactions and the subunit orientation in both states. Note also the major reorganization of the baseplate. (From A. Aksyuk et al., *EMBO J.* 28:821–829, 2009. With permission from John Wiley & Sons.)

the rigid tail-tube through the bacterial envelope, the tail-sheath shortens to 45 nm. The short tail-fibers act as stays that prevent the phage particle from being pushed away from the bacterial surface.

This process is closely regulated. Unlike most protein filaments, the length of the tail is uniquely specified. Assembly of the tail-sheath is templated on the underlying tail-tube (see Figure 8.2B), whose length, in turn, is specified by a *tape-measure* protein. If baseplates cannot form, through mutation of a key component, gp18 accumulates in the infected cell and eventually polymerizes into **polysheath**, which has the same subunit packing as in the contracted tail-sheath but is of indefinite length. The reason tail-sheaths normally assemble instead of polysheaths appears to be that the baseplate–tail-tube complex nucleates polymerization of gp18 at a concentration that is lower than the critical concentration for polysheath assembly.

Tail-sheath contraction has been shown to be exothermic; that is, it releases energy in the form of heat. As ascertained from an X-ray structure for gp18 and from cryo-EM reconstructions of tails before and after contraction, the transition involves rigid-body rotations of the gp18, shifting inter-subunit interactions to different molecular surfaces (**Figure 14.71**). From start to finish, each gp18 subunit moves radially outward by ~50 Å and rotates through 45°; however, the subunits remain in continuous contact throughout the contraction process.

Fertilization by *Limulus* sperm involves uncoiling a long bundle of cross-linked actin filaments

The eggs of certain marine invertebrates are protected by a jelly-like coating. To achieve fertilization, sperm cells must traverse this layer. They do so by uncoiling a 50 μm-long filamentous structure called the **acrosomal process** (AP) that pushes through the jelly layer and fuses at its tip with the underlying ovum (**Figure 14.72**). The AP consists of a highly ordered bundle of actin filaments cross-linked by a protein called scruin, enclosed within a membrane. Actin filaments in the AP are captured in a slightly different state of twist (by ~0.23° per subunit) than in F-actin free in solution, and the inferred torsional strain has been proposed to be the source of energy that drives the fertilization reaction. The trigger for uncoiling the AP appears to be the release of stored Ca^{2+}, which induces a conformational change in scruin. Although the local structural changes in the actin filaments are slight, they accomplish a major rearrangement of the overall structure of the AP and do so within a few seconds. This scenario is reminiscent of the hand-changing switch in bacterial

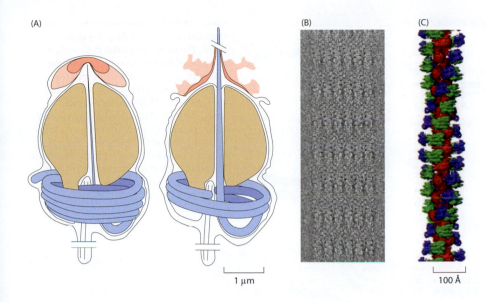

(A)   (B)   (C)

1 µm   100 Å

Figure 14.72 Uncoiling of the acrosomal process of *Limulus* sperm. (A) When uncoiling is triggered, the leading end of the process (blue tip) is thrust outward via a central channel through the sperm cell. (B, C) Cryo-electron micrograph (B) and reconstruction (C) of an actin filament (red) decorated with scruin at a 1:1 stoichiometry. Scruin has two domains, 50 kDa (blue) and 60 kDa (green). (A, adapted from D.J. DeRosier et al., *J. Cell Biol*. 93:324–337, 1982. With permission from Rockefeller University Press; B and C, from M.F. Schmid et al., *Nature* 431:104–107, 2004. With permission from Macmillan Publishers Ltd.)

flagella, which is also slight on the local level (see Figure 14.82) but drastic overall; it is thus qualitatively different from the large rigid-body rotations undergone by protein subunits in a contracting phage tail (see above).

Tensed macromolecular springs drive other reactions from insect hopping to membrane fusion

Another example of such a reaction is offered by the case of vorticelids, protozoa that attach to substrates via a stalk containing a contractile fibril. Here the motile reaction involves contraction (as in the phage tail), rather than straightening (as in the acrosome). The mechanism appears to involve the state of aggregation of a protein called **spasmin**, and the trigger may again be Ca^{2+}.

The only known example of a multi-cycle reaction of this kind involves a mechanism that has evolved in certain jumping insects. By storing strain energy in elastic body parts containing the protein **resilin**, they are able to release large amounts of energy in a short time in a series of powerful hops. Although the number of distinct reactions involving protein polymers with hundreds or thousands of subunits is limited, the same principle of controlled release of conformational free energy appears to drive membrane fusion in the infection process of enveloped viruses (Section 8.7) and perhaps also other membrane-scission events.

14.11 CHEMOTAXIS I: THE BACTERIAL FLAGELLUM

As long ago as 1683, motile bacteria were observed by Antoni van Leeuwenhoek, using his self-constructed compound microscope. Then as now (when viewed in a standard light microscope), cells were seen to move quite rapidly in seemingly random directions and to change direction with no obvious purpose. The advent of chemical stains revealed long filaments extending from the bacteria. With the introduction of negative-staining EM in the late 1950s, these filaments, called *flagella*, were seen to be ~25 nm in diameter. Dark-field light microscopy, differential interference contrast microscopy, and fluorescence microscopy of bacteria expressing fusions of the protein *flagellin,* the main constituent of flagella, with green fluorescent protein have demonstrated that they undergo rapid rotational movement, as shown with *E. coli*. These observations established that flagella are the motile organelle and that the resulting motion is not entirely random but, rather, the bacterium zigzags in a series of linear path segments. A change of direction arises from a change in the sense of flagellar rotation, whereby the cell 'tumbles' and then heads off in a new direction. In this way, bacteria perform biased Brownian walks toward sources of nutrients. The process is called **chemotaxis**.

The navigation system of chemotactic bacteria is based on clusters of membrane-traversing receptor molecules (*chemoreceptors*) that detect ligands in the external milieu and respond

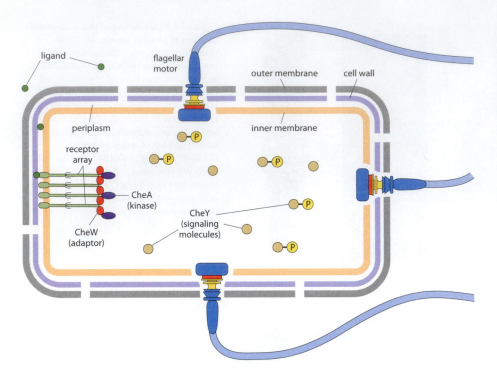

Figure 14.73 Communication between chemoreceptors and flagellar motors in *E. coli*. The receptors cluster in an array at one pole of the cell. An adaptor, a kinase, and other regulators associate with this array. These proteins are designated Che (for <u>che</u>motaxis) followed by a capital letter. The level of phosphorylation of CheY (when phosphorylated, CheYp) is regulated by modulation of CheA kinase activity in response to ligands binding to the receptors. The probability of a motor switching from clockwise to counter-clockwise rotation is determined by the amount of CheYp.

by initiating a cascade of intracellular reactions (**Figure 14.73**). The pathway by which signals are detected, transformed, and relayed to the flagellar motor that is embedded in the bacterial cell wall and controls the direction of rotation of the flagellum is probably the simplest known behavioral response.

The rotary motor that drives flagellar motility is reversible and powered by ion gradients

The flagellum consists of a long filament attached to the cell by its basal body, a structure that spans the inner membrane, cell wall, and outer membrane. Flagellated bacteria are classified on the basis of the number of flagella and their distribution over the cell surface (**Figure 14.74**). Two bacteria on which many of the fundamental studies of flagellar motility have been performed, *Escherichia coli* and *Salmonella typhimurium*, belong to the peritrichous class, which have flagella projecting in all directions that bundle together and rotate in concert. An additional class is exemplified by the *spirochetes*, long thin bacteria whose flagella, called *axial filaments*, are confined to the periplasm, where their rotation propels the cell forward in a corkscrew-like motion.

The rotary nature of the flagellar motor was demonstrated by tethering bacteria to glass slides via their flagella, whereupon the cell body was observed to rotate around the point of attachment. Being larger and therefore subject to greater viscous drag, the cell rotates more slowly than the flagellum would. Unlike linear molecular motors, which are quite widespread, rotary motors are rare. The only other one on record is the F_1F_0 ATPase (Section 15.4).

The motor that causes the filament to rotate, part of the basal body, is powered by the electrical potential of the proton gradient (pH gradient) that is normally present across the inner membrane. Alternatively, some species of marine or alkylophilic bacteria such as *Vibrio* or *Bacillus* use sodium gradients to power their motors. In either case, it is the flow of protons or sodium ions through channels in the membrane that drives the motor. It has been shown that sodium channels from *Vibrio* are able to replace the proton channels in *E. coli* and generate motor rotation by using a sodium gradient as energy source. It follows that the same mechanism is used to generate torque, regardless of whether the ion flux is provided by H^+ or Na^+.

Much like a conventional electric motor, the flagellar motor has components that are fixed relative to the cell body (stator complexes) as well as moving parts (the rotor and attached

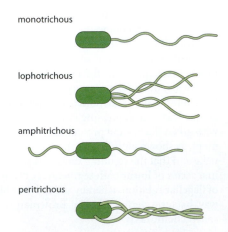

Figure 14.74 Four classes of flagellated bacteria.

filament). The filament is a long helical polymer of protein subunits that is supercoiled, giving it a corkscrew-like form. As the filament rotates, it diverts the viscous drag so as to generate a vectorial force that pushes the bacterium forward; that is, away from the pole to which the flagellum is attached. Thus the filament acts as a supercoiled helical propeller.

The flagellar motor further resembles an electric motor in being reversible. In this case, reversal occurs not through a change in direction of the current (proton flow) but through conformational changes in the rotor and/or stators—a mechanism comparable to a gear shift.

The flagellum has a modular architecture built from 24 different proteins

More than 40 genes are known to be involved in motility in *S. typhimurium* and *E. coli*. The nomenclature for the flagellar and chemotaxis genes has been standardized to apply to homologs from all species. It is based on the phenotypes of mutants; thus, mutations in *mot* genes are nonmotile (but have flagella), and cells with defects in *fla* genes lack flagella. The more than 26 proteins with fla phenotypes have been divided into subgroups—flg, flh, fli, and flj—based on how the genes cluster into operons. These 40-plus gene products direct the assembly of the flagellum and regulate its activity. As many as 24 of them are physical components of the flagellum. Their locations are mapped in **Figure 14.75**.

The filamentous portion of the flagellum has three distinct segments: the rod, 25 nm long, which is coupled directly to the basal body and functions as a drive shaft; the hook, 55 nm long, which is pliable and serves as a universal joint; and the filament proper, which is up to 10 μm long (**Figure 14.76A**). The fixed lengths of the rod and hook are specified by

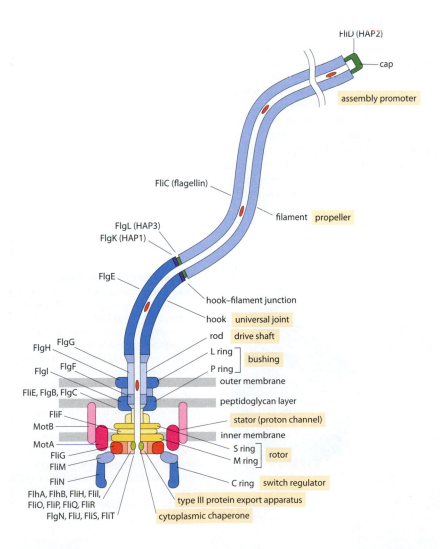

Figure 14.75 Protein components of the *E. coli* flagellum. (Courtesy of Keiichi Namba.)

(A)

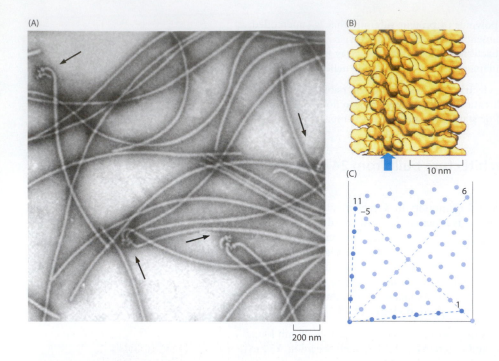

(B)

(C)

200 nm

10 nm

Figure 14.76 Structure of bacterial flagella. (A) Electron micrograph of negatively stained flagella detached from *E. coli*. Arrows point to basal body/hook complexes. (B) Cryo-EM reconstruction of the flagellum from a strain of *S. typhimurium* with straight R-type flagella; the 11 long-pitch helices or protofilaments (one is marked with a blue arrow) are right-handed. (C) Radial projection of the helical lattice. One long-pitch helix is marked with blue dashes and an 11; also marked are members of the 1-fold, 5-fold, and 6-fold families of helices (minus, as in −5, means left-handed). (A, courtesy of Dennis Thomas; B and C, from Y. Mimori-Kinouye et al., *Proc. Natl Acad. Sci. USA* 93:15108–15113, 1996 National Academy of Sciences. With permission from PNAS.)

tape-measure proteins. The rod, hook, and filament are polymers of different proteins but all three structures are arranged on a common helical lattice with 11 protofilaments (see Figure 14.73B,C).

The *basal body* is a barrel-like structure comprising two sets of rings: the MS rings (<u>m</u>embrane-associated and <u>s</u>upra-membrane-associated) and the L and P rings (associated with the <u>l</u>ipid outer membrane and <u>p</u>eptidoglycan cell wall; **Figure 14.77**). The MS rings form part of the rotor. The C ring is located in the <u>c</u>ytoplasm in close association with the proton channels and the MS ring. It is likely that the C ring also rotates, because it incorporates part of the torque-generating protein FliG and presents, on FliM, the binding site for the signaling molecule, CheY, that controls the direction of rotation. MS rings are polymorphic; the number of subunits varies from 24 to 26 within the same bacterium. C rings vary even more widely, from 31-fold to 38-fold (**Figure 14.78**). C-ring symmetry appears not to correlate with MS-ring symmetry, and their polymorphisms have been observed both in native preparations and when ring proteins have been overexpressed. The order of symmetry of the LP rings is not known, although they appear to resemble MS rings in this respect.

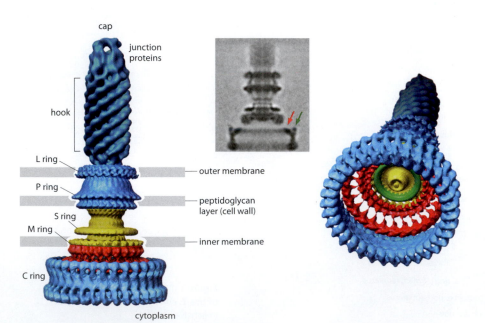

cap

junction proteins

hook

L ring — outer membrane

P ring — peptidoglycan layer (cell wall)

S ring

M ring — inner membrane

C ring

cytoplasm

Figure 14.77 Cryo-EM of the basal body/hook complex. Middle panel, an averaged image in grayscale. The inner lobe of the C ring (red arrow) is part of FliG and has the MS ring symmetry. The outer lobe (green arrow) is FliM/N and has the same symmetry as the rest of the C ring. The other two panels show two views of the complex, with the various rings distinguished. This is a composite combining several reconstructions of subcomplexes. The green annulus under the M ring is likely to be export proteins, which do not follow any symmetry and so appear cylindrically symmetric when averaged in the reconstruction. See also Figure 14.79. (Adapted from D.J. DeRosier, *Curr. Biol.* 16:R928–R930, 2006. With permission from Elsevier.)

Figure 14.78 C-ring polymorphism. Cryo-EM images enhanced by averaging show axial and side views for the 36-fold (top pair) and 34-fold (bottom pair) polymorphic variants. The axial views are from particles isolated from an overexpressing strain of *S. typhimurium*. The red bar is the same length in both panels and is intended to bring out the small difference in inner diameter. The side views are from a strain lacking the filament but expressing normal motors with the rod and hook. (From H.S. Young et al., *Biophys. J.* 84:571–577, 2003 and D.J. Thomas et al., *J. Bacteriol.* 188:7039–7048, 2006. With permission from Elsevier and American Society for Microbiology.)

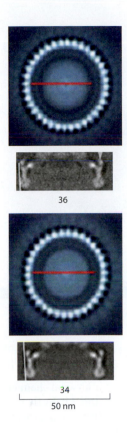

The number of proton channels (stators) associated with the motor appears not to be fixed. When the genes for stator proteins were expressed from a plasmid in a strain of *E. coli* lacking them, quantized increases in torque and motor speed were observed, 8–10 in number, suggesting that there are 8–10 stators per motor. Stator complexes have been visualized *in situ* by cryo-electron tomography of *Borrelia burgdorferi*. This bacterium has a ring of 16 stator complexes (**Figure 14.79**). This representation of the motor complex reveals the relative positioning of the C ring, the bushings, the rod, and other motor components, as well as their disposition relative to the inner membrane.

In summary, then, the flagellar apparatus has two main parts: the stators, which are fixed in the cell envelope; and the basal body surmounted by the hook and filament, which rotates as a single integrated entity. A notable feature of flagellar architecture is the existence of junctions between structurally differentiated segments. These junctions incur symmetry mismatches: between the C ring and the MS rings, and between the rotationally symmetric rotor and the helically symmetric rod. In principle, these junctions are weak points because of the paucity of bonds, but in practice they are strong enough to withstand the torque generated by the motor. At the rod–hook and hook–filament junctions, there is geometric continuity of the helical lattice but discontinuity in the kind of subunit arrayed on it.

When a flagellum is synthesized, the innermost components (C ring and MS rings) are assembled first and the distal components are fed out sequentially through them. As amino acid sequence data accumulated, it became apparent that there is a close similarity between flagellar proteins and some gene products associated with bacterial virulence. These virulence factors, now known as the *Type III secretion system*, form a complex, the *injectisome*, that resembles the basal body. Overall, the flagellar assembly pathway has two Type III-like secretory segments and one SecA-like segment. These export pathways are described in Section 10.7.

The flagellar filament is a hollow tube with a backbone of packed α helices

The filament is composed of up to 20,000 copies of flagellin (55 kDa), whose subunit has four domains: D0, D1, D2, and D3 (**Figure 14.80**). Both terminal regions, which contain heptad repeats, are disordered in solution. In the filament, they form an antiparallel coiled coil (D0). The structure of the *S. typhimurium* filament has been determined by Keiichi Namba and associates, using two approaches. First, a fragment of flagellin lacking the terminal regions was solved by X-ray crystallography and fitted into cryo-EM density maps of both left-handed and right-handed filaments. Later, the right-handed filament was solved by cryo-EM at a resolution high enough for an atomic model to be built. It revealed the arrangement of D0 domains in the filament backbone in two concentric rings of α helices,

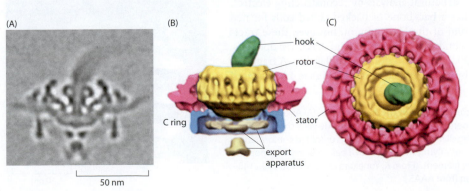

(A) (B) (C)

hook
rotor
stator
C ring
export apparatus

50 nm

Figure 14.79 Cryo-tomographic reconstruction of the *Borrelia burgdorferi* basal body *in situ*. (A) Central section. (B, C) Surface renderings of a side view ((B), partly cut away) and an axial view seen from the periplasmic side (C). To obtain this density map, basal bodies from many tomograms were averaged but not rotationally symmetrized. Note, nevertheless, the 16-fold symmetry of the stator complexes in (C). (Adapted from J. Liu et al., *J. Bact.* 191:5026–5036, 2009.)

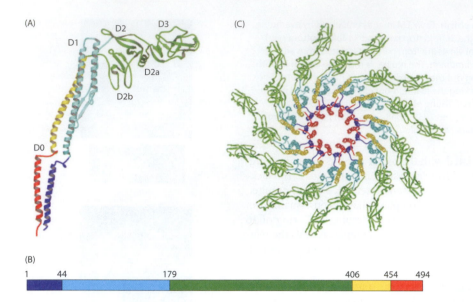

(A)

D1
D2
D3
D2a
D2b
D0

(B)

1   44   179   406   454   494

(C)

Figure 14.80 Flagellar filament structure.
(A) The flagellin subunit (FliC). (B) Map of the polypeptide chain, with segments color-coded as in (A). (C) Axial view of the filament from the distal end. Eleven subunits are displayed. (From K. Yonekura et al., *Nature* 424:643–650, 2003. With permission from Macmillan Publishers Ltd.)

whose tight packing confers a stiffness that is advantageous in the role of the filament as a propeller. D2 and D3 are rich in β sheet structures.

The filament has an axial channel, 2.5 nm wide. In a growing filament, flagellin monomers migrate along this channel, to fold and be incorporated at the outer tip, under the cap complex. The cap is a pentameric ring of fliD subunits, mounted at the tip of an 11-protofilament filament—yet another symmetry mismatch (**Figure 14.81**). The cavity under the cap is thought to provide a compartment conducive to folding of the flagellin subunits, akin to the 'Anfinsen cage' of cytosolic *chaperones* (Section 6.3).

The filament is supercoiled into a corkscrew whose hand and helical pitch depend on the conformational states of its protofilaments (**Figure 14.82**). The supercoil switches hand when the direction of motor rotation changes. When a switch takes place, the change initiates in the cell body and propagates outward. Its basis appears to lie in two alternative conformations for the flagellin subunits, with all of the subunits in a given protofilament assuming the same conformation. The observed variability in pitch and hand can be explained by having different combinations of protofilaments in the left-handed and right-handed states; for example, with four protofilaments in the right-handed conformation and seven in the left-handed conformation ('semi-coil' in Figure 14.82), the filament is left-handed with a pitch of ~0.5 μm. Locally, the difference between the two states is very small, amounting to a subunit-to-subunit shift of less than 1 Å, but they add up to a large and functionally important change in the overall structure of the filament.

The hook functions as a universal joint and the rod serves as the drive shaft of the flagellar motor

The hook is naturally curved to a greater degree than the filament but straightens out at low temperature, a property that facilitated structural analysis by reconstructing electron micrographs. Like the filament, the hook has a backbone of packed coiled coils formed from heptad-containing tracts at both termini of the FlgE subunit; however, these tracts are shorter than in flagellin, so there is no second ring of α helices. FlgE has three domains: D0, D1, and D2, roughly equivalent to the D0, D2, and D3 domains of flagellin. A fragment consisting of D1 and D2 has been solved by X-ray crystallography (**Figure 14.83**). Both

(A)

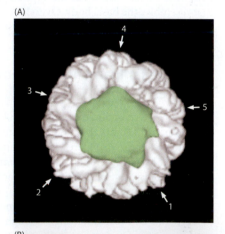

(B)

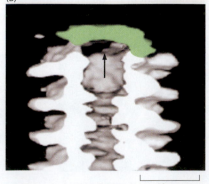

10 nm

Figure 14.81 Cryo-EM reconstruction of the flagellar cap. (A) Axial view of the filament, showing the pentameric cap. (B) Cut-away side view showing the axial lumen along which flagellin subunits migrate in an unfolded conformation, to fold in the wider region underlying the cap (arrow) before being incorporated into the growing end of the filament. (From K. Yonekura et al., *Science* 290:2148–2152, 2000. With permission from AAAS.)

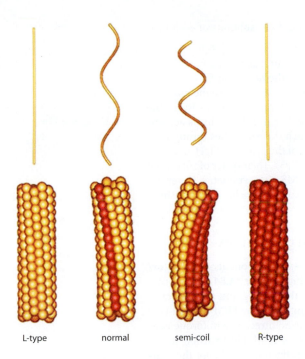

L-type   normal   semi-coil   R-type

Figure 14.82 Different supercoiling states of the flagellar filament. Changes are made between these states when different combinations of protofilaments are in one or other of two conformations. The conformations are color-coded as gold and orange. (From K. Namba and F. Vonderviszt, *Q. Rev. Biophys.* 30:1–65, 1997. With permission from Cambridge University Press.)

domains consist mainly of antiparallel β sheets. In a model obtained by fitting this structure into cryo-EM density maps, D1 packs against the inward-facing surface of D2 on the next subunit along the same protofilament. The flexible linker between D1 and D2 allows the domains to remain in contact even as the spacing between D2 domains expands and contracts on opposing sides of the curved hook.

Two hook-associated proteins, FlgK and FlgL, couple the hook to the filament. Their function is to transmit torque and changes in motor rotation—and, hence, changes in the hand

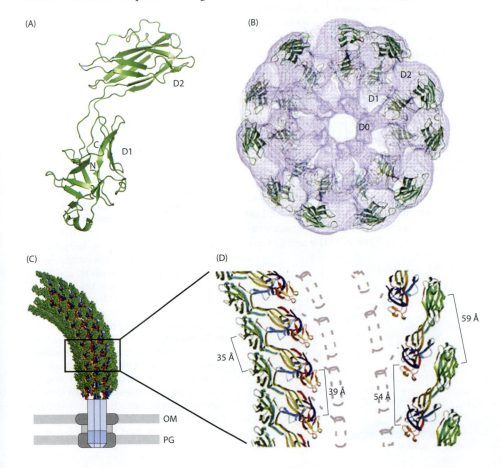

Figure 14.83 Hook structure and modeling of hook flexibility. (A) Crystal structure of D1 and D2 domains of the hook protein FlgE. The domains are connected by a flexible linker region. (PDB 2BGY) (B) Axial view of an EM density map with subunits fitted in. The near-axial ring of density is assigned to the coiled coils (D0 domains). (C, D) Model of a curved hook. (D) Enlargement of a central slab with the D1 domains now in various colors. OM, outer membrane; PG, peptidoglycan layer. (B, from T.R. Shaikh et al., *Proc. Natl Acad. Sci. USA* 102:1023–1028, 2005. With permission from National Academy of Sciences; C and D, from F.A. Samatey et al., *Nature* 431:1062–1068, 2004. With permission from Macmillan Publishers Ltd.)

of the filament. It appears that only one helical turn, involving 6–11 subunits of each protein, constitutes the junction.

The embedding of the rod within the basal body has made structural analysis difficult. Nevertheless, it appears to follow the same helical organization as the hook and filament, and its subunits also have N-terminal and C-terminal tracts of heptad repeats. Like the motor of an automobile, the flagellar drive shaft (rod) needs to transmit torque to the outside of the cell without allowing its contents to leak out, and it must be free to rotate. This is accomplished by deployment of a bushing system by means of which the rod passes through the cell wall and outer membrane. It comprises two hollow rings of the proteins, FlgH and FlgI, inserted into the cell wall and outer membrane (see Figure 14.75). Their order of rotational symmetry is thought to be 25-fold, matching the commonest MS-ring symmetry. The entire basal body of the flagellar motor of *Borrelia burgdorferi,* presumably including its bushing apparatus, has been visualized *in situ* by cryo-electron tomography (see Figure 14.79).

Six basal-body proteins contribute to torque generation

Four of the basal-body proteins—FliF, FliG, FliM, and FliN—are components of the **rotor/switch complex**, which comprises the C ring and the MS ring (**Figure 14.84**). The C ring protrudes into the cytoplasm but is tightly associated with the M ring and also interacts with the stator. C-ring protomers consist of four FliN subunits and one FliM subunit (the C-terminal domain of FliM is homologous to FliN). FliN has no direct role in torque generation but is required for secretion and as a binding partner for FliM. FliM participates in torque generation by interacting with FliG. FliM also contains the binding site for the signaling molecule CheY (see Figure 14.1 and Section 14.12), in the interaction that switches the sense of rotation of the rotor. The outer proximal lobe of the C ring would be a logical location for the CheY-binding domain of FliM, but this has yet to be demonstrated.

FliG, the protein most directly involved in torque generation, is partly in the C ring and partly in the MS ring. MS protomers consist of one subunit of FliF and part of a FliG subunit. FliF is a membrane protein that serves as the mechanical mount for FliG; it forms the S ring and part of the M ring. (The S ring resides in the periplasm, whereas the M ring traverses the inner membrane and is partly exposed to the cytoplasm.) The FliG moiety is located on the cytoplasmic face of the M ring, with its C-terminal region extending out to interact with both the stator and the C ring (Figure 14.84A). Also associated with the MS ring are several other proteins that are needed for export of other flagellar components (see Figures 14.77 and 14.79).

The flagellar motor is a powerful and efficient stepping motor

The torque that the motor generates and its speed under load are proportional to the number of stators present. Proton-driven motors rotate at speeds up to 300 Hz, generating a torque of 1300 pN.nm. Some 1200 protons are passed per revolution. Measurements of rotation have been performed for *E. coli*, using a chimeric motor in which the *E. coli* rotor was paired with stators from *Vibrio* sp. (Na⁺-driven motors are faster, rotating at speeds of up to 1200 Hz). The chimeric motor was found to rotate at *Vibrio* speeds, attesting to the dominant role of the stator in specifying motor performance. Under conditions of low sodium-motive force, 26 discrete steps per rotation were seen. At limiting Na⁺ concentrations, occasional backward steps of ~1/35 of a revolution were observed. The results were

Figure 14.84 The rotor/switch complex surrounded by stators. (A) Diagram showing MotA/B stator complexes distributed around the rotor and contacting the peptidoglycan layer (PG). OM, outer membrane; IM, inner membrane. FliG is shown as spanning both lobes of the C ring next to the MS ring. For clarity, the MS ring, which comprises the membrane-embedded M ring and the periplasmic S ring, is shown as a single entity (red). (B) Topological diagram of MotA and MotB. MotA has a cytoplasmic domain that interacts with FliG. Most of MotB, including its PG-binding domain (PBD), resides in the periplasm. Amino acids important to motor function are labeled; those in FliG (blue) are in its C-terminal domain in the C ring. The only amino acid residue known to be indispensable for motor function is Asp32 in MotB. (From S. Kojima and D.F. Blair, *Int. Rev. Cytol.* 233:93–134, 2004. With permission from Elsevier.)

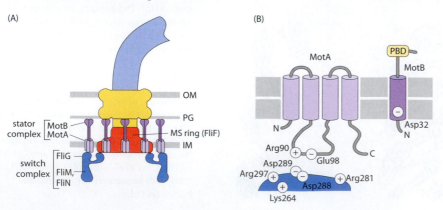

consistent with known structural properties of the rotor: 26 is a typical MS-ring symmetry and 35 is a common C-ring symmetry. These experiments established the flagellar motor as a stepping motor.

The key question of motor mechanism has yet to be fully answered. The proton channels consist of two proteins, MotA and MotB. There are four MotA and two MotB subunits per stator complex, housing two channels. The protonation or deprotonation of Asp32 near the cytoplasmic end of MotB in the channel (Figure 14.84B) drives a conformational change in MotA that moves FliG along. Torque–speed curves are smooth until high speeds are reached, which suggests a mechanism in which energy is derived continuously as protons move down the pH gradient. The motor can be driven backward without resistance, suggesting that there is not a bound state waiting for protons. Switching, when it occurs, is rapid (~1 ms) and involves a cooperative change in conformation that causes the stator to drive FliG in the opposite direction.

14.12 CHEMOTAXIS II: SIGNALING BY CHEMORECEPTOR ARRAYS

Chemoreceptors are transmembrane proteins that cluster at one pole of the bacterial cell (**Figure 14.85**). The clusters vary in size and shape, depending on the strain and growth conditions, but are typically 0.2–0.4 μm across and contain up to 5000 or so receptor molecules. Receptors with different specificities are found in the same cluster: Tsr for serine, Tar for aspartate, Tap for dipeptides and pyrimidines, Trg for ribose and galactose, and Aer for redox potential. Each receptor also recognizes other ligands; for example, Tar responds, in

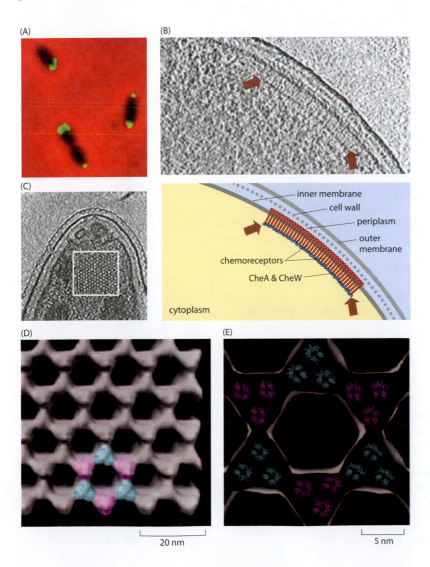

Figure 14.85 Clusters of chemotaxis receptors. (A) Light micrograph of *Rhodobacter* expressing a chemoreceptor fused to GFP (green fluorescent protein). A cluster is detected at one pole, except in dividing cells, which also have a nascent cluster at the other pole. (B) Slice 5 nm thick from a cryo-electron tomogram of an *E. coli* cell (an interpretive diagram is shown below). Positive density is dark. The edges of the cluster are marked with brown arrows. The layer of density midway between the inner and outer membranes is the peptidoglycan cell wall. (C) Tangential slice from a cryo-electron tomogram of a *Caulobacter crescentus* cell, showing the honeycomb lattice of a chemoreceptor cluster. The white box is 160 nm on a side. (D) Surface rendering of the cluster, enhanced by averaging; six densities at positions of local 3-fold symmetry are colored (three in turquoise, three in magenta). (E) Trimers of dimers placed into the trigonal densities. (B, from G.L. Hazelbauer et al., *Trends Biochem. Sci.* 33:9–19, 2007. With permission from Elsevier; C–E, from C.M. Khursigara et al., *J. Bact.* 190:6805–6810, 2008. With permission from American Society for Microbiology.)

addition to aspartate, to maltose (via a maltose-binding protein), Ni^{2+} (a repellent), hydrogen ions, and temperature. This multifunctional characteristic accounts for the wide range of attractants and repellents sensed by bacteria; for instance, *E. coli* responds appropriately to some 50 distinct chemical signals. The sensing system is highly sensitive and has a vast dynamic range: aspartate can be detected at concentrations below 10 nM and is still effective above 1 mM.

Chemoreceptors modulate the activity of a kinase, CheA

The kinase CheA associates with the cluster with the help of an adaptor protein, CheW. CheA transfers phosphate groups from ATP to a small protein, CheY (see below). Its level of activity reflects the conformational state of the receptor(s) to which CheA is bound; this activity is decreased by attractant ligands binding to their receptor and stimulated by repellents. Phosphorylated CheY (CheYp) diffuses to the flagellar motors, where its binding switches their direction of rotation. The higher the concentration of CheYp, the more likely a motor is to turn clockwise and generate a tumble.

Two other enzymes mediate an *adaptation* response: CheR methylates receptors, and the phosphorylated form of CheB, CheBp, demethylates them. Methylation increases the tendency of the receptor to be in an active state—that is, to stimulate CheA kinase. In a typical chemotactic response, encounter with an attractant inactivates receptors and decreases CheA activity. The intracellular concentration of CheYp falls, reducing the tendency of the cell to tumble. The cell continues longer on its present path—an appropriate response, because it has located nutrients. Adaptation then ensues on a slower time course and, as methyl groups are added to the receptor, its activity returns to baseline. After some minutes (depending on the size of the stimulus), the bacterium reverts to the pattern of swimming it had before it encountered the attractant, and is ready to respond to further stimuli.

Chemoreceptors have globular periplasmic domains and coiled-coil cytoplasmic domains

Tsr and Tar, the most abundant receptors in *E. coli*, are homodimers of 60 kDa subunits. Crystal structures have been determined for the principal domains of receptors. A composite molecular model is shown in **Figure 14.86**. The cytoplasmic domains consist of two intramolecular antiparallel coiled coils that pair to form a four-helix bundle.

Sensing is performed by the periplasmic domain(s). Binding of attractant induces a small conformational change that is transmitted along the molecule. Although there are two binding sites per dimer, single occupancy produces a full physiological response. Next is the transmembrane region, followed by the HAMP domain (see **Guide**), an α-helical motif of ~60 aa that is widespread among bacterial <u>h</u>istidine kinases, <u>a</u>denyl cyclases, <u>m</u>ethyl-accepting proteins, and <u>p</u>hosphatases. After the HAMP domain comes the methylation region of the cytoplasmic domain, which is responsible for adaptation reactions. Distributed along this region are glutamate residues—8 per Tar dimer and 10 per Tsr dimer—that are subject to methylation by CheR. In this reaction, reactive methyl groups are transferred from *S*-adenosylmethionine to the γ-carboxylic acids of these glutamates to form methyl esters. The concomitant loss of negative charge is thought to bias the receptor to its active conformation.

The distal portion of the molecule consists of the signaling domain containing the regions that communicate the current conformational state of the receptor to CheA and CheW. In addition, this domain mediates interactions with neighboring receptors in the cluster.

A short segment at the C terminus of the receptor, located approximately halfway along the cytoplasmic tail, is *natively unfolded*. This segment ends in a pentapeptide that is recognized by both CheR and CheB and appears to act as a tether for both enzymes. Such an arrangement increases their local concentrations and hence their activity in the receptor cluster.

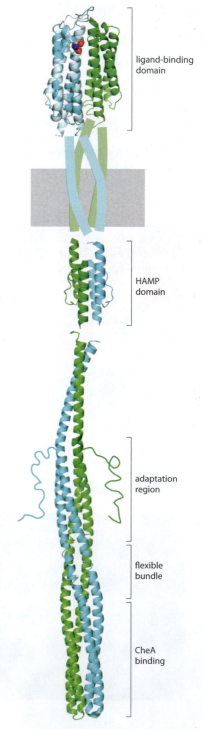

Figure 14.86 Structure of a chemoreceptor dimer. A composite model assembled from crystal structures of the ligand-binding domain of the aspartate receptor Tar (PDB 1VLT) and of most of the cytoplasmic domain of the serine receptor Tsr (PDB 1QU7), and an NMR structure for an archaeal HAMP domain (PDB 2ASX). (Adapted from G.L. Hazelbauer et al., *Trends Biochem. Sci.* 33:9–19, 2007. With permission from Elsevier.)

ligand-binding domain

HAMP domain

adaptation region

flexible bundle

CheA binding

Chemotactic signals are conveyed by allosteric switches

In this respect, bacterial chemosensing resembles other intracellular signaling pathways (Chapter 12). Each protein involved, from receptor to flagellin, can adopt more than one conformational state, and specific states are stabilized by such events as the binding of a small molecule, the attachment of a modifying group, or interaction with a neighboring protein. Waves of chemical reactions thereby propagate from the initial detection event in the periplasm, across the membrane, along the receptor, and finally across the cytoplasm to the flagellar motors (see Figure 14.73).

Chemosensing begins when an attractant or a repellent binds to a receptor. Ligands such as aspartate and serine stabilize the periplasmic domains of their receptors in the 'inactive' state that inhibits tumbling. Switching into this conformation is relayed across the membrane by means of shifts in membrane-spanning segments, the critical change being a small (1–2 Å) extrusion of one helix into the cytoplasm. This piston-like event is communicated to the HAMP domain and thence to the signaling domain, resulting in a change in CheA activity.

The HAMP domain dimer contains two pairs of short α helices, making a parallel four-helix bundle (**Figure 14.87**). An NMR structure of the HAMP domain from an archaeal protein revealed a non-canonical packing of α helices, called da-x packing, whereby the helices are rotated by 26° around an axis perpendicular to the plane of the membrane, relative to the conformations that they would assume in canonical 'knobs into holes' packing. This observation led to the proposal that these two states may represent the active and inactive states of the receptor with respect to its regulation of CheA, with switching to the active conformation transmitting an allosteric signal to the CheA-binding region at the tip of the signaling domain.

The simple nature of this communication system was demonstrated by experiments in which the periplasmic domain of Tar was spliced onto the signaling portion of the human insulin receptor. The activity of the chimeric receptor as a *tyrosine kinase* (the insulin receptor activity) could be stimulated by aspartate, as though the periplasmic domain was sending messages across the membrane, like a puppeteer pulling strings.

CheA is the engine that powers the chemical signaling events of the chemotaxis pathway

The CheA kinase is continually active, phosphorylating itself by adding the terminal phosphate of ATP to a key histidine residue. The high-energy phosphoryl group thus created is then transferred to either CheY or CheB.

CheA is a homodimer, with each subunit having five domains with distinct functions (**Figure 14.88**):

- The substrate domain, P1, contains the His residue that is phosphorylated by the autokinase reaction.
- The CheY/B-binding domain, P2, although non-essential, optimizes phosphotransfer from P1 to CheY or CheB.

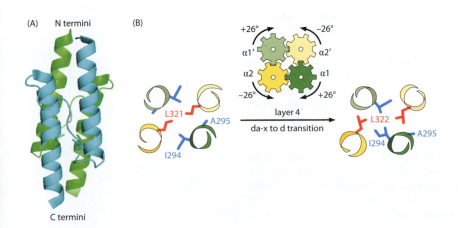

Figure 14.87 Signaling-related conformational change of a HAMP domain. (A) Structure solved by NMR reveals a dimeric four-chain parallel coiled coil. (PDB 2ASX) (B) Switching between the observed da-x arrangement for packing of side chains and canonical packing may represent an activating conformational transition. (From M.F. Hulko et al., *Cell* 126:929–940, 2006. With permission from Elsevier.)

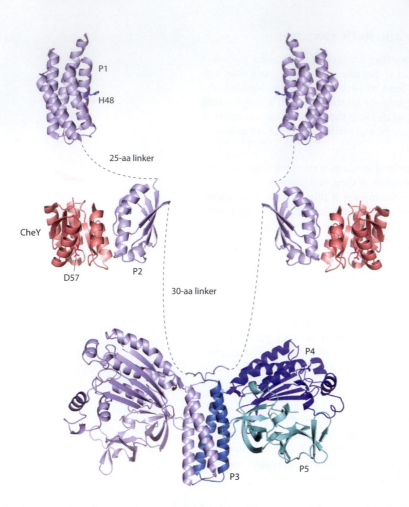

Figure 14.88 Structure of the CheA kinase dimer. The subunit has five domains, P1–P5. This model was assembled from crystal structures of the P3/P4/P5 dimer (PDB 1B3Q), the P2/CheY complex (PDB 1UOS), and P1 (PDB 1I5N). Extended linkers between P1 and P2 and between P2 and P3 confer mobility. The catalytic domain P4 transfers a phosphate from ATP to His48 on P1, whence it is transferred to P2-bound CheY (pink).

- The dimerization domain, P3, is a helical hairpin that pairs to form a stable four-helix bundle.
- The catalytic domain, P4, has the ATP-binding site as well as residues that catalyze phosphotransfer from ATP to P1.
- The regulatory domain, P5, binds CheW and the receptor and mediates their control of the kinase activity.

These domains are separated by flexible linkers. The resulting freedom of movement allows the autophosphorylation reaction of CheA to operate via a *trans* mechanism in which the P4 domain of one subunit phosphorylates the histidine residue on the P1 domain of the other subunit. In theory, this mobility could also permit neighboring CheA molecules to influence each other's activity.

The receptor cluster is a plate-like assembly of 'trimers-of-dimers'

Receptors cluster in a loosely ordered hexagonal lattice. Viewed in a transverse slice (see Figure 14.85B), a layer of density at its base marks the location of CheA and CheW, as was confirmed by immuno-gold labeling EM. A second, fainter, layer midway between the base of the plate and the inner membrane is thought to represent CheR and CheB. The lattice is honeycomb-like, with triangular densities of appropriate size to accommodate a trimer of dimers (see Figure 14.85C–E). In the trimers, different kinds of dimer are mixed, apparently at random. The residues that effect inter-dimer contacts are closely conserved, so that a given trimer can contain more than one kind of dimer. Another protein located mainly in the cluster is CheZ, a phosphatase that removes phosphoryl residues from CheYp.

The receptor cluster represents a specialized compartment of the cell. Not only do receptors detect molecules of interest to the bacterium, but they also act as a processing unit. The receptor cluster maintains a short-term memory of ligand concentrations encountered over the previous few seconds. By comparing past and present conditions, the bacterium is able to decide whether things are getting better (so it should continue on its present course) or worse (so it should change direction by tumbling). The receptor cluster also amplifies signals so that the binding of a few attractant molecules is able to trigger a cascade of chemical events involving a large number of kinase molecules and, ultimately, a motile response by the bacterium.

14.13 SUMMARY

Motility is essential for life, accomplishing the directed movement of individual cells and multicellular organisms, the division of growing cells, and intracellular transport of cargoes between compartments, among other processes. The forces employed derive from only a few chemical sources: of these, the most prevalent involves the hydrolysis of ATP or GTP, but other sources involve coupling to the flow of protons across a membrane, or the release of conformational free energy from a metastable complex. Macromolecular machines with moving parts transduce the energy thus released into diverse forms of motion, which may include

- stepping, as by motor proteins translocating their cargoes along cytoskeletal filaments;
- pushing, as in polymerizing protein filaments impinging on a membrane, or a contractile bacteriophage tail penetrating a host cell;
- gliding, as in single cells migrating across a substrate;
- pulling, as in the separation of dividing chromosomes by depolymerizing microtubules;
- swimming, as by bacteria or eukaryotic cells propelled by beating flagella.

The cytoskeletal proteins actin and tubulin are central players in motility. Both generate forces by two distinct mechanisms: first, polymerization/depolymerization reactions, as managed by regulatory proteins, impose forces on structures to which these filaments are coupled; and second, they provide linear tracks along which motor proteins—myosins in the case of actin, and kinesins and dyneins in the case of tubulin—move cargoes. Muscle contraction represents a specialized and highly coordinated form of the general mechanism of motorized transport.

'Single-molecule' biophysical experiments have allowed the measurement of forces exerted by individual motile macromolecules as generally in the range of 1–5 pN, and occasionally as high as 50 pN; their step-lengths, generally 3–10 nm, and occasionally as high as 35 nm; and their efficiencies, which can approach 50%.

Sections of this chapter are based on the following contributions

14.2–14.4   Kenneth C. Holmes, Max Planck Institute for Medical Research, Heidelberg, Germany.
14.5   Roger Craig, University of Massachusetts Medical School, Worcester, USA.
14.6–14.8   Linda A. Amos, MRC Laboratory of Molecular Biology, Cambridge, UK.
14.11   Dennis R. Thomas, Max Planck Institute for Biochemistry, Martinsried, Germany.
14.12   Dennis Bray, University of Cambridge, UK.

REFERENCES

14.1 Introduction

Bray D (2001) Cell Movements: From Molecules to Motility, 2nd ed. Garland Science.
Howard J (2001) Mechanics of Motor Proteins and the Cytoskeleton. Sinauer.
Squire J & Parry DAD (eds) (2005) Fibrous Proteins: Muscle and Molecular Motors. Elsevier Inc./Academic Press.

14.2 Actin filaments and associated proteins

Dominguez R (2009) Actin filament nucleation and elongation factors: structure–function relationships. *Crit Rev Biochem Mol Biol* 44:351–366.
Dominguez R & Holmes KC (2011) Actin structure and function. *Annu Rev Biophys* 40:169–186.
Oosawa F & Asakura S (1975) Thermodynamics of the Polymerization of Protein. Academic Press.

Pederson T & Aebi U (2002) Actin in the nucleus: what form and what for? *J Struct Biol* 140:3–9.
Pollard TD & Cooper JA (2009) Actin, a central player in cell shape and movement. *Science* 326:1208–1212.

14.3 The myosin motor proteins

Coluccio LM (2008) Myosins, a Superfamily of Molecular Motors. Springer.
Fischer S, Windshugel B, Horak D et al (2005) Structural mechanism of the recovery stroke in the myosin molecular motor. *Proc Natl Acad Sci USA* 102:6873–6878.
Foth BJ, Goedecke MC & Soldati D (2006) New insights into myosin evolution and classification. *Proc Natl Acad Sci USA* 103:3681–3686.
Holmes KC, Angert I, Kull FJ et al (2003) Electron cryo-microscopy shows how strong binding of myosin to actin releases nucleotide. *Nature* 425:423–427.

Rayment I, Holden HM, Whittaker M et al (1993) Structure of the actin–myosin complex and its implications for muscle contraction. *Science* 261:58–65.

Vale RD & Milligan RA (2000) The way things move: looking under the hood of molecular motor proteins. *Science* 288:88–95.

14.4   Force generation in muscle

Craig R & Padron R (2004) Molecular structure of the sarcomere. In Myology (Engel AG, Franzini-Armstrong C eds), pp 129–166. McGraw-Hill.

Geeves MA & Holmes KC (2005) The molecular mechanism of muscular contraction. *Adv Prot Chem* 71:161–193.

Granzier HL & Labeit S (2005) Titin and its associated proteins: the third myofilament system of the sarcomere. *Adv Prot Chem* 71:89–119.

Hamilton SL & Serysheva I (2008) Ryanodine receptor structure: progress and challenges. *J Biol Chem* 284:4047–4051.

Squire J (1981) The Structural Basis of Muscular Contraction. Plenum.

Puchner EM, Alexandrovich A, Kho AL et al (2008) Mechanoenzymatics of titin kinase. *Proc Natl Acad Sci USA* 105:13385–13390.

Reconditi M (2006) Recent improvements in small angle x-ray diffraction for the study of muscle physiology. *Rep Prog Phys* 69:2709–2759.

14.5   Myosin filaments

Bullard B, Linke WA & Leonard K (2002) Varieties of elastic protein in invertebrate muscles. *J Muscle Res Cell Motil* 23:435–447.

Craig R & Woodhead JL (2006) Structure and function of myosin filaments. *Curr Opin Struct Biol* 16:204–212.

Flashman E, Redwood C, Moolman-Smook J et al (2004) Cardiac myosin binding protein C: its role in physiology and disease. *Circ Res* 94:1279–1289.

Squire JM (1973) General model of myosin filament structure. 3. Molecular packing arrangements in myosin filaments. *J Mol Biol* 77:291–323.

Tskhovrebova L & Trinick J (2003) Titin: properties and family relationships. *Nature Rev Mol Cell Biol* 4:679–689.

14.6   Microtubules and associated proteins

Amos LA & Schlieper D (2005) Microtubules and MAPs. *Adv Prot Chem* 71:257–298.

Des Georges A, Katsuki M, Drummond DR et al (2008) Mal3, the *Schizosaccharomyces pombe* homolog of EB1, changes the microtubule lattice. *Nature Struct Mol Biol* 15:1102–1108.

Kikkawa M, Ishikawa T, Nakata T et al (1994) Direct visualization of the microtubule lattice seam both *in vitro* and *in vivo*. *J Cell Biol* 127:1965–1971.

McIntosh JR, Grishchuk EL & West RR (2002) Chromosome-microtubule interactions during mitosis. *Annu Rev Cell Dev Biol* 18:193–219.

Nogales E, Wolf SG & Downing KH (1998) Structure of the αβ tubulin dimer by electron crystallography. *Nature* 391:199–203.

Ravelli RBG, Gigant B, Curmi PA et al (2004) Insight into tubulin regulation from a complex with colchicine and a stathmin-like domain. *Nature* 428:198–202.

14.7   Motorized transport along microtubules

Block SM (1995) Nanometres and piconewtons: the macromolecular mechanics of kinesin. *Trends Cell Biol* 5:169–175.

Howard J & Hyman AA (2003) Dynamics and mechanics of the microtubule plus ends. *Nature* 422:753–758.

Kikkawa M, Sablin EP, Okada Y et al (2001) Switch-based mechanism of kinesin motors. *Nature* 411:439–445.

Kon T, Oyama T, Shimo-Kon R et al (2012) The 2.8 Å crystal structure of the dynein motor domain. *Nature* 484:345–350.

Miki H, Okada Y & Hirokawa N (2005) Analysis of the kinesin superfamily: insights into structure and function. *Trends Cell Biol* 15:467–476.

14.8   Motile organelles built from microtubules

Aoyama S & Kamiya R (2005) Cyclical interactions between two outer doublet microtubules in split flagellar axonemes. *Biophys J* 89:3261–3268. Supplementary movie.

Brokaw CJ (2002) Computer simulation of flagellar movement. VIII. Coordination of dynein by local curvature control can generate helical bending waves. *Cell Motil Cytoskel* 53:103–124.

Ishikawa T (2012) Structural biology of cytoplasmic and axonemal dyneins. *J Struct Biol* 179:229–234.

King S (ed) (2011) Dyneins: Structure, Biology and Disease. Elsevier Inc./Academic Press.

Nicastro D, Schwartz C, Pierson J et al (2006) The molecular architecture of axonemes revealed by cryoelectron tomography. *Science* 313:944–948.

14.9   Polymerization/depolymerization machines

Astumian RD (1997) Thermodynamics and kinetics of a Brownian motor. *Science* 276:917–922.

Hill TL & Kirschner MW (1982) Bioenergetics and kinetics of microtubule and actin filament assembly-disassembly. *Int Rev Cytol* 78:1–125.

Michie KA & Lowe J (2006) Dynamic filaments of the bacterial cytoskeleton. *Annu Rev Biochem* 75:467–492.

Mogilner A & Oster G (2003) Polymer motors: pushing out the front and pulling up the back. *Curr Biol* 13:R721–R733.

Pollard TD (2007) Regulation of actin filament assembly by Arp2/3 complex and formins. *Annu Rev Biophys Biomol Struct* 36:451–477.

Theriot JA (2000) The polymerization motor. *Traffic* 1:19–28.

14.10   Motility powered by supramolecular springs

Burrows M, Shaw SR & Sutton GP (2008) Resilin and chitinous cuticle form a composite structure for energy storage in jumping by froghopper insects. *BMC Biol* 6:41.

Kostyuchenko VA, Chipman PR, Leiman PG et al (2005) The tail structure of bacteriophage T4 and its mechanism of contraction. *Nature Struct Mol Biol* 12:810–813.

Leiman PG, Chipman PR, Kostyuchenko VA et al (2004) Three-dimensional rearrangement of proteins in the tail of bacteriophage T4 on infection of its host. *Cell* 118:419–429.

Mahadevan L & Matsudaira P (2000) Motility powered by supramolecular springs and ratchets. *Science* 288:95–100.

14.11   Chemotaxis I: the bacterial flagellum

Berg HC (2003) The rotary motor of bacterial flagella. *Annu Rev Biochem* 72:19–54.

Berg HC (2008) Bacterial flagellar motor. *Curr Biol* 18:R689–R691.

Macnab RM (2003) How bacteria assemble flagella. *Annu Rev Microbiol* 57:77–100.

Macnab RM (2004) Type III flagellar protein export and flagellar assembly. *Biochim Biophys Acta* 1694:207–217.

Namba K & Vonderviszt F (1997) Molecular architecture of bacterial flagellum. *Q Rev Biophys* 30:1–65.

14.12   Chemotaxis II: signaling by chemoreceptor arrays

Baker MD, Wolanin PM & Stock JB (2005) Signal transduction in bacterial chemotaxis. *BioEssays* 28:9–22.

Bray D & Duke T (2004) Conformational spread: the propagation of allosteric states in large multiprotein complexes. *Annu Rev Biophys Biomol Struct* 33:53–73.

Gloor SL & Falke JJ (2009) Thermal domain motions of CheA kinase in solution: disulfide trapping reveals the motional constraints leading to trans-autophosphorylation. *Biochemistry* 48:3631–3644.

Kim KK, Yokota H & Kim S-H (1999) Four-helical-bundle structure of the cytoplasmic domain of a serine chemotaxis receptor. *Nature* 400:787–792.

Sourjik V & Berg HC (2004) Functional interactions between receptors in bacterial chemotaxis. *Nature* 426:437–441.

4 H^+

intermembrane
space

Q → QH_2

mitochondrial
matrix

Q → QH_2

e^-

e^-

½O_2 4 H^+ H_2O

succinate fumarate

2 H^+

Bioenergetics

15.1 INTRODUCTION

To survive, grow and divide, even to die by apoptosis, cells need a source of energy. The chemical energy currency of all cells commonly is ATP. This can be generated in two basic ways: oxidation of nutrients, such as carbohydrates, fats, and the amino acids from proteins; or photosynthesis, whereby organisms capture energy from light.

Aerobic organisms oxidize substrates and, in doing so, consume molecular oxygen, an exergonic (energy-releasing) process akin to the burning of fossil fuels. A widespread example is the series of reactions that brings about the oxidation of glucose, depicted schematically in **Figure 15.1**. The first part is termed glycolysis (in which a 6C molecule is converted to two 3C molecules) and the second is the citric acid cycle (or tricarboxylic acid cycle), two turns of which convert the two 3C molecules to $6CO_2$. The sum of the individual reactions, namely

$$C_6H_{12}O_6 + 6O_2 \rightarrow 6CO_2 + 6H_2O$$

can be regarded as two part-reactions:

$$C_6H_{12}O_6 + 6H_2O \longrightarrow 6CO_2 + 24H^+ + 24e^-$$

$$24H^+ + 24e^- + 6O_2 \rightarrow 12H_2O$$

The CO_2 is released (breathed out by animals) and the 24 protons and 24 electrons are 'captured' as 10 $NADH_2$ (that is, 10 NADH + 10 H^+ at pH 7) plus 2 $FADH_2$ (**Box 15.1**). Of the ten NADH for each glucose molecule oxidized, two are generated in glycolysis, two by the pyruvate dehydrogenase (PDH) multienzyme complex reaction (Chapter 9), which links glycolysis and the citric acid cycle, and the remaining six together with the two $FADH_2$ in the two turns of the citric acid cycle. The 10 NADPH and 2 $FADH_2$ are reoxidized to NAD^+ and FAD, and in the process 6 O_2 are reduced to 12 H_2O.

Note that in glycolysis (6C → 2 × 3C) two ATP are used by kinases, whereas four ATP are generated in the later reactions, giving a net gain of two ATP. In addition, two GTP, which are equivalent to two ATP, are generated when succinyl-CoA is converted to succinate in the subsequent two turns (2 × 3C → $6CO_2$) of the citric acid cycle. In all these instances, the ATP or GTP is generated by phosphoryl group transfer from a phosphorylated intermediate to ADP or GDP. This is termed **substrate-level phosphorylation** and it relies on coupled chemical reactions (Section 1.6). It requires no molecular machines and is not considered further in this chapter.

In both prokaryotes and eukaryotes, the molecular machinery for reoxidizing NADH and $FADH_2$ is a chain of protein complexes (the *respiratory chain* or *electron transport chain*) embedded in a membrane: the plasma membrane of a prokaryotic cell and the inner membrane of the mitochondria in eukaryotic cells (Section 1.9). The ultimate electron acceptor is molecular oxygen, which is reduced to H_2O. As this succession of complexes acts on NADH and $FADH_2$, protons are selectively translocated across the membrane, from the cytoplasm into the periplasmic space in the case of bacteria and from the matrix into the intermembrane space in the case of mitochondria. This creates an *electrochemical proton gradient* across the membrane. The gradient stores the free energy liberated by the reactions catalyzed by the enzyme complexes, and this can be used to drive other reactions. In particular, ATP is generated by allowing the protons to flow back across the membrane

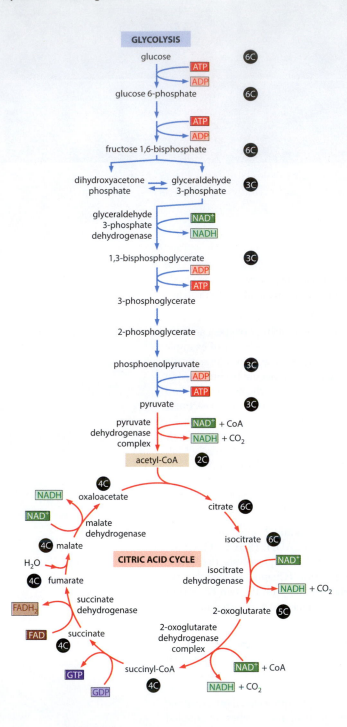

Figure 15.1 An outline of the intracellular reactions that oxidize glucose to CO$_2$ and water. Glycolysis is a widespread series of reactions (depicted in blue) that converts one molecule of glucose (six carbon atoms, 6C) to two molecules of pyruvate (3C). The pyruvate is oxidatively decarboxylated by the pyruvate dehydrogenase (PDH) multienzyme complex (Chapter 9), and the product, acetyl-CoA (2C), enters the citric acid cycle (reactions depicted in red). The cycle has to turn over twice to oxidize the two acetyl-CoA molecules derived from one glucose molecule, releasing four molecules of CO$_2$ in the process. The number of C atoms in each intermediate is shown in a black circle.

through another molecular machine (*ATP synthase*) embedded in the membrane, forcing it to catalyze a reaction that is the reverse of ATP hydrolysis:

$$ADP + P_i \rightarrow ATP + H_2O$$

This mode of coupling chemical reactions (the **chemiosmotic theory**) is radically different from the conventional coupling of reactions through chemical intermediates, as observed in substrate-level phosphorylation. It was first proposed by Peter Mitchell and, though at first highly controversial, is now universally accepted. The generation of ATP from the free energy stored in the proton gradient is generally known as **respiratory chain phosphorylation**, and the overall process of ATP synthesis from the oxidation of a substrate (substrate-level plus respiratory chain phosphorylation) is termed **oxidative phosphorylation**.

The generation of ATP by photosynthesis is different from oxidative phosphorylation but has some important similarities. In certain bacteria (examples are found in purple sulfur, purple non-sulfur, and green sulfur bacteria), molecular machines embedded in the cell membrane use light energy to extract high-energy electrons from donors such as hydrogen, sulfide ions, or Fe^{2+} ions. These electrons then flow down a free energy gradient through other membrane-embedded machines that deposit protons selectively on one side of the membrane. The electrochemical proton gradient thus created, as in the chemiosmotic theory of respiratory chain phosphorylation, stores the free energy needed for an ATP synthase to drive the synthesis of ATP from ADP and P_i (photophosphorylation).

In cyanobacteria and plants, evolution has taken the photosynthetic process a crucial step further. A further set of molecular machines remarkably uses water as the electron donor, with the concomitant formation and release of oxygen:

$$2H_2O \rightarrow 4H^+ + 4e^- + O_2$$

Some of the reducing power thus generated can be used to reduce $NADP^+$ to NADPH, required by the organism for various biosynthetic reactions. This is termed *oxygenic* photosynthesis, as opposed to the use of other electron donors (*anoxygenic* photosynthesis) by bacterial systems.

In plants, the photosynthetic machinery is found in the thylakoid membranes of chloroplasts, specialized plastid organelles that resemble mitochondria (Box 1.5) in many ways. Crucially, the NADPH generated in oxygenic photosynthesis is used in the synthesis of carbohydrate, for example glucose from CO_2, according to the equation

$$6CO_2 + 12NADPH + 12H^+ \rightarrow C_6H_{12}O_6 + 12NADP^+ + 6H_2O$$

This is an endergonic (energy-requiring) process and is thus the converse of the exergonic reactions of biological oxidation, as in mitochondria. The biosynthetic reactions are independent of light and are traditionally called *dark reactions*. Oxygenic plant photosynthesis is the ultimate source of almost all the oxygen and carbon compounds—fixation of CO_2—in the biosphere. It accounts for $\sim 10^{18}$ kJ of free energy each year stored on Earth, corresponding to more than 10^{10} tons of carbon in the form of carbohydrate and other organic matter. This in turn is what aerobic organisms require for the oxidative phosphorylation that supplies their own energy needs, generating CO_2 in the process. Life on Earth depends on the delicate balance between the generation of oxygen and fixation of CO_2 and the oxidation of carbon compounds and release of CO_2.

From fossil evidence it is likely that photosynthetic organisms appeared around 3.4 billion years ago, and that oxygenic photosynthesis in cyanobacteria emerged shortly afterward. Importantly, this led to the buildup of oxygen in the atmosphere after the vast amounts of Fe^{2+} in the early seas had been oxidized to the Fe^{3+} form. This created conditions for the further evolution of organisms capable of profiting from the large amounts of energy released by oxidative metabolism of the carbonaceous material derived from CO_2 fixation.

The molecular machinery of respiratory chain phosphorylation is described in Section 15.2, that of photosynthesis is covered in Section 15.3, and the ATP synthase, a molecular motor driven by chemiosmosis, is explained in Section 15.4.

15.2 BIOLOGICAL OXIDATION AND THE RESPIRATORY CHAIN

Biological oxidation generally consists of the removal of electrons or hydrogen atoms from a substrate and their transfer to an acceptor (see Box 15.1). In aerobic organisms, the NADH and $FADH_2$ generated by the oxidative metabolism of, say, glucose (see Figure 15.1) or fatty acids (Chapter 9) are reoxidized with molecular oxygen as the ultimate electron acceptor. The reducing equivalents are passed to O_2 along an **electron transport chain** (**ETC**) composed of a series of large multi-subunit protein complexes. These complexes are vectorially embedded in a membrane: the cell membrane in bacteria, or the inner membrane of the mitochondria in eukaryotes.

In the uncompartmentalized cells of bacteria, the enzymes of carbohydrate and fatty acid oxidation are all located in the cytosol. In contrast, the enzymes of glycolysis in eukaryotes are in the cytoplasm but the enzymes of the citric acid cycle (including the PDH complex)

Box 15.1   Cofactors of biological oxidation

Biological oxidation normally requires the removal of two H atoms or of one or more electrons from a designated substrate: typically, say, the oxidation of an alcohol to a ketone ($R_1R_2CHOH \rightarrow R_1R_2C=O + 2H$) or the oxidation of the Fe ion in a cytochrome ($Fe^{2+} \rightarrow Fe^{3+} + e^-$). In oxidoreductases (dehydrogenases), the coenzyme most often required is either NAD^+ (nicotinamide adenine dinucleotide) or $NADP^+$ (nicotinamide adenine dinucleotide phosphate); in most instances, NAD^+ is used if the physiological reaction proceeds in the oxidative direction (degradative, or *catabolic*), whereas NADPH is used for reductions (biosynthetic, or *anabolic*). The removal of two H atoms can also be regarded as equivalent to the removal of two protons and two electrons or one proton and one hydride ion ($H^- \equiv H^+ + 2e^-$). With NAD^+ or $NADP^+$ acting as coenzyme at physiologically neutral pH, one H atom is transferred as a hydride ion and the complementary proton enters solution (**Figure 15.1.1**). Reduction is the reverse.

The hydride transfer to NAD^+ or $NADP^+$ is stereospecific. When substrates with deuterium substituted for hydrogen at given positions were studied, it became apparent that some dehydrogenases transfer the hydride to one side of the nicotinamide ring, designated A, whereas other dehydrogenases transfer it to the opposite side, designated B. In some instances, enzymes responsible for steps in the same metabolic pathway have been found to use complementary sides; for example, in glycolysis, glyceraldehyde-3-phosphate dehydrogenase generates NADH of class B stereospecificity, whereas lactate dehydrogenase, which uses NADH to reduce pyruvic acid to lactate in the last step of anaerobic glycolysis (the source of lactic acid in muscle during exercise), acts on NADH with class A stereospecificity. This initially suggested that the complementary stereo-specificity might facilitate the use of the NADH produced by glyceraldehyde-3-phosphate dehydrogenase in the reduction catalyzed by lactate dehydrogenase.

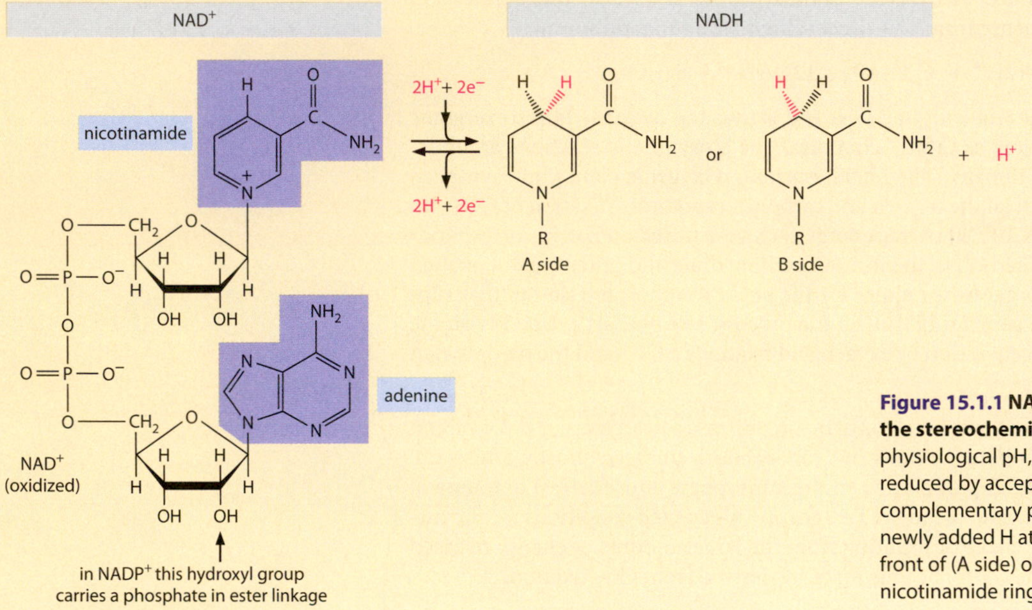

Figure 15.1.1 NAD⁺ and NADP⁺ and the stereochemistry of reduction. At physiological pH, NAD^+ and $NADP^+$ are reduced by accepting a hydride ion and the complementary proton enters solution. The newly added H atom (in red) can project in front of (A side) or behind (B side) the planar nicotinamide ring.

and fatty acid oxidation (Chapter 9) are confined to the inner matrix of the mitochondrion (**Figure 15.2**). Thus in eukaryotes the relevant substrates, for example the pyruvate and NADH derived from glycolysis or the fatty acids, have to enter the mitochondrial matrix to be oxidized. Conversely, ATP is required for use outside the mitochondrion and must be exported, to be replaced with ADP and P_i entering the organelle from the cytosol. The mitochondrial inner membrane is generally impermeable to the passage of molecules and ions, and specific transport systems are needed for the various substrates and products to enter and exit (**Box 15.2**). The protons that are pumped out of the mitochondrion by the action of the ETC are retained in the intermembrane space and flow back through the ATP synthase.

The free energy released in the oxidation of NADH and FADH₂ is stored as an electrochemical proton gradient

The overall reaction for the oxidation of NADH is

$$NADH + H^+ + \tfrac{1}{2}O_2 \rightarrow NAD^+ + H_2O$$

Box 15.1 *continued*

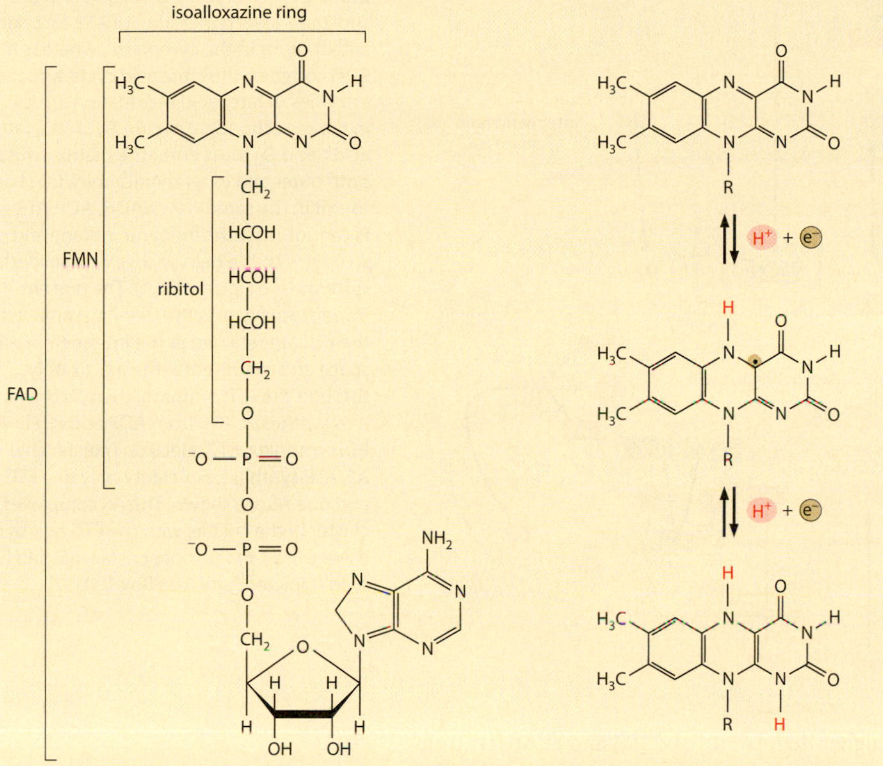

Figure 15.1.2 FMN and FAD. Left, FMN is effectively half of FAD, lacking the AMP moiety. Right, like ubiquinone (see Figure 15.4), FAD can exist in one-electron (FADH·, one H atom added) or two-electron (FADH$_2$, two H atoms added) reduced forms. This is due to a stable free radical (semiquinone) form of the isoalloxazine ring in FADH·. The added H atoms are shown in red and depicted as H$^+$ + e$^-$. FMN behaves similarly.

This suggests that there is a kind of channeling of the coenzyme between the two enzymes (Chapter 9)—but no evidence for this has been found. An interesting structural correlation, however, is that class A dehydrogenases bind NAD$^+$ with the nicotinamide ring in the *anti*-conformation about the glycosidic bond, whereas in class B enzymes the NAD$^+$ is bound with the ring in the *syn* conformation.

FMN (flavin mononucleotide) and FAD (flavin adenine dinucleotide) are also carriers of reducing equivalents in redox reactions. Unlike NAD$^+$ and NADP$^+$, which are colorless, FMN and FAD are named in part for their yellow color (Latin *flavus*, yellow). They can accept two H atoms, thereby being converted to FMNH$_2$ or FADH$_2$, but differ from the nicotinamide coenzymes in that they also exist in one-electron reduced forms; this is because they are able to sustain a stable free radical intermediate (**Figure 15.1.2**). They differ from NAD(P)$^+$ also in that they are prosthetic groups, normally bound firmly into their redox proteins and passing on their reducing equivalents to other electron carriers, such as ubiquinone (see Figure 15.4), that interact with them *in situ*. In some cases, the isoalloxazine ring is covalently linked to the protein.

For the two half-reactions,

$$NADH + H^+ \rightarrow NAD^+ + 2H^+ + 2e^-$$

and

$$\tfrac{1}{2}O_2 + 2H^+ + 2e^- \rightarrow H_2O$$

at pH 7, the NADH/NAD$^+$ couple has a strong tendency to donate electrons ($E'_o = -320$ mV), whereas the H$_2$O/$\tfrac{1}{2}$O$_2$ couple is avid for electrons ($E'_o = +820$ mV). The large difference in redox potential for the overall reaction (+1140 mV, corresponding to a $\Delta G° = -220$ kJ/mol) means that electron flow down the ETC is thermodynamically very favorable.

The free energy for the synthesis of ATP is governed by the equation

$$\Delta G = \Delta G° + RT\ln [ATP]/[ADP][P_i]$$

Under standard conditions, where ATP, ADP, and P$_i$ are all at concentrations of 1 M, $\Delta G = \Delta G° = 31$ kJ/mol, meaning that in theory it should be possible to synthesize multiple copies of ATP from the oxidation of 1 mol of NADH (a 2e$^-$ transfer). In fact, under normal

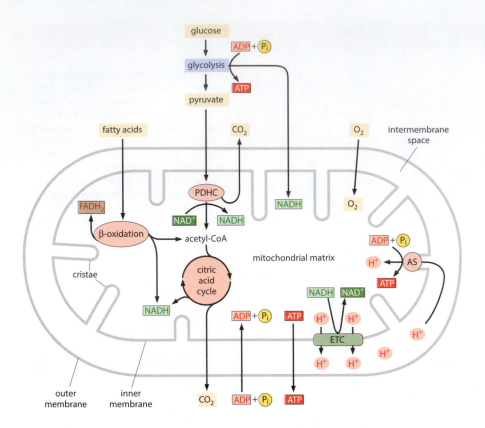

Figure 15.2 Schematic diagram of the reactions of oxidative phosphorylation in the cytoplasm and mitochondrial matrix. In eukaryotes, glycolysis and the substrate-level generation of ATP associated with it occur in the cytoplasm, whereas the PDH complex, the citric acid cycle, and the enzymes of fatty acid β-oxidation are found in the mitochondrial matrix. Pyruvate, fatty acids, and O_2 must enter the matrix across both outer and inner membranes; CO_2 has to exit in the same way. NADH, ADP, ATP, and P_i cannot cross the inner membrane and do so only with the help of specific transport systems (see Figure 15.2.1). The protons translocated across the inner membrane by the ETC are retained in the intermembrane space and can re-enter the matrix only through the ATP synthase, thereby driving it to synthesize ATP from ADP and P_i. PDHC, PDH complex; ETC, electron transfer chain; AS, ATP synthase. For clarity only one ETC and one AS are shown. The AS complexes cluster in the cristae, with the ETC nearby; there will be many more cristae, AS, and ETC than depicted here (see Box 1.5).

cellular conditions in which [ATP] is much higher than [ADP] and [P_i] and all are far from 1 M, the free energy of ATP synthesis is much higher (estimated at 46–55 kJ/mol). This implies that fewer ATP are synthesized per NADH oxidized. However, as we shall see later, because the overall reaction is broken down into smaller steps and the free energy released is stored in a proton gradient across the membrane, the efficiency of energy conversion is very high.

In bacteria and the mitochondria of eukaryotes, the ETC comprises four membrane-bound multi-subunit enzyme complexes. Intermediary two-electron and one-electron carriers, including small-molecule quinones, protein-bound cytochromes, and a soluble cytochrome, are involved. The spectra of the different cytochrome components were first identified by David Keilin in 1925. Since then, a combination of biochemistry, time-resolved spectroscopy, and X-ray crystallography (see **Methods**) of membrane proteins (the latter pioneered by Johann Deisenhofer, Robert Huber, and Hartmut Michel with the protein complexes involved in photosynthesis; see Section 15.3), has led to outstanding progress in mechanistic understanding. A schematic overview of the system of respiratory chain complexes and the parts they play in pumping protons across the membrane in which they are embedded is shown in **Figure 15.3**.

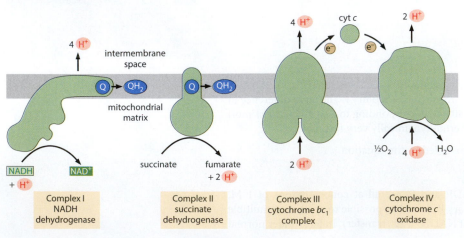

Figure 15.3 Sequential action of the complexes in the respiratory chain. The membrane-embedded complexes of the ETC are depicted schematically and their relative contributions to proton translocation across the membrane are indicated. NADH feeds in at Complex I, whereas the reducing equivalents of $FADH_2$, from the oxidation of succinate in the citric acid cycle or from fatty acid oxidation, enter as QH_2. Cyt c is the water-soluble cytochrome that transfers electrons from Complex III to Complex IV. The membrane shown is the inner mitochondrial membrane; in bacteria it would be the inner membrane.

Box 15.2   Shuttle systems exist to transfer the reducing equivalents of NADH into the mitochondrion

The bacterial cell lacks cytoplasmic compartments, and the oxidative and reductive reactions that form part of intermediary metabolism take place side by side. In eukaryotic cells, much of the oxidative part of metabolism (for example the citric acid cycle and fatty acid oxidation) is confined to the mitochondrial matrix—mitochondria have been called the 'powerhouses' of the cell—whereas reductive metabolism (normally requiring NADPH) takes place largely in the cytosol. The NADH and $FADH_2$ generated in the mitochondrial matrix feed directly into the ETC, whereas the NADH generated by glycolysis in the cytosol must be transported into the matrix for aerobic oxidation (see Figure 15.2). Cytoplasmic NADH can enter the mitochondrial intermembrane space, but the matrix membrane is not permeable to NADH nor indeed to many compounds or ions. However, its reducing equivalents can be transferred by means of the **malate shuttle**, a complementary series of reactions on either side of the matrix membrane (**Figure 15.2.1A**). There are specific transporters that allow malate and aspartate to pass in and out of the mitochondrial matrix; the net effect is

to oxidize cytoplasmic NADH to NAD^+ and reduce matrix NAD^+ to NADH.

A similar shuttle, prominent in insect flight muscle and the so-called *brown adipose tissue* of mammals, uses NADH to reduce dihydroxyacetone phosphate, an intermediate of glycolysis, to 3-phosphoglycerol. In turn, 3-phosphoglycerol reduces the FAD cofactor of a flavoprotein dehydrogenase in the inner mitochondrial membrane (Figure 15.2.1B). The NAD^+ remains extra-mitochondrial and can participate again in glycolysis, whereas the $FADH_2$ is reoxidized by the ETC. Because electrons from $FADH_2$ enter the ETC only at Complex III, electrons coming from NADH via the **glycerol phosphate shuttle** do not pass through Complex I. Thus they cannot contribute to the part of the electrochemical proton gradient that is generated by Complex I, and correspondingly less ATP is synthesized. The 'lost' energy appears as heat, but this is an advantage in warming the flight muscle of insects to permit flight, and also in non-shivering thermogenesis in mammals.

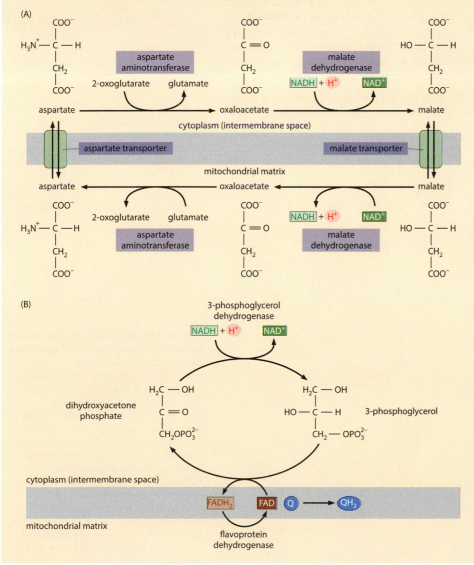

Figure 15.2.1 The malate and glycerol phosphate shuttles. (A) The malate shuttle. In the cytoplasm, a malate dehydrogenase uses NADH to reduce oxaloacetate to malate. Malate, unlike oxaloacetate, has a specific transporter in the inner mitochondrial membrane that allows it to pass into the matrix, where a mitochondrial malate dehydrogenase reverses the reaction, generating NADH. Oxaloacetate is converted to aspartate by means of a mitochondrial aminotransferase that catalyzes the transfer of an amino group from glutamate. Aspartate also has a specific transporter in the inner mitochondrial membrane that allows it to pass into the cytoplasm, where a cytoplasmic aminotransferase transfers the amino group to 2-oxoglutarate to form glutamate and regenerate oxaloacetate, ready for a repeat of the shuttle. (B) The glycerol phosphate shuttle. Cytosolic NADH is used to reduce dihydroxyacetone phosphate to 3-phosphoglycerol, which is a substrate for an FAD-dependent dehydrogenase in the inner mitochondrial membrane. This regenerates dihydroxyacetone phosphate, and the $FADH_2$ produced can enter the ETC at the level of Complex III.

Complex I (NADH–ubiquinone oxidoreductase) is the entry point for oxidation of NADH

NADH feeds reducing equivalents into the electron transport chain through Complex I, an NADH–ubiquinone oxidoreductase. The oxidation of NADH is coupled to the reduction of the membrane-soluble quinone **ubiquinone** (**Q**; **Figure 15.4**), according to the reaction

$$NADH + H^+ + Q \rightarrow NAD^+ + QH_2$$

This reaction is very favorable ($\Delta E'_o = 360$ mV, $\Delta G° = -69.5$ kJ/mol). The reduced ubiquinone (QH_2, a **quinol**) can diffuse laterally in the membrane as a substrate for subsequent components of the ETC. Complex I is the largest electron transport complex and in mitochondria is composed of ~45 protein subunits with a total molecular mass of ~1 MDa. In electron micrographs the complex has a characteristic L shape (**Figure 15.5A**); one arm is composed of numerous hydrophobic subunits and lies in the plane of the membrane, whereas the other arm (also multi-subunit) is hydrophilic and projects more than 100 Å into the mitochondrial matrix at an angle of about 100°. Complex I of prokaryotes is much smaller (~550 kDa), usually consisting of only 14 subunits, but is similarly shaped and correspondingly projects from the cell membrane into the cytoplasm. These 14 subunits, conserved from bacteria to humans, contain all of the redox cofactors and proton pumping machinery and appear to constitute a modular core from which the larger eukaryotic complexes are descended.

The structure of the hydrophilic part of Complex I from a thermophilic bacterium, *Thermus thermophilus*, is shown in Figure 15.5B. It contains eight protein subunits that together harbor the noncovalently bound cofactor FMN and nine **iron–sulfur (Fe–S) clusters** (the mitochondrial enzyme contains only eight distinct FeS clusters, all of which are conserved in the thermophilic bacterium). In FMN, as in FAD, the flavin moiety is stable in the one-electron reduced form, a **free radical** or **semiquinone**, but capable of a further one-electron reduction to $FMNH_2$ or $FADH_2$ (see Box 15.1, second part).

The structures of typical iron–sulfur [2Fe–2S] and [4Fe–4S] clusters are depicted in **Figure 15.6**). In each case the Fe ions are coordinated with an equal number of sulfide (S^{2-}) ions.

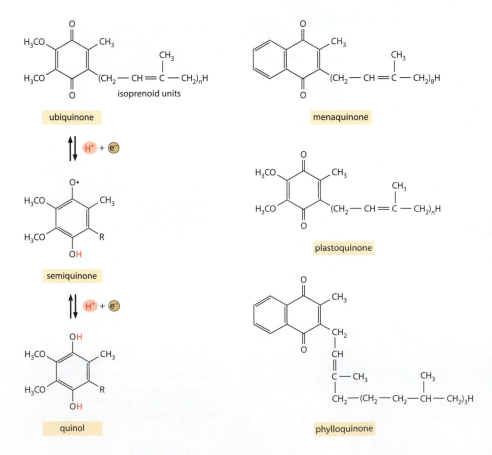

Figure 15.4 Quinones as electron carriers in the membrane. Ubiquinone (*ubi*quitous quinone), designated CoQ or just Q, is widespread. It can carry either one electron and one proton in a stable half-reduced form (semiquinone, a free radical), or two electrons and two protons in its fully reduced form (quinol). The long hydrophobic side chain composed of repeated isoprenoid units confines ubiquinone to the membrane, enabling it to diffuse laterally between complexes as an intermediate connecting one- and two-electron donors and acceptors. In most mammalian species, Q has 10 isoprenoid units. In bacteria, menaquinone has 8; in plant chloroplasts, plastoquinone has 6–10 depending on the source, and phylloquinone has 3.

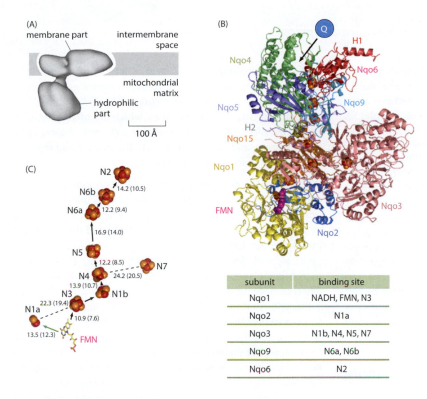

subunit	binding site
Nqo1	NADH, FMN, N3
Nqo2	N1a
Nqo3	N1b, N4, N5, N7
Nqo9	N6a, N6b
Nqo6	N2

Figure 15.5 Structure of Complex I and pathway of electron transfer. (A) A model of bovine Complex I from cryoelectron microscopy. (B) Crystal structure of the hydrophilic part of Complex I of *T. thermophilus*. The orientation is the same as in (A), and the eight protein subunits (Nqo1, Nqo2, and so on) are depicted in different colors. The FMN is shown in a space-filling form in purple, and the 9 Fe–S clusters, two [2Fe–2S], and seven [4Fe–4S], also in space-filling form, are in red (Fe) and yellow (S). The binding cavity for Q is indicated. (C) The arrangement of FMN (depicted in stick model) and the various Fe–S clusters in the hydrophilic part. The arrows represent the path taken by electrons from the FMN to [Fe–S] N2. The green arrow to [Fe–S] N1a represents a likely side-path involved in preventing the release of reactive oxygen species. The center-to-center distances between redox centers are given in Å, with the shortest edge-to-edge distances (also in Å) in parentheses. [Fe–S]N7 is too far from other clusters to be involved in efficient electron transfer, is not conserved, and may be an evolutionary remnant. The table lists the assignments of individual subunits in binding NADH, FMN, and Fe–S clusters. Q, ubiquinone. (PDB 3I9V) (A, adapted from N. Grigorieff, *J. Mol. Biol.* 277:1033–1046, 1998. With permission from Elsevier; B and C, adapted from L.A. Sazanov, *Biochemistry* 46:2275–2288, 2007. With permission from American Chemical Society.)

In iron–sulfur proteins, also called **non-heme iron proteins**, the Fe–S clusters are covalently bound and are normally coordinated by the thiol groups of four Cys residues. The Fe–S clusters are capable of reversible one-electron reduction, the negative charge becoming delocalized over the conjugated Fe ions (partly Fe^{2+} and Fe^{3+}). Thus they are able to pass just one electron at a time.

The electron transfer pathway in Complex I (Figure 15.5C) can be inferred from the X-ray structure starting from the NADH-binding site in the 51 kDa peripheral subunit Nqo1 near the tip of the hydrophilic arm. Two electron-equivalents are transferred from NADH to the FMN redox center in the same subunit (NADH + H^+ + FMN → NAD^+ + FMNH$_2$) and are then released one by one (FMNH$_2$ → FMNH· → FMN) to enter an electron 'wire' of seven successive Fe–S clusters. This begins with cluster N3 and terminates at cluster N2 in subunit Nqo6 close to the membrane-embedded segment, ~95 Å from the FMN cofactor. The bound NADH, FMN, and seven Fe–S centers are not in direct contact but are all separated by distances of less than 14 Å, the maximum distance permissible for efficient electron transfer by quantum tunneling (**Box 15.3**). From the N2 Fe–S cluster, the first electron is transferred to Q, accompanied by the uptake of a proton from the mitochondrial matrix, to generate the stable semiquinone QH· (see Figure 15.4). The same sequence of events leading to the transfer of a second electron and uptake of a proton generates the quinol (QH· → QH$_2$).

Of the four Cys ligands to Fe–S cluster N2 provided by subunit Nqo6, two (Cys45 and Cys46) are highly unusual in coming from adjacent residues in the polypeptide chain, forcing a sub-optimal bonding geometry. On reduction of the hydrophilic part of Complex I by NADH, conformational changes take place. As judged by the lack of expected electron density, Cys46 appears to become 'disconnected' as a ligand. If so, and it were to become protonated, this could be part of a route for the translocation of one proton across the membrane, as discussed further below.

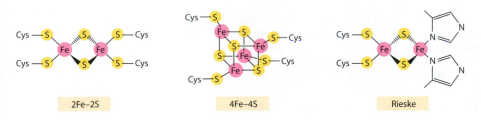

Figure 15.6 Structures of iron–sulfur clusters. The sulfides, S^{2-}, are denoted in yellow and the Fe ions in purple. The 2Fe–2S and 4Fe–4S clusters are conjugated by four Cys residues from the surrounding protein, whereas in the Rieske Fe–S clusters, two of the Cys residues are replaced by His residues.

Box 15.3 Quantum mechanical tunneling by electrons and protons

Enzyme-catalyzed reactions usually proceed through a transition state in which chemical bonds are being made or broken over distances of up to 2 Å (Section 1.6). In the reactions associated with biological oxidation (see, for example, Chapter 9), the transfer of an electron or hydride ion can be direct from one substrate or cofactor to another. However, the redox centers in some electron transfer proteins are found to be as much as 15–20 Å apart. This is widely observed in the protein complexes involved with respiratory chain phosphorylation or photosynthesis. In such instances electron transfer proceeds by **quantum mechanical tunneling**; that is, the electron tunnels through a transition state barrier over which it classically could not pass. Electron tunneling has been known in physics for decades and is explained in terms of the wave–particle duality of matter. An electron on the left-hand side of the barrier has a nonzero probability of tunneling through the barrier and being found on the right-hand side, even if its energy is less than the barrier height. Compared with transition state theory, tunneling is a function of barrier width and height, but width dominates.

Any particle has associated with it a wavelength λ given by the de Broglie equation

$$\lambda = \sqrt{(h^2/2\pi m k T)}$$

where h is Planck's constant, k is the Boltzmann constant, m is the particle mass, and T is the temperature. Thus, in reality, only a particle of low mass can tunnel, with distance and the ΔG of the overall transfer as major factors. As the mass increases, the wavelength becomes so small in comparison with the macroscopic dimensions of the particle that its wave properties are undetectable (the classical situation with respect to familiar objects). For an electron at rest, $\lambda = 27$ Å from the above equation, and in biology its tunneling distances are generally in the range of 4–14 Å. The rate of electron transfer is rapid; it is $\sim 10^{13}/s$

at the shortest distance but falls off by a factor of about 10 for every 1.7 Å extra of separation between donor and receptor. Thus, it occurs faster than the millisecond turnover times that are characteristic of most enzymes, which are often related to proton transfers, protein conformational changes, ligand binding or release, and can accommodate the picosecond reaction times common in photosynthesis.

Hydrogen is the only other particle of low enough mass to be capable of tunneling, but its mass (1800 times that of the electron) means that the tunneling distance is limited to ~0.6 Å. In some enzymes that catalyze the breaking of the C–H bond (methylamine dehydrogenase, lipooxygenase, and a thermophilic alcohol dehydrogenase, among others), exceptionally large **kinetic isotope effects** (**KIEs**), k_H/k_D, have been observed when comparing the rate of hydrogen transfer (protium, mass 1 Da) with that of the heavier deuterium (mass 2 Da). With some enzymes there is evidence of a possible tunneling component below the saddle-point of the potential surface (**Figure 15.3.1A,B**), but this is not always a sufficient explanation of the KIE. Quantum tunneling should be independent of temperature but, although the measured KIEs are close to temperature-independent, the rate of the enzyme-catalyzed reaction may increase significantly as the temperature is raised. It has been suggested that raising the temperature increases the inherent and rapid (nanosecond to millisecond) dynamical motions in the enzyme protein, fostering a family of states in which the donor and acceptor in the H transfer (AH + B $\rightarrow$ A + BH) are brought closer together in the enzyme–substrate complex (Figure 15.3.1C). Local and much faster (perhaps picosecond) motions may also contribute. Together they transiently compress the width of the potential energy barrier, creating essentially degenerate quantum states and increasing the probability of hydrogen tunneling. *(Box continues on next page.)*

Structure of intact Complex I and a possible proton pumping mechanism

Measurements indicate that Complex I pumps four protons out of the mitochondrial matrix for every two electrons transferred from NADH to Q, leading to a net reaction

$$NADH + 5H^+_{in} + Q \rightarrow NAD^+ + QH_2 + 4H^+_{out}$$

The structure of the intact Complex I from *T. thermophilus* (**Figure 15.7A**) shows the hydrophobic membrane-embedded part to be about 180 Å long, comprising seven protein subunits with a total of 64 transmembrane helices. It is slightly curved, both in the plane of the lipid bilayer and perpendicular to it. The additional subunit, Nqo16, in the hydrophilic part (compare Figure 15.5C) appears to be required for crystallization (perhaps as an assembly factor) but is not essential for catalytic activity. Loops from Nqo4 and Nqo6 of the hydrophilic part and Nqo8 of the hydrophobic part are found at the interface between the parts, and most of the interactions involve Nqo8. The three largest subunits (Nqo12, Nqo13, and Nqo14) in the hydrophobic part have amino acid sequences and three-dimensional structures similar to known Na⁺ or K⁺/H⁺ antiporters (see Section 16.5), suggesting that they may be participants in proton pumping. The Fe–S cluster N2 is located about 30 Å outside the lipid bilayer but is near a narrow cavity, about 30 Å long, at the interface of Nqo4, Nqo6, and Nqo8, where the two parts of Complex I abut. X-ray studies of crystals of Complex I in the presence of Q analogs indicate that this cavity is the Q-binding site (menaquinone

Box 15.3   *continued*

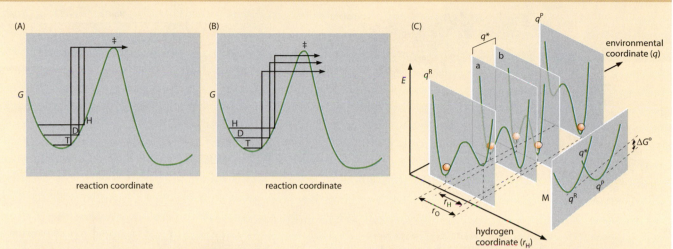

Figure 15.3.1 Electron and proton tunneling in enzyme-catalyzed reactions. (A) Energy profile of a H-transfer reaction across a transition barrier, in which the differences in rate (transfer of H is faster than that of D, which is faster than that of T) arise from the differences in ground-state vibrational levels. (B) Energy profile where tunneling occurs below the top of the transition state barrier. The tunneling correction is expected to decrease from H to D to T as the mass increases (H = 1 Da, D = 2 Da, and T = 3 Da). (C) Energy profiles of a H-transfer reaction in which H-tunneling occurs similar to that of the electron. Tunneling is predominantly from the vibrational ground state of the wavefunction of the hydrogen nucleus, represented by the yellow spheres. Reaction is along the q axis, the environmental coordinate. E, energy; q^R, energy profile for initial reactant; q^*, transition state; q^P, energy profile for product. The two profiles designated **a** and **b** represent essentially degenerate quantum states in the double well from which the proton is able to tunnel. In profile **b**, the distance between the two energy wells on the hydrogen coordinate axis (r_H) is less than the equilibrium value (r_O) in profile **a**, owing to rapid (picosecond) thermally induced fluctuations in the enzyme protein. M is a Marcus-like free-energy profile, indicating $\Delta G°$ between reactant and product. (Adapted from M.J. Sutcliffe et al., *Phil. Trans. R. Soc. B* 361:1375–1386, 2006. With permission from The Royal Society.)

In addition to explaining electron or proton transfer in some enzyme reactions, quantum tunneling has been proposed to account for mutations in DNA, whereby a hydrogen atom in a hydrogen bond tunnels through the barrier from the electronegative atom of one base in a pair to the electronegative atom of the other base in the pair.

References

Klinman JP (2014) The power of integrating kinetic isotope effects into the formalism of the Michaelis–Menten equation. *FEBS J* 281:489–497.

Moser CC, Chobot SE, Page CC & Dutton PL (2008) Distance metrics for heme protein electron tunneling. *Biochim Biophys Acta* 1777:1032–1037.

in vivo). Although it is above the membrane surface, the cavity is enclosed from solvent and has only a narrow entry point for the quinone. The narrowness of the cavity restricts the long isoprenoid tail of the quinone to an extended conformation, which seals the reaction chamber. It places the quinone ring ~12 Å (center to center) away from the Fe–S cluster N2, sufficiently close for efficient electron transfer but suggesting that slight structural rearrangements are needed for the quinone to move in and out.

In each of the three antiporter-like subunits Nqo12, Nqo13, and Nqo14, key transmembrane helices are broken in the middle by flexible peptide regions surrounded by charged and polar amino acid residues. This allows for an N-terminal and C-terminal half-channel in each protein (Figure 15.7B). Subunit Nqo8, despite showing only ~11–18% sequence identity to antiporters, similarly contains an N-terminal half-channel, although its transmembrane helices are tilted by up to 45° from the more usual position of roughly normal to the plane of the membrane. It also contains a 'funnel' of charged residues leading from the Q-binding site to a cluster of charged residues deep in the membrane: four Glu residues from Nqo8 and an Asp from Nqo10. On the other side of the break in the transmembrane helix of Nqo10, a conserved Tyr residue of Nqo10 interacts with an essential Glu of Nqo11, and these are part of another half-channel. Taken together, these elements make up four potential H⁺-translocating channels, as depicted in Figure 15.7B. A possible proton-pumping mechanism is thus that electrons travel from NADPH to Fe–S cluster N2 and thence to menaquinone in its binding cavity. This initiates a conformational change in the proton channel in Nqo8, Nqo10, and Nqo11, which is propagated through 'antiporter' Nqo14 to

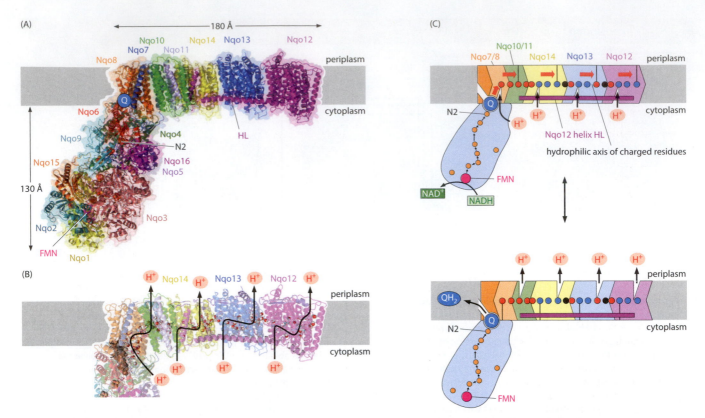

'antiporters' Nqo13 and Nqo12 (Figure 15.7C). The first half-channel in each case closes to the cytoplasm and a proton is donated to the second half-channel, from there to be ejected into the periplasm when the complex returns to the oxidized state. A total of four H^+ are translocated during each turnover.

The funnel of charged residues in the Q-binding site, plus the charged and polar residues (a central hydrophilic axis) surrounding the flexible breaks in the relevant transmembrane helices are crucial to the lateral cascade of conformational change. Coordination is aided by (1) the long amphipathic helix HL at the C-terminal end of Nqo12, which lies in the membrane on the cytoplasmic side, parallel to the central hydrophilic axis, and makes contact with both Nqo13 and Nqo14, and (2) β-hairpin loops interacting with helices in each of Nqo14, Nqo13, and Nqo12 on the periplasmic side. The reduced menaquinone (QH_2) is released into the membrane, retained there by its long hydrophobic tail, and is replaced by an oxidized Q ready for another turnover of the complex.

The mitochondrial Complex I resembles the bacterial Complex I but has many more subunits

Comparison of a lower-resolution X-ray structure of the intact Complex I of the aerobic yeast *Yarrowia lipolytica* with that of the *T. thermophilus* complex suggests that it closely resembles its bacterial counterpart in terms of its mechanistic core. This eukaryotic complex (947 kDa) contains 40 subunits, and the 26 additional subunits (up to 31 for the bovine complex) are arranged around the 14 core subunits in ways that have yet to be determined in detail. In the mitochondrial Complex I of most species, 7 of the 14 core subunits are encoded in the mitochondrial DNA and form part of the membrane arm of the complex, whereas the other subunits (38 for the bovine complex) are encoded in the nuclear DNA (**Table 15.1**).

The functions of the ~31 non-core (sometimes called *supernumerary*) subunits are largely unknown, but some have sequence similarities to other recognizable proteins. For example, the bovine mitochondrial Complex I contains a subunit with sequence similarity to the acyl-carrier protein of fatty acid biosynthesis (Section 9.5). Another subunit is the protein GRIM-19 (Gene associated with Retinoid-IFN induced Mortality), thought to be involved in cell death (Section 13.6). Thus, in addition to its function in electron transfer and proton pumping in the respiratory chain, Complex I may turn out to have other roles.

Figure 15.7 Structure of Complex I and a possible proton pumping mechanism. (A) Structure of the entire Complex I of *T. thermophilus*. The various subunits are depicted in different colors. The Q-binding cavity and Fe–S cluster N2 are indicated. A long amphipathic helix (HL) at the C-terminal end of Nqo12 lies parallel to the membrane and makes contact with Nqo13 and Nqo14. (B) A view of the membrane part of the complex to illustrate the four putative H^+-translocating channels from cytoplasm to periplasm. (C) A schematic representation of a possible proton pumping mechanism. The color coding is the same as in (A) and (B). Charged residues in the central hydrophilic axis are depicted as red (Glu) and blue (Lys/His) circles. Black circles represent conserved Pro residues at key helix break points. The red arrows in the reduced form indicate the direction of the transmission of conformational changes across the hydrophobic part of the complex, emanating from the Q-binding site. (PDB 4HE8 and 4HEA) (Adapted from R. Baradaran et al., *Nature* 494:443–448, 2013. With permission from Macmillan Publishers Ltd.)

Table 15.1 Genes and proteins involved in structure and assembly of Complexes I–IV of human mitochondria.

Location of genes	Complex I	Complex II	Complex III	Complex IV
Subunits encoded in mitochondrial DNA	7	0	1	3
Subunits encoded in nuclear DNA	~38	4	10	10
Nuclear-encoded assembly proteins	~11	~2	~9	~30

Complex II (succinate–ubiquinone oxidoreductase) is a succinate dehydrogenase and not a proton pump

Complex II provides a second entry point for electrons into the ETC. It is a succinate dehydrogenase, catalyzing the oxidation of succinate to fumarate in the citric acid cycle (see Figure 15.1). It uses FAD as a covalently bound cofactor and brings about the reduction of Q to QH_2. Complex II is the only membrane-bound enzyme of the citric acid cycle and the only complex in the respiratory chain that is not a proton pump.

The reaction catalyzed is

$$FADH_2 + Q \rightarrow FAD + QH_2$$

for which $\Delta E^{\circ\prime} = 85$ mV, with a corresponding $\Delta G^{\circ\prime}$ of -16.4 kJ/mol, much the smallest of the four complexes in the ETC. A structurally homologous enzyme, quinol–fumarate reductase, is used by bacteria such as *Escherichia coli* growing anaerobically to reduce fumarate to succinate as a terminal electron acceptor in place of oxygen.

Complex II (~125 kDa) is much smaller than Complex I and has only four types of subunit. In mammalian mitochondria it is the only respiratory complex in which all of the subunits are encoded by nuclear genes. The enzyme from *E. coli* is a homotrimer, and in each of the three protomers the hydrophilic portion is composed of two subunits; the larger (SdhA) contains the substrate-binding site and the FAD, whereas the smaller subunit (SdhB) contains three Fe–S clusters (**Figure 15.8**). These two subunits are bound to two small hydrophobic membrane-spanning subunits (SdhC, SdhD) that together ligate a *b* heme cofactor (**Figure 15.9**). Heme *b* may partake in electron transfer but is not essential for function; it may help stabilize the structure of the membrane-bound complex and prevent the leakage of electrons to oxygen and the formation of dangerous *reactive oxygen species* (see below). The binding site for Q utilizes amino acid residues from the Fe–S subunit and both membrane-spanning subunits.

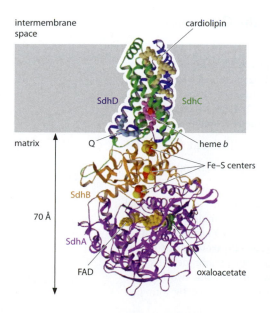

Figure 15.8 The structure of Complex II. Only one protomer of the trimeric complex from *E. coli* is shown for simplicity. FAD in SdhA is in gold, Fe–S centers in SdhB are in red and yellow, and heme *b* is in magenta; all are in space-filling form. Also shown are a molecule of oxaloacetate (a reversible inhibitor bound as a result of the crystallization conditions) in dark green, and a molecule of cardiolipin (a membrane lipid) in beige. A molecule of bound ubiquinone (Q), in blue, marks the exit point of the product QH_2. (PDB 1NEN)

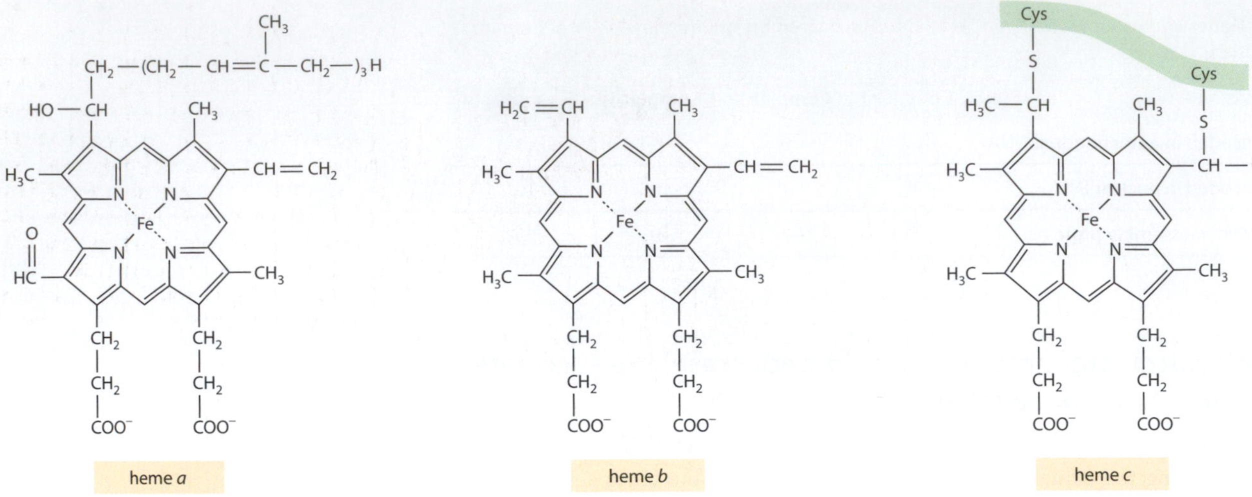

heme *a*

heme *b*

heme *c*

Figure 15.9 The heme groups of cytochromes *a*, *b*, and *c*. The heme group is a planar tetrapyrrole ring (or porphyrin) at the center of which is an Fe ion, coordinated with the four N atoms. In cytochromes the heme is mostly buried and the Fe has two axial ligands, one on either side of the plane. In cytochromes *a* and *b* both axial ligands are His residues; in contrast, in cytochrome *c* they are the imidazole N of a His residue and the sulfur of a Met residue, and the heme group is covalently linked to the protein by two thioether bonds from Cys residues. Heme *a* uniquely carries a hydrophobic chain of repeating isoprenoid units. Cytochromes become oxidized and reduced as the Fe ion reversibly switches between Fe^{3+} and Fe^{2+}. The cytochromes have characteristic absorption spectra characterized by three main peaks: α (present only in the reduced species), β, and γ (the most intense and often called **Soret**), in order of decreasing wavelength. They arise from the conjugated array of single and double bonds in the heme groups, and the spectrum changes when the heme becomes reduced. The wavelength of the α band is used to distinguish between cytochromes, for example, the α band of cytochrome *a* has a wavelength of 600 nm, that of cytochrome *c* is 550 nm, and that of cytochrome c_1 is 554 nm. Cytochromes differ markedly from the oxygen carriers, hemoglobin and myoglobin. The heme group, protoporphyrin IX, in hemoglobin is the same as that in cytochrome *b*, but there is only one axial ligand (a His) from the protein and the Fe must remain Fe^{2+} for hemoglobin to be able to bind O_2 as the other ligand.

As in Complex I, the three well-spaced Fe–S clusters ([2Fe–2S]–[4Fe–4S]–[3Fe–4S]) in Complex II provide a linear 'wire' of more than 40 Å in length to convey electrons through to the site of reduction of the Q (FADH$_2$ → FADH· → FAD and, correspondingly, Q → QH·→ QH$_2$) The FAD-containing subunit comprises two domains, termed a capping domain and flavin domain. The succinate-binding site is sandwiched between them, and the capping domain appears to rotate to allow entry of substrate (succinate) and exit of product (fumarate). These conformational changes are also affected by the presence of electrons on the Fe–S clusters, but there is no proton translocation.

Complex III (cytochrome *bc*$_1$ complex) oxidizes the QH$_2$ produced by Complexes I and II

Complexes I and II both generate ubiquinol (QH$_2$), which diffuses laterally in the membrane. Other sources of QH$_2$ include glycerol phosphate dehydrogenase (see Box 15.2) and the β-oxidation of fatty acids in mitochondria brought about by the electron-transfer flavoprotein (Section 9.5), all of which use FAD as cofactor. Complex III catalyzes the oxidation of QH$_2$ and the reduction of soluble cytochrome *c* according to the reaction

$$QH_2 + 2 \text{ cytochrome } c \text{ (oxidized, } Fe^{3+}) \rightarrow Q + 2 \text{ cytochrome } c \text{ (reduced, } Fe^{2+}) + 2H^+$$

For this reaction, $\Delta E^{\circ\prime} = 190$ mV ($\Delta G^{\circ\prime} = -37$ kJ/mol), which is large and favorable.

Complex III from mammalian mitochondria is a dimer in which each protomer (~240 kDa) contains 11 subunits (**Figure 15.10A**). Only three of these subunits have bacterial homologs, and between them they carry all the *redox centers* of the complex. One of the three is cytochrome *b*, which has eight transmembrane helices and two *b* hemes. The other two are an iron–sulfur protein (ISP) carrying a **Rieske-type [2Fe–2S] iron–sulfur center** and cytochrome c_1, each anchored to the membrane by a single transmembrane helix. The Rieske [2Fe–2S] center is unusual in that it is coordinated by two Cys and two His side chains, unlike the more common [2Fe–2S] centers, which are coordinated by four Cys side chains (see Figure 15.6). The eight accessory subunits in eukaryotic cytochrome bc_1 complexes have uncertain roles. Two of them (core 1 and core 2) are homologous to the Zn^{2+}-dependent mitochondrial metalloendopeptidases, which cleave presequences from proteins after targeted protein import. Complex III, like Complex I, may have other functions in addition to its direct role in the respiratory chain.

Proton pumping by Complex III depends on a Q cycle

Complex III functions by means of a **Q cycle**, as originally proposed by Peter Mitchell and later modified by Anthony Crofts (**Figure 15.11**). The presence of two active sites in the cytochrome bc_1 complex is an essential feature. The inhibitor myxothiazol binds in or near the Q_o site, where QH_2 is oxidized, located near the outside of the inner mitochondrial membrane, between the Rieske ISP and the low-potential heme (b_L, b_{566}) of cytochrome b; in contrast, the inhibitor antimycin binds in or near the Q_i site toward the other side of the membrane close to the high-potential heme (b_H, b_{562}) and facing the mitochondrial matrix (see Figure 15.10A).

The two electrons donated by QH_2 (releasing $2H^+$) can follow one or other of two distinct pathways through Complex III. The first electron taken from QH_2 at the Q_o site is

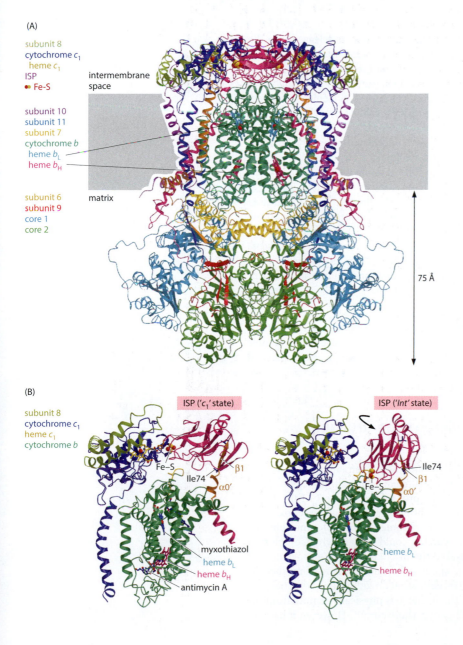

(A)

subunit 8
cytochrome c_1
  heme c_1
ISP
  • Fe–S

intermembrane space

subunit 10
subunit 11
subunit 7
cytochrome b
  heme b_L
  heme b_H

subunit 6   matrix
subunit 9
core 1
core 2

75 Å

(B)

subunit 8
cytochrome c_1
  heme c_1
cytochrome b

ISP ('c_1' state)

ISP ('*Int*' state)

Fe–S
Ile74
β1
α0'

Ile74
β1
Fe–S
α0'

myxothiazol
heme b_L
heme b_H
antimycin A

heme b_L
heme b_H

Figure 15.10 The structure of Complex III. (A) The bovine mitochondrial complex is a dimer and contains 11 subunits in each protomer. The three key subunits in the mechanism, with counterparts in bacterial complexes, are cytochrome b, cytochrome c_1, and the ISP. The heme c_1 protein ligands are His41 and Met160, and the heme is covalently linked to Cys27 and Cys40. Subunit 9 is the cleaved mitochondrial import sequence of the ISP and may not be a functional protein subunit as such. (PDB 1BGY) (B) The structures around the ISP in two different crystal forms of the protein. The position of the membrane is omitted for clarity. Left, in the $P6_522$ form, the ISP interacts with residues of cytochrome c_1 and is in position for electron transfer between the Fe–S center and heme c_1. Right, in the $P6_5$ form, the ISP is positioned differently, with its Fe–S center closer to the heme b_L of cytochrome b. This is made possible by conformational changes in the linking region involving the helix designated α0' and the β strand designated β1 (shown in orange), centered round Ile74 (depicted in blue), where the ISP transmembrane anchor helix is emerging from the membrane. (Adapted from S. Iwata et al., *Science* 281:64–71, 1998. With permission from AAAS.)

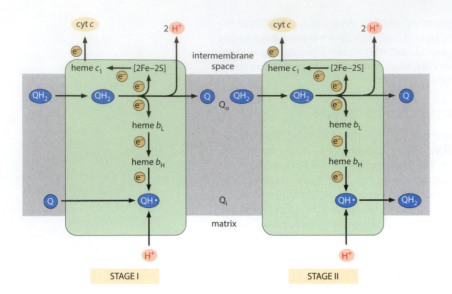

Figure 15.11 The two-stage mechanism of the Q cycle. In stage 1, QH_2 is oxidized to Q at the Q_o site, and Q is reduced to QH· at the Q_i site. In the second stage, another QH_2 is oxidized to Q at the Q_o site, and the QH· at the Q_i site is reduced to QH_2. At each stage, one H^+ is taken up from the matrix and two H^+ are pumped into the intermembrane space. Two QH_2 are oxidized and one QH_2 is replenished after the completion of both stages. The diagram is schematic only, and for simplicity the cytochrome bc_1 complex is not depicted as a dimer.

transferred to the Rieske Fe–S center in the ISP. A proton from QH_2 is released into the intermembrane space and the semiquinone QH· remains bound. On the basis of chemical and X-ray crystallographic evidence, it appears that the reduced Rieske Fe–S domain can then undergo a 35° rotation and swing by ~20 Å, to bring the Fe–S center close to the heme group of cytochrome c_1 (Figure 15.10B). The cytochrome c_1 heme is reduced, and from there the electron is passed to the heme group of a soluble cytochrome c that makes transient contact with Complex III. The reduced cytochrome c is free to move in the intermembrane space and makes transient contact with another complex, the terminal cytochrome oxidase (Complex IV), to which the electron is finally transferred. The second electron, coming from the QH· left bound at the Q_o site, follows a different path. It is passed across the membrane through a lower-potential chain consisting of the b_L and b_H hemes of cytochrome b and is used to reduce a Q at the Q_i site to QH·. The now fully oxidized Q at the Q_o site can dissociate and join the Q pool. A second QH_2 replaces it and undergoes oxidation at the Q_o site, exactly as before, the first electron being used to produce another reduced ISP (and ultimately reduced cytochrome c) with the second electron passed across the membrane through the b_L and b_H hemes to the Q_i site. The QH· waiting at the Q_i site is reduced to QH_2, the proton required being taken from the mitochondrial matrix. Thus, for every two QH_2 that deliver electrons into the Q cycle at the Q_o site, one QH_2 is regenerated at the Q_i site.

To summarize, in the overall reaction of the Q cycle, the oxidation of one molecule of QH_2 delivers two electrons via the ISP and cytochrome c_1 to reduce two molecules of soluble cytochrome c. In parallel, two H^+ are taken up from the matrix and four H^+ are dispatched into the intermembrane space:

$$QH_2 + 2 \text{ cytochrome } c \text{ (Fe}^{3+}) + 2H^+_{in} \rightarrow 2 \text{ cytochrome } c \text{ (Fe}^{2+}) + 4H^+_{out}$$

The ISPs in the Complex III dimer undergo large structural movements

After it accepts the first electron from QH_2 at the Q_o site, the movement of the [2Fe–2S] domain of the ISP is made possible by a conformational change in a hinge region at the beginning of the transmembrane helix that anchors the protein (see Figure 15.10B). In the Complex III dimer, the transmembrane helix is in one protomer, whereas the redox active [2Fe–2S] domain is catalytically engaged with the cytochrome c_1 and ubiquinol-binding site provided by the other protomer. Such an interaction could explain why Complex III is dimeric. Bifurcated electron transfer is essential to the mechanism, and a movement of the reduced [2Fe–2S] domain away from the Q_o site may be critical in ensuring that the second electron, coming from the residual QH·, travels instead down the low-potential path through the b_L and b_H hemes of cytochrome b. The b_L heme is physically remote from the cytochrome c_1 heme, which obviates any possibility of a 'short-circuit.' However, it has also been argued that the bifurcated electron transport is a result of the first electron transfer from QH_2 to the oxidized ISP being the rate-determining step in the overall process, the [2Fe–2S] domain then dissociating from the Q_o site and rapidly moving away, leaving the second electron, from the bound $QH\cdot$, to pass to the b_L/b_H hemes.

Cytochrome c is a soluble protein. Its structure is highly conserved and the heme is essentially buried, but a cluster of Lys residues surrounds the heme crevice where one of the two thioether bridges covalently binding the heme is exposed (**Figure 15.12**). Electrostatic interactions with negatively charged side chains on cytochrome c_1 may be important in orienting cytochrome c to promote rapid electron transfer during its transient interaction with other proteins.

Complex IV (cytochrome *c* oxidase) reduces molecular oxygen to water

The final step in the respiratory chain is the reduction of molecular oxygen to water, according to the reaction

$$2\ \text{cytochrome } c\ (Fe^{2+}) + 2H^+ + \tfrac{1}{2}O_2 \rightarrow 2\ \text{cytochrome } c\ (Fe^{3+}) + H_2O$$

This reaction is catalyzed by Complex IV (cytochrome c oxidase, CcO); $\Delta E^{\circ\prime}$ = 580 mV ($\Delta G^{\circ\prime}$ = –112 kJ/mol), which is very large and favorable.

Each protomer (~410 kDa) of the dimeric CcO of mammalian mitochondria has 13 subunits, whereas that of the bacterial enzymes has fewer, usually only four. The three large subunits (I, II, and III) at the core of the structure (**Figure 15.13A**) are encoded by the mitochondrial DNA and are homologous to those of bacterial CcOs. The other 10 subunits are encoded in the nuclear DNA and appear to be involved with stabilizing and assembling the eukaryotic complex. CcO does not appear to undergo the substantial conformational changes associated with Complexes I–III, but subtle changes are likely to be essential for electron transport and for opening and closing the proton channels involved in pumping protons across the membrane.

On the concave surface of ox heart mitochondrial CcO that faces the intermembrane space, there are negatively charged amino acid side chains complementing the positively charged Lys side chains of an incoming cytochrome c and facilitating the transient contact. The

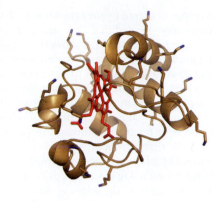

Figure 15.12 The structure of cytochrome *c*. The structure shown is that of horse heart cytochrome c. The heme is depicted in red and the lysine side chains are shown with the N^6-NH_2 groups in blue. (PDB 1HRC)

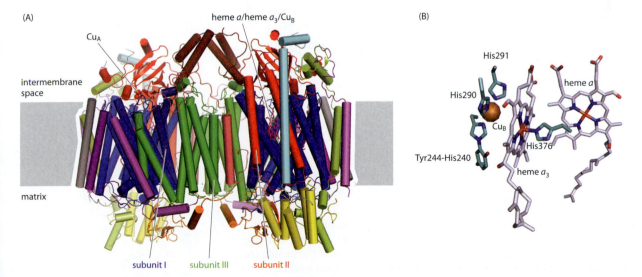

Figure 15.13 The structure of Complex IV. (A) The structure shown is that of the CcO from ox heart mitochondria. The 13 subunits in each protomer of the dimer are all colored differently, with helices as cylinders for simplicity. The Cu atoms are shown in red and the heme groups in purple. There are several lipid molecules bound (not shown) in the crystal structure, including phosphatidylethanolamine, cardiolipin, and phosphatidylglycerol; their hydrophobic tails are directed inward into the hydrophobic region of the membrane, and their hydrophilic phosphates are in the membrane surface: two on the cytosolic side and six on the matrix side. (B) The structure surrounding the heme a_3/Cu$_B$ binuclear center, illustrating the nearby Tyr residue covalently linked to a His. The remainder of the protein and the membrane are omitted for simplicity. The structure is viewed in the same orientation in both panels. (PDB 1V54 and 2OCC)

electron delivered by the cytochrome c first reduces a pair of copper ions (Cu^+/Cu^{2+}) in the Cu_A center located in a 10-β-stranded extramembrane domain of subunit II. The two Cu ions are held about 7.5 Å above the membrane surface and are like a [2Fe–2S] center (see Figure 15.6) in that they are coordinated with the sulfur atoms of two Cys residues, Cys196 and Cys200. The electron passes, in turn, to a low-spin heme a, from which it is transferred to a binuclear heme a_3/Cu_B center. Both hemes are bound with their rings perpendicular to the plane of the membrane in a cluster of transmembrane helices of subunit I. The distance of the nearer Cu in Cu_A to the Fe of heme a is 21 Å and to the Fe of heme a_3 is 23 Å. This is too large for efficient through-space electron transfer. In keeping with the spectroscopic evidence that heme a is reduced before heme a_3, it appears that the electron could travel to heme a by a fast through-bond route ($t_{\frac{1}{2}} = 0.1$ ms) involving His204 of subunit II, a peptide bond between Arg438 and Arg439 of subunit I, and the propionate side chain (see Figure 15.9) of heme a. The edges of the two hemes are close (4.7 Å), which would then allow very fast electron transfer laterally from heme a to heme a_3. The Cu_B ion is coordinated by three His residues, one of which (His240) is covalently linked to a Tyr residue (Tyr244), as illustrated in Figure 15.13B. At the heme a_3/Cu_B center, the electron meets with a proton coming from the matrix on the other side of the membrane and is ready to participate in the reduction of molecular oxygen to water.

Reduction of molecular oxygen is a four-electron reaction

Four electrons (and four protons) are required to reduce one molecule of oxygen to two molecules of water ($O_2 + 4H^+ + 4e^- \rightarrow 2H_2O$). This makes it a difficult reaction to bring about and it requires a number of steps. All four electrons needed are supplied from reduced cytochrome c, and the overall eight protons involved (four for reduction of O_2 and four to be pumped) are supplied from the matrix side of the membrane. The mechanism is still somewhat speculative, but spectroscopic data are consistent with the following series of reactions at the heme a_3/Cu_B center, where TyrOH represents Tyr244, which is covalently linked to His240 (Figure 15.13B)

The first two electrons prime the active site and each also pumps a proton:

$$Fe^{3+}\ ^-OH\ HO^-Cu^{2+}\ TyrOH + (e^- + 2H^+_{in}) \rightarrow Fe^{3+}H_2O\ HO^-Cu^+\ TyrOH + H^+_{out}$$

$$Fe^{3+}H_2O\ HO^-Cu^+\ TyrOH + (e^- + 2H^+_{in}) \rightarrow Fe^{2+}H_2O\ H_2OCu^+\ TyrOH + H^+_{out}$$

It is thought that O_2 binds transiently to this doubly reduced heme a_3–Cu_B center. Within this primed active site, the TyrOH contributes an electron and becomes the radical TyrO·. Two more electrons are provided by the Fe^{2+} becoming Fe^{4+} (the *ferryl* state) and a fourth electron by the Cu^+ becoming Cu^{2+}. With these four electrons, the O–O bond can be broken:

$$Fe^{2+}H_2O\ H_2OCu^+\ TyrOH + O_2 \rightarrow Fe^{4+}{=}O^{2-}\ HO^-Cu^{2+}\ TyrO· + 2H_2O$$

The heme a_3/Cu_B center now has to return to its starting position. For this to happen, two more electrons are provided by reduced cytochrome c molecules; one is used to reduce TyrO· back to TyrOH and the other to reduce the Fe^{4+} back to Fe^{3+}. In doing so, two more protons are pumped:

$$Fe^{4+}{=}O^{2-}\ HO^-Cu^{2+}\ TyrO· + (e^- + 2H^+_{in}) \rightarrow Fe^{4+}{=}O^{2-}\ HO^-Cu^{2+}\ TyrOH + H^+_{out}$$

$$Fe^{4+}{=}O^{2-}\ HO^-Cu^{2+}\ TyrOH + (e^- + 2H^+_{in}) \rightarrow Fe^{3+}\ ^-OH\ HO^-Cu^{2+}\ TyrOH + H^+_{out}$$

The overall reaction is thus

$$8H^+_{in} + 4e^- + O_2 \rightarrow 2H_2O + 4H^+_{out}$$

For each turnover of CcO, four protons are pumped out of the matrix (or bacterial cytoplasm) and four protons are taken up in the reduction of O_2 to water. The reaction may have several parts to it, but it takes place in ~1 ms. A proton pumped across the membrane during electron transfer is often called a **pumped proton**, whereas a proton transferred to the active site in the reduction of oxygen is referred to as a substrate or **chemical proton**.

Protons enter and leave the membrane-embedded CcO through gated pathways

There appear to be two proton uptake channels that connect the buried active site of CcO to the mitochondrial matrix; these are termed D and K after conserved Asp (D) and Lys

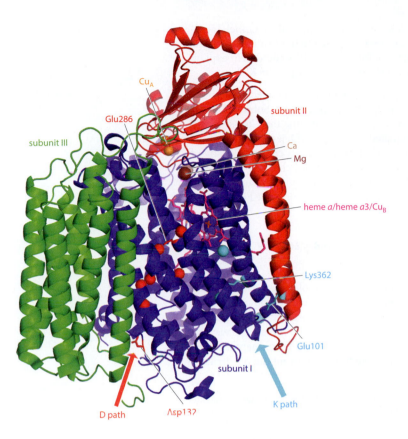

Figure 15.14 The D and K proton channels in the C*c*O of *Rhodobacter sphaeroides*. The proposed D path is shown in red (hydrogen-bonded waters as red spheres) and the K path in blue (hydrogen-bonded waters as blue spheres). The residue numbering is that of the *Rb. sphaeroides* complex, with the corresponding conserved residues in subunits I and II of the bovine enzyme in parentheses, as follows: $Glu286_I (Glu242_I)$, $Tyr288_I (Tyr240_I)$, $Asp132_I (Asp91_I)$, $Lys362_I (Lys319_I)$. Glu101 is provided by subunit II. The rest of the protein and the position of the membrane are omitted for clarity. (PDB 1M56) (Adapted from J.P. Hosler, S. Ferguson-Miller and D.A. Mills, *Annu. Rev. Biochem.* 75:165–187, 2006.)

(K) residues, respectively, found within them. Mutagenesis and other experiments suggest that the K pathway is used to transfer two of the four matrix protons required to reduce an O_2 molecule to water at the active site, whereas the D pathway transfers a total of six protons, the further two required as substrate protons and the four that are pumped across the membrane. It has also been suggested that a third pathway, designated H and located near heme *a*, may be involved in pumping protons across the membrane. Entrances to the D and K channels can be discerned in the X-ray structures of C*c*O and its bacterial homologs, but it is less clear where the translocated protons exit on the other side of the membrane and how the molecule of O_2 reaches the active site. No H pathway has yet been noted in the bacterial enzymes.

The four-subunit bacterial C*c*Os have been particularly well studied because of the relative ease of their preparation and experimental (including genetic) manipulation. The conserved Asp (Asp132) at the entry to the D channel of *Rhodobacter sphaeroides* C*c*O (**Figure 15.14**) is connected via a chain of hydrogen-bonded water molecules to a conserved Glu (Glu286). This corresponds to Glu242 in the bovine sequence $H_{240}PE_{242}VY_{244}$ (in the one-letter code). This spans a distance of some 26 Å, and Glu242 is close to but approximately equidistant from the two hemes (*a* and a_3). A proton-hopping mechanism (Box 9.2) is possible, and the pK_a and protonation state of Glu242 (continuing with the bovine sequence) and other local ionizable residues appears to be sensitive to redox changes. From Glu242 the substrate protons flow on through another short series of water molecules to Cu_B, a distance of 10–12 Å. In contrast, the K pathway has only two visible ordered water molecules (see Figure 15.14) but there is strong hydrogen bonding between the OH of Tyr242 and the OH of the side chain farnesyl group of heme a_3 in the oxidized form of the enzyme.

High-resolution X-ray structures of the C*c*O in the reduced form reveal a displacement of heme a_3 plus small nearby protein movements, including residues of subunit I in the K channel, and these changes are reversed on reoxidation. The hydrogen bond between Tyr242-OH and heme a_3 farnesyl-OH is lost, and additional ordered water molecules appear close to the binuclear center. It is possible that these changes could affect the connection of Glu242 to the active site or to a proton exit path, providing a conformational control over alternating openings of the D and K paths for proton transport. A further suggestion is that the changes in the conformation of the side chain of Glu242 and the disposition of surrounding water molecules allows Glu242 to act as a one-way valve to prevent a backward leak of protons, which could occur if there were not some sort of gate in the channel. Ionizable amino acid side chains from subunit III of CcO are juxtaposed to the Asp residue at the entrance to the D channel and may act as an antenna for protons. This could help explain why a majority of protons are recruited to pass through the D rather than the K channel.

Studies of the CcO of *T. thermophilus*, a three-subunit enzyme, have revealed its similarity to the aa_3 CcOs described above but with some significant differences. Subunit I binds heme b and contains a heme a_3/Cu_B catalytic site; subunit II contains the Cu_A site; and subunit IIa is a single transmembrane helix. Designated a ba_3-type CcO, it contains just one identifiable proton channel, corresponding to the K channel of the aa_3 enzymes. Many of the mechanistic ideas about CcOs, especially of proton pumping and channeling, are still tentative and much remains to be established. High-resolution structures are increasingly revealing conserved binding sites for specific lipids (including cardiolipin—see Figure 15.17 below) and steroids, and these too may be important.

The oxidation of QH₂ in some prokaryotes is catalyzed by specialized quinol oxidases

For eukaryotes the most common electron donors are organic molecules; NADH is the common coenzyme and O_2 is the ultimate electron acceptor. However, some prokaryotes (known as *lithotrophs*, from the Greek for 'rock-eaters') can use inorganic molecules (H_2, CO, S, S^{2-}, and Fe^{2+} among them) as electron donors and, growing anaerobically, have specific terminal *reductases* whose synthesis is induced according to the environmental milieu and which reduce particular electron acceptors (fumarate to succinate, nitrate or nitrite to dinitrogen, and so on).

If growing anaerobically, bacteria such as *E. coli* can make use of a terminal fumarate reductase whereas, if growing aerobically, the QH_2 produced by Complex II in the ETC or by other enzymes is oxidized by quinol oxidases. *E. coli* does not have a Complex III (cytochrome bc_1); instead, the quinol oxidases (cytochrome bo_3 oxidase and cytochrome bd oxidase) catalyze the two-electron oxidation of QH_2 and four-electron reduction of O_2 to H_2O. The genes encoding cytochrome bo_3 are expressed when O_2 is plentiful, and those for cytochrome bd when O_2 is limiting. Cytochrome bo_3 is a proton pump, thus resembling CcO. However, unlike the mitochondrial CcO, the quinol oxidase contains no Cu_A site. Instead, electrons are passed from QH_2 to an enzyme-bound Q molecule and thence to the b heme. From there they pass to a binuclear site containing an o_3 heme (like heme a but with a $-CH_3$ group replacing the $-CHO$ on one of the pyrrole rings; see Figure 15.9) and a mononuclear copper (Cu_B), where O_2 is reduced to H_2O. In contrast, cytochrome bd does not pump protons, although it does contribute to the electrochemical proton gradient by depositing the protons from the oxidation of QH_2 into the periplasm and taking up the protons for the four-electron reduction of O_2 to H_2O from the cytoplasm. It shows no sequence homology with the heme–copper oxidases; it does not contain copper but has three heme groups (two b and one d), the d heme (structurally related to b heme) being the site of the reduction of O_2 to H_2O.

Similar situations are encountered in many prokaryotes. The energy coupling provided by the oxidation of QH_2 in the eukaryotic ETC is superior to that of the quinol oxidases, but there may be compensating advantages in the way in which prokaryotes can manipulate their oxidative phosphorylation in the face of changing environmental conditions, low oxygen tension, the presence of inhibitors, and so on. Nitric oxide (NO), a key player in a wide range of eukaryotic physiological processes, including host immunity to microbial infection, is a potent reversible inhibitor of CcOs. Cytochrome bd is also reversibly inhibited by NO, but the off-rate for NO is much higher. A lower sensitivity to NO inhibition is associated with virulence in pathogens, and the lack of eukaryotic homologs suggests that this enzyme is a potential antibacterial drug target. The low sensitivity of cytochrome bd to poisoning by carbon monoxide, together with its high intrinsic catalase activity offering protection against H_2O_2 as a *reactive oxygen species* (see below), may also be significant.

Respiratory chain complexes are assembled by modular pathways

It is reasonable to expect that complexes as complicated as those of the ETC will be put together along assembly pathways. In humans, more than 50 nuclear genes encode protein subunits or assembly factors required for the complexes (Table 15.1). Many mitochondria-related diseases involve malfunctions in the assembly pathways and have thrown light on their existence. Some are due to defects in the proteins responsible for assembling the Fe–S clusters (there are 12 altogether in Complexes I–III). In yeasts, proteins have been identified that are responsible for the insertion of copper into Complex IV. Additional factors are also needed for the assembly and insertion of the heme *a* of Complex IV and the heme *b* of Complex III.

The assembly of mitochondrial Complexes I, III, and IV is additionally complicated by the necessity for interplay between genes in mitochondrial DNA and others in the nuclear DNA. A common theme is modularity. The evidence suggests that building blocks that had evolved separately in microorganisms came together to generate the multi-subunit assemblies found today. In the metabolism of certain bacteria, hydrogenases are important in catalyzing the reversible reaction $H_2 \leftrightarrow 2H^+ + 2e^-$ with a number of different electron acceptors or donors. For the hydrophilic part of mitochondrial Complex I, an assembly intermediate composed of 30 kDa and 49 kDa subunits, which are homologs of hydrogenase subunits, has been identified. It is also suggested that an NADH dehydrogenase module (51 kDa and 24 kDa subunits) assembles independently and later associates with the hydrogenase module. The membrane-embedded part of Complex I appears to assemble separately, and other proteins may be necessary to join the assembled modules together. Many more factors are likely to be required and remain to be identified.

What we know of the assembly of yeast Complex III from modules in the inner mitochondrial membrane is informative (**Figure 15.15**). First, a central core, composed of the cytochrome *b* and two smaller accessory subunits (Qcr7p and Qcr8p), is formed. This joins with another module formed by association of Core Proteins 1 and 2 and the cytochrome c_1 to make a 500 kDa subcomplex, which apparently acts as a scaffold to which subunits Qcr9p,

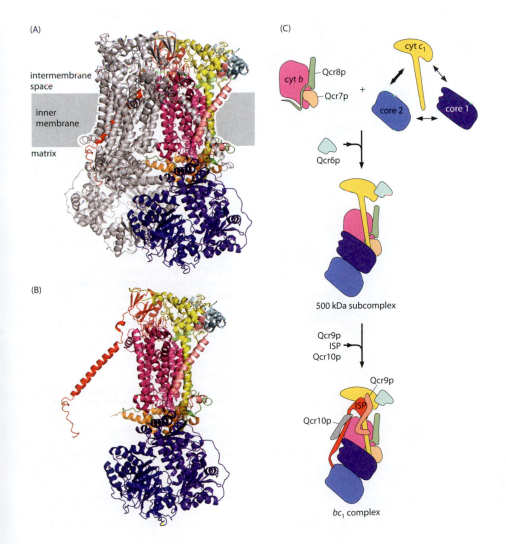

(A)

intermembrane space

inner membrane

matrix

(B)

(C)

cyt c_1

cyt *b*

Qcr8p

Qcr7p

core 2   core 1

+

Qcr6p

500 kDa subcomplex

Qcr9p
ISP
Qcr10p

Qcr9p

ISP

Qcr10p

bc_1 complex

Figure 15.15 A pathway for the modular assembly of Complex III. The *Saccharomyces cerevisiae* complex contains 10 different proteins in each protomer. Only cytochrome *b* is encoded in the mitochondrial DNA; all the other subunits are synthesized in the cytoplasm and imported into the mitochondrion. (A) Structure of the cytochrome bc_1 dimer, with all the subunits in one protomer shown in silver. The subunits in the other protomer are as follows: cytochrome *b* (385 aa, pink), cytochrome c_1 (248 aa, yellow), Rieske iron–sulfur protein (185 aa, red), core protein 1 (431 aa, light blue), core protein 2 (352 aa, dark blue), Qcr6p (122 aa, cyan), Qcr7p (126 aa, orange), Qcr8p (92 aa, green), and Qcr9p (66 aa, salmon). Subunit 10 (76 aa) is not in the crystal structure. (B) Structure of a protomer of the cytochrome bc_1 dimer, with the same color coding. (C) Schematic assembly pathway based on the analysis by polyacrylamide gel electrophoresis of complexes from detergent-solubilized mitochondrial membranes of wild-type and mutant yeast strains. (PDB 1EZV) (Adapted from V. Zara, L. Conte and B.L. Trumpower, *Biochim. Biophys. Acta* 1793:89–96, 2009. With permission from Elsevier.)

Qcr10p, and the Rieske-type ISP can then bind. It may be that the ISP and Qcr9p associate with each other before binding the 500 kDa subcomplex.

Respiratory chain complexes may come together in higher-order structures

The respiratory chain complexes have so far been described as distinct entities that are presumably free to diffuse laterally within the membrane and interact on the basis of random collision. However, delicate nondenaturing gel electrophoresis of the complexes solubilized with digitonin and isolated gently from mitochondria indicates that they adopt a kind of supramolecular organization (sometimes termed a 'respirasome'). Complexes I, III, and IV have been described as forming a variety of supercomplexes, such as $I_1III_2IV_1$, I_2III_2, III_2IV_4, $I_1III_2IV_4$, I_1III_2, depending on the source and method of preparation.

In cryoelectron microscopy, such supercomplexes can plausibly be reconciled with the known X-ray structures of their constituents (**Figure 15.16**). Respirasomes might facilitate electron transfer through the respiratory chain by a kind of solid-state transfer or by limiting the distance that cytochrome c has to diffuse between Complexes III and IV to no more than ~100 Å (only three times its own diameter). In Figure 15.16, the overall distance between the NADH-binding site in Complex I and the O_2-reduction site in Complex IV is about 40 nm. The respiratory chain may best be viewed as a collection of various functional conglomerates of catalytic complexes, and it is thought that cardiolipin, an unusual phospholipid prevalent in the mitochondrial inner membrane (about 20% of the total lipid), may

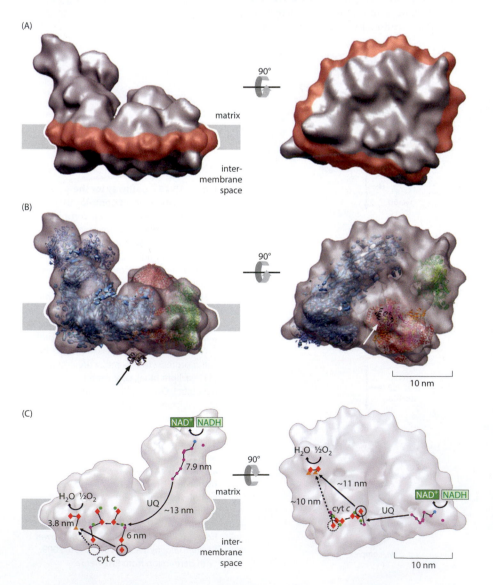

Figure 15.16 A bovine mitochondrial supercomplex. (A) Structure of a supercomplex ($I_1III_2IV_1$) from cryo-EM viewed (left) in the plane of the inner mitochondrial membrane, and (right) from the intermembrane space. Respirasomes were prepared by gentle solubilization of bovine mitochondria with digitonin, followed by sucrose density gradient centrifugation and polyacrylamide gel electrophoresis. The digitonin was replaced with *amphipol*, a polyacrylate-based carbohydrate polymer with positively and negatively charged side chains that interacts with the hydrophobic surface areas of membrane proteins and stabilizes them in soluble form. The amphipol layer is depicted in brown. (B) Docking of X-ray structures of Complex I (blue), Complex III dimer (red), and Complex IV (green) into the structures shown in (A). The small black structure (arrowed) is a cytochrome c molecule bound to the cytochrome c_1 subunit in one of the monomers in the dimer of Complex III, facing the intermembrane space. (C) Proposed electron transfer pathway in the supercomplex. The supercomplex is viewed (left) from the side, rotated by 180° compared with the images in (A) and (B), and (right) from the matrix. Cofactors are indicated by colored symbols: FMN in Complex I (blue), Fe–S clusters (purple), quinols (green), hemes (red), and Cu atoms (orange). UQ, ubiquinol; cyt c, cytochrome c. The straight solid and dotted arrows in (C) represent the shortest distances between the two cytochrome c-binding sites on the dimer of Complex III and the site of O_2 reduction in Complex IV. The dotted symbol and arrow refer to the cytochrome c-binding site unfilled in the Complex III dimer in the supercomplex structure. (Adapted from T. Althoff et al., *EMBO J.* 30:4652–4664, 2011. With permission from John Wiley & Sons.)

Figure 15.17 The structure of cardiolipin. Cardiolipin is an unusual phospholipid in that it has three glycerol moieties, two of which with pendant fatty acyl groups (a total of four) are in phosphodiester linkage with the third.

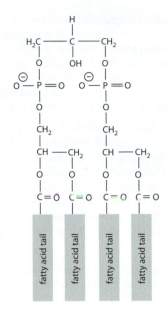

act as a 'glue' to help hold supercomplexes together (**Figure 15.17**). Together, and importantly in conjunction with the F_1F_o-ATP synthase (see below), they also help shape the organization and efficacy of the inner mitochondrial membrane (Section 15.4).

Respiration can produce dangerous side-products, which are eliminated by protective enzymes

Oxygen is essential to aerobic life, but as an element avid for electrons it is also dangerous. If it picks up an electron it can form the **superoxide** anion ($O_2 + e^- \rightarrow O_2^-\cdot$) and thence peroxide ($O_2^-\cdot + e^- \rightarrow O_2^{2-}$). As a free radical, superoxide itself is dangerous, but it can also lead to the generation of other even more **reactive oxygen species**; for example, if protonated it becomes $HO_2^-\cdot$, and if it reacts with hydrogen peroxide it can form the very damaging hydroxyl radical:

$$H_2O_2 + O_2^-\cdot \rightarrow H_2O + O_2 + OH\cdot$$

The respiratory chain complexes very rarely leak these potentially dangerous intermediates during catalysis, but small amounts are inevitably formed. The danger is diminished by an enzyme superoxide dismutase, capable of operating at close to a diffusion-controlled rate (Box 1.4), which eliminates superoxide radicals by the dismutation reaction

$$2O_2^-\cdot + 2H+ \rightarrow H_2O_2 + O_2$$

Hydrogen peroxide is degraded by a second enzyme, *catalase*, concentrated in peroxisomes and also operating at close to a diffusion-controlled rate, which turns it into water and the less harmful molecular oxygen ($2H_2O_2 \rightarrow 2H_2O + O_2$). It is also a substrate for *glutathione peroxidase*, which catalyzes the oxidation of the thiol group in the tripeptide glutathione (GSH, γ-Glu-Cys-Gly) to form water plus oxidized glutathione (GSSG):

$$H_2O_2 + 2GSH \rightarrow 2H_2O + GSSG$$

The importance of superoxide dismutase and glutathione peroxidase is underlined by their ubiquity in mitochondria. It has also been argued that some at least of the additional (supernumerary) subunits in the electron transfer complexes of eukaryotic mitochondria compared with their bacterial counterparts may be an evolutionary response to limiting the leak of reactive oxygen species. The leak has been associated with damage to mitochondrial DNA, with aging and with diseases such as cardiovascular and neurodegenerative disorders and tumor formation.

The respiratory chain is very efficient in the capture of energy as a proton-motive force

For each pair of electrons from NADH that is passed through the respiratory chain and used to reduce molecular oxygen, 10 protons are pumped out of the matrix of the mitochondrion (4 by Complex I, 4 by Complex III, and 2 by Complex IV). The protons cannot diffuse back through the membrane, and the resulting proton gradient stores energy (**electrochemical potential** or **proton-motive force**) in two ways: as chemical potential energy because protons are unequally distributed between the two sides of the membrane (ΔpH), and as electrical potential energy because protons have positive charge and, in being pumped across the membrane without a counter-ion, leave an electrical imbalance, measured as a membrane potential ($\Delta\psi$), negative on the inside and positive on the outside.

The change in free energy created by an ion pump is given by

$$\Delta G = RT \ln (C_2/C_1) + ZF \Delta\psi$$

where C_2 and C_1 represent the concentrations of ions on opposite sides of the membrane, Z is the electrical charge on the ion (1 for H^+), and F is the Faraday constant (the charge per mole of electrons = 96,485 coulombs/mol). In this instance,

$$\ln(C_2/C_1) = 2.303 \log_{10} ([H^+_{out}]/\log_{10} [H^+_{in}]) = 2.303(pH_{in} - pH_{out}) = 2.303\Delta pH$$

The pH in the intermembrane space is difficult to measure and may vary under different cell growth conditions. However, it is thought generally to be ~0.75 pH units below that of the mitochondrial matrix, and $\Delta\psi$ is ~170 mV. Substituting these values into the above equation and assuming that $T = 298K$, when 1 mol of H^+ is pumped across the inner membrane, $\Delta G = 4300 + 14{,}500 = 18{,}800$ joules, or 18.8 kJ/mol. Because there are 10 protons pumped across the membrane per NADH oxidized, 188 kJ/mol is stored, a very high return on the $\Delta G^{\circ\prime}$ of 220 kJ/mol released in the oxidation.

These calculations should be viewed with a certain amount of caution. The value of ΔpH can vary, and the standard free energies ($\Delta G^{\circ\prime}$) assigned to the various reactions described above are based on assumptions such as equal concentrations (1 M) of reactants, whereas we know that, in actively respiring mitochondria, $[NADH]/[NAD^+]$ is well above 1.0 and therefore that the true value of ΔG would be substantially more (that is, more negative) than $\Delta G^{\circ\prime} = -220$ kJ/mol. Nonetheless, it is clear that the respiratory chain is very efficient in capturing the free energy released in biological oxidation and storing it as an electrochemical potential. How the proton-motive force is used to drive the synthesis of ATP is the topic of Section 15.4.

In the mitochondrial inner membrane, notably in the mitochondria of brown adipose tissue, there are also specific *uncoupling proteins* related to the anionic ADP/ATP transporters (see Figure 15.2). Protons can leak back into the matrix through these proteins without passing through the F_1F_0-ATP synthase. The energy thus dissipated appears as heat, and uncoupling is particularly important for thermogenesis in newborn infants and for hibernating animals living off fat stores. Poisons such as dinitrophenol are also uncoupling agents, once thought of as means of promoting weight loss but now discredited. Uncoupling proteins are activated by free fatty acids and, among other things, by reactive oxygen species. It is thought that 'mild uncoupling' of the latter kind might help diminish the damage caused by oxygen stress.

Damaging mitochondrial disorders are found in more than 1 in 5000 persons, notably in the nervous system and skeletal muscle, degenerative diseases, aging, and cancer. The disorders can be traced in some instances to problems with the respiratory chain, overproduction of reactive oxygen species and oxidative stress, lowered ATP production, and compromised Ca^{2+} homeostasis. In some instances, mutations have been identified in the mitochondrial DNA, in which 11 genes encode protein subunits in Complexes I, III, and IV and two encode subunits in the ATP synthase (see below). Interestingly, more than half the deleterious mutations occur in the 22 tRNA genes in the mitochondrial DNA. In other instances, mutations in nuclear genes are responsible. Thus, in humans, mutations in the nuclear-only genes responsible for Complex II have been found to be associated with neurodegeneration and early death (mainly alterations in SdhA) and with tumorigenesis (mainly alterations in other subunits and assembly factors). The reasons for the different manifestations of disease are not yet fully apparent. This is developing into a fertile area of biomedical research.

15.3 PHOTOSYNTHETIC REACTION CENTERS AND LIGHT-HARVESTING COMPLEXES

Photosynthetic organisms derive their energy from light rather than from the oxidation of nutrients. A series of reactions based on two-electron and one-electron steps is responsible, beginning with a light-induced one-electron charge separation that is the defining feature of photosynthesis. As in the ETC (Section 15.2), the enzymes that make this possible are giant membrane-embedded protein complexes. They are termed *reaction centers* and *light-harvesting antenna complexes* and require light-absorbing pigments as noncovalently bound cofactors in the reactions. The electron transfers are accompanied by the translocation of protons across the membrane in which the photosynthetic enzyme complexes reside, and the energy stored in the proton gradient is available to drive ATP synthesis (Section 15.4).

In plants the photosynthetic machinery is located in organelles called chloroplasts

In plants the photosynthetic reactions take place in the **chloroplast** (**Figure 15.18**), an organelle that resembles the mitochondrion in many ways (see Section 1.9) but is significantly larger (~5 μm). It too has a circular DNA, sometimes found as multiple copies joined

(A)

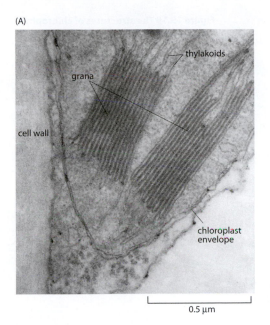

(B)

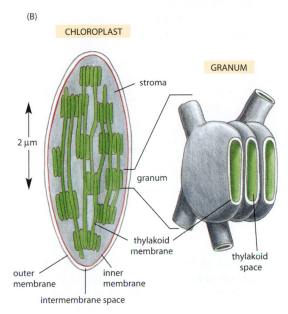

Figure 15.18 The chloroplast. (A) Electron micrograph of a thin section of part of a plant chloroplast, showing the outer and inner membrane, the stroma contained by the inner membrane, and, within the stroma, two grana. The grana are generated from the thylakoids, stacked flattened membranous structures that interconnect and generate the thylakoid space (or lumen). (B) A schematic drawing of the internal structure of a chloroplast. (Adapted from B. Alberts et al., Molecular Biology of the Cell, 6th ed. New York: Garland Science, 2015.)

end to end, but one that is generally larger than mitochondrial DNA. For example, that of *Arabidopsis thaliana* (154,478 bp) has about 120 genes (encoding 4 ribosomal RNAs, 37 tRNAs, and potentially 87 proteins, among them an RNA polymerase and some involved in photosynthesis). Many of the chloroplast protein complexes have subunits encoded in both chloroplast and nuclear DNA. Chloroplast ribosomes closely resemble those of bacteria, and it is likely that chloroplasts arose from engulfment of an O_2-producing photosynthetic bacterium resembling a latter-day cyanobacterium. Transport proteins in the chloroplast inner membrane control entry into the interior space, called the **stroma**. However, the inner membrane is not folded into cristae; instead, the photosynthetic apparatus and the ATP synthase are located in the **thylakoids**, a distinct membranous system within the stroma. Thylakoids are essentially flattened disk-like sacs, which form local stacks called **grana**; the interior spaces (lumens) of the thylakoids interconnect, thereby creating a third compartment called the thylakoid space.

In mitochondria the electron transport chains in the inner membrane pump protons across the membrane from the mitochondrial matrix into the space between the inner and outer membranes. They flow back into the matrix through the ATP synthase, whose F_1 component is in the matrix (Section 15.4). In the thylakoid system, protons are pumped out of the stroma into the thylakoid space and then flow back through an ATP synthase, whose F_1 component projects into the stroma.

Photosynthesis depends on the photochemical capabilities of light-absorbing pigments

Chlorophylls, carotenoids, and phycobilins are the main light-absorbing pigments in photosynthesis. Chlorophyll (and bacteriochlorophyll) are made up of two parts: the chlorin rings and the phytol tails (**Figure 15.19**). The chlorins have a Mg^{2+} ion bound at the center of four linked pyrrole rings, like the Fe^{2+} ion at the center of the heme in hemoglobin or the Fe^{2+}/Fe^{3+} ion in the cytochromes (see Figure 15.9). However, the Mg^{2+} ion is incapable of changing its valence state and does not participate directly in any electron transfer reactions. The hydrophobic phytol tails keep the chlorophylls attached to the membranes and membrane–protein complexes, but take no part in the photochemistry. The photochemical reactions of the chlorophylls and their ability to absorb light derive from the extended system of

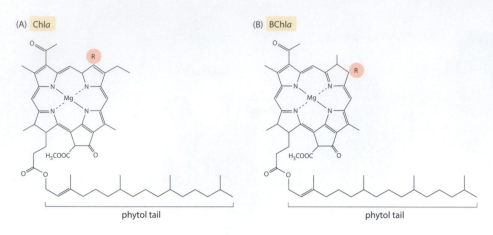

(A) Chla

(B) BChla

Figure 15.19 The structures of chlorophyll and bacteriochlorophyll. (A) In chlorophyll *a*, R is –CH₃; in chlorophyll *b*, R is –CHO. (B) In bacteriochlorophyll *a*, R is –CH₂CH₃; in bacteriochlorophyll *b*, R is =CHCH₃.

conjugated carbon–carbon double bonds in the chlorin rings. The absorption spectra of chlorophyll *a* (Chl*a*) and Chl*b* are shown in **Figure 15.20A**. Chl*a* and Chl*b* absorb strongly in the blue and red regions of the spectrum, and in solution in an organic solvent they appear green (bacteriochlorophyll *a* absorbs strongly in the near ultraviolet and in the near infrared, and a solution of it therefore appears blue). The dependence of photosynthetic activity on the absorption of light by the chlorophylls is clear from the action spectrum depicted in Figure 15.20B.

When a chlorophyll molecule absorbs a photon it becomes energetically excited and an electron is promoted from its ground state. The molecule very rapidly finds itself in the first excited **singlet state** (**Box 15.4**). If a more highly excited singlet state is initially formed, for example by the absorption of a blue (higher-frequency) photon, it quickly converts internally to the first excited singlet state of lowest energy, and the excess energy is lost as heat. All the subsequent photochemistry takes place from this first excited singlet state. The electron can fall back to the ground state, with the excess energy dissipated either as heat (*radiationless decay*) or by the emission of a photon of lower energy and thus longer wavelength (*fluorescence*). The excited singlet state can also internally convert to form a **triplet state**, the spin of the excited electron 'flipping' in a process called *intersystem crossing*.

Two additional fates are possible, depending on the type of protein complex in which the chlorophyll is located. In a *reaction center*, an electron can be transferred from the first excited singlet state to a suitable acceptor, initiating a chain of further electron transfers and a *charge separation*. In contrast, in a light-harvesting complex, molecules of chlorophyll sufficiently close together can transfer energy from a donor to an acceptor molecule—**exciton transfer** (**resonance energy transfer**)—but without any transfer of an electron.

The first excited singlet state of Chl*a* persists for only about 1 ns (10^{-9} s), and productive electron transfer and energy transfer in photosynthesis must therefore take place in much less than 1 ns. The triplet states of chlorophyll, however, last much longer, typically for microseconds or milliseconds. This is long enough for them to react with molecular oxygen and produce singlet oxygen (see Box 15.4). Singlet oxygen is a very powerful and destructive oxidizing agent, a reactive oxygen species (Section 15.2) that rapidly kills cells. However, as described below, carotenoids can act to quench singlet oxygen and avert the danger.

(A)

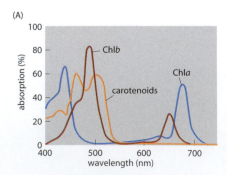

(B)

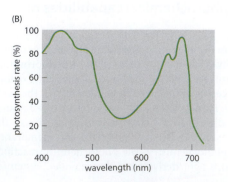

Figure 15.20 The absorption spectra of chlorophylls and carotenoids. (A) Spectra over the range 400–700 nm, the biologically important part of the solar spectrum. (B) An action spectrum of a chloroplast, showing how the rate of photosynthesis varies with the wavelength of the incident light. (Adapted from J. Whitmarsh and Govindjee, Encyclopedia of Applied Physics, vol. 13, pp. 513–532. Stuttgart: VCH Publishers Inc., 1995.)

Box 15.4 Singlet and triplet states and intersystem crossing

The electron has an intrinsic magnetic moment, most simply accounted for by ascribing to it a property of *spin*. In the presence of a magnetic field, the spin axis can take up only two orientations. For two electrons, each with a spin quantum number of ½ (+½ or −½), each particle can be *spin up* or *spin down*, which means that there are four states in all for such an electron pair:

$$\uparrow\uparrow, \uparrow\downarrow, \downarrow\uparrow, \downarrow\downarrow$$

The Pauli Exclusion Principle requires that if the spatial quantum numbers are the same, the spin quantum numbers must be different (in other words, if the energy level is the same, the spins must be opposite).

There is one state with a total angular momentum of 0: $(\uparrow\downarrow - \downarrow\uparrow)/\sqrt{2}$. This is a *singlet*, in which all electron spins are paired. There are three states with total angular momentum of 1: $\uparrow\uparrow$, $(\uparrow\downarrow + \downarrow\uparrow)/\sqrt{2}$, and $\downarrow\downarrow$. The three energy levels may not be identical (although the differences may be small), and the group forms a *triplet*. According to Hund's Rule, if two electrons occupy different orbitals, the triplet state will be at a lower energy level than the singlet.

If a molecule with a singlet ground state is excited by, say, absorption of a photon, an electron is promoted to a higher energy level, and this generates either an excited singlet state or an excited triplet state. In the triplet state, the excited electron is no longer paired with the ground state electron and has the same spin. Given that this requires a 'forbidden' spin transition for the promoted electron, it is more likely that the excited state will be singlet rather than triplet.

If the energy levels in the excited singlet and triplet states are very close, and those of the triplet state lie just below the excited singlet state, a process of *intersystem crossing* can take place (**Figure 15.4.1**). In this process the singlet state decays without radiation to a triplet state, with the spin of one of the electrons flipping to generate an unpaired electron.

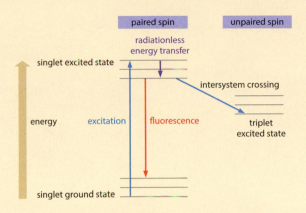

Figure 15.4.1 The process of intersystem crossing. A singlet ground state is raised to a singlet excited state and can then be converted to a triplet excited state of similar or slightly lower energy.

Molecular oxygen is important in both respiration and oxygenic photosynthesis. The ground state of O_2 is a triplet state (**Figure 15.4.2**). The triplet state has two unpaired electrons (shown in red), which occupy two degenerate molecular orbitals. The spin angular momenta of the two electrons in the $\pi_g 2p$ orbital add together, thereby explaining the paramagnetism and magnetic moment of O_2. The higher-energy singlet state of O_2 has the two electron spins opposed, as depicted in the two electronic configurations predicted by molecular orbital theory. The $O_2(b^1\Sigma_g^+)$ state is very short lived and relaxes quickly to the lower lying excited state, $O_2(a^1\Delta_g)$, which is therefore commonly referred to as singlet oxygen.

The triplet state of O_2 is stable and relatively unreactive; in contrast, the singlet state is highly reactive and very dangerous to biological matter. The way in which this is overcome in biological photosynthesis is discussed in the text.

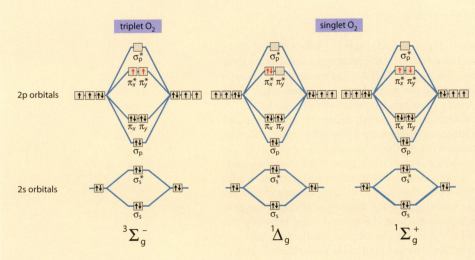

Figure 15.4.2 The electronic configurations of the triplet and singlet states of molecular oxygen.

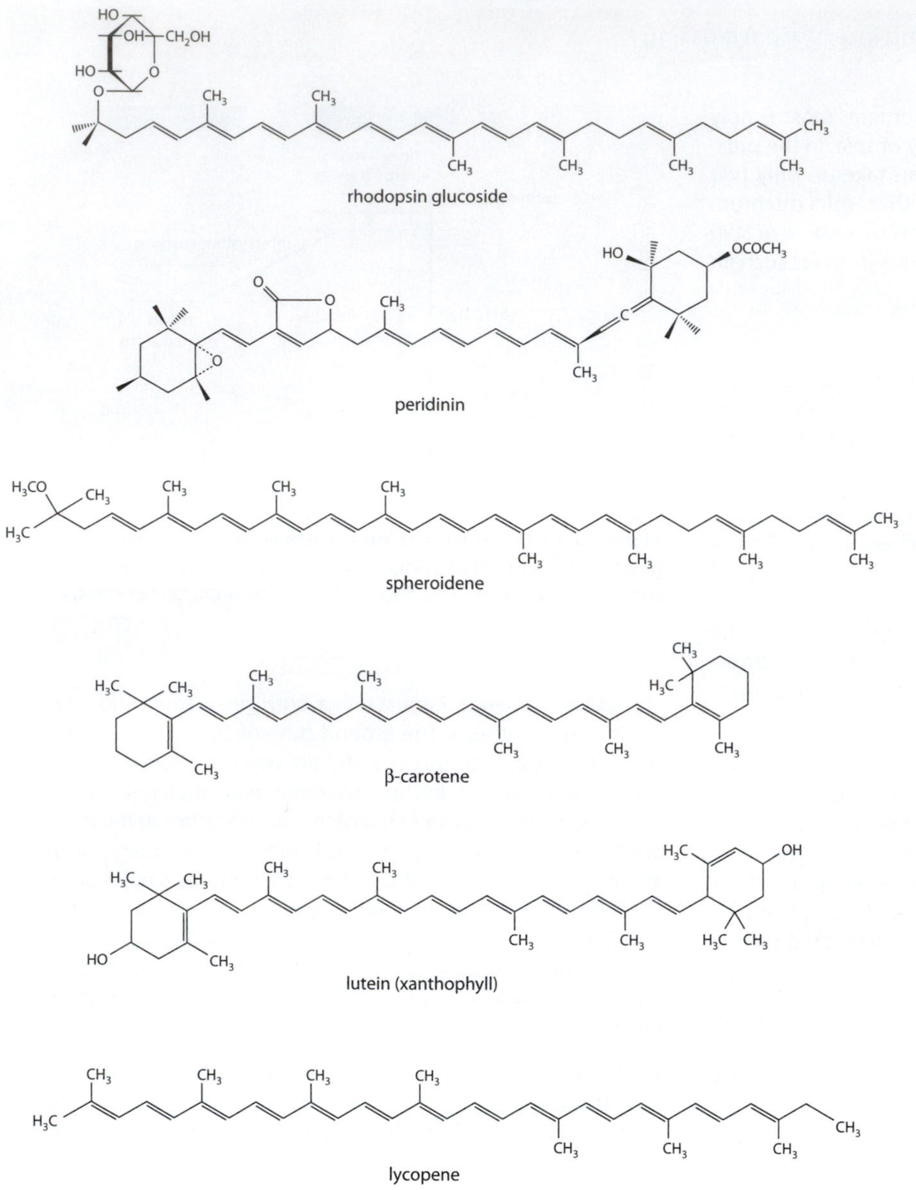

Figure 15.21 The structures of carotenoids. The name carotene is derived from carrot, the orange-colored vegetable, and carotenes are in part responsible for the reds and yellows of the leaves of deciduous trees in the fall.

rhodopsin glucoside

peridinin

spheroidene

β-carotene

lutein (xanthophyll)

lycopene

The **carotenoids** are typically C_{40} polyenes (**Figure 15.21**), substantially different from chlorophylls. The extended system of carbon–carbon double bonds again leads to a strong absorption of radiation in the visible region (see Figure 15.20A). The main, strong absorption band represents the electronic transition from the ground state to the very short-lived second excited singlet state, which lasts for only a few hundred femtoseconds (fs; 10^{-15} s). It rapidly converts internally to the first excited singlet state, which decays back to the ground state in a few picoseconds (ps; 10^{-12} s). Carotenoids also have a low-lying triplet state but this is not usually formed by direct excitation because the donor-excited singlet states are too short-lived. However, if a carotenoid molecule is sufficiently close to a chlorophyll molecule, a newly formed triplet chlorophyll can generate a carotenoid triplet by triplet–triplet exchange. This is a very important reaction because the energy of the triplet state of a carotenoid is below that of singlet oxygen and cannot induce the formation of singlet oxygen. Instead, the carotenoid triplet decays harmlessly back to the ground state, releasing the excess energy as heat. Thus carotenoids, which are found in all photosynthetic organisms, are able to quench chlorophyll triplets before the latter can lead to the formation of dangerous singlet oxygen.

The **phycobilins** are tetrapyrroles, but are open chains not closed rings. They are related to the bile pigment, bilirubin, and can appear red, orange, and blue; uniquely among photosynthetic pigments they are bonded to water-soluble proteins (*phycobiliproteins*). Like carotenoids they can capture light energy (especially red, yellow, and green wavelengths) and pass it on to chlorophylls; they are found in cyanobacteria but not in algae or higher plants.

Chlorophylls and carotenoids are crucial components of reaction centers and light-harvesting (antenna) complexes

In **photosynthetic reaction centers** (**RCs**), a small number of noncovalently bound chlorophyll molecules, together with carotenoids and other compounds such as (bacterio)pheophytin [(bacterio)Chl$a$ with the central Mg^{2+} ion replaced by two H^+ ions] and quinones (see Figure 15.4), make use of light energy to drive a series of reactions that transfer electrons across membranes and create the electrochemical potential for the synthesis of ATP and NADPH.

However, most of the chlorophylls and carotenoids are noncovalently bound in **light-harvesting (antenna) complexes** (**LHCs**), which capture solar energy and rapidly transfer it (*exciton transfer*) to the RCs. Exciton transfer in an antenna system is *singlet–singlet* energy transfer; the electron in the first excited singlet state in a donor chlorophyll falls back to the ground state and an electron that was in the ground state in an acceptor chlorophyll is raised to its first excited singlet state. The shorter the lifetime of the donor's excited singlet state, the less time there is for energy transfer to occur before it decays. Other factors are also important. Energy transfer falls off rapidly with distance ($1/r^6$) and depends on the relative angles between the transition dipole moments of the donor and acceptor molecules (parallel favorable, orthogonal unfavorable). Likewise, it increases with the spectral overlap between the emission band of the donor and the absorption band of the acceptor (reflecting the energy gap between the excited states involved in the energy transfer reaction).

LHCs surround a RC in a notional **photosynthetic unit** (**PSU**). In a typical plant, there are about 200–250 chlorophyll molecules in a PSU. The clustering of LHCs around RCs in the photosynthetic membrane of purple bacteria is illustrated in **Figure 15.22**. The number of chlorophyll molecules in a single RC protein complex is relatively small, and the probability of a chlorophyll receiving a productive direct 'hit' from a photon is low. It has been estimated that, even on a sunny day, it would be only ~1 photon per second. However, the numerous LHCs surrounding a RC increase the cross-sectional area for productive photon capture by at least two orders of magnitude.

Carotenoids absorb light in different regions of the spectrum from chlorophylls and extend the range of wavelengths of light available to power photosynthesis. Higher plants use Chl$a$ and Chl$b$ as their major light-absorbing pigments; these absorb the blue light and the red light up to about 750 nm. Purple photosynthetic bacteria, which are anaerobic and typically live near the bottom of ponds and streams, survive on either green light or red light of wavelengths beyond 750 nm. They use carotenoids to harvest the green light, and bacteriochlorophyll a (BChl$a$) to absorb near-infrared light with wavelengths between 800 and 900 nm. Dinoflagellates, a simple eukaryote living in the sea at depths where most of the available light is in the blue-green region of the spectrum, have evolved a special LHC in which the major pigments are carotenoids whose absorption bands are in this spectral range.

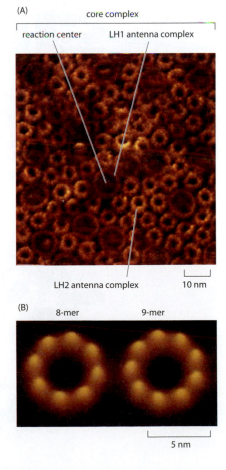

Figure 15.22 Assemblies of reaction centers and light-harvesting complexes. Images of a high-light-adapted membrane of the bacterium *Rhodospirillum photometricum* obtained by atomic force microscopy (AFM). (A) In this AFM topogram, a reaction center (RC) is surrounded by the light-harvesting (antenna) complex LH1 to form a core complex, with large numbers of LH2 antenna complexes clustered round it. The RC/LH1 core complex is slightly elliptical. The LH2 complexes have nine α/β heterodimeric proteins in the ring. In this membrane patch the overall ratio of LH2 to RC/LH1 is about 3.5 : 1 and the complexes are clustered together in what appear to be lipid-depleted regions of the membrane but not in any regular array. (B) The LH2 complexes of *R. photometricum* can vary in size. Image analysis reveals polymorphism; averaged images for the 8-mer and 9-mer are shown at higher magnification. (Adapted from S. Scheuring and J.N. Sturgis, *Photosynthesis Res.* 102:197–211, 2009. With permission from Springer Science and Business Media.)

Two types of LHC serve different purposes in purple bacteria

Two different types of LHC, namely LH1 and LH2, are found in purple bacteria: the LH1 complex surrounds a RC and forms with it an LH1/RC core complex, whereas multiple copies of the LH2 complex cluster round each LH1/RC core complex in the photosynthetic cell membrane. Both types of complex are circular or elliptical structures (see Figure 15.22), constructed from heterodimers (α/β) of small proteins, typically only 50–60 aa, and numerous photosynthetic pigments.

The crystal structure of the LH2 complex from the purple bacterium *Rhodopseudomonas acidophila* is shown in **Figure 15.23**. The inner wall of the complex is made up of the α helices of nine α-apoproteins, and the outer wall consists of the α helices of nine β-apoproteins. The polypeptides fold over at the top and bottom of the structure, effectively enclosing the pigments between the transmembrane helices. There are 27 BChl*a* molecules in an LH2 complex. One group of nine BChl*a* molecules, one in each α/β apoprotein pair, have their long-wavelength absorption band at 800 nm (*B800 BChla*). Their (bacterio)chlorin rings lie parallel to the plane of the membrane, at right-angles to the transmembrane helices. The central Mg^{2+} ion in each chlorin ring is coordinated to a carbonyl group from an unusual source, a carboxylated α-amino group belonging to the N-terminal Met residue (^-OCONH-Met-) of an α subunit. The other 18 BChl*a* molecules (*B850 BChla*) comprise the second group, two associated with each α/β apoprotein pair, with their chlorin rings parallel to the transmembrane helices. The central Mg^{2+} ion in each chlorin is coordinated to two His residues, one from an α subunit and one from a β subunit. The B850 BChl*a* molecules form a tightly coupled ring and behave essentially as a single unit in passing on energy by exciton transfer.

There are also nine carotenoid molecules in each LH2 antenna complex, one associated with each α/β subunit pair. In *Rps. acidophila* the carotenoid is rhodopsin-glucoside (see Figure 15.21) in the all-*trans* configuration. It starts on one side of the complex in one protein pair,

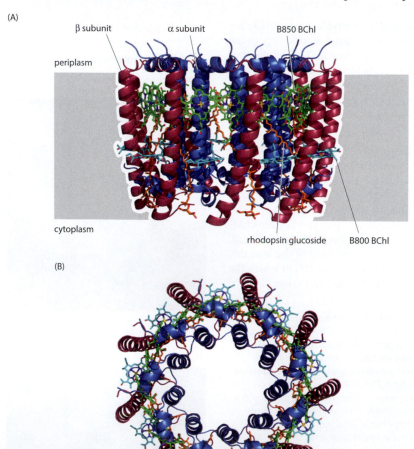

(A)

β subunit   α subunit   B850 BChl

periplasm

cytoplasm

rhodopsin glucoside   B800 BChl

(B)

Figure 15.23 The structure of the LH2 antenna complex from *Rps. acidophila*. (A) The monomeric complex in the plane of the membrane. Color code: green, B850 BChl*a* molecules; cyan, B800 BChl*a* molecules; orange, rhodopsin glucoside. The phytol tails of the BChl*a* molecules have been omitted for clarity. (B) The LH2 complex viewed from above. (PDB 1KZU)

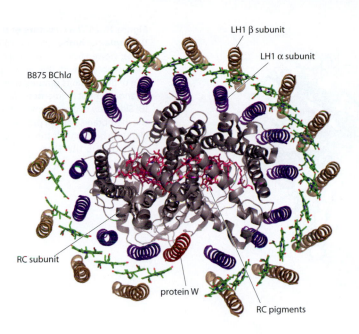

LH1 β subunit

LH1 α subunit

B875 BChl*a*

RC subunit

protein W

RC pigments

Figure 15.24 The structure of the LH1/RC complex from *Rps. palustris*. View of the LH1/RC core complex as seen from the periplasmic side of the membrane. The polypeptides of the RC are depicted in gray, and the pigments in the RC responsible for charge separation are shown in magenta. The B875 BChl*a* molecules have their long-wavelength absorption band at 875 nm. The membrane is omitted for clarity. (PDB 1PYH)

then crosses over into the next pair and finishes on the other side of the complex (see Figure 15.23). In this way it 'glues' neighboring protein pairs together; in its absence, LH2 is not a stable assembly. The carotenoid makes van der Waals contact with both groups of BChl*a* molecules and, as described below, this is essential for proper function. LH2 complexes can vary somewhat in size; for example the LH2 complexes in membranes of *Rhodospirillum photometricum* mostly have nine α/β protein pairs in a ring, but eightfold and tenfold rings have also been reported (see Figure 15.22). The structure of an LH2 complex from *Rhodospirillum molischianum* with eight rather than nine α/β protein pairs shows that it is almost identically constructed. It is not known what governs ring size and how it might affect function.

The LH1/RC core complexes come either as monomers, as in *Rhodopseudomonas palustris*, or as dimers, as in *Rhodobacter sphaeroides*. In a crystal structure of the monomeric LH1/RC complex from *Rps. palustris*, despite relatively low resolution, important structural features can be identified (**Figure 15.24**). The LH1 complex is elliptical and the RC lies centrally within it, as in the atomic force microscopy images of intact photosynthetic membranes (see Figure 15.22). In the *Rps. palustris* LH1, there are 15 α/β apoprotein pairs, leaving a gap in the ring in which a protein called W is found. As with the LH2 complex, some structural variability is observed: in some *Rhodobacter* species an LH1 assembly takes the shape of an S rather than a closed ring, and a RC is associated with each loop of the S. It is unclear how these different structures come to be assembled and whether there is any biological significance.

The LHCs of plants and cyanobacteria differ from those of purple bacteria

The major LHC of plants, LHC-II, is embedded in the chloroplast thylakoid membrane. It differs from the purple bacterial LH2 in being a trimer of subunits with molecular masses of between 24 and 29 kDa encoded by a family of *Lhcb* genes. Despite the mixed protein compositions, the similarity in their amino acid sequences has made it possible to crystallize spinach LHC-II and determine its structure (**Figure 15.25**). Each monomer contains three membrane-spanning α helices connected by either short loops or short amphipathic helices, plus eight Chl*a* molecules, six Chl*b* molecules, and four carotenoids. The carotenoid composition is rather variable, but two central lutein molecules (see Figure 15.21) interact strongly at the center of each monomer with the two longest transmembrane α helices. The chlorophylls are arranged essentially in two layers, one (five Chl*a* molecules and three Chl*b* molecules) nearer the stromal surface, and the other (three Chl*a* molecules and three Chl*b* molecules) nearer the luminal surface; the carotenoids are all in van der Waals contact with at least one chlorophyll molecule, and the Chl*b* molecules are organized into clusters. These are important features for the functioning of LHC-II.

(A)

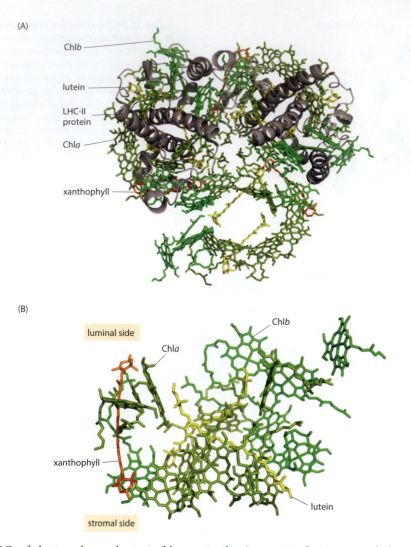

(B)

Figure 15.25 The structure of the LHC-II complex of higher plants. (A) The trimer viewed from the stromal side of the membrane, with the protein part of one of the subunits omitted to make the arrangement of pigments easier to see. (B) Side view of a monomer in the plane of the membrane, with all protein omitted for clarity. In both panels the membrane is also omitted. (PDB 1RWT)

The RCs of plants and cyanobacteria (blue-green algae) occur in what are termed photosystems I and II. In both instances they are surrounded by LHCs, which operationally resemble the LH1 surrounding the RC of purple bacteria. They are described in detail later (see Figure 15.31 and Figure 15.36 below).

Energy transfer reactions in LHCs are very fast and very efficient

Photosynthetic energy transfer has been best elucidated for the purple bacteria because their different pigment groups can be independently excited and their reactions are spectrally well resolved. This is much more difficult in plants as the spectra of the different chlorophylls overlap. As described above, the first excited singlet states of both Chl$a$ and BChl$a$ persist for only about 1 ns. However, ultra-short laser flashes can be generated and, when a BChl molecule is excited by light, the creation of the first excited singlet state results in bleaching of the long-wavelength absorption band (see Figure 15.20A). The recovery in absorbance can then be monitored to follow the process of energy transfer.

The light-harvesting BChl molecules (B800 and B850) in the peripheral LH2 complexes absorb more to the blue (800 and 850 nm; shorter wavelength, higher energy) than those in the LH1 complexes (B875, absorbing at 875 nm; longer wavelength, lower energy), which are in direct contact with the RCs where *charge separation* takes place. Energy travels down an energy gradient by exciton transfer, from the B800 group to the B850 group and then to the B875 group of chlorophylls. From there it passes finally into the RC, as illustrated schematically in **Figure 15.26**.

If the nine B800 BChl$a$ molecules in LH2 are excited with a 200 fs laser pulse, it takes only 0.9 ps for the excitation energy to be transferred to the B850 BChl$a$ molecules. The reaction time is largely independent of temperature and only slows down to about 2 ps at 4 K. This

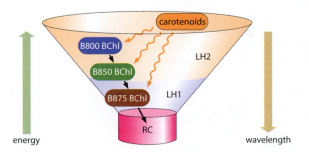

Figure 15.26 Downhill exciton energy transfer from LHCs to a RC. In purple photosynthetic bacteria, energy passes from the B800 BChl to the B850 BChl molecules in the LH2 complexes. From there it passes to the B875 BChl molecules in the LH1 antenna complex that surrounds a reaction center.

demonstrates that no molecular motion is required and that the pigments are organized very effectively for exciton transfer. Subsequent energy transfer from the LH2 complexes to the ring of B875 BChls in LH1 surrounding a RC takes place in 3–5 ps. The 18 B850 BChl molecules behave as an extended interactive assembly over which the excited state is delocalized and from which energy can be transferred out from any B850 BChl*a* molecule with equal probability. The ring of B875 BChl molecules in LH1 behaves similarly, but the final energy transfer to the RC takes 30–50 ps, the slowest step in the process because it involves the longest distance (~40 Å). These energy transfer reactions are illustrated in a structural context in **Figure 15.27**.

The distance between the BChl molecules in LH1 and those in the RC of an LH1/RC core complex is much more than 14 Å, the maximum distance for efficient electron transfer (see Box 15.3). Thus only energy transfer can be expected to occur. This is vital because, if a single BChl molecule of in the B875 ring of LH1 were to lose an electron, it would act as a 'quencher' and reduce the lifetime of the B875 excited singlet state so severely that it could no longer function in light harvesting. Overall, the transit time from the periphery of the PSU into the RC is just a few tens of picoseconds, crucially much less than the nanosecond lifetime of the chlorophyll excited singlet state, and the overall efficiency of energy transfer is close to 100%.

The structural blueprint of the RC of purple photosynthetic bacteria is conserved throughout photosynthetic organisms

The structure of the RC from *Rb. sphaeroides* is depicted in **Figure 15.28**. The RC is embedded in the plasma membrane surrounded by the elliptical LH1 complex (see Figure 15.24) amid multiple LH2 antenna complexes (see Figure 15.22). It consists of three subunits. The L (31 kDa) and M (34 kDa) subunits each contain five membrane-spanning α helices. Their overall folds are very similar, suggesting a common evolutionary origin. The H subunit (29 kDa) is largely located at the cytoplasmic surface of the RC and has only a single transmembrane helix. The RC contains four molecules of BChl*a*, two molecules of bacteriopheophytin (Bphe), two molecules of ubiquinone (Q), one carotenoid (spheroidene), and one Fe^{2+} (non-heme) ion, all of which are noncovalently bound to subunits L and M.

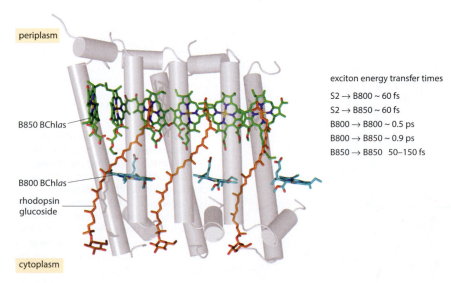

exciton energy transfer times

S2 → B800 ~ 60 fs
S2 → B850 ~ 60 fs
B800 → B800 ~ 0.5 ps
B800 → B850 ~ 0.9 ps
B850 → B850 50–150 fs

Figure 15.27 Time course for exciton energy transfer between pigments in the LH2 antenna complex. A side view of one-third (three α/β pairs) of the nonameric ring of the LH2 antenna complex of *Rhodopseudomonas acidophila* in the same orientation as that in Figure 15.23. The transfer times for exciton energy between B800 and B850 BChl*a* molecules and the carotenoid rhodopsin glucoside are indicated. S2 refers to the singlet state of the carotenoid from which energy transfer occurs. The membrane is omitted for clarity. (PDB 1KZU)

The pigments are arranged pseudo-symmetrically in two branches, A (L subunit) and B (M subunit). On the periplasmic side of the RC, two BChl*a* molecules are found close together (Mg–Mg distance ~7Å) as the so-called 'special pair,' designated P870. Each branch then proceeds via an accessory BChl, a monomeric Bphe and finally a ubiquinone molecule. The ferrous ion (Fe^{2+}) is located between the two ubiquinone molecules (Q_A and Q_B at the end of the A and B branches, respectively) and lies on the two-fold axis of pseudo-symmetry that separates the two branches. The carotenoid, unusually in a 15–15' *cis* configuration, is located near the accessory BChl molecule on the B branch. When the RC is excited, usually by exciton energy transfer from an excited LH1 complex, the 'special pair' P870 of BChl*a* molecules adopts the excited singlet state P870*. Electron transfer is initiated by P870* losing an electron to the accessory BChl on the A branch, and in doing so becoming P870⁺. This transfer, a process of **charge separation** in the RC, takes about 2 ps. The electron then moves onto the Bphe on the A branch in about 1 ps and finally to the ubiquinone Q_A also on the A branch in about 200 ps. Spectroscopic studies with polarized light on crystals of RC show that only the A branch is active in electron transport. The result is the half-reduction of Q_A (Q_A to Q_A^{-}·). It is not clear why the B branch is inactive in electron transport.

(A)

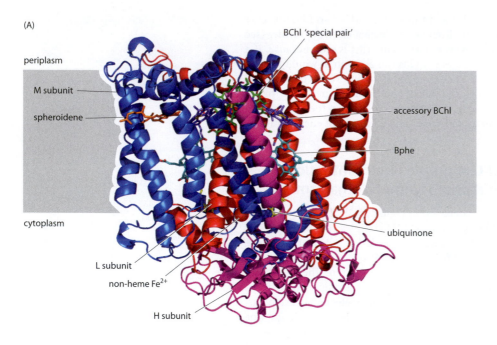

(B)

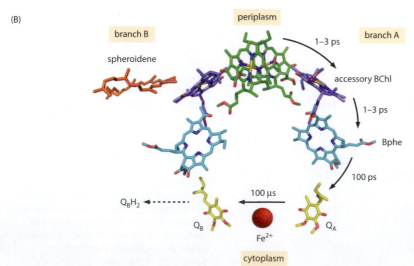

Figure 15.28 The structure of the RC of *Rb. sphaeroides*. (A) The protein complex viewed in the plane of the membrane. The phytol tails of the chlorophylls have been omitted for clarity. (PDB 2RCR) (B) Organization of the pigments and times for electron transfer in the complex. Only branch A is active in electron transfer. The Q_BH_2 generated dissociates from the RC and diffuses out laterally in the membrane. The color code is as in (A).

The semiquinone $Q_A^-\cdot$ is oxidized back to Q_A by transferring an electron to the more solvent-exposed ubiquinone Q_B (generating $Q_B^-\cdot$), and $P870^+$ is converted back to P870 by taking up an electron from a reduced cytochrome c_2, a soluble c-type cytochrome that is free to move in the periplasmic space. Another photon capture, another $P870^* \rightarrow P870^+$, and a second turnover of the RC, generates $Q_A^-\cdot$ again. This donates an electron to $Q_B^-\cdot$, thereby producing Q_B^{2-} and re-forming Q_A. The non-heme Fe does not appear to participate in these electron transfers from Q_A to Q_B. Q_B^{2-} takes up two protons from the cytoplasm to become the fully reduced Q_BH_2. The Q_BH_2 dissociates from the RC and is then free to diffuse out through the elliptical LH1 in the gap where protein W is located (see Figure 15.24) into the plane of the plasma membrane. There it is able to deliver its reducing equivalents ($2H^+$ and $2e^-$) to a membrane cytochrome bc_1 complex, which can reduce (transfer electrons back to) cytochrome c_2 in the periplasmic space. This ensures that a reduced cytochrome c_2 is always available to donate an electron to the $P870^+$ form of the BChl 'special pair' for another turnover of the light-driven reaction. The oxidized quinone (Q_B) joins the Q_B pool in the membrane and a fresh Q_B diffuses into the binding site in a RC, readying it for another round of reaction.

The sequence of electron carriers in the RC, each with a progressively more positive redox potential, is important for the efficiency of the process. As the electron travels down the A branch, each of the forward reactions is fast and, with each downward hop, the subsequent back-reaction is slower by three orders of magnitude. Thus, by the time the RC is in the state $P870^+/Q_A^-\cdot$, the back-reaction has slowed to a timescale of microseconds. This is long enough for the reaction to be made irreversible by transfer of an electron to $P870^+$ from the reduced cytochrome c_2, which takes place on a timescale of microseconds.

Unlike LHCs, different RCs turn out to have very similar structures. This is probably because the efficiency of electron transfer falls off sharply with distance. Energy transfer is more tolerant so long as the donor and acceptor pigments are reasonably well orientated relative to each other, and LHCs can differ much more.

In purple bacteria, the cytochrome bc_1 complex generates a proton gradient that drives ATP synthesis

The cytochrome bc_1 complex in the photosynthetic membranes of purple bacteria resembles the proton-translocating Complex III of the mitochondrial respiratory chain (Section 15.2). It, too, utilizes a two-stage Q cycle (see Figure 15.11) in transferring reducing equivalents from Q_BH_2 to a soluble periplasmic cytochrome c_2, which means that four protons are transferred from cytoplasm to periplasm for every two electrons transferred from Q_BH_2 to cytochrome c_2.

As described above, there is no net oxidation or reduction in the photosynthetic reactions of the purple bacterial RC. No external electron donor or acceptor is required. The electron donated by $P870^*$ to Q_A, and thence to Q_B, passes through cytochrome bc_1 and ultimately is taken up by periplasmic cytochrome c_2, which then returns it to $P870^+$ to regenerate P870. In this way the LHCs and RCs combine with cytochrome bc_1 to create a continuous **cyclic electron transport** pathway (**Figure 15.29**). The action of the cytochrome bc_1 complex sets up a proton gradient across the bacterial cell membrane; the energy stored in the proton gradient, originating from the absorbed photons, can then be used to drive the synthesis of ATP (Section 15.4).

In cyanobacteria and plant chloroplasts, two RCs work in series with water as electron donor

The reactions of photosynthesis in plants and oxygenic cyanobacteria differ significantly from those in purple bacteria. They require an extrinsic electron donor; water acts as that donor and, in a remarkable series of reactions, it is split, molecular oxygen is evolved, and $NADP^+$ is reduced to NADPH:

$$2H_2O \rightarrow 4H^+ + 4e^- + O_2$$

$$NADP^+ + H^+ + 2e^- \rightarrow NADPH$$

Water does not readily let go of electrons—the redox potential for the reaction $H_2O \rightarrow 2H^+ + 2e^- + \frac{1}{2}O_2$ is +820 mV. The redox potential for the reaction

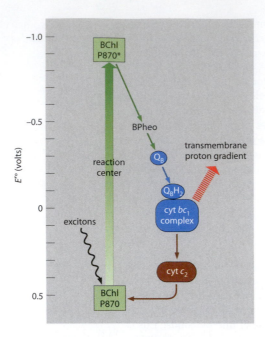

Figure 15.29 Cyclic electron transport in purple sulfur bacteria generates a transmembrane proton gradient. Exciton energy transfer from LH1 drives the BChl 'special pair' P870 in the RC to the excited singlet state P870*. Charge separation and electron transfer in the RC reduce ubiquinone, Q_B, to Q_B^{2-}. Q_B^{2-} picks up $2H^+$ from the cytoplasm and the Q_BH_2 diffuses in the plane of the membrane to the cytochrome bc_1 (cyt bc_1) complex, which reoxidizes Q_BH_2 to Q_B and concomitantly reduces soluble cytochrome c_2 in the periplasmic space. In so doing the cytochrome bc_1 complex pumps $4H^+$ across the membrane into the periplasmic space for each Q_BH_2 oxidized (see Figure 15.11). The reduced cytochrome c_2 donates its electron to reduce the BChl P870$^+$ back to BChl P870, ready for a further photon-driven excitation to BChl P870*. Reactions depicted in green take place in the RC, reactions in blue take place in the cell membrane, and reactions in brown take place in the periplasmic space.

$NADP^+ + H^+ + 2e^- \rightarrow NADPH$ is +320 mV. Thus the net reaction has a standard redox potential of +1140 mV. To bring about such a thermodynamically uphill reaction, the thylakoid membranes of cyanobacteria (foldings of the cell membranes) and of plants (inside the chloroplasts; see Figure 15.18) employ two types of RC, termed **photosystem I** and **photosystem II**, working in series in a so-called Z-scheme, named for its shape (**Figure 15.30**). In the first, which for historical reasons is called photosystem II, the redox potential at which an electron (coming from water) is provided to the 'special pair' of chlorophylls (P680) is much more positive than that at which it is provided by cytochrome c_2 to the 'special pair' (P870) in the RC of purple bacteria (see Figure 15.28). The second light-driven step (photosystem I) is then able to start at a higher energy level than the first. A transmembrane proton gradient is generated, and this is used to drive the synthesis of ATP.

Photosystem II oxidizes water and reduces a quinone

The structure of photosystem II from the cyanobacterium *Thermosynechococcus elongatus* is shown in **Figure 15.31**. In structure and mechanism, its RC closely resembles that of purple bacteria (see Figure 15.29) except that photosystem II is dimeric, with a two-fold axis of symmetry. Each protomer contains 19 subunits, 14 of which are embedded in the membrane. Five of the latter are of major mechanistic importance. Polypeptides CP43 and CP47 constitute a (core antenna) LHC. They have similar protein folds, each with six membrane-spanning α helices and arranged on opposite sides of the polypeptides D1 and D2, which constitute the RC. CP47 has 16 bound chlorophyll molecules and CP43 has 13; most of these chlorophylls are arranged in two layers, displaced toward different sides of the membrane, and thus matching the organization of chlorophyll molecules in the LHC-II complexes described above (see Figure 15.25). The chlorophyll molecules in the antenna complex (CP43 and CP47) are well separated from those in the RC (D1 and D2) and, as in the LH1/RC complex of purple bacteria (see Figure 15.24), this prevents the antenna chlorophylls from entering into destructive electron transfer reactions. In addition, CP43 and CP47 contain three and five carotenoids, respectively. Their compositions can vary but they are thought normally to be all-*trans* β-carotene, and all are in van der Waals contact with at least one chlorophyll molecule.

The structures of D1 and D2 closely resemble those of the L and M subunits of the purple bacterial RC; each contains five membrane-spanning helices, and the electron transport components are arranged in two branches. The fifth major player is cytochrome b_{559}. On the luminal side of the membrane, photosystem II has three extrinsic proteins: PbsO (also called the 33 kDa protein), PbsU (12 kDa), and PbsV (cytochrome c_{550}). These subunits together cover the region of the complex where a specialized Mn cluster (see below) is bound; their removal destabilizes the Mn cluster.

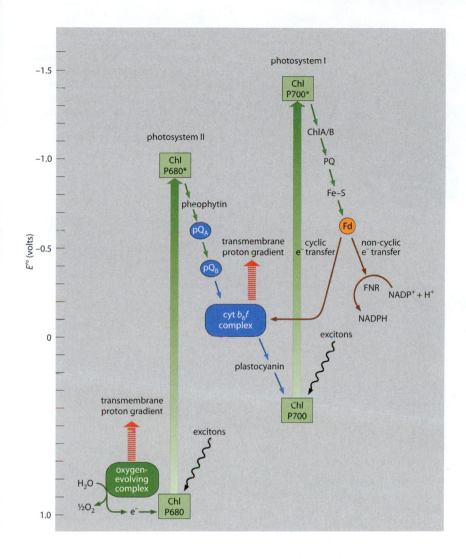

Figure 15.30 The sequential operation of photosystems I and II. Electrons abstracted from water by the oxygen-evolving complex (OEC) in photosystem II replace electrons ejected by photon excitation from the Chl special pair (P680/680*) and which pass through a series of carriers including pheophytin to reduce plastoquinone (pQ). The quinol, pQ$_B$H$_2$, diffuses laterally in the membrane to cytochrome b_6f, where it is reoxidized and a soluble lumen protein, plastocyanin, is reduced. Reduced plastocyanin feeds electrons to the Chl$a$ special pair (P700/P700*) of photosystem I, to replace electrons ejected by photon excitation into a series of carriers including electron-transfer Chl$a$ molecules (ChlA/B), phylloquinone (PQ), and various Fe–S centers in photosystem I, culminating in the reduction of a soluble protein, ferredoxin (Fd), in the stroma. Both the OEC and the cytochrome b_6f complex pump protons into the thylakoid space as part of their activity. The reduced ferredoxin has two possible fates: in the first, non-cyclic electron transfer, it is reoxidized by a stromal enzyme, ferredoxin–NADP$^+$ reductase (FNR), with NADPH as the ultimate product; in the second (a cyclic process of electron transfer), electrons flow back into the cytochrome b_6f complex, through which they can re-enter photosystem I with further protons being pumped into the thylakoid space. The cyclic process is very similar to that of purple sulfur bacteria (Figure 15.29). Reactions depicted in green take place in photosystems I and II, those in blue are due to the cytochrome b_6f complex and the stromal protein plastocyanin, and those in brown are the stromal reactions involving Fd and FNR.

Upon excitation with light, the special pair of Chl$a$ molecules (P680, P$_{D1}$P$_{D2}$) in the RC of photosystem II donates an electron to the chlorophyll on the D1 side (**Figure 15.32**). From there the electron is transferred via the pheophytin on the D1 side to a **plastoquinone** (pQ$_A$), a quinone specific to plants (see Figure 15.4). The semiquinone, pQ$_A^-\cdot$, then transfers the electron to pQ$_B$. These electron transfers resemble those in the purple bacterial RC; they take place very rapidly and again proceed only on one side, in this case the D1 side. After two rounds of photosynthetic electron transfer, pQ$_B$ is fully reduced (pQ$_B^{2-}$) and, on taking up two protons from the stroma, it dissociates from the RC as pQ$_B$H$_2$. The overall reaction catalyzed by photosystem II can thus be summarized as

$$2pQ_B + 2H_2O \longrightarrow 2pQ_BH_2 + O_2$$

Having donated an electron to the photosynthetic reactions, P680 is a very powerful oxidizing species. It has a redox potential of more than 1 V, sufficient to oxidize water (in other words, to remove four H atoms from 2H$_2$O), but it is also dangerous. It is able to oxidize amino acids and to generate destructive reactive oxygen species. Indeed, during normal functioning *in vivo*, D1 is continually being damaged and is typically replaced every 30–60 min.

The oxidation of water is a four-electron process catalyzed by a specialized Mn cluster in photosystem II

A molecule of oxygen (O$_2$) can be released only when four protons and four electrons have been removed from two molecules of H$_2$O. Four successive one-electron removals by the RC must be coupled in some way, and correspondingly four positive charges must be stored. In photosystem II this is carried out by the **oxygen-evolving complex** (**OEC**) located on the

(A)

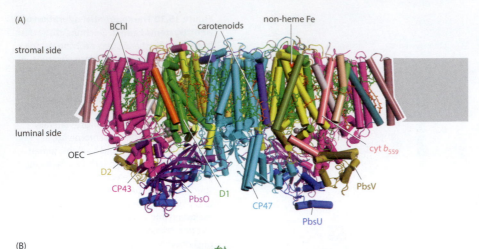

Figure 15.31 The structure of photosystem II from the cyanobacterium *T. elongatus*. (A) A view in the plane of the bacterial cell (thylakoid) membrane. The antenna complex in each protomer of the symmetric dimer is provided by subunits CP43 and CP47. D1 and D2 are part of the RC. (B) A view from the stromal side of the membrane. The membrane is omitted for clarity. (PDB 3BZ1 and 3BZ2)

(B)

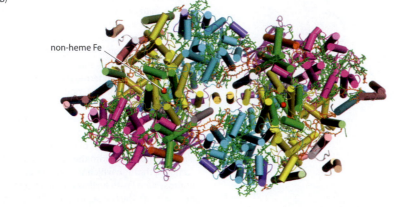

luminal surface as part of the D1 subunit. With each turnover of photosystem II, the 'hole' on P680 (P680$^+$) left by the electron donated to Chl$_{D1}$ is filled by transfer of an electron from a tyrosine residue, Tyr161 on D1, called Tyr$_Z$ (see Figure 15.32). This leaves Tyr$_Z$ as a radical (TyrO·) that can be detected by electron spin resonance; each time, Tyr$_Z$ is regenerated by transfer of an electron to TyrO· from the so-called **S-state system**. The S-state system comprises a cluster of four Mn ions (the redox state is still uncertain) and a Ca^{2+} ion. Each one-electron turnover in the RC causes the S-state cluster to accumulate one more

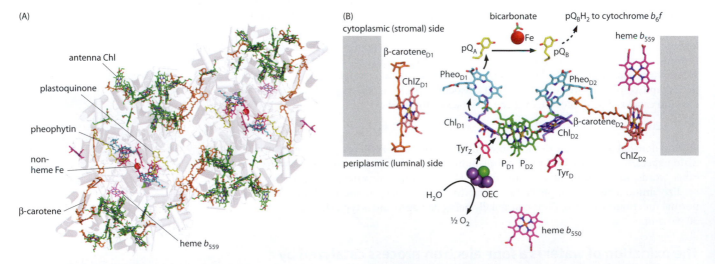

Figure 15.32 The organization of pigments in photosystem II.
(A) The dimeric *T. elongatus* complex is viewed from the periplasmic side of the bacterial cell (thylakoid) membrane (thylakoid space in plants). The pigments are shown with the protein scaffold of the dimer in faint gray and the phytol tails of the chlorophylls removed for clarity. (B) The electron transfer pathway through one protomer of the RC, in which subscripts D1 and D2 refer to subunits D1 and D2. Arrows show the direction of electron transfer from the OEC to the special pair (P$_{D1}$, P$_{D2}$) and thence through additional electron carriers to pQ$_B$. The fully reduced pQ$_B$H$_2$ diffuses laterally through the membrane to the cytochrome b_6f complex, where it becomes reoxidized and plastocyanin is reduced. Pheo, pheophytin. (PDB 3BZ1 and 3BZ2)

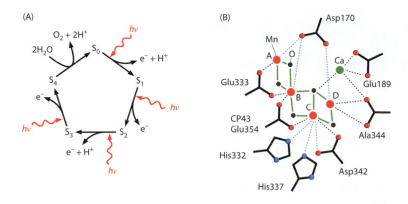

positive charge (**Figure 15.33A**). Eventually, four electrons have been transferred and a total of four positive charges have been accumulated. Four electrons can then be removed from two water molecules in a concerted reaction, leaving four protons and a molecule of oxygen to be released into the thylakoid space.

There are now several high-resolution structures of photosystem II, but the detailed mechanism of oxygen evolution remains unclear. X-rays at the intensity required for X-ray crystallographic analysis interact with the Mn cluster of the S-system, induce redox changes in it, and cause damage. As a result, the calculated X-ray structure is still not well resolved in the region of the Mn cluster. However, a reasonable hypothesis based on extended X-ray absorption fine structure (EXAFS) analysis is shown in Figure 15.33B.

Figure 15.33 A schematic mechanism for the OEC of photosystem II. (A) Each of four successive light-driven reactions in the RC of photosystem II is supplied by an electron donated by a cluster of four Mn^{2+} ions and a Ca^{2+} ion (possibly a Mn_4CaO_5 complex) in the S-state system (S_0 to S_4). When four positive charges have been accumulated, four electrons can be removed from two water molecules in a concerted reaction to generate one molecule of oxygen ($S_4 \rightarrow S_0$). Overall, $4H^+$ are pumped into the thylakoid space. (B) A possible structure of the Mn/Ca cluster, derived from EXAFS and X-ray crystallography. The four Mn ions are depicted as filled red circles and labeled A–D, bridging O atoms are depicted as filled black circles, the Ca ion is shown as a filled green circle, and bonds in the S cluster are shown as green lines. The local ligands are all amino acid side chains (N atoms in blue, O atoms in red) from subunit D1, with the exception of Glu354 of subunit CP43. (B, adapted from J. Yano et al., *Science* 314:821–825, 2006. With permission from AAAS.)

A cytochrome b_6f complex links photosystem II to photosystem I and generates a proton gradient

The quinol pQ_BH_2 generated by photosystem II dissociates from the RC and migrates in the plane of the thylakoid membrane where it can pass two electrons, one at a time, to a **cytochrome b_6f complex** (see Figure 15.30). The quinone pQ joins the pool of pQ in the membrane and a fresh pQ molecule enters the RC. The structure of the cytochrome b_6f complex of the cyanobacterium *Mastigocladus laminosus*, a dimer, is shown in **Figure 15.34**. The resemblance to the proton-translocating cytochrome bc_1 complexes of mitochondria (see Figure 15.10) and purple bacteria is clear, but there are differences. Each protomer (~109 kDa) contains eight subunits (**Table 15.2**).

There are also some cofactors not represented in mitochondrial Complex III, among them a molecule of Chl*a*, a β-carotene, and a novel heme designated heme *x* (sometimes called heme c_n) covalently linked by a single thioether bond to Cys35 of cytochrome b_6 and located on the stromal side of the photosynthetic membrane near heme b_L.

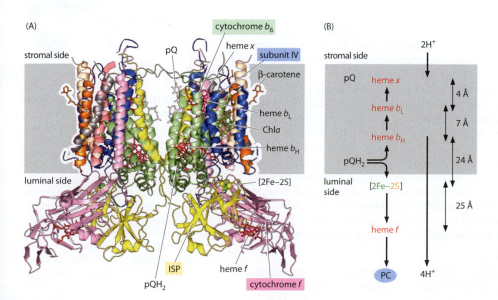

Figure 15.34 The structure of the cytochrome b_6f complex from *M. laminosus*. (A) A view of the dimer in the plane of the membrane. Each protomer (109 kDa) contains eight subunits, with a total of 13 transmembrane helices. The four major subunits are cytochrome b_6, subunit IV, cytochrome *f*, and an iron–sulfur protein (ISP). The four small (3–4 kDa) subunits Pet G, L, M, and N are colored salmon, wheat, orange, and gray, respectively. (PDB 1VF5) (B) A schematic view of the electron transfer pathways through the cytochrome b_6f complex in the oxidation of pQH_2 and the reduction of the soluble protein plastocyanin (PC) in the periplasm. The double-headed arrows indicate the edge-to-edge distances between the various heme groups/[2Fe–2S] centers through which the electrons pass. For each pQH_2 oxidized, $4H^+$ are pumped across the membrane by a Q cycle mechanism similar to that for the cytochrome bc_1 complex of mitochondria.

Table 15.2 Subunit composition of the cytochrome b_6f complex from *M. laminosus* and the cytochrome bc_1 Complex III of bovine mitochondria.

Cytochrome b_6f complex	Features	Counterpart in cytochrome bc_1 complex
Cytochrome b_6	24.7 kDa heme b_H heme b_L heme x	N-terminal half of cytochrome b (no heme x)
Subunit IV	17.5 kDa	C-terminal half of cytochrome b
Cytochrome f	32.3 kDa heme f	cytochrome c_1, but which is structurally unrelated
Rieske iron–sulfur protein	19.3 kDa 2Fe–2S cluster	Rieske ISP
Pet G	4.1 kDa	none
Pet L	3.5 kDa	none
Pet M	3.8 kDa	none
Pet N	3.3 kDa	none

The mechanism appears to be similar to that of cytochrome bc_1, with electrons reaching the ISP being used to reduce the heme f instead of c_1, from which they pass not to a soluble cytochrome c but to **plastocyanin** (**PC**), a small (11 kDa) and soluble copper-containing protein found in the lumen, with an intense blue color in its oxidized form (**Figure 15.35**). The net reaction is thus

$$pQ_BH_2 + 2PC\text{-}Cu^{2+} \rightarrow pQ_B + 2PC\text{-}Cu^+ + 2H^+$$

The reduced plastocyanin is free to diffuse and convey reducing equivalents to photosystem I. The cytochrome b_6f complex functions by means of a Q cycle like that of the proton-translocating cytochrome bc_1 complex of mitochondria and purple bacteria (see Figure 15.11). This causes four protons to be pumped across the membrane, from stroma into the thylakoid space (lumen) for chloroplasts, for each plastoquinol oxidized (that is, for every two PC molecules reduced).

Photosystem I generates a reductant powerful enough to reduce CO_2 to carbohydrate

The structure of photosystem I, a trimer, from the cyanobacterium *T. elongatus* is shown in **Figure 15.36**. Each protomer (356 kDa) is made up of 12 protein subunits, 9 transmembrane and 3 located on the stromal side of the membrane, together with >100 pigments and cofactors. The main integral membrane subunits are the homologous proteins PsaA (83 kDa) and PsaB (81 kDa). They can be regarded as equivalent to a combination of either D1 and CP43 or D2 and CP47 of photosystem II, with core antenna domains and RC domains linked together in single polypeptide chains. The core antenna domains contain 90 Chl$a$ molecules and 22 carotenoids. The RC domains are similar to both the D1 and D2 subunits of photosystem II and the L and M subunits of the purple bacterial RC. On the stromal surface of the complex, three extrinsic proteins—PsaC, PsaD, and PsaE—are bound. The PsaC protein contains two 4Fe–4S clusters, F_A and F_B. A third 4Fe–4S cluster, F_X, is coordinated by four Cys residues, two from each of PsaA and PsaB, and lies on the pseudo-two-fold axis that relates PsaA and PsaB. PsaC is the site of interaction with **ferredoxin** (**Fd**), in cyanobacteria a small (6 kDa) soluble protein with two 4Fe–4S clusters, which becomes reduced by receiving an electron as the terminal acceptor of photosystem I. The structure of subunit PsaC resembles that of bacterial ferredoxin.

The organization of the light-harvesting chlorophylls and carotenoids in the core antenna domains is similar to that seen in CP43 and CP47 of photosystem II, with most of the

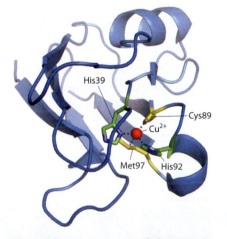

Figure 15.35 The structure of plastocyanin. The 106-residue protein from the cyanobacterium *Anabaena variabilis* has a single Cu^{2+} ion liganded by His39, Cys89, His92, and Met97. The liganding is a distorted tetrahedral geometry, and the Cu^{2+} ion can alternate between Cu^{2+} (oxidized) and Cu^+ (reduced). Four ligands to Cu^{2+} ions normally adopt a square planar geometry, and the observed distortion toward the tetrahedral geometry favored by Cu^+ ions may account for the unusually high redox potential (~0.37 V) of PC compared with that (0.16 V) for the normal Cu^{2+}/Cu^+ half-reaction. (PDB 2GIM)

(A)

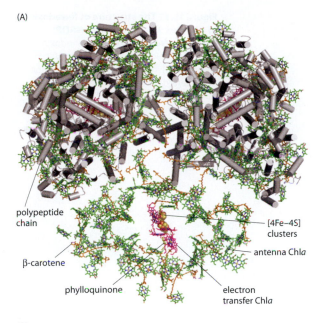

polypeptide
chain

β-carotene

phylloquinone

[4Fe–4S]
clusters

antenna Chl*a*

electron
transfer Chl*a*

(B)

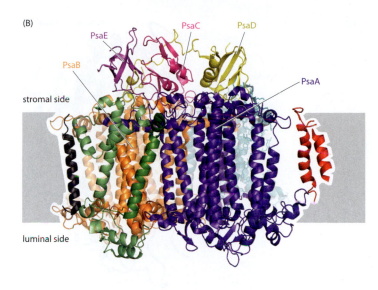

PsaE

PsaC

PsaD

PsaB

stromal side

PsaA

luminal side

(C)

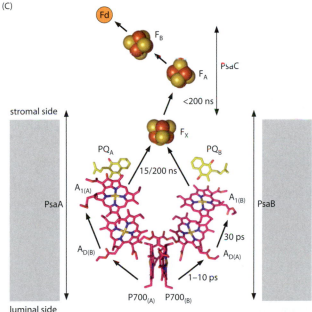

Figure 15.36 The structure of photosystem I from the cyanobacterium
T. elongatus. (A) A view of the trimer is shown from the luminal side of the
membrane, with the subunits of one of the protomers omitted for clarity.
Color code: green, chlorophyll *a*; orange, β-carotene; magenta, electron
transfer chlorophylls; yellow, phylloquinone; yellow/orange spheres, Fe–S
centers. Proteins PsaA and PsaB are related by a two-fold axis of pseudo-
symmetry. (PDB 1 JB0) (B) A protomer viewed in the plane of the membrane,
to illustrate the positions of the transmembrane proteins PsaA and PsaB,
and the extrinsic proteins PsaC, PsaD, and PsdE. (C) The organization of
the pigments and times taken for electron transfer in the photosystem I
complex. $A_{D(A)}$ and $A_{1(A)}$ are electron-transfer chlorophylls liganded to
PsaA, whereas $A_{D(B)}$ and $A_{1(B)}$ are electron-transfer chlorophylls liganded to
PsaB (designated ChlA/B in Figure 15.30). PQ_A and PQ_B are phylloquinone
molecules bound to PsaA and PsaB, respectively. F_X, F_A, and F_B are [4Fe–4S]
clusters bound in transmembrane subunits PsaA/B and extrinsic subunit
PsaC. Ferredoxin (Fd) is the ultimate electron acceptor of photosystem I. The
phytol tails of the chlorophylls are omitted for clarity.

chlorophylls arranged in two main groups, one toward each side of the membrane. The pigments involved in the electron transfer reactions (none more than 14 Å apart) are arranged
in two branches (Figure 15.36C). Cofactors with subscript A are bound by PsaA, and those
with subscript B are bound by PsaB. The arrangement resembles the reaction centers of
purple bacteria and photosystem II, but there are some interesting differences. First, the
two chlorophylls in the P700 'special pair,' which act in the same way as P870 in the purple
bacterial reaction center, are nonidentical: $P700_{(A)}$ is a stereoisomer, the C13′ epimer, of
$P700_{(B)}$. Second, spectroscopy indicates that electron transfer occurs down both branches,
albeit perhaps at different rates. From the special pair, both branches pass electrons through
two intermediate chlorophylls, $A_{D(B)}$ and $A_{1(A)}$ on one side and $A_{D(A)}$ and $A_{1(B)}$ on the
other, and thence to a molecule of **phylloquinone** (see Figure 15.4) but then join up to
transfer their electrons to the iron–sulfur complex, F_X. F_X in turn reduces F_A/F_B, which then
reduces ferredoxin. The ferredoxin (11 kDa) of plant chloroplasts (**Figure 15.37**) contains a
single 2Fe–2S cluster coordinated by four Cys residues.

Photosystem I of higher plants is monomeric, but otherwise appears to be very similar,
albeit with some additional antenna proteins. The donor side of photosystem I is generally

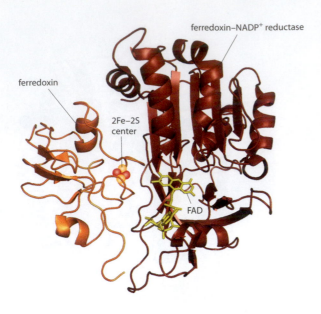

ferredoxin–NADP⁺ reductase

ferredoxin

2Fe–2S
center

FAD

Figure 15.37 The structure of ferredoxin in complex with ferredoxin–NADP⁺ reductase from maize leaf. Ferredoxin (98 aa) is depicted in orange, the monomeric FNR (314 aa) in brown. The 2Fe–2S cluster of the ferredoxin is positioned close to the FAD (yellow) of FNR. NADP⁺, not seen bound in this complex, is the electron acceptor required to complete the reaction. (PDB 1GAQ)

similar to that of the purple bacterial RC, except that P700⁺ is replenished by electrons supplied by the reduced plastocyanin from photosystem II/cytochrome b_6f rather than by the cyclic transfer illustrated for P870⁺ in Figure 15.29. P870 and P700 have very similar standard redox potentials (~+500 mV). However, the acceptor side of P700 is very different (see Figure 15.36). It operates at very low redox potentials and uses Fe–S centers rather than just quinones in electron transfers. The redox potential of the Fe–S center F_X is about –700 mV, whereas the redox potential of Q_A in photosystem II is only about –50 mV. This means that the reduced ferredoxin is able to bring about the reduction of NADP⁺ to NADPH, which is sufficiently powerful as a reducing agent to be able to reduce CO_2 to carbohydrate. This is an outcome of profound biological importance.

The reduction of NADP⁺ by reduced ferredoxin is catalyzed by a stromal enzyme, ferredoxin–NADP⁺ reductase, a monomeric 35 kDa FAD-containing protein (see Figure 15.37). The bound FAD can accept two electrons, one at a time from reduced ferredoxin, passing through the semiquinone form, before NADP⁺ is reduced:

$$2Fd(e^-) + NADP^+ + H^+ \rightarrow 2Fd + NADPH$$

In green sulfur bacteria, which are obligate anaerobes, the electrons required for the photosynthetic reduction of NADP⁺ do not come from H_2O but from H_2S, leaving elemental S rather than molecular O_2 as the by-product. Their RCs, like photosystem I, are able to generate reduced ferredoxin, which by a comparable non-cyclic pathway is used to reduce NAD⁺ to NADH for metabolic purposes. Both arms of the RC electron pathway also function in green sulfur bacteria. Overall it is remarkable how highly conserved the photosynthetic machinery is from purple sulfur and green sulfur bacteria to higher plants. As with mitochondria (Section 1.9), this supports the idea that chloroplasts are derived from the 'capture' of a bacterial partner, in this case a cyanobacterial ancestor, by a progenitor eukaryotic cell more than a billion years ago.

ATP can be generated by cyclic and non-cyclic electron transport

In anoxygenic bacteria, the LHCs and RCs turn light energy into an electrochemical proton gradient by virtue of protons pumped across the bacterial cell membrane by a cytochrome bc_1 complex in a process of cyclic electron transport (see Figure 15.29). In the oxygenic photosynthesis conducted in plants and cyanobacteria, photosystems II and I combine with an analogous cytochrome b_6f complex to create an electrochemical proton gradient, but the electron transport is non-cyclic and leads instead to the reduction of NADP⁺ to NADPH (see Figure 15.30). In both instances, the free energy stored in the electrochemical proton gradient can be used to drive ATP s means of an ATP synthase (Section 15.4). However, in oxygenic photosynthesis, as outlined above, the NADPH generated is utilized for the biosynthesis of carbohydrates from CO_2 by a set of chemical reactions, the **dark reactions**, so named because they are able to proceed in the absence of light. However,

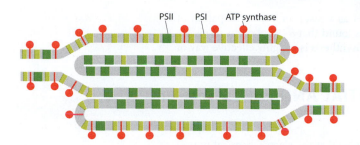

PSII   PSI   ATP synthase

Figure 15.38 Clustering of photosystems I and II in chloroplast thylakoid membranes. Photosystems I and II (PSI and PSII) exhibit lateral heterogeneity in the thylakoid membranes. PSII (dark green) is largely found in the appressed membranes of the stacked grana, whereas PSI (light green) is restricted to the membrane domains facing the stroma. The chloroplast ATP synthase (red) is more evenly distributed with the extra-membrane part (F_1) projecting into the stroma. (Adapted from J.M. Anderson, *FEBS Lett.* 124:1–10, 1981. With permission from Elsevier.)

overall these biosynthetic reactions are endergonic and require coupling to ATP hydrolysis, and circumstances can arise in which different amounts of ATP relative to NADPH are required. Any need for additional ATP is met by cyclic electron transport, in which electrons from reduced ferredoxin, the product of photosystem I, are diverted from NADP and returned to cytochrome b_6f for recycling through photosystem I (see Figure 15.30). This is comparable with the process of cyclic electron transfer in purple bacteria. Photosystem II is not involved, no O_2 is formed, and no NADPH is produced. The balance struck between cyclic and non-cyclic phosphorylation is finely controlled in response to metabolic circumstances, light intensities, and so on.

Biochemical fractionation, electron microscopy (see **Methods**), and atomic force microscopy have revealed that photosystems I and II cluster in the chloroplast thylakoid membrane—but they do so separately: photosystem II/LHC-II supercomplexes plus additional LHC-II are predominantly in cylindrical thylakoids in granal stacks, whereas photosystem I and ATP synthase are in thylakoid domains exposed to the stroma (**Figure 15.38**). Cytochrome b_6f complexes are evenly and reversibly distributed between the stacked and unstacked regions and appear to rely on the lateral diffusion of pQH_2. Their physical separation prevents exciton energy transfer from photosystem II (P680, shorter wavelength, higher energy) to photosystem I (P700, longer wavelength, lower energy). In higher plants, the ratio of granal to stromal domains and the amounts of b_6f complexes vary widely according to growth under high or low light conditions. The situation is also dynamic in the shorter term. Partitioning of LHC-II is subject to cycles of phosphorylation and dephosphorylation. When LHC-II is associated with photosystem II, it favors the production of pQH_2 and thus of reduced plastocyanin, the source of electrons for photosystem I. An excess of reduced plastocyanin triggers a protein kinase to phosphorylate specific Thr residues in LHC-II and it migrates laterally from stacked grana, diminishing their number, to associate with photosystem I and enhance its activity. When the level of oxidized plastocyanin rises, LHC-II becomes dephosphorylated and migrates back to photosystem II.

Photosynthesis is very efficient in the capture of solar energy

It has been estimated that more than 95% of photons absorbed give rise to a charge separation at a reaction center. In oxygenic organisms, two photons are required by photosystem II to reduce one molecule of Q to QH_2, and a total of four to release one molecule of O_2 from H_2O. Four protons are discharged into the thylakoid space by the OEC of photosystem II for each molecule of O_2 formed. Four protons are also pumped across the membrane for each QH_2 oxidized by the passage of electrons through the cyanobacterial cytochrome bc_1 or plant cytochrome b_6f complexes. In cyclic photophosphorylation in oxygenic organisms, this means that 4 photons are absorbed and 12 protons enter the thylakoid space for each O_2 molecule evolved. This pumping of protons causes the pH in the thylakoid space to fall some 3 pH units below that of the stroma, and the energy stored in this electrochemical proton gradient is used to drive ATP synthesis.

However, the thylakoid membrane is quite permeable to certain other ions, such as Cl^- and Mg^{2+}, the movement of which maintains electrical neutrality and dissipates most of the electrical potential ($\Delta\psi$) in the proton gradient. Thus, chloroplasts differ from mitochondria in that the electrochemical potential arises almost entirely from the ΔpH, a total of 17 kJ/mol ($\Delta G = 2.303RT \times \Delta pH$). For the movement of 12 H^+ per O_2, the free energy change is therefore ~204 kJ/mol, enough in principle to cover the synthesis of up to about six molecules of ATP under standard conditions ($\Delta G^{\circ\prime} = 31$ kJ/mol). As with the mitochondrial respiratory chain, these numbers need to be viewed with caution, and the additional energy stored in the reduction of $NADP^+$ to NADPH as a result of the participation of photosystem I has been neglected. Experimentally it is found that 8–10 photons of visible light are needed to generate one molecule of O_2. Photosynthesis is a highly effective way of capturing solar energy.

15.4   ELECTROCHEMICAL POTENTIAL AND THE BIOSYNTHESIS OF ATP

In 1961 Peter Mitchell put forward the revolutionary idea that the electrochemical potential in an ion gradient across an impermeable membrane could store energy. In particular, Mitchell proposed that it was in the form of a proton gradient across the mitochondrial inner membrane that the respiratory chain captured the energy from the oxidation of NADH and $FADH_2$, and it was this gradient that was used to drive the synthesis of ATP. This *chemiosmotic theory* met with considerable resistance. There was great reluctance to relinquish the idea of 'high-energy' chemical intermediates, but the experimental evidence piled up and, as we have seen in Sections 15.2 and 15.3, the outcome of respiration in mitochondria and of photosynthesis is indeed a proton gradient across the membrane in which the relevant molecular machines are embedded.

ATP synthase is the name given to the machine that is driven by the proton gradient and synthesizes ATP from ADP and P_i. The reaction ($ADP + P_i \rightarrow ATP + H_2O$) is more commonly thought of in reverse, namely ATP hydrolysis coupled to make otherwise endergonic reactions possible (Section 1.6). Early biochemical studies together with electron micrographs of negatively stained mitochondrial membranes demonstrated that the seat of ATP synthesis was in an assembly (about 90 Å in diameter) of multiple polypeptide chains protruding from the inner membrane into the matrix on a stalk ~30 Å long (**Figure 15.39A**). This part was termed the F_1 *component*, and the part in the membrane to which it is attached was designated the F_0 *component* (because it was where the antibiotic <u>o</u>ligomycin bound and inhibited ATP synthesis). Later structures determined by cryoelectron microscopy revealed the presence of a *peripheral stalk* in addition to the *central stalk* (Figure 15.39B). Membrane fragments (still capable of respiratory chain respiration) catalyzed ATP hydrolysis, not synthesis. The F_1 component, which could be reversibly detached from the membrane by gentle treatment with urea, also catalyzed only ATP hydrolysis. For these reasons, F_1–F_0 and the F_1 component came to be known as the **F_1F_0-ATPase** and **F_1-ATPase**, respectively. However, under physiological conditions, it normally functions as an ATP synthase, driven by the electrochemical potential of the proton gradient across the intact membrane in which it resides. Hence it has been given the physiologically relevant name **F_1F_0-ATP synthase**.

The F_1F_0-ATP synthase is constructed from two rotary motors

The F_1F_0-ATP synthase is composed of two multi-subunit rotary motors, one in F_1 and one in F_0 (**Figure 15.40**). In bacterial and mitochondrial ATP synthases, F_1 has the subunit composition $\alpha_3\beta_3\gamma_1\delta_1\epsilon_1$. The α and β subunits form an $\alpha_3\beta_3$ ring, into the central cavity of which enters the γ subunit projecting from F_0. The ϵ subunit is a regulatory subunit, at least in bacteria and chloroplasts, binding to the exposed part of the γ subunit. Together the γ and ϵ subunits form the central stalk between F_1 and F_0. The δ subunit is also a connector subunit between F_1 and F_0. [Confusingly, the bacterial δ subunit is the homolog of the OSCP (<u>o</u>ligomycin <u>s</u>ensitivity <u>c</u>onferring <u>p</u>rotein) of the mitochondrial F_1, and the bacterial ϵ subunit is the homolog of the mitochondrial subunit designated δ. The bacterial enzyme contains no counterpart of the mitochondrial subunit designated ϵ.]

In bacterial synthases, F_0 has the subunit composition $a_1b_2c_{10-15}$ and is embedded in the cell membrane. The c-subunits, which range in number from 10–15 in different organisms, form a ring structure that binds tightly to the γ and ϵ subunits and from which they protrude toward F_1. When driven by the flow of protons through a channel in F_0, the c-ring (and with it the γ and ϵ subunits) rotates. However, parts of subunit a and the b_2 dimer make up the peripheral stalk that projects from the membrane and is tightly bound to both $\alpha_3\beta_3$ and δ, thereby preventing the outer part of F_1 from rotating also. The F_0 of mitochondria is similar, but it has additional subunits whose function is still not fully understood. Thus, the flow of protons through F_0 from the 'outside' (the periplasmic space of bacteria or the intermembrane space of mitochondria and chloroplast thylakoids) to the 'inside' (bacterial cytoplasm, mitochondrial matrix, or chloroplast stroma) drives the rotation of the c-ring.

(A)

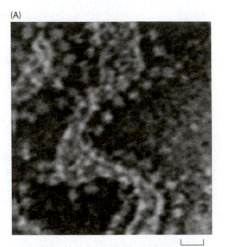

|_____| 100 Å

(B)

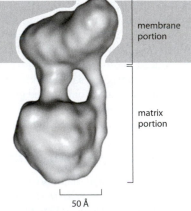

membrane portion

matrix portion

|____| 50 Å

Figure 15.39 Electron microscope images of ATP synthases in mitochondrial membranes. (A) Electron micrograph of negatively stained cristae of bovine heart mitochondria, showing the 'balls on stalks' (subsequently identified as F_1 components of ATP synthases) projecting from the inner membrane into the mitochondrial matrix. (B) Structure from cryoelectron microscopy of bovine mitochondrial F_1F_0-ATPase solubilized in a detergent micelle. The matrix portion is joined to the membrane portion by the central and peripheral stalks. (A, adapted from H. Fernandez-Moran at al., *J. Cell Biol.* 22:63–100, 1964; B, adapted from L.A. Baker et al., *Proc. Natl. Acad. Sci. USA* 109:11675–11680, 2012. With permission from National Academy of Sciences.)

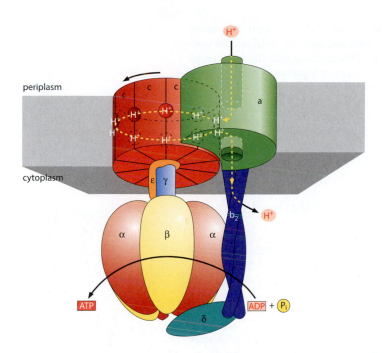

Figure 15.40 Schematic diagram of the structure of the F₁Fₒ-ATP synthase of _E. coli._ The peripheral stalk comprises proteins a and b₂ (a dimer) embedded at one end in the cell membrane; the c-ring, through which protons pass to cause it to rotate in the plane of the membrane about a vertical axis, is shown in red; and the F₁ component (α₃β₃γδε), the site of ATP synthesis, projects into the cytoplasm. The γ and ε subunits (central stalk) rotate with the c-ring, while the α₃β₃δ part is held stationary by interaction with the peripheral stalk. The inferred position of the membrane is depicted in gray.

The accompanying rotation of the γ subunit within the static $\alpha_3\beta_3$ ring is what in turn drives ATP synthesis in F_1.

A schematic comparison of the bacterial and mitochondrial F_1F_o-ATP synthases is shown in **Figure 15.41**. The bacterial F_1F_o is reversible and, if required, the rotation of the γ subunit can be reversed and the free energy of hydrolysis of ATP in F_1 used instead to pump protons through F_o. It is therefore designated an ATP synthase/ATPase. In contrast, the mitochondrial enzyme contains a small additional protein, IF1 (84 aa in the bovine ATP synthase), that forms a long α helix which dimerizes and, at pH values below about 7.0 (high [H⁺]) in the mitochondrial matrix, jams the reverse rotation of the γ subunit and blocks ATPase activity. At higher pH values (the lower [H⁺] encouraging proton flow into the mitochondrial matrix), IF1 is ejected and the dimers associate and are rendered inactive. Forward rotation of the γ subunit can then resume and ATP synthesis ensues. If and when

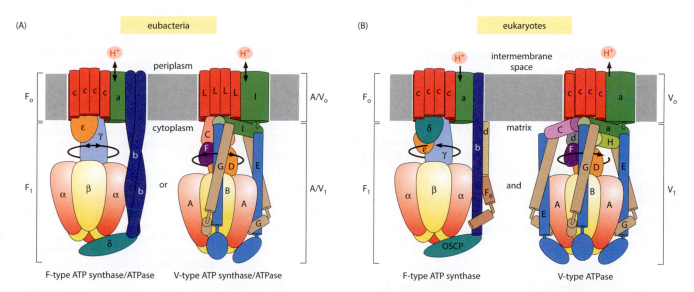

Figure 15.41 Schematic structures of F- and V-ATPases from prokaryotes and eukaryotes. (A) F-type and V-type ATPase/synthases from eubacteria. Both are reversible. If H⁺ ions are driven into the cytoplasm by virtue of the electrochemical proton gradient, the system acts as a synthase; if ATP hydrolysis is the more powerful source of free energy, the system operates in reverse and protons are expelled. (B) F-type ATP synthases and V-type ATPases from eukaryotes. Neither is reversible. The V-type ATPase may operate with other cations, for example Na⁺, in certain instances. (Adapted from A.G. Stewart et al., _BioArchitecture_ 3:2–12, 2013. With permission from Taylor & Francis.)

the internal pH falls again, IF1 dimers jam the reverse rotation of the γ subunit once more. In addition to these F-type ATP synthases, eukaryotic cells and some bacterial species contain a related ATPase, termed a vacuolar or V-ATPase. This machine uses the free energy of ATP hydrolysis to pump cations across membranes, for example pumping protons in the acidification of intracellular compartments such as lysosomes or creating electrochemical ion gradients that drive symporters or antiporters in selectively moving ions across internal or plasma membranes (Section 16.5). In eubacteria, the V-type ATPase (V_1V_o) is reversible (an ATP synthase/ATPase) like the F_1F_o, but has two peripheral stalks, whereas the V_1V_o of eukaryotes is irreversible (an ATPase only) and has three peripheral stalks (see Figure 15.41). In archaea, the enzyme complexes (A_1A_o) are reversible and have two peripheral stalks, like those of eubacteria.

The F_1 component functions by means of a chemical binding change mechanism

There are three active sites in the F_1 component, principally in the β subunits and located at the α/β interfaces. Inactivation of any one of the three inhibits most or all catalytic activity. If ADP and P_i are incubated with isolated F_1 in the presence of $H_2^{18}O$, three or four atoms of ^{18}O are rapidly incorporated into the P_i. This indicates that ATP is being reversibly synthesized and broken down ($ADP^{2-} + HPO_4^{2-} + H^+ \leftrightarrow ATP^{3-} + H_2O$) by the isolated F_1 component. Moreover, the rate constants for the process indicate that the equilibrium constant for the reaction on the surface of the enzyme is close to 1.0; that is, $\Delta G^{o'} = 0$, compared with a value of 31 kJ/mol in free solution. These and other experiments indicate that during net synthesis a tightly bound ATP cannot be released from one active site until ADP and P_i have bound at another site, and that it must be the energy stored in the proton gradient *in vivo* that causes release of the bound ATP from F_1F_o.

A chemical mechanism for F_1-catalyzed ATP hydrolysis can be formulated as one of successive binding change, as first proposed by Paul Boyer (**Figure 15.42**). Each of the three active sites in F_1 cycles through three successive states: the T state binds ATP tightly and hydrolyzes it to ADP and P_i; on conversion to the L state, the subunit is catalytically inactive and ADP and P_i are more loosely bound; and finally, in the O state (originally open), the catalytic site is still inactive and binds ligands very loosely or is empty. The cycle is then ready to repeat, with three ATP molecules having been hydrolyzed (one each per active site). The conformational changes necessary for the cycle of changes in each subunit are made possible by the free energy of hydrolysis of the ATP. However, if driven to act in reverse, three molecules of ATP will be synthesized, with the free energy required for the reversed cycle of conformational changes then coming from the electrochemical proton gradient, as we shall see.

The structure of the F_1 component reveals the rotary mechanism of ATP synthase

This chemistry fits with the crystal structure of the ox heart mitochondrial F_1-ATPase (371 kDa) established by John Walker and colleagues (**Figure 15.43A**). The three α and three β subunits alternate in a ring around a helical extension of the γ subunit (the δ and ε subunits are not visible here and may be missing). The α and β subunits have nearly identical folds despite a sequence identity of only 20%, but each of the three α/β pairs adopts a different conformation. If crystallized in the presence of ADP and a non-hydrolyzable ATP analog, 5′-adenylyl-β,γ-imidodiphosphate (AMP-PNP), one group of three interface regions, each formed mainly from residues contributed by the β subunit, can be seen to represent three stages of a catalytic cycle: one site ($β_{empty}$) appears to be unoccupied, one ($β_{ATP}$) contains an AMP-PNP mimicking ATP, and the third ($β_{ADP}$) is occupied by ADP plus an azide anion at the γ-phosphate binding site derived from the crystallization buffer.

Figure 15.42 A binding change mechanism for ATP hydrolysis by the F_1 component. There are three α/β pairs in the F_1 component, but they form three nonidentical catalytic sites: T (tight), L (loose), and O (open), viewed here along the central stalk from the F_o part in the membrane. At any one time, only the T state is catalytically active, binding ATP and hydrolyzing it to ADP and P_i, which are retained. On conversion of this subunit to the L state, the ADP and P_i, are bound much less tightly. Concomitantly, a neighboring subunit in the O state binds ATP and converts to the T state, and the third subunit, in the L state, changes to the O state and releases ADP and P_i. The process is then ready to repeat, meaning that three ATP molecules are hydrolyzed for a complete cycle of changes in which each subunit goes successively through T, L, and O states. The free energy required for the conformational changes derives from the hydrolysis of the ATP. The blue arrow in the center represents the position of the central stalk at each stage, as it rotates in a counterclockwise direction. If the cycle of changes is driven in the reverse direction by a stronger opposing source of free energy, the system will synthesize ATP and become an ATP synthase.

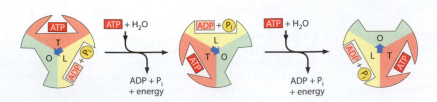

(A)

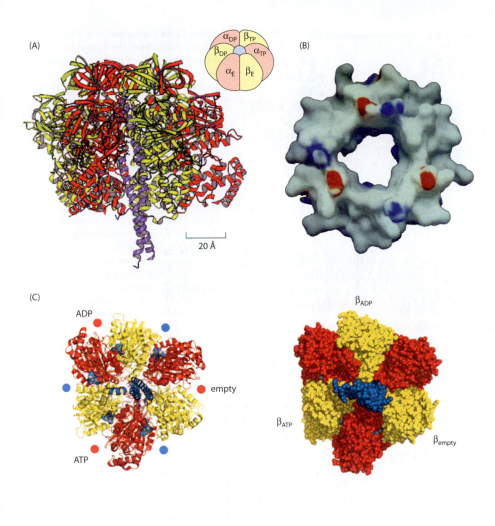

20 Å

(B)

(C)

ADP

empty

ATP

β_{ADP}

β_{ATP}

β_{empty}

(D)

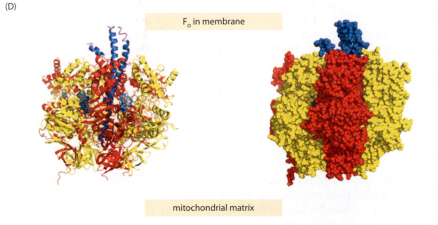

F$_o$ in membrane

mitochondrial matrix

Figure 15.43 The structure of the F$_1$ component of ox heart mitochondrial ATP synthase. (A) The α subunits are shown in red, the β subunits in yellow. The γ subunit is shown in blue. A schematic interpretation of the structure, viewed from above, is included at the top right. Bound nucleotides are portrayed in black in a 'ball-and-stick' representation. (PDB 1BMF) (B) An electrostatic surface potential of the 'sleeve' region contributed by the three α and three β subunits surrounding the double-helical 'shaft' of the γ subunit (calculated with GRASP from the coordinates of α287–294 and β274–281). The sleeve is essentially hydrophobic (white) with little in the way of negative (red) or positive (blue) charges. This is matched by the hydrophobic nature of the γ subunit in this region where it passes through the sleeve. (C) Two views (ribbon, left, and space-filling, right) of the F$_1$ component viewed down the γ subunit (blue) from the F$_o$ part in the membrane. All three α subunits (red) are found with AMP-PNP bound (inactive), depicted as blue circles; the three β subunits (yellow) are found as β_{ADP} (L), β_{ATP} (T), and β_{empty} (O), depicted as red circles. (PDB 1E79) (D) A side view of the images in (C), projecting down into the matrix from F$_o$ above.

In the absence of azide, the catalytic site of β_{ADP} is occupied by AMP-PNP, implying that β_{ADP} represents an intermediate state of ATP synthesis (or hydrolysis).

Much of the γ subunit in the form of an α helix lies in the cavity inside the $\alpha_3\beta_3$ ring. The C-terminal end appears in a dimple ~15 Å deep on the 'top' of the structure, whereas the major part adopts a left-handed coiled-coil structure that runs approximately 90 Å up the center of the $\alpha_3\beta_3$ ring but projects out as a stem about 30 Å long (the central stalk) at the 'bottom.' It passes through an essentially hydrophobic sleeve made up from the three α and three β subunits surrounding it (Figure 15.43B). The conformations of the three α subunits are similar, as are those of the β_{ATP} and β_{ADP} subunits (Figure 15.43C,D). However, the β_{empty} subunit has undergone a substantial conformational change. A displacement of the

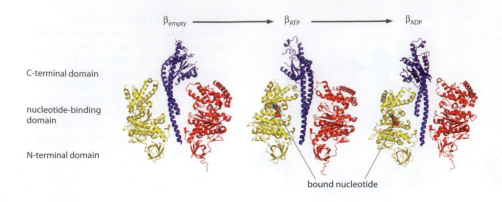

C-terminal domain

nucleotide-binding
domain

N-terminal domain

$\beta_{empty} \longrightarrow \beta_{ATP} \longrightarrow \beta_{ADP}$

bound nucleotide

Figure 15.44 Conformational change in the β subunits of the F₁ component induced by rotation of the γ subunit. Three views of an α/β pair contributing an active site, with the β subunit going through its succession of conformational changes as the γ subunit rotates. The β subunits are in yellow, the α subunits are in red, and the γ subunit is in blue. The successive changes in the structure of the β subunit are attributed to interaction with the γ subunit as it rotates about a vertical axis. The α and β subunits are composed of an N-terminal domain, a nucleotide-binding domain, and a C-terminal domain (from bottom to top). β_{empty} has an open conformation in which the α-helical C-terminal domain has been rotated upward, thereby opening the cleft of the nucleotide-binding pocket. Both β_{ATP} and β_{ADP} have a closed conformation, trapping the nucleotide within the closed pocket. All α subunits are present in an open conformation.

coiled-coil structure of the γ subunit away from the central axis of the $\alpha_3\beta_3$ ring appears to have forced what is best described as a rigid-body rotation of the C-terminal domain of the β_{empty} subunit outward by some 30° from the same central axis. This prevents the β_{empty} subunit from adopting the conformation of the β_{ATP} and β_{ADP} subunits (**Figure 15.44**).

The obvious inference is that the γ subunit rotates inside the $\alpha_3\beta_3$ ring and forces the β_{ATP} subunit to adopt the β_{empty} conformation and discharge its ATP, whereupon the next β subunit in the direction of rotation changes from β_{ADP} to β_{ATP} and ATP is synthesized from the now tightly bound ADP and P_i. Concomitantly, the preceding β subunit changes from β_{empty} to β_{ADP} and binds ADP and P_i loosely. Each rotational movement of the γ subunit by 120° will produce one molecule of ATP for export from the complex, and a complete rotation will therefore lead to the synthesis of three ATP molecules.

A likely reaction mechanism for ATP hydrolysis (ATP synthesis would be expected to be the reverse) can be inferred from the schematic view of residues in the active site of the β_{ATP} subunit shown in **Figure 15.45**. Critical residues are provided by the P-loop (GXXXXGKT/S in the one-letter code) commonly found in ATPases (see Box 3.1). A water molecule, potentially activated by hydrogen bonding to the carboxylate of Glu188, is poised for an in-line nucleophilic attack on the γ-phosphate of the ATP. (Another view, however, is that Glu188 acts as acceptor of the proton released upon ATP cleavage.) The negative charge on a penta-coordinated transition state would be stabilized by the cluster of basic residues (Lys162, Arg189, and αArg373 provided by the neighboring α subunit). Collapse of the transition state and release of a proton would leave ADP and P_i.

Single-molecule experiments demonstrate rotation of the γ subunit in F₁

The rotation inferred from the chemistry and X-ray structure has been elegantly demonstrated in single-molecule experiments with a *Bacillus* PS3 F₁-ATPase ($\alpha_3\beta_3\gamma$, from the relevant genes expressed in *E. coli*). To make this possible, a run of 10 His residues (a His-tag) was introduced at the N terminus of the β subunits in the ATPase by means of directed mutagenesis, and a surface Cys residue was similarly introduced into the γ subunit. The complex was treated with a maleimide that had been covalently linked to biotin (see Section 9.2). The maleimide reacts with the new Cys thiol group on the γ subunit, thereby attaching the biotin in an exposed position. An actin filament about 2.5 μm long (Section 14.2) was prepared with a similarly biotinylated actin subunit at one end and made fluorescent by reacting the surface amino groups of all the actin subunits with a fluorescent isothiocyanate. In a parallel experiment, a microscope coverslip was coated with a carrier protein, horseradish peroxidase, that had been conjugated with Ni^{2+}-nitriloacetic acid, Ni^{2+}-$N(CH_2COOH)_3$. When exposed to a solution of biotinylated F₁-ATPase, the Ni^{2+} binds tightly to the His-tag at the N termini of the β subunits, thereby immobilizing the F₁-ATPase. Given that the N termini of the α and β subunits are on the 'matrix' side of the $\alpha_3\beta_3$ ring in the orientation in Figure 15.43, the F₁-ATPase therefore becomes immobilized, with the exposed stalk of the γ subunit with its biotin 'tag' on top. The fluorescent actin filament could then be attached specifically to the γ subunit by a bridging molecule of streptavidin, a bacterial protein that has four very-high-affinity binding sites for biotin, two at each end of the molecule. This is depicted in **Figure 15.46**.

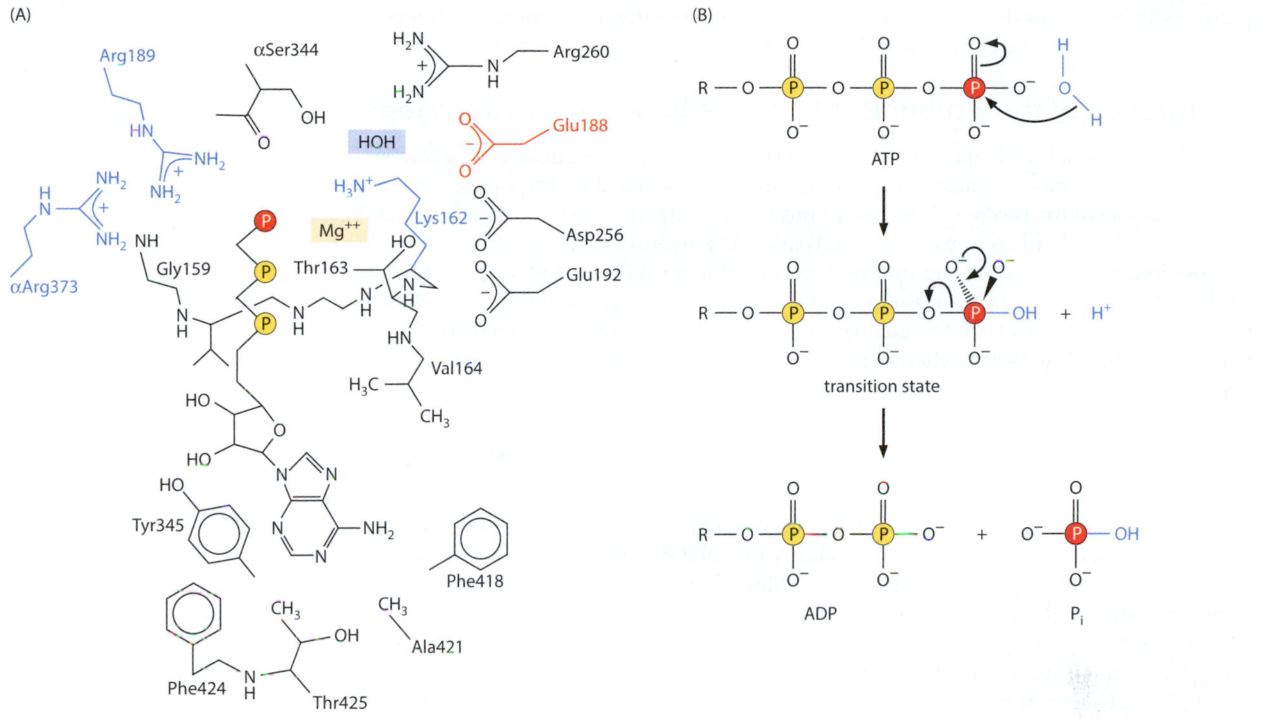

(A)

(B)

Figure 15.45 A likely reaction mechanism for ATP-hydrolysis by the F$_1$ component. (A) All the catalytically important residues come from the β subunit with the exception of αArg373. The phosphate groups of the ATP shown bound in the active site are in close contact with residues Gly159–Thr163, including Lys162, provided by the P-loop of the β subunit. The adenine ring lies in a hydrophobic pocket consisting of Tyr345, Phe418, and Phe424. An ordered water molecule, activated by hydrogen bonding to Glu188, is positioned for an in-line nucleophilic attack on the ATP γ-phosphate (depicted in red). A cluster of basic residues (depicted in blue) is well placed to stabilize the penta-coordinated transition state. (B) In-line attack of ordered H$_2$O activated by Glu188 leads to the release of a proton and the formation of a penta-coordinated transition state whose collapse generates ADP and P$_i$. In the corresponding site of the α subunit, there is a glutamine residue (Gln208), which explains why the α subunits are catalytically inactive. The biosynthesis of ATP is this reaction in reverse. (A, adapted from A.G.W. Leslie and J.E. Walker, *Phil. Trans. R. Soc. Lond. B* 355:465–472, 2000. With permission from The Royal Society.)

The assembly was viewed in a flow chamber under a fluorescence microscope. No movement was detected in the absence of ATP, but when 2 mM ATP was infused into the chamber, rotation of the actin filaments in a significant proportion of the immobilized F$_1$-ATPases was observed. They all rotated in only one direction, and the direction (β$_{ATP}$ → β$_{ADP}$ → β$_{empty}$) was consistent with the rotary mechanism for ATP hydrolysis inferred from the crystal structure. The rotation rate was at most ~4 rev/s, far below the rate of 52 rev/s calculated from the measured rate of ATP consumption in solution (assuming 3 ATP to be consumed per revolution), and it decreased with increasing size of the actin filament. This was

Figure 15.46 Direct demonstration of the rotation of the F$_1$ component of an ATP synthase driven by ATP hydrolysis. A His-tag was introduced at the N-terminal end of each of the three β subunits in the F$_1$ component, and these interact with a covering of Ni^{2+}-nitriloacetic acid (Ni^{2+}-NTA) applied to a glass microscope coverslip, rendering the α$_3$β$_3$γ complex immobile with the projecting stalk of the γ subunit on the upper surface. A fluorescent actin filament was tightly but noncovalently bound to the γ subunit by a bridging molecule of streptavidin (a tetramer) that interacts with biotin tags strategically placed on both partners, as shown. The direction of rotation of the actin filament (and thus of the γ subunit) that takes place during the hydrolysis of exogenous ATP is indicated. (Adapted from H. Noji et al., *Nature* 386:299–302, 1997. With permission from Macmillan Publishers Ltd.)

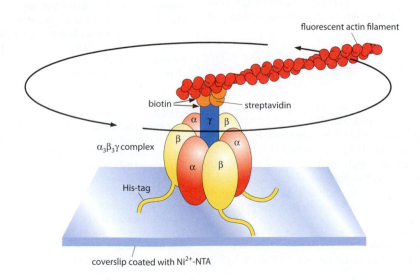

presumably because of the viscous drag on the filament, but the rotary mechanism was vividly confirmed. Moreover, it was clear that the δ and ε subunits are not essential for rotation.

The rotation of the γ subunit in F₁ can be broken down into steps

By replacing the actin filament with a biotinylated 40 nm gold particle, the viscous drag on the rotation is much reduced. Rotation is thus much speeded up but can be visualized in laser dark-field microscopy. It is then found to comprise discrete steps and these can be correlated with individual steps in the catalysis of ATP hydrolysis. Three distinct pauses in the rotation, each 120° apart, are interpreted as waiting for ATP to bind, and the 120° step has been further resolved into substeps of 80° and 40°. Kinetic analyses of the substeps using various concentrations of ATP or analogs indicate that the 80° substep is triggered by ATP binding (the **binding dwell**), whereas that the 40° substep is initiated after two consecutive reactions, each having a time constant of approximately 1 ms (the **catalytic dwell**). The use of site-directed and temperature-sensitive mutants of F_1, together with a slowly hydrolyzed analog of ATP (ATPγS), points to the catalytic dwell reactions as being the hydrolysis of ATP and then release of the product P_i.

During one revolution of the γ subunit, F_1 consumes one molecule of ATP in each step of 120°. Further insight has come from the ability to make simultaneous observations of the rotation of the γ subunit and of the binding/unbinding of a fluorescently labeled ATP. These show that, after binding ATP, F_1 retains it for a rotation of 240°, and releases it as ADP between 240° and 320°, probably at 240°. By using a hybrid F_1 constructed with a single copy of a mutant β subunit that has a slow rate of ATP hydrolysis, it was possible to identify the angular position at which ATP hydrolysis occurs as being +200° (that is, 120° + 80°) round from the binding angle of the ATP, defined as 0°. The steps in a complete rotation of the γ subunit, as currently understood, are summarized in **Figure 15.47**. Each β subunit completes a single turnover of ATP hydrolysis during the course of one revolution of the γ subunit; it begins with ATP binding at 0°, hydrolysis of the ATP at 200°, release of the product ADP at 240°, and finally release of P_i at 320°.

To which catalytic state in the coupling scheme does the crystal structure of F_1 (see Figure 15.43) correspond? This question has been approached in a number of ways. A pair of Cys residues, one in the γ subunit (Leu276Cys) and another in the α subunit (Glu284Cys), was introduced into the F_1F_0 ATPase of *E. coli*. The mutations were chosen on the basis of the crystal structure. By chemically oxidizing the two Cys thiol groups, a disulfide bridge is formed and the rotating γ subunit is arrested in the same orientation as in the crystal structure. Correlation of the locked structure with the angle that has been reached in a rotation shows that the crystal structure correlates not with the ATP-waiting dwell, as might first be supposed, but with the catalytic dwell. A similar result has been inferred from experiments in which an F_1-ATPase of *Bacillus* PS3a was constructed containing a β subunit that was modified by attachment of a fluorophore to its C-terminal helix. Changes in the fluorescence allowed changes at the β subunit to be correlated with simultaneous observation of the substep in the rotation of the γ subunit. These observations identify $β_{ATP}$ as corresponding to the 80° state, $β_{ATP}$ to the 200° state, and $β_{ATP}$ to the 320° state (see Figure 15.47). They do not invalidate any of the mechanistic inferences but are important in helping to interpret a snapshot of a rotary motion.

The c-subunits form a ring in F₀ that rotates against a stator complex

The structural information for F_0 (a membrane complex) is limited in comparison with that available for F_1. The amino acid sequence of the *E. coli* a-protein (30 kDa) indicates that it has five transmembrane helices, whereas the b-protein (18 kDa) has only one transmembrane helix, in its N-terminal region. This anchors an extended polar domain that protrudes from the membrane, and its primary structure suggests a right-handed coiled coil (much rarer than left-handed coiled-coils) in a b_2 dimer. This is the peripheral stalk. At its distal end, b_2 binds tightly to $α_3β_3$, both directly and through the δ subunit. Together these proteins ($a_1b_2δα_3β_3$) create the **stator** in which the $α_3β_3$ ring is held stationary in the F_1F_0-ATP synthase, illustrated schematically in Figure 15.40.

A structure of a partial complex from bovine heart mitochondria (F_1 still attached to the c-subunits from F_0; the other subunits were lost during purification and crystallization) is

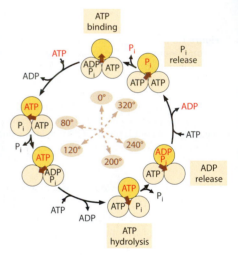

Figure 15.47 Discrete steps in the rotation mechanism of F₁. Each of the three catalytic sites in an F_1 component is represented by a beige disk, within which is indicated the chemical status of that active site at any point. The radial arrows (brown) represent the angular position of the γ subunit in its rotation counterclockwise, starting from 0°, depicted at the top of the circular reaction cycle. One catalytic site is highlighted in yellow to make it easier to follow an individual α/β pair through the reaction sequence. Each catalytic site retains the bound nucleotide as ATP (picked out in red for the highlighted site) until the γ subunit has rotated 200° from the binding angle at 0° for that particular catalytic site. At this point, the bound ATP is hydrolyzed to ADP and P_i, after which the ADP is released at 240° and the P_i is released at 320°. (Adapted from R. Watanabe, R. Iino and H. Noji, *Nat. Chem. Biol.* 6:814–820, 2010. With permission from Macmillan Publishers Ltd.)

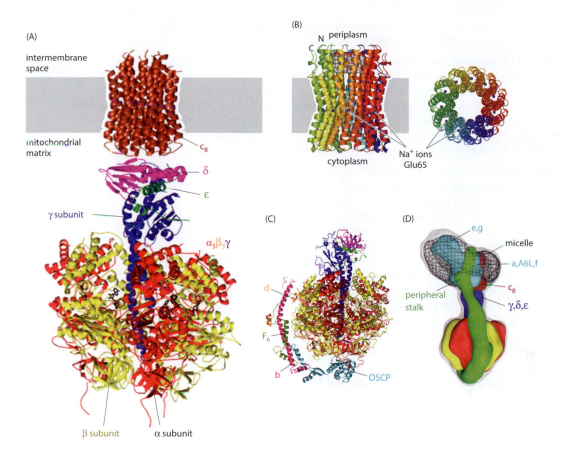

Figure 15.48 Structures of the c-ring and stator complex of the F₁F₀-ATP synthase. (A) The structure of a partial ATP synthase from bovine heart mitochondria showing the interaction of the F₁ component with the c-ring. Eight c-subunits, transmembrane helical hairpins, make up the c₈-ring (brown). The subunits of the central stalk (bovine complex nomenclature) are δ (purple), ε (green), and γ (blue). (PDB 2XND) (B) Structure of the c₁₁-ring of the Na⁺-translocating ATP synthase of *I. tartaricus*. Each of the 11 c-subunits is depicted in a different color, and Na⁺ ions are shown as black dots. Left, a view in the plane of the membrane; right, a view from the cytoplasmic side (membrane omitted for clarity). (PDB 1YCE) (C) A structure of the bovine mitochondrial F₁–stator complex, with the peripheral stalk on the left. The color-coding of the F₁ component is as in (A). The proteins of the peripheral stalk are colored as follows: teal, residues 1–146 and 169–189 of OSCP; pink, residues 122–207 of subunit b; green, residues 5–25 and 35–57 of F₆; orange, residues 30–40, 65–74, and 85–91 of subunit d. (PDB 2WSS). (D) Likely arrangement of subunits in the cryo-EM model of the intact F₁F₀-ATPase (Figure 15.39B), shown here with the peripheral stalk in front. The central stalk (γ,δ,ε) is in blue, the peripheral stalk is depicted in green, the c₈-ring is in brown; α and β subunits are in red and yellow, and other subunits are as indicated. The detergent micelle replacing the F₀ membrane is a gray mesh. The regions of electron density in the membrane into which proteins have yet to be fitted illustrate the present lack of knowledge about the remainder of the F₀ part. (A, adapted from I.N. Watt et al., *Proc. Natl Acad. Sci. USA* 107:16823–16827, 2010; B, adapted from T. Meier et al., *Science* 308:659–662, 2005. With permission from AAAS; C, from D.M. Rees, A.G. Leslie and J.E. Walker, *Proc. Natl Acad. Sci. USA* 106:21597–21601, 2009; D, adapted from L.A. Baker et al., *Proc. Natl Acad. Sci. USA* 109:11675–11680, 2012. A, C, and D with permission from National Academy of Sciences.)

shown in **Figure 15.48A**. Eight c-subunits, each (75 aa) in the shape of a helical hairpin, form a ring. The eight C-terminal helices (41 Å long) make up an outer ring closely packed against the eight N-terminal helices (52 Å long), which form an inner ring, leaving a central hole ~40 Å in diameter that is probably filled with phospholipids. The loop regions between the helices are on the matrix side and make extensive interactions with the γ, δ, and ε subunits (mitochondrial ATP synthase terminology) at the proximal end of the central stalk. The structure of the 11-subunit (each 89 aa) c-ring from the Na⁺-translocating F-type ATP synthase of the Gram-negative bacterium *Ilyobacter tartaricus* (which uses the electrochemical potential of a Na⁺ gradient to generate ATP) is similar (Figure 15.48B). In this instance, a Na⁺ ion can be seen at the central bend in each of the 11 subunits, sandwiched in a site formed from one N-terminal helix and two C-terminal helices but with access from the outside. A negatively charged residue, Glu65, forms part of the Na⁺-binding site. A negatively charged residue (Glu58) is at the same relative position in the bovine c-subunit. Indeed, the amino acid sequences of c-subunits (where known) are identical in almost all vertebrates and highly conserved more generally. In the c-subunits of animal ATP synthases, the N^6-NH₂ group of Lys43 is trimethylated and probably marks the binding site of cardiolipin molecules essential for c-ring stability and full proton pumping. However, the trimethylation is not found in rings with 10 or more c-subunits observed in other species (see below).

In mitochondrial ATP synthases, the peripheral stalk consists of single copies of OSCP, b, d, and F_6 subunits (Figure 15.48C). The peripheral stalk binds tightly to the surface of the $\alpha_3\beta_3$ ring, as in the simpler *E. coli* complex, holding it stationary while the γ subunit rotates within it to drive ATP synthesis. In the cryo-EM (see **Methods**) model of the intact synthase, subunit b at its base is embedded in the membrane region (the only peripheral stalk protein to be so), the a-subunit appears to abut the c_8-ring, and other membrane subunits of F_o (including A6L, e, f, and g) have also been placed nearby (Figure 15.48D).

Rotation of the c-ring of an intact F_1F_o-ATPase has been observed by using techniques similar to those described above for F_1 alone. The $\alpha_3\beta_3$ hexamer in the F_1 component was immobilized on a surface, and a fluorescent actin filament was attached to a subunit in the c-ring in F_o rather than to the γ subunit. In the presence of ATP, rotation of the c-ring was observed while ATP was being hydrolyzed in F_1 and causing the γ subunit to rotate. A similar result was achieved by engineering disulfide bridges between Cys residues introduced by directed mutagenesis to crosslink the γ, ϵ, and c-subunits into a single entity in the *E. coli* F_1F_o-ATPase. Remarkably, covalently clamping these three components together had no effect on the ATP hydrolysis, proton translocation, or sensitivity to inhibitors. ATP synthesis is the reverse of ATP hydrolysis and these experiments make it clear that the c-ring, like the γ subunit to which it is tightly bound, rotates as part of the mechanism. A stepping of 36° in the rotation of the c-ring (commensurate with its 10 c-subunits) has been detected for the *E. coli* ATP synthase (see below).

Proton transport across the membrane by the c-ring makes it rotate

How does proton translocation by F_o force the c-ring to rotate and thereby drive ATP synthesis in F_1? Biochemical analysis and directed mutagenesis of bacterial enzymes indicate that the a- and c-subunits both contain functional groups that are essential for activity. A model that couples rotation of the c-ring (and with it the γ subunit within the $\alpha_3\beta_3$ hexamer) to proton translocation is depicted in **Figure 15.49**. It envisages two separate and offset half channels formed where the c-ring abuts the a-subunit in the membrane; one channel connects with the 'outside' (bacterial periplasm or intermembrane space of mitochondria) and the other with the 'inside' (bacterial cytoplasm or mitochondrial matrix). Halfway across the span of the c-subunit, in its C-terminal helix, is a highly conserved acidic residue, Asp61 in *E. coli*, corresponding to Glu58 and Glu65 in the *I. tartaricus* and bovine c-rings, respectively (see Figure 15.48). A proton entering the 'outside' channel encounters Asp61, which is unprotonated (and negatively charged) in the local aqueous environment. Protonation of Asp61 makes it electrically neutral and the c-ring can therefore rotate a fraction of a turn with the protonated (and uncharged) Asp61 now in contact with the membrane. This part-rotation correspondingly brings a protonated Asp61 in another c-subunit out of the membrane milieu and into position at the entrance to the hydrophilic exit channel between the c-ring and the a-subunit. The change from hydrophobic to aqueous environment causes Asp61 to deprotonate, possibly promoted by a nearby invariant and positively charged Arg

Figure 15.49 A schematic model of c-ring rotation driven by proton translocation. A proton from the high proton concentration 'outside' (periplasm or mitochondrial intermembrane space) enters a channel in the a-subunit, and then protonates a conserved Asp (or Glu) residue in a c-subunit (Asp61 in *E. coli*). The proton remains bound until, by rotation of the c-ring, the Asp of this c-subunit appears at the entrance to a separate 'exit' channel in the a-subunit, offset from the first, where a protonated Arg residue (Arg210 in the a-subunit of *E. coli*) awaits. At this point, the Asp carboxyl group deprotonates and the proton is discharged into the inside (bacterial cytoplasm or mitochondrial matrix). Each c-subunit in turn undergoes the same process, one rotation of the c-ring thereby translocating as many protons as there are c-subunits in the ring. The c-ring is depicted here with 10 subunits, as in *E. coli* F_1F_o-ATP synthase.

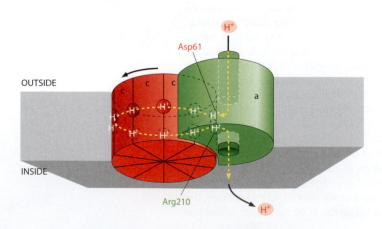

residue (Arg210, halfway across the fourth transmembrane helix in the *E. coli* a-subunit), and the proton released passes down the exit channel into the inside. At the next part-rotation of the c-ring, this newly unprotonated Asp61 will present itself at the receiving end of the proton entrance channel and will become protonated. The process then repeats.

This mechanism is supported by the presence of Na^+ ions bound to the c-subunits all round the *I. tartaricus* c-ring (see Figure 15.48B), reflecting its physiological function of converting an electrochemical potential stored in a Na^+ (not H^+) gradient into the synthesis of ATP. In principle, the c-ring could undergo rotational movements in either direction driven by Brownian motion. However, the rotation of the c-ring is made unidirectional by the protonation of the Asp (or Glu) residue in the c-subunit, enabling it to enter the hydrophobic environment of the membrane, after which it can only be deprotonated when rotation of the c-ring ultimately brings it opposite the protonated Arg residue in the exit channel of the a-subunit. Thus the ring can go round in only one direction, a *Brownian ratchet*. This direction is clockwise, as viewed from the outside, the direction in which the $\gamma(\varepsilon)$ subunit must rotate to drive ATP synthesis in F_1. A similar argument applies to the V-type ion-transporting ATPase.

The number of c-subunits in the c-ring varies widely from organism to organism: 8 in mammalian mitochondria, 10 in yeast and *E. coli*, 11 in *I. tartaricus*, 13 in the thermoalkalophilic *Bacillus* sp strain TA2A1, 14 in chloroplasts, and 15 in the alkaliphilic cyanobacterium *Spirulina platensis*. Until the discovery of the latter it was thought that a symmetry mismatch between the c-ring and the pseudo-three-fold symmetry of the $\alpha_3\beta_3$ ring in F_1 was required to avoid rotations with integral ratios. It was argued that these might trap the rotor complex in a deep potential energy well if the minima of F_1 and F_o should coincide. Although widespread, however, a symmetry mismatch is evidently not essential.

The transport of protons can be correlated with the efficiency of ATP synthesis

One revolution of the $\gamma(\varepsilon)$ subunit/c-ring generates three molecules of ATP from ADP and P_i. The passage of two electrons from NADH through the respiratory chain in mitochondria causes 10 electrogenic proton transfers across the mitochondrial membrane (Section 15.2). If the c-ring of the F_1F_o-ATPase contains 10 c-subunits, as it does in *E. coli* and yeast, the ratio of ATP synthesized for every 10 electrons translocated back again (enough for one complete revolution of the c-ring) would be 3.0. For the oxidation of $FADH_2$, which enters the respiratory chain after Complex I and is coupled to the translocation of only six protons, the number of ATP synthesized would be limited to 1.8. In mammalian mitochondria, these values would rise to 3.8 (NADH) and 2.3 ($FADH_2$), respectively, given that there are only eight subunits in a c-ring.

There will inevitably be some nonspecific and unproductive proton leakage, and the important action of uncoupling proteins in mitochondria will diminish the proton gradient for thermogenesis and other reasons (Section 15.2). Moreover, ATP is made in the mitochondrial matrix and must be translocated across the inner membrane into the cytoplasm for the many ATP-requiring reactions located there (see Figure 15.2). The translocation is catalyzed by an ADP/ATP antiporter (*adenine nucleotide transferase*) in the inner membrane,

$$ADP^{3-}_{cytosol} + ATP^{4-}_{matrix} \rightarrow ATP^{4-}_{cytosol} + ADP^{3-}_{matrix}$$

which causes net transfer of one negative charge from the matrix per ATP translocated. This is driven by the membrane potential (positive outside, negative inside) that forms part of the proton-motive force. For the synthesis of ATP from the transported ADP, P_i also has to be transferred into the matrix from the cytoplasm. This is made possible by a *phosphate carrier* that catalyzes the symport of P_i and H^+, driven by the ΔpH component of the proton-motive force and at a consequent cost of one proton per ATP delivered to the cytoplasm.

Taking these numbers into account, for a mammalian ATP synthase with eight subunits in the c-ring, the number of ATP molecules made per O atom ($\frac{1}{2}O_2$) consumed would be about 2.7 for the oxidation of NADH and 1.6 for $FADH_2$. In respiring mitochondria, experimental values of these **P/O ratios** are commonly about 2.5 and 1.5, respectively. The coupling of the proton-motive force to ATP synthesis is well above 80% efficient.

In mitochondria the proton-motive force resides less in ΔpH and more in $\Delta\psi$ (Section 15.2), whereas in chloroplasts it is mostly in ΔpH (Section 15.3). The presence of 14 c-subunits in a chloroplast c-ring means that the number of H^+ translocated per ATP synthesized should be 4.67 if coupling were 100% efficient. However, in experiments with chloroplasts, a ratio of 4.0 is commonly observed. At the moment it is difficult to understand an apparent efficiency of more than 100%. In alkaliphilic bacteria, the interior pH is maintained at about 1.5–2 pH units below that of the external environment, yet bacterial growth continues despite this adverse ΔpH. It has been argued that the larger c-ring in organisms such as *Sp. platensis* enables the torque necessary to drive a rotation of the γ subunit to be spread over more steps of the c-ring. A c-ring with 15 c-subunits would generate a 1.5-fold higher torque than one with only 10 c-subunits under a given electrochemical potential.

There is still much to be understood before the advantages or disadvantages in various c-ring sizes can be satisfactorily rationalized. Nevertheless, it is clear that the ATP synthases are exceptionally efficient molecular machines, capable of running at 5000 rev/min or more, and able to meet the heavy demand for ATP needed to maintain life (for example, an estimated daily turnover of ~50 kg of ATP in the human body).

ATP synthase dimers help shape the mitochondrial cristae and form respiratory supercomplexes

Cryoelectron tomography of mitochondria and mitochondrial membranes from a wide range of organisms has revealed that ATP synthases are found as dimers in extensive rows along the highly curved ridges of the cristae (**Figure 15.50A,B**). The F_1 heads, identified as 10 nm spherical densities on stalks 5 nm above the membranes, are consistently 28 nm apart, and the angle between them is always 86°. Atomic models of the ATP synthase can be fitted to the dimers (Figure 15.50C), and the angle is consistent with the local curvature in the micelle in the cryo-EM model of the ATP synthase monomer (see Figure 15.48D). The distance between adjacent dimers in a given row varies from 12 nm to >20 nm, indicating that the dimers do not interact directly. It is likely that the local deformations in the lipid bilayer induced by the bends associated with the ATP synthase dimers are partly relieved by the dimers lining up in rows. In so doing, they help shape the cristae. In yeast, certain membrane-associated proteins are found only with the dimers and, in mutants lacking them, the dimers do not form. In such mutants, wild-type lamellar cristae with tightly curved ridges are not observed and individual ATP synthase monomers are found more randomly scattered over smoother balloon-shaped membranes.

Figure 15.50 ATP synthase dimers and respiratory chain complexes in mitochondrial cristae. (A) Tomographic slice through a mitochondrion of the fungus *Podospora anserina*. Yellow arrowheads highlight rows of ATP synthase dimers on ridges of cristae. Ribosomes are visible in the matrix as dark gray particles. OM, outer membrane; IM, inner membrane. (B) Surface-rendered volumes of mitochondrial membranes from *S. cerevisiae*. Left, a typical flat lamellar crista; right, a tubular crista. Membranes are depicted in gray, ATP synthase dimers in yellow. (C) Averaged ATP synthase dimer from a cryoelectron tomogram of *S. cerevisiae* mitochondrial membranes, to which known atomic models have been fitted. (D) Schematic model of possible arrangements of ATP synthase dimers (yellow) and respiratory chain complexes (in green) in mitochondrial cristae. Protons are represented by red dots, pumped out of the matrix by the respiratory chain complexes to create the electrochemical proton gradient and returned to the matrix via the ATP synthases. (A and D, adapted from K.M. Davies et al., *Proc. Natl Acad. Sci. USA* 108:14121–14126, 2011; B and C, from K.M. Davies et al., *Proc. Natl Acad. Sci. USA* 109:13602–13607, 2012. All with permission from National Academy of Sciences.)

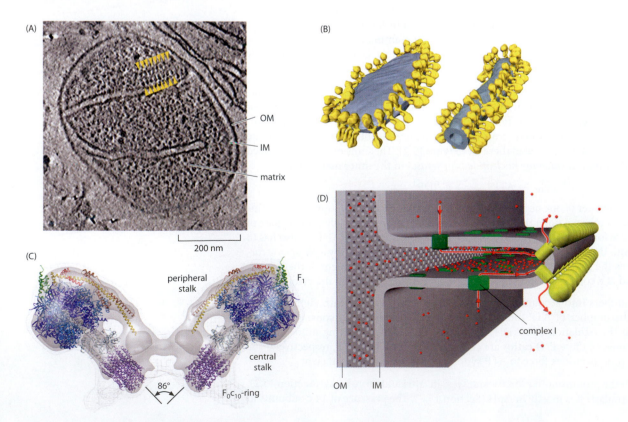

(A)

OM

IM

matrix

200 nm

(B)

(C)

peripheral stalk

F_1

central stalk

86°

F_0c_{10}-ring

(D)

complex I

OM   IM

The respiratory complexes in the inner mitochondrial membrane also form supercomplexes (see Figure 15.16). For each Complex I there is an average of three to four ATP synthase monomers and, although the experimental results are not unambiguous, electron tomography suggests that Complex I occupies flatter parts of the inner membrane, as opposed to the ridges in the cristae in which the rows of ATP synthase dimers are located. This may have implications for the efficiency of energy conversions in that a higher concentration of pumped protons might thus exist locally in the neighborhood of the ATP synthases. Protons move ~10 times faster along membrane surfaces compared with diffusion into bulk solvent, and their preferential diffusion toward the rows of ATP synthase dimers would enhance ATP synthesis, as illustrated schematically in Figure 15.50D.

15.5 SUMMARY

Cells need energy to survive, grow, and divide. The energy currency of the cell is ATP, which is generated either by the oxidation of nutrients or by photosynthesis. In both processes, membrane-embedded molecular machines translocate protons across a membrane. The free energy stored in the resulting electrochemical proton gradient is then used by another molecular machine to drive ATP synthesis. In eukaryotes, oxidative phosphorylation takes place in mitochondria, whereas photosynthesis, in relevant organisms, occurs in chloroplasts.

Oxidation of nutrients generates NADH and $FADH_2$, which are then used to reduce molecular oxygen to water. This process is catalyzed by a succession of four membrane complexes that form a respiratory (electron transport) chain. In the course of their activity, protons are pumped across the membrane, from the matrix into the intermembrane space in the mitochondria of eukaryotes or out of the cytoplasm in prokaryotes. Complex I (NADH–ubiquinone oxidoreductase) is the entry point for the oxidation of NADH. In this complex of up to 45 protein subunits, a flavin mononucleotide (FMN) redox center accepts reducing equivalents from NADH ($2H \approx 2H^+ + 2e^-$) and passes electrons along a 'wire' composed of nine iron–sulfur (Fe–S) clusters in a succession of subunits to ubiquinone (Q). In so doing, four protons are pumped across the membrane per electron pair from NADH. Complex II (succinate–ubiquinone oxidoreductase) is the entry point for oxidation of $FADH_2$. Again, electrons are passed through Fe–S clusters to Q but, unlike the other respiratory chain complexes, Complex II is not a proton pump. Complex III oxidizes the ubiquinol (QH_2) released into the membrane by Complexes I and II. Three core subunits, cytochromes b and c_1 and a Rieske-type iron–sulfur protein (ISP), combine to pump four more protons across the membrane and reduce two molecules of a soluble cytochrome c in the intermembrane space. Complex IV (cytochrome c oxidase), another multi-subunit structure, uses four molecules of reduced cytochrome c to deliver the four electrons (equivalent to two pairs) required to reduce molecular oxygen (O_2) to water, and a further four protons are moved into the intermembrane space. Thus a total of 10 protons are pumped across the membrane for each NADH oxidized.

Photosynthesis harnesses the photochemical capabilities of multiple light-absorbing pigments. Light is captured by the chlorophylls and carotenoids in the reaction centers of large membrane-bound multi-subunit protein complexes embedded in the thylakoid membranes in chloroplasts (plants) or folded regions of the cell membrane in photosynthetic bacteria. They are surrounded by giant light-harvesting complexes, which greatly enhance energy capture by very fast and efficient exciton energy transfer to the reaction centers. From the initial charge separation in chlorophylls at a photochemical reaction center, electrons are used to reduce a quinone Q to QH_2. In purple bacteria, the quinol is the substrate for a cytochrome bc_1 complex that resembles the cytochrome bc_1 complex (Complex III) of the respiratory chain. The reoxidation of QH_2 generates a reduced cytochrome c_2 in the periplasmic space and, as with the respiratory cytochrome bc_1 complex, protons are simultaneously pumped across the membrane. The reduced cytochrome c_2 finally passes its electron back to the chlorophylls in the reaction center, making this a cyclic process.

In cyanobacteria and plants, two reaction centers act in series. The first (photosystem II) passes electrons to a quinone and thence through a cytochrome b_6/f complex to a small protein, plastocyanin, instead of a cytochrome c, and protons are correspondingly pumped across the membrane in the process. In a second light-driven step, photosystem I accepts electrons from reduced plastocyanin and passes them on through a quinone and iron–sulfur complex to a small soluble protein, ferredoxin. The reduced ferredoxin can return its electron to the cytochrome b_6/f complex on the donor side of photosystem I, a form of cyclic transfer resembling that of purple bacteria. Alternatively, the reduced ferredoxin can be used by an enzyme, ferredoxin–$NADP^+$ reductase, to reduce $NADP^+$ to NADPH, a **non-cyclic electron** transfer. In cyanobacteria and plants, photosynthesis is oxygenic; the source of the electrons entering photosystem II in non-cyclic transfer is H_2O (abstraction of $2H^+ + 2e^-$), thereby releasing O_2. This is energetically a difficult reaction to bring about: it depends on the light-driven reactions in the photosystem II reaction center, and it too is accompanied by proton pumping across the membrane. NADPH is required for biosynthetic reactions, notably carbohydrate synthesis, and the balance between cyclic and non-cyclic electron transfer can be adjusted according to metabolic need and light intensity.

The energy stored in the proton gradient across the membrane generated by respiration and photosynthesis is used by another membrane-embedded macromolecular machine, F_1F_0-ATP synthase, to generate ATP. The synthase is constructed from two rotary motors; the passage of protons back across the membrane in response to the proton gradient drives the rotation of one of them (F_o) against the other (F_1), which is held stationary. The rotation of F_o enforces sequential conformational changes in F_1, which in turn drives ATP synthesis from ADP and P_i in each of its three active sites. One complete rotation of F_o is thus responsible for the synthesis of three molecules of ATP. In mitochondria the enzymes of the respiratory chain and the F_1F_0-ATP synthase appear to associate in supramolecular complexes in the inner membrane. This helps shape the sharp membrane folds in the cristae and may enhance the already high efficiency of energy conversion to ATP.

Sections of this chapter are based on the following contributions

15.2   Gary Cecchini, The Veteran's Health Research Institute, San Francisco, USA.
15.3   Richard J. Cogdell, University of Glasgow, UK.
15.4   Hiroyuki Noji, University of Tokyo, Japan and Ryota Iino, National Institutes of Natural Sciences, Japan.

REFERENCES

General

Blankenship RE (2002) Molecular Mechanisms of Photosynthesis. Blackwell Scientific.
Keilin D (1966) The History of Cell Respiration and Cytochromes. Cambridge University Press.
Nicholls DG & Ferguson SJ (2013) Bioenergetics, 4th ed. Academic Press.

15.2 Biological oxidation and the respiratory chain

Acin-Perez R, Fernandez-Silva P, Peleato ML et al (2008) Respiratory active mitochondrial supercomplexes. *Mol Cell* 32:529–539.
Borisov VB, Forte E, Davletshin A et al (2013) Cytochrome *bd* oxidase from *Escherichia coli* displays high catalase activity: an additional defense against oxidative stress. *FEBS Lett* 587:2214–2218.
Carroll J, Fearnley IM, Skehel JM et al (2006) Bovine complex I is a complex of 45 different subunits. *J Biol Chem* 281:32724–32727.
Crofts AR, Hong S, Wilson C et al (2013) The mechanism of ubihydroquinone oxidation at the Q_o-site of the cytochrome bc_1 complex. *Biochim Biophys Acta* 1827:1362–1377.
Darrouzet K, Moser CC, Dutton PL & Daldal F (2001) Large scale domain movement in cytochrome bc_1: a new device for electron transfer in proteins. *Trends Biochem Sci* 26:445–451.
Efremov RG & Sazanov LA (2011) Respiratory complex I: 'steam engine' of the cell? *Curr Opin Struct Biol* 21:532–540.
Ferguson-Miller S, Hiser C & Liu J (2012) Gating and regulation of the cytochrome *c* oxidase proton pump. *Biochim Biophys Acta* 1817:489–494.
Genova ML & Lenaz G (2014) Functional role of mitochondrial respiratory supercomplexes. *Biochim Biophys Acta* 1837:427–443.
Heinemeyer J, Braun H-P, Boekema EJ & Kouřil R (2007) Structural model of the cytochrome *c* reductase/oxidase supercomplex from yeast mitochondria. *J Biol Chem* 282:12240–12248.

Hunte C, Zickermann V & Brandt U (2010) Functional modules and structural basis of conformational coupling in mitochondrial complex I. *Science* 329:448–451.
Iverson TM, Maklashina E & Cecchini G (2012) Structural basis for malfunction in Complex II. *J Biol Chem* 287:35430–35438.
Johansson AL, Högbom M, Carlsson J et al (2013) Role of aspartate 132 at the orifice of a proton pathway in cytochrome *c* oxidase. *Proc Natl Acad Sci USA* 110:8912–8917.
Johnson D, Dean DR, Smith AD & Johnson MK (2005) Structure, function and formation of biological iron–sulfur clusters. *Annu Rev Biochem* 74:247–281.
Kaila VRI, Johansson MP, Sundholm D et al (2009) The chemistry of the Cu_B site in cytochrome *c* oxidase and the importance of the unique His–Tyr bond. *Biochim Biophys Acta* 1787:221–233.
Moser CC, Chobot SE, Page CC & Dutton PL (2008) Distance metrics for heme protein electron tunneling. *Biochim Biophys Acta* 1777:1032–1037.
Rich PR & Maréchal A (2010) The mitochondrial respiratory chain. *Essays Biochem* 47:1–23.
Schon EA, DiMauro S & Hirano M (2012) Human mitochondrial DNA: roles of inherited and somatic mutations. *Nat Rev Genet* 13:878–890.
Smirnova I, Chang H-Y, von Ballmoos C et al (2013) Single mutations that redirect internal proton transfer in the ba_3 cytochrome oxidase from *Thermus thermophilus*. *Biochemistry* 52:7022–7030.
Solmaz SR & Hunte C (2008) Structure of complex III with bound cytochrome *c* in reduced state and definition of a minimal core interface for electron transfer. *J Biol Chem* 283:17542–17549.
Sun F, Hyo X, Zhai Y et al (2006) Crystal structure of mitochondrial respiratory membrane complex I. *Cell* 121:1043–1057.
Xia D, Esser L, Tang W-K et al (2013) Structural analysis of cytochrome bc_1 complexes: implications to the mechanism of function. *Biochim Biophys Acta* 1827:1278–1294.

Yap LL, Lin MT, Ouyang H et al (2010) The quinone-binding sites of the cytochrome bo_3 ubiquinol oxidase of *Escherichia coli*. *Biochim Biophys Acta* 1797:1924–1932.

Yoshikawa S, Muramoto K, Shinzawa-Itoh K & Mochizuki M (2012) Structural studies on bovine heart cytochrome *c* oxidase. *Biochim Biophys Acta* 1817:579–589.

15.3 Photosynthetic reaction centers and light-harvesting complexes

Amunts A, Toporik H, Borovikova A & Nelson N (2010) Structure determination and improved model of plant photosystem I. *J Biol Chem* 285:3478–3486.

Anderson JM, Horton P, Kim E-H & Chow WS (2012) Towards elucidation of dynamic structural changes of plant thylakoid architecture. *Phil Trans R Soc B* 367:3515–3524.

Barber J (2012) Photosystem II: the water-splitting enzyme of photosynthesis. *Cold Spring Harb Symp Quant Biol* 77:295–307.

Cramer WA, Zhang H, Yan J et al (2006) Transmembrane traffic in the cytochrome b_6f complex. *Annu Rev Biochem* 75:769–790.

Deisenhofer J, Epp O, Miki K et al (1985) Structure of the protein subunits in the photosynthetic reaction center of *Rhodopseudomonas viridis* at 3 Å resolution. *Nature* 318:618–624.

Fromme P (ed.) (2008) Photosynthetic Protein Complexes: A Structural Approach. Wiley VCH.

Grotjohann I & Fromme P (2005) Structure of cyanobacterial photosystem I. *Photosynth Res* 85:51–72.

Hofmann E, Wrench PM, Sharples FP et al (1996) Structural basis of light-harvesting by carotenoids: peridinin-chlorophyll-protein from *Amphidinium carterae*. *Science* 272:1788–1791.

Hohmann-Marriott MF & Blankenship RE (2011) Evolution of photosynthesis. *Annu Rev Plant Biol* 62:515–548.

Pagliano C, Nield J, Marsano F et al (2013) Proteomic characterization and three-dimensional electron microscopy of PSII–LHCII supercomplexes from higher plants. *Biochim Biophys Acta* 9:1454–1462 (DOI:10.1016/j.bbabio.2013.11.004).

Pagliano C, Saracco G & Barber J (2013) Structural, functional and auxiliary proteins of photosystem II. *Photosynth Res* 116:167–188.

Passarini F, Wientjes E, Hienewadel R & Croce R (2009) Molecular basis of light harvesting and photoprotection in CP24. Unique features of the most recent antenna complex. *J Biol Chem* 284:29536–29546.

Rochaix J-D, Lemeille S, Shapiguzov A et al (2012) Protein kinases and phosphatases involved in the acclimation of the photosynthetic apparatus to a changing light environment. *Phil Trans R Soc B* 367:3466–3474.

Ruban AV, Solovieva S, Lee PJ et al (2006) Plasticity in the composition of the light harvesting antenna of higher plants preserves structural integrity and biological function. *J Biol Chem* 281:14981–14990.

Standfuss J, van Scheltinga ACT, Lamborghini M & Kühlbrandt W (2005) Mechanisms of photoprotection and nonphotochemical quenching in pea light-harvesting complex at 2.5 Å resolution. *EMBO J* 24:919–928.

15.4 Electrochemical potential and the biosynthesis of ATP

Abrahams JP, Leslie AGW, Lutter R & Walker JE (1994) Structure at 2.8 Å resolution of F_1-ATPase from bovine heart mitochondria. *Nature* 370:621–668.

Adachi K, Oiwa K, Nishizaka T et al (2007) Coupling of rotation and catalysis in F_1-ATPase revealed by single-molecule imaging and manipulation. *Cell* 130:309–321.

Bowler MW, Montgomery MG, Leslie AG & Walker JE (2007) Ground state structure of F_1-ATPase from bovine heart mitochondria at 1.9 Å resolution. *J Biol Chem* 282:14238–14242.

Boyer PD (1997) The ATP synthase—a splendid molecular machine. *Annu Rev Biochem* 66:717–749.

Fillingame RH & Steed PR (2014) Half channels mediating H^+ transport and the mechanism of gating in the F_o sector of *Escherichia coli* F_1F_o ATP synthase. *Biochim Biophys Acta* 7:1063–1068.

Grüber G, Manimekalai MSS, Mayer F & Müller M (2014) ATP synthases from archaea: the beauty of a molecular motor. *Biochim Biophys Acta* 1837:940–952.

Hayashi S, Ueno H, Shaikh AR et al (2012) Molecular mechanism of ATP hydrolysis in F_1-ATPase revealed by molecular simulations and single-molecule observations. *J Am Chem Soc* 134:8447–8454.

Junge W, Sielaff H & Engelbrecht S (2009) Torque generation and elastic power transmission in the rotary F_oF_1-ATPase. *Nature* 459:364–370.

Mitchell P (1961) Coupling of phosphorylation to electron and hydrogen transfer by a chemi-osmotic type of mechanism. *Nature* 191:144–148.

Muench SP, Trinick J & Harrison MA (2011) Structural divergence of the rotary ATPases. *Quart Rev Biophys* 44:311–356.

Okuno D, Fujisawa R, Iino R et al (2008) Correlation between the conformational states of F_1-ATPase as determined from its crystal structure and single-molecule rotation. *Proc Natl Acad Sci USA* 105:20722–20727.

Pogoryelov D, Yu J, Meier T et al (2005) The c_{15} ring of the *Spirulina platensis* F-ATP synthase: F_1/F_o symmetry mismatch is not obligatory. *EMBO Rep* 6:1040–1044.

Strauss M, Hofhaus G, Schröder RR & Kühlbrandt W (2008) Dimer ribbons of ATP synthase shape the inner mitochondrial membrane. *EMBO J* 27:1154–1160.

von Ballmoos C, Wiedenmann A & Dimroth P (2009) Essentials for ATP synthesis by F_1F_o ATP synthases. *Annu Rev Biochem* 78:649–672.

Watt IN, Montgomery MG, Runswick MJ et al (2010) Bioenergetic cost of making an adenosine triphosphate molecule in animal mitochondria. *Proc Natl Acad Sci USA* 107:16823–16827.

Wehrle F, Kaim G & Dimroth P (2002) Molecular mechanism of the ATP synthase's F_o motor probed by mutational analyses of subunit a. *J Mol Biol* 322:369–381.

Yasuda R, Noji H, Yoshida M et al (2001) Resolution of distinct rotational substeps by submillisecond kinetic analysis of F_1-ATPase. *Nature* 410:898–904.

Membrane Channels and Transporters

16

16.1 INTRODUCTION

Biological membranes consist of lipid bilayers with proteins bound to or embedded in them. By itself, the bilayer is fluid, an electrical insulator, and almost totally impermeable to ions but mildly permeable to water and small uncharged molecules. The controlled movement of ions, water, and small molecules across the cell's membranes is essential for life, and these functions are carried out by specialized proteins. When ions (Na^+, K^+, Ca^{2+}, Mg^{2+}, Cl^-, HCO^{3-}, and HPO_4^{2-}) are selectively driven across a membrane, the resulting gradients can provide an energy source to transport nutrients into cells, to generate electrical signals, and to promote secretion. Regulation of water transport across a cell membrane maintains a constant osmotic pressure and governs the cell volume.

The proteins that enable ions and small molecules to cross the membrane barrier fall into two classes—channels and transporters. **Channels** have membrane-spanning water-filled pores through which ions or water (substrates) may diffuse down their electrochemical gradient. Many channels are gated and ion fluxes take place only when the gate is open. Each channel contains a **selectivity filter** in its pore where substrates bind with specific interactions. *Transporters* move ions and small molecules against their electrochemical gradient by coupling the transport process to an energy source such as ATP hydrolysis or the movement of another ion or substrate molecule down its concentration gradient. Transporters undergo cyclic conformational changes that are linked to substrate or ion binding. This allows access to the binding site from one side of the membrane, followed by a conformational change that leads to the substrate or ion being occluded from both sides of the membrane. A second conformational change delivers the substrate or ion to the other side of the membrane. This process is known as the **alternating access mechanism**.

There are differences in the rates at which the ions or molecules cross the membrane. The flux through ion channels is usually high, often close to the diffusion limit (typically 10^5–10^8/s) (**Box 16.1**). The flux through transporters is slower (typically 1–10^3/s). These rates reflect the differing energy barriers of the limiting steps in the two processes: low barriers to diffusion when all gates are open, and high barriers when conformational changes are required for alternating access mechanisms. This distinction is not absolute: there are also low-flux channels and high-flux transporters. For example, the CLC family of chloride channels (Section 16.2) includes both channels and Cl^-/H^+ exchange transporters.

In this chapter, we summarize the molecular architecture of selected channels (Sections 16.2–16.4) and transporters (Sections 16.5–16.6). We describe how selectivity is engineered, the structural basis for gating in ion channels, and the alternating access model in transporters.

16.2 ION CHANNELS

The flow of ions across the cell membrane is important in many biological processes. In the nervous system, **ion channels** are central to all fast communication, with voltage-gated channels propagating electrical signals along neuronal membranes, and ligand-gated channels mediating the transmission of electrical signals across the synapse. They also have key roles in muscle contraction, where the signal to contract is delivered through changes in electrical potential across the cell membrane, and in insulin secretion, where β cells in the pancreas are stimulated to release insulin by the closing of potassium channels. Ion

The ion flux (that is, the number of ions crossing per unit area per unit time) is proportional to the difference in ion concentration between the two sides of the membrane (Δc) and inversely proportional to the distance traveled (l). Thus the flux (J) may be calculated from the simple diffusion equation

$$J = D\Delta c/l$$

where D is the ion diffusion constant (typically $\sim 2 \times 10^9$ nm^2/s). Assuming a typical value for Δc (~ 100 mM or $\sim 6 \times 10^{-2}$ molecules/nm^3) and a step length of ~ 5 nm, then $J = 2 \times 10^7$ nm^{-2}/s. Given a pore diameter of 1.6 nm, wide enough to allow the passage of hydrated ions, the cross-sectional area is 2 nm^2, and hence the number of ions crossing per unit area is $\sim 4 \times 10^7$/s. This value is probably an upper limit, and diffusion will depend on the size of the pore and a decrease in D from restricted pore diffusion compared with diffusion in bulk solution.

The charge carried by a univalent ion is 1.602×10^{-19} coulombs (C). For current, 1 amp (A) = 1 C/s. Hence the current arising from the flow of 4×10^7 univalent ions such as K$^+$ or Na$^+$ per second is 4×10^7/s $\times 1.602 \times 10^{-19}$ A = 6.4 pA. If the ion concentrations on either side of the membrane are not very different, the current flowing through the open channel is given by Ohm's law:

$$I = V/R$$

where I is the current in amps, V is the voltage gradient across the membrane in volts, and R is the resistance in ohms. Conductance is the reciprocal of resistance, and its unit is the siemens (S) = 1/ohm. Hence a channel with a voltage difference of 100 mV and a current of 6.4 pA has a conductance of 64 pS.

When the concentrations of permeant ions are the same on both sides of the membrane, the membrane potential is zero. A change in membrane potential to a more positive value is known as depolarization and a change to a more negative value as hyperpolarization.

channels are **gated** so that they only open briefly, allowing ions to flow in response to physiological signals, and then close again. The principal stimuli that cause ion channels to open are a change in voltage across the membrane (**voltage-gated channels**), mechanical stress (**mechanically gated channels**), and binding of ligands (**ligand-gated channels**). Ligands include neurotransmitters, intracellular ions, G proteins (see Section 12.2), nucleotides, phosphoinositol 4,5-bisphosphate, and pH. Some representative ion channels are summarized in **Table 16.1**.

Ion channels have pores that are highly selective and gated

Ion and water channels contain selectivity filters to which the ion in question or water bind specifically. When an ion enters the filter it has to shed its hydration shell at a cost in free energy (for example, the solvation free energy for K$^+$ is ~ 320 kJ/mol). Groups that line the filter provide polar interactions that substitute for the water lost, decreasing the energy penalty from dehydration. These interacting groups provide selectivity for different ions, as described below.

For ion channels, a hydrophobic constriction in the pore at a separate region from the selectivity filter acts as a barrier to permeation. In gating, this constriction opens, allowing ions to flow. Molecular dynamics simulations have shown that a short hydrophobic pore is closed to water if it is narrower than 9 Å, and closed to K$^+$ if it is narrower than 13 Å. Thus, conduction can be blocked by hydrophobic gates with diameters appreciably larger than those of the diffusing molecules (2.8 Å for water; 2.66 Å for K$^+$, 1.90 Å for Na$^+$, augmented by hydration). Conformational changes are needed to open these gates.

The KcsA channel allows a rapid and highly selective flux of potassium ions

Na$^+$ and K$^+$ are the most abundant cations in biological systems. Outside the cell, Na$^+$ is present at high concentrations whereas K$^+$ is present at high concentrations inside the cell (**Table 16.2**). Gradients of these ions provide the energy source for action potentials generated by pore opening. K$^+$ channels belong to a single protein family found in all kingdoms of life. They have a conserved sequence motif, TV/IGYG, which forms the selectivity filter that prevents the passage of Na$^+$ but allows K$^+$ to flow at rates approaching the diffusion limit (up to 10^8 ions/s). The difference in atomic radius between K$^+$ and Na$^+$ is just 0.38 Å; nevertheless, K$^+$ channels manage to select for K$^+$ over Na$^+$ by a factor of more than 1000. Moreover, selectivity for K$^+$ is achieved without compromising the rate of conduction.

Table 16.1 Properties of some ion channels.		
Channel class	**Ion**	**Regulation**
[a]Voltage-gated K$^+$ channels (Kv)	K$^+$	Voltage
Ca^{2+}-activated K$^+$ channels (K$_{Ca}$)	K$^+$	Ca^{2+}
[a]Inward-rectifying K$^+$ channels (Kir)	K$^+$	Mg^{2+}, spermine, pH, ATP, GTP and G proteins, phosphorylation
Voltage-gated Na$^+$ channels (Nav)	Na$^+$	Voltage
Voltage-gated Ca^{2+} channels	Ca^{2+}	Wide range of cytoplasmic modulators including neurotransmitters and phosphorylation
[a]Cl$^-$ channels (transporters)	Cl$^-$	H$^+$
Cyclic-nucleotide-gated channels	Na$^+$ and K$^+$	cGMP, cAMP
Cystic fibrosis transmembrane conductance regulator (CFTR)	Cl$^-$	ATP
[a]Epithelial Na$^+$ channel (ENaC)	Na$^+$	cAMP, α-spectrin, pH, ATP
Ligand-gated Ca^{2+} channels	Ca^{2+}	
Ryanodine receptor		Ca^{2+}, ATP, cyclic ADP-ribose
Inositol 1,4,5-trisphosphate receptor		Inositol 1,4,5-trisphosphate, ATP
[b]Nicotinic acetylcholine receptors (nAChR)	Na$^+$	Acetylcholine
[a]Ionotropic glutamate receptors		
AMPA receptors	Na$^+$, K$^+$	Glutamate, AMPA
NMDA receptors	Na$^+$, K$^+$, Ca^{2+}	Glutamate, NMDA, external Mg^{2+}
Glycine receptors (GlyR)	Cl$^-$	Glycine, phosphorylation
[b]GABA receptors (GABA$_A$R)	Cl$^-$	GABA

[a]A representative of this class of ion channels is discussed in this Section or in Section 16.3[b]. (Data from F.M. Ashcroft, Ion Channels and Disease. San Diego: Academic Press, 2000.)

The first crystal structure of a K$^+$ channel, solved by Roderick MacKinnon and colleagues in 1998, was that of KcsA, a prokaryotic channel that is similar in sequence to several vertebrate and invertebrate K$^+$ channels. It comprises four identical subunits surrounding a central pore (**Figure 16.1A**). Each subunit has a transmembrane α helix (M1, the outer helix; that is, further from the pathway), a tilted pore helix (P) that spans half the membrane, followed by a loop and a second transmembrane helix (M2, the inner helix). The pore helix directs its partly negatively charged C-terminal end toward the ion pathway. The KcsA pore has a wide water-filled cavity facing the cytoplasm, which crosses about two-thirds of the bilayer, ending at the selectivity filter. The cavity allows ions to diffuse rapidly from the cytoplasm (where K$^+$ concentration is high) to the selectivity filter, which forms the remainder of the pathway (12 Å) across the lipid bilayer.

The selectivity filter is lined by four copies of the signature motif TVGYG. Transiting K$^+$ encounters four evenly spaced layers of carbonyl oxygen atoms and a single layer of Thr hydroxyl oxygen atoms, which create four binding sites (Figure 16.1B). At each of these sites, K$^+$ binds in a dehydrated state, surrounded by eight oxygen atoms from the protein, four 'above' and four 'below' each ion. This arrangement is very similar to that of water molecules around a hydrated K$^+$ (Figure 16.1C). The size of the cavity is a good fit for K$^+$, with a mean K$^+$–O distance of 2.8 Å. Thus, K$^+$ is readily able to diffuse into the selectivity filter.

Table 16.2 Ion concentrations and equilibrium potentials.

Ion	Extracellular concentration (mM)	Intracellular concentration (mM)	Equilibrium potential (mV)	Ionic radius (Å)
Na^+	135–145	12	+66	0.95
K^+	3.5–5	140	–93	1.33
Ca^{2+}	2.2–2.5	10^{-4} (free)	+135	1.14
Cl^-	115	2.5–50	–42	1.81
pH	7.37–7.42 (arterial)	7.1–7.2	–15	

Extracellular concentrations refer to the range typically found in human blood. Intracellular concentrations are given for a typical mammalian cell. The equilibrium potential (E) is calculated from E (volts) $= RT/zF \ln([X]_o/[X]_i)$, where R is the gas constant (8.314 J K^{-1} mol^{-1}), T is the temperature in kelvins, z is the valency (charge) of the ion, F is the Faraday constant (96,500 C mol^{-1}), and $[X]_o$ and $[X]_i$ are respectively the extracellular and intracellular concentrations of ion X. This equation can be simplified to E (mV) $= 58/z \log_{10}([X]_o/[X]_i)$. The resting potential in most cells is –60 to –100 mV.

(From F.M. Ashcroft, Ion Channels and Disease. San Diego: Academic Press, 2000.)

The key factor discriminating against Na^+ is the size and geometry of the filter. Sodium ions do not enter, even when present in vast excess. This is because Na^+ in its hydrated state is too large, and in its dehydrated state Na^+ prefers to be six-coordinated; it is therefore too small to make contact with the carbonyl oxygens and cannot overcome the energy barrier of dehydration.

High selectivity and high conduction rates present an apparent paradox. In principle, the precise coordination required for high selectivity could result in the ions' binding too tightly, slowing diffusion through the pore. However, the selectivity filter accommodates more than one ion, and electrostatic repulsion between closely spaced ions helps to decrease the affinity of each ion for its binding site. Across the four positions of the selectivity filter there are, on average, two K^+ present at any one time. They reside predominantly in two arrangements: 1 and 3 or 2 and 4 (Figure 16.1D). Molecular dynamics simulations suggest that as an ion enters the queue from one side of the filter, another ion exits from the opposite side. The direction of flow is determined by the electrochemical gradient for K^+.

Gating is mediated through extra domains attached to the pore unit

Diversity in the K^+ channel family is related mainly to differing gating domains attached to the conserved pore unit. Ligand-gated K^+ channels may have cytoplasmic or extracellular domains that bind ligands. Voltage-gated K^+ channels have integral membrane domains for sensing voltage differences.

KcsA was crystallized in a closed conformation in which the KcsA inner helices are straight and form a bundle near the cytoplasmic surface of the membrane (the activation gate) (Figure 16.1B). At this point the pore narrows and, lined with hydrophobic amino acids, creates a barrier to the flow of K^+. Another prokaryotic channel protein, MthK, a Ca^{2+}-gated K^+ channel, was crystallized in the presence of Ca^{2+} at concentrations that open its pore. In this open channel, the inner helices are bent at a hinge point where there is a conserved Gly residue (also present in KcsA), opening the central cavity to the cytoplasm (Figure 16.1E).

Inward rectifier K^+ (Kir) channels are regulated by cytoplasmic gating domains

Insight into how cytoplasmic domains regulate ion flux has come from studies of Kir channels. The term 'inward rectifier' indicates that, under physiological conditions, these channels support high flux rates for K^+ entering the cell when the membrane's voltage potential is negative (inside the cell), but near-zero flux rates for K^+ flowing out of the cell when the membrane voltage potential is zero or positive. This is because Mg^{2+} and polyamines, which

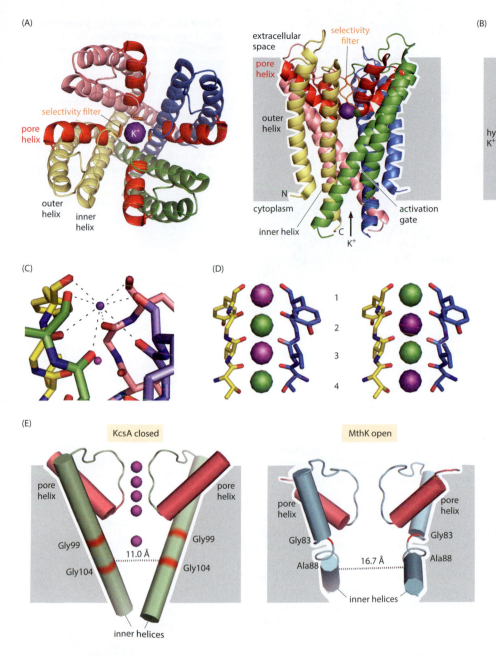

(A)

selectivity filter

pore helix

outer helix

inner helix

K⁺

extracellular space

selectivity filter

pore helix

outer helix

N

cytoplasm

inner helix

activation gate

C

K⁺

(B)

hydrated K⁺

(C)

(D)

1

2

3

4

(E)

KcsA closed

pore helix

pore helix

Gly99

Gly99

11.0 Å

Gly104

Gly104

inner helices

MthK open

pore helix

pore helix

Gly83

Gly83

Ala88

16.7 Å

Ala88

inner helices

Figure 16.1 The potassium channel KcsA.
(A) The channel viewed axially from the extracellular side (left) and in side view (right). This construct lacked the cytoplasmic domain. The four subunits are in different colors, except for the pore helices (red). (PDB 1K4C) (B) The K⁺ conduction pore, showing two of the four subunits. In the selectivity filter are four dehydrated K⁺ at positions 1–4 (external to internal) and there is a hydrated K⁺ below the filter. (C) The K⁺ at position 1 showing eight-coordination by main-chain carbonyl oxygens. (D) Density observed in the region of the selectivity filter is an average of two arrangements (left and right) of K⁺ (magenta) alternating with water (green). The TVGY motif is shown (bottom to top) for two subunits. (E) Gate opening occurs mainly through bending of the inner helix. The closed KcsA structure (residues 62–114, left) is compared with the open MthK structure (residues 46–96, right). Only two subunits are shown for clarity. The MthK inner helix is bent at Gly83, widening the gate to 16.7 Å at the narrowest point (Ala88). In KcsA, the corresponding distance is 11 Å at Gly104, which is too small to allow a flow of K⁺. Helices are shown as cylinders. (PDB 1K4C (KcsA) and 1LNQ (MthK)) (B, adapted from R. MacKinnon, *FEBS Lett.* 555:62–65, 2003. With permission from Elsevier.)

are abundant inside the cell, block the pore when the electrochemical gradient favors the outward flow of K⁺. Consequently, Kir channels operate near the resting membrane potential (–60 to –100 mV), allowing a flow of K⁺ into the cell, and are silenced when voltage-dependent Na⁺ or Ca²⁺ channels depolarize the membrane.

In mammals, Kir channels have several important roles: Kir1 channels regulate electrolyte flow across kidney epithelial cells; Kir2 channels control the resting potential that is important for electrical activity in muscle and neurons; Kir3 channels are responsible for heart rate control in cardiac cells; and Kir6 channels mediate insulin secretion from pancreatic β cells. Their cytoplasmic domains extend the length of the pore and provide the components responsible for gating by Mg²⁺ and polyamines and for modulation of activity by βγ G-protein subunits, phosphatidylinositol 4,5-bisphosphate, ATP, Na⁺, and pH.

In the Kir3.1 channel that was crystallized, three-quarters of the mammalian pore domain was replaced with the pore of a prokaryotic channel (KirBac3.1), whereas the cytoplasmic domain and membrane interfacial regions came from mouse Kir3.1 (99% identical to the human protein). The resulting structure had two Kir3.1 channels per asymmetric unit that were identical in their pore regions but differed in their cytoplasmic domains, one of which

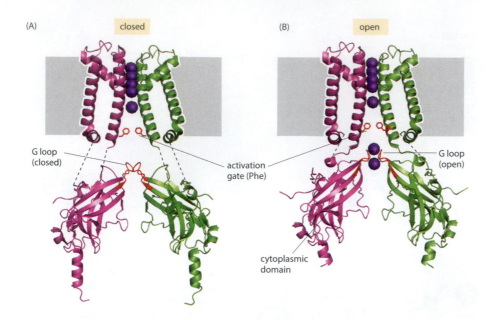

(A) closed   (B) open

G loop (closed)

activation gate (Phe)

G loop (open)

cytoplasmic domain

Figure 16.2 The conduction pore of the Kir3.1–prokaryotic chimera. Only two of the four subunits are shown (in green and magenta), for clarity. Breaks in electron density are marked by dashes. (A) Closed conformation. The constriction site in the cytoplasmic domain is created by the four G loops (residues 302–309). (B) Open conformation. Sites 1–7 (numbered from the top) at which Rb$^+$ (as a mimic for K$^+$) binds are shown in purple. The sites are probably not all occupied simultaneously. (PDB 2QKS)

was designated to be open and the other closed (**Figure 16.2**). Soaking experiments with Rb$^+$ as a substitute for K$^+$ showed seven sites for the open state. Four ions occupied the selectivity filter and a fifth occupied the central cavity, as in KcsA. Two ions positioned in sites 6 and 7 in the cytoplasmic pore were unique to the open state.

The pore has two constrictions. One corresponds to the inner helix gate where four hydrophobic residues (Phe for Kir3.1) come together, as described above for KcsA. The other is in the cytoplasmic domain where the four subunits associate in the region of the G (gating) loops. In the closed state, the G-loop residue Met308 blocks the pore. In the open state, a change in conformation of residues Thr306 and Gly307 exposes oxygens to line the pore, creating a pathway wide enough to allow passage of a hydrated K$^+$ (Figure 16.2B).

Binding studies with the Kir2.2 channel from chicken, comparing Rb$^+$ and Sr^{2+} as mimics of monovalent and divalent ions respectively, provided a structural basis for inward rectification. Sr^{2+} bound at two major sites, close to sites 6 and 7 in the Kir3.1 structure (Figure 16.2B). Each site contains a ring of acidic amino acids and, although the diameters of the sites are just too wide (~9 Å) to allow direct coordination of Rb$^+$, ions at the center can engage through strong electrostatic interactions. Thus, conducting ions (K$^+$ or Rb$^+$) compete with blocking ions (Mg^{2+} or Sr^{2+}) for the same sites in the cytoplasmic domain pore. When extracellular K$^+$ concentrations are high, K$^+$ can flow in and occupy these sites, but a depolarizing voltage will drive the blocking ions into the pore from the cytoplasm to replace the conducting K$^+$.

Mutations in Kir channels or defects in their regulation result in neuronal degeneration, failure of salt absorption, and defective insulin secretion. Some of these mutations are in the G loop or the pore loop, and all affect the ability of the gate to be open or closed.

Voltage gating requires a movement of charge through the membrane

In resting nerve cells, the electrical potential is negative inside the cell relative to outside, typically by –80 mV. When a cell becomes depolarized (less negative inside), voltage-activated ion channels sense the potential across its membrane and react by opening or closing. Voltage-gated channels exhibit a nearly switch-like dependence on membrane voltage in changing from 'off' to 'on.' Crystal structures of voltage-gated channels have provided insight into how Nature uses electric charge to open channels.

Voltage-gated K$^+$ (Kv) channels are composed of an α-subunit ion channel (of which there are 12 different classes) and a nonconducting β subunit. In 1987, a mutation in *Drosophila melanogaster* was found to cause the insect to shake uncontrollably when exposed to ether (used as an anesthetic when examining flies). The defect was traced to a gene coding for a K$^+$ channel that became named the 'shaker' ion channel. In the mid-1990s, a part of the

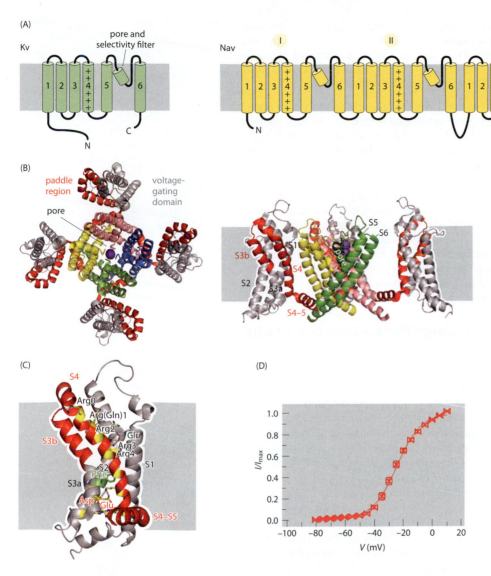

Figure 16.3 Voltage-gated channels: regulation by a transmembrane domain. (A) Membrane topology of the α subunits for the Kv and Nav channels. (B) The chimeric K⁺ channel Kv1.2–Kv2.1–paddle in the open conformation viewed axially (left) and in side view (right). Each pore domain is in a different color, with K⁺ in the pore (purple ball). In the right-hand panel, one subunit has been removed for clarity. The inner helix S6 is bent in this open conformation. (PDB 2R9R) (C) The voltage sensor domain of the Kv1.2–Kv2.1–paddle chimera. Side chains of positively charged residues on S4 (Arg3 and Arg4) form hydrogen bonds with negatively charged residues. (Arg1 is replaced by a Gln in Kv2.1.) The Phe (in green) lines an occluded binding site for Lys5 (not visible) that interacts with Asp and Glu (in red). (D) Voltage activation curve for the wild-type (Phe233) shaker channel. Measurements of the fraction of maximum activatable current (I/I_{max}) are plotted as a function of the depolarization voltage (V). (Adapted from X. Tao et al., *Science* 328:67–73, 2010. With permission from AAAS.)

channel termed the S4 region was identified as a likely source of charge sensitivity because it has a motif of five Arg or Lys residues, each separated by three residues. Mutational studies and measurements of ion currents allowed an estimate of total gating charge to be converted into a charge per channel. In the shaker Kvα1.2 channel, the charge transferred across the membrane in each cycle corresponded to 12–14 elementary charges per channel, or three to four charges through each voltage-sensing subunit.

Voltage-gated Na⁺ (Nav) channels also have a pore-forming α subunit and a β subunit. Whereas Kv α subunits have six transmembrane helices and form a tetramer to create the channel, the Nav α subunit (~2000 aa) comprises four homologous domains (I–IV), each with six transmembrane helices (**Figure 16.3A**). It is expected that these four domains form the pore from their helices S5 and S6 (equivalent to the outer and inner helices of KcsA) and the connecting pore helices.

The crystal structure of the Kv channel Kv1.2 (a chimera in which the so-called S4 paddle region was replaced by that of Kv2.1) showed an open conformation. The channel is composed of four subunits, each with a characteristic ion-conducting pore domain, linked to an N-terminal voltage-sensor domain (Figure 16.3B). The voltage-sensor domains, each composed of four helices, surround the pore. Such domains are also found in other channels and are conserved from archaea to mammals. When spliced to KcsA, a voltage-sensor domain can confer voltage sensitivity for the conduction of K⁺. These domains are able to provide voltage regulation simply by being coupled to a pore domain.

The voltage-sensing 'paddle' region comprises helices S3b and S4, where the positively charged amino acids reside (Figure 16.3C). Labeling studies, in which biotin was attached at several positions on S4 and its accessibility to avidin (its binding partner) was tested from both sides of the membrane, found that accessibility depended on whether the channel was in the open or the closed conformation. From these and other labeling and mutagenesis studies, it was concluded—somewhat surprisingly—that the charged amino acids move 15–20 Å across the membrane when the channel switches from closed to open.

The positively charged residues are shown as Arg0, Arg1, Arg2, Arg3, and Arg4 in Figure 16.3C. In the open state, Arg0, Arg1, and Arg2 are positioned to interact favorably with the lipid head-groups and extracellular water, while Arg3 and Arg4 form H-bonds with negatively charged residues within the membrane. Lys5 (hidden in the view shown in Figure 16.3C) is shielded from the aqueous surface by the side chain of the conserved Phe233. In the open conformation, Lys5 forms ion pairs with Glu and Asp residues on helix S2 (Figure 16.3C). It is proposed that in the open conformation, Lys5 occupies the occluded Glu/Asp-binding site, but in the fully closed conformation, Arg1 moves into the occluded site, a shift of 21 Å along the S4 helix and 18 Å perpendicular to the membrane. In the closed conformation, all five charged residues reach a cytoplasm-accessible location, resulting in the displacement of three to four charges per voltage sensor. These studies imply that hydrophobic and electrostatic forces can balance each other so that the delocalized charge on an Arg residue can be drawn into the membrane at the protein/lipid interface. Figure 16.3D shows a plot of current against voltage that indicates that the pore opens between −40 and +20 mV.

The glutamate receptor combines a large extracellular region with a conventional K$^+$-selective pore

In the nervous system, cell-to-cell signaling occurs at specialized synapses where information is transferred by *neurotransmitters*. Glutamate is an excitatory neurotransmitter that binds to and activates ionotropic glutamate receptors. (Ionotropic receptors are ligand-gated ion channels and distinct from metabotropic receptors, which operate by activating a second messenger cascade.) These receptors have been implicated in most aspects of the development and function of the nervous system. Traditionally they have been classified, according to their preferred agonist, into AMPA (α-amino-3-hydroxy-5-methyl-4-isoxazole propionic acid), kainite, and NMDA (*N*-methyl-D-aspartate) subtypes. (An agonist binds to a receptor, inducing a response; an antagonist blocks the activity of an agonist). The structure of the rat AMPA-subtype glutamate receptor, composed of four three-domain GluA2 subunits, is shown in **Figure 16.4A**. On the extracellular side are the amino-terminal domain (ATD) and the ligand-binding domain (LBD), followed by the transmembrane domain (TMD), which forms the pore. An unusual feature of this structure is the way in which the extracellular domains interact. At the ATD region, subunit A interacts with subunit B (and C with D), whereas at the LBD region, subunit A interacts mostly with subunit D (and C with B); in the membrane, the TMDs associate with 4-fold symmetry.

The channel was crystallized in the presence of an antagonist that binds at the glutamate site in the LBD, and so the channel is in its closed conformation. In the TMD, the three helices M1, M2, and M3 share structural similarity with the K$^+$ channels (for example KcsA), where M1 is equivalent to the outer helix, M2 to the pore helix, and M3 to the inner helix (see Figure 16.1A). However, in the AMPA-receptor channel (Figure 16.4B), the arrangement is upside down relative to KcsA: its transmembrane region starts with the N terminus of M1 and ends with the C terminus of M3 both in the extracellular space, whereas in KcsA the outer helix M1 starts and the inner helix M3 ends in the cytoplasm. Thus, in the AMPA receptor, the selectivity filter (disordered in the crystal structure) is closer to the cytoplasm. After the M3 helix, the polypeptide chain completes part of the extracellular domain and then ends with a fourth transmembrane helix, M4, which associates with the TMD of an adjacent subunit. Comparison of isolated LBD structures with bound agonists and antagonists suggests that the linkers between the TMD and the LBD are key to how the binding of an agonist to the LBD induces conformational changes that result in separation of the linkers and opening of the gate.

The trimeric acid-sensing ion channel (ASIC) imposes Na$^+$ selectivity by its pore architecture

ASICs are Na$^+$-selective, voltage-independent, ligand-gated ion channels that give a highly cooperative response to changes in pH, switching from closed to open over a narrow pH

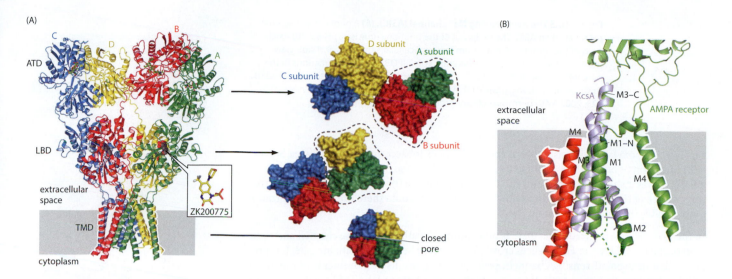

Figure 16.4 Regulation of the AMPA glutamate receptor by extracellular domains. (A) The receptor is shown in side view (left) and top views at three levels (right). The competitive antagonist ZK200775 bound in the LBD domain is shown as gray spheres; its structure is shown in the inset (yellow, carbon; blue, nitrogen; red, oxygen; gray, fluorine; orange, phosphorus). The top views illustrate the switch in oligomerization pattern between the LBDs and the ATDs. (PDB 3KGC)
(B) The transmembrane helices M1, M2, M3 (residues 521–586 with breaks between M1 and M2 and between M2 and M3 shown as dashes) and M4 (residues 789–817) for the A subunit in green and the B subunit in red. Part of the green subunit LBD domain is also shown. The view is similar to that in panel A. Part of a KcsA subunit in gray is superimposed on the green subunit to show the close correspondence between their triplets of membrane helices, although with the AMPA receptor inverted with respect to KcsA. (Thus helix M1 of AMPA appears as gray, not green). The selectivity filter between M2 and M3 is disordered in the AMPA channel structure (dashed region). (A, adapted from A.I. Slobolevsky et al., *Nature* 462:745–756, 2009. With permission from Macmillan Publishers Ltd.)

range (6.8–6.6) (**Figure 16.5A**). These trimeric proteins are present in multicellular eukaryotes and are implicated in the perception of pain, mechanosensing, learning, memory, and strokes due to blood clots. ASICs are related in sequence to the epithelial Na^+ channels (ENaCs) (Table 16.1).

The chicken ASIC1 channel was crystallized at pH 6.5 in a closed state despite the presence of a bound ligand, a state known as **desensitized** (Figure 16.5B). Each subunit has a large extracellular domain connected by a 'wrist' region to two transmembrane helices (TM1 and TM2), followed by a disordered cytoplasmic region. The extracellular domain is funnel-shaped but there is no continuous pore along the 3-fold axis. Openings near the extracellular membrane surface at the 'wrist' region provide a possible route for ions to access the pore.

The pore is mostly lined with residues from TM2. It has an hourglass shape, being ~8 Å wide at the extracellular side, ~15 Å wide at the cytoplasmic side, and narrowing to a constriction created from three symmetry-related Asp residues (Asp433). To locate the binding sites for monovalent cations, the crystals were soaked in solutions containing Cs^+, which is heavier than Na^+ and thus easier to detect. The difference in ionic radii (0.95 Å for Na^+, versus 1.69 Å for Cs^+) means that the positions of cesium ions do not define the Na^+ site unambiguously but they are likely to be important because the channel is able to conduct Cs^+ weakly. Cs^+ binds at two sites within the pore, both located on the 3-fold symmetry axis (Figure 16.5C). The sites are 3.4 Å apart and are probably not occupied simultaneously. At site 1, side-chain oxygens from Asp433 coordinate the Cs^+. At site 2, the Cs^+ is coordinated by three side-chain oxygens from Asp433 and three main-chain oxygens from Gly432. These six interactions provide a perfect octahedral arrangement to coordinate a monovalent cation.

It is expected that a pH-sensitive channel would use amino acid side chains that change their state of ionization in the pH range over which the channel is sensitive. From the ASIC structure, possible proton-binding sites were identified, including several acidic pairs (Asp/Asp or Glu/Asp) in the extracellular domain. The pK_a of Asp and Glu carboxyl groups is usually about 5; however, if two are forced together at neutral pH, one is protonated and the other remains charged; that is, they form a carboxyl/carboxylate pair. (It is noteworthy that

(A)

Figure 16.5 The acid-sensing Na⁺ channel (ASIC). (A) Plot of current against pH for chicken ASIC. The midpoint of the titration curve is at pH 6.7. (B) Axial view of the trimeric channel from the extracellular side (top), and side view (bottom). Each subunit is in a different color. (PDB 3IJ4) (C) Interactions in the pore with Cs⁺ at sites 1 and 2. Cs⁺ is shown with smaller than actual ionic radius, to show interactions. (PDB 3IJ4) (A, adapted from J. Jasti et al., *Nature* 449:316–323, 2007. With permission from Macmillan Publishers Ltd.)

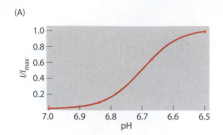

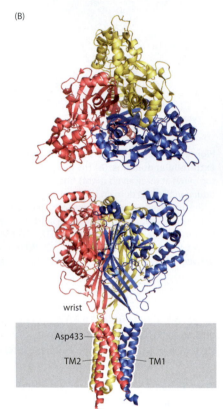

carboxyl/carboxylate pairs also govern the assembly of tobacco mosaic virus—see Section 8.3.) Mutation of one of these residues to Asn led to a decrease in pH sensitivity of the channel: there was a shift to lower pH for full opening (pH 6.2–6.0), consistent with the inferred importance of these pairs. These and other acidic residues could therefore act as the proton-binding sites that make this channel acid-sensitive.

In the low-pH desensitized state, there is no continuous path through the channel and there is an occlusion where the TM2 helices cross. This seems to be the desensitizing gate. It is proposed that in the acid-sensing channels, proton binding results in movements of the extracellular domains that open the activation gate through rotation of the TM2 helices, allowing the pore to open and Asp433 and carbonyl oxygens to participate in ion conduction.

Chloride channels (CLCs) have conserved pore regions and diverse extracellular and cytoplasmic regions

CLCs form a large family of membrane proteins found in both prokaryotes and eukaryotes. In eukaryotes, their functions include: stabilization of the resting membrane potential and regulation of excitability in skeletal muscle; regulation of fluid and electrolyte transport in the kidney; and regulation of pH in intracellular compartments through coupled H⁺/Cl⁻ exchange. CLC proteins may be channels or transporters. As channels they allow Cl⁻ to diffuse down an electrochemical gradient, and as transporters they couple Cl⁻ movement in one direction to H⁺ movement in the opposite direction at a stoichiometry of two Cl⁻ per H⁺. The flux rate (10^5 s⁻¹) of the *Escherichia coli* CLC transporter is low for a channel but high for a transporter, suggesting that conformational changes between states are small and readily accomplished. The sequences of CLCs are conserved in their integral membrane regions, implying that the channels and transporters are likely to have similar structures.

Cl⁻ is the most abundant halide ion in biological systems. CLCs are faced with the modest challenge of selecting Cl⁻ over other negatively charged ions such as the larger PO_4^{2-}. Single channel analysis (of patch-clamp records) of the CLC-0 channel of the *Torpedo* ray electric organ indicated that it contains two pores. This finding has been supported by EM and crystallographic studies. The two pores act independently, but in some channels they turn on and off simultaneously.

Two prokaryotic CLC transporters and a eukaryotic CLC transporter have similar transmembrane architectures but differ in their extracellular and cytoplasmic regions. The prokaryotic CLC transporters are homodimers with a 2-fold axis of symmetry perpendicular to the membrane (**Figure 16.6A**). Each subunit has its own pore and selectivity filter, consistent with the electrophysiological observations, and contains 18 α helices (labeled A to R) with a pseudo-2-fold axis relating helices B–I and J–Q (Figure 16.6B). This axis lies in the plane of the membrane so that the second pseudo-repeat is inverted relative to the first one.

The Cl⁻ selectivity filter is located at the neck of an hourglass-shaped ion pathway near the middle of the membrane. At its narrowest point, conserved groups from four different parts of the protein surround a Cl⁻ and share their protons to neutralize the ion's charge (Figure 16.6C). The hydroxyl from Tyr445 shares its proton because the aromatic ring is able to delocalize the excess negative charge left on the oxygen atom; the Ser107 hydroxyl shares its proton because the oxygen is H-bonded to a main-chain amide; and finally, main-chain amides direct their protons toward the Cl⁻. The Cl⁻-binding site is at the N-terminal ends of α helices. Thus, the partial positive charge of the helix dipole is directed to the anion-binding site.

Just above the selectivity filter on the extracellular side, the pore is blocked by Glu148 (Figure 16.6C), the residue thought to regulate gating. Electrophysiological studies of CLC-0

(B)

(C)

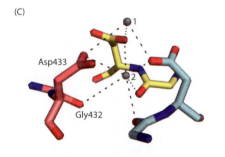

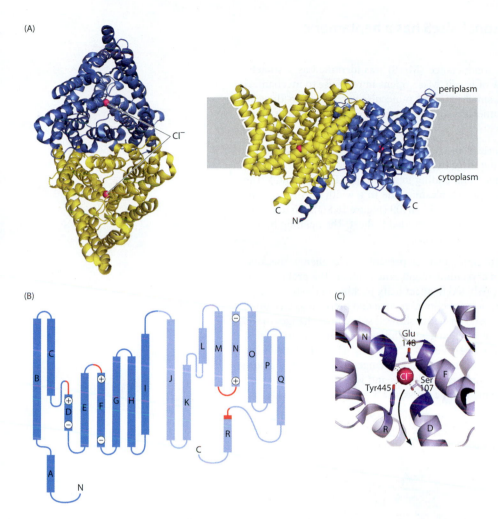

Figure 16.6 The CLC chloride channel/transporter. (A) The prokaryotic CLC dimer viewed axially from the extracellular side (left) and in side view (right). (PDB 1OTS) (B) Schematic diagram of one CLC subunit showing the arrangement of α helices and the internal pseudo-2-fold symmetry. The two halves are shown in different shades of blue. Regions forming the Cl⁻ pore are in red. (C) Close-up view of the Cl⁻-binding site. The polar groups of Ser107, Tyr445, and main-chain nitrogens of Ile356 and Phe357 at the start of helix N contact the Cl⁻. Access and exit routes are shown with curved arrows. In the closed state, Glu148 blocks access. The N-terminal ends of α helices D, F, and N are directed toward the anion-binding site and are colored dark blue. (PDB 1OTS) (B, from R. Dutzler et al., *Nature* 415:287–294, 2002. With permission from Macmillan Publishers Ltd.)

mutants Glu148Ala and Glu148Gln showed the gate to be open. Crystallographic analysis of the corresponding mutants in *E. coli* CLC showed that the side chain is displaced and an additional Cl⁻ replaces its carboxylate group of the Glu. Hence it was deduced that when the Glu148 carboxyl is ionized, the pore is closed; when the carboxylate is protonated, the pore opens, allowing Cl⁻ to pass.

Studies of eukaryotic and prokaryotic transporters gave differing views of Cl⁻ occupancy in the pore. In the eukaryotic CLC, the gating Glu occupied the central site, and two halogen ions (Br⁻ was used) were located at internal and external sites, ~10 Å apart (**Figure 16.7**). In the Glu148Gln mutant *E. coli* CLC, three Cl⁻ were observed bound at internal, central, and external sites (internal to central, 7 Å; central to external, 4 Å), while in the wild-type protein, two Cl⁻ were observed at the internal and central sites. These structures captured snapshots along the ion conduction pathway.

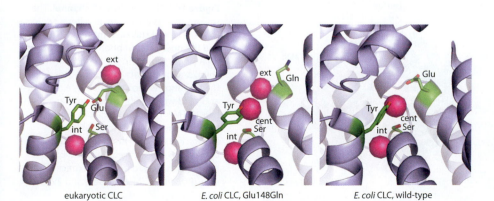

eukaryotic CLC | *E. coli* CLC, Glu148Gln | *E. coli* CLC, wild-type

Figure 16.7 Chloride channel pores (CLC) in three different states. Left: a eukaryotic CLC (from a thermophilic red alga) with two 'Cl⁻' (pink spheres) bound at the external ('ext') and internal ('int') sites and Glu in the position of the central site (labeled 'cent' in the middle panel). (PDB 3ORG) Middle: the *E. coli* Glu148Gln mutant CLC has the Gln external (mimicking a protonated Glu) and all three Cl⁻ sites occupied. (PDB 1OTU) Right: the *E. coli* wild-type protein has the ionized Glu occupying the external site and the central and internal sites occupied by Cl⁻. (PDB 1OTS)

The mechanosensitive ion channel MscS has a heptameric structure

The mechanosensitive channel of small conductance (MscS) was identified as a stretch-activated channel in the inner membrane of *E. coli*. MscS opens in response to mechanical stress applied to the membrane, so that small ions and molecules can exit from the cell, preserving the osmotic pressure. The channel is relatively unselective but has a slight preference for anions.

The channel structure was first characterized in its closed state. It has 7-fold symmetry (**Figure 16.8A,D**) with each subunit having three transmembrane helices (TM1, TM2, and TM3) followed by a cytoplasmic αβ domain and ending in a β strand (Figure 16.8C). The TM3 helices line a pore that narrows on the cytoplasmic side to a width of 5 Å, as constricted by the hydrophobic side chains of Leu105 and Leu109 (Figure 16.8D). On the cytoplasmic side, the C-terminal β strands form a seven-stranded β sheet. The opening here is only 3.2 Å across, further obstructing the passage of ions.

Obtaining crystals of MscS in an open conformation presented a challenge, because mechanical stress is difficult to apply to a crystalline membrane protein. The problem was solved by creating a mutant, Ala106Val, predicted to affect helix packing in the transmembrane domain (Figure 16.8B). The mutant was functional *in vivo* and behaved as a channel in which the open state was favored. Its crystal structure showed substantial changes in the packing of the transmembrane helices but no changes in the cytoplasmic region (Figure 16.8C). Relative to the closed state, helices TM1 and TM2 rotate by 45° and change their tilt. TM3a pivots around Gly113, whereas TM3b remains unchanged. As a result, the pore diameter increases to 13 Å, just wide enough to conduct ions and water molecules (Figure 16.8E). It is suggested that, in response to membrane stress, the TM1 and TM2 helices shift

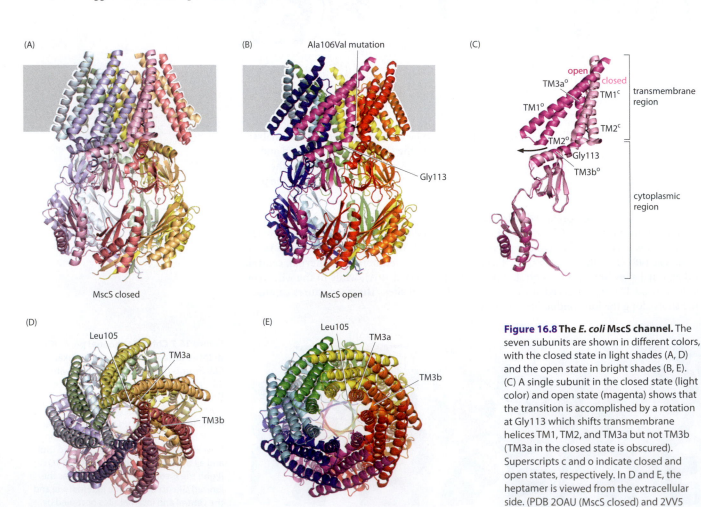

(A) MscS closed

(B) Ala106Val mutation — Gly113 — MscS open

(C) open TM3aᵒ / closed TM1ᶜ — transmembrane region; TM1ᵒ; TM2ᶜ; TM2ᵒ; Gly113; TM3bᵒ — cytoplasmic region

(D) Leu105 — TM3a — TM3b — MscS closed

(E) Leu105 — TM3a — TM3b — MscS open

Figure 16.8 The *E. coli* MscS channel. The seven subunits are shown in different colors, with the closed state in light shades (A, D) and the open state in bright shades (B, E). (C) A single subunit in the closed state (light color) and open state (magenta) shows that the transition is accomplished by a rotation at Gly113 which shifts transmembrane helices TM1, TM2, and TM3a but not TM3b (TM3a in the closed state is obscured). Superscripts c and o indicate closed and open states, respectively. In D and E, the heptamer is viewed from the extracellular side. (PDB 2OAU (MscS closed) and 2VV5 (MscS open))

to a more buried state, taking TM3a with them in a pore-opening movement. Movement of helices is a frequently observed mechanism for pore opening in membrane transporters (Section 16.5).

16.3 NICOTINIC ACETYLCHOLINE RECEPTOR

Nerve cells communicate with their neighbors through specialized junctions called synapses (**Figure 16.9**). The arrival of an electrical impulse (an **action potential**) at the presynaptic terminal triggers the release of **neurotransmitter** molecules, which diffuse across the synaptic cleft (a distance of 20–30 nm) and bind to ligand-gated ion channels in the postsynaptic membrane of the target cell. The channels respond by opening an ion-conducting pathway through that membrane, thereby changing its potential. The channels exist in a ligand-free closed state and a ligand-bound open state. The acetylcholine receptor translates a chemical signal (acetylcholine, ACh) into an electrical signal (ion flow). For millions of years, survival of animals against predators has depended on the robustness and speed of this receptor.

The neurotransmitter ACh (146 Da) binds to nicotinic ACh receptors exposed on the surface of nerve or muscle cells. These receptors can also be activated by binding nicotine and are different from the muscarinic ACh receptors found in heart muscle, which regulate the rate and strength of muscle contraction in response to ACh binding and can also be activated by the fungal alkaloid muscarine. The muscle nicotinic ACh receptor was the first ion channel to be solubilized from membranes and characterized biochemically. Its function has been studied by biophysical measurements of purified ion channels inserted into synthetic membranes. The electric organ of the ray *Torpedo* has had a vital role in many of these studies. The muscle-derived cells from this organ (electrocytes) contain extensive postsynaptic membranes, densely packed with ACh receptors (~20,000 μm^{-2}). Vesicles rich in ACh receptors can be extracted simply and quickly from these cells, and detergent-solubilized receptors can be purified in high yield.

The ACh receptor is highly selective for cations

These proteins belong to the 'Cys-loop' family of ion channels, which can be cation- or anion-selective. The family also includes the 5HT$_3$ (5-hydroxytryptamine or serotonin), GABA$_A$ (γ-aminobutyric acid) and glycine receptors. All are pentamers of homologous (and sometimes identical) subunits arranged around a central ion pathway. Each subunit has a 13-aa loop flanked by a conserved motif—the Cys-loop, which gives the family its name. Each family member can draw on a large pool of similar subunits (more than 16 for ACh receptors). Variations in subunit type provide a wealth of functional diversity, presumably to meet a wide range of physiological needs.

The muscle ACh receptor is assembled from two copies of the α subunit and one each of the β, γ, and δ subunits. The overall structure, in the closed state, has been determined from cryo-EM (see **Methods**) images of *Torpedo* postsynaptic membranes. The isolated membranes readily form tubular vesicles with receptors arranged on a helical lattice (**Figure 16.10**).

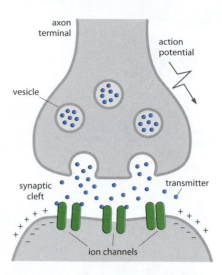

Figure 16.9 Arrival of an action potential at an axon terminal. When the action potential arrives, synaptic vesicles fuse with the nerve terminal, releasing neurotransmitters into the synaptic cleft. These molecules diffuse rapidly across the cleft and activate ligand-gated ion channels in the postsynaptic membrane of the target cell.

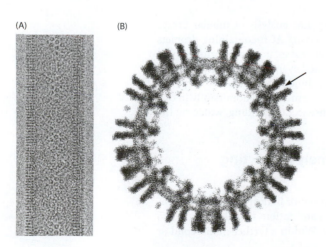

Figure 16.10 Acetylcholine receptor arrays. (A) Cryo-EM image of a tubular vesicle with ACh receptors packed on a helical lattice. The diameter of the tube is 78 nm. (B) Cross section of the tube, calculated by image reconstruction. The receptor protein projects from either side of the membrane, for which the two leaflets are resolved. The arrow points to the axial channel of a single centrally sectioned receptor. (Courtesy of Nigel Unwin.)

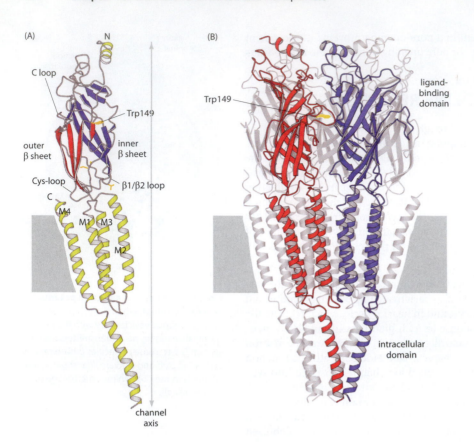

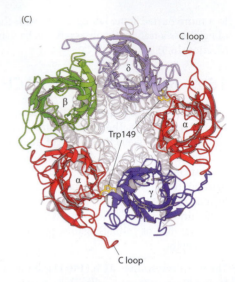

Figure 16.11 The nicotinic acetylcholine receptor. (A) Side view of a single α subunit. The β-sheet core of the extracellular ligand-binding domain is in red (outer β sheet) and blue (inner β sheet). There is a break in the electron density between the end of helix M3 and the start of the intracellular helix before the C-terminal helix M4. (B) Side view of the whole receptor. One α subunit is in red and the γ subunit in blue; other subunits are in gray. One ACh ligand-binding site (marked by Trp149) lies between the α and γ subunits. (C) Axial view from the extracellular side. (PDB 2BG9)

The helical symmetry presents views of the receptor from all directions, enabling it to be reconstructed in three dimensions. The local packing is similar to that found *in vivo*.

The subunits are long rods with three parts: an N-terminal extracellular portion organized around a β-sandwich core; a membrane-spanning portion composed of four α helices; and a single intracellular α helix (**Figure 16.11A**). Together, the five subunits make a ~160 Å long, 290 kDa tube with three functionally distinct regions: one containing the two ACh-binding sites; a membrane-spanning domain containing the channel gate; and an intracellular domain, which interacts with proteins that help to localize the receptor to the synapse (Figure 16.11B).

The ion-conducting pathway consists of a narrow pore with large open vestibules on either side. The vestibules have charged groups on their inner walls, which influence ion selectivity. The ACh receptor is cation-selective and when open allows only small cations (Na^+ and K^+) to pass. The charged groups are mostly negative and located strategically on the helices framing the lateral openings in the intracellular domain. Unlike the situation with K^+ channels (Section 16.2), the pore itself does not have a major role in ion discrimination, but contributes by restricting the size of the permeant ion.

Insight into the gating mechanism has been obtained by EM analysis of tubular arrays in which receptors were captured in the open state by spraying ACh-containing droplets on to tube-bearing microscope grids, ~5 ms before near-instantaneous freezing. These experiments showed that binding of ACh initiates two related events in the ligand-binding domain. One is a local disturbance in the region of the ACh-binding sites, and the other is a larger-scale conformational change, involving rotational movements, predominantly in one of the α subunits. The inner M2 helices also shift in response to ACh binding, widening the pore at the middle of the membrane.

The ACh receptor has evolved to be extremely fast-acting and efficient

ACh released into the synaptic cleft diffuses rapidly to its receptor in the postsynaptic membrane. Within ~10 μs of ligand binding, the channel opens and cations flow through, down their electrochemical gradients (mainly an influx of Na^+). Within a further millisecond or so, acetylcholinesterase molecules resident in the synaptic cleft deplete ACh, the bound ACh dissociates from the receptor, and the channel closes. The speed of these opening and closing events is crucial in achieving rapid initiation and termination of the synaptic response, so that closely spaced impulses are communicated with high fidelity.

After the release process triggered by a single nerve impulse, the local concentration of ACh in the synaptic cleft rises briefly to ~0.5 mM. The resulting opening of many channels depolarizes the membrane from its resting potential of ~−90 mV to ~0 mV, allowing an action potential to be generated, which causes the muscle to contract. It is important that the channels have a low probability of opening when the ACh concentration is low, and a high probability of opening when ACh is present in saturating amounts (more than 100 μM).

Electrophysiological recordings at the single-channel level have indicated how channels respond to ACh and other agonists. Most characteristically, they exhibit an all-or-none response. The channel is normally closed. If it is open, it conducts ions swiftly (~2.5×10^7/s; current 4 pA) and at the same rate, whatever concentration of ACh (or other agonist) is used (**Figure 16.12**). When ACh is present in saturating amounts (so that both binding sites are occupied), the probability of the channel being open approaches unity (~0.98). These observations point to the existence of two distinct conformations, closed and open, with ACh acting to shift the equilibrium between the two. Evidently, this receptor is an off–on switch that has been fine-tuned through evolution.

The two ligand-binding sites have different affinities

Both sites are ~35 Å from the membrane surface, at the interfaces of the α subunits with their γ and δ neighbors. Crystal structures of an ACh-binding protein (AChBP)—a homopentameric homolog of the ligand-binding domain—have shown that each site is in an aromatic cleft shaped largely by a projecting loop (C loop in Figure 16.11C) and the body of the α subunit (**Figure 16.13A**). Furthermore, a structure of AChBP complexed with carbamylcholine (an ACh analog, Figure 16.13B) has revealed how bound ACh is coordinated with the surrounding side chains. The highly conserved Trp149 (Figure 16.11B,C) and two Tyr residues on loops A, B, and C of the α-subunit-equivalent contact the bound ligand (Figure 16.13A). Several more variable residues on the adjacent subunit (corresponding to γ or δ of the receptor) also participate in these interactions. The AChBP/carbamylcholine complex is thought to resemble a nonconducting desensitized conformation, as would be obtained with receptors subjected for long periods (more than 50 ms) to ACh. However, desensitization is of little consequence for normal neuromuscular transmission, in which the channels are exposed to ACh for no more than a few milliseconds

Although detailed structures of the receptor binding sites in the closed- and open-channel forms are not yet available, estimates have been obtained of their affinities for ACh. This is important in terms of the activation mechanism, a central tenet of allosteric theory

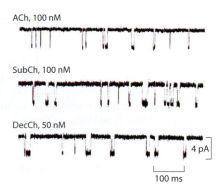

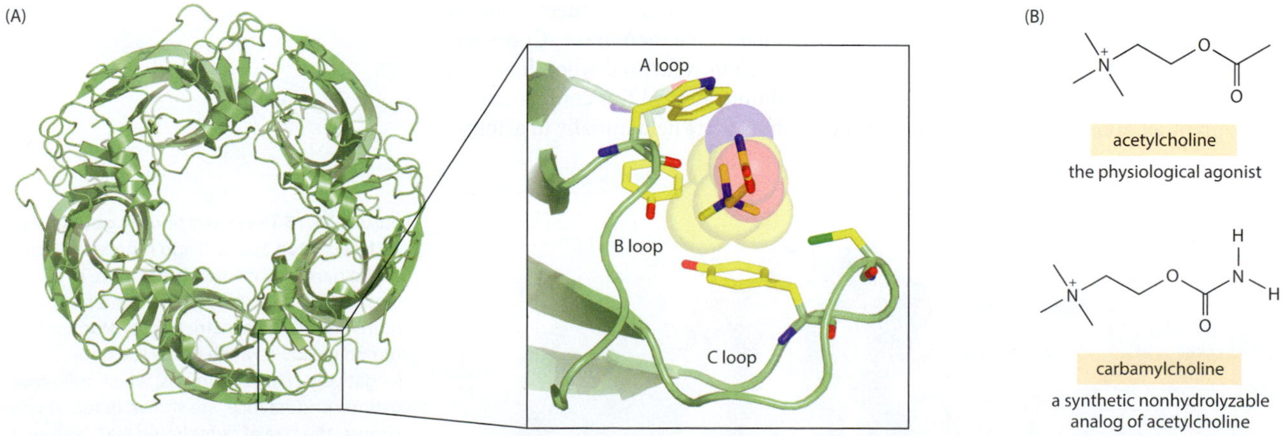

ACh, 100 nM

SubCh, 100 nM

DecCh, 50 nM

4 pA

100 ms

Figure 16.12 Electrophysiological single-channel recordings from ACh receptors. The records show responses of receptors to ACh and other agonists. The receptors were isolated from postsynaptic membranes of frog muscle. The 'blips' occurring at irregular intervals represent the transmembrane current generated when a channel opens. There are only two conductance levels (open and closed), regardless of the ACh concentration or the analog used (SubCh, suberyldicholine; DecCh, decan-1,10-dicarboxylic acid dicholine ester). (Adapted from D. Colquhoun and B. Sakmann, *J. Physiol.* 369:501–557, 1985. With permission from John Wiley & Sons.)

(A)

A loop

B loop

C loop

(B)

acetylcholine

the physiological agonist

H

carbamylcholine

a synthetic nonhydrolyzable analog of acetylcholine

Figure 16.13 The ACh-binding protein bound to the ACh analog carbamylcholine. (A) Axial view of the AChBP pentamer corresponding to that of the receptor in Figure 16.11; one of the regions responsible for ligand binding is boxed. The enlargement shows the arrangement of conserved aromatic side chains and the Cys residue that coordinate with the bound ligand. (PDB 1UV6) (B) Carbamylcholine differs from ACh only in the replacement of the terminal NH_2 group with a CH_3 group. (A, adapted from P.H. Celie et al., *Neuron* 41:907–914, 2004. With permission from Elsevier.)

(Section 1.7) being that agonists bind to the inactive state of a protein (closed channel) less tightly than to the active state (open channel), and that this difference in binding energy drives the activation event. For the *Torpedo* receptor, electrophysiological and biochemical studies have shown that the closed-state affinity of site α_γ for ACh ($K_d \sim 100$ μM) is about a 100-fold weaker than that of α_δ ($K_d \sim 1$ μM), whereas the open-state affinities are more similar and in the nanomolar range. The low affinity of one site may facilitate rapid termination of the synaptic response, while the higher affinity of the other site may accelerate activation. There is no evidence that the two binding sites are required to provide a cooperative response, which is often the *raison d'être* for multiple binding sites in allosteric proteins.

The gate is a hydrophobic girdle in the middle of the membrane

When the channel is open, it can conduct ions at a rate close to the diffusion limit (see Box 1.4); when closed, it gives no measurable ion flow. These contrasting permeation properties are determined by a narrow pore ~40 Å long. **Figure 16.14** shows the 5-fold arrangement of transmembrane domains that makes up this pore. Each subunit contributes four 40 Å-long α helices that protrude slightly beyond the membrane surface on the extracellular side. Together they form an inner ring of helices (M2) lining the pore, and an outer ring of helices (M1, M3, and M4) facing the lipids. The M2 helices only make significant side-to-side contacts near the middle of the bilayer.

The M2 helices curve radially outward toward their extracellular ends, presenting a tapering path for ions. In the closed state, the pore is narrowest near the middle of the bilayer, where it is surrounded by hydrophobic side chains projecting from the surrounding helices. The most constricting portion, shaped by conserved Leu and Val side chains, is 6–7 Å in diameter and ~8 Å long (**Figure 16.15**). This bore is too narrow for Na^+ or K^+ to pass through while retaining their hydration shells, and the ions cannot readily shed these shells in the absence of polar surfaces that would substitute for water. The tight hydrophobic girdle creates a gate that presents an energetic barrier to ion permeation. Hydrophobic gating—where there is a constricting hydrophobic girdle, rather than a physical occlusion, to prevent the flow of ions—is a recurring theme with ion channels (Section 16.2). Molecular dynamics simulations have shown that just a small increase in pore size can transform a hydrophobic constriction from being robustly impermeable to one that permits a full flow of hydrated ions (**Figure 16.16**). For the ACh receptor, this means an increase in radius of ~2 Å.

Gating involves movements in the ligand-binding and membrane-spanning domains

The gate is more than 50 Å from the ACh-binding sites. How is the binding of ACh communicated almost instantly over such a large distance to open the gate? Although we do not yet have a detailed answer, the structure of the closed channel provides some clues. In it, the two α subunits have a slightly different conformation from the other three subunits in terms of the arrangement of their β-sandwich cores. In particular, their inner β sheet (blue in Figure 16.11A) is rotated slightly about an axis perpendicular to the membrane. Comparison with AChBP suggests that the non-α-subunit conformation is favored when ligand is present, and hence that the binding of ACh drives the α subunits toward this conformation. By analogy with other allosteric proteins, it is as if the α subunits are held initially in a 'tense'

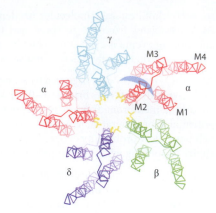

Figure 16.14 Membrane pore of the ACh receptor. A near-perfect 5-fold arrangement of α helices encircling the heteromeric pore, viewed from the extracellular side. The blue crescent identifies a water-filled space behind the upper portions of the pore-lining M2 helices. The two rings of conserved hydrophobic residues (Val and Leu) forming the gate are in yellow.

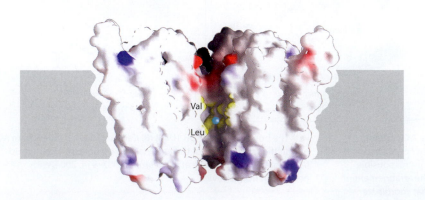

Figure 16.15 The hydrophobic gate of the ACh receptor. Space-filling representation of the narrow transmembrane pore in the closed state with the front subunit removed, highlighting the constricting hydrophobic region (yellow), which forms the gate. Negatively charged regions, which influence cation conductance, are shown in red. A cyan sphere, the size of dehydrated Na^+, is shown for scale. A hydrated Na^+ would be too large to enter this site in the closed state. (Adapted from A. Miyazawa et al., *Nature* 423:949–955, 2003. With permission from Macmillan Publishers Ltd.)

to a more buried state, taking TM3a with them in a pore-opening movement. Movement of helices is a frequently observed mechanism for pore opening in membrane transporters (Section 16.5).

16.3 NICOTINIC ACETYLCHOLINE RECEPTOR

Nerve cells communicate with their neighbors through specialized junctions called synapses (**Figure 16.9**). The arrival of an electrical impulse (an **action potential**) at the presynaptic terminal triggers the release of **neurotransmitter** molecules, which diffuse across the synaptic cleft (a distance of 20–30 nm) and bind to ligand-gated ion channels in the postsynaptic membrane of the target cell. The channels respond by opening an ion-conducting pathway through that membrane, thereby changing its potential. The channels exist in a ligand-free closed state and a ligand-bound open state. The acetylcholine receptor translates a chemical signal (acetylcholine, ACh) into an electrical signal (ion flow). For millions of years, survival of animals against predators has depended on the robustness and speed of this receptor.

The neurotransmitter ACh (146 Da) binds to nicotinic ACh receptors exposed on the surface of nerve or muscle cells. These receptors can also be activated by binding nicotine and are different from the muscarinic ACh receptors found in heart muscle, which regulate the rate and strength of muscle contraction in response to ACh binding and can also be activated by the fungal alkaloid muscarine. The muscle nicotinic ACh receptor was the first ion channel to be solubilized from membranes and characterized biochemically. Its function has been studied by biophysical measurements of purified ion channels inserted into synthetic membranes. The electric organ of the ray *Torpedo* has had a vital role in many of these studies. The muscle-derived cells from this organ (electrocytes) contain extensive postsynaptic membranes, densely packed with ACh receptors (~20,000 μm^{-2}). Vesicles rich in ACh receptors can be extracted simply and quickly from these cells, and detergent-solubilized receptors can be purified in high yield.

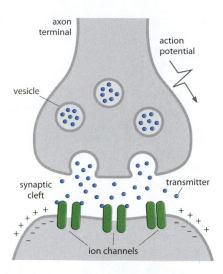

Figure 16.9 Arrival of an action potential at an axon terminal. When the action potential arrives, synaptic vesicles fuse with the nerve terminal, releasing neurotransmitters into the synaptic cleft. These molecules diffuse rapidly across the cleft and activate ligand-gated ion channels in the postsynaptic membrane of the target cell.

The ACh receptor is highly selective for cations

These proteins belong to the 'Cys-loop' family of ion channels, which can be cation- or anion-selective. The family also includes the 5HT$_3$ (5-hydroxytryptamine or serotonin), GABA$_A$ (γ-aminobutyric acid) and glycine receptors. All are pentamers of homologous (and sometimes identical) subunits arranged around a central ion pathway. Each subunit has a 13-aa loop flanked by a conserved motif—the Cys-loop, which gives the family its name. Each family member can draw on a large pool of similar subunits (more than 16 for ACh receptors). Variations in subunit type provide a wealth of functional diversity, presumably to meet a wide range of physiological needs.

The muscle ACh receptor is assembled from two copies of the α subunit and one each of the β, γ, and δ subunits. The overall structure, in the closed state, has been determined from cryo-EM (see **Methods**) images of *Torpedo* postsynaptic membranes. The isolated membranes readily form tubular vesicles with receptors arranged on a helical lattice (**Figure 16.10**).

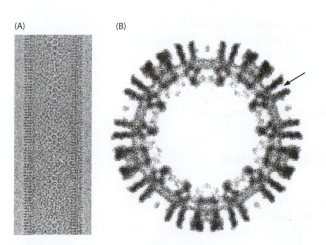

(A)   (B)

Figure 16.10 Acetylcholine receptor arrays. (A) Cryo-EM image of a tubular vesicle with ACh receptors packed on a helical lattice. The diameter of the tube is 78 nm. (B) Cross section of the tube, calculated by image reconstruction. The receptor protein projects from either side of the membrane, for which the two leaflets are resolved. The arrow points to the axial channel of a single centrally sectioned receptor. (Courtesy of Nigel Unwin.)

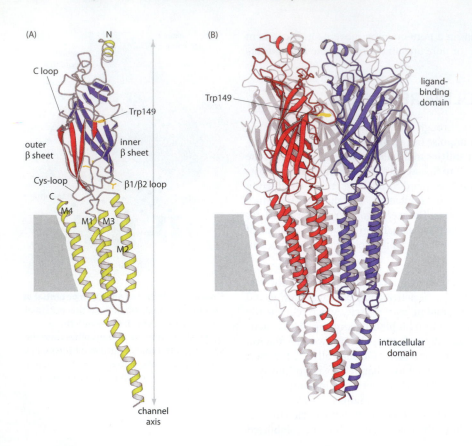

(A)

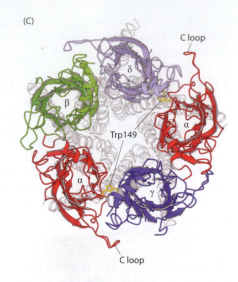

(C)

Figure 16.11 The nicotinic acetylcholine receptor. (A) Side view of a single α subunit. The β-sheet core of the extracellular ligand-binding domain is in red (outer β sheet) and blue (inner β sheet). There is a break in the electron density between the end of helix M3 and the start of the intracellular helix before the C-terminal helix M4. (B) Side view of the whole receptor. One α subunit is in red and the γ subunit in blue; other subunits are in gray. One ACh ligand-binding site (marked by Trp149) lies between the α and γ subunits. (C) Axial view from the extracellular side. (PDB 2BG9)

The helical symmetry presents views of the receptor from all directions, enabling it to be reconstructed in three dimensions. The local packing is similar to that found *in vivo*.

The subunits are long rods with three parts: an N-terminal extracellular portion organized around a β-sandwich core; a membrane-spanning portion composed of four α helices; and a single intracellular α helix (**Figure 16.11A**). Together, the five subunits make a ~160 Å long, 290 kDa tube with three functionally distinct regions: one containing the two ACh-binding sites; a membrane-spanning domain containing the channel gate; and an intracellular domain, which interacts with proteins that help to localize the receptor to the synapse (Figure 16.11B).

The ion-conducting pathway consists of a narrow pore with large open vestibules on either side. The vestibules have charged groups on their inner walls, which influence ion selectivity. The ACh receptor is cation-selective and when open allows only small cations (Na^+ and K^+) to pass. The charged groups are mostly negative and located strategically on the helices framing the lateral openings in the intracellular domain. Unlike the situation with K^+ channels (Section 16.2), the pore itself does not have a major role in ion discrimination, but contributes by restricting the size of the permeant ion.

Insight into the gating mechanism has been obtained by EM analysis of tubular arrays in which receptors were captured in the open state by spraying ACh-containing droplets on to tube-bearing microscope grids, ~5 ms before near-instantaneous freezing. These experiments showed that binding of ACh initiates two related events in the ligand-binding domain. One is a local disturbance in the region of the ACh-binding sites, and the other is a larger-scale conformational change, involving rotational movements, predominantly in one of the α subunits. The inner M2 helices also shift in response to ACh binding, widening the pore at the middle of the membrane.

The ACh receptor has evolved to be extremely fast-acting and efficient

ACh released into the synaptic cleft diffuses rapidly to its receptor in the postsynaptic membrane. Within ~10 μs of ligand binding, the channel opens and cations flow through, down their electrochemical gradients (mainly an influx of Na^+). Within a further millisecond or

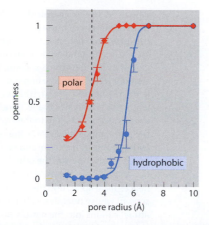

Figure 16.16 Hydrophobic gating simulated by molecular dynamics. The calculations show that very small changes in the diameter (or polarity) of a hydrophobic pore suffice to change it from impermeable to a fully open water-filled pathway. The figure plots pore 'openness' against radius for model hydrophobic (blue) and polar (red) pores. The vertical line corresponds to the radius of the hydrophobic girdle in the closed ACh receptor (yellow in Figure 16.15). (Adapted from O. Beckstein et al., *FEBS Lett.* 555:85–90, 2003. With permission from Elsevier.)

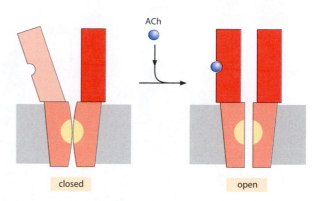

Figure 16.17 An allosteric model for ACh binding. ACh binding to the 'tense' α subunits (light red) triggers a global conformational change, making all five subunits in the ligand-binding domain nearly equivalent (red). Only two are shown, for clarity. As a result, the balance between the two pore conformations (closed or open) is shifted in favor of open.

(distorted) state by their interactions with the neighboring subunits, and the energy of ACh binding allows the subunits to switch to the 'relaxed' form (**Figure 16.17**).

Rotation of the inner β sheets of the α subunits toward the favored non-α-subunit positions would give rise to small displacements of the β1/β2 loops at the interface between the ligand-binding and membrane-spanning domains. For the α subunits, these loops are important because they communicate directly with the pore-lining M2 helices through a Val side chain, which makes a pin-into-socket interaction with their ends. To effect gating, the β1/β2 displacements must destabilize the weak lateral interactions of the pore-lining helices that keep the channel closed, so they can adopt an alternative conformation, making the pore ion-permeable (**Figure 16.18**). This mechanism has parallels in other membrane proteins, such as the ABC transporter (Section 16.5) and P-type Ca^{2+}-ATPase (Section 16.6).

The special construction of the membrane-spanning domain enables small movements of the pore-lining M2 helices to occur independently of the outer helices and away from the influence of lipids. It is likely that the pore-lining helices have distinct 'closed' and 'open' conformations of nearly equal stability, allowing controlled movements in the ligand-binding domain to tip the balance toward one or the other. Small cooperative displacements or flexure of helices taking place on a nanosecond timescale would provide a suitably fast means of achieving all-or-none opening of the pore.

16.4 AQUAPORINS

Aquaporins are channels that allow water or small uncharged molecules to move across membranes, independently of the transport of ions. Transfer of water across the plasma membrane is essential to maintain the correct osmotic pressure in the cell and to maintain cellular integrity, but the lipid bilayer presents an impermeable barrier to the efficient transfer of water. Early biophysical studies indicated that aqueous channels exist in a variety of cells. These led to the discovery of the first water channel, aquaporin-1 (AQP1) from the red blood cell, by Peter Agre and colleagues in 1992.

Aquaporins are present in all forms of life. In mammals, 13 have been recognized, all of clinical importance (**Table 16.3**). Aquaporins regulate water homeostasis in the brain, concentrate urine in the kidney, maintain lens transparency in the eye, control the secretion of sweat from the skin and tears from the eye, govern glycerol concentration for fat metabolism, and facilitate cell migration during angiogenesis. Some aquaporins allow the permeation of water only, whereas others, the aquaglyceroporins (AQGPs), also allow the transport of other uncharged solutes. Cells and organs vary in their requirements for the regulation

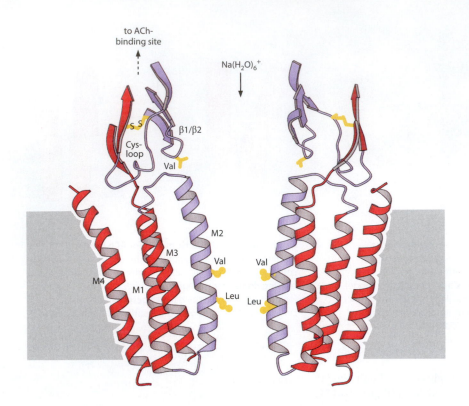

Figure 16.18 The chemical-to-electrical transduction machinery of the ACh receptor. The main functional components are the ligand-binding domain (just part of its β-sheet core is shown), the element that controls gating movements in the membrane; the β1/β2 loop, which communicates the ACh-induced conformational change to the ends of the pore-lining helices (M2); the central hydrophobic girdle of Val and Leu residues, which forms the gate; the outer helices (M1, M3, and M4), which shield the pore-lining helices from lipids; and the vestibules (not shown) whose negatively charged surfaces attract cations into the narrow pore. For clarity, portions of only the two α subunits are shown.

Table 16.3 Human aquaporins.

Aquaporin	Permeability	Tissue distribution	Primary cellular distribution
AQP0	Water (low)	Lens	Plasma membrane
AQP1	Water (high)	Red blood cell, kidney, lung, vascular endothelium, brain, eye	Plasma membrane
AQP2	Water (high)	Kidney, vas deferens	Apical plasma membrane, intracellular vesicles
AQP3	Water (high), glycerol (high), urea (moderate)	Kidney, skin, lung, eye, colon	Basolateral plasma membrane
AQP4	Water (high)	Brain, muscle, kidney, lung, stomach, small intestine	Basolateral plasma membrane
AQP5	Water (high)	Salivary gland, lacrimal gland, sweat gland, lung, cornea	Apical plasma membrane
AQP6	Water (low), anions ($NO_3^- > Cl^-$)	Kidney	Intracellular vesicles
AQP7	Water (high), glycerol (high), urea (high)	Adipose tissue, kidney, testis	Plasma membrane
AQP8	Water (high)	Testis, kidney, liver, pancreas, small intestine, colon	Plasma membrane, intracellular vesicles
AQP9	Water (low), glycerol (high), urea (high)	Liver, leukocytes, brain, testis	Plasma membrane
AQP10	Water (low), glycerol (high), urea (high)	Small intestine	Intracellular vesicles
AQP11	Water	Testis, liver, kidney, brain	Intracellular vesicles
AQP12	Water	Acinar cells of pancreas	Intracellular vesicles

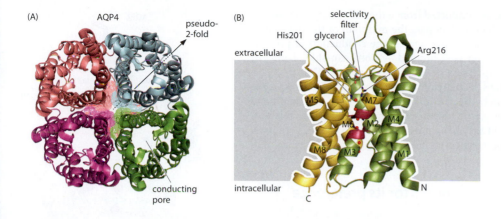

Figure 16.19 Structure of human aquaporin 4 (AQP4). (A) Top view of the AQP4 tetramer from the extracellular side. The van der Waals surfaces of residues that block the region along the 4-fold axis are shaded. The pseudo-2-fold axis that relates the two tandem repeats in the monomer is shown for the cyan subunit. (PDB 3GD8) (B) Side view of an AQP4 subunit. The tandem repeats, relatively inverted, are in green (helices M1–M4) and yellow (helices M5–M8). M3 and M7 meet in the middle of the membrane; the NPA motifs at their N termini are in magenta. Seven water molecules (small red balls) are bound in the pore. A glycerol molecule binds at the entrance but does not enter.

of water transport across membranes, and the diverse aquaporins meet these needs. The bacterium *E. coli* has a water channel AqpZ and a glycerol facilitator GlpF, representatives of the two aquaporin groups. Plants, which are particularly dependent on water balance to maintain cell turgor (rigidity deriving from osmotic pressure applied by intracellular water) and to transport water to the tips of tall trees, are especially rich in aquaporins (for example, there are 35 in *Arabidopsis thaliana*).

All aquaporins are tetramers of 33-kDa subunits. Within each subunit there are two sequence repeats, each with the conserved motif Asn-Pro-Ala (NPA). The first structures were described in 2000, by X-ray diffraction for detergent-solubilized *E. coli* GlpF and by electron crystallography for the human red-cell AQP1 in reconstituted membranes. They showed that each subunit has six transmembrane helices with a water-conducting pore at their center (**Figure 16.19A**). Two additional short helices, each containing the NPA motif, are directed into the pore: one reaches to the cytoplasmic side and the other to the extracellular side of the membrane. Structures are now available for AQP0, APQ1, APQ4, APQ5, and several plant, yeast, archaeal, and bacterial aquaporins.

The structure of aquaporin 4 shows the basis for water conductance

AQP4 is primarily responsible for water homeostasis in the brain, where it localizes to astrocytes. Because the brain is enclosed within the rigid skull, any increase in intracranial pressure can result in brain injury unless the secretion and uptake of fluid are controlled. Astrocytes have plasma membrane projections that surround the blood capillaries and separate them from the neurons. Brain endothelial cells are joined by *tight junctions,* creating a *blood/brain barrier* (Section 11.3 and Box 11.1) that keeps the environment in the brain stable and prevents dangerous substances from entering the brain from the bloodstream.

AQP4 has a C-terminal sequence that binds to the PDZ domain (see **Guide**) of α-syntrophin, a component of the dystrophin complex that links AQP4 to the actin cytoskeleton in the so-called foot processes of astrocytes.

The human AQP4 channel has a high selectivity for water over glycerol. In the AQP4 tetramer (and some other aquaporins), there is no central pore: rather, there are four independent pores. Each subunit has eight α helices (Figure 16.19B). The two tandem repeats per subunit have an antiparallel arrangement with a pseudo-2-fold axis of symmetry, as also observed in the chloride transporter (Section 16.2), although the structures are quite different. Two helices, M3 and M7, extend just halfway across the membrane and meet at their N-terminal ends, where the conserved NPA sequences lie. The pore down the center of each subunit narrows to a 2.8 Å constriction at a point ~8 Å from the extracellular side of the bilayer. This region constitutes the *selectivity filter*. Two residues—His201 and Arg216 in AQP4—create the selectivity filter; these are signature residues for water-selective AQP channels.

In its open state, the pore, which is 25 Å long, contains a line of water molecules (**Figure 16.20**). The pore is amphipathic, as it is in all AQPs: the hydrophobic wall is formed from side chains of Phe (not shown), Ile, Leu, and Val, whereas on the hydrophilic wall, main-chain carbonyl oxygens act as H-bond acceptors for the seven water molecules transiting the pore. This arrangement allows water to be conducted from either side. The Asn residues of the NPA motifs orient the water molecules as they pass the midpoint of the pore: their amide side chains act as H-bond donors that insulate against the conduction of protons or ions. In AQP1 and AqpZ, the central water is H-bonded by both Asn side chains. In this arrangement, the water cannot rotate to accept or exchange its proton. Thus, H_3O^+ is excluded from this site, which explains the selectivity of the pore for water. Other ions are excluded because of the energetic cost of dehydration and because the filter is lined on one side by hydrophobic groups. The conserved Arg at the filter entrance contributes to selectivity by repelling positively charged ions, and the channel is too narrow to allow Cl^- to pass. This combination of pore size and charge restrictions underlies the unique permeability characteristics of aquaporins.

The wider selectivity filter of GlpF accounts for its preference for glycerol

Glycerol is a component of about two-thirds of all lipids and is an important metabolite. In the cell, it is phosphorylated by glycerol kinase; this lowering of its concentration creates an inward gradient that drives the uptake of glycerol. The *E. coli* glycerol-conducting protein GlpF has a much higher conductance for glycerol than for water. Its glycerol influx is 100–1000 times greater than is expected for a transporter and is nonsaturable up to more than 200 mM, indicating that the protein is acting as a channel and not a transporter.

GlpF is similar in structure to water-conducting aquaporins, as expected from sequence similarity. The biggest difference is in the selectivity filter. The constriction in GlpF has a diameter of 3.4–3.8 Å, which is 0.6–1 Å wider than in AQP4 and sufficient to allow passage of glycerol (**Figure 16.21**). In water-conducting aquaporins, the selectivity filter is composed of Arg and His residues: in GlpF and other AGQPs, it is more hydrophobic. The Arg is conserved but the other residues are aromatics: a Trp and Phe in GlpF. The side chains of these residues interact favorably with the carbon backbone of glycerol (**Figure 16.22A**).

Three glycerol molecules bind within the GlpF pore (Figure 16.22B). The first, G1, is outside the filter and makes contacts at the extracellular vestibule. The other two, G2 and G3, are located in the pore that extends ~28 Å to the cytoplasmic surface and is only wide enough to accommodate a glycerol molecule with its long axis oriented parallel to the pore. G2 and G3 are linked by a H-bonded water, suggesting that the movements of water and glycerol are coupled. At the G2 site, the alkyl part of glycerol packs against the two aromatic residues in the selectivity filter. The O1 and O2 hydroxyl groups are H-bonded to an Arg residue and to main-chain carbonyl oxygens. G3 is H-bonded through its O1 and O2 hydroxyl groups to the Asn residues of both NPA motifs, which act as H-bond donors. The H-bonds to both glycerols compensate for dehydration on entering the filter. As in water-conducting AQPs, the amphipathic nature of the pore matches the separation of CH–OH groups of glycerol with polar contacts that are only possible on one side.

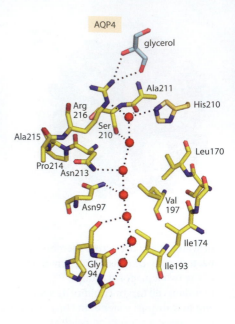

Figure 16.20 The pore of an aquaporin. Some of the residues that line the pore of human AQP4 are shown in stick representation, together with the seven water molecules (red balls) and a glycerol.

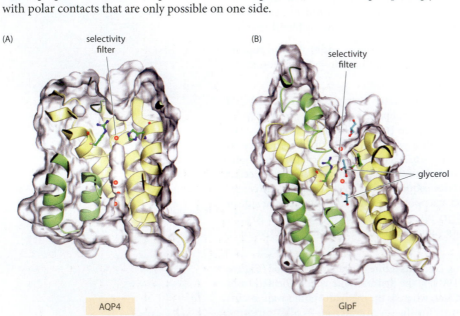

Figure 16.21 Aquaporin selectivity filters. Cutaway diagrams exposing the molecular surfaces of the two types of pore. (A) In AQP4, the pore is occluded at the selectivity filter but allows water molecules (red balls) to bind. (B) In GlpF, the selectivity filter is wider, allowing glycerol to bind. The two aquaporins are shown in slightly differing orientations. The N-terminal domains are in green and the C-terminal domains in pale yellow. (PDB 3GD8 and 1FX8)

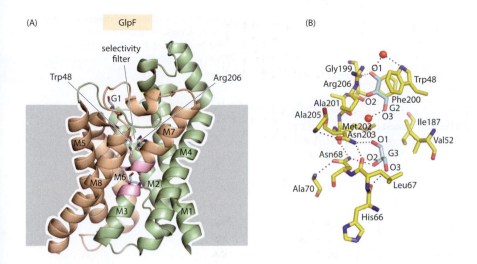

Figure 16.22 The *E. coli* glycerol facilitator GlpF. (A) Side view of a GlpF monomer. The selectivity filter comprises Trp48, Phe200 (obscured by M8), and Arg206. Three glycerol molecules (G1, G2, and G3) are bound, with G2 and G3 hidden in the pore. The two NPA motifs are in pink (compare with Figure 16.19). (PDB 1FX8) (B) Details of the interactions of G2 and G3 within the filter, showing the polar groups on the left-hand side and the mostly nonpolar groups on the right-hand side. The side chains of Asn68 and Asn203 are positioned by H-bonds that ensure that only their NH$_2$ groups are available to bind glycerol G3.

The lower conductivity of GlpF for water may be explained by the hydrophobic nature of the selectivity filter. Mutagenesis experiments that converted the selectivity filter of the *E. coli* water channel AqpZ to that of GlpF decreased the conductance rate for water but not for glycerol. Thus, it seems that other structural elements also have a role in imposing selectivity.

16.5 TRANSPORTERS

Transporters move small molecules and ions across the membrane against a concentration gradient. Unlike channels, transporters have substrate-binding sites that can only be accessed from one side of the membrane at a time. They use an alternating access mechanism, cycling between a conformation in which the substrate-binding site is accessible from outside the membrane (outward-facing) and one in which it is inward-facing (**Figure 16.23**).

Transporter activity requires energy that is usually provided by ATP hydrolysis (primary active transport) or by tapping into concentration gradients of protons or other ions (secondary active transport). Other transporters mediate the movement of molecules down a concentration gradient, which occurs spontaneously (passive transport). Here we describe the structures and modes of action of selected examples of primary and secondary transporters.

ABC transporters employ an alternating access mechanism driven by ATP hydrolysis

The ATP-binding cassette (ABC) transporters are primary transporters and can be divided into two major classes: exporters and importers. Exporters transport molecules out of cells or into organelles and are found in all organisms. The human genome encodes about 48 of them. ABC importers transport nutrients into cells and are found only in prokaryotes.

All ABC transporters have a common core architecture (**Figure 16.24**), consisting of a dimer of transmembrane domains (TMDs), each with a nucleotide-binding domain (NBD)

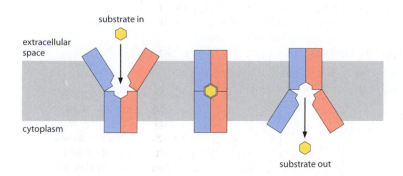

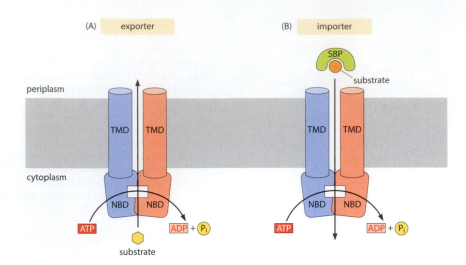

Figure 16.24 Schematic diagram of ABC transporters. (Adapted from K.P. Locher, *Phil. Trans. R. Soc. Lond. B* 364:239–245, 2008. With permission from The Royal Society.)

on the cytoplasmic side. The core may be built from one (for example P-glycoprotein), two (for example MsbA), or four (for example MalFGK$_2$) subunits and elaborated with regulatory domains. Importers have an associated substrate-binding protein (SBP) in the periplasm, which delivers substrate to the TMDs (Figure 16.24B).

MsbA is an exporter found in Gram-negative bacteria that transports lipid A and lipopolysaccharides from the cytoplasm to the periplasm en route to the outer membrane. Binding and hydrolysis of ATP promotes dimerization of the NBDs, accompanied by a conformational change in the transmembrane helices, whereby the molecule shifts from an inward-facing to an outward-facing conformation (**Figure 16.25A,B**).

The bacterial maltose transporter MalFGK$_2$ (Figure 16.25C,D), an importer of maltose oligosaccharides, has a periplasmic maltose-binding protein (MBP), two integral membrane proteins (MalF and MalG), and two copies of the ABC cassette, MalK. The crystal structure of MBP by itself shows a cleft between two lobes in which the oligosaccharide is completely engulfed. In the intact transporter, the MBP delivers the substrate to the TMD and stimulates ATPase activity, ensuring that one molecule of substrate is transported in each cycle of ATP hydrolysis. Figure 16.25C shows the transporter without both nucleotide and MBP in an inward-facing conformation. Figure 16.25D shows an intermediate nucleotide-bound state in which maltose is bound to the TMD in a large solvent-filled cavity shielded from bulk solvent by MalF and MalG. Two aromatic residues stack against the carbohydrate rings, as is frequently observed at sugar-binding sites in proteins. The MBP does not have maltose bound and is in an open conformation.

The transmembrane domains exhibit a variety of helical arrangements

All TMDs are built from membrane-spanning α helices, but their number and topology vary. Their key property is their ability to switch between inward- and outward-facing conformations. Exporters such as MsbA have six transmembrane helices per TMD (Figure 16.25A,B). They are organized as two 'wings,' each consisting of two helices (TM1 and TM2) from one TMD, intertwined with four helices (TM3–TM6) from the other TMD. The TMD helices extend 25 Å into the cytoplasm, where they interact with the NBDs. In the nucleotide-bound state, the two wings diverge toward the periplasmic side of the membrane, creating a hydrophilic cavity that affords a pathway for substrates.

Importers show more variation in their TMDs, which have from five to ten α helices. In MalFGK$_2$, the transmembrane region consists of two subunits: MalF with eight TM helices and MalG with six TM helices. Helices 1–3 of MalF contact helices 2–4 of MalG, and helix 1 of MalG packs against the bundle of helices 4–8 of MalF (Figure 16.25D). MalF has an extended periplasmic domain inserted between TM helices 3 and 4 (Figure 16.25D). MalF helices 3–8 and MalG helices 1–6 have similar topology and are related by an approximate 2-fold axis perpendicular to the membrane.

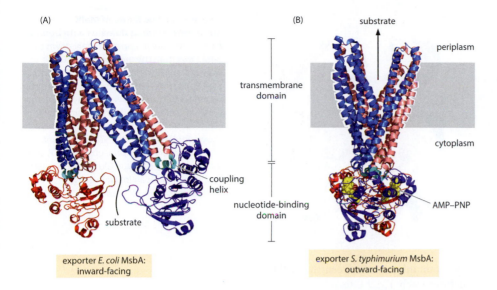

(A)

transmembrane
domain

nucleotide-binding
domain

coupling
helix

substrate

exporter *E. coli* MsbA:
inward-facing

(B)

substrate

periplasm

cytoplasm

AMP–PNP

exporter *S. typhimurium* MsbA:
outward-facing

Figure 16.25 Structures of ABC transporters. (A) MsbA from *E. coli* is a dimer; its two subunits are shown in red and blue, with the coupling helices in cyan. In the absence of nucleotide, the substrate-binding site is inward-facing. (PDB 3B5W) (B) The same protein from *Salmonella typhimurium* with bound nucleotide 5′-adenylyl-β,γ-imidodiphosphate (AMP-PNP, in yellow). The two NBDs associate tightly and the exporter is outward-facing. (PDB 3B60) (C) *E. coli* maltose transporter MalFGK$_2$ has four subunits: MalF (pink), MalG (blue), and two MalKs (NBDs, in red and darker blue) in a nucleotide-free state. (PDB 3FH6) (D) MalFGK$_2$ with ATP bound (yellow). This structure represents an intermediate state with the maltose released from the open maltose-binding protein (MBP, yellow) and bound to the TMD. The TM helices in MalF and MalG are numbered. (PDB 2R6G)

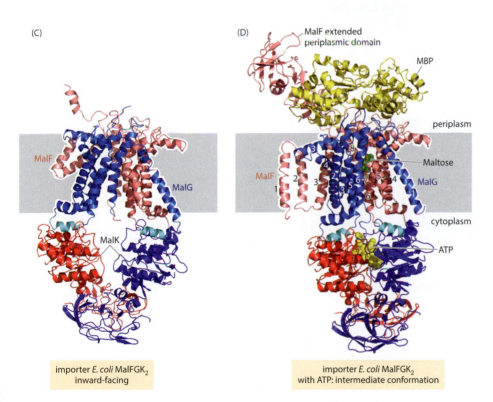

(C)

MalF

MalG

MalK

importer *E. coli* MalFGK$_2$
inward-facing

(D)

MalF extended
periplasmic domain

MBP

periplasm

Maltose

MalF   MalG

cytoplasm

ATP

importer *E. coli* MalFGK$_2$
with ATP: intermediate conformation

The nucleotide-binding domains form dimers with ATP

The NBDs are conserved throughout the ABC transporter family. These domains have been called the 'engine' of the transporter, because it is here that ATP binding, hydrolysis, and release generate the energy needed to move molecules up a concentration gradient. NBDs have two subdomains: a catalytic core domain and an α-helical domain (**Figure 16.26A**). The core domain has several conserved features including a Walker A motif (P-loop) that interacts with the phosphate groups of the nucleotide, and a Walker B motif (see Box 3.1 for definition of Walker motifs and their role in ATP hydrolysis). The α-helical domain contains a motif Leu-Ser-Gly-Gly-Gln (LSGGQ) known as the ABC signature motif (or Q loop), which is involved in nucleotide binding. Each ATP-binding site is formed from the Walker A and Walker B motifs of one subunit and the signature motif of the other subunit, with the ATP binding in the dimer interface (Figure 16.26B).

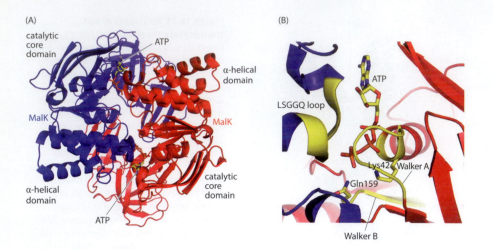

Figure 16.26 The dimer of MalK nucleotide-binding domains with bound ATP. (A) As viewed approximately down the 2-fold axis from the membrane. (B) Close-up view of the ATP-binding site, showing how the LSGGQ loop (yellow) is provided by one subunit (blue) and the Walker A (yellow) and Walker B (yellow) motifs are provided by the other subunit (red). The side chains are also shown for Lys42 from the Walker A motif, which contacts the triphosphate of ATP, and for Gln159. The catalytic Glu159 was mutated to Gln to provide an inactive enzyme amenable to crystallography. (PDB 2R6G)

Coupling of transport to ATP binding requires communication between the NBDs and the TMDs. Despite the variability in their structures, each TMD interacts with an NBD in similar fashion through an α helix known as the coupling helix (Figures 16.25A and **Figure 16.27**). Although this helix is structurally conserved, there are transporter-specific variations in its sequence and position. In the exporter Sav1866 and the importer ModBC, both of which have six TM helices, the coupling helix is between TM4 and TM5. In MalFGK$_2$, the coupling helices are between TM6 and TM7 for MalF and between TM4 and TM5 for MalG.

ABC exporters employ an 'ATP switch' mechanism (Figure 16.27). The substrate first binds to protein in the inward-facing conformation, triggering a conformational change that brings the NBDs closer together. ATP binding then causes another conformational change that switches the protein to be outward-facing. This change is transmitted through the coupling helices. The substrate can then dissociate. ATP hydrolysis and the release of ADP and phosphate return the protein to its inward-facing conformation. Intracellular concentrations of ATP are ~10 times the K_m of most ABC transporters and so the nucleotide-binding sites are readily saturated once they have been correctly formed. A similar mechanism is envisaged for importers, but instead of interacting directly with the TMDs, the substrate first binds to a periplasmic binding protein that relays it to the transporter (for example, MalFGK$_2$).

Major facilitator superfamily proteins tap into concentration gradients to transport their substrates

MFS proteins can either transfer a substrate down a concentration gradient (**uniporter**) or move substrates up this gradient by coupling the process to a movement of H$^+$, Na$^+$, or solutes, either in the same direction (**symporter**) or in the opposite direction (**antiporter**). Although family members vary substantially in mechanism, they are structurally similar as illustrated by two bacterial transporters, a proton symporter (LacY) and a solute antiporter (GlpT).

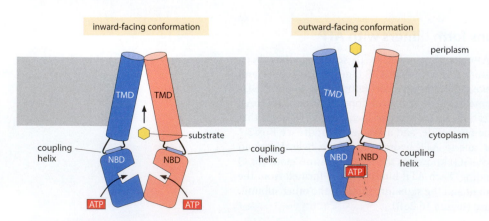

Figure 16.27 Proposed mechanism for ABC exporters. First, substrate binds to the protein in an inward-facing conformation. This triggers a conformational change that brings the NBDs closer together. ATP binding then causes another conformational change that switches the protein to outward-facing. This change is transmitted through the coupling helices. The substrate can then dissociate. (Adapted from K.J. Linton and C.F. Higgins, *Pflügers Arch.* 453:555–567, 2007. With permission from Springer Science and Business Media.)

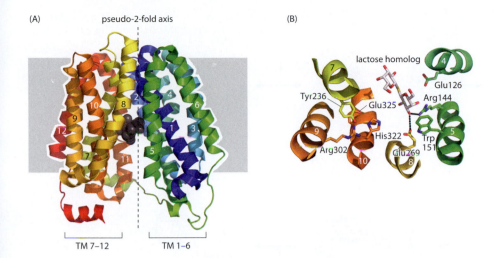

(A)
pseudo-2-fold axis
TM 7–12   TM 1–6

(B)
lactose homolog
Tyr236
Glu325
Arg302
His322
Glu269
Trp 151
Glu126
Arg144

Figure 16.28 Structure of lactose permease (LacY) from *E. coli*. (A) Side view. This inward-facing conformation was stabilized for crystallographic studies by a substitution, Cys154Gly, that allows the permease to bind substrate but not to transport it. The two lobes (TM helices 1–6, colored blue to green, and TM helices 7–12, colored light green through yellow to red) are related by a pseudo-2-fold axis. The bound substrate (lactose homolog TDG, β-D-galactopyranosyl-1-thio-β-D-galactopyranoside) is shown in dark gray. (PDB 1PV7) (B) Substrate-binding site, viewed from above, showing the sugar and residues important for binding and transport.

Lactose permease (LacY) transports galactosides such as lactose into the cell. The energy needed is obtained from a simultaneous flux of H^+ into the cell. LacY has two domains, each consisting of six transmembrane helices, related by a pseudo-2-fold axis (**Figure 16.28A**). In this inward-facing conformation, the protein has a V shape open to the cytoplasmic side. The substrate-binding site (Figure 16.28B) is situated at the interdomain interface in a hydrophilic cavity within the membrane. The involvement of a Trp or other aromatic side chain in a carbohydrate-binding site (as here) is a common feature in carbohydrate-binding proteins.

This structure suggested a mechanism in which the two domains rock against one another, allowing access to the binding site to alternate between the two sides of the membrane through an intermediate state that is closed to both sides. Mutagenesis experiments identified six residues as essential for active transport: Glu126 and Arg144 are crucial for substrate binding; Glu269 for substrate binding and H^+ translocation; and Arg302, His322, and Glu325 for H^+ translocation. On the basis of these assignments and a wealth of other evidence, a six-step mechanism has been proposed for galactoside uptake (**Figure 16.29A**). In the absence of substrate, the outward-facing conformation of the protein is unstable (step 1) and Glu269 is rapidly protonated (step 2). The galactoside then binds, guided by Trp151 and the protonated Glu269 (step 3). Upon its binding, the salt bridge between Glu126 and Arg144 (in the N-terminal domain, marked in step 1) is broken as Arg144 interacts with the galactoside and forms a new salt bridge with Glu269 in the C-terminal domain (step 3). For this bridge to be made, the H^+ is released from Glu269 and transferred to His322, leading to a rapid transition to the inward-facing conformation, with the H^+ shared by His322 and Glu325. Glu325 is in a hydrophobic environment, which would favor its protonated state (step 4). The substrate is then released into the cytoplasm, and Arg144 reestablishes its interaction with Glu126 (step 5). Because of a decrease in the pK_a of Glu325 caused by this conformational change, H^+ is also released into the cytoplasm (step 6). The final step is the transition to the outward-facing conformation (step 1).

GlpT is an antiporter that uses the gradient of inorganic phosphate (P_i) to drive the uptake of glycerol 3-phosphate. It is very similar in structure to LacY. In the proposed alternating access mechanism (Figure 16.29B), the same binding site is used to transport P_i out of the cell and to transport glycerol 3-phosphate in. The cycle starts from the stable conformation of the inward-facing unliganded protein (step 1). P_i binds to the protein, interacting with Arg45 on the NTD and Arg269 on the CTD (step 2) and pulling the two Arg residues closer together. At the same time, the nearby His165 is protonated, strengthening its interaction with the P_i (step 3). The binding energy thus released lowers the activation energy sufficiently to allow the protein to switch conformation so that the substrate-binding site faces the periplasm (step 3). Deprotonation of His165 weakens the interaction with P_i, which then dissociates (step 4). The transporter is now able to take up glycerol 3-phosphate to be moved into the cytoplasm by a similar mechanism (step 4). The direction of transport is governed by concentrations and binding constants. P_i is at a higher concentration than glycerol 3-phosphate in the cytoplasm and so binds preferentially to the inward-facing

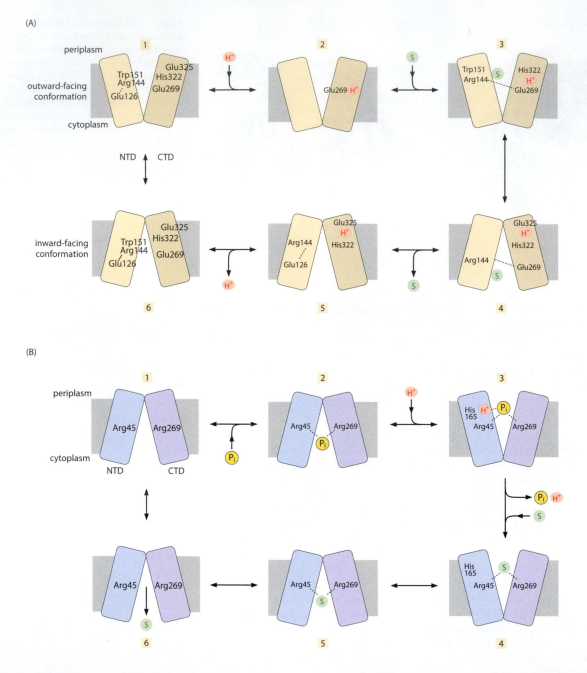

Figure 16.29 Proposed mechanisms for import of substrates by a symporter and an antiporter. Only the amino acids critical at each step are marked. (A) The proton symporter LacY for galactoside import. S, substrate. The pathway was deduced from crystal structures of the inward-facing conformation with and without substrate (steps 4 and 5) and modeling of conformations for the other stages. (B) The antiporter GlpT for glycerol 3-phosphate import and phosphate export, based on a crystal structure for an unliganded inward-facing state. The ligand-binding site was deduced from inspection of the pore between the two domains. In the mid-membrane region, it involves two Arg residues binding the negatively charged groups of P_i and glycerol 3-phosphate. However, their guanidinium groups are nearly 10 Å apart in the solved structure and must shift to bind ligand. (A, adapted from O. Mirza et al., *EMBO J.* 25:1177–1183, 2006; B, adapted from C. J. Law et al., *Annu. Rev. Microbiol.* 62:289–305, 2008. With permission from Annual Reviews.)

conformation. Glycerol 3-phosphate has a higher affinity than P_i for the binding site, so it binds preferentially to the outward-facing conformation. The protein then switches to expose the site to the cytoplasm (step 5, the reverse of step 3), whereupon the substrate dissociates (step 6). Substrate binding is proposed to facilitate interconversion between the inward-facing and outward-facing conformations of GlpT, allowing the phosphate gradient to drive transport of glycerol 3-phosphate.

(A)

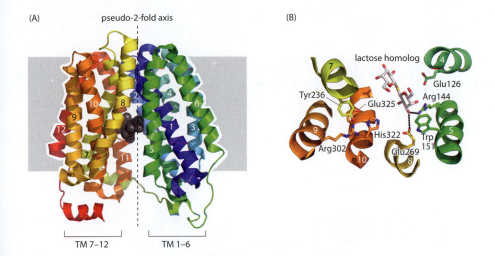

Figure 16.28 Structure of lactose permease (LacY) from *E. coli*. (A) Side view. This inward-facing conformation was stabilized for crystallographic studies by a substitution, Cys154Gly, that allows the permease to bind substrate but not to transport it. The two lobes (TM helices 1–6, colored blue to green, and TM helices 7–12, colored light green through yellow to red) are related by a pseudo-2-fold axis. The bound substrate (lactose homolog TDG, β-D-galactopyranosyl-1-thio-β-D-galactopyranoside) is shown in dark gray. (PDB 1PV7) (B) Substrate-binding site, viewed from above, showing the sugar and residues important for binding and transport.

Lactose permease (LacY) transports galactosides such as lactose into the cell. The energy needed is obtained from a simultaneous flux of H^+ into the cell. LacY has two domains, each consisting of six transmembrane helices, related by a pseudo-2-fold axis (**Figure 16.28A**). In this inward-facing conformation, the protein has a V shape open to the cytoplasmic side. The substrate-binding site (Figure 16.28B) is situated at the interdomain interface in a hydrophilic cavity within the membrane. The involvement of a Trp or other aromatic side chain in a carbohydrate-binding site (as here) is a common feature in carbohydrate-binding proteins.

This structure suggested a mechanism in which the two domains rock against one another, allowing access to the binding site to alternate between the two sides of the membrane through an intermediate state that is closed to both sides. Mutagenesis experiments identified six residues as essential for active transport: Glu126 and Arg144 are crucial for substrate binding; Glu269 for substrate binding and H^+ translocation; and Arg302, His322, and Glu325 for H^+ translocation. On the basis of these assignments and a wealth of other evidence, a six-step mechanism has been proposed for galactoside uptake (**Figure 16.29A**). In the absence of substrate, the outward-facing conformation of the protein is unstable (step 1) and Glu269 is rapidly protonated (step 2). The galactoside then binds, guided by Trp151 and the protonated Glu269 (step 3). Upon its binding, the salt bridge between Glu126 and Arg144 (in the N-terminal domain, marked in step 1) is broken as Arg144 interacts with the galactoside and forms a new salt bridge with Glu269 in the C-terminal domain (step 3). For this bridge to be made, the H^+ is released from Glu269 and transferred to His322, leading to a rapid transition to the inward-facing conformation, with the H^+ shared by His322 and Glu325. Glu325 is in a hydrophobic environment, which would favor its protonated state (step 4). The substrate is then released into the cytoplasm, and Arg144 reestablishes its interaction with Glu126 (step 5). Because of a decrease in the pK_a of Glu325 caused by this conformational change, H^+ is also released into the cytoplasm (step 6). The final step is the transition to the outward-facing conformation (step 1).

GlpT is an antiporter that uses the gradient of inorganic phosphate (P_i) to drive the uptake of glycerol 3-phosphate. It is very similar in structure to LacY. In the proposed alternating access mechanism (Figure 16.29B), the same binding site is used to transport P_i out of the cell and to transport glycerol 3-phosphate in. The cycle starts from the stable conformation of the inward-facing unliganded protein (step 1). P_i binds to the protein, interacting with Arg45 on the NTD and Arg269 on the CTD (step 2) and pulling the two Arg residues closer together. At the same time, the nearby His165 is protonated, strengthening its interaction with the P_i (step 3). The binding energy thus released lowers the activation energy sufficiently to allow the protein to switch conformation so that the substrate-binding site faces the periplasm (step 3). Deprotonation of His165 weakens the interaction with P_i, which then dissociates (step 4). The transporter is now able to take up glycerol 3-phosphate to be moved into the cytoplasm by a similar mechanism (step 4). The direction of transport is governed by concentrations and binding constants. P_i is at a higher concentration than glycerol 3-phosphate in the cytoplasm and so binds preferentially to the inward-facing

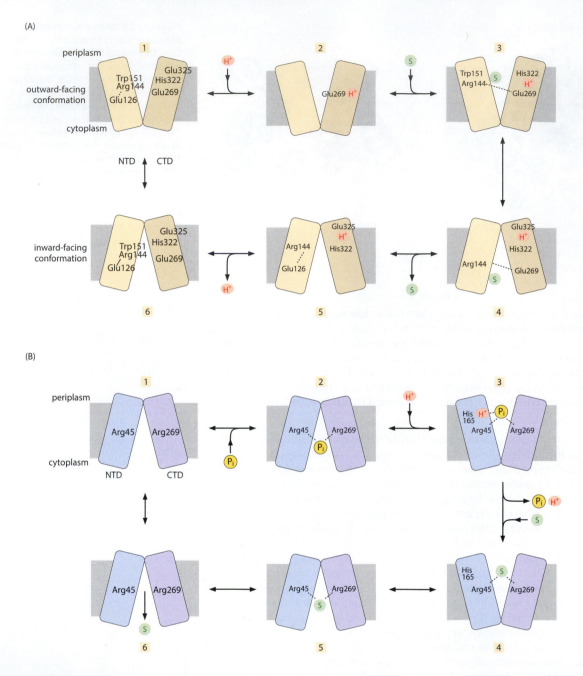

Figure 16.29 Proposed mechanisms for import of substrates by a symporter and an antiporter. Only the amino acids critical at each step are marked. (A) The proton symporter LacY for galactoside import. S, substrate. The pathway was deduced from crystal structures of the inward-facing conformation with and without substrate (steps 4 and 5) and modeling of conformations for the other stages. (B) The antiporter GlpT for glycerol 3-phosphate import and phosphate export, based on a crystal structure for an unliganded inward-facing state. The ligand-binding site was deduced from inspection of the pore between the two domains. In the mid-membrane region, it involves two Arg residues binding the negatively charged groups of P_i and glycerol 3-phosphate. However, their guanidinium groups are nearly 10 Å apart in the solved structure and must shift to bind ligand. (A, adapted from O. Mirza et al., *EMBO J.* 25:1177–1183, 2006; B, adapted from C. J. Law et al., *Annu. Rev. Microbiol.* 62:289–305, 2008. With permission from Annual Reviews.)

conformation. Glycerol 3-phosphate has a higher affinity than P_i for the binding site, so it binds preferentially to the outward-facing conformation. The protein then switches to expose the site to the cytoplasm (step 5, the reverse of step 3), whereupon the substrate dissociates (step 6). Substrate binding is proposed to facilitate interconversion between the inward-facing and outward-facing conformations of GlpT, allowing the phosphate gradient to drive transport of glycerol 3-phosphate.

The transport mechanism of nucleobase cation symport 1 exploits pseudosymmetry

Members of the NCS1 family, found in bacteria and yeast, transport nucleobases, compounds that include the purines and pyrimidines of RNA and DNA. They share a fold based on a core of 10 transmembrane helices, which can be divided into two sets of five related by a pseudo-2-fold axis (**Figure 16.30A–C**). Instead of the separate domains seen in MFS transporters, these two sets are intertwined. The substrate-binding sites are in the middle of the transmembrane region.

Mhp1, a bacterial protein, exploits the Na^+ gradient created by other transporters that pump Na^+ out, to drive solutes into the cell. Its structure has been solved in three conformations: outward-open, in which Na^+ but not substrate is bound; occluded, in which both substrate and Na^+ are present; and inward-open, in which neither is bound. These structures have allowed the formulation of a mechanism for substrate transport into the cell. This is diagrammed in Figure 16.30D–F, in which the helices have been regrouped into three functional units and recolored. When substrate binds to Mhp1 in the outward-open conformation, helix 10 closes over it, with helices 3 and 8 barring passage into the cell. To go from this occluded conformation to the inward-open conformation, the block of helices 3, 4 and 8, 9 rotates through ~30°. This opens a pathway for substrate to move into the cell, while blocking its exit to the exterior. Helix 5, the N-terminal equivalent to helix 10, also flexes to further open the import pathway for substrate.

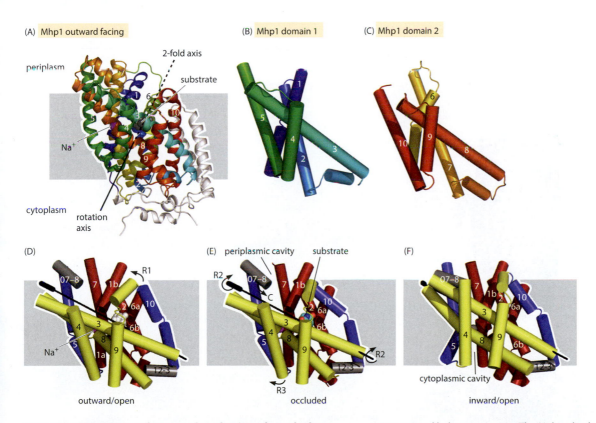

Figure 16.30 Structure and proposed mechanism of a nucleobase cation symport-1 protein. (A) Hydantoin permease from *Microbacterium liquefaciens* (Mhp1) in the occluded state with both Na^+ and substrate bound. The core of 10 helices, in rainbow coloring, comprises two tandem repeats of five helices related by a pseudo-2-fold axis in the plane of the membrane. Additional nonconserved helices are in gray. The binding of Na^+ (purple sphere) and substrate (gray spheres) is shown. (PDB 2JLN) (B, C) The five-helix repeats of Mhp1. (B) Helices 1–5 as in A. (C) Helices 6–10 rotated around the pseudo-2-fold axis to orient them as in B. The topologies of the two sets of helices match. (D–F) Proposed transport mechanism. The helices are shown as cylinders and have been recolored in three functionally linked groups (yellow, red, and blue). Additional nonconserved helices are in gray. The Na^+ and substrate are shown in white (D), where they are not present in the crystal structure. (D) The outward-facing state with Na^+ bound. The view is slightly rotated relative to panels A and B. On binding of substrate (benzylhydantoin), the occluded state is reached (E). The N-terminal part of helix 10 bends and packs over the substrate (movement R1). A 30° rotation of helices 3, 4 and 8, 9 around the shown rotation axis (movement R2) then switches the protein into the inward-facing conformation (F). Accompanying this change, helix O7–8 moves into the extracellular cavity, and the N-terminal part of helix 5 (complementary to the N-terminal part of helix 10) shifts to further open the intracellular cavity (movement R3). (Adapted from T. Shimamura et al., *Science* 328:470–473, 2010. With permission from AAAS.)

Transport of the substrate hydantoin correlates with Na$^+$ flux. Sodium increases the affinity of Mhp1 for hydantoin about tenfold, and hydantoin increases the affinity for Na$^+$ tenfold. The Na$^+$-binding site lies between helices 1 and 8 (Figure 16.30A,E). In the outward-facing conformation, this site is well defined; however, in the inward-facing conformation, the distances from Na$^+$ to its protein contacts are greater, indicative of a weaker binding site. This suggests that Na$^+$ stabilizes the outward-facing conformation. The switch to the inward-facing conformation occurs only upon substrate binding, and this prevents Na$^+$ from being transported without the concomitant movement of substrate. Once the protein is in the inward-facing conformation, Na$^+$ and substrate can both dissociate.

A multidrug resistance protein employs a rotary version of the alternating access mechanism

<u>A</u>criflavin <u>r</u>esistance protein B (AcrB) confers drug resistance in Gram-negative bacteria by actively exporting drugs from the cell. To do this, AcrB in the inner membrane cooperates with an outer-membrane channel TolC and a periplasmic adapter protein, AcrA. The structures of the three components were determined separately. TolC and AcrB can be fitted together to give a continuous pore from the inner membrane across the periplasm and through the outer membrane (**Figure 16.31**). AcrB is a trimer of very large subunits (1049 aa). Each subunit has three layers: a transmembrane domain, a pore domain, and a

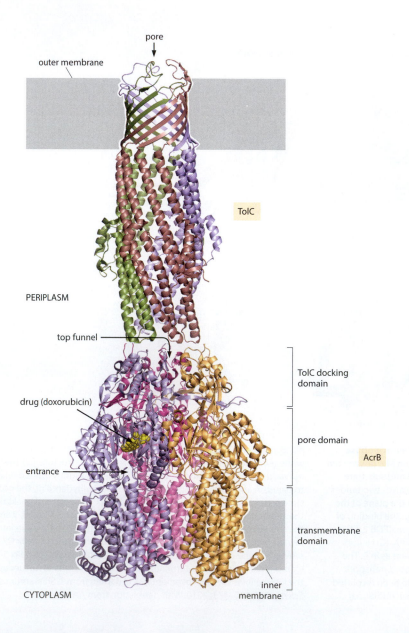

Figure 16.31 The AcrB/TolC efflux pump.
The three subunits of TolC and of AcrB are shown in different colors. The complex spans the distance from the cytoplasmic surface of the bacterial inner membrane to the extracellular surface of the outer membrane. (PDB 1EK9 (TolC) and 2DR6 (AcrB))

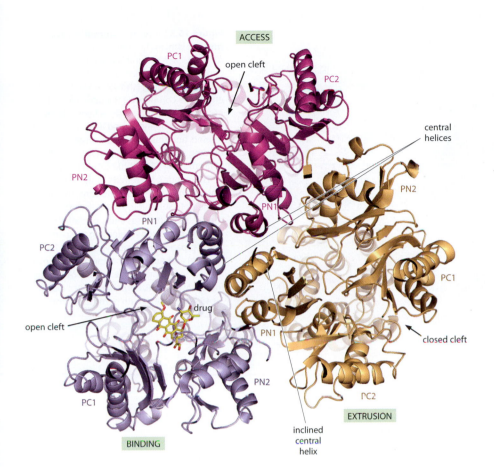

Figure 16.32 The AcrB pore domains.
A view from the periplasm, looking inward from the TolC-docking domain. The three subunits are colored purple (access subunit), blue (binding subunit), and gold (extrusion subunit), as in Figure 16.31. The drug doxorubicin, in yellow, is attached to the binding subunit. (PDB 2DR6)

TolC-docking domain. TolC, also a trimer, forms a tube of α helices leading to a β-barrel domain that provides an export pathway through the outer membrane. AcrA (not shown) surrounds TolC and AcrB, completing the complex.

AcrB determines substrate specificity and uses a H^+ gradient for energy. The three subunits enclose a cavity which is constricted by three central helices in the region where the pore domains meet the TolC docking domains (**Figure 16.32**). A pore domain consists of four subdomains named PN1, PN2, PC1, and PC2, each based on two β–α–β folds. These domains house the substrate-binding pocket, which is lined with aromatic hydrophobic residues and is large enough (~5000 $Å^3$) to accommodate a variety of molecules. AcrB takes in substrates from the periplasm through an opening in the pore domain at the membrane/periplasm boundary and extrudes them from the top funnel into the TolC pore.

A key feature of the AcrB trimer is that its three subunits are not *equivalent* (Figure 16.32). Drug binds to only one subunit at a time—the binding subunit—not to the extrusion subunit nor the access subunit. The central helix of the extrusion subunit inclines by nearly 15° toward the binding subunit, blocking exit from the binding site into the central cavity. In contrast, the vacant binding site of the extrusion subunit is open toward the exit tunnel. The access subunit also has a vacant binding site but has an upright central helix and represents an intermediate state between extrusion and binding, awaiting the next drug-binding event.

AcrB uses a rotary conformational change mechanism in which each subunit cycles through the conformations for access, binding, and extrusion (**Figure 16.33**). In the access state, the cleft is open for drug entry but the binding site is closed. In the binding state, the binding site expands (domains PN2 and PC1 move away from PN1 and PC2) to form the hydrophobic pocket that accommodates the drug, and the cleft remains open. Exit from the binding site to the pore is blocked by the central helix. In the extrusion state, the access cleft closes and the binding site shrinks (PN2 and PC1 move back toward PN1 and PC2). The central helix shifts and the drug can exit via the axial pore. The drug is thereby squeezed from the binding site to the pore and thence into the top funnel.

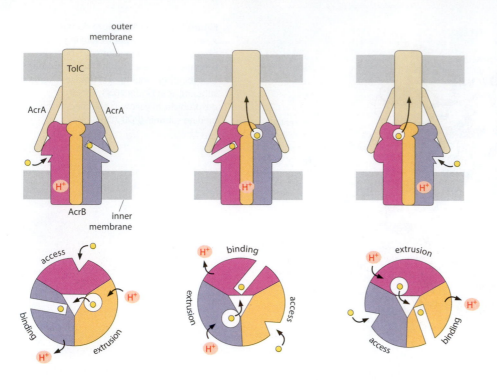

Figure 16.33 A rotary alternating access mechanism. In this mechanism proposed for AcrB, each subunit is in a different conformation and functional state. In a first step, the drug (shown as a yellow ball) enters the complex via the open cleft of the access subunit (magenta). A hydrophobic pocket then opens up and allows substrate to bind, and this subunit becomes the binding subunit (middle panel: this is the conformation of the blue subunit in step 1). This subunit then shifts to the extrusion state (right panel: this is the conformation of the gold subunit in step 1). In this state, the access cleft closes and a pocket opens near the pore funnel, releasing the substrate. The top panel shows side views, with TolC (hollow) and AcrA. The bottom panel shows an axial view from the periplasm. (Adapted from S. Murakami, *Curr. Opin. Struct. Biol.* 18:459–465, 2008. With permission from Elsevier.)

Opening and closing of the pore entrances and of the gate formed by the central helices is coupled to rearrangements in the proton-translocation site in the transmembrane (TM) region. This site involves the only charged residues in the middle of the TM region; these have been shown by mutagenesis studies to be essential for AcrB function. The transitions are driven by protonation and deprotonation of Asp residues and the making and breaking of salt bridges. As a result, conformational change is induced in a TM helix that in turn triggers conformational changes in the pore domain.

In contrast to the MFS transporters, in which H^+ and substrate are translocated via the same pathway, in AcrB the drug-binding sites and the proton translocation sites are separated. The mechanism of AcrB is reminiscent of that of F_1 ATPase (see Chapter 15): both are trimeric assemblies and operate by coupling a binding change mechanism to proton transport.

16.6   THE P-TYPE ATPASE PUMPS

The **P-type ATPases** are so called because the transporter is autophosphorylated by ATP during the reaction cycle. The Na^+/K^+-ATPase, the first P-type ATPase to be identified, was discovered by Jens Christian Skou in 1957. This ATPase pumps Na^+ out of the cell against its gradient and pumps K^+ in, thus keeping the concentration of K^+ 10–30 times greater inside the cell than outside. The family also includes the H^+/K^+-ATPases, responsible for acidifying the stomach as part of the digestive system, and the Ca^{2+}-ATPase, responsible for keeping Ca^{2+} levels low in the cytoplasm. Here we focus on the Ca^{2+}-ATPase and the Na^+/K^+-ATPase, two systems for which structural data have clarified the underlying mechanisms.

Calcium ions are key intracellular signaling molecules. In resting cells, cytoplasmic Ca^{2+} is kept low (<1 μM) by pumping Ca^{2+} out into the extracellular space or an internal compartment (the ER or, in muscle, the sarcoplasmic reticulum (SR)), in which levels of Ca^{2+} are high (> 1 mM) (Table 16.2). These channels are stimulated to admit Ca^{2+} by voltage changes or ligand binding, for example in the stimulation of muscle contraction (Section 14.4). When the stimulus for opening ceases, the Ca^{2+} channels close and transporter proteins restore the cytoplasmic Ca^{2+} concentration to a low level. The most important component is the Ca^{2+}-ATPase pump located in the plasma membrane and in the ER and SR (the latter is also called the <u>s</u>arco(<u>end</u>o)plasmic <u>r</u>eticulum <u>C</u>a^{2+}-<u>A</u>TPase or SERCA). Muscle and nerve cells also have a second pump, the Ca^{2+} transport protein, which couples the efflux of Ca^{2+} to the influx of Na^+ across the plasma membrane.

The SERCA ATPase pumps Ca^{2+} out of the cytoplasm

Early EM (see **Methods**) studies showed that the SR membranes of fast-twitch muscles are packed with Ca^{2+}-ATPases in ordered arrays. This abundance facilitated biochemical characterization and further structural studies. SERCAs have turnover rates of ~100/s, but influx of Ca^{2+} through Ca^{2+} channels in response to signals is much faster (~10^6/s). Accordingly, large numbers of ATPases are needed to clear Ca^{2+} rapidly from the cytoplasm. SERCA enzymes are estimated to account for ~25% of the total ATP turnover in working muscle.

The challenge facing the Ca^{2+}-ATPase is to bind Ca^{2+} with high affinity at the cytoplasmic face of the SR membrane, translocate it across the membrane against a concentration gradient, and release it into the lumen. High-affinity cytoplasm-facing Ca^{2+}-binding sites must be converted into low-affinity lumen-facing sites. The Ca^{2+}-ATPase transports counter-ions (H$^+$) in the opposite direction: two Ca^{2+} are pumped out of the cytoplasm and two or three H$^+$ are exchanged per ATP hydrolyzed.

Biochemical studies have distinguished two states of the enzyme—E1 and E2—in the delivery cycle (**Figure 16.34**). E1 states have a high affinity for cytoplasmic Ca^{2+}, whereas E2 states have low affinity for lumenal Ca^{2+}. Step 1 involves the binding of cytoplasmic Ca^{2+} and ATP, yielding the E1.2Ca^{2+}.ATP state. Transfer of the γ-phosphate from ATP to an Asp residue (Asp351) brings the enzyme into the E1P.2Ca^{2+} state, in which the bound Ca^{2+} are occluded at their intramembrane binding sites (step 2). There follows a conformational shift from the E1P to the E2P state, which delivers the Ca^{2+} to the SR lumen. At the same time, two or three H$^+$ access the now vacant cation sites to protonate acidic residues, forming the E2P.nH$^+$ state (step 3). H$^+$ binding is coupled with hydrolysis of the phosphoryl-aspartate and release of the phosphate, giving the E2.nH$^+$ state (step 4). A further conformational shift brings the enzyme back to the E1 state, releasing H$^+$ ions to the cytoplasm and allowing access of Ca^{2+}, thereby permitting formation of the E1.2Ca^{2+} state.

ATP can bind in all conformational states, and at physiological concentrations (2–10 mM), the enzyme is likely to be saturated with ATP. ATP has either a catalytic or a modulatory role, depending on the state of the enzyme. The Ca^{2+}-ATPase is only phosphorylated from ATP in its catalytic role, when the two cytoplasmic Ca^{2+} are bound. Its functions of Ca^{2+} translocation, enzyme dephosphorylation, and release of H$^+$ depends on ATP in its modulatory role. For simplicity, the modulatory roles have been omitted from Figure 16.34.

The binding sites for Ca^{2+} and ATP in Ca^{2+}-ATPase are 50 Å apart

Rabbit skeletal muscle SR Ca^{2+}-ATPase is a 110 kDa integral membrane protein. The first crystal structure depicted the E1.2Ca^{2+} state. It revealed 10 transmembrane helices and a large cytoplasmic domain made up of three subdomains: the P-domain (phosphorylation) adjacent to the TM helices and including the phosphorylation site Asp351; the N-domain (nucleotide-binding); and the A-domain (actuator), which undergoes large rotations in switching between different conformational states (**Figure 16.35A,B**). The P-domain,

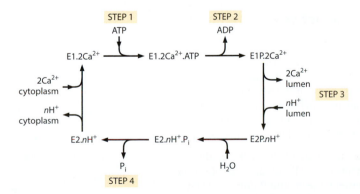

Figure 16.34 Reaction cycle of the Ca^{2+}-ATPase. There are two conformational states, E1 and E2. E1.2Ca^{2+} is E1 with two Ca^{2+} bound, E1P is E1 with phosphate bound to Asp351, and so on. Mg^{2+} (not shown) is bound to the ATP-bound and phosphorylated states. The number of protons, *n*, is two or three. (Adapted from T. Shinoda et al., *Nature* 459:446–450, 2009. With permission from Macmillan Publishers Ltd.)

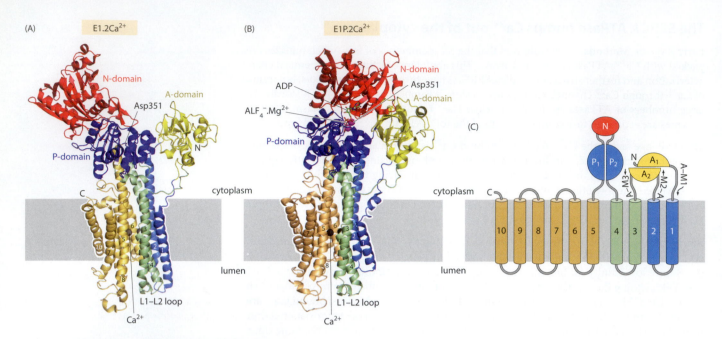

Figure 16.35 Structure of the Ca²⁺-ATPase in two E1 states. (A) The E1.2Ca²⁺ state in the presence of an inactive ATP analog, adenosine 5′-(β,γ-methylene)triphosphate (AMPPCP). TM helices 1–10 are labeled. See (C) for color-coding. Two Ca²⁺ (gray balls) are bound in the TM region. (PDB 1SU4) (B) The E1P.2Ca²⁺ state, as determined from the ATPase in the presence of Ca²⁺ and ADP.AlF₄⁻.2Mg²⁺. Rotations of the N-domain and A-domain bring Asp351, the phosphorylation site, close to AlF₄⁻, the group that mimics the γ-phosphate of ATP. (PDB 1T5T) (C) Schematic diagram of the connectivity of TM helices and cytoplasmic domains.

combining a seven-stranded parallel β sheet with eight short helices, resembles the nucleotide-binding Rossmann fold.

From both the A-domain and the P-domain, the polypeptide chain makes excursions into the TM helices (Figure 16.35C). Thus, there is a direct line of communication between the cytoplasmic domains and the TM region. The two Ca²⁺-binding sites are in the middle of the TM region. Site I is located between helices TM5 and TM6; this Ca²⁺ is chelated by side-chain oxygens. Site II is 5.7 Å away and contacts residues from TM4 and TM6; the coordination of Ca²⁺ is mostly to main-chain oxygens (**Figure 16.36A**). Access to the binding pocket from the cytoplasm can occur via a route between TM1 and TM2.

The two functional sites, ATP-binding and Ca²⁺-binding, are more than 50 Å apart, and Asp351 is more than 25 Å from the ATP-binding site (Figure 16.35B). Thus, major conformational changes are needed to connect the phosphorylation of Asp351 with Ca²⁺ transport.

Conformational changes switch the enzyme between different functional states

Crystal structures have been determined for the Ca²⁺-ATPase in its E1P.2Ca²⁺ state (Figure 16.36B) and E1.2Ca²⁺.ATP state (not shown). The two structures are very similar and differ from the E1.2Ca²⁺ state (Figure 16.36A). E1P.2Ca²⁺ was solved in the presence of AlF₄⁻, a planar molecule that mimics the ATPase transition state in which the γ-phosphate of ATP has moved toward Asp351. During transfer, the phosphate group undergoes inversion, and in the transition state it should be planar (see Box 3.1 for phosphoryl transfer mechanisms).

In the E1P.2Ca²⁺ and E1.2Ca²⁺.ATP states, the cytoplasmic part of the ATPase adopts a compact structure, achieved by rotating the N-domain toward the P-domain and a 45° rotation of the A-domain to make contact with a crevice between the N- and P-domains (Figure 16.35A,B). The A-domain rotation is accompanied by a shift of the TM2 helix by 9 Å and sliding of the L1–L2 loop toward the membrane. The N-terminal part of TM1 becomes kinked, lying almost parallel to the membrane surface. As a result of these movements, the gate to the Ca²⁺ sites becomes closed by a hydrophobic plug, sequestering the bound Ca²⁺ (Figure 16.36B).

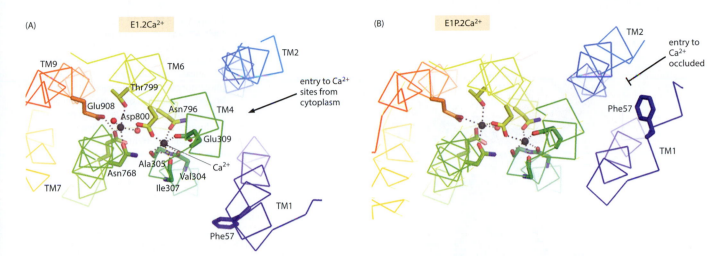

Figure 16.36 The Ca²⁺-binding sites in the Ca²⁺-ATPase transmembrane region. The view is perpendicular to the membrane plane, from the cytoplasmic side. The TM helices are colored as in Figure 16.35. The residues chelating the two Ca²⁺ (gray spheres) are labeled. Water molecules are shown as red balls. (A) The E1.2Ca²⁺ state. There is an open route to the Ca²⁺ sites from the cytoplasm between helices TM1 and TM2. (B) The E1P.2Ca²⁺ state. Movement of TM1 relative to TM2 occludes the site, as shown by the shift of Phe57. (PDB 1SU4 and 1T5T)

In the E1.2Ca²⁺.ADP.AlF₄⁻ state, the N- and P-domains are brought together so that the AlF₄⁻ interacts with Asp351 and the β-phosphate of ADP in a near-linear arrangement (**Figure 16.37A**). The structure is stabilized by two Mg²⁺.

The conformational change from E1P to E2P delivers Ca²⁺ to the lumen

To visualize the E2 state, the Ca²⁺-ATPase was crystallized in the presence of BeF₃⁻, a compound that covalently modifies Asp351 and binds Mg²⁺ to simulate the phosphorylated state E2P.nH⁺. Biochemical studies characterized this form of the enzyme as an E2P state with low affinity for Ca²⁺. The structure showed large changes from the E1P.2Ca²⁺ state. The residues at the Ca²⁺-binding sites shift, opening a pathway to the lumen (**Figure 16.38**). How does phosphorylation trigger formation of this exit pathway? In the E2P conformation, the tight contact between the A-domain and the N-domain relaxes and the interactions between the A-domain and the Asp351 phosphorylation site are altered. A loop from the A-domain with the conserved signature motif Thr¹⁸¹-Gly-Glu-Ser¹⁸⁴ (TGES) replaces

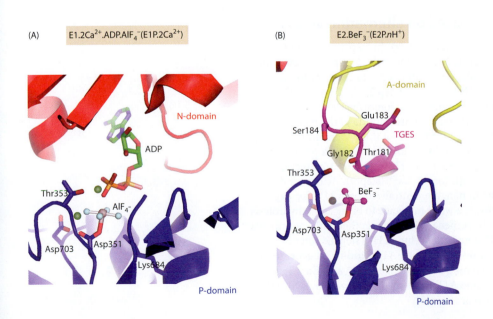

Figure 16.37 The Ca²⁺-ATPase phosphorylation site. (A) The E1P state (as in the complex with ADP.Mg²⁺ and AlF₄⁻). The adenine and ribose of ADP bind to the N-domain, while the β-phosphate is directed toward the P-domain. The AlF₄⁻ is covalently bonded to Asp351 and mimics the transition state for phosphoryl transfer. Key residues from the P-domain are shown. The two Mg²⁺ are shown as green balls. (B) The E2P. nH⁺ state from the complex with BeF₃⁻. The A-domain has displaced the N domain and the TGES loop is in the site previously occupied by ADP. ATP remains bound to the N-domain, which has shifted. The BeF₃⁻ forms a covalent bond to Asp351, mimicking the E2P state. The Mg²⁺ is represented by a gray ball. (PDB 1T5T and 3B9B)

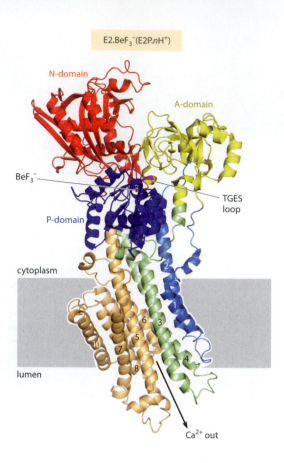

E2.BeF$_3^-$(E2P.nH$^+$)

N-domain

A-domain

BeF$_3^-$

P-domain

TGES
loop

cytoplasm

lumen

Ca^{2+} out

Figure 16.38 The E2P state of the Ca^{2+}-ATPase. The A-domain has shifted significantly from the E1P state, accompanied by a rearrangement of the TM helices. The TGES loop now contacts the phosphorylation site marked by the BeF$_3^-$ bound to Asp351. A path opens between TM3 and TM5 from the Ca^{2+}-binding sites to the lumen, and the Ca^{2+}-binding residues are displaced. (PDB 3B9B)

ADP and makes extensive interactions with the Mg^{2+}-bound aspartyl phosphoanhydride mimic (Figure 16.37B). The N-domain shifts but still binds ATP in its modulatory role. Movement of the TGES motif in the A-domain correlates with shifts in the TM1–TM2 and TM3–TM4 helices, which spread out and separate from the TM5–TM6 helices. As a consequence of these shifts, three of the Ca^{2+}-binding residues (Glu309, Glu771, and Asn796) become exposed to the lumen, and the affinity for Ca^{2+} is lowered. The Ca^{2+} sites are coupled to the phosphorylation sites through linkage of the A-domain to helices TM1 and TM3, which controls the opening and closing of the lumenal pathway.

A mechanism for Ca^{2+} transport is based on alternating access

This account of the functional cycle is shown in **Figure 16.39**. In step 1, ATP binds in its modulatory role and the Ca^{2+} sites are accessible (E1.ATP). Once the pump has been loaded with Ca^{2+}, the conformation at the catalytic site alters, allowing the reaction to proceed. Asp351 becomes phosphorylated, giving the E1P.2Ca^{2+} state (step 2). At this stage, the Ca^{2+} is sequestered. The phosphorylated enzyme then undergoes conformational changes that lead to the release of Ca^{2+} (and the uptake of H$^+$) on the lumenal side of the membrane in the E2P.nH$^+$ state (step 3). The main driver is a 120° rotation of the A-domain that brings the TGES loop close to the phosphorylation site. As the A-domain rotates, it pulls on helix pairs TM1–TM2 and TM3–TM4, separating them from TM5–TM10. This distorts the Ca^{2+}-binding sites and creates a lumenal exit path for the Ca^{2+} (step 3). Acidic residues at the cation-binding site, now exposed to the lumen, are neutralized with H$^+$. For dephosphorylation, residues in the TGES motif activate a water molecule to bring about hydrolysis of the phosphorylated Asp351. Closure of the lumenal pathway together with movements of the A- and P-domains lead to the E2.nH$^+$ state with occluded H$^+$ (step 4). After dephosphorylation, the cycle completes, stimulated by modulatory ATP binding to generate the E1.ATP state, with a cytoplasmic pathway opening for exchange of H$^+$ and Ca^{2+} (back to step 1). The net reaction for this cycle entails one ATP molecule hydrolyzed to ADP and P$_i$, and two Ca^{2+} transported in exchange for two or three H$^+$.

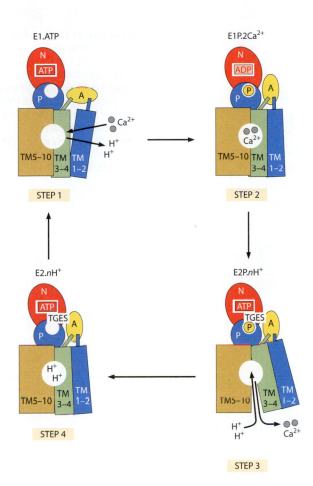

Figure 16.39 Schematic representation of the Ca^{2+}-ATPase reaction cycle. Step 1: the Ca^{2+} site is readily accessible. The N-domain binds ATP at the modulatory site. Step 2: transfer of the γ-phosphate from ATP at the catalytic site to an Asp residue yields the covalently modified phospho-Ca^{2+}-ATPase in the E1P.$2Ca^{2+}$ state in which two bound Ca^{2+} are occluded at their transmembrane binding sites. Step 3: a conformational switch from the E1P state to the E2P state delivers Ca^{2+} to the SR lumen. At the same time, two or three H^+ access the now vacant cation sites to protonate acidic residues, giving E2P.nH^+. Step 4: hydrolysis of the phosphoryl-Asp catalyzed by the TGES loop results in the release of inorganic phosphate and the formation of E2.nH^+. A further conformational shift brings the enzyme back to the E1 state, releasing the H^+ to the cytoplasm and allowing access of a Ca^{2+} to produce the E1.$2Ca^{2+}$ state. (Adapted from C. Olesen et al., *Nature* 450:1036–1044, 2007. With permission from Macmillan Publishers Ltd.)

The Na^+/K^+-ATPase regulates the cellular concentrations of Na^+ and K^+

This enzyme is present in the plasma membranes of almost all animal cells and is a major contributor to the electrochemical gradient. It transports two K^+ inward for every three Na^+ it pumps out. The Na^+ gradient thus produced drives the transport of nutrients into cells and regulates cytoplasmic pH. The Na^+/K^+-ATPase also controls the concentration of solutes inside the cell and contributes to the osmolarity.

In the Na^+/K^+-ATPase, an Asp residue is transiently phosphorylated in the presence of Na^+ during its transport cycle. The phosphoryl-Asp intermediate is hydrolyzed if K^+ is present. Two states, E1 and E2, have been recognized, as with the Ca^{2+}-ATPase (**Figure 16.40**). The E1.ATP state accepts three Na^+ from the cytoplasm, leading to the phosphorylation of Asp376 and formation of the phospho-intermediate E1P.$3Na^+$. After the E1→E2 transition, the Na^+ are delivered to the extracellular space and two K^+ become bound, giving the E2P.$2K^+$ state. The phospho-intermediate is then hydrolyzed, yielding the E2P.$2K^+P_i$ state and then the E2.$2K^+$ state. In the presence of ATP, the E2 state equilibrates to the E1 state, delivers two K^+ to the cytoplasm, and is ready to take up Na^+. Both the E1 and E2 states can adopt conformations in which the ions are occluded.

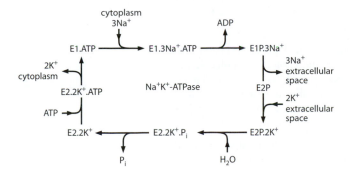

Figure 16.40 Reaction cycle of the Na^+/K^+-ATPase. There are two conformational states, E1 and E2. The E1.ATP state accepts three Na^+ from the cytoplasm, and conversion to the E2P state delivers three Na^+ to the extracellular space. The E2P state accepts two K^+ from the extracellular space, and subsequent conversion to the E1.ATP state delivers these to the cytoplasm. Mg^{2+} is bound to the ATP and phosphorylated states but is not shown. (Adapted from T. Shinoda et al., *Nature* 459:446–450, 2009. With permission from Macmillan Publishers Ltd.)

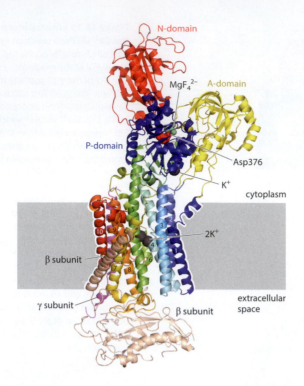

Figure 16.41 The E2.2K⁺.Pᵢ state of the Na⁺/K⁺-ATPase. The view is similar to Figure 16.38 for the Ca²⁺-ATPase. The β subunit is composed of one TM helix and a large extracellular domain. The γ subunit (purple) is partly obscured by the TM9 helix. Two K⁺ in the TM region and a third K⁺ at a site at the boundary between the TM and the P-domain are shown as gray spheres. The MgF₄⁻ mimics the phospho group and is close to Asp376. (PDB 2ZXE)

The structure of Na⁺/K⁺-ATPase differs from Ca²⁺-ATPase at the cation-binding sites

The Na⁺/K⁺-ATPase from shark renal gland consists of a 113 kDa α subunit and a 35 kDa β subunit. The α subunit contains the binding sites for Na⁺, K⁺, and ATP and shares 31% sequence identity with the rabbit Ca²⁺-ATPase over structurally similar regions (~80% of the molecule). The β subunit serves to locate the α subunit at the plasma membrane and define the K⁺-binding conformation. A third component, the γ subunit, is a small (10 kDa) intrinsic membrane protein that appears to regulate pumping in a tissue-specific manner.

Crystallography of the Na⁺/K⁺-ATPase in the E2.2K⁺.Pᵢ state revealed structural similarities to the Ca²⁺-ATPase (**Figure 16.41**). However, the A- and N-domains of the Na⁺/K⁺-ATPase hardly interact and the N-domain is 22° farther from the P-domain than in the Ca²⁺-ATPase. ATP can be docked at the nucleotide-binding site by analogy with the Ca²⁺-ATPase.

The potassium ions bind at two sites (I and II) in the transmembrane region, at positions similar but not identical to the Ca²⁺-binding sites of the Ca²⁺-ATPase. The K⁺ are coordinated as shown in **Figure 16.42**. In the central part of TM5, a Pro residue causes the TM5 helix to kink so that the main-chain oxygen of Thr779 becomes available for K⁺ coordination. The two bound K⁺ (ionic radius 1.33 Å) are only 4.1 Å apart and share contacts to Asn783 and Asp811. At the Ca²⁺-binding sites in the E1 Ca²⁺-ATPase, the Ca²⁺ are 5.7 Å apart (compare Figure 16.36). Despite this, the Na⁺/K⁺-ATPase has a larger binding cavity for K⁺. The positions of the coordinating residues are very similar in the E2.Pᵢ states of the two ATPases (except for Asn783), and the Na⁺/K⁺-ATPase has been shown to bind H⁺ as well as K⁺. The Ca²⁺-ATPase is only able to bind H⁺ in the E2P state. A third K⁺ site is observed in the P-domain and may be involved in regulation.

The TM helix of the β subunit is in contact with TM7 and TM10 (Figure 16.41) and helps to protect the unwound portion of TM7. There is also a bound cholesterol molecule (not shown) that shields the unwound part of TM7 from lipid. Na⁺/K⁺-ATPase activity is dependent on the presence of cholesterol. TM7 interacts with TM5 and thus the β subunit stabilizes the K⁺-bound state. The γ subunit has a transmembrane segment that interacts with TM9 and also with the β subunit, engaging in interactions that stabilize the α/β complex.

The structure of the Na⁺/K⁺-ATPase in the E2.2K⁺.Pᵢ state explains its high-affinity K⁺ binding. Mutagenesis experiments and analysis of this structure indicate that the unwinding of

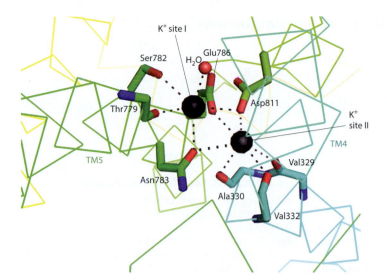

Figure 16.42 The two K⁺-binding sites in the transmembrane region of the Na⁺/K⁺-ATPase. The view is perpendicular to the membrane from the cytoplasmic side. The TM helices are colored as in Figure 16.41. Water molecules are shown as red spheres. (PDB 2ZXE)

the TM7 helix and its contacts that stabilize TM5 and the C-terminal region of TM10 are likely to be important in the E1P–E2P transition.

16.7 SUMMARY

The proteins that convey ions and small polar molecules across the hydrophobic barrier presented by a membrane can be divided into two classes: channels and transporters. In channels, ions or water (substrates) move by diffusion through gated pores; in transporters, the freight is moved by coupling transport to an energy source.

Channel proteins have transmembrane regions composed of α helices, often in a symmetrical arrangement around a central pore, as in the pentameric acetylcholine receptor. In the dimeric chloride channels and tetrameric aquaporins, each subunit has an independent pore. Channels have selectivity filters within their pores whose geometry allows their polar groups to substitute for the hydration shells of the ions being transported, reducing the energy penalty associated with dehydration. This filter is matched with the coordination geometry of the ion in question. Electrostatic repulsion between the multiple bound ions and conformational changes induced in the pore by ion binding help to keep binding weak enough to permit rapid flux through the pore.

Additional structural regions provide a 'gate' to the pore that allows regulation of ion flux by ligand binding or voltage sensing. Ligand-binding domains may be on the intracellular or extracellular side, whereas voltage sensing is performed by integral membrane domains. In the mechanosensitive MscS channel, transmembrane helices move in response to mechanical stress, opening the central pore. The acetylcholine receptor acts as an allosterically controlled 'on-off' switch, whereby the binding of ligand (acetylcholine) to two extracellular sites induces a conformational change in the gating region of the pore, ~50 Å distant, transiently opening it to allow the passage of ions.

Transporters bind substrate on one side of the membrane and change conformation, first so that the substrate is then occluded from both sides of the membrane, and then again, releasing the substrate on the other side of the membrane. This is known as the alternating access mechanism. Active transporters, such as the ABC transporters and the Ca²⁺-ATPase, harness the hydrolysis of ATP to transport solutes or ions across membranes. Other transporters, such as lactose permease, use concentration gradients of ions as energy source. In addition to transmembrane domains, transporters have substrate-binding domains and nucleotide-binding domains that function as the engines of these transporter machines. During its pumping cycle, the cytoplasmic domains of the Ca²⁺-ATPase undergo large rotations. The multidrug resistance protein AcrB shows a variation on the alternating access mechanism: each of its three homomeric subunits cycles through three conformational states, sequentially allowing substrate access, binding, and extrusion.

Sections of this chapter are based on the following contributions

16.3   Nigel Unwin, MRC Laboratory of Molecular Biology, Cambridge, UK.

16.4   Robert M. Stroud, University of California, San Francisco, USA.

16.5   Alex Cameron, Diamond Light Source, Didcot, UK and Konstantinos Beis, Imperial College London, UK.

16.6   Thomas Sørensen, Diamond Light Source, Didcot, UK.

REFERENCES

General

Ashcroft F, Gadsby D & Miller C (2009) The blurred boundary between channels and transporters. *Phil Trans R Soc B* 364:145–147.

Colquhoun D, Lape R & Sivilotti L (2009). Single ion channels. In Single Molecule Biology (Knight AE ed.), pp 223–251. San Diego: Academic Press.

Sonoda Y, Cameron A, Newstead S et al (2010) Tricks of the trade used to accelerate high-resolution structure determination of membrane proteins. *FEBS Lett* 584:2539–2547.

16.2 Ion channels

Accardi A & Miller C (2004) Secondary active transport mediated by a prokaryotic homologue of CLC Cl⁻ channels. *Nature* 427:803–807.

Ashcroft FM (2000) Ion Channels and Disease. London: Academic Press.

Bass RB, Strop P, Barclay M et al (2002) Crystal structure of *Escherichia coli* MscS, a voltage-modulated and mechanosensitive channel. *Science* 298:1582–1587.

Beckstein O & Sansom MSP (2004) The influence of geometry, surface character, and flexibility on the permeation of ions, and water through biological pores. *Phys Biol* 1:42–52.

Doyle DA, Morais Cabral J, Pfuetzner RA et al (1998) The structure of the potassium channel: molecular basis of K⁺ conduction and selectivity. *Science* 280:69–77.

Dutzler R, Campbell EB & MacKinnon R (2003) Gating the selectivity filter in CLC chloride channels. *Science* 300:108–112.

Gonzales EB, Kawate T & Gouaux E (2009) Pore architecture and ion sites in acid-sensing ion channels and P2X receptors. *Nature* 460:599–604.

Gouaux E & Mackinnon R (2005) Principles of selective ion transport in channels and pumps. *Science* 310:1461–1465.

Jiang Y, Lee A, Chen J et al (2002) Crystal structure and mechanism of a calcium-gated potassium channel. *Nature* 417:515–522.

Jiang Y, Lee A, Chen J et al (2002) The open pore conformation of potassium channels. *Nature* 417:523–526.

Long SB, Tao X, Campbell EB et al (2007) Atomic structure of a voltage-dependent K⁺ channel in a lipid membrane-like environment. *Nature* 450:376–382.

Nishida M, Cadene M, Chait BT et al (2007) Crystal structure of a Kir3.1-prokaryotic Kir channel chimera. *EMBO J* 26:4005–4015.

Sobolevsky AI, Rosconi MP & Gouaux E (2009) X-ray structure, symmetry and mechanism of an AMPA-subtype glutamate receptor. *Nature* 462:745–756.

Tao X, Avalos JL, Chen J et al (2009) Crystal structure of the eukaryotic strong inward-rectifier K⁺ channel Kir2.2 at 3.1 Å resolution. *Science* 326:1668–1674.

Wang W, Black SS, Edwards MD et al (2008) The structure of an open form of an *E. coli* mechanosensitive channel at 3.45 Å resolution. *Science* 321:1179–1183.

16.3 Nicotinic acetylcholine receptor

Albuquerque EX, Pereira EFR, Alkondon M et al (2009) Mammalian nicotinic acetylcholine receptors: from structure to function. *Physiol Rev* 89:72–120.

Brejc K, van Dijk WJ, Klaassen RV et al (2001) Crystal structure of an ACh-binding protein reveals the ligand-binding domain of nicotinic receptors. *Nature* 411:269–276.

Dellisanti CD, Yao Y, Stroud JC et al (2007) Crystal structure of the extracellular domain of nAChR α1 bound to α-bungarotoxin at 1.94 Å resolution. *Nature Neurosci* 10:953–962.

Hanek AP, Lester HA & Dougherty DA (2008) A stereochemical test of a proposed structural feature of the nicotinic acetylcholine receptor. *J Am Chem Soc* 130:13216–13218.

Miyazawa A, Fujiyoshi Y & Unwin N (2003) Structure and gating mechanism of the acetylcholine receptor pore. *Nature* 423:949–955.

Sine SM & Engel AG (2006) Recent advances in Cys-loop receptor structure and function. *Nature* 440:448–455.

Unwin N (2005) Refined structure of the nicotinic acetylcholine receptor at 4 Å resolution. *J Mol Biol* 346:967–989.

Unwin N & Fujiyoshi Y (2012) Gating movement of acetylcholine receptor caught by plunge-freezing. *J Mol Biol* 422:617–934.

16.4 Aquaporins

Engel A, Fujiyoshi Y, Gonen T et al (2008) Junction-forming aquaporins. *Curr Opin. Struct Biol* 18:229–235.

Ho JD, Yeh R, Sandstrom A et al (2009) Crystal structure of human aquaporin 4 at 1.8 Å and its mechanism of conductance. *Proc Natl Acad Sci USA* 106:7437–7442.

King LS, Kozono D & Agre P (2004) From structure to disease: the evolving tale of aquaporin biology. *Nature Rev* 5:687–698.

Murata K, Mitsuoka K, Hirai T et al (2000) Structural determinants of water permeation through aquaporin-1. *Nature* 407:599–605.

Savage DF, O'Connell JD, Miercke LJ et al (2010) Structural context shapes the aquaporin selectivity filter. *Proc Natl Acad Sci USA* 107:17164–17169.

Stroud RM, Savage D, Miercke LJ et al (2003) Selectivity and conductance among the glycerol and water conducting aquaporin family of channels. *FEBS Lett* 555:79–84.

16.5 Transporters

Abramson J, Smirnova I, Kasho V et al (2003) Structure and mechanism of the lactose permease of *Escherichia coli*. *Science* 301:610–615.

Aller SG, Yu J, Ward A et al (2009) Structure of P-glycoprotein reveals a molecular basis for poly-specific drug binding. *Science* 323:1718–1722.

Faham S, Watanabe A, Mercado Besserer G et al (2008) The crystal structure of a sodium galactose transporter reveals mechanistic insights into Na⁺/sugar symport. *Science* 321:810–814.

Guan L & Kaback HR (2006) Lessons from lactose permease. *Annu Rev Biophys Biomol Struct* 35:67–91.

Higgins CF (2007) Multiple molecular mechanisms for multidrug resistance transporters. *Nature* 446:749–757.

Hollenstein K, Dawson RJ & Locher KP (2007) Structure and mechanism of ABC transporter proteins. *Curr Opin Struct Biol* 17:412–418.

Huang Y, Lemieux MJ, Song J et al (2003) Structure and mechanism of the glycerol-3-phosphate transporter from *Escherichia coli*. *Science* 301:616–620.

Law CJ, Maloney PC & Wang DN (2008) Ins and outs of major facilitator superfamily antiporters. *Annu Rev Microbiol* 62:289–305.

Locher KP (2009) Structure and mechanism of ATP-binding cassette transporters. *Phil Trans R Soc* 364:239–245.

Murakami S (2008) Multidrug efflux transporter, AcrB—the pumping mechanism. *Curr Opin Struct Biol* 18:459–465.

Rees DC, Johnson E & Lewinson O (2009) ABC transporters: the power to change. *Nature Rev Mol Cell Biol* 10:218–227.

Shimamura T, Weyand S, Beckstein O et al (2010) Molecular basis of alternating access membrane transport by the sodium-hydantoin transporter Mhp1. *Science* 328:470–473.

Weyand S, Shimamura T, Yajima S et al (2008) Structure and molecular mechanism of a nucleobase-cation-symport-1 family transporter. *Science* 322:709–713.

Yamashita A, Singh SK, Kawate T et al (2005) Crystal structure of a bacterial homologue of Na^+/Cl^--dependent neurotransmitter transporters. *Nature* 437:215–223.

16.6 The P-type ATPase pumps

Bublitz M, Poulsen H, Morth JP et al (2010) In and Out of the cation pumps: P-type ATPase structure revisited. *Curr Opin Struct Biol* 20:431–439.

Olesen C, Picard M, Winther AM et al (2007) The structural basis of calcium transport by the calcium pump. *Nature* 450:1036–1042.

Shinoda T, Ogawa H, Cornelius F et al (2009) Crystal structure of the sodium–potassium pump at 2.4 Å resolution. *Nature* 459:446–450.

Skou JC (1957) The influence of some cations on an adenosine triphosphatase from peripheral nerves. *Biochim Biophys Acta* 23:394–401.

Sørensen TL, Møller JV & Nissen P (2004) Phosphoryl transfer and calcium ion occlusion in the calcium pump. *Science* 304:1672–1675.

Toyoshima C, Nakasako M, Nomura H et al (2000) Crystal structure of the calcium pump of sarcoplasmic reticulum at 2.6 Å resolution. *Nature* 405:647–655.

Toyoshima C, Nomura H & Tsuda T (2004) Lumenal gating mechanism revealed in calcium pump crystal structures with phosphate analogues. *Nature* 432:361–368.

Complexes of the Immune System

17.1 INTRODUCTION

Our bodies are constantly exposed to microorganisms present in the environment, some of which have the potential to cause disease. The first line of defense is the layer of epithelial cells, held together by tight junctions (Section 11.3), which covers the outer surfaces of the body and lines the gastrointestinal, respiratory, and urogenital tracts. Infection occurs when microorganisms cross this barrier. In addition, two major defense systems enable us to actively resist infection. The **innate immune system**, evolutionarily the older, is found in organisms ranging from animals to fungi and plants. This system mounts a generic defense by recognizing molecules on the surfaces of the invading microorganisms and stimulating phagocytes that engulf and digest microorganisms and by producing an inflammatory response (**Box 17.1**). The **adaptive immune system**, occurring only in higher animals (vertebrates), recognizes specific pathogens and mobilizes neutralizing responses against them. Innate immunity provides an immediate response but does not confer long-term protection. The adaptive response takes several days to mature. During this time, there is a clonal expansion of B cells and T cells that produces the antigen-specific effector cells that confer long-lasting immunity.

The innate immune defense depends on phagocytes, cells that include macrophages and neutrophils. They detect pathogens by means of cell surface receptors that recognize particular structural motifs displayed by pathogens. The principal receptors are the **Toll-like receptors** (TLRs). Activation of TLRs triggers the production of pro-inflammatory *cytokines* and *chemokines* and the expression of co-stimulatory molecules. Innate immunity also employs a specialized set of proteins, the **complement system**, that recognizes pathogens and is rapidly activated to help destroy invaders. Complement comprises a large network of proteins synthesized mainly in the liver and circulating in blood plasma. Most of these proteins are inactive until activated by infection. Their name derives from the discovery of a heat-sensitive factor in serum that 'complemented' antibodies in the killing of bacteria. Through complement, the innate system is also able to enhance the adaptive immune response.

The adaptive immune system identifies foreign **antigens** via two pathways: the antibody-mediated pathway and the T-cell-mediated pathway. **Antibodies** are extracellular proteins of the immunoglobulin (Ig) family secreted into the bloodstream by *B cells* in the bone marrow. Antibodies bind specifically to foreign antigens (for example, toxic proteins, viruses, bacteria), leading to their neutralization or elimination. *T cells*, produced in the thymus, distinguish between self and foreign antigens by means of specialized machineries. These involve ingestion of the pathogen by dendritic cells through endocytosis, proteolytic degradation of the antigens to peptides, and presentation of the peptides at the target cell surface, where they are recognized by T cells. The infected cell is then destroyed. A combination of positive and negative selection processes operates during T cell development to shape the T cell receptor repertoire. These processes help to ensure that only T cells with potentially useful receptors survive and mature, while all others die by *apoptosis* (see Section 13.6).

Progress has been made toward understanding the macromolecular assemblies of both the innate and the adaptive immune systems, providing a platform for new therapeutic treatments to combat infectious diseases.

Inflammation is a condition marked by reddening, swelling, heating, and sensitivity or pain in the affected tissue. Inflammation represents the body's response to injury or to infection by a microbe or a virus. The symptoms of inflammation are generally induced by the host's immune system, primarily the innate immune system, to promote healing and combat the infection. Cells present at the site of infection (macrophages, dendritic cells, mast cells, and others) recognize the intruder and release substances such as cytokines that promote the inflammatory response. One consequence is vasodilation, which leads to enhanced blood flow into the affected area (hence the reddening and swelling) and a chemotactically driven influx of leukocytes, some of which ingest infecting bacterial cells by phagocytosis while others release hydrolytic enzymes to degrade invaders. Pain follows if the swelling impinges on nerve ends. Inflammation is a feature of many clinical conditions and can involve a wide variety of cellular and molecular mechanisms. If the signaling system that mobilizes inflammation is impaired in some way, such as by mutation, the symptoms of inflammation may be elicited in the absence of infection, giving rise to autoinflammatory and autoimmune diseases.

17.2   TOLL-LIKE RECEPTORS AND INFLAMMASOMES

This section focuses on two assemblies of the innate immune system: the Toll-like receptors (TLRs) and the recently discovered inflammasomes. These both recognize pathogens and foreign surfaces, triggering signaling events that lead to an inflammatory response.

Toll-like receptors invoke an inflammatory response that also assists the adaptive immune response

TLRs are named after a family of genes called Toll (German for 'amazing'). They were first identified in 1984 in a genetic screen of *Drosophila melanogaster*, where they were found to direct dorsal–ventral patterning in embryos. By the mid-1990s, it was established that insect TLRs also function in the innate immune response, giving these proteins a wider context than insect development. From genome sequence projects, similar proteins were identified in vertebrates. There are 10 such proteins in humans (the number varies in other vertebrates). All have a common organization that comprises a large N-terminal extracellular domain containing multiple repeats of a leucine-rich repeat (LRR) motif (see **Guide**), followed by a single transmembrane helix and an intracellular C-terminal domain that is related to the interleukin-1 receptor and is known as the Toll/interleukin-1 receptor (TIR) domain. (Interleukin-1 is a cytokine that is part of the acute-phase inflammatory response.)

Some mammalian TLRs act as cell surface receptors, whereas others are found inside cells in endosomal membranes, where they sense the presence of pathogens and pathogen components taken into the cell by endocytosis. Each human TLR recognizes a distinct set of macromolecular patterns found on microorganisms but not, normally, on vertebrate cells (**Table 17.1**). Some receptors form heterodimers while others act as homodimers. TLR4 is an exception in that it requires a co-receptor, the myeloid differentiation factor 2 (MD2), and ligand recognition is assisted by transient association with the peripheral membrane protein CD14, a protein that binds lipopolysaccharides (LPS), thereby detecting the presence of Gram-negative bacteria (Section 1.9). TLR dimerization is promoted by ligand binding. Dimerization in turn triggers the recruitment of adaptor proteins to the TIR domains to initiate signaling.

A schematic diagram for TLR4 signaling in response to LPS recognition is given in **Figure 17.1**. TLR dimerization signals through the membrane to bring the TIR domains together. TIR dimerization promotes recognition by the adaptors MyD88-adaptor-like (Mal) and TRIF-related adaptor molecule (TRAM). The Mal/TIR complex stimulates activities of the interleukin receptor kinase (IRAK) in association with myeloid differentiation primary response protein 88 (MyD88), and together with IκB kinase (IKK) (not shown) these complexes activate the transcription factor NFκB. NFκB induces the expression of a variety of genes including those encoding pro-inflammatory cytokines and chemokines. Through the TRIF protein (TIR domain-containing adaptor protein inducing interferon-β), the TRAM/TIR complex stimulates the interferon response factor 3 (IRF3) to induce interferon-β and interferon (IFN)-inducible genes. Other TLRs use similar signaling pathways either through Mal and MyD88 or through TRAM and TRIF.

Table 17.1 Innate immune recognition by Toll-like receptors.		
Human Toll-like receptor	Localization	Ligand
TLR1/TLR2	Plasma membrane	Lipoproteins or lipopeptides
TLR2/TLR6	Plasma membrane	Lipoproteins or lipopeptides
(TLR3)$_2$	Endosome	Viral double-stranded RNA
(TLR4)$_2$ plus MD2 and CD14	Plasma membrane	Lipopolysaccharide
(TLR5)$_2$	Plasma membrane	Flagellin
TLR6/TLR2	Plasma membrane	Diacyl lipopeptides
(TLR8)$_2$	Endosome	Single-stranded RNA; guanosine analogs and synthetic imidazolquinolines
(TLR9)$_2$	Endosome	Microbial DNA; unmethylated CpG DNA
(TLR10)$_2$	Endosome	Ligand not identified
TLR11	Plasma membrane	Profilin-like molecule

Adapted from O. Takeuchi and S. Akira, *Cell* 140:805–820, 2010.

Monophosphoryl lipid A (MPLA) preferentially induces TRAM signaling and is used as an adjuvant whereby it can drive an adaptive response. Thus there are links between the innate immune pathway and the adaptive immune response.

Toll-like receptors recognize a variety of ligands within a common framework

The extracellular domains of all TLRs contain 14–28 LRRs in addition to domains called LRRNT and LRRCT at their N and C termini. These domains do not have the LRR motif but

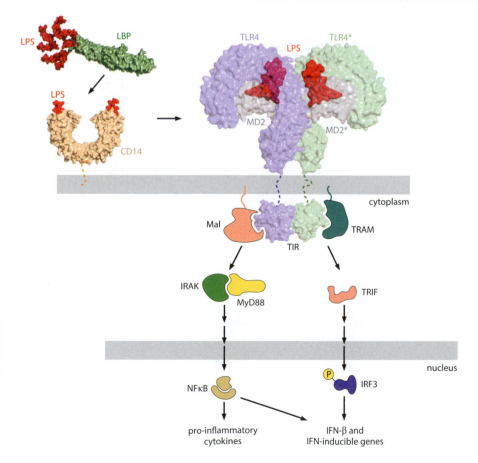

Figure 17.1 The TLR4 signaling pathway. Lipopolysaccharide (LPS, red) is recognized by the lipid-binding protein (LBP) and transferred to CD14, which then transiently associates with and assists LPS binding by TLR4/MD2. LPS binding results in dimerization of TLR4/MD2 and intracellular signaling that activates the inflammatory response through the transcription factor NFκB and interferon-β and interferon (IFN)-inducible genes. Molecules are not drawn to scale. (Adapted from B.S. Park et al., *Nature* 458:1191–1196, 2009. With permission from Macmillan Publishers Ltd.)

have several cysteine residues that engage in disulfide bonds. The structures of some TLR extracellular domains with bound ligands are shown in **Figure 17.2**. The TLR2/TLR1 heterodimer recognizes bacterial lipoproteins. Its crystal structure was determined associated

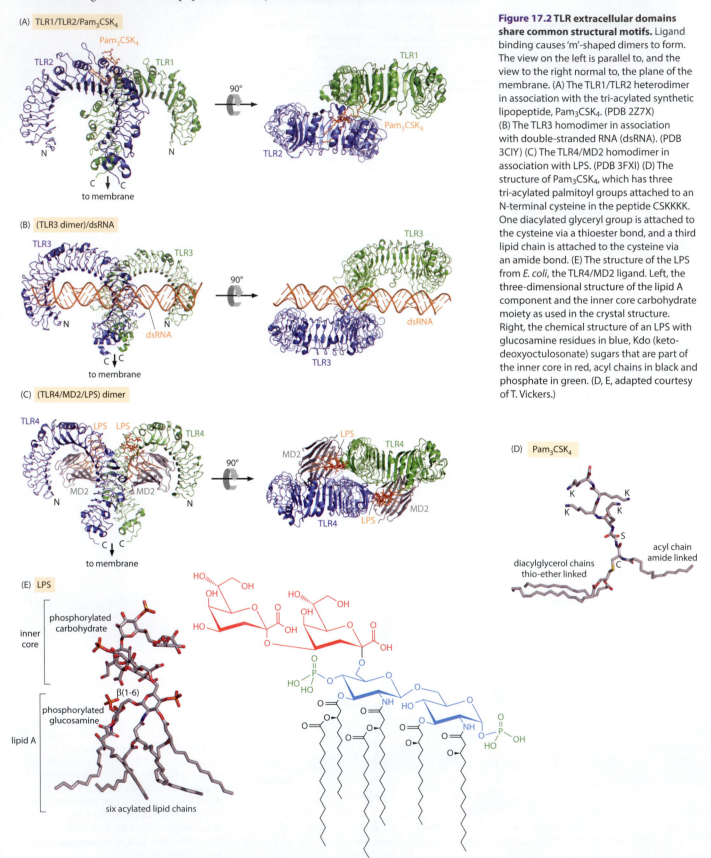

Figure 17.2 TLR extracellular domains share common structural motifs. Ligand binding causes 'm'-shaped dimers to form. The view on the left is parallel to, and the view to the right normal to, the plane of the membrane. (A) The TLR1/TLR2 heterodimer in association with the tri-acylated synthetic lipopeptide, Pam$_3$CSK$_4$. (PDB 2Z7X) (B) The TLR3 homodimer in association with double-stranded RNA (dsRNA). (PDB 3CIY) (C) The TLR4/MD2 homodimer in association with LPS. (PDB 3FXI) (D) The structure of Pam$_3$CSK$_4$, which has three tri-acylated palmitoyl groups attached to an N-terminal cysteine in the peptide CSKKKK. One diacylated glyceryl group is attached to the cysteine via a thioester bond, and a third lipid chain is attached to the cysteine via an amide bond. (E) The structure of the LPS from *E. coli*, the TLR4/MD2 ligand. Left, the three-dimensional structure of the lipid A component and the inner core carbohydrate moiety as used in the crystal structure. Right, the chemical structure of an LPS with glucosamine residues in blue, Kdo (keto-deoxyoctulosonate) sugars that are part of the inner core in red, acyl chains in black and phosphate in green. (D, E, adapted courtesy of T. Vickers.)

with a synthetic tri-acylated lipopeptide called Pam$_3$CSK$_4$ (Figure 17.2D). The dimer has a characteristic 'm' shape with the two C-terminal domains meeting at the center (Figure 17.2A). The acyl chains of the LPS ligand bridge the two TLRs: two chains insert into a pocket on TLR2 and the remaining amide-bound chain inserts into a channel on TLR1. These interactions are predominantly hydrophobic. The ligand-binding site on the convex surface of the horseshoe-shaped subunit is unusual in that most LRR proteins have their ligand-binding sites on the concave surface.

TLR3 recognizes double-stranded RNA (dsRNA), a replication intermediate found in cells infected with RNA viruses. In most cell types, TLR3 resides in the endosome, which has an acidic pH; TLR3 requires a low pH for ligand binding. The low pH and sequestration to the endosome prevent the host's own nucleic acids from generating an immune response. The TLR3 subunit has a flat horseshoe shape that is more similar to that of a canonical LRR than those of TLR1/TLR2 (compare Figure 17.2A,B). Its ligand-binding mode is also different from that of TLR1/TLR2. The dsRNA-binding sites are positioned in two regions near the N and C termini of TLR3. The complex is 140 Å long, thus explaining why TLR3 needs a minimum of 40–50 base pairs (40×3.4 Å $= 136$ Å) for recognition and activation. The interactions between TLR3 and dsRNA are mostly electrostatic and hydrogen bonds to the phosphate and the ribose, and there is no specificity for nucleotide bases. The structure of the extracellular domain of TLR3 does not change on dimer formation, thus supporting a mechanism in which the signaling pathway is induced by ligand-dependent dimerization that brings together the two cytoplasmic TIR domains.

TLR4 in association with cofactor MD2 mediates responses to the LPS of Gram-negative bacteria, as described above. *E. coli* LPS is composed of a hydrophobic lipid A attached to a branched-chain sugar core and an O antigen (Figure 17.2E). The lipid A portion, with four to seven acyl chains and phosphorylated di-glucosamine, triggers the immune response. The structure of LPS-bound TLR4/MD2 shows TLR4 to have the characteristic horseshoe form with MD2, a β-cup composed of two anti-parallel β sheets forming the major lipid-binding site through its large hydrophobic pocket (Figure 17.2C). LPS binds to MD2 through five of its lipid chains and contacts the TLR4* of the dimer through the sixth lipid chain. (The asterisk denotes the other TLR4 subunit of the dimer.) MD2 makes contact with the central portions of TLR4. The phosphate groups of lipid A bind to positively charged groups on TLR4, TLR4*, and MD2. LPS thus mediates dimerization.

An accessory protein called CD14 assists in LPS binding. It has been proposed that the CD14/LPS complex makes direct but transient contact with the TLR4/MD2 complex to transfer the ligand. CD14 is also an LRR protein with 11 repeats per subunit and dimerizes through interaction of the two C termini. Its LPS-binding site is located near the N-terminal region in a crevice on the convex surface.

LPS recognition by TLR4/MD2 can lead to *fatal septic shock syndrome* if the inflammatory response is over-amplified and uncontrolled. Knowledge of the TLR4/MD2/LPS dimer structure has led to the design of several soluble MD2-binding 'decoy' proteins based on the LRR scaffold. These have been shown to be effective in mice in attenuating the LPS cytokine response, thus opening the way for a possible new treatment of bacterial sepsis.

The cytosolic TIR domains of Toll-like receptors promote downstream signaling

TLR dimerization, triggered by the binding of ligand to the extracellular domains, brings the intracellular TIR domains together, and this leads to downstream signaling. The TIR domains of TLR1, TLR2, and TLR10 have a common fold, comprising a five-stranded β sheet surrounded by five α helices (**Figure 17.3**). The BB loop between βB and αB has been implicated, from mutational evidence, in dimerization and in signaling to adaptors. Modeling and mutational studies have implicated regions near the BB loops as the recognition sites for adaptor molecules.

Returning to the TLR4/MD2/LPS system, an adaptor called MyD88 is centrally involved. In addition to a C-terminal TIR domain, MyD88 has a short intermediate domain and an N-terminal death domain (DD) (a six-helix bundle; see **Guide**). Through its DD, MyD88 interacts with IRAKs, protein kinases that also have an N-terminal DD, connected in this case to a C-terminal Ser/Thr kinase domain. Despite the apparent simplicity of the MyD88/IRAK complex (illustrated schematically in Figure 17.1), these proteins assemble into more

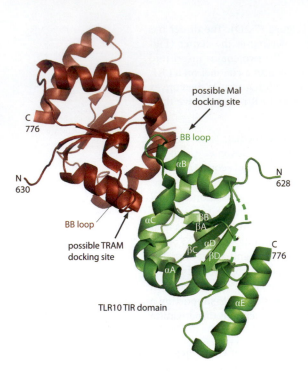

Figure 17.3 The TIR domain dimer of human TLR10. Disordered regions between αA and βB and between αB and βD are indicated by dashed lines on the green subunit. Possible docking sites for the Mal and TRAM adaptors were deduced from modeling and mutational studies for TLR4 TIR domain. In the intact protein, the N termini connect to the transmembrane region via linkers of 20–30 aa. The N termini are 58 Å apart and it is not clear how the connection to the transmembrane region is made. (PDB 2J67)

elaborate structures. A complex called the **myddosome** comprises a total of 14 DDs, six from MyD88, four from IRAK4, and four from IRAK2. The subunits are arranged in 3.5 turns of a left-handed helix, nucleated on the MyD88 DDs, to which IRAK4 is added, and finally IRAK2 (**Figure 17.4**). The C termini of the IRAK DDs are on the outside of the myddosome, suggesting that with the full-length proteins, appended kinase domains should cluster in this region, an arrangement conducive to activation by autophosphorylation. Although many details remain to be determined, the model shown in Figure 17.4C gives a plausible overall scheme for this ligand-responsive signaling machine.

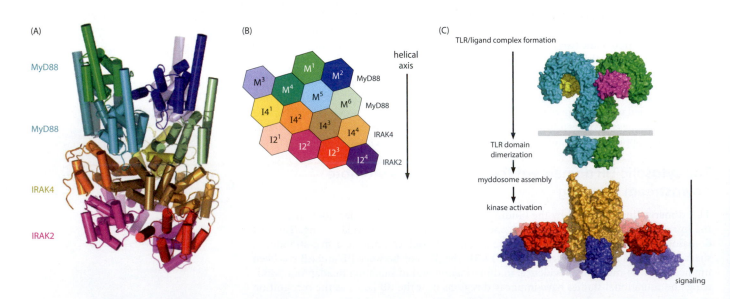

Figure 17.4 The myddosome is a 14-subunit complex implicated in transducing signals from the TLR4/MD2 dimer. (A) Structure of the myddosome, which comprises six MyD88 DD domains (cold colors), four IRAK4 DD domains (earth tones), and four IRAK2 DD domains (warm colors); α helices are shown as cylinders. (B) Schematic radial projection of the helical arrangement colored as in A. The 14 subunits make 3.5 turns of a single-start left-handed helix. (C) A model of initial events on the signaling pathway. Binding of LPS to the extracellular portions of TLR4/MD2 promotes their dimerization (see Figure 17.1) and, consequently, that of their TIR domains. Either directly or indirectly (with additional adaptors involved), this event leads to the assembly of a helical array of 14 DDs (a myddosome, beige), which clusters the kinase domains of IRAK2 (blue) and IRAK4 (red), promoting activation by phosphorylation. (From S.-C. Lin et al., *Nature* 465:885–890, 2010. With permission from Macmillan Publishers Ltd.)

Pattern-recognition receptors promote the assembly of inflammasomes

The innate immune system repertoire includes other classes of receptors that also recognize molecular patterns present on microbes and other challengers. These include the family of **Nod-like receptors** (**NLRs**), and they promote the assembly of signaling complexes whose activity leads ultimately to an inflammatory response. Accordingly, these complexes are termed **inflammasomes**. They are multisubunit structures assembled from several different kinds of proteins. The subunits are compiled from a base set of domains, concatenated in specific different combinations. Although there is as yet little information on the structures of fully assembled inflammasomes, a recurring theme seems to be the pairing, domino-style, of like domains, to make both homodimers and heterodimers, which associate into larger structures. The immediate consequence of inflammasome assembly is not phosphorylation but proteolytic conversion of the inactive zymogen pro-caspase-1 to the active protease caspase-1 (see Section 13.6).

One relatively simple system is the AIM2 (absent in melanoma) inflammasome, which responds to the presence of DNA in the cytosol. Host DNA is nuclear material, so cytosolic DNA is treated as a foreign substance. Its presence may signify infection by a dsDNA virus such as vaccinia virus (see Figure 8.57), which replicates its genome in the cytoplasm. Cytosolic DNA is recognized by the AIM2 protein (not itself an NLR) (**Figure 17.5A**), which binds DNA cooperatively via its HIN (hemopoietic, interferon inducibility, nuclear localization) domain (Figure 17.5C). Consisting of two tandem subdomains with the OB fold (see **Guide**), HIN binds to both strands on the outside of the DNA duplex, indicating specificity for dsDNA but in a sequence-independent manner. This binding initiates assembly of an inflammasome in a process involving the adaptor protein ASC (apoptosis-associated speck-like protein containing a CARD domain; see **Guide**) (Figure 17.5B). This complex binds and processes pro-caspase-1, releasing the activated protease. In turn, caspase-1 processes the precursor to the cytokine interleukin-1β, which is then secreted from the cell.

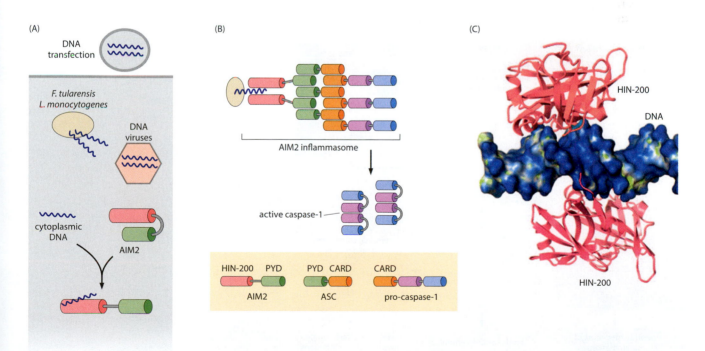

Figure 17.5 Activation of the AIM2 inflammasome by cytosolic DNA. (A) AIM2 functions as a DNA sensor, detecting DNA introduced by transfection or infection with a bacterial pathogen such as *Francisella tularensis*, *Listeria monocytogenes* or a DNA virus. (B) Binding of multiple copies of AIM2 to DNA promotes assembly of the inflammasome, shown schematically here. The actual numbers of subunits involved have not been determined, but ~80 base pairs of DNA are required to maximize IL-1β production, and a HIN domain covers 8–9 base pairs. (C) Crystal structure of two HIN domains complexed with a DNA duplex. (A and B, adapted from P. Broz and D.M. Monack, *Immunol. Rev.* 243:174–190, 2011. With permission from John Wiley & Sons; C, adapted from T. Jin et al., *Immunity* 36:561–571, 2012. With permission from Elsevier.)

17.3 THE COMPLEMENT SYSTEM

Like the TLRs, the complement system recognizes foreign organisms through molecules displayed on their surface. However, unlike the TLRs, which signal through the membrane to produce an inflammatory response, complement uses soluble extracellular proteins to tag the surfaces of foreign particles, marking them for destruction. Complement pathways exhibit considerable complexity, mainly reflecting regulatory requirements to keep the system inactive until it is stimulated by a pathogen and then to confine the destructive activity to the pathogen, sparing the host cells and the plasma proteins.

An overview of the steps involved is shown in **Figure 17.6** and a schematic representation in **Figure 17.7**. There are three ways in which complement activation is initiated, but these three pathways converge on the central and most abundant protein component of complement, the C3 convertase (an enzyme that can convert C3 to fragments C3a and C3b). The *classical pathway* and the *lectin pathway* recognize molecular signals on foreign particles: charge clusters and antibody/antigen complexes on the classical pathway, and neutral sugars on the lectin pathway. These recognition events lead to the activation of C4 and C2 through limited proteolysis. In the *alternative pathway*, C3 undergoes spontaneous auto-activation at a low rate, a process that is amplified after infection and recognition of a foreign pathogen.

The activated complement system invokes a proteolytic cascade that generates protein fragments with strong immunological activities. Upon proteolytic activation, each enzyme cleaves and activates many molecules, leading to a large amplification of the C3 activation step. This proteolytic amplification produces protein fragments among which are **anaphylatoxins** (C3a and C5a), and also the C3 convertases (C3bBb and C4bC2a), which generate C3b **opsonins**. (Anaphylatoxins are complement proteins that elicit an inflammatory response. Opsonins are molecules that bind to the surfaces of foreign particles such as bacteria, marking them for phagocytosis.) C3 activation leads to assembly of the proteins C5, C6, C7, C8, and C9 to form the large **membrane attack complex** (**MAC**) that causes cell lysis. Other fragments of complement proteins also bind to infecting pathogens, promoting their recognition by the adaptive immune response.

Opsonic fragments (mainly C3b, derived from C3) also bind in small numbers to host cells, but the mammalian system has developed a number of protective regulator molecules that

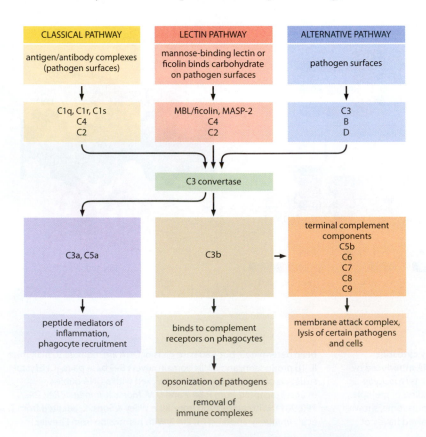

Figure 17.6 Overview of the complement system. (Adapted from K. Murphy, Janeway's Immunobiology, 6th ed. New York: Garland Science, 2012.)

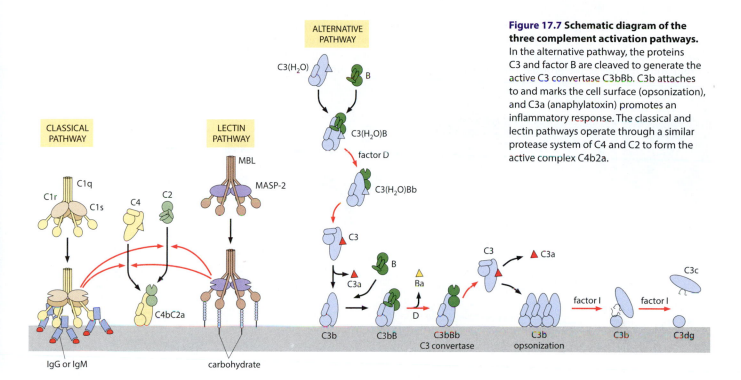

Figure 17.7 Schematic diagram of the three complement activation pathways. In the alternative pathway, the proteins C3 and factor B are cleaved to generate the active C3 convertase C3bBb. C3b attaches to and marks the cell surface (opsonization), and C3a (anaphylatoxin) promotes an inflammatory response. The classical and lectin pathways operate through a similar protease system of C4 and C2 to form the active complex C4b2a.

block both the activation of C3 and the action of the MAC, thereby protecting the host's own healthy cells. Several viruses and bacteria are able to evade attack by complement by expressing surface proteins similar to these regulator proteins or by expressing proteins that capture host regulators and bind them to their own surfaces.

The nomenclature system for complement proteins is described in **Box 17.2**.

The classical and lectin pathways use the multimeric complexes C1, mannose-binding lectin (MBL), and the ficolins for pathogen recognition

On the classical pathway, C1 recognizes many targets, including bacteria, apoptotic cells, antigen/antibody complexes, and *amyloids*. The 770-kDa C1 complex has three components: C1q, C1r, and C1s. C1q is a 430-kDa hexamer, in which each protomer is assembled from three homologous subunits A, B, and C. As shown by EM (see **Methods**) and biochemical studies in the 1970s, their N-terminal regions form collagen-like triple helices (see Section 11.7) that splay apart at the ends where they join the C-terminal globular heads, giving the appearance of a bunch of tulips (**Figure 17.8A**). Each head consists of three domains, one each from the A, B, and C chains, and binds to targets mainly through the recognition of charge clusters. C1r and C1s are serine proteases (Section 7.2) arranged as a C1s/C1r/C1r/C1s tetramer that associates with the collagen-like stalks of C1q. In the unactivated system, C1r and C1s are present as inactive zymogens. When C1q binds to a target, conformational changes take place that lead to the activation of C1r through auto-proteolysis. C1r then cleaves C1s, activating it. Activated C1s then hydrolyzes C4 and C2.

Box 17.2 Complement nomenclature

The proteins of the classical pathway, the first pathway to be discovered, are designated C1–C9. Proteins of the alternative pathway are called factors and designated by letters, such as factors B, H, and I. Properdin is an exception to this practice. The proteins of the lectin pathway are MBL (mannose-binding

lectin), the ficolins (L-, H-, and M-ficolin), and the MBL-associated proteases (MASPs). After proteolytic activation, the fragments formed are denoted by the protein name followed by lower case letters; for example, C3 is cleaved to a small fragment C3a and a large fragment C3b.

C1r and C1s are homologous molecules that consist of two CUB domains (Complement C1r/C1s, Uegf, Bmp1) (see **Guide**) separated by an EGF domain (epidermal growth factor-like) (see **Guide**) and followed by two CCP domains (complement control protein) (see **Guide**) and then the serine protease (SP) domain (Figure 17.8B). C1r dimerizes through

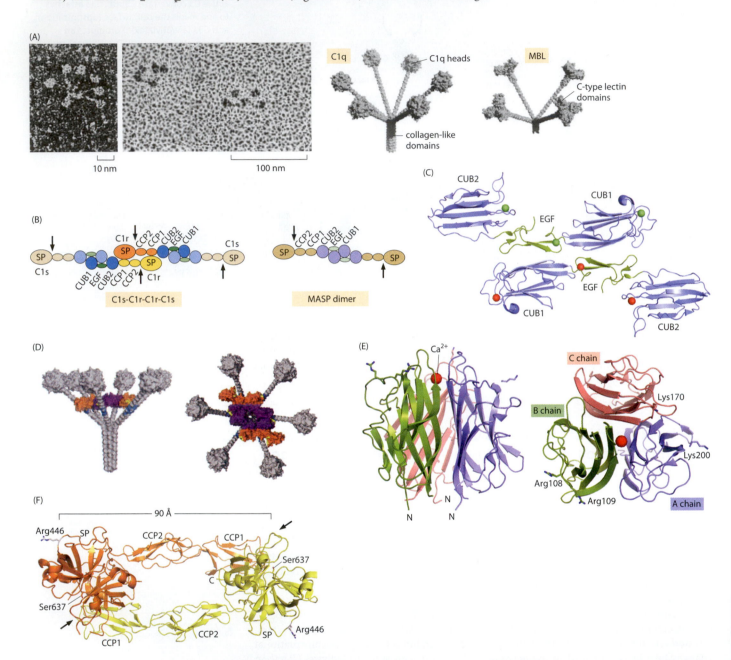

Figure 17.8 The C1 and MBL/MASP components of complement.
(A) Electron micrographs of C1q (negatively stained) and MBL (rotary shadowed with platinum) with models for the arrangements of globular heads and collagen-like stalks. (B) Domain organization of the C1rs tetramer and the MASP homodimer. Arrows indicate the cleavage sites for activation. (C) Arrangement of the CUB1–EGF–CUB2 domains in MASP showing dimerization through contacts between the CUB1 of one chain and the EGF of the other chain. Ca^{2+} is shown as red spheres in one chain and green spheres in the other. (PDB 3DEM) (D) Model derived from mutational, binding, and modeling studies for C1 as a C1q hexamer binding a C1rs tetramer. Only the CUB1–EGF–CUB2 domains of C1r (purple) and C1s (orange) are shown. The CCP1–CCP2–SP domains would be inside the complex in inactive C1. Binding sites for C1q are marked in blue, and those for C1rs in yellow. (E) Side and top views of a trimeric C1q head; Ca^{2+} is shown as a red sphere. In the side view, the three N termini (N) would connect to the collagen-like stalks. Residues that have been implicated in binding to IgG and IgM (Lys200 (A chain), Lys170 (C chain), and Arg108 and Arg109 (B chain) are indicated. (PDB 1PK6) (F) The CCP1–CCP2–SP dimer of the inactive C1r zymogen. One chain is in yellow, the other in orange. The contacts at the dimer interface (arrows) are between the CCP1 of one chain and the SP domain of the other chain. Ser637 marks the catalytic site. Arg446 marks the substrate site for autocatalytic activation. In the crystallographic studies, Arg446 was mutated to Gln to prevent activation. (PDB 1GPZ) Residues are numbered as in the crystal structure omitting the signal peptide. (A, from H.R. Knobel et al., *Eur. J. Immunol.* 5:78–82, 1975, and J. Lu et al., *J. Immunol.* 144:2287–2294, 1990; D, from R. Wallis et al., *Immunobiology* 215:1–11, 2010. With permission from Elsevier.)

interactions between the SP domain of one subunit and the CCP1 domain of its partner. In the presence of Ca^{2+}, each C1r binds to C1s via the CUB1-EGF domains to form the C1s/C1r/C1r/C1s tetramer (~340 kDa).

The lectin pathway was discovered in the late 1980s, much later than the classical pathway. The mannose-binding lectin (MBL) binds carbohydrates on target surfaces to activate the MBL-associated serine proteases (MASPs) (Figure 17.8A,B). MBL resembles C1q in having collagen-like and globular regions but exists in several oligomeric states, of which the tetramer is the most common. The globular heads each contain three C-type lectin CRD domains (Ca^{2+}-dependent carbohydrate recognition domains). These bind N-acetylglucosamine, fucose, and mannose, sugars that are common on microbial surfaces. The ficolins, of which there are three in humans (L-, H-, and M-, or ficolins 1, 3, and 2 respectively), also have collagen-like and globular regions and form mainly tetramers, but their heads consist of three *fibrinogen*-like domains. Ficolins bind to apoptotic cells and microbial targets, but their specificity is not well established. L-ficolin binds to molecules bearing acetyl groups, including N-acetyl amino sugars and some sialic acids.

MBL and the ficolins associate with MASP proteases, which are homologous to C1r and C1s. There are three MASPs, named MASP-1, -2, and -3, each a homodimer. MASP-2 acts like C1s, activating C4 and C2. The natural substrates of MASP-1 and MASP-3 are not clearly identified. In the MASP-2 dimer, the subunits are linked through their CUB1-EGF domains (Figure 17.8C). One MASP-2 dimer interacts with one MBL or ficolin tetramer and is auto-activated when the complex binds to a target surface.

No structure has yet been determined for intact C1, but there is a wealth of structural data on its components. Surface plasmon resonance and mutational binding studies indicate that C1rs tetramers have six binding sites for C1q located on their CUB domains, two from each C1r chain and one from each C1s chain, and these contact the C1q collagen-like stalk. Similarly, the MASP-2 dimer has four sites, two on each of its CUB domains. Models for these interactions are shown in Figure 17.8D. Ca^{2+} is important in mediating the interactions at conserved binding sites on the CUB and EGF domains between C1r and C1s and also between C1rs and C1q and between MBL and MASP. EM and neutron scattering experiments indicate that the C1rs tetramers exhibit flexibility when free in solution but become more constrained on binding to C1q. Likewise, unliganded C1q is flexible and becomes more rigid when bound to C1rs tetramers.

The first step in C1 activation involves C1q recognition of the target and auto-activation of C1r

Each interaction of C1q with a target surface is relatively weak, but high-avidity binding is achieved through multiple interactions. The heads have a 10-stranded β-sandwich conformation with jelly roll topology (Figure 17.8E). The three subunits (A, B, and C chains) exhibit distinct patterns of charged and hydrophobic residues that are important for ligand recognition. Some of the residues involved in binding IgG, IgM, and bacterial LPS have been mapped by mutagenesis. After C1q binds to a target, it communicates with the C1rs tetramer through conformational changes in the stalk region to stimulate activation. The intimate connection between the stalks and the proteases provides a route for communication.

The C1r protease cleaves after Arg residues but its substrates are limited to its own auto-activation and to C1s, indicating that it requires a special mechanism for substrate selection. A striking feature of the structures of both the zymogen and the activated C1r dimer is the wide separation of the two SP domains (Figure 17.8F). The Ser in the catalytic site on one chain is ~90 Å from the Arg at the cleavage site on the other chain. During auto-activation, a direct physical contact must be achieved so that the bond to be cleaved is presented to the catalytic site of the other subunit. Contacts observed in the crystal structure of the activated C1r dimer suggest that this might be arranged in the C1q/C1rs complex by the dissociation and reassociation of C1r dimers. The catalytic sites of C1r face outward and are available to cleave C1s.

C1 inhibitor (C1-INH) is a *serpin*, a member of a family of inhibitory proteins that control the activities of a number of blood proteases, including C1r, C1s, and MASP-1 and -2. The primary role of C1-INH seems to be to limit the lifetime of active C1s and MASP-2 and thus to regulate C3 cleavage (see below). C1-INH is a 100-kDa protein that acts as an auto-inhibitory substrate. Binding to the protease results in inhibitor cleavage followed by a massive conformational change that leads to a stable conformation in which the protease catalytic site is distorted. After activation, the three complexes—C1, MBL/MASP-2, or ficolin/MASP-2—proteolytically activate the proteins C4 and C2 (Figure 17.6 and Figure 17.7), as discussed further below.

Loss of C1-INH regulation through genetic or acquired deficiencies leads to the disease angioedema, a recurrent tissue swelling. Human C1-INH has been approved for administration in the treatment of hereditary angioedema, and several peptidomimetic C1s inhibitors are in clinical trials.

The alternative pathway and the classical and lectin pathways converge on the C3 convertase

The proteins C3 and factor B are initially inactive. On the alternative pathway, C3 is activated spontaneously by hydrolysis of a thioester bond (see below), yielding a form, named $C3(H_2O)$, that has a low level of activity. This version of C3 cannot bind to target surfaces but can bind factor B. A plasma protease called factor D then cleaves factor B to Ba and Bb. Bb remains part of the soluble $C3(H_2O)Bb$ complex. Although only present in small amounts, this complex is able to cleave C3 to C3a and C3b through the serine protease activity of Bb. C3b attaches to the surface of pathogens via a thioester bond; otherwise it is rapidly hydrolyzed. Surface-bound C3b recruits factor B, stimulating factor D cleavage of factor B, yielding the C3 convertase complex C3bBb. This complex then amplifies the response by proteolyzing more C3 molecules, generating more of the anaphylatoxin C3a and the opsonin C3b (Figure 17.7).

This mechanism has been duplicated during evolution. The two multidomain proteins C4 (205 kDa) and C2 (100 kDa) of the classical and lectin pathways are homologous to C3 (187 kDa) and factor B (90 kDa), respectively. The C1 complex and the MBL/MASP-2 complex cleave C4 to C4a and the opsonin C4b. Like C3b, C4b binds covalently to surfaces through a thioester group. Next, C2 binds C4b, forming the C4bC2 complex. C1s and MASP2 both cleave C2 into C2a and C2b, generating a complex that is homologous to C3bBb of the alternative pathway. This complex is referred to as C4bC2a (where C2b refers to the N-terminal part and C2a to the larger protease fragment of C2—an unfortunate inconsistency in terminology). Cleavage of C3 by this convertase generates C3a and C3b, giving an amplification of C3b.

Association of C3b with C3 convertases (forming C3bBbC3b or C4bC2aC3b, the so-called C5 convertases) shifts the substrate specificity from C3 to C5. C5 is homologous to C3 but lacks a thioester and hence cannot bind to surfaces. Cleavage of C5 by these convertases generates a potent anaphylatoxin and chemotactic factor, C5a, and a C5b molecule.

The anaphylatoxins (the three homologous proteins C3a, C4a, and C5a) induce an inflammatory response, which is associated with the contraction of smooth muscle, the degranulation of mast cells and basophils, and the release of histamine and other substances that increase capillary permeability. They also increase blood flow and make it easier for plasma proteins and cells to pass out of the blood into the site of infection. C5a is a potent chemotactic factor for neutrophils.

The pathway is terminated by the plasma protease factor I, which cleaves C3b to iC3b and then to C3dg and C3c, thus inactivating the C3 convertase (Figure 17.7). Factor I activity is promoted by the regulator, factor H (see below), which displaces factor B from C3b and facilitates cleavage by factor I. Likewise, factor I cleaves C4b and this reaction is helped by the C4-binding protein (C4BP) that displaces C2a.

Cleavage of C3 causes major structural changes that lead to activation

C3 is the central protein of the complement system. Cleavage of C3 yields C3a and C3b. Comparison of the structures of C3 and C3b illustrates the dramatic consequences of this cleavage. Native C3, derived from its precursor pro-C3 by an unidentified protease, consists of a β subunit and an α subunit connected by a disulfide bond. It is organized in an intricate arrangement of 13 domains (**Figure 17.9A**). Its core consists of eight macroglobulin (MG) domains (Figure 17.9B), which have a fibronectin type 3-like fold (see **Guide**). Domain MG6

(A)

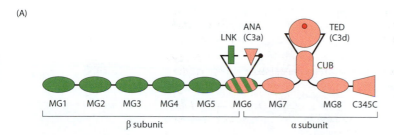

(B)

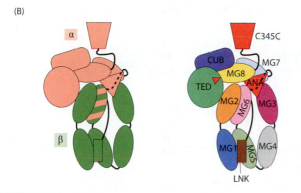

(C)

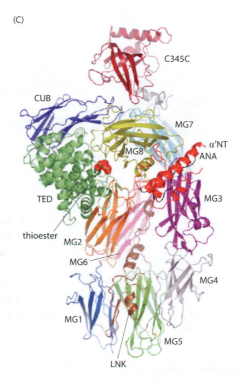

contains β strands from both the β and α chains, while the other MG domains are either entirely from α or from β. These domains are not apparent from sequence alignments or from an examination of intron/exon boundaries, indicating that C3 and α₂-macroglobulin are ancient molecules that evolved from a common ancestor through many gene duplications, mutations, insertions, and so on, so that sequence similarity has been lost while structural similarity of the domains has endured.

Other functional domains are added to the MG core of C3. Inserted into MG6 are a linker (LNK) domain at the end of the β subunit and the anaphylatoxin domain (ANA) at the start of the α subunit (Figure 17.9A). A processing site consisting of four consecutive Arg residues connects the linker and the ANA domain, which is removed in the conversion of pro-C3 to the native two-chain C3 molecule. In C3, the C terminus of the β subunit and the N terminus of the α subunit are ~50 Å apart, and hence some rearrangement must take place when pro-C3 is converted to C3. A linker α′NT connects the ANA domain back to MG6.

Two domains are inserted between MG7 and MG8: a CUB domain and, inserted into it, the all α-helical thioester-containing domain (TED). Finally, at the C terminus is a domain common to C3, C4, and C5, called the C345C domain. This domain is attached to MG8 through a short anchoring region and is essential for binding factor B. The overall structure of C3 is shown in Figure 17.9C.

C3 is activated by cleavage (by C4bC2a or C3bBb) at a site just after the ANA domain (**Figure 17.10A**). Cleavage releases the ANA domain C3a and induces massive conformational changes in the remaining C3b molecule (Figure 17.10B). The MG7 and MG8 domains swivel, while the CUB and TED domains shift by up to 95 Å. In this way, C3b exposes the binding sites for a number of proteins (for instance, factor B, C5, and complement regulators). Moreover, conformational changes in the TED domain bring a His and Glu close to the thioester (Figure 17.10C,D). In native C3, a reactive thioester is formed by residues Cys988 and Gln991 and is protected from hydrolysis by water by hydrophobic groups from the MG8 and TED domains. The thioester, when exposed, is reactive with nucleophilic groups. In C3b, the TED changes conformation and the thioester reacts with His1104, forming a cysteine thiolate ion and an acylimidazole intermediate that is stabilized by Glu1106 (Figure 17.10D). The reactive intermediate forms covalent bonds with hydroxyls (or amino groups) on target surfaces when C3b opsonizes the surface. About 90% of activated C3 just reacts with water, the most abundant nucleophile in biological systems. These C3 molecules in which the acylimidazole is hydrolyzed do not participate further in the reactions of the complement system. Rapid reaction with water prevents the activated C3 from diffusing far enough to react with bystanders, such as particles or host cells near the target.

Figure 17.9 The C3 complement protein. (A) Domain organization of C3 β subunit (green; residues 1–645) and α subunit (pink; residues 650–1641) which are linked by a disulfide bridge. (B) Schematic diagram showing the organization of the α and β subunits (left) and of the individual domains (right). (C) Structure of C3 with the individual domains color-coded as in B. (PDB 2A73) (C, from B.J.C. Janssen et al., *Nature* 437:505–511, 2005. With permission from Macmillan Publishers Ltd.)

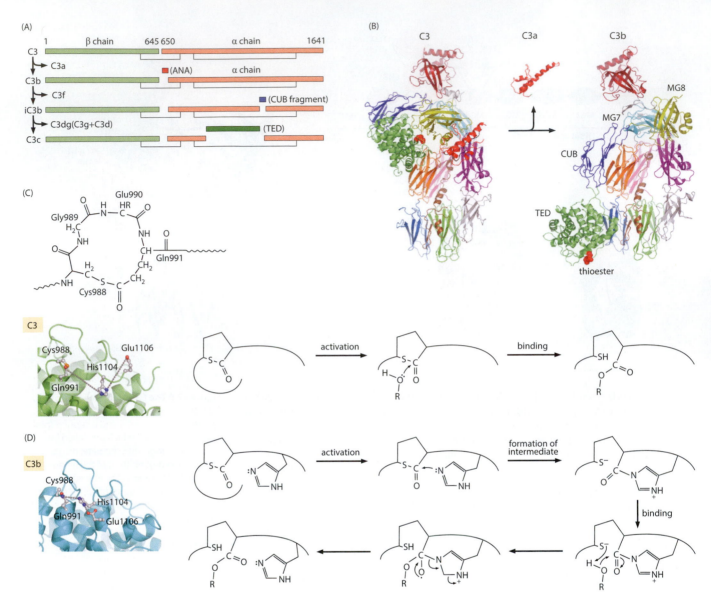

Figure 17.10 Conversion of C3 to C3b. (A) Schematic diagram of the factor D-mediated proteolytic cleavage of C3 between Arg726 and Ser727 to give C3a (the ANA domain) and C3b. A disulfide bridge that links the α and β chains and one within the α chain are indicated, but other disulfides and the glycosylation sites are omitted for clarity. Factor I promotes further cleavage of C3b to iC3b and then to C3dg and C3c. (B) Structures of C3 and of C3b after release of C3a, showing the drastic rearrangements including that of the TED domain and in the position of the thioester. The MG7 and MG8 domains (cyan and yellow) swivel, while the CUB and TED domains (blue and green) move by up to 95 Å. (C) The thioester linkage of C3. Top, schematic drawing of the thioester between Cys988 and Gln991 in C3. Below, the local arrangement of residues in C3. Right, the reaction of the thioester with a hydroxyl-containing group (HOR). (D) The thioester environment in C3b. The thioester reacts with His1104 to form an acylimidazole intermediate that is stabilized by Glu1106. Right, the reaction mechanism that leads to formation of the acylimidazole intermediate and the reaction of the intermediate with hydroxyl-containing groups. (PDB 2I07) (From B.J.C. Janssen et al., *Nature* 444:213–216, 2006. With permission from Macmillan Publishers Ltd.)

Amplification of the response involves the C3 convertase, C3bBb

Surface-bound C3b is a starting point for producing short-lived C3 convertases, which amplify the complement response by cleaving additional copies of C3 into C3a and C3b close to the target surface. This C3 convertase forms in two steps (**Figure 17.11A**). Factor B binds C3b in a Mg^{2+}-dependent manner, yielding the C3bB complex. Next, factor D excises fragment Ba, yielding an active protease complex consisting of C3b and fragment Bb (C3bBb). This complex activates more C3 molecules before dissociating irreversibly into C3b and the now inactive fragment Bb. Without Ba, Bb cannot reassociate with C3b. Surface-associated assembly gives a brief local burst of C3b molecules that opsonize the target surface, marking it for phagocytosis.

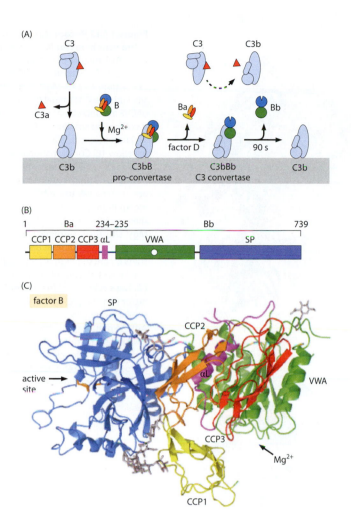

Figure 17.11 Factor B: function and structure. (A) Schematic diagram showing C3b forming a complex with factor B, with domains color-coded as in B. The C3bB complex is proteolyzed by factor D to form the C3bBb complex (the C3 convertase), which converts more C3 to C3b, amplifying the response. (B) Domain map of factor B. The white circle in the VWA domain marks the Mg²⁺-binding site. (C) Structure of factor B with domains colored as in (B). αL is in magenta. Sites with attached carbohydrates are also shown. (PDB 2OK5) (Adapted from F.J. Milder et al., *Nature Struct. Mol. Biol.* 14:224–228, 2007. With permission from Macmillan Publishers Ltd.)

Factor B has five domains: three complement control protein (CCP) domains connected by a linker to a von Willebrand factor type A (VWA) domain (see **Guide**) and a C-terminal chymotrypsin-like serine protease (SP) domain (Figure 17.11B). After cleavage by factor D, the CCP domains and the linker form fragment Ba, and the VWA and SP domains form fragment Bb. The overall fold of intact factor B is shown in Figure 17.11C.

Insights into the recognition of factor B by C3b have come from a crystal structure of the cobra venom factor (CVF) complexed with factor B. CVF is 46% identical in sequence to C3 but lacks the TED domain. It is processed by proteases in the venom gland of cobras into a three-subunit molecule with C3b-like properties. The CVF-containing convertase (CVFBb) is more stable (half-life ~7 hours) than the C3bBb convertase (half-life ~90 seconds) and cleaves C3 to destroy the complement components of the cobra prey. The long-lived venom convertase may find therapeutic applications in the use of humanized CVF to treat conditions in which tissue damage caused by inappropriate complement activity could be alleviated by complement depletion.

CVF binds factor B at two distinct sites: one on fragment Ba and one on fragment Bb (**Figure 17.12A**). Both fragments contribute to the binding surface, but Ba (comprising the three CCP domains and the helix αL) provides 73% of the total buried surface area and is essential for loading Bb onto CVF (or C3b). The N-terminal CCP domains of Ba bind to several domains of CVF. These C3b domains rearrange on the conversion of C3 to C3b to form the factor B-binding site. The structure shows why only C3b, and not C3, can bind factor B. Domains CCP2–3 of factor B make a critical contact to the CUB domain of CVF (or C3b). These contacts in C3b would be lost when further proteolytic processing of C3b by factor I yields fragments iC3b, C3c, and C3d that cannot form C3 convertases

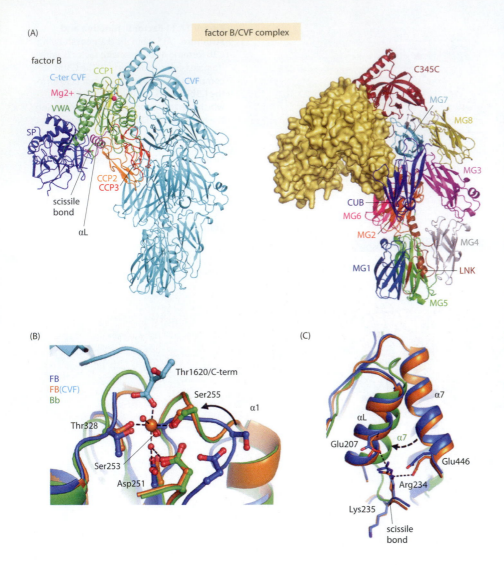

(A)

factor B/CVF complex

factor B

C-ter CVF   CCP1   CVF

Mg2+

VWA

SP

CCP2
CCP3

scissile
bond

αL

C345C

MG7

MG8

MG3

CUB

MG6

MG2   MG4

MG1   LNK

MG5

(B)

FB
FB(CVF)
Bb

Thr1620/C-term

Ser255

α1

Thr328

Ser253

Asp251

(C)

α7

αL

α7

Glu207

Glu446

Arg234

Lys235

scissile
bond

Figure 17.12 Recognition of factor B by cobra venom factor (CVF), a mimic of C3b. (A) Structure of CVF in complex with factor B. Left, factor B is colored by domain as in Figure 17.11B. Mg2+ marks the MIDAS site with the C-terminal (C-term) carboxylate of CVF in contact with the metal. Right, CVF is colored by domain (as for C3 in Figure 17.7), and factor B (FB) is surface-rendered in yellow. (B) Conformational rearrangement around the MIDAS site of the VWA domain. In free factor B (dark blue), Asp251 and Ser255 are too far away to contribute to the Mg2+ recognition site. In the complex with CVF, these residues shift, creating a high-affinity Mg2+ site; the C-terminal carboxylate of CVF also contributes. In Bb, corresponding shifts create the high-affinity Mg2+ site. (C) The helices αL and α7 of the VWA domain of FB. The structures of free factor B (blue) and factor B in complex with CVF (orange) show similar arrangements of these helices, and Arg234 of the scissile bond is in contact with two glutamate residues. The peptide bond is not available for cleavage. Activation of factor B to Bb (green) results in a shift of α7 (dashed arrow) that takes the place of αL, which is removed as part of Ba.

(Figure 17.10A). Fragment Bb binds CVF through the Mg2+ of a so-called metal-ion-dependent adhesion site (MIDAS) in the VWA domain (Figure 17.12B). Similar Mg2+-binding sites were first described for the homologous 'inserted' (I) domains in *integrins* (~50% of integrin α subunits—see Figure 11.39—have an I domain located above their β-propeller domain). In cell/cell adhesion, ligand binding to Mg2+ in the MIDAS of integrins induces large rearrangements that signal cell/cell adhesion across the cell membrane (Sections 11.6 and 11.7). In the CVF/factor B complex, the C-terminal carboxylate of the α chain (Thr1620 in CVF, equivalent to Asn1641 in C3b) binds the Mg2+ in the MIDAS of factor B (Figure 17.12B).

In unbound factor B, the Arg234 residue of the scissile bond between Ba and Bb forms salt bridges to helix α7 of the VWA domain and helix αL (residues 202–213) of the linker of the Ba segment (Figure 17.12C). Helix αL (Ba region) occupies a groove in the VWA domain (Bb region) and helix α7 is displaced, thereby 'locking' free factor B in an inactive state. In the CVF complex with B, there is a similar arrangement. In convertase formation, factor D (a single SP domain) circulates in plasma and encounters C3b-bound factor B, which it cleaves, removing Ba. In the final complex, C3bBb (deduced from the structure of C3bBb in complex with a bacterial inhibitor and the isolated Bb structure), helix α7 occupies a groove taking the place of αL (which has been cleaved off as Ba). The conformations of α7 and the MIDAS site are, by analogy to the integrins, those that correspond to a high-affinity ligand-binding state. The high-affinity state is associated with slow off kinetics, which may explain both the 90-second half-life and the irreversible dissociation of the C3 convertase formed on the target surface.

Complement activation leads to four major consequences: cell lysis, an inflammatory response, phagocytosis, and B cell stimulation

On complement activation, the five complement proteins C5, C6, C7, C8, and C9 form the membrane attack complex (MAC) that causes cytolysis of the infecting pathogen. Cleavage of C5 by the C5 convertase yields C5b, which is similar to C3b but lacks a reactive thioester moiety. C5b initiates the **terminal pathway** of the complement system. Assembly of the MAC proceeds by formation of a C5b/C6 complex, which adds C7 to form the C5b/C6/C7 complex. After binding C8, the complex recruits multiple copies of C9 to complete the MAC (**Figure 17.13A**). The MAC associates with the pathogen membrane, penetrates the lipid bilayer, and forms pores. Early EM studies on complement-lysed membranes showed pores of ~10 nm internal diameter (Figure 17.13B).

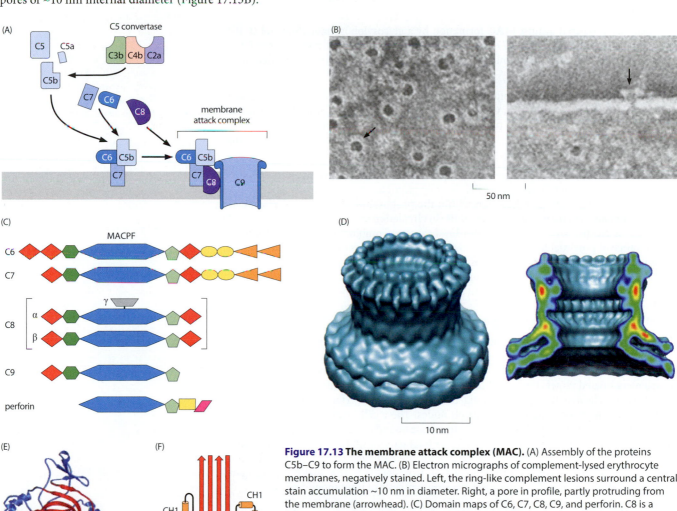

Figure 17.13 The membrane attack complex (MAC). (A) Assembly of the proteins C5b–C9 to form the MAC. (B) Electron micrographs of complement-lysed erythrocyte membranes, negatively stained. Left, the ring-like complement lesions surround a central stain accumulation ~10 nm in diameter. Right, a pore in profile, partly protruding from the membrane (arrowhead). (C) Domain maps of C6, C7, C8, C9, and perforin. C8 is a heterotrimer of C8α, C8β, and the lipocalin protein C8γ. All except C8γ have a MAC-perforin (MACPF) domain flanked by small regulatory domains: thrombospondin type I (TSP1; red diamonds), low-density lipoprotein receptor class A (LDLa; green hexagons), epidermal growth factor (EGF; green pentagons), complement control protein (CCP; yellow ovals), and factor I MAC (FIMAC; orange triangles). (D) Surface and cutaway views of a cryo-EM reconstruction of a perforin pore with 20-fold symmetry. The cutaway plane is color-coded by density, with red representing high density. (E) Structure of a perforin subunit. The protein was mutated at one site to prevent oligomerization. The central sheet of the MACPF domain is red, the CH1 and CH2 helix bundles are orange, the remainder of the domain is blue, the EGF domain green, the C2 Ca²⁺-binding Ig domain yellow, and the C-terminal region (CTR) magenta. (PDB 3NSJ) (F) Model derived from cholesterol-dependent cytolysin (CDC) for the unwinding of CH1 and CH2 helices to β-hairpins to form the pore. (A, adapted from M.J. Walport, *New Engl. J. Med.* 344:1058–1066, 2001; B, from J. Tranum-Jensen and S. Bhakdi *J. Cell Biol.* 97:618–626, 1983. With permission from Rockefeller University Press; D–F, from R.H.P. Law et al., *Nature* 468:447–451, 2010. With permission from Macmillan Publishers Ltd.)

C6 through C9 are large homologous proteins that have a central 40 kDa MAC-perforin (MACPF) domain flanked by regulatory domains (Figure 17.13C). The homologous protein, perforin, is secreted by cytotoxic T cells when killing infected target cells. Cryo-EM (see **Methods**) of human perforin inserted into liposomes showed pores with a variable number of subunits: 19–24 per pore with pore diameters ranging from 13 to 20 nm. A pore with 20-fold symmetry is shown in Figure 17.13C. An explanation of how the water-soluble MACPF-containing proteins rearrange into membrane-penetrating molecules came from crystal structures of C8α, perforin, and a distantly related cholesterol-dependent cytolysin (CDC) that is secreted by Gram-positive bacteria. In them (Figure 17.13E), the MACPF domain contains a central four-stranded β sheet flanked by two helical regions, CH1 and CH2, and other regions. Spectroscopic studies of CDCs suggested that on membrane insertion, CH1 and CH2 refold into amphipathic β hairpins which insert into the membrane, forming a β-barrel pore (Figure 17.13F). It is proposed that there is a similar restructuring of perforin.

Host cells are protected against MAC formation by a glycosylphosphatidylinositol (GPI)-anchored surface protein CD59 (77 aa) that is present on almost all human cells. CD59 binds C8α and C9 in the transmembrane β-hairpin regions and thereby blocks membrane insertion and pore formation.

The inflammatory response that follows complement activation is raised by the small anaphylatoxic protein fragments C3a, C4a, and the highly potent C5a. These 10 kDa helical proteins bind to their respective G-protein-coupled receptors, the C3a receptor (which also binds C4a) and the C5a receptor on leukocytes, producing a chemotactic response that brings leukocytes to the site of infection.

When complement proteins bind to a foreign particle (such as a bacterium or an apoptotic cell), several of them act as opsonins. That is, they promote adherence to phagocytic cells and subsequent ingestion and digestion within the phagosome. The principal phagocytic cells, macrophages (in tissues) and neutrophils (in circulation and tissue fluids), have a wide repertoire of receptors by which they can bind targets. Combinations of these receptors allow a range of potential targets to be recognized. The major opsonin is derived from C3, and adherence of particles to phagocytes requires many ligand/receptor interactions. C3b binds to complement receptor 1 (CR1) on phagocytic cells, as does C4b. CR1 (also known as CD35) consists of 30 CCP domains and is also, as discussed below, a complement regulator. The C3b breakdown product, iC3b, is a more effective opsonin than C3b, because it is bound by the more abundant macrophage receptors, the integrins CR3 and CR4.

The adjuvant activity of complement that promotes the adaptive immune response has similarities to opsonic activity and relies on the interaction between complement-coated particles and cells. A range of complement/receptor interactions promotes the binding of complement-coated targets (antigens) to dendritic cells, the major antigen-presenting cells. The dendritic cells may transport the antigen from the periphery (for example, close to the skin) via the lymphatic system to the lymph nodes. These cells also ingest and digest the antigen, and present antigen peptides on their surface bound to HLA class II (Section 17.4). Another mechanism of adjuvant effect is a direct interaction with the CR2 (CD21) receptor (a homolog of CR1) on B lymphocytes. CR2 binds the pathogen surface-bound opsonins iC3b and C3dg, the proteolytic products of C3b from factor I. The dual receptor/ligand interactions stimulate B cell proliferation, causing synthesis of immunoglobulins.

Host cells are protected by regulator proteins

To protect host cells and tissues against opsonization, the host expresses a family of 'regulators' that consist of long strings of CCP domains (from 4 to 30 in number). These proteins are either expressed on the cell surface or present in blood. Decay-accelerating factor (DAF, also known as CD55), membrane cofactor protein (MCP, also known as CD46) and complement receptor 1 (CR1, also known as CD35) are all membrane-bound regulators. The soluble regulators, factor H and C4bp, bind to negative charge clusters on host cells, such as sialic acid or highly sulfated glycosaminoglycans (GAGs). Regulators work through two mechanisms, referred to as **decay accelerating activity** and **cofactor activity**, that stop the local production of C3b. Decay accelerating activity breaks down surface-bound C3 convertases, while cofactor activity assists in binding protease factor I, which cleaves C3b and C4b into inactivated species (iC3b and C4d and C4c) (see Figure 17.7). Some regulators, such as the soluble regulator factor H, exhibit both activities; others, such as the surface-bound DAF, have a single activity. They may act on both C4b and C3b (for example, CR1),

(A)

cell surface
and fHbp
binding

C3b binding

cell surface
binding

factor H (1)(2)(3)(4)(5)(6)(7)(8)(9)(10)(11)(12)(13)(14)(15)(16)(17)(18)(19)(20)

(B)

factor H (CCP1–4)

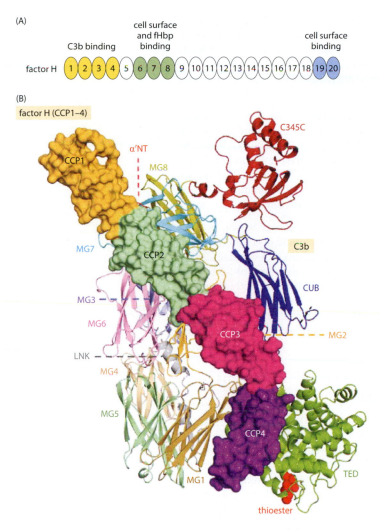

Figure 17.14 Factor H regulation of C3b.
(A) The domain structure of factor H.
(B) Structure of domains CCP1–4 in complex
with C3b. (PDB 2WII). The binding site
partly overlaps the Bb binding site to C3b.
(Adapted from J. Wu et al., *Nature Immunol.*
10:728–734, 2009. With permission from
Macmillan Publishers Ltd.)

or they may act only on C3b (for example, factor H) or only on C4b (for example, C4bp). The importance of these regulators is underscored by various mutations that affect the C3b/ regulator interactions, causing loss of protection and complement-mediated tissue damage in patients.

Factor H is the most abundant soluble regulator, with a concentration in plasma of 200– 800 mg/l. This 155 kDa protein consists of 20 CCP domains, each of ~60 aa (**Figure 17.14A**). Factor H binds to one face of C3b through CCP domains 1–4 and to the opposite face of C3b through domains 19–20. It also binds to GAGs through domains 6–8 and 19–20. The roles of factor H are to prevent soluble C3b from forming convertases and, by interacting with host cells, to diminish convertase formation on the host. Localization of factor H is crucial for regulation. When the alternative pathway is activated on a surface that does not have GAGs or when the classical or lectin pathways increase the rate at which C3b is generated, the effects of factor H can be overridden.

The crystal structure of CCP domains 1–4 complexed with C3b revealed an extended arrangement of these domains, yielding an interface 100 Å long (Figure 17.14B). Their binding has a K_d of ~11 μM, indicating a weak interaction, but *in vivo* the high concentration of factor H favors complex formation. The factor H/C3b interface consists of four contact regions, which include the α'NT, MG7, CUB, and TED domains. These domains change their conformation during the conversion of C3 to C3b, and factor H can only bind to the conformation adopted in C3b.

Domains CCP1–4 of factor H are able to dissociate Bb from the active convertase C3bBb. The structure indicates that CCP1–2 compete with the binding of Bb to C3b. Moreover, the surfaces of factor H and Bb are both negatively charged and electrostatic repulsion probably contributes to the destabilization of the C3bBb complex. Factor H is able to recruit factor I, which then degrades C3b. Domains CCP2–3 are close to the C3b CUB domain

(Figure 17.14B), which houses the cleavage sites for factor I; mutagenesis data have implicated these domains in factor I binding. The specific but weak interactions between C3b and factor H(CCP1–4) allow fast and selective regulation but also render factor H liable to functional impairment by small modifications in its sequence. Several diseases (such as age-related *macular degeneration*) are associated with mutations in the gene that encodes factor H.

The membrane-bound regulators also prevent complement attack on host cells. Almost all human cells express DAF or MCP, or both, on their surfaces. Membrane-bound DAF accelerates the decay of the C3 convertase-releasing Bb. DAF has four CCP domains followed by a heavily glycosylated stalk and a GPI anchor that tethers it to the plasma membrane. Functional studies suggest that DAF interacts with C3bBb through its CCP2 and CCP4 domains. DAF has no cofactor activity and does not interact with factor I. MCP, a homolog of similar size to DAF, has cofactor but no decay accelerating activity. CR1 has both activities.

Properdin is a different type of regulator from those described above, having positive instead of negative activities. Properdin is an oligomer of two to five 50 kDa subunits, each containing six thrombospondin (TSR) domains. The role of properdin is to stabilize C3bBb. It binds to both C3b and Bb, and increases the half-life of C3bBb more than 10-fold. Properdin is also a recognition protein for the alternative pathway. It binds directly to several complement-activating bacteria, via charge clusters. Once bound, it may serve as an assembly point for C3b (generated as described above by $C3(H_2O)$) and then for factor B, forming a C3bB/properdin complex on the bacterial surface, which is converted into the stabilized convertase, C3bBb/properdin.

Several viruses and bacteria use host protection mechanisms to evade complement-mediated clearance

Vaccinia, variola (smallpox virus), and Kaposi's sarcoma-associated herpesvirus express homologs of regulators on their surfaces. These molecules comprise four CCP domains with decay accelerating and cofactor activities. Several bacteria, including *Neisseria meningitidis* and *Borrelia burgdorferi*, causative agents of sepsis and Lyme disease respectively, have another mechanism for avoiding detection: they express factor H-binding proteins on their surface that hijack the host's factor H, thus acquiring decay accelerating and cofactor activities that enable them to avoid opsonization.

The structure of factor H-binding protein (fHbp), a 27 kDa surface lipoprotein from *Neisseria meningitidis*, in complex with factor H domains CCP6–7 (**Figure 17.15A**) has shown how bacteria employ protein/protein interactions instead of sialic acid or GAG recognition to recruit factor H to their cell surfaces and hence engineer protection from the host complement system. The K_d for the fHbp/factor H (domains CCP6–7) interaction is 5 nM, indicative of tight binding. This is achieved by extensive interactions between the double β-barrel structure of fHbp and factor H domain CCP6. The structure of factor H domains CCP6–8 in complex with sucrose octasulfate, an analog of the GAGs found on the surface of endothelial cells, has defined the region of factor H that interacts with cell surfaces (Figure 17.15B). Comparison of these structures shows that fHbp and the GAGs both bind to the same part of the surface of factor H. Thus the bacterium mimics the mechanism by which host cells recruit factor H for protection and does so with high specificity, binding only human factor H and not factor H from other species. Although human and primate factor H are 90% identical in sequence, the sequence differences cluster to the fHbp-interacting region in the CCP6 domain. The structures suggest how an fHbp vaccine might be raised to abolish factor H binding and hence allow the complement system to recognize and destroy this pathogen.

17.4   T-CELL-MEDIATED IMMUNITY

The adaptive immune responses—the **antibody response** and the **T-cell-mediated immune response**—are carried out by specialized cells derived from lymphocytes. **B cells** from bone marrow produce antibodies; **T cells**, generated in the thymus, produce T-cell-mediated immunity. Here we focus on the macromolecular assemblies that orchestrate T-cell-mediated immunity. A brief description of antibodies is given in **Box 17.3**. In T-cell-mediated immune responses, activated T cells recognize a foreign antigen presented on the surface of

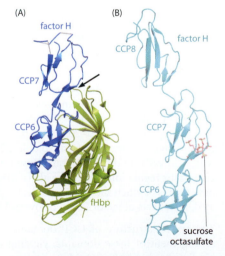

Figure 17.15 *Neisseria meningitidis* recruits factor H using protein mimicry of host glycosaminoglycans (GAGs). (A) Structure of factor H binding protein (fHbp) complexed with domains CCP6–7 of factor H. (PDB 2W81) (B) Structure of factor H domains CCP6–8 in complex with sucrose octasulfate, a mimic of cell surface GAGs. (PDB 2V8E) The GAG-binding site on domain CCP7 is at a similar position to that recognized by fHbp, marked by the arrow in A.

Box 17.3 Immunoglobulins

Antibodies belong to the IgG class of **immunoglobulins** (Igs), proteins synthesized by plasma cells derived from B cells in response to the presence of a foreign substance. This substance, the antigen, can be protein, carbohydrate, or glycoprotein. Any given antibody recognizes and binds to only a small part of the antigen surface, known as its *epitope* or antigenic determinant.

An IgG is made up of four polypeptide chains, two light chains L (each 25 kDa) and two heavy chains H (each 50 kDa) (**Figure 17.3.1**). Each chain has a constant region (C) and a variable region (V). The variable regions are responsible for recognizing specific antigens. Early biochemical studies showed that the IgG molecule (150 kDa) can be cleaved by the protease papain to yield the Fab fragments (ab for antigen-binding) (2×50 kDa) and the Fc fragment (c for crystallizable) (1×50 kDa). The Fc region is responsible for dimerization and for downstream signaling to the complement system (Section 17.3) and the uptake of antibody/antigen complexes by macrophages.

The L and H chains are composed of subdomains. The L chain has one variable subdomain (V_L) and one constant subdomain (C_L), whereas the H chain has one variable subdomain (V_H) and three constant subdomains: C_H1, C_H2, and C_H3. The disulfide bond that links the two heavy chains is between residues in the hinge region between C_H1 (Fab) and C_H2 (Fc). The hinge region allows flexibility in the arrangement of the two Fabs.

Each of these subdomains has the Ig fold (see **Guide**), which consists of two β sheets, each composed of antiparallel β strands with a defined topology. Many hydrophobic residues pack between the sheets, which are linked by a disulfide bond. The Ig fold recurs in many other proteins of the immune system (for example, MHC class I and II molecules, T cell receptors, and CD3, CD4, CD8, CD28, and B7 proteins) and also in many non-immune proteins.

The region that recognizes antigen comprises six loops from the variable chains, three each from the V_H and V_L chains. These loops are known as **complementarity-determining regions** (**CDRs**; **Figure 17.3.2**). Through their sequence variability, CDRs can recognize antigens with a wide variety of shapes and differing chemical properties. The epitope may be a single loop on the surface of the antigen (linear epitope) or a cluster of several motifs from different parts of the polypeptide chain (discontinuous epitope). The interactions involve numerous van der Waals and hydrophobic interactions and hydrogen bonds that, together with shape complementarity, create a tight binding interaction (typically, $K_d \approx 20$ nM).

In all, there are five classes of Igs (**Figure 17.3.3**). There are two types of L chains (κ and λ), which are the same in each of the five classes, and five types of H chains (γ for IgG, α for IgA, μ for IgM, δ for IgD, and ε for IgE). The different Igs have

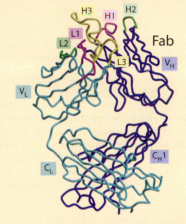

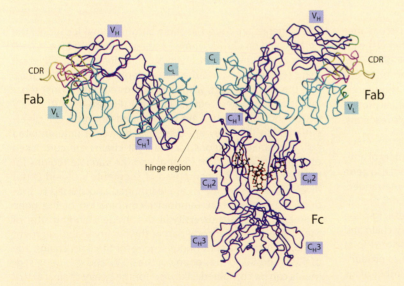

Figure 17.3.1 Structure of IgG b12, an anti-HIV antibody. The two light chains (L) and two heavy chains (H) are shown in cyan and blue, respectively. The 12 subdomains of the Fab and Fc domains are labeled. C_H1 is the first constant subdomain of the heavy chain; V_L is the variable subdomain of the light chain, and so on. The Fabs are connected to the Fc by flexible hinges that allow each Fab arm to rotate freely. (PDB 1HZH)

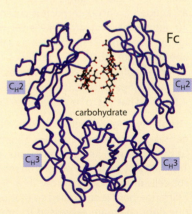

Figure 17.3.2 The Fab and Fc fragments. The Fab fragment (top) is made of one light chain (cyan) and the N-terminal half of a heavy chain (blue). The Fab recognizes antigen via its CDR loops: L1, L2, and L3 from the light chain, and H1, H2, and H3 from the heavy chain. The Fc fragment (bottom) consists of the C-terminal halves of two heavy chains. The Fc dimer has carbohydrate chains attached in an inner cavity.

Box 17.3 **Immunoglobulins** *(continued)*

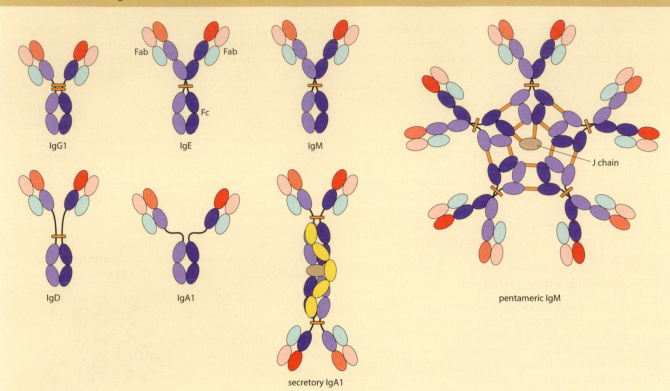

Figure 17.3.3 The five classes of immunoglobulins. These schematics show the overall domain structures. Each Ig domain is represented by an oval. The constant domains of the heavy chains are in two shades of blue, and that of the light chain in cyan. The variable domains are in two shades of red (heavy chain) and pink (light chain). Secretory IgA and pentameric IgM both contain a J chain (beige), and secretory IgA also contains a secretory component chain (yellow). Disulfide bonds within the hinge regions are shown in orange. Flexible hinge regions connect Fab to Fc in the IgG1, IgA1, and IgD molecules. The IgE and IgM molecules lack a hinge region but have an extra C domain in their heavy chains.

different functions. IgM is the first Ig to appear in serum after an antigen is encountered. It is a pentamer (five dimers) with 10 antigen-binding sites. IgG is the major Ig in blood serum. IgA is found in secretions such as saliva, tears, and bronchial and intestinal mucus, and provides the first line of antibody defense against bacteria and viruses. IgE helps confer protection against parasites but is also implicated in allergic reactions. A function has yet to be assigned to IgD.

The ability of Igs to distinguish between large numbers of antigens is generated by a complex genetic process. For an Ig gene to be expressed, its gene segments must first be rearranged. This takes place in the development of B cells from B precursor cells in the bone marrow. In a developing B cell, Ig-gene rearrangement is controlled so that only one heavy chain and one light chain are expressed to provide a monoclonal Ig of unique specificity. As B cells may be programmed to produce any combination of light and heavy chains, the number of Igs with distinct specificities is very large. The Ig is then presented on the B cell surface as a receptor, which can recognize its antigen. The binding of antigen to the Ig of a naive B cell triggers the cell's proliferation and differentiation, and secretion of the Ig. The next time the host encounters this antigen, there are numerous specific Igs available so that the intruder is recognized and effectively intercepted.

The Ig genes are found at three chromosomal locations: the heavy chain on chromosome 14, the κ light chain on chromosome 2, and the λ light chain on chromosome 22. Different gene segments encode the leader peptide, the variable part, and the constant part, with the coding regions (exons) separated by noncoding regions (introns) (**Figure 17.3.4**). The V genes and the C genes are translocated during B cell differentiation so that they are close together. In a differentiated cell, the variable regions for the light chains are composed of segments V and J (for joining) and the variable regions for the heavy chains are composed of segments V, J, and D (for diversity). These segments are rearranged and spliced to produce the final protein. Random recombination of the gene segments generates diversity in the antigen-binding sites of the Ig. The J genes encode part of the hypervariable segment CDR3. In forming a continuous V gene for the light chains, any of 40 V_κ and 5 J_κ genes or any of the 30 V_λ and 4 J_λ gene segments can be combined. Likewise, for the heavy chains, any of the 65 V_H, 27 D, and 6 J_H segments can be combined through somatic recombination to give 320 light chains and 10,530 heavy chains. Beyond that, recombinase enzymes create further diversity at the junctions between the gene segments for the CDR3 regions of both heavy and light chains, adding a factor of $\sim 10^7$ in overall diversity.

Box 17.3 Immunoglobulins *(continued)*

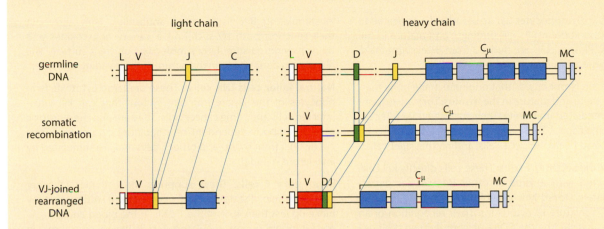

light chain   heavy chain

Figure 17.3.4 Immunoglobulin diversity. The light-chain V regions are constructed from two segments; a variable segment V and a joining segment J from the genomic DNA. Introns (thin black lines) are spliced out during mRNA processing (Section 5.4). The exon for the leader sequence (L) is also shown. The heavy-chain V regions are constructed from three regions: V, D, and J. The heavy chain for IgM μ is shown. The membrane-coding (MC) exons are also shown. The J segment is distinct from the J chain, which is part of the IgA and IgE assemblies shown in Figure 17.3.3. (Adapted from P. Parham, The Immune System, 4th ed. New York: Garland Science, 2015.)

a host cell, referred to as an *antigen-presenting cell* (APC). (Note: the same initials, APC, are used to denote the anaphase-promoting complex in Chapters 12 and 13.) T cells either kill the infected cell or help it to eliminate the microbe. In contrast to innate immune responses that are activated mainly at sites of infection or injury, adaptive immune responses are activated in peripheral organs such as lymph nodes and spleen. The response to infection starts with dendritic cells at the site of infection. These cells recognize pathogens and take them up by phagocytosis; they produce antigens by *proteasomal* proteolysis (Section 7.4) of microbial proteins, yielding peptide fragments. The dendritic cells migrate to the peripheral lymphoid organs, where they become APCs. They use protein complexes called **major histocompatibility complexes** (**MHCs**) to bind the peptide fragments and convey them to the cell surface, where T cells can recognize them. When T cells encounter a foreign antigen displayed on an APC, they are activated to proliferate and differentiate into effector cells. This clonal expansion of T cells confers protection from future infection by the same pathogen.

Major histocompatibility complexes (MHCs) are glycoproteins that present foreign and self-antigens at the surface of antigen-presenting cells (APCs)

The primary event governing the cellular immune response is the recognition by T cells of antigenic peptides presented in the context of MHC molecules. There are two types of MHCs, termed class I (MHC-I) and class II (MHC-II). In humans, the best characterized MHC molecules are human leukocyte-associated antigens HLA-A, -B, and -C (class I) and HLA-DP, DQ, and DR (class II). In mice, the class I molecules are called histocompatibility-2 (H-2) D, L, and K, and the class II molecules are called H-2A and H-2E. MHC-I molecules are recognized by cytotoxic (CD8+) T cells, whereas MHC-II molecules are recognized by helper (CD4+) T cells. (T cell nomenclature is given in **Box 17.4**.) Because of extensive polymorphism in the MHC genes, it is rare for two persons (except for identical twins) to have the same set of MHC alleles. This MHC-related uniqueness is the primary cause of organ transplant rejection.

While MHC-I and MHC-II are termed classical MHCs, other MHC-like molecules also participate in the cellular immune system. These nonclassical MHCs resemble the classical ones structurally but differ in their functions. They include ligands for natural killer (NK) receptor cells, such as MICA (MHC-I-related chain A), as well as ligands for γδ T cells, such as T10 and T22.

Box 17.4 A summary of the cells involved in the immune response

B cells: B cells are lymphocytes (white blood cells) dedicated to making immunoglobulins. B cells originate and mature in the bone marrow. Plasma cells are derived from B cells and make and secrete large amounts of soluble antibodies.

T cells: T cells are lymphocytes that originate in the bone marrow and migrate to and mature in the thymus. They are responsible for cell-mediated immunity. Their surface antigen receptor is the T cell receptor. There are three main classes of T cells: effector cytotoxic T cells, helper T cells, and natural killer cells.

Effector cytotoxic (CD8+) T cells: These cells directly kill cells that are infected with a virus or other intracellular pathogen. They contain stored lytic granules, a modified lysosome that contains a mixture of cytotoxins. Their recognition of an infected cell is assisted by the CD8+ glycoprotein on their cell surface. Cytotoxic CD8+ T cells migrate in the blood to sites of infection where they recognize, through their T cell receptors, the MHC-I/peptide complexes presented by infected cells. They secrete the contents of lytic granules onto the surface of the infected cell, promoting its apoptosis (orderly cell death).

Helper T (CD4+) cells: When activated by an antigen-presenting cell, some T cells differentiate into helper CD4+ T cells in the lymph node. These cells help stimulate other cells. They express the co-receptor CD4 and recognize MHC-II/peptide complexes. CD4+ helper T cells of the class T_H1 help activate tissue macrophages to engulf and kill extracellular pathogens. CD4+ helper T cells of the class T_H2 are involved

in stimulating B cells to make antibodies. They carry out their function both by secreting a variety of cytokines and by displaying co-stimulatory molecules on their surface. The response is dependent on the nature of the invading pathogen.

Natural killer cells (NK cells): These are large cytotoxic lymphocytes of the innate immune system that can kill virus-infected cells and some cancer cells. They are also the cytotoxic cells of antibody-dependent cell-mediated cytotoxicity. They are distinct from B cells and other T cells and are stimulated by interferons α and β.

Regulatory T cells: These suppress the activities of other cells and help distinguish self from nonself. Although they make up less than 10% of the T cells in the blood and peripheral lymphoid organs, regulatory T cells have a crucial role in immunological self-tolerance by suppressing the activities of self-reactive cytotoxic and helper T cells. They also express the co-receptor CD4.

Thymocytes: These are developing T cells in the thymus, which is located in the chest just above the heart. T cells migrate in the blood from bone marrow to the thymus, where they undergo development. Mature T cells flow from the thymus to lymphoid tissues and then return to the blood. A combination of positive and negative selection processes shape the T cell repertoire. They ensure that T cells with potentially useful receptors survive and mature, while others die by apoptosis. Helper and cytotoxic T cells that could react with self-antigens are eliminated.

MHC-I is a heterodimer of an α subunit (also called the heavy chain, ~33 kDa) and a highly conserved β subunit (also called β_2-microglobulin, β_2M, ~12 kDa). The α subunit is composed of three globular domains (α1, α2, α3), a single-pass transmembrane segment, and a small cytoplasmic domain. Their ectodomains (extracellular regions) are shown in **Figure 17.16A,B**. The overall architecture of MHC-II resembles that of MHC-I but it is a heterodimer of two similar subunits (α and β), each with two globular domains (α1, α2 and β1, β2), a transmembrane segment, and a short cytoplasmic tail (Figure 17.16C,D). The α3 domains and β_2M in class I, and α2 and β2 in class II, adopt Ig folds.

The first crystal structure of a MHC-I molecule, determined in 1987 by Pamela Bjorkman and Don Wiley, revealed unexpected electron density representing an unknown peptide or mixture of peptides, bound in a long closed groove at the membrane-distal region of the molecule. This observation identified the antigen-binding site. The MHC-I ectodomain has a peptide-binding core, which has a β-sheet floor made up of seven anti-parallel β strands, with two anti-parallel α helices forming walls on either side of the peptide-binding groove. This core is formed by the α1 and α2 domains in class I, and by α1 and β1 in class II (Figure 17.16A,C). The residues that line the peptide-binding groove confer specificity to particular regions or 'pockets' within the groove. Variations in the size, shape, and charge of these pockets mediate peptide recognition by the MHC (**Figure 17.17**), while the side chains of residues of the bound peptide, protruding from the groove, contact the T cell receptor (TCR) (discussed below).

The peptide-binding groove of MHC-I is closed at both ends, limiting the length of bound peptides to 8–10 residues (Figure 17.16A and Figure 17.17A). Longer peptides can be accommodated if the peptide bulges out in the middle of the groove. In MHC-II, the ends of the groove are open, thus accommodating longer peptides, typically of 13–15 residues (Figure 17.16C and Figure 17.17B).

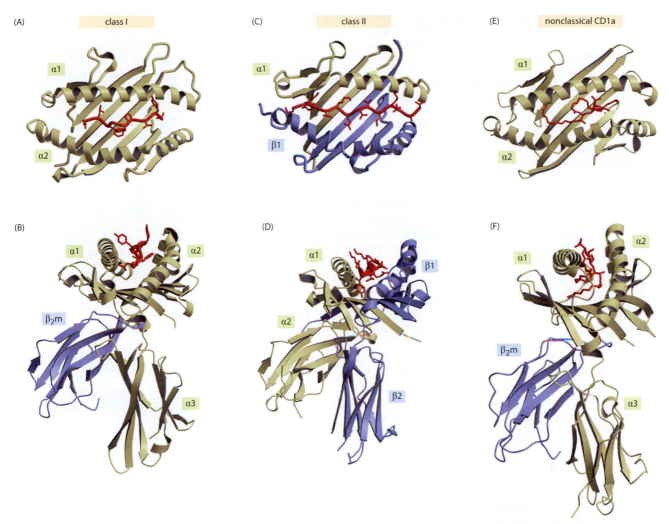

Figure 17.16 Class I, class II, and nonclassical MHC extracellular domains. (A) Top view of MHC-I HLA-A2 complexed with a nine-residue peptide from the human T lymphotropic virus (red). (PDB 1AO7) The α1 and α2 domains of the α chain are visible in this view, but not the α3 and β2M subunits. (B) Side view of MHC-I HLA-A2, looking down the peptide-binding groove. (C) Top view of the MHC-II HLA-DR1 in complex with a 13-residue peptide from influenza virus hemagglutinin. (PDB 1FYT) (D) Side view of HLA-DR1. (E) Top view of CD1α with bound sulfatide (a sulfated galactosyl lipid: 3'-sulfated β-1-D-galatosylceramide) ligand in red. (PDB 1ONQ) (F) Side view of CD1α, looking down the ligand-binding groove. This hydrophobic ligand binds much more deeply than the peptide ligands to MHC-I or MHC-II.

Peptides are loaded on to MHC-I and MHC-II via different pathways

MHC-I molecules mainly present peptides derived from the proteolysis of cytosolic proteins, whereas MHC-II molecules present peptides derived from imported extracellular proteins proteolyzed in endosomal compartments. These pathways reflect the respective roles of the two kinds of receptors in the cellular immune system.

For MHC-I, peptide loading occurs along with MHC folding in the ER, as shown in **Figure 17.18A**. The nascent heavy chain is secreted into the ER and anchored in the ER membrane (1) and associates (2) with the ER-resident chaperone calnexin, in a carbohydrate-dependent interaction. The MHC-I/calnexin complex associates with β2M, and calnexin is displaced by calreticulin (3). The disulfide-isomerase ERP57 associates with nascent MHC-I. The ternary complex then associates with the TAP molecule (transporter associated with antigen processing), a membrane-spanning ABC transporter (Section 16.5), through interaction with the chaperone tapasin (4). Polypeptides in the cytosol are digested by the proteasome, producing peptides of both endogenous ('self') and antigenic ('nonself') origin. Peptides thus produced are transported into the ER lumen by the TAP complex and loaded into the peptide-binding groove of MHC-I through interaction with tapasin, producing peptide-bound MHC-1 (pepMHC-I) molecules that contain both antigenic and

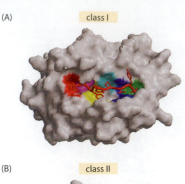

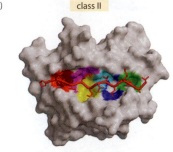

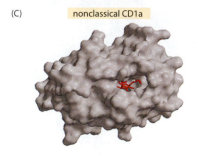

Figure 17.17 Binding grooves for MHC molecules. (A) MHC-I HLA-A201 in complex with the Tax peptide (LLFGYPVYV). The specificity pockets of the binding groove are colored in red, magenta, yellow, cyan, blue, and green. The following Tax peptide residues are pointing into the pockets: L1 (red), L2 (magenta), F3 (yellow), P6 (cyan), and V9 (green). (B) MHC-II HLA-DR1 in complex with the influenza HA peptide (PKYVKQNTLKLAT). The following residues are pointing into the pockets: Y3 (red), K5 (magenta), Q6 (yellow), T8 (cyan), L9 (blue), and L11 (green). Pockets colored as in A. (C) Nonclassical MHC CD1α in complex with a sulfatide ligand. The binding groove is more hydrophobic and deeper, with a lid over the ligand that partly occludes the antigen except for the carbohydrate head-group. CD1 does not have as many well-defined specificity pockets as MHC-I and MHC-II.

non-antigenic peptides (5). Mature pepMHC-I is then transported (6) to the cell surface (7), via the *trans* Golgi network. Most peptides presented by MHC-I are from the host cell and hence are not antigenic, except in rare instances when tolerance is broken and autoimmunity arises. However, upon viral infection or tumor growth, proteolysis of viral or tumor-derived polypeptides in the cytosol leads to the production of antigenic (nonself) peptides that are loaded onto MHC-I for presentation to the T cell receptor at the cell surface.

MHC-II employs several pathways for peptide loading as shown in Figure 17.18B. In the major pathway, nascent class II α/β dimers associate with a chaperone called the invariant chain (Ii) within the ER. α and β subunits are secreted (1) into the ER lumen and associate (2) to form nascent MHC-II, which is rapidly bound (3) by Ii, to prevent association with endogenous peptide fragments from within the ER. The MHC-II/Ii complex is then transported (4) from the ER, through the Golgi, to the endocytic pathway, where cathepsin proteases degrade Ii (5, 6), leaving only a small fragment (class II-associated invariant-chain peptide, or CLIP) in the peptide-binding groove (7). Protein antigens are endocytosed by the antigen-presenting cells and digested to antigenic peptides by proteases in early and late endosomes (top right). These peptides are loaded onto nascent MHC-II molecules within late endosomes (MHII compartments, or MIICs), where they replace the CLIP peptide in the peptide-binding groove in a reaction catalyzed by the MHC-II-like molecule HLA-DM (8). The mature pepMHC-II complex is then transported (9) to the cell surface.

CD1 and other nonclassical MHC molecules may have functions other than antigen presentation

Among the best-studied nonclassical MHCs is the CD1 family. CD1 molecules bear strong structural similarity to MHC-I (Figure 17.16E,F), but present self and nonself lipids such as bacterial cell envelope components, rather than peptides, at the surface of APCs. To accommodate these more hydrophobic ligands, which are usually but not always glycolipids, the CD1 ligand-binding grooves are more hydrophobic and often have 'lids' covering most of the groove (Figure 17.17C). Like MHC-I, CD1 molecules have an α chain with α1, α2, and α3 domains and a β₂M chain. Humans have five CD1 genes (CD1A–E), whereas mice have just two CD1D genes. CD1 can be recognized by either αβ or γδ T cell receptors, whereas CD1δ is primarily recognized by natural killer (NK) T cells.

The T cell receptor (TCR) is the primary antigen-recognition molecule on the surface of T cells

The TCR is central to the function of the cellular immune system and is a highly specific antigen receptor, similar in topology and genetic diversity to the antigen-binding Fabs of antibodies (Box 17.3). However, antigen recognition by the TCR differs from that of antibodies in two important ways. First, the TCR is a membrane-bound receptor, involved in T cell activation together with other T cell surface co-receptors and co-stimulatory molecules. Second, the TCR only recognizes antigens presented in the context of MHC molecules.

TCRs are made up of either one α and one β subunit (αβTCR), or one γ and one δ subunit (γδTCR). αβTCRs recognize peptides displayed by MHC molecules. The ligands recognized by γδTCRs are not as well understood but include both intact proteins and non-peptidic ligands presented by MHCs or MHC-like molecules. The enormous sequence diversity of TCRs is generated by a similar mechanism to that of antibodies—that is, somatic recombination of different V, D, and J genes (Figure 17.3.4 in Box 17.3). Each

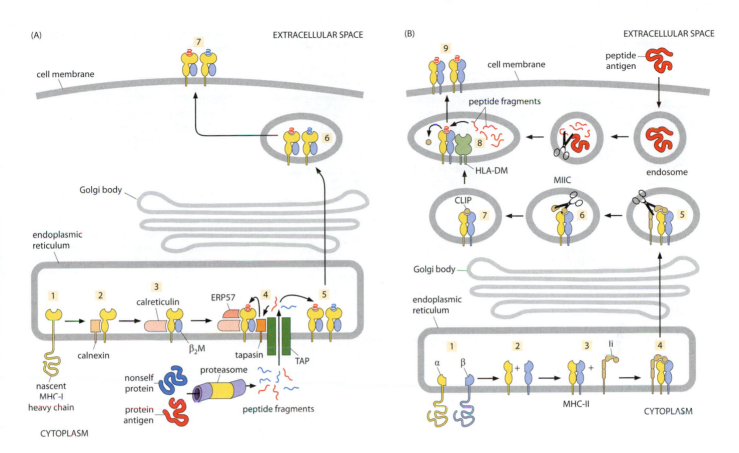

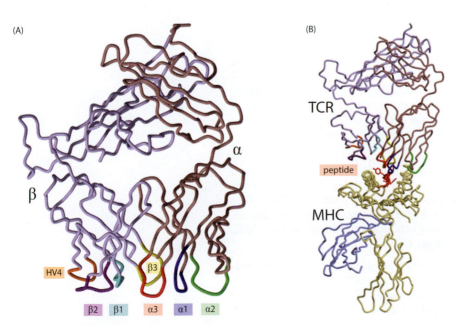

subunit comprises one variable (N-terminal) and one constant (C-terminal) Ig domain followed by a single-pass transmembrane domain. The two subunits combine within the ER to form a disulfide-linked heterodimer. As with antibodies, the variable domains contain hypervariable sequences and highly conserved constant domains. Together, the variable domains of the TCR muster seven hypervariable loops at the membrane-distal region of the molecule. Six of these correspond to the CDR loops of antibodies (Box 17.3), with an additional loop in the β subunit, termed hypervariable 4 (HV4) (**Figure 17.19A**). Thus, three CDR loops come from the α subunit (CDR1α, CDR2α, CDR3α) and four from the β subunit (CDR1β, CDR2β, CDR3β, HV4). As in antibodies, the CDR loops vary in length and sequence and in their flexibility.

Figure 17.18 Peptide loading of MHC-I and MHC-II. (A) Loading of MHC-I occurs during protein folding within the ER and the mature pepMHC-I is then transported to the cell surface via the *trans* Golgi network. (B) Loading of MHC-II with antigenic peptides takes place in specialized late-endosomal compartments, known as MHC-II compartments (MIICs), from which they are transported to the cell surface.

Figure 17.19 Structure and function of the αβTCR. (A) Mouse αβTCR 2C. (PDB 1TCR) The α subunit is shown in brown and the β subunit in light blue. CDR loops α1, α2, α3, β1, β2, β3, and HV4 are shown in different colors. In these diagrams only the trace of the polypeptide chain is shown for simplicity. (B) Human αβTCR/pepMHC complex. The TCR is B7; the MHC is class I HLA-A 0201, and the peptide is from human T lymphotropic virus. (PDB 1BD2) Coloring as in A and Figure 17.16.

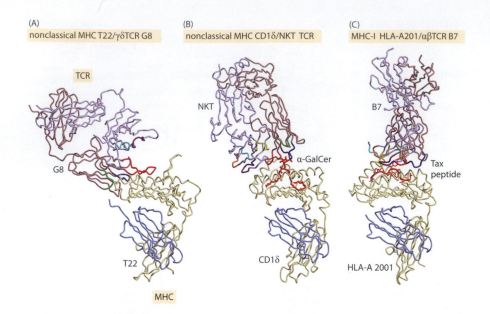

(A) nonclassical MHC T22/γδTCR G8

TCR

G8

T22

MHC

(B) nonclassical MHC CD1δ/NKT TCR

NKT

α-GalCer

CD1δ

(C) MHC-I HLA-A201/αβTCR B7

B7

Tax peptide

HLA-A 2001

Figure 17.20 Comparison of TCR–MHC recognition. (A) Nonclassical MHC T22 bound to γδTCR G8. (PDB 1PYZ) In this complex there is no peptide bound to the non-classical T22 and recognition is directly via the G8 TCR. (B) Nonclassical MHC CD1δ bound to the NKT TCR. (PDB 2PO6) (C) Classical MHC-I HLA-A201 bound to αβTCR B7. (PDB 1BD2) The MHC-I is bound to the Tax peptide as in Figure 17.17A. The structures have been aligned on their MHC components. The α and β subunits of the TCR each provide ~50% of the binding surface to the MHC. The G8-T22 binding mode shown in A is different; the mode of interaction is at an acute angle, and the TCR β subunit makes most of the contacts to the MHC. Peptide ligands are in red.

TCR molecules bind to the peptide-bound MHC complex (pepMHC) via their CDR loops (Figure 17.19B). The α subunit usually contacts the N-terminal side of the peptide, and the β subunit the C-terminal side. CDRs α3 and β3 generally have the most contact with the peptide, whereas the other CDRs contact the relatively conserved MHC residues along the two helices bordering the peptide-binding groove. Variations in the CDR loops of the TCR optimize the recognition of pepMHC complexes. Comparison of different TCR/MHC complexes has indicated that TCRs bind to the pepMHC in a rather uniform manner, with the TCR oriented diagonally across the peptide but with slight variations in the angle of the TCR relative to the MHC (**Figure 17.20**). However, some TCRs position themselves toward one or other end of the groove and hence are centered asymmetrically atop the pepMHC.

The CD3 complex initiates and transmits signals in the T cell interior

Antigen recognition at the T cell surface is mediated by sustained interaction between the pepMHC and the hypervariable (V) domains of a clone-specific TCR. The TCR itself has no substantial cytoplasmic domain, and intracellular signaling originates through associated CD3 proteins (see below). Upon TCR/pepMHC engagement, intracellular kinases rapidly phosphorylate the CD3 complex, initiating a sequence of events that leads to the specific effector responses of T cells. These events are assisted by the stimulatory co-receptor molecules, CD8+ of cytotoxic T cells and CD4+ of helper T cells that interact with the TCR/pepMHC complex (see below).

The CD3 complex comprises four invariant subunits, CD3ε, -γ, -δ, and -ζ, which combine with the TCR to produce a functional signaling complex. In mammals, the CD3ε, CD3δ, and CD3γ ectodomains (**Figure 17.21**) each have a single constant-type Ig ectodomain connected, via a short stalk of 5–10 aa and a single-pass transmembrane segment, to an intracellular cytoplasmic domain of ~55 aa. In contrast, the majority of CD3ζ is intracellular; it has a small ectopeptide (~9 aa), a transmembrane segment, and a cytoplasmic domain of ~113 aa.

In the ER, CD3ε binds CD3δ or CD3γ to form heterodimers. Interactions between paired ectodomains have been characterized in structural analyses of CD3εδ and CD3εγ. The ectodomains interact through parallel association of their C-terminal β strands (labeled G in Figure 17.21) to give a side-by-side conformation. Additional interactions involve electrostatic complementarity in the subunits' mainly hydrophobic transmembrane regions. CD3ζ, in contrast, forms a covalently associated homodimer, CD3ζζ.

The intracellular portions of CD3 subunits are phosphorylated at conserved tyrosine residues, known as immunoreceptor tyrosine-based activation motifs (ITAMs) with a consensus sequence of YXXL/IX$_{6-12}$YXXL/I (where X is any amino acid). The distribution and function of ITAMs are highly conserved. The intracellular domains of CD3ε, -δ, and -γ each

(A)

(B)

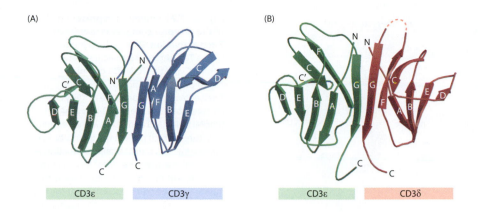

CD3ε	CD3γ

CD3ε	CD3δ

Figure 17.21 Ectodomains of the CD3 complex. (A) Human CD3εγ ectodomain. (PDB 1SY6) (B) Human CD3εδ ectodomain. (PDB 1XIW) CD3γ and CD3δ contact the corresponding region of CD3ε. The overall topologies of the CD3εδ and CD3εγ heterodimers are similar. The ectodomains of both dimers are thought to contact the invariant domains of the TCR, although the molecular basis for this interaction has yet to be determined.

have a single ITAM, whereas that of CD3ζ has three. CD3ζζ, with its six ITAMs, is considered to be the principal signaling component of the TCR/CD3 complex, despite its lack of a substantial ectodomain (**Figure 17.22**). The state of phosphorylation of CD3 subunits is the net outcome of the opposing activities of tyrosine kinases, such as the Src-family kinases Lck and Fyn, and phosphatases such as CD45.

The TCR/CD3 complex: expression of the TCR at the cell surface is accompanied by co-expression of CD3

The large number of distinct subunits (TCRα, -β; CD3ε, -δ, -γ, -ζ) in the mature TCR/CD3 complex reflects the level of complexity manifested in TCR-mediated signal transduction. Measurements of their relative abundances at the cell surface suggested that an individual complex comprises a single TCRαβ plus one CD3εδ, one CD3εγ, and one CD3ζζ. The TCR/CD3 complex is assembled stepwise in the ER, starting with dimerization of CD3ε with CD3δ or CD3γ (Figure 17.22). Then, CD3εδ associates with TCRα and CD3εγ with TCRβ. These components unite to form a TCRαβ/CD3εδ/CD3εγ complex, which finally binds CD3ζζ to form the mature TCR/CD3 complex. Addition of the CD3ζζ homodimers to the CD3 complex is sensitive to correct folding and assembly of the other components, and CD3ζζ is thereby thought to have a role in quality control in the assembly of TCR/CD3 complexes.

The arrangement of subunits in these complexes at the surface of mature T cells has implications for the mechanism of TCR-mediated signal transduction and T cell activation; however, this mechanism remains a contentious issue. The primary mode of association between CD3 subunits and the TCR is thought to be via complementary charge interactions in the transmembrane region, whereby conserved basic residues of the TCR interact with conserved acidic residues of CD3ζζ, CD3εδ, and CD3εγ. However, the presence of charged amino acids within the hydrophobic transmembrane region would be unfavorable unless these charged groups were able to make complementary interactions. Additional interactions between the TCR constant domains and CD3 ectodomain dimers have been proposed but have yet to be substantiated.

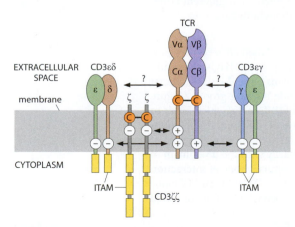

Figure 17.22 Interacting elements within the TCR/CD3 complex. The mature TCR/CD3 complex comprises eight subunits, organized as four dimers. Cys residues involved in inter-subunit disulfide bonds are designated by C–C. The CD3 subunits form CD3εδ and CD3εγ heterodimers through interactions involving the Ig ectodomains. CD3ζ lacks an Ig ectodomain and forms a homodimer through disulfide bonding in the transmembrane region. Assembly of the TCR/CD3 complex seems to involve electrostatic interactions between the transmembrane domains of the TCR and CD3 dimers. Three basic (+) residues on the TCR interact with a pair of acidic (–) residues on each CD3 dimer, respectively, to form a three-way interaction. The basis for interaction between the ectodomains of the TCR, CD3εδ, and CD3εγ has yet to be determined, but it has been demonstrated that CD3εδ interacts with the DE loop of the TCRα constant domain (Cα) and that CD3εγ interacts with the CC' loop of the TCRβ constant domain (Cβ). The mechanism of TCR/CD3 interaction is common to CD4+ (helper) and CD8+ (cytotoxic) T cells.

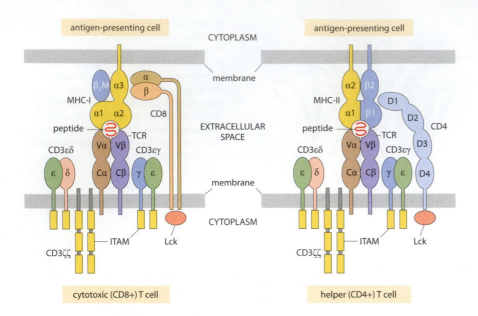

Figure 17.23 Schematic representation of the receptor complexes required for T cell activation. Peptide recognition by the TCR, through interaction between pepMHC and the TCR/CD3 complex, gives rise to phosphorylation of CD3 ITAMs. T cell activation requires the activity of co-receptors CD8 (pepMHC-I) or CD4 (pepMHC-II). CD8 and CD4 have extended structures, which project the ligand-binding domains across the intercellular junction to contact the membrane-distal region of the pepMHC. Upon TCR/pepMHC/co-receptor interaction, the co-receptors recruit the Lck tyrosine kinase to interact with intracellular components of the CD3 complex, thereby promoting phosphorylation of the ITAMs. The curved lines from CD8α and β to the membrane and Lck represent the extended CD8 stalk, which is 21–30 aa long, rich in Pro, Ser, and Thr, but of currently unknown structure.

TCR co-receptor molecules are required for T cell development and activation

The co-receptors CD8 and CD4 serve to extend the reach of the TCR across the intercellular junction to contact the membrane-proximal region of pepMHC at the surface of the APC. These co-receptors help direct helper and cytotoxic T cells to appropriate targets by recognizing an invariant part of the relevant MHC complex. The co-receptors interact with the TCR/CD3/pepMHC complex and stimulate antigen-specific T cell activation (**Figure 17.23**). CD8 at the surface of cytotoxic T cells (CTLs) interacts with pepMHC-I. CD4 at the surface of helper T cells interacts with pepMHC-II. Both interact with the membrane-proximal region of pepMHC at a site distinct from the TCR-binding site and independent of the nature of the associated peptide antigen. During TCR engagement, the co-receptors bring intracellular tyrosine kinases into close proximity with the CD3 ITAMs, promoting signal transduction in response to TCR-mediated antigen recognition.

The co-receptors CD4 and CD8 have strikingly different structures (Figure 17.23 and **Figure 17.24A,B**). CD4 has an extended ectodomain of four tandem Ig domains, a transmembrane segment, and a small intracellular domain that associates with the tyrosine kinase Lck. CD8 has distinct α and β subunits, which form disulfide-linked dimers, CD8αβ and CD8αα. Both subunits have an Ig domain tethered to the membrane by a long flexible stalk, a transmembrane domain, and a 28-aa cytoplasmic domain that is responsible for interaction with Lck. CD8β has less than 20% sequence identity with CD8α and its stalk is 10–13 aa shorter. CD8αβ is the abundant isoform on the surface of thymocytes and mature cytotoxic T cells and is critical for TCR-mediated activation, development, and differentiation. CD8αα does not function as a TCR co-receptor, although it is thought to have a regulatory role in T cell function.

Characteristics of the interactions between pepMHC complexes and their co-receptors have been revealed in crystal structures of CD4/pepMHC-II and CD8αα/pepMHC-I. CD4 and CD8 both interact with the invariant membrane-proximal domains of pepMHC, such that they bind beneath and perpendicular to the side of the peptide-binding groove (Figure 17.24C,D). CD4 contacts pepMHC-II via its membrane-distal D1 domain, which binds to the α2 and β2 domains. The Ig domain dimer of CD8αα employs the CDR-equivalent loops between the BC, C'C'', and FG strands of each subunit to contact conserved regions of pepMHC-I in an antibody-like manner. CD8αα interacts primarily with the α3 domain of pepMHC-I, but also contacts the α1, α2, and β2M domains. A similar mode of interaction has been proposed for the CD8αβ heterodimer. In both the CD4/pepMHC-II and CD8αα/pepMHC-I complexes, the peptide-presentation domain of the pepMHC is not contacted by the co-receptor.

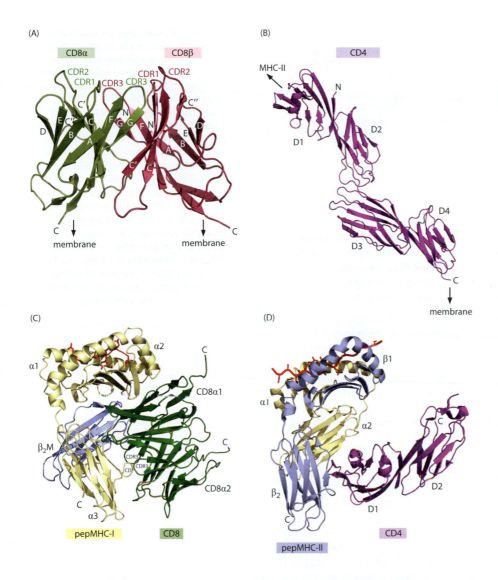

Figure 17.24 The TCR co-receptors CD8 and CD4. CD8 and CD4 have analogous functions but strikingly different structures. (A) The mouse CD8αβ Ig heterodimer (PDB 2ATP) has two V-type Ig domains, paired along the C' (β subunit, mauve) and G (α subunit, green) β strands of the respective subunits, such that the Ig dimer resembles the antigen-binding (Fv) region of antibodies (see Box 17.3). CDR-like loops are formed between the BC (CDR1), CC' (CDR2), and FG (CDR3) strands of each subunit. Both subunits are attached to the membrane through stalks 20–30 aa long. (B) The mouse CD4 ectodomain (PDB 1WIP) has four tandem Ig domains (D1–D4) connected to a single-pass transmembrane motif C-terminal to D4. The elongated shape of CD4 allows the D1 domain to reach across the intracellular junction, where it contacts the membrane-proximal region of pepMHC-II. (C) In the mouse pepMHC-I/CD8αα complex (PDB 1BQH), CD8 (green) binds pepMHC-I (beige) by interacting with the α1, α2, and α3 domains, as well as with β2M (light blue). The CDR3-like loops of both CD8 subunits clamp around the exposed CD loop of the pepMHC-I α3 domain. (D) The mouse pepMHC-II/CD4 complex showing the D1 and D2 domains of CD4 (PDB 1JL4). The D1 domain contacts the α2 and β2 domains of pepMHC-II, whereas domains D2–D4 do not participate in the interaction.

Antigen recognition by the TCR initiates intracellular signaling via the associated CD3 complex

Models for T cell activation can be assigned to three categories: aggregation, segregation, and conformational change. In aggregation models, distinct TCR/CD3 complexes are brought together by interaction with pepMHC at the surface of an APC, leading to local enrichment of CD3 ITAMs and effective co-receptor/kinase function. Segregation models envisage that phosphorylation of CD3 ITAMs occurs as a result of the exclusion of phosphatases with large ectodomains from the spatially restricted intercellular contact zone. Conformational change models suggest that the TCR/pepMHC interactions give rise to changes in the structure and/or organization of the TCR/CD3 complex that are transmitted across the T cell membrane and induce the phosphorylation of CD3 ITAMs. The latter models imply that interaction between the TCR and CD3 is influenced by pepMHC interaction, and that activation involves changes in the association between the TCR and CD3 subunits. All three models have supporting experimental evidence, and it is likely that TCR-mediated activation employs elements of all three mechanisms.

Phosphorylation of CD3 leads to activation of T cells

In the lymph node, when a naive T cell encounters a specific antigen presented on an APC for the first time, it is stimulated to differentiate into an effector T cell. The T cells use as co-stimulatory molecules the cell surface B7 proteins CD80 and CD86 from the APC and the CD28 protein on the T cell surface. These molecules help facilitate cellular adhesion and enhance T cell receptor signaling. The proliferation and differentiation of a naive T cell to

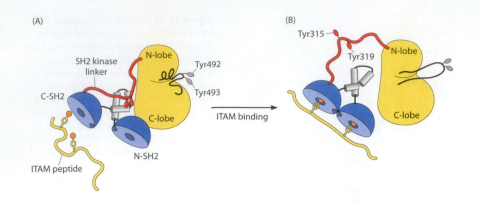

Figure 17.25 Signaling via Zap70 after T cell activation. In this schematic diagram, Zap70 consists of an N-terminal SH2 domain (blue), an inter-SH2 linker (gray), a C-terminal SH2 domain (blue), an SH2-kinase linker (red), and the kinase domain (yellow); the doubly phosphorylated ITAM peptide is in yellow. (A) In the autoinhibited state of Zap70, the N-SH2 unit is docked onto the catalytic domain, and Tyr315 and Tyr319 in the C-SH2-kinase linker are buried at the interface. (B) ITAM binding stabilizes a conformation of the tandem SH2 unit that is inconsistent with its docking onto the kinase domain. Tyr315 and Tyr319 are exposed for phosphorylation by Lck. Tyr492 and Tyr493 on the activation segment (black ribbon in the kinase domain) become available for phosphorylation and activation of Zap70. (Adapted from B.B. Au-Yeung et al., *Immunol. Rev.* 228:41–57, 2009. With permission from John Wiley & Sons.)

an activated antigen-specific effector T cell is initiated by the phosphorylation of CD3. The TCR signaling pathways eventually lead to the activation of transcription factors for interleukin-2, and expression of genes for cytoskeleton remodeling, and integrin cell adhesion molecules (Section 11.6). The signaling pathways are complex and diverse. Here we give a simplified account of how TCR signaling to the nucleus promotes gene expression.

Upon recognition of the antigenic peptide presented by the MHC complex to the TCR, the T cell is activated. The kinase Lck, recruited by CD4 or CD8, phosphorylates the ITAM motifs present on the CD3δ, -γ, -ε, and -ζ subunits. Phosphorylation of the dual ITAM motifs of CD3ζ subunits recruits the kinase Zap70 (zeta-associated protein of 70 kDa). Zap70 binds to phosphorylated ITAMs through its two tandem SH2 domains, which relieves the kinase inhibitory interactions of the SH2 domains (**Figure 17.25**). Zap70 is then recruited to the plasma membrane, where it is activated by Lck phosphorylation on two tyrosines in the linker between the SH2 and the kinase domains (Tyr315 and Tyr319) and on two tyrosines of the activation segment.

Active Zap70 subsequently phosphorylates at least two adaptor proteins, LAT and SLP76 (abbreviations given in the legend to **Figure 17.26**). LAT lacks enzymatic activity but acts as a scaffold for the assembly of signaling complexes. Zap70-phosphorylated LAT recruits SH2 domain-containing adaptors and effector proteins, including GADS and GADS-bound SLP76. In turn, the LAT/SLP76 complex recruits PLCγ and an activating kinase ITK. PLCγ hydrolyzes phosphatidylinositol bisphosphate (PIP_2) to diacylglycerol (DAG), a membrane-associated lipid, and inositol triphosphate (IP_3). DAG recruits RasGRP, a guanine nucleotide exchange factor (GEF), and signals through the small GTPase Ras, to activate the Raf-MEK1/2-Erk pathway. Phosphorylated Erk enters the nucleus to activate the transcription factor Fos. Ras is also activated by the GEF SOS, which is recruited to LAT via the adaptor protein Grb2. DAG recruits to the membrane PKCθ, which associates with the complex MALT1/Bcl10/CARMA1. This complex activates MEKK, which in turn activates the IKK complex that phosphorylates and dissociates the inhibitor IκB to release NFκB. This transcription factor then enters the nucleus.

SLP76 also recruits the adaptor proteins ADAP, NCK, and VAV. VAV is a guanine nucleotide exchange factor GEF that activates the small GTPases Rac and Cdc42 (Figure 17.26). Rac/Cdc42 promotes the first stage in cytoskeleton remodeling through Wiscott–Aldrich syndrome protein (WASP): see Section 11.6. This remodeling process involves actin polymerization, generating a lamellipodial sheet, and integrin-mediated adhesion molecules that facilitate the establishment of a long-lived T cell/APC contact. Rac/Cdc42 signals through the MAP kinase pathway via MEKK and either MKK3 or MKK6 to activate p38, which activates the transcription factor Fos, or through MKK4 or MKK7 to activate JNK, which in turn activates the transcription factor Jun. Together, Fos and Jun form the heterodimerTCR transcription factor activation protein 1 (AP1).

The other product of PIP_2 hydrolysis, IP_3, stimulates the release of Ca^{2+} that activates the phosphatase calcineurin through calmodulin (see **Guide**) (Figure 17.26). The transcription factor NFAT is phosphorylated and inactivated by the kinase GSK3 when Ca^{2+} levels are low. When Ca^{2+} levels are high, calcineurin dephosphorylates NFAT and allows it to enter the nucleus.

These factors promote the transcription of the cytokine *interleukin-2* (IL-2) gene and other genes. IL-2 is synthesized and secreted by the T cell. It then binds to IL-2 receptors at the

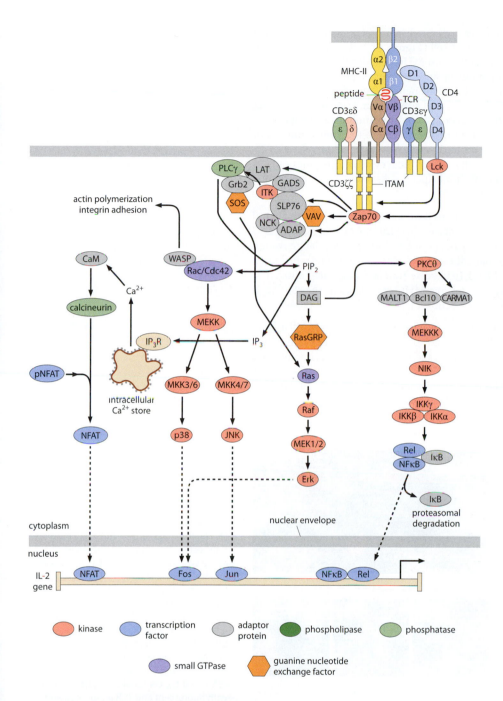

Figure 17.26 T cell receptor signaling to the nucleus. The pathways that involve actin polymerization and integrin adhesion, the co-stimulatory receptor CD28, and the coupling to the immunological synapse are not shown. Abbreviations: ADAP, adhesion and degranulation-promoting adaptor protein; AP1, activator protein 1; Bcl10, B cell lymphoma 10; CaM, calmodulin; CARMA1, caspase recruitment domain-containing membrane-associated guanylate kinase protein 1; DAG, diacylglycerol; Erk, extracellular signal-regulated kinase; GADS, Grb2-related adaptor protein downstream of Shc; Grb2, growth-factor receptor bound protein; GSK3, glycogen synthase kinase 3; IκB, inhibitor of NFκB; IKK, IκB kinase; IP$_3$, inositol 1,4,5-trisphosphate; IP$_3$R, IP$_3$ receptor; ITK, interleukin-2-inducible T cell kinase; JNK, Jun kinase; LAT, linker for the activation of T cells; Lck, leukocyte-specific tyrosine kinase; MALT1, mucosa-associated lymphoid tissue lymphoma translocation gene 1; MAPK, mitogen-activated protein kinase; MEK, MAPK or Erk kinase; MEKK, MEK kinase; MEKKK, MEK kinase kinase; MKK, MAPK kinase; NCK, non-catalytic region of tyrosine kinase containing SH3 and SH2 domains; NFAT, nuclear factor of activated T cells; NIK, NFκB-inducing kinase; NFκB, nuclear factor κB; p38, protein of 38 kDa; PIP$_2$, phosphatidylinositol 4,5-bisphosphate; PKCθ, protein kinase Cθ; PLCγ, phospholipase Cγ; RasGRP, Ras guanyl nucleotide releasing protein; SLP76, SH2-domain-containing leucocyte protein of 76 kDa; SOS, Son of Sevenless; WASP, Wiscott–Aldrich syndrome protein. (Adapted from Cell Signaling Technology and M.J. Huse, *Cell Sci.* 122:1269–1273, 2009.)

T cell surface to drive clonal expansion of the activated T cell. There ensues an enormous proliferation, with thousands of cells originating from a single T cell clone.

Feedback regulation at several points in the pathways permits different outcomes, depending on cell type and environment. The pathways are turned off by specific phosphatases and inhibitory kinases. The tyrosine phosphatase SHP1 dephosphorylates and inactivates Zap70 and Lck. The tyrosine kinase Csk inhibits Lck by phosphorylating a C-terminal tyrosine residue. The E3 ubiquitin ligase Cbl targets several proteins, including Lck, Zap70, and VAV, for proteasome degradation.

The immunological synapse mediates the cytolytic machinery of cytotoxic T cells

Once inside the cell, a pathogen is inaccessible to the immune system proteins but can be eliminated through efforts of the cell itself or through direct attack by effector cytotoxic T cells (CTLs). These cells kill the target cell by apoptosis (Section 13.6), induced either by ligand interactions or by cytotoxic enzymes contained in secretory lytic vesicles that are stored in the CTL and released on to the target cell (**Figure 17.27A**). Contact is established when CTLs recognize target cells presenting, on MHC-I molecules, antigenic peptides derived from pathogens. CTL activation during an immune response begins with the formation of a stable junction between CTL TCRs and the peptide-bound MHC (pepMHC) on the APC. Soon after contact, there is rearrangement of adhesion and signaling molecules on both plasma membranes, resulting in the formation of an **immunological synapse** (IS). IS formation leads rapidly (within a few minutes) to destruction of the virally infected or tumorigenic APC. The CTL may then move on to destroy other infected target cells recognized by the TCR. In this way, effector T cells deliver their cytotoxic enzymes only to infected cells. A similar synapse forms when effector helper T cells interact with their target B cells and activate them to produce antibodies.

Fluorescence microscopy studies with labeled proteins have shown that the IS has a central zone called the central supramolecular activation cluster (cSMAC), which is rich in TCR/pepMHC complexes. This zone is surrounded by the peripheral supramolecular activation cluster (pSMAC), which is rich in junctional adhesion proteins (Section 11.4), including ICAM-1 (on the APC) and the integrin LFA-1 (leukocyte function-associated antigen 1) on the T cell (Figure 17.27B). The buildup of TCR/pepMHC complexes in the cSMAC is accompanied by localized actin polymerization (Section 14.2).

Once bound to its target cell, an effector CTL can employ one of two strategies to kill it, both of which operate by inducing the target cell to undergo apoptosis (Section 13.6). In killing a target cell, the CTL releases a pore-forming protein called perforin, which is homologous to complement component C9 (see Figure 17.13). The CTL stores perforin in secretory vesicles and releases it by exocytosis at the point of contact with the target cell. Perforin then assembles into transmembrane pores in the plasma membrane of the target cell. The

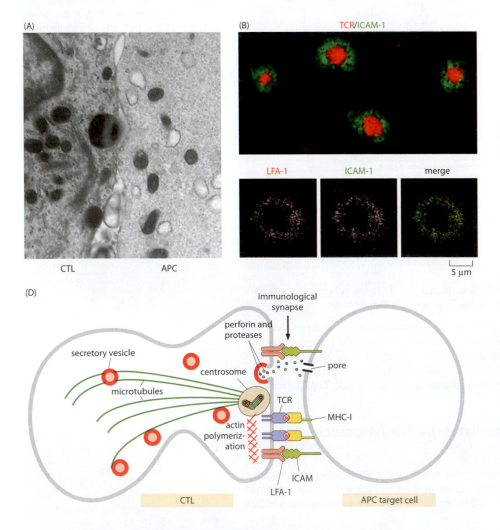

Figure 17.27 The immunological synapse. (A) Electron micrograph showing a transverse thin section through an IS. A granule is shown at the synapse between the CTL cytoplasm and the inter-cell interface, where it may be discharging from the CTL. Other smaller dark organelles are mitochondria, which are numerous on both sides of the IS. (B) An artificial IS between T cells interacting with planar lipid bilayers containing pepMHC and ICAM-1, mimicking an APC. Top, confocal image showing the separation of cSMAC (at the center of the IS, labeled red with fluorescent anti-TCR antibodies) and pSMAC zones (at the periphery, labeled green with anti-ICAM-1 antibodies). Bottom, co-localization of ICAM-1 with LFA-1 in the pSMAC, detected by fluorescently labeled antibodies. (C) Schematic drawing showing the immunological synapse between a CTL and a target APC illustrating the inferred migration of secretory vesicles on microtubules to the IS. (A, from G. Bossi et al., *Immunol Rev.* 189:152–160, 2002. With permission from John Wiley & Sons; B, from Y. Kaizuka et al., *Proc. Natl Acad. Sci. USA* 104:20296–20301, 2007 National Academy of Sciences, USA. With permission from PNAS; C, adapted from M.R. Jenkins and G.M. Griffiths, *Curr. Opin. Immunol.* 22:1–6, 2010. With permission from Elsevier.)

secretory vesicles also contain serine proteases, which enter the target cell through the perforin pores and promote apoptosis. In the second killing strategy, the CTL activates a death-inducing caspase cascade in the target cell through the homotrimeric protein, Fas death ligand, on the CTL surface that binds to transmembrane receptor Fas on the target cell. The binding stimulates the apoptotic extrinsic pathway (Section 13.6).

IS formation is accompanied by a cascade of signaling events on both sides of the synapse that lead to the delivery of secretory vesicles to the synapse. These vesicles are formed in the CTL cytosol after the precursor CD8+ T cell has been primed by prior interaction with an APC. On formation of the IS, a microtubule-organizing center (centrosome; Section 13.5) moves from its usual location on the periphery of the nucleus to the synapse region of the CTL; from the centrosome, microtubules grow out. Cytolytic vesicles are transported toward the synapse along the microtubules, presumably by minus end-directed molecular motors (Section 14.7). From there, the vesicles are transferred into a space between the cSMAC on the CTL and the pSMAC on the APC, where they discharge their contents (Figure 17.27C).

17.5 SUMMARY

The immune system, which protects organisms from infection, has two parts: innate immunity and adaptive immunity. Innate immunity is evolutionarily the older and is immediately available to resist infection. It has two components: (1) Toll-like receptors and inflammasomes; (2) the complement system. Adaptive immunity is found only in higher animals (vertebrates) and tailors its response specifically to the infecting pathogen (or foreign substance)—a process that takes several days. This system also has two components: the antibody response and the T cell-mediated response.

Toll-like receptors (TLRs) recognize ligands ('molecular patterns') present on pathogenic microorganisms. A variety of ligands are recognized within a common framework. Ligands bind to the TLR extracellular domains. TLRs signal via ligand-induced dimerization, with the ligand acting as a bridge connecting two monomers. In this way, the intracellular TIR domains are brought together in a scaffold that can bind post-receptor adaptor proteins, leading into downstream signaling by phosphorylation. Inflammasomes are signaling platforms assembled when foreign substances are detected inside the cell. Their subunits represent different combinations drawn from the same base set of domains; they assemble in domino-like fashion by homomeric interactions. This results in proteolytic activation of the protease caspase-1. Like TLRs, inflammasomes elicit an inflammatory response (hence, the name).

Complement consists of a protein network circulating in blood plasma. Foreign organisms are recognized through molecules displayed on their surfaces. Their recognition activates a proteolytic cascade that generates protein fragments, some of which (opsonins) mark foreign cells and particles for phagocytosis, while others (anaphylatoxins) have inflammatory activity. There are three initiation pathways—classical, lectin, and alternative—which all converge on the C3 convertase. Cleavage of C3 to C3a and C3b (conversion) is accompanied by major structural changes that lead to activation. Host cells are protected by regulator proteins, which inactivate the complement convertases, a system that has been mimicked by several viruses and bacteria to evade complement-mediated clearance.

Antibodies are (mainly) extracellular proteins circulating in the bloodstream. They consist of multiple copies of subunits with the immunoglobulin fold (hence the name), since found in many other kinds of proteins. Foreign entities ('antigens') are recognized by an array of six juxtaposed loops on the antibody ('complementarity-determining regions'), three each from the 'variable' regions of two subunits. Their variety is generated by recombination. The immune complexes formed by the antigen-antibody interaction lead to elimination of the foreign substance by a variety of mechanisms.

The T cell-mediated immune response is driven by the T cell receptor (TCR)/CD3 complex, a multifunctional machine that controls the activation of T cells and restricts the cellular response to a specific population of target cells. The TCRs recognize antigens bound to major histocompatibility complexes (MHC) on the surface of antigen-presenting cells (APCs). The antigens are peptides derived from the infecting organism bound in a groove on the MHC surface (pepMHC). Recognition by the TCR leads to intracellular signal transduction via the associated CD3 complex, initiated by phosphorylation of intracellular CD3 domains and depending also on activity of the co-receptor proteins, CD4 or CD8. Other accessory molecules at the T cell surface provide co-stimulatory or co-inhibitory signals that modulate the outcome of the TCR/CD3/pepMHC interaction and represent a secondary level of control of T cell activation. On forming an immunological synapse with an APC, the lytic machinery of a cytotoxic T cell is actuated.

Sections of this chapter are based on the following contributions

17.3   Piet Gros, University of Utrecht, The Netherlands.
17.4   David A. Shore, Robyn L. Stanfield, and Ian A. Wilson, The Scripps Research Institute, La Jolla, USA.

REFERENCES

General

Murphy K (2016) Janeway's Immunobiology, 9th ed. New York: Garland Science.

Owen J, Punt J & Stanford S (2013) Kuby Immunology, 7th ed. New York: WH Freeman.

Parham P (2015) The Immune System, 4th ed. New York: Garland Science.

17.2 Toll-like receptors

Bryant CE, Spring DR, Gangloff M et al (2010) The molecular basis of the host response to lipopolysaccharide. *Nature Rev Microbiol* 8:8–14.

Gay NJ & Gangloff M (2007) Structure and function of Toll receptors and their ligands. *Annu Rev Biochem* 76:141–165.

Gross O, Thomas CJ, Guarda G et al (2011) The inflammasome: an integrated view. *Immunol Rev* 243:136–151.

Jin MS & Lee JO (2008) Structures of the Toll-like receptor family and its ligand complexes. *Immunity* 29:182–191.

Jin MS, Kim SE, Heo JY et al (2007) Crystal structure of the TLR1-TLR2 heterodimer induced by binding of a tri-acylated lipopeptide. *Cell* 130:1071–1082.

Jung K, Lee JE, Kim HZ et al (2009) Toll-like receptor 4 decoy, TOY, attenuates Gram-negative bacterial sepsis. *PloS ONE* 4:e7403.

Liu L, Botos I, Wang Y et al (2008) Structural basis of Toll-like receptor 3 signaling with double-stranded RNA. *Science* 320:379–381.

Lu A & Wu H (2015) Structural mechanisms of inflammasome assembly. *FEBS J* 282:435–444.

Nyman T, Stenmark P, Flodin S et al (2008) The crystal structure of the human toll-like receptor 10 cytoplasmic domain reveals a putative signaling dimer. *J Biol Chem* 283:11861–11865.

Park BS, Song DH, Kim HM et al (2009) The structural basis of lipopolysaccharide recognition by the TLR4–MD-2 complex. *Nature* 458:1191–1195.

Yin Q, Fu TM, Li J et al (2015) Structural biology of innate immunity. *Annu Rev Immunol* 11:393–416.

17.3 Complement

Arlaud GJ, Barlow PN, Gaboriaud C et al (2007) Deciphering complement mechanisms: the contributions of structural biology. *Mol Immunol* 44:3809–3822.

Gal P, Dobo J, Zavodszky P et al (2009) Early complement proteases: C1r, C1s and MASPs. A structural insight into activation and functions. *Mol Immunol* 46:2745–2752.

Gros P, Milder FJ & Janssen BJ (2008) Complement driven by conformational changes. *Nature Rev* 8:48–58.

Janssen BJ & Gros P (2007) Structural insights into the central complement component C3. *Mol Immunol* 44:3–10.

Janssen BJ, Christodoulidou A, McCarthy A et al (2006) Structure of C3b reveals conformational changes that underlie complement activity. *Nature* 444:213–216.

Janssen BJ, Gomes L, Koning RI et al (2009) Insights into complement convertase formation based on the structure of the factor B-cobra venom factor complex. *EMBO J* 28:2469–2478.

Janssen BJ, Huizinga EG, Raaijmakers HC et al (2005) Structures of complement component C3 provide insights into the function and evolution of immunity. *Nature* 437:505–511.

Kardos J, Harmat V, Pallo A et al (2008) Revisiting the mechanism of the autoactivation of the complement protease C1r in the C1 complex: structure of the active catalytic region of C1r. *Mol Immunol* 45:1752–1760.

Law RHP, Lukoyanova N, Voskoboinik I et al (2010) The structural basis for membrane binding and pore formation by lymphocyte perforin. *Nature* 468:447–451.

Milder FJ, Gomes L, Schouten A et al (2007) Factor B structure provides insights into activation of the central protease of the complement system. *Nature Struct Mol Biol* 14:224–228.

Ponnuraj K, Xu Y, Macon K et al (2004) Structural analysis of engineered Bb fragment of complement factor B: insights into the activation mechanism of the alternative pathway C3-convertase. *Mol Cell* 14:17–28.

Qu H, Ricklin D & Lambris JD (2009) Recent developments in low molecular weight complement inhibitors. *Mol Immunol* 47:185–195.

Rooijakkers SH, Wu J, Ruyken M et al (2009) Structural and functional implications of the alternative complement pathway C3 convertase stabilized by a staphylococcal inhibitor. *Nature Immunol* 10:721–727.

Rosado CJ, Buckle AM, Law RH et al (2007) A common fold mediates vertebrate defense and bacterial attack. *Science* 317:1548–1551.

Schneider MC, Prosser BE, Caesar JJ et al (2009) *Neisseria meningitidis* recruits factor H using protein mimicry of host carbohydrates. *Nature* 458:890–893.

Tilley SJ & Saibil HR (2006) The mechanism of pore formation by bacterial toxins. *Current Opin Struct Biol* 16:230–236.

Torreira E, Tortajada A, Montes T et al (2009) 3D structure of the C3bB complex provides insights into the activation and regulation of the complement alternative pathway convertase. *Proc Natl Acad Sci USA* 106:882–887.

Torreira E, Tortajada A, Montes T et al (2009) Coexistence of closed and open conformations of complement factor B in the alternative pathway C3bB(Mg^{2+}) proconvertase. *J Immunol* 183:7347–7351.

Wallis R, Mitchell DA, Schmid R et al (2010) Paths reunited: initiation of the classical and lectin pathways of complement activation. *Immunobiology* 215:1–11.

Wu J, Wu YQ, Ricklin D et al (2009) Structure of complement fragment C3b-factor H and implications for host protection by complement regulators. *Nature Immunol* 10:728–733.

17.4 T-cell-mediated immunity

Adams EJ, Chien YH & Garcia KC (2005) Structure of a γδ T cell receptor in complex with the nonclassical MHC T22. *Science* 308:209–210.

Arnett KL, Harrison SC & Wiley DC (2004) Crystal structure of a human CD3εδ dimer in complex with a UCHT1 single-chain antibody fragment. *Proc Natl Acad Sci USA* 101:16268–16273.

Au-Yeung BB, Deindl S, Hsu L-Y et al (2009) The structure, regulation and function of ZAP-70. *Immunol Rev* 228:41–57.

Bjorkman PJ, Saper MA, Samraoui B et al (1987) The foreign antigen binding site and T cell recognition regions of class I histocompatibility antigens. *Nature* 329:512–518.

Borg NA, Wun KS, Kjer-Nielsen L et al (2007) CD1δ-lipid-antigen recognition by the semi-invariant NKT T-cell receptor. *Nature* 448:44–49.

Bryant P & Ploegh H (2004) Class II MHC peptide loading by the professionals. *Curr Opin Immunol* 16:96–102.

Call ME, Pyrdol J & Wucherpfennig KW (2004) Stoichiometry of the T-cell receptor–CD3 complex and key intermediates assembled in the endoplasmic reticulum. *EMBO J* 23:2348–2357.

Call ME, Pyrdol J, Wiedmann M et al (2002) The organizing principle in the formation of the T cell receptor–CD3 complex. *Cell* 111:967–979.

Chang HC, Tan K, Ouyang J et al (2005) Structural and mutational analyses of a CD8αβ heterodimer and comparison with the CD8αα homodimer. *Immunity* 23:661–671.

Ding YH, Smith KJ, Garboczi DN et al (1998) Two human T cell receptors bind in a similar diagonal mode to the HLA-A2/Tax peptide complex using different TCR amino acids. *Immunity* 8:403–411.

Dustin ML (2014) The immunological synapse. *Cancer Immunol Res* 2:1023–1033.

Fremont DH, Matsumura M, Stura EA et al (1992) Crystal structures of two viral peptides in complex with murine MHC class I H-2K^b. *Science* 257:919–927.

Gao GF, Tormo J, Gerth UC et al (1997) Crystal structure of the complex between human CD8 and HLA-A2. *Nature* 387:630–634.

Garcia KC, Degano M, Stanfield RL et al (1996) An αβ T cell receptor structure at 2.5 Å and its orientation in the TCR–MHC complex. *Science* 274:176–177.

Garcia KC, Scott CA, Brunmark A et al (1996) CD8 enhances formation of stable T-cell receptor/MHC class I molecule complexes. *Nature* 384:577–581.

Hennecke J, Carfi A & Wiley DC (2000) Structure of a covalently stabilized complex of a human αβ T-cell receptor, influenza HA peptide and MHC class II molecule, HLA-DR1. *EMBO J* 19:5611–5624.

Huse M (2009) The T-cell receptor signaling network. *J Cell Sci* 122:1269–1273.

Kaizuka Y, Douglass AD, Vrma R et al (2007) Mechanisms for segregating T-cell receptor and adhesion molecules during immunological synapse formation in Jurkat T cells. *Proc Natl Acad Sci USA* 104:20296–20301.

Kern PS, Teng MK, Smolyar A et al (1998) Structural basis of CD8 coreceptor function revealed by crystallographic analysis of a murine CD8αα ectodomain fragment in complex with H-2K^b. *Immunity* 9:519–530.

Kjer-Nielsen L, Dunstone MA, Kostenko L et al (2004) Crystal structure of the human T cell receptor CD3εγ heterodimer complexed to the therapeutic mAb OKT3. *Proc Natl Acad Sci USA* 101:7675–7680.

Koretzky GA, Abtahian F & Silverman MA (2006) SLP76 and SLP65: complex regulation of signaling in lymphocytes and beyond. *Nature Rev Immunol* 6:67–78.

Purcell AW & Elliott T (2008) Molecular machinations of the MHCI peptide loading complex. *Curr Opin Immunol* 20:75–81.

Ritter AT, Angus KL & Griffiths GM (2013) The role of the cytoskeleton at the immunological synapse. *Immunol Rev* 256:107–117.

Rudolph MG, Stanfield RL & Wilson IA (2006) How TCRs bind MHCs, peptides, and co-receptors. *Annu Rev Immunol* 24:419–466.

Saphire EO, Stanfield RL, Crispin MD et al (2003) Crystal structure of an intact human IgG: antibody asymmetry, flexibility, and a guide for HIV-1 vaccine design. *Adv Exp Med Biol* 535:55–66.

Sun ZJ, Kim KS, Wagner G et al (2001) Mechanisms contributing to T cell receptor signaling and assembly revealed by the solution structure of an ectodomain fragment of the CD3 εγ heterodimer. *Cell* 105:913–923.

Sun ZY, Kim ST, Kim IC et al (2004) Solution structure of the CD3εδ ectodomain and comparison with CD3εγ as a basis for modeling T cell receptor topology and signaling. *Proc Natl Acad Sci USA* 101:16867–16872.

van der Merwe PA & Davis SJ (2003) Molecular interactions mediating T cell antigen recognition. *Annu Rev Immunol* 21:659–684.

Wang JH, Meijers R, Xiong Y et al (2001) Crystal structure of the human CD4 N terminal two-domain fragment complexed to a class II MHC molecule. *Proc Natl Acad Sci USA* 98:10799–10804.

Wu H, Kwong PD & Hendrickson WA (1997) Dimeric association and segmental variability in the structure of human CD4. *Nature* 387:527–530.

Zajonc DM, Elsliger MA, Teyton L et al (2003) Crystal structure of CD1a in complex with a sulfatide self antigen at a resolution of 2.15 Å. *Nature Immunol* 4:808–815.

Glossary

α-actinin
Rodlike α-helical protein that cross-links actin filaments within the Z lines of myofibrils.

α-satellite DNA
A 171 bp DNA sequence found in repeated arrays in human centromeres.

abasic
See AP site.

accessory domains
Additional protein domains that modify the core function of a protein.

acrosomal process
Coiled filamentous structure that straightens, propelling the sperm of marine invertebrates through the jelly-like coating of the egg.

actin-binding protein (ABP)
Proteins that regulate the assembly, disassembly, and network organization of actin filaments.

actin filament
Helical protein filament. A major constituent of the cytoskeleton of all eukaryotic cells and part of the contractile apparatus of skeletal muscle (Figure 14.2).

action potential
Rapid, transient, self-propagating electrical excitation in the plasma membrane of a cell such as a neuron or muscle cell; makes possible long-distance signaling in the nervous system (Figure 16.9).

activation energy
The extra energy that must be acquired by atoms or molecules in addition to their ground-state energy in order to reach the transition state required for them to undergo a particular chemical reaction.

active site
Region of an enzyme surface to which a substrate molecule binds in order to undergo a catalyzed reaction.

active site-directed inhibitor
A molecule resembling the substrate of an enzyme that selectively reacts with a functional group on the enzyme, causing irreversible inhibition of the active site.

activity coefficient
The ratio of the effective concentration of a macromolecule (its **thermodynamic activity**) and its physical concentration. Used to correct for variations from ideal thermodynamic behavior.

acyl carrier protein (ACP)
Protein that binds phosphopantetheine and aids the addition of acyl groups to growing fatty acid chains.

adaptive immune system
Defense system in vertebrates that recognizes specific pathogens and mobilizes neutralizing responses against them. It consists of B lymphocytes (B cells), which secrete antibodies that bind specifically to the pathogen or its products, and T lymphocytes (T cells), which can either directly kill cells infected with the pathogen or produce secreted or cell-surface signal proteins that stimulate other host cells to help eliminate the pathogen.

adaptor
General term for a protein that functions solely to link two or more different proteins together in an intracellular signaling pathway or protein complex.

adherens junction
One of the structures that coordinate cell–cell adhesion in epithelial tissues. Consists of a double array of cadherin proteins at the cell surface that links the cells together, while their endodomains connect to the actin cytoskeleton.

adhesin
A component of a bacterial pilus, conferring binding specificity and often located at its tip.

adhesome
The catalog of the proteins and their interactions that make up focal adhesions.

adrenaline
See epinephrine.

adrenergic receptors
Class of G-protein-coupled receptors that are targets of catecholamine hormones, including epinephrine.

affinity
The strength of binding of a molecule to its ligand at a single binding site.

agonist
Chemical that binds to a receptor, causing it to adopt the conformation of its active state.

allostery
Change in a protein's conformation brought about by the binding of a regulatory ligand or other interaction. The resulting conformational change alters the activity of the protein (Figures 1.32, 1.33, and 1.34).

alternating access mechanism
Cycling of a transporter between one conformation in which the substrate-binding site is accessible from the outer surface of the membrane and another in which it is inward facing (Figure 16.23).

aminoacyl tRNA synthetases
Enzymes that charge transfer RNAs with the appropriate amino acid.

amyloid
Fibrillar aggregates of misfolded proteins that are insoluble, protease-resistant, thermostable, and have cross-β structure.

anaphase
Stage of mitosis during which sister chromatids separate and move away from each other (Figure 13.1).

anaphylatoxin
Peptides produced as part of the complement system that cause smooth muscle contraction, histamine release from mast cells, and enhanced vascular permeability.

anoikis
A form of apoptosis triggered when a cell loses adhesion to the extracellular matrix.

anomalous diffusion
A diffusion process with a nonlinear relationship to time.

antagonist
Chemical that binds to a receptor, causing it to adopt the conformation of its inactive state.

antibody
Protein secreted by activated B cells in response to a pathogen or foreign molecule. Binds tightly to the pathogen or foreign molecule, inactivating it or marking it for destruction by phagocytosis or complement-induced lysis.

antigen
Molecule that can induce an adaptive immune response or that can bind to an antibody or T-cell receptor.

antiporter
Carrier protein that transports two different ions or small molecules across a membrane in opposite directions, either simultaneously or in sequence (Figure 16.29).

AP site
A site within DNA that has neither a purine nor a pyrimidine base.

APC/C (anaphase-promoting complex)
Ubiquitin ligase that catalyzes the ubiquitylation and destruction of securin and M- and S-cyclins, initiating the separation of sister chromatids in the metaphase-to-anaphase transition during mitosis.

apoptosis
Form of programmed cell death, in which a 'suicide' program is activated within an animal cell, leading to rapid cell death mediated by intracellular proteolytic enzymes called caspases.

apoptosome
Complex of Apaf1 proteins and cytochrome *c* that forms on activation of the intrinsic apoptotic pathway; it recruits and activates initiator caspases that subsequently activate downstream executioner caspases to induce apoptosis.

aquaporin
Channel protein embedded in the plasma membrane that greatly increases the cell's permeability to water, allowing water, but not ions, to cross the membrane at a high rate.

Archaea
A kingdom of single-celled organisms without a nucleus. Together with bacteria, they comprise the prokaryotes. Although superficially similar to bacteria, their genetic machinery more closely resembles that of eukaryotes (Figures 1.38 and 1.40).

arginine finger
A structural motif of AAA+ ATPases, positioning an arginine residue that interacts with the nucleotide γ-phosphate.

Arp2/3 complex
Complex of proteins that nucleates the growth of actin filaments as side branches of a 'mother' actin filament.

ASCE (additional strand conserved E)
Superfamily of ATPases with a common topology of five β strands sandwiched between α helices, and with a conserved catalytic Glu residue. These ATPases are active only as dimers or higher order oligomers.

aspartyl protease
Type of protease that has a reactive aspartic acid in the active site.

autocatalytic (reaction)
A chemical reaction in which at least one of the reactants is also a product.

automaton
A self-propagating machine.

autonomously replicating sequence (ARS)
The origin of replication in the *Saccharomyces cerevisiae* genome.

autotransporter
Type 5 secretion system of bacteria employed for the secretion of toxins and virulence factors (Figure 10.56).

avidity
Accumulated strength of multiple noncovalent binding affinities between one protein or protein complex and another.

axoneme
Bundle of microtubules and associated proteins that forms the core of a cilium or a flagellum in eukaryotic cells and generates motility.

β arcade
A polymeric protein structure consisting of in-register stacking of strand-turn-strand motifs (Figure 6.46).

β helix
A type of β solenoid protein structure with three β strands per turn.

β-oxidation
Catabolic process that breaks down fatty acids; a cycle of four enzymic reactions shortens fatty acids by two carbons each cycle.

β sandwich
A type of β solenoid protein structure with two β strands per turn.

β serpentines
Type of β sheet protein structure in which the polypeptide chain zigzags in the planar fold, with the backbone interactions along the fibril axis and intrasheet interactions parallel to the fibril axis (Figure 6.46).

β solenoid
Structural motif found in amyloid and other proteins comprising short β strands coiled into a single-start helix (Figure 6.46).

B cells
Lymphocytes produced in bone marrow, which make antibodies.

bacteria
Kingdom of single-celled organisms without a nucleus. Together with the domain Archaea, they comprise the prokaryotes (Figures 1.38 and 1.40).

basal body
Short cylindrical array of microtubules and their associated proteins found at the base of a eukaryotic cell cilium or flagellum, which nucleates the growth of the axoneme. The term is also used for the structure at the base of a bacterial flagellum.

base-excision repair (BER)
DNA repair pathway in which single faulty bases are removed from the DNA helix and replaced.

Bcl-2
Archetypal member of a family of proteins of the outer mitochondrial membrane that binds and inhibits other pro-apoptotic family members and prevents inappropriate activation of the intrinsic pathway of apoptosis.

biofilm
Multicellular aggregate of bacteria embedded in a matrix of secreted proteins and polysaccharides.

biopanning
A laboratory technique used to identify short sequences displayed on the surface of bacteriophages that bind to a particular target.

biotin
Vitamin that serves as a prosthetic group for the attachment of catalytic intermediates to swinging arms in carboxylases.

breathing behavior (macromolecules)
Dynamic conformational fluctuations of a macromolecule or macromolecular complex in solution.

cadherin
Member of a large superfamily of transmembrane adhesion proteins whose ectodomains form intercellular contacts while their endodomains connect, via adaptor proteins, to the underlying cytoskeleton.

calcium release unit
Tetrad array of dihydropyridine receptors and ryanodine receptors in skeletal muscle cells that facilitates the release of calcium from internal stores to initiate muscle contraction.

calmodulin
Ubiquitous intracellular Ca^{2+}-binding protein that undergoes a large conformational change when it binds Ca^{2+}, allowing it to regulate the activity of many target proteins

cAMP
Cyclic AMP: a nucleotide generated from ATP by adenylyl cyclase in response to various extracellular signals and which acts as a small intracellular signaling molecule.

capping (microtubules)
The blocking of polymerization of protein filaments by end-binding.

capsid
Protein shell or 'coat' of a virus, formed by assembly of one or more types of protein subunit into a geometrically regular structure.

capsomers
Protein oligomers, usually ringlike, that make up a viral capsid.

capsule
The complex polysaccharide coating of some bacteria.

carbohydrate-binding module (CBM)
Protein module that targets cellulosomes to their substrates.

carboxysome
A protein-bound microcompartment in a bacterial cell that houses the CO_2-fixing enzyme ribulose-1,5-bisphosphate carboxylase.

carotenoids
Pigment molecules found in chloroplasts, which capture light energy and pass it on to chlorophylls and also protect chlorophylls from photodamage (Figures 15.20 and 15.21).

caspases
Intracellular proteases involved in mediating the intracellular events of apoptosis.

catalytic antibodies
Antibodies produced by immunization with a molecule resembling the transition state of a chemical reaction, which can catalyze the reaction represented by the transition state analog.

catalytic rate constant (k_{cat})
Constant that describes the rate-limiting step in enzyme catalysis, also known as the **turnover number**.

catalytic triad
Cluster of three amino acids (Ser/His/Asp) found in the active sites of some proteases.

catastrophe (microtubule disassembly)
Rapid disassembly of microtubules caused by the loss of GTP-bound β subunits at the plus end.

cathepsins
Family of proteases found predominantly in lysosomes that use cysteine and histidine in the active site.

cell cycle
The orderly sequence of events by which a eukaryotic cell duplicates its chromosomes and, usually, the other cell contents, and divides into two.

cell migration
Movement of a cell across a solid substrate using forces generated by systems of intracellular protein filaments.

cellulosomes
Large extracellular assemblies of hydrolytic enzymes generated by anaerobic microorganisms and used to break down plant cell walls.

centrioles
Short cylindrical array of microtubules, a pair of which is usually found at the center of a centrosome in animal cells.

centromere
Constricted region of a mitotic chromosome that holds sister chromatids together; also the site on the DNA where the kinetochore forms so as to capture microtubules from the mitotic spindle.

centrosomes
Centrally located organelle of animal cells that is the primary microtubule-organizing center (MTOC) and acts as the spindle pole during mitosis. In most animal cells it contains a pair of centrioles.

channel
Transmembrane protein complex that allows inorganic ions or other small molecules to diffuse across the lipid bilayer.

chaperone
Protein that helps guide proper folding of other proteins or helps them avoid misfolding.

chaperone/usher pathway
Assembly pathway for some bacterial pili, in which secreted pilin subunits are bound by a periplasmic chaperone until transferred into the growing pili (Figures 11.42 and 11.44).

chaperonins
Class of heat shock proteins (Hsp60) that help guide the proper folding of other proteins, or help them avoid misfolding.

charge separation
In photosynthesis, the light-induced transfer of a high-energy electron from chlorophyll to an acceptor molecule resulting in the formation of a positive charge on the chlorophyll and a negative charge on a mobile electron carrier.

checkpoints (cell cycle)
Transition points in the cell cycle where, if the preceding step is imperfectly executed, progression to the next phase can be blocked.

chemical proton
Proton transferred to the active site of cytochrome c oxidase during the reduction of molecular oxygen at the end of the respiratory chain.

chemiosmotic theory
Theory postulated by Peter Mitchell to describe ATP synthesis using the electrochemical proton gradient generated by the electron transport chain.

chemotaxis
Directed movement of a cell towards or away from some diffusible chemical.

chiral
Describes an object (atom or molecule) that is asymmetric in the sense of not being superposable on its mirror image.

chloroplast
Organelle in green algae and plants that contains chlorophyll and carries out photosynthesis. A specialized form of plastid.

chromatin
Complex of DNA, histones, and non-histone proteins found in the nucleus of a eukaryotic cell; the material of which chromosomes are made.

chromatin remodeling complex
Protein complexes that alter the presence, position, or structure of nucleosomes on chromosomal DNA.

chromonema fibers
Supercondensed form of chromatin, 130 nm in diameter.

cilium (plural: cilia)
Hairlike extension of a eukaryotic cell containing a core bundle of microtubules.

clamp loader
Protein complex that utilizes ATP hydrolysis to load the sliding clamp on to a primer–template junction in the process of DNA replication (Figures 3.23, 3.24, and 3.25).

clathrin
Protein that assembles into a polyhedral cage on the cytosolic side of a membrane, leading to the budding off of vesicles that can transport cargo to endosomes.

claudin
A family of intrinsic membrane proteins that are the main components of tight junctions.

coenzymes
Organic molecules that are required by some enzymes to carry out catalysis.

co-factor
A (non-protein) chemical compound that is required by some enzymes to carry out catalysis.

co-factor activity (innate immune system)
Mechanism to protect host cells from opsonization, in which regulators on the host cell bind protease factor I, which cleaves and inactivates complement proteins.

cohesin (bacterial)
Protein module of the non-catalytic core of cellulosomes that serves, to attach enzymatic subunits via dockerin modules.

colchicine
Drug, extracted from crocus plants, that promotes the disassembly of microtubules.

competitive inhibitor
A molecule that closely resembles the substrate of an enzyme and which binds reversibly to the active site, thereby blocking the reaction.

complement system
System of blood proteins that can be activated by antibody/antigen complexes or pathogens to help eliminate the pathogens, by directly causing their lysis, by promoting their phagocytosis, or by activating an inflammatory response (Figures 17.6 and 17.7).

complementarity
Of interacting molecules: a measure of how well two molecular surfaces fit together (Figure 1.9).

complementarity-determining regions (CDRs)
Triads of loops on the surface of antibodies that recognize the shapes and chemical properties of specific antigens.

complexin
Small soluble protein that plays a regulatory role in Ca^{2+}-triggered neurotransmitter release.

congression
Alignment of sister chromatids at the center of the mitotic spindle at metaphase.

conjugation
Intercellular transfer of DNA between bacterial cells.

connector
See portal protein.

connexin
Transmembrane protein that forms the hexameric channels which make up gap junctions (Figure 11.19).

connexon
Water-filled pore in the plasma membrane formed by a ring of six connexin protein subunits, comprising one-half of a gap junction (Figure 11.19).

cooperativity
Phenomenon in which the binding of one ligand molecule to a target molecule promotes the subsequent binding of additional ligand molecules.

COP9 signalosome (CSN)
Protein complex that catalyzes the hydrolysis of NEDD8 protein from the cullin subunit of Cullin-RING ubiquitin ligases (deneddylation).

CpG islands
Regions of the genome rich in cytosine nucleotides adjacent to guanosine nucleotides, which often occur near the promoter.

Creuzfeldt–Jakob disease (CJD)
A fatal, degenerative disease affecting the central nervous system of humans. The sporadic form affects elderly people and the variant form is caused by the consumption of prion-infected bovine products.

crista (plural: cristae)
A specialized invagination of the inner mitochondrial membrane.

cross β structure
Fibrillar protein structure rich in β sheets whose strands run perpendicular to the fibril axis (Figure 6.42).

cubic symmetry
A form of point group symmetry (Figure 1.16).

cullins
Family of hydrophobic proteins providing a scaffold for E3 ubiquitin ligases.

curli
Family of secreted bacterial proteins, two of which form fibrils that contribute to the formation of biofilms (Figure 6.49).

cyclic electron transport
Cyclical light-driven flow of electrons through the photosynthetic reaction center of purple sulfur bacteria, which generates a transmembrane proton gradient (Figure 15.29).

cyclic symmetry
A form of point group symmetry, in which the subunits form closed rings (Figure 1.15).

cyclin
A protein whose concentration periodically rises and falls in step with the eukaryotic cell cycle, activating protein kinases and thereby helping to control progression from one stage of the cell cycle to the next.

cyclin-dependent kinase (CDK)
Protein kinase that has to be complexed with a cyclin protein in order to act and that triggers a specific step in the cell cycle by phosphorylating specific target proteins.

cyclosome
See APC/C (anaphase-promoting complex).

cytochrome b_6f complex
Protein complex that passes electrons from photosystem II to photosystem I, generating a proton gradient across the thylakoid membrane (Figure 15.34).

cytokine
Extracellular signaling protein or peptide that acts as a local mediator in cell–cell communication through the JAK–STAT pathway.

cytokine receptor
Cell-surface receptor that binds a specific cytokine or hormone and acts through the JAK–STAT signaling pathway.

cytokinesis
Division of the cytoplasm of a plant or animal cell into two, which follows the associated division of its nucleus (mitosis).

cytoskeleton
System of protein filaments in the cytoplasm of a eukaryotic cell that gives the cell shape and the capacity for directed movement. Its most abundant components are actin filaments, microtubules, and intermediate filaments.

cytosol
Contents of the cytoplasm, excluding membrane-delimited organelles such as endoplasmic reticulum, Golgi, mitochondria, etc.

dark reactions
Chemical reactions that use NADPH generated by oxygenic photosynthesis to synthesize carbohydrates from CO_2.

death effector domains (DEDs)
Protein–protein interaction domains found in some proteins that make up the DISC. Homotypic interactions of these domains allow initiator capases to interact and activate, in the extrinsic pathway of apoptosis.

death receptors
Transmembrane receptor protein that can signal the cell to undergo apoptosis when it binds its extracellular ligand.

death-inducing signaling complex (DISC)
Activation complex in which initiator caspases interact and are activated following binding of extracellular ligands to cell-surface death receptors in the extrinsic pathway of apoptosis (Figure 13.40).

decay accelerating activity
Mechanism to protect host cells from opsonization, which breaks down surface-bound complement C3 convertases, interrupting the complement cascade.

decoding center (DC)
Area on the ribosome small subunit where the codon and anticodon are assessed for complementarity (Figure 6.3).

degradation motif (degron)
Recognition sequence or site on a protein that marks it for proteolytic degradation.

denaturation
A process by which macromolecules lose their folded structures, disrupting the multiple weak forces that combine to stabilize a folded protein or nucleic acid.

desensitized
Adjustment of sensitivity following repeated stimulation. The mechanism that allows a cell to react to small changes in stimuli even against a high background level of stimulation.

desmin
A type III intermediate filament protein found in muscle cells.

desmocollin
A type of cadherin with a short cytoplasmic tail, found in desmosomes.

desmoglein
A type of cadherin with a long cytoplasmic tail, found in desmosomes.

desmosome
Junctional complex that coordinates cell–cell adhesion in epithelial tissues; links to intermediate filaments within the interacting cells.

diastereoisomerism
A form of stereoisomerism for molecules that contain two or more chiral atoms (stereocenters).

diffusion
Random motion of molecules in solution.

dihedral symmetry
A form of **cyclic symmetry**, in which a twofold axis is added at right-angles to the n-fold axis (Figure 1.15)

dissociation constant (K_d)
Measure of the tendency of a complex to dissociate. The inverse of affinity constant (K_a).

DNA ligase
Enzyme that joins the ends of two strands of DNA together with a phosphodiester bond to make a continuous DNA strand (Figure 3.26).

DNA photolyase
Enzyme that reverses the damage to DNA caused by ultraviolet light.

DNA polymerase
Enzyme that synthesizes DNA by joining nucleotides together using a DNA template (Figures 3.15 and 3.16).

DNA primase
Enzyme that synthesizes a short strand of RNA on a DNA template, producing a primer for DNA synthesis.

DNA replication
Process by which a copy of a DNA molecule is made. The two strands of the double helix separate to expose nucleobases in each strand, each of which then acts as a template for the generation of a two-stranded daughter molecule.

DNA supercoiling
The over- or under-winding of a DNA strand.

dockerin
A protein module of enzymes that serves to attach the enzymes to the catalytic core of cellulosomes.

docking
The apposition of two membranes, prior to their fusion.

domains (protein domains)
An independently folding region of protein (~50 aa to ~250 aa) that forms either a subunit or part of a larger protein.

donor strand exchange
In the assembly of bacterial pili, the insertion of a β strand from one pilin subunit into the Ig domain of the next subunit, displacing the β strand of the chaperone.

dynactin
Protein complex that anchors a dynein motor to the cell membrane, facilitating the intracellular movement of organelles or vesicles.

dynamic instability
Sudden transition from growth to shrinkage, and vice versa, in a protein filament such as a microtubule.

dynamin
Cytosolic GTPase that binds to the neck of a clathrin-coated vesicle that is in the process of budding from the membrane (Figure 10.13).

dynamin-like proteins
GTPases that resemble dynamin and which serve similar functions in the scission of newly formed vesicles from membranes.

dynasore
Inhibitor of the dynamin GTPase, which leads to accumulation of coated pits at the beginning and end of neck formation.

dyneins
Class of motor protein that uses the energy of ATP hydrolysis to move along microtubules, toward the minus ends.

ectodomain
Extracellular domain of an intrinsic membrane protein, i.e. one outwith the hydrophobic phase of the lipid bilayer.

elastin
Extracellular protein that forms extensible fibers (elastic fibers) in connective tissues.

electrochemical potential
See proton-motive force.

electron transport chain (ETC)
Series of multi-subunit protein complexes embedded in a cell membrane which transfer electrons from donors to acceptors via redox reactions, and couple this electron transfer with the transfer of protons across the membrane (Figure 15.3).

elongation complex (EC)
Complex comprising RNA polymerase, the DNA being copied, and the newly synthesized RNA (Figure 5.5).

elongation factors
Proteins that escort tRNAs into and within the ribosome (Figure 6.12).

enantiomers
Stereoisomers that are mirror images of each other and are non-superposable.

endocytosis
Uptake of material into a cell by an invagination of the plasma membrane and its internalization in a membrane-enclosed vesicle.

endodomain
Cytoplasmic domain of an intrinsic membrane protein.

endoplasmic reticulum (ER)
Labyrinthine membrane-delimited compartment in the cytoplasm of eukaryotic cells, where lipids are synthesized and membrane-bound proteins and secretory proteins are made.

endoprotease
An enzyme that recognizes and hydrolyzes peptide bonds that are not at the termini of a substrate protein.

endosome
Membrane-delimited organelle containing materials newly ingested into the cell and destined for fusion into lysosomes where their protein cargo is degraded.

endosymbiont theory
Theory that organelles of eukaryotes arose by the internalization of one (prokaryotic) organism by another, followed by their symbiotic association.

enhanceosome
Protein complex assembled at the enhancer site of a target eukaryotic gene, facilitating its transcription (Figure 5.18).

enhancer
Regulatory DNA sequence to which gene regulatory proteins bind, increasing the rate of transcription of a gene that can be thousands of base pairs away.

enthalpy
The heat content of a system at constant pressure.

entropy
Thermodynamic quantity that measures the degree of disorder or randomness in a system.

envelope (viral)
Outer layer of some viruses, containing viral glycoproteins embedded in a host-derived membrane.

epigenetics
The study of heritable changes in phenotype caused by mechanisms that alter gene expression without changing the underlying DNA sequence.

epinephrine
Hormone produced by the adrenal glands that prepares mammals for 'fight or flight'.

ERK2
See MAP kinase.

euchromatin
Region of an interphase chromosome that stains diffusely; 'normal' chromatin, as opposed to the more condensed **heterochromatin**.

Eukarya
The kingdom of organisms that contain a nucleus and other membrane-delimited organelles (Figures 1.38 and 1.40).

eukaryotes
Organisms whose cells contain a nucleus and other membrane-delimited organelles (Figures 1.38 and 1.40).

exciton transfer (resonance energy transfer)
Transfer of energy from a donor to an acceptor molecule without transfer of an electron. Occurs within a light-harvesting complex.

excluded volume effect
Effective increase in concentration arising from the mutual exclusion of molecules in the system from the space that they occupy. Also known as **macromolecular crowding**.

exocytosis
Excretion of material from the cell by vesicle fusion with the plasma membrane; can occur constitutively or be regulated.

exon
Segment of a eukaryotic gene that remains present in the final mature RNA produced from that gene. Encodes amino acids in a protein-coding gene.

exoproteases
Enzymes that detach residues from one end of a polypeptide chain.

exosome
RNA-degrading machine with an interior rich in 3′ to 5′ RNA exonucleases; important for RNA quality control. The same term is also used for extracellular vesicles.

exportins
Proteins that mediate transport from the nucleus to the cytoplasm through the nuclear pore complex.

extracellular matrix
An elaborate extracellular assembly of glycoproteins and polysaccharides which coats cells and mediates cell–cell interactions.

extremophile
An organism that thrives in geochemically or physically extreme environments, such as hot springs or salt lakes.

F-actin (filamentous actin)
See actin filament.

F₁-ATPase
ATPase that, when associated with a membrane-bound F_o component, can synthesize ATP using the electrochemical potential of the proton gradient.

F₁F₀-ATPase
See F_1F_o-ATP synthase.

F₁F₀-ATP synthase
Membrane-bound ATP synthase that uses the electrochemical potential of the proton gradient to generate ATP (Figure 15.40).

F-box
Substrate receptor subunit of the SCF complex, an E3 ubiquitin ligase.

fast retrograde axonal transport
Rapid movement of vesicles from the cell body toward nerve termini in axons.

feedback inhibition
The process in which a product of a reaction feeds back to inhibit a previous reaction in the same pathway.

ferredoxin (Fd)
Soluble membrane iron–sulfur protein found in chloroplasts which mediates electron transfer in photosystem I.

fibrillum
An extended structure at the tip of bacterial pili that is composed of several different proteins and which determines binding specificity.

fimbriae
Another name for pili, the extracellular protein filaments of bacteria.

flagellum
Long, whiplike protrusion whose undulations drive a cell through a fluid medium (Figure 14.57).

focal adhesion
Anchoring cell junction, forming a patch on the surface of a fibroblast or other cell that is anchored to the extracellular matrix. Attachment is mediated by transmembrane proteins such as integrins linked through adaptors to actin filaments in the cytoplasm.

focal complex
Small structures that form at the cell edge and which mature into focal adhesions, anchoring a cell to the extracellular matrix.

folding cage
Compartment in the GroES chaperone that transiently accommodates a substrate protein, allowing it to fold (Figure 6.26).

free energy (G) (Gibbs free energy)
The energy that can be extracted from a system to drive reactions.

free energy landscape
Three-dimensional map whose local minima represent the multiple intermediate and final folding states of a protein (Figure 1.12).

free radical
Atom or molecule that has a single unpaired electron.

FRET (fluorescence resonance energy transfer)
Technique for monitoring the closeness of two fluorescently labeled molecules (and thus their interaction) in cells.

Fullerene
Three-dimensional structure composed of 12 pentagons and 20 or more hexagons; represents a geometrical generalization of the dodecahedron; named for the architect Buckminster Fuller.

functional amyloids
Insoluble protein fibrils (*see* amyloid) that function in normal physiological processes.

fusion peptide
Portion of a viral envelope glycoprotein that, after a conformational change, is inserted into the host membrane.

fusogenic
Facilitating fusion, particularly of viruses and their host cells.

γ-tubulin ring complex (γTurc)
Protein complex found in centrosomes that acts as a nucleation point for microtubule assembly.

G-actin
Monomeric globular protein that polymerizes to form helical filamentous actin (**F-actin**).

G0 phase
Quiescent stage, when cells are not actively dividing (Figure 13.1).

G1 phase
Long period of the cell cycle between mitosis and S phase (Figure 13.1).

G2 phase
Long period of the cell cycle between S phase and mitosis (Figure 13.1).

G proteins
Family of guanine nucleotide-binding proteins with intrinsic GTPase activity that couple transmembrane receptors to either enzymes or ion channels (Box 12.1).

G-protein-coupled receptor (GPCR)
Seven-pass membrane receptor that, when activated by an extracellular ligand, activates a G protein, which then activates either an enzyme or ion channel in the plasma membrane (Figure 12.6).

G-quadruplex
A four-stranded DNA structure generated by the stacking of G-tetrads, formed from four G nucleotides held together by Hoogsteen base pairing (Figure 3.31).

gap junction
Channel-forming cell–cell junction present in most animal tissues that allows ions and small molecules to pass from the cytoplasm of one cell to the cytoplasm of an adjacent cell (*see* connexon).

gating (of ion channels)
The opening and closing of ion channels in response to a stimulus such as membrane potential, ligand binding or mechanical deformation.

general transcription factors
Proteins whose assembly at all promoters of a given type is required for the binding and activation of RNA polymerase and the initiation of transcription (Table 5.2).

genetic strain
A genetic variant or subtype of a microorganism, e.g. yeast.

geometric isomerism
A property of molecules containing a double bond, which allows alternative dispositions of the same atoms across the double bond.

giant proteases
Large homomultimeric enzymes that present multiple cavities where oligopeptides are degraded.

glycerol phosphate shuttle
A biochemical system for translocating electrons produced during glycolysis in the cytoplasm across the mitochondrial matrix membrane, utilizing FAD, with the 'lost' energy appearing as heat.

glycosylases (DNA glycosylases)
Enzymes that remove damaged DNA bases, as the first step in base-excision repair.

Goldberg diagram
Hexagonal lattice that specifies the triangulation (*T*) numbers (integers) compatible with quasi-equivalence for icosahedral viral capsids.

Golgi apparatus (Golgi complex)
Complex organelle in eukaryotic cells, centered on a stacked system of flattened vesicles in which proteins from the **endoplasmic reticulum** are modified and sorted (Figure 1.35).

grana
Stacked membrane discs (thylakoids) in chloroplasts that contain chlorophyll and are the site of light-trapping reactions of photosynthesis (Figure 15.18).

Grotthuss-like mechanism
Mechanism by which a proton diffuses through the hydrogen bond network of water molecules (Figure 9.2.1E).

GTPase-activating protein (GAP)
Protein that binds to a GTPase and stimulates its intrinsic activity, causing the enzyme to hydrolyze its bound GTP to GDP.

guanine exchange factor (GEF)
Protein that binds to and activates a GTPase, stimulating it to release its tightly bound GDP and thereby allowing it to bind GTP.

head-full packaging
Type of viral assembly in which a physical mechanism signals that the capsid contains sufficient genomic DNA.

heat shock proteins
Molecular chaperones that promote correct protein folding and are synthesized in elevated amounts under stressful conditions such as higher temperatures.

hemidesmosome
A junctional structure that connects cells to the extracellular matrix; it contains integrins which link to intermediate filaments in the cytoplasm.

heterochromatin
Chromatin that is highly condensed even in interphase; generally transcriptionally inactive (compare with **euchromatin**).

heterokaryon incompatibility
Inability of two fungal strains to undergo fusion of their vegetative cells.

heteromeric (assembly)
Protein complex containing more than one type of subunit.

heterotrimeric G proteins
Trimeric guanine nucleotide-binding proteins with more than one kind of subunit, having intrinsic GTPase activity that couples the activation of transmembrane receptors to intracellular responses.

heterotropic (enzyme)
A type of allosteric enzyme in which the binding of a small molecule other than the substrate triggers the allosteric reflection.

heterotypic (complexes)
Multiprotein complexes comprising two or more different proteins.

hexon
Trimer of adenovirus coat proteins that form a pseudo-hexamer.

histones
A group of small abundant proteins, rich in arginine and lysine residues, that combine to form the nucleosome cores around which DNA is wrapped in eukaryotic chromosomes.

histone chaperones
Proteins that bind free histones, releasing them once they have been incorporated into newly replicated chromatin.

histone code
Hypothesis that proposes an epigenetic marking system using different combinations of histone PTM patterns to regulate functional outputs of eukaryotic genomes.

Holliday junction (HJ)
X-shaped structure observed in DNA undergoing recombination, in which the two DNA duplexes are held together at the crossover site.

homeostasis
Property of a system in which parameters such as protein concentrations are regulated so that the stability of the system is maintained when external conditions change.

homologous recombination (HR)
Genetic exchange between a pair of identical or very similar DNA sequences, typically those located on two copies of the same chromosome. Also a DNA repair mechanism for double-strand breaks.

homomeric (assembly)
Protein complex containing just one type of subunit.

homotropic (enzyme)
A type of allosteric enzyme in which the binding of a substrate at a site remote from the active site triggers the allosteric reflection.

homotypic (complexes)
Multi-subunit complexes with a single type of subunit.

hydrogen bond
Noncovalent bond formed when a hydrogen atom attached to an electronegative atom approaches another electronegative atom and there is a partial sharing of the proton (Figure 1.5).

hydrophobic effects
The observed tendency of nonpolar groups to associate with each other, excluding water molecules. May be measured in terms of the free energy change when a nonpolar group is transferred from water into a nonpolar solvent (Figure 1.7 and Table 1.1).

icosahedron
A regular polyhedron with 5-3-2 point group symmetry.

immunoglobulin
An antibody molecule. There are five classes, each with a different role in adaptive immune responses (Box 17.3).

immunological synapse
Complex of adhesion and signaling molecules on the surfaces of a cytotoxic T cell and its target antigen-presenting cell (APC), which leads to the rapid destruction of the APC (Figure 17.27).

immunoproteasome
Large protein complex that recognizes and degrades foreign proteins within eukaryotic cells, producing peptides that are subsequently exposed at the surface of antigen-presenting cells.

importins
Proteins that mediate transport into the nucleus through the nuclear pore complex.

imprinting
Phenomenon in which a gene is either expressed or not expressed in the offspring depending on which parent it is inherited from.

inflammasome
Protein complex formed after activation of cytoplasmic NOD-like receptors with adaptor proteins; contains a caspase enzyme that cleaves pro-inflammatory cytokines from their precursor proteins.

initiation factors
Proteins that help load initiator tRNA onto the ribosome, thus initiating translation (Figure 6.11).

injectisomes
Molecular machines that enable bacteria to deliver virulence factors into eukaryotic host cells.

innate immune system
Defense system found in both plants and animals helping them to resist invading pathogens; it includes both antimicrobial molecules and phagocytic cells.

innexin
Connexin-like gap junction protein found in invertebrates.

inside-out signaling
Signals transmitted from the intracellular tails of transmembrane proteins, influencing their extracellular conformation and binding affinity (Figures 11.26 and 11.39).

insulin receptor
Transmembrane receptor which, when activated by ligand binding, regulates lipid, protein, and carbohydrate metabolism.

integrin
Transmembrane adhesion protein involved in the attachment of cells to the extracellular matrix; found in hemidesmosomes and focal adhesions (Figures 11.2 and 11.26).

interaction patch
Region on the surface of a protein that binds to a complementary patch on another protein via multiple weak interactions.

interphase
Long period of the cell cycle between one mitosis and the next; includes G1 phase, S phase, and G2 phase (Figure 13.1).

introns
Segment of a eukaryotic gene that is removed by RNA splicing during maturation to yield the final RNA product of the gene.

ion channel
Transmembrane protein complex that forms a water-filled channel across the lipid bilayer through which specific ions can diffuse down their electrochemical gradient.

iron–sulfur (Fe–S) proteins
Proteins that contain ensembles of iron and sulfur centers that function as oxidation–reduction centers in electron transport chains. The sulfur atoms are normally coordinated by cysteine residues.

isologous (interaction)
Face-to-face interaction of two identical units.

JAK (Janus kinase)
Cytoplasmic tyrosine kinase associated with cytokine receptors, which phosphorylates and activates transcription regulators called STATs.

junction
Region where two cells meet and their plasma membranes become closely apposed.

karyopherins
Family of nucleocytoplasmic transport factors.

katanin
Member of the AAA+ family of ATPases that is responsible for severing microtubules, triggering their disassembly.

kinesins
Class of motor proteins that use the energy of ATP hydrolysis to move along microtubules, toward the plus ends.

kinetic isotope effect (KIE)
The change in the rate of a chemical reaction when one of the atoms in the reactants is substituted with one of its isotopes (Box 15.3).

kinetochore
Large protein complex that connects the centromere of a chromosome to microtubules of the mitotic spindle.

lagging strand
One of the two newly synthesized strands of DNA found at a replication fork. The lagging strand is made in discontinuous lengths that are later joined covalently (Figure 3.13).

lamin
A type of intermediate filament found in the nucleus.

laminin
Extracellular matrix fibrous protein found in basal laminae, where it forms a sheetlike network (Figures 11.31 and 11.37).

last universal common ancestor (LUCA)
The most recent primordial organism from which all cellular organisms now living on Earth are descended.

leading strand
One of the two newly synthesized strands of DNA found at a replication fork. The leading strand is made by continuous synthesis in the 5′ to 3′ direction (Figure 3.13).

lever arm (myosin)
Extended protein domain that links the axis of rotation of the myosin motor to the site at which force is applied on the actin filament.

ligand-gated channel
A transmembrane ion channel that opens in response to the binding of a chemical messenger.

ligase core complexes
Protein complexes that contain, together with other proteins, an E3 ubiquitin ligase that ubiquitylates substrate endoreticulum-associated proteins destined for proteolytic degradation.

light-harvesting (antenna) complex (LHC)
Collection of chlorophylls and carotenoids surrounding a photosynthetic reaction center (RC), which capture solar energy and rapidly transfer it to the RC.

line symmetry
A symmetry in which structure on one side of a straight line is matched by a reflected version of the same structure on the other side of the line (Figure 1.15).

linking number (Lk)
Parameter that describes the intertwining of two strands of DNA. The sum of the factors twist and writhe.

lipid raft
Small region of a membrane locally enriched in sphingolipids and cholesterol.

lipoic acid
Enzyme co-factor that serves as a prosthetic group for the attachment of catalytic intermediates to swinging arms in oxidative decarboxylases.

lipoyl domain
Mobile domain on the surface of oxidative decarboxylases that serves as a swinging arm, moving catalytic intermediates between catalytic sites.

lock-and-key hypothesis
Emil Fischer's proposition that an enzyme and its substrate possess complementary geometric shapes (*see* complementarity).

lysis
Bursting open of a cell's plasma membrane leading to its death; often caused by a virus or enzyme.

lysogeny
The process whereby a viral genome is incorporated into the genome of a host cell where it is replicated passively with the host genome.

lysosome
A membrane-delimited organelle in eukaryotic cells that contains enzymes that degrade foreign materials taken into cells by **endocytosis**.

M phase
See mitosis.

macromolecular crowding
Situation in which macromolecules are concentrated to the point that their combined volumes account for a significant fraction of the total volume available; this affects the rate of diffusion-limited reactions. Also known as **excluded volume effect** (Figure 1.35).

major capsid protein (MCP)
Principal building block of a viral capsid; often co-assembled with other (minor) structural proteins.

major histocompatibility complexes (MHCs)
Proteins found in antigen-presenting cells which bind peptides derived by proteolysis of foreign proteins and convey them to the cell surface where they are recognized by T cells.

malate shuttle
A biochemical system for translocating electrons produced during glycolysis in the cytoplasm across the mitochondrial matrix membrane (Box 15.2).

MAP kinase (mitogen-activated kinase)
Protein kinase at the end of a three-component signaling module involved in relaying signals from the plasma membrane to the nucleus.

MAP kinase kinase
Protein kinase in the center of a three-component signaling module involved in relaying signals from the plasma membrane to the nucleus.

maturation (viral)
Triggering of conformational changes in a virion that render it infectious; often controlled by the activity of a protease.

mechanically gated channel
Type of transmembrane ion channel that opens in response to a mechanically induced switch.

mechanotransduction
A biological response or reaction elicited by the application of mechanical force.

Mediator
Multiprotein complex that links gene-specific transcription factors to the pre-initiation complex for eukaryotic transcription (Figures 5.1 and 5.19).

membrane attack complex (MAC)
Complex formed as part of the complement system, which associates with the pathogen membrane, forming pores that lead to cytolysis of the pathogen.

membrane potential
Voltage difference across a membrane due to a slight excess of positive ions on one side and of negative ions on the other.

messenger RNA
RNA molecule that specifies the amino acid sequence of a protein. Produced in eukaryotes by processing of an RNA molecule made by RNA polymerase. It is translated into protein in a process catalyzed by ribosomes.

metabolic control analysis
A mathematical framework based on the idea that each step in a metabolic pathway contributes in some degree to the control of flux through that pathway.

metaphase
The third phase of mitosis, at which chromosomes are firmly attached to the mitotic spindle at the equator but are not yet segregated toward opposite poles (Figure 13.1).

Michaelis constant (K_m)
The concentration of substrate at which an enzyme works at half its maximum rate (Figure 1.27).

microhomology-mediated end joining (MMEJ)
Pathway for the repair of double-strand breaks in DNA, in which damaged and blocked ends are resected before rejoining by DNA ligation (Figure 4.23).

microtubule (MT)
Long hollow cylindrical structure composed of polymerized tubulin protein (a heterodimer). A major constituent of the cytoskeleton in all eukaryotic cells, providing the tracks along which motor proteins run (Figure 14.1).

microtubule-associated protein (MAP)
Any protein that binds to microtubules and modifies their properties. [Not to be confused with the 'MAP' (mitogen-activated protein kinase) or 'MAP kinase'].

microtubule-organizing center (MTOC)
Region of the cytoplasm occupied by a structure out of which microtubules grow, such as a centrosome or basal body.

mini-chromosome maintenance (MCM) assembly
Hexameric protein complex that is bound to the DNA at origins of replication in eukaryotic chromosomes. Its ATPase activity unwinds the duplex DNA (Figure 3.4).

mismatch repair (MMR)
Pathway for the repair of mismatched bases arising from errors in DNA replication.

mitochondria
Membrane-delimited organelles that carry out oxidative phosphorylation and produce most of the ATP in eukaryotic cells (Figure 1.5.1 in Box 1.5).

mitochondrial matrix
Large internal compartment of the mitochondrion.

mitosis
Division of a nucleus of a eukaryotic cell, involving condensation of DNA into chromosomes, and separation of the duplicated chromosomes into two identical sets (Figure 13.1).

mitotic checkpoint complex (MCC)
Protein complex that holds the anaphase-promoting complex in check until kinetochores have attained bi-orientation and are ready to separate the duplicated chromosomes (Figure 13.22).

mitotic spindle
Bipolar array of microtubules and associated molecules that forms between the opposite poles of a eukaryotic cell during mitosis and serves to move the duplicated chromosomes apart.

mobile genetic elements
Tracts of DNA that can be moved around the genome through the process of site-specific recombination.

molecular chaperone
See chaperone.

molecular tape-measure
A molecule that specifies the length of a polymer; for example, the genomic nucleic acid of helical viruses determines the length of the virion.

molecular tunnel
Tunnel through an enzyme complex linking two active sites; allows the channeling of catalytic intermediates.

molten globule
An intermediate conformational state (partially folded) between the unfolded and native (fully folded) states of a globular protein.

motor protein
Protein that uses energy derived from nucleoside triphosphate hydrolysis to propel itself along a linear track (protein filament or other polymeric molecule).

multienzyme complexes
Assemblies containing two or more enzymes of a catalytic pathway.

murein
See peptidoglycan.

muscle fiber
A bundle of many myofibrils, each containing a bundle of actin and myosin filaments, within a single cell.

mutation
Heritable change in a nucleotide sequence.

myddosome
Protein complex implicated in transducing signals from Toll-like receptor 4 (TLR4), eliciting the production of pro-inflammatory cytokines and chemokines (Figure 17.4).

myofibril
Long, highly organized bundle of actin filaments, myosin filaments, and other proteins in muscle cells that contracts by a sliding filament mechanism.

myosin
A motor protein that uses the energy of ATP hydrolysis to move along actin filaments.

N-end rule
Particular amino acids at the N terminus of a protein that determine its likelihood of being degraded.

nanoinjectors
See injectisomes.

natively unfolded proteins
Proteins that under physiological conditions have no specific structure.

nebulin
Giant protein that serves as a molecular tape-measure to specify the length of actin filaments.

Nedd8 (neural precursor cell expressed, developmentally down-regulated 8)
Small ubiquitin-like protein that is covalently attached to larger proteins to modify their function in processes such as cell cycle progression and cytoskeletal regulation.

neurofilament
A type of intermediate filament found in nerve cells (Figure 11.4).

neurotransmitter
Small signal molecule secreted by the presynaptic nerve cell at a chemical synapse to relay the signal to the postsynaptic cell.

nexin
Cross-linking protein with elastic properties that forms part of the axome of eukaryotic flagella.

non-cyclic electron transport
Linear pathway of light-induced electron transport in plants and cyanobacteria that leads to the reduction of $NADP^+$ to NADPH (Figure 15.30).

non-equivalence
Form of virus capsid structure that exhibits multiple sets of quite different interactions between subunits (Figure 8.12).

non-heme iron proteins
See iron–sulfur proteins.

nonhomologous end joining (NHEJ)
A DNA repair mechanism for double-strand breaks in which the broken ends of DNA are brought together and rejoined by DNA ligation, generally with the loss of one or more nucleotides at the site of joining.

NSF (*N*-ethylmaleimide sensitive factor)
Protein with ATPase activity that disassembles SNARE complexes, allowing the recycling of their components for another round of fusion.

nuclear envelope
Double lipid bilayer that surrounds the genetic material and nucleolus in eukaryotic cells.

nuclear pore complex (NPC)
Large multiprotein structure forming an aqueous channel through the nuclear envelope that allows selected molecules to move between the nucleus and cytoplasm.

nucleocapsid
Viral nucleic acid surrounded by a regular polymer of one or more protein subunits.

nucleoid
The region of a bacterial cell that contains the chromosome.

nucleoporins
Family of proteins that make up the nuclear pore complex (Figures 10.32 and 10.33).

nucleosome
Beadlike structure in eukaryotic chromatin, composed of a short length of DNA wrapped around an octameric core of histone proteins. The fundamental structural unit of chromatin.

nucleotide-excision repair (NER)
Type of DNA repair that corrects damage of the DNA double helix, such as that caused by chemicals or UV light, by cutting out the damaged region on one strand and resynthesizing it using the undamaged strand as template.

nucleus
The double membrane-delimited organelle of eukaryotic cells that contains DNA organized into chromosomes.

occludin
A transmembrane protein that forms part of tight junctions between cells.

Okazaki fragments
Series of short DNA fragments synthesized from the lagging strands during DNA synthesis (Figure 3.13).

opsonin
Molecules that bind to the surfaces of foreign particles such as bacteria, marking them for phagocytosis.

origin recognition complex (ORC)
Large protein complex that is bound to the DNA at origins of replication in eukaryotic chromosomes throughout the cell cycle (Figure 3.4).

outside-in signaling
Ligand binding to transmembrane proteins that transmits a signal into the cells (Figures 11.26 and 11.39).

oxidative phosphorylation
Process in bacteria and mitochondria in which ATP formation is

driven by the transfer of electrons through the electron transport chain to molecular oxygen. Involves the intermediate generation of an electrochemical proton gradient across a membrane and a chemiosmotic coupling of that gradient to ATP synthase (Figure 15.2).

oxygen-evolving complex (OEC)
Enzyme complex associated with photosystem II that is responsible for the photo-oxidation of water during the light reactions of photosynthesis.

P/O ratios
Ratio of ATP molecules made per oxygen atom, a measure of the efficiency of the coupling of the proton-motive force to ATP synthesis.

P-type ATPase
Transmembrane transporter protein that is autophosphorylated by ATP during its reaction cycle.

packaging signal
Motif in a viral genome that is recognized by the viral capsid protein, leading to assembly of the virion.

paired helical filaments
Two protofibrils forming an amyloid fibril.

pannexin
Connexin-like protein found in vertebrates that forms transmembrane ATP release channels.

pantothenic acid
Vitamin precursor that serves as a prosthetic group for the attachment of catalytic intermediates to swinging arms of fatty acid synthases.

parallel superpleated β structures
See β serpentines.

pause (microtubules)
Unexpected halt to microtubule disassembly.

pentasymmetron
Sub-assembly of large viral capsids comprising pentagonal clusters of hexons arranged symmetrically around a vertex (Figure 8.47).

peptidoglycan
A sheetlike polymer of the bacterial cell envelope formed by linear chains of two amino sugars, *N*-acetyl glucosamine and *N*-acetyl muramic acid, linked by a β-(1,4) glycosidic bond and cross-linked by short (4–5 aa) peptides.

peptidyl transfer center (PTC)
Region of the ribosome large subunit where amino acids are covalently added to the nascent protein (Figure 6.3).

pericentriolar material (PCM)
Material surrounding the centriole.

periplasm
Matrix contained between the inner cytoplasmic membrane and the outer membrane of Gram-negative bacteria.

periplasmic space
See periplasm.

permselectivity
The property of gap junction channels to discriminate according to molecular size, charge, or shape.

peroxisome
Small membrane-delimited organelle that uses molecular oxygen to oxidize organic molecules. Contains some enzymes that produce and others that degrade hydrogen peroxide.

phage display
A laboratory technique in which the gene for a protein of interest is inserted a bacteriophage coat protein gene and the protein is displayed on the surface of the bacteriophage virion.

phase diagram
Graphical representation of an equilibrium of different states of a system as a function of two condition parameters.

phase variation
The use of conservative site-specific recombination to control the expression of particular genes. The ability of *Neisseria gonorrhoeae* bacterium, for example, to insert a variant genetic cassette into pilin genes to alter the antigenic character of the pilus (Figure 11.45).

phosphopantetheine
Vitamin that serves as a prosthetic group for the attachment of catalytic intermediates to swinging arms of fatty acid synthases.

photophosphorylation
Process in plants in which ATP formation is driven by the transfer of electrons through an electron transport chain initiated by the absorption of light energy. Involves the intermediate generation of an electrochemical proton gradient across a membrane and a chemiosmotic coupling of that gradient to ATP synthase.

photosynthetic reaction centers (RCs)
Complex of proteins, pigments, and co-factors that together make use of light energy to create the electrochemical potential for the synthesis of ATP and NADPH.

photosynthetic unit (PSU)
Photosynthetic reaction center and its surrounding light-harvesting complex of antenna chlorophylls and carotenoids.

photosystem I
Class of photosynthetic reaction center containing chlorophyll P700, which accepts electrons from photosystem II in cyanobacteria and higher plants.

photosystem II
Class of photosynthetic reaction center containing chlorophyll P680, which passes electrons on to photosystem I and oxidizing water in cyanobacteria and higher plants.

phycobilins
Open-chain tetrapyrrole pigments found in chloroplasts and cyanobacteria that capture light energy and pass it on to chlorophylls.

phylloquinone
Membrane-soluble molecule that acts as an electron carrier in photosystem II (Figure 15.4).

picornaviruses
A family of small RNA viruses whose members include poliovirus.

pili
Extracellular protein filaments on the surface of bacteria.

pilin
Protein subunit of a bacterial pilus.

plakin repeat
Repeated protein motif found in cytolinkers that bind directly to intermediate filaments (Figure 11.8C).

plakoglobin
Adaptor protein that links cadherin proteins to actin filaments in both desmosomes and adherens junctions (Figure 11.14).

plakophilin-1
Adaptor protein containing ARM-repeat domains that forms part of the desmosome (Figure 11.14).

plaque
Hexagonal arrays of channels forming a gap junction (Figures 11.18 and 11.21).

plastocyanin (PC)
Small soluble copper-containing protein that plays a role in photosynthetic electron transfer between photosystems II and I.

plastoquinone
A plant-specific quinone that plays a role in photosynthetic electron transfer in photosystem II (Figure 15.4).

pleiomorphic
Describes a substance that exists in a continuously variable set of related structures.

point mutation
The alteration of a single nucleotide base within DNA or RNA (Figure 4.2).

polyadenylation
Addition of a long sequence of A nucleotides (the poly-A tail) to the 3′ end of a nascent RNA molecule.

polyhead
Virus structure forming an open-ended tube.

polymerization engine
Force exerted within a cell by the polymerization of protein filaments.

polymorphic
Describes a substance that coexists as two or more discrete forms.

polyprotein
A single protein cleaved after synthesis to form two or more proteins.

polyribosomes (polysomes)
Messenger RNA to which are attached multiple ribosomes engaged in protein synthesis (Figure 6.19).

polysheath
Abnormal assembly of bacteriophage tail-sheath proteins that forms in the absence of a bacteriophage baseplate.

porins
Channel-forming proteins of the outer membranes of bacteria, mitochondria, and chloroplasts.

portal protein
Specialized capsid protein inserted in some viruses to allow entry and exit of DNA or attachment of a tail.

power stroke (muscle contraction)
The stroke of a myosin motor that generates force (Figure 14.12).

pregenome
RNA packaged with a reverse transcriptase that gives rise to the viral DNA genome within the capsid.

preinitiation complex (PIC)
Multiprotein complex comprising general transcription factors and RNA polymerase, which assembles on a promoter sequence close to the transcription start site (Figures 5.14 and 5.16).

pre-replicative complex (pre-RC)
Multiprotein complex that is assembled at origins of replication during late mitosis and early G phases of the cell cycle; a prerequisite to license the assembly of a preinitiation complex, and the subsequent initiation of DNA replication.

primase
Enzyme that synthesizes a short strand of RNA on a DNA template, producing a primer for DNA synthesis.

prion
An infectious, abnormally folded form of an endogenous protein.

procapsid
Precursor viral protein that assembles and combines with the viral genome before maturation into the infectious virion.

pro-enzymes
Inactive precursors of enzymes; the active enzyme is released by the action of a protease.

progenote
A hypothetical entity with an organizational level simpler than that of prokaryotic cells.

programmed necrosis
A form of cell death that involves the formation of amyloid fibrils known as necrosomes and the subsequent rupture of cell membranes.

prokaryotes
Single-celled organisms that lack a well-defined, membrane-enclosed nucleus. Either a bacterium or an archaeon.

prometaphase
Second stage of mitosis, during which the nuclear envelope breaks down (Figure 13.1).

promoter
Nucleotide sequence in DNA to which RNA polymerase binds to begin transcription.

propeptide
Peptide extension that keeps an enzyme in an inactive conformation until its removal.

prophase
First stage of mitosis, during which the chromosomes condense but are not yet attached to a mitotic spindle (Figure 13.1).

prosthetic groups
A type of enzyme co-factor that is bound very tightly, or even covalently, to the parent protein.

26S proteasome
Large protein complex that degrades unneeded or damaged proteins by ATP-dependent proteolysis within eukaryotic cells.

proteasome activators
Regulatory protein complexes that bind to the ends of proteosome core particles.

protein kinase
Enzyme that transfers the terminal phosphate group of ATP to one or more specific amino acids (serine, threonine, or tyrosine) of a target protein.

protein kinase A
Enzyme that phosphorylates target proteins in response to a rise in intracellular cyclic AMP.

protein quality control
The elimination of misfolded, foreign, and otherwise aberrant proteins.

proteoglycan
Molecule consisting of one or more glycosaminoglycan chains attached to a core protein, commonly found in the extracellular matrix (Figure 11.31).

protofibril
Single protein filament that forms part of an amyloid fibril.

protofilaments
Linear strings of tubulin proteins joined end to end, which associate laterally to construct the hollow tubes of microtubules.

protomers
The structural unit of an oligomeric protein.

proton-motive force
The force exerted by the electrochemical proton gradient that moves protons across a membrane.

pumped proton
A proton pumped across the membrane during electron transfer (compare with **chemical proton**).

pyridoxal phosphate (PLP)
Prosthetic group needed for the enzyme tryptophan synthase.

Q cycle
Series of reactions involving the sequential oxidation and reduction of ubiquinone/ubiquinol that results in the net pumping of protons across a lipid bilayer (Figure 15.11).

quantum mechanical tunneling
Passage of an electron or protein through a transition state barrier (Box 15.3).

quasi-equivalence
Principle that accounts for the formation of icosahedral viral capsids of greater than 60 subunits.

quinol
Reduced form of ubiquinone, a membrane-soluble molecule that acts as an electron carrier in oxidative phosphorylation.

Rab
Small GTPase present in membranes and involved in conferring specificity on vesicle docking.

racemic mixture
A mixture of equal amounts of two **enantiomers** of a **chiral** molecule.

Ran
Small GTPase required for the active transport of macromolecules into and out of the nucleus through nuclear pore complexes.

random coil
An ensemble of unfolded structures of a polymer (e.g. protein or nucleic acid).

Ras
Archetypal monomeric GTPase that helps to relay signals from cell-surface receptor tyrosine kinase receptors to the nucleus, frequently in response to signals that stimulate cell division.

rate-limiting step
The slowest step in a chemical reaction, that determines the rate at which the overall reaction proceeds.

reactive oxygen species
Chemically reactive molecules containing oxygen. Examples include HO_2^- and OH•.

RecA
Prototype for a class of DNA-binding proteins with ATPase activity that catalyze synapsis of DNA strands during genetic recombination.

receptor tyrosine kinases (RTKs)
Family of cell-surface receptors with an extracellular ligand-binding domain and an intracellular kinase domain that phosphorylate signaling proteins on tyrosine residues in response to ligand binding.

recovery stroke (muscle contraction)
Movement of the lever arm of a molecular motor in preparation for the power stroke that will generate force (Figure 14.12).

release factors
Proteins that bind to a stop codon within mRNA and release the newly synthesized protein from the ribosome (Figure 6.16).

remodeler
See chromatin remodeling complex.

replication fork
Y-shaped region of a replicating DNA molecule at which the two strands of the DNA are being separated and the daughter strands are being formed (Figures 3.1 and 3.3).

replicative DNA helicase
Enzyme responsible for unwinding duplex DNA to enable it to be copied.

replicon
Unit of DNA replication, comprising a replication origin, a replisome and the replication forks traveling in opposite directions (Figure 3.12).

replisome
Complex molecular machine that carries out DNA replication (Figure 3.12).

resilin
Protein found in some jumping insects that stores strain energy that is released quickly and in large amounts.

respirasome
Supramolecular complex of respiratory chain complexes in the mitochondrial membrane.

respiratory chain phosphorylation
The generation of ATP from the free energy stored in a proton gradient.

reverse transcription
The transfer of information from single-stranded RNA to double-stranded DNA.

rhodopsin
Seven-span membrane protein of the GPCR family that acts as a light sensor in rod photoreceptor cells in the vertebrate retina. Contains the light-sensitive prosthetic group retinol (Figure 12.3).

ribosome
Particle composed of rRNAs and ribosomal proteins that catalyzes the synthesis of protein using information provided by mRNA (Figure 6.3).

ribosome recycling factor (RRF)
Protein that, together with EF-G, separates the two subunits of the ribosome after termination, so that they can engage with other mRNAs for a new round of translation (Figure 6.17).

Rieske-type [2Fe–2S] iron–sulfur center
Type of iron–sulfur center in which two of the coordinating cysteine residues are replaced with histidine residues.

rigor complex
Rigid complex formed between actin filaments and myosin in the absence of ATP (Figure 14.2C).

RING (really interesting new gene) domain
Type of zinc-binding finger domain often found in E3 ligases, which mediates ubiquitination by facilitating the direct transfer of ubiquitin from E2 enzymes to lysine residues on the target substrate.

Rossmann fold
A folding motif commonly found in proteins that bind nucleotides (NADP, FAD, etc.), as pointed out by the crystallographer Michael G. Rossmann in 1973. The basic Rossmann fold is βαβ with a two-stranded parallel β sheet and an α helix, and the sheet may be extended with additional β strands.

RNA-dependent RNA polymerase
An enzyme that catalyzes the synthesis of an RNA molecule from an RNA template.

RNA polymerases
Enzymes that catalyze the synthesis of an RNA molecule on a DNA template from ribonucleoside triphosphate precursors (Figure 5.3).

RNA primer
A short strand of RNA (~10 bp) that serves as a starting point for DNA synthesis (Figure 3.18).

rolling circle
Form of DNA or RNA replication that produces multiple copies of circular genomes and plasmids.

rotor/switch complex
Cytoplasm (C) ring and membrane (M) ring of the bacterial flagellum (Figure 14.84).

S-layer
Regular arrays of glycoproteins on the outer surface of many Archaea and bacteria.

S-layer homology (SLH)
Protein domain that serves to attach cellulosomes to the bacterial cell wall.

S phase
Period of a eukaryotic cell cycle in which DNA is replicated.

S-state system
Cluster of four manganese ions and one calcium ion that accumulates four successive positive charges from electron transfers by photosystem II, leading to the oxidation of water to molecular oxygen.

sarcomere
Repeating unit of a myofibril in a muscle cell, composed of an array of overlapping myosin and actin filaments between two adjacent Z lines (Figure 14.20).

saturation kinetics
The leveling off of the rate of an enzymatic reaction as the substrate concentration is increased.

scaffoldin
Non-catalytic core of cellulosomes, large extracellular assemblies of hydrolytic enzymes as found in anaerobic microorganisms.

SCF complex
Multi-subunit RING E3 ligase controlling transitions between G1/S and G2/M phases of the cell cycle.

scrapie
A fatal, degenerative disease affecting the central nervous system of sheep and goats, involving a prion protein.

Sec system
Bacterial system for the intracellular translocation and secretion of proteins.

second messengers
Small intracellular mediators that are formed or released in response to an extracellular signal and which help to relay the signal inside the cell.

secretion systems (SS)
Specialized bacterial systems that secrete effector proteins that interact with host cells.

selectivity filter
The part of an ion channel structure that determines which ions it can transport.

self-assembly
The formation of an organized quaternary structure from preexisting subunits.

self-compartmentalization
Principle of sequestration of active sites in the interior of large proteolytic complexes such as the 20S proteasome.

semiquinone
Stable, one-electron reduced form of ubiquinone, a membrane-soluble molecule that acts as an electron carrier in oxidative phosphorylation.

sensor 1
Polar residue that is one of the signature motifs of AAA+ ATPases.

sensor 2
Arginine residue that is one of the signature motifs of AAA+ ATPases.

septate junction
A form of tight junction between cells of invertebrates (Figure 11.9).

serine/threonine kinases
Enzymes that phosphorylate specific proteins on serine or threonine residues.

sigma (σ)
Prokaryotic general transcription factor that binds to promoter sequences upstream of a transcription start site (Figure 5.13).

signal recognition particle (SRP)
Ribonucleoprotein particle that binds an ER signal sequence on a partially synthesized polypeptide chain and directs the polypeptide and its attached ribosome to the endoplasmic reticulum (Figure 6.18).

silk (proteins)
A type of functional amyloid secreted by spiders.

single-stranded DNA-binding protein
Protein that binds to the single strands of the opened-up DNA double helix, preventing helical structures from reforming while the DNA is being replicated.

singlet state
A quantum state of a system with a spin of 0 (i.e. all electron spins are paired). Occurs when a chlorophyll molecule absorbs a photon.

sister chromatids
Tightly linked pair of chromosomes that arise from chromosome duplication during S phase, and which separate during M phase and segregate into two daughter cells.

site-specific recombination
A type of DNA recombination that takes place between short, specific sequences of DNA and occurs without the gain or loss of nucleotides.

sliding filament mechanism
Mechanism of muscle movement, in which actin and myosin filaments slide past each other to generate contractile force (Figure 14.22).

SMADs
Family of latent transcription regulators that are phosphorylated and activated by receptor serine/threonine kinases and carry the signal from the cell surface to the nucleus (Figure 12.35).

small heat shock proteins (sHsps)
Proteins of 15–30 kDa that bind to partly folded or aggregation-prone proteins, thereby preventing or delaying their aggregation. They are synthesized in increased amounts under stressful conditions such as raised temperature.

SNAPs
Proteins that, together with the ATPase NSF, disassemble SNARE complexes, allowing the recycling of their components for another round of fusion.

SNARE proteins
Members of a large family of transmembrane proteins present in organelle membranes and the vesicles derived from them, which mediate the many membrane fusion events in cells, including neurotransmitter release.

solvent-accessible surface
The accessible surface area (ASA) over which a protein and solvent make contact. Measured by computationally rolling a spherical solvent molecule over the van der Waals surface of the protein (Figure 1.8).

SOS response
A global response to DNA damage in bacteria, in which there is controlled expression of genes for DNA repair, cell cycle arrest, and mutagenesis.

spasmin
Contractile protein of the calmodulin superfamily found in vorticelids and protozoa.

spastin
Protein with ATPase activity that promotes microtubule disassembly by interacting with the C termini of tubulin subunits (Figure 14.44).

special pair
Two bacterial chlorophyll *a* molecules that together form P870 in the photosynthetic reaction center of purple sulfur bacteria.

species barriers
The natural mechanisms that prevent a virus or disease spreading from one species to another.

specificity constant
See Michaelis constant.

spindle assembly checkpoint (SAC)
Regulatory system that operates during mitosis to ensure that all chromosomes are properly attached to the spindle before sister chromatid separation starts.

spliceosome
Large assembly of RNA and protein molecules that performs pre-mRNA splicing in eukaryotic cells (Figure 5.22).

Src
Archetypal member of a family of cytoplasmic tyrosine kinases that associate with the cytoplasmic domains of some enzyme-linked cell-surface receptors that lack intrinsic tyrosine kinase activity. They transmit a signal onward by phosphorylating the receptor itself and specific intracellular signaling proteins.

STAT (signal transducer and activator of transcription)
Latent gene regulatory protein that is activated by phosphorylation by JAK kinases and enters the nucleus in response to signaling from cytokine receptors.

stathmin
Tubulin-sequestering protein that alters the curvature of tubulin dimers and caps their ends, thus interfering with microtubule assembly.

stator
Stationary component of a rotatory motor such as F_1F_o ATP synthase.

stereoisomers
Molecules that have the same molecular formula and sequence of bonded atoms but which differ in their geometric orientation (Box 1.1).

steric zipper
Interdigitation of β sheets in the structure of amyloid fibrils.

stress fiber
Cortical fibers of contractile actin–myosin II bundles that connect the cell to the extracellular matrix or adjacent cells through focal adhesions or a circumferential belt and adherens junctions.

striated muscle
Muscle fibers in which the bundles of myofibrils are aligned in register, forming sarcomeres that are visible as stripes under a microscope (Figure 14.19)

stroma
Interior space of a chloroplast, containing enzymes that incorporate CO_2 into sugars (Figure 15.18).

structural isomerism
Molecules that are made of the same atoms but with different bond arrangements; for example, the sugars glucose and fructose (Box 1.1).

substrate channeling
The passing of a catalytic intermediate from one active site to the next within a multienzyme complex.

substrate-level phosphorylation
The transfer of a phosphoryl group from a phosphorylated intermediate to ADP or GDP.

SUMO (small ubiquitin-like modifier)
Small ubiquitin-like protein that is covalently attached to larger proteins to modify their function in processes such as nuclear–cytosolic transport, transcription regulation, and cell cycle progression.

superoxide
Oxygen molecule with one additional unpaired electron (O_2^-).

swinging arm
An elongated side chain that binds a catalytic intermediate via a prosthetic group, moving it between enzymic catalytic sites in multienzyme complexes.

swinging lever arm hypothesis (muscle contraction)
Hypothesis of muscle contraction which suggests that a lever arm on myosin 'rows' the actin filament past the myosin tail.

symporter
Carrier protein that transports two types of solute across the membrane in the same direction (Figure 16.29).

synapse
Communicating cell–cell junction that allows signals to pass from a nerve cell to another cell.

synaptic vesicle
Small neurotransmitter-filled secretory vesicle found at the axon terminals of nerve cells. Its contents are released into the synaptic cleft by exocytosis when an action potential reaches the axon terminal.

synaptotagmin
Protein that acts as a Ca^{2+} sensor for synchronous neurotransmitter release.

T cell
Lymphocytes produced in the thymus, which discriminate between self and foreign peptides and serve a variety of roles in the adaptive immune system.

tail-sheath
Stacked hexameric protein rings surrounding bacteriophage tail fibers that can contract, propelling the tail tube through the bacterial envelope.

talin
An adaptor protein that links integrins to actin filaments in focal adhesions (Figures 11.28 and 11.39).

tape-measure (molecular)
Molecule that limits the size of a molecular assembly.

TATA box
Sequence in the promoter region of many eukaryotic genes that binds a general transcription factor (TFIID) and hence specifies the position at which transcription is initiated.

Tau
Microtubule-associated protein abundant in neurons that is implicated in several neurodegenerative diseases including Alzheimer's disease.

telomerase
Enzyme that elongates telomere sequences in DNA, which occur at the ends of eukaryotic chromosomes.

telomere
End of a chromosome, associated with a characteristic DNA sequence that is replicated in a special way. Counteracts the tendency of the chromosome otherwise to shorten with each round of replication.

telophase
Final stage of mitosis in which the two sets of separated chromosomes decondense and become enclosed by nuclear envelopes (Figure 13.1).

terminal pathway (complement system)
Final part of the complement system, leading to the assembly of the membrane attack complex that causes cytolysis of the infecting pathogen.

terminase
Motor protein that translocates viral DNA into the procapsid, cleaving it from the concatameric DNA.

termination factors
Ancillary proteins that bring about the end of mRNA translation.

tethering
See docking.

thermodynamic activity
The effective concentration of a macromolecule in a reaction, found by multiplying its physical concentration by its **activity coefficient.**

thermosome
Chaperonin protein complex found in Archaea (Figure 6.28).

thylakoid
Flattened sac of membrane in a chloroplast that contains chlorophyll and other pigments and carries out the light-trapping reactions of photosynthesis (Figure 15.18).

tight junction
Cell–cell junction that seals adjacent epithelial cells together, preventing the passage of most dissolved molecules from one side of the epithelial sheet to the other (Figure 11.2).

titin
Giant protein that serves as a molecular tape-measure to specify the length of myosin filaments, and as a stress sensor during muscle contraction (Figure 14.20).

Toll-like receptor
Family of pattern recognition receptors on or in cells of the innate immune system. They recognize pathogen-associated immunostimulants associated with microbes and promote downstream signaling.

tonofilaments
Intermediate filaments forming bundles (tonofibrils) in epithelial cells.

topoisomerase
Enzyme that binds to DNA and reversibly breaks a phosphodiester bond in one or both strands, thus resolving supercoiling, catenation, and knotting of DNA.

traffic ATPases
Family of proteins containing the bacterial PilT protein and which is part of the secretion superfamily of NTPases (Figure 11.46).

transcription
Copying of one strand of DNA into a complementary RNA sequence by the enzyme RNA polymerase.

transcription factor
Any protein required to initiate or regulate eukaryotic transcription. Includes general and gene-specific proteins.

transducin
G protein that is activated by a light-induced conformation change in rhodopsin, and transmits the signal within the cell by activating phosphodiesterase.

transfer RNA (tRNA)
Set of small RNA molecules used in protein synthesis as an interface (adaptor) between mRNA and amino acids. Each type of tRNA molecule is covalently linked to a particular amino acid (Figure 6.2).

transformation
The uptake of DNA into a (bacterial) cell.

transforming growth factor α (TGFα)
Ligand for transmembrane receptor tyrosine kinases that activates a signaling pathway for cell proliferation, differentiation, and development.

transforming growth factor β (TGFβ)
Family of secreted hormones that regulate cell proliferation and differentiation, transmitting their signals through transmembrane serine/threonine kinase receptors (Figure 12.35).

transition state
Structure that forms transiently in the course of a chemical reaction and has the highest free energy of any reaction intermediate.

translesion synthases (TLS)
DNA polymerases that are capable of reading past damaged DNA bases.

translocon system
Pore-forming protein complex within a membrane that assists with the co- or post-translational translocation of newly synthesized proteins (Figure 6.18).

transporter
Membrane transport protein that binds to a solute on one side of a membrane and transports it to the other side against a concentration gradient by undergoing a series of conformational changes.

treadmilling
Process by which a polymeric protein filament is maintained at constant length by addition of protein subunits at one end and loss of subunits at the other (Figure 14.3).

triangulation number (T)
Number of basic triangles in each triangular facet of an icosahedral viral capsid shell, $T = h^2 + hk + k^2$

triplet state
A quantum state of a system with a spin of 1, with one unpaired electron spin, which sometimes occurs when a chlorophyll molecule absorbs a photon.

triskelion
Spidery trimer protein that is the building block of clathrin coats.

trisymmetron
Sub-assembly of large viral capsids comprising triangular plates of hexons (Figure 8.47).

tropism
Movement in response to a stimulus.

tubulin
Protein subunit of microtubules.

turnover number
The number of substrate molecules converted to product per second at an enzyme's active site, also known as the **catalytic rate constant.**

twin-arginine translocation (Tat) system
Protein complex responsible for the translocation of folded proteins across the bacterial inner membrane.

twist (Tw)
Number of turns of one DNA strand about the other in B-DNA.

twitching motility
A form of bacterial locomotion across a substrate surface, mediated by pili.

two-partner system (TPS)
Type 5 secretion system of bacteria employed for the secretion of toxins and virulence factors across the outer membrane of Gram-negative bacteria.

tyrosine kinases
Enzymes that phosphorylate specific proteins on tyrosines.

ubiquinone
One of a family of membrane-soluble molecules that act as electron carriers in oxidative phosphorylation. Also known as coenzyme Q.

ubiquitin
Small, highly conserved protein present in all eukaryotic cells that becomes covalently attached to lysines of other proteins, marking these proteins for intracellular destruction by a proteasome or otherwise determining their fate.

ubiquitylation
Process for covalently attaching a small protein to lysine residues, which marks a protein for intracellular destruction by a proteosome.

unfoldase-assisted proteases
Large protein complex capable of unfolding proteins prior to their degradation.

uniporter
Carrier protein that transports a single solute from one side of the membrane to the other.

uroplakin
Membrane protein that forms hexagonal arrays on the surfaces of epithelial cells lining the urinary tract (Figure 11.41).

van der Waals interactions
Type of (individually weak) noncovalent bond that is formed at close range between nonpolar atoms (Figure 1.3).

vimentin
Type of intermediate filament found in fibroblasts (Figures 11.1, 11.4, and 11.5).

vinculin
An adaptor protein that links cadherins to actin filaments in focal adhesions (Figures 11.17, 11.28, and 11.39).

vinnexin
Connexin-like gap junction protein found in viruses.

viral envelope
Membrane that surrounds the nucleocapsid of some viruses.

virus-like particle (VLP)
Structure comprising viral proteins but lacking the viral genome that can be used as a vaccine, or as a delivery system for drugs or nucleic acids into cells.

virus surfing
Movement of a virus from its first point of contact with a host cell to its point of entry.

voltage-gated channel
Type of ion channel found in the membranes of electrically excitable cells (such as nerve, endocrine, egg, and muscle cells). Opens in response to a shift in membrane potential past a threshold value.

Walker-A (or P-loop) motif
Protein sequence (G/AXXXXGKT/S) characteristic of enzymes that bind ATP and which coordinates the phosphates of the bound ATP.

Walker-B motif
Protein sequence ($\Phi\Phi\Phi\Phi D/E$, where Φ is any hydrophobic amino acid) characteristic of enzymes that bind ATP and which coordinates the phosphates of bound Mg^{2+}.

weak forces
Noncovalent interactions that govern protein folding and oligoisomerizations. Includes electrostatics, hydrogen bonds, van der Waals interactions, and hydrophobic interactions.

writhe (Wr)
Number of superhelical twists of an under- or overwound DNA helix.

X-inactivation
Inactivation of one copy of the X chromosome in the somatic cells of female mammals.

Z-lines
Molecular scaffold for actin filaments at the ends of a sarcomere, running perpendicular to the myofibril axis.

zymogens
See pro-enzymes.

Index

Note: abbreviations following page numbers are: B, box; F, figure; and T, table. Prefixes are ignored in the alphabetical sequence – thus β-Catenin will be found under the letter C.